TABLE 1.7 Approximate Physical Properties of Some Common Gases at Standard Atmospheric Pressure (BG Units)

Gas	Temperature (°F)	Density, ρ (slugs/ft³)	Specific Weight, γ (lb/ft³)	Dynamic Viscosity, μ (lb·s/ft²)	Kinematic Viscosity, ν (ft²/s)	Gas Constant,[a] R (ft·lb/slug·°R)	Specific Heat Ratio,[b] k
Air (standard)	59	2.38 E − 3	7.65 E − 2	3.74 E − 7	1.57 E − 4	1.716 E + 3	1.40
Carbon dioxide	68	3.55 E − 3	1.14 E − 1	3.07 E − 7	8.65 E − 5	1.130 E + 3	1.30
Helium	68	3.23 E − 4	1.04 E − 2	4.09 E − 7	1.27 E − 4	1.242 E + 4	1.66
Hydrogen	68	1.63 E − 4	5.25 E − 3	1.85 E − 7	1.13 E − 4	2.466 E + 4	1.41
Methane (natural gas)	68	1.29 E − 3	4.15 E − 2	2.29 E − 7	1.78 E − 4	3.099 E + 3	1.31
Nitrogen	68	2.26 E − 3	7.28 E − 2	3.68 E − 7	1.63 E − 4	1.775 E + 3	1.40
Oxygen	68	2.58 E − 3	8.31 E − 2	4.25 E − 7	1.65 E − 4	1.554 E + 3	1.40

[a] Values of the gas constant are independent of temperature.

[b] Values of the specific heat ratio depend only slightly on temperature.

TABLE 1.8 Approximate Physical Properties of Some Common Gases at Standard Atmospheric Pressure (SI Units)

Gas	Temperature (°C)	Density, ρ (kg/m³)	Specific Weight, γ (N/m³)	Dynamic Viscosity, μ (N·s/m²)	Kinematic Viscosity, ν (m²/s)	Gas Constant,[a] R (J/kg·K)	Specific Heat Ratio k
Air (standard)	15	1.23 E + 0	1.20 E + 1	1.79 E − 5	1.46 E − 5	2.869 E + 2	1.40
Carbon dioxide	20	1.83 E + 0	1.80 E + 1	1.47 E − 5	8.03 E − 6	1.889 E + 2	1.30
Helium	20	1.66 E − 1	1.63 E + 0	1.94 E − 5	1.15 E − 4	2.077 E + 3	1.66
Hydrogen	20	8.38 E − 2	8.22 E − 1	8.84 E − 6	1.05 E − 4	4.124 E + 3	1.41
Methane (natural gas)	20	6.67 E − 1	6.54 E + 0	1.10 E − 5	1.65 E − 5	5.183 E + 2	1.31
Nitrogen	20	1.16 E + 0	1.14 E + 1	1.76 E − 5	1.52 E − 5	2.968 E + 2	1.40
Oxygen	20	1.33 E + 0	1.30 E + 1	2.04 E − 5	1.53 E − 5	2.598 E + 2	1.40

[a] Values of the gas constant are independent of temperature.

[b] Values of the specific heat ratio depend only slightly on temperature.

FUNDAMENTALS
OF
FLUID MECHANICS

FUNDAMENTALS OF FLUID MECHANICS

Bruce R. Munson

Donald F. Young
Department of Engineering Science and Mechanics

Theodore H. Okiishi
Department of Mechanical Engineering
Iowa State University
Ames, Iowa, USA

John Wiley & Sons

New York   Chichester   Brisbane   Toronto   Singapore

Library of Congress Cataloging-in-Publication Data:
Munson, Bruce Roy, 1940–
Fundamentals of fluid mechanics/Bruce R. Munson, Donald F.
Young, Theodore H. Okiishi.
Includes bibliographical references.
ISBN 0-471-85526-X
1. Fluid mechanics.   I. Young, Donald F.   II. Okiishi, T. H.
(Theodore Hisao), 1939–   III. Title.
TA357.M86   1990
620.1'06—dc20   89-38368
                                                              CIP

ABOUT THE AUTHORS

Bruce R. Munson, Professor of Engineering Science and Mechanics at Iowa State University since 1974, received his B.S. and M.S. degrees from Purdue University and his Ph.D. degree from the Aerospace Engineering and Mechanics Department of the University of Minnesota in 1970.

From 1970 to 1974, Dr. Munson was on the mechanical engineering faculty of Duke University. From 1964 to 1966, he worked as an engineer in the jet engine fuel control department of Bendix Aerospace Corporation, South Bend, Indiana.

Dr. Munson's main professional activity has been in the area of fluid mechanics education and research. He has been responsible for the development of many fluid mechanics courses for studies in civil engineering, mechanical engineering, engineering science, and agricultural engineering and is the recipient of an Iowa State University Superior Engineering Teacher Award.

He has authored and coauthored many theoretical and experimental technical papers on hydrodynamic stability, low Reynolds number flow, secondary flow, and the applications of viscous incompressible flow. He is a member of The American Society of Mechanical Engineers, The American Physical Society, and The American Society for Engineering Education.

Donald F. Young, Anson Marston Distinguished Professor in Engineering, is a faculty member in the Department of Engineering Science and Mechanics at Iowa State University. Dr. Young received his B.S. degree in mechanical engineering, his M.S. and Ph.D. degrees in theoretical and applied mechanics from Iowa State, and has taught both undergraduate and graduate courses in fluid mechanics for many years. In addition to being named a Distinguished Professor in the College of Engineering, Dr. Young has also received the Standard Oil Foundation Outstanding Teacher Award and the Iowa State University Alumni Association Faculty Citation. He has been engaged in fluid mechanics research for more than 35 years, with special interests in similitude and modeling and the interdisciplinary field of biomedical fluid mechanics. Dr. Young has contributed to many technical publications and is the author or coauthor of two textbooks on applied mechanics. He is a Fellow of The American Society of Mechanical Engineers.

Theodore H. Okiishi, Professor of Mechanical Engineering at Iowa State University, has taught fluid mechanics courses there since 1967. He received his undergraduate and graduate degrees at Iowa State.

From 1965 to 1967, Dr. Okiishi served as a U.S. Army officer with duty assignments at the National Aeronautics and Space Administration Lewis Research Center, Cleveland, Ohio, where he participated in rocket nozzle heat transfer research, and at the Combined Intelligence Center, Saigon, Republic of South Vietnam, where he studied seasonal river flooding problems.

Professor Okiishi is active in research on turbomachinery fluid dynamics. He and his graduate students have written a number of journal articles based on their studies. Some of these projects have involved significant collaboration with government and industrial laboratory researchers.

Dr. Okiishi has received several awards for teaching. He has developed undergraduate and graduate courses in classical fluid dynamics as well as the fluid dynamics of turbomachines.

He is a licensed professional engineer. His technical society activities include being a member of the board of directors of The American Society of Mechanical Engineers (ASME) International Gas Turbine Institute and serving as an associate editor for gas turbines of the ASME *Journal of Turbomachinery* and the *Journal of Engineering for Gas Turbines and Power*.

__ PREFACE __

This book is intended for junior and senior engineering students who are interested in learning some fundamental aspects of fluid mechanics. This area of mechanics is mature, and a complete coverage of all aspects of it obviously cannot be accomplished in a single volume. We developed this text to be used as a first course. The principles considered are classical and have been well established for many years. However, fluid mechanics education has improved with experience in the classroom, and we have brought to bear in this book our own ideas about the teaching of this interesting and important subject.

One of our aims is to represent fluid mechanics as it really is—an exciting and useful discipline. To this end, we include analyses of numerous everyday examples of fluid flow phenomena to which students and faculty can easily relate. More than 160 examples are presented that provide detailed solutions to a variety of problems. Also, a generous set of homework problems in each chapter stresses the practical application of principles. Those problems that can be best worked with a programmable calculator or a computer, about 10% of the problems, are so identified. The examples and homework problems illustrate the considerable versatility of fluid mechanical analyses.

Our message to students is that fluid motion is consistent with well-established physical laws. The mathematical statements, or equations that represent these laws and thus describe fluid behavior, form the basis for problem solving. In some instances, the solution of these fundamental equations results in the answers sought. Often, however, experimental data-based correlations and dimensional analysis are required in addition to basic equations for solution closure.

Since this is an introductory text, we have designed the presentation of material to allow for the gradual development of student confidence in fluid mechanics problem solving. Each important concept or notion is considered in terms of simple and easy-to-understand circumstances before more complicated features are introduced.

Two systems of units are used throughout the text: the British Gravitational System (pounds, slugs, feet, and seconds), and the International System of Units (newtons, kilograms, meters, and seconds). Both systems are widely used, and we believe that students need to be knowledgeable and comfortable with both systems. Approximately one half of the examples and homework problems use the British System; the other half is based on the International System.

In the first four chapters, the student is made aware of some fundamental aspects of fluid motion, including important fluid properties, regimes of flow, pressure variations in fluids at rest and in motion, fluid kinematics, and methods of flow description and analysis. The Bernoulli equation is introduced in Chapter 3 to draw attention, early on, to some of the interesting effects of fluid motion on the distribution of pressure in a flow field. We believe that this timely consideration of elementary fluid dynamics will increase student enthusiasm for the more complicated material that follows. In Chapter 4, we convey the essential elements of kinematics, including Eulerian and Lagrangian mathematical descriptions of flow phenomena, and indicate the vital relationship between the two views. For teachers who wish to consider kinematics in detail before the material on elementary fluid mechanics, Chapters 3 and 4 can be interchanged without loss of continuity.

Chapters 5, 6, and 7 expand on the basic analysis methods generally used to solve or to begin solving fluid mechanics problems. Emphasis is placed on understanding how flow phenomena are described mathematically and on when and how to use infinitesimal and finite control volumes. The effects of fluid friction on pressure and velocity distributions are also considered in some detail. A formal course in thermodynamics is not required to understand the various portions of the text that consider some elementary aspects of the thermodynamics of fluid flow. Experiments or tests must be relied on when mathematical analysis alone is inadequate to solve a problem. The advantages of using dimensional analysis and similitude for organizing test data and for planning experiments and the basic techniques involved are featured in Chapter 7.

Chapters 8 to 11 offer students opportunities for the further application of the principles learned earlier in the text. Also, where appropriate, additional important notions such as boundary layers, transition from laminar to turbulent flow, and flow separation are introduced. Practical concerns such as pipe flow, open-channel flow, flow measurement, drag and lift, and the effects of compressibility are discussed.

Students who study this text and who solve a representative set of the exercises provided should acquire a useful knowledge of the fundamentals of fluid mechanics. Faculty who use this text are provided with numerous topics to select from to meet the objectives of their own courses. More material than can be reasonably covered in one term is included. All are reminded of the fine collection of supplementary material. Where appropriate, we have cited throughout the text the articles and books that are available for enrichment.

We wish to express our thanks to the many colleagues who have helped in the development of this text. We are indebted to the following reviewers for their comments and suggestions: Willem F. Brutsaert, University of Maine; Robert R. Faddick, Colorado School of Mines; James A. Liburdy, Clemson University; Robert A. Medrow, University of Missouri-Rolla; Charles L. Merkle, Pennsylvania State University; Edgar A. O'Rear, University of Oklahoma; and Laura L. Pauley, Pennsylvania State University. We also appreciate the help provided by Dennis Cronin, Nikolaos Stergiopulos, and Mark Kroneman relative to the homework problems and solutions manual. We wish to express our gratitude to the many persons who supplied the photographs used throughout the text and to Milton Van Dyke for his help in this effort. We acknowledge the superb typing of the manuscript by Carolyn Taylor. Finally, we thank our families for their continued encouragement during the writing of this text.

Working with students over the years has taught us much about fluid mechanics education. We have tried in earnest to draw from this experience for the benefit of users of this book. Obviously we are still learning, and we thus welcome any suggestions and comments from you.

Bruce R. Munson
Donald F. Young
Theodore H. Okiishi

Publisher's Note. There are two important supplements that are available to professors who adopt this book for classroom use. The first is an Instructor's Manual containing complete solutions to all of the problems in the text and more than 39 transparency masters. The second is IBM-PC compatible software which consists of programs that allow students to solve the computer problems in the book. Both supplements may be obtained directly from the publisher.

CONTENTS

**CHAPTER SIX
DIFFERENTIAL ANALYSIS
OF FLUID FLOW** /319

FUNDAMENTALS
OF
FLUID MECHANICS

Liquid jet breakup: A jet of liquid breaks up into a series of drops because of surface tension effects. The number, size, and spacing of the drops is a function of various fluid properties including surface tension, density, and viscosity. (Water in air.) (Photograph courtesy of G. J. Jameson.)

ONE

INTRODUCTION

Fluid mechanics is that discipline within the broad field of applied mechanics concerned with the behavior of liquids and gases at rest or in motion. This field of mechanics obviously encompasses a vast array of problems that may vary from the study of blood flow in the capillaries (which are only a few microns in diameter) to the flow of crude oil across Alaska through an 800-mile-long 4-ft-diameter pipe. Fluid mechanics principles are needed to explain why airplanes are made streamlined with smooth surfaces for the most efficient flight, whereas golf balls are made with rough surfaces (dimpled) to increase their efficiency. Numerous interesting questions can be answered by using relatively simple fluid mechanics ideas. For example:

- How can a rocket generate thrust without having any air to push against in outer space?
- Why can't you hear a supersonic airplane until it has gone past you?
- How can a river flow downstream with a significant velocity even though the slope of the surface is so small that it could not be detected with an ordinary level?
- How can information obtained from model airplanes be used to design the real thing?
- Why does a stream of water from a faucet sometimes appear to have a smooth surface, but sometimes a rough surface?
- How much greater gas mileage can be obtained by improved aerodynamic design of cars and trucks?

The list of applications and questions goes on and on—but you get the point; fluid mechanics is a very important, practical subject. It is very likely that during your career as an engineer you will be involved in the analysis and design of systems that require a good understanding of fluid mechanics. It is hoped that this introductory text will provide a sound foundation in the fundamental aspects of fluid mechanics.

1.1 SOME CHARACTERISTICS OF FLUIDS

One of the first questions we need to explore is—what is a fluid? Or we might ask—what is the difference between a solid and a fluid? We have a general, vague idea of the difference. A solid is "hard" and not easily deformed, whereas a fluid is "soft" and is easily deformed (we can readily move through air). Although quite descriptive, these casual observations of the differences between solids and fluids are not very satisfactory from a scientific or engineering point of view. A closer look at the molecular structure of materials reveals that matter that we commonly think of as a solid (steel, concrete, etc.) has densely spaced molecules with large intermolecular cohesive forces that allow the solid to maintain its shape, and to not be easily deformed. However, for matter that we normally think of as a liquid (water, oil, etc.), the molecules are spaced further apart and the intermolecular forces are smaller than for solids, and the molecules have more freedom of movement. Thus, liquids can be easily deformed (but not easily compressed), and can be poured into containers or forced through a tube. Gases (air, oxygen, etc.) have even greater molecular spacing and freedom of motion with negligible cohesive intermolecular forces and as a consequence are easily deformed (and compressed) and will completely fill the volume of any container in which they are placed.

Although the differences between solids and fluids can be explained qualitatively on the basis of molecular structure, a more specific distinction is based on how they deform under the action of an external load. Specifically, *a fluid is defined as a substance that deforms continuously when acted on by a shearing stress of any magnitude.* A shearing stress (force per unit area) is created whenever a tangential force acts on a surface. When common solids such as steel or other metals are acted on by a shearing stress, they will initially deform (usually a very small deformation), but they will not continuously deform (flow). However, common fluids such as water, oil, and air satisfy the definition of a fluid—that is, they will flow when acted on by a shearing stress. Some materials, such as slurries, tar, putty, and so on, are not easily classified since they will behave as a solid if the applied shearing stress is small, but if the stress exceeds some critical value, the substance will flow. The study of such materials is called *rheology,* and does not fall within the province of classical fluid mechanics. Thus, all the fluids we will be concerned with in this text will conform to the definition of a fluid given previously.

Although the molecular structure of fluids is important in distinguishing one fluid from another, it is not possible to study the behavior of individual molecules when trying to describe the behavior of fluids at rest or in motion. Rather, we characterize the behavior by considering the average, or macroscopic, value of the quantity of interest, where the average is evaluated over a small volume containing a large number of molecules. Thus, when we say that the velocity at a certain point in a fluid is so much we are really indicating the average velocity of the molecules in a small volume surrounding the point. The volume is small compared with the physical dimensions of the system of interest, but large compared with the average distance between molecules. Is this a reasonable way to describe the behavior of a fluid? The answer is generally yes, since the spacing between molecules is typically very small. For gases at normal pressures and temperatures, the spacing is on the order of 10^{-6} mm, and for liquids it is on the order of 10^{-7} mm. The number of molecules per cubic millimeter is on the order of 10^{18} for gases and 10^{21} for liquids. It is thus clear that the number of molecules in a very tiny volume is huge and the idea of using average values taken over this volume is certainly reasonable. We thus assume that all the fluid characteristics we are interested in (pressure, velocity, etc.) vary continuously throughout the fluid—that is, we treat the fluid as a *continuum*. This concept will certainly be valid for all the circumstances considered in this text. One area of fluid mechanics for which the continuum concept breaks down is in the study of rarefied gases such as would be encountered at very high altitudes. In this case the spacing between air molecules can become large and the continuum concept is no longer acceptable.

1.2 DIMENSIONS, DIMENSIONAL HOMOGENEITY, AND UNITS

Since in our study of fluid mechanics we will be dealing with a variety of fluid characteristics, it is necessary to develop a system for describing these characteristics both *qualitatively* and *quantitatively*. The qualitative aspect serves to identify the nature, or type, of the characteristics (such as, length, time, stress, and velocity), whereas the quantitative aspect provides a numerical measure of the characteristics. The quantitative description requires both a number and a standard by which various quantities can be compared. A standard for a length might be a meter or foot, for time an hour or second, and for mass a slug or kilogram. Such standards are called *units*, and several systems of units are in common use as described in the following section. The qualitative description is conveniently given in terms of certain *primary quantities*, such as length, L, time, T, mass, M, and temperature, Θ. These primary quantities can then be used to provide a qualitative description of any other *secondary quantity*, for example, area $\doteq L^2$, velocity $\doteq LT^{-1}$, density $\doteq ML^{-3}$, and so on, where the symbol $\doteq$ is used to indicate the *dimensions* of the secondary quantity in terms of the primary quantities. Thus, to describe qualitatively a velocity, V, we would write

$$V \doteq LT^{-1}$$

and say that "the dimensions of a velocity equals length divided by time." The primary quantities are also referred to as *basic dimensions*.

For a wide variety of problems involving fluid mechanics, only the three basic dimensions, L, T, and M are required. Alternatively, L, T, and F could be used, where F is the basic dimension of force. Since Newton's law states that force is equal to mass times acceleration, it follows that $F \doteq MLT^{-2}$ or $M \doteq FL^{-1}T^2$. Thus, secondary quantities expressed in terms of M can be expressed in terms of F through the relationship above. For example, stress, σ, is a force per unit area, so that $\sigma \doteq FL^{-2}$, but an equivalent dimensional equation is $\sigma \doteq ML^{-1}T^{-2}$. Table 1.1 provides a list of dimensions for a number of common physical quantities.

All theoretically derived equations are *dimensionally homogeneous*—that is, the dimensions of the left side of the equation must be the same as those on the right side, and all additive separate terms must have the same dimensions. We accept as a fundamental premise that all equations describing physical phenomena must be dimensionally homogeneous. If this were not true, we would be attempting to equate or add unlike physical quantities, which would not make sense. For example, the equation for the velocity, V, of a uniformly accelerated body is

$$V = V_0 + at \tag{1.1}$$

where V_0 is the initial velocity, a the acceleration, and t the time interval. In terms of dimensions the equation is

$$LT^{-1} \doteq LT^{-1} + LT^{-1}$$

TABLE 1.1   Dimensions Associated with Common Physical Quantities

	FLT System	MLT System
Acceleration	LT^{-2}	LT^{-2}
Angle	$F^0L^0T^0$	$M^0L^0T^0$
Angular acceleration	T^{-2}	T^{-2}
Angular velocity	T^{-1}	T^{-1}
Area	L^2	L^2
Density	$FL^{-4}T^2$	ML^{-3}
Energy	FL	ML^2T^{-2}
Force	F	MLT^{-2}
Frequency	T^{-1}	T^{-1}
Heat	FL	ML^2T^{-2}
Length	L	L
Mass	$FL^{-1}T^2$	M
Modulus of elasticity	FL^{-2}	$ML^{-1}T^{-2}$
Moment of a force	FL	ML^2T^{-2}
Moment of inertia (area)	L^4	L^4
Moment of inertia (mass)	FLT^2	ML^2
Momentum	FT	MLT^{-1}
Power	FLT^{-1}	ML^2T^{-3}
Pressure	FL^{-2}	$ML^{-1}T^{-2}$
Specific heat	$L^2T^{-2}\Theta^{-1}$	$L^2T^{-2}\Theta^{-1}$
Specific weight	FL^{-3}	$ML^{-2}T^{-2}$
Strain	$F^0L^0T^0$	$M^0L^0T^0$
Stress	FL^{-2}	$ML^{-1}T^{-2}$
Surface tension	FL^{-1}	MT^{-2}
Temperature	Θ	Θ
Time	T	T
Torque	FL	ML^2T^{-2}
Velocity	LT^{-1}	LT^{-1}
Viscosity (dynamic)	$FL^{-2}T$	$ML^{-1}T^{-1}$
Viscosity (kinematic)	L^2T^{-1}	L^2T^{-1}
Volume	L^3	L^3
Work	FL	ML^2T^{-2}

and thus Eq. 1.1 is dimensionally homogeneous.

Some equations that are known to be valid contain constants having dimensions. The equation for the distance, d, traveled by a freely falling body can be written as

$$d = 16.1t^2 \tag{1.2}$$

and a check of the dimensions reveals that the constant must have the dimensions of LT^{-2} if the equation is to be dimensionally homogeneous. Actually, Eq. 1.2 is a special form of the well-known equation from physics for freely falling bodies,

$$d = \frac{gt^2}{2} \tag{1.3}$$

in which g is the acceleration of gravity. Equation 1.3 is dimensionally homogeneous and valid in any system of units. For $g = 32.2$ ft/s^2 the equation reduces to Eq. 1.2 and thus Eq. 1.2 is valid only for the system of units using feet and seconds. Equations that are restricted to a particular system of units can be denoted as *restricted homogeneous equations,* as opposed to equations valid in any system of units, which are *general homogeneous equations.* The preceding discussion indicates one rather elementary, but important, use of the concept of dimensions: the determination of one aspect of the generality of a given equation simply based on a consideration of the dimensions of the various terms in the equation. The concept of dimensions also forms the basis for the powerful tool of *dimensional analysis,* which is considered in detail in Chapter 7.

EXAMPLE 1.1

A commonly used equation for determining the volume rate of flow, Q, of a liquid through an orifice located in the side of a tank is

$$Q = 0.61 A \sqrt{2gh}$$

where A is the area of the orifice, g is the acceleration of gravity, and h is the height of the liquid above the orifice. Investigate the dimensional homogeneity of this formula.

SOLUTION

The dimensions of the various terms in the equation are

$$Q = \text{volume/time} \doteq L^3 T^{-1}$$

$$A = \text{area} \doteq L^2$$

$$g = \text{acceleration of gravity} \doteq LT^{-2}$$

$$h = \text{height} \doteq L$$

These terms, when substituted into the equation, yield the dimensional form:

$$(L^3 T^{-1}) \doteq (0.61)(L^2)(\sqrt{2})(LT^{-2})^{1/2}(L)^{1/2}$$

or

$$(L^3 T^{-1}) \doteq [(0.61)\sqrt{2}](L^3 T^{-1})$$

It is clear from this result that the equation is dimensionally homogeneous (both sides of the formula have the same dimensions of $L^3 T^{-1}$), and the numbers (0.61 and $\sqrt{2}$) are dimensionless.

If we were going to use this relationship repeatedly we might be tempted to simplify it by replacing g with its standard value of 32.2 ft/s^2 and rewriting the formula as

$$Q = 4.90 A \sqrt{h} \tag{1}$$

A quick check of the dimensions reveals that

$$L^3 T^{-1} \doteq (4.90)(L^{5/2})$$

and, therefore, the equation expressed as Eq. 1 can only be dimensionally correct if the number, 4.90, has the dimensions of $L^{1/2} T^{-1}$. Whenever a number appearing in an equation or formula has dimensions it means that the specific value of the number will depend on the system of units used. Thus, for the case being considered with feet and seconds used as units, the number 4.90 has units of ft$^{1/2}$/s. Equation (1) will only give the correct value for Q (in ft^3/s) when A is expressed in square feet and h in feet. Thus, Eq. (1) is a *restricted* homogeneous equation, whereas the original equation is a *general* homogeneous equation that would be valid for any consistent system of units. A quick check of the dimensions of the various terms in an equation is a useful practice and will often be helpful in eliminating errors—that is, as noted previously, all physically meaningful equations must be dimensionally homogeneous. We have briefly alluded to units in this example, and this important topic will be considered in more detail in the next section.

1.2.1 Systems of Units

In addition to the qualitative description of the various quantities of interest, it is generally necessary to have a quantitative measure of any given quantity. For example, if we measure the width of this page in the book and say that it is 10 units wide, the statement has no meaning until the unit of length is defined. If we indicate that the unit of length is a meter, and define the meter as some standard length, a unit system for length has been established (and a numerical value can be given to the page width). In addition to length, a unit must be established for each of the remaining basic quantities (force, mass, time, and temperature). There are several systems of units in use and we shall consider three systems that are commonly used in engineering.

British Gravitational (BG) System In the BG system the unit of length is the foot (ft), the time unit is the second (s), the force unit is the pound (lb), and the temperature unit is the degree Fahrenheit (°F), or the absolute temperature unit is the degree Rankine (°R), where

$$°R = °F + 459.67$$

The mass unit, called the *slug,* is defined from Newton's second law (force = mass × acceleration) as

$$1 \text{ lb} = (1 \text{ slug})(1 \text{ ft/s}^2)$$

This relationship indicates that a 1-lb force acting on a mass of 1 slug will give the mass an acceleration of 1 ft/s^2.

The weight, W (which is the force due to gravity, g) of a mass, m, is given by the equation

$$W = mg$$

and in BG units

$$W \text{ (lb)} = m \text{ (slugs) } g \text{ (ft/s}^2)$$

Since the earth's standard gravity is taken as $g = 32.174$ ft/s^2 (commonly approximated as 32.2 ft/s^2) it follows that a mass of 1 slug weighs 32.2 lb under standard gravity.

International System (SI) In 1960 the Eleventh General Conference on Weights and Measures, the international organization responsible for maintaining precise uniform standards of measurements, formally adopted the *International System of Units* as the international standard. This system, commonly termed SI, has been widely adopted worldwide and is widely used (although certainly not exclusively) in the United States. It is expected that the long-term trend will be for all countries to accept SI as the accepted standard and it is imperative that engineering students become familiar with this system. In SI the unit of length is the meter (m), the time unit is the second (s), the mass unit is the kilogram (kg), and the temperature unit is the kelvin (K). Note that there is no degree symbol used when expressing a temperature in kelvin units. The Kelvin temperature scale is an absolute scale and is related to the Celsius (centigrade) scale (°C) through the relationship

$$K = {}^\circ C + 273.15$$

Although the Celsius scale is not in itself part of SI, it is common practice to specify temperatures in degrees Celsius when using SI units.

The force unit, called the newton (N), is defined from Newton's second law as

$$1 \text{ N} = (1 \text{ kg})(1 \text{ m/s}^2)$$

Thus, a 1-N force acting on a 1-kg mass will give the mass an acceleration of 1 m/s². Standard gravity in SI is 9.807 m/s² (commonly approximated as 9.81 m/s²) so that a 1-kg mass weighs 9.81 N under standard gravity. Note that weight and mass are different, both qualitatively and quantitatively! The unit of *work* in SI is the joule (J) which is the work done when the point of application of a 1-N force is displaced through a 1-m distance in the direction of a force. Thus,

$$1 \text{ J} = 1 \text{ N} \cdot \text{m}$$

The unit of *power* is the watt (W) defined as a joule per second. Thus,

$$1 \text{ W} = 1 \text{ J/s} = 1 \text{ N} \cdot \text{m/s}$$

Prefixes for forming multiples and fractions of SI units are given in Table 1.2. For example, the notation kN would be read as "kilonewtons" and stands for 10^3 N. Similarly, mm would be read as "millimeters" and stands for 10^{-3} m. The centimeter is not an accepted unit of length in the SI system, and for most problems in fluid mechanics in which SI units are used, lengths will be expressed in millimeters or meters.

English Engineering (EE) System   In the EE system units for force *and* mass are defined independently; thus special care must be exercised when using this system in

TABLE 1.2 Prefixes for SI Units

Factor by Which Unit Is Multiplied	Prefix	Symbol
10^{12}	tera	T
10^9	giga	G
10^6	mega	M
10^3	kilo	k
10^2	hecto	h
10	deka	da
10^{-1}	deci	d
10^{-2}	centi	c
10^{-3}	milli	m
10^{-6}	micro	μ
10^{-9}	nano	n
10^{-12}	pico	p
10^{-15}	femto	f
10^{-18}	atto	a

conjunction with Newton's second law. The basic unit of mass is the pound mass (lbm), the unit of force is the pound (lb),[1] the unit of length is the foot (ft), the unit of time is the second (s), and the absolute temperature scale is the degree Rankine (°R). To make the equation expressing Newton's second law dimensionally homogeneous we write it as

$$\mathbf{F} = \frac{m\mathbf{a}}{g_c} \tag{1.4}$$

where g_c is a constant of proportionality which allows us to define units for both force and mass. For the BG system only the force unit was prescribed and the mass unit defined in a consistent manner such that $g_c = 1$. Similarly, for SI the mass unit was prescribed and the force unit defined in a consistent manner such that $g_c = 1$. For the EE system, a 1-lb force is defined as that force which gives a 1 lbm a standard acceleration of gravity which is taken as 32.174 ft/s². Thus, for Eq. 1.4 to be both numerically and dimensionally correct

$$1 \text{ lb} = \frac{(1 \text{ lbm})(32.174 \text{ ft/s}^2)}{g_c}$$

so that

$$g_c = \frac{(1 \text{ lbm})(32.174 \text{ ft/s}^2)}{(1 \text{ lb})}$$

With the EE system weight and mass are related through the equation

$$W = \frac{mg}{g_c}$$

where g is the local acceleration of gravity. Under conditions of standard gravity ($g = g_c$) the weight in pounds and the mass in pound mass are numerically equal. Also, since a 1-lb force gives a mass of 1 lbm an acceleration of 32.174 ft/s², and a mass of 1 slug an acceleration of 1 ft/s², it follows that

$$1 \text{ slug} = 32.174 \text{ lbm}$$

In this text we will primarily use the BG system and SI for units. The EE system is used very sparingly, and only in those instances where convention dictates its use. Approximately one-half the problems and examples are given in BG units and one-half in SI units. We cannot overemphasize the importance of paying close attention to units when solving problems. It is very easy to introduce huge errors into problem solutions through the use of incorrect units. Get in the habit of using a *consistent* system of units throughout a given solution. It really makes no difference which system you use as long as you are consistent; for example, don't mix slugs and newtons. If problem data are specified in SI units, then use SI units throughout the solution. If the data are specified in BG units, then use BG units throughout the solution. Tables 1.3 and 1.4 provide conversion factors for some quantities that are commonly encountered in fluid mechanics. For convenient reference these tables are also reproduced on the inside of the back cover. Note that in these tables (and others) the numbers are expressed by using computer exponential notation. For example, the number 5.154 E + 2 is equivalent to 5.154×10^2 in scientific notation, and the number 2.832 E − 2 is equivalent to 2.832×10^{-2}. More extensive tables of conversion factors for a large variety of unit systems can be found in Appendix A.

---

[1] It is also common practice to use the notation, lbf, to indicate pound force.

EXAMPLE 1.2

A tank of water having a total mass of 36 kg rests on the floor of an elevator. Determine the force (in newtons) that the tank exerts on the floor when the elevator is accelerating upward at 7 ft/s^2.

SOLUTION

A free-body diagram of the tank is shown in Fig. E1.2 where W is the weight of the tank and water, and F_f is the reaction of the floor on the tank. Application of Newton's second law of motion to this body gives

$$\sum \mathbf{F} = m\mathbf{a}$$

or

$$F_f - W = ma \tag{1}$$

where we have taken upward as the positive direction. Since $W = mg$, Eq. 1 can be written as

$$F_f = m(g + a) \tag{2}$$

Before substituting any number into Eq. 2 we must decide on a system of units, and then be sure all of the data are expressed in these units. Since we want F_f in newtons we will use SI units so that

$$F_f = 36 \text{ kg } [9.81 \text{ m/s}^2 + (7 \text{ ft/s}^2)(0.3048 \text{ m/ft})]$$

$$= 430 \text{ kg} \cdot \text{m/s}^2$$

Since $1 \text{ N} = 1 \text{ kg} \cdot \text{m/s}^2$ it follows that

$$F_f = 430 \text{ N} \quad \text{(downward on floor)} \tag{Ans}$$

The direction is downward since the force shown on the free-body diagram is the force of the floor *on the tank* so that the force the tank exerts *on the floor* is equal in magnitude but opposite in direction.

As you work through a large variety of problems in this text, you will find that units play an essential role in arriving at a numerical answer. Be careful! It is easy to mix units and cause large errors. If in the above example the elevator acceleration had been left as 7 ft/s^2 with m and g expressed in SI units, we would have calculated the force as 605 N and the answer would have been 41% too large!

FIGURE E1.2

1.3 ANALYSIS OF FLUID BEHAVIOR

The study of fluid mechanics involves the same fundamental laws you have encountered in physics and other mechanics courses. These laws include Newton's laws of motion, conservation of mass, and the first and second laws of thermodynamics. Thus, there are strong similarities between the general approach to fluid mechanics and to rigid-body and deformable-body solid mechanics. This is indeed helpful since many of the concepts and techniques of analysis used in fluid mechanics will be ones you have encountered before in other courses.

The broad subject of fluid mechanics can be generally subdivided into *fluid statics,* in which the fluid is at rest, and *fluid dynamics,* in which the fluid is moving. In the following chapters we will consider both of these areas in detail. Before we can proceed, however, it will be necessary to define and discuss certain fluid *properties* that are intimately related to fluid behavior. It is obvious that different fluids can have grossly different characteristics. For example, gases are light and compressible, whereas liquids are heavy (by comparison) and relatively incompressible. A syrup flows slowly from a container, but water flows rapidly when poured from the same container. To quantify these differences certain fluid properties are used. In the following several sections the properties that play an important role in the analysis of fluid behavior are considered.

TABLE 1.3 Conversion Factors from BG and EE Units to SI Units[a]

	To Convert from	to	Multiply by
Acceleration	ft/s^2	m/s^2	3.048 E − 1
Area	ft^2	m^2	9.290 E − 2
Density	lbm/ft^3	kg/m^3	1.602 E + 1
	slugs/ft^3	kg/m^3	5.154 E + 2
Energy	Btu	J	1.055 E + 3
	ft·lb	J	1.356
Force	lb	N	4.448
Length	ft	m	3.048 E − 1
	in.	m	2.540 E − 2
	mile	m	1.609 E + 3
Mass	lbm	kg	4.536 E − 1
	slug	kg	1.459 E + 1
Power	ft·lb/s	W	1.356
	hp	W	7.457 E + 2
Pressure	in. Hg (60 °F)	N/m^2	3.377 E + 3
	lb/ft^2 (psf)	N/m^2	4.788 E + 1
	lb/in.^{2} (psi)	N/m^2	6.895 E + 3
Specific weight	lb/ft^3	N/m^3	1.571 E + 2
Temperature	°F	°C	$T_C = (5/9)(T_F - 32°)$
	°R	K	5.556 E − 1
Velocity	ft/s	m/s	3.048 E − 1
	mi/hr (mph)	m/s	4.470 E − 1
Viscosity (dynamic)	lb·s/ft^2	N·s/m^2	4.788 E + 1
Viscosity (kinematic)	ft^2/s	m^2/s	9.290 E − 2
Volume flowrate	ft^3/s	m^3/s	2.832 E − 2
	gal/min (gpm)	m^3/s	6.309 E − 5

[a] If more than four-place accuracy is desired, refer to Appendix A.

FIGURE 1.1 Density of water as a function of temperature.

1.4 MEASURES OF FLUID MASS AND WEIGHT

1.4.1 Density

The *density* of a fluid, designated by the Greek symbol ρ (rho), is defined as its *mass* per unit volume. Density is typically used to characterize the mass of a fluid system. In the BG system ρ has units of slugs/ft³ and in SI the units are kg/m³.

The value of density can vary widely between different fluids, but for liquids, variations in pressure and temperature generally have only a small effect on the value of ρ. The small change in the density of water with large variations in temperature is illustrated in Fig. 1.1. Tables 1.5 and 1.6 list values of density for several common

TABLE 1.4 Conversion Factors from SI Units to BG and EE Units[a]

	To Convert from	to	Multiply by
Acceleration	m/s²	ft/s²	3.281
Area	m²	ft²	1.076 E + 1
Density	kg/m³	lbm/ft³	6.243 E − 2
	kg/m³	slugs/ft³	1.940 E − 3
Energy	J	Btu	9.478 E − 4
	J	ft·lb	7.376 E − 1
Force	N	lb	2.248 E − 1
Length	m	ft	3.281
	m	in.	3.937 E + 1
	m	mile	6.214 E − 4
Mass	kg	lbm	2.205
	kg	slug	6.852 E − 2
Power	W	ft·lb/s	7.376 E − 1
	W	hp	1.341 E − 3
Pressure	N/m²	in. Hg (60 °F)	2.961 E − 4
	N/m²	lb/ft² (psf)	2.089 E − 2
	N/m²	lb/in.² (psi)	1.450 E − 4
Specific weight	N/m³	lb/ft³	6.366 E − 3
Temperature	°C	°F	$T_F = 1.8\,T_C + 32°$
	K	°R	1.800
Velocity	m/s	ft/s	3.281
	m/s	mi/hr (mph)	2.237
Viscosity (dynamic)	N·s/m²	lb·s/ft²	2.089 E − 2
Viscosity (kinematic)	m²/s	ft²/s	1.076 E + 1
Volume flowrate	m³/s	ft³/s	3.531 E + 1
	m³/s	gal/min (gpm)	1.585 E + 4

[a] If more than four-place accuracy is desired, refer to Appendix A.

13

TABLE 1.5 Approximate Physical Properties of Some Common Liquids (BG Units)

Liquid	Temperature (°F)	Density, ρ (slugs/ft³)	Specific Weight, γ (lb/ft³)	Dynamic Viscosity, μ (lb·s/ft²)	Kinematic Viscosity, ν (ft²/s)	Surface Tension,[a] σ (lb/ft)	Vapor Pressure, p_v [lb/in.² (abs)]	Bulk Modulus,[b] E_v (lb/in.²)
Carbon tetrachloride	68	3.09	99.5	2.00 E − 5	6.47 E − 6	1.84 E − 3	1.9 E + 0	1.91 E + 5
Ethyl alcohol	68	1.53	49.3	2.49 E − 5	1.63 E − 5	1.56 E − 3	8.5 E − 1	1.54 E + 5
Gasoline[c]	60	1.32	42.5	6.5 E − 6	4.9 E − 6	1.5 E − 3	8.0 E + 0	1.9 E + 5
Glycerin	68	2.44	78.6	3.13 E − 2	1.28 E − 2	4.34 E − 3	2.0 E − 6	6.56 E + 5
Mercury	68	26.3	847	3.28 E − 5	1.25 E − 6	3.19 E − 2	2.3 E − 5	4.14 E + 6
SAE 30 oil[c]	60	1.77	57.0	8.0 E − 3	4.5 E − 3	2.5 E − 3	—	2.2 E + 5
Seawater	60	1.99	64.0	2.51 E − 5	1.26 E − 5	5.03 E − 3	2.56 E − 1	3.39 E + 5
Water	60	1.94	62.4	2.34 E − 5	1.21 E − 5	5.03 E − 3	2.56 E − 1	3.12 E + 5

[a] In contact with air.
[b] Isentropic bulk modulus calculated from speed of sound.
[c] Typical values. Properties of petroleum products vary.

TABLE 1.6 Approximate Physical Properties of Some Common Liquids (SI Units)

Liquid	Temperature (°C)	Density, ρ (kg/m³)	Specific Weight, γ (kN/m³)	Dynamic Viscosity, μ (N·s/m²)	Kinematic Viscosity, ν (m²/s)	Surface Tension,[a] σ (N/m)	Vapor Pressure, p_v [N/m² (abs)]	Bulk Modulus,[b] E_v (N/m²)
Carbon tetrachloride	20	1,590	15.6	9.58 E − 4	6.03 E − 7	2.69 E − 2	1.3 E + 4	1.31 E + 9
Ethyl alcohol	20	789	7.74	1.19 E − 3	1.51 E − 6	2.28 E − 2	5.9 E + 3	1.06 E + 9
Gasoline[c]	15.6	680	6.67	3.1 E − 4	4.6 E − 7	2.2 E − 2	5.5 E + 4	1.3 E + 9
Glycerin	20	1,260	12.4	1.50 E + 0	1.19 E − 3	6.33 E − 2	1.4 E − 2	4.52 E + 9
Mercury	20	13,600	133	1.57 E − 3	1.15 E − 7	4.66 E − 1	1.6 E − 1	2.85 E + 10
SAE 30 oil[c]	15.6	912	8.95	3.8 E − 1	4.2 E − 4	3.6 E − 2	—	1.5 E + 9
Seawater	15.6	1,030	10.1	1.20 E − 3	1.17 E − 6	7.34 E − 2	1.77 E + 3	2.34 E + 9
Water	15.6	999	9.80	1.12 E − 3	1.12 E − 6	7.34 E − 2	1.77 E + 3	2.15 E + 9

[a] In contact with air.
[b] Isentropic bulk modulus calculated from speed of sound.
[c] Typical values. Properties of petroleum products vary.

liquids. The density of water at 60 °F is 1.94 slugs/ft³ or 999 kg/m³. The large difference between these two values illustrates the importance of paying attention to units! Unlike liquids, the density of a gas is strongly influenced by both pressure and temperature, and this difference will be discussed in the next section.

The *specific volume, v*, is the *volume* per unit mass and is therefore the reciprocal of the density—that is,

$$v = \frac{1}{\rho} \tag{1.5}$$

This property is not commonly used in fluid mechanics but is used in thermodynamics.

1.4.2 Specific Weight

The *specific weight* of a fluid, designated by the Greek symbol γ (gamma), is defined as its *weight* per unit volume. Thus, specific weight is related to density through the equation

$$\gamma = \rho g \tag{1.6}$$

where g is the local acceleration of gravity. Just as density is used to characterize the mass of a fluid system, the specific weight is used to characterize the weight of the system. In the BG system γ has units of lb/ft³ and in SI the units are N/m³. Under conditions of standard gravity (g = 32.174 ft/s² = 9.807 m/s²) water at 60 °F has a specific weight of 62.4 lb/ft³ and 9.80 kN/m³. Tables 1.5 and 1.6 list values of specific weight for several common liquids (based on standard gravity). More complete tables for water can be found in Appendix B (Tables B.1 and B.2).

1.4.3 Specific Gravity

The *specific gravity* of a fluid, designated as SG, is defined as the ratio of the density of the fluid to the density of water at some specified temperature. Usually the specified temperature is taken as 4 °C (39.2 °F), and at this temperature the density of water is 1.94 slugs/ft³ or 1000 kg/m³. In equation form specific gravity is expressed as

$$SG = \frac{\rho}{\rho_{H_2O@4°C}} \tag{1.7}$$

and since it is the *ratio* of densities, the value of SG does not depend on the system of units used. For example, the specific gravity of mercury at 20 °C is 13.55 and the density of mercury can thus be readily calculated in either BG or SI units through the use of Eq. 1.7 as

$$\rho_{Hg} = (13.55)(1.94 \text{ slugs/ft}^3) = 26.3 \text{ slugs/ft}^3$$

or

$$\rho_{Hg} = (13.55)(1000 \text{ kg/m}^3) = 13.6 \times 10^3 \text{ kg/m}^3$$

It is clear that density, specific weight, and specific gravity are all interrelated, and from a knowledge of any one of the three the others can be calculated.

1.5 IDEAL GAS LAW

Gases are highly compressible in comparison to liquids, with changes in gas density directly related to changes in pressure and temperature through the equation

$$p = \rho RT \tag{1.8}$$

TABLE 1.7 Approximate Physical Properties of Some Common Gases at Standard Atmospheric Pressure (BG Units)

Gas	Temperature (°F)	Density, ρ (slugs/ft³)	Specific Weight, γ (lb/ft³)	Dynamic Viscosity, μ (lb·s/ft²)	Kinematic Viscosity, ν (ft²/s)	Gas Constant,[a] R (ft·lb/slug·°R)	Specific Heat Ratio,[b] k
Air (standard)	59	2.38 E − 3	7.65 E − 2	3.74 E − 7	1.57 E − 4	1.716 E + 3	1.40
Carbon dioxide	68	3.55 E − 3	1.14 E − 1	3.07 E − 7	8.65 E − 5	1.130 E + 3	1.30
Helium	68	3.23 E − 4	1.04 E − 2	4.09 E − 7	1.27 E − 4	1.242 E + 4	1.66
Hydrogen	68	1.63 E − 4	5.25 E − 3	1.85 E − 7	1.13 E − 4	2.466 E + 4	1.41
Methane (natural gas)	68	1.29 E − 3	4.15 E − 2	2.29 E − 7	1.78 E − 4	3.099 E + 3	1.31
Nitrogen	68	2.26 E − 3	7.28 E − 2	3.68 E − 7	1.63 E − 4	1.775 E + 3	1.40
Oxygen	68	2.58 E − 3	8.31 E − 2	4.25 E − 7	1.65 E − 4	1.554 E + 3	1.40

[a] Values of the gas constant are independent of temperature.
[b] Values of the specific heat ratio depend only slightly on temperature.

TABLE 1.8 Approximate Physical Properties of Some Common Gases at Standard Atmospheric Pressure (SI Units)

Gas	Temperature (°C)	Density, ρ (kg/m³)	Specific Weight, γ (N/m³)	Dynamic Viscosity, μ (N·s/m²)	Kinematic Viscosity, ν (m²/s)	Gas Constant,[a] R (J/kg·K)	Specific Heat Ratio,[b] k
Air (standard)	15	1.23 E + 0	1.20 E + 1	1.79 E − 5	1.46 E − 5	2.869 E + 2	1.40
Carbon dioxide	20	1.83 E + 0	1.80 E + 1	1.47 E − 5	8.03 E − 6	1.889 E + 2	1.30
Helium	20	1.66 E − 1	1.63 E + 0	1.94 E − 5	1.15 E − 4	2.077 E + 3	1.66
Hydrogen	20	8.38 E − 2	8.22 E − 1	8.84 E − 6	1.05 E − 4	4.124 E + 3	1.41
Methane (natural gas)	20	6.67 E − 1	6.54 E + 0	1.10 E − 5	1.65 E − 5	5.183 E + 2	1.31
Nitrogen	20	1.16 E + 0	1.14 E + 1	1.76 E − 5	1.52 E − 5	2.968 E + 2	1.40
Oxygen	20	1.33 E + 0	1.30 E + 1	2.04 E − 5	1.53 E − 5	2.598 E + 2	1.40

[a] Values of the gas constant are independent of temperature.
[b] Values of the specific heat ratio depend only slightly on temperature.

where p is the absolute pressure, ρ the density, T the absolute temperature,[2] and R is a gas constant. Equation 1.8 is commonly termed the *ideal* or *perfect gas law*, or the *equation of state* for an ideal gas. It is known to closely approximate the behavior of real gases under normal conditions when the gases are not approaching liquefaction.

Pressure in a fluid at rest is defined as the normal force per unit area exerted on a plane surface (real or imaginary) immersed in a fluid, and is created by the bombardment of the surface with the fluid molecules. From the definition, pressure has the dimension of FL^{-2}, and in BG units is expressed as lb/ft² (psf) or lb/in.² (psi) and in SI units as N/m². In SI, 1 N/m² is defined as a *pascal*, abbreviated as Pa, and pressures are commonly specified in pascals. The pressure in the ideal gas law must be expressed as an *absolute* pressure, which means that it is measured relative to absolute zero pressure (a pressure that would only occur in a perfect vacuum). Standard sea-level atmospheric pressure (by international agreement) is 14.696 psi (abs) or 101.33 kPa (abs). For most calculations these pressures can be rounded to 14.7 psi and 101 kPa, respectively. In engineering it is common practice to measure pressure relative to the local atmospheric pressure, and when measured in this fashion it is called *gage* pressure. Thus, the absolute pressure can be obtained from the gage pressure by adding the value of the atmospheric pressure. For example, a pressure of 30 psi (gage) in a tire is equal to 44.7 psi (abs) at standard atmospheric pressure. Pressure is a particularly important fluid characteristic and it will be discussed more fully in the next chapter.

The gas constant, R, which appears in Eq. 1.8, depends on the particular gas and is related to the molecular weight of the gas, M_{gas}, through the relationship

$$R = \frac{\lambda}{M_{gas}}$$

where λ is a universal constant ($\lambda = 49{,}720$ ft·lb/slug·°R $= 8314$ J/kg·K). Values of the gas constant for several common gases are listed in Tables 1.7 and 1.8. Also in these tables the gas density and specific weight are given for standard atmospheric pressure and gravity and for the temperature listed. More complete tables for air at standard atmospheric pressure can be found in Appendix B (Tables B.3 and B.4).

EXAMPLE 1.3

A compressed air tank has a volume of 0.84 ft³. When the tank is filled with air at a gage pressure of 50 psi, determine the density of the air and the weight of air in the tank. Assume the temperature is 70 °F and the atmospheric pressure is 14.7 psi (abs).

SOLUTION

The air density can be obtained from the ideal gas law (Eq. 1.8) expressed as

$$\rho = \frac{p}{RT}$$

so that

$$\rho = \frac{(50 \text{ lb/in.}^2 + 14.7 \text{ lb/in.}^2)(144 \text{ in.}^2/\text{ft}^2)}{(1716 \text{ ft·lb/slug·°R})[(70 + 460)°R]} = 0.0102 \text{ slugs/ft}^3 \qquad \text{(Ans)}$$

---

[2] We will use T to represent temperature in thermodynamic relationships although T is also used to denote the basic dimension of time.

Note that both the pressure and temperature were changed to absolute values.

The weight, W, of the air is equal to

$$W = \rho g \times (\text{volume})$$
$$= (0.0102 \text{ slugs/ft}^3)(32.2 \text{ ft/s}^2)(0.84 \text{ ft}^3)$$

so that

$$W = 0.276 \text{ lb} \qquad\qquad (\text{Ans})$$

since 1 lb = 1 slug·ft/s².

1.6 VISCOSITY

The properties of density and specific weight are measures of the "heaviness" of a fluid. It is clear, however, that these properties are not sufficient to uniquely characterize how fluids behave since two fluids (such as water and oil) can have approximately the same value of density but behave quite differently when flowing. There is apparently some additional property that is needed to describe the "fluidity" of the fluid.

To determine this additional property consider a hypothetical experiment in which a material is placed between two very wide parallel plates as shown in Fig. 1.2a. The bottom plate is rigidly fixed, but the upper plate is free to move. If a solid, such as steel, were placed between the two plates and loaded with the force P as shown, the top plate would be displaced through some small distance, δa (assuming the solid was mechanically attached to the plates). The vertical line AB would be rotated through the small angle, $\delta\beta$, to the new position AB'. We note that to resist the applied force, P, a shearing stress, τ, would be developed at the plate–material interface, and for equilibrium to occur $P = \tau A$ where A is the effective upper plate area (Fig. 1.2b). It is well known that for elastic solids, such as steel, the small angular displacement, $\delta\beta$, (called the shearing strain) is proportional to the shearing stress, τ, that is developed in the material.

What happens if the solid is replaced with a fluid such as water? We would immediately notice a major difference. When the force P is applied to the upper plate it will move continuously with a velocity, U, as illustrated in Fig. 1.3. This behavior is consistent with the definition of a fluid—that is, if a shearing stress is applied to a fluid it will deform continuously. A closer inspection of the fluid motion between the

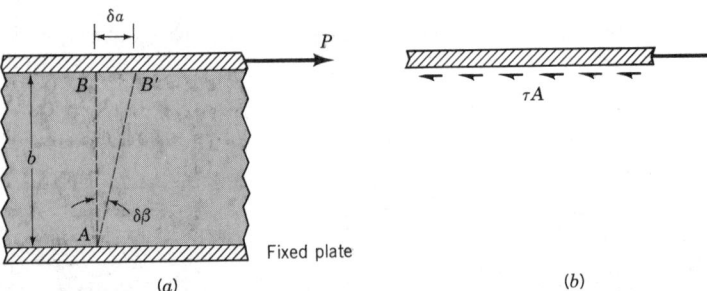

FIGURE 1.2 (a) Deformation of material placed between two parallel plates. (b) Forces acting on upper plate.

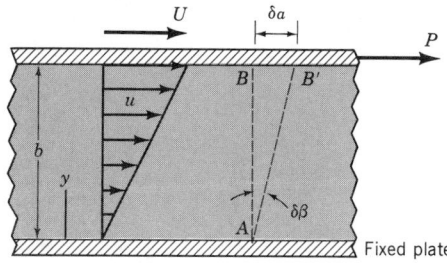

FIGURE 1.3 Behavior of a fluid placed between two parallel plates.

two plates would reveal that the fluid in contact with the upper plate moves with the plate velocity, U, and the fluid in contact with the bottom fixed plate has a zero velocity. The fluid between the two plates moves with velocity $u = u(y)$ that would be found to vary linearly as illustrated in Fig. 1.3. Thus, a *velocity gradient, du/dy,* is developed in the fluid between the plates. In this particular case the velocity gradient is a constant since $du/dy = U/b$, but in more complex flow situations this would not be true. The experimental observation that the fluid "sticks" to the solid boundaries is a very important one in fluid mechanics and is usually referred to as the *no-slip condition*. All fluids, both liquids and gases, satisfy this condition.

In a small time increment, δt, an imaginary vertical line AB in the fluid would rotate through an angle, $\delta\beta$, so that

$$\tan \delta\beta \approx \delta\beta = \frac{\delta a}{b}$$

Since $\delta a = U\,\delta t$ it follows that

$$\delta\beta = \frac{U\,\delta t}{b}$$

We note that in this case, $\delta\beta$ is a function not only of the force P (which governs U) but also of time. Thus, it is not reasonable to attempt to relate the shearing stress, τ, to $\delta\beta$ as is done for solids. Rather, we consider the *rate* at which $\delta\beta$ is changing, and define the *rate of shearing strain, $\dot{\gamma}$,* as

$$\dot{\gamma} = \lim_{\delta t \to 0} \frac{\delta\beta}{\delta t}$$

which in this instance is equal to

$$\dot{\gamma} = \frac{U}{b} = \frac{du}{dy}$$

A continuation of this experiment would reveal that as the shearing stress, τ, is increased by increasing P (recall that $\tau = P/A$) the rate of shearing strain is increased in direct proportion—that is

$$\tau \propto \dot{\gamma}$$

or

$$\tau \propto \frac{du}{dy}$$

Shearing stress, τ

Crude oil (60 °F)

μ

1

Water (60 °F)

Water (100 °F)

Air (60 °F)

Rate of shearing strain, $\dfrac{du}{dy}$

FIGURE 1.4 Linear variation of shearing stress with rate of shearing strain for common fluids.

This result indicates that for common fluids such as water, oil, gasoline, and air the shearing stress and rate of shearing strain (velocity gradient) can be related with a relationship of the form

$$\tau = \mu \frac{du}{dy} \tag{1.9}$$

where the constant of proportionality is designated by the Greek symbol μ (mu) and is called the *absolute viscosity*, *dynamic viscosity*, or simply the *viscosity* of the fluid. In accordance with Eq. 1.9, plots of τ versus du/dy should be linear with the slope equal to the viscosity as illustrated in Fig. 1.4. The actual value of the viscosity depends on the particular fluid, and for a particular fluid the viscosity is also highly dependent on temperature as illustrated in Fig. 1.4 with the two curves for water. Fluids for which the shearing stress is *linearly* related to the rate of shearing strain (also referred to as rate of angular deformation) are designated as *Newtonian fluids*. Fortunately most common fluids, both liquids and gases, are Newtonian, but there are important exceptions. Blood, paint, sludges, and certain plastics, for example, do not follow the linear stress versus rate of strain relationships. These fluids are designated as *non-Newtonian fluids*. It is beyond the scope of this book to consider the behavior of such fluids and we will only be concerned with Newtonian fluids. A more general formulation of Eq. 1.9 which applies to more complex flows of Newtonian fluids is given in Section 6.8.1.

From Eq. 1.9 it can be readily deduced that the dimensions of viscosity are FTL^{-2}. Thus, in BG units viscosity is given as lb·s/ft^2 and in SI units as N·s/m^2. Values of viscosity for several common liquids and gases are listed in Tables 1.5 through 1.8. A quick glance at these tables reveals the wide variation in viscosity among fluids. Viscosity is only mildly dependent on pressure and the effect of pressure is usually neglected. However, as previously mentioned, and as illustrated in Fig. 1.5, viscosity is very sensitive to temperature. For example, as the temperature of water changes from 60 to 100 °F the density decreases by less than 1% but the viscosity decreases

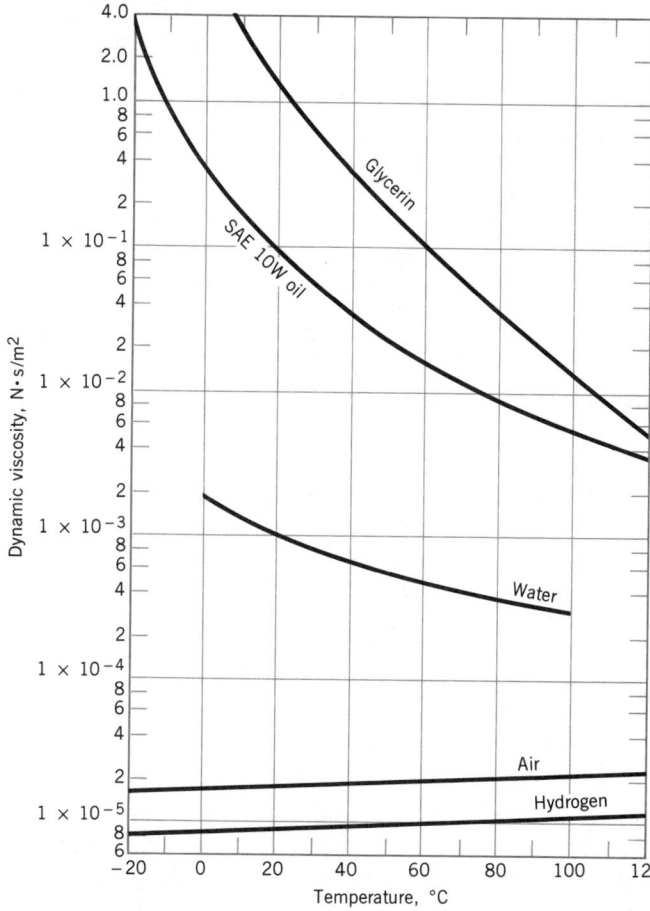

FIGURE 1.5 Dynamic (absolute) viscosity of some common fluids as a function of temperature.

by about 40%. It is thus clear that particular attention must be given to temperature when determining viscosity.

Figure 1.5 shows in more detail how the viscosity varies from fluid to fluid and how for a given fluid it varies with temperature. It is to be noted from this figure that the viscosity of liquids decreases with an increase in temperature, whereas for gases an increase in temperature causes an increase in viscosity. This difference in the effect of temperature on the viscosity of liquids and gases can again be traced back to the difference in molecular structure. The liquid molecules are closely spaced, with strong cohesive forces between molecules, and the resistance to relative motion between adjacent layers of fluid is related to these intermolecular forces. As the temperature increases, these cohesive forces are reduced with a corresponding reduction in resistance to motion. Since viscosity is an index of this resistance, it follows that the viscosity is reduced by an increase in temperature. In gases, however, the molecules are widely spaced and intermolecular forces negligible. In this case resistance to relative motion arises due to the exchange of momentum of gas molecules between adjacent layers. As molecules are transported by random motion from a region of low bulk velocity to mix with molecules in a region of higher bulk velocity (and vice versa), there is an effective momentum exchange which resists the relative motion between the layers. As the temperature of the gas increases, the random molecular activity increases with a corresponding increase in viscosity.

The effect of temperature on viscosity can be closely approximated using two empirical formulas. For gases the *Sutherland equation* can be expressed as

$$\mu = \frac{CT^{3/2}}{T + S} \tag{1.10}$$

where C and S are empirical constants, and T is absolute temperature. Thus, if the viscosity is known at two temperatures, C and S can be determined. Or, if more than two viscosities are known, the data can be correlated with Eq. 1.10 by using some type of curve-fitting scheme.

For liquids an empirical equation that has been used is

$$\mu = De^{B/T} \tag{1.11}$$

where D and B are constants and T is absolute temperature. This equation is often referred to as *Andrade's equation.* As was the case for gases, the viscosity must be known at least for two temperatures so the two constants can be determined. A more detailed discussion of the effect of temperature on fluids can be found in Ref. 1.

Quite often viscosity appears in fluid flow problems combined with the density in the form

$$\nu = \frac{\mu}{\rho}$$

This ratio is called the *kinematic viscosity* and is denoted with the Greek symbol ν (nu). The dimensions of kinematic viscosity are L^2/T, and the BG units are ft^2/s and SI units are m^2/s. Values of kinematic viscosity for some common liquids and gases are given in Tables 1.5 through 1.8. More extensive tables giving both the dynamic and kinematic viscosities for water and air can be found in Appendix B (Tables B.1 through B.4), and graphs showing the variation in both dynamic and kinematic viscosity with temperature for a variety of fluids are also provided in Appendix B (Figs. B.1 and B.2).

Although in this text we are primarily using BG and SI units, dynamic viscosity is often expressed in the metric CGS (centimeter-gram-second) system with units of dyne·s/cm^2. This combination is called a *poise*, abbreviated P. In the CGS system, kinematic viscosity has units of cm^2/s, and this combination is called a *stoke*, abbreviated St.

EXAMPLE 1.4

A dimensionless combination of variables that is important in the study of viscous flow through pipes is called the *Reynolds number*, Re, defined as $\rho VD/\mu$ where ρ is the fluid density, V the mean fluid velocity, D the pipe diameter, and μ the fluid viscosity. A Newtonian fluid having a viscosity of 0.38 N·s/m^2 and a specific gravity of 0.91 flows through a 25-mm-diameter pipe with a velocity of 2.6 m/s. Determine the value of the Reynolds number using (a) SI units, and (b) BG units.

SOLUTION

(a) The fluid density is calculated from the specific gravity as

$$\rho = SG\ \rho_{H_2O@4°C} = 0.91\ (1000\ kg/m^3) = 910\ kg/m^3$$

and from the definition of the Reynolds number

$$Re = \frac{\rho VD}{\mu} = \frac{(910\ kg/m^3)(2.6\ m/s)(25\ mm)(10^{-3}\ m/mm)}{0.38\ N·s/m^2}$$

$$= 156\ (kg·m/s^2)/N$$

However, since $1\ N = 1\ kg·m/s^2$ it follows that the Reynolds number is unitless—that is,

$$Re = 156 \tag{Ans}$$

The value of any dimensionless quantity does not depend on the system of units used if all variables that make up the quantity are expressed in a consistent set of units. To check this we will calculate the Reynolds number using BG units.

(b) We first convert all the SI values of the variables appearing in the Reynolds number to BG values by using the conversion factors from Table 1.4. Thus,

$$\rho = (910\ kg/m^3)(1.940 \times 10^{-3}) = 1.77\ slugs/ft^3$$

$$V = (2.6\ m/s)(3.281) = 8.53\ ft/s$$

$$D = (0.025\ m)(3.281) = 8.20 \times 10^{-2}\ ft$$

$$\mu = (0.38\ N·s/m^2)(2.089 \times 10^{-2}) = 7.94 \times 10^{-3}\ lb·s/ft^2$$

and the value of the Reynolds number is

$$Re = \frac{(1.77\ slugs/ft^3)(8.53\ ft/s)(8.20 \times 10^{-2}\ ft)}{7.94 \times 10^{-3}\ lb·s/ft^2}$$

$$= 156\ (slug·ft/s^2)/lb = 156 \tag{Ans}$$

since $1\ lb = 1\ slug·ft/s^2$. The values from part (a) and part (b) are the same, as expected. Dimensionless quantities play an important role in fluid mechanics and the significance of the Reynolds number as well as other important dimensionless combinations will be discussed in detail in Chapter 7. It should be noted that in the Reynolds number it is actually the ratio μ/ρ that is important, and this is the property that we have defined as the kinematic viscosity.

EXAMPLE 1.5

The velocity distribution for the flow of a Newtonian fluid between two wide, parallel plates (see Fig. E1.5) is given by the equation

$$u = \frac{3V}{2}\left[1 - \left(\frac{y}{h}\right)^2\right]$$

where V is the mean velocity. The fluid has a viscosity of 0.04 lb·s/ft². When $V = 2$ ft/s and $h = 0.2$ in. determine: (a) the shearing stress acting on the bottom wall, and (b) the shearing stress acting on a plane parallel to the walls and passing through the centerline (midplane).

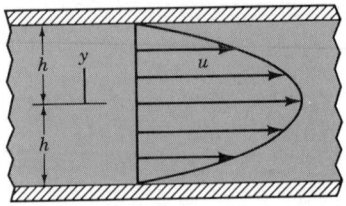

FIGURE E1.5

SOLUTION

For this type of parallel flow the shearing stress is obtained from Eq. 1.9,

$$\tau = \mu \frac{du}{dy} \tag{1}$$

Thus, if the velocity distribution, $u = u(y)$ is known, the shearing stress can be determined at all points by evaluating the velocity gradient, du/dy. For the distribution given

$$\frac{du}{dy} = -\frac{3Vy}{h^2} \tag{2}$$

(a)   Along the bottom wall $y = -h$ so that (from Eq. 2)

$$\frac{du}{dy} = \frac{3V}{h}$$

and therefore the shearing stress is

$$\tau_{\substack{\text{bottom}\\\text{wall}}} = \mu \left(\frac{3V}{h} \right) = \frac{(0.04 \text{ lb·s/ft}^2)(3)(2 \text{ ft/s})}{(0.2 \text{ in.})(1 \text{ ft}/12 \text{ in.})}$$

$$= 14.4 \text{ lb/ft}^2 \text{ (in direction of flow)} \qquad \text{(Ans)}$$

This stress creates a drag on the wall. Since the velocity distribution is symmetrical, the shearing stress along the upper wall would have the same magnitude and direction.

(b)   Along the midplane where $y = 0$ it follows from Eq. 2 that

$$\frac{du}{dy} = 0$$

and thus the shearing stress is

$$\tau_{\text{midplane}} = 0 \qquad \text{(Ans)}$$

From Eq. 2 we see that the velocity gradient (and therefore the shearing stress) varies linearly with y and in this particular example varies from 0 at the center of the channel to 14.4 lb/ft^2 at the walls. For the more general case the actual variation will, of course, depend on the nature of the velocity distribution.

1.7   COMPRESSIBILITY OF FLUIDS

1.7.1   Bulk Modulus

An important question to answer when considering the behavior of a particular fluid is how easily can the volume (and thus the density) of a given mass of the fluid be changed when there is a change in pressure? That is, how compressible is the fluid?

A property that is commonly used to characterize compressibility is the *bulk modulus*, E_v, defined as

$$E_v = -\frac{dp}{d\Psi/\Psi} \tag{1.12}$$

where dp is the differential change in pressure needed to create a differential change in volume, $d\Psi$, of a volume Ψ. The negative sign is included since an increase in pressure will cause a decrease in volume. Since a decrease in volume of a given mass, $m = \rho\Psi$, will result in an increase in density, Eq. 1.12 can also be expressed as

$$E_v = \frac{dp}{d\rho/\rho} \tag{1.13}$$

The bulk modulus (also referred to as the *bulk modulus of elasticity*) has dimensions of pressure, FL^{-2}. In BG units values for E_v are usually given as lb/in.^{2} (psi) and in SI units as N/m^2 (Pa). Large values for the bulk modulus indicate that the fluid is relatively incompressible—that is, it takes a large pressure change to create a small change in volume. As expected, values of E_v for common liquids are large (see Tables 1.5 and 1.6). For example, at atmospheric pressure and a temperature of 60 °F it would require a pressure of 3120 psi to compress a unit volume of water 1%. This result is representative of the compressibility of liquids. Since such large pressures are required to effect a change in volume, we conclude that liquids can be considered as *incompressible* for most practical engineering applications. As liquids are compressed the bulk modulus increases, but the bulk modulus near atmospheric pressure is usually the one of interest. The use of bulk modulus as a property describing compressibility is most prevalent when dealing with liquids, although the bulk modulus can also be determined for gases.

1.7.2 Compression and Expansion of Gases

When gases are compressed (or expanded) the relationship between pressure and density depends on the nature of the process. If the compression or expansion takes place under constant temperature conditions (*isothermal process*), then

$$\frac{p}{\rho} = \text{constant} \tag{1.14}$$

If the compression or expansion is frictionless and no heat is exchanged with the surroundings (*isentropic process*), then

$$\frac{p}{\rho^k} = \text{constant} \tag{1.15}$$

where k is the ratio of the specific heat at constant pressure, c_p, to the specific heat at constant volume, c_v (i.e., $k = c_p/c_v$). The two specific heats are related to the gas constant, R, through the equation $R = c_p - c_v$. As was the case for the ideal gas law, the pressure in both Eqs. 1.14 and 1.15 must be expressed as an absolute pressure. Values of k for some common gases are given in Tables 1.7 and 1.8, and for air over a range of temperatures, in Appendix B (Tables B.3 and B.4).

With explicit equations relating pressure and density the bulk modulus for gases can be determined by obtaining the derivative $dp/d\rho$ from Eq. 1.14 or 1.15 and substituting the results into Eq. 1.13. It follows that for an isothermal process

$$E_v = p \tag{1.16}$$

and for an isentropic process

$$E_v = kp \tag{1.17}$$

Note that in both cases the bulk modulus varies directly with pressure. For air under standard atmospheric conditions with $p = 14.7$ psi (abs) and $k = 1.40$, the isentropic bulk modulus is 20.6 psi. A comparison of this figure with that for water under the same conditions ($E_v = 312{,}000$ psi) shows that air is approximately 15,000 times as compressible as water. It is thus clear that in dealing with gases greater attention will need to be given to the effect of compressibility on fluid behavior. However, as will be discussed further in later sections, gases can often be treated as incompressible fluids if the changes in pressure are small.

EXAMPLE 1.6

A cubic foot of helium at an absolute pressure of 14.7 psi is compressed isentropically to $\frac{1}{2}$ ft^3. What is the final pressure?

SOLUTION

For an isentropic compression

$$\frac{p_i}{\rho_i^k} = \frac{p_f}{\rho_f^k}$$

where the subscripts i and f refer to initial and final states, respectively. Since we are interested in the final pressure, p_f, it follows that

$$p_f = \left(\frac{\rho_f}{\rho_i}\right)^k p_i$$

As the volume is reduced by one half, the density must double, since the mass of the gas remains constant. Thus,

$$p_f = (2)^{1.66}(14.7 \text{ psi}) = 46.5 \text{ psi (abs)} \tag{Ans}$$

1.7.3 Speed of Sound

Another important consequence of the compressibility of fluids is that disturbances introduced at some point in the fluid propagate at a finite velocity. For example, if a fluid is flowing in a pipe and a valve at the outlet is suddenly closed (thereby creating a localized disturbance) the effect of the valve closure is not felt instantaneously upstream. It takes a finite time for the increased pressure created by the valve closure to propagate to an upstream location. Similarly, a loud speaker diaphragm causes a localized disturbance as it vibrates, and the small change in pressure created by the motion of the diaphragm is propagated through the air with a finite velocity. The velocity at which these small disturbances propagate is called the *acoustic velocity* or the *speed of sound*, c. It will be shown in Chapter 11 that the speed of sound is related to changes in pressure and density of the fluid medium through the equation

$$c = \sqrt{\frac{dp}{d\rho}} \tag{1.18}$$

or in terms of the bulk modulus defined by Eq. 1.13

$$c = \sqrt{\frac{E_v}{\rho}} \tag{1.19}$$

Since the disturbance is small and moves rapidly there is negligible heat transfer and the process is assumed to be isentropic. Thus, the pressure–density relationship used in Eq. 1.18 should be that for an isentropic process.

For gases undergoing an isentropic process $E_v = kp$ (Eq. 1.17) so that

$$c = \sqrt{\frac{kp}{\rho}}$$

and making use of the ideal gas law, it follows that

$$c = \sqrt{kRT} \tag{1.20}$$

Thus, for ideal gases the speed of sound is proportional to the square root of the absolute temperature. For example, for air at 60 °F with $k = 1.40$ and $R = 1716$ ft·lb/slug·°R it follows that $c = 1117$ ft/s. The speed of sound in air at various temperatures can be found in Appendix B (Tables B.3 and B.4). Equation 1.19 is also valid for liquids and values of E_v can be used to determine the speed of sound in liquids. For water at 20 °C, $E_v = 2.19$ GN/m² and $\rho = 998.2$ kg/m³ so that $c = 1481$ m/s or 4860 ft/s. Note that the speed of sound in water is much higher than in air. If a fluid were truly incompressible ($E_v = \infty$) the speed of sound would be infinite. The speed of sound in water for various temperatures can be found in Appendix B (Tables B.1 and B.2).

EXAMPLE 1.7

A jet aircraft flies at a speed of 550 mph at an altitude of 35,000 ft, where the temperature is -66 °F. Determine the ratio of the speed of the aircraft, V, to that of the speed of sound, c, at the specified altitude. Assume $k = 1.40$.

SOLUTION

From Eq. 1.20 the speed of sound can be calculated as

$$c = \sqrt{kRT} = \sqrt{(1.40)(1716 \text{ ft·lb/slug·°R})(-66 + 460) \text{ °R}}$$
$$= 973 \text{ ft/s}$$

Since the air speed is

$$V = \frac{(550 \text{ mi/hr})(5280 \text{ ft/mi})}{(3600 \text{ s/hr})} = 807 \text{ ft/s}$$

the ratio is

$$\frac{V}{c} = \frac{807 \text{ ft/s}}{973 \text{ ft/s}} = 0.829 \tag{Ans}$$

This ratio is called the *Mach number*, Ma. If Ma < 1.0 the aircraft is flying at *subsonic* speeds, whereas for Ma > 1.0 it is flying at *supersonic* speeds. The Mach number is an important dimensionless parameter used in the study of the flow of gases at high speeds and will be further discussed in Chapters 7 and 11.

1.8   VAPOR PRESSURE

It is a common observation that liquids such as water and gasoline will evaporate if they are simply placed in a container open to the atmosphere. Evaporation takes place because some liquid molecules at the surface have sufficient momentum to overcome the intermolecular cohesive forces and escape into the atmosphere. If the container is closed with a small air space left above the surface, and this space evacuated to form a vacuum, a pressure will develop in the space as a result of the vapor that is formed by the escaping molecules. When an equilibrium condition is reached so that the number of molecules leaving the surface is equal to the number entering, the vapor is said to be saturated and the pressure the vapor exerts on the liquid surface is termed the *vapor pressure*. Since the development of a vapor pressure is closely associated with molecular activity, the value of vapor pressure for a particular liquid depends on temperature. Values of vapor pressure for water at various temperatures can be found in Appendix B (Tables B.1 and B.2), and values of vapor pressure for several common liquids at room temperatures are given in Tables 1.5 and 1.6.

Boiling, which is the formation of vapor bubbles within a fluid mass, is initiated when the absolute pressure in the fluid reaches the vapor pressure. As commonly observed in the kitchen, water at standard atmospheric pressure will boil when the temperature reaches 212 °F (100 °C)—that is, the vapor pressure of water at 212 °F is 14.7 psi (abs). However, if we attempt to boil water at a higher elevation, say 10,000 ft above sea level, where the atmospheric pressure is 10.1 psi (abs), we find that boiling will start when the temperature is about 193 °F. At this temperature the vapor pressure of water is 10.1 psi (abs). Thus, boiling can be induced at a given pressure acting on the fluid by raising the temperature, or at a given fluid temperature by lowering the pressure.

An important reason for our interest in vapor pressure and boiling lies in the common observation that in flowing fluids it is possible to develop very low pressure due to the fluid motion, and if the pressure is lowered to the vapor pressure, boiling will occur. For example, this phenomenon may occur in flow through the irregular, narrowed passages of a valve or pump. When vapor bubbles are formed in a flowing fluid they are swept along into regions of higher pressure where they suddenly collapse with sufficient intensity to actually cause structural damage. The formation and subsequent collapse of vapor bubbles in a flowing fluid, called *cavitation*, is an important fluid flow phenomenon to be given further attention in Chapters 3 and 7.

1.9   SURFACE TENSION

At the interface between a liquid and a gas, or between two immiscible liquids, forces develop in the liquid surface which cause the surface to behave as if it were a "skin" or "membrane" stretched over the fluid mass. Although such a skin is not actually present, this conceptual analogy allows us to explain several commonly observed phenomena. For example, a steel needle will float on water if placed gently on the surface because the tension developed in the hypothetical skin supports the needle. Small droplets of mercury will form into spheres when placed on a smooth surface because the cohesive forces in the surface tend to hold all the molecules together in a compact shape. Similarly, discrete water droplets will form when placed on a newly waxed surface. These various types of surface phenomena are due to the unbalanced cohesive forces acting on the liquid molecules at the fluid surface. Molecules in the

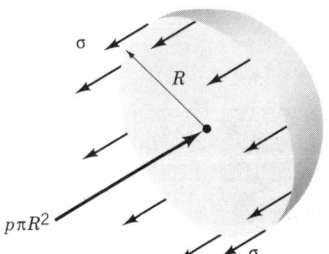

FIGURE 1.6 Forces acting on one-half of a liquid drop.

interior of the fluid mass are surrounded by molecules that are attracted to each other equally. However, molecules along the surface are subjected to a net force toward the interior. The apparent physical consequence of this unbalanced force along the surface is to create the hypothetical skin or membrane. A tensile force may be considered to be acting in the plane of the surface along any line in the surface. The intensity of the molecular attraction per unit length along any line in the surface is called the *surface tension* and is designated by the Greek symbol σ (sigma). Surface tension is a property of the liquid and depends on temperature as well as the other fluid it is in contact with at the interface. The dimensions of surface tension are FL^{-1} with BG units of lb/ft and SI units of N/m. Values of surface tension for some common liquids (in contact with air) are given in Tables 1.5 and 1.6 and in Appendix B (Tables B.1 and B.2) for water at various temperatures. The value of the surface tension decreases as the temperature increases.

The pressure inside a drop of fluid can be calculated using the free-body diagram in Fig. 1.6. If the spherical drop is cut in half (as shown) the force developed around the edge due to surface tension is $2\pi R\sigma$. This force must be balanced by the pressure, p, acting over the circular area, πR^2. Thus,

$$2\pi R\sigma = p\pi R^2$$

or

$$p = \frac{2\sigma}{R} \tag{1.21}$$

where p is the difference between the pressure inside the drop and the external pressure. It is apparent from this result that the pressure inside the drop is greater than the pressure surrounding the drop. (Would the pressure on the inside of a bubble of water be the same as that on the inside of a drop of water of the same diameter and at the same temperature?)

Among common phenomena associated with surface tension is the rise (or fall) of a liquid in a capillary tube. If a small open tube is inserted into water, the water level in the tube will rise above the water level outside the tube as is illustrated in Fig. 1.7*a*. In this situation we have a liquid–gas–solid interface. For the case illustrated there is an attraction (adhesion) between the wall of the tube and liquid molecules which is strong enough to overcome the mutual attraction (cohesion) of the molecules and pull them up the wall. Hence, the liquid is said to *wet* the solid surface.

The height, h, is governed by the value of the surface tension, σ, the tube radius, R, the specific weight of the liquid, γ, and the *angle of contact*, θ, between the fluid and tube. From the free-body diagram of Fig. 1.7*b* we see that the vertical force due

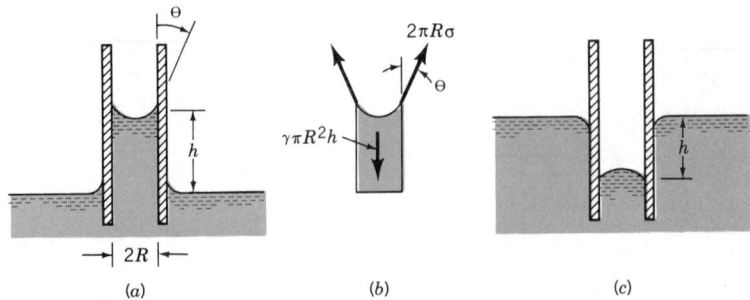

FIGURE 1.7 Effect of capillary action in small tubes. (*a*) Rise of column for a liquid that wets the tube. (*b*) Free-body diagram for calculating column height. (*c*) Depression of column for a nonwetting liquid.

to the surface tension is equal to $2\pi R\sigma \cos\theta$ and the weight is $\gamma\pi R^2 h$ and these two forces must balance for equilibrium. Thus,

$$\gamma\pi R^2 h = 2\pi R\sigma \cos\theta$$

so that the height is given by the relationship

$$h = \frac{2\sigma \cos\theta}{\gamma R} \tag{1.22}$$

The angle of contact is a function of both the liquid and the surface. For water in contact with clean glass $\theta \approx 0°$. It is clear from Eq. 1.22 that the height is inversely proportional to the tube radius, and therefore the rise of a liquid in a tube as a result of capillary action becomes increasingly pronounced as the tube radius is decreased.

EXAMPLE 1.8

Pressures are sometimes determined by measuring the height of a column of liquid in a vertical tube. What diameter of clean glass tubing is required so that the rise of water at 20 °C in a tube due to capillary action is less than 1.0 mm?

SOLUTION

From Eq. 1.22

$$h = \frac{2\sigma \cos\theta}{\gamma R}$$

so that

$$R = \frac{2\sigma \cos\theta}{\gamma h}$$

For water at 20 °C (from Table B.2) $\sigma = 0.0728$ N/m and $\gamma = 9.789$ kN/m^3. Since $\theta \approx 0°$ it follows that for $h = 1.0$ mm,

$$R = \frac{2(0.0728 \text{ N/m})(1)}{(9.789 \times 10^3 \text{ N/m}^3)(1.0 \text{ mm})(10^{-3} \text{ m/mm})} = 0.0149 \text{ m}$$

and the minimum required tube diameter, D, is

$$D = 2R = 0.0298 \text{ m} = 29.8 \text{ mm} \qquad \text{(Ans)}$$

If adhesion of molecules to the solid surface is weak compared to the cohesion between molecules, the liquid will not wet the surface and the level in a tube placed in a nonwetting liquid will actually be depressed as shown in Fig. 1.7c. Mercury is a good example of a nonwetting liquid when it is in contact with a glass tube. For nonwetting liquids the angle of contact is greater than 90°, and for mercury in contact with clean glass $\theta \approx 130°$.

Surface tension effects play a role in many fluid mechanics problems including the movement of liquids through soil and other porous media, flow of thin films, formation of drops and bubbles, and the breakup of liquid jets. Surface phenomena associated with liquid–gas, liquid–liquid, liquid–gas–solid interfaces are exceedingly complex, and a more detailed and rigorous discussion of them is beyond the scope of this text. Fortunately, in many fluid mechanics problems surface phenomena, as characterized by surface tension, are not important, since inertial, gravitational, and viscous forces are much more dominant.

1.10   A BRIEF LOOK BACK IN HISTORY

Before proceeding with our study of fluid mechanics, we should pause for a moment to consider the history of this important engineering science. As is true of all basic scientific and engineering disciplines, their actual beginnings are only faintly visible through the haze of early antiquity. But, we know that interest in fluid behavior dates back to the ancient civilizations. Through necessity there was a practical concern about the manner in which spears and arrows could be propelled through the air, in the development of water supply and irrigation systems, and in the design of boats and ships. These developments were of course based on trial and error procedures without any knowledge of mathematics or mechanics. However, it was the accumulation of such empirical knowledge that formed the basis for further development during the emergence of the ancient Greek civilization and the subsequent rise of the Roman Empire. Some of the earliest writings that pertain to modern fluid mechanics are those of Archimedes (287–212 B.C.), a Greek mathematician and inventor who first expressed the principles of hydrostatics and flotation. Elaborate water supply systems were built by the Romans during the period from the fourth century B.C. through the early Christian period, and Sextus Julius Frontinus (A.D. 40–103), a Roman engineer, described these systems in detail. However, for the next 1000 years during the Middle Ages (also referred to as the Dark Ages), there appears to have been little added to further understanding of fluid behavior.

Beginning with the Renaissance period (about the fifteenth century) a rather continuous series of contributions began that forms the basis of what we consider to be the science of fluid mechanics. Leonardo da Vinci (1452–1519) described through sketches and writings many different types of flow phenomena. The work of Galileo Galilei (1564–1642) marked the beginning of experimental mechanics. Following the early Renaissance period and during the seventeenth and eighteenth centuries, numerous significant contributions were made. These include theoretical and mathematical advances associated with the famous names of Newton, Bernoulli, Euler, and d'Alembert. Experimental aspects of fluid mechanics were also advanced during this period, but unfortunately the two different approaches, theoretical and experimental, developed along separate paths. *Hydrodynamics* was the term associated with the theoretical or mathematical study of idealized, frictionless fluid behavior, with the term *hydraulics* being used to describe the applied or experimental aspects of real fluid behavior, particularly the behavior of water. Further contributions and refinements were made to both theoretical hydrodynamics and experimental hydraulics during the nineteenth century, with the general differential equations describing fluid motions that are used in modern fluid mechanics being developed in this period. Experimental hydraulics became more of a science, and many of the results of experiments performed during the nineteenth century are still used today.

At the beginning of the twentieth century both the fields of theoretical hydrodynamics and experimental hydraulics were highly developed, and attempts were being made to unify the two. In 1904 a classic paper was presented by a German professor, Ludwig Prandtl (1857–1953), who introduced the concept of a "fluid boundary layer," which laid the foundation for the unification of the theoretical and experimental aspects of fluid mechanics. Prandtl's idea was that for flow next to a solid boundary a thin fluid layer (boundary layer) develops in which friction is very important, but outside this layer the fluid behaves very much like a frictionless fluid. This relatively simple concept provided the necessary impetus for the resolution of the conflict between the hydrodynamicists and the hydraulicists. Prandtl is generally accepted as the founder of modern fluid mechanics.

Also, during the first decade of the twentieth century, powered flight was first successfully demonstrated with the subsequent vastly increased interest in *aerodynamics*. Because the design of aircraft required a degree of understanding of fluid flow and an ability to make accurate predictions of the effect of air flow on bodies, the field of aerodynamics provided a great stimulus for the many rapid developments in fluid mechanics that have taken place during the twentieth century.

As we proceed with our study of the fundamentals of fluid mechanics, we will continue to note the contributions of many of the pioneers in the field. Table 1.9 provides a chronological listing of some of these contributors and reveals the long journey that makes up the history of fluid mechanics. This list is certainly not comprehensive with regard to all of the past contributors, but includes those who are mentioned in this text. As mention is made in succeeding chapters of the various individuals listed in Table 1.9, a quick glance at this table will reveal where they fit into the historical chain.

It is, of course, impossible to summarize the rich history of fluid mechanics in a few paragraphs. Only a brief glimpse is provided, and we hope it will stir your interest. References 2 to 5 are good starting points for further study, and in particular Ref. 2 provides an excellent, broad, easily read history. Try it—you might even enjoy it!

REFERENCES

1. Reid, R. C., Prausnitz, J. M., and Sherwood, T. K., *The Properties of Gases and Liquids,* 3rd Ed., McGraw-Hill, New York, 1977.

2. Rouse, H. and Ince, S., *History of Hydraulics,* Iowa Institute of Hydraulic Research, Iowa City, 1957, Dover, New York, 1963.

3. Tokaty, G. A., *A History and Philosophy of Fluidmechanics,* G. T. Foulis and Co., Ltd., Oxfordshire, Great Britain, 1971.

4. Rouse, H., *Hydraulics in the United States 1776–1976,* Iowa Institute of Hydraulic Research, Iowa City, Iowa, 1976.

5. Garbrecht, G., ed., *Hydraulics and Hydraulic Research—A Historical Review,* A. A. Balkema, Rotterdam, Netherlands, 1987.

TABLE 1.9 Chronological Listing of Some Contributors to the Science of Fluid Mechanics Noted in Text[a]

ARCHIMEDES (287–212 B.C.)
Established elementary principles of buoyancy and flotation.

SEXTUS JULIUS FRONTINUS
(40–103 A.D.)
Wrote treatise on Roman methods of water distribution.

LEONARDO da VINCI (1452–1519)
Expressed elementary principle of continuity; observed and sketched many basic flow phenomena; suggested designs for hydraulic machinery.

GALILEO GALILEI (1564–1642)
Indirectly stimulated experimental hydraulics; revised Aristotelian concept of vacuum.

EVANGELISTA TORRICELLI
(1608–1647)
Related barometric height to weight of atmosphere, and form of liquid jet to trajectory of free fall.

BLAISE PASCAL (1623–1662)
Finally clarified principles of barometer, hydraulic press, and pressure transmissibility.

ISAAC NEWTON (1642–1727)
Explored various aspects of fluid resistance—inertial, viscous, and wave; discovered jet contraction.

HENRI de PITOT (1695–1771)
Constructed double-tube device to indicate water velocity through differential head.

DANIEL BERNOULLI (1700–1782)
Experimented and wrote on many phases of fluid motion, coining name "hydrodynamics"; devised manometry technique and adapted primitive energy principle to explain velocity–head indication; proposed jet propulsion.

LEONHARD EULER (1707–1783)
First explained role of pressure in fluid flow; formulated basic equations of motion and so-called Bernoulli theorem; introduced concept of cavitation, and principle of centrifugal machinery.

JEAN le ROND d'ALEMBERT
(1717–1783)
Originated notion of velocity and acceleration components, differential expression of continuity, and paradox of zero resistance to steady nonuniform motion.

ANTOINE CHEZY (1718–1798)
Formulated similarity parameter for predicting flow characteristics of one channel from measurements on another.

GIOVANNI BATTISTA VENTURI
(1746–1822)
Performed tests on various forms of mouthpieces—in particular, conical contractions and expansions.

LOUIS MARIE HENRI NAVIER
(1785–1836)
Extended equations of motion to include "molecular" forces.

AUGUSTIN LOUIS de CAUCHY
(1789–1857)
Contributed to the general field of theoretical hydrodynamics and to the study of wave motion.

GOTTHILF HEINRICH LUDWIG HAGEN (1797–1884)
Conducted original studies of resistance in and transition between laminar and turbulent flow.

JEAN LOUIS POISEUILLE (1799–1869)
Performed meticulous tests on resistance of flow through capillary tubes.

HENRI PHILIBERT GASPARD DARCY
(1803–1858)
Performed extensive tests on filtration and pipe resistance; initiated open-channel studies carried out by Bazin.

JULIUS WEISBACH (1806–1871)
Incorporated hydraulics in treatise on engineering mechanics, based on original experiments; noteworthy for flow patterns, nondimensional coefficients, weir, and resistance equations.

TABLE 1.9 (continued)

WILLIAM FROUDE (1810–1879)
Developed many towing-tank techniques, in particular the conversion of wave and boundary layer resistance from model to prototype scale.

ROBERT MANNING (1816–1897)
Proposed several formulas for open-channel resistance.

GEORGE GABRIEL STOKES (1819–1903)
Derived analytically various flow relationships ranging from wave mechanics to viscous resistance—particularly that for the settling of spheres.

ERNST MACH (1838–1916)
One of the pioneers in the field of supersonic aerodynamics.

OSBORNE REYNOLDS (1842–1912)
Described original experiments in many fields—cavitation, river model similarity, pipe resistance—and devised two parameters for viscous flow; adapted equations of motion of a viscous fluid to mean conditions of turbulent flow.

JOHN WILLIAM STRUTT, LORD RAYLEIGH (1842–1919)
Investigated hydrodynamics of bubble collapse, wave motion, jet instability, laminar flow analogies, and dynamic similarity.

VINCENZ STROUHAL (1850–1922)
Investigated the phenomenon of "singing wires."

EDGAR BUCKINGHAM (1867–1940)
Stimulated interest in the United States in the use of dimensional analysis.

MORITZ WEBER (1871–1951)
Emphasized the use of the principles of similitude in fluid flow studies and formulated a capillarity similarity parameter.

LUDWIG PRANDTL (1875–1953)
Introduced concept of the boundary layer and is generally considered to be the father of present-day fluid mechanics.

LEWIS FERRY MOODY (1880–1953)
Provided many innovations in the field of hydraulic machinery. Proposed a method of correlating pipe resistance data which is widely used.

THEODOR VON KÁRMÁN (1881–1963)
One of the recognized leaders of twentieth century fluid mechanics. Provided major contributions to our understanding of surface resistance, turbulence, and wake phenomena.

PAUL RICHARD HEINRICH BLASIUS (1883–1970)
One of Prandtl's students who provided an analytical solution to the boundary layer equations. Also, demonstrated that pipe resistance was related to the Reynolds number.

---

[a] Adapted from Ref. 2; used by permission of the Iowa Institute of Hydraulic Research, The University of Iowa.

PROBLEMS

Note: Unless specific values of required fluid properties are given in the statement of the problem, use the values found in the tables on the inside of the front cover. Problems designated with an (*) are intended to be solved with the aid of a programmable calculator or a computer.

1.1 Determine the dimensions, in both the *FLT* system and the *MLT* system, for **(a)** the product of force times acceleration, **(b)** the product of force times velocity divided by area, and **(c)** momentum divided by volume.

1.2 Verify the dimensions, in both the *FLT* and *MLT* systems, of the following quantities which appear in Table 1.1: **(a)** angular velocity, **(b)** energy, **(c)** moment of inertia (area), **(d)** power, and **(e)** pressure.

1.3 Verify the dimensions, in both the *FLT* system and the *MLT* system, of the following quantities which appear in Table 1.1: **(a)** frequency, **(b)** stress, **(c)** strain, **(d)** torque, and **(e)** work.

1.4 If P is a force and x a length, what are the dimensions (in the FLT system) of **(a)** dP/dx, **(b)** d^3P/dx^3, and **(c)** $\int P \, dx$?

1.5 If u is a velocity, x a length, and t a time, what are the dimensions (in the MLT system) of **(a)** $\partial u/\partial t$, **(b)** $\partial^2 u/\partial x \partial t$, and **(c)** $\int (\partial u/\partial t) \, dx$?

1.6 If V is a velocity, ℓ a length, and v a fluid property having dimensions of $L^2 T^{-1}$, which of the following combinations are dimensionless: **(a)** $V\ell v$, **(b)** $V\ell/v$, **(c)** $V^2 v$, **(d)** $V/\ell v$?

1.7 Dimensionless combinations of quantities (commonly called dimensionless parameters) play an important role in fluid mechanics. Make up five possible dimensionless parameters by using combinations of some of the quantities listed in Table 1.1.

1.8 If V is a velocity, ℓ a length, W a weight, and μ a fluid property having dimensions of $FL^{-2}T$, determine the dimensions of: **(a)** $V\ell W/\mu$, **(b)** $W\mu\ell$, **(c)** $V\mu/\ell$, and **(d)** $V\ell^2\mu/W$.

1.9 The force, P, that is exerted on a spherical particle moving slowly through a liquid is given by the equation

$$P = 3\pi\mu DV$$

where μ is a fluid property (viscosity) having dimensions of $FL^{-2}T$, D is the particle diameter, and V is the particle velocity. What are the dimensions of the constant, 3π? Would you classify this equation as a general homogeneous equation?

1.10 Determine the dimensions of the coefficients A and B which appear in the dimensionally homogeneous equation

$$\frac{d^2x}{dt^2} + A\frac{dx}{dt} + Bx = 0$$

where x is a length and t is time.

1.11 The pressure difference, Δp, across a partial blockage in an artery (called a *stenosis*) is approximated by the equation

$$\Delta p = K_v \frac{\mu V}{D} + K_u \left(\frac{A_0}{A_1} - 1\right)^2 \rho V^2$$

where V is the blood velocity, μ the blood viscosity ($FL^{-2}T$), ρ the blood density (ML^{-3}), D the artery diameter, A_0 the area of the unobstructed artery, and A_1 the area of the stenosis. Determine the dimensions of the constants K_v and K_u. Would this equation be valid in any system of units?

1.12 During a study of a certain flow system the following equation relating the pressures p_1 and p_2 at two points was developed:

$$p_2 = p_1 + \frac{f\ell V}{Dg}$$

In this equation V is a velocity, ℓ the distance between the two points, D a diameter, g the acceleration of gravity, and f a dimensionless coefficient. Is the equation dimensionally consistent?

1.13 A formula to estimate the volume rate of flow, Q, flowing over a dam of length, B, is given by the equation

$$Q = 3.09BH^{3/2}$$

where H is the depth of the water above the top of the dam (called the head). This formula gives Q in ft^3/s when B and H are in feet. Is the constant, 3.09, dimensionless? Would this equation be valid if units other than feet and seconds were used?

1.14 Cite an example of a restricted homogeneous equation contained in a technical article found in an engineering journal in your field of interest. Define all terms in the equation, explain why it is a restricted equation, and provide a complete journal citation (title, date, etc.).

1.15 Make use of Table 1.3 to express the following quantities in SI units: **(a)** 10.2 in./min, **(b)** 4.81 slugs, **(c)** 3.02 lb, **(d)** 73.1 ft/s^2, **(e)** 0.0234 lb·s/ft^2.

1.16 Make use of Table 1.4 to express the following quantities in BG units: **(a)** 14.2 km, **(b)** 8.14 N/m^3, **(c)** 1.61 kg/m^3, **(d)** 0.0320 N·m/s, **(e)** 5.67 mm/hr.

1.17 Make use of Appendix A to express the following quantities in SI units: **(a)** 160 acre, **(b)** 742 Btu, **(c)** 240 miles, **(d)** 79.1 hp, **(e)** 60.3 °F.

1.18 Make use of Table 1.4 to express the following quantities in BG units: **(a)** 465 W, **(b)** 92.1 J, **(c)** 536 N/m^2, **(d)** 85.9 mm^3, **(e)** 386 kg/m^2.

1.19 For Table 1.3 verify the conversion relationships for: **(a)** area, **(b)** density, **(c)** velocity, and **(d)** specific weight. Use the basic conversion relationships: 1 ft = 0.3048 m; 1 lb = 4.4482 N; and 1 slug = 14.594 kg.

1.20 For Table 1.4 verify the conversion relationships for: **(a)** acceleration, **(b)** density, **(c)** pressure, and **(d)** volume flowrate. Use the basic conversion relationships: 1 m = 3.2808 ft; 1 N = 0.22481 lb; and 1 kg = 0.068521 slug.

1.21   A person weighs 180 lb at the earth's surface. Determine the person's mass in slugs, kilograms, and pounds mass.

1.22   Water flows from a large drainage pipe at a rate of 1200 gal/min. What is this volume rate of flow in m^3/s and in liters/min?

1.23   A tank of oil has a mass of 20 slugs. **(a)** Determine its weight in pounds and in newtons at the earth's surface. **(b)** What would be its mass (in slugs) and its weight (in pounds) if located on the moon's surface where the gravitational attraction is approximately one-sixth that at the earth's surface?

1.24   A certain object weighs 400 N at the earth's surface. Determine the mass of the object (in kilograms) and its weight (in newtons) when located on a planet with an acceleration of gravity equal to 4.0 ft/s^2.

1.25   An important dimensionless parameter in certain types of fluid flow problems is the *Froude number* defined as $V/\sqrt{g\ell}$, where V is a velocity, g the acceleration of gravity, and ℓ a length. Determine the value of the Froude number for $V = 10$ ft/s, $g = 32.2$ ft/s^2, and $\ell = 2$ ft. Recalculate the Froude number using SI units for V, g, and ℓ. Explain the significance of the results of these calculations.

1.26   The specific weight of a certain liquid is 70.3 lb/ft^3. Determine its density and specific gravity.

1.27   The density of a certain type of jet fuel is 775 kg/m^3. Determine its specific gravity and specific weight.

1.28   Make use of Fig. 1.1 to determine the specific gravity of water at 22 and 89 °C. What is the specific volume of water at these two temperatures?

1.29   The specific gravity of mercury at 80 °C is 13.4. Determine its density and specific weight at this temperature. Express your answer in both BG and SI units.

1.30   A liquid when poured into a graduated cylinder is found to weigh 6 N when occupying a volume of 500 ml (milliliters). Determine its specific weight, density, and specific gravity.

1.31   A beaker contains 10 $in.^3$ of pure glycerin. If 2 $in.^3$ of water is added to the glycerine, what is the specific gravity of the mixture?

1.32*   The variation in the density of water, ρ, with temperature, T, in the range 20 °C $\leq T \leq$ 60 °C, is given in the following table.

| Density (kg/m^3) | 998.2 | 997.1 | 995.7 | 994.1 | 992.2 | 990.2 | 988.1 |
| Temperature (°C) | 20 | 25 | 30 | 35 | 40 | 45 | 50 |

Use these data to determine an empirical equation of the form $\rho = c_1 + c_2T + c_3T^2$ which can be used to predict the density over the range indicated. Compare the predicted values with the data given. What is the density of water at 42.1 °C?

1.33   A 1-ft-diameter cylindrical tank that is 5 ft long weighs 125 lb and is filled with a liquid having a specific weight of 69.6 lb/ft^3. Determine the vertical force required to give the tank an upward acceleration of 12 ft/s^2.

1.34   Calculate the density and specific weight of air at a gage pressure of 100 psi and a temperature of 100 °F. Assume standard atmospheric pressure.

1.35   The density of oxygen contained in a tank is 2.0 kg/m^3 when the temperature is 25 °C. Determine the gage pressure of the gas if the atmospheric pressure is 99 kPa.

1.36   Nitrogen is compressed to a density of 4 kg/m^3 under an absolute pressure of 400 kPa. Determine the temperature in degrees Celsius.

1.37   The molecular weight of carbon monoxide is 28.01. Determine the density of this gas when the absolute pressure is 50 psi and the temperature is 150 °F.

1.38   The molecular weight of chlorine gas is 70.9. Calculate the specific weight of this gas at a gage pressure of 10 psi and a temperature of 60 °F.

1.39   A tire having a volume of 3 ft^3 contains air at a gage pressure of 26 psi and a temperature of 70 °F. Determine the density of the air and the weight of the air contained in the tire.

1.40   A compressed air tank contains 5 kg of air at a temperature of 80 °C. A gage on the tank reads 300 kPa. Determine the volume of the tank.

1.41   A large dirigible having a volume of 90,000 m^3 contains helium under standard atmospheric conditions [pressure = 101 kPa (abs) and temperature = 15 °C]. Determine the density and total weight of the helium.

1.42*   Develop a computer program for calculating the density of an ideal gas when the gas pressure in pascals (abs), the temperature in degrees Celsius, and the gas constant in J/kg·K are specified.

1.43*   Repeat Problem 1.42 for the case in which the pressure is given in psi (gage), the temperature in degrees Fahrenheit, and the gas constant in ft·lb/slug·°R.

1.44   What is the dynamic viscosity of glycerin at 85 °F? Express your answer in both SI and BG units.

1.45   Determine the ratio of the dynamic viscosity of water to air at a temperature of 60 °C. Compare this value with the corresponding ratio of kinematic viscosities. Assume the air is at standard atmospheric pressure.

1.46   The kinematic viscosity and specific gravity of a liquid are 3.5×10^{-4} m²/s and 0.82, respectively. What is the dynamic viscosity of the liquid in SI units?

1.47   A liquid has a specific weight of 52 lb/ft³ and a dynamic viscosity of 2.75 lb·s/ft². Determine its kinematic viscosity.

1.48   The viscosity of a certain fluid is 4×10^{-4} poise. Determine its viscosity in both SI and BG units.

1.49   The kinematic viscosity of oxygen at 20 °C and a pressure of 150 kPa (abs) is 0.104 stokes. Determine the dynamic viscosity of oxygen at this temperature and pressure.

1.50   Determine from Figs. B.1 and B.2 in Appendix B the dynamic and kinematic viscosity of kerosene and mercury at 40 °C. Express your answer in both SI and BG units.

1.51   SAE 30 oil at 60 °F flows through a 2-in.-diameter pipe with a mean velocity of 7 ft/s. Determine the value of the Reynolds number (see Example 1.4).

1.52   Calculate the Reynolds numbers for the flow of water and for air through a 3-mm-diameter tube, if the mean velocity is 2 m/s and the temperature is 30 °C in both cases (see Example 1.4). Assume the air is at standard atmospheric pressure.

1.53   For air at standard atmospheric pressure the values of the constants that appear in the Sutherland equation (Eq. 1.10) are $C = 1.458 \times 10^{-6}$ kg/(m·s·K$^{1/2}$) and $S = 110.4$ K. Use these values to predict the viscosity of air at 10 °C and 90 °C and compare with values given in Table B.4 in Appendix B.

1.54*   Use the values of viscosity of air given in Table B.4 at temperatures of 0, 20, 40, 60, 80, and 100 °C to determine the constants C and S which appear in the Sutherland equation (Eq. 1.10). Compare your results with the values given in Problem 1.53. (*Hint*: Rewrite the equation in

the form

$$\frac{T^{3/2}}{\mu} = \left(\frac{1}{C}\right)T + \frac{S}{C}$$

and plot $T^{3/2}/\mu$ versus T. From the slope and intercept of this curve C and S can be obtained.)

1.55   For a certain liquid $\mu = 7.1 \times 10^{-5}$ lb·s/ft² at 40 °F and $\mu = 1.9 \times 10^{-5}$ lb·s/ft² at 150 °F. Make use of these data to determine the constants D and B which appear in Andrade's equation (Eq. 1.11). What would be the viscosity at 80 °F?

1.56*   Use the value of the viscosity of water given in Table B.2 at temperatures of 0, 20, 40, 60, 80, and 100 °C to determine the constants D and B which appear in Andrade's equation (Eq. 1.11). Calculate the value of the viscosity at 50 °C and compare with the value given in Table B.2. (*Hint*: Rewrite the equation in the form

$$\ln \mu = (B)\frac{1}{T} + \ln D$$

and plot $\ln \mu$ versus $1/T$. From the slope and intercept of this curve B and D can be obtained. If a nonlinear curve fitting program is available the constants can be obtained directly from Eq. 1.11 without rewriting the equation.)

1.57   Crude oil having a viscosity of 9.52×10^{-4} lb·s/ft² is contained between parallel plates. The bottom plate is fixed and upper plate moves when a force P is applied (see Fig. 1.3). If the distance between the two plates is 0.2 in. what value of P is required to translate the plate with a velocity of 3 ft/s? The effective area of the upper plate is 200 in.²

1.58   A Newtonian fluid having a specific gravity of 0.92 and a kinematic viscosity of 5×10^{-4} m²/s flows past a fixed surface. The velocity profile near the surface is shown in Fig. P1.58. Determine the magnitude and direction of the shearing stress developed on the plate. Express your answer in terms of U and δ, with U and δ expressed in units of meters per second and meters, respectively.

$$\frac{u}{U} = \frac{3}{2}\frac{y}{\delta} - \frac{1}{2}\left(\frac{y}{\delta}\right)^3$$

FIGURE P1.58

1.59 Solve Problem 1.58 if the velocity profile is given by the equation

$$\frac{u}{U} = \sin\left(\frac{\pi}{2}\frac{y}{\delta}\right)$$

1.60 For a parallel plate arrangement of the type shown in Fig. 1.3 it is found that when the distance between plates is 2 mm, a shearing stress of 150 Pa develops at the upper plate when it is pulled at a velocity of 1 m/s. Determine the viscosity of the fluid between the plates. Express your answer in SI units.

1.61 A 10-kg block slides down a smooth inclined surface as shown in Fig. P1.61. Determine the terminal velocity of the block if the 0.1-mm gap between the block and the surface contains SAE 30 oil at 60 °F. Assume the velocity distribution in the gap is linear, and the area of the block in contact with the oil is 0.1 m².

FIGURE P1.61

1.62 A solid cube measuring 0.5 ft per side and weighing 100 lb slides down a smooth surface which is inclined 30° from the horizontal. The block slides on a film of oil having a viscosity of 1.71×10^{-2} lb·s/ft². If the terminal velocity of the block is 1.5 ft/s, what is the film thickness? Assume a linear velocity distribution in the film.

1.63 A layer of water flows down an inclined fixed surface with the velocity profile shown in Fig. P1.63. Determine the magnitude and direction of the shearing stress that the water exerts on the fixed surface for $U = 2$ m/s and $h = 0.1$ m.

FIGURE P1.63

1.64 A large movable plate is located between two large fixed plates as shown in Fig. P1.64. Two Newtonian fluids having the viscosities indicated are contained between the plates. Determine the magnitude and direction of the shearing stresses that act on the fixed walls when the moving plate has a velocity of 4 m/s as shown. Assume that the velocity distribution between the plates is linear.

FIGURE P1.64

1.65* Standard air flows past a flat surface and velocity measurements near the surface indicate the following distribution:

y (ft)	0.005	0.01	0.02	0.04	0.06	0.08
u (ft/s)	0.74	1.51	3.03	6.37	10.21	14.43

The coordinate y is measured normal to the surface and u is the velocity parallel to the surface. **(a)** Assume the velocity distribution is of the form

$$u = C_1 y + C_2 y^3$$

and use a standard curve-fitting technique to determine the constants C_1 and C_2. **(b)** Make use of the results of part (a) to determine the magnitude of the shearing stress at the wall ($y = 0$) and at $y = 0.05$ ft.

1.66 The viscosity of liquids can be measured through the use of a *rotating cylinder viscometer* of the type illustrated in Fig. P1.66. In this device

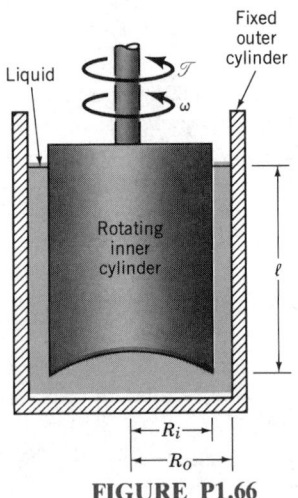

FIGURE P1.66

the outer cylinder is fixed and the inner cylinder is rotated with an angular velocity, ω. The torque $\mathcal{T}$ required to develop ω is measured and the viscosity is calculated from these two measurements. Develop an equation relating μ, ω, $\mathcal{T}$, ℓ, R_o and R_i. Neglect end effects and assume the velocity distribution in the gap is linear.

1.67 The space between two 6-in. long concentric cylinders is filled with glycerin (viscosity $= 8.5 \times 10^{-3}$ lb·s/ft²). The inner cylinder has a radius of 3 in. and the gap width between cylinders is 0.1 in. Determine the torque required to rotate the inner cylinder at 180 rev/min. The outer cylinder is fixed. Assume the velocity distribution in the gap to be linear.

1.68 Determine the torque required to rotate a 50-mm-diameter vertical cylinder at a constant angular velocity of 30 rad/s inside a fixed outer cylinder that has a diameter of 50.2 mm. The gap between the cylinders is filled with SAE 10 oil at 20 °C. The length of the inner cylinder is 200 mm. Neglect bottom effects and asume the velocity distribution in the gap is linear. If the temperature of the oil increases to 80 °C what will be the percentage change in the torque?

1.69* The following torque–angular velocity data were obtained with a rotating cylinder viscometer of the type described in Problem 1.66.

Torque (ft·lb)	13.1	26.0	39.5	52.7	64.9	78.6
Angular velocity (rad/s)	1.0	2.0	3.0	4.0	5.0	6.0

For this viscometer $R_o = 2.50$ in., $R_i = 2.45$ in., and $\ell = 5.00$ in. Make use of these data and a standard curve-fitting program to determine the viscosity of the liquid contained in the viscometer.

1.70 A 25-mm-diameter shaft is pulled through a cylindrical bearing as shown in Fig. P1.70. The lubricant that fills the 0.3-mm gap between the shaft and bearing is an oil having a kinematic viscosity of 8.0×10^{-4} m²/s and a specific gravity of 0.91. Determine the force P required to pull the shaft at a velocity of 3 m/s. Assume the velocity distribution in the gap is linear.

FIGURE P1.70

1.71 A 1-m³ volume of water is contained in a rigid container. Estimate the change in the volume of the water when a piston applies a pressure of 35 MPa.

1.72 Estimate the increase in pressure (in psi) required to decrease a unit volume of mercury by 0.1%.

1.73 In a test to determine the bulk modulus of a liquid it was found that as the absolute pressure was changed from 15 to 3000 psi the volume decreased from 10.240 to 10.138 in.³ Determine the bulk modulus for this liquid.

1.74 Calculate the speed of sound in m/s for **(a)** gasoline, **(b)** mercury, and **(c)** seawater.

1.75 Air is enclosed by a rigid cylinder containing a piston. A pressure gage attached to the cylinder indicates an initial reading of 20 psi. Determine the reading on the gage when the piston has compressed the air to one-third its original volume. Assume the compression process to be isothermal and the local atmospheric pressure to be 14.7 psi.

1.76 Repeat Problem 1.75 if the compression process takes place without friction and without heat transfer (isentropic process).

1.77 Oxygen at 30 °C and 300 kPa absolute pressure expands isothermally to an absolute pressure of 170 kPa. Determine the final density of the gas.

1.78 Natural gas at 60 °F and standard atmospheric pressure of 14.7 psi is compressed isentropically to a new absolute pressure of 60 psi. Determine the final density and temperature of the gas.

1.79 Compare the isentropic bulk modulus of air at 101 kPa (abs) with that of water at the same pressure.

1.80 What is the isothermal bulk modulus of nitrogen at a temperature of 80 °F and an absolute pressure of 5600 lb/ft²?

1.81* Develop a computer program for calculating the final gage pressure of gas when the initial gage pressure, initial and final volumes, atmospheric pressure, and the type of process (isothermal or isentropic) are specified. Use BG units. Check your program against the results obtained for Problem 1.75.

1.82 Determine the speed of sound at 20 °C in **(a)** air, **(b)** helium, and **(c)** natural gas. Express your answer in m/s.

1.83 Compare the speed of sound in mercury and oxygen at 20 °C.

1.84   At a certain altitude it was found that water boils at 90 °C. What is the atmospheric pressure at this altitude?

1.85   When a fluid flows through a sharp bend low pressures may develop in localized regions of the bend. Estimate the minimum absolute pressure (in psi) that can develop without causing cavitation if the fluid is water at 140 °F.

1.86   Estimate the minimum absolute pressure (in pascals) that can be developed at the inlet of a pump to avoid cavitation if the fluid is ethyl alcohol at 20 °C.

1.87   At what atmospheric pressure will water boil at 35 °C? Express your answer in both SI and BG units.

1.88   A partially filled closed tank contains ethyl alcohol at 68 °F. If the air above the alcohol is evacuated what is the minimum absolute pressure that develops in the evacuated space?

1.89   Estimate the excess pressure inside a rain drop having a diameter of 2 mm.

1.90   Small droplets of carbon tetrachloride at 68 °F are formed with a spray nozzle. If the average diameter of the droplets is 200 μm what is the difference in pressure between the inside and outside of the droplets?

1.91   What is the difference between the pressure inside a soap bubble and atmospheric pressure for a 3-in.-diameter bubble? Assume the surface tension of the soap film to be 70% of that of water at 80 °F.

1.92   Determine the height water at 60 °F will rise due to capillary action in a clean, $\frac{1}{4}$-in.-diameter tube. What will be the height if the diameter is reduced to 0.01 in.?

1.93   Two vertical, parallel, clean, glass plates are spaced a distance of 2 mm apart. If the plates are placed in water how high will the water rise between the plates due to capillary action?

1.94   An open, clean glass tube, having a diameter of 4 mm, is inserted vertically into a dish of mercury at 20 °C. How far will the column of mercury in the tube be depressed?

1.95   An open 2-mm-diameter tube is inserted into a pan of ethyl alcohol and a similar 4-mm-diameter tube is inserted into a pan of water. In which tube will the height of the rise of the fluid column due to capillary action be the greatest? Assume the angle of contact is the same for both tubes.

1.96*   The capillary rise in a tube depends on the cleanliness of both the fluid and the tube. Typically, values of h are less than those predicted by Eq. 1.22 using values of σ and θ for clean fluids and tubes. Some measurements of the height, h, a water column rises in a vertical open tube of diameter, d, are given below. The water was tap water at a temperature of 60 °F and no particular effort was made to clean the glass tube. Fit a curve to these data and estimate the value of the product $\sigma \cos \theta$. If it is assumed that σ has the value given in Table 1.5 what is the value of θ? If it is assumed that θ is equal to 0° what is the value of σ?

d (in.)	0.3	0.25	0.20	0.15	0.10	0.05
h (in.)	0.133	0.165	0.198	0.273	0.421	0.796

$v = 0.1 \, m/s$

$\rightarrow F$

$\rightarrow F$

$\Sigma F = 0 = F = 2 F_s$

$F_s = \tau A$

$\tau = \mu \dfrac{dv}{dy} \approx \mu \dfrac{\Delta v}{\Delta y}$

$= (6.81 \times 10^{-3}) \left(\dfrac{.100}{.001} \right)$

$\tau = 0.681 \, N/m^2$

$F = 2 (0.681)(0.25)^2$

An image of hurricane Allen viewed via satellite: although there is considerable motion and structure to a hurricane, the pressure variation in the vertical direction is approximated by the pressure-depth relationship for a static fluid. (Visible and infrared image pair from a NOAA satellite using a technique developed at NASA/GSPC.) (Photograph courtesy of A. F. Hasler, [Ref. 7].)

TWO

FLUID
STATICS

In this chapter we will consider an important class of problems in which the fluid is either at rest or moving in such a manner that there is no relative motion between adjacent particles. In both instances there will be no shearing stresses in the fluid, and the only forces that develop on the surfaces of the particles will be due to the pressure. Thus, our principal concern is to investigate pressure and its variation throughout a fluid, and the effect of pressure on submerged surfaces. The absence of shearing stresses greatly simplifies the analysis, and as we will see, allows us to obtain relatively simple solutions to many important practical problems.

2.1 PRESSURE AT A POINT

As we briefly discussed in Chapter 1, the term pressure is used to indicate the normal force per unit area at a given point acting on a given plane within the fluid mass of interest. A question that immediately arises is how the pressure at a point varies with the orientation of the plane passing through the point. To answer this question consider the free-body diagram, illustrated in Fig. 2.1, that was obtained by removing a small triangular wedge of fluid from some arbitrary location within a fluid mass. Since we are considering the situation in which there are no shearing stresses, the only external forces acting on the wedge are due to the pressure and the weight. For simplicity the forces in the x direction are not shown, and the z axis is taken as the vertical axis so the weight acts in the negative z direction. Although we are primarily interested in fluids at rest, to make the analysis as general as possible, we will allow the fluid element to have accelerated motion. The assumption of zero shearing stresses will still be valid so long as the fluid element moves as a rigid body; that is, there is no relative motion between adjacent elements.

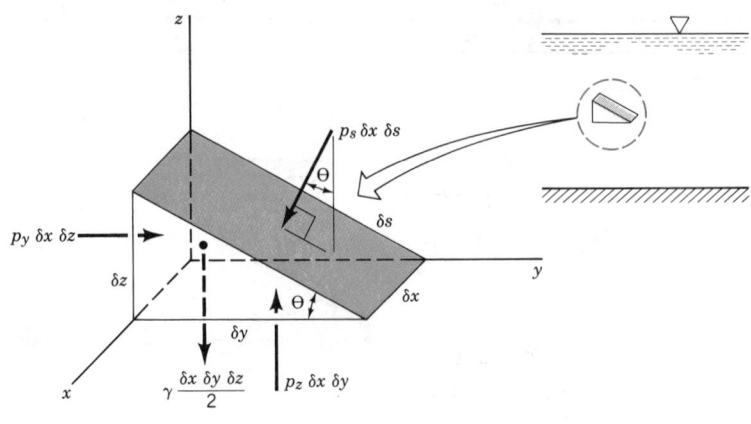

FIGURE 2.1
Forces on an arbitrary wedged-shaped element of fluid.

The equations of motion (Newton's second law, $\mathbf{F} = m\mathbf{a}$) in the y and z directions are, respectively,

$$\sum F_y = p_y \, \delta x \, \delta z - p_s \, \delta x \, \delta s \sin \theta = \rho \frac{\delta x \, \delta y \, \delta z}{2} a_y$$

$$\sum F_z = p_z \, \delta x \, \delta y - p_s \, \delta x \, \delta s \cos \theta - \gamma \frac{\delta x \, \delta y \, \delta z}{2} = \rho \frac{\delta x \, \delta y \, \delta z}{2} a_z$$

where p_s, p_y, and p_z are the average pressures on the faces, γ and ρ are the fluid specific weight and density, respectively, and a_y, a_z the accelerations. Note that a pressure must be multiplied by an appropriate area to obtain the force generated by the pressure. It follows from the geometry that

$$\delta y = \delta s \cos \theta \qquad \delta z = \delta s \sin \theta$$

so that the equations of motion can be rewritten as

$$p_y - p_s = \rho a_y \frac{\delta y}{2}$$

$$p_z - p_s = (\rho a_z + \gamma) \frac{\delta z}{2}$$

Since we are really interested in what is happening at a point, we take the limit as δx, δy, and δz, approach zero (while maintaining the angle θ), and it follows that

$$p_y = p_s \qquad p_z = p_s$$

or $p_s = p_y = p_z$. The angle θ was arbitrarily chosen so we can conclude that *the pressure at a point in a fluid at rest, or in motion, is independent of direction as long as there are no shearing stresses present.* This important result is known as *Pascal's law* named in honor of Blaise Pascal (1623–1662), a French mathematician who made important contributions in the field of hydrostatics. In Chapter 6 it will be shown that for moving fluids in which there is relative motion between particles (so that shearing stresses develop) the normal stress at a point, which corresponds to pressure in fluids at rest, is not necessarily the same in all directions. In such cases the pressure is defined as the *average* of any three mutually perpendicular normal stresses at the point.

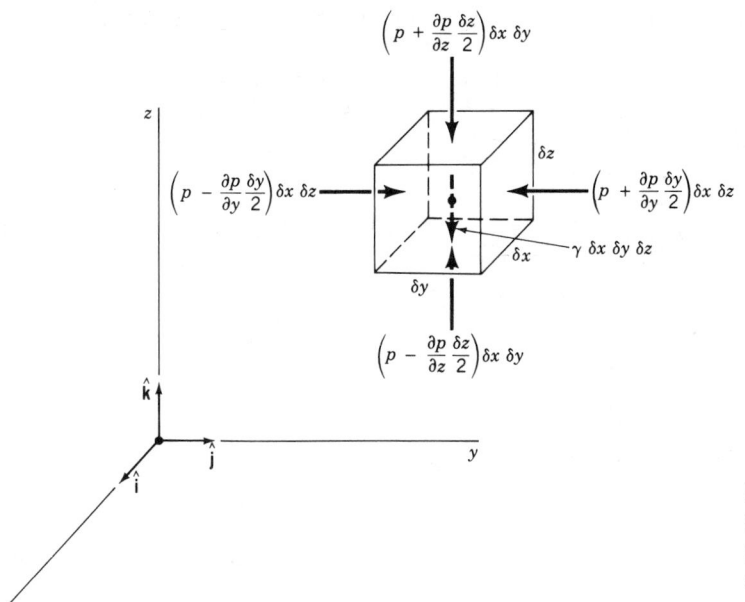

FIGURE 2.2
Surface and body
forces acting on
small fluid ele-
ment.

2.2 BASIC EQUATION FOR PRESSURE FIELD

Although we have answered the question of how the pressure at a point varies with direction, we are now faced with an equally important question—how does the pressure in a fluid in which there are no shearing stresses vary from point to point? To answer this question consider a small rectangular element of fluid removed from some arbitrary position within the mass of fluid of interest as illustrated in Fig. 2.2. There are two types of forces acting on this element: *surface forces* due to the pressure, and a *body force* equal to the weight of the element. Other possible types of body forces, such as those due to magnetic fields, will not be considered in this text.

If we let the pressure at the center of the element be designated as p then the average pressure on the various faces can be expressed in terms of p and its derivatives as shown in Fig. 2.2. We are actually using a Taylor series expansion of the pressure at the element center to approximate the pressures a short distance away and neglecting higher order terms that will vanish as we let δx, δy, and δz approach zero. For simplicity the surface forces in the x direction are not shown. The resultant surface force in the y direction is

$$\delta F_y = \left(p - \frac{\partial p}{\partial y} \frac{\delta y}{2} \right) \delta x\, \delta z - \left(p + \frac{\partial p}{\partial y} \frac{\delta y}{2} \right) \delta x\, \delta z$$

or

$$\delta F_y = -\frac{\partial p}{\partial y} \delta x\, \delta y\, \delta z$$

Similarly, for the x and z directions the resultant surface forces are

$$\delta F_x = -\frac{\partial p}{\partial x} \delta x\, \delta y\, \delta z \qquad \delta F_z = -\frac{\partial p}{\partial z} \delta x\, \delta y\, \delta z$$

The resultant surface force acting on the element can be expressed in vector form as

$$\delta \mathbf{F}_s = \delta F_x \hat{\mathbf{i}} + \delta F_y \hat{\mathbf{j}} + \delta F_z \hat{\mathbf{k}}$$

or

$$\delta \mathbf{F}_s = -\left(\frac{\partial p}{\partial x} \hat{\mathbf{i}} + \frac{\partial p}{\partial y} \hat{\mathbf{j}} + \frac{\partial p}{\partial z} \hat{\mathbf{k}} \right) \delta x\, \delta y\, \delta z \qquad (2.1)$$

where $\hat{\mathbf{i}}$, $\hat{\mathbf{j}}$, and $\hat{\mathbf{k}}$ are the unit vectors along the coordinate axes shown in Fig. 2.2. The group of terms in parentheses in Eq. 2.1 represents in vector form the *pressure gradient* and can be written as

$$\frac{\partial p}{\partial x} \hat{\mathbf{i}} + \frac{\partial p}{\partial y} \hat{\mathbf{j}} + \frac{\partial p}{\partial z} \hat{\mathbf{k}} = \nabla p$$

where

$$\nabla(\ ) = \frac{\partial(\ )}{\partial x} \hat{\mathbf{i}} + \frac{\partial(\ )}{\partial y} \hat{\mathbf{j}} + \frac{\partial(\ )}{\partial z} \hat{\mathbf{k}}$$

and the symbol ∇ is the *gradient* or "del" vector operator. Thus, the resultant surface force per unit volume can be expressed as

$$\frac{\delta \mathbf{F}_s}{\delta x\, \delta y\, \delta z} = -\nabla p$$

Since the z axis is vertical, the weight of the element is

$$\delta \mathbf{W} = -\gamma\, \delta x\, \delta y\, \delta z\, \hat{\mathbf{k}}$$

where the negative sign indicates that the force due to the weight is downward (in the negative z direction). Newton's second law, applied to the fluid element, can be expressed as

$$\sum \delta \mathbf{F} = \delta m\, \mathbf{a}$$

where $\sum \delta \mathbf{F}$ represents the resultant force acting on the element, $\mathbf{a}$ is the acceleration of the element, and δm is the element mass, which can be written as $\rho\, \delta x\, \delta y\, \delta z$. It follows that

$$\sum \delta \mathbf{F} = \delta \mathbf{F}_s + \delta \mathbf{W} = \delta m\, \mathbf{a}$$

or

$$-\nabla p\, \delta x\, \delta y\, \delta z - \gamma\, \delta x\, \delta y\, \delta z\, \hat{\mathbf{k}} = \rho\, \delta x\, \delta y\, \delta z\, \mathbf{a}$$

and, therefore,

$$-\nabla p - \gamma \hat{\mathbf{k}} = \rho \mathbf{a} \qquad (2.2)$$

Equation 2.2 is the general equation of motion for a fluid in which there are no shearing stresses. We will use this equation in Section 2.12 when we consider the pressure distribution in a moving fluid. For the present, however, we will restrict our attention to the special case of a fluid at rest.

2.3 PRESSURE VARIATION IN A FLUID AT REST

For a fluid at rest $\mathbf{a} = 0$ and Eq. 2.2 reduces to

$$\nabla p + \gamma \hat{\mathbf{k}} = 0$$

or in component form

$$\frac{\partial p}{\partial x} = 0 \qquad \frac{\partial p}{\partial y} = 0 \qquad \frac{\partial p}{\partial z} = -\gamma \tag{2.3}$$

These equations show that the pressure does not depend on x or y. Thus, as we move from point to point in a horizontal plane (any plane parallel to the x–y plane) the pressure does not change. Since p depends only on z, the last of Eqs. 2.3 can be written as the ordinary differential equation

$$\frac{dp}{dz} = -\gamma \tag{2.4}$$

Equation 2.4 is the fundamental equation for fluids at rest and can be used to determine how pressure changes with elevation. This equation indicates that the pressure gradient in the vertical direction is negative; that is, the pressure decreases as we move upward in a fluid at rest. There is no requirement that γ be a constant. Thus, it is valid for fluids with constant specific weight, such as liquids, as well as fluids whose specific weight may vary with elevation, such as air or other gases. However, to proceed with the integration of Eq. 2.4 it is necessary to stipulate how the specific weight varies with z.

2.3.1 Incompressible Fluid

Since the specific weight is equal to the product of fluid density and acceleration of gravity ($\gamma = \rho g$), changes in γ are caused either by a change in ρ or g. For most engineering applications the variation in g is negligible, so our main concern is with the possible variation in the fluid density. For liquids the variation in density is negligible, even over large vertical distances, so that the assumption of constant specific weight when dealing with liquids is a good one. For this instance, Eq. 2.4 can be directly integrated

$$\int_{p_1}^{p_2} dp = -\gamma \int_{z_1}^{z_2} dz$$

to yield

$$p_2 - p_1 = -\gamma(z_2 - z_1)$$

or

$$p_1 - p_2 = \gamma(z_2 - z_1) \tag{2.5}$$

where p_1 and p_2 are pressures at the vertical elevations z_1 and z_2, as is illustrated in Fig. 2.3.

Equation 2.5 can be written in the compact form

$$p_1 - p_2 = \gamma h \tag{2.6}$$

or

$$p_1 = \gamma h + p_2 \tag{2.7}$$

FIGURE 2.3 Notation for pressure variation in a fluid at rest with a free surface.

where h is the distance, $z_2 - z_1$, which is the depth of fluid measured downward from the location of p_2. This type of pressure distribution is commonly called a *hydrostatic distribution*, and Eq. 2.7 shows that in an incompressible fluid at rest the pressure varies linearly with depth.

It can also be observed from Eq. 2.6 that the pressure difference between two points can be specified by the distance, h, since

$$h = \frac{p_1 - p_2}{\gamma}$$

In this case h is called the *pressure head* and is interpreted as the height of a column of fluid of specific weight, γ, required to give a pressure difference, $p_1 - p_2$. For example, a pressure difference of 10 psi can be specified in terms of pressure head as 23.1 ft of water ($\gamma = 62.4$ lb/ft^3), or 518 mm of Hg ($\gamma = 133$ kN/m^3).

When one works with liquids there is often a free surface, as is illustrated in Fig. 2.3, and it is convenient to use this surface as a reference plane. The reference pressure p_0 would correspond to the pressure acting on the free surface (which would frequently be atmospheric pressure), and thus if we let $p_2 = p_0$ in Eq. 2.7 it follows that the pressure p at any depth h below the free surface is given by the equation:

$$p = \gamma h + p_0 \tag{2.8}$$

As is demonstrated by Eqs. 2.7 or 2.8, the pressure in a homogeneous, incompressible fluid at rest depends on the depth of the fluid relative to some reference plane, and it is *not* influenced by the *size* or *shape* of the tank or container in which the fluid is held. Thus, in Fig. 2.4 the pressure is the same at all points along the line AB even though the container may have the very irregular shape shown in the figure. The actual value of the pressure along AB depends only on the depth, h, the surface pressure, p_0, and the specific weight, γ, of the liquid in the container.

FIGURE 2.4 Fluid equilibrium in a container of arbitrary shape.

EXAMPLE 2.1

Because of a leak in a buried gasoline storage tank, water has seeped in to the depth shown in Fig. E2.1. If the specific gravity of the gasoline is SG = 0.68, determine the pressure at the gasoline–water interface, and at the bottom of the tank. Express the pressure in units of lb/ft², lb/in.², and as a pressure head in feet of water.

Open

Gasoline

(1)

(2)

Water

17 ft

3 ft

FIGURE E2.1

SOLUTION

Since we are dealing with liquids at rest, the pressure distribution will be hydrostatic, and therefore the pressure variation can be found from the equation:

$$p = \gamma h + p_0$$

With p_0 corresponding to the pressure at the free surface of the gasoline, then the pressure at the interface is

$$p_1 = (SG)(\gamma_{H_2O})(h) + p_0$$
$$= (0.68)(62.4 \text{ lb/ft}^3)(17 \text{ ft}) + p_0$$
$$= 721 + p_0 \text{ (lb/ft}^2)$$

If we measure the pressure relative to atmospheric pressure (gage pressure), it follows that $p_0 = 0$, and therefore

$$p_1 = 721 \text{ lb/ft}^2 \qquad \text{(Ans)}$$

$$p_1 = \frac{721 \text{ lb/ft}^2}{144 \text{ in.}^2/\text{ft}^2} = 5.01 \text{ lb/in.}^2 \qquad \text{(Ans)}$$

$$\frac{p_1}{\gamma_{H_2O}} = \frac{721 \text{ lb/ft}^2}{62.4 \text{ lb/ft}^3} = 11.6 \text{ ft} \qquad \text{(Ans)}$$

It is noted that a rectangular column of water 11.6 ft tall and 1 ft² in cross section weighs 721 lb. A similar column with a 1-in.² cross section weighs 5.01 lb.

We can now apply the same relationship to determine the pressure at the tank bottom; that is,

$$p_2 = \gamma_{H_2O}h_{H_2O} + p_1$$
$$= (62.4 \text{ lb/ft}^3)(3 \text{ ft}) + 721 \text{ lb/ft}^2$$
$$= 908 \text{ lb/ft}^2 \qquad \text{(Ans)}$$

$$p_2 = \frac{908 \text{ lb/ft}^2}{144 \text{ in.}^2/\text{ft}^2} = 6.31 \text{ lb/in.}^2 \qquad \text{(Ans)}$$

$$\frac{p_2}{\gamma_{H_2O}} = \frac{908 \text{ lb/ft}^2}{62.4 \text{ lb/ft}^3} = 14.6 \text{ ft} \qquad \text{(Ans)}$$

Observe that if we wish to express these pressures in terms of *absolute* pressure, we would have to add the local atmospheric pressure (in appropriate units) to the previous results. A further discussion of gage and absolute pressure is given in Section 2.5.

The required equality of pressures at equal elevations throughout a system is important for the operation of hydraulic jacks, lifts, and presses, as well as hydraulic controls on aircraft and other types of heavy machinery. The fundamental idea behind such devices and systems is demonstrated in Fig. 2.5. A piston located at one end of a closed system filled with a liquid, such as oil, can be used to change the pressure throughout the system, and thus transmit an applied force F_1 to a second piston where the resulting force is F_2. Since the pressure p acting on the faces of both pistons is the same (the effect of elevation changes is usually negligible for this type of hydraulic device), it follows that $F_2 = (A_2/A_1)F_1$. The piston area A_2 can be made much larger than A_1 and therefore a large mechanical advantage can be developed; that is, a small force applied at the smaller piston can be used to develop a large force at the larger piston. The applied force could be created manually through some type of mechanical device, such as a hydraulic jack, or through compressed air acting directly on the surface of the liquid, as is done in hydraulic lifts commonly found in service stations.

2.3.2 Compressible Fluid

We normally think of gases such as air, oxygen, and nitrogen as being compressible fluids since the density of the gas can change significantly with changes in pressure and temperature. Thus, although Eq. 2.4 applies at a point in a gas, it is necessary to consider the possible variation in γ before the equation can be integrated. However, as was discussed in Chapter 1, the specific weights of common gases are small when compared with those of liquids. For example, the specific weight of air at sea level and 60 °F is 0.0763 lb/ft^3, whereas the specific weight of water under the same conditions is 62.4 lb/ft^3. Since the specific weights of gases are comparatively small, it follows from Eq. 2.4 that the pressure gradient in the vertical direction is correspondingly small, and even over distances of several hundred feet the pressure will remain essentially constant for a gas. This means we can neglect the effect of elevation changes on the pressure in gases in tanks, pipes, and so forth in which the distances involved are small.

$F_1 = pA_1$ $F_2 = pA_2$

FIGURE 2.5 Transmission of fluid pressure.

For those situations in which the variations in heights are large, on the order of thousands of feet, attention must be given to the variation in the specific weight. As is described in Chapter 1, the equation of state for an ideal (or perfect) gas is

$$p = \rho R T$$

where p is the absolute pressure, R is the gas constant, and T is the absolute temperature. This relationship can be combined with Eq. 2.4 to give

$$\frac{dp}{dz} = -\frac{gp}{RT}$$

and by separating variables

$$\int_{p_1}^{p_2} \frac{dp}{p} = \ln \frac{p_2}{p_1} = -\frac{g}{R} \int_{z_1}^{z_2} \frac{dz}{T} \tag{2.9}$$

where g and R are assumed to be constant over the elevation change from z_1 to z_2. Although the acceleration of gravity, g, does vary with elevation, the variation is very small (see Tables C.1 and C.2 in Appendix C), and g is usually assumed constant at some average value for the range of elevation involved.

Before completing the integration, one must specify the nature of the variation of temperature with elevation. For example, if we assume that the temperature has a constant value T_0 over the range z_1 to z_2 (*isothermal* conditions), it then follows from Eq. 2.9 that

$$p_2 = p_1 \exp\left[-\frac{g(z_2 - z_1)}{R T_0} \right] \tag{2.10}$$

This equation provides the desired pressure-elevation relationship for an isothermal layer. For nonisothermal conditions a similar procedure can be followed if the temperature–elevation relationship is known, as is discussed in the following section.

EXAMPLE 2.2

The Empire State Building in New York City, one of the tallest buildings in the world, rises to a height of approximately 1250 ft. Estimate the ratio of the pressure at the top of the building to the pressure at its base, assuming the air to be at a common temperature of 59 °F. Compare this result with that obtained by assuming the air to be incompressible with $\gamma = 0.0765$ lb/ft³ at 14.7 psi(abs) (values for air at standard conditions).

SOLUTION

For the assumed isothermal conditions, and treating air as a compressible fluid, Eq. 2.10 can be applied to yield

$$\frac{p_2}{p_1} = \exp\left[-\frac{g(z_2 - z_1)}{R T_0} \right]$$

$$= \exp\left\{ -\frac{(32.2 \text{ ft/s}^2)(1250 \text{ ft})}{(1716 \text{ ft·lb/slug·°R})[(59 + 460)°\text{R}]} \right\} = 0.956 \tag{Ans}$$

If the air is treated as an incompressible fluid we can apply Eq. 2.5. In this case

$$p_2 = p_1 - \gamma(z_2 - z_1)$$

or

$$\frac{p_2}{p_1} = 1 - \frac{\gamma(z_2 - z_1)}{p_1}$$

$$= 1 - \frac{(0.0765 \text{ lb/ft}^3)(1250 \text{ ft})}{(14.7 \text{ lb/in.}^2)(144 \text{ in.}^2/\text{ft}^2)} = 0.955 \qquad \text{(Ans)}$$

Note that there is little difference between the two results. Since the pressure difference between the bottom and top of the building is small, it follows that the variation in fluid density is small and, therefore, the compressible fluid and incompressible fluid analyses yield essentially the same result.

We see that for both calculations the pressure decreases by less than 5% as we go from ground level to the top of this tall building. It does not require a very large pressure difference to support a 1250 ft tall column of fluid as light as air. This result supports the earlier statement that the changes in pressures in air and other gases due to elevation changes are very small, even for distances of hundreds of feet. Thus, the pressure differences between the top and bottom of a horizontal pipe carrying a gas, or a gas storage tank, are negligible since the distances involved are very small.

2.4   STANDARD ATMOSPHERE

An important application of Eq. 2.9 relates to the variation in pressure in the earth's atmosphere. Ideally, we would like to have measurements of pressure versus altitude over the specific range for the specific conditions (temperature, reference pressure) for which the pressure is to be determined. However, this type of information is usually not available. Thus, a "standard atmosphere" has been determined that can be used in the design of aircraft, missiles, and spacecraft, and in comparing their performance under standard conditions. The concept of a standard atmosphere was first developed in the 1920s and since that time many national and international committees and organizations have pursued the development of such a standard. The currently accepted standard atmosphere is based on a report published in 1962, and updated in 1976 (see Refs. 1 and 2), defining the so-called *U.S. standard atmosphere*, which is an idealized representation of middle-latitude, year-round mean conditions of the earth's atmosphere. Several important properties for standard atmospheric conditions at *sea level* are listed in Table 2.1, and Fig. 2.6 shows the temperature profile for the U.S. standard atmosphere. As is shown in this figure the temperature decreases with altitude in the region nearest the earth's surface (*troposphere*), then

TABLE 2.1   Properties of U.S. Standard Atmosphere at Sea Level[a]

Property	SI Units	BG Units
Temperature, T	288.15 K (15 °C)	518.67 °R (59.00 °F)
Pressure, p	101.33 kPa (abs)	2116.2 lb/ft² (abs)
		[14.696 lb/in.² (abs)]
Density, ρ	1.225 kg/m³	0.002377 slugs/ft³
Specific weight, γ	12.014 N/m³	0.07647 lb/ft³
Viscosity, μ	1.789×10^{-5} N·s/m²	3.737×10^{-7} lb·s/ft²

[a] Acceleration of gravity at sea level = 9.807 m/s² = 32.174 ft/s².

FIGURE 2.6 Variation of temperature with altitude in the U.S. standard atmosphere.

becomes essentially constant in the next layer (*stratosphere*), and subsequently starts to increase in the next layer.

Since the temperature variation is represented by a series of linear segments, it is possible to integrate Eq. 2.9 to obtain the corresponding pressure variation. For example, in the troposphere, which extends to an altitude of about 11 km (~36,000 ft), the temperature variation is of the form

$$T = T_a - \beta z \tag{2.11}$$

where T_a is the temperature at sea level ($z = 0$) and β is the *lapse rate* (the rate of change of temperature with elevation). For the standard atmosphere in the troposphere $\beta = 0.00650$ K/m or 0.00357 °R/ft.

Equation 2.11 used with Eq. 2.9 yields

$$p = p_a\left(1 - \frac{\beta z}{T_a}\right)^{g/R\beta} \tag{2.12}$$

where p_a is the absolute pressure at $z = 0$. With p_a, T_a, and g obtained from Table 2.1, and with the gas constant $R = 286.9$ J/kg·K or 1716 ft·lb/slug·°R, the pressure variation throughout the troposphere can be determined from Eq. 2.12. This calculation shows that at the outer edge of the troposphere, where the temperature is -56.5 °C, the absolute pressure is about 23 kPa (3.3 psia). It is to be noted that modern jetliners cruise at approximately this altitude. Pressures at other altitudes are shown in Fig. 2.6, and tabulated values for temperature, acceleration of gravity, pressure, density, and viscosity for the U.S. standard atmosphere are given in Tables C.1 and C.2 in Appendix C.

2.5 MEASUREMENT OF PRESSURE

Since pressure is a very important characteristic of a fluid field, it is not surprising that numerous devices and techniques are used in its measurement. As is noted briefly

FIGURE 2.7 Graphical representation of gage and absolute pressure.

in Chapter 1, the pressure at a point within a fluid mass will be designated as either an *absolute* pressure or a *gage* pressure. Absolute pressure is measured relative to a perfect vacuum (absolute zero pressure), whereas gage pressure is measured relative to the local atmospheric pressure. Thus, a gage pressure of zero corresponds to a pressure that is equal to the local atmospheric pressure. Absolute pressures are always positive, but gage pressures can be either positive or negative depending on whether the pressure is above atmospheric pressure (a positive value) or below atmospheric pressure (a negative value). A negative gage pressure is also referred to as a *suction* or *vacuum* pressure. For example, 10 psi (abs) could be expressed as -4.7 psi (gage), if the local atmospheric pressure is 14.7 psi, or alternatively 4.7 psi suction or 4.7 psi vacuum. The concept of gage and absolute pressure is illustrated graphically in Fig. 2.7 for two typical pressures located at points 1 and 2.

In addition to the reference used for the pressure measurement, the *units* used to express the value are obviously of importance. As is described in Section 1.5, pressure is a force per unit area, and the units in the BG system are lb/ft^2 or $lb/in.^2$, commonly abbreviated psf or psi, respectively. In the SI system the units are N/m^2; this combination is called the pascal and written as Pa $(1 \ N/m^2 = 1 \ Pa)$. As noted earlier, pressure can also be expressed as the height of a column of liquid. Then, the units will refer to the height of the column (in., ft, mm, m, etc.), and in addition, the liquid in the column must be specified (H_2O, Hg, etc.). For example, standard atmospheric pressure can be expressed as 760 mm Hg (abs). *In this text, pressures will be assumed to be gage pressures unless specifically designated absolute,* for example, 10 psi or 100 kPa would be gage pressures, whereas 10 psia or 100 kPa (abs) would refer to absolute pressures. It is to be noted that *pressure differences* are independent of the reference, so that no special notation is required in this case.

The measurement of atmospheric pressure is usually accomplished with a mercury *barometer,* which in its simplest form, consists of a glass tube closed at one end with the open end immersed in a container of mercury as shown in Fig. 2.8. The tube is initially filled with mercury (inverted with its open end up) and then turned upside down (open end down) with the open end in the container of mercury. The column of mercury will come to an equilibrium position where its weight plus the force due to the vapor pressure (which develops in the space above the column) balances the force due to the atmospheric pressure. Thus,

$$p_{atm} = \gamma h + p_{vapor} \tag{2.13}$$

where γ is the specific weight of mercury. For most practical purposes the contribution

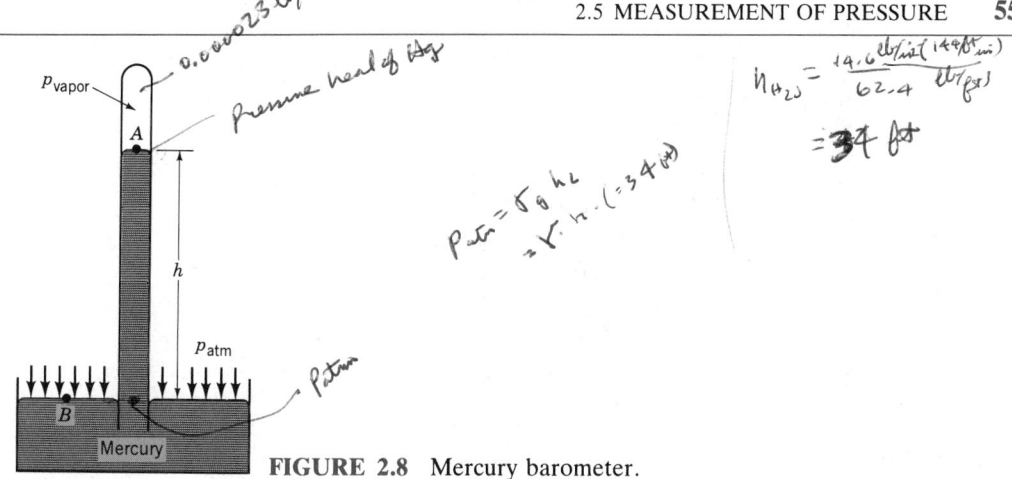

FIGURE 2.8   Mercury barometer.

of the vapor pressure can be neglected since it is very small [for mercury, p_{vapor} = 0.000023 lb/in.² (abs) at a temperature of 68 °F] so that $p_{atm} \approx \gamma h$. It is conventional to specify atmospheric pressure in terms of the height, h, in millimeters or inches of mercury. Note that if water were used instead of mercury, the height of the column would have to be approximately 34 ft rather than 29.9 in. of mercury for an atmospheric pressure of 14.7 psia! The concept of the mercury barometer is an old one, with the invention of this device attributed to Evangelista Torricelli in about 1644.

EXAMPLE 2.3

A mountain lake has an average temperature of 10 °C and a maximum depth of 40 m. For a barometric pressure of 598 mm Hg, determine the absolute pressure (in pascals) at the deepest part of the lake.

SOLUTION

The pressure in the lake at any depth, h, is given by the equation

$$p = \gamma h + p_0$$

where p_0 is the pressure at the surface. Since we want the absolute pressure, p_0 will be the local barometric pressure expressed in a consistent system of units; that is

$$\frac{p_{barometric}}{\gamma_{Hg}} = 598 \text{ mm} = 0.598 \text{ m}$$

and for γ_{Hg} = 133 kN/m³

$$p_0 = (0.598 \text{ m})(133 \text{ kN/m}^3) = 79.5 \text{ kN/m}^2$$

From Table B.2, γ_{H_2O} = 9.804 kN/m³ at 10 °C and therefore

$$p = (9.804 \text{ kN/m}^2)(40 \text{ m}) + 79.5 \text{ kN/m}^2$$
$$= 392 \text{ kN/m}^2 + 79.5 \text{ kN/m}^2 = 472 \text{ kPa (abs)} \qquad \text{(Ans)}$$

This simple example illustrates the need for close attention to the units used in the calculation of pressure; that is, be sure to use a *consistent* unit system, and be careful not to add a pressure head (m) to a pressure (Pa).

FIGURE 2.9   Piezometer tube.

2.6   MANOMETRY

A standard technique for measuring pressure involves the use of liquid columns in vertical or inclined tubes. Pressure measuring devices based on this technique are called *manometers*. The mercury barometer is an example of one type of manometer, but there are many other configurations possible, depending on the particular application. Three common types of manometers include the piezometer tube, the U-tube manometer, and the inclined-tube manometer.

2.6.1   Piezometer Tube

The simplest type of manometer consists of a vertical tube, open at the top, and attached to the container in which the pressure is desired, as illustrated in Fig. 2.9. Since manometers involve columns of fluids at rest, the fundamental equation describing their use is Eq. 2.8

$$p = \gamma h + p_0$$

which gives the pressure at any elevation within a homogeneous fluid in terms of a reference pressure p_0 and the vertical distance h between p and p_0. Remember that in a fluid at rest pressure will *increase* as we move *downward,* and will decrease as we move *upward.* Application of this equation to the piezometer tube of Fig. 2.9 indicates that the pressure p_A can be determined by a measurement of h_1 through the relationship

$$p_A = \gamma_1 h_1$$

where γ_1 is the specific weight of the liquid in the container. Note that since the tube is open at the top, the pressure p_0 can be set equal to zero (we are now using gage pressure), with the height h_1 measured from the meniscus at the upper surface to point (1). Since point (1) and point A within the container are at the same elevation, $p_A = p_1$.

Although the piezometer tube is a very simple and accurate pressure measuring device, it has several disadvantages. It is only suitable if the pressure in the container is greater than atmospheric pressure (otherwise air would be sucked into the system), and the pressure to be measured must be relatively small so the required height of

Open

γ_1

A

(1)

h_1

(2)

γ_2
(gage fluid)

h_2

(3)

p_A

$p_A = p_1$

$p_2 = p_A + h_1 \gamma_1$

$p_3 = p_2$

$p_3 = p_A + h_1 \gamma_1$

$p_4 = p_3 - \gamma_2 h_2$

$p_A + h_1 \gamma_1 - h_2 \gamma_2 = p_4 = p_{atm}$

$p_A =$

FIGURE 2.10 Simple U-tube manometer.

the column is reasonable. Also, the fluid in the container in which the pressure is to be measured must be a liquid rather than a gas.

2.6.2 U-Tube Manometer

To overcome the difficulties noted previously, another type of manometer which is widely used consists of a tube formed into the shape of a U as is shown in Fig. 2.10. The fluid in the manometer is called the *gage fluid*. To find the pressure p_A in terms of the various column heights, we start at one end of the system and work our way around to the other end, simply utilizing Eq. 2.8. Thus, for the U-tube manometer shown in Fig. 2.10, we will start at point A and work around to the open end. The pressure at points A and (1) are the same, and as we move from point (1) to (2) the pressure will increase by $\gamma_1 h_1$. The pressure at point (2) is equal to the pressure at point (3), since the pressures at equal elevations in a continuous mass of fluid at rest must be the same. Note that we could not simply "jump across" from point (1) to a point at the same elevation in the right-hand tube since these would not be points within the same continuous mass of fluid. With the pressure at point (3) specified we now move to the open end where the pressure is zero. As we move vertically upward the pressure decreases by an amount $\gamma_2 h_2$. In equation form these various steps can be expressed as

$$p_A + \gamma_1 h_1 - \gamma_2 h_2 = 0$$

and, therefore, the pressure p_A can be written in terms of the column heights as

$$p_A = \gamma_2 h_2 - \gamma_1 h_1 \qquad (2.14)$$

A major advantage of the U-tube manometer lies in the fact that the gage fluid can be different from the fluid in the container in which the pressure is to be determined. For example, the fluid in A in Fig. 2.10 can be either a liquid or a gas. If A does contain a gas, the contribution of the gas column, $\gamma_1 h_1$, is almost always negligible so that $p_A \approx p_2$ and in this instance Eq. 2.14 becomes

$$p_A = \gamma_2 h_2$$

Thus, for a given pressure the height, h_2, is governed by the specific weight, γ_2, of the gage fluid used in the manometer. If the pressure p_A is large, then a heavy gage fluid, such as mercury, can be used and a reasonable column height (not too long) can still be maintained. Alternatively, if the pressure p_A is small, a lighter gage fluid, such as water, can be used so that a relatively large column height (which is easily read) can be achieved.

EXAMPLE 2.4

A closed tank contains compressed air and oil ($SG_{oil} = 0.90$) as is shown in Fig. E2.4. A U-tube manometer using mercury ($SG_{Hg} = 13.6$) is connected to the tank as shown. For column heights $h_1 = 36$ in., $h_2 = 6$ in., and $h_3 = 9$ in., determine the pressure reading (in psi) of the gage.

FIGURE E2.4

SOLUTION

Following the general procedure of starting at one end of the manometer system and working around to the other, we will start at the air–oil interface in the tank and proceed to the open end where the pressure is zero. The pressure at level (1) is

$$p_1 = p_{air} + \gamma_{oil}(h_1 + h_2)$$

This pressure is equal to the pressure at level (2), since these two points are at the same elevation in a homogeneous fluid at rest. As we move from level (2) to the open end, the pressure must decrease by $\gamma_{Hg}h_3$, and at the open end the pressure is zero. Thus, the manometer equation can be expressed as

$$p_{air} + \gamma_{oil}(h_1 + h_2) - \gamma_{Hg}h_3 = 0$$

or

$$p_{air} + (SG_{oil})(\gamma_{H_2O})(h_1 + h_2) - (SG_{Hg})(\gamma_{H_2O})h_3 = 0$$

For the values given

$$p_{air} + (0.9)(62.4 \text{ lb/ft}^3)\left(\frac{36 + 6}{12} \text{ ft}\right) - (13.6)(62.4 \text{ lb/ft}^3)\left(\frac{9}{12} \text{ ft}\right) = 0$$

so that

$$p_{air} = 440 \text{ lb/ft}^2$$

Since the specific weight of the air above the oil is much smaller than the specific weight of the oil, the gage should read the pressure we have calculated; that is,

$$p_{gage} = \frac{440 \text{ lb/ft}^2}{144 \text{ in.}^2/\text{ft}^2} = 3.06 \text{ psi} \qquad \text{(Ans)}$$

$$P_{p_5} \approx P_B$$

$$P_{\gamma_5} + \gamma_3\, h_3 + \gamma_2\, h_2 - \gamma_1\, h_1 = P_A$$

$$P_2 \approx P_3$$

FIGURE 2.11   Differential U-tube manometer.

The U-tube manometer is also widely used to measure the *difference* in pressure between two containers or two points in a given system. Consider a manometer connected between containers A and B as is shown in Fig. 2.11. The difference in pressure between A and B can be found by again starting at one end of the system and working around to the other end. For example, at A the pressure is p_A, which is equal to p_1, and as we move to point (2) the pressure increases by $\gamma_1 h_1$. The pressure at p_2 is equal to p_3, and as we move upward to point (4) the pressure decreases by $\gamma_2 h_2$. Similarly, as we continue to move upward from point (4) to (5) the pressure decreases by $\gamma_3 h_3$. Finally, $p_5 = p_B$, since they are at equal elevations. Thus,

$$p_A + \gamma_1 h_1 - \gamma_2 h_2 - \gamma_3 h_3 = p_B$$

and the pressure difference is

$$p_A - p_B = \gamma_2 h_2 + \gamma_3 h_3 - \gamma_1 h_1$$

When the time comes to substitute in numbers, be sure to use a consistent system of units!

Capillarity due to surface tension at the various fluid interfaces in the manometer is usually not considered, since for a simple U-tube with a meniscus in each leg, the capillary effects cancel (assuming the surface tensions and tube diameters are the same at each meniscus), or we can make the capillary rise negligible by using relatively large bore tubes (with diameters of about 0.5 in. or larger). Two common gage fluids are water and mercury. Both give a well-defined meniscus, a very important characteristic for a gage fluid, and their properties are well known. Of course, the gage fluid must be immiscible with respect to the other fluids in contact with it. For highly accurate measurements, special attention should be given to temperature since the various specific weights of the fluids in the manometer will vary with temperature.

EXAMPLE 2.5

As will be discussed in Chapter 3, the volume rate of flow, Q, through a pipe can be determined by means of a flow nozzle located in the pipe as illustrated in Fig. E2.5. The nozzle creates a pressure drop, $p_A - p_B$, along the pipe which is related to the flow through the equation, $Q = K\sqrt{p_A - p_B}$, where K is a constant depending on the pipe and nozzle size. The pressure drop is frequently measured with a differential U-tube manometer of the type illustrated. (a) Determine an equation for $p_A - p_B$ in terms of the specific weight of the flowing fluid, γ_1,

FIGURE E2.5

the specific weight of the gage fluid, γ_2, and the various heights indicated. (b) For $\gamma_1 =$ 9.80 kN/m³, $\gamma_2 = 15.6$ kN/m³, $h_1 = 1.0$ m, and $h_2 = 0.5$ m what is the value of the pressure drop, $p_A - p_B$?

SOLUTION

(a) Although the fluid in the pipe is moving, the fluids in the columns of the manometer are at rest so that the pressure variation in the manometer tubes is hydrostatic. If we start at point A and move vertically upward to level (1), the pressure will decrease by $\gamma_1 h_1$ and will be equal to the pressure at (2) and at (3). We can now move from (3) to (4) where the pressure has been further reduced by $\gamma_2 h_2$. The pressures at levels (4) and (5) are equal, and as we move from (5) to B the pressure will increase by $\gamma_1(h_1 + h_2)$. Thus, in equation form

$$p_A - \gamma_1 h_1 - \gamma_2 h_2 + \gamma_1(h_1 + h_2) = p_B$$

or

$$p_A - p_B = h_2(\gamma_2 - \gamma_1) \qquad \text{(Ans)}$$

It is to be noted that the only column height of importance is the differential reading, h_2. The differential manometer could be placed 0.5 or 5.0 m above the pipe ($h_1 = 0.5$ m or $h_1 = 5.0$ m) and the value of h_2 would remain the same.

(b) The specific value of the pressure drop for the data given is

$$p_A - p_B = (0.5 \text{ m})(15.6 \text{ kN/m}^3 - 9.80 \text{ kN/m}^3)$$

$$= 2.90 \text{ kPa} \qquad \text{(Ans)}$$

2.6.3 Inclined-Tube Manometer

To measure small pressure changes, a manometer of the type shown in Fig. 2.12 is frequently used. One leg of the manometer is inclined at an angle θ, and the differential reading ℓ_2 is measured along the inclined tube. The difference in pressure $p_A - p_B$ can be expressed as

$$p_A + \gamma_1 h_1 - \gamma_2 \ell_2 \sin \theta - \gamma_3 h_3 = p_B$$

or

$$p_A - p_B = \gamma_2 \ell_2 \sin \theta + \gamma_3 h_3 - \gamma_1 h_1 \qquad (2.15)$$

where it is to be noted that the pressure difference between points (1) and (2) is due to the *vertical* distance between the points, which can be expressed as $\ell_2 \sin \theta$. Thus,

FIGURE 2.12 Inclined tube manometer.

for relatively small angles the differential reading along the inclined tube can be made large even for small pressure differences. The inclined-tube manometer is often used to measure small differences in gas pressures so that if pipes A and B contain a gas then

$$p_A - p_B = \gamma_2 \ell_2 \sin \theta$$

or

$$\ell_2 = \frac{p_A - p_B}{\gamma_2 \sin \theta} \tag{2.16}$$

where the contributions of the gas columns h_1 and h_3 have been neglected. Equation 2.16 shows that the differential reading ℓ_2 (for a given pressure difference) of the inclined-tube manometer can be increased over that obtained with a conventional U-tube manometer by the factor $1/\sin \theta$. Recall that $\sin \theta \to 0$ as $\theta \to 0$.

2.7   MECHANICAL AND ELECTRONIC PRESSURE MEASURING DEVICES

Although manometers are widely used, they are not well suited for measuring very high pressures, or pressures that are changing rapidly with time. In addition, they require the measurement of one or more column heights, which, although not particularly difficult, can be time consuming. To overcome some of these problems numerous other types of pressure-measuring instruments have been developed. Most of these make use of the idea that when a pressure acts on an elastic structure the structure will deform, and this deformation can be related to the magnitude of the pressure. Probably the most familiar device of this kind is the *Bourdon* pressure gage, which is shown in Fig. 2.13*a*. The essential mechanical element in this gage is the hollow, elastic curved tube (Bourdon tube) which is connected to the pressure source as shown in Fig. 2.13*b*. As the pressure within the tube increases the tube tends to straighten, and although the deformation is small, it can be translated into the motion of a pointer on a dial as illustrated. Since it is the difference in pressure between the outside of the tube (atmospheric pressure) and the inside of the tube that causes the movement of the tube, the indicated pressure is gage pressure. The Bourdon gage must be calibrated so that the dial reading can directly indicate the pressure in suitable units such as psi, psf, or pascals. A zero reading on the gage indicates that the measured pressure is equal to the local atmospheric pressure. This type of gage can be used to measure a negative gage pressure (vacuum) as well as positive pressures.

(a)   *(b)*

FIGURE 2.13   (*a*) Liquid-filled Bourdon pressure gages for various pressure ranges. (*b*) Internal elements of Bourdon gages. The "C-shaped" Bourdon tube is shown on the left, and the "coiled spring" Bourdon tube for high pressures of 1000 psi and above is shown on the right. (Photographs courtesy of Weiss Instruments, Inc.)

The *aneroid* barometer is another type of mechanical gage that is used for measuring atmospheric pressure. Since atmospheric pressure is specified as an absolute pressure, the conventional Bourdon gage is not suitable for this measurement. The common aneroid barometer contains a hollow, closed, elastic element which is evacuated so that the pressure inside the element is near absolute zero. As the external atmospheric pressure changes, the element deflects, and this motion can be translated into the movement of an attached dial. As with the Bourdon gage, the dial can be calibrated to give atmospheric pressure directly, with the usual units being millimeters or inches of mercury.

For many applications in which pressure measurements are required, the pressure must be measured with a device that converts the pressure into an electrical output. For example, it may be desirable to continuously monitor a pressure that is changing with time. This type of pressure measuring device is called a *pressure transducer,* and many different designs are used. One possible type of transducer is one in which a Bourdon tube is connected to a linear variable differential transformer (LVDT), as is illustrated in Fig. 2.14. The core of the LVDT is connected to the free end of the Bourdon so that as a pressure is applied the resulting motion of the end of the tube moves the core through the coil and an output voltage develops. This voltage is a

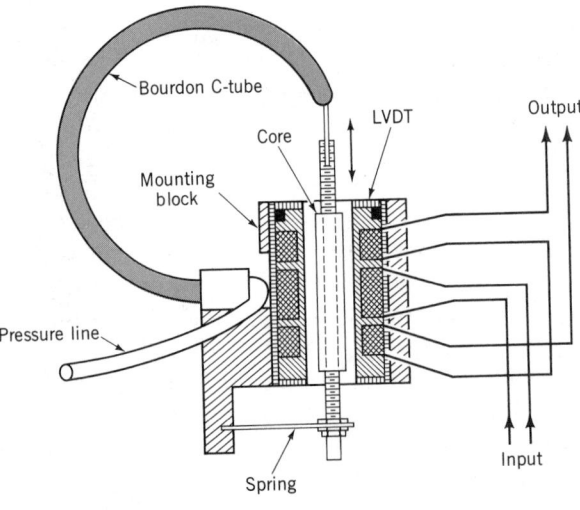

FIGURE 2.14   Pressure transducer which combines a linear variable differential transformer (LVDT) with a Bourdon gage. (From Ref. 4, used by permission.)

FIGURE 2.15
(*a*) Two different sized strain-gage pressure transducers (Spectramed Models P10EZ and P23XL) commonly used to measure physiological pressures. Plastic domes are filled with fluid and connected to blood vessels through a needle or catheter. (Photograph courtesy of Spectramed, Inc.)
(*b*) Schematic diagram of the P23XL transducer with the dome removed. Deflection of the diaphragm due to pressure is measured with a silicon beam on which strain gages and an associated bridge circuit have been deposited.

linear function of the pressure and could be recorded on an oscillograph, or digitized for storage or processing on a computer.

One disadvantage of a pressure transducer using a Bourdon tube as the elastic sensing element is that it is limited to the measurement of pressures that are static or only changing slowly (quasistatic). Because of the relatively large mass of the Bourdon tube it cannot respond to rapid changes in pressure. To overcome this difficulty a different type of transducer is used in which the sensing element is a thin, elastic diaphragm which is in contact with the fluid. As the pressure changes, the diaphragm deflects, and this deflection can be sensed and converted into an electrical voltage. One way to accomplish this is to locate strain gages either on the surface of the diaphragm not in contact with the fluid, or on an element attached to the diaphragm. These gages can accurately sense the small strains induced in the diaphragm and provide an output voltage proportional to pressure. This type of transducer is capable of measuring accurately both small and large pressures, as well as both static and dynamic pressures. For example, strain-gage pressure transducers of the type shown in Fig. 2.15 are used to measure arterial blood pressure, which is a relatively small pressure that varies periodically with a fundamental frequency of about 1 Hz. The transducer is usually connected to the blood vessel by means of a liquid-filled, small diameter tube called a pressure catheter.

Although the strain-gage type of transducers can be designed to have very good frequency response (up to approximately 10 kHz), they become less sensitive at the higher frequencies since the diaphragm must be made stiffer to achieve the higher frequency response. As an alternative the diaphragm can be constructed of a piezo-electric crystal to be used as both the elastic element and the sensor. When a pressure is applied to the crystal a voltage develops because of the deformation of the crystal. This voltage is directly related to the applied pressure. Depending on the design, this type of transducer can be used to measure both very low and high pressures (up to approximately 100,000 psi) at high frequencies. Additional information on pressure transducers can be found in Refs. 3, 4, and 5.

2.8 HYDROSTATIC FORCE ON A PLANE SURFACE

When a surface is submerged in a fluid, forces develop on the surface due to the fluid. The determination of these forces is important in the design of storage tanks, ships, dams, and other hydraulic structures. For fluids at rest we know that the force must be *perpendicular* to the surface, since there are no shearing stresses present. We also know that the pressure will vary linearly with depth if the fluid is incompressible. For a horizontal surface, such as the bottom of a liquid-filled tank (Fig. 2.16), the magnitude of the resultant force is simply $F_R = pA$, where p is the uniform pressure on the bottom and A is the area of the bottom. For the open tank shown $p = \gamma h$. Note that if atmospheric pressure acts on both sides of the bottom, as is illustrated, the *resultant* force on the bottom is simply due to the liquid in the tank. Since the pressure is constant and uniformly distributed over the bottom, the resultant force acts through the centroid of the area as shown in Fig. 2.16.

For the more general case in which a submerged plane surface is inclined, as is illustrated in Fig. 2.17, the determination of the resultant force acting on the surface is more involved. For the present we will assume that the fluid surface is open to the atmosphere. Let the plane in which the surface lies intersect the free surface at 0 and make an angle θ with this surface as in Fig. 2.17. The area can have an arbitrary shape as shown. We wish to determine the direction, location, and magnitude of the resultant force acting on one side of this area due to the liquid in contact with the area. At any given depth, h, the force acting on dA (the differential area of Fig. 2.17) is $dF = \gamma h \, dA$ and is perpendicular to the surface. Thus, the magnitude of the resultant

FIGURE 2.16 Pressure and resultant hydrostatic force developed on the bottom of an open tank.

FIGURE 2.17
Notation for hy-
drostatic force on
an inclined plane
surface of arbi-
trary shape.

force can be found by summing these differential forces over the entire surface. In equation form

$$F_R = \int_A \gamma h \, dA = \int_A \gamma y \sin \theta \, dA$$

where $h = y \sin \theta$. For constant γ and θ

$$F_R = \gamma \sin \theta \int_A y \, dA \qquad (2.17)$$

The integral appearing in Eq. 2.17 is the *first moment of the area* with respect to the x axis, so we can write

$$\int_A y \, dA = y_c A$$

where y_c is the y coordinate of the centroid measured from the x axis which passes through 0. Equation 2.17 can thus be written as

$$F_R = \gamma A y_c \sin \theta$$

or more simply as

$$F_R = \gamma h_c A \qquad (2.18)$$

where h_c is the vertical distance from the fluid surface to the centroid of the area. Note that the magnitude of the force is independent of the angle θ and depends only on the specific weight of the fluid, the total area, and the depth of the centroid of the area below the surface. In effect, Eq. 2.18 indicates that the magnitude of the resultant force is equal to pressure at the centroid of the area multiplied by the total area. Since all the differential forces that were summed to obtain F_R are perpendicular to the surface, the resultant F_R must also be perpendicular to the surface.

Although our intuition might suggest that the resultant force should pass through the centroid of the area, this is not actually the case. The y coordinate, y_R, of the resultant force can be determined by summation of moments around the x axis. That is, the moment of the resultant force must equal the moment of the distributed pressure force, or

$$F_R y_R = \int_A y \, dF = \int_A \gamma \sin \theta \, y^2 \, dA$$

and, therefore, since $F_R = \gamma A y_c \sin \theta$

$$y_R = \frac{\int_A y^2 \, dA}{y_c A}$$

The integral in the numerator is the *second moment of the area* (*moment of inertia*), I_x, with respect to an axis formed by the intersection of the plane containing the surface and the free surface (x axis). Thus, we can write

$$y_R = \frac{I_x}{y_c A}$$

Use can now be made of the parallel axis theorem to express I_x as

$$I_x = I_{xc} + A y_c^2$$

where I_{xc} is the second moment of the area with respect to an axis passing through its *centroid* and parallel to the x axis. Thus,

$$y_R = \frac{I_{xc}}{y_c A} + y_c \tag{2.19}$$

Equation 2.19 clearly shows that the resultant force does not pass through the centroid but is always *below* it, since $I_{xc}/y_c A > 0$.

The x coordinate, x_R, for the resultant force can be determined in a similar manner by summing moments about the y axis. Thus,

$$F_R x_R = \int_A \gamma \sin \theta \, xy \, dA$$

and, therefore,

$$x_R = \frac{\int_A xy \, dA}{y_c A} = \frac{I_{xy}}{y_c A}$$

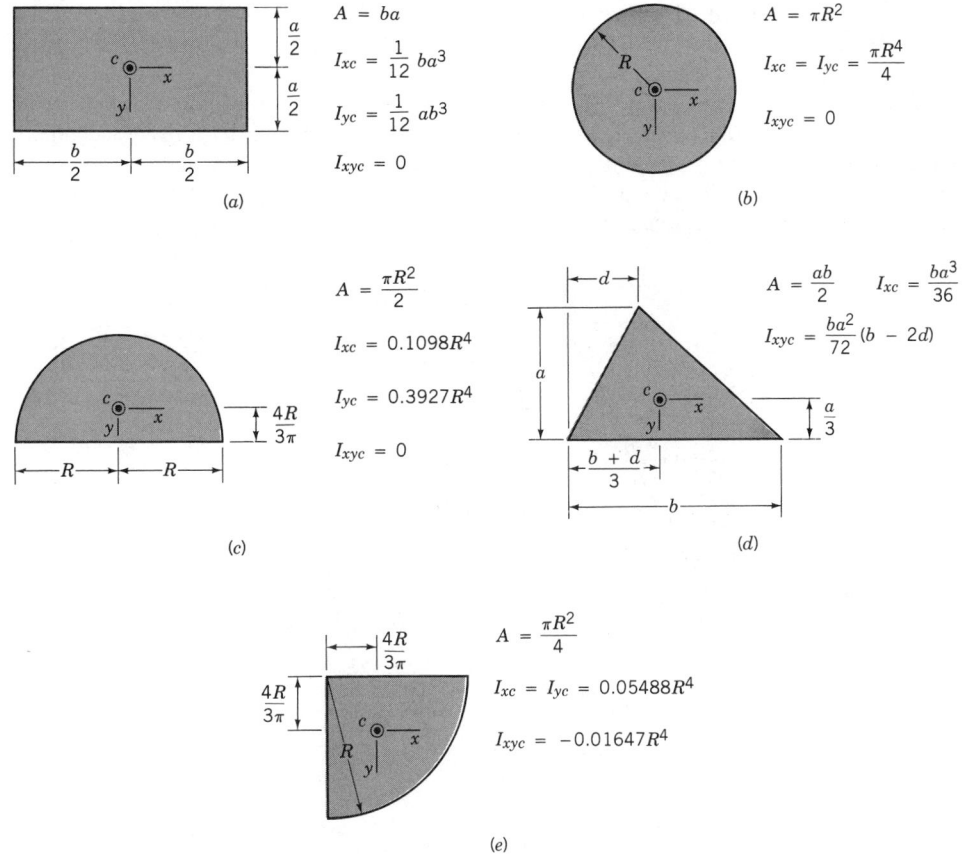

FIGURE 2.18 Geometric properties of some common shapes.

where I_{xy} is the product of inertia with respect to the x and y axes. Again, using the parallel axis theorem,[1] we can write

$$x_R = \frac{I_{xyc}}{y_c A} + x_c \tag{2.20}$$

where I_{xyc} is the product of inertia with respect to an orthogonal coordinate system passing through the *centroid* of the area, and formed by a translation of the xy coordinate system. If the submerged area is symmetrical with respect to an axis passing through the centroid and parallel to either the x or y axes, the resultant force must lie along the line $x = x_c$, since I_{xyc} is identically zero in this case. The point through which the resultant force acts is called the *center of pressure*. It is to be noted from Eqs. 2.19 and 2.20 that as the depth of submergence increases, y_c increases and the center of pressure moves closer to the centroid of the area. Centroidal coordinates and moments of inertia for some common areas are given in Fig. 2.18.

---

[1] Recall that the parallel axis theorem for the product of inertia of an area states that the product of inertia with respect to an orthogonal set of axes (xy coordinate system) is equal to the product of inertia with respect to an orthogonal set of axes parallel to the original set and passing through the centroid of the area, plus the product of the area and the x and y coordinates of the centroid of the area. Thus, $I_{xy} = I_{xyc} + Ax_c y_c$.

EXAMPLE 2.6

The 4-m-diameter circular gate of Fig. E2.6a is located in the inclined wall of a large reservoir containing water ($\gamma = 9.80\ \text{kN/m}^3$). The gate is mounted on a shaft along its horizontal diameter. For a water depth of 10 m above the shaft determine: (a) the magnitude and location of the resultant force exerted on the gate by the water, and (b) the moment that would have to be applied to the shaft to overcome the moment due to the water.

FIGURE E2.6

SOLUTION

(a)   To find the magnitude of the force of the water we can apply Eq. 2.18,

$$F_R = \gamma h_c A$$

and since the vertical distance from the fluid surface to the centroid of the area is 10 m it follows that

$$F_R = (9.80 \times 10^3\ \text{N/m}^3)(10\ \text{m})(4\pi\ \text{m}^2)$$

$$= 1230 \times 10^3\ \text{N} = 1.23\ \text{MN} \qquad \text{(Ans)}$$

To locate the point (center of pressure) through which F_R acts, we use Eqs. 2.19 and 2.20,

$$x_R = \frac{I_{xyc}}{y_c A} + x_c \qquad y_R = \frac{I_{xc}}{y_c A} + y_c$$

For the coordinate system shown $x_R = 0$ since the area is symmetrical, and the center of pressure must lie along the diameter A–A. To obtain y_R, we have from Fig. 2.18

$$I_{xc} = \frac{\pi R^4}{4}$$

and y_c is shown in Fig. E2.6b. Thus,

$$y_R = \frac{(\pi/4)(2\ \text{m})^4}{(10\ \text{m/sin}\ 60°)(4\pi\ \text{m}^2)} + \frac{10\ \text{m}}{\sin\ 60°}$$

$$= 0.0866\ \text{m} + 11.55\ \text{m} = 11.6\ \text{m}$$

and the distance (along the gate) below the shaft to the center of pressure is

$$y_R - y_c = 0.0866 \text{ m} \qquad \text{(Ans)}$$

We can conclude from this analysis that the force on the gate due to the water has a magnitude of 1.23 MN and acts through a point along its diameter A–A at a distance of 0.0866 m (along the gate) below the shaft. The force is perpendicular to the gate surface as shown.

(b) The moment required to open the gate can be obtained with the aid of the free-body diagram of Fig. E2.6c. In this diagram W is the weight of the gate and O_x and O_y are the horizontal and vertical reactions of the shaft on the gate. We can now sum moments about the shaft

$$\sum M_c = 0$$

and, therefore,

$$\begin{aligned} M &= F_R(y_R - y_c) \\ &= (1230 \times 10^3 \text{ N})(0.0866 \text{ m}) \\ &= 1.07 \times 10^5 \text{ N·m} \qquad \text{(Ans)} \end{aligned}$$

EXAMPLE 2.7

A large fish-holding tank contains seawater ($\gamma = 64.0 \text{ lb/ft}^3$) to a depth of 10 ft as shown in Fig. E2.7a. To repair some damage to one corner of the tank, a triangular section is replaced with a new section as illustrated. Determine the magnitude and location of the force of the seawater on this triangular area.

(a)

(b)

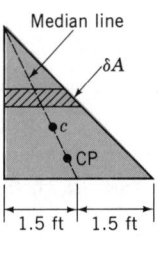

(c)

FIGURE E2.7

SOLUTION

The various distances needed to solve this problem are shown in Fig. E2.7b. Since the surface of interest lies in a vertical plane, $y_c = h_c = 9$ ft and from Eq. 2.18 the magnitude of the force is

$$F_R = \gamma h_c A$$

$$= (64.0 \text{ lb/ft}^3)(9 \text{ ft})(9/2 \text{ ft}^2) = 2590 \text{ lb} \qquad \text{(Ans)}$$

Note that this force is independent of the tank length. The result is the same if the tank is 0.25 ft, 25 ft, or 25 miles long. The y coordinate of the center of pressure (CP) is found from Eq. 2.19,

$$y_R = \frac{I_{xc}}{y_c A} + y_c$$

and from Fig. 2.18

$$I_{xc} = \frac{(3 \text{ ft})(3 \text{ ft})^3}{36} = \frac{81}{36} \text{ ft}^4$$

so that

$$y_R = \frac{81/36 \text{ ft}^4}{(9 \text{ ft})(9/2 \text{ ft}^2)} + 9 \text{ ft}$$

$$= 0.0556 \text{ ft} + 9 \text{ ft} = 9.06 \text{ ft} \qquad \text{(Ans)}$$

Similarly, from Eq. 2.20

$$x_R = \frac{I_{xyc}}{y_c A} + x_c$$

and from Fig. 2.18

$$I_{xyc} = \frac{(3 \text{ ft})(3 \text{ ft})^2}{72}(3 \text{ ft}) = \frac{81}{72} \text{ ft}^4$$

so that

$$x_R = \frac{81/72 \text{ ft}^4}{(9 \text{ ft})(9/2 \text{ ft}^2)} + 0 = 0.0278 \text{ ft} \qquad \text{(Ans)}$$

Thus, we conclude that the center of pressure is 0.0278 ft to the right of and 0.0556 ft below the centroid of the area. If this point is plotted, we find that it lies on the median line for the area as illustrated in Fig. E2.7c. Since we can think of the total area as consisting of a number of small rectangular strips of area δA (and the fluid force on each of these small areas acts through its center), it follows that the resultant of all these parallel forces must lie along the median.

2.9 PRESSURE PRISM

An informative and useful graphical interpretation can be made for the force developed by a fluid acting on a plane area. Consider the pressure distribution along a vertical wall of a tank of width b, which contains a liquid having a specific weight γ. Since the pressure must vary linearly with depth, we can represent the variation as is shown in Fig. 2.19a, where the pressure is equal to zero at the upper surface and equal to γh at the bottom. It is apparent from this diagram that the average pressure occurs at the depth $h/2$, and therefore the resultant force acting on the rectangular area $A = bh$ is

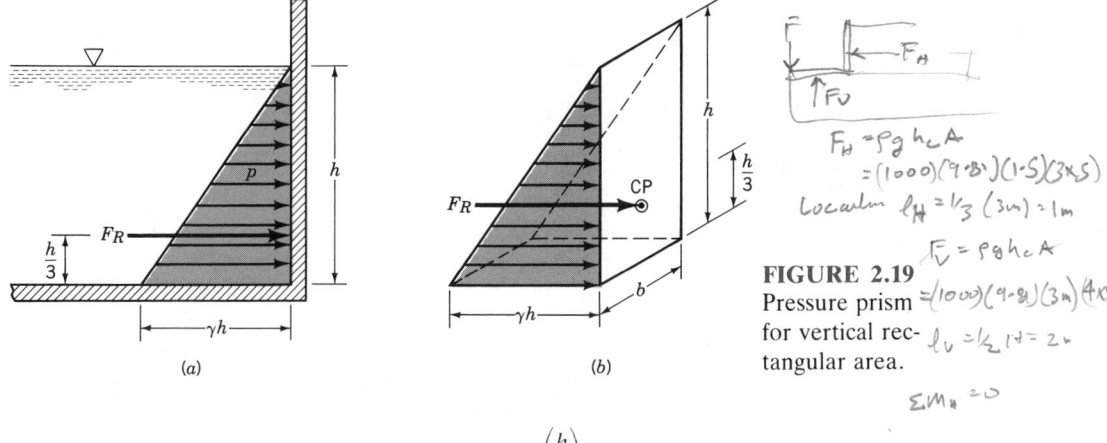

FIGURE 2.19 Pressure prism for vertical rectangular area.

$$F_R = p_{av} A = \gamma \left(\frac{h}{2}\right) A$$

which is the same result as obtained from Eq. 2.18. The pressure distribution shown in Fig. 2.19a applies across the vertical surface so we can draw the three-dimensional representation of the pressure distribution as shown in Fig. 2.19b. The base of this "volume" in pressure–area space is the plane surface of interest, and its altitude at each point is the pressure. This volume is called the *pressure prism*, and it is clear that the magnitude of the resultant force acting on the surface is equal to the volume of the pressure prism. Thus, for the prism of Fig. 2.19b the fluid force is

$$F_R = \text{volume} = \frac{1}{2}(\gamma h)(bh) = \gamma \left(\frac{h}{2}\right) A$$

where bh is the area of the rectangular surface, A.

The resultant force must pass through the *centroid* of the pressure prism. For the volume under consideration the centroid is located along the vertical axis of symmetry of the surface, and at a distance of $h/3$ above the base (since the centroid of a triangle is located at $h/3$ above its base). This result can readily be shown to be consistent with that obtained from Eqs. 2.19 and 2.20.

This same graphical approach can be used for plane surfaces that do not extend up to the fluid surface as illustrated in Fig. 2.20a. In this instance, the cross section of the pressure prism is trapezoidal. However, the resultant force is still equal in

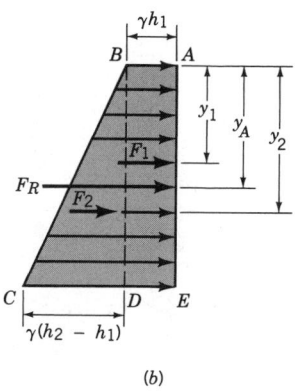

FIGURE 2.20 Graphical representation of hydrostatic forces on a vertical rectangular surface.

FIGURE 2.21   Pressure variation along an inclined plane area.

magnitude to the volume of the pressure prism, and it passes through the centroid of the volume. Specific values can be obtained by decomposing the pressure prism into two parts, $ABDE$ and BCD, as shown in Fig. 2.20b. Thus,

$$F_R = F_1 + F_2$$

where the components can readily be determined by inspection for rectangular surfaces. The location of F_R can be determined by summing moments about some convenient axis, such as one passing through A. In this instance

$$F_R y_A = F_1 y_1 + F_2 y_2$$

and y_1 and y_2 can be determined by inspection.

For inclined plane surfaces the pressure prism can still be developed, and the cross section of the prism will generally be trapezoidal as is shown in Fig. 2.21. Although it is usually convenient to measure distances along the inclined surface, the pressures developed depend on the vertical distances as illustrated.

The use of pressure prisms for determining the force on submerged plane areas is convenient if the area is rectangular so the volume and centroid can be easily determined. However, for other nonrectangular shapes, integration would generally be needed to determine the volume and centroid. In these circumstances it is more convenient to use the equations developed in the previous section, in which the necessary integrations have been made and the results presented in a convenient and compact form that is applicable to submerged plane areas of any shape.

The effect of atmospheric pressure on a submerged area has not yet been considered, and we may ask how this pressure will influence the resultant force. If we again consider the pressure distribution on a plane vertical wall, as is shown in Fig. 2.22a, the pressure varied from zero at the surface to γh at the bottom. Since we are setting the surface pressure equal to zero, we are using atmospheric pressure as our datum, and thus the pressure used in the determination of the fluid force is gage pressure. If

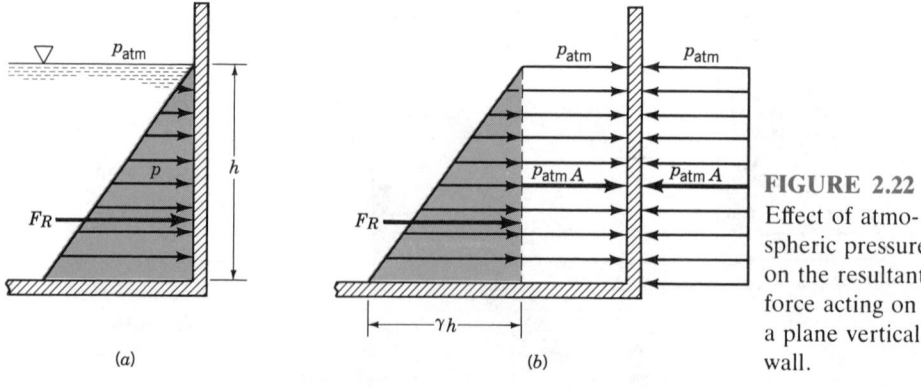

FIGURE 2.22 Effect of atmospheric pressure on the resultant force acting on a plane vertical wall.

we wish to include atmospheric pressure, the pressure distribution will be as is shown in Fig. 2.22b. We note that in this case the force on one side of the wall now consists of F_R as a result of the hydrostatic pressure distribution, plus the contribution of the atmospheric pressure, $p_{atm}A$, where A is the area of the surface. However, if we are going to include the effect of atmospheric pressure on one side of the wall we must realize that this same pressure acts on the outside surface (assuming it is exposed to the atmosphere), so that an equal and opposite force will be developed as illustrated in the figure. Thus, we conclude that the *resultant* fluid force on the surface is that due only to the gage pressure contribution of the liquid in contact with the surface— the atmospheric pressure does not contribute to this resultant. Of course, if the surface pressure of the liquid is different from atmospheric pressure (such as might occur in a closed tank), the resultant force acting on a submerged area, A, will be changed in magnitude from that caused simply by hydrostatic pressure by an amount $p_s A$, where p_s is the gage pressure at the liquid surface (the outside surface is assumed to be exposed to atmospheric pressure).

EXAMPLE 2.8

A pressurized tank contains oil ($SG = 0.90$) and has a square, 0.6 m by 0.6 m plate bolted to its side, as is illustrated in Fig. E2.8a. When the pressure gage on the top of the tank reads 50 kPa, what is the magnitude and location of the resultant force on the attached plate? The outside of the tank is at atmospheric pressure.

(a)

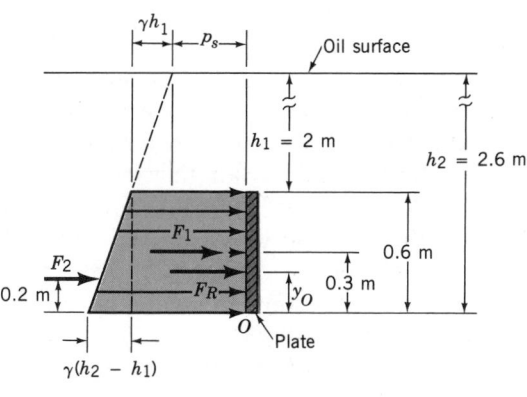

(b)

FIGURE E2.8

SOLUTION

The pressure distribution acting on the inside surface of the plate is shown in Fig. E2.8b. The pressure at a given point on the plate is due to the air pressure, p_s, at the oil surface, and the pressure due to the oil, which varies linearly with depth as is shown in the figure. The resultant force on the plate (having an area A) is due to the components, F_1 and F_2, with

$$F_1 = (p_s + \gamma h_1)A$$
$$= [50 \times 10^3 \text{ N/m}^2 + (0.90)(9.81 \times 10^3 \text{ N/m}^3)(2 \text{ m})](0.36 \text{ m}^2)$$
$$= 24.4 \times 10^3 \text{ N}$$

and F_2

$$F_2 = \gamma \left(\frac{h_2 - h_1}{2}\right) A$$
$$= (0.90)(9.81 \times 10^3 \text{ N/m}^3) \left(\frac{0.6 \text{ m}}{2}\right) (0.36 \text{ m}^2)$$
$$= 0.954 \times 10^3 \text{ N}$$

The magnitude of the resultant force, F_R, is therefore

$$F_R = F_1 + F_2 = 25.4 \times 10^3 \text{ N} = 25.4 \text{ kN} \qquad \text{(Ans)}$$

The vertical location of F_R can be obtained by summing moments around an axis through point O so that

$$F_R y_O = F_1(0.3 \text{ m}) + F_2(0.2 \text{ m})$$

or

$$(25.4 \times 10^3 \text{ N}) y_O = (24.4 \times 10^3 \text{ N})(0.3 \text{ m}) + (0.954 \times 10^3 \text{ N})(0.2 \text{ m})$$

$$y_O = 0.296 \text{ m} \qquad \text{(Ans)}$$

Thus, the force acts at a distance of 0.296 m above the bottom of the plate along the vertical axis of symmetry.

Note that the air pressure used in the calculation of the force was gage pressure. Atmospheric pressure does not affect the resultant force (magnitude or location), since it acts on both sides of the plate, thereby canceling its effect.

2.10   HYDROSTATIC FORCE ON A CURVED SURFACE

The equations developed in Section 2.8 for the magnitude and location of the resultant force acting on a submerged surface only apply to plane surfaces. However, many surfaces of interest (such as those associated with dams, pipes, and tanks) are non-planar. Although the resultant fluid force can be determined by integration, as was done for the plane surfaces, this is generally a rather tedious process and no simple, general formulas can be developed. As an alternative approach we will consider the equilibrium of the fluid volume enclosed by the curved surface of interest and the horizontal and vertical projections of this surface.

For example, consider the curved section BC of the open tank of Fig. 2.23a. We wish to find the resultant fluid force acting on this section, which has a unit length perpendicular to the plane of the paper. We first isolate a volume of fluid that is bounded by the surface of interest, in this instance section BC, and the horizontal

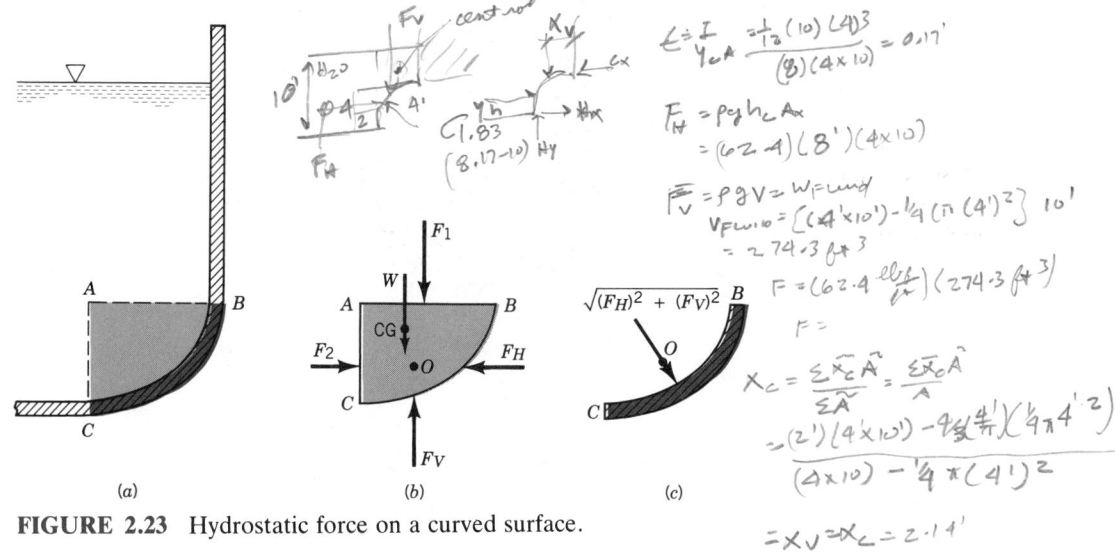

FIGURE 2.23 Hydrostatic force on a curved surface.

plane surface AB and the vertical plane surface AC. The free-body diagram for this volume is shown in Fig. 2.23b. The magnitude and location of forces F_1 and F_2 can be determined from the relationships for planar surfaces. The weight, W, is simply the specific weight of the fluid times the enclosed volume and acts through the center of gravity (CG) of the mass of fluid contained within the volume. The forces F_H and F_V represent the components of the force that the tank *exerts on the fluid.*

In order for this force system to be in equilibrium, the horizontal component F_H must be equal in magnitude and collinear with F_2, and the vertical component F_V equal in magnitude and collinear with the resultant of the vertical forces F_1 and W. This follows since the three forces acting on the fluid mass (F_2, the resultant of F_1 and W, and the resultant force that the tank exerts on the mass) must form a *concurrent* force system. That is, from the principles of statics, it is known that when a body is held in equilibrium by three nonparallel forces they must be concurrent (their lines of action intersect at a common point), and coplanar. Thus,

$$F_H = F_2$$

$$F_V = F_1 + W$$

and the magnitude of the resultant is obtained from the equation

$$F_R = \sqrt{(F_H)^2 + (F_V)^2}$$

The resultant F_R passes through the point O, which can be located by summing moments about an appropriate axis. The resultant force of the fluid acting *on the curved surface BC* is equal and opposite in direction to that obtained from the free-body diagram of Fig. 2.23b. The desired fluid force is shown in Fig. 2.23c.

EXAMPLE 2.9

The 6-ft-diameter drainage conduit of Fig. E2.9a is half full of water at rest. Determine the magnitude and line of action of the resultant force that the water exerts on a 1-ft length of the curved section BC of the conduit wall.

(a)

(b)

(c)

FIGURE E2.9

SOLUTION

We first isolate a volume of fluid bounded by the curved section BC, the horizontal surface AB, and the vertical surface AC, as shown in Fig. E2.9b. The volume has a length of 1 ft. The forces acting on the volume are the horizontal force, F_1, which acts on the vertical surface AC, the weight, W, of the fluid contained within the volume, and the horizontal and vertical components of the force of the conduit wall on the fluid, F_H and F_V, respectively.

The magnitude of F_1 is found from the equation

$$F_1 = \gamma h_c A = (62.4 \text{ lb/ft}^3)(\tfrac{3}{2} \text{ ft})(3 \text{ ft}^2) = 281 \text{ lb}$$

and this force acts 1 ft above C as shown. The weight, W, is

$$W = \gamma \text{ vol} = (62.4 \text{ lb/ft}^3)(9\pi/4 \text{ ft}^2)(1 \text{ ft}) = 441 \text{ lb}$$

and acts through the center of gravity of the mass of fluid, which according to Fig. 2.18 is located 1.27 ft to the right of AC as shown. Therefore, to satisfy equilibrium

$$F_H = F_1 = 281 \text{ lb}$$

$$F_V = W = 441 \text{ lb}$$

and the magnitude of the resultant force is

$$F_R = \sqrt{(F_H)^2 + (F_V)^2}$$
$$= \sqrt{(281 \text{ lb})^2 + (441 \text{ lb})^2} = 523 \text{ lb} \qquad \text{(Ans)}$$

The force the water exerts *on* the conduit wall is equal, but *opposite in direction*, to the forces F_H and F_V shown in Fig. E2.9b. Thus, the resultant force *on the conduit wall* is shown in Fig. E2.9c. This force acts through the point O at the angle shown.

An inspection of this result will show that the line of action of the resultant force passes through the center of the conduit. In retrospect, this is not a surprising result, since at each point on the curved surface of the conduit the elemental force due to the pressure is normal to the surface, and each line of action must pass through the center of the conduit. It therefore follows that the resultant of this concurrent force system must also pass through the center of concurrence of the elemental forces that make up the system.

This same general approach can also be used for determining the force on curved surfaces of pressurized, closed tanks. If these tanks contain a gas, the weight of the gas is usually negligible in comparison with the forces developed by the pressure. Thus, the forces (such as F_1 and F_2 in Fig. 2.23b) on horizontal and vertical projections of the curved surface of interest can simply be expressed as the internal pressure times the appropriate projected area.

2.11 BUOYANCY, FLOTATION, AND STABILITY

2.11.1 Archimedes' Principle

When a body is completely submerged in a fluid, or floating so that it is only partially submerged, the resultant fluid force acting on the body is called the *buoyant force*. A net upward vertical force results because pressure increases with depth and the pressure forces acting from below are larger than the pressure forces acting from above. This force can be determined through an approach similar to that used in the previous article for forces on curved surfaces. Consider a body of arbitrary shape, having a volume V, that is immersed in a fluid as illustrated in Fig. 2.24a. We enclose the body in a parallelepiped and draw a free-body diagram of the parallelepiped with the body removed as shown in Fig. 2.24b. Note that the forces F_1, F_2, F_3, and F_4 are

(a)

(b)

(c)

(d)

FIGURE 2.24
Buoyant force on submerged and floating bodies.

simply the forces exerted on the plane surfaces of the parallelepiped (for simplicity the forces in the x direction are not shown), W is the weight of the shaded fluid volume (parallelepiped minus body), and F_B is the force the body is exerting *on the fluid*. The forces on the vertical surfaces, such as F_3 and F_4, are all equal and cancel so the equilibrium equation of interest is in the z direction and can be expressed as

$$F_B = F_2 - F_1 - W \tag{2.21}$$

If the specific weight of the fluid is constant, then

$$F_2 - F_1 = \gamma(h_2 - h_1)A$$

where A is the horizontal area of the upper (or lower) surface of the parallelepiped, and Eq. 2.21 can be written as

$$F_B = \gamma(h_2 - h_1)A - \gamma[(h_2 - h_1)A - \forall]$$

Simplifying, we arrive at the desired expression for the buoyant force

$$F_B = \gamma\forall \tag{2.22}$$

where γ is the specific weight of the fluid and $\forall$ is the volume of the body. The direction of the buoyant force, which is the force of the fluid *on the body,* is opposite to that shown on the free-body diagram. Therefore, the buoyant force has a magnitude equal to the weight of the fluid displaced by the body, and is directed vertically upward. This result is commonly referred to as *Archimedes' principle* in honor of Archimedes (287–212 B.C.), a Greek mechanician and mathematician who first enunciated the basic ideas associated with hydrostatics.

The location of the line of action of the buoyant force can be determined by summing moments of the forces shown on the free-body diagram in Fig. 2.24*b* with respect to some convenient axis. For example, summing moments about an axis perpendicular to the paper through point D we have

$$F_B y_c = F_2 y_1 - F_1 y_1 - W y_2$$

and on substitution for the various forces

$$\forall y_c = \forall_T y_1 - (\forall_T - \forall) y_2 \tag{2.23}$$

where $\forall_T$ is the total volume $(h_2 - h_1)A$. The right-hand side of Eq. 2.23 is the first moment of the displaced volume $\forall$ with respect to the xz plane so that y_c is equal to the y coordinate of the centroid of the volume $\forall$. In a similar fashion it can be shown that the x coordinate of the buoyant force coincides with the x coordinate of the centroid. Thus, we conclude that the *buoyant force passes through the centroid of the displaced volume* as shown in Fig. 2.24*c*. The point through which the buoyant force acts is called the *center of buoyancy*.

These same results apply to floating bodies which are only partially submerged, as illustrated in Fig. 2.24*d*, if the specific weight of the fluid above the liquid surface is very small compared with the liquid in which the body floats. Since the fluid above the surface is usually air, for practical purposes this condition is satisfied.

In the derivations presented above, the fluid is assumed to have a constant specific weight, γ. If a body is immersed in a fluid in which γ varies with depth, such as in a layered fluid, the magnitude of the buoyant force remains equal to the weight of the displaced fluid. However, the buoyant force does not pass through the centroid of the displaced volume, but rather, it passes through the center of gravity of the displaced volume.

EXAMPLE 2.10

A spherical buoy has a diameter of 1.5 m, weighs 8.50 kN, and is anchored to the sea floor with a cable as is shown in Fig. E2.10a. Although the buoy normally floats on the surface, at certain times the water depth increases so that the buoy is completely immersed as illustrated. For this condition what is the tension of the cable?

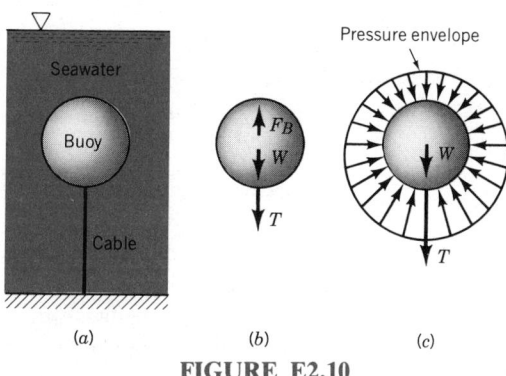

FIGURE E2.10

SOLUTION

We first draw a free-body diagram of the buoy as is shown in Fig. E2.10b, where F_B is the buoyant force acting on the buoy, W is the weight of the buoy, and T is the tension in the cable. For equilibrium it follows that

$$T = F_B - W$$

From Eq. 2.22

$$F_B = \gamma \forall$$

and for seawater with $\gamma = 10.1$ kN/m³ and $\forall = \pi d^3/6$ then

$$F_B = (10.1 \times 10^3 \text{ N/m}^3)[(\pi/6)(1.5 \text{ m})^3] = 1.785 \times 10^4 \text{ N}$$

The tension in the cable can now be calculated as

$$T = 1.785 \times 10^4 \text{ N} - 0.850 \times 10^4 \text{ N} = 9.35 \text{ kN} \qquad \text{(Ans)}$$

Note that we replaced the effect of the hydrostatic pressure force on the body by the buoyant force, F_B. Another correct free-body diagram of the buoy is shown in Fig. E2.10c. The net effect of the pressure forces on the surface of the buoy is equal to the upward force of magnitude, F_B (the buoyant force).

2.11.2   Stability

Another interesting and important problem associated with submerged or floating bodies is concerned with the stability of the bodies. A body is said to be in a *stable equilibrium* position if, when displaced, it returns to its equilibrium position. Conversely, it is in an *unstable equilibrium* position if, when displaced (even slightly), it moves to a new equilibrium position. Stability considerations are particularly important for submerged or floating bodies since the centers of buoyancy and gravity do not necessarily coincide. A small rotation can result in either a restoring or overturning

FIGURE 2.25 Stability of a completely immersed body—center of gravity below centroid.

FIGURE 2.26 Stability of a completely immersed body—center of gravity above centroid.

couple. For example, for the *completely* submerged body shown in Fig. 2.25, which has a center of gravity below the center of buoyancy, a rotation from its equilibrium position will create a restoring couple formed by the weight, W, and the buoyant force, F_B, which causes the body to rotate back to its original position. Thus, for this configuration the body is stable. It is to be noted that as long as the center of gravity falls *below* the center of buoyancy, this will always be true; that is, the body is in a *stable equilibrium* position with respect to small rotations. However, as is illustrated in Fig. 2.26, if the center of gravity is above the center of buoyancy, the resulting couple formed by the weight and the buoyant force will cause the body to overturn and move to a new equilibrium position. Thus, a completely submerged body with its center of gravity *above* its center of buoyancy is in an *unstable equilibrium* position.

For *floating* bodies the stability problem is more complicated, since as the body rotates the location of the center of buoyancy (which passes through the centroid of the displaced volume) may change. As is shown in Fig. 2.27, a floating body such as a barge that rides low in the water can be stable even though the center of gravity lies above the center of buoyancy. This is true since as the body rotates the buoyant force, F_B, shifts to pass through the centroid of the newly formed displaced volume and, as illustrated, combines with the weight, W, to form a couple which will cause the body to return to its original equilibrium position. However, for the relatively tall, slender body shown in Fig. 2.28, a small rotational displacement can cause the buoyant force and the weight to form an overturning couple as illustrated.

It is clear from these simple examples that the determination of the stability of submerged or floating bodies can be difficult since the analysis depends in a complicated fashion on the particular geometry and weight distribution of the body. The problem can be further complicated by the necessary inclusion of other types of

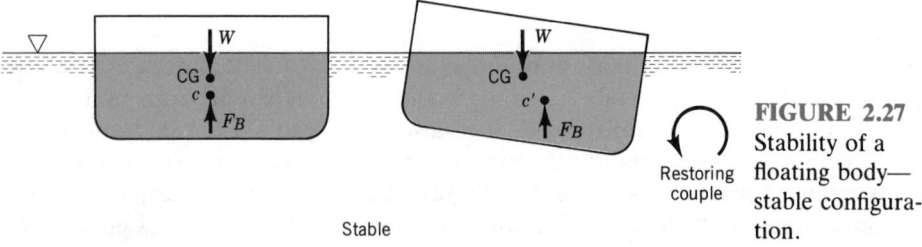

FIGURE 2.27 Stability of a floating body—stable configuration.

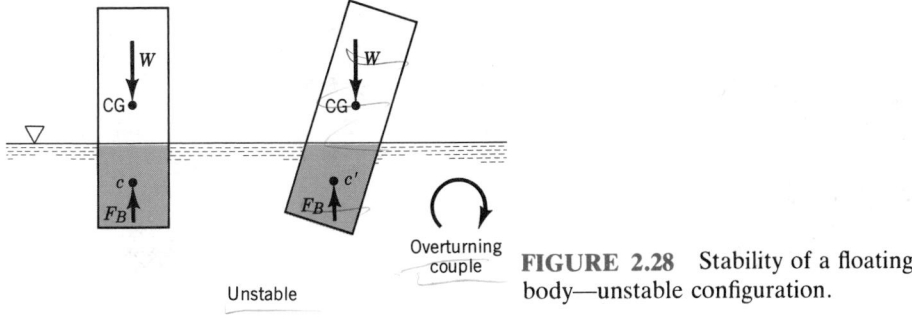

FIGURE 2.28 Stability of a floating body—unstable configuration.

external forces such as those induced by wind gusts or currents. Stability considerations are obviously of great importance in the design of ships, submarines, bathyscapes, and so forth, and such considerations play a significant role in the work of naval architects (see, for example, Ref. 6).

2.12 PRESSURE VARIATION IN A FLUID WITH RIGID-BODY MOTION

Although in this chapter we have been primarily concerned with fluids at rest, the general equation of motion (Eq. 2.2)

$$-\nabla p - \gamma \hat{\mathbf{k}} = \rho \mathbf{a}$$

was developed for both fluids at rest and fluids in motion, with the only stipulation being that there were no shearing stresses present. Equation 2.2 in component form, based on rectangular coordinates with the positive z axis being vertically upward, can be expressed as

$$-\frac{\partial p}{\partial x} = \rho a_x$$

$$-\frac{\partial p}{\partial y} = \rho a_y \qquad\qquad (2.24)$$

$$-\frac{\partial p}{\partial z} = \gamma + \rho a_z$$

A general class of problems involving fluid motion in which there are no shearing stresses occurs when a mass of fluid undergoes rigid-body motion. For example, if a container of fluid accelerates along a straight path, the fluid will move as a rigid mass (after the initial sloshing motion has died out) with each particle having the same acceleration. Since there is no deformation, there will be no shearing stresses and, therefore, Eq. 2.2 applies. Similarly, if a fluid is contained in a tank that rotates about a fixed axis, the fluid will simply rotate with the tank as a rigid body, and again Eq. 2.2 can be applied to obtain the pressure distribution throughout the moving fluid. Specific results for these two cases (rigid-body uniform motion and rigid-body rotation) are developed in the following two sections. Although problems relating to fluids having rigid-body motion are not, strictly speaking, "fluid statics" problems, they are included in this chapter because, as we will see, the analysis and resulting pressure relationships are similar to those for fluids at rest.

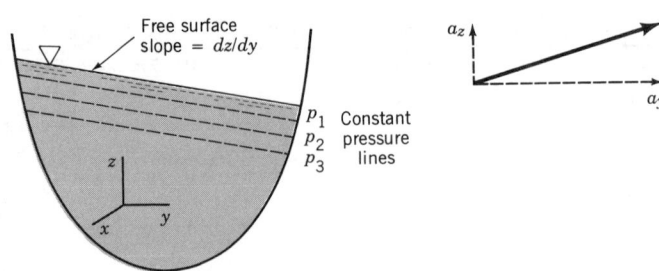

FIGURE 2.29 Linear acceleration of a liquid with a free surface.

2.12.1   Linear Motion

We first consider an open container of a liquid that is translating along a straight path with a constant acceleration **a** as illustrated in Fig. 2.29. Since $a_x = 0$ it follows from the first of Eqs. 2.24 that the pressure gradient in the x direction is zero $(\partial p / \partial x = 0)$. In the y and z directions

$$\frac{\partial p}{\partial y} = -\rho a_y \tag{2.25}$$

$$\frac{\partial p}{\partial z} = -\rho(g + a_z) \tag{2.26}$$

The change in pressure between two closely spaced points located at y, z, and $y + dy$, $z + dz$ can be expressed as

$$dp = \frac{\partial p}{\partial y} dy + \frac{\partial p}{\partial z} dz$$

or in terms of the results from Eqs. 2.25 and 2.26

$$dp = -\rho a_y \, dy - \rho(g + a_z) \, dz \tag{2.27}$$

Along a line of *constant* pressure, $dp = 0$, and therefore from Eq. 2.27 it follows that the slope of this line is given by the relationship

$$\frac{dz}{dy} = -\frac{a_y}{g + a_z} \tag{2.28}$$

Along a free surface the pressure is constant, so that for the accelerating mass shown in Fig. 2.29 the free surface will be inclined if $a_y \neq 0$. In addition, all lines of constant pressure will be parallel to the free surface as illustrated.

For the special circumstance in which $a_y = 0$, $a_z \neq 0$, which corresponds to the mass of fluid accelerating in the vertical direction, Eq. 2.28 indicates that the fluid surface will be horizontal. However, from Eq. 2.26 we see that the pressure distribution is not hydrostatic, but is given by the equation:

$$\frac{dp}{dz} = -\rho(g + a_z)$$

For fluids of constant density this equation shows that the pressure will vary linearly with depth, but the variation is due to the combined effects of gravity and the externally induced acceleration, $\rho(g + a_z)$, rather than simply the specific weight ρg. Thus, for example, the pressure along the bottom of a liquid-filled tank which is resting on the floor of an elevator that is accelerating upward will be increased over that which exists when the tank is at rest (or moving with a constant velocity). It is to be noted that for a *freely falling* fluid mass ($a_z = -g$), the pressure gradients in all three coordinate directions are zero, which means that if the pressure surrounding the mass is zero, the pressure throughout will be zero.

EXAMPLE 2.11

The cross section for the fuel tank of an experimental vehicle is shown in Fig. E2.11. The rectangular tank is vented to the atmosphere, and a pressure transducer is located in its side as illustrated. During testing of the vehicle, the tank is subjected to a constant linear acceleration, a_y. (a) Determine an expression that relates a_y and the pressure (in lb/ft²) at the transducer for a fuel with a $SG = 0.65$. (b) What is the maximum acceleration that can occur before the fuel level drops below the transducer?

FIGURE E2.11

SOLUTION

(a)   For a constant horizontal acceleration the fuel will move as a rigid body, and from Eq. 2.28 the slope of the fuel surface can be expressed as

$$\frac{dz}{dy} = -\frac{a_y}{g}$$

since $a_z = 0$. Thus, for some arbitrary a_y, the change in depth, z_1, of liquid on the right side of the tank can be found from the equation

$$-\frac{z_1}{0.75 \text{ ft}} = -\frac{a_y}{g}$$

or

$$z_1 = (0.75 \text{ ft}) \left(\frac{a_y}{g}\right)$$

Since there is no acceleration in the vertical, z direction, the pressure along the wall varies hydrostatically as shown by Eq. 2.26. Thus, the pressure at the transducer is given by the relationship

$$p = \gamma h$$

where h is the depth of fuel above the transducer, and therefore

$$p = (0.65)(62.4 \text{ lb/ft}^3)[0.5 \text{ ft} - (0.75 \text{ ft})(a_y/g)]$$

$$= 20.3 - 30.4 \frac{a_y}{g} \qquad \text{(Ans)}$$

for $z_1 \geq 0.5$ ft. As written, p would be given in lb/ft^2.

(b)   The limiting value for a_y (when the fuel level reaches the transducer) can be found from the equation

$$0.5 \text{ ft} = (0.75 \text{ ft}) \left[\frac{(a_y)_{max}}{g} \right]$$

or

$$(a_y)_{max} = \frac{2g}{3}$$

and for standard acceleration of gravity

$$(a_y)_{max} = \tfrac{2}{3} (32.2 \text{ ft/s}^2) = 21.5 \text{ ft/s}^2 \qquad \text{(Ans)}$$

Note that the pressure in horizontal layers is not constant in this example since $\partial p / \partial y = -\rho a_y \neq 0$. Thus, for example, $p_1 \neq p_2$.

2.12.2   Rigid-Body Rotation

After an initial "start-up" transient, a fluid contained in a tank that rotates with a constant angular velocity ω about an axis as is shown in Fig. 2.30 will rotate with the tank as a rigid body. It is known from elementary particle dynamics that the acceleration of a fluid particle located at a distance r from the axis of rotation is equal in magnitude to $r\omega^2$, and the direction of the acceleration is toward the axis of rotation as is illustrated in the figure. Since the paths of the fluid particles are circular, it is convenient to use cylindrical polar coordinates r, θ, and z, defined in the insert in Fig. 2.30. It will be shown in Chapter 6 that in terms of cylindrical coordinates the pressure gradient ∇p can be expressed as

$$\nabla p = \frac{\partial p}{\partial r} \, \hat{\mathbf{e}}_r + \frac{1}{r} \frac{\partial p}{\partial \theta} \, \hat{\mathbf{e}}_\theta + \frac{\partial p}{\partial z} \, \hat{\mathbf{e}}_z \qquad (2.29)$$

Thus, in terms of this coordinate system

$$\mathbf{a}_r = -r\omega^2 \, \hat{\mathbf{e}}_r, \qquad \mathbf{a}_\theta = 0 \qquad \mathbf{a}_z = 0$$

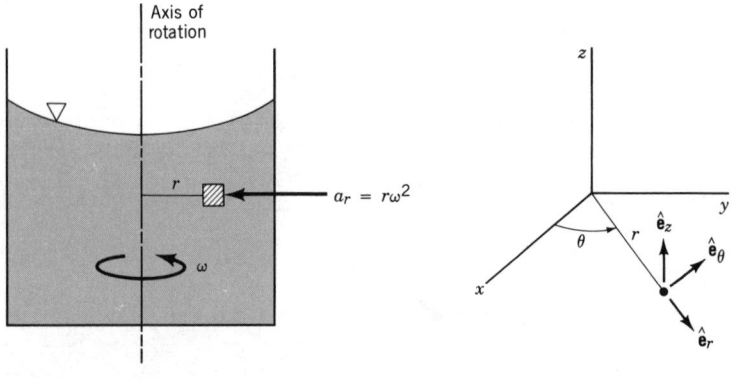

FIGURE 2.30
Rigid-body rotation of a liquid in a tank.

and from Eq. 2.2

$$\frac{\partial p}{\partial r} = \rho r \omega^2$$

$$\frac{\partial p}{\partial \theta} = 0$$

$$\frac{\partial p}{\partial z} = -\gamma \qquad (2.30)$$

These results show that for this type of rigid-body rotation, the pressure is a function of two variables r and z, and therefore the differential pressure is

$$dp = \frac{\partial p}{\partial r} dr + \frac{\partial p}{\partial z} dz$$

or

$$dp = \rho r \omega^2 dr - \gamma dz \qquad (2.31)$$

Along a surface of constant pressure, such as the free surface, $dp = 0$, so that from Eq. 2.31

$$\frac{dz}{dr} = \frac{r \omega^2}{g}$$

and, therefore, the equation for surfaces of constant pressure is

$$z = \frac{\omega^2 r^2}{2g} + \text{constant} \qquad (2.32)$$

This equation reveals that these surfaces of constant pressure are parabolic as illustrated in Fig. 2.31.

Integration of Eq. 2.31 yields

$$\int dp = \rho \omega^2 \int r \, dr - \gamma \int dz$$

or

$$p = \frac{\rho \omega^2 r^2}{2} - \gamma z + \text{constant} \qquad (2.33)$$

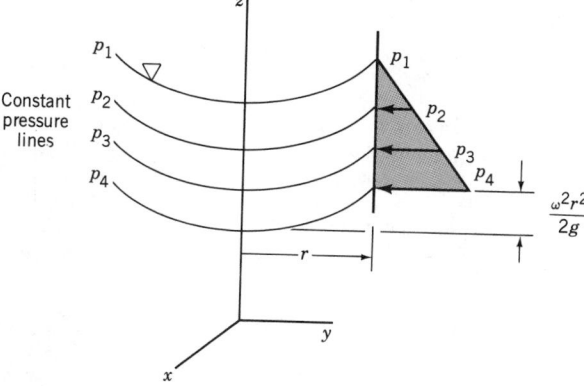

FIGURE 2.31 Pressure distribution in a rotating liquid.

where the constant of integration can be expressed in terms of a specified pressure at some arbitrary point r_0, z_0. This result shows that the pressure varies with the distance from the axis of rotation, but at a fixed radius, the pressure varies hydrostatically in the vertical direction as shown in Fig. 2.31.

EXAMPLE 2.12

It has been suggested that the angular velocity, ω, of a rotating body or shaft can be measured by attaching an open cylinder of liquid, as shown in Fig. E2.12a, and measuring with some type of depth gage the change in the fluid level, $H - h_0$, caused by the rotation of the fluid. Determine the relationship between this change in fluid level and the angular velocity.

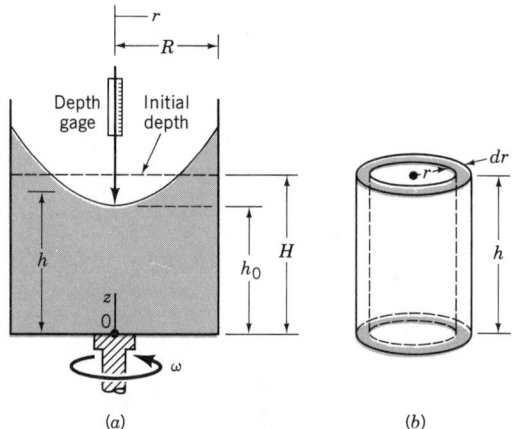

FIGURE E2.12

SOLUTION

The height, h, of the free surface above the tank bottom can be determined from Eq. 2.32, and it follows that

$$h = \frac{\omega^2 r^2}{2g} + h_0$$

The initial volume of fluid in the tank, V_i, is equal to

$$V_i = \pi R^2 H$$

The volume of the fluid with the rotating tank can be found with the aid of the differential element shown in Fig. E2.12b. This cylindrical shell is taken at some arbitrary radius, r, and its volume is

$$dV = 2\pi r h\, dr$$

The total volume is, therefore

$$V = 2\pi \int_0^R r\left(\frac{\omega^2 r^2}{2g} + h_0\right) dr$$

or

$$V = \frac{\pi\omega^2 R^4}{4g} + \pi R^2 h_0$$

Since the volume of the fluid in the tank must remain constant (assuming that none spills over the top) it follows that

$$\pi R^2 H = \frac{\pi \omega^2 R^4}{4g} + \pi R^2 h_0$$

or

$$H - h_0 = \frac{\omega^2 R^2}{4g} \qquad \text{(Ans)}$$

This is the relationship we were looking for. It shows that the change in depth could indeed be used to determine the rotational speed, although the relationship between the change in depth and speed is not a linear one.

REFERENCES

1. *The U.S. Standard Atmosphere, 1962,* U.S. Government Printing Office, Washington, D.C., 1962.

2. *The U.S. Standard Atmosphere, 1976,* U.S. Government Printing Office, Washington, D.C., 1976.

3. Benedict, R. P., *Fundamentals of Temperature, Pressure, and Flow Measurements,* 3rd Ed., Wiley, New York, 1984.

4. Dally, J. W., Riley, W.F., and McConnell, K. G., *Instrumentation for Engineering Measurements,* Wiley, New York, 1983.

5. Holman, J. P., *Experimental Methods for Engineers,* 4th Ed., McGraw-Hill, New York, 1983.

6. Comstock, J. P., ed., *Principles of Naval Architecture,* Society of Naval Architects and Marine Engineers, New York, 1967.

7. Hasler, A. F., Pierce, H., Morris, K. R., and Dodge, J. "Meteorological Data Fields 'In Perspective' ", Bulletin of the American Meteorological Society, Vol. 66, No. 7, July 1985.

PROBLEMS

Note: Unless otherwise indicated use the values of fluid properties found in the tables on the inside of the front cover. Problems designated with an (*) are intended to be solved with the aid of a programmable calculator or a computer.

2.1 The water level in an open standpipe is 80 ft above the ground. What is the static pressure at a fire hydrant that is connected to the standpipe and located at ground level? Express your answer in psi.

2.2 How high a column of SAE 30 oil would be required to give the same pressure as 700 mm Hg?

2.3 What pressure, expressed in pascals, will a skin diver be subjected to at a depth of 50 m in seawater?

2.4 A closed tank is partially filled with glycerin. If the air pressure in the tank is 8 lb/in.^{2} and the depth of glycerin is 10 ft, what is the pressure in lb/ft^2 at the bottom of the tank?

2.5 Compare the column heights of water, carbon tetrachloride, and mercury, corresponding to a pressure of 50 kPa. Express your answer in meters.

FIGURE P2.6

FIGURE P2.10

2.6   The closed tank of Fig. P2.6 is filled with water and is 5 ft long. The pressure gage on the tank reads 5 psi. Determine: **(a)** the height, h, in the open water column, **(b)** the gage pressure acting on the bottom tank surface AB, and **(c)** the absolute pressure of the air in the top of the tank if the local atmospheric pressure is 14.7 psia.

2.7   Bathyscaphes are capable of submerging to great depths in the ocean. What is the pressure at a depth of 6 km, assuming that seawater has a constant specific weight of 10.1 kN/m³? Express your answer in pascals and psi.

2.8   For the great depths that may be encountered in the ocean the compressibility of seawater may become an important consideration. **(a)** Assume that the bulk modulus for seawater is constant and derive a relationship between pressure and depth which takes into account the change in fluid density with depth. **(b)** Make use of part (a) to determine the pressure at a depth of 6 km assuming seawater has a bulk modulus of 2.3×10^9 Pa, and a density of 1030 kg/m³ at the surface. Compare this result with that obtained by assuming a constant density of 1030 kg/m³.

2.9   Blood pressure is usually given as the ratio of the maximum pressure (systolic pressure) to the minimum pressure (diastolic pressure). For example, a typical value for this ratio for a human would be 120/70, where the pressures are in mm Hg. What would these pressures be in pascals and in psi?

2.10   Two hemispherical shells are bolted together as shown in Fig. P2.10. The resulting spherical container, which weighs 360 lb, is filled with mercury and supported by a cable as shown. The container is vented at the top. If eight bolts are symmetrically located around the circumference, what is the vertical force that each bolt must carry?

2.11   Develop an expression for the pressure variation in a liquid in which the specific weight increases with depth, h, as $\gamma = Kh + \gamma_0$, where K is a constant and γ_0 is the specific weight at the free surface.

2.12*   In a certain liquid at rest, measurements of the specific weight at various depths show the following variation:

h (ft)	γ (lb/ft³)
0	70
10	76
20	84
30	91
40	97
50	102
60	107
70	110
80	112
90	114
100	115

The depth, $h = 0$, corresponds to a free surface at atmospheric pressure. Determine, through numerical integration of Eq. 2.4, the corresponding variation in pressure, and show the results on a plot of pressure (in psf) versus depth (in feet).

2.13   The basic elements of a hydraulic press are shown in Fig. P2.13. The plunger has an area

FIGURE P2.13

of 1 in.², and a force, F_1, can be applied to the plunger through a lever mechanism having a mechanical advantage of 8 to 1. If the large piston has an area of 150 in.², what load, F_2, can be raised by a force of 20 lb applied to the lever?

2.14   A 0.3-m-diameter pipe is connected to a 0.02-m-diameter pipe and both are rigidly held in place. Both pipes are horizontal with pistons at each end. If the space between the pistons is filled with water, what force will have to be applied to the larger piston to balance a force of 90 N applied to the smaller piston? Neglect friction.

2.15   A barometric pressure of 29.4 in. Hg corresponds to what value of atmospheric pressure in psia, and in pascals?

2.16   What would be the barometric pressure reading, in mm Hg, at an elevation of 4 km in the U.S. standard atmosphere? (Refer to Table C.2 in Appendix C.)

2.17   An absolute pressure of 9 psia corresponds to what gage pressure for standard atmospheric pressure of 14.7 psia?

2.18   A Bourdon pressure gage attached to the outside of a tank containing air reads 64.0 psi when the local atmospheric pressure is 760 mm Hg. What will be the gage reading if the atmospheric pressure increases to 773 mm Hg?

2.19   On the suction side of a pump a Bourdon pressure gage reads 40-kPa vacuum. What is the corresponding absolute pressure if the local atmospheric pressure is 100 kPa (abs)?

2.20   For an atmospheric pressure of 101 kPa (abs) determine the heights of the fluid columns in barometers containing one of the following liquids: **(a)** mercury, **(b)** water, and **(c)** ethyl alcohol. Calculate the heights including the effect of vapor pressure, and compare the results with those obtained neglecting vapor pressure. Do these results support the widespread use of mercury for barometers? Why?

2.21   Aneroid barometers can be used to measure changes in altitude. If a barometer reads 30.1 in. Hg at one elevation, what has been the change in altitude in meters when the barometer reading is 28.3 in. Hg? Assume a standard atmosphere, and that Eq. 2.12 is applicable over the range of altitudes of interest.

2.22   Pikes Peak near Denver, Colorado has an elevation of 14,110 ft. **(a)** Determine the pressure at this elevation, based on Eq. 2.12. **(b)** If the air is assumed to have a constant specific weight of 0.07647 lb/ft³, what would the pressure be at this altitude? **(c)** If the air is assumed to

have a constant temperature of 59 °F what would the pressure be at this elevation? For all three cases assume standard atmospheric conditions at sea level (see Table 2.1).

2.23   Equation 2.12 provides the relationship between pressure and elevation in the atmosphere for those regions in which the temperature varies linearly with elevation. Derive this equation and verify the value of the pressure given in Table C.2 in Appendix C for an elevation of 5 km.

2.24   As shown in Fig. 2.6 for the U.S. standard atmosphere, the troposphere extends to an altitude of 11 km where the pressure is 22.6 kPa (abs). In the next layer, called the stratosphere, the temperature remains constant at -56.5 °C. Determine the pressure and density in this layer at an altitude of 15 km. Assume $g = 9.77$ m/s² in your calculations. Compare your results with those given in Table C.2 in Appendix C.

2.25*   Under normal conditions the temperature of the atmosphere decreases with increasing elevation. In some situations, however, a temperature inversion may exist so that the air temperature increases with elevation. A series of temperature probes on a mountain give the elevation–temperature data shown in the table below. If the barometric pressure at the base of the mountain is 12.1 psia, determine by means of numerical integration the pressure at the top of the mountain.

Elevation (ft)	Temperature (°F)
5000	50.1 (base)
5500	55.2
6000	60.3
6400	62.6
7100	67.0
7400	68.4
8200	70.0
8600	69.5
9200	68.0
9900	67.1 (top)

2.26   A U-tube manometer is connected to a closed tank as shown in Fig. P2.26. The air pressure in the tank is 0.50 psi and the liquid in the tank is oil ($\gamma = 54.0$ lb/ft³). The pressure at point A is 2.00 psi. Determine: **(a)** the depth of oil, z, and **(b)** the differential reading, h, on the manometer.

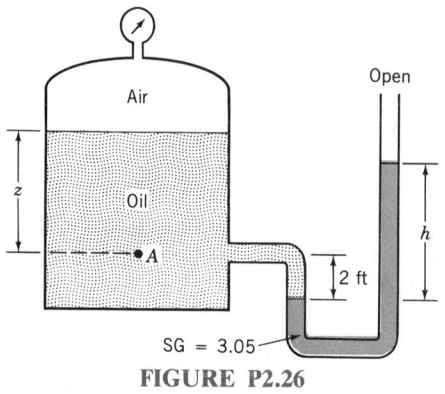

FIGURE P2.26

2.27   A tank is constructed of a series of cylinders having diameters of 0.30, 0.25, and 0.15
m as shown in Fig. P2.27. The tank contains oil,
water, and glycerin and a mercury manometer is
attached to the bottom as illustrated. Calculate
the manometer reading, h.

FIGURE P2.27

2.28   In Fig. P2.28 pipe A contains carbon tetrachloride ($SG = 1.60$) and the closed storage
tank B contains a salt brine ($SG = 1.15$). Determine the air pressure in tank B if the pressure
in pipe A is 20 psi.

FIGURE P2.28

2.29   A U-tube manometer contains oil, mercury, and water as shown in Fig. P2.29. For the
column heights indicated what is the pressure differential between pipes A and B?

FIGURE P2.29

2.30   The inverted U-tube manometer of Fig.
P2.30 contains oil and water as shown. The
pressure differential between pipes A and B,
$p_A - p_B$, is -5 kPa. Determine the differential
reading, h.

FIGURE P2.30

2.31   For the configuration shown in Fig.
P2.31 what must be the value of the specific

FIGURE P2.31

weight of the unknown fluid? Express your answer in lb/ft³.

2.32 A flowrate measuring device is installed in a horizontal pipe through which water is flowing. A U-tube manometer is connected to the pipe through pressure taps located 3 in. on either side of the device. The gage fluid in the manometer has a specific weight of 112 lb/ft³. Determine the differential reading of the manometer corresponding to a pressure drop between the taps of 0.5 lb/in.².

2.33 The mercury manometer of Fig. P2.33 indicates a differential reading of 0.30 m when the pressure in pipe A is 25 mm Hg vacuum. Determine the pressure in pipe B.

FIGURE P2.33

2.34 For the inclined-tube manometer of Fig. P2.34 the pressure in pipe A is 0.8 psi. The fluid in both pipes A and B is water, and the gage fluid in the manometer has a specific gravity of 2.6. What is the pressure in pipe B corresponding to the differential reading shown?

FIGURE P2.34

2.35 The differential mercury manometer of Fig. P2.35 is connected to pipe A containing gasoline ($SG = 0.65$), and to pipe B containing water. Determine the differential reading, h, corresponding to a pressure in A of 20 kPa and a vacuum of 150 mm Hg in B.

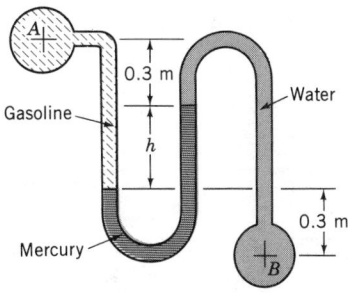

FIGURE P2.35

2.36 Small differences in gas pressures are commonly measured with a *micromanometer* of the type illustrated in Fig. P2.36. This device consists of two large reservoirs each having a cross-sectional area, A_r, which are filled with a liquid having a specific weight, γ_1, and connected by a U-tube of cross-sectional area, A_t, containing a liquid of specific weight, γ_2. When a differential gas pressure, $p_1 - p_2$, is applied a differential reading, h, develops. It is desired to have this reading sufficiently large (so that it can be easily read) for small pressure differentials. Determine the relationship between h and $p_1 - p_2$ when the area ratio A_t/A_r is small, and show that the differential reading, h, can be magnified by making the difference in specific weights, $\gamma_2 - \gamma_1$, small. Assume that initially (with $p_1 = p_2$) the fluid levels in the two reservoirs are equal.

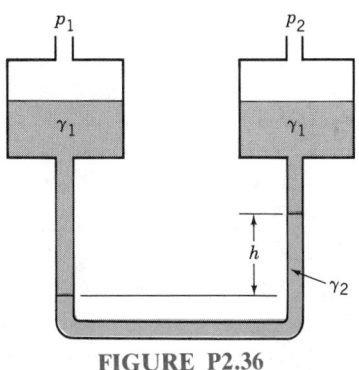

FIGURE P2.36

2.37 An inverted U-tube manometer containing oil ($SG = 0.8$) is located between two reservoirs as shown in Fig. P2.37. The reservoir on the left, which contains carbon tetrachloride, is closed and pressurized to 8 psi. The reservoir on the right contains water and is open to the atmosphere. With the given data, determine the depth of water, h, in the right reservoir.

FIGURE P2.37

2.38 A mercury manometer is used to measure the pressure difference in the two pipelines of Fig. P2.38. Fuel oil (specific weight = 53.0 lb/ft³) is flowing in A and SAE 30 lube oil (specific weight = 57.0 lb/ft³) is flowing in B. An air pocket has become entrapped in the lube oil as indicated. Determine the pressure in pipe B if the pressure in A is 15.3 psi.

FIGURE P2.38

2.39 A piston having a cross-sectional area of 0.07 m² is located in a cylinder containing water as shown in Fig. P2.39. An open U-tube manometer is connected to the cylinder as shown. For $h_1 = 60$ mm and $h = 100$ mm, what is the value of the applied force, P, acting on the piston? The weight of the piston is negligible.

FIGURE P2.39

2.40 Two pipes are connected by a manometer as shown in Fig. P2.40. Determine the pressure difference, $p_A - p_B$, between the pipes.

FIGURE P2.40

2.41 An air-filled, hemispherical shell is attached to the ocean floor at a depth of 10 m as shown in Fig. P2.41. A mercury barometer located inside the shell reads 765 mm Hg, and a mercury U-tube manometer designed to give the outside water pressure indicates a differential reading of 735 mm Hg as illustrated. Based on these data what is the atmospheric pressure at the ocean surface?

FIGURE P2.41

2.42* Both ends of the U-tube mercury manometer of Fig. P2.42 are initially open to the atmosphere and under standard atmospheric pressure. When the valve at the top of the right leg is open the level of mercury below the valve is h_i. After the valve is closed, air pressure is applied to the left leg. Determine the relationship between the differential reading on the manometer and the applied gage pressure, p_g. Show on a plot how the differential reading varies with p_g for h_i = 25, 50, 75, and 100 mm over the range $0 \le p_g \le 300$ kPa. Assume that the temperature of the trapped air remains constant.

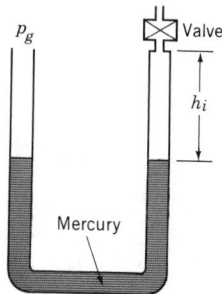

FIGURE P2.42

2.43 In Fig. P2.43 pipe A contains gasoline ($SG = 0.7$), pipe B contains oil ($SG = 0.9$), and the manometer fluid is mercury. Determine the new differential reading if the pressure in pipe A is decreased 25 kPa, and the pressure in pipe B remains constant. The initial differential reading is 0.30 m as shown.

FIGURE P2.43

2.44 A 6-in.-diameter piston is located within a cylinder which is connected to a $\frac{1}{2}$-in.-diameter inclined-tube manometer as shown in Fig. P2.44. The fluid in the cylinder and the manometer is oil (specific weight = 59 lb/ft³). When a weight W is placed on the top of the cylinder the fluid level in the manometer tube rises from point (1) to (2). How heavy is the weight? Assume that the change in position of the piston is negligible.

FIGURE P2.44

2.45 The manometer fluid in the manometer of Fig. P2.45 has a specific gravity of 3.46. Pipes A and B both contain water. If the pressure in pipe A is decreased by 1.3 psi and the pressure in pipe B increases by 0.8 psi, determine the new differential reading of the manometer.

FIGURE P2.45

2.46 Determine the ratio of areas, A_1/A_2, of the two manometer legs of Fig. P2.46 if a change in pressure in pipe B of 0.5 psi gives a corresponding change of 1 in. in the level of the mercury in the right leg. The pressure in pipe A does not change.

FIGURE P2.46

2.47 The inclined differential manometer of Fig. P2.47 contains carbon tetrachloride. Initially the pressure differential between pipes A and B, which contain a brine ($SG = 1.1$), is zero as illustrated in the figure. It is desired that the manometer give a differential reading of 12 in. (measured along the inclined tube) for a pressure differential of 0.1 psi. Determine the required angle of inclination, θ.

FIGURE P2.47

2.48 Determine the new differential reading along the inclined leg of the mercury manometer of Fig. P2.48, if the pressure in pipe A is decreased 12 kPa and the pressure in pipe B remains unchanged. The fluid in A has a specific gravity of 0.9 and the fluid in B is water.

FIGURE P2.48

2.49 Determine the change in the elevation of the mercury in the left leg of the manometer of Fig. P2.49 as a result of an increase in pressure

FIGURE P2.49

of 5 psi in pipe A while the pressure in pipe B remains constant.

2.50* Water initially fills the funnel and its connecting tube as shown in Fig. P2.50. Oil ($SG = 0.85$) is poured into the funnel until it reaches a level $h > H/2$ as indicated. Determine and plot the value of the rise in the water level in the tube, ℓ, as a function of h for $H/2 \le h \le H$, with $H = D = 2$ ft and $d = 0.1$ ft.

FIGURE P2.50

2.51 A swimming pool is 18 m long and 7 m wide. Determine the magnitude and location of the resultant force of the water on the vertical end of the pool where the depth is 2.5 m.

2.52 A square gate (4 m by 4 m) is located on the 45° face of a dam. The top edge of the gate lies 6 m below the water surface. Determine the force of the water on the gate and the point through which it acts.

2.53 A vertical rectangular gate is 8 ft wide and 10 ft long and weighs 6000 lb. The gate slides in vertical slots in the side of a reservoir containing water. The coefficient of friction between the slots and the gate is 0.03. Determine the minimum vertical force required to lift the gate when the water level is 4 ft above the top edge of the gate.

2.54 A 3-ft-diameter circular plate is located in the vertical side of an open tank containing gasoline. The resultant force that the gasoline exerts on the plate acts 3.1 in. below the centroid of the plate. What is the depth of the liquid above the centroid?

2.55 A rectangular gate having a width of 5 ft is located in the sloping side of a tank as shown in Fig. P2.55. The gate is hinged along its top edge and is held in position by the force P. Friction at the hinge and the weight of the gate can be neglected. Determine the required value of P.

FIGURE P2.55

FIGURE P2.60

2.56 A long, vertical wall separates seawater from freshwater. If the seawater stands at a depth of 7 m, what depth of freshwater is required to give a zero resultant force on the wall? When the resultant force is zero will the moment due to the fluid forces be zero? Explain.

2.57 An area in the form of an isosceles triangle with a base width of 6 ft and an altitude of 8 ft lies in the plane forming one wall of a tank which contains a liquid having a specific weight of 79.8 lb/ft³. The side slopes upward making an angle of 60° with the horizontal. The base of the triangle is horizontal and the vertex is above the base. Determine the resultant force the fluid exerts on the area when the fluid depth is 20 ft above the base of the triangular area. Show, with the aid of a sketch, where the center of pressure is located.

2.58 Solve Problem 2.57 if the isosceles triangle is replaced with a right triangle having the same base width and altitude.

2.59 A horizontal 2-m-diameter conduit is half filled with a liquid ($SG = 1.6$) and is capped at both ends with plane vertical surfaces. The air pressure in the conduit above the liquid surface is 200 kPa. Determine the resultant force of the fluid acting on one of the end caps, and locate this force relative to the bottom of the conduit.

2.60 A rectangular gate that is 2 m wide is located in the vertical wall of a tank containing water as shown in Fig. P2.60. It is desired to have the gate open automatically when the depth of water above the top of the gate reaches 10 m. **(a)** At what distance, d, should the frictionless horizontal shaft be located? **(b)** What is the magnitude of the force on the gate when it opens?

2.61 A tank containing water has a homogeneous rectangular gate in one side as shown in Fig. P2.61. The gate, weighing 200 lb, pivots about a horizontal axis located 10 ft below the surface. If the gate can only open inwardly, what will be the reaction of the gate (in a direction perpendicular to gate) against the bottom edge of the tank opening? The width of the gate is 2 ft, and friction at the axis is negligible.

FIGURE P2.61

2.62 A vertical plane area having the shape shown in Fig. P2.62 is immersed in an oil bath (specific weight = 8.75 kN/m³). Determine the magnitude of the resultant force acting on one side of the area as a result of the oil.

FIGURE P2.62

2.63*   A 200-lb homogeneous gate of 10-ft width and 5-ft length is hinged at point A and held in place by a 12-ft-long brace as shown in Fig. P2.63. As the bottom of the brace is moved to the right, the water level remains at the top of the gate. The line of action of the force that the brace exerts on the gate is along the brace. **(a)** Plot the magnitude of the force exerted on the gate by the brace as a function of the angle of the gate, θ, for $0 \le \theta \le 90°$. **(b)** Repeat the calculations for the case in which the weight of the gate is negligible. Comment on the results as $\theta \to 0$.

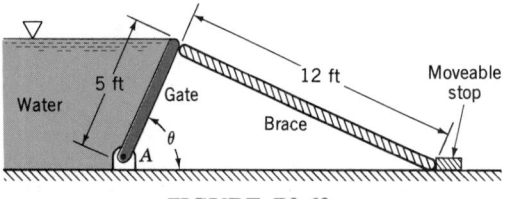

FIGURE P2.63

2.64   An open rectangular tank is 2 m wide and 4 m long. The tank contains water to a depth of 2 m and oil ($SG = 0.8$) on top of the water to a depth of 1 m. Determine the magnitude and location of the resultant fluid force acting on one end of the tank.

2.65   The closed vessel of Fig. P2.65 contains water with an air pressure of 10 psi at the water surface. One side of the vessel contains a spout that is closed by a 6-in.-diameter circular gate that is hinged along one side as illustrated. The horizontal axis of the hinge is located 10 ft below the water surface. Determine the minimum torque that must be applied at the hinge to hold the gate shut. Neglect the weight of the gate and friction at the hinge.

FIGURE P2.65

2.66   A gate having the cross section shown in Fig. P2.66 closes an opening 5 ft wide and 4 ft high in a water reservoir. The gate weighs 500 lb

and its center of gravity is 1 ft to the left of AC and 2 ft above BC. Determine the horizontal reaction that is developed on the gate at C.

FIGURE P2.66

2.67   The vertical cross section of a 7-m-long closed storage tank is shown in Fig. P2.67. The tank contains ethyl alcohol and the air pressure is 30 kPa. Determine the magnitude of the resultant fluid force acting on one end of the tank.

FIGURE P2.67

2.68   The rectangular gate CD of Fig. P2.68 is 1.8 m wide and 2.0 m long. Assuming the material of the gate to be homogeneous and neglecting friction at the hinge C, determine the weight of the gate necessary to keep it shut until the water level rises to 2.0 m above the hinge.

FIGURE P2.68

2.69 An open rectangular container contains a liquid that has a specific weight that varies according to the equation $\gamma = c_1 + c_2h$, where c_1 and c_2 are constants and h is a vertical coordinate measured downward from the free surface. Derive an equation for the magnitude of the liquid force exerted on one wall of the container having a width, b, and height, H, and an equation that gives the vertical coordinate of this force.

2.70* An open rectangular settling tank contains a liquid suspension that at a given time has a specific weight that varies approximately with depth according to the following data:

h (m)	γ (kN/m^3)
0	10.0
0.4	10.1
0.8	10.2
1.2	10.6
1.6	11.3
2.0	12.3
2.4	12.7
2.8	12.9
3.2	13.0
3.6	13.1

The depth $h = 0$ corresponds to the free surface. Determine, by means of numerical integration, the magnitude and location of the resultant force that the liquid suspension exerts on a vertical wall of the tank that is 6 m wide. The depth of fluid in the tank is 3.6 m.

2.71 The inclined face AD of the tank of Fig. P2.71 is a plane surface containing a gate ABC, which is hinged along line BC. The shape of the gate is shown in the plan view. If the tank contains water, determine the magnitude of the force that the water exerts on the gate.

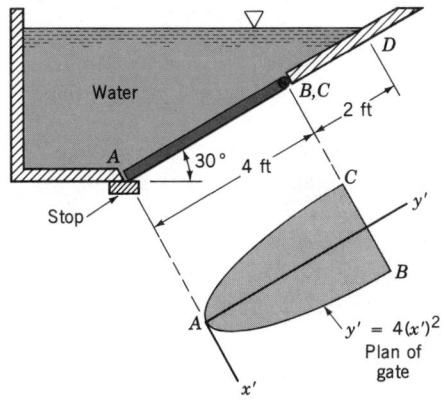

FIGURE P2.71

2.72 The concrete dam of Fig. P2.72 weighs 23.6 kN/m^3 and rests on a solid foundation. Determine the minimum coefficient of friction between the dam and the foundation required to keep the dam from sliding at the water depth shown. Assume no fluid uplift pressure along the base. Base your analysis on a unit length of the dam.

FIGURE P2.72

2.73* Water backs up behind a concrete dam as shown in Fig. P2.73. Leakage under the foundation gives a pressure distribution under the dam as indicated. If the water depth, h, is too great, the dam will topple over about its toe (point A). For the dimensions given, determine the maximum water depth for the following widths of the dam: $\ell = 20$, 30, 40, 50, and 60 ft. Base your analysis on a unit length of the dam. The specific weight of the concrete is 150 lb/ft^3.

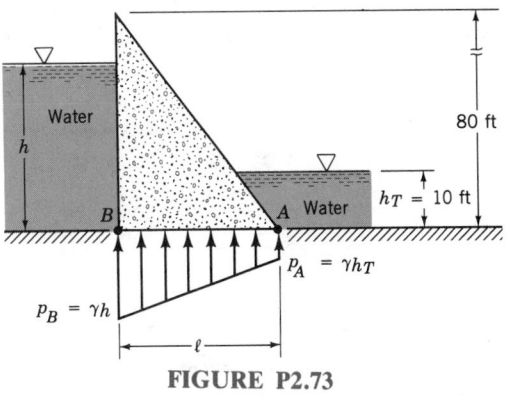

FIGURE P2.73

2.74 A 3-m-long curved gate is located in the side of a reservoir containing water as shown in Fig. P2.74. Determine the magnitude of the horizontal and vertical components of the force of the water on the gate. Will this force pass through point A? Explain.

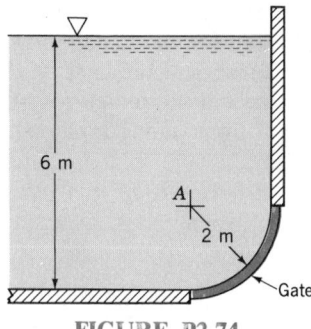

FIGURE P2.74

2.75 A 3-m-diameter open cylindrical tank contains water and has a hemispherical bottom as shown in Fig. P2.75. Determine the magnitude, line of action, and direction of the force of the water on the curved bottom.

FIGURE P2.75

2.76 The 18-ft-long gate of Fig. P2.76 is a quarter circle and is hinged at H. Determine the horizontal force, P, required to hold the gate in place. Neglect friction at the hinge and the weight of the gate.

FIGURE P2.76

2.77 A plug in the bottom of a pressurized tank is conical in shape as shown in Fig. P2.77. The air pressure is 50 kPa and the liquid in the tank has a specific weight of 27 kN/m³. Determine the magnitude, direction, and line of action of the force exerted on the curved surface of the cone within the tank due to the 50-kPa pressure and the liquid.

FIGURE P2.77

2.78 A 12-in.-diameter pipe contains a gas under a pressure of 120 psi. If the pipe wall thickness is $\frac{1}{4}$-in., what is the average circumferential stress developed in the pipe wall?

2.79 The concrete (specific weight = 150 lb/ft³) seawall of Fig. P2.79 has a curved surface and restrains seawater at a depth of 24 ft. The trace of the surface is a parabola as illustrated. Determine the moment of the fluid force (per unit length) with respect to an axis through the toe (point A).

FIGURE P2.79

2.80 A cylindrical tank with its axis horizontal has a diameter of 2.0 m and a length of 4.0 m. The ends of the tank are vertical planes. A vertical, 0.1-m-diameter pipe is connected to the top of the tank. The tank and the pipe are filled with ethyl alcohol to a level of 1.5 m above the top of the tank. Determine the resultant force of the alcohol on one end of the tank and show where it acts.

2.81 If the tank ends in Problem 2.80 are hemispherical, what is the magnitude of the resultant horizontal force of the alcohol on one of the curved ends?

2.82 Imagine the tank of Problem 2.80 split by a horizontal plane. Determine the magnitude of the resultant force of the alcohol on the bottom half of the tank.

2.83   A gate in the form of a partial cylindrical surface (called a *Tainter gate*) holds back water on top of a dam as shown in Fig. P2.83. The radius of the surface is 22 ft, and its length is 36 ft. The gate can pivot about point A, and the pivot point is 10 ft above the seat, C. Determine the magnitude of the resultant water force on the gate. Will the resultant pass through the pivot? Explain.

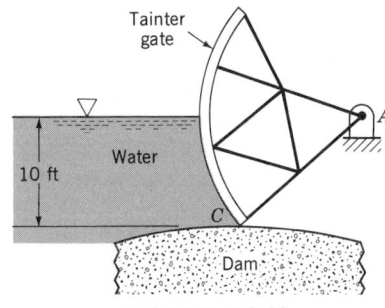

FIGURE P2.83

2.84   The 9-ft-long cylinder of Fig. P2.84 floats in oil and rests against a wall. Determine the horizontal force the cylinder exerts on the wall at the point of contact, A.

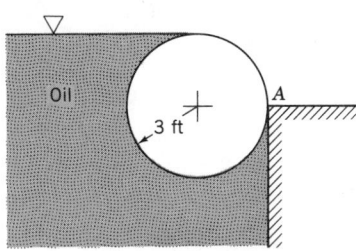

FIGURE P2.84

2.85   A tank wall has the shape shown in Fig. P2.85. Determine the horizontal and vertical components of the force of the water on a 1-ft length of the curved section AB.

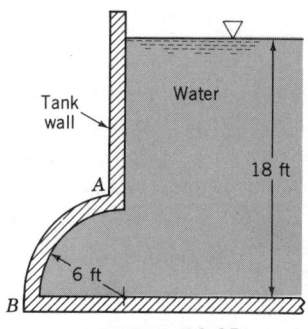

FIGURE P2.85

2.86   A conical plug is located in the side of a tank as shown in Fig. P2.86. **(a)** Show that the horizontal component of the force of the water on the plug does not depend on h. **(b)** For the depth indicated, what is the magnitude of this component?

FIGURE P2.86

2.87   An irregularly shaped piece of a solid material weighs 8.05 lb in air and 5.26 lb when completely submerged in water. Determine the density of the material.

2.88   The homogeneous timber AB of Fig. P2.88 is 0.15 m by 0.30 m in cross section. Determine the specific weight of the timber and the tension in the rope.

FIGURE P2.88

2.89   A 2-ft-thick block constructed of wood ($SG = 0.6$) is submerged in oil ($SG = 0.8$), and has a 2-ft-thick aluminum (specific weight = 168 lb/ft³) plate attached to the bottom as indicated in Fig. P2.89. Determine completely the force required to hold the block in the position shown. Locate the force with respect to point A.

FIGURE P2.89

2.90 The homogeneous wooden block A of Fig. P2.90 is 0.7 m by 0.7 m by 1.3 m and weighs 2.4 kN. The concrete block B (specific weight = 23.6 kN/m³) is suspended from A by means of the slender cable causing A to float in the position indicated. Determine the volume of B.

FIGURE P2.90

2.91 A river barge, whose cross section is approximately rectangular, carries a load of grain. The barge is 28 ft wide and 90 ft long. When unloaded its draft (depth of submergence) is 5 ft, and with the load of grain the draft is 7 ft. Determine: **(a)** the unloaded weight of the barge, and **(b)** the weight of the grain.

2.92 A hot-air balloon weighs 500 lb, including the weight of the balloon, the basket, and one person. The air outside the balloon has a temperature of 80 °F, and the heated air inside the balloon has a temperature of 150 °F. Assume the inside and outside air to be at standard atmospheric pressure of 14.7 psia. Determine the required volume of the balloon to support the weight. If the balloon had a spherical shape, what would be the required diameter?

2.93 The hydrometer shown in Fig. P2.93 has a mass of 0.045 kg and the cross-sectional area of its stem is 320 mm². Determine the distance between graduations (on the stem) for specific gravities of 1.00 and 0.90.

FIGURE P2.93

2.94 A 1-m-diameter cylindrical mass, M, is connected to a 2-m-wide rectangular gate as shown in Fig. P2.94. The gate is to open when the water level, h, drops below 2.5 m. Determine the required value for M. Neglect friction at the gate hinge and the pulley.

FIGURE P2.94

2.95 The 3-ft-wide homogeneous rectangular gate of Fig. P2.95 is hinged at A and an 8-ft³ block of concrete (specific weight = 150 lb/ft³) is connected to the gate through its center of gravity. Determine the minimum force P required to hold the gate closed. The weight of the gate can be neglected and the hinge is smooth.

FIGURE P2.95

2.96 The thin-walled, 1-m-diameter tank of Fig. P2.96 is closed at one end and has a mass of 90 kg. The open end of the tank is lowered into the water and held in the position shown by a steel block having a density of 7840 kg/m³. Assume that the air that is trapped in the tank is compressed at a constant temperature. Determine: **(a)** the reading on the pressure gage at the top of the tank, and **(b)** the volume of the steel block.

FIGURE P2.96

2.97* An inverted hollow cone is pushed into the water as is shown in Fig. P2.97. Determine the distance, ℓ, that the water rises in the cone as a function of the depth, d, of the lower edge of the cone. Plot the results for $0 \le d \le H$, when H is equal to 1 m. Assume the temperature of the air within the cone remains constant.

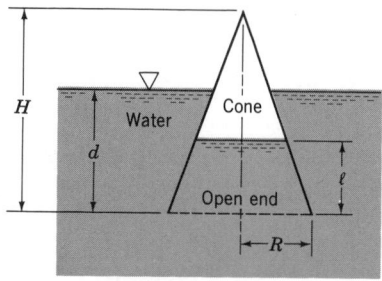

FIGURE P2.97

2.98 An open container of oil rests on the flatbed of a truck that is traveling along a horizontal road at 65 mi/hr. As the truck slows uniformly to a complete stop in 5 s, what will be the slope of the oil surface during the period of constant deceleration?

2.99 A 5-gal, cylindrical open container with a bottom area of 100 in.² is filled with glycerin and rests on the floor of an elevator. **(a)** Determine the fluid pressure at the bottom of the container when the elevator has an upward acceleration of 3 ft/s². **(b)** What resultant force does the container exert on the floor of the elevator during this acceleration? The weight of the container is negligible. (Note: 1 gal = 231 in.³)

2.100 An open rectangular tank 1 m wide and 2 m long contains gasoline to a depth of 1 m. If the height of the tank sides is 1.5 m, what is the maximum horizontal acceleration (along the long axis of the tank) that can develop before the gasoline would begin to spill?

2.101 If the tank of Problem 2.100 slides down a frictionless plane that is inclined at 30° with the horizontal, determine the angle the free surface makes with the horizontal.

2.102 A closed cylindrical tank that is 6 ft in diameter and 24 ft long is completely filled with gasoline. The tank, with its long axis horizontal, is pulled by a truck along a horizontal surface. Determine the pressure difference between the ends (along the long axis of the tank) when the truck undergoes an acceleration of 5 ft/s².

2.103 A container that is partially filled with water is pulled with a constant acceleration along a plane horizontal surface. With this acceleration the water surface slopes downward at an angle of 40° with respect to the horizontal. Determine the acceleration. Express your answer in m/s².

2.104 The open U-tube of Fig. P2.104 is partially filled with a liquid. When this device is accelerated with a horizontal acceleration, a, a differential reading, h, develops between the manometer legs which are spaced a distance ℓ apart. Determine the relationship between a, ℓ, and h.

FIGURE P2.104

2.105 An open 1-m-diameter tank contains water at a depth of 0.5 m when at rest. As the tank is rotated about its vertical axis the center of the fluid surface is depressed. At what angular velocity will the bottom of the tank first be exposed? No water is spilled from the tank.

2.106   The U-tube of Fig. P2.106 is partially filled with water and rotates around the axis *a–a*. Determine the angular velocity that will cause the water to start to vaporize at the bottom of the tube (point *A*).

FIGURE P2.106

2.107   An open, 2-ft-diameter tank contains water to a depth of 3 ft when at rest. If the tank is rotated about its vertical axis with an angular velocity of 180 rev/min, what is the minimum height of the tank walls to prevent water from spilling over the sides?

2.108   The U-tube of Fig. P2.108 contains mercury and rotates about the off-center axis *a–a*. At rest, the depth of mercury in each leg is 150 mm as illustrated. Determine the angular velocity for which the difference in heights between the two legs is 75 mm.

FIGURE P2.108

2.109   A closed, 0.4-m-diameter cylindrical tank is completely filled with oil (*SG* = 0.9) and rotates about its vertical longitudinal axis with an angular velocity of 40 rad/s. Determine the difference in pressure just under the vessel cover between a point on the circumference and a point on the axis.

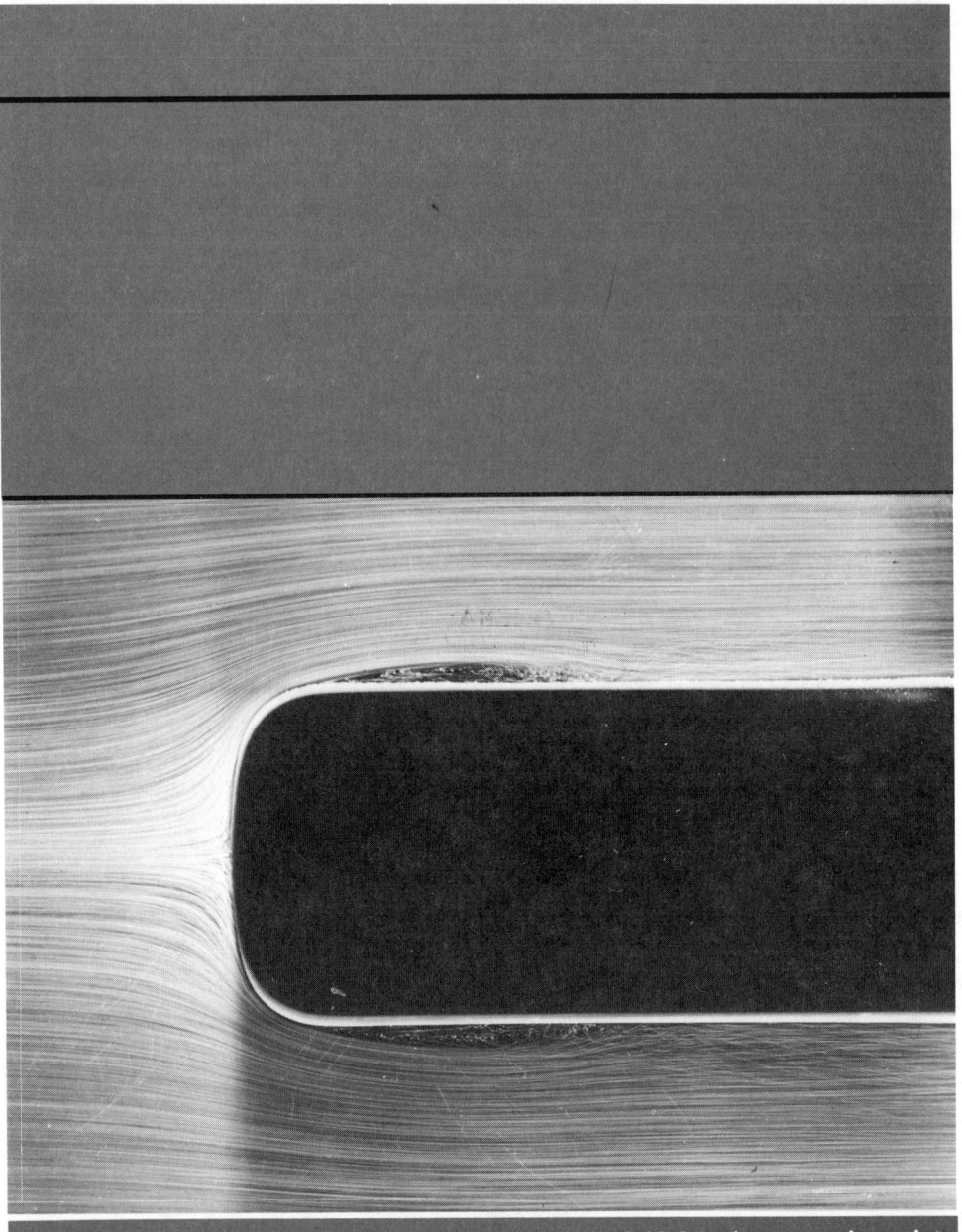

Flow past a blunt body: On any object placed in a moving fluid there is a stagnation point on the front of the object where the velocity is zero. This location has a relatively large pressure and divides the flow field into two portions—one flowing over the body, and one flowing under the body. (Air bubbles in water) (Photograph courtesy of ONERA, France).

___ THREE ___

ELEMENTARY FLUID DYNAMICS—
THE BERNOULLI EQUATION

As was discussed in the previous chapter, there are many situations involving fluids in which the fluid can be considered as stationary. In general, however, the use of fluids involves motion of some type. In fact, a dictionary definition of the word "fluid" is "free to change in form." In this chapter we investigate some typical fluid motions (fluid dynamics) in an elementary way.

To understand the interesting phenomena associated with fluid motion, one must consider the fundamental laws that govern the motion of fluid particles. Such considerations include the concepts of force and acceleration. We will discuss in some detail the use of Newton's second law ($\mathbf{F} = m\mathbf{a}$) as it is applied to fluid particle motion that is "ideal" in some sense. We will obtain the celebrated Bernoulli equation and apply it to various flows. Although this equation is one of the oldest in fluid mechanics and the assumptions involved in its derivation are numerous, it can be effectively used to predict and analyze a variety of flow situations. However, if the equation is applied without proper respect for its restrictions, serious errors can arise. Indeed, the Bernoulli equation is appropriately called "the most used and the most abused equation in fluid mechanics."

A thorough understanding of the elementary approach to fluid dynamics involved in this chapter will be useful on its own. It also provides a good foundation for the material in the following chapters where some of the present restrictions are removed and "more nearly exact" results are presented.

3.1 NEWTON'S SECOND LAW

As a fluid particle moves from one location to another, it usually experiences an acceleration or deceleration. According to Newton's second law of motion, the net force acting on the fluid particle under consideration must equal its mass times its acceleration,

$$\mathbf{F} = m\mathbf{a}$$

In this chapter we consider the motion of inviscid fluids. That is, the fluid is assumed to have zero viscosity. If the viscosity is zero then the thermal conductivity of the fluid is also zero and there can be no heat transfer (except by radiation).

In practice there are no inviscid fluids, since every fluid supports shear stresses when it is subjected to a rate of strain displacement. For many flow situations the viscous effects are relatively small compared with other effects. As a first approximation for such cases it is often possible to ignore viscous effects. For example, often the viscous forces developed in flowing water may be several orders of magnitude smaller than forces due to other influences, such as gravity or pressure differences. For other water flow situations, however, the viscous effects may be the dominant ones. Similarly, the viscous effects associated with the flow of a gas are often negligible, although in some circumstances they are very important.

We assume that the fluid motion is governed by pressure and gravity forces only and examine Newton's second law as it applies to a fluid particle in the form:

(Net pressure force on a particle) + (net gravity force on particle) =
(particle mass) × (particle acceleration)

The results of the interaction between the pressure, gravity, and acceleration provide numerous useful applications in fluid mechanics.

To apply Newton's second law to a fluid (or any other object), we must define an appropriate coordinate system in which to describe the motion. In general the motion will be three-dimensional and unsteady so that three space coordinates and time are needed to describe it. There are numerous coordinate systems available, including the most often used rectangular (x, y, z) and cylindrical (r, θ, z) systems. Usually the specific flow geometry dictates which system would be most appropriate.

In this chapter we will be concerned with two-dimensional motion like that confined to the x–z plane as is shown in Fig. 3.1a. Clearly we could choose to describe the flow in terms of the components of acceleration and forces in the x and z coordinate directions. The resulting equations are frequently referred to as a two-dimensional form of the *Euler equations* of motion in rectangular Cartesian coordinates. This approach will be discussed in Chapter 6.

As is done in the study of dynamics (Ref. 1), the motion of each fluid particle is described in terms of its velocity vector, $\mathbf{V}$, which is defined as the time rate of change

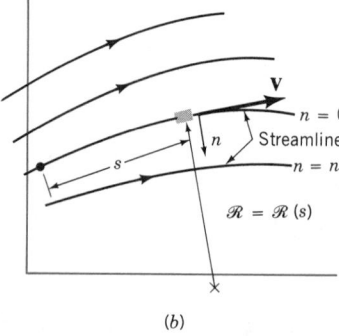

(a)

(b)

FIGURE 3.1 (*a*) Flow in the *x*–*z* plane. (*b*) Flow in terms of streamline and normal coordinates.

of the position of the particle. The particle's velocity is a vector quantity with a magnitude (the speed, $V = |\mathbf{V}|$) and direction. As the particle moves about it follows a particular path, the shape of which is governed by the velocity of the particle. The location of the particle along the path is a function of where the particle started at the initial time and its velocity along the path. If the flow is steady (i.e., nothing changes with time at a given location in the flow field), each successive particle that passes through a given point [such as point (1) in Fig. 3.1a] will follow the same path. For such cases the path is a fixed line in the x–z plane. Neighboring particles that pass on either side of point (1) follow their own paths, which may be of different shape than the one passing through (1). The entire x–z plane is filled with such paths.

For steady flows each particle slides along its path, and its velocity vector is everywhere tangent to the path. The lines that are tangent to the velocity vectors throughout the flow field are called *streamlines*. For many situations it is easiest to describe the flow in terms of the "streamline" coordinates based on the streamlines as are illustrated in Fig. 3.1b. The particle motion is described in terms of its distance, $s = s(t)$, along the streamline from some convenient origin and the local radius of curvature of the streamline, $\mathcal{R} = \mathcal{R}(s)$. The distance along the streamline is related to the particle's speed by $V = ds/dt$, and the radius of curvature is related to shape of the streamline. In addition to the coordinate along the streamline, s, the coordinate normal to the streamline, n, as is shown in Fig. 3.1b, will be of use.

To apply Newton's second law to a particle flowing along its streamline, we must write the particle acceleration in terms of the streamline coordinates. By definition, the acceleration is the time rate of change of the velocity of the particle, $\mathbf{a} = d\mathbf{V}/dt$. For two-dimensional flow in the x–z plane, the acceleration has two components—one along the streamline, a_s, the streamwise acceleration, and one normal to the streamline, a_n, the normal acceleration.

The streamwise acceleration results from the fact that the speed of the particle generally varies along the streamline, $V = V(s)$. For example, in Fig. 3.1a the speed may be 100 ft/s at point (1) and 50 ft/s at point (2). Thus, by use of the chain rule of differentiation, the s component of the acceleration is given by $a_s = dV/dt = (\partial V/\partial s)(ds/dt) = (\partial V/\partial s)V$. We have used the fact that $V = ds/dt$. The normal component of acceleration is the centrifugal acceleration given in terms of the particle speed and the radius of curvature of its path. Thus, $a_n = V^2/\mathcal{R}$, where both V and $\mathcal{R}$ may vary along the streamline. These equations for the acceleration should be familiar from the study of particle motion in physics (Ref. 2) or dynamics (Ref. 1). A more complete derivation and discussion of these topics can be found in Chapter 4.

Thus, the components of acceleration in the s and n directions, a_s and a_n, are given by

$$a_s = V\frac{\partial V}{\partial s}, \qquad a_n = \frac{V^2}{\mathcal{R}} \tag{3.1}$$

where $\mathcal{R}$ is the local radius of curvature of the streamline, and s is the distance measured along the streamline from some arbitrary initial point. In general there is acceleration along the streamline (because the particle speed changes along its path, $\partial V/\partial s \neq 0$) and acceleration normal to the streamline (because the particle does not flow in a straight line, $\mathcal{R} \neq \infty$). To produce this acceleration there must be a net, nonzero force on the fluid particle.

To determine the forces necessary to produce a given flow (or conversely, what

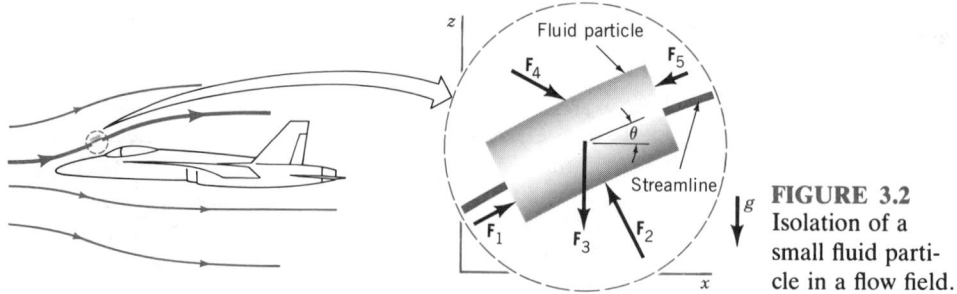

FIGURE 3.2 Isolation of a small fluid particle in a flow field.

flow results from a given set of forces), we consider the free-body diagram of a small fluid particle as is shown in Fig. 3.2. The particle of interest is removed from its surroundings, and the reactions of the surroundings on the particle are indicated by the appropriate forces present, $\mathbf{F}_1$, $\mathbf{F}_2$, and so forth. For the present case, the important forces are assumed to be gravity and pressure. Other forces, such as viscous forces and surface tension effects, are assumed negligible. The acceleration of gravity, g, is assumed to be constant and acts vertically, in the negative z direction, at an angle θ relative to the normal to the streamline.

3.2 F = ma ALONG A STREAMLINE

Consider the small fluid particle of size δs by δn in the plane of the figure and δy normal to the figure as shown in the free-body diagram of Fig. 3.3. Unit vectors along and normal to the streamline are denoted by $\hat{s}$ and $\hat{n}$, respectively. For steady flow, the component of Newton's second law along the streamline direction, s, can be written as

$$\sum \delta F_s = \delta m \, a_s = \delta m \, V \frac{\partial V}{\partial s} = \rho \, \delta \Psi \, V \frac{\partial V}{\partial s} \tag{3.2}$$

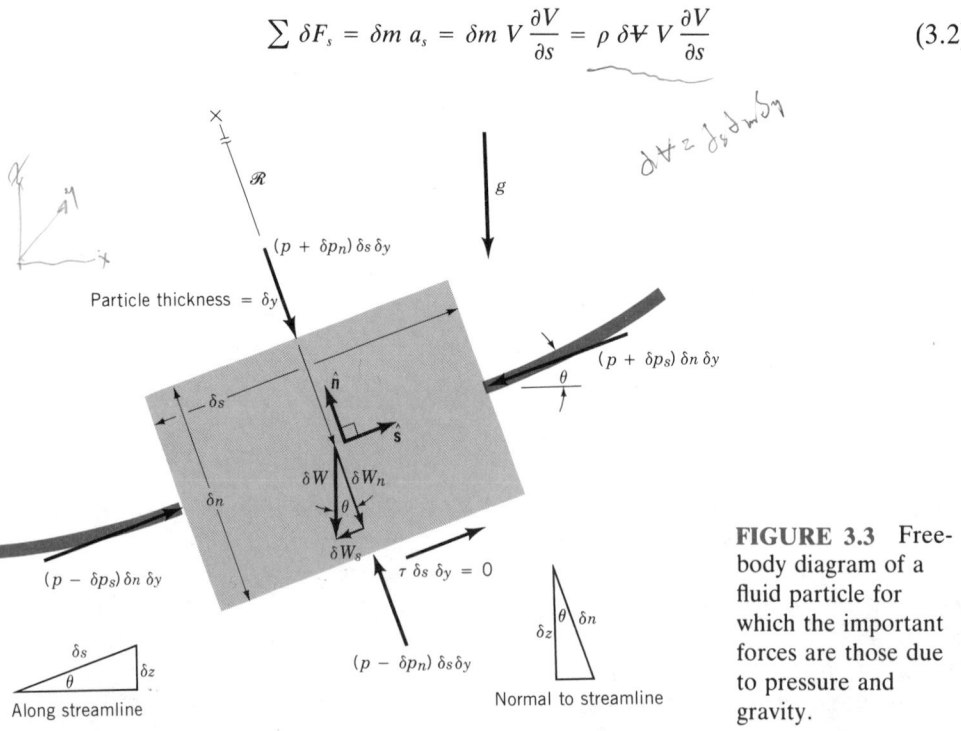

FIGURE 3.3 Free-body diagram of a fluid particle for which the important forces are those due to pressure and gravity.

where $\Sigma \, \delta F_s$ represents the sum of the s components of all the forces acting on the particle, which has mass $\delta m = \rho \, \delta \Psi$, and $V \, \partial V / \partial s$ is the acceleration in the s direction. Here, $\delta \Psi = \delta s \, \delta n \, \delta y$ is the particle volume. Equation 3.2 is valid for both compressible and incompressible fluids. That is, the density need not be constant throughout the flow field.

The gravity force (weight) on the particle can be written as $\delta W = \gamma \, \delta \Psi$, where $\gamma = \rho g$ is the specific weight of the fluid. Hence, the component of the weight force in the direction of the streamline is

$$\delta W_s = -\delta W \sin \theta = -\gamma \, \delta \Psi \sin \theta$$

If the streamline is horizontal at the point of interest, then $\theta = 0$, and there is no component of particle weight along the streamline to contribute to its acceleration in that direction.

As is indicated in Chapter 2, the pressure is not constant throughout a stationary fluid ($\nabla p \neq 0$) because of the fluid weight. Likewise, in a flowing fluid the pressure is usually not constant. In general, for steady flow, $p = p(s, n)$. If the pressure at the center of the particle shown in Fig. 3.3 is denoted as p, then its average value on the two end faces that are perpendicular to the streamline are $p + \delta p_s$ and $p - \delta p_s$. Since the particle is "small," we can use a one-term Taylor series expansion for the pressure field (as was done in Chapter 2 for the pressure forces in static fluids) to obtain

$$\delta p_s \approx \frac{\partial p}{\partial s} \frac{\delta s}{2}$$

Thus, if δF_{ps} is the net pressure force on the particle in the streamline direction, it follows that

$$\delta F_{ps} = (p - \delta p_s) \, \delta n \, \delta y - (p + \delta p_s) \, \delta n \, \delta y = -2 \, \delta p_s \, \delta n \, \delta y$$

$$= -\frac{\partial p}{\partial s} \delta s \, \delta n \, \delta y$$

That is,

$$\delta F_{ps} = -\frac{\partial p}{\partial s} \delta \Psi$$

Note that the actual level of the pressure, p, is not important. What produces a net pressure force is the fact that the pressure is not constant throughout the fluid. The nonzero pressure gradient, $\nabla p = \partial p / \partial s \, \hat{\mathbf{s}} + \partial p / \partial n \, \hat{\mathbf{n}}$, is what provides a net pressure force on the particle. Viscous forces, represented by $\tau \, \delta s \, \delta y$, are zero, since the fluid is inviscid.

Thus, the net force acting in the streamline direction on the particle shown in Fig. 3.3 is given by

$$\sum \delta F_s = \delta W_s + \delta F_{ps} = \left(-\gamma \sin \theta - \frac{\partial p}{\partial s} \right) \delta \Psi \tag{3.3}$$

By combining Eqs. 3.2 and 3.3 we obtain the following equation of motion along the streamline direction:

$$-\gamma \sin \theta - \frac{\partial p}{\partial s} = \rho V \frac{\partial V}{\partial s} \tag{3.4}$$

We have divided out the common particle volume factor, $\delta \Psi$, that appears in both the force and the acceleration portions of the equation. This is a representation of the fact that it is the fluid density (mass per unit volume), not the mass, per se, of the fluid particle that is important.

The physical interpretation of Eq. 3.4 is that a change in fluid particle speed is accomplished by the appropriate combination of pressure gradient and particle weight along the streamline. For fluid static situations this balance between pressure and gravity forces is such that no change in particle speed is produced—the right-hand side of Eq. 3.4 is zero, and the particle remains stationary. In a flowing fluid the pressure and weight forces do not necessarily balance—the force unbalance provides the appropriate acceleration and, hence, particle motion.

EXAMPLE 3.1

Consider the inviscid, incompressible, steady flow along the horizontal streamline A–B in front of the sphere of radius a as shown in Fig. E3.1a. From a more advanced theory of flow past a sphere, the fluid velocity along this streamline is

$$V = V_0 \left(1 + \frac{a^3}{x^3} \right)$$

(a)

(b)

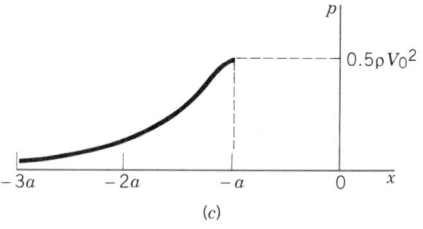

(c)

FIGURE E3.1

Determine the pressure variation along the streamline from point A far in front of the sphere ($x_A = -\infty$ and $V_A = V_0$) to point B on the sphere ($x_B = -a$ and $V_B = 0$).

SOLUTION

Since the flow is steady and inviscid, Eq. 3.4 is valid. In addition, since the streamline is horizontal, $\sin\theta = \sin 0 = 0$ and the equation of motion along the streamline reduces to

$$\frac{\partial p}{\partial s} = -\rho V \frac{\partial V}{\partial s} \tag{1}$$

With the given velocity variation along the streamline, the acceleration term is

$$V\frac{\partial V}{\partial s} = V\frac{\partial V}{\partial x} = V_0\left(1 + \frac{a^3}{x^3}\right)\left(-\frac{3V_0 a^3}{x^4}\right)$$

$$= -3V_0^2\left(1 + \frac{a^3}{x^3}\right)\frac{a^3}{x^4}$$

where we have replaced s by x, since the two coordinates are identical along streamline A–B. It follows that $V\,\partial V/\partial s < 0$ along the streamline. The fluid slows down from V_0 far ahead of the sphere to zero velocity on the "nose" of the sphere ($x = -a$).

Thus, according to Eq. 1, to produce the given motion the pressure gradient along the streamline is

$$\frac{\partial p}{\partial x} = \frac{3\rho a^3 V_0^2(1 + a^3/x^3)}{x^4} \tag{2}$$

This variation is indicated in Fig. E3.1b. It is seen that the pressure increases in the direction of flow ($\partial p/\partial x > 0$) from point A to point B. The maximum pressure gradient ($0.610\,\rho V_0^2/a$) occurs just slightly ahead of the sphere ($x = -1.205a$). It is the pressure gradient that slows the fluid down from $V_A = V_0$ to $V_B = 0$.

The pressure distribution along the streamline can be obtained by integrating Eq. 2 from $p = 0$(gage) at $x = -\infty$ to pressure p at location x. The result, plotted in Fig. E3.1c, is

$$p = -\rho V_0^2\left[\left(\frac{a}{x}\right)^3 + \frac{(a/x)^6}{2}\right] \tag{Ans}$$

The pressure at B, a stagnation point since $V_B = 0$, is the highest pressure along the streamline ($p_B = \rho V_0^2/2$). Note that the pressure gradient and pressure are directly proportional to the density of the fluid, a representation of the fact that the fluid inertia is proportional to its mass.

Equation 3.4 can be rearranged and integrated as follows. First, we note from Fig. 3.3 that along the streamline $\sin\theta = dz/ds$. Also, we can write $V\,dV/ds = \frac{1}{2}d(V^2)/ds$. Finally, along the streamline the value of n is constant ($dn = 0$) so that $dp = (\partial p/\partial s)\,ds + (\partial p/\partial n)\,dn = (\partial p/\partial s)\,ds$. Hence, along the streamline $\partial p/\partial s = dp/ds$. These ideas combined with Eq. 3.4 give the following result valid along a streamline

$$-\gamma\frac{dz}{ds} - \frac{dp}{ds} = \frac{1}{2}\rho\frac{d(V^2)}{ds}$$

This simplifies to

$$dp + \frac{1}{2}\rho d(V^2) + \gamma\,dz = 0 \qquad \text{(along a streamline)} \tag{3.5}$$

which can be integrated to give

$$\int \frac{dp}{\rho} + \frac{1}{2} V^2 + gz = C \qquad \text{(along a streamline)} \qquad (3.6)$$

where C is a constant of integration to be determined by the conditions at some point on the streamline.

In general it is not possible to integrate the pressure term because the density may not be constant and, therefore, cannot be removed from under the integral sign. To carry out this integration we must know specifically how the density varies with pressure. This is not always easily determined. For example, for a perfect gas the density, pressure, and temperature are related according to $p = \rho RT$, where R is the gas constant. To know how the density varies with pressure, we must also know the temperature variation. For now we will assume that the density is constant (incompressible flow). The justification for this assumption and the consequences of compressibility will be considered further in Section 3.8.1 and more fully in Chapter 11.

With the additional assumption that the density remains constant (a very good assumption for liquids and also for gases if the speed is "not too high"), Eq. 3.6 assumes the following simple representation for steady, inviscid, incompressible flow.

$$p + \tfrac{1}{2} \rho V^2 + \gamma z = \text{constant along streamline} \qquad (3.7)$$

This is the celebrated *Bernoulli equation*—a very powerful tool in fluid mechanics. In 1738 Daniel Bernoulli (1700–1782) published his *Hydrodynamics* in which an equivalent of this famous equation first appeared. To use it correctly we must constantly remember the basic assumptions used in its derivation: (1) viscous effects are assumed negligible; (2) the flow is assumed to be steady; (3) the flow is assumed to be incompressible; (4) the equation is applicable along a streamline. In the derivation of Eq. 3.7, we assumed that the flow takes place in a plane (the x–z plane). In general, this equation is valid for both planar and nonplanar (three-dimensional) flows, provided it is applied along the streamline.

If any of the above assumptions are violated, the use of the Bernoulli equation will give incorrect results. We will provide many examples to illustrate the correct use of the Bernoulli equation and will show how a violation of the basic assumptions can lead to erroneous conclusions. The constant of integration in the Bernoulli equation can be evaluated if sufficient information about the flow is known at one location along the streamline.

EXAMPLE 3.2

Consider the flow of air around a bicyclist moving through still air with velocity V_0 as is shown in Fig. E3.2. Determine the difference in the pressure between points (1) and (2).

FIGURE E3.2

SOLUTION

In a coordinate system fixed to the bike, it appears as though the air is flowing steadily toward the bicyclist with speed V_0. If the assumptions of Bernoulli's equation are valid (steady, incompressible, inviscid flow), Eq. 3.7 can be applied as follows along the streamline that passes through (1) and (2)

$$p_1 + \tfrac{1}{2}\rho V_1^2 + \gamma z_1 = p_2 + \tfrac{1}{2}\rho V_2^2 + \gamma z_2$$

We consider (1) to be in the free stream so that $V_1 = V_0$ and (2) to be at the tip of the bicyclist's nose and assume that $z_1 = z_2$ and $V_2 = 0$ (both of which, as is discussed in Section 3.4, are reasonable assumptions). It follows that the pressure at (2) is greater than that at (1) by an amount

$$p_2 - p_1 = \tfrac{1}{2}\rho V_1^2 \qquad\qquad\text{(Ans)}$$

A similar result was obtained in Example 3.1 by integrating the pressure gradient, which was known because the velocity distribution along the streamline, $V(s)$, was known. The Bernoulli equation is a general integration of $F = ma$. To determine $p_2 - p_1$, knowledge of the detailed velocity distribution is not needed—only the "boundary conditions" at (1) and (2) are required. Of course, knowledge of the value of V along the streamline is needed to determine the pressure at points between (1) and (2). Note that if we measure $p_2 - p_1$ we can determine the speed, V_0.

The difference in fluid velocity between two points in a flow field, V_1 and V_2, can often be controlled by appropriate geometric constraints of the fluid. For example, a garden hose nozzle is designed to give a much higher velocity at the exit of the nozzle than at its entrance where it is attached to the hose. As is shown by the Bernoulli equation, the pressure within the hose must be larger than that at the exit (for constant elevation, an increase in velocity requires a decrease in pressure if Eq. 3.7 is valid). It is this pressure drop that accelerates the water through the nozzle. Similarly, an airfoil is designed so that the fluid velocity over its upper surface is greater (on the average) than that along its lower surface. From the Bernoulli equation, therefore, the average pressure on the lower surface is greater than on the upper surface. A net upward force, the lift, results.

3.3 $F = ma$ NORMAL TO A STREAMLINE

In this section we will consider application of Newton's second law in a direction normal to the streamline. In many flows the streamlines are relatively straight, the flow is essentially one-dimensional, and variations in parameters across streamlines (in the normal direction) can often be neglected when compared to the variations along the streamline. However, in numerous other situations valuable information can be obtained from considering $F = ma$ normal to the streamlines. For example, the devastating low pressure region at the center of a tornado can be explained by applying Newton's second law across the nearly circular streamlines of the tornado.

We again consider the force balance on the fluid particle shown in Fig. 3.3. This time, however, we consider components in the normal direction, $\hat{n}$, and write Newton's second law in this direction as

$$\sum \delta F_n = \frac{\delta m\, V^2}{\mathcal{R}} = \frac{\rho\, \delta V\!\!\!\!- V^2}{\mathcal{R}} \qquad\qquad (3.8)$$

where $\Sigma\ \delta F_n$ represents the sum of n components of all the forces acting on the particle. We assume the flow is steady with a normal acceleraton $a_n\ =\ V^2/\mathcal{R}$, where $\mathcal{R}$ is the local radius of curvature of the streamlines. This acceleration is produced by the change in direction of the particle's velocity as it moves along a curved path.

We again assume that the only forces of importance are pressure and gravity. The component of the weight (gravity force) in the normal direction is

$$\delta W_n\ =\ -\delta W \cos\theta\ =\ -\gamma\ \delta\rlap{\forall}V \cos\theta$$

If the streamline is vertical at the point of interest, $\theta = 90°$, and there is no component of the particle weight normal to the direction of flow to contribute to its acceleration in that direction.

If the pressure at the center of the particle is p, then its values on the top and bottom of the particle are $p + \delta p_n$ and $p - \delta p_n$, where $\delta p_n = (\partial p/\partial n)(\delta n/2)$. Thus, if δF_{pn} is the net pressure force on the particle in the normal direction, it follows that

$$\delta F_{pn}\ =\ (p\ -\ \delta p_n)\ \delta s\ \delta y\ -\ (p\ +\ \delta p_n)\ \delta s\ \delta y\ =\ -2\ \delta p_n\ \delta s\ \delta y$$

$$=\ -\frac{\partial p}{\partial n}\ \delta s\ \delta n\ \delta y$$

or

$$\delta F_{pn}\ =\ -\frac{\partial p}{\partial n}\ \delta\rlap{\forall}V$$

Hence, the net force acting in the normal direction on the particle shown in Fig. 3.3 is given by

$$\Sigma\ \delta F_n\ =\ \delta W_n\ +\ \delta F_{pn}\ =\ \left(-\gamma \cos\theta\ -\ \frac{\partial p}{\partial n}\right)\delta\rlap{\forall}V \tag{3.9}$$

By combining Eqs. 3.8 and 3.9 and using the fact that along a line normal to the streamline $\cos\theta = dz/dn$ (see Fig. 3.3), we obtain the following equation of motion along the normal direction

$$-\gamma\frac{dz}{dn}\ -\ \frac{\partial p}{\partial n}\ =\ \frac{\rho V^2}{\mathcal{R}} \tag{3.10}$$

The physical interpretation of Eq. 3.10 is that a change in the directon of flow of a fluid particle ($\mathcal{R} < \infty$) is accomplished by the appropriate combination of pressure gradient and particle weight normal to the streamline. A larger speed or density or a smaller radius of curvature of the motion requires a larger force unbalance to produce the motion. For example, if gravity is neglected (as is commonly done for gas flows) or if the flow is in a horizontal ($dz/dn = 0$) plane, Eq. 3.10 becomes

$$\frac{\partial p}{\partial n}\ =\ -\frac{\rho V^2}{\mathcal{R}}$$

This indicates that the pressure increases with distance away from the center of curvature ($\partial p/\partial n$ is negative since $\rho V^2/\mathcal{R}$ is positive—the positive n direction points toward the "inside" of the curved streamline). Thus, the pressure outside a tornado (typical atmospheric pressure) is larger than it is near the center of the tornado (where an often dangerously low partial vacuum may occur). This pressure difference is needed to balance the centrifugal acceleration associated with the curved streamlines of the fluid motion.

EXAMPLE 3.3

Shown in Figs. E3.3a,b are two flow fields with circular streamlines. The velocity distributions are

$$V(r) = C_1 r \quad \text{for case } (a)$$

and

$$V(r) = \frac{C_2}{r} \quad \text{for case } (b)$$

where C_1 and C_2 are constants. Determine the pressure distributions, $p = p(r)$, for each.

(a)   (b)

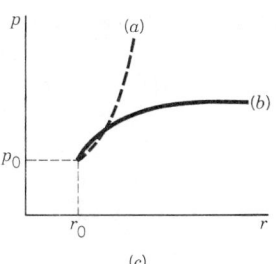

(c)

FIGURE E3.3

SOLUTION

We assume the flows are steady, inviscid, and incompressible with streamlines in the horizontal plane ($dz/dn = 0$). Since the streamlines are circles, the coordinate n points in a direction opposite of that of the radial coordinate, $\partial/\partial n = -\partial/\partial r$, and the radius of curvature is given by $\mathcal{R} = r$. Hence, Eq. 3.10 becomes

$$\frac{\partial p}{\partial r} = \frac{\rho V^2}{r}$$

For case (a) this gives

$$\frac{\partial p}{\partial r} = \rho C_1^2 r$$

while for cases (b) it gives

$$\frac{\partial p}{\partial r} = \frac{\rho C_2^2}{r^3}$$

For either case the pressure decreases as r decreases. Integration of these equations with respect to r, starting with a known pressure $p = p_0$ at $r = r_0$, gives

$$p = \frac{1}{2} \rho C_1^2 (r^2 - r_0^2) + p_0 \qquad \text{(Ans)}$$

for case (a) and

$$p = \frac{1}{2} \rho C_2^2 \left(\frac{1}{r_0^2} - \frac{1}{r^2} \right) + p_0 \qquad \text{(Ans)}$$

for case (b). These pressure distributions are sketched in Fig. E3.3c. The pressure distributions needed to balance the centrifugal accelerations in cases (a) and (b) are not the same because the velocity distributions are different. In fact for case (a) the pressure increases without bound as $r \to \infty$, while for case (b) the pressure approaches a finite value as $r \to \infty$. The streamline patterns are the same for each case, however.

Physically, case (a) represents rigid body rotation (as obtained in a can of water on a turntable after it has been "spun up") and case (b) represents a free vortex (an approximation to a tornado or the swirl of water in a drain, the "bathtub vortex").

If we multiply Eq. 3.10 by dn, use the fact that $\partial p / \partial n = dp/dn$ if s is constant, and integrate across the streamline (in the n direction) we obtain

$$\int \frac{dp}{\rho} + \int \frac{V^2}{\mathcal{R}}\, dn + gz = \text{constant across the streamline} \qquad (3.11)$$

To complete the indicated integrations we must know how the density varies with pressure and how the fluid speed and radius of curvature vary with n. For incompressible flow the density is constant and the integration involving the pressure term gives simply p/ρ. We are still left, however, with the integration of the second term in Eq. 3.11. Without knowing the n dependence in $V = V(s, n)$ and $\mathcal{R} = \mathcal{R}(s, n)$ this integration cannot be completed.

Thus, the final form of Newton's second law applied across the streamlines for steady, inviscid, incompressible flow is

$$p + \rho \int \frac{V^2}{\mathcal{R}}\, dn + \gamma z = \text{constant across the streamline} \qquad (3.12)$$

As with the Bernoulli equation, we must be careful that the assumptions involved in the derivation of this equation are not violated when it is used.

3.4   PHYSICAL INTERPRETATION

In the previous two sections we developed the basic equations governing fluid motion under a fairly stringent set of restrictions. In spite of the numerous assumptions imposed on these flows, a variety of flows can be readily analyzed with them. A physical interpretation of the equations will be of help in understanding the processes involved. To this end, we rewrite Eqs. 3.7 and 3.12 here and interpret them physically. Application of $\mathbf{F} = m\mathbf{a}$ along and normal to the streamline results in

$$p + \tfrac{1}{2}\rho V^2 + \gamma z = \text{constant along the streamline} \qquad (3.13)$$

and

$$p + \rho \int \frac{V^2}{\mathcal{R}} \, dn + \gamma z = \text{constant across the streamline} \qquad (3.14)$$

The following basic assumptions were made to obtain these equations: The flow is steady and the fluid is inviscid and incompressible. In practice none of these assumptions is exactly true.

A violation of one or more of the above assumptions is a common cause for obtaining an incorrect match between the "real world" and solutions obtained by use of the Bernoulli equation. Fortunately, many "real world" situations are adequately modeled by the use of Eqs. 3.13 and 3.14 because the flow is nearly steady and incompressible and the fluid behaves as if it were nearly inviscid.

The Bernoulli equation was obtained by integration of the equation of motion along the "natural" coordinate direction of the streamline. To produce an acceleration, there must be an unbalance of the resultant forces, of which only pressure and gravity were considered to be important. Thus, there are three processes involved in the flow—mass times acceleration (the $\rho V^2/2$ term), pressure (the p term), and weight (the γz term).

Integration of the equation of motion to give Eq. 3.13 actually corresponds to the work–energy principle often used in the study of dynamics [see any standard dynamics text (Ref. 1)]. This principle results from a general integration of the equations of motion for an object in a way very similar to that done for the fluid particle in Section 3.2. With certain assumptions, a statement of the work–energy principle may be written as follows:

The work done on a particle by all forces acting on the particle is equal to the change of the kinetic energy of the particle.

The Bernoulli equation is a mathematical statement of this principle.

As the fluid particle moves, both gravity and pressure forces do work on the particle. Recall that the work done by a force is equal to the product of the distance the particle travels times the component of force in the direction of travel (i.e., work = $\mathbf{F} \cdot \mathbf{d}$). The terms γz and p in Eq. 3.13 are related to the work done by the weight and pressure forces, respectively. The remaining term, $\rho V^2/2$, is obviously related to the kinetic energy of the particle. In fact, an alternate method of deriving the Bernoulli equation is to use the first and second laws of thermodynamics (the energy and entropy equations), rather than Newton's second law. With the appropriate restrictions, the general energy equation reduces to the Bernoulli equation. This alternate approach is discussed in Section 5.4.

EXAMPLE 3.4

Consider the flow of water from the syringe shown in Fig. E3.4. A force applied to the plunger will produce a pressure greater than atmospheric at point (1) **within the syringe**. The water flows from the needle, point (2), with relatively high velocity and **coasts up** to point (3) at the top of its trajectory. Discuss the energy of the fluid at points (1), (2), and (3) by using the Bernoulli equation.

FIGURE E3.4

SOLUTION

If the assumptions (steady, inviscid, incompressible flow) of the Bernoulli equation are approximately valid, it then follows that the flow can be explained in terms of the partition of the total energy of the water. According to Eq. 3.13 the sum of the three types of energy (kinetic, potential, and pressure) must remain constant. The following table indicates the relative magnitude of each of these energies at the three points shown in the figure.

| | Energy Type | | |
| | Kinetic | Potential | Pressure |
Point	$\rho V^2/2$	γz	p
1	Small	Zero	Large
2	Large	Small	Zero
3	Zero	Large	Zero

(Ans)

The motion results in (or is due to) a change in the magnitude of each type of energy as the fluid flows from one location to another. An alternate way to consider this flow is as follows. The pressure gradient between (1) and (2) produces an acceleration to eject the water from the needle. Gravity acting on the particle between (2) and (3) produces a deceleration to cause the water to come to a momentary stop at the top of its flight.

If friction (viscous) effects were important, there would be an energy loss between (1) and (3) and for the given p_1 the water would not be able to reach the height indicated in the figure. Such friction may arise in the needle (see Chapter 8, pipe flow) or between the water stream and the surrounding air (see Chapter 9, external flow).

A net force is required to accelerate any mass. For steady flow the acceleration can be interpreted as arising from two distinct occurrences—a change in speed along the streamline and a change in direction because the streamline is not straight. Integration of the equation of motion along the streamline accounts for the change in speed (kinetic energy change) and results in the Bernoulli equation. Integration of the equation of motion normal to the streamline accounts for the centrifugal acceleration ($V^2/\mathcal{R}$) and results in Eq. 3.14.

When a fluid particle travels along a curved path, a net force directed toward the center of curvature is required. Under the assumptions valid for Eq. 3.14 this force may be either gravity or pressure, or a combination of both. In many instances the streamlines are nearly straight so that centrifugal effects are negligible ($\mathcal{R} = \infty$) and the pressure variation across the streamlines is merely hydrostatic (because of gravity alone), even though the fluid is in motion.

EXAMPLE 3.5

Consider the inviscid, incompressible, steady flow shown in Fig. E3.5. From section A to B the streamlines are straight, while from C to D they follow circular paths. Describe the pressure variation between points (1) and (2) and points (3) and (4).

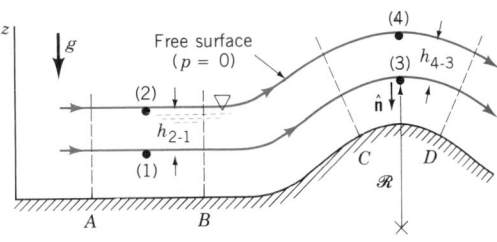

FIGURE E3.5

SOLUTION

With the above assumptions and the fact that $\mathcal{R} = \infty$ for the portion from A to B, Eq. 3.14 becomes

$$p + \gamma z = \text{constant}$$

The constant can be determined by evaluating the known variables at the two locations using $p_2 = 0$ (gage), $z_1 = 0$, and $z_2 = h_{2-1}$ to give

$$p_1 = p_2 + \gamma(z_2 - z_1) = p_2 + \gamma h_{2-1} \qquad \text{(Ans)}$$

Note that since the radius of curvature of the streamline is infinite, the pressure variation in the vertical direction is the same as if the fluid were stationary.

However, if we apply Eq. 3.14 between points (3) and (4) we obtain (using $dn = -dz$)

$$p_4 + \rho \int_{z_3}^{z_4} \frac{V^2}{\mathcal{R}} (-dz) + \gamma z_4 = p_3 + \gamma z_3$$

With $p_4 = 0$ and $z_4 - z_3 = h_{4-3}$ this becomes

$$p_3 = \gamma h_{4-3} - \rho \int_{z_3}^{z_4} \frac{V^2}{\mathcal{R}} \, dz \qquad \text{(Ans)}$$

To evaluate the integral we must know the variation of V and $\mathcal{R}$ with z. Even without this detailed information we note that the integral has a positive value. Thus, the pressure at (3) is less than the hydrostatic value, γh_{4-3}, by an amount equal to $\rho \int_{z_3}^{z_4} (V^2/\mathcal{R}) \, dz$. This lower pressure, caused by the curved streamline, is necessary to accelerate the fluid around the curved path.

Note that we did not apply the Bernoulli equation (Eq. 3.13) across the streamlines from (1) to (2) or (3) to (4). Rather we used Eq. 3.14. As is discussed in Section 3.6, application of the Bernoulli equation across streamlines (rather than along them) may lead to serious errors.

3.5   STATIC, STAGNATION (TOTAL), AND DYNAMIC PRESSURE

A useful concept associated with the Bernoulli equation deals with the stagnation and dynamic pressures. These pressures arise from the conversion of kinetic energy in a flowing fluid into a "pressure rise" as the fluid is brought to rest (as in Example 3.2). In this section we explore various results of this process. Each term of the Bernoulli equation, Eq. 3.13, has the dimensions of force per unit area—psi, lb/ft², N/m². The first term, p, is the actual thermodynamic pressure of the fluid as it flows. To measure its value, one could move along with the fluid, thus being "static" relative to the moving fluid. Hence, it is normally termed the *static pressure*. Another way to measure the static pressure would be to drill a hole in a flat surface and fasten a piezometer tube as indicated by the location of point (3) in Fig. 3.4. As we saw in Example 3.5, the pressure in the flowing fluid at (1) is $p_1 = \gamma h_{3-1} + p_3$, the same as if the fluid were static. From the manometer considerations of Chapter 2, we know that $p_3 = \gamma h_{4-3}$. Thus, since $h_{3-1} + h_{4-3} = h$ it follows that $p_1 = \gamma h$.

The third term in Eq. 3.13, γz, is termed the *hydrostatic pressure,* in obvious regard to the hydrostatic pressure variation discussed in Chapter 2. It is not actually a pressure, but does represent the change in pressure possible due to potential energy variations of the fluid as a result of elevation changes.

The second term in the Bernoulli equation, $\rho V^2/2$, is termed the *dynamic pressure.* Its interpretation can be seen in Fig. 3.4 by considering the pressure at the end of a small tube inserted into the flow and pointing upstream. After the initial transient motion has died out, the liquid will fill the tube to a height of H as shown. The fluid in the tube, including that at its tip, (2), will be stationary. That is, $V_2 = 0$, or point (2) is a *stagnation point.*

If we apply the Bernoulli equation between points (1) and (2), using $V_2 = 0$ and assuming that $z_1 = z_2$, we find that

$$p_2 = p_1 + \tfrac{1}{2}\rho V_1^2$$

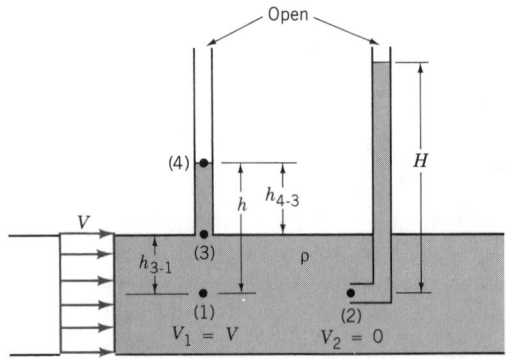

FIGURE 3.4 Measurement of static and stagnation pressures.

(a)   (b)

Stagnation point

FIGURE 3.5 Stagnation points on bodies in flowing fluids.

Hence, the pressure at the stagnation point is greater than the static pressure, p_1, by an amount $\rho V_1^2/2$, the dynamic pressure.

It can be shown that there is a stagnation point on any stationary body that is placed into a flowing fluid. Some of the fluid flows "over" and some "under" the object. The dividing line (or surface for two-dimensional flows) is termed the *stagnation streamline* and terminates at the stagnation point on the body. For symmetrical objects (such as a sphere) the stagnation point is clearly at the tip or front of the object as shown in Fig. 3.5a. For nonsymmetrical objects such as the airplane shown in Fig. 3.5b, the location of the stagnation point is not always obvious.

If elevation effects are neglected, the *stagnation pressure, $p + \rho V^2/2$*, is the largest pressure obtainable along a given streamline. It represents the conversion of all of the kinetic energy into a pressure rise. The sum of the static pressure, hydrostatic pressure, and dynamic pressure is termed the *total pressure, p_T*. The Bernoulli equation is a statement that the total pressure remains constant along a streamline. That is,

$$p + \tfrac{1}{2}\rho V^2 + \gamma z = p_T = \text{constant along a streamline} \qquad (3.15)$$

Again, we must be careful that the assumptions used in the derivation of this equation are appropriate for the flow being considered.

Knowledge of the values of the static and stagnation pressures in a fluid implies that the fluid speed can be calculated. This is the principle on which the *Pitot-static tube* is based. As shown in Fig. 3.6, two concentric tubes are attached to two pressure gages (or a differential gage) so that the values of p_3 and p_4 (or the difference $p_3 - p_4$) can be determined. The center tube measures the stagnation pressure at its open tip. If elevation changes are negligible,

$$p_3 = p + \tfrac{1}{2}\rho V^2$$

where p and V are the pressure and velocity of the fluid upstream of point (2). The outer tube is made with several small holes at an appropriate distance from the tip

FIGURE 3.6 The Pitot-static tube.

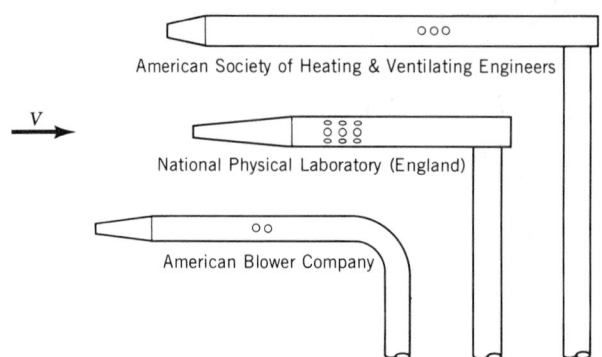

FIGURE 3.7   Typical Pitot-static tube designs.

so that they measure the static pressure. If the elevation difference between (1) and (4) is negligible, then

$$p_4 = p_1 = p$$

By combining these two equations we see that

$$p_3 - p_4 = \tfrac{1}{2}\rho V^2$$

which can be rearranged to give

$$V = \sqrt{2(p_3 - p_4)/\rho} \qquad\qquad (3.16)$$

The actual shape and size of Pitot-static tubes vary considerably. Some of the more common types are shown in Fig. 3.7.

EXAMPLE 3.6

An airplane flies 100 mi/hr at an elevation of 10,000 ft in a standard atmosphere as shown in Fig. E3.6. Determine the pressure at point (1) far ahead of the airplane, the pressure at the stagnation point on the nose of the airplane, point (2), and the pressure difference indicated by a Pitot-static probe attached to the fuselage.

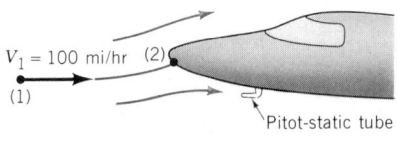

FIGURE E3.6

SOLUTION

From Table C.1 we find that the static pressure at the altitude given is

$$p_1 = 1456 \text{ lb/ft}^2 \text{ (abs)} = 10.11 \text{ psia} \qquad\qquad \text{(Ans)}$$

Also, the density is $\rho = 0.001756$ slug/ft^3.

   If the flow is steady, inviscid, and incompressible and elevation changes are neglected, Eq. 3.13 becomes

$$p_2 = p_1 + \frac{\rho V_1^2}{2}$$

With $V_1 = 100$ mi/hr $= 146.7$ ft/s and $V_2 = 0$ (since the coordinate system is fixed to the airplane) we obtain

$$p_2 = 1456 \text{ lb/ft}^2 + (0.001756 \text{ slugs/ft}^3)(146.7^2 \text{ ft}^2/\text{s}^2)/2$$

$$= (1456 + 18.9) \text{ lb/ft}^2 \text{ (abs)}$$

Hence, in terms of gage pressure

$$p_2 = 18.9 \text{ lb/ft}^2 = 0.1313 \text{ psi} \tag{Ans}$$

Thus, the pressure difference indicated by the Pitot-static tube is

$$p_2 - p_1 = \frac{\rho V_1^2}{2} = 0.1313 \text{ psi} \tag{Ans}$$

Note that it is very easy to obtain incorrect results by using improper units. Do not add lb/in.^2 and lb/ft^2. Recall that $(\text{slug/ft}^3)(\text{ft}^2/\text{s}^2) = (\text{slug·ft/s}^2)/(\text{ft}^2) = \text{lb/ft}^2$.

The Pitot-static tube provides a simple, relatively inexpensive way to measure fluid speed. Its use depends on the ability to measure the static and stagnation pressures. Care is needed to obtain these values accurately. For example, an accurate measurement of static pressure requires that none of the fluid's kinetic energy be converted into a pressure rise at the point of measurement. This requires a smooth hole with no burrs or imperfections. As indicated in Fig. 3.8, such imperfections can cause the measured pressure to be greater or less than the actual static pressure.

Also, the pressure along the surface of an object varies from the stagnation pressure at its stagnation point to values that may be less than the free stream static pressure. A typical pressure variation for a Pitot-tube is indicated in Fig. 3.9. Clearly it is important that the pressure taps be properly located to ensure that the pressure measured is actually the static pressure.

FIGURE 3.8 Incorrect and correct design of static pressure taps.

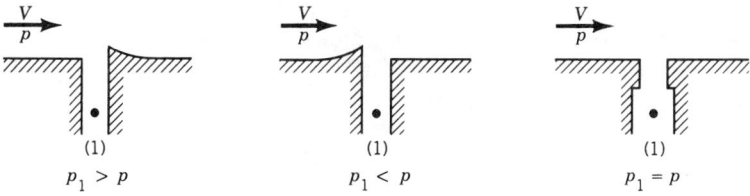

FIGURE 3.9 Typical pressure distribution along a Pitot-static tube.

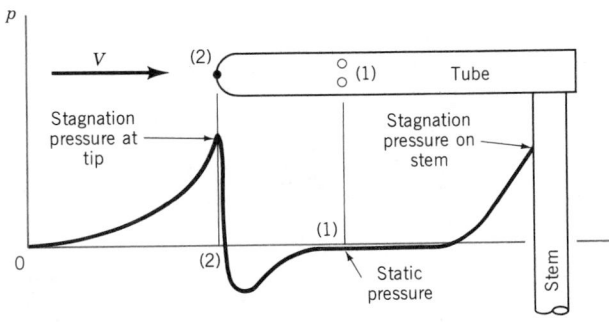

If $\theta = 0$

$p_1 = p_3 = p$

$p_2 = p + \frac{1}{2}\rho V^2$

FIGURE 3.10   Cross section of a directional finding Pitot-static tube.

In practice it is often difficult to align the Pitot-static tube directly into the flow direction. Any misalignment will produce a nonsymmetrical flow field that may introduce errors. Typically, yaw angles up to 12 to 20° (depending on the particular probe design) give results that are less than 1% in error from the perfectly aligned results. Generally it is more difficult to measure static pressure than stagnation pressure.

One method of determining the flow direction and its speed (thus the velocity) is to use a directional finding Pitot tube as is illustrated in Fig. 3.10. Three pressure taps are drilled into a small circular cylinder, fitted with small tubes, and connected to three pressure transducers. The cylinder is rotated until the pressures in the two side holes are equal, thus indicating that the center hole points directly upstream. The center tap then measures the stagnation pressure. The two side holes are located at a specific angle ($\beta = 29.5°$) so that they measure the static pressure. The speed is then obtained from $V = [2(p_2 - p_1)/\rho]^{1/2}$.

The above discussion is valid for incompressible flows. At high speeds compressibility becomes important (the density is not constant) and other phenomena occur. Some of these ideas are discussed in Section 3.8, while others (such as shockwaves for supersonic Pitot-tube applications) are discussed in Chapter 11.

The concepts of static, dynamic, stagnation, and total pressure are useful in a variety of flow problems. These ideas will be used more fully in the remainder of the book.

3.6   EXAMPLES OF USE OF THE BERNOULLI EQUATION

In this section we illustrate various additional applications of the Bernoulli equation. Between any two points, (1) and (2), on a streamline in steady, inviscid, incompressible flow the Bernoulli equation can be applied in the form

$$p_1 + \tfrac{1}{2}\rho V_1^2 + \gamma z_1 = p_2 + \tfrac{1}{2}\rho V_2^2 + \gamma z_2 \qquad \textit{along a streamline} \qquad (3.17)$$

Obviously if five of the six variables are known, the remaining one can be determined. In many instances it is necessary to introduce other equations, such as the conservation of mass for uniform incompressible flow. Such considerations will be discussed briefly in this section and in more detail in Chapter 5.

3.6.1   Free Jets

One of the oldest equations in fluid mechanics deals with the flow of a liquid from a large reservoir as is shown in Fig. 3.11. A jet of liquid of diameter d flows from the nozzle with velocity V as shown. Application of Eq. 3.17 between points (1) and (2) on the streamline shown gives

$$\gamma h = \tfrac{1}{2}\rho V^2$$

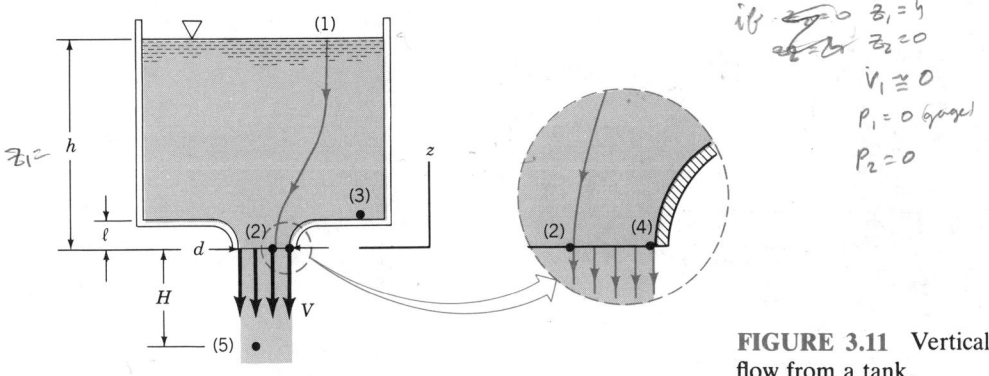

FIGURE 3.11 Vertical flow from a tank.

We have used the facts that $z_1 = h$, $z_2 = 0$, the reservoir is large ($V_1 \cong 0$), open to the atmosphere ($p_1 = 0$ gage), and the fluid leaves as a "free jet" ($p_2 = 0$). Thus, we obtain

$$V = \sqrt{2\frac{\gamma h}{\rho}} = \sqrt{2gh} \qquad (3.18)$$

which is the modern version of a result obtained in 1643 by Torricelli (1608–1647), an Italian physicist.

The fact that the exit pressure equals the surrounding pressure ($p_2 = 0$) can be seen by applying $\mathbf{F} = m\mathbf{a}$, as given by Eq. 3.14, across the streamlines between (2) and (4). If the streamlines at the tip of the nozzle are straight ($\mathcal{R} = \infty$), it follows that $p_2 = p_4$. Since (4) is on the surface of the jet, in contact with the atmosphere, we have $p_4 = 0$. Thus, $p_2 = 0$ also. Since (2) is an arbitrary point in the exit plane of the nozzle, it follows that the pressure is atmospheric across this plane. Physically, since there is no component of the weight force or acceleration in the normal (horizontal) direction, the pressure is constant in that direction.

Once outside the nozzle, the stream continues to fall as a free jet with zero pressure throughout ($p_5 = 0$) and as seen by applying Eq. 3.17 between points (1) and (5), the speed increases according to

$$V = \sqrt{2g(h + H)}$$

where H is the distance the fluid has fallen outside the nozzle.

Equation 3.18 could also be obtained by writing the Bernoulli equation between points (3) and (4) using the fact that $z_4 = 0$, $z_3 = \ell$. Also, $V_3 = 0$ since it is far from the nozzle, and from hydrostatics, $p_3 = \gamma(h - \ell)$.

Recall from physics or dynamics that any object dropped from rest through a distance h in a vacuum will obtain the speed $V = \sqrt{2gh}$, the same as the liquid leaving the nozzle. This is consistent with the fact that all of the particle's potential energy is converted to kinetic energy, provided viscous (friction) effects are negligible. Recall that for the case shown in Fig. 3.11 the pressure is the same (atmospheric) at points (1) and (2).

For the horizontal nozzle of Fig. 3.12, the velocity of the fluid at the centerline, V_2, will be slightly greater than that at the top, V_1, and slightly less than that at the bottom, V_3, due to the differences in elevation. In general $d \ll h$ and we can safely use the centerline velocity as a reasonable "average velocity."

FIGURE 3.12 Horizontal flow from a tank.

FIGURE 3.13 Vena contracta effect for a sharp edged orifice.

If the exit is not a smooth, well-contoured nozzle, but rather a flat plate as shown in Fig. 3.13, the diameter of the jet, d_j, will be less than the diameter of the hole, d_h. This phenomenon, called a *vena contracta* effect, is a result of the inability of the fluid to turn the sharp 90° corner indicated by the dotted lines in the figure.

Since the streamlines in the exit plane are curved ($\mathcal{R} < \infty$), the pressure across them is not constant. It would take an infinite pressure gradient across the streamlines

FIGURE 3.14 Typical flow patterns and contraction coefficients for various round exit configurations.

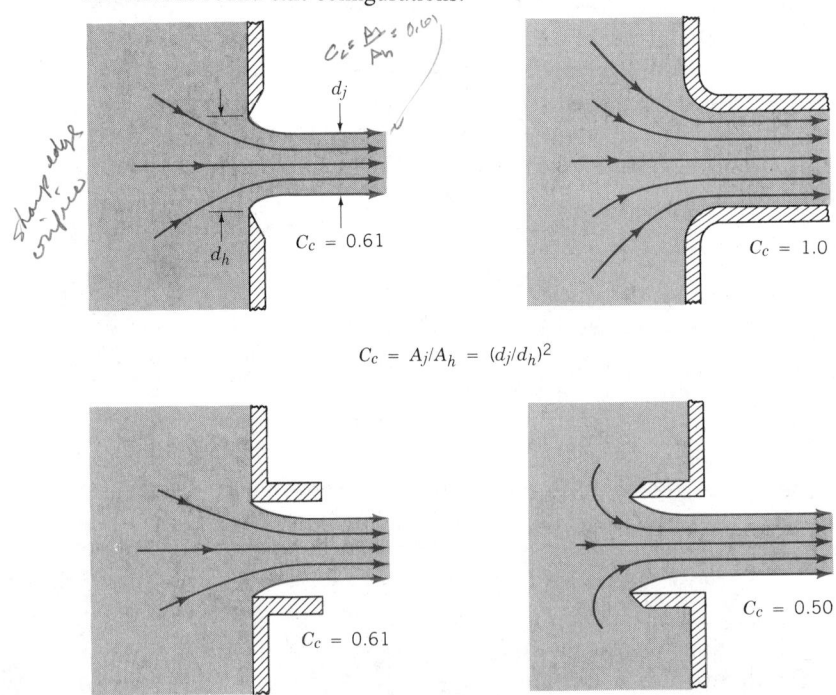

$$C_c = A_j/A_h = (d_j/d_h)^2$$

$C_c = 0.61$

$C_c = 1.0$

$C_c = 0.61$

$C_c = 0.50$

FIGURE 3.15 Steady flow into and out of a tank.

to cause the fluid to turn a "sharp" corner ($\mathcal{R} = 0$). The highest pressure occurs along the centerline at (2) and the lowest pressure, $p_1 = p_3 = 0$, is at the edge of the jet. Thus, the assumption of uniform velocity with straight streamlines and constant pressure is not valid at the exit plane. It is valid, however, in the plane of the vena contracta, section a–a. The uniform velocity assumption is valid at this section provided $d_j \ll h$, as is discussed for the flow from the nozzle shown in Fig. 3.12.

The vena contracta effect is a function of the geometry of the outlet. Some typical configurations are shown in Fig. 3.14 along with typical values of the experimentally obtained *contraction coefficient*, $C_c = A_j/A_h$, where A_j and A_h are the areas of the jet at the vena contracta and the area of the hole, respectively.

In many instances it is necessary to use the concept of conservation of mass (the continuity equation) along with the Bernoulli equation. The derivation and use of this equation is discussed in detail in Chapters 4 and 5. For the needs of this chapter we can use a simplified form of the continuity equation obtained from the following intuitive arguments. Consider a fluid flowing through a fixed volume (such as a tank) that has one inlet and one outlet as shown in Fig. 3.15. If the flow is steady so that there is no additional accumulation of fluid within the volume, the rate at which the fluid flows into the volume must equal the rate at which it flows out of the volume (otherwise mass would not be conserved).

The *mass flowrate* from an outlet, $\dot{m}$ (slugs/s or kg/s), is given by $\dot{m} = \rho Q$, where Q (ft³/s or m³/s) is the *volume flowrate*. If the outlet area is A and the fluid flows across this area (normal to the area) with an average velocity V, then the volume of the fluid crossing this area in a time interval δt is $VA\ \delta t$, equal to that in a volume of length $V\ \delta t$ and cross-sectional area A (see Fig. 3.15). Hence, the volume flowrate (volume per unit time) is $Q = VA$. Thus, $\dot{m} = \rho VA$. To conserve mass, the inflow rate must equal the outflow rate. If the inlet is designated as (1) and the outlet as (2), it follows that $\dot{m}_1 = \dot{m}_2$, or $\rho_1 A_1 V_1 = \rho_2 A_2 V_2$. If the density remains constant, $\rho_1 = \rho_2$, this reduces to $A_1 V_1 = A_2 V_2$. For example, if the outlet flow area is one-half the size of the inlet flow area, it follows that the outlet velocity is twice that of the inlet velocity, since $V_2 = A_1 V_1/A_2 = 2V_1$. The use of the Bernoulli equation and the flowrate equation (continuity equation) is demonstrated by Example 3.7.

EXAMPLE 3.7

A stream of water of diameter $d = 0.1$ m flows steadily from a tank of diameter $D = 1.0$ m as shown in Fig. E3.7a. Determine the flowrate, Q, needed from the inflow pipe if the water depth remains constant, $h = 2.0$ m.

(a)

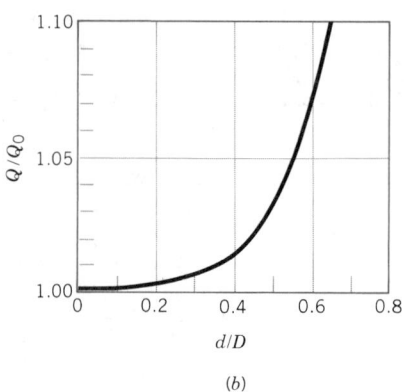

(b)

FIGURE E3.7

SOLUTION

For steady, inviscid, incompressible flow the Bernoulli equation applied between points (1) and (2) is

$$p_1 + \tfrac{1}{2}\rho V_1^2 + \gamma z_1 = p_2 + \tfrac{1}{2}\rho V_2^2 + \gamma z_2 \tag{1}$$

With the assumptions that $p_1 = p_2 = 0$, $z_1 = h$, and $z_2 = 0$, Eq. 1 becomes

$$\tfrac{1}{2}V_1^2 + gh = \tfrac{1}{2}V_2^2 \tag{2}$$

Although the water level remains constant (h = constant), there is an average velocity, V_1, across section (1) because of the flow from the tank. For steady incompressible flow, conservation of mass requires $Q_1 = Q_2$, where $Q = AV$. Thus,

$$\frac{\pi}{4}D^2V_1 = \frac{\pi}{4}d^2V_2$$

or

$$V_1 = \left(\frac{d}{D}\right)^2 V_2 \tag{3}$$

Equations 1 and 3 can be combined to give

$$V_2 = \sqrt{\frac{2gh}{1 - (d/D)^4}}$$

Thus, with the given data

$$V_2 = \sqrt{\frac{2(9.81 \text{ m/s}^2)(2.0 \text{ m})}{1 - (0.1\text{m}/1\text{m})^4}} = 6.26 \text{ m/s}$$

and

$$Q = A_1 V_1 = A_2 V_2 = \frac{\pi}{4}(0.1 \text{ m})^2(6.26 \text{ m/s}) = 0.0492 \text{ m}^3/\text{s} \qquad \text{(Ans)}$$

In this example we have not neglected the kinetic energy of the water in the tank ($V_1 \neq 0$). If the tank diameter is large compared to the jet diameter ($D \gg d$), Eq. 3 indicates that $V_1 \ll V_2$ and the assumption that $V_1 \approx 0$ would be reasonable. The error associated with this assumption can be seen by calculating the ratio of the flowrate assuming $V_1 \neq 0$, denoted Q, to that assuming $V_1 = 0$, denoted Q_0. This ratio, written as

$$\frac{Q}{Q_0} = \frac{V_2}{V_2|_{D=\infty}} = \frac{\sqrt{2gh/[1 - (d/D)^4]}}{\sqrt{2gh}} = \frac{1}{\sqrt{1 - (d/D)^4}}$$

is plotted in Fig. E3.7b. With $0 < d/D < 0.4$ it follows that $1 < Q/Q_0 \lesssim 1.01$ and the error in assuming $V_1 = 0$ is less than 1%. Thus, it is often reasonable to assume $V_1 = 0$.

3.6.2 Confined Flows

In many cases the fluid is physically constrained within a device so that its pressure cannot be prescribed a priori as was done for the free jet examples above. Such cases include nozzles and pipes of variable diameter for which the fluid velocity changes because the flow area is different from one section to another. This kinetic energy change is accompanied by a change in pressure as shown by Example 3.8.

EXAMPLE 3.8

Air flows steadily from a tank, through a hose of diameter $D = 0.03$ m and exits to the atmosphere from a nozzle of diameter $d = 0.01$ m as shown in Fig. E3.8. The pressure in the tank remains constant at 3.0 kPa (gage) and the atmospheric conditions are standard temperature and pressure. Determine the flowrate and the pressure in the hose.

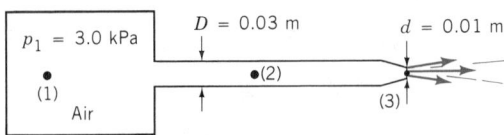

FIGURE E3.8

SOLUTION

If the flow is assumed steady, inviscid, and incompressible, we can apply the Bernoulli equation along the streamline shown as

$$p_1 + \tfrac{1}{2}\rho V_1^2 + \gamma z_1 = p_2 + \tfrac{1}{2}\rho V_2^2 + \gamma z_2$$
$$= p_3 + \tfrac{1}{2}\rho V_3^2 + \gamma z_3$$

With the assumption that $z_1 = z_2 = z_3$ (horizontal hose), $V_1 = 0$ (large tank), and $p_3 = 0$ (free jet) this becomes

$$V_3 = \sqrt{\frac{2p_1}{\rho}}$$

and

$$p_2 = p_1 - \tfrac{1}{2}\rho V_2^2 \tag{1}$$

The density of the air in the tank is obtained from the perfect gas law, using absolute pressure and temperature, as

$$\rho = \frac{p_1}{RT_1}$$
$$= [(3.0 + 101) \text{ kN/m}^2]$$
$$\times \frac{10^3 \text{ N/kN}}{(286.9 \text{ N·m/kg·K})(15 + 273)\text{K}}$$
$$= 1.26 \text{ kg/m}^3$$

Thus, we find that

$$V_3 = \sqrt{\frac{2(3.0 \times 10^3 \text{ N/m}^2)}{1.26 \text{ kg/m}^3}} = 69.0 \text{ m/s}$$

or

$$Q = A_3 V_3 = \frac{\pi}{4} d^2 V_3 = \frac{\pi}{4} (0.01 \text{ m})^2(69.0 \text{ m/s})$$
$$= 0.00542 \text{ m}^3/\text{s} \tag{Ans}$$

Note that the value of V_3 is determined strictly by the value of p_1 (and the assumptions involved in the Bernoulli equation), independent of the "shape" of the nozzle. Although we used gage pressure in the Bernoulli equation ($p_3 = 0$), we had to use absolute pressure in the perfect gas law when calculating the density.

The pressure within the hose can be obtained from Eq. 1 and the continuity equation

$$A_2 V_2 = A_3 V_3$$

Hence,

$$V_2 = A_3 V_3 / A_2 = \left(\frac{d}{D}\right)^2 V_3 = \left(\frac{0.01 \text{ m}}{0.03 \text{ m}}\right)^2 (69.0 \text{ m/s})$$
$$= 7.67 \text{ m/s}$$

and

$$p_2 = 3.0 \times 10^3 \text{ N/m}^2 - \tfrac{1}{2}(1.26 \text{ kg/m}^3)(7.67 \text{ m/s})^2$$
$$= (3000 - 37.1)\text{N/m}^2 = 2963 \text{ N/m}^2 \tag{Ans}$$

In the absence of viscous effects the pressure throughout the hose is constant, and equal to p_2. Physically, the decreases in pressure from p_1 to p_2 to p_3 accelerate the air and increase its kinetic energy from zero in the tank to an intermediate value in the hose and finally to its maximum value at the nozzle exit. Since the air velocity in the nozzle exit is nine times that in the hose, most of the pressure drop occurs across the nozzle ($p_1 = 3000 \text{ N/m}^2$, $p_2 = 2963$ N/m² and $p_3 = 0$).

Since the pressure change from (1) to (3) is not too great [that is, $(p_1 - p_3)/p_1 = 3.0/101 = 0.03$], it follows from the perfect gas law that the density change is also not significant. Hence, the incompressibility assumption is reasonable for this problem.

In many situations the combined effects of kinetic energy, pressure, and gravity are important. Example 3.9 illustrates this.

EXAMPLE 3.9

Water flows through a pipe reducer as is shown in Fig. E3.9. The static pressures at (1) and (2) are measured by the inverted U-tube manometer containing oil of specific gravity, SG, less than one. Determine the manometer reading, h.

FIGURE E3.9

SOLUTION

With the assumptions of steady, inviscid, incompressible flow, the Bernoulli equation can be written as

$$p_1 + \tfrac{1}{2}\rho V_1^2 + \gamma z_1 = p_2 + \tfrac{1}{2}\rho V_2^2 + \gamma z_2$$

The continuity equation provides a second relationship between V_1 and V_2 if we assume the velocity profiles are uniform at those two locations:

$$Q = A_1 V_1 = A_2 V_2$$

By combining these two equations we obtain

$$p_1 - p_2 = \gamma(z_2 - z_1) + \tfrac{1}{2}\rho V_2^2[1 - (A_2/A_1)^2] \tag{1}$$

This pressure difference is measured by the manometer and can be determined by using the pressure–depth ideas developed in Chapter 2. Thus,

$$p_1 - \gamma(z_2 - z_1) - \gamma\ell - \gamma h + SG\,\gamma h + \gamma\ell = p_2$$

or

$$p_1 - p_2 = \gamma(z_2 - z_1) + (1 - SG)\gamma h \tag{2}$$

As discussed in Chapter 2, this pressure difference is neither merely γh nor $\gamma(h + z_1 - z_2)$.
Equations 1 and 2 can be combined to give the desired result as follows

$$(1 - SG)\gamma h = \frac{1}{2}\rho V_2^2\left[1 - \left(\frac{A_2}{A_1}\right)^2\right]$$

or since $V_2 = Q/A_2$

$$h = (Q/A_2)^2\frac{1 - (A_2/A_1)^2}{2g(1 - SG)} \tag{Ans}$$

The difference in elevation, $z_1 - z_2$, was not needed because the change in elevation term in the Bernoulli equation exactly cancels the elevation term in the manometer equation. The pressure difference, $p_1 - p_2$, depends on the angle θ, because of the elevation, $z_1 - z_2$, in Eq. 1. Thus, for a given flowrate, the pressure difference, $p_1 - p_2$, as measured by a pressure gage would vary with θ. However, the manometer reading, h, would be independent of θ.

In general, an increase in velocity is accompanied by a decrease in pressure. For example, the velocity of the air flowing over the top surface of an airplane wing is, on the average, faster than that flowing under the bottom surface. Thus, the net pressure force is greater on the bottom than on the top—the wing generates a lift.

If the differences in velocity are considerable, the differences in pressure can also be considerable. For flows of gases, this may introduce compressibility effects as discussed in Section 3.8 and Chapter 11. For flows of liquids, this may result in cavitation, a potentially dangerous situation that results when the liquid pressure is reduced to the vapor pressure and the liquid "boils."

As discussed in Chapter 1, the vapor pressure, p_v, is the pressure at which vapor bubbles form in a liquid. It is the pressure at which the liquid starts to boil. Obviously this pressure depends on the type of liquid and its temperature. For example, water, which boils at 212 °F at standard atmospheric pressure, 14.7 psia, boils at 80 °F if the pressure is 0.507 psia. That is, $p_v = 0.507$ psia at 80 °F and $p_v = 14.7$ psia at 212 °F. (See Tables B.1 and B.2.)

One way to produce cavitation in a flowing liquid is noted from the Bernoulli equation. If the fluid velocity is increased (for example, by a reduction in flow area

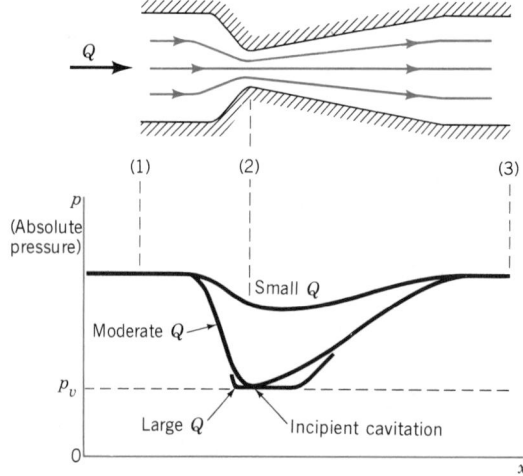

FIGURE 3.16 Pressure variation and cavitation in a variable area pipe.

FIGURE 3.17 Tip cavitation from a propeller. (Photograph courtesy of Garfield Thomas Water Tunnel, Pennsylvania State University)

as shown in Fig. 3.16) the pressure will decrease. This pressure decrease (needed to accelerate the fluid through the constriction) can be large enough so that the pressure in the liquid is reduced to its vapor pressure. A simple example of cavitation can be demonstrated with an ordinary garden hose. If the hose is "kinked," a restriction in the flow area in some ways analogous to that shown in Fig. 3.16 will result. The water velocity through this restriction will be relatively large. With a sufficient amount of restriction the sound of the flowing water will change—a definite "hissing" sound is produced. This sound is a result of cavitation.

In such situations boiling occurs (though the temperature need not be high), vapor bubbles form, and then collapse as the fluid moves into a region of higher pressure (lower velocity). This process can produce dynamic effects (imploding) that cause very large pressure transients in the vicinity of the bubbles. Pressures as large as 100,000 psi (690 MPa) are believed to occur. If the bubbles collapse close to a physical boundary they can, over a period of time, cause damage to the surface in the cavitation area. Tip cavitation from a propeller is shown in Fig. 3.17. In this case the high-speed rotation of the propeller produced a corresponding low pressure on the propeller. Obviously, proper design and use of equipment is needed to eliminate cavitation damage.

EXAMPLE 3.10

Water at 60 °F is siphoned from a large tank through a constant diameter hose as shown in Fig. E3.10. Determine the maximum height of the hill, H, over which the water can be siphoned without cavitation occurring. The end of the siphon is 5 ft below the bottom of the tank. Atmospheric pressure is 14.7 psia.

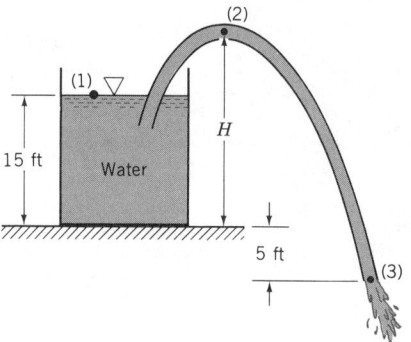

FIGURE E3.10

SOLUTION

If the flow is steady, inviscid, and incompressible we can apply the Bernoulli equation along the streamline from (1) to (2) to (3) as follows

$$p_1 + \tfrac{1}{2}\rho V_1^2 + \gamma z_1 = p_2 + \tfrac{1}{2}\rho V_2^2 + \gamma z_2 = p_3 + \tfrac{1}{2}\rho V_3^2 + \gamma z_3 \tag{1}$$

With the tank bottom as the datum, we have $z_1 = 15$ ft, $z_2 = H$, and $z_3 = -5$ ft. Also, $V_1 = 0$ (large tank), $p_1 = 0$ (open tank), $p_3 = 0$ (free jet), and from the continuity equation $A_2V_2 = A_3V_3$, or because the hose is constant diameter, $V_2 = V_3$. Thus, the speed of the fluid in the hose is determined from Eq. 1 to be

$$V_3 = \sqrt{2g(z_1 - z_3)} = \sqrt{2(32.2 \text{ ft/s}^2)[15 - (-5)]} \text{ ft}$$
$$= 35.9 \text{ ft/s} = V_2$$

Use of Eq. 1 between points (1) and (2) then gives the pressure p_2 at the top of the hill as

$$p_2 = p_1 + \tfrac{1}{2}\rho V_1^2 + \gamma z_1 - \tfrac{1}{2}\rho V_2^2 - \gamma z_2 = \gamma(z_1 - z_2) - \tfrac{1}{2}\rho V_2^2 \tag{2}$$

From Table B.1, the vapor pressure of water at 60 °F is 0.256 psia. Hence, for incipient cavitation the lowest pressure in the system will be $p = 0.256$ psia. Careful consideration of Eq. 2 and Fig. E3.10 will show that this lowest pressure will occur at the top of the hill. Since we have used gage pressure at point (1) ($p_1 = 0$), we must use gage pressure at point (2) also. Thus, $p_2 = 0.256 - 14.7 = -14.4$ psi and Eq. 2 gives

$$(-14.4 \text{ lb/in.}^2)(144 \text{ in.}^2/\text{ft}^2) = (62.4 \text{ lb/ft}^3)(15 - H)\text{ft} - \tfrac{1}{2}(1.94 \text{ slugs/ft}^3)(35.9 \text{ ft/s})^2$$

or

$$H = 28.2 \text{ ft} \tag{Ans}$$

For larger values of H, vapor bubbles will form at point (2) and the siphon action may stop.

Note that we could have used absolute pressure throughout [$p_2 = 0.256$ psia and $p_1 = 14.7$ psia] and obtained the same result. The lower the elevation of point (3), the larger the flowrate and, therefore, the smaller the value of H allowed.

We could also have used the Bernoulli equation between (2) and (3), with $V_2 = V_3$, to obtain the same value of H. In this case it would not have been necessary to determine V_2 by use of the Bernoulli equation between (1) and (3).

The above results are independent of the diameter and length of the hose (provided viscous effects are not important). Proper design of the hose (or pipe) is needed to ensure that it will not collapse due to the large pressure difference (vacuum) across it.

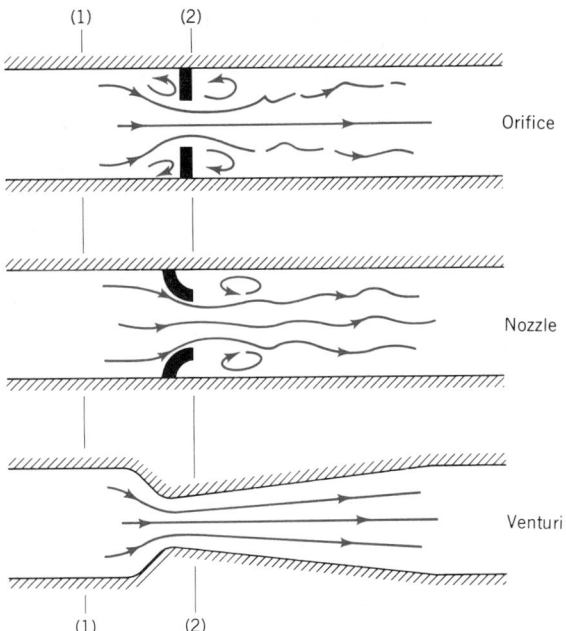

FIGURE 3.18 Typical devices for measuring flowrate in pipes.

3.6.3   Flowrate Measurement

Many types of devices using principles involved in the Bernoulli equation have been developed to measure fluid velocities and flowrates. The Pitot-static tube discussed in Section 3.5 is an example. Other examples discussed below include devices to measure flowrates in pipes and conduits and devices to measure flowrates in open channels. In this chapter we will consider "ideal" flowmeters—those devoid of viscous, compressibility, and other "real world" effects. Corrections for these effects are discussed in Chapters 8 and 10. Our goal here is to understand the basic operating principles of these simple flowmeters.

An effective way to measure the flowrate through a pipe is to place some type of restriction within the pipe as shown in Fig. 3.18 and to measure the pressure difference between the low-velocity, high-pressure upstream section (1), and the high-velocity, low-pressure downstream section (2). Three commonly used types of flowmeters are illustrated: the *orifice meter,* the *nozzle meter,* and the *Venturi meter.* The operation of each is based on the same physical principles—an increase in velocity causes a decrease in pressure. The differences between them is a matter of cost, accuracy, and how closely their actual operation obeys the idealized flow assumptions.

We assume the flow is horizontal ($z_1 = z_2$), steady, inviscid, and incompressible between points (1) and (2). The Bernoulli equation becomes

$$p_1 + \tfrac{1}{2}\rho V_1^2 = p_2 + \tfrac{1}{2}\rho V_2^2$$

(The effect of nonhorizontal flow can be incorporated easily by including the change in elevation, $z_1 - z_2$, in the Bernoulli equation.)

If we assume the velocity profiles are uniform at sections (1) and (2), the continuity equation can be written as

$$Q = A_1 V_1 = A_2 V_2$$

where A_2 is the small $(A_2 < A_1)$ flow area at section (2). Combination of these two equations results in the following theoretical flowrate

$$Q = A_2 \sqrt{\frac{2(p_1 - p_2)}{\rho[1 - (A_2/A_1)^2]}}$$ (3.19)

Thus, for a given flow geometry (A_1 and A_2) the flowrate can be determined if the pressure difference, $p_1 - p_2$, is measured. The actual measured flowrate, Q_{actual}, will be smaller than this theoretical result because of various differences between the "real world" and the assumptions used in the derivation of Eq. 3.19. These differences (which are quite consistent and may be as small as 1 to 2% or as large as 40% depending on the geometry used) are discussed in Chapter 8.

EXAMPLE 3.11

Kerosene ($SG = 0.85$) flows through the Venturi meter shown in Fig. E3.11 with flowrates between 0.005 and 0.050 m³/s. Determine the range in pressure difference, $p_1 - p_2$, needed to measure these flowrates.

Kerosene, $SG = 0.85$

$D_1 = 0.1$ m

(2)● $D_2 = 0.06$ m

(1)

Q

0.005 m³/s ≤ Q ≤ 0.050 m³/s

FIGURE E3.11

SOLUTION

If the flow is assumed to be steady, inviscid, and incompressible, the relationship between flowrate and pressure is given by Eq. 3.19. This can be rearranged to give

$$p_1 - p_2 = \frac{Q^2 \rho [1 - (A_2/A_1)^2]}{2A_2^2}$$

With a density of the flowing fluid of

$$\rho = SG \, \rho_{H_2O} = 0.85(1000 \text{ kg/m}^3) = 850 \text{ kg/m}^3$$

the pressure difference for the smallest flowrate is

$$p_1 - p_2 = (0.005 \text{ m}^3/\text{s})^2(850 \text{ kg/m}^3) \frac{[1 - (0.06 \text{ m}/0.10 \text{ m})^4]}{2[(\pi/4)(0.06 \text{ m})^2]^2}$$

$$= 1160 \text{ N/m}^2 = 1.16 \text{ kPa}$$

Likewise, the pressure difference for the largest flowrate is

$$p_1 - p_2 = (0.05)^2(850) \frac{[1 - (0.06/0.10)^4]}{2[(\pi/4)(0.06)^2]^2}$$

$$= 1.16 \; 10^5 \text{ N/m}^2$$

$$= 116 \text{ kPa}$$

Thus,

$$1.16 \text{ kPa} \le p_1 - p_2 \le 116 \text{ kPa} \qquad \text{(Ans)}$$

These values represent the pressure differences for inviscid, steady, incompressible conditions.

It is seen from Eq. 3.19 that the flowrate varies as the square root of the pressure difference. Hence, as indicated by the numerical results, a tenfold increase in flowrate requires a one-hundredfold increase in pressure difference. This nonlinear relationship can cause difficulties when measuring flowrates over a wide range of values. Such measurements would require pressure transducers with a wide range of operation. An alternative is to use two flowmeters in parallel—one for the larger and one for the smaller flowrate ranges.

Other flowmeters based on the Bernoulli equation are used to measure flowrates in open channels such as flumes and irrigation ditches. Two of these devices, the *sluice gate* and the *sharp crested weir,* are discussed below under the assumption of steady, inviscid, incompressible flow. These and other open-channel flow devices are discussed in more detail in Chapter 10.

A sluice gate as shown in Fig. 3.19 is often used to regulate and measure the flowrate in an open channel. The flowrate, Q, is a function of the water depth upstream, z_1, the width of the gate, b, and the gate opening, a. Application of the Bernoulli equation and continuity equation between points (1) and (2) can provide a good approximation to the actual flowrate obtained. We assume the velocity profiles are uniform sufficiently far upstream and downstream of the gate.

Thus, we apply the Bernoulli and continuity equations between points on the free surfaces at (1) and (2) to give

$$p_1 + \tfrac{1}{2}\rho V_1^2 + \gamma z_1 = p_2 + \tfrac{1}{2}\rho V_2^2 + \gamma z_2$$

and

$$Q = bV_1 z_1 = bV_2 z_2$$

With the fact that $p_1 = p_2 = 0$, these equations can be combined and rearranged to give the flowrate as

$$Q = z_2 b \sqrt{\frac{2g(z_1 - z_2)}{1 - (z_2/z_1)^2}} \tag{3.20}$$

In the limit of $z_1 \gg z_2$ this result simply becomes

$$Q = z_2 b \sqrt{2gz_1} \tag{3.21}$$

FIGURE 3.19 Sluice gate geometry.

This limiting result represents the fact that if the depth ratio, z_1/z_2, is large, the kinetic energy of the fluid upstream of the gate is negligible and the fluid velocity after it has fallen a distance $(z_1 - z_2) \approx z_1$ is approximately $V_2 = \sqrt{2gz_1}$.

The results of Eq. 3.20 could also be obtained by using the Bernoulli equation between points (3) and (4) and the fact that $p_3 = \gamma z_1$ and $p_4 = \gamma z_2$ since the streamlines at these sections are straight. In this formulation, rather than the potential energies at (1) and (2), we have the pressure contributions at (3) and (4).

The downstream depth, z_2, not the gate opening, a, was used to obtain the result of Eq. 3.20. As was discussed relative to flow from an orifice (Fig. 3.14), the fluid cannot turn a sharp 90° corner. A vena contracta results with a contraction coefficient, $C_c = z_2/a$, less than 1. Typically C_c is approximately 0.61 over the depth ratio range of $0 < a/z_1 < 0.2$. For larger values of a/z_1 the value of C_c increases rapidly.

EXAMPLE 3.12

Water flows under the sluice gate shown in Fig. E3.12. Determine the approximate flowrate per unit width of the channel.

6.0 m

5.0 m

0.8 m

FIGURE E3.12

SOLUTION

Under the assumptions of steady, inviscid, incompressible flow, we can apply Eq. 3.20 to obtain Q/b, the flowrate per unit width, as

$$\frac{Q}{b} = z_2 \sqrt{\frac{2g(z_1 - z_2)}{1 - (z_2/z_1)^2}}$$

In this instance $z_1 = 5.0$ m and $a = 0.80$ m so the ratio $a/z_1 = 0.16 < 0.20$, and we can assume that the contraction coefficient is approximately $C_c = 0.61$. Thus, $z_2 = C_c a = 0.61$ (0.80 m) $= 0.488$ m and we obtain the flowrate

$$\frac{Q}{b} = (0.488 \text{ m}) \sqrt{\frac{2(9.81 \text{ m/s}^2)(5.0 \text{ m} - 0.488 \text{ m})}{1 - (0.488 \text{ m}/5.0 \text{ m})^2}}$$

$$= 4.61 \text{ m}^2/\text{s} \qquad\qquad \text{(Ans)}$$

If we consider $z_1 \gg z_2$ and neglect the kinetic energy of the upstream fluid, Eq. 3.21 would be valid so that

$$\frac{Q}{b} = 0.488 \text{ m} \sqrt{2(9.81 \text{ m/s}^2)(5.0 \text{ m})} = 4.83 \text{ m}^2/\text{s}$$

In this case the difference in Q with or without including V_1 is not too significant because the depth ratio is fairly large ($z_1/z_2 = 5.0/0.488 = 10.2$). Thus, it is often reasonable to neglect the kinetic energy upstream from the gate compared to that downstream of it.

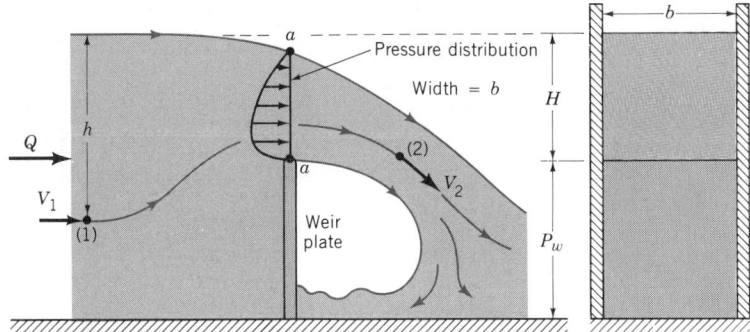

FIGURE 3.20   Rectangular, sharp crested weir geometry.

Another device used to measure flow in an open channel is a *weir*. A typical rectangular, sharp crested weir is shown in Fig. 3.20. For such devices the flowrate of liquid over the top of the weir plate is dependent on the weir height, P_w, the width of the channel, b, and the head, H, of the water above the top of the weir. Application of the Bernoulli equation can provide a simple approximation of the flowrate expected for these situations, even though the actual flow is quite complex.

Between points (1) and (2) the pressure and gravitational fields cause the fluid to accelerate from velocity V_1 to velocity V_2. At (1) the pressure is $p_1 = \gamma h$, while at (2) the pressure is essentially atmospheric, $p_2 = 0$. Across the curved streamlines directly above the top of the weir plate (section a–a) the pressure changes from atmospheric on the top surface to some maximum value within the fluid stream and then to atmospheric again at the bottom surface. This distribution is indicated in Fig. 3.20. Such a pressure distribution, combined with the streamline curvature and gravity, produces a rather nonuniform velocity profile across this section. This velocity distribution can be obtained from experiments or a more advanced theory.

For now, we will take a very simple approach and assume that the weir flow is similar in many respects to an orifice type flow with a free streamline. In this instance we would expect the average velocity across the top of the weir to be proportional to $\sqrt{2gH}$ and the flow area for this rectangular weir to be proportional to Hb. Hence, it follows that

$$Q = C_1 Hb \sqrt{2gH} = C_1 b \sqrt{2g} \, H^{3/2}$$

where C_1 is a constant to be determined.

Simple use of the Bernoulli equation has provided a method to analyze the relatively complex flow over a weir. The correct functional dependence of Q on H has been obtained ($Q \sim H^{3/2}$), but the value of the coefficient C_1 is unknown. Even a more advanced analysis cannot predict its value accurately. As is discussed in Chapter 10, experiments are used to determine the value of C_1.

EXAMPLE 3.13

Water flows over a triangular weir as is shown in Fig. E3.13. Based on a simple analysis using the Bernoulli equation, determine the dependence of the flowrate on the depth H. If the flowrate is Q_0 when $H = H_0$, estimate the flowrate when the depth is increased to $H = 3H_0$.

FIGURE E3.13

SOLUTION

With the assumption that the flow is steady, inviscid, and incompressible, it is reasonable to assume from Eq. 3.18 that the average speed of the fluid over the triangular notch in the weir plate is proportional to $\sqrt{2gH}$. Also, the flow area for a depth of H is proportional to $H \, [H \tan (\theta/2)]$. The combination of these two ideas gives

$$Q = C_2 \tan \frac{\theta}{2} \sqrt{2g} \, H^{5/2} \qquad \text{(Ans)}$$

where C_2 is an unknown constant to be determined experimentally.

Thus, an increase in the depth by a factor of three (from H_0 to $3H_0$) results in an increase of the flowrate by a factor of

$$\frac{Q_{3H_0}}{Q_{H_0}} = \frac{C_2 \tan(\theta/2) \sqrt{2g} \, (3H_0)^{5/2}}{C_2 \tan(\theta/2) \sqrt{2g} \, (H_0)^{5/2}} \qquad \text{(Ans)}$$

$$= 15.6$$

Note that for a triangular weir the flowrate is proportional to $H^{5/2}$, whereas for the rectangular weir discussed above, it is proportional to $H^{3/2}$. The triangular weir can be accurately used over a wide range of flowrates.

3.7   THE ENERGY LINE AND THE HYDRAULIC GRADE LINE

As was discussed in Section 3.4, the Bernoulli equation is actually an energy equation representing the partitioning of energy for an inviscid, incompressible, steady flow. The sum of the various energies of the fluid remains constant as the fluid flows from one section to another. A useful interpretation of the Bernoulli equation can be obtained through the use of the concepts of the *hydraulic grade line* (HGL) and the *energy line* (EL). These ideas represent a geometrical interpretation of a flow and can often be effectively used to better grasp the fundamental processes involved.

For steady, inviscid, incompressible flow the total energy remains constant along a streamline. The concept of "head" can be introduced by dividing each term in Eq. 3.7 by the specific weight, $\gamma = \rho g$, to give the Bernoulli equation in the following form

$$\frac{p}{\gamma} + \frac{V^2}{2g} + z = \text{constant on a streamline} = H \qquad (3.22)$$

Each of the terms in this equation has the units of length (feet or meters) and represents a certain type of head.

FIGURE 3.21   Representation of the energy line and the hydraulic grade line.

The elevation term, z, is related to the potential energy of the particle and is called the *elevation head*. The pressure term, p/γ, is called the *pressure head* and represents the height of a column of the fluid that is needed to produce the pressure p. The velocity term, $V^2/2g$, is the *velocity head* and represents the vertical distance needed for the fluid to fall freely if it is to reach velocity V from rest. The Bernoulli equation states that the sum of the pressure head, the velocity head, and the elevation head is constant along a streamline. This constant is called the *total head, H*.

The energy line is a line that represents the total head available to the fluid. As shown in Fig. 3.21, the elevation of the energy line can be obtained by measuring the stagnation pressure with a Pitot tube. (A Pitot tube is the portion of a Pitot-static tube that measures the stagnation pressure. See Section 3.5.) The stagnation point at the end of the Pitot tube provides a measurement of the total head (or energy) of the flow. The static pressure tap connected to the piezometer tube shown, on the other hand, measures the sum of the pressure head and the elevation head, $p/\gamma + z$. This sum is often called the *piezometric head*. The static pressure tap does not measure the velocity head.

According to Eq. 3.22 the total head remains constant along the streamline (provided the assumptions of the Bernoulli equation are valid). Thus, a Pitot tube at another location in the flow will measure the same total head, as is shown in the figure. The elevation head and pressure head may be different along the streamline, however.

The locus of elevations provided by a series of Pitot tubes is termed the energy line, EL. That provided by a series of piezometer taps is termed the hydraulic grade line, HGL. Under the assumptions of the Bernoulli equation, the energy line is horizontal. If the fluid velocity changes along the streamline, the hydraulic grade line will not be horizontal. If viscous effects are important (as they often are in pipe flows) the total head does not remain constant due to a loss in energy as the fluid flows along its streamline. This means that the energy line is no longer horizontal. Such viscous effects are discussed in Chapter 8.

FIGURE 3.22 The energy line and hydraulic grade line for flow from a tank.

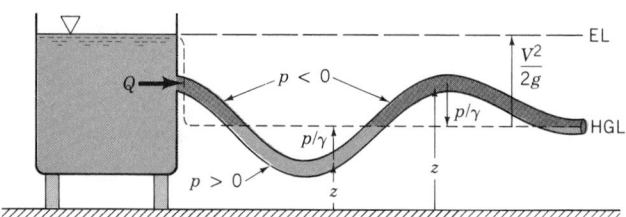

FIGURE 3.23 Use of the energy line and the hydraulic grade line.

The energy line and hydraulic grade line for flow from a large tank are shown in Fig. 3.22. If the flow is steady, incompressible, and inviscid, the energy line is horizontal and at the elevation of the liquid in the tank (since the fluid velocity in the tank and the pressure on the surface are zero). The hydraulic grade line lies a distance of one velocity head, $V^2/2g$, below the energy line. Thus, a change in fluid velocity due to a change in the pipe diameter results in a change in the elevation of the hydraulic grade line. At the pipe outlet the pressure head is zero (gage) so the pipe elevation and the hydraulic grade line coincide.

The distance from the pipe to the hydraulic grade line indicates the pressure within the pipe as is shown in Fig. 3.23. If the pipe lies below the hydraulic grade line the pressure within the pipe is positive (above atmospheric). If the pipe lies above the hydraulic grade line the pressure is negative (below atmospheric). Thus, a scale drawing of a pipeline and the hydraulic grade line can be used to readily indicate regions of positive or negative pressure within a pipe.

EXAMPLE 3.14

Water is siphoned from the tank shown in Fig. E3.14 through a hose of constant diameter. A small hole is found in the hose at location (1) as indicated. When the siphon is used, will water leak out of the hose, or will air leak into the hose, thereby possibly causing the siphon to malfunction?

FIGURE E3.14

SOLUTION

Whether air will leak into or water will leak out of the hose depends on whether the pressure within the hose at (1) is less than or greater than atmospheric. Which happens can be easily determined by using the energy line and hydraulic grade line concepts as follows. With the assumption of steady, incompressible, inviscid flow it follows that the total head is constant—thus, the energy line is horizontal.

Since the hose diameter is constant, it follows from the continuity equation ($AV =$ constant) that the water velocity in the hose is constant throughout. Thus, the hydraulic grade line is a constant distance, $V^2/2g$, below the energy line as shown in Fig. E3.14. Since the pressure at the end of the hose is atmospheric, it follows that the hydraulic grade line is at the same elevation as the end of the hose outlet. The fluid within the hose at any point above the hydraulic grade line will be at less than atmospheric pressure.

Thus, air will leak into the hose through
the hole at point (1).   (Ans)

In practice, viscous effects may be quite important, making this simple analysis (horizontal energy line) incorrect. However, if the hose is "not too small diameter," "not too long," and the flowrate "not too large," the above result may be very accurate. If the end of the hose were closed so the flowrate were zero, the hydraulic grade line would coincide with the energy line ($V^2/2g = 0$ throughout), the pressure at (1) would be greater than atmospheric, and water would leak through the hole at (1).

The above discussion of the hydraulic grade line and the energy line is restricted to ideal situations involving inviscid, incompressible flows. Another restriction is that there are no "sources" or "sinks" of energy within the flow field. That is, there are no pumps, turbines, or head exchangers involved. Alterations in the energy line and hydraulic grade line concepts due to these devices are discussed in Chapter 8.

3.8   RESTRICTIONS ON THE USE OF THE BERNOULLI EQUATION

Proper use of the Bernoulli equation requires close attention to the assumptions used in its derivation. In this section we review some of these assumptions and consider the consequences of incorrect use of the equation.

3.8.1 Compressibility Effects

One of the main assumptions is that the fluid is incompressible. Although this is reasonable for most liquid flows, it can, in certain instances, introduce considerable errors for gases.

In the previous section we saw that the stagnation pressure is greater than the static pressure by an amount $\rho V^2/2$, provided that the density remains constant. If this dynamic pressure is not too large compared with the static pressure, the density change between two points is not very large and the flow can be considered incompressible. However, since the dynamic pressure varies as V^2, the error associated with the assumption that a fluid is incompressible increases with the square of the velocity of the fluid. To account for compressibility effects we must return to Eq. 3.6 and properly integrate the term $\int dp/\rho$ when ρ is not constant.

A simple, although specialized, case of compressible flow occurs when the temperature of a perfect gas remains constant along the streamline—isothermal flow. Thus, we consider $p = \rho RT$, where T is constant. For steady, inviscid, isothermal flow Eq. 3.6 becomes

$$RT \int \frac{dp}{p} + \frac{1}{2} V^2 + gz = \text{constant}$$

where we have used $\rho = p/RT$. The pressure term is easily integrated and the constant of integration evaluated if z_1, p_1, and V_1 are known at some location on the streamline. The result is

$$\frac{V_1^2}{2g} + z_1 + \frac{RT}{g} \ln \left(\frac{p_1}{p_2} \right) = \frac{V_2^2}{2g} + z_2 \tag{3.23}$$

Equation 3.23 is the inviscid, isothermal analog of the incompressible Bernoulli equation. In the limit of small pressure difference, $p_1/p_2 = 1 + (p_1 - p_2)/p_2 = 1 + \varepsilon$, with $\varepsilon \ll 1$, Eq. 3.23 reduces to the standard incompressible Bernoulli equation. This can be shown by use of the approximation $\ln(1 + \varepsilon) \approx \varepsilon$ for small ε. The use of Eq. 3.23 in practical applications is restricted by the inviscid flow assumption, since (as is discussed in Section 11.5) most isothermal flows are accompanied by viscous effects.

A much more common compressible flow condition is that of isentropic (constant entropy) flow of a perfect gas. Such flows are reversible adiabatic processes—"no friction or heat transfer"—and are closely approximated in many physical situations. As discussed fully in Chapter 11, for isentropic flow of a perfect gas the density and pressure are related by $p/\rho^k = C$, where k is the specific heat ratio and C is a constant. Hence, the $\int dp/\rho$ integral of Eq. 3.6 can be evaluated as follows. The density can be written in terms of the pressure as $\rho = p^{1/k}C^{-1/k}$ so that Eq. 3.6 becomes

$$C^{1/k} \int p^{-1/k} \, dp + \frac{1}{2} V^2 + gz = \text{constant}$$

The pressure term can be integrated between points (1) and (2) on the streamline and the constant C evaluated at either point ($C^{1/k} = p_1^{1/k}/\rho_1$ or $C^{1/k} = p_2^{1/k}/\rho_2$) to give the following

$$C^{1/k} \int_{p_1}^{p_2} p^{-1/k} \, dp = C^{1/k} \left(\frac{k}{k-1} \right) [p_2^{(k-1)/k} - p_1^{(k-1)/k}]$$

$$= \left(\frac{k}{k-1} \right) \left(\frac{p_2}{\rho_2} - \frac{p_1}{\rho_1} \right)$$

Thus, the final form of Eq. 3.6 for compressible, isentropic, steady flow of a perfect gas is

$$\left(\frac{k}{k-1}\right)\frac{p_1}{\rho_1} + \frac{V_1^2}{2} + gz_1 = \left(\frac{k}{k-1}\right)\frac{p_2}{\rho_2} + \frac{V_2^2}{2} + gz_2 \tag{3.24}$$

The similarities between the results for compressible isentropic flow (Eq. 3.24) and incompressible isentropic flow (the Bernoulli equation, Eq. 3.7) are apparent. The only difference is the factor of $[k/(k-1)]$ that multiplies the pressure terms. In the limit of "low-speed flow" the two results are exactly the same as is seen by the following.

We consider the stagnation point flow of Section 3.5 to illustrate the difference between the incompressible and compressible results. As is shown in Chapter 11, Eq. 3.24 can be written in dimensionless form as

$$\frac{p_2 - p_1}{p_1} = \left[\left(1 + \frac{k-1}{2}\text{Ma}_1^2\right)^{k/k-1} - 1\right] \text{ (compressible)} \tag{3.25}$$

where (1) denotes the upstream conditions and (2) the stagnation conditions. We have assumed $z_1 = z_2$ and have denoted $\text{Ma}_1 = V_1/c_1$ as the upstream *Mach number*—the ratio of the fluid velocity to the speed of sound, $c_1 = \sqrt{kRT_1}$.

A comparison between this compressible result and the incompressible result is perhaps most easily seen if we write the incompressible flow result in terms of the pressure ratio and the Mach number. Thus, we divide each term in the Bernoulli equation, $\rho V_1^2/2 + p_1 = p_2$, by p_1 and use the perfect gas law, $p_1 = \rho RT_1$, to obtain

$$\frac{p_2}{p_1} = 1 + \frac{V_1^2}{2RT_1}$$

Since $\text{Ma}_1 = V_1/\sqrt{kRT_1}$ this can be written as

$$\frac{p_2 - p_2}{p_1} = \frac{k\text{Ma}_1^2}{2} \quad \text{(incompressible)} \tag{3.26}$$

Equations 3.25 and 3.26 are plotted in Fig. 3.24. In the low speed limit of $\text{Ma}_1 \to 0$ both of the results are the same. This can be seen by denoting $(k-1)\text{Ma}_1^2/2 = \tilde{\varepsilon}$ and using the binomial expansion, $(1 + \tilde{\varepsilon})^n = 1 + n\tilde{\varepsilon} + n(n-1)\tilde{\varepsilon}^2/2 + \cdots$, where $n = k/(k-1)$, to write Eq. 3.25 as

$$\frac{p_2 - p_1}{p_1} = \frac{k\text{Ma}_1^2}{2}\left(1 + \frac{1}{4}\text{Ma}_1^2 + \frac{2-k}{24}\text{Ma}_1^4 + \cdots\right) \quad \text{(compressible)}$$

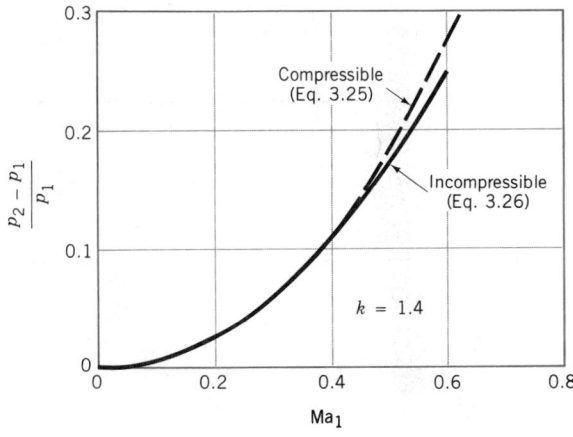

FIGURE 3.24 Pressure ratio as a function of Mach number for incompressible and compressible (isentropic) flow.

For $Ma_1 \ll 1$ this compressible flow result agrees with Eq. 3.26. The incompressible and compressible equations agree to within about 2% up to a Mach number of approximately $Ma_1 = 0.3$. For larger Mach numbers the disagreement between the two results increases.

Thus, a "rule of thumb" is that the flow of a perfect gas may be considered as incompressible provided the Mach number is less than about 0.3. In standard air $(T_1 = 59\ °F, c_1 = \sqrt{kRT_1} = 1117\ ft/s)$ this corresponds to a speed of $V_1 = c_1 Ma_1 = 0.3\ (1117\ ft/s) = 335\ ft/s = 228\ mi/hr$. At higher speeds, compressibility may become important.

EXAMPLE 3.15

Consider flow past the airplane of Example 3.6, which flies 100 mi/hr at an altitude of 10,000 ft. Determine the stagnation pressure if the flow is compressible rather than incompressible as was considered previously.

SOLUTION

For the conditions given, the upstream temperature (see Tables 1.7 and C.1) is $T_1 = 23.36\ °F$, $k = 1.4$, and $R = 1716\ ft\cdot lb/slug\cdot°R$ so that the upstream Mach number is

$$Ma_1 = \frac{V_1}{c_1} = \frac{V_1}{\sqrt{kRT_1}}$$

$$= \frac{100\ mi/hr\ (88\ ft/s)/(60\ mi/hr)}{\sqrt{1.4(1716\ ft\cdot lb/slug\cdot°R)(23.36 + 460)°R}}$$

$$= 0.1361$$

Thus, for steady, inviscid isentropic flow with negligible elevation changes, Eq. 3.25 gives

$$\frac{p_2 - p_1}{p_1} = \left\{ \left[1 + \frac{1.4 - 1}{2} (0.1361)^2 \right]^{1.4/(1.4-1)} - 1 \right\}$$

$$= 0.01303$$

or

$$p_2 - p_1 = 0.01303 p_1 = 0.01303(10.1\ psia) = 0.1316\ psi \qquad \text{(Ans)}$$

As expected (since $Ma_1 < 0.3$) the result is essentially the same as that obtained under the assumption of constant density flow given in Example 3.6 ($p_2 - p_1 = 0.1313\ psi$). Note that we must use absolute pressure, not gage pressure, when we use Eqs. 3.25 or 3.26. This is a result of having used the perfect gas law (for which the pressure is the absolute pressure) in the derivation of these equations.

EXAMPLE 3.16

A Boeing 747 flies at Mach 0.82 at an altitude of 10 km in a standard atmosphere. Determine the stagnation pressure on the leading edge of its wing if the flow is incompressible; if the flow is compressible isentropic.

SOLUTION

From Tables 1.8 and C.2 we find that $p_1 = 26.5$ kPa (abs), $T_1 = -49.9$ °C, $\rho = 0.414$ kg/m^3, and $k = 1.4$. Thus, if we assume incompressible flow Eq. 3.26 gives

$$\frac{p_2 - p_1}{p_1} = \frac{k\mathrm{Ma}_1^2}{2} = 1.4 \frac{(0.82)^2}{2} = 0.471$$

or

$$p_2 - p_1 = 0.471 \ (26.5 \ \text{kPa}) = 12.5 \ \text{kPa} \qquad \text{(Ans)}$$

On the other hand, if we assume isentropic flow, Eq. 3.25 gives

$$\frac{p_2 - p_1}{p_1} = \left\{ \left[1 + \frac{(1.4 - 1)}{2} (0.82)^2 \right]^{1.4/(1.4-1)} - 1 \right\} = 0.555$$

or

$$p_2 - p_1 = 0.555 \ (26.5 \ \text{kPa}) = 14.7 \ \text{kPa} \qquad \text{(Ans)}$$

We see that at Mach 0.82 compressibility effects are of importance. The pressure (and, to a first approximation, the lift and drag on the airplane) is approximately $14.7/12.5 = 1.18$ times greater according to the compressible flow calculations. This may be very significant. As discussed in Chapter 11, for Mach numbers greater than 1 (supersonic flow) the differences between incompressible and compressible results are often not only quantitative but also qualitative.

3.8.2 Unsteady Effects

Another restriction of the Bernoulli equation (Eq. 3.7) is the assumption that the flow is steady. For such flows the velocity along the streamline is a function of only s, the location along the streamline. That is, along a streamline $V = V(s)$. For unsteady flows the velocity is also a function of time, so that along a streamline $V = V(s, t)$. Thus, when taking the time derivative of the velocity to obtain the streamwise acceleration, we obtain $a_s = \partial V/\partial t + V \, \partial V/\partial s$ rather than just $a_s = V \, \partial V/\partial s$ as is true for steady flow. For steady flows the acceleration is due to the change in velocity resulting from a change in position of the particle (the $V \, \partial V/\partial s$ term), whereas for unsteady flow there is an additional contribution to the acceleration resulting from a change in velocity with time at a fixed location (the $\partial V/\partial t$ term). These effects are discussed in detail in Chapter 4. The net effect is that the inclusion of the unsteady term, $\partial V/\partial t$, does not allow the equation of motion to be easily integrated (as was done to obtain the Bernoulli equation) unless additional assumptions are made.

The Bernoulli equation was obtained by integrating the component of Newton's second law (Eq. 3.5) along the streamline. When integrated, the acceleration contribution to this equation, the $\frac{1}{2}\rho d(V^2)$ term, gave rise to the kinetic energy term in the Bernoulli equation. If the steps leading to Eq. 3.5 are repeated with the inclusion of the unsteady effect ($\partial V/\partial t \neq 0$) the following is obtained

$$\rho \frac{\partial V}{\partial t} ds + dp + \frac{1}{2} \rho d(V^2) + \gamma \, dz = 0 \qquad \text{(along a streamline)}$$

For incompressible flow this can be easily integrated between points (1) and (2) to give

$$p_1 + \frac{1}{2} \rho_1 V_1^2 + \gamma z_1 = \rho \int_{s_1}^{s_2} \frac{\partial V}{\partial t} ds + p_2 + \frac{1}{2} \rho V_2^2 + \gamma z_2 \qquad \text{(along a streamline)} \qquad (3.27)$$

FIGURE 3.25   Oscillation of a liquid column in a U-tube.

Equation 3.27 is an unsteady form of the Bernoulli equation valid for unsteady, incompressible, inviscid flow. Except for the integral involving the local acceleration, $\partial V/\partial t$, it is identical to the steady Bernoulli equation. In general, it is not easy to evaluate this integral because the variation of $\partial V/\partial t$ along the streamline is not known. In some situations the concepts of "irrotational flow" and the "velocity potential" can be used to simplify this integral. These topics are discussed in Chapter 6.

A simple flow in which unsteady effects are dominant is that of the oscillation of a column of liquid in a U-tube as is shown in Fig. 3.25. When released from a nonequilibrium position, the liquid column will oscillate at a specific frequency that can be calculated by use of Eq. 3.27 as follows. Let points (1) and (2) be at the air–water interfaces of the two columns of the tube and $z = 0$ correspond to the equilibrium position of these interfaces. Hence, $p_1 = p_2 = 0$ and if $z_1 = z$, then $z_2 = -z$. In general, z is a function of time, $z = z(t)$. For a constant diameter tube, at any instant in time the fluid speed is constant throughout the tube, $V_1 = V_2 = V$, and the integral representing the unsteady effect in Eq. 3.27 can be written as

$$\int_{s_1}^{s_2} \frac{\partial V}{\partial t}\, ds = \frac{dV}{dt} \int_{s_1}^{s_2} ds = \ell \frac{dV}{dt}$$

where ℓ is the total length of the liquid column as shown in the figure. Thus, Eq. 3.27 can be written as

$$\gamma(-z) = \rho\ell \frac{dV}{dt} + \gamma z$$

Since $V = dz/dt$ and $\gamma = \rho g$, this can be written as the second order differential equation describing simple harmonic motion

$$\frac{d^2 z}{dt^2} + \frac{2g}{\ell} z = 0$$

which has the solution $z(t) = C_1 \sin(\sqrt{2g/\ell}\, t) + C_2 \cos(\sqrt{2g/\ell}\, t)$. The values of the constants C_1 and C_2 depend on the initial state (velocity and position) of the liquid at $t = 0$. Thus, the liquid oscillates in the tube with a frequency $\omega = \sqrt{2g/\ell}$ that depends on the length of the columns and the acceleration of gravity (in a manner very similar to the oscillation of a pendulum). The period of this oscillation (the time required to complete on oscillation) is $t_0 = 2\pi \sqrt{\ell/2g}$.

In a few unsteady flow cases the flow can be made steady by an appropriate selection of the coordinate system. Example 3.17 illustrates this.

EXAMPLE 3.17

A submarine moves through the seawater ($SG = 1.03$) at a depth of 50 m with velocity $V_0 = 5.0$ m/s as shown in Fig. E3.17. Determine the pressure at the stagnation point (2).

FIGURE E3.17

SOLUTION

In a coordinate system fixed to the ground, the flow is unsteady. For example, the water velocity at (1) is zero with the submarine in its initial position, but at the instant when the nose, (2), reaches point (1) the velocity there becomes $\mathbf{V}_1 = -V_0\,\hat{\mathbf{i}}$. Thus, $\partial\mathbf{V}_1/\partial t \neq 0$ and the flow is unsteady. Application of the steady Bernoulli equation between (1) and (2) would give the incorrect result that "$p_1 = p_2 + \rho V_0^2/2$." According to this result the static pressure is greater than the stagnation pressure—an incorrect use of the Bernoulli equation.

We can either use an unsteady analysis for the flow (which is outside the scope of this text) or redefine the coordinate system so that it is fixed on the submarine, giving steady flow with respect to this system. The correct method would be

$$p_2 = \frac{\rho V_1^2}{2} + \gamma h = [(1.03)(1000)\ \text{kg/m}^3]\,(5.0\ \text{m/s})^2/2$$

$$+\ (9.80 \times 10^3\ \text{N/m}^3)(1.03)(50\ \text{m})$$

$$= (12{,}900 + 505{,}000)\ \text{N/m}^2 = 518\ \text{kPa} \qquad \text{(Ans)}$$

similar to that discussed in Example 3.2.

If the submarine were accelerating, $\partial V_0/\partial t \neq 0$, the flow would be unsteady in either of the above coordinate systems and we would be forced to use an unsteady form of the Bernoulli equation.

Some unsteady flows may be treated as "quasisteady" and solved approximately by using the steady Bernoulli equation. In these cases the unsteadiness is "not too great" (in some sense), and the steady flow results can be applied at each instant in time as though the flow were steady. The draining of a tank filled with liquid provides an example of this type of flow.

EXAMPLE 3.18

A stream of liquid of diameter d drains from a circular tank of diameter D as is shown in Fig. E3.18. The depth of the water was h_0 at time $t = 0$. Determine the water depth as a function of time, $h = h(t)$.

FIGURE E3.18

SOLUTION

Clearly this is an unsteady flow—the deeper the water, the faster it flows from the tank. However, if the hole in the tank is not too big, the water will drain slowly, and the unsteady effect, $\partial V/\partial t$, at any point in the flow will be smaller than the steady effect, $V\, \partial V/\partial s$. Under these conditions it is reasonable to consider the flow as "quasisteady" and to apply the steady Bernoulli equation as follows.

As was shown in Section 3.6, the velocity of the water leaving the tank can be written as

$$V_2 = \sqrt{\frac{2gh}{1 - (d/D)^4}}$$

Hence, by equating the flowrate from the tank, $V_2 A_2$, and the rate at which the amount of water in the tank changes with time, $-(dh/dt)\, A_1$, we obtain

$$\frac{dh}{dt} = -\frac{V_2 A_2}{A_1} = -\left(\frac{d}{D}\right)^2 \sqrt{\frac{2gh}{1 - (d/D)^4}} \tag{1}$$

This result can be integrated from the initial time and depth, $t = 0$ when $h = h_0$, to an arbitrary time and depth as follows.

$$\int_{h_0}^{h} \frac{1}{\sqrt{h}}\, dh = -\left(\frac{d}{D}\right)^2 \sqrt{\frac{2g}{1 - (d/D)^4}} \int_0^t dt$$

or

$$2(\sqrt{h} - \sqrt{h_0}) = -\left(\frac{d}{D}\right)^2 \sqrt{\frac{2g}{1 - (d/D)^4}}\, t$$

This can be arranged into the form

$$\frac{h}{h_0} = \left[1 - \frac{t\sqrt{g/2h_0}}{\sqrt{(D/d)^4 - 1}}\right]^2 \tag{2 \quad (Ans)}$$

The results of Eq. 2 correlate quite well with experiments, provided d/D is not too large, even though we have used a steady flow analysis for an unsteady flow. This is another way of saying $\partial V/\partial t \ll V\, \partial V/\partial s$. For larger values of d/D the unsteady Bernoulli equation gives a nonlinear, second order differential equation that, unlike Eq. 1, is not easy to integrate.

3.8.3 Rotational Effects

Another of the restrictions of the Bernoulli equation is that it is applicable along the streamline. Application of the Bernoulli equation across streamlines (i.e., from a point on one streamline to a point on another streamline) can lead to considerable errors, depending on the particular flow conditions involved. In general, the Bernoulli constant varies from streamline to streamline. However, under certain restrictions this constant is the same throughout the entire flow field. Example 3.19 illustrates this fact.

EXAMPLE 3.19

Consider the uniform flow in the channel shown in Fig. E3.19a. Discuss the use of the Bernoulli equation between points (1) and (2), points (3) and (4), and points (4) and (5). The liquid in the vertical piezometer tube is stationary.

FIGURE E3.19

SOLUTION

If the flow is steady, inviscid, and incompressible, Eq. 3.7 written between points (1) and (2) gives

$$p_1 + \tfrac{1}{2}\rho V_1^2 + \gamma z_1 = p_2 + \tfrac{1}{2}\rho V_2^2 + \gamma z_2 = \text{constant} = C_{12}$$

Since $V_1 = V_2 = V_0$ and $z_1 = z_2 = 0$, it follows that $p_1 = p_2 = p_0$ and the Bernoulli constant for this streamline, C_{12}, is given by

$$C_{12} = \tfrac{1}{2}\rho V_0^2 + p_0$$

Along the streamline from (3) to (4) we note that $V_3 = V_4 = V_0$ and $z_3 = z_4 = h$. As was shown in Example 3.5, application of $\mathbf{F} = m\mathbf{a}$ across the streamline (Eq. 3.12) gives $p_3 = p_1 - \gamma h$ because the streamlines are straight and horizontal. The above facts combined with the Bernoulli equation applied between (3) and (4) show that $p_3 = p_4$ and that the Bernoulli constant along this streamline is the same as that along the streamline between (1) and (2). That is, $C_{34} = C_{12}$, or

$$p_3 + \tfrac{1}{2}\rho V_3^2 + \gamma z_3 = p_4 + \tfrac{1}{2}\rho V_4^2 + \gamma z_4 = C_{34} = C_{12}$$

Similar reasoning shows that the Bernoulli constant is the same for any streamline in Fig. E3.19. Hence,

$$p + \tfrac{1}{2}\rho V^2 + \gamma z = \text{constant throughout the flow in the channel}$$

Again from Example 3.5 we recall that

$$p_4 = p_5 + \gamma H = \gamma H$$

If we apply the Bernoulli equation across streamlines from (4) to (5) we obtain the incorrect result "$H = p_4/\gamma + V_4^2/2g$." The correct result is $H = p_4/\gamma$.

From the above we see that we can apply the Bernoulli equation across streamlines (1)–(2) and (3)–(4) (that is, $C_{12} = C_{34}$) but not across streamlines from (4) to (5). The reason for this is that while the flow in the channel is "irrotational," it is "rotational" between the flowing fluid in the channel and the stationary fluid in the piezometer tube. Because of the uniform velocity profile across the channel, it is seen that the fluid particles do not rotate or "spin" as they move. The flow is "irrotational." However, as seen in Fig. E3.19b, there is a very thin shear layer between (4) and (5) in which adjacent fluid particles interact and rotate or "spin". This produces a "rotational" flow. A more complete analysis would show that the Bernoulli equation cannot be applied across streamlines if the flow is "rotational" (see Chapter 6).

As is indicated by Example 3.19, if the flow is "irrotational" (that is, the fluid particles do not "spin" as they move), it is appropriate to use the Bernoulli equation across streamlines. However, if the flow is "rotational" (fluid particles "spin"), use of the Bernoulli equation is restricted to flow along a streamline. Thus, it is not appropriate to apply the Bernoulli equation between points (1) and (2) for the flow in Fig. 3.26a, but such use is permitted between points (3) and (4) in Fig. 3.26b. The distinction between irrotational and rotational flow is often a very subtle and confusing one. These topics are discussed in more detail in Chapter 6.

3.8.4   Other Restrictions

Another restriction on the Bernoulli equation is that the flow is inviscid. As is discussed in Section 3.4, the Bernoulli equation is actually a first integral of Newton's second law along a streamline. This general integration was possible because, in the absence of viscous effects, the fluid system considered was a conservative system. The total energy of the system remains constant. If viscous effects are important the system is nonconservative and energy losses occur. A more detailed analysis is needed for these cases. Such material is presented in Chapter 8.

The final basic restriction on use of the Bernoulli equation is that there are no mechanical devices (pumps or turbines) in the system between the two points along the streamline for which the equation is applied. These devices represent sources or sinks of energy. Since the Bernoulli equation is actually one form of the energy equation, it must be altered to include pumps or turbines, if these are present. The inclusion of pumps and turbines is covered in Chapter 5.

FIGURE 3.26   Rotational and irrotational flows.

Rotational flow

(a)

Irrotational flow

(b)

In this chapter we have spent considerable time investigating fluid dynamic situations governed by a relatively simple analysis for steady, inviscid, incompressible flows. Many flows can be adequately analyzed by use of these ideas. However, because of the rather severe restrictions imposed, many others cannot. An understanding of these basic ideas will provide a firm foundation for the remainder of the topics in this book.

REFERENCES

1. Meriam, J. L., and Kraige, L. G., *Engineering Mechanics, Vol. 2: Dynamics*, Wiley, New York, 1986.

2. Tipler, P. A., *Physics*, Worth, New York, 1982.

PROBLEMS

Note: Unless otherwise indicated use the values of fluid properties found in the tables on the inside of the front cover. Problems designated with an (*) are intended to be solved with the aid of a programmable calculator or a computer.

3.1 Water flows steadily through the variable area horizontal pipe shown in Fig. P3.1. The velocity is given by $\mathbf{V} = 10(1 + x)\mathbf{\hat{i}}$ ft/s, where x is in feet. Viscous effects are neglected. **(a)** Determine the pressure gradient, $\partial p/\partial x$, (as a function of x) needed to produce this flow. **(b)** If the pressure at section (1) is 50 psi, determine the pressure at (2) by: (i) integration of the pressure gradient obtained in **(a)**; (ii) application of the Bernoulli equation.

FIGURE P3.1

3.2 Repeat Problem 3.1 if the pipe is vertical with the flow downward.

3.3 An incompressible fluid flows steadily past a circular cylinder as shown in Fig. P3.3. The fluid velocity along the dividing streamline ($-\infty \leq x \leq -a$) is found to be $V = V_0 (1 - a^2/x^2)$, where a is the radius of the cylinder and V_0 is the upstream velocity. **(a)** Determine the pressure gradient along this streamline. **(b)** If the upstream

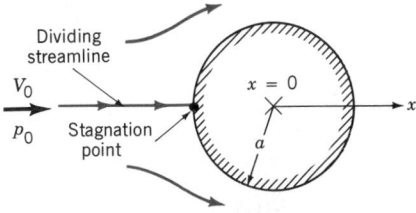

FIGURE P3.3

pressure is p_0, integrate the pressure gradient to obtain the pressure $p(x)$ for $-\infty \leq x \leq -a$. **(c)** Show from the result of part **(b)** that the pressure at the stagnation point ($x = -a$) is $p_0 + \rho V_0^2/2$, as expected from the Bernoulli equation.

3.4 What pressure gradient along the streamline, dp/ds, is required to accelerate water in a horizontal pipe at a rate of 10 m/s²?

3.5 What pressure gradient along the streamline, dp/ds, is required to accelerate air at standard temperature and pressure in a horizontal pipe at a rate of 300 ft/s²?

3.6 What pressure gradient along the streamline, dp/ds, is required to accelerate water upward in a vertical pipe at a rate of 20 ft/s²? What is the answer if the flow is downward?

3.7 Consider a compressible fluid for which the pressure and density are related by $p/\rho^n = C_0$, where n and C_0 are constants. Integrate the equation of motion along the streamline, Eq. 3.6, to obtain the "Bernoulli equation" for this compressible flow as $[n/(n - 1)]p/\rho + V^2/2 + gz = $ constant.

3.8* A wind of velocity V_0 blows past a 5.0-ft-diameter smokestack. Based on the velocity distribution given in Problem 3.3, plot the pressure distribution from a distance 30 ft ahead of the smokestack to the stagnation point of the smokestack for wind speeds of $V_0 = 0, 10, 20, 30, 40$, and 50 mph.

3.9   It can be shown that if viscous and gravitational effects are neglected, the fluid velocity along the surface of a circular cylinder of radius a is $V = 2V_0 \sin \theta$, where V_0 is the upstream velocity and $s = a\theta$ is the distance measured along the streamline that coincides with the cylinder (see Fig. P3.9). For a fluid of density ρ, determine the pressure gradient in the radial direction, $\partial p/\partial r$, on the surface of the cylinder. Assume the axis of the cylinder is vertical. Is $\partial p/\partial r$ positive or negative? Explain physically. For what θ is $\partial p/\partial r$ the maximum? Explain why.

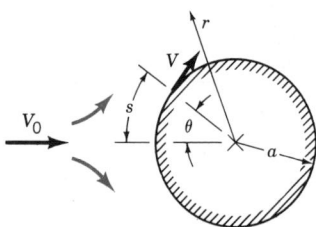

FIGURE P3.9

3.10   Water flows around the vertical two-dimensional bend with circular streamlines and constant velocity as shown in Fig. P3.10. If the pressure is 40 kPa at point (1), determine the pressures at points (2) and (3). Assume that the velocity profile is uniform as indicated.

FIGURE P3.10

3.11   An incompressible, inviscid fluid flows steadily with circular streamlines around a horizontal bend as shown in Fig. P3.11. The radial variation of the velocity profile is given by $rV = $

r_0V_0, where V_0 is the velocity at the inside of the bend which has radius $r = r_0$. Determine the pressure variation across the bend in terms of V_0, r_0, ρ, r, and p_0, where p_0 is the pressure at $r = r_0$. Plot the pressure distribution, $p = p(r)$, if $r_0 = 1.2$ m, $r_1 = 1.8$ m, $V_0 = 12$ m/s, $p_0 = 20$ kN/m^2, and the fluid is water. Neglect gravity.

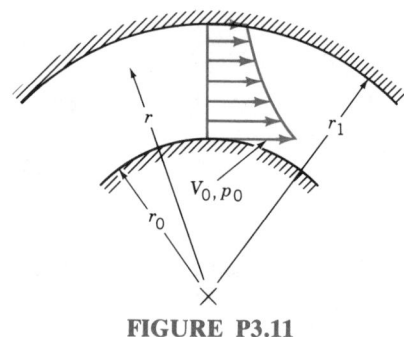

FIGURE P3.11

3.12   Repeat Problem 3.10 if the velocity profile is parabolic with zero velocity at points (1) and (3) and a maximum velocity of $V = 10$ m/s at point (2).

3.13*   Water flows around a vertical two-dimensional bend with circular streamlines as is shown in Fig. P3.13. The pressure at point (1) is measured to be $p_1 = 25$ psi and the velocity across section a–a is as indicated in the table. Calculate and plot the pressure across section a–a of the channel [$p = p(z)$ for $0 \le z \le 2$ ft].

z (ft)	V (ft/s)
0	0
0.2	8.0
0.4	14.3
0.6	20.0
0.8	19.5
1.0	15.6
1.2	8.3
1.4	6.2
1.6	3.7
1.8	2.0
2.0	0

FIGURE P3.13

FIGURE P3.15

3.14   Some animals have learned to take advantage of the Bernoulli effect without having read a fluid mechanics book. For example, a typical prairie dog burrow contains two entrances—a flat front door, and a mounded back door as shown in Fig. P3.14. When the wind blows with velocity V_0 across the front door, the average velocity across the back door is greater than V_0 because of the mound. Assume the air velocity across the back door is $1.07V_0$. For a wind velocity of 6 m/s, what pressure differences, $p_1 - p_2$, is generated to provide a fresh air flow within the burrow?

3.16   Several holes are punched into a tin can as shown in Fig. P3.16. Which of the figures represents the variation of the water velocity as it leaves the holes? Justify your choice.

FIGURE P3.16

3.17   Observations show that it is not possible to blow the table tennis ball from the funnel shown in Fig. P3.17a. In fact, the ball can be kept in an inverted funnel, Fig. P3.17b, by blowing through it. The harder one blows through the funnel, the harder the ball is held within the funnel. Explain this phenomenon.

FIGURE P3.14

3.15   Water flows from the faucet on the first floor of the building shown in Fig. P3.15 with a maximum velocity of 15 ft/s. For steady inviscid flow, determine the maximum water velocity from the basement faucet and from the faucet on the second floor (assume each floor is 12 ft tall).

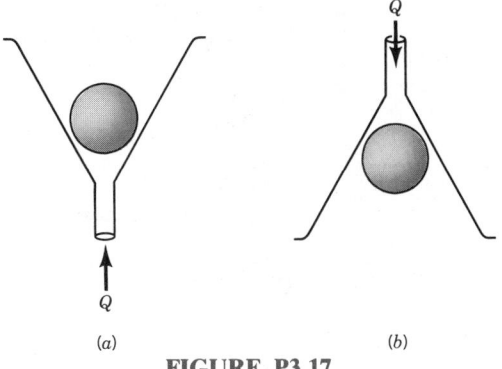

FIGURE P3.17

3.18 Water flows from a garden hose nozzle with a velocity of 15 m/s. What is the maximum height above the nozzle that it can reach?

3.19 The pressure in domestic water pipes is typically 70 psi above atmospheric. If viscous effects are neglected determine the height reached by a jet of water through a small hole in the top of the pipe.

3.20 A fire hose nozzle has a diameter of $1\frac{1}{8}$ in. According to some fire codes, the nozzle must be capable of delivering at least 300 gal/min. If the nozzle is attached to a 3-in.-diameter hose, what pressure must be maintained just upstream of the nozzle to deliver this flowrate?

3.21 Air flows through the two-dimensional conduit of 2 in. width shown in Fig. P3.21 at a rate of 1.32 ft³/s. One wall of the channel is flat, but the other wall has considerable curvature to it. Pressure measurements obtained from a set of static pressure taps located along each wall are shown in the table. Assume the flow is one-dimensional (that is, neglect curvature effects) and determine the theoretical pressure distribution along the channel. Neglect gravity. Explain why the measured pressure distributions shown in the figure are different than the theoretical one.

x (in.)	b (in.)
0.75	2.00
2.50	2.00
4.00	1.28
4.62	1.05
5.38	1.05
8.25	1.29
10.75	1.54
13.22	1.77
15.80	2.00
21.86	2.00

FIGURE P3.21

3.22* An inviscid liquid drains from a large tank through a square duct of width b as shown in Fig. P3.22. The velocity of the fluid at the outlet is not precisely uniform because of the difference in elevation across the outlet. If $b \ll h$, this difference in velocity is negligible. For given b and h, determine v as a function of x and integrate the results to determine the average velocity, $V = Q/b^2$. Plot the velocity distribution, $v = v(x)$, across the outlet if $h = 1$ and $b = 0.1, 0.2, 0.4, 0.6, 0.8,$ and 1.0 m. How small must b be if the centerline velocity, v at $x = b/2$, is to be within 3% of the average velocity?

FIGURE P3.22

3.23* Water flows from a pipe of diameter 20 mm with a flowrate Q as shown in Fig. P3.23. Plot the diameter of the water stream, d, as a function of distance below the faucet, h, for values of $0 \le h \le 1$ m and $0 \le Q \le 0.004$ m³/s. Discuss the validity of the one-dimensional assumption used to calculate $d = d(h)$, noting, in particular, the conditions of small h and small Q.

FIGURE P3.23

3.24 A person holds her hand out of an open car window while the car drives through still air at 55 mph. Under standard atmospheric conditions, what is the maximum pressure on her hand? What would be the maximum pressure if the "car" were an Indy 500 racer traveling 220 mph?

3.25 A person thrusts his hand into the water while traveling 3 m/s in a motor boat. What is the maximum pressure on his hand?

3.26 When an airplane is flying 200 mph at 5000-ft altitude in a standard atmosphere, the air velocity at a certain point on the wing is 273 mph relative to the airplane. What suction pressure is developed on the wing at that point? What is the pressure at the leading edge (a stagnation point) of the wing?

3.27 Air at standard temperature and pressure is drawn through the carburetor of a car's engine as is shown in Fig. P3.27. If the velocity of the air through the throat of the carburetor is approximately 17 m/s and viscous effects are negligible, determine the pressure in the carburetor throat.

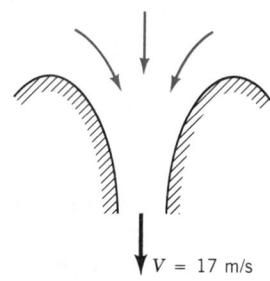

$V = 17$ m/s

FIGURE P3.27

3.28 A 4-in.-diameter pipe carries 300 gal/min of water at a pressure of 50 psi. Determine: **(a)** the pressure head in feet of water, **(b)** the velocity head, and **(c)** the total head with reference to a datum plane 20 ft below the pipe.

3.29 A hang glider soars through standard sea level air with an airspeed of 10 m/s. What is the gage pressure at a stagnation point on the structure?

3.30 Air is drawn into a small open-circuit wind tunnel as shown in Fig. P3.30. Atmospheric pressure is 98.7 kPa (abs) and the temperature is 27 °C. If viscous effects are negligible, determine the pressure at the stagnation point on the nose of the airplane. Also determine the manometer reading, h, for the manometer attached to the static pressure tap within the test section of the wind tunnel if the air velocity within the test section is 60 m/s.

FIGURE P3.30

3.31 A loon is a diving bird equally at home "flying" in the air or water. What swimming velocity under water will produce a dynamic pressure equal to that when it flies in the air at 40 mph?

3.32 Niagara Falls is approximately 167 ft high. If the water flows over the crest of the falls with

a velocity of 8 ft/s and viscous effects are neglected, with what velocity does the water strike the rocks at the bottom of the falls? What is the maximum pressure of the water on the rocks? Repeat the calculations for the 1430-ft-high Upper Yosemite Falls in Yosemite National Park. Is it reasonable to neglect viscous effects for these falls? Explain.

3.33 Water flows through the pipe contraction shown in Fig. P3.33. For the given 0.2-m difference in manometer level, determine the flowrate as a function of the diameter of the small pipe, D.

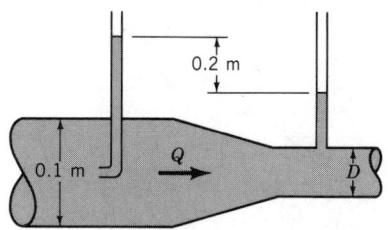

0.2 m

0.1 m | Q | D

FIGURE P3.33

3.34 Water flows through the pipe contraction shown in Fig. P3.34. For the given 0.2-m difference in the manometer level, determine the flowrate as a function of the diameter of the small pipe, D.

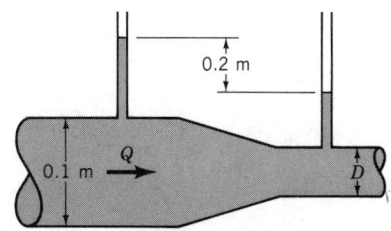

0.2 m

0.1 m Q | D

FIGURE P3.34

3.35 Water flows through the pipe contraction shown in Fig. P3.35. For the given 0.2-m difference in manometer level, determine the flowrate as a function of the diameter of the small pipe, D.

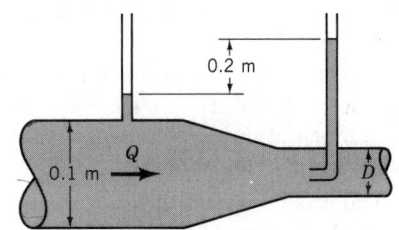

0.2 m

0.1 m Q | D

FIGURE P3.35

3.36 A Pitot-static tube is used to measure the velocity of helium in a pipe. The temperature and pressure are 40 °F and 25 psia. A water manometer connected to the Pitot-static tube indicates a reading of 2.3 in. Determine the helium velocity. Is it reasonable to consider the flow as incompressible? Explain.

3.37 Water flows from the large open tank shown in Fig. P3.37. If viscous effects are neglected, determine the heights, h_1, h_2, and h_3, to which the three streams rise.

Diameter $D/2$ Diameter D

FIGURE P3.37

3.38 Streams of water from two tanks impinge upon each other as shown in Fig. P3.38. If viscous effects are negligible and point A is a stagnation point, determine the height h.

FIGURE P3.38

3.39 Carbon tetrachloride flows in a pipe of variable diameter with negligible viscous effects. At point A in the pipe the pressure and velocity are 20 psi and 30 ft/s, respectively. At location B the pressure and velocity are 23 psi and 14 ft/s. Which point is at the higher elevation and by how much?

3.40 A 0.15-m-diameter pipe discharges into a 0.10-m-diameter pipe. Determine the velocity head in each pipe if they are carrying 0.12 m³/s of kerosene.

3.41 Water flows upward through a variable area pipe with a constant flowrate, Q, as shown in Fig. P3.41. If viscous effects are negligible,

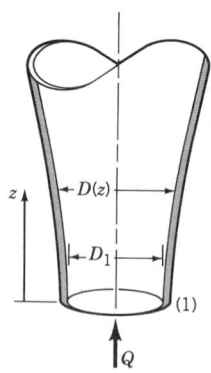

FIGURE P3.41

determine the diameter, $D(z)$, in terms of D_1 if the pressure is to remain constant throughout the pipe. That is, $p(z) = p_1$.

3.42 Repeat Problem 3.41 if the flow were down rather than up.

3.43 The circular stream of water from a faucet is observed to taper from a diameter of 20 mm to 10 mm in a distance of 40 cm. Determine the flowrate.

3.44 Water is siphoned from the tank shown in Fig. P3.44. The water barometer indicates a reading of 30.2 ft. Determine the maximum value of h allowed without cavitation occurring. Note that the pressure of the vapor in the closed end of the barometer equals the vapor pressure.

FIGURE P3.44

3.45 Water is siphoned from a tank as shown in Fig. P3.45. Determine the flowrate and the pressure at point A, a stagnation point.

FIGURE P3.45

3.46 Water flows from a nozzle of triangular cross section as shown in Fig. P3.46. After it has fallen a distance of 2.7 ft its cross section is circular (because of surface tension effects) with a diameter $D = 0.11$ ft. Determine the flowrate, Q.

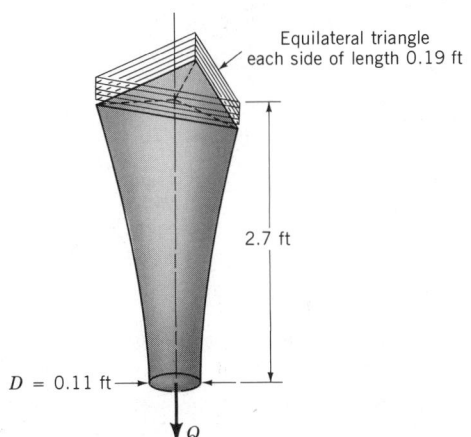

Equilateral triangle each side of length 0.19 ft

2.7 ft

$D = 0.11$ ft

Q

FIGURE P3.46

3.47 An inviscid fluid flows steadily through the contraction shown in Fig. P3.47. Derive an expression for the fluid velocity at (2) in terms of D_1, D_2, ρ, ρ_m, and h if the flow is assumed incompressible.

ρ

D_1

Q

D_2

h

Density ρ_m

FIGURE P3.47

3.48 Repeat Problem 3.47 if the fluid is air that is assumed to be isothermal (i.e., the density changes, but the temperature remains constant). Assume the pressure p_2 is given. Compare the isothermal and incompressible results and show

that in the limit of small pressure difference between (1) and (2) (hence, small h), the two results coincide.

3.49 Water from a faucet fills a 16-oz glass (volume = 28.9 in.³) in 20 s. If the diameter of the jet leaving the faucet is 0.60 in., what is the diameter of the jet when it strikes the water surface in the glass which is positioned 14 in. below the faucet?

3.50 A garden hose with an inside diameter of 15 mm is used to drain a wading pool as is shown in Fig. P3.50. If viscous effects are neglected, what is the flowrate from the pool?

0.2 m

0.23 m

FIGURE P3.50

3.51 Gasoline flows from a 0.3-m-diameter pipe in which the pressure is 300 kPa into a 0.15-m-diameter pipe in which the pressure is 120 kPa. If the pipes are horizontal and viscous effects are negligible determine the flowrate.

3.52 Carbon dioxide flows at a rate of 1.5 ft³/s from a 3-in. pipe in which the pressure and temperature are 20 psi (gage) and 30 °F into a 1.5-in. pipe. If viscous effects are neglected and incompressible conditions are assumed, determine the pressure in the smaller pipe.

3.53 Oil of specific gravity 0.83 flows in the pipe shown in Fig. P3.53. If viscous effects are neglected, what is the flowrate?

4 in.

Water

3 in.

$SG = 0.83$

Q

4 in.

FIGURE P3.53

3.54 Water flows without viscous effects from the nozzle shown in Fig. P3.54. Determine the flowrate and the height, h, to which the water can flow.

FIGURE P3.54

3.55 Determine the flowrate through the pipe in Fig. P3.55.

FIGURE P3.55

3.56 A plastic tube of 50-mm-diameter is used to siphon water from the large tank shown in Fig. P3.56. If the pressure on the outside of the tube is more than 30 kPa greater than the pressure within the tube, the tube will collapse and the siphon will stop. If viscous effects are negligible, determine the minimum value of h allowed without the siphon stopping.

FIGURE P3.56

3.57 For the pipe enlargement shown in Fig. P3.57, the pressures at sections (1) and (2) are 46.3 and 49.2 psi, respectively. Determine the weight flow rate (lb/s) of the gasoline in the pipe.

FIGURE P3.57

3.58 Water is pumped from a lake through an 8-in. pipe at a rate of 10 ft^3/s. If viscous effects are negligible what is the pressure in the suction pipe (the pipe between the lake and the pump) at an elevation 3.5 ft above the lake?

3.59 Air flows through the device shown in Fig. P3.59. If the flowrate is large enough, the pressure within the constriction will be low enough to draw the water up into the tube. Determine the flowrate, Q, and the pressure needed at section (1) to draw the water into section (2). Neglect compressibility and viscous effects.

FIGURE P3.59

3.60 Helium flows through a 0.30-m-diameter horizontal pipe with a temperature of 20 °C and a pressure of 200 kPa (abs) at a rate of 0.30 kg/s. If the pipe reduces to 0.25-m-diameter determine the pressure difference between these two sections. Assume incompressible, inviscid flow.

3.61* Use the results of Problems 3.47 and 3.48 to plot a graph of flowrate, Q, as a function of manometer reading, h, for both the incompressible and isothermal cases. Assume $D_1 = 5$ in., $D_2 = 3$ in., $p_1 = 20$ psia, $T_1 = 90$ °F and $\rho_m = 1.94$ slugs/ft^3. Consider flowrates from $Q = 0.1$ to 2.0 ft^3/s and assume three different fluids—air, helium, and hydrogen.

3.62* The device shown in Fig. P3.62 is used to spray an appropriate mixture of water and insecticide. The flowrate from tank A is to be $Q_A = 0.02$ gal/min when the water flowrate through the hose is $Q = 1$ gal/min. Determine the pressure needed at point (1) and the diameter, D, of the device. For the diameter determined above, plot the ratio of insecticide flowrate to water flowrate as a function of water flowrate, Q, for $0.1 \leq Q \leq 1$ gal/min. Can this device be used to provide a reasonably constant ratio of insecticide to water regardless of the water flow-rate? Explain.

FIGURE P3.62

3.63*   Water flows from a pressurized tank through six equally spaced outlets on the vertical spray tower shown in Fig. P3.63. The diameter of the lowest outlet is 13 mm. Determine the diameters of the other five outlets if the flowrate is to be the same from each of the outlets when the pressure in the tank is $p_1 = 200$ kPa. With outlets of these diameters, the flowrates from each outlet will not be equal to each other if the pressure in the tank is not equal to 200 kPa. On one graph plot each of the six flowrates as a function of p_1 for 100 kPa $< p_1 <$ 300 kPa. What range in p_1 is acceptable if it is required that the flowrate from each outlet must be within 10% of the flowrate from any of the other outlets?

FIGURE P3.63

3.64*   Water flows from a large tank and through a pipe of variable area as shown in Fig. P3.64. The area of the pipe is given by $A = A_0 [1 - x(1 - x/\ell)/2\ell)]$, where A_0 is the area at the beginning ($x = 0$) and end ($x = \ell$) of the pipe. Plot graphs of the pressure within the pipe as a function of distance along the pipe for water depths of $h = 1, 4, 10$, and 25 m.

FIGURE P3.64

3.65*   Air flows through a horizontal pipe of variable diameter, $D = D(x)$, at a rate of 1.5 ft^3/s. The static pressure distribution obtained from a set of 12 static pressure taps along the pipe wall is as shown below. Plot the pipe shape, $D(x)$, if the diameter at $x = 0$ is 1, 2, or 3 in. Neglect viscous and compressibility effects.

x (in.)	p (in. H$_2$O)	x (in.)	p (in. H$_2$O)
0	1.00		
1	0.72	7	0.44
2	0.16	8	0.51
3	−0.96	9	0.65
4	−0.31	10	0.78
5	0.27	11	0.90
6	0.39	12	1.00

3.66   The vent on the tank shown in Fig. P3.66 is closed and the tank pressurized to increase the flowrate. What pressure, p_1, is needed to produce twice the flowrate of that when the vent is open?

FIGURE P3.66

3.67   Water flows steadily through the large tanks shown in Fig. P3.67. Determine the water depth, h_A.

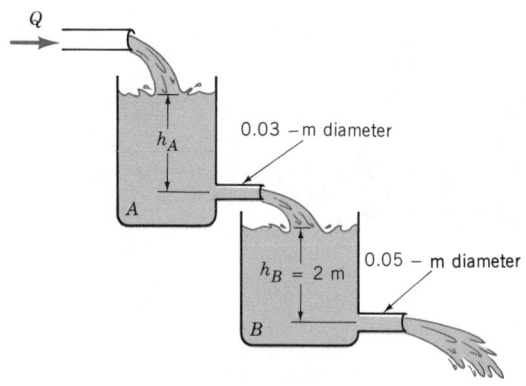

FIGURE P3.67

3.68 Air at 60 °F and 14.2 psia flows into the tank shown in Fig. P3.68. Determine the flowrate in ft³/s, lb/s, and slugs/s. Assume incompressible flow.

FIGURE P3.68

3.69 Water flows into a large tank at a rate of 0.011 m³/s as shown in Fig. P3.69. The water leaves the tank through 20 holes in the bottom of the tank, each of which produces a stream of 10-mm diameter. Determine the equilibrium height, h, for steady state operation.

FIGURE P3.69

3.70 Water flows from a large tank as shown in Fig. P3.70. Atmospheric pressure is 14.5 psia and the vapor pressure is 0.62 psia. If viscous effects are neglected, at what height, h, will cavitation begin? To avoid cavitation should the value of D_1 be increased or decreased? To avoid cavi-

tation should the value of D_2 be increased or decreased? Explain.

FIGURE P3.70

3.71 If viscous effects are neglected and the tank is large, determine the flowrate from the tank shown in Fig. P3.71.

FIGURE P3.71

3.72 Water flows from the tank shown in Fig. P3.72. If viscous effects are negligible determine the value of h in terms of H and the specific gravity, SG, of the manometer fluid.

FIGURE P3.72

3.73 What pressure, p_1, is needed to produce a flowrate of 0.07 ft³/s from the tank shown in Fig. P3.73?

FIGURE P3.73

3.74 Water collects in the bottom of a rectangular oil tank as shown in Fig. P3.74. How long will it take for the water to drain from the tank through a 0.02-m-diameter drain hole in the bottom of the tank? Assume quasisteady flow.

FIGURE P3.74

3.75 A 0.3-m-diameter soccer ball, pressurized to 20 kPa, develops a small leak with an area equivalent to 0.008 mm². If viscous effects are neglected and the air is assumed to be incompressible, determine the flowrate through the hole. Would the ball become noticeably softer during a 1-hr soccer game? Explain. Is it reasonable to assume incompressible flow for this situation? Explain.

3.76 Water is siphoned from the tank shown in Fig. P3.76. Determine the flowrate from the tank and the pressures at points (1), (2), and (3) if viscous effects are negligible.

FIGURE P3.76

3.77 Redo Problem 3.76 if a 1-in.-diameter nozzle is placed at the end of the tube.

3.78 Determine the manometer reading, h, for the flow shown in Fig. P3.78.

FIGURE P3.78

3.79 The specific gravity of the manometer fluid shown in Fig. P3.79 is 1.15. Determine the volume flowrate, Q, if the flow is inviscid and incompressible and the flowing fluid is **(a)** water, **(b)** gasoline, or **(c)** air at standard conditions.

FIGURE P3.79

3.80 JP-4 fuel ($SG = 0.77$) flows through the Venturi meter shown in Fig. P3.80 with a velocity of 15 ft/s in the 6-in. pipe. If viscous effects are negligible determine the elevation, h, of the fuel in the open tube connected to the throat of the Venturi meter.

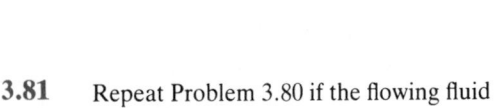

$V = 15$ ft/s

FIGURE P3.80

3.81   Repeat Problem 3.80 if the flowing fluid is water rather than JP-4 fuel.

3.82   Air at standard conditions flows through the cylindrical drying stack shown in Fig. P3.82. If viscous effects are negligible and the inclined water-filled manometer reading is 30 mm as indicated, determine the flowrate.

FIGURE P3.82

3.83   Oil flows through the system shown in Fig. P3.83 with negligible losses. Determine the flowrate.

FIGURE P3.83

3.84   Determine the flowrate through the submerged orifice shown in Fig. P3.84 if the contraction coefficient is $C_c = 0.68$.

FIGURE P3.84

3.85   Determine the flowrate through the Venturi meter shown in Fig. P3.85 if ideal conditions exist.

FIGURE P3.85

3.86   For what flowrate through the Venturi meter of Prob. 3.85 will cavitation begin if $p_1 = 275$ kPa gage, atmospheric pressure is 101 kPa (abs), and the vapor pressure is 5.2 kPa (abs)?

3.87   What diameter orifice hole, d, is needed if under ideal conditions the flowrate through the orifice meter of Fig. P3.87 is to be 30 gal/min of seawater with $p_1 - p_2 = 1.37$ lb/in.²? The contraction coefficient is assumed to be 0.63.

FIGURE P3.87

3.88 An ancient device for measuring time is shown in Fig. P3.88. The axisymmetric vessel is shaped so that the water level falls at a constant rate. Determine the shape of the vessel, $R = R(z)$, if the water level is to decrease at a rate of 0.10 m/hr and the drain hole is 5.0 mm in diameter. The device is to operate for 12 hr without needing refilling. Make a scale drawing of the shape of the vessel.

FIGURE P3.88

3.89 A long water trough of triangular cross section is formed from two planks as is shown in Fig. P3.89. A gap of 0.1 in. remains at the junction of the two planks. If the water depth initially was 2 ft, how long a time does it take for the water depth to reduce to 1 ft.?

FIGURE P3.89

3.90* Salt water flows from tank A through a hole in diameter d as indicated in Fig. P3.90. Initially (time $t = 0$) tank A contains concentrated salt water with a specific gravity of 1.15 and tank B contains pure water. Each tank has a diameter of 2 m and an initial liquid depth of 3 m. For $d = 25, 50, 75,$ and 100 mm, plot the specific gravity of the salt water in the lower tank as a function of time—from $t = 0$ to when tank A is drained. Assume the liquid in tank B is mixed continuously so that the specific gravity is uniform throughout the tank. Also assume that the flow is quasisteady.

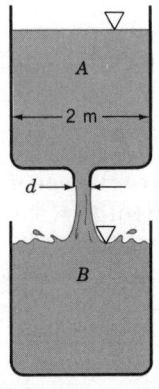

FIGURE P3.90

3.91* A spherical tank of diameter D has a drain hole of diameter d at its bottom. A vent at the top of the tank maintains atmospheric pressure within the tank. The flow is quasisteady and inviscid and the tank is full of water initially. Determine the water depth as a function of time, $h = h(t)$, and plot graphs of $h(t)$ for tank diameters of 1, 5, 10, and 20 ft if $d = 1$ in.

3.92* An inexpensive timer is to be made from a funnel as indicated in Fig. P3.92. The funnel is filled to the top with water and the plug is removed at time $t = 0$ to allow the water to run out. Marks are to be placed on the wall of the funnel indicating the time in 15-s intervals, from 0 to 3 min (at which time the funnel becomes empty). If the funnel outlet has a diameter of $d = 0.1$ in., draw to scale the funnel with the timing marks for funnels with angles of $\theta = 30, 45,$ and 60°. Repeat the problem if the diameter is changed to 0.05 in.

FIGURE P3.92

FIGURE P3.94

3.93* The surface area, A, of the pond shown in Fig. P3.93 varies with the water depth, h, as shown in the table. At time $t = 0$ a valve is opened and the pond is allowed to drain through a pipe of diameter D. If viscous effects are negligible and quasisteady conditions are assumed, plot the water depth as a function of time from when the valve is opened ($t = 0$) until the pond is drained for pipe diameters of $D = 0.5, 1.0, 1.5, 2.0, 2.5,$ and 3.0 ft. Assume $h = 18$ ft at $t = 0$.

h (ft)	A [acres (1 acre = 43,560 ft^2)]
0	0
2	0.3
4	0.5
6	0.8
8	0.9
10	1.1
12	1.5
14	1.8
16	2.4
18	2.8

3.95 An agricultural spraying apparatus has a boom that serves as a manifold for the eight equally spaced, equal diameter spray nozzles as shown in Fig. P3.95. To produce the desired flow conditions, it is necessary to maintain the same pressure at each nozzle inlet. Determine the diameter of the boom, $D = D(x)$, if the pressure remains constant at 30 lb/in.^{2} and viscous effects are negligible.

FIGURE P3.95

3.96 Water flows from a large tank through a large pipe that splits into two smaller pipes as shown in Fig. P3.96. If viscous effects are negligible, determine the flowrate from the tank and the pressure at point (1).

FIGURE P3.93

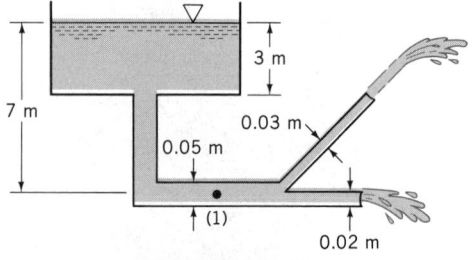

FIGURE P3.96

3.94 Water flows through a horizontal branching pipe as shown in Fig. P3.94. Determine the pressure at section (3).

3.97 Water flows through the horizontal Y-fitting shown in Fig. P3.97. If the flowrate and pressure in pipe (1) are $Q_1 = 2.3$ ft^3/s and $p_1 = 63$ lb/in.^{2}, determine the pressures, p_2 and p_3, in pipes (2) and (3) under the assumption that the flowrate divides evenly between pipes (2) and (3).

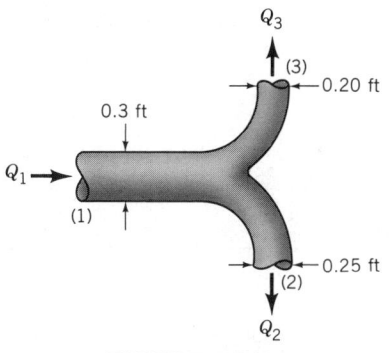

FIGURE P3.97

3.98   Water flows from the pipe shown in Fig. P3.98 as a free jet and strikes a circular flat plate. The flow geometry shown is axisymmetrical. Determine the flowrate and the manometer reading, H.

FIGURE P3.98

3.99   Water flows past the hydrofoil shown in Fig. P3.99 with an upstream velocity of V_0. A more advanced analysis indicates that the maximum velocity of the water in the entire flow field occurs at point B and is equal to $1.1V_0$. Calculate the velocity, V_0, at which cavitation will begin if the atmospheric pressure is 101 kPa (abs) and the vapor pressure of the water is 3.2 kPa (abs).

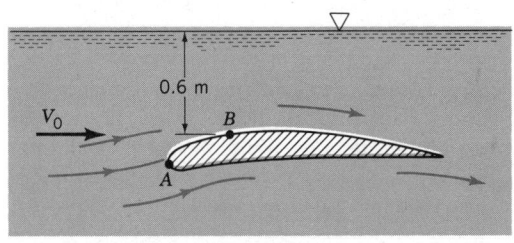

FIGURE P3.99

3.100   Air, assumed incompressible and inviscid, flows into the outdoor cooking grill through nine holes of 0.40-in. diameter as shown in Fig. P3.100. If a flowrate of 30 in.^{3}/s into the grill is required to maintain the correct cooking conditions, determine the pressure within the grill near the holes.

FIGURE P3.100

3.101   A conical plug is used to regulate the air flow from the pipe shown in Fig. P3.101. The air leaves the edge of the cone with a uniform thickness of 0.02 m. If viscous effects are negligible and the flowrate is 0.50 m^3/s, determine the pressure within the pipe.

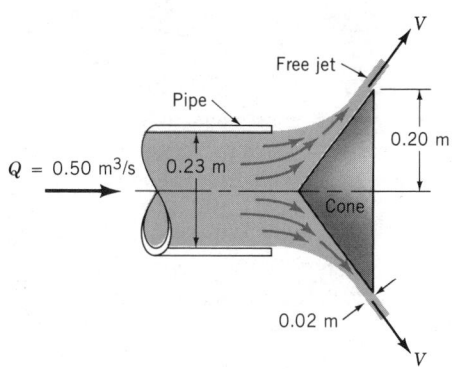

FIGURE P3.101

3.102   An air cushion vehicle is supported by forcing air into the chamber created by a skirt around the periphery of the vehicle as shown in Fig. P3.102. The air escapes through the 3-in. clearance between the lower end of the skirt and the ground (or water). Assume the vehicle weighs 10,000 lb and is essentially rectangular in shape, 30 by 65 ft. The volume of the chamber is large enough so that the kinetic energy of the air within the chamber is negligible. Determine the flowrate, Q, needed to support the vehicle. If the ground clearance were reduced to 2 in., what flowrate would be needed? If the vehicle weight were reduced to 5000 lb and the ground clearance maintained at 3 in., what flowrate would be needed?

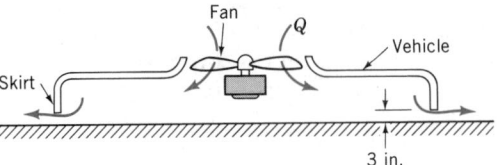

FIGURE P3.102

3.103 Air flows from a hole of diameter 0.03 m in a flat plate as shown in Fig. P3.103. A circular disk of diameter D is placed a distance h from the lower plate. The pressure in the tank is maintained at 1 kPa. Determine the flowrate as a function of h if viscous effects and elevation changes are assumed negligible and the flow exits radially from the circumference of the circular disk with uniform velocity.

FIGURE P3.103

3.104 An automatic boat bailer is shown in Fig. P3.104. The "bump" on the bottom of the hull is designed to increase the velocity at point (1) to approximately $1.2V_0$, where V_0 is the velocity of the boat through the water. If the pressures at (1) and (2) are essentially equal, what minimum velocity is needed to initiate the bailing action?

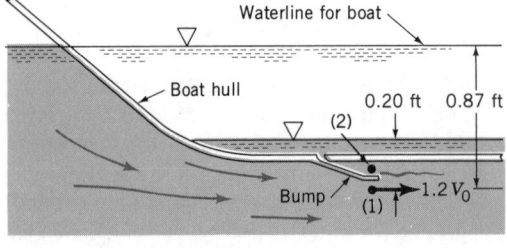

FIGURE P3.104

3.105 Water flows over the weir plate which has a semicircular opening as shown in Fig. P3.105. Determine the functional dependence of the flowrate on the head, $Q = Q(H)$, for $0 \le H \le R$. Illustrate this dependence with a

graph of Q/Q_{max} as a function of H/R for $0 \le H/R \le 1$, where Q_{max} is the flowrate when $H = R$.

FIGURE P3.105

3.106 A weir of trapezoidal cross section is used to measure the flowrate in a channel as shown in Fig. P3.106. If the flowrate is Q_0 when $H = \ell/2$, what flowrate is expected when $H = \ell$?

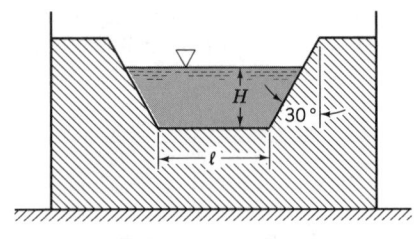

FIGURE P3.106

3.107 Water flows over the spillway shown in Fig. P3.107. If the velocity is uniform at sections (1) and (2) and viscous effects are negligible, determine the flowrate per unit width of the spillway.

FIGURE P3.107

3.108 Water flows down the sloping ramp shown in Fig. P3.108 with negligible viscous effects. The flow is uniform at sections (1) and (2). For the conditions given show that three solutions for the downstream depth, h_2, are obtained by use of the Bernoulli and continuity equations. However, show that only two of these solutions are realistic. Determine these values.

FIGURE P3.108

3.109   The flowrate in a water channel is sometimes determined by use of a device called a Venturi flume. As shown in Fig. P3.109, this device consists simply of a hump on the bottom of the channel. If the water surface dips a distance of 0.07 m for the conditions shown, what is the flowrate per width of the channel? Assume the velocity is uniform and viscous effects are negligible.

FIGURE P3.109

3.110   Water flows in a rectangular channel that is 0.5 m wide as shown in Fig. P3.110. The upstream depth is 70 mm. The water surface rises 40 mm as it passes over a portion where the channel bottom rises 10 mm. If viscous effects are negligible, what is the flowrate?

FIGURE P3.110

3.111   Water flows down the ramp shown in the channel of Fig. P3.111. The channel width decreases from 15 ft at section (1) to 9 ft at section (2). For the conditions shown, determine the flowrate.

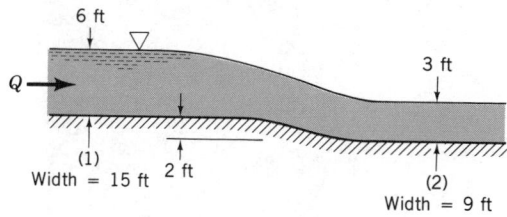

FIGURE P3.111

3.112   Water flows under the sluice gate shown in Fig. P3.112. Determine the flowrate if the gate is 4.6 ft wide.

FIGURE P3.112

3.113   Water flows under the sluice gate shown in Fig. P3.113. Determine the depth downstream, z_2, if the flowrate is 3.5 m³/s per meter width of the gate.

FIGURE P3.113

3.114   Water flows in a vertical pipe of 0.15-m diameter at a rate of 0.2 m³/s and a pressure of 200 kPa at an elevation of 25 m. Determine the velocity head and pressure head at elevations of 20 and 55 m.

3.115   Draw the energy line and the hydraulic grade line for the flow of Problem 3.70.

3.116   Draw the energy line and hydraulic grade line for the flow shown in Problem 3.76.

3.117   Draw the energy line and hydraulic grade line for the flow shown in Problem 3.77.

3.118*   Water flows up the ramp shown in Fig. P3.118 with negligible viscous losses. The upstream depth and velocity are maintained at $h_1 = 0.3$ m and $V_1 = 6$ m/s. Plot a graph of the downstream depth, h_2, as a function of the ramp height, H, for $0 \le H \le 2$ m. Note that for each value of H there are three solutions, not all of which are realistic.

FIGURE P3.118

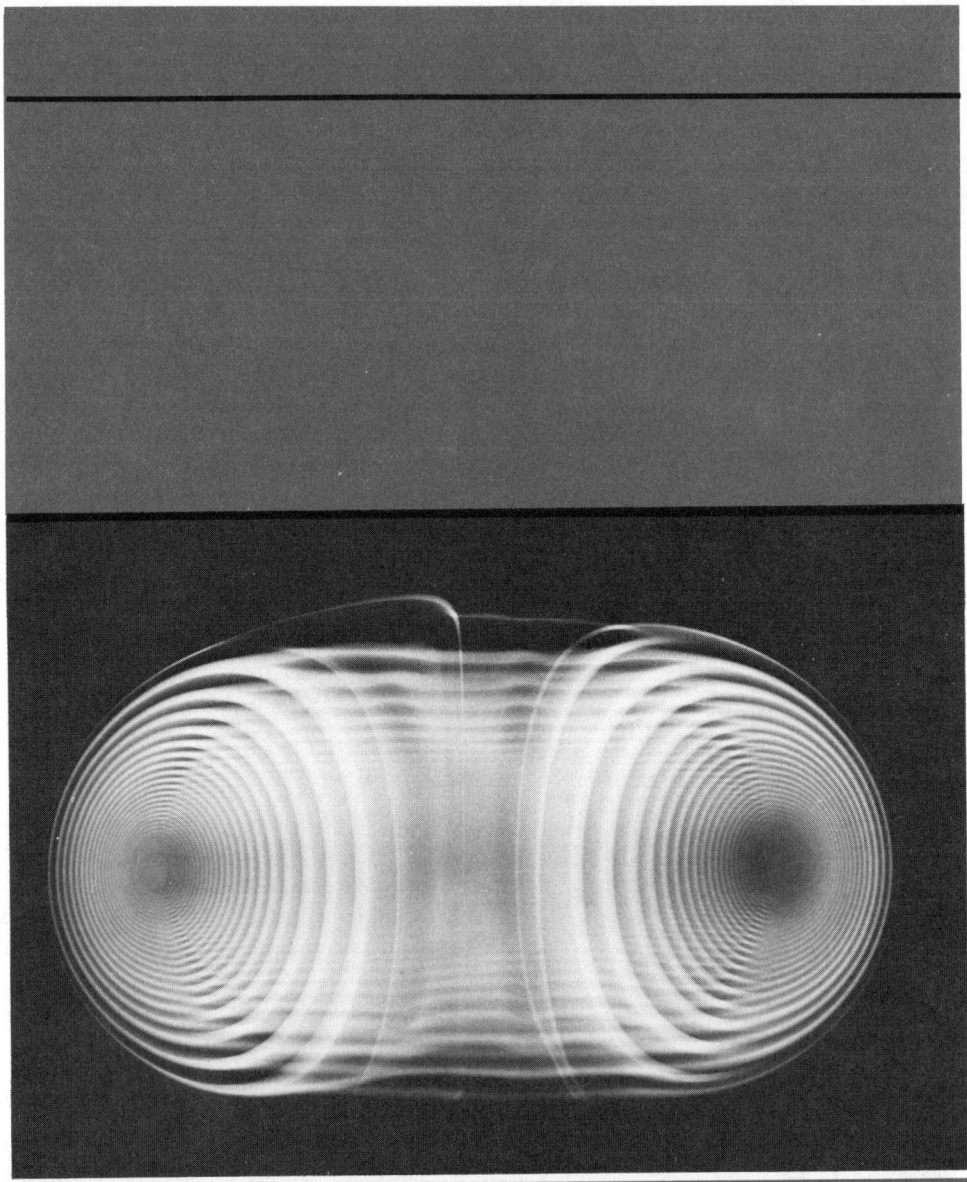

A vortex ring: The complex, three-dimensional structure of a smoke ring is indicated in this cross-sectional view. (Smoke in air.) (Photograph courtesy of R. H. Magarvey and C. S. MacLatchy, Ref. 4.)

FOUR

FLUID KINEMATICS

In the previous three chapters we have defined some basic properties of fluids and have considered various situations involving fluids that are either at rest or are moving in a rather elementary manner. We have seen that there are numerous situations in which a fluid is either stationary or its motion is small enough (in some sense) to be negligible. In general, however, fluids have a well-known tendency to move or flow. It is very difficult to "tie down" a fluid and restrain it from moving. The slightest of shear stresses will cause the fluid to move. Similarly, an appropriate imbalance of normal stresses (pressure) will cause fluid motion.

In this chapter we will discuss various aspects of fluid motion without being concerned with the actual forces necessary to produce the motion. That is, we will consider the *kinematics* of the motion—the velocity and acceleration of the fluid, and the description and visualization of its motion. The analysis of the specific forces necessary to produce the motion (the *dynamics* of the motion) will be discussed in detail in the following chapters. A wide variety of useful information can be gained from a thorough understanding of fluid kinematics. Such an understanding of how to describe and observe fluid motion is an essential step to the complete understanding of fluid dynamics.

We have all observed fascinating fluid motions like those associated with the smoke emerging from a chimney or the flow of the atmosphere as indicated by the motion of clouds. The motion of waves on a lake or the mixing of paint in a bucket provide other common, although quite different, examples of flow visualization. Considerable insight into these fluid motions can be gained by considering the kinematics of such flows without being concerned with the specific force that drives them.

4.1 THE VELOCITY FIELD

In general, fluids flow. That is, there is a net motion of the molecules of the fluid from one point in space to another point as a function of time. As is discussed in Chapter 1, a typical portion of fluid contains so many molecules that it becomes totally unrealistic (except in special cases) for us to attempt to account for the motion of individual molecules. Rather, we employ the continuum hypothesis and consider fluids to be made up of fluid particles that interact with each other and with their surroundings. Each particle contains numerous molecules. Thus, we can describe the flow of a fluid in terms of the motion of fluid particles rather than individual molecules. This motion can be described in terms of the velocity and acceleration of the fluid particles.

The infinitesimal particles of a fluid are tightly packed together (as is implied by the continuum assumption). Thus, at a given instant in time, a description of any fluid property (such as density, pressure, velocity, and acceleration) may be given as a function of the fluid's location. This representation of fluid parameters as functions of the spatial coordinates is termed a *field representation* of the flow. Of course, the specific field representation may be different at different times, so that to describe a fluid flow we must determine the various parameters not only as a function of the spatial coordinates (x, y, z, for example) but also as a function of time, t. Thus, to completely specify the temperature, T, in a room we must specify the temperature field, $T = T(x, y, z, t)$, throughout the room (from floor to ceiling and wall to wall) at any time of the day or night.

One of the most important fluid variables is the *velocity field,*

$$\mathbf{V} = u(x, y, z, t)\hat{\mathbf{i}} + v(x, y, z, t)\hat{\mathbf{j}} + w(x, y, z, t)\hat{\mathbf{k}}$$

where u, v, and w are the x, y, and z components of the velocity vector. By definition, the velocity of a particle is the time rate of change of the position vector for that particle. As is illustrated in Fig. 4.1, the position of particle A relative to the coordinate system is given by its *position vector*, $\mathbf{r}_A$, which (if the particle is moving) is a function of time. The time derivative of this position gives the *velocity* of the particle, $\mathbf{V}_A$. By writing the velocity for all of the particles we can obtain the field description of the velocity vector $\mathbf{V} = \mathbf{V}(x, y, z, t)$.

Since the velocity is a vector, it has both a direction and a magnitude. The magnitude of $\mathbf{V}$, denoted $V = |\mathbf{V}| = (u^2 + v^2 + w^2)^{1/2}$, is the speed of the fluid. (It is very

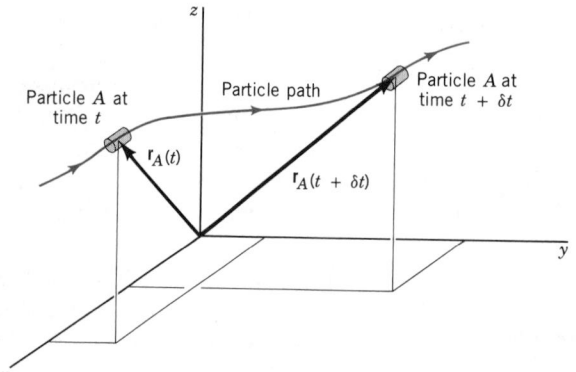

FIGURE 4.1 Particle location in terms of its position vector.

common in practical situations to call V velocity rather than speed, i.e., "the velocity of the fluid is 12 m/s.") As is discussed in the next section, a change in velocity results in an acceleration. This acceleration may be due to a change in speed and/or direction.

4.1.1   Eulerian and Lagrangian Flow Descriptions
There are two general approaches in analyzing fluid mechanics problems (or problems in other branches of the physical sciences, for that matter). The first method, called the *Eulerian method,* uses the field concept introduced above. In this case, the fluid motion is given by completely prescribing the necessary properties (pressure, density, velocity, etc.) as functions of space and time. From this method we obtain information about the flow in terms of what happens at fixed points in space as the fluid flows past those points.

The second method, called the *Lagrangian method,* involves following individual fluid particles as they move about and determining how the fluid properties associated with these particles change as a function of time. That is, the fluid particles are "tagged" or identified, and their properties determined as they move.

The difference between the two methods of analyzing fluid flow problems can be seen in the example of smoke discharging from a chimney, as is shown in Fig. 4.2. In the Eulerian method one may attach a temperature-measuring device to the top of the chimney (point 0) and record the temperature at that point as a function of time. At different times there are different fluid particles passing by the stationary device. Thus, one would obtain the temperature, T, for that location ($x = x_0$, $y = y_0$, and $z = z_0$) as a function of time. That is, $T = T(x_0, y_0, z_0, t)$. The use of numerous temperature measuring devices fixed at various locations would provide the temperature field, $T = T(x, y, z, t)$. The temperature of a particle as a function of time would not be known unless the location of the particle were known as a function of time.

In the Lagrangian method, one would attach the temperature-measuring device to a particular fluid particle (particle A) and record that particle's temperature as it moves about. Thus, one would obtain that particle's temperature as a function of time, $T_A = T_A(t)$. The use of many such measuring devices moving with various fluid particles would provide the temperature of these fluid particles as a function of time. The temperature would not be known as a function of position unless the location of each particle were known as a function of time. If enough information in Eulerian

FIGURE 4.2 Eulerian and Lagrangian description of temperature of a flowing fluid.

form is available, Lagrangian information can be derived from the Eulerian data—and vice versa.

In fluid mechanics it is usually easier to use the Eulerian method to describe a flow—in either experimental or analytical investigations. There are, however, certain instances in which the Lagrangian method is more convenient. For example, some numerical fluid mechanics calculations are based on determining the motion of individual fluid particles (based on the appropriate interactions among the particle), thereby describing the motion in Lagrangian terms. Similarly, in some experiments individual fluid particles are "tagged" and are followed throughout their motion, providing a Lagrangian description. Oceanographic measurements obtained from devices that flow with the ocean currents provide this information. Similarly, by using X-ray opaque dyes it is possible to trace blood flow in arteries and to obtain a Lagrangian description of the fluid motion. A Lagrangian description may also be useful in describing fluid machinery (such as pumps and turbines) in which fluid particles gain or lose energy as they move along their flow paths.

Another illustration of the difference between the Eulerian and Lagrangian descriptions can be seen in the following biological example. Each year thousands of birds migrate between their summer and winter habitats. Ornithologists study these migrations to obtain various types of important information. One set of data obtained is the rate at which birds pass a certain location on their migration route (birds per hour). This corresponds to an Eulerian description—"flowrate" at a given location as a function of time. Individual birds need not be followed to obtain this information. Another type of information is obtained by "tagging" certain birds with radio transmitters and following their motion along the migration route. This corresponds to a Lagrangian description—"position" of a given particle as a function of time.

EXAMPLE 4.1

A velocity field is given by $\mathbf{V} = (V_0/\ell)(x\hat{\mathbf{i}} - y\hat{\mathbf{j}})$, where V_0 and ℓ are constants. At what location in the flow field is the speed equal to V_0? Make a sketch of the velocity field in the first quadrant ($x \geq 0$, $y \geq 0$) by drawing arrows representing the fluid velocity at representative locations.

SOLUTION

The x, y, and z components of the velocity are given by $u = V_0 x/\ell$, $v = -V_0 y/\ell$, and $w = 0$ so that the fluid speed, V, is

$$V = (u^2 + v^2 + w^2)^{1/2} = \frac{V_0}{\ell}(x^2 + y^2)^{1/2} \tag{1}$$

The speed is $V = V_0$ at any location on the circle of radius ℓ centered at the origin as shown in Fig. E4.1a.　(Ans)

The direction of the fluid velocity relative to the x axis is given in terms of $\theta = $ arctan (v/u) as shown in Fig. E4.1b. For this flow,

$$\tan \theta = \frac{v}{u} = \frac{-V_0 y/\ell}{V_0 x/\ell} = \frac{-y}{x}.$$

Thus, along the x axis ($y = 0$) we see that $\tan \theta = 0$, so that $\theta = 0°$ or $\theta = 180°$. Similarly, along the y axis ($x = 0$) we obtain $\tan \theta = \pm\infty$ so that $\theta = 90°$ or $\theta = 270°$. Also, for $y = 0$ we find $\mathbf{V} = (V_0 x/\ell)\hat{\mathbf{i}}$, while for $x = 0$ we have $\mathbf{V} = (-V_0 y/\ell)\hat{\mathbf{j}}$, indicating (if $V_0 > 0$) that

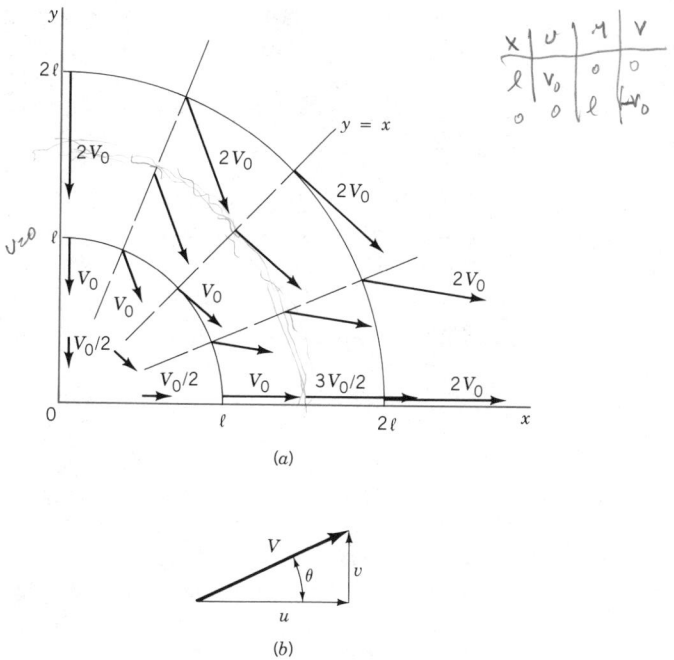

(a)

(b)

FIGURE E4.1

the flow is directed toward the origin along the y axis and away from the origin along the x axis as shown in Fig. E4.1a.

By determining $\mathbf{V}$ and θ for other locatons in the x–y plane, the velocity field can be sketched as shown in the figure. For example, on the line $y = x$ the velocity is at a $-45°$ angle relative to the x axis ($\tan\theta = v/u = -y/x = -1$). At the origin $x = y = 0$ so that $\mathbf{V} = 0$. This point of stationary fluid is called a stagnation point. The farther from the origin the fluid is, the faster it is flowing (as seen from Eq. 1). By careful consideration of the velocity field it is possible to determine considerable information about the flow.

The above provides an Eulerian description of the flow. For a Lagrangian description we would need to determine the velocity as a function of time for each particle as it flows along from one point to another.

4.1.2 One-, Two-, and Three-Dimensional Flows

Generally, a fluid flow is a rather complex three-dimensional, time-dependent phenomenon—$\mathbf{V} = \mathbf{V}(x, y, z, t)$. In many situations, however, it is possible to make simplifying assumptions that allow a much easier understanding of the problem without sacrificing needed accuracy. One of these simplifications involves approximating a real flow as a simpler one- or two-dimensional flow.

In almost any flow situation, the velocity field actually contains all three velocity components (u, v, and w, for example) and each is a function of all three space coordinates (x, y, and z, for example). In many situations the *three-dimensional flow* characteristics are important in terms of the physical effects they produce. For these situations it is necessary to analyze the flow in its complete three-dimensional character. Neglect of one or two of the velocity components in these cases would lead to considerable misrepresentation of the effects produced by the actual flow.

The flow of air past an airplane wing provides an example of a complex three-dimensional flow. A feel for the three-dimensional structure of such flows can be obtained by studying Fig. 4.3, which is a photograph of the flow past a model airfoil; the flow has been made visible by using a flow visualization technique.

In many situations one of the velocity components may be small (in some sense) relative to the two other components. In situations of this kind it may be reasonable to neglect the smaller component and assume *two-dimensional flow*. That is, $\mathbf{V} = u\hat{\mathbf{i}} + v\hat{\mathbf{j}}$.

It is sometimes possible to further simplify a flow analysis by assuming that two of the velocity components are negligible, leaving the velocity field to be approximated as a *one-dimensional flow* field. That is, $\mathbf{V} = u\hat{\mathbf{i}}$. As we will learn from examples throughout the remainder of the book, although there are very few, if any, flows that are truly one-dimensional, there are many flow fields for which the one-dimensional flow assumption provides a reasonable approximation. There are also many flow situations for which use of a one-dimensional flow field assumption will give completely erroneous results.

An example of the one-, two-, or three-dimensional flow approximations is given by the flow of a meandering river as shown in Fig. 4.4. Most rivers tend to produce curved or meandering stream beds as are indicated in the photograph. The actual velocity field in a river is very complex and three-dimensional. However, for certain analysis it is possible to approximate the velocity field as a simple one-dimensional flow given by $\mathbf{V} = V_0\hat{\mathbf{i}}$, where the x direction points "downstream," and V_0 is the "average velocity" of the river. As we will see in Chapter 10, such a one-dimensional approximation is not strictly valid, since the water speed is not constant across the river cross section. Thus, a better approximation would be $\mathbf{V} = u(y, z)\hat{\mathbf{i}}$, where the y coordinate is normal to the riverbank. A typical velocity contour map is shown in Fig. 4.4b.

In addition, if the river is curved, the flow will not be in the x direction only. Because of centrifugal effects a secondary flow will be produced in the plane of the cross section. That is, the flow field is truly three-dimensional, $\mathbf{V} = u\hat{\mathbf{i}} + v\hat{\mathbf{j}} + w\hat{\mathbf{k}}$. This complex flow is illustrated in Fig. 4.4c. In fact, it is this secondary flow (the three-dimensional character represented by v and w) that causes the meandering of rivers. The cross flow (secondary flow) removes material from the bank on the outer edge of the curve and deposits it on the bank on the inner edge of the curve. The meandering grows (becomes more pronounced) because of this three-dimensional

FIGURE 4.3 Flow visualization of the complex three-dimensional flow past a model airfoil. (Photograph by M. R. Head.)

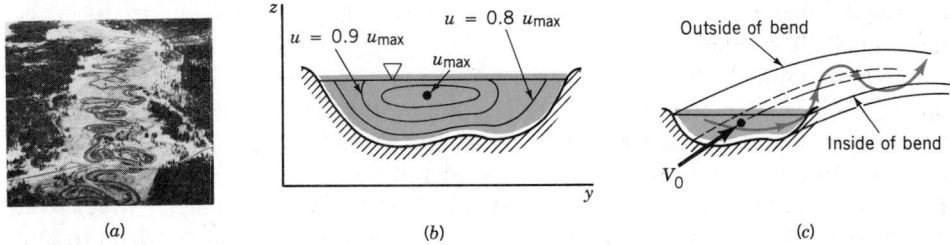

FIGURE 4.4 (*a*) The meandering Deschutes River near Bend, Oregon. (Photograph courtesy of J. S. Shelton.) (*b*) Typical velocity distribution in a river. (*c*) Meandering produced by the secondary flow.

effect of the flow. If one wishes to study the meandering of the river, it is necessary to include three-dimensional effects. If one only wishes to know the flowrate in the river, a simpler one-dimensional analysis may be sufficient.

It is not always easy to decide whether a flow should be considered as one-, two-, or three-dimensional. We must continually keep in mind what is involved in these various assumptions. In fact, sometimes the velocity or a velocity component of a certain motion may be so small that it is not easy to decide if the effects are important enough to warrant its inclusion in an analysis. We must be careful to realize what "small" means. In some situations a "small" effect can have considerable consequences if it acts over a long period of time. The drift velocity of the South American continent relative to the African continent is "very, very small"—on the order of 1 cm/yr. Over a period of millions of years the result has been nonnegligible. The continents that were once connected are now thousands of miles apart!

4.1.3 Steady and Unsteady Flows

In the above discussion we have neglected the effect that time has on the velocity field. That is, we have assumed *steady flow*—the velocity at a given point in space does not vary with time, $\partial \mathbf{V}/\partial t = 0$. In reality, almost all flows are unsteady in some sense. That is, the velocity does vary with time. It is not difficult to believe that *unsteady flows* are usually more difficult to analyze (and to investigate experimentally) than are steady flows. Hence, considerable simplicity often results if one can make the assumption of steady flow without compromising the usefulness of the results. Among the various types of unsteady flows are nonperiodic flow, periodic flow, and truly random flow. Whether or not unsteadiness of one or more of these types must be included in an analysis is not always immediately obvious.

An example of a nonperiodic, unsteady flow is that produced by turning off a faucet to stop the flow of water. Usually this unsteady flow process is quite mundane and the forces developed as a result of the unsteady effects need not be considered. However, if the water is turned off suddenly (as with an electrically operated valve in a dishwasher), the unsteady effects can become important [as in the "water hammer" effects made apparent by the loud banging of the pipes under such conditions (Ref. 1)].

In other flows the unsteady effects may be periodic, occurring time after time in basically the same manner. The periodic injection of the air–gasoline mixture into the cylinder of an automobile engine is such an example. The unsteady effects are quite regular and repeatable in a regular sequence. They are very important in the operation of the engine.

In many situations the unsteady character of a flow is quite random. That is, there is no repeatable sequence or regular variation to the unsteadiness. This behavior occurs in *turbulent flow* and is absent from *laminar flow*. The "smooth" flow of highly viscous syrup onto a pancake represents a "deterministic" laminar flow. It is quite different from the turbulent flow observed in the "irregular" splashing of water from a faucet onto the sink below it. The "irregular" gustiness of the wind represents another random turbulent flow. The differences between these types of flows will be discussed in considerable detail in Chapters 8 and 9.

It must be understood that the definition of steady or unsteady flow pertains to the behavior of a fluid property as observed at a fixed point in space. For steady flow, the values of all fluid properties (velocity, temperature, density, etc.) at any fixed point are independent of time. However, the value of those properties for a given fluid particle may change with time as the particle flows along, even in steady flow. Thus, the temperature of the exhaust at the exit of a car's exhaust pipe may be constant for several hours, but the temperature of a fluid particle that left the exhaust pipe 5 min ago is lower now than it was when it left the pipe, even though the flow is steady.

4.1.4   Streamlines, Streaklines, and Pathlines

Although fluid motion can be quite complicated, there are various concepts that can be used to help in the visualization and analysis of flow fields. To this end we discuss the use of streamlines, streaklines, and pathlines in flow analysis. The streamline is often used in analytical work while the streakline and pathline are often used in experimental work.

A *streamline* is a line that is everywhere tangent to the velocity field. If the flow is steady, nothing at a fixed point (including the velocity direction) changes with time, so the streamlines are fixed lines in space. For unsteady flows the streamlines may change shape with time. Streamlines are obtained analytically by integrating the equations defining lines tangent to the velocity field. For two-dimensional flows the slope of the streamline, dy/dx, must be equal to the tangent of the angle that the velocity vector makes with the x axis, or

$$\frac{dy}{dx} = \frac{v}{u} \tag{4.1}$$

If the velocity field is known as a function of x and y (and t if the flow is unsteady), this equation can be integrated to give the equation of the streamlines.

For unsteady flow there is no easy way to produce streamlines experimentally in the laboratory. As is discussed below, the observation of dye, smoke, or some other tracer injected into a flow can provide useful information, but for unsteady flows it is not necessarily information about the streamlines.

EXAMPLE 4.2

Determine the streamlines for the two-dimensional steady flow discussed in Example 4.1, $\mathbf{V} = (V_0/\ell)(x\hat{\mathbf{i}} - y\hat{\mathbf{j}})$.

SOLUTION

Since $u = (V_0/\ell)x$ and $v = -(V_0/\ell)y$ it follows that streamlines are given by solution of the equation

$$\frac{dy}{dx} = \frac{v}{u} = \frac{-(V_0/\ell)y}{(V_0/\ell)x} = -\frac{y}{x}$$

in which variables can be separated and the equation integrated to give

$$\int \frac{dy}{y} = -\int \frac{dx}{x}$$

or

$$\ln y = -\ln x + \text{constant}$$

Thus, along the streamline

$$xy = C, \qquad \text{where } C \text{ is a constant} \tag{Ans}$$

By using different values of the constant C, we can plot various lines in the x–y plane—the streamlines. The usual notation for a streamline is $\psi = $ constant on a streamline. Thus, the equation for the streamlines of this flow are

$$\psi = xy$$

As is discussed more fully in Chapter 6, the function $\psi = \psi(x, y)$ is called the *stream function*. The streamlines in the first quadrant are plotted in Fig. E4.2. A comparison of this figure with Fig. E4.1a illustrates the fact that streamlines are lines parallel to the velocity field.

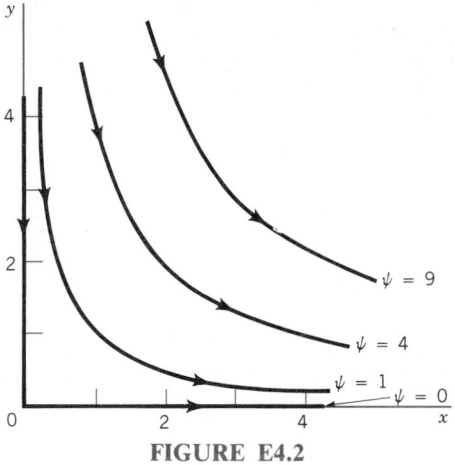

FIGURE E4.2

A *streakline* consists of all particles in a flow that have previously passed through a common point. Streaklines are more of a laboratory tool than an analytical tool. They can be obtained by taking instantaneous photographs of marked particles that all passed through a given location in the flow field at some earlier time. Such a line can be produced by continuously injecting marked fluid (neutrally buoyant smoke in air, or dye in water) at a given location (Ref. 2). If the flow is steady, each successively injected particle follows precisely behind the previous one, forming a steady streakline that is exactly the same as the streamline through the injection point.

For unsteady flows, particles injected at the same point at different times need not follow the same path. An instantaneous photograph of the marked fluid would show the streakline at that instant, but it would not necessarily coincide with the streamline through the point of injection at that particular time nor with the streamline through the same injection point at a different time.

The third method used for visualizing and describing flows involves the use of *pathlines*. A pathline is the line traced out by a given particle as it flows from one point to another. The pathline is a Lagrangian concept that can be produced in the laboratory by marking a fluid particle (dying a small fluid element) and taking a time exposure photograph of its motion.

If the flow is steady, the path taken by a marked particle (a pathline) will be the same as the line formed by all other particles that previously passed through the point of injection (a streakline). For such cases these lines are tangent to the velocity field. Hence, pathlines, streamlines, and streaklines are the same for steady flows. For unsteady flows none of these three types of lines need be the same (Ref. 3). Often one sees pictures of "streamlines" made visible by the injection of smoke or dye into a flow as is shown in Fig. 4.3. Actually, such pictures show streaklines rather than streamlines. However, for steady flows the two are identical; only the nomenclature is incorrectly used.

EXAMPLE 4.3

Water flowing from the oscillating slit shown in Fig. E4.3a produces a velocity field given by $\mathbf{V} = u_0 \sin[\omega(t - y/v_0)]\hat{\mathbf{i}} + v_0\hat{\mathbf{j}}$, where u_0, v_0, and ω are constants. Thus, the y component of velocity remains constant ($v = v_0$) and the x component of velocity at $y = 0$ coincides with the velocity of the oscillating sprinkler head [$u = u_0 \sin(\omega t)$ at $y = 0$].

(a) Determine the streamline that passes through the origin at $t = 0$; at $t = \pi/2\omega$. (b) Determine the pathline of the particle that was at the origin at $t = 0$; at $t = \pi/2\omega$. (c) Discuss the shape of the streakline that passes through the origin.

SOLUTION

(a)   Since $u = u_0 \sin[\omega(t - y/v_0)]$ and $v = v_0$ it follows from Eq. 4.1 that streamlines are given by the solution of

$$\frac{dy}{dx} = \frac{v}{u} = \frac{v_0}{u_0 \sin[\omega(t - y/v_0)]}$$

in which the variable can be separated and the equation integrated to give

$$u_0 \int \sin\left[\omega\left(t - \frac{y}{v_0}\right)\right] dy = v_0 \int dx,$$

or

$$u_0(v_0/\omega) \cos\left[\omega\left(t - \frac{y}{v_0}\right)\right] = v_0 x + C \tag{1}$$

where C is a constant. For the streamline at $t = 0$ that passes through the origin ($x = y = 0$), the value of C is obtained from Eq. 1 as $C = u_0 v_0/\omega$. Hence, the equation for this streamline is

$$x = \frac{u_0}{\omega}\left[\cos\left(\frac{\omega y}{v_0}\right) - 1\right] \tag{2} \quad \text{(Ans)}$$

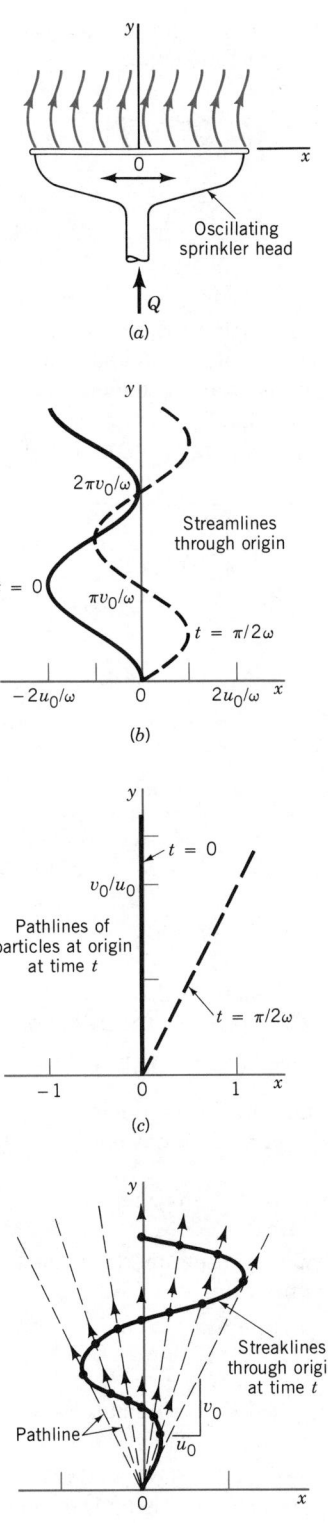

FIGURE E4.3

Similarly, for the streamline at $t = \pi/2\omega$ that passes through the origin, Eq. 1 gives $C = 0$. Thus, the equation for this streamline is

$$x = \frac{u_0}{\omega} \cos \left[\omega \left(\frac{\pi}{2\omega} - \frac{y}{v_0} \right) \right] = \frac{u_0}{\omega} \cos \left(\frac{\pi}{2} - \frac{\omega y}{v_0} \right)$$

or

$$x = \frac{u_0}{\omega} \sin \left(\frac{\omega y}{v_0} \right) \qquad \qquad (3) \quad \text{(Ans)}$$

These two streamlines, plotted in Fig. E4.3b, are not the same because the flow is unsteady. For example, at the origin ($x = y = 0$) the velocity is $\mathbf{V} = v_0 \hat{\mathbf{j}}$ at $t = 0$ and $\mathbf{V} = u_0 \hat{\mathbf{i}} + v_0 \hat{\mathbf{j}}$ at $t = \pi/2\omega$. Thus, the angle of the streamline passing through the origin changes with time. Similarly, the shape of the entire streamline is a function of time.

(b)   The pathline of a particle (the location of the particle as a function of time) can be obtained from the velocity field and the definition of the velocity. Since $u = dx/dt$ and $v = dy/dt$ we obtain

$$\frac{dx}{dt} = u_0 \sin \left[\omega \left(t - \frac{y}{v_0} \right) \right] \quad \text{and} \quad \frac{dy}{dt} = v_0$$

The y equation can be integrated (since $v_0 = $ constant) to give the y coordinate of the pathline as

$$y = v_0 t + C_1 \qquad \qquad (4)$$

where C_1 is a constant. With this known $y = y(t)$ dependence, the x equation for the pathline becomes

$$\frac{dx}{dt} = u_0 \sin \left[\omega \left(t - \frac{v_0 t + C_1}{v_0} \right) \right] = -u_0 \sin \left(\frac{C_1 \omega}{v_0} \right)$$

This can be integrated to give the x component of the pathline as

$$x = - \left[u_0 \sin \left(\frac{C_1 \omega}{v_0} \right) \right] t + C_2 \qquad \qquad (5)$$

where C_2 is a constant. For the particle that was at the origin ($x = y = 0$) at time $t = 0$, Eqs. 4 and 5 give $C_1 = C_2 = 0$. Thus, the pathline is

$$x = 0 \quad \text{and} \quad y = v_0 t \qquad \qquad (6) \quad \text{(Ans)}$$

Similarly, for the particle that was at the origin at $t = \pi/2\omega$, Eqs. 4 and 5 give $C_1 = -\pi v_0/2\omega$ and $C_2 = -\pi u_0/2\omega$. Thus, the pathline for this particle is

$$x = u_0 \left(t - \frac{\pi}{2\omega} \right) \quad \text{and} \quad y = v_0 \left(t - \frac{\pi}{2\omega} \right) \qquad (7)$$

The pathline can be drawn by plotting the locus on $x(t)$, $y(t)$ values for $t \geq 0$ or by eliminating the parameter t from Eq. 7 to give

$$y = \frac{v_0}{u_0} x \qquad \qquad (8) \quad \text{(Ans)}$$

The pathline given by Eqs. 6 and 8, shown in Fig. 4.3c, are straight lines from the origin (rays). The pathlines and streamlines do not coincide because the flow is unsteady.

(c)   The streakline through the origin at time $t = 0$ is the locus of particles at $t = 0$ that previously ($t < 0$) passed through the origin. The general shape of the streaklines can be seen as follows. Each particle that flows through the origin travels in a straight line (pathlines are rays from the origin), the slope of which lies between $\pm v_0/u_0$ as shown in Fig. E4.3d. Particles passing through the origin at different times are located on different rays from the origin and at different distances from the origin. The net result is that a stream of dye continually injected at the origin (a streakline) would have the shape shown in Fig. E4.3*d*. Because of the unsteadiness, the streakline will vary with time, although it will always have the oscillating, sinuous character shown. Similar streaklines are given by the stream of water from a garden hose nozzle that oscillates back and forth in a direction normal to the axis of the nozzle.

In this example neither the streamlines, pathlines, nor streaklines coincide. If the flow were steady all of these lines would be the same.

4.2   THE ACCELERATION FIELD

As indicated in the previous section, we can describe fluid motion by either (1) following individual particles (Lagrangian description), or (2) remaining fixed in space and observing different particles as they pass by (Eulerian description). In either case, to apply Newton's second law ($\mathbf{F} = m\mathbf{a}$) we must be able to describe the particle acceleration in an appropriate fashion. For the infrequently used Lagrangian method, we describe the fluid acceleration just as is done in solid body dynamics—$\mathbf{a} = \mathbf{a}(t)$ for each particle. For the Eulerian description we describe the *acceleration field* as a function of position and time without actually following any particular particle. This is analogous to describing the flow in terms of the velocity field, $\mathbf{V} = \mathbf{V}(x, y, z, t)$, rather than the velocity for particular particles. In this section we will discuss how to obtain the acceleration field if the velocity field is known.

The acceleration of a particle is the time rate of change of its velocity. For unsteady flows the velocity at a given point in space (occupied by different particles) may vary with time, giving rise to a portion of the fluid acceleration. In addition, a fluid particle may experience an acceleration because its velocity changes as it flows from one point to another in space. For example, water flowing through a garden hose nozzle under steady conditions (constant number of gallons per minute from the hose) will experience an acceleration as it changes from its relatively low velocity in the hose to its relatively high velocity at the tip of the nozzle.

4.2.1   The Material Derivative

Consider a fluid particle moving along its pathline as is shown in Fig. 4.5. In general, the particle's velocity, denoted $\mathbf{V}_A$ for particle A, is a function of its location and the

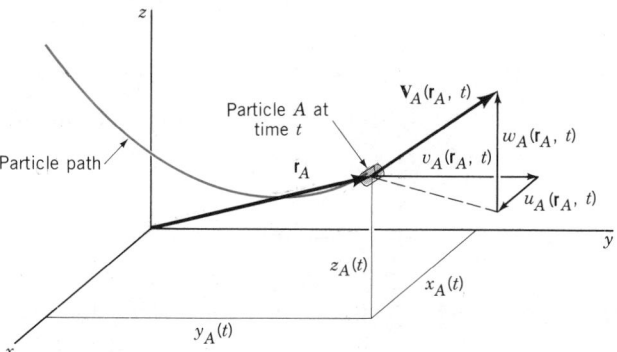

FIGURE 4.5 Velocity and position of particle A at time t.

time. That is,

$$\mathbf{V}_A = \mathbf{V}_A(\mathbf{r}_A, t) = \mathbf{V}_A[x_A(t), y_A(t), z_A(t), t]$$

where $x_A = x_A(t)$, $y_A = y_A(t)$, and $z_A = z_A(t)$ define the location of the moving particle. By definition, the acceleration of a particle is the time rate of change of its velocity. Since the velocity may be a function of both position and time, its value may change because of the change in time as well as a change in the particle's position. Thus, we use the chain rule of differentiation to obtain the acceleration of particle A, $\mathbf{a}_A$, as

$$\mathbf{a}_A(t) = \frac{d\mathbf{V}_A}{dt} = \frac{\partial \mathbf{V}_A}{\partial t} + \frac{\partial \mathbf{V}_A}{\partial x}\frac{dx_A}{dt} + \frac{\partial \mathbf{V}_A}{\partial y}\frac{dy_A}{dt} + \frac{\partial \mathbf{V}_A}{\partial z}\frac{dz_A}{dt} \tag{4.2}$$

Using the fact that the particle velocity components are given by $u_A = dx_A/dt$, $v_A = dy_A/dt$, and $w_A = dz_A/dt$, Eq. 4.2 becomes

$$\mathbf{a}_A = \frac{\partial \mathbf{V}_A}{\partial t} + u_A \frac{\partial \mathbf{V}_A}{\partial x} + v_A \frac{\partial \mathbf{V}_A}{\partial y} + w_A \frac{\partial \mathbf{V}_A}{\partial z}$$

Since the above is valid for any particle, we can drop the reference to particle A and obtain the acceleration field from the velocity field as

$$\mathbf{a} = \frac{\partial \mathbf{V}}{\partial t} + u \frac{\partial \mathbf{V}}{\partial x} + v \frac{\partial \mathbf{V}}{\partial y} + w \frac{\partial \mathbf{V}}{\partial z} \tag{4.3}$$

This is a vector result whose scalar components can be written as

$$a_x = \frac{\partial u}{\partial t} + u \frac{\partial u}{\partial x} + v \frac{\partial u}{\partial y} + w \frac{\partial u}{\partial z}$$

$$a_y = \frac{\partial v}{\partial t} + u \frac{\partial v}{\partial x} + v \frac{\partial v}{\partial y} + w \frac{\partial v}{\partial z} \tag{4.4}$$

and

$$a_z = \frac{\partial w}{\partial t} + u \frac{\partial w}{\partial x} + v \frac{\partial w}{\partial y} + w \frac{\partial w}{\partial z}$$

where a_x, a_y, and a_z are the x, y, and z components of the acceleration.

The above result is often written in shorthand notation as

$$\mathbf{a} = \frac{D\mathbf{V}}{Dt}$$

where the operator

$$\frac{D(\ )}{Dt} \equiv \frac{\partial(\ )}{\partial t} + u \frac{\partial(\ )}{\partial x} + v \frac{\partial(\ )}{\partial y} + w \frac{\partial(\ )}{\partial z} \tag{4.5}$$

is termed the *material derivative* or *substantial derivative*. An often used shorthand notation for the material derivative operator is

$$\frac{D(\ )}{Dt} = \frac{\partial(\ )}{\partial t} + (\mathbf{V} \cdot \nabla)(\ ) \tag{4.6}$$

The dot product of the velocity vector, $\mathbf{V}$, and the gradient operator, $\nabla(\ ) = \partial(\ )/\partial x\,\hat{\mathbf{i}} + \partial(\ )/\partial y\,\hat{\mathbf{j}} + \partial(\ )/\partial z\,\hat{\mathbf{k}}$ (a vector operator) provides a convenient notation for the spatial derivative terms appearing in the Cartesian coordinate representation of the material derivative. Note that the notation $\mathbf{V} \cdot \nabla$ represents the operator $(\mathbf{V} \cdot \nabla(\ ) = u\partial(\ )/\partial x + v\partial(\ )/\partial y + w\partial(\ )/\partial z$.

The material derivative concept is very useful in analysis involving various fluid parameters, not just the acceleration. The material derivative of any variable is the rate at which that variable changes with time for a given particle (as seen by one moving along with the fluid—the Lagrangian description). For example, consider a temperature field $T = T(x, y, z, t)$ associated with a given flow, like that shown in Fig. 4.2. It may be of interest to determine the time rate of change of temperature of a fluid particle (particle A) as it moves through this temperature field. If the velocity, $\mathbf{V} = \mathbf{V}(x, y, z, t)$, is known, we can apply the chain rule to determine the rate of change of temperature as

$$\frac{dT_A}{dt} = \frac{\partial T_A}{\partial t} + \frac{\partial T_A}{\partial x}\frac{dx_A}{dt} + \frac{\partial T_A}{\partial y}\frac{dy_A}{dt} + \frac{\partial T_A}{\partial z}\frac{dz_A}{dt}$$

This can be written as

$$\frac{DT}{Dt} = \frac{\partial T}{\partial t} + u\frac{\partial T}{\partial x} + v\frac{\partial T}{\partial y} + w\frac{\partial T}{\partial z} = \frac{\partial T}{\partial t} + \mathbf{V} \cdot \nabla T$$

As in the determination of the acceleration, the material derivative operator, $D(\ )/Dt$, appears.

EXAMPLE 4.4

An incompressible, inviscid fluid flows steadily past a sphere of radius a as shown in Fig. E4.4a. According to a more advanced analysis of the flow, the fluid velocity along streamline A–B is given by

$$\mathbf{V} = u(x)\hat{\mathbf{i}} = V_0\left(1 + \frac{a^3}{x^3}\right)\hat{\mathbf{i}}$$

(a)

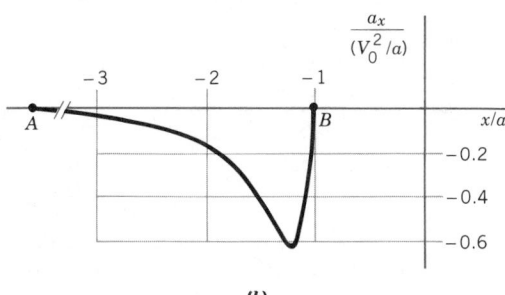

(b)

FIGURE E4.4

where V_0 is the upstream velocity far ahead of the sphere. Determine the acceleration experienced by fluid particles as they flow along this streamline.

SOLUTION

Along streamline $A–B$ there is only one component of velocity ($v = w = 0$) so that from Eq. 4.3

$$\mathbf{a} = \frac{\partial \mathbf{V}}{\partial t} + u \frac{\partial \mathbf{V}}{\partial x} = \left(\frac{\partial u}{\partial t} + u \frac{\partial u}{\partial x} \right) \hat{\mathbf{i}}$$

or

$$a_x = \frac{\partial u}{\partial t} + u \frac{\partial u}{\partial x}, \qquad a_y = 0, \qquad a_z = 0$$

Since the flow is steady the velocity at a given point in space does not change with time. Thus, $\partial u / \partial t = 0$. With the given velocity distribution along the streamline the acceleration becomes

$$a_x = u \frac{\partial u}{\partial x} = V_0 \left(1 + \frac{a^3}{x^3} \right) V_0 [a^3(-3x^{-4})]$$

or

$$a_x = -3(V_0^2/a) \frac{1 + (a/x)^3}{(x/a)^4} \tag{Ans}$$

Along streamline $A–B$ ($-\infty \le x \le -a$ and $y = 0$) the acceleration has only an x component and it is negative (a deceleration). Thus, the fluid slows down from its upstream velocity of $\mathbf{V} = V_0\hat{\mathbf{i}}$ at $x = -\infty$ to its stagnation point velocity of $\mathbf{V} = 0$ at $x = -a$, the "nose" of the sphere. The variation of a_x along streamline $A–B$ is shown in Fig. E4.4b. It is the same result as is obtained in Example 3.1 by using the streamwise component of the acceleration, $a_s = V \, \partial V / \partial s$. The maximum deceleration occurs at $x = -1.205a$ and has a value of $a_x = -0.610V_0^2/a$.

In general, for fluid particles on streamlines other than $A–B$, all three components of the acceleration (a_x, a_y, and a_z) will be nonzero. For example, for this steady flow the y component of acceleration is given by

$$a_y = u \frac{\partial v}{\partial x} + v \frac{\partial v}{\partial y} + w \frac{\partial v}{\partial z} \tag{1}$$

For particles that travel along streamline $A–B$, $v = 0$ and $w = 0$. A more complex analysis of the flow would show that along streamline $A–B$ $\partial v / \partial x = 0$, although $\partial v / \partial y$ and $\partial v / \partial z$ are not zero. Hence, from Eq. 1, $a_y = 0$. Similarly, the z component of acceleration also involves three terms ($u \, \partial w / \partial x$, $v \, \partial w / \partial y$, and $w \, \partial w / \partial z$), each of which along streamline $A–B$ is zero. Thus, $a_z = 0$ also. At an arbitrary point in a general flow field, none of the nine terms involved in the acceleration for steady flow (Eq. 4.4 with $\partial u / \partial t = \partial v / \partial t = \partial w / \partial t = 0$) are zero. A description of the complete velocity field is needed if the entire acceleration field is to be determined.

Fairly large accelerations (or decelerations) often occur in fluid flows. Consider air flowing past a baseball of radius $a = 0.14$ ft with a velocity of $V_0 = 100$ mi/hr $= 147$ ft/s. According to the results of Example 4.4, the maximum deceleration of an air particle approaching the stagnation point along the streamline in front of the ball is

$$|a_x|_{max} = |a_x|_{x = -0.168 ft} = \frac{0.610(147 \text{ ft/s})^2}{0.14 \text{ ft}}$$

$$= 94.2 \times 10^3 \text{ ft/s}^2$$

This is a deceleration of approximately 3000 times that of gravity. In some situations the acceleration or deceleration experienced by fluid particles may be very large. An extreme case involves flow through shock waves that can occur in supersonic flow past objects (see Chapter 11). In such circumstances the fluid particles may experience decelerations hundreds of thousands of times greater than gravity. Large forces are obviously needed to produce such accelerations.

4.2.2 Unsteady Effects

As is seen from Eq. 4.5, the material derivative formula contains two types of terms—those involving the time derivative $[\partial(\ )/\partial t]$ and those involving spatial derivatives $[\partial(\ )/\partial x, \partial(\ )/\partial y,$ and $\partial(\ )/\partial z]$. The time derivative portions are denoted as the *local derivative*. They represent effects of the unsteadiness of the flow. If the parameter involved is the acceleration, that portion given by $\partial \mathbf{V}/\partial t$ is termed the *local acceleration*. For steady flow the time derivative is zero throughout the flow field $[\partial(\ )/\partial t \equiv 0]$, and the local effect vanishes. Physically, there is no change in flow parameters at a fixed point in space if the flow is steady. There may be a change of those parameters for a fluid particle as it moves about, however.

If a flow is unsteady its parameter values (velocity, temperature, density, etc.) at any location may change with time. For example, an unstirred ($\mathbf{V} = 0$) cup of coffee will cool down in time because of heat transfer to its surroundings. That is, $DT/Dt = \partial T/\partial t + \mathbf{V} \cdot \nabla T = \partial T/\partial t < 0$. Similarly, a fluid particle may have nonzero acceleration as a result of the unsteady effect of the flow. Consider flow in a constant diameter pipe as is shown in Fig. 4.6. The flow is assumed to be spatially uniform throughout the pipe. That is, $\mathbf{V} = V_0(t)\,\hat{\mathbf{i}}$ at all points in the pipe. The value of the acceleration depends on whether V_0 is being increased, $\partial V_0/\partial t > 0$, or decreased, $\partial V_0/\partial t < 0$. Unless V_0 is independent of time ($V_0 \equiv$ constant) there will be an acceleration because of the local acceleration term. Thus, the acceleration field, $\mathbf{a} = \partial V_0/\partial t\,\hat{\mathbf{i}}$, is uniform throughout the entire flow, although it may vary with time ($\partial V_0/\partial t$ need not be constant). The acceleration due to the spatial variations of velocity ($u\,\partial u/\partial x, v\,\partial v/\partial y,$ etc.) vanishes automatically for this flow, since $\partial u/\partial x = 0$ and $v = w = 0$. That is,

$$\mathbf{a} = \frac{\partial \mathbf{V}}{\partial t} + u\frac{\partial \mathbf{V}}{\partial x} + v\frac{\partial \mathbf{V}}{\partial y} + w\frac{\partial \mathbf{V}}{\partial z} = \frac{\partial \mathbf{V}}{\partial t} = \frac{\partial V_0}{\partial t}\,\hat{\mathbf{i}}$$

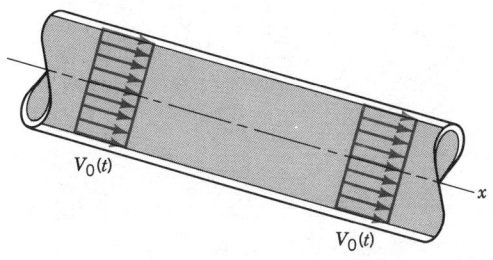

FIGURE 4.6 Uniform, unsteady flow in a constant diameter pipe.

FIGURE 4.7 Steady state operation of a water heater.

4.2.3 Convective Effects

The portion of the material derivative (Eq. 4.5) represented by the spatial derivatives is termed the *convective derivative*. It represents the fact that a flow property associated with a fluid particle may vary because of the motion of the particle from one point in space where the parameter has one value to another point in space where its value is different. This contribution to the time rate of change of the parameter for the particle can occur whether the flow is steady or unsteady. It is due to the convection, or motion, of the particle through space in which there is a gradient $[\nabla(\ ) = \partial(\ )/\partial x\ \hat{\mathbf{i}} + \partial(\ )/\partial y\hat{\mathbf{j}} + \partial(\ )/\partial z\ \hat{\mathbf{k}}]$ in the parameter value. That portion of the acceleration given by the term $(\mathbf{V} \cdot \nabla)\mathbf{V}$ is termed the *convective acceleration*.

As is illustrated in Fig. 4.7, the temperature of a water particle changes as it flows through a water heater. The water entering the heater is always the same cold temperature and the water leaving the heater is always the same hot temperature. The flow is steady. However, the temperature, T, of each water particle increases as it passes through the heater—$T_{\text{out}} > T_{\text{in}}$. Thus, $DT/Dt \neq 0$ because of the convective term in the total derivative of the temperature. That is, $\partial T/\partial t = 0$, but $u\ \partial T/\partial x \neq 0$ (where x is directed along the streamline), since there is a nonzero temperature gradient along the streamline. A fluid particle traveling along this nonconstant temperature path ($\partial T/\partial x \neq 0$) at a specified speed ($u$) will have its temperature change with time at a rate of $DT/Dt = u\ \partial T/\partial x$ even though the flow is steady ($\partial T/\partial t = 0$).

The same type of processes are involved with fluid accelerations. Consider flow in a variable area pipe as shown in Fig. 4.8. It is assumed that the flow is steady and one-dimensional with velocity that increases and decreases in the flow direction as indicated. As the fluid flows from section (1) to section (2), its velocity increases from V_1 to V_2. Thus, even though $\partial \mathbf{V}/\partial t = 0$, fluid particles experience an acceleration given by $a_x = u\ \partial u/\partial x$. For $x_1 < x < x_2$ it is seen from Fig. 4.8 that $\partial u/\partial x > 0$ so that $a_x > 0$—the fluid accelerates. For $x_2 < x < x_3$, it is seen that $\partial u/\partial x < 0$ so that $a_x < 0$—the fluid decelerates. If $V_1 = V_3$, the amount of acceleration precisely balances the amount of deceleration even though the distances between x_2 and x_1 and x_3 and x_2 are not the same.

FIGURE 4.8 Uniform, steady flow in a variable area pipe.

The concepts of the material derivative, local acceleration, and convective acceleration are very useful in various applications throughout the study of fluid mechanics.

EXAMPLE 4.5

Consider the steady, two-dimensional flow field discussed in Example 4.2. Determine the acceleration field for this flow.

SOLUTION

The velocity is given by $\mathbf{V} = (V_0/\ell)(x\hat{\mathbf{i}} - y\hat{\mathbf{j}})$ so that $u = (V_0/\ell)x$ and $v = -(V_0/\ell)y$. For steady $[\partial(\ )/\partial t = 0]$, two-dimensional ($w = 0$ and $\partial(\ )/\partial z = 0$) flow, Eq. 4.3 becomes

$$\mathbf{a} = u\frac{\partial \mathbf{V}}{\partial x} + v\frac{\partial \mathbf{V}}{\partial y} = \left(u\frac{\partial u}{\partial x} + v\frac{\partial u}{\partial y}\right)\hat{\mathbf{i}} + \left(u\frac{\partial v}{\partial x} + v\frac{\partial v}{\partial y}\right)\hat{\mathbf{j}}$$

Hence, for this flow the acceleration is given by

$$\mathbf{a} = \left[\left(\frac{V_0}{\ell}\right)(x)\left(\frac{V_0}{\ell}\right) + \left(\frac{V_0}{\ell}\right)(y)(0)\right]\hat{\mathbf{i}} + \left[\left(\frac{V_0}{\ell}\right)(x)(0) + \left(\frac{-V_0}{\ell}\right)(y)\left(\frac{-V_0}{\ell}\right)\right]\hat{\mathbf{j}}$$

or

$$a_x = \frac{V_0^2 x}{\ell^2}, \qquad a_y = \frac{V_0^2 y}{\ell^2} \tag{Ans}$$

The fluid experiences an acceleration in both the x and y directions. Since the flow is steady, there is no local acceleration—the fluid velocity at any given point is constant in time. However, there is a convective acceleration due to the change in velocity from one point on the particle's pathline to another. Recall that the velocity is a vector—it has both a magnitude and a direction. In this flow both the fluid speed (magnitude) and flow direction change with location (see Fig. E4.1a).

For this flow the magnitude of the acceleration is constant on circles centered at the origin, as is seen from the fact that

$$|\mathbf{a}| = (a_x^2 + a_y^2 + a_z^2)^{1/2} = \left(\frac{V_0}{\ell}\right)^2 (x^2 + y^2)^{1/2} \tag{1}$$

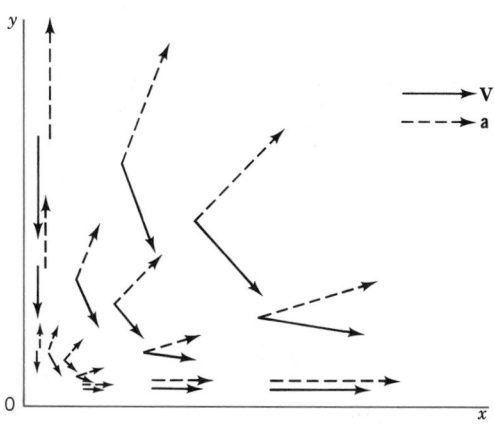

FIGURE E4.5

Also, the acceleration vector is oriented at an angle θ from the x axis, where

$$\tan \theta = \frac{a_y}{a_x} = \frac{y}{x}$$

This is the same angle as that formed by a ray from the origin to point (x, y). Thus, the acceleration is directed along rays from the origin and has a magnitude proportional to the distance from the origin. Typical acceleration vectors (from Eq. 1) and velocity vectors (from Example 4.1) are shown in Fig. E4.5 for the flow in the first quadrant. Note that **a** and **V** are not parallel except along the x and y axes, (a fact that is responsible for the curved pathlines of the flow) and that both the acceleration and velocity are zero at the origin ($x = y = 0$). An infinitesimal fluid particle placed precisely at the origin will remain there, but its neighbors (no matter how close they are to the origin) will drift away.

The concept of the material derivative can be used to determine the time rate of change of any parameter associated with a particle as it moves about. Its use is not restricted to fluid mechanics alone. The basic ingredients needed to use the material derivative concept are the field description of the parameter, $P = P(x, y, z, t)$, and the rate at which the particle moves through that field, $\mathbf{V} = \mathbf{V}(x, y, z, t)$.

EXAMPLE 4.6

A manufacturer produces a perishable product in a factory located at $x = 0$ and sells the product along the distribution route $x > 0$. The selling price of the product, P, is a function of the length of time after it was produced, t, and the location at which it is sold, x. That is, $P = P(x, t)$. At a given location the price of the product decreases in time (it is perishable) according to $\partial P/\partial t = -C_1$, where C_1 is a positive constant. In addition, because of shipping costs the price increases with distance from the factory according to $\partial P/\partial x = C_2$, where C_2 is a positive constant. If the manufacturer wishes to sell the product for the same price anywhere along the distribution route, determine how fast he must travel along the route.

SOLUTION

For a given batch of the product (Lagrangian description), the time rate of change of the price can be obtained by using the material derivative

$$\frac{DP}{Dt} = \frac{\partial P}{\partial t} + \mathbf{V} \cdot \nabla P = \frac{\partial P}{\partial t} + u\frac{\partial P}{\partial x} + v\frac{\partial P}{\partial y} + w\frac{\partial P}{\partial z} = \frac{\partial P}{\partial t} + u\frac{\partial P}{\partial x}$$

We have used the fact that the motion is one-dimensional with $\mathbf{V} = u\hat{\mathbf{i}}$, where u is the speed at which the product is convected along its route. If the price is to remain constant as the product moves along the distribution route, then

$$\frac{DP}{Dt} = 0 \quad \text{or} \quad \frac{\partial P}{\partial t} + u\frac{\partial P}{\partial x} = 0$$

Thus, the correct delivery speed is

$$u = -\frac{\partial P/\partial t}{\partial P/\partial x} = \frac{C_1}{C_2} \tag{Ans}$$

With this speed, the decrease in price because of the local effect ($\partial P/\partial t$) is exactly balanced by the increase in price due to the convective effect ($u\,\partial P/\partial x$). A faster delivery speed will cause the price of the given batch of the product to increase in time ($DP/Dt > 0$; it is rushed to distant markets before it spoils), while a slower delivery speed will cause its price to decrease ($DP/Dt < 0$; the increased costs due to distance from the factory is more than offset by reduced costs due to spoilage.)

4.2.4 Streamline Coordinates

In many flow situations it is convenient to use a coordinate system defined in terms of the streamlines of the flow. An example for steady, two-dimensional flows is illustrated in Fig. 4.9. Such flows can be described either in terms of the usual x, y Cartesian coordinate system (or some other system such as the r, θ polar coordinate system) or the streamline coordinate system. In the streamline coordinate system the flow is described in terms of one coordinate along the streamlines, denoted s, and the second coordinate normal to the streamlines, denoted n. Unit vectors in these two directions are denoted by $\hat{s}$ and $\hat{n}$ as shown in the figure. Care is needed not to confuse the coordinate distance s (a scalar) with the unit vector along the streamline direction, $\hat{s}$.

The flow plane is therefore covered by an orthogonal curved net of coordinate lines. At any point the s and n directions are perpendicular, but the lines of constant s or constant n are not necessarily straight. Without knowing the actual velocity field (hence, the streamlines) it is not possible to construct this flow net. In many situations appropriate simplifying assumptions can be made so that this lack of information does not present an insurmountable difficulty. One of the major advantages of using the streamline coordinate system is that the velocity is always tangent to the s direction. That is,

$$\mathbf{V} = V\hat{s}$$

This allows simplifications in describing the fluid particle acceleration and in solving the equations governing the flow.

For steady, two-dimensional flow we can determine the acceleration as

$$\mathbf{a} = \frac{D\mathbf{V}}{Dt} = a_s\hat{s} + a_n\hat{n}$$

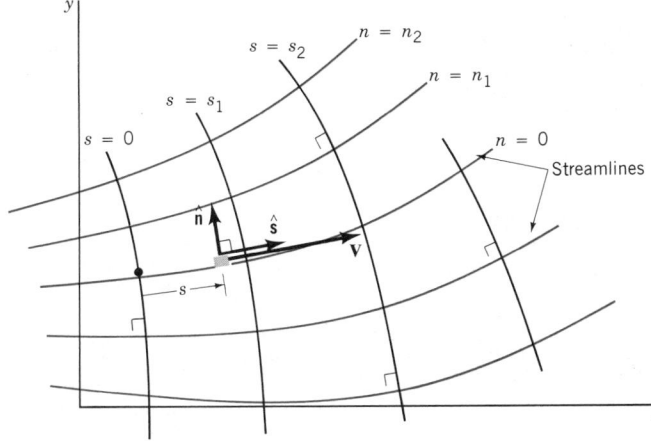

FIGURE 4.9 Streamline coordinate system for two-dimensional flow.

where a_s and a_n are the streamline and normal components of acceleration, respectively. We use the material derivative because by definition the acceleration is the time rate of change of the velocity of a given particle as it moves about. If the streamlines are curved, both the speed of the particle and its direction of flow may change from one point to another. In general, for steady flow both the speed and the flow direction are a function of location—$V = V(s, n)$ and $\hat{s} = \hat{s}(s, n)$. For a given particle, the value of s changes with time, but the value of n remains fixed because the particle flows along a streamline defined by $n =$ constant. (Recall that streamlines and pathlines coincide in steady flow.) Thus, application of the chain rule gives

$$\mathbf{a} = \frac{D(V\hat{s})}{Dt} = \frac{DV}{Dt}\hat{s} + V\frac{D\hat{s}}{Dt}$$

or

$$\mathbf{a} = \left(\frac{\partial V}{\partial t} + \frac{\partial V}{\partial s}\frac{ds}{dt} + \frac{\partial V}{\partial n}\frac{dn}{dt}\right)\hat{s} + V\left(\frac{\partial \hat{s}}{\partial t} + \frac{\partial \hat{s}}{\partial s}\frac{ds}{dt} + \frac{\partial \hat{s}}{\partial n}\frac{dn}{dt}\right)$$

This can be simplfied by using the fact that for steady flow nothing changes with time at a given point so that both $\partial V/\partial t$ and $\partial \hat{s}/\partial t$ are zero. Also, the velocity along the streamline is $V = ds/dt$ and the particle remains on its streamline ($n =$ constant) so that $dn/dt = 0$. Hence,

$$\mathbf{a} = \left(V\frac{\partial V}{\partial s}\right)\hat{s} + V\left(V\frac{\partial \hat{s}}{\partial s}\right)$$

The quantity $\partial \hat{s}/\partial s$ represents the limit as $\delta s \to 0$ of the change in the unit vector along the streamline, $\delta \hat{s}$, per change in distance along the streamline, δs. The magnitude of $\hat{s}$ is constant ($|\hat{s}| = 1$; it is a unit vector), but its direction is variable if the streamlines are curved. From Fig. 4.10 it is seen that the magnitude of $\partial \hat{s}/\partial s$ is equal to the inverse of the radius of curvature of the streamline, $\mathcal{R}$, at the point in question. This follows because the two triangles shown (AOB and $A'O'B'$) are similar triangles so that $\delta s/\mathcal{R} = |\delta \hat{s}|/|\hat{s}| = |\delta \hat{s}|$, or $|\delta \hat{s}/\delta s| = 1/\mathcal{R}$. Similarly, in the limit $\delta s \to 0$, the direction of $\delta \hat{s}/\delta s$ is seen to be normal to the streamline. That is,

$$\frac{\partial \hat{s}}{\partial s} = \lim_{\delta s \to 0} \frac{\delta \hat{s}}{\delta s} = \frac{\hat{n}}{\mathcal{R}}$$

Hence, the acceleration for steady, two-dimensional flow can be written in terms of

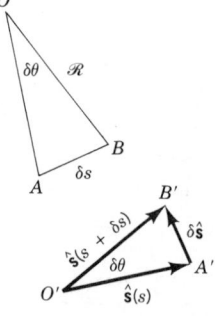

FIGURE 4.10 Relationship between the unit vector along the streamline, $\hat{s}$, and the radius of curvature of the streamline, $\mathcal{R}$.

its streamwise and normal components in the form

$$\mathbf{a} = V\frac{\partial V}{\partial s}\hat{\mathbf{s}} + \frac{V^2}{\mathcal{R}}\hat{\mathbf{n}} \quad \text{or} \quad a_s = V\frac{\partial V}{\partial s}, \qquad a_n = \frac{V^2}{\mathcal{R}} \tag{4.7}$$

The first term, $a_s = V\,\partial V/\partial s$, represents the convective acceleration along the streamline and the second term, $a_n = V^2/\mathcal{R}$, represents centrifugal acceleration (one type of convective acceleration) normal to the fluid motion. These components can be noted in Fig. E4.5 by resolving the acceleration vector into its components along and normal to the velocity vector. Note that the unit vector $\hat{\mathbf{n}}$ is directed from the streamline toward the center of curvature. These forms of the acceleration are probably familiar from previous dynamics or mechanics considerations.

EXAMPLE 4.7

An incompressible fluid flows through the converging duct shown in Fig. E4.7a with velocity V_0 at the entrance. Measurements indicate that the actual velocity of the fluid near the wall of the duct along streamline A–F is as is shown in Fig. E4.7b. Sketch the component of acceleration along this streamline, a_s, as a function of s.

(a)

(b)

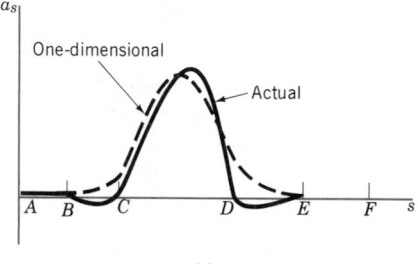

(c)

FIGURE E4.7

SOLUTION

The streamwise acceleration is obtained from Eq. 4.7 as $a_s = V \, \partial V / \partial s$. Its value can be determined from the graph of V versus s and its slope, $\partial V / \partial s$, which can also be obtained from that graph. Since the original data, $V(s)$, is in graphical form, some type of graphical or numerical method must be used to determine a_s. It is seen that the value of $\partial V / \partial s$ is negative along the streamline from B to C and from D to E; it is essentially zero from A to B and from E to F; and it is positive from C to D. Thus, the streamwise acceleration has the character shown in Fig. E4.7c.

In traveling through the contraction along the streamline A–F, the fluid initially decelerates slightly, then accelerates considerably, and finally decelerates slightly again. The net result is an increase in velocity from V_0 to $1.5V_0$ (the net acceleration is greater than the deceleration). If the flow were assumed to be one-dimensional with a uniform velocity profile across any section of the duct, the velocity profile and acceleration would be as shown by the dashed lines in Figs. E4.7b and E4.7c. The actual flow field is apparently more complex than that implied by a one-dimensional assumption.

4.3 CONTROL VOLUME AND SYSTEM REPRESENTATIONS

As is discussed in Chapter 1, a fluid is a type of matter that is relatively free to move and interact with its surroundings. As with any matter, a fluid's behavior is governed by a set of fundamental physical laws which are approximated by an appropriate set of equations. The application of laws such as the conservation of mass, Newton's laws of motion, and the laws of thermodynamics form the foundation of fluid mechanics analyses. There are various ways that these governing laws can be applied to a fluid, including the system approach and the control volume approach. By definition, a *system* is a collection of matter of fixed identity (always the same atoms or fluid particles), which may move, flow, and interact with its surroundings. A *control volume*, on the other hand, is a volume in space (a geometric entity, independent of mass) through which fluid may flow. The control volume is not unique—for a given flow situation there is a wide variety of control volumes that one may use.

A system is a specific, identifiable quantity of matter. It may consist of a relatively large amount of mass (such as all of the air in the earth's atmosphere), or it may be of infinitesimal size (such as a single fluid particle). In any case, the molecules making up the system are "tagged" in some fashion (dyed red, either actually or only in your mind) so that they can be continually identified as they move about. The system may interact with its surroundings by various means (by the transfer of heat or the exertion of a pressure force, for example). It may continually change size and shape, but it always contains the same mass.

A mass of air drawn into an air compressor can be considered as a system. It changes shape and size (it is compressed), its temperature may change, and it is eventually expelled through the outlet of the compressor. The matter associated with the original air drawn into the compressor remains as a system, however. The behavior of this material could be investigated by applying the appropriate governing equations to this system.

One of the important concepts used in the study of statics and dynamics is that of the free-body diagram. That is, we identify an object, isolate it from its surroundings, replace its surroundings by the equivalent actions that they put on the object, and apply Newton's laws of motion. The body in such cases is our system—an identified portion of matter that we follow during its interactions with its surroundings. In fluid mechanics, it is often quite difficult to identify and keep track of a specific quantity of matter. A finite portion of a fluid contains an uncountable number of fluid particles that move about quite freely, unlike a solid that may deform but usually remains relatively easy to identify. For example, we cannot as easily follow a specific portion of water flowing in a river as we can follow a branch floating on its surface.

We may often be more interested in determining the forces put on a fan, airplane, or automobile by air flowing past the object than we are in the information obtained by following a given portion of the air (a system) as it flows along. For these situations we often use the control volume approach. We identify a specific volume in space (a volume associated with the fan, airplane, or automobile, for example) and analyze the fluid flow within, through, or around that volume. In general, the control volume can be a moving volume, although for most situations considered in this book we will use only fixed, nondeformable control volumes. The matter within a control volume may change with time as the fluid flows through it. Similarly, the amount of mass within the volume may change with time. The control volume itself is a specific geometric entity, independent of the flowing fluid.

Examples of control volumes and *control surfaces* (the surface of the control volume) are shown in Fig. 4.11. For case (*a*), fluid flows through a pipe. The fixed control surface consists of the inside surface of the pipe, the outlet end at section (2), and a section across the pipe at (1). One portion of the control surface is a physical surface (the pipe), while the remainder is simply a surface in space (across the pipe). Fluid flows across part of the control surface, but not across all of it.

Another control volume is the rectangular volume surrounding the jet engine shown in Fig. 4.11*b*. If the airplane to which the engine is attached is sitting still on the runway, air flows through this control volume because of the action of the engine within it. The air that was within the engine itself at time $t = t_1$ (a system) has passed through the engine and is partially outside of the control volume at a later time $t = t_2$ as indicated. At this later time other air (a different system) is within the engine. If the airplane is moving, the control volume is fixed relative to an observer on the airplane, but it is a moving control volume relative to an observer on the ground. In either situation air flows through and around the engine as indicated.

The deflating balloon shown in Fig. 4.11*c* provides an example of a deforming control volume. As time increases the control volume, whose surface is the inner surface of the balloon, decreases in size. If we do not hold onto the balloon, it becomes a moving, deforming control volume as it darts about the room. The majority of the problems we will analyze can be solved by using a fixed, nondeforming control volume. In some instances, however, it will be advantageous, in fact necessary, to use a moving, deforming control volume.

FIGURE 4.11   Typical control volumes: (*a*) fixed control volume, (*b*) fixed or moving control volume, (*c*) deforming control volume.

In many ways the relationship between a system and a control volume is similar to the relationship between the Lagrangian and Eulerian description of a flow. In the system or Lagrangian description we follow the fluid and observe its behavior as it moves about. In the control volume or Eulerian description we remain stationary and observe the fluid's behavior at a fixed location. (If a moving control volume is used, it virtually never moves with the system—the system flows through the control volume.) These ideas are discussed in more detail in the next section.

All of the laws governing the motion of a fluid are stated in their basic form in terms of a system approach. For example, "the mass of a system remains constant," or "the time rate of change of momentum of a system is equal to the sum of all the forces acting on the system." Note the word system, not control volume, in these statements. To use the governing equations in a control volume approach to problem solving, we must rephrase the laws in an appropriate manner. To this end we introduce the Reynold's transport theorem in the following section.

4.4   THE REYNOLDS TRANSPORT THEOREM

We are sometimes interested in what happens to a particular part of the fluid as it moves about. Other times we may be interested in what effect the fluid has on a particular object or volume in space as fluid interacts with it. Thus, we need to describe the laws governing fluid motion using both system concepts (consider a given mass of the fluid) and control volume concepts (consider a given volume). To do this we need an analytical tool to shift from one representation to the other. The *Reynolds transport theorem* provides this tool.

All physical laws are stated in terms of various physical parameters. Velocity, acceleration, mass, temperature, and momentum are but a few of the more common parameters. Let B represent any of these (or other) fluid parameters and b represent the amount of that parameter per unit mass. That is,

$$B = mb$$

where m is the mass of the portion of fluid of interest. For example, if $B = m$, the mass, it follows that $b = 1$. (The mass per unit mass is unity.) If $B = mV^2/2$, the kinetic energy of the mass, then $b = V^2/2$, the kinetic energy per unit mass. The parameters B and b may be scalars or vectors. Thus, if $\mathbf{B} = m\mathbf{V}$, the momentum of the mass, then $\mathbf{b} = \mathbf{V}$. (The momentum per unit mass is the velocity.)

The parameter B is termed an *extensive property* and the parameter b is termed an *intensive property*. The value of B is directly proportional to the amount of the mass being considered, whereas the value of b is independent of the amount of mass. The amount of an extensive property that a system possesses at a given instant, B_{sys}, can be determined by adding up the amount associated with each fluid particle in the system. For infinitesimal fluid particles of size δV and mass $\rho \, \delta V$, this summation (in the limit of $\delta V \to 0$) takes the form of an integration over all the particles in the system and can be written as

$$B_{sys} = \lim_{\delta V \to 0} \sum_i b_i(\rho_i \, \delta V_i) = \int_{sys} \rho b \, dV$$

The limits of integration cover the entire system—a (usually) moving volume. We have used the fact that the amount of B in a fluid particle of mass $\rho \, \delta \mathcal{V}$ is given in terms of b by $\delta B = b\rho \, \delta \mathcal{V}$.

Most of the laws governing fluid motion involve the time rate of change of an extensive property of a fluid system—the rate at which the momentum of a system changes with time, the rate at which the mass of a system changes with time, and so on. Thus, we often encounter terms such as

$$\frac{dB_{sys}}{dt} = \frac{d\left(\int_{sys} \rho b \, d\mathcal{V}\right)}{dt} \tag{4.8}$$

To formulate the laws into a control volume approach, we must obtain an expression for the time rate of change of an extensive property within a control volume, B_{cv}, not within a system. This can be written as

$$\frac{dB_{cv}}{dt} = \frac{d\left(\int_{cv} \rho b \, d\mathcal{V}\right)}{dt} \tag{4.9}$$

where the limits of integration, denoted by cv, cover the control volume of interest. Although Eqs. 4.8 and 4.9 may look very similar, the physical interpretation of each is quite different. Mathematically, the difference is represented by the difference in the limits of integration. Recall that the control volume is a volume in space (in most cases stationary, although if it moves it need not move with the system). On the other hand, the system is an identifiable collection of mass that moves with the fluid (indeed it is a specified portion of the fluid). We will learn that even for those instances when the control volume and the system momentarily occupy the same volume in space, the two quantities dB_{sys}/dt and dB_{cv}/dt need not be the same. The Reynolds transport theorem provides the relationship between the time rate of change of an extensive property for a system and that for a control volume—the relationship between Eqs. 4.8 and 4.9.

EXAMPLE 4.8

Fluid flows from the fire extinguisher tank shown in Fig. E4.8. Discuss the differences between dB_{sys}/dt and dB_{cv}/dt if B represents mass.

FIGURE E4.8

SOLUTION

With $B = m$, the system mass, it follows that $b = 1$ and Eqs. 4.8 and 4.9 can be written as

$$\frac{dB_{\text{sys}}}{dt} \equiv \frac{dm_{\text{sys}}}{dt} = \frac{d\left(\int_{\text{sys}} \rho \, d\mathcal{V}\right)}{dt}$$

and

$$\frac{dB_{\text{cv}}}{dt} \equiv \frac{dm_{\text{cv}}}{dt} = \frac{d\left(\int_{\text{cv}} \rho \, d\mathcal{V}\right)}{dt}$$

Physically these represent the time rate of change of mass within the system and the time rate of change of mass within the control volume, respectively. We choose our system to be the fluid within the tank at the time the valve was opened ($t = 0$) and the control volume to be the tank itself. A short time after the valve is opened, part of the system has moved outside of the control volume as is shown in Fig. E4.8b. The control volume remains fixed. The limits of integration are fixed for the control volume; they are a function of time for the system.

Clearly, if mass is to be conserved (one of the basic laws governing fluid motion), the mass of the fluid in the system is constant, so that

$$\frac{d\left(\int_{\text{sys}} \rho \, d\mathcal{V}\right)}{dt} = 0$$

On the other hand, it is equally clear that some of the fluid has left the control volume through the nozzle on the tank. Hence, the amount of mass within the tank (the control volume) decreases with time, or

$$\frac{d\left(\int_{\text{cv}} \rho \, d\mathcal{V}\right)}{dt} < 0$$

The actual numerical value of the rate at which the mass in the control volume decreases will depend on the rate at which the fluid flows through the nozzle (that is, the size of the nozzle and the speed and density of the fluid). Clearly the meaning of dB_{sys}/dt and dB_{cv}/dt are different. For this example $dB_{\text{cv}}/dt < dB_{\text{sys}}/dt$. Other situations may have $dB_{\text{cv}}/dt \geq dB_{\text{sys}}/dt$.

4.4.1   Derivation of the Reynolds Transport Theorem

A simple version of the Reynolds transport theorem relating system concepts to control volume concepts can be obtained easily for the one-dimensional flow through a fixed control volume as is shown in Fig. 4.12a. We consider the control volume to be that stationary volume within the pipe or duct between sections (1) and (2) as indicated. The system that we consider is that fluid occupying the control volume at some initial time t. A short time later, at time $t + \delta t$, the system has moved slightly to the right. The fluid particles that coincided with section (2) of the control surface at time t have moved a distance $\delta l_2 = V_2 \, \delta t$ to the right, where V_2 is the velocity of the fluid as it passes section (2). Similarly, the fluid initially at section (1) has moved a distance $\delta l_1 = V_1 \, \delta t$, where V_1 is the fluid velocity at section (1). We assume the

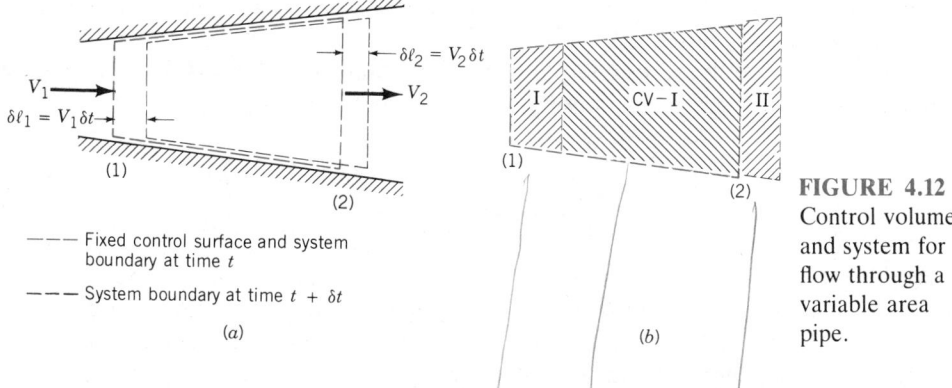

--- Fixed control surface and system
boundary at time t

--- System boundary at time $t + \delta t$

(a)

(b)

FIGURE 4.12
Control volume
and system for
flow through a
variable area
pipe.

fluid flows across sections (1) and (2) in a direction normal to these surfaces and that V_1 and V_2 are constant across sections (1) and (2).

As is shown in Fig. 4.12b, the outflow from the control volume from time t to $t + \delta t$ is denoted as volume II, the inflow as volume I, and the control volume itself as CV. Thus, the system at time t consists of the fluid in section CV ("SYS = CV" at time t), while at time $t + \delta t$ the system consists of the same fluid that now occupies sections (CV − I) + II. That is, "SYS = CV − I + II" at time $t + \delta t$. The control volume remains as section CV for all time.

If B is an extensive parameter of the system, then the value of it for the system at time t is

$$B_{sys}(t) = B_{cv}(t)$$

since the system and the fluid within the control volume coincide at this time. Its value at time $t + \delta t$ is

$$B_{sys}(t + \delta t) = B_{cv}(t + \delta t) - B_1(t + \delta t) + B_{II}(t + \delta t)$$

Thus, the change in the amount of B in the system in the time interval δt divided by this time interval is given by

$$\frac{\delta B_{sys}}{\delta t} = \frac{B_{sys}(t + \delta t) - B_{sys}(t)}{\delta t}$$

$$= \frac{B_{cv}(t + \delta t) - B_1(t + \delta t) + B_{II}(t + \delta t) - B_{sys}(t)}{\delta t}$$

By using the fact that at the initial time t we have $B_{sys}(t) = B_{cv}(t)$, this ungainly expression may be rearranged as follows.

$$\frac{\delta B_{sys}}{\delta t} = \frac{B_{cv}(t + \delta t) - B_{cv}(t)}{\delta t} - \frac{B_1(t + \delta t)}{\delta t} + \frac{B_{II}(t + \delta t)}{\delta t} \qquad (4.10)$$

In the limit as $\delta t \to 0$ the left-hand side of Eq. 4.10 is equal to the time rate of change of B for the system and is denoted as DB_{sys}/Dt. We use the material derivative notation, $D(\ )/Dt$, to denote this time rate of change to emphasize the Lagrangian character of this term. (Recall from Section 4.2.1 that the material derivative, DP/Dt, of any quantity P represents the time rate of change of that quantity associated with a given fluid particle as it moves along.) Similarly, the quantity DB_{sys}/Dt represents the time rate of change of property B associated with a system (a given portion of fluid) as it moves along.

In the limit $\delta t \to 0$, the first term on the right-hand side of Eq. 4.10 is seen to be the time rate of change of the amount of B within the control volume

$$\lim_{\delta t \to 0} \frac{B_{\text{cv}}(t + \delta t) - B_{\text{cv}}(t)}{\delta t} = \frac{\partial B_{\text{cv}}}{\partial t} = \frac{\partial \left(\int_{\text{cv}} \rho b \, d\Psi \right)}{\partial t} \tag{4.11}$$

The third term on the right-hand side of Eq. 4.10 represents the rate at which the extensive parameter B flows from the control volume, across the control surface. This can be seen from the fact that the amount of B within region II, the outflow region, is its amount per unit volume, ρb, times the volume $\delta\Psi_{\text{II}} = A_2 \, \delta\ell_2 = A_2(V_2 \, \delta t)$. Hence,

$$B_{\text{II}}(t + \delta t) = (\rho_2 b_2)(\delta\Psi_{\text{II}}) = \rho_2 b_2 A_2 V_2 \, \delta t$$

where b_2 and ρ_2 are the constant values of b and ρ across section (2). Thus, the rate at which this property flows from the control volume, $\dot{B}_{\text{out}}$, is given by

$$\dot{B}_{\text{out}} = \lim_{\delta t \to 0} \frac{B_{\text{II}}(t + \delta t)}{\delta t} = \rho_2 A_2 V_2 b_2 \tag{4.12}$$

Similarly, the inflow of B into the control volume across section (1) during the time interval δt corresponds to that in region I and is given by the amount per unit volume times the volume, $\delta\Psi_{\text{I}} = A_1 \, \delta\ell_1 = A_1(V_1 \, \delta t)$. Hence,

$$B_{\text{I}}(t + \delta t) = (\rho_1 b_1)(\delta\Psi_{\text{I}}) = \rho_1 b_1 A_1 V_1 \, \delta t$$

where b_1 and ρ_1 are the constant values of b and ρ across section (1). Thus, the rate on inflow of the property B into the control volume, $\dot{B}_{\text{in}}$, is given by

$$\dot{B}_{\text{in}} = \lim_{\delta t \to 0} \frac{B_{\text{I}}(t + \delta t)}{\delta t} = \rho_1 A_1 V_1 b_1 \tag{4.13}$$

If we combine Eqs. 4.10, 4.11, 4.12, and 4.13 we see that the relationship between the time rate of change of B for the system and that for the control volume is given by

$$\frac{DB_{\text{sys}}}{Dt} = \frac{\partial B_{\text{cv}}}{\partial t} + \dot{B}_{\text{out}} - \dot{B}_{\text{in}} \tag{4.14}$$

or

$$\frac{DB_{\text{sys}}}{Dt} = \frac{\partial B_{\text{cv}}}{\partial t} + \rho_2 A_2 V_2 b_2 - \rho_1 A_1 V_1 b_1 \tag{4.15}$$

This is a version of the Reynolds transport theorem valid under the restrictive assumptions associated with the flow shown in Fig. 4.12—fixed control volume with one inlet and one outlet having uniform properties (density, velocity) across the inlet and outlet with the velocity normal to sections (1) and (2). Note that the time rate of change of B for the system (the left-hand side of Eq. 4.15 or the quantity in Eq. 4.8) is not necessarily the same as the rate of change of B within the control volume (the first term on the right-hand side of Eq. 4.15 or the quantity in Eq. 4.9). This is true because the inflow rate $(b_1 \rho_1 V_1 A_1)$ and the outflow rate $(b_2 \rho_2 V_2 A_2)$ of the property B for the control volume need not be the same.

EXAMPLE 4.9

Consider again the flow from the fire extinguisher shown in Fig. E4.8. Let the extensive property of interest be the system mass ($B = m$, the system mass, or $b = 1$), and write the appropriate form of the Reynolds transport theorem for this flow.

SOLUTION

Again we take the control volume to be the fire extinguisher, and the system to be the fluid within it at time $t = 0$. For this case there is no inlet, section (1), across which the fluid flows into the control volume ($A_1 = 0$). There is, however, an outlet, section (2). Thus, the Reynolds transport theorem, Eq. 4.15, can be written as

$$\frac{Dm_{sys}}{Dt} = \frac{\partial \left(\int_{cv} \rho \, d\Psi \right)}{\partial t} + \rho_2 A_2 V_2 \qquad (1) \quad (\text{Ans})$$

If we proceed one step further and use the basic law of conservation of mass, we may set the left-hand side of this equation equal to zero (the amount of mass in a system is constant) and rewrite Eq. 1 in the form:

$$\frac{\partial \left(\int_{cv} \rho \, d\Psi \right)}{\partial t} = -\rho_2 A_2 V_2 \qquad (2)$$

The physical interpretation of this result is that the rate at which the mass in the tank decreases in time is equal in magnitude but opposite to the rate of flow of mass from the exit, $\rho_2 A_2 V_2$. Note the units for the two terms of Eq. 2 (kg/s or slugs/s). We will deal with the consequences of mass conservation in considerably more detail in following chapters.

Equation 4.15 is a simplified version of the Reynolds transport theorem. We will now derive it for much more general conditions. A general, fixed control volume with fluid flowing through it is shown in Fig. 4.13. The flow field may be quite

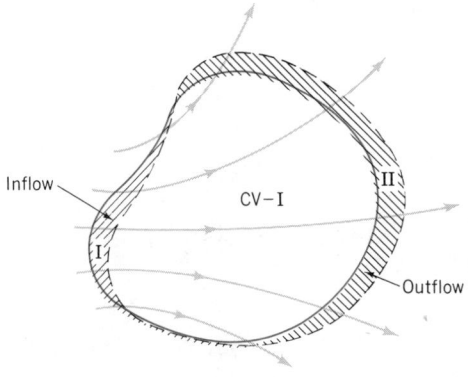

Inflow

CV–I

II

I

Outflow

———— Fixed control surface and system boundary at time t

— — — System boundary at time $t + \delta t$

FIGURE 4.13 Control volume and system for flow through an arbitrary fixed control volume.

simple (as in the above one-dimensional flow considerations), or it may involve a quite complex, unsteady, three-dimensional situation. In any case we again consider the system to be the fluid within the control volume at the initial time t. A short time later a portion of the fluid (region II) has exited from the control volume and additional fluid (region I, not part of the original system) has entered the control volume.

We consider an extensive fluid property B and seek to determine how the rate of change of B associated with the system is related to the rate of change of B within the control volume at any instant. By repeating the exact steps that we did for the simplified control volume shown in Fig. 4.12, we see that Eq. 4.14 is valid for the general case also, provided that we give the correct interpretation to the terms $\dot{B}_{\text{out}}$ and $\dot{B}_{\text{in}}$. In general, the control volume may contain more (or less) than one inlet and one outlet. A typical pipe system may contain several inlets and outlets as are shown in Fig. 4.14. In such instances we think of all inlets grouped together (I = $I_a + I_b + I_c + \cdots$) and all outlets grouped together (II = $II_a + II_b + II_c + \cdots$), at least conceptually.

The term B_{out} represents the net flowrate of the property B from the control volume. Its value can be thought of as arising from the addition (integration) of the contributions through each infinitesimal area element of size δA on the portion of the control surface dividing region II and the control volume. This surface is denoted CS_{out}. As is indicated in Fig. 4.15, in time δt the volume of fluid that passes across each area element is given by $\delta \Psi = \delta \ell_n \, \delta A$, where $\delta \ell_n = \delta \ell \cos \theta$ is the height (normal to the base, δA) of the small volume element, and θ is the angle between the velocity vector and the outward pointing normal to the surface, $\hat{\mathbf{n}}$. Thus, since $\delta \ell = V \, \delta t$, the amount of the property B carried across the area element δA in the time interval δt is given by

$$\delta B = b\rho \, \delta \Psi = b\rho (V \cos \theta \, \delta t) \, \delta A$$

The rate at which B is carried out of the control volume across the small area element δA, denoted $\delta \dot{B}_{\text{out}}$, is

$$\delta \dot{B}_{\text{out}} = \lim_{\delta t \to 0} \frac{\rho b \, \delta \Psi}{\delta t} = \lim_{\delta t \to 0} \frac{(\rho b V \cos \theta \, \delta t) \, \delta A}{\delta t} = \rho b V \cos \theta \, \delta A$$

FIGURE 4.14
Typical control volume with more than one inlet and outlet.

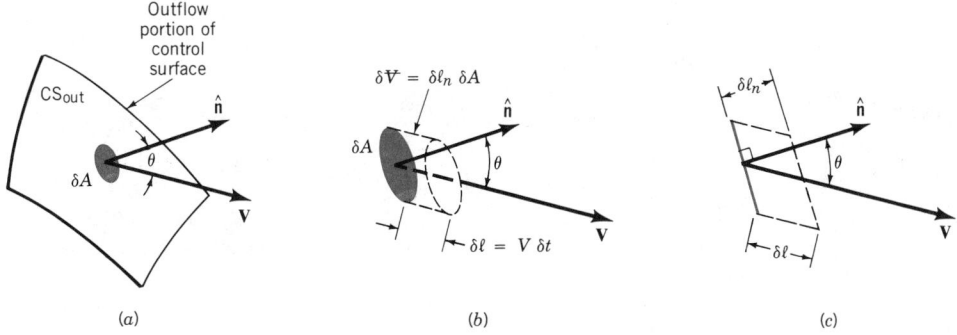

FIGURE 4.15 Outflow across a typical portion of the control surface.

By integrating over the entire outflow portion of the control surface, CS_{out}, we obtain

$$\dot{B}_{out} = \int_{cs_{out}} d\dot{B}_{out} = \int_{cs_{out}} \rho b V \cos \theta \, dA$$

The quantity $V \cos \theta$ is the component of the velocity normal to the area element δA and can be written as $V \cos \theta = \mathbf{V} \cdot \hat{\mathbf{n}}$. Hence, an alternate form of the outflow rate is

$$\dot{B}_{out} = \int_{cs_{out}} \rho b \mathbf{V} \cdot \hat{\mathbf{n}} \, dA \qquad (4.16)$$

In a similar fashion, by considering the inflow portion of the control surface, CS_{in}, as shown in Fig. 4.16, we find that the inflow rate of B into the control volume is

$$\dot{B}_{in} = -\int_{cs_{in}} \rho b V \cos \theta \, dA = -\int_{cs_{in}} \rho b \mathbf{V} \cdot \hat{\mathbf{n}} \, dA \qquad (4.17)$$

We use the standard notation that the unit normal vector to the control surface, $\hat{\mathbf{n}}$, points out from the control volume. Thus, as is shown in Fig. 4.17, $-90° < \theta < 90°$ for outflow regions (the normal component of $\mathbf{V}$ is positive; $\mathbf{V} \cdot \hat{\mathbf{n}} > 0$). For inflow regions $90° < \theta < 270°$ (the normal component of $\mathbf{V}$ is negative; $\mathbf{V} \cdot \hat{\mathbf{n}} < 0$). The

FIGURE 4.16 Inflow across a typical portion of the control surface.

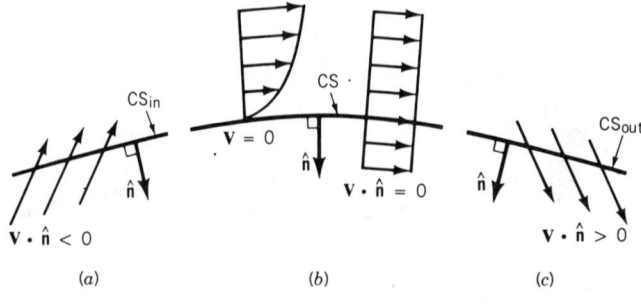

FIGURE 4.17 Possible velocity configurations on portions of the control surface: (a) inflow, (b) no flow across the surface, (c) outflow.

value of cos θ is, therefore, positive on the CV_{out} portions of the control surface and negative on the CV_{in} portions. Over the remainder of the control surface there is no inflow or outflow, leading to $\mathbf{V} \cdot \hat{\mathbf{n}} = V \cos \theta = 0$ on those portions. On such portions either $V = 0$ (the fluid "sticks" to the surface) or $\cos \theta = 0$ (the fluid "slides" along the surface without crossing it) (see Fig. 4.17). Therefore, the net flux (flowrate) of parameter B across the entire control surface is

$$\dot{B}_{out} - \dot{B}_{in} = \int_{cs_{out}} \rho b \mathbf{V} \cdot \hat{\mathbf{n}} \, dA - \left(-\int_{cs_{in}} \rho b \mathbf{V} \cdot \hat{\mathbf{n}} \, dA \right)$$

$$= \int_{cs} \rho b \mathbf{V} \cdot \hat{\mathbf{n}} \, dA \qquad (4.18)$$

where the integration is over the entire control surface.

By combining Eqs. 4.14 and 4.18 we obtain

$$\frac{DB_{sys}}{Dt} = \frac{\partial B_{cv}}{\partial t} + \int_{cs} \rho b \mathbf{V} \cdot \hat{\mathbf{n}} \, dA$$

This can be written in a slightly different form by using $B_{cv} = \int_{cv} \rho b \, d\mathcal{V}$ so that

$$\frac{DB_{sys}}{Dt} = \frac{\partial}{\partial t} \int_{cv} \rho b \, d\mathcal{V} + \int_{cs} \rho b \, \mathbf{V} \cdot \hat{\mathbf{n}} \, dA \qquad (4.19)$$

Equation 4.19 is the general form of the Reynolds transport theorem for a fixed, nondeforming control volume. Its interpretation and use are discussed in the following sections.

4.4.2 Physical Interpretation

The Reynolds transport theorem as given in Eq. 4.19 is widely used in fluid mechanics (and other areas as well). At first it appears to be a rather formidable mathematical expression—perhaps one to be steered clear of if possible. However, a physical understanding of the concepts involved will show that it is a rather straightforward, relatively easy to use tool. Its purpose is to provide a link between control volume ideas and system ideas.

The left side of Eq. 4.19 is the time rate of change of an arbitrary extensive parameter of a system. This may represent the rate of change of mass, momentum, energy, or angular momentum of the system, depending on the choice of the parameter B.

Because the system is moving and the control volume is stationary, the time rate of change of the amount of B within the control volume is not necessarily equal to that of the system. The first term on the right side of Eq. 4.19 represents the rate of change of B within the control volume as the fluid flows through it. Recall that b is the amount of B per unit mass, so that $\rho b \, d\mathcal{V}$ is the amount of B in a small volume $d\mathcal{V}$. Thus, the time derivative of the integral of ρb throughout the control volume is the time rate of change of B within the control volume at a given time.

The last term in Eq. 4.19 (an integral over the control surface) represents the net flowrate of the parameter B across the entire control surface. Over a portion of the control surface this property is being carried out of the control volume ($\mathbf{V} \cdot \hat{\mathbf{n}} > 0$); over other portions it is being carried into the control volume ($\mathbf{V} \cdot \hat{\mathbf{n}} < 0$). Over the remainder of the control surface there is no transport of B across the surface since $b\mathbf{V} \cdot \hat{\mathbf{n}} = 0$, because either $b = 0$, $\mathbf{V} = 0$, or $\mathbf{V}$ is parallel to the surface at those locations. The mass flow rate through area element δA, given by $\rho \mathbf{V} \cdot \hat{\mathbf{n}} \, \delta A$, is positive for outflow (efflux) and negative for inflow (influx). Each fluid particle or fluid mass carries a certain amount of B with it, as given by the product of B per unit mass, b, and the mass. The rate at which this B is carried across the control surface is given by the area integral term of Eq. 4.19. This net rate across the entire control surface may be negative, zero, or positive depending on the particular situation involved.

The Reynolds transport theorem can also be useful in areas other than fluid mechanics. The main ingredients needed for its use are a system, a control volume, a parameter of interest, and some type of flux or flow involved in a process.

EXAMPLE 4.10

Cars are driven into and out of a certain neighborhood of a city as is indicated in Fig. E4.10a. Use the Reynolds transport theorem to discuss the time rate of change of the number of cars within this neighborhood.

SOLUTION

We consider the neighborhood to be a control volume (a two-dimensional control volume since it has no height) and let N represent the number of cars. Thus, N_{sys} is the number of cars in a system of cars (a collection of cars that moves about and coincides with the cars in the neighborhood at some initial time t_0), and N_{cv} is the number of cars in the neighborhood. When $t = t_0$, $N_{\text{sys}} = N_{\text{cv}}$. According to the Reynolds transport theorem (Eq. 4.15 for this case), the relationship between the number of cars in the system and the number of cars in the neighborhood is

$$\frac{DN_{\text{sys}}}{Dt} = \frac{\partial N_{\text{cv}}}{\partial t} + \dot{N}_{\text{out}} - \dot{N}_{\text{in}} \tag{1}$$

where $\dot{N}_{\text{out}}$ and $\dot{N}_{\text{in}}$ represent the net rate at which cars leave and enter the neighborhood (cars per hour). There may be many streets crossing the control surface along which cars travel into and out of the control volume. Thus, the total efflux and influx of cars is given by

$$\dot{N}_{\text{out}} = \sum_i \dot{N}_{i_{\text{out}}} \quad \text{and} \quad \dot{N}_{\text{in}} = \sum_i \dot{N}_{i_{\text{in}}} \tag{2}$$

where $\dot{N}_i$ represents the rate of cars along the ith street crossing the boundary of the neighborhood.

Similarly, the number of cars within the control volume can be written in terms of the density of the cars throughout the neighborhood, $n = n(x, y, t)$. The value of n, the number of cars

(a)

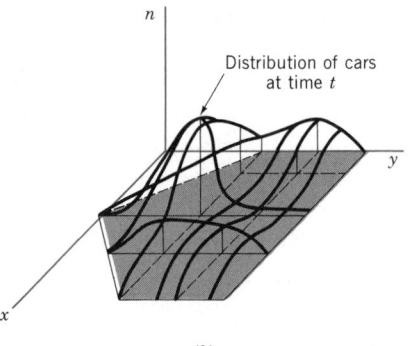

(b)

FIGURE E4.10

per acre, varies with location (there are fewer cars per acre in front yards than there are in parking lots) and with time. Such a distribution may look as is shown in Fig. E4.10b. We can add up the number of cars in each portion of the control volume to obtain the total number of cars in the control volume at any time

$$N_{cv} = \int_{cv} n \, dA$$

The parameters N and n coincide with the extensive and intensive parameters B and b in the usual notation. Thus, the time rate of change of the number of cars in the neighborhood is

$$\frac{\partial N_{cv}}{\partial t} = \frac{\partial}{\partial t} \int_{cv} n \, dA \tag{3}$$

By combining Eqs. 1, 2, and 3 we obtain

$$\frac{DN_{sys}}{Dt} = \frac{\partial}{\partial t} \int_{cv} n \, dA + \sum_i \dot{N}_{i_{out}} - \sum_i \dot{N}_{i_{in}} \tag{4}$$

Physically, this states that the rate at which the number of cars in the system changes with time is equal to the rate at which the number of cars in the neighborhood changes with time, plus the net rate at which cars cross the neighborhood boundary. This is the Reynolds transport theorem relationship for our situation.

To proceed further we must apply a basic law governing the system of cars. For example, we could assume that the number of cars in the system remains constant in time, $DN_{sys}/Dt = 0$. Equation 4 then provides the means to convert from this basic system-oriented law to the same physical situation translated into control volume ideas. Therefore, if N_{sys} = constant, Eq.

4 gives

$$\frac{\partial}{\partial t} \int_{cv} n \, dA = \sum_i \dot{N}_{i_{\text{in}}} - \sum_i \dot{N}_{i_{\text{out}}}$$

The physical interpretation is that if cars are neither created nor destroyed ($DN_{\text{sys}}/Dt = 0$), then the rate of increase of cars within a given neighborhood equals the difference between the rate at which they are driven into the neighborhood and the rate that they are driven out. If the basic law governing the number of cars in a system is different than what we have assumed (i.e., if $N_{\text{sys}} \neq$ constant), we can provide the new law (a relationship describing the rate at which cars in the system are created or destroyed) and use the Reynolds transport theorem, Eq. 4, with a nonzero left-hand side.

4.4.3   Relationship to Material Derivative

In Section 4.2.1 we discussed the concept of the material derivative $D(\ )/Dt = \partial(\ )/\partial t + \mathbf{V} \cdot \nabla(\ ) = \partial(\ )/\partial t + u \, \partial(\ )/\partial x + v \, \partial(\ )/\partial y + w \, \partial(\ )/\partial z$. The physical interpretation of this derivative is that it provides the time rate of change of a fluid property (temperature, velocity, etc.) associated with a particular fluid particle as it flows. The value of that parameter for that particle may change because of unsteady effects [the $\partial(\ )/\partial t$ term] or because of effects associated with the particle's motion [the $\mathbf{V} \cdot \nabla(\ )$ term].

Careful consideration of Eq. 4.19 indicates the same type of physical interpretation for the Reynolds transport theorem. The term involving the time derivative of the control volume integral represents unsteady effects associated with the fact that values of the parameter within the control volume may change with time. For steady flow this effect vanishes—fluid flows through the control volume but the amount of any property, B, within the control volume is constant in time. The term involving the control surface integral represents the convective effects associated with the flow of the system across the fixed control surface. The sum of these two terms gives the rate of change of the parameter B for the system. This corresponds to the interpretation of the material derivative, $D(\ )/Dt = \partial(\ )/\partial t + \mathbf{V} \cdot \nabla(\ )$, in which the sum of the unsteady effect and the convective effect gives the rate of change of a parameter for a fluid particle. As is discussed in Section 4.2, the material derivative operator may be applied to scalars (such as temperature) or vectors (such as velocity). This is also true for the Reynolds transport theorem. The particular parameters of interest, B and b, may be scalars or vectors.

Thus, both the material derivative and the Reynolds transport theorem equations represent ways to transfer from the Lagrangian viewpoint (follow a particle or follow a system) to the Eulerian viewpoint (observe the fluid at a given location in space or observe what happens in the fixed control volume). The material derivative (Eq. 4.5) is essentially the infinitesimal (or derivative) equivalent of the finite size (or integral) Reynolds transport theorem (Eq. 4.19).

4.4.4   Steady Effects

Consider a steady flow [$\partial(\ )/\partial t \equiv 0$] so that Eq. 4.19 reduces to

$$\frac{DB_{\text{sys}}}{Dt} = \int_{cs} \rho b \mathbf{V} \cdot \hat{\mathbf{n}} \, dA \qquad (4.20)$$

In such cases if there is to be a change in the amount of B associated with the system (nonzero left-hand side), there must be a net difference in the rate that B flows into the control volume compared with the rate that it flows out of the control volume. That is, the integral of $\rho b \mathbf{V} \cdot \hat{\mathbf{n}}$ over the inflow portions of the control surface would not be equal and opposite to that over the outflow portions of the surface.

Consider steady flow through the "black box" control volume that is shown in Fig. 4.18. If the parameter B is the mass of the system, the left-hand side of Eq. 4.20 is zero (conservation of mass for the system as discussed in detail in Section 5.1). Hence, the flowrate of mass into the box must be the same as the flowrate of mass out of the box because the right-hand side of Eq. 4.20 represents the net flowrate through the control surface. On the other hand, assume the parameter B is the momentum of the system. The momentum of the system need not be constant. In fact, according to Newton's second law the time rate of change of the system momentum equals the net force, $\mathbf{F}$, acting on the system. In general, the left-hand side of Eq. 4.20 will therefore be nonzero. Thus, the right-hand side, which then represents the net flux of momentum across the control surface, will be nonzero. The flowrate of momentum into the control volume need not be the same as the flux of momentum from the control volume. We will investigate these concepts much more fully in Chapter 5. They are the basic principles describing the operation of such devices as jet or rocket engines.

For steady flows the amount of the property B within the control volume does not change with time. The amount of the property associated with the system may or may not change with time, depending on the particular property considered and the flow situation involved. The difference between that associated with the control volume and that associated with the system is determined by the rate at which B is carried across the control surface—the term $\int_{cs} \rho b \mathbf{V} \cdot \hat{\mathbf{n}} \, dA$.

4.4.5 Unsteady Effects

Consider unsteady flow $[\partial(\ )/\partial t \neq 0]$ so that all terms in Eq. 4.19 must be retained. When they are viewed from a control volume standpoint, the amount of parameter B within the system may change because the amount of B within the fixed control volume may change with time [the $\partial(\int_{cv} \rho b \, d\mathcal{V})/\partial t$ term] and because there may be a net non-zero flow of that parameter across the control surface (the $\int_{cs} \rho b \mathbf{V} \cdot \hat{\mathbf{n}} \, dA$ term).

For the special unsteady situations in which the rate of inflow of parameter B is exactly balanced by its rate of outflow, it follows that $\int_{cs} \rho b \mathbf{V} \cdot \hat{\mathbf{n}} \, dA = 0$, and Eq. 4.19 reduces to

$$\frac{DB_{\text{sys}}}{Dt} = \frac{\partial}{\partial t} \int_{cv} \rho b \, d\mathcal{V} \tag{4.21}$$

For such cases, any rate of change in the amount of B associated with the system is

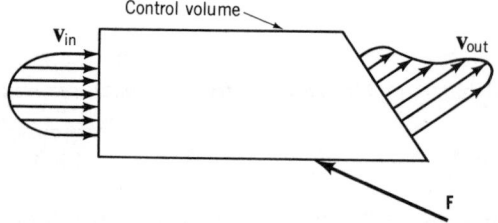

FIGURE 4.18 Steady flow through a control volume.

FIGURE 4.19 Unsteady flow through a constant diameter pipe.

------ Control surface

equal to the rate of change of B within the control volume. This can be illustrated by considering flow through a constant diameter pipe as is shown in Fig. 4.19. The control volume is as shown, and the system is the fluid within this volume at time t_0. We assume the flow is one-dimensional with $\mathbf{V} = V_0\hat{\mathbf{i}}$ and that $V_0(t)$ is a function of time. At any instant in time, all particles in the system have the same velocity. We let $\mathbf{B}$ = system momentum = $m\mathbf{V}$ = $mV_0\hat{\mathbf{i}}$, where m is the system mass, so that $\mathbf{b} = \mathbf{B}/m = \mathbf{V} = V_0\hat{\mathbf{i}}$, the fluid velocity. The magnitude of the momentum efflux across the outlet [section (2)] is the same as the magnitude of the momentum influx across the inlet [section (1)]. However, the sign of the efflux is opposite to that of the influx since $\mathbf{V} \cdot \hat{\mathbf{n}} > 0$ for the outflow and $\mathbf{V} \cdot \hat{\mathbf{n}} < 0$ for the inflow. Note that $\mathbf{V} \cdot \hat{\mathbf{n}} = 0$ along the sides of the control volume. Thus, with $\mathbf{V} \cdot \hat{\mathbf{n}} = -V_0$ on section (1), $\mathbf{V} \cdot \hat{\mathbf{n}} = V_0$ on section (2), and $A_1 = A_2$ we obtain

$$\int_{cs} \rho b \mathbf{V} \cdot \hat{\mathbf{n}}\, dA = \int_{cs} \rho(V_0\hat{\mathbf{i}})(\mathbf{V} \cdot \hat{\mathbf{n}})\, dA$$
$$= \int_{(1)} \rho(V_0\hat{\mathbf{i}})(-V_0)\, dA + \int_{(2)} \rho(V_0\hat{\mathbf{i}})(V_0)\, dA$$
$$= -\rho V_0^2 A_1 \hat{\mathbf{i}} + \rho V_0^2 A_2 \hat{\mathbf{i}} = 0$$

It is seen that for this special case Eq. 4.21 is valid. The rate at which the momentum of the system changes with time is the same as the rate of change of momentum within the control volume. If V_0 is constant in time, there is no rate of change of momentum of the system and for this special case each of the terms in the Reynolds transport theorem is zero by itself.

Consider the flow through a variable area pipe shown in Fig. 4.20. In such cases the fluid velocity is not the same at section (1) as it is at (2). Hence, the efflux of momentum from the control volume is not equal to the influx of momentum, so that the convective term in Eq. 4.20 [the integral of $\rho\mathbf{V}(\mathbf{V} \cdot \hat{\mathbf{n}})$ over the control surface] is not zero. These topics will be discussed in considerably more detail in Chapter 5.

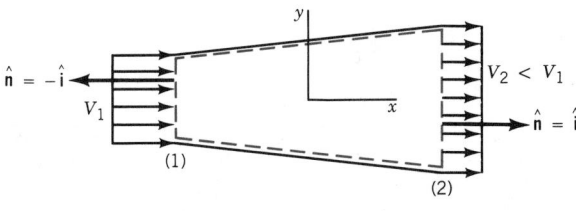

FIGURE 4.20 Flow through a variable area pipe.

------ Control surface

4.4.6 Moving Control Volumes

For most problems in fluid mechanics, the control volume may be considered as a fixed volume through which the fluid flows. There are, however, situations for which the analysis is simplified if the control volume is allowed to move or deform. The most general situation would involve a control volume that moves, accelerates, and deforms. As one might expect, the use of these control volumes can become fairly complex.

A number of important problems can be most easily analyzed by using a nonde-forming control volume that moves with a constant velocity. Such an example is shown in Fig. 4.21 in which a stream of water with velocity $\mathbf{V}_1$ strikes a vane that is moving with constant velocity $\mathbf{V}_0$. It may be of interest to determine the force, $\mathbf{F}$, that the water puts on the vane. Such problems frequently occur in turbines where a stream of fluid (water or steam, for example) strikes a series of blades that move past the nozzle. To analyze such problems it is advantageous to use a moving control volume. We will obtain the Reynolds transport theorem for such control volumes.

We consider a control volume that moves with a constant velocity as is shown in Fig. 4.22. The shape, size, and orientation of the control volume do not change with time. The control volume merely translates with a constant velocity, $\mathbf{V}_{cv}$, as shown. In general, the velocity of the control volume and the fluid are not the same, so that there is a flow of fluid through the moving control volume just as in the stationary control volume cases discussed in Section 4.4.2. The main difference between the fixed and the moving control volume cases is that it is the *relative velocity*, $\mathbf{V}_r$, that carries fluid across the moving control surface, whereas it is the *absolute velocity*, $\mathbf{V}$, that carries the fluid across the fixed control surface. The relative velocity is the fluid velocity relative to the moving control volume—the fluid velocity seen by an observer riding along on the control volume. The absolute velocity is the fluid velocity as seen by a stationary observer in a fixed coordinate system.

The difference between the absolute and relative velocities is the velocity of the control volume, $\mathbf{V}_{cv} = \mathbf{V} - \mathbf{V}_r$, or

$$\mathbf{V} = \mathbf{V}_r + \mathbf{V}_{cv} \tag{4.22}$$

Since the velocity is a vector, we must use vector addition as is shown in Fig. 4.23 to obtain the relative velocity if we know the absolute velocity and the velocity of the control volume. Thus, if the water leaves the nozzle in Fig. 4.21 with a velocity of $\mathbf{V}_1 = 100\hat{\mathbf{i}}$ ft/s and the vane has a velocity of $\mathbf{V}_0 = 20\hat{\mathbf{i}}$ ft/s (the same as the control volume), it appears to an observer riding on the vane that the water approaches the vane with a velocity of $\mathbf{V}_r = \mathbf{V} - \mathbf{V}_{cv} = 80\hat{\mathbf{i}}$ ft/s. In general, the absolute velocity,

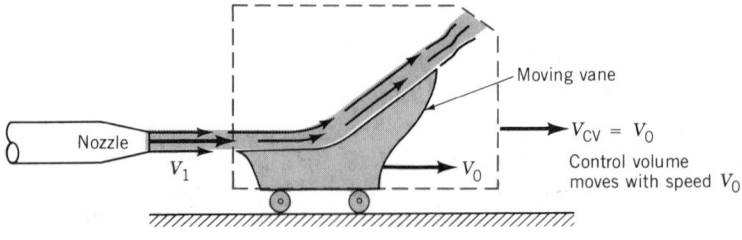

FIGURE 4.21 Example of a moving control volume.

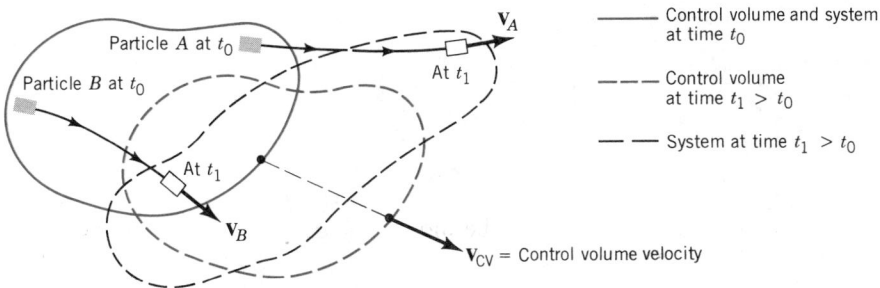

FIGURE 4.22 Typical moving control volume and system.

$\mathbf{V}$, and the control volume velocity, $\mathbf{V}_{cv}$, will not be in the same direction so that the relative and absolute velocities will have different directions (see Fig. 4.23).

The Reynolds transport theorem for a moving, nondeforming control volume can be derived in the same manner that it was obtained for a fixed control volume. As is indicated in Fig. 4.24, the only difference that needs be considered is the fact that relative to the moving control volume the fluid velocity observed is the relative velocity, not the absolute velocity. An observer fixed to the moving control volume may or may not even know that he or she is moving relative to some fixed coordinate system. If we follow the derivation that led to Eq. 4.19 (the Reynolds transport theorem for a fixed control volume), we note that the corresponding result for a moving control volume can be obtained by simply replacing the absolute velocity, $\mathbf{V}$, in that equation by the relative velocity, $\mathbf{V}_r$. Thus, the Reynolds transport theorem for a control volume moving with constant velocity is given by

$$\frac{DB_{\text{sys}}}{Dt} = \frac{\partial}{\partial t} \int_{cv} \rho b \, d\mathcal{V} + \int_{cs} \rho b \, \mathbf{V}_r \cdot \hat{\mathbf{n}} \, dA \tag{4.23}$$

where the relative velocity is given by Eq. 4.22.

4.4.7 Selection of a Control Volume

Any volume in space can be considered as a control volume. It may be of finite size or it may be infinitesimal in size, depending on the type of analysis to be carried out. In most of our cases, the control volume will be a fixed, nondeforming volume. In some situations we will consider control volumes that move with constant velocity. In either case it is important that considerable thought go into the selection of the specific control volume to be used.

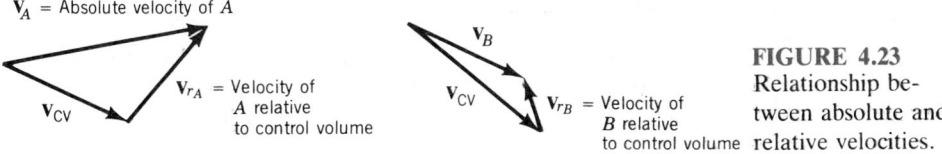

FIGURE 4.23 Relationship between absolute and relative velocities.

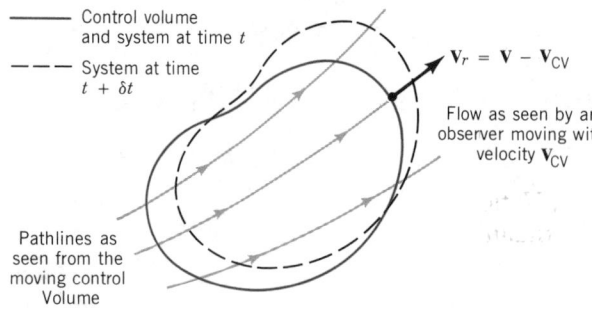

FIGURE 4.24 Control volume and system as seen by an observer moving with the control volume.

The selection of an appropriate control volume in fluid mechanics is very similar to the selection of an appropriate free-body diagram in dynamics or statics. In dynamics we select the body in which we are interested, represent the object in a free-body diagram, and then apply the appropriate governing laws to that body. The ease of solving a given dynamics problem is often very dependent upon the specific free-body diagram drawn. Similarly, the ease of solving a given fluid mechanics problem is often very dependent upon the choice of the control volume used.

In later chapters we will often use the Reynolds transport theorem, Eq. 4.19, in solving problems associated with finite sized control volumes. Use of this equation can often be simplified by the thoughtful selection of the control volume. Only by practice can we develop skill at selecting the "best" control volume. None are "wrong," but some are "much better" than others.

Solution of a typical problem will involve determining parameters such as velocity, pressure, and force at some point in the flow field. It is usually best to ensure that this point is located on the control surface, not "buried" within the control volume. The unknown will then appear in the convective term (the surface integral) of the Reynolds transport theorem. If possible, the control surface should be normal to the fluid velocity so that the angle θ ($\mathbf{V} \cdot \hat{\mathbf{n}} = V \cos \theta$) in the flux terms of Eq. 4.19 will be 0 or 180°. This will usually simplify the solution process.

Figure 4.25 illustrates three possible control volumes associated with flow through a pipe. If the problem is to determine the pressure at point (1), the selection of the control volume (a) is better than that of (b) because point (1) lies on the control surface. Similarly, control volume (a) is better than (c) because the flow is normal to the inlet and exit portions of the control volume. Neither control volume is wrong—(a) will be easier to use. Proper control volume selection will become much clearer in Chapter 5 where the Reynolds transport theorem is used to transform the governing equations from the system formulation into the control volume formulation and numerous examples using control volume ideas are discussed.

(a)

(b)

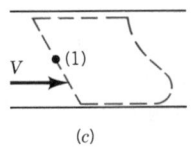
(c)

FIGURE 4.25 Various control volumes for flow through a pipe.

REFERENCES

1. Streeter, V. L., and Wylie, E. B., *Fluid Mechanics, Eighth Edition*, McGraw-Hill, New York, 1985.
2. Goldstein, R. J., *Fluid Mechanics Measurements*, Hemisphere, New York, 1983.
3. Kline, S. J., *Flow Visualization*, National Committee for Fluid Mechanics Films, distributed by Encyclopaedia Britannica Educational Corp., Chicago, 1972.
4. Magarvey, R. H., and MacLatchy, C. S., Canadian Journal of Physics, Vol. 42, 1964.

PROBLEMS

Note: Unless otherwise indicated use the values of fluid properties found in the tables on the inside of the front cover. Problems designated with an (*) are intended to be solved with the aid of a programmable calculator or a computer.

4.1 The velocity field of a flow is given by $V = (3y + 2)\hat{i} + (x - 4)\hat{j} + 5z\hat{k}$ ft/s, where x, y, and z are in feet. Determine the fluid speed at the origin ($x = y = z = 0$) and on the y axis ($x = z = 0$).

4.2 The velocity field of a flow is given by $V = 2x^2t\hat{i} + [4y(t - 1) + 2x^2t]\hat{j}$ m/s, where x and y are in meters and t is in seconds. For fluid particles on the x axis, determine the speed and direction of flow.

4.3 The velocity field of a flow is given by $V = 10y/(x^2 + y^2)^{1/2}\hat{i} - 10x/(x^2 + y^2)^{1/2}\hat{j}$ ft/s, where x and y are in feet. Determine the fluid speed at points along the x axis; along the y axis. What is the angle between the velocity vector and the x axis at points $(x, y) = (5, 0)$, $(5, 5)$, and $(0, 5)$?

4.4 A balloon rises through the atmosphere with a vertical velocity component of $w = 2e^{-t/10,000}$ m/s, where t is in seconds. At the same time the wind blows horizontally with a constant speed of 3 m/s. Determine the speed of the balloon relative to the ground when $t = 0$, 5000, and ∞. At what time is the balloon's speed a maximum; a minimum?

4.5 The x and y components of velocity for a two-dimensional flow are $u = 6y$ ft/s and $v = 3$ ft/s, where y is in feet. Determine the equation for the streamlines and sketch representative streamlines in the upper half plane.

4.6 A velocity field is given by $V = x\hat{i} + x(x - 1)(y + 1)\hat{j}$, where u and v are in ft/s and x and y are in feet. Plot the streamline that passes through $x = 0$ and $y = 0$. Compare this streamline with the streakline through the origin.

4.7 A velocity field is given by $u = y - 1$ and $v = y - 2$, where u and v are in m/s and x and y are in meters. Plot the streamline that passes through point $(x, y) = (2, 3)$. Compare this streamline with the streakline through the point $(x, y) = (2, 3)$.

4.8 Show that the streamlines for a flow whose velocity components are $u = c(x^2 - y^2)$ and $v = -2cxy$, where c is a constant, are given by the equation $x^2y - y^3/3 = $ constant. At which point (points) is the flow parallel to the x axis? At which point (points) is the fluid stationary?

4.9 The velocity field of a flow is given by $u = -V_0y/(x^2 + y^2)^{1/2}$ and $v = V_0x/(x^2 + y^2)^{1/2}$, where V_0 is a constant. Where in the flow field is the speed equal to V_0? Determine equation of the streamlines and discuss the various characteristics of this flow.

4.10 A flow in the x–y plane is given by the following velocity field: $u = 3$ and $v = 6$ m/s for $0 < t < 20$ s; $u = -4$ and $v = 0$ m/s for $20 < t < 40$ s. Dye is released at the origin ($x = y = 0$) for $t = 0$. **(a)** Draw the pathline at $t = 30$ s for two particles that were released from the origin—one released at $t = 0$ and the other released at $t = 20$ s. **(b)** On the same graph draw the streamlines at times $t = 10$ s and $t = 30$ s.

4.11 From time $t = 0$ to $t = 5$ hr radioactive steam is released from a nuclear power plant accident located at $x = -1$ mile and $y = 3$ miles.

The following wind conditions are expected: $\mathbf{V} = 10\hat{\mathbf{i}} - 5\hat{\mathbf{j}}$ mph for $0 < t < 3$ hr, $\mathbf{V} = 15\hat{\mathbf{i}} + 8\hat{\mathbf{j}}$ mph for $3 < t < 10$ hr, and $\mathbf{V} = 5\hat{\mathbf{i}}$ mph for $t > 10$ hr. Draw to scale the expected streakline of the steam for $t = 3$, 10, and 15 hr.

4.12   A weather balloon is released from point 0 at time $t = 0$ as shown in Fig. P4.12 and travels upward with a vertical component of velocity of 8 ft/s. From ground level to 2000-ft altitude the wind velocity is $\mathbf{V} = 3\hat{\mathbf{i}} + 2\hat{\mathbf{j}}$ ft/s, while from 2000- to 10,000-ft altitude the wind velocity is $\mathbf{V} = 5\hat{\mathbf{j}}$ ft/s. Above 10,000 ft the velocity is $\mathbf{V} = -5\hat{\mathbf{i}} + 10\hat{\mathbf{j}}$ ft/s. Plot the projection in the x–y plane of the balloon's path for $0 \le t \le 25$ min.

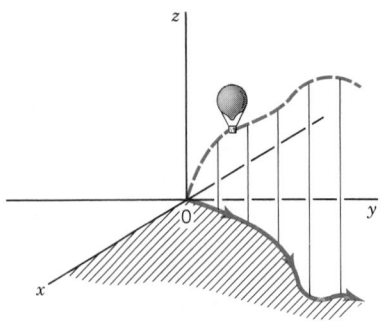

FIGURE P4.12

4.13   Repeat Problem 4.12 if the vertical component of velocity of the balloon is not constant but is given by $w = w_0 e^{-at}$, where $w_0 = 8$ ft/s and $a = 4 \times 10^{-4}$/s. What is the maximum altitude reached by the balloon?

4.14   In addition to the customary horizontal velocity components of the air in the atmosphere (the "wind"), there often are vertical air currents (thermals) caused by buoyant effects due to uneven heating of the air as indicated in Fig. P4.14. Assume that the velocity field in a certain region is approximated by $u = u_0$, $v = v_0 (1 - y/h)$ for $0 < y < h$, and $u = u_0$, $v = 0$ for $y > h$. Plot

FIGURE P4.14

the shape of the streamline that passes through the origin for values of $u_0/v_0 = 0.5$, 1, and 2.

4.15*   Repeat Problem 4.14 using the same information except that $u = u_0 y/h$ for $0 \le y \le h$ rather than $u = u_0$. Use values of $u_0/v_0 = 0$, 0.1, 0.2, 0.4, 0.6, 0.8, and 1.0.

4.16   A velocity field is given by $u = cx^2$ and $v = cy^2$, where c is a constant. Determine the x and y components of the acceleration. At what point (points) in the flow field is the acceleration zero?

4.17   A three-dimensional velocity field is given by $u = x$, $v = -y$, and $w = z$. Determine the acceleration vector.

4.18   Water flows in a pipe so that its velocity triples every 20 s. At $t = 0$ it has $u = 5$ ft/s. That is, $\mathbf{V} = u(t)\hat{\mathbf{i}} = 5 \, (3^{t/20})\hat{\mathbf{i}}$ ft/s. Determine the accleration when $t = 0$, 10, and 20 s.

4.19   The velocity of the water in the pipe shown in Fig. P4.19 is given by $V_1 = 0.50t$ m/s and $V_2 = 1.0t$ m/s, where t is in seconds. Determine the local acceleration at points (1) and (2). Is the average convective acceleration between these two points negative, zero, or positive? Explain.

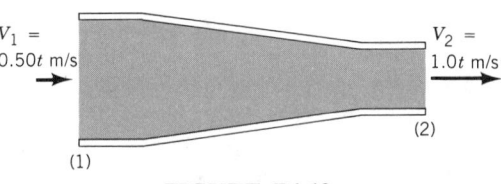

FIGURE P4.19

4.20   Water flows through a constant diameter pipe with a uniform velocity given by $\mathbf{V} = (10/t + 5)\hat{\mathbf{j}}$ m/s, where t is in seconds. Determine the acceleration at time $t = 1$, 2, and 10 s.

4.21   Air flows through a pipe with a uniform velocity of $\mathbf{V} = 5 \, t^2\hat{\mathbf{i}}$ ft/s, where t is in seconds. Determine the acceleration at time $t = -1$, 0, and 1 s.

4.22*   Water flows through a pipe with $\mathbf{V} = u(t)\hat{\mathbf{i}}$ where the approximate measured values of $u(t)$ are shown in the table. Plot the acceleration as a function of time for $0 \le t \le 20$ s. Plot the acceleration as a function of time if all of the values of $u(t)$ are increased by a factor of 2; by a factor of 5.

t (s)	u (ft/s)	t (s)	u (ft/s)
0	0	11.2	8.1
1.8	1.7	12.3	8.4
3.1	3.2	13.9	8.3
4.0	3.8	15.0	8.1
5.5	4.6	16.4	7.9
6.9	5.8	17.5	7.0
8.1	6.3	18.4	6.6
10.0	7.1	20.0	5.7

4.23 The fluid velocity along the x axis shown in Fig. P4.23 changes from 12 m/s at point A to 36 m/s at point B. It is also known that the velocity is a linear function of distance along the streamline. Determine the acceleration at points A, B, and C. Assume steady flow.

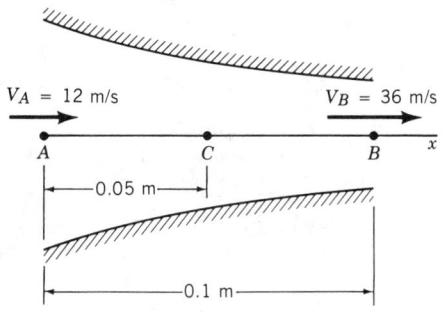

FIGURE P4.23

4.24 When a fluid flows into a round pipe as shown in Fig. P4.24 viscous effects may cause the velocity profile to change from a uniform profile ($\mathbf{V} = V_0\hat{\mathbf{i}}$) at the entrance of the pipe to a parabolic profile $\{\mathbf{V} = 2V_0[1 - (r/R)^2]\hat{\mathbf{i}}\}$ at the end of the entrance region ($x = \ell$). Velocity profiles for various values of x are as indicated in the figure. Use this graph to show that a fluid particle moving along the centerline ($r = 0$) experiences an acceleration, but a particle close to the edge of the pipe ($r \approx R$) experiences a deceleration. Does a particle traveling along the line $r = 3R/4$ experience an acceleration or deceleration, or both? Explain.

(a)

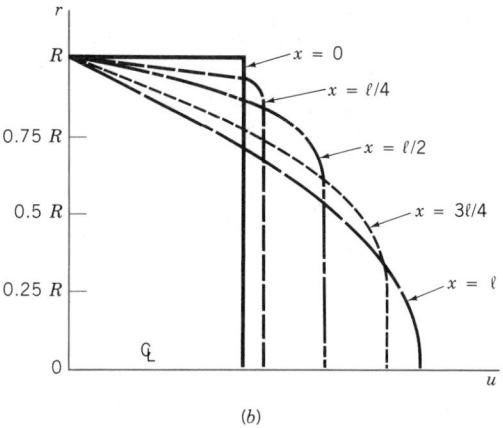

(b)

FIGURE P4.24

4.25 As a valve is opened, water flows through the diffuser shown in Fig. P4.25 at an increasing flowrate so that the velocity along the centerline is given by $\mathbf{V} = u\hat{\mathbf{i}} = V_0(1 - e^{-ct})(1 - x/\ell)\hat{\mathbf{i}}$, where u_0, c, and ℓ are constants. Determine the acceleration as a function of x and t. If $V_0 = 10$ ft/s and $\ell = 5$ ft, what value of c (other than $c = 0$) is needed to make the acceleration zero for any x at $t = 2$ s? Explain how the acceleration can be zero if the flowrate is increasing with time.

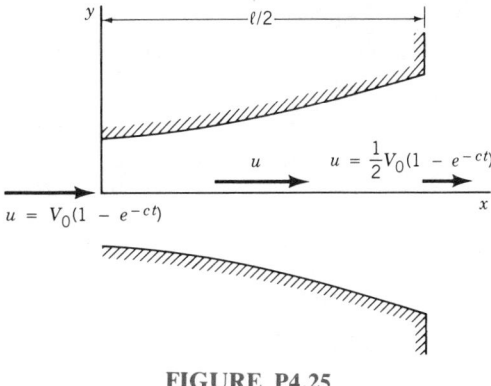

FIGURE P4.25

4.26 A fluid flows along the x axis with a velocity given by $\mathbf{V} = (x/t)\hat{\mathbf{i}}$, where x is in feet and t in seconds. **(a)** Plot the speed for $0 \le x \le 10$ ft and $t = 3$ s. **(b)** Plot the speed for $x = 5$ ft and $2 \le t \le 4$ s. **(c)** Determine the local and convective acceleration. **(d)** Show that the acceleration of any fluid particle in the flow is zero. **(e)** Explain physically how the velocity of a particle in this unsteady flow remains constant throughout its motion.

4.27 A shock wave is a very thin layer (thickness $= \ell$) in a high-speed (supersonic) gas flow across which the flow properties (velocity, density, pressure, etc.) change from state (1) to state (2) as shown in Fig. P4.27. If $V_1 = 2000$ fps, $V_2 = 750$ fps, and $\ell = 10^{-4}$ in., estimate the average deceleration of the gas as it flows across the shock wave. How many g's deceleration does this represent?

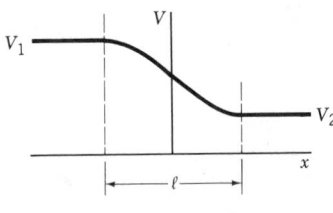

FIGURE P4.27

4.28 A hydraulic jump is a rather sudden change in depth of a liquid layer as it flows in an open channel as shown in Fig. P4.28. In a relatively short distance (thickness $= \ell$) the liquid depth changes from z_1 to z_2, with a corresponding change in velocity from V_1 to V_2. If $V_1 = 5$ m/s, $V_2 = 1$ m/s, and $\ell = 0.3$ m, estimate the average deceleration of the liquid as it flows across the hydraulic jump. How many g's deceleration does this represent?

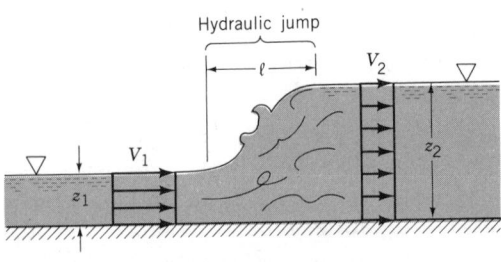

FIGURE P4.28

4.29 In the conical nozzle shown in Fig. P4.29 the streamlines are essentially radial lines emanating from point A and the fluid velocity is given

approximately by $V = C/r^2$, where C is a constant. The fluid velocity is 2 m/s along the centerline at the beginning of the nozzle ($x = 0$). Determine the acceleration along the nozzle centerline as a function of x. What is the value of the acceleration at $x = 0$ and $x = 0.3$ m?

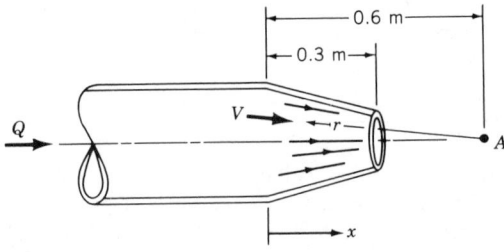

FIGURE P4.29

4.30 A nozzle is designed to accelerate the fluid from V_1 to V_2 in a linear fashion. That is, $V = ax + b$, where a and b are constants. If the flow is constant with $V_1 = 5$ m/s at $x_1 = 0$ and $V_2 = 15$ m/s at $x_2 = 1$ m, determine the local acceleration, the convective acceleration, and the acceleration of the fluid at points (1) and (2).

4.31 Repeat Problem 4.30 if $V = ax^2 + b$, where a and b are constants.

4.32 Repeat Problem 4.30 with the assumption that the flow is not steady, but at the time when $V_1 = 5$ m/s and $V_2 = 15$ m/s, it is known that $\partial V_1/\partial t = 20$ m/s^2 and $\partial V_2/\partial t = 60$ m/s^2.

4.33 Air flows along the vertical streamline shown in Fig. P4.33. The temperature and location of two particles, A and B, along this streamline are listed in the table. Estimate the time rate of change of temperature of the particle, DT/Dt, from estimated values of $\partial T/\partial t$ and $\partial T/\partial y$. Compare your result with the estimate of the rate at which the temperature of particle A changes with time by considering T_A at the two times indicated.

Particle	Time (s)	Location, y (ft)	T (°F)
A	0	0	80
A	1	1	85
B	1	0	82

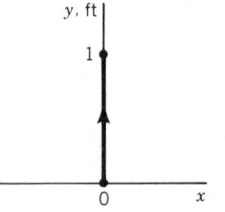

FIGURE P4.33

4.34* Air flows steadily through a variable area pipe with a velocity of $\mathbf{V} = u(x)\hat{\mathbf{i}}$ ft/s, where the approximate measured values of $u(x)$ are given in the table. Plot the acceleration as a function of x for $0 \le x \le 12$ in. Plot the acceleration if the flowrate is increased by a factor of N (i.e., the values of u are increased by a factor of N), for $N = 2, 4, 10$.

x (in.)	u (ft/s)	x (in.)	u (ft/s)
0	10.0	7	20.1
1	10.2	8	17.4
2	13.0	9	13.5
3	20.1	10	11.9
4	28.3	11	10.3
5	28.4	12	10.0
6	25.8	13	10.0

4.35 Assume the temperature of the exhaust in an exhaust pipe can be approximated by $T = T_0(1 + ae^{-bx})[1 + c\cos(\omega t)]$, where $T_0 = 100\,°C$, $a = 3$, $b = 0.03\,m^{-1}$, $c = 0.05$, and $\omega = 100$ rad/s. If the exhaust speed is a constant 3 m/s, determine the time rate of change of temperature of the fluid particles at $x = 0$ and $x = 4$ m when $t = 0$.

4.36* As is indicated in Fig. P4.36, the speed of exhaust in a car's exhaust pipe varies in time and distance because of the periodic nature of the engine's operation and the damping effect with distance from the engine. Assume that the speed is given by $V = V_0[1 + ae^{-bx}\sin(\omega t)]$, where $V_0 = 8$ fps, $a = 0.05$, $b = 0.2\,ft^{-1}$, and $\omega = 50$ rad/s. Calculate and plot the fluid acceleration at $x = 0, 1, 2, 3, 4$, and 5 ft for $0 \le t \le \pi/25$ s.

FIGURE P4.36

4.37 The pressure in the pipe near the discharge of a reciprocating pump fluctuates according to $p = [200 + 40\sin(8t)]$ kPa, where t is in seconds. If the fluid speed in the pipe is 5 m/s, determine the maximum rate of change of pressure experienced by a fluid particle.

4.38 Water flows over the crest of a dam with speed V as shown in Fig. P4.38. Determine the speed if the magnitude of the normal acceleration at point (1) is to equal the acceleration of gravity, g.

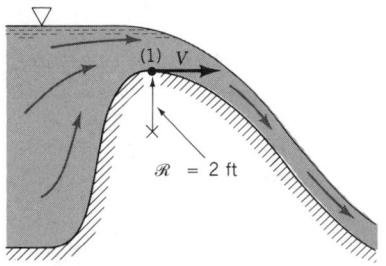

FIGURE P4.38

4.39 Water flows under the sluice gate shown in Fig. P4.39. If $V_1 = 3$ m/s, what is the normal acceleration at point (1)?

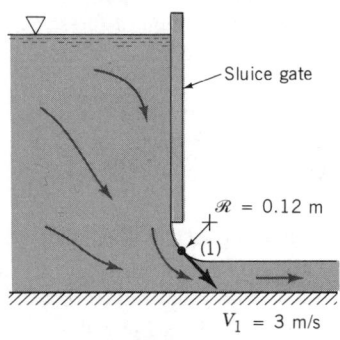

FIGURE P4.39

4.40 Water flows down the face of the dam shown in Fig. P4.40. The face of the dam consists of two circular arcs with radii of 10 and 20 ft as shown. If the speed of the water along streamline A–B is approximately $V = (2gh)^{1/2}$, where the distance h is as indicated, plot the normal acceleration as a function of distance along the streamline, $a_n = a_n(s)$.

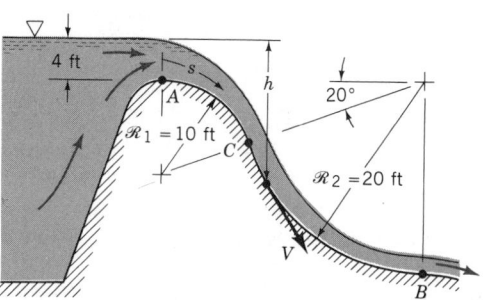

FIGURE P4.40

4.41 For the flow given in Problem 4.40, plot the streamwise acceleration, a_s, as a function of distance, s, along the surface of the dam from A to C.

4.42 A fluid flows past a sphere with an upstream velocity of $V_0 = 30$ m/s as shown in Fig. P4.42. From a more advanced theory it is found that the speed of the fluid along the front part of the sphere is $V = \frac{3}{2}V_0 \sin \theta$. Determine the streamwise and normal components of acceleration at point A if the radius of the sphere is $a = 0.20$ m.

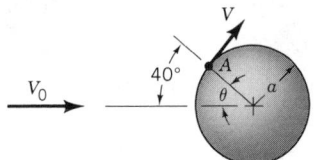

FIGURE P4.42

4.43* For flow past a sphere as discussed in Problem 4.42, plot a graph of the streamwise acceleration, a_s, the normal acceleration, a_n, and the magnitude of the acceleration as a function of θ for $0 \le \theta \le 90°$ with $V_0 = 50$ ft/s and $a = 0.1$, 1.0, and 10 ft. Repeat for $V_0 = 5$ ft/s. At what point is the acceleration a maximum; a minimum?

4.44 A fluid flows past a circular cylinder of radius a with an upstream speed of V_0 as shown in Fig. P4.44. A more advanced theory indicates that if viscous effects are negligible, the velocity of the fluid along the surface of the cylinder is given by $V = 2V_0 \sin \theta$. Determine the streamline and normal components of acceleration on the surface of the cylinder as a function of V_0, a, and θ.

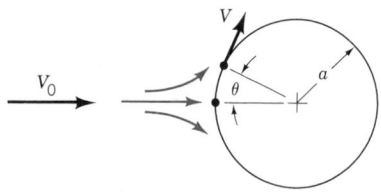

FIGURE P4.44

4.45* Use the results of Problem 4.44 to plot graphs of a_s and a_n for $0 \le \theta \le 90°$ with $V_0 = 10$ m/s and $a = 0.01$, 0.10, 1.0, and 10.0 m.

4.46 Determine the x and y components of acceleration for the flow given in Problem 4.8. If $c > 0$, is the particle at point $x = x_0 > 0$ and $y = 0$ accelerating or decelerating? Explain. Repeat if $x_0 < 0$.

4.47 A fluid flows steadily along the streamline as shown in Fig. P4.47. Determine the acceleration at point A. At point A what is the angle between the acceleration and the x axis? At point A what is the angle between the acceleration and the streamline?

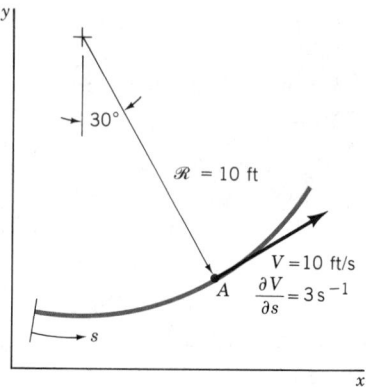

FIGURE P4.47

4.48 Water flows steadily through the funnel shown in Fig. P4.48. Throughout most of the funnel the flow is approximately radial (along rays from O) with a velocity of $V = c/r^2$, where r is the radial coordinate and c is a constant. If the velocity is 0.6 m/s when $r = 0.1$ m, determine the acceleration at points A and B.

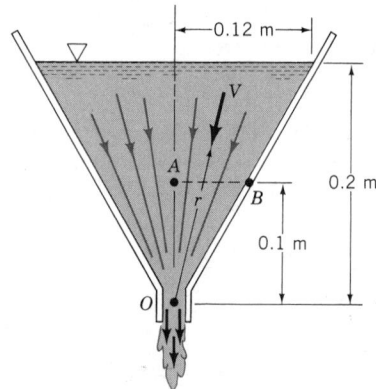

FIGURE P4.48

4.49 Water flows through the slit at the bottom of a two-dimensional water trough as shown in Fig. P4.49. Throughout most of the trough the flow is approximately radial (along rays from O) with a velocity of $V = c/r$, where r is the radial coordinate and c is a constant. If the velocity is 0.06 m/s when $r = 0.1$ m, determine the acceleration at points A and B.

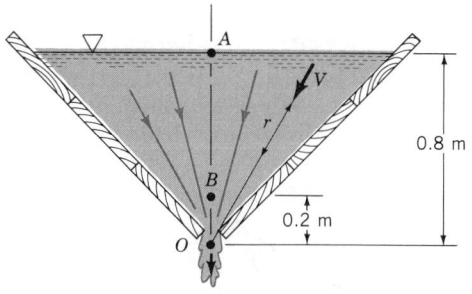

FIGURE P4.49

4.50 Air flows from a pipe into the region between two parallel circular disks as shown in Fig. P4.50. The fluid velocity in the gap between the disks is closely approximated by $V = V_0 R/r$, where R is the radius of the disk, r is the radial coordinate, and V_0 is the fluid velocity at the edge of the disk. Determine the acceleration for $r = 1, 2,$ or 3 ft if $V_0 = 5$ ft/s and $R = 3$ ft.

FIGURE P4.50

4.51 Air flows from a pipe into the region between a circular disk and a cone as shown in Fig. P4.51. The fluid velocity in the gap between the disk and the cone is closely approximated by $V = V_0 R^2/r^2$, where R is the radius of the disk, r is the radial coordinate, and V_0 is the fluid velocity at the edge of the disk. Determine the acceleration for $r = 1$ and 2 ft if $V_0 = 5$ ft/s and $R = 2$ ft.

FIGURE P4.51

4.52 Water flows through a duct of square cross section as shown in Fig. P4.52 with a constant, uniform velocity of $V = 20$ m/s. Consider fluid particles that lie along line A–B at time $t = 0$. Determine the position of these particles, denoted by line A'–B', when $t = 0.10$ s. Use the volume of fluid in the region between lines A–B and A'–B' to determine the flowrate in the duct. Repeat the problem for fluid particles originally along line C–D; along line E–F. Compare your three answers.

FIGURE P4.52

4.53 Repeat Problem 4.52 if the velocity profile is linear from 0 to 20 m/s across the duct as shown in Fig. P4.53.

FIGURE P4.53

4.54 In the region just downstream of a sluice gate, the water may develop a reverse flow region as is indicated in Fig. P4.54. The velocity profile is assumed to consist of two uniform regions, one with velocity $V_a = 10$ fps and the other with $V_b = 3$ fps. Determine the net flowrate of water across the portion of the control surface at section (2) if the channel is 20 ft wide.

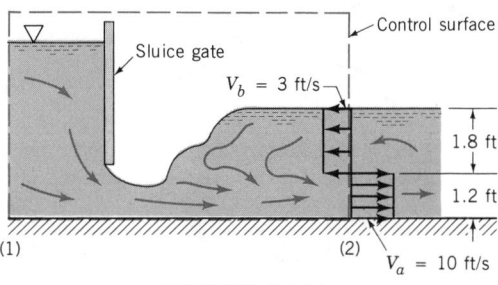

FIGURE P4.54

4.55 At time $t = 0$ the valve on an initially empty (perfect vacuum, $\rho = 0$) tank is opened and air rushes in. If the tank has a volume of V_0 and the density of air within the tank increases as $\rho = \rho_\infty(1 - e^{-bt})$, where b is a constant, determine the time rate of change of mass within the tank.

4.56 A plunger is pushed into a closed, air-filled pipe of diameter D with velocity V_p as shown in Fig. P4.56. The flow within the control volume consists of two distinct regions—one of zero velocity and constant density ρ_1; and the other of uniform velocity $\mathbf{V} = V_p\hat{\mathbf{i}}$ and density ρ_2. The two regions are separated by an interface (a shock wave) that travels to the right with constant speed, V_s (different than the fluid speed; $V_s > V_p$). Determine the mass within the control volume as a function of time from $t = 0$ (when the shock wave is at $x = 0$ as indicated) until the shock wave reaches the end of the pipe ($x = \ell$).

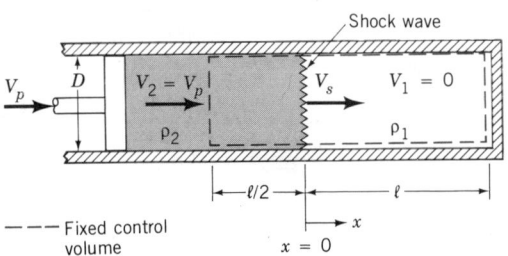

FIGURE P4.56

4.57 A sanding operation injects 10^4 particles/s into the air in a room as shown in Fig. P4.57. The amount of dust in the room is maintained at a constant level by a ventilating fan that draws clean air into the room at section (1) and expels dusty air at section (2). Consider a control volume whose surface is the interior surface of the room (excluding the sander) and a system consisting of the material within the control volume at time $t = 0$. **(a)** If N is the number of particles, discuss the physical meaning of and evaluate the terms DN_{sys}/Dt and $\partial N_{cv}/\partial t$. **(b)** Use the Reynolds transport theorem to determine the concentration of particles (particles/m³) in the exhaust air for steady state conditions.

FIGURE P4.57

4.58 Consider Problem 4.57 for the case that the ventilating fan fails to work so that $V_1 =$

$V_2 = 0$. Evaluate and discuss the meaning of the terms DN_{sys}/Dt and $\partial N_{cv}/\partial t$ for this case.

4.59 Air flows over a flat plate with a velocity profile given by $\mathbf{V} = u(y)\hat{\mathbf{i}}$, where $u = 2y$ ft/s for $0 \le y \le 0.5$ ft and $u = 1$ ft/s for $y > 0.5$ ft as shown in Fig. P4.59. The fixed rectangular control volume $ABCD$ coincides with the system at time $t = 0$. Make a sketch to indicate **(a)** the boundary of the system at time $t = 0.5$ s, **(b)** the fluid that moved out of the control volume in the interval $0 \le t \le 0.5$ s, and **(c)** the fluid that moved into the control volume during that time interval.

FIGURE P4.59

4.60 Water flows in the branching pipe shown in Fig. P4.60 with uniform velocity at each inlet and outlet. The fixed control volume indicated coincides with the system at time $t = 20$ s. Make a sketch to indicate **(a)** the boundary of the system at time $t = 20.1$ s, **(b)** the fluid that left the control volume during that 0.1-s interval, and **(c)** the fluid that entered the control volume during that time interval.

FIGURE P4.60

4.61 Two plates are pulled in opposite directions with speeds of 1.0 ft/s as shown in Fig. P4.61. The oil between the plates moves with a velocity given by $\mathbf{V} = 10\,y\hat{\mathbf{i}}$ ft/s, where y is in feet. The fixed control volume $ABCD$ coincides with

with the system at time $t = 0$. Make a sketch to indicate **(a)** the system at time $t = 0.2$ s and **(b)** the fluid that has entered and exited the control volume in that time period.

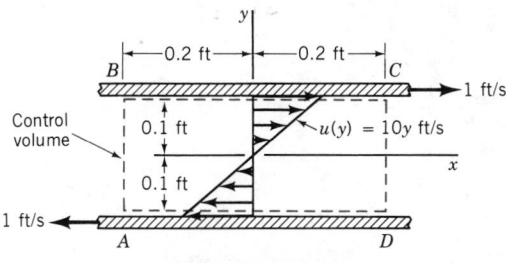

FIGURE P4.61

4.62 Water is squirted from a syringe with a speed of $V = 5$ m/s by pushing in the plunger with a speed of $V_p = 0.03$ m/s as shown in Fig. P4.62. The surface of the deforming control vol-

ume consists of the sides and end of the cylinder and the end of the plunger. The system consists of the water in the syringe at $t = 0$ when the plunger is at section (1) as shown. Make a sketch to indicate the control surface and the system when $t = 1$ s.

4.63 Water flows from a nozzle with a speed of $V = 10$ m/s and is collected in a container that moves toward the nozzle with a speed of $V_{cv} = 2$ m/s as shown in Fig. P4.63. The moving control surface consists of the inner surface of the container. The system consists of the water in the container at time $t = 0$ and the water between the nozzle and the tank in the constant diameter stream at $t = 0$. At time $t = 0.1$ s what volume of the system remains outside of the control volume? How much water has entered the control volume during this time period? Repeat the problem for $t = 0.3$ s.

FIGURE P4.62

FIGURE P4.63

A jet of water injected into stationary water: Upon emerging from the slit at the left, the jet of fluid loses some of its momentum to the surrounding fluid. This causes the jet to slow down and its width to increase (air bubbles in water). (Photograph courtesy of ONERA, France.)

FIVE

FINITE CONTROL VOLUME ANALYSIS

Many practical problems in fluid mechanics require analysis of the behavior of the contents of a finite region in space (a control volume). For example, we may be asked to calculate the anchoring force required to hold a jet engine in place during a test. Or, we could be called on to determine the amount of time to allow for complete filling of a large storage tank. An estimate of how much power it would take to move water from one location to another at a higher elevation and several miles away may be sought. As you will learn by studying the material in this chapter, these and many other important questions can be readily answered with finite control volume analyses. The bases of this analysis method are some fundamental principles of physics, namely, conservation of mass, Newton's second law of motion, and the first and second[1] laws of thermodynamics. Thus, as one might expect, the resultant techniques are powerful and applicable to a wide variety of fluid mechanical circumstances that require engineering judgment. Furthermore, the finite control volume formulas are easy to interpret physically and thus are not difficult to use.

The control volume formulas are derived from the equations representing basic laws applied to a collection of mass (a system). The system statements are probably more familiar to you presently. However, in fluid mechanics, the control volume or Eulerian view is generally less complicated and, therefore, more convenient to use than the system or Lagrangian view. The notion of a control volume and system occupying the same region of space at an instant (coincident condition) and the

---

[1] The section (Section 5.4) on the second law of thermodynamics may be omitted without loss of continuity in the text material.

Reynolds transport theorem (Eqs. 4.19 and 4.23) are key elements in the derivation of the control volume equations.

Integrals are used throughout the chapter for generality. Volume integrals can accommodate spatial variations of the material properties of the contents of a control volume. Control surface area integrals allow for surface distributions of flow variables. However, in this chapter, for simplicity we often assume that flow variables are uniformly distributed over cross section areas where fluid enters or leaves the control volume. This uniform flow is called one-dimensional flow. Later on (in Chapters 8, 9, and 10), when we discuss velocity profiles and other flow variable distributions, the effects of nonuniformities will be covered in more detail.

Mostly steady flows are considered. However, some simple examples of unsteady flow analyses are introduced.

Although fixed, nondeforming control volumes are emphasized in this chapter, a few examples of moving, nondeforming control volumes and deforming control volumes are also included.

5.1   CONSERVATION OF MASS—THE CONTINUITY EQUATION

5.1.1   Derivation of the Continuity Equation

A system is defined as a collection of unchanging contents, so the conservation of mass principle for a system is simply stated as

time rate of change of the system mass = 0

or

$$\frac{DM_{sys}}{Dt} = 0 \tag{5.1}$$

where the system mass, M_{sys}, is more generally expressed as

$$M_{sys} = \int_{sys} \rho \, d\mathcal{V} \tag{5.2}$$

where the integration is over the volume of the system. In words, Eq. 5.2 states that the system mass is equal to the sum of all the density–volume element products for the contents of the system.

For a system and a fixed, nondeforming control volume that are coincident at an instant of time as are illustrated in Fig. 5.1, the Reynolds transport theorem (Eq. 4.19) allows us to state that

$$\frac{D}{Dt} \int_{sys} \rho \, d\mathcal{V} = \frac{\partial}{\partial t} \int_{cv} \rho \, d\mathcal{V} + \int_{cs} \rho \mathbf{V} \cdot \hat{\mathbf{n}} \, dA \tag{5.3}$$

or

$$
\begin{array}{c}
\text{time rate of change} \\
\text{of the mass of the} \\
\text{coincident system}
\end{array}
=
\begin{array}{c}
\text{time rate of change} \\
\text{of the mass of the} \\
\text{contents of the coin-} \\
\text{cident control volume}
\end{array}
+
\begin{array}{c}
\text{net rate of flow} \\
\text{of mass through} \\
\text{the control} \\
\text{surface}
\end{array}
$$

In Eq. 5.3, we express the time rate of change of the system mass as the sum of two control volume quantities, the time rate of change of the mass of the contents of the

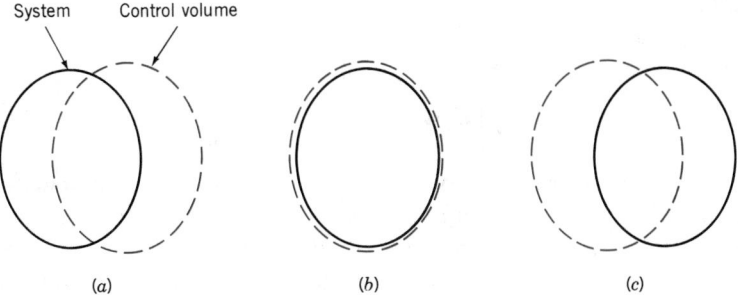

System   Control volume

(a)   (b)   (c)

FIGURE 5.1 System and control volume at three different instances of time. (a) System and control volume at time $= t - \delta t$. (b) System and control volume at time $= t$; coincident condition. (c) System and control volume at time $= t + \delta t$.

control volume,

$$\frac{\partial}{\partial t} \int_{cv} \rho \, d\mkern-2mu\rlap{-}V$$

and the net rate of mass flow through the control surface,

$$\int_{cs} \rho \mathbf{V} \cdot \hat{\mathbf{n}} \, dA$$

When a flow is steady, all field properties including density remain constant with time and the time rate of change of the mass of the contents of the control volume is zero, that is,

$$\frac{\partial}{\partial t} \int_{cv} \rho \, d\mkern-2mu\rlap{-}V = 0$$

The integrand, $\mathbf{V} \cdot \hat{\mathbf{n}} \, dA$, in the mass flowrate integral represents the product of the component of velocity, $\mathbf{V}$, perpendicular to the small portion of control surface, area, dA, and the differential area. Thus, $\mathbf{V} \cdot \hat{\mathbf{n}} \, dA$ is the volume flowrate through dA and $\rho \mathbf{V} \cdot \hat{\mathbf{n}} \, dA$ is the mass flowrate through dA. Furthermore, the sign of the dot product $\mathbf{V} \cdot \hat{\mathbf{n}}$ is "$+$" for flow *out* of the control volume and "$-$" for flow *into* the control volume since $\hat{\mathbf{n}}$ is considered positive when it points out of the control volume. When all of the differential quantities, $\rho \mathbf{V} \cdot \hat{\mathbf{n}} \, dA$, are summed over the entire control surface, as indicated by the integral

$$\int_{cs} \rho \mathbf{V} \cdot \hat{\mathbf{n}} \, dA$$

the result is the net mass flowrate through the control surface, or

$$\int_{cs} \rho \mathbf{V} \cdot \hat{\mathbf{n}} \, dA = \sum \dot{m}_{out} - \sum \dot{m}_{in} \tag{5.4}$$

where $\dot{m}$ is the mass flowrate. If the integral in Eq. 5.4 is positive, the net flow is out of the control volume; if the integral is negative, the net flow is into the control volume.

The control volume expression for *conservation of mass*, which is commonly called the *continuity equation,* for a fixed, nondeforming control volume is obtained by combining Eqs. 5.1, 5.2, and 5.3 to obtain

$$\frac{\partial}{\partial t} \int_{cv} \rho \, d\mathbf{\Psi} + \int_{cs} \rho \mathbf{V} \cdot \hat{\mathbf{n}} \, dA = 0 \tag{5.5}$$

In words, Eq. 5.5 states that the time rate of change of the mass of the contents of the control volume plus the net rate of mass flow through the control surface must equal zero to conserve mass. Actually, the same result could have been obtained more directly by equating the rates of mass flow into and out of the control volume to the rates of accumulation and depletion of mass within the control volume. It is reassuring, however, to see that the Reynolds transport theorem works for this simple-to-understand case. This confidence will serve us well as we develop control volume expressions for other important principles.

An expression for mass flowrate, $\dot{m}$, through a section of control surface having area A that is often used in solving fluid mechanics problems is

$$\dot{m} = \rho Q = \rho A V \tag{5.6}$$

where ρ is the fluid density, Q is the volume flowrate, and V is the component of fluid velocity perpendicular to area A. Since

$$\dot{m} = \int_A \rho \mathbf{V} \cdot \hat{\mathbf{n}} \, dA$$

application of Eq. 5.6 involves the use of *representative* values of fluid density, ρ, and fluid velocity, V. For incompressible flows, ρ is uniformly distributed over area A. For compressible flows, we will normally consider a uniformly distributed fluid density at each section of flow and allow density changes to occur only from section to section. The appropriate fluid velocity to use in Eq. 5.6 is the average value of the component of velocity normal to the section area involved. This average value can be defined as

$$\overline{V} = \frac{\displaystyle\int_A \rho \mathbf{V} \cdot \hat{\mathbf{n}} \, dA}{\rho A} \tag{5.7}$$

If the velocity is considered uniformly distributed (one-dimensional flow) over the section area, A, then

$$\overline{V} = \frac{\displaystyle\int_A \rho \mathbf{V} \cdot \hat{\mathbf{n}} \, dA}{\rho A} = V \tag{5.8}$$

and the bar notation is not necessary (as in Example 5.1). When the flow is not uniformly distributed over the flow cross section area, the bar notation reminds us that an average velocity is being used (as in Examples 5.2 and 5.4).

5.1.2   Fixed, Nondeforming Control Volume
In many applications of fluid mechanics, an appropriate control volume to use is fixed and nondeforming. Several example problems that involve the continuity equation for fixed, nondeforming control volumes (Eq. 5.5) follow.

EXAMPLE 5.1

Seawater flows steadily through a simple conical shaped nozzle at the end of a firehose as illustrated in Fig. E5.1. If the nozzle exit velocity must be at least 20 m/s, determine the minimum pumping capacity required in m^3/s.

Section (1) (pump discharge)

Section (2) (nozzle exit)

Flow

$V_2 = 20$ m/s
$D_2 = 40$ mm

Control volume

FIGURE E5.1

SOLUTION

The pumping capacity sought is the volume flowrate delivered by the fire pump to the hose and nozzle. Since we desire knowledge about the pump discharge flowrate and we have information about the nozzle exit flowrate, we link these two flowrates with the control volume designated with the dashed line in Fig. E5.1. This control volume contains, at any instant, seawater that is within the hose and nozzle from the pump discharge to the nozzle exit plane.

Equation 5.5 is applied to the contents of this control volume to give

$$\frac{\partial}{\partial t}\int_{cv} \rho\, d\cancel{V} + \int_{cs} \rho \mathbf{V} \cdot \hat{\mathbf{n}}\, dA = 0 \qquad (1)$$

0 (flow is steady)

The time rate of change of the mass of the contents of this control volume is zero because the flow is steady. From Eq. 5.4, we see that the control surface integral in Eq. 1 involves mass flowrates at the pump discharge, section (1), and at the nozzle exit, section (2), or

$$\int_{cs} \rho \mathbf{V} \cdot \hat{\mathbf{n}}\, dA = \dot{m}_2 - \dot{m}_1 = 0$$

so that

$$\dot{m}_2 = \dot{m}_1 \qquad (2)$$

Since the mass flowrate is equal to the product of fluid density, ρ, and volume flowrate, Q, (see Eq. 5.6) we obtain from Eq. 2

$$\rho_2 Q_2 = \rho_1 Q_1 \qquad (3)$$

Liquid flow at low speeds, as in this example, may be considered incompressible. Therefore

$$\rho_2 = \rho_1 \qquad (4)$$

and from Eqs. 3 and 4,

$$Q_2 = Q_1 \qquad (5)$$

The pumping capacity is equal to the volume flowrate at the nozzle exit. If, for simplicity, the velocity distribution at the nozzle exit plane, section (2), is considered uniform (one-dimensional), then from Eqs. 5, 5.6, and 5.8,

$$Q_1 = Q_2 = V_2 A_2$$

$$= V_2 \frac{\pi}{4} D_2^2 = (20 \text{ m/s}) \frac{\pi}{4} \left(\frac{40 \text{ mm}}{1000 \text{ mm/m}} \right)^2 = 0.0251 \text{ m}^3/\text{s} \qquad \text{(Ans)}$$

EXAMPLE 5.2

Air flows steadily between two sections in a long, straight portion of 4-in.-inside diameter pipe as indicated in Fig. E5.2. The uniformly distributed temperature and pressure at each section are given. If the average air velocity (nonuniform velocity distribution) at section (2) is 1000 ft/s, calculate the average air velocity at section (1).

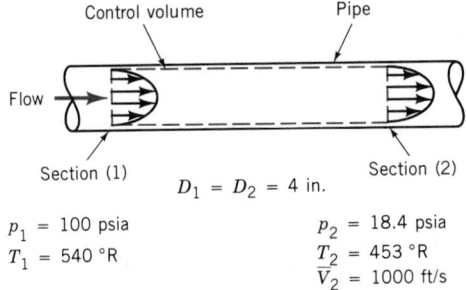

$$D_1 = D_2 = 4 \text{ in.}$$

$p_1 = 100 \text{ psia}$
$T_1 = 540 \text{ °R}$

$p_2 = 18.4 \text{ psia}$
$T_2 = 453 \text{ °R}$
$\overline{V}_2 = 1000 \text{ ft/s}$

FIGURE E5.2

SOLUTION

The average fluid velocity at any section is that velocity which yields the section mass flowrate when multiplied by the section average fluid density and section area (Eq. 5.7). We relate the flows at sections (1) and (2) with the control volume designated with a dashed line in Fig. E5.2.

Equation 5.5 is applied to the contents of this control volume to obtain

$$\overset{0 \text{ (flow is steady)}}{\frac{\partial}{\partial t} \int_{cv} \rho \, d\mathcal{V}} + \int_{cs} \rho \mathbf{V} \cdot \hat{\mathbf{n}} \, dA = 0$$

The time rate of change of the mass of the contents of this control volume is zero because the flow is steady. The control surface integral involves mass flowrates at sections (1) and (2) so that from Eq. 5.4 we get

$$\int_{cs} \rho \mathbf{V} \cdot \hat{\mathbf{n}} \, dA = \dot{m}_2 - \dot{m}_1 = 0$$

or

$$\dot{m}_1 = \dot{m}_2 \qquad (1)$$

and from Eqs. 1, 5.6, and 5.7 we obtain

$$\rho_1 A_1 \overline{V}_1 = \rho_2 A_2 \overline{V}_2 \qquad (2)$$

or

$$\overline{V}_1 = \frac{\rho_2}{\rho_1} \overline{V}_2 \qquad (3)$$

since $A_1 = A_2$.

Air at the pressures and temperatures involved in this example problem behaves like an ideal gas. The ideal gas equation of state (Eq. 1.8) is

$$\rho = \frac{p}{RT} \tag{4}$$

Thus, combining Eqs. 3 and 4 we obtain

$$\overline{V}_1 = \frac{p_2 T_1 \overline{V}_2}{p_1 T_2}$$

$$= \frac{(18.4 \text{ psia})(540 \text{ °R})(1000 \text{ ft/s})}{(100 \text{ psia})(453 \text{ °R})} = 219 \text{ ft/s} \tag{Ans}$$

We learn from this example that the continuity equation (Eq. 5.5) is valid for compressible as well as incompressible flows. Also, nonuniform velocity distributions can be handled with the average velocity concept.

EXAMPLE 5.3

Moist air (a mixture of dry air and water vapor) enters a dehumidifier at the rate of 22 slugs/ hr. Liquid water drains out of the dehumidifier at a rate of 0.5 slug/hr. Determine the mass flowrate of the dry air and the water vapor leaving the dehumidifier. A simplified sketch of the process is provided in Fig. E5.3.

FIGURE E5.3

SOLUTION

The unknown mass flowrate at section (2) is linked with the known flowrates at sections (1) and (3) with the control volume designated with a dashed line in Fig. E5.3. The contents of the control volume are the air and water vapor mixture and the condensate (liquid water) in the dehumidifier at any instant.

Not included in the control volume are the fan and its motor, and the condenser coils and refrigerant. Even though the flow in the vicinity of the fan blade is unsteady, it is unsteady in a cyclical way. Thus, the flowrates at sections (1), (2), and (3) appear steady and the time rate of change of the mass of the contents of the control volume may be considered equal to zero on a time-average basis. The application of Eqs. 5.4 and 5.5 to the control volume contents results in

$$\int_{cs} \rho \mathbf{V} \cdot \hat{\mathbf{n}} \, dA = -\dot{m}_1 + \dot{m}_2 + \dot{m}_3 = 0$$

or

$$\dot{m}_2 = \dot{m}_1 - \dot{m}_3 = 22 \text{ slugs/hr} - 0.5 \text{ slugs/hr} = 21.5 \text{ slugs/hr} \qquad \text{(Ans)}$$

Note that the continuity equation (Eq. 5.5) can be used when there is more than one stream of fluid flowing through the control volume.

EXAMPLE 5.4

Incompressible, laminar water flow develops in a straight pipe having radius R as indicated in Fig. E5.4. At section (1), the velocity profile is uniform; the velocity is equal to a constant value U and is parallel to the pipe axis everywhere. At section (2), the velocity profile is axisymmetric and parabolic, with zero velocity at the pipe wall and a maximum value of u_{max} at the centerline. How are U and u_{max} related? How are the average velocity at section (2), $\overline{V}_2$, and u_{max} related?

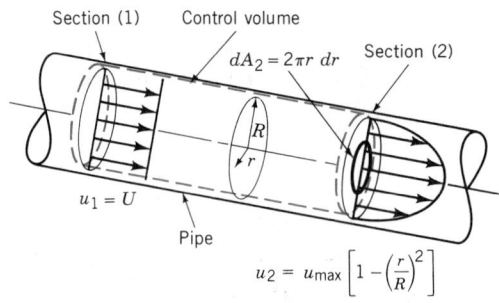

Section (1)   Control volume   Section (2)

$dA_2 = 2\pi r \, dr$

R

r

$u_1 = U$

Pipe

$u_2 = u_{max}\left[1 - \left(\dfrac{r}{R}\right)^2\right]$

FIGURE E5.4

SOLUTION

An appropriate control volume is sketched (dashed lines) in Fig. E5.4. The application of Eq. 5.5 to the contents of this control volume yields

$$\int_{cs} \rho \mathbf{V} \cdot \hat{\mathbf{n}} \, dA = 0$$

The surface integral is evaluated at sections (1) and (2) to give

$$-\rho_1 A_1 U + \int_A \rho \mathbf{V} \cdot \hat{\mathbf{n}} \, dA_2 = 0 \qquad (1)$$

or, since the component of velocity, $\mathbf{V}$, perpendicular to the area at section (2) is u_2, and the element of cross section area, dA_2, is equal to $2\pi r \, dr$ (see shaded area element in Fig. E.5.4), Eq. 1 becomes

$$-\rho_1 A_1 U + \rho_2 \int_0^R u_2 2\pi r \, dr = 0 \qquad (2)$$

Since the flow is considered incompressible, $\rho_1 = \rho_2$. The parabolic velocity relationship for flow through section (2) is used in Eq. 2 to yield

$$-A_1 U + 2\pi u_{max} \int_0^R \left[1 - \left(\frac{r}{R} \right)^2 \right] r \, dr = 0 \tag{3}$$

Integrating, we get from Eq. 3

$$-\pi R^2 U + 2\pi u_{max} \left(\frac{r^2}{2} - \frac{r^4}{4R^2} \right)_0^R = 0$$

or

$$u_{max} = 2U \tag{Ans}$$

Since this flow is incompressible, we conclude from Eq. 5.8 that U is the average velocity at all sections of the control volume. Thus, the average velocity at section (2), $\overline{V}_2$, is one-half the maximum velocity, u_{max}, there, or

$$\overline{V}_2 = \frac{u_{max}}{2} \tag{Ans}$$

EXAMPLE 5.5

A bathtub is being filled with water from a faucet. The rate of flow from the faucet is steady at 9 gal/min. The tub volume is approximated by a rectangular space as indicated in Fig. E5.5. Estimate the time rate of change of the depth of water in the tub, $\partial h/\partial t$, in in./min at any instant.

FIGURE E5.5

SOLUTION

We will see later (Example 5.9) that this problem can also be solved with a deforming control volume that includes only the water in the tub at any instant. We presently use the fixed, nondeforming control volume outlined with a dashed line in Fig. E5.5. This control volume includes in it, at any instant, the water accumulated in the tub, some of the water flowing from the faucet into the tub, and some air. Application of Eqs. 5.4 and 5.5 to these contents of the control volume results in

$$\frac{\partial}{\partial t} \int_{\substack{air \\ volume}} \rho_{air} \, d\mathcal{V}_{air} + \frac{\partial}{\partial t} \int_{\substack{water \\ volume}} \rho_{water} \, d\mathcal{V}_{water} - \dot{m}_{water} + \dot{m}_{air} = 0$$

Note that the time rate of change of air mass and water mass are each not zero. Recognizing, however, that the air mass must be conserved, we stated that the time rate of change of the mass of air in the control volume must be equal to the rate of air mass flow out of the control volume. For simplicity, we disregard any water evaporation that occurs. Thus, applying Eqs. 5.4 and 5.5 to the air only and to the water only, we obtain

$$\frac{\partial}{\partial t} \int_{\substack{air \\ volume}} \rho_{air} \, d\mathcal{V}_{air} + \dot{m}_{air} = 0$$

for air, and

$$\frac{\partial}{\partial t} \int_{\substack{water \\ volume}} \rho_{water} \, d\mathcal{V}_{water} = \dot{m}_{water} \qquad (1)$$

for water. For the water,

$$\int_{\substack{water \\ volume}} \rho_{water} \, d\mathcal{V}_{water} = \rho_{water} \left[h(2 \text{ ft})(5 \text{ ft}) + (1.5 \text{ ft} - h)A_j \right] \qquad (2)$$

where A_j is the cross section area of the water flowing from the faucet into the tub. Combining Eqs. 1 and 2, we obtain

$$\rho_{water} (10 \text{ ft}^2 - A_j) \frac{\partial h}{\partial t} = \dot{m}_{water}$$

and, thus

$$\frac{\partial h}{\partial t} = \frac{Q_{water}}{(10 \text{ ft}^2 - A_j)}$$

For $A_j \ll 10 \text{ ft}^2$ we can conclude that

$$\frac{\partial h}{\partial t} = \frac{Q_{water}}{(10 \text{ ft}^2)}$$

or

$$\frac{\partial h}{\partial t} = \frac{(9 \text{ gal/min})(12 \text{ in./ft})}{(7.48 \text{ gal/ft}^3)(10 \text{ ft}^2)} = 1.44 \text{ in./min} \qquad \text{(Ans)}$$

The preceding example problems illustrate some important results of applying the conservation of mass principle to the contents of a fixed, nondeforming control volume. The dot product $\mathbf{V} \cdot \hat{\mathbf{n}}$ is considered "+" for flow out of the control volume and "−" for flow into the control volume. Thus, mass flowrate out of the control volume is "+" and mass flowrate in is "−". When the flow is steady, the time rate of change of the mass of the contents of the control volume,

$$\frac{\partial}{\partial t} \int_{cv} \rho \, d\mathcal{V}$$

is zero and the net amount of mass flowrate, $\dot{m}$, through the control surface is therefore also zero;

$$\sum \dot{m}_{out} - \sum \dot{m}_{in} = 0 \qquad (5.9)$$

If the steady flow is also incompressible, the net amount of volume flowrate, Q, through the control surface is also zero;

$$\sum Q_{out} - \sum Q_{in} = 0 \qquad (5.10)$$

An unsteady, but cyclical flow can be considered steady on a time-average basis. When the flow is unsteady, the instantaneous time rate of change of the mass of the contents of the control volume is not zero and can be an important variable. When the value of

$$\frac{\partial}{\partial t} \int_{cv} \rho \, d\mathcal{V}$$

is " + ," the mass of the contents of the control volume is increasing. When it is " − ," the mass of the contents of the control volume is decreasing.

When the flow is uniformly distributed over the opening in the control surface (one-dimensional flow),

$$\dot{m} = \rho A V$$

where V is the uniform value of velocity component normal to the section area A. When the velocity is nonuniformly distributed over the opening in the control surface,

$$\dot{m} = \rho A \overline{V} \tag{5.11}$$

where $\overline{V}$ is the average value of the component of velocity normal to the section area A as defined by Eq. 5.7.

For steady flow involving only one stream of a specific fluid flowing through the control volume,

$$\dot{m} = \rho_1 A_1 \overline{V}_1 = \rho_2 A_2 \overline{V}_2 \tag{5.12}$$

and for incompressible flow,

$$Q = A_1 \overline{V}_1 = A_2 \overline{V}_2 \tag{5.13}$$

For steady flow involving more than one stream of a specific fluid or more than one specific fluid flowing through the control volume,

$$\sum \dot{m}_{in} = \sum \dot{m}_{out}$$

The variety of example problems solved above should give the correct impression that the fixed, nondeforming control volume is versatile and useful.

5.1.3 Moving, Nondeforming Control Volume

It is sometimes necessary to use a nondeforming control volume attached to a moving reference frame. Examples include control volumes containing a gas turbine engine on an aircraft in flight, the exhaust stack of a ship at sea, and the gasoline tank of an automobile passing by.

As we discussed earlier in Section 4.4.6, when a moving control volume is used, the fluid velocity relative to the moving control volume (relative velocity) is an important flow field variable. The relative velocity, $\mathbf{V}_r$, is the fluid velocity seen by an observer moving with the control volume. The control volume velocity, $\mathbf{V}_{cv}$, is the velocity of the control volume as seen from a fixed coordinate system. The absolute velocity, $\mathbf{V}$, is the fluid velocity seen by a stationary observer in a fixed coordinate system. These velocities are related to each other by the vector equation

$$\mathbf{V} = \mathbf{V}_r + \mathbf{V}_{cv} \tag{5.14}$$

which is the same as Eq. 4.22, introduced earlier.

For a system and a moving, nondeforming control volume that are coincident at an instant of time, the Reynolds transport theorem (Eq. 4.23) for a moving control

volume leads to

$$\frac{DM_{sys}}{Dt} = \frac{\partial}{\partial t} \int_{cv} \rho \, d\!\!\!\forall + \int_{cs} \rho \mathbf{V}_r \cdot \hat{\mathbf{n}} \, dA \qquad (5.15)$$

From Eqs. 5.1 and 5.15, we can get the control volume expression for conservation of mass (the continuity equation) for a moving, nondeforming control volume, namely,

$$\frac{\partial}{\partial t} \int_{cv} \rho \, d\!\!\!\forall + \int_{cs} \rho \mathbf{V}_r \cdot \hat{\mathbf{n}} \, dA = 0 \qquad (5.16)$$

Some examples of the application of Eq. 5.16 follow.

EXAMPLE 5.6

An airplane moves forward at a speed of 971 km/hr as shown in Fig. E5.6a. The frontal intake area of the jet engine is 0.80 m² and the entering air density is 0.736 kg/m³. A stationary observer estimates that relative to the earth, the jet engine exhaust gases move away from the engine with a speed of 1050 km/hr. The engine exhaust area is 0.558 m² and the exhaust gas density is 0.515 kg/m³. Estimate the mass flowrate of fuel into the engine in kg/hr.

(a)

(b)

FIGURE E5.6

SOLUTION

The control volume, which moves with the airplane (see Figs E5.6a and E5.6b), surrounds the engine and its contents and includes all fluids involved at an instant. The application of Eq. 5.16 to these contents of the control volume yields

0 (flow relative to moving control
volume is considered steady on a
time-average basis)

$$\frac{\partial}{\partial t} \int_{cv} \rho \, d\!\!\!\forall + \int_{cs} \rho \, \mathbf{V}_r \cdot \hat{\mathbf{n}} \, dA = 0 \qquad (1)$$

Assuming one-dimensional flow, we evaluate the surface integral in Eq. 1 and get

$$-\dot{m}_{\substack{\text{fuel} \\ \text{in}}} - \rho_1 A_1 V_{r,1} + \rho_2 A_2 V_{r,2} = 0$$

or

$$\dot{m}_{\substack{\text{fuel} \\ \text{in}}} = \rho_2 A_2 V_{r,2} - \rho_1 A_1 V_{r,2} \tag{2}$$

We consider the intake velocity, $V_{r,1}$ relative to the moving control volume, as being equal in magnitude to the speed of the airplane, 971 km/hr. The exhaust velocity, $V_{r,2}$, also needs to be as measured relative to the moving control volume. Since a fixed observer noted that the exhaust gases were moving away from the engine at a speed of 1050 km/hr, the speed of the exhaust gases relative to the moving control volume, $V_{r,2}$, is determined as follows by using Eq. 5.14

$$V_2 = V_{r,2} + V_{\text{plane}}$$

or

$$V_{r,2} = V_2 - V_{\text{plane}} = 1050 \text{ km/hr} + 971 \text{ km/hr} = 2021 \text{ km/hr}$$

and is shown in Fig. E5.6b.

From Eq. 2,

$$\dot{m}_{\substack{\text{fuel} \\ \text{in}}} = (0.515 \text{ kg/m}^3)(0.558 \text{ m}^2)(2021 \text{ km/hr})(1000 \text{ m/km})$$

$$- (0.736 \text{ kg/m}^3)(0.80 \text{ m}^2)(971 \text{ km/hr})(1000 \text{ m/km})$$

$$\dot{m}_{\substack{\text{fuel} \\ \text{in}}} = 9050 \text{ kg/hr} \tag{Ans}$$

EXAMPLE 5.7

Water enters a rotating lawn sprinkler through its base at the steady rate of 100 ml/s as sketched in Fig. E5.7. If the exit area of each of the two nozzles is 30 mm², determine the average speed of the water leaving each nozzle, relative to the nozzle, if **(a)** the rotary sprinkler head is stationary; **(b)** the sprinkler head rotates at 600 rpm; **(c)** the sprinkler head accelerates from 0 to 600 rpm.

FIGURE E5.7

SOLUTION

We specify a control volume that contains the water in the rotary sprinkler head at any instant. This control volume is nondeforming, but it moves (rotates) with the sprinkler head. The application of Eq. 5.16 to the contents of this control volume for situation **(a)**, **(b)**, or **(c)** of the problem results in the same expression, namely,

0 flow is steady or the control volume is filled with an incompressible fluid.

$$\frac{\partial}{\partial t}\int_{cv} \rho\, d\mathcal{V} + \int_{cs} \rho\, \mathbf{V}_r \cdot \hat{\mathbf{n}}\, dA = 0$$

The flow is steady in the control volume reference frame when the control volume is stationary [(part (a)] and when it moves [parts (b) and (c)]. Also, the control volume is filled with water. Thus, the time rate of change of the mass of the water in the control volume is zero. The control surface integral has a nonzero value only where water enters and leaves the control volume; thus,

$$\int_{cv} \rho\mathbf{V}_r \cdot \mathbf{n}\, dA = -\dot{m}_{in} + \dot{m}_{out} = 0$$

or

$$\dot{m}_{out} = \dot{m}_{in} \tag{1}$$

Since

$$\dot{m}_{out} = 2\rho A_2 \overline{V}_{r,2}$$

and

$$\dot{m}_{in} = \rho Q$$

it follows from Eq. 1 that

$$\overline{V}_{r,2} = \frac{Q}{2A_2}$$

or

$$V_{r,2} = \frac{(1000\ \text{ml/s})(0.001\ \text{m}^3/\text{liter})(10^6\ \text{mm}^2/\text{m}^2)}{(1000\ \text{ml/liter})(2)(30\ \text{mm}^2)} = 16.7\ \text{m/s} \tag{Ans}$$

The value of $V_{r,2}$ is independent of the speed of rotation of the sprinkler head and represents the average velocity of the water exiting from each nozzle with respect to the nozzle for cases **(a)**, **(b)**, **(c)**. The velocity of water discharging from each nozzle, when viewed from a stationary reference (i.e. V_2), will vary as the rotation speed of the sprinkler head varies since from Eq. 5.14,

$$V_2 = V_{r,2} - U$$

where $U = \omega R$ is the speed of the nozzle and ω and R are the angular velocity and radius of the sprinkler head.

When a moving, nondeforming control volume is used, the dot product sign convention used earlier for fixed, nondeforming control volume application is still valid. Also, if the flow within the moving control volume is steady, or steady on a time-average basis, the time rate of change of the mass of the contents of the control volume is zero. Velocities seen from the control volume reference frame (relative velocities) must be used in the continuity equation. Relative and absolute velocities are related by a vector equation (Eq. 5.14), which also involves the control volume velocity.

5.1.4 Deforming Control Volume

Occasionally, a deforming control volume can simplify the solution of a problem. A deforming control volume involves changing volume size and control surface movement. Thus, the Reynolds transport theorem for a moving control volume can be used for this case, and Eqs. 4.23 and 5.1 lead to

$$\frac{DM_{sys}}{Dt} = \frac{\partial}{\partial t}\int_{cv} \rho \, dV + \int_{cs} \rho \mathbf{V}_r \cdot \hat{\mathbf{n}} \, dA = 0 \qquad (5.17)$$

The time rate of change term in Eq. 5.17,

$$\frac{\partial}{\partial t}\int_{cv} \rho \, dV$$

is usually nonzero and must be carefully evaluated because the extent of the control volume varies with time. The mass flowrate term in Eq. 5.17,

$$\int_{cs} \rho \mathbf{V}_r \cdot \hat{\mathbf{n}} \, dA$$

must be determined with the relative velocity, $\mathbf{V}_r$, the velocity referenced to the control surface. Since the control volume is deforming, the control surface velocity is not necessarily uniform and identical to the control volume velocity, $\mathbf{V}_{cv}$, as was true for moving, nondeforming control volumes. For the deforming control volume,

$$\mathbf{V} = \mathbf{V}_r + \mathbf{V}_{cs} \qquad (5.18)$$

where $\mathbf{V}_{cs}$ is the velocity of the control surface as seen by a fixed observer. The relative velocity, $\mathbf{V}_r$, must be ascertained with care wherever fluid crosses the control surface. Two example problems that illustrate the use of the continuity equation for a deforming control volume, Eq. 5.17, follow.

EXAMPLE 5.8

A syringe (Fig. E5.8) is used to inoculate a cow. The plunger has a face area of 500 mm². If the liquid in the syringe is to be injected steadily at a rate of 300 cm³/min, at what speed should the plunger be advanced? The leakage rate past the plunger is 0.10 times the volume flowrate out of the needle.

FIGURE E5.8

SOLUTION

The control volume selected for solving this problem is the deforming one illustrated in Fig. E5.8. Section (1) of the control surface moves with the plunger. The surface area of section (1), A_1, is considered equal to the circular area of the face of the plunger, A_p although this is not strictly true, since leakage occurs. The difference is small, however. Thus,

$$A_1 = A_p \tag{1}$$

Liquid leaves the needle through section (2), which involves fixed area A_2. The application of Eq. 5.17 to the contents of this control volume gives

$$\frac{\partial}{\partial t} \int_{cv} \rho \, d\Psi + \dot{m}_2 + \rho Q_{\text{leak}} = 0 \tag{2}$$

Even though Q_{leak} and the flow through section area A_2 are steady, the time rate of change of the mass of liquid in the shrinking control volume is not zero because the control volume is getting smaller. To evaluate the first term of Eq. 2, we note that

$$\int_{cv} \rho \, d\Psi = \rho(\ell A_1 + \Psi_{\text{needle}}) \tag{3}$$

where ℓ is the changing length of the control volume (see Fig. E5.8) and Ψ_{needle} is the volume of the needle. From Eq. 3, we obtain

$$\frac{\partial}{\partial t} \int_{cv} \rho \, d\Psi = \rho A_1 \frac{\partial \ell}{\partial t} \tag{4}$$

Note that

$$-\frac{\partial \ell}{\partial t} = V_p \tag{5}$$

where V_p is the speed of the plunger sought in the problem statement. Combining Eqs. 2, 4 and 5 we obtain

$$-\rho A_1 V_p + \dot{m}_2 + \rho Q_{\text{leak}} = 0 \tag{6}$$

However, from Eq. 5.6, we see that

$$\dot{m}_2 = \rho Q_2 \tag{7}$$

and Eq. 6 becomes

$$-\rho A_1 V_p + \rho Q_2 + \rho Q_{\text{leak}} = 0 \tag{8}$$

Solving Eq. 8 for V_p yields

$$V_p = \frac{Q_2 + Q_{\text{leak}}}{A_1} \tag{9}$$

Since $Q_{\text{leak}} = 0.1 Q_2$, Eq. 9 becomes

$$V_p = \frac{Q_2 + 0.1 Q_2}{A_1} = \frac{1.1 Q_2}{A_1}$$

and

$$V_p = \frac{(1.1)(300 \text{ cm}^3/\text{min})}{(500 \text{ mm}^2)} \left(\frac{1000 \text{ mm}^3}{\text{cm}^3} \right) = 660 \text{ mm/min} \tag{Ans}$$

EXAMPLE 5.9

Solve the problem of Example 5.5 using a deforming control volume that includes only the water accumulating in the bath tub.

SOLUTION

For this deforming control volume, Eq. 5.17 leads to

$$\frac{\partial}{\partial t} \int_{\substack{\text{water} \\ \text{volume}}} \rho \, d\mathcal{V} + \int_{cs} \rho \mathbf{V}_r \cdot \hat{\mathbf{n}} \, dA = 0 \tag{1}$$

The first term of Eq. 1 can be evaluated as

$$\frac{\partial}{\partial t} \int_{\substack{\text{water} \\ \text{volume}}} \rho \, d\mathcal{V} = \frac{\partial}{\partial t} \left[\rho h(2 \text{ ft})(5 \text{ ft}) \right]$$

$$= \rho (10 \text{ ft}^2) \frac{\partial h}{\partial t} \tag{2}$$

The second term of Eq. 1 can be evaluated as

$$\int_{cs} \rho \, \mathbf{V}_r \cdot \hat{\mathbf{n}} \, dA = -\rho \left(V_j + \frac{\partial h}{\partial t} \right) A_j \tag{3}$$

where A_j and V_j are the cross section area and velocity of the water flowing from the faucet into the tube. Thus, from Eqs. 1, 2, and 3 we obtain

$$\frac{\partial h}{\partial t} = \frac{V_j A_j}{(10 \text{ ft}^2 - A_j)} = \frac{Q}{(10 \text{ ft}^2 - A_j)}$$

or for $A_j \ll 10 \text{ ft}^2$

$$\frac{\partial h}{\partial t} = \frac{9(\text{gal/min})(12 \text{ in./ft})}{(7.48 \text{ gal/ft}^3)(10 \text{ ft}^2)} = 1.44 \text{ in./min} \tag{Ans}$$

Note that these results using a deforming control volume are the same as that obtained in Example 5.5 with a fixed control volume.

The conservation of mass principle is easily applied to the contents of a control volume. The appropriate selection of a specific kind of control volume (for example, fixed and nondeforming, moving and nondeforming, or deforming) can make the solution of a particular problem less complicated. In general, where fluid flows through the control surface, it is advisable to make the control surface perpendicular to the flow. We shall learn, in the sections ahead, that the conservation of mass principle is usually used in combination with other important laws to solve problems.

5.2   NEWTON'S SECOND LAW—THE LINEAR MOMENTUM AND MOMENT-OF-MOMENTUM EQUATIONS

5.2.1   Derivation of the Linear Momentum Equation
Newton's second law of motion for a system is

> time rate of change of the = sum of external forces
> linear momentum of the system   acting on the system

or

$$\frac{D}{Dt} \int_{\text{sys}} \mathbf{V} \rho \, d\mathcal{V} = \sum \mathbf{F}_{\text{sys}} \tag{5.19}$$

Any reference or coordinate system for which this statement is true is called *inertial*. A fixed coordinate system is inertial. A coordinate system that moves in a straight line with constant velocity and is thus without acceleration is also inertial. We proceed to develop the control volume formula for this important law. When a control volume is coincident with a system at an instant of time, the forces acting on the system and the forces acting on the contents of the coincident control volume (see Fig. 5.2) are instantaneously identical, that is

$$\sum \mathbf{F}_{\text{sys}} = \sum \mathbf{F}_{\substack{\text{contents of the} \\ \text{coincident control volume}}} \tag{5.20}$$

Furthermore, for a system and the contents of a coincident control volume that is fixed and nondeforming, the Reynolds transport theorem (Eq. 4.19) allows us to conclude that

$$\frac{D}{Dt} \int_{\text{sys}} \mathbf{V}\rho \, d\forall = \frac{\partial}{\partial t} \int_{\text{cv}} \mathbf{V}\rho \, d\forall + \int_{\text{cs}} \mathbf{V}\rho \mathbf{V} \cdot \hat{\mathbf{n}} \, dA \tag{5.21}$$

or

| time rate of change of the linear momentum of the system | = | time rate of change of the linear momentum of the contents of the control volume | + | net rate of flow of linear momentum through the control surface |

Equation 5.21 states that the time rate of change of system linear momentum is expressed as the sum of two control volume quantities, the time rate of change of the *linear momentum of the contents of the control volume,* and the net rate of *linear momentum flow through the control surface.* As particles of mass move into or out of a control volume through the control surface, they carry linear momentum in or out. Thus, linear momentum flow should seem no more unusual than mass flow.

For a control volume that is fixed (inertial) and nondeforming, Eqs. 5.19, 5.20, and 5.21 suggest that an appropriate mathematical statement of Newton's second law of motion is

$$\frac{\partial}{\partial t} \int_{\text{cv}} \mathbf{V}\rho \, d\forall + \int_{\text{cs}} \mathbf{V}\rho \mathbf{V} \cdot \hat{\mathbf{n}} \, dA = \sum \mathbf{F}_{\substack{\text{contents of the} \\ \text{control volume}}} \tag{5.22}$$

We call Eq. 5.22 the *linear momentum equation.*

In our application of the linear momentum equation, we initially confine ourselves to fixed, nondeforming control volumes for simplicity. Subsequently, we discuss the use of a moving but inertial, nondeforming control volume. We do not consider

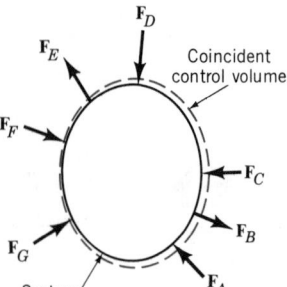

FIGURE 5.2 External forces acting on system and coincident control volume.

deforming control volumes and accelerating (noninertial) control volumes. If a control volume is noninertial, the acceleration components involved (for example, translation acceleration, Coriolis acceleration, and centrifugal acceleration) require consideration.

The forces involved in

$$\sum \mathbf{F}_{\substack{\text{contents of the} \\ \text{control volume}}}$$

are body and surface forces that act on what is contained in the control volume. The only body force we consider in this chapter is the one associated with the action of gravity. We experience this body force as weight. The surface forces are basically exerted on the contents of the control volume by material just outside the control volume in contact with material just inside the control volume. For example, a wall in contact with fluid can exert a reaction surface force on the fluid it bounds. Similarly, fluid just outside the control volume can push on fluid just inside the control volume at a common interface, usually an opening in the control surface through which fluid flow occurs. An immersed object can resist fluid motion with surface forces.

The linear momentum terms in the momentum equation deserve careful explanation. We clarify their physical significance in the following sections.

5.2.2 Application of the Linear Momentum Equation

The linear momentum equation for an inertial control volume is a vector equation (Eq. 5.22). In engineering applications, components of this vector equation resolved along orthogonal coordinates, for example, x, y, and z (rectangular coordinate system) or r, θ, and x (cylindrical coordinate system), will normally be used. A simple example involving one-dimensional, steady, incompressible flow is considered first.

EXAMPLE 5.10

Determine the anchoring force required to hold in place a conical nozzle attached to the end of a laboratory sink faucet (see Fig. E5.10a) when the water flowrate is 0.6 liter/s. The nozzle mass is 0.1 kg. The nozzle inlet and exit diameters are 16 mm and 5 mm, respectively. The nozzle axis is vertical and the axial distance between sections (1) and (2) is 30 mm. The pressure at section (1) is 464 kPa.

SOLUTION

The anchoring force sought is the reaction force between the faucet and nozzle threads. To evaluate this force we select a control volume that includes the entire nozzle and the water contained in the nozzle at an instant as is indicated in Figs. E5.10a and E5.10b. All of the vertical forces acting on the contents of this control volume are identified in Fig. E5.10b. The action of atmospheric pressure cancels out in every direction and is not shown. Gage pressure forces do not cancel out in the vertical direction and are shown. Application of the vertical or z direction component of Eq. 5.22 to the contents of this control volume leads to

$$\underbrace{\frac{\partial}{\partial t} \int_{cv} w\rho \, d\mathcal{V}}_{0 \text{ (flow is steady)}} + \int_{cs} w\rho \mathbf{V} \cdot \hat{\mathbf{n}} \, dA = F_A - W_n - p_1 A_1 - W_w + p_2 A_2 \qquad (1)$$

where w is the z direction component of fluid velocity.

Note that the positive direction is considered "up" for the forces. We will use this same

FIGURE E5.10a

sign convention for the fluid velocity, w, in Eq. 1. In Eq. 1, the dot product, $\mathbf{V} \cdot \hat{\mathbf{n}}$, is "+" for flow out of the control volume and "−" for flow into the control volume. For this particular example

$$\mathbf{V} \cdot \hat{\mathbf{n}} \, dA = \pm |w| \, dA \tag{2}$$

with the "+" used for flow out of the control volume and "−" used for flow in. To evaluate the control surface integral in Eq. 1, we need to assume a distribution for fluid velocity, w, and fluid density, ρ. For simplicity, we assume that w is uniformly distributed or constant, with magnitudes of w_1 and w_2 over cross section areas A_1 and A_2. Also, this flow is incompressible so the fluid density, ρ, is constant throughout. Proceeding further we obtain for Eq. 1

$$(-|w_1|)\rho(-|w_1|A_1) + (-|w_2|)\rho(+|w_2|A_2)$$
$$= F_A - W_n - p_1 A_1 - W_w + p_2 A_2 \tag{3}$$

Note that $-|w_1|$ and $-|w_2|$ are used because both of these velocities are "down." Also, $-|w_1|A_1$ is used because it is associated with flow into the control volume. Similarly, $+|w_2|A_2$ is used because it is associated with flow out of the control volume. The absolute value notation is used above for clarity. However, this notation is dropped in the remainder of this example with the understanding that absolute values of velocities are being used. Solving Eq. 3 for the anchoring force, F_A, we obtain

$$F_A = \rho w_1^2 A_1 - \rho w_2^2 A_2 + W_n + p_1 A_1 + W_w - p_2 A_2 \tag{4}$$

From the conservation of mass equation, Eq. 5.12, we obtain

$$\dot{m} = \rho w_1 A_1 = \rho w_2 A_2 \tag{5}$$

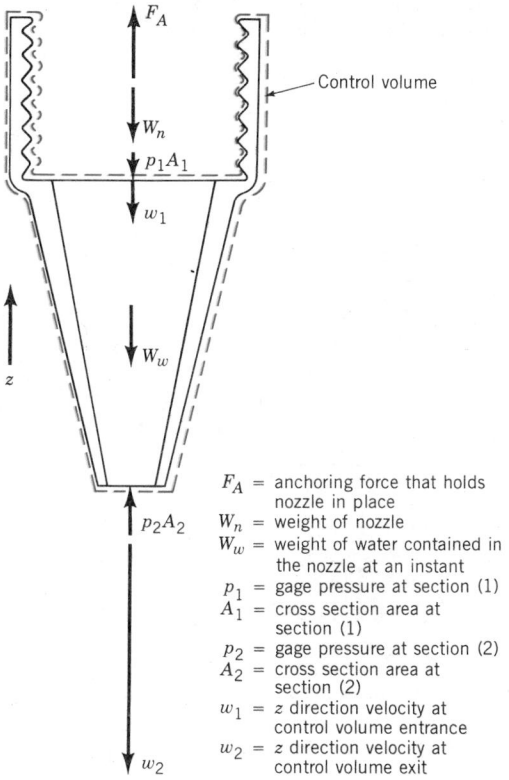

F_A = anchoring force that holds
       nozzle in place
W_n = weight of nozzle
W_w = weight of water contained in
       the nozzle at an instant
p_1 = gage pressure at section (1)
A_1 = cross section area at
       section (1)
p_2 = gage pressure at section (2)
A_2 = cross section area at
       section (2)
w_1 = z direction velocity at
       control volume entrance
w_2 = z direction velocity at
       control volume exit

FIGURE E5.10b

which when combined with Eq. 4 gives

$$F_A = \dot{m}(w_1 - w_2) + W_n + p_1 A_1 + W_w - p_2 A_2 \tag{6}$$

It is instructive to note how the anchoring force is affected by the different actions involved. As expected, the nozzle weight, W_n, the water weight, W_w, and gage pressure force at section (1), $p_1 A_1$, all increase the anchoring force, while the gage pressure force at section (2), $p_2 A$, acts to decrease the anchoring force. The change in the vertical momentum flowrate, $\dot{m}(w_1 - w_2)$, will, in this instance, decrease the anchoring force because this change is negative ($w_2 > w_1$).

To complete this example we use quantities given in the problem statement to quantify the terms on the right-hand side of Eq. 6. From Eq. 5.6,

$$\dot{m} = \rho w_1 A_1 = \rho Q = (999 \text{ kg/m}^3)(0.6 \text{ liter/s})(10^{-3} \text{ m}^3/\text{liter}) = 0.599 \text{ kg/s} \tag{7}$$

and

$$w_1 = \frac{Q}{A_1} = \frac{Q}{\pi(D_1^2/4)} = \frac{(0.6 \text{ liter/s})(10^{-3} \text{ m}^3/\text{liter})}{\pi(16 \text{ mm})^2/4(1000^2 \text{ mm}^2/\text{m}^2)} = 2.98 \text{ m/s} \tag{8}$$

Also from Eq. 5.6,

$$w_2 = \frac{Q}{A_2} = \frac{Q}{\pi(D_2^2/4)} = \frac{(0.6 \text{ liter/s})(10^{-3} \text{ m}^3/\text{liter})}{\pi(5 \text{ mm})^2/4(1000^2 \text{ mm}^2/\text{m}^2)} = 30.6 \text{ m/s} \tag{9}$$

The weight of the nozzle, W_n, can be obtained from the nozzle mass, m_n, with

$$W_n = m_n g = (0.1 \text{ kg})(9.81 \text{ m/s}^2) = 0.981 \text{ N} \tag{10}$$

The weight of the water in the control volume, W_w, can be obtained from the water density, ρ, and the volume of water, $\mathcal{V}_w$, in the truncated cone of height h. That is,

$$W_w = \rho \mathcal{V}_w g = \rho \tfrac{1}{12} \pi h (D_1^2 + D_2^2 + D_1 D_2) g$$

Thus,

$$W_w = (999 \text{ kg/m}^3) \frac{1}{12} \pi \frac{(30 \text{ mm})}{(1000 \text{ mm/m})}$$

$$\times \left[\frac{(16 \text{ mm})^2 + (5 \text{ mm})^2 + (16 \text{ mm})(5 \text{ mm})}{(1000^2 \text{ mm}^2/\text{m}^2)} \right] (9.81 \text{ m/s}^2) = 0.0278 \text{ N} \qquad (11)$$

The gage pressure at section (2), p_2, is zero since, as discussed in Section 3.6.1, when a subsonic flow discharges to the atmosphere as in the present situation, the discharge pressure is essentially atmospheric. The anchoring force, F_A, can now be determined from Eqs. 6 through 11 with

$$F_A = (0.599 \text{ kg/s})(2.98 \text{ m/s} - 30.6 \text{ m/s}) + 0.981 \text{ N}$$

$$+ (464 \text{ kPa})(1000 \text{ Pa/kPa}) \frac{\pi (16 \text{ mm})^2}{4(1000^2 \text{ mm}^2/\text{m}^2)}$$

$$+ 0.0278 \text{ N} + 0$$

or

$$F_A = -16.5 \text{ N} + 0.981 \text{ N} + 93.3 \text{ N} + 0.0278 \text{ N} = 77.8 \text{ N} \qquad \text{(Ans)}$$

Since the anchoring force, F_A, is positive, it acts upward in the z direction.

Several important generalities about the application of the linear momentum equation (Eq. 5.22) are apparent in the example just considered.

1. When the flow is uniformly distributed over a section of the control surface where flow into or out of the control volume occurs, the integral operations are simplified. Thus, one-dimensional flows are easier to work with than flows involving nonuniform velocity distributions.

2. Linear momentum is directional; it can have components in as many as three orthogonal coordinate directions. Furthermore, along any one coordinate, the linear momentum of a fluid particle can be in the positive or negative direction and thus be considered as a positive or a negative quantity. In Example 5.10, only the linear momentum in the z direction was considered (all of it was in the negative z direction and was hence treated as being negative).

3. The flow of positive or negative linear momentum *into* a control volume involves a negative $\mathbf{V} \cdot \hat{\mathbf{n}}$ product. Momentum flow *out* of the control volume involves a positive $\mathbf{V} \cdot \hat{\mathbf{n}}$ product. The correct algebraic sign ($+$ or $-$) to assign to momentum flow ($\mathbf{V}\rho\mathbf{V} \cdot \hat{\mathbf{n}} \, dA$) will depend on the sense of the linear momentum ($+$ in positive coordinate direction, $-$ in negative coordinate direction) and the $\mathbf{V} \cdot \hat{\mathbf{n}}$ product ($+$ for flow out of the control volume, $-$ for flow into the control volume). In Example 5.10, the momentum flow into the control volume past section (1) was a positive ($+$) quantity while the momentum flow out of the control volume at section (2) was a negative ($-$) quantity.

4. The time rate of change of the linear momentum of the contents of a nondeforming control volume (i.e., $\partial/\partial t \int_{cv} \mathbf{V}\rho \, d\mathcal{V}$) is zero for steady flow. The momentum problems considered in this text all involve steady flow.

5. If the control surface is selected so that it is perpendicular to the flow where fluid

enters or leaves the control volume, the surface force exerted at these locations by fluid outside the control volume on fluid inside will be due to pressure. Furthermore, when subsonic flow exits from a control volume into the atmosphere, atmospheric pressure prevails at the exit cross section. In Example 5.10, the flow was subsonic and so we set the exit flow pressure at the atmospheric level. The continuity equation (Eq. 5.12) allowed us to evaluate the fluid flow velocities w_1 and w_2 at sections (1) and (2).

6. The forces due to atmospheric pressure on the control surface cancel each other and do not require further consideration if gage pressure is used.

7. The external forces have algebraic sign, positive if the force is in the assigned positive coordinate direction and negative otherwise.

8. Only external forces acting on the contents of the control volume are considered in the linear momentum equation (Eq. 5.22). If the fluid alone is included in a control volume, reaction forces between the fluid and the surface or surfaces in contact with the fluid [wetted surface(s)] will need to be in Eq. 5.22. If the fluid and the wetted surface or surfaces are within the control volume, the reaction forces between fluid and wetted surface(s) do not appear in the linear momentum equation (Eq. 5.22) because they are internal, not external forces. The anchoring force that holds the wetted surface(s) in place is an external force, however, and must therefore be in Eq. 5.22.

9. The force required to anchor an object will generally exist in response to surface pressure and/or shear forces acting on the control surface, to a change in linear momentum flow through the control volume containing the object, and to the weight of the object and the fluid contained in the control volume. In Example 5.10 the nozzle anchoring force was required mainly because of pressure forces and partly because of a change in linear momentum flow associated with accelerating the fluid in the nozzle. The weight of the water and the nozzle contained in the control volume influenced the size of the anchoring force only slightly.

To further demonstrate the use of the linear momentum equation (Eq. 5.22), we consider another one-dimensional flow example before moving on to other facets of this important equation.

EXAMPLE 5.11

Water flows through a horizontal, 180° pipe bend as illustrated in Fig. E5.11a. The flow cross section area is constant at a value of 0.1 ft² through the bend. The flow velocity everywhere in the bend is axial and 50 ft/s. The absolute pressures at the entrance and exit of the bend are 30 psia and 24 psia, respectively. Calculate the horizontal (x and y) components of the force exerted by the bend on the flowing water.

SOLUTION

Since we want to evaluate components of the reaction force between the pipe bend and the fluid, an appropriate control volume (see dashed line in Fig. E5.11a) contains only the water in the bend at an instant. The horizontal forces acting on the contents of this control volume are identified in Fig. E5.11b. Note that the weight of the water is vertical (in the negative z direction) and does not contribute to the x and y components of the reaction force. All of the horizontal normal and tangential forces exerted on the flowing fluid by the wetted wall of the

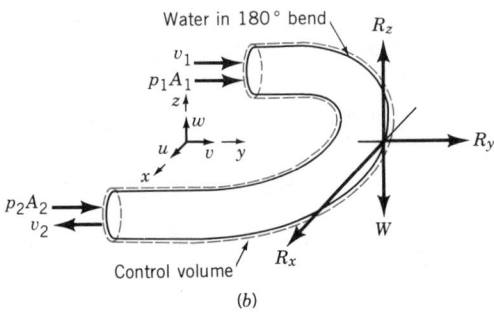

FIGURE E5.11

pipe bend are resolved and combined into the two resultant components, R_x and R_y. These two forces act on the control volume contents, and thus for the x direction, Eq. 5.22 leads to

$$\int_{cs} u\rho \mathbf{V} \cdot \hat{\mathbf{n}} \, dA = R_x \tag{1}$$

At sections (1) and (2), the flow is in the y direction and therefore $u = 0$ at both cross sections. There is no x direction momentum flow into or out of the control volume and we conclude from Eq. 1 that $R_x = 0$.

For the y direction, we get from Eq. 5.22

$$\int_{cs} v\rho \mathbf{V} \cdot \hat{\mathbf{n}} \, dA = R_y + p_1 A_1 + p_2 A_2 \tag{2}$$

For one-dimensional flow, the surface integral in Eq. 2 is easy to evaluate and Eq. 2 becomes

$$(+v_1)\rho_1(-v_1 A_1) + (-v_2)\rho_2(+v_2 A_2) = R_y + p_1 A_1 + p_2 A_2 \tag{3}$$

Note that the velocity in the y direction is positive at section (1) but is negative at section (2). Also, the $\mathbf{V} \cdot \hat{\mathbf{n}}$ product is negative at section (1) (flow in) and is positive at section (2) (flow out). From the continuity equation (Eq. 5.12), we get

$$\dot{m} = \rho_1 A_1 v_1 = \rho_2 A_2 v_2 \tag{4}$$

and thus Eq. 3 can be written as

$$-\dot{m}(v_1 + v_2) = R_y + p_1 A_1 + p_2 A_2 \tag{5}$$

Solving Eq. 5 for R_y we obtain

$$R_y = -\dot{m}(v_1 + v_2) - p_1 A_1 - p_2 A_2 \tag{6}$$

From the given data we can calculate $\dot{m}$ from Eq. 5.6 as

$$\dot{m} = \rho_1 A_1 v_1 = (1.94 \text{ slugs/ft}^3)(0.1 \text{ ft}^2)(50 \text{ ft/s}) = 9.70 \text{ slugs/s}$$

If we want to include the atmospheric pressure effect in the reaction force components, we should use absolute pressures for p_1 and p_2. If we want to allow the atmospheric pressure forces everywhere on the control surface to cancel out, then gage pressure should be used for p_1 and p_2. Throughout this book we will let the atmospheric pressure forces cancel and use gage pressure forces only. Thus, by substituting numerical values of variables into Eq. 6, we get

$$R_y = -(9.70 \text{ slugs/s})(50 \text{ ft/s} + 50 \text{ ft/s})\{1 \text{ lb/[slug} \cdot (\text{ft/s}^2)]\}$$
$$- (30 \text{ psia} - 14.7 \text{ psia})(144 \text{ in.}^2/\text{ft}^2)(0.1 \text{ ft}^2)$$
$$- (24 \text{ psia} - 14.7 \text{ psia})(144 \text{ in.}^2/\text{ft}^2)(0.1 \text{ ft}^2)$$
$$R_y = -970 \text{ lb} - 220 \text{ lb} - 134 \text{ lb} = -1324 \text{ lb} \qquad \text{(Ans)}$$

The negative sign for R_y is interpreted as meaning that the y component of the reaction force is actually in the negative y direction, not the positive y direction as originally indicated in Fig. E5.11b.

In Example 5.11, the direction of flow entering the control volume was different by 180° from the direction of flow leaving the control volume. This change in flow direction only (the flow speed remained constant) resulted in a large portion of the reaction force exerted by the pipe on the water. This is in contrast to the small contribution of fluid acceleration to the anchoring force of Example 5.10. From Examples 5.10 and 5.11 we see that changes in flow speed and/or direction result in a reaction force. Other types of problems that can be solved with the linear momentum equation (Eq. 5.22) are illustrated in the following examples.

EXAMPLE 5.12

Air flows steadily between two cross sections in a long, straight portion of 4 in.-inside diameter pipe as indicated in Fig. E5.12, where the uniformly distributed temperature and pressure at each cross section are given. If the average air velocity at section (2) is 1000 ft/s, we found in Example 5.2 that the average air velocity at section (1) must be 219 ft/s. Assuming uniform velocity distributions at sections (1) and (2), determine the frictional force exerted by the pipe wall on the air flow between sections (1) and (2).

FIGURE E5.12

SOLUTION

The control volume of Example 5.2 is appropriate for this problem. The forces acting on the air between sections (1) and (2) are identified in Fig. E5.12. The weight of air is considered negligibly small. The reaction force between the wetted wall of the pipe and the flowing air, R_x, is the frictional force sought. Application of the axial component of Eq. 5.22 to this control volume yields

$$\int_{cs} u\rho \mathbf{V} \cdot \hat{\mathbf{n}} \, dA = -R_x + p_1 A_1 - p_2 A_2 \qquad (1)$$

The positive x direction is set as being to the right. Furthermore, for uniform velocity distributions (one-dimensional flow), Eq. 1 becomes

$$(+u_1)\rho_1(-u_1A_1) + (+u_2)\rho_2(+u_2A_2) = -R_x + p_1A_1 - p_2A_2 \tag{2}$$

From conservation of mass (Eq. 5.12) we get

$$\dot{m} = \rho_1A_1u_1 = \rho_2A_2u_2 \tag{3}$$

so that Eq. 2 becomes

$$\dot{m}(u_2 - u_1) = -R_x + A_2(p_1 - p_2) \tag{4}$$

Solving Eq. 4 for R_x, we get

$$R_x = A_2(p_1 - p_2) - \dot{m}(u_2 - u_1) \tag{5}$$

The equation of state gives

$$\rho_2 = \frac{p_2}{RT_2} \tag{6}$$

and the equation for area A_2 is

$$A_2 = \frac{\pi D_2^2}{4} \tag{7}$$

Thus, from Eqs. 3, 6, and 7

$$\dot{m} = \left(\frac{p_2}{RT_2}\right)\left(\frac{\pi D_2^2}{4}\right)u_2 = \frac{(18.4 \text{ psia})(144 \text{ in.}^2/\text{ft}^2)}{(1716 \text{ ft·lb/slug·°R})(453 \text{ °R})}$$

$$\times \frac{\pi(4 \text{ in.})^2}{4(144 \text{ in.}^2/\text{ft}^2)} (1000 \text{ ft/s}) = 0.297 \text{ slugs/s} \tag{8}$$

Thus, from Eqs. 5 and 8

$$R_x = \frac{\pi(4 \text{ in.})^2}{4} (100 \text{ psia} - 18.4 \text{ psia})$$

$$- (0.297 \text{ slugs/s})(1000 \text{ ft/s} - 219 \text{ ft/s})[1 \text{ lb/(slug·ft/s}^2)]$$

or

$$R_x = 1025 \text{ lb} - 232 \text{ lb} = 793 \text{ lb} \tag{Ans}$$

EXAMPLE 5.13

If the flow of Example 5.4 is vertically upward, develop an expression for the fluid pressure drop that occurs between section (1) and section (2).

SOLUTION

A control volume (see dashed lines in Fig. E5.4) that includes only fluid from section (1) to section (2) is selected. The forces acting on the fluid in this control volume are identified in Fig. E5.13. The application of the axial component of Eq. 5.22 to the fluid in this control volume results in

$$\int_{cs} w\rho\mathbf{V} \cdot \hat{\mathbf{n}} \, dA = p_1A_1 - R_z - W - p_2A_2 \tag{1}$$

where R_z is the resultant force of the wetted pipe wall on the fluid. Further, for uniform flow

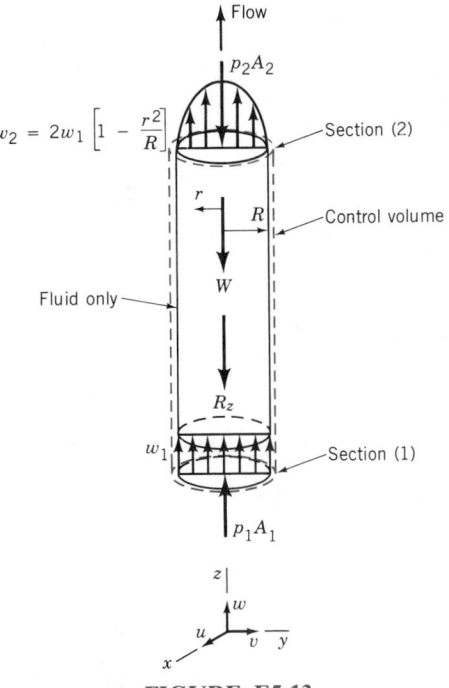

FIGURE E5.13

at section (1), and because the flow at section (2) is out of the control volume, Eq. 1 becomes

$$(+w_1)\rho(-w_1 A_1) + \int_{A_2} (+w_2)\rho(+w_2 \, dA_2) = p_1 A_1 - R_z - W - p_2 A_2 \tag{2}$$

The positive direction is considered up. The surface integral over the cross section area at section (2), A_2, is evaluated by using the parabolic velocity profile obtained in Example 5.4, $w_2 = 2w_1[1 - (r/R)^2]$, as

$$\int_{A_2} w_2 \rho w_2 \, dA_2 = \rho \int_0^R w_2^2 \, 2\pi r \, dr = 2\pi\rho \int_0^R (2w_1)^2 \left[1 - \left(\frac{r}{R}\right)^2\right]^2 r \, dr$$

or

$$\int_{A_2} w_2 \rho w_2 \, dA_2 = 4 \, \pi\rho w_1^2 \frac{R^2}{3} \tag{3}$$

Combining Eqs. 2 and 3 we obtain

$$-w_1^2 \rho\pi R^2 + w_1^2 \tfrac{4}{3}\rho\pi R^2 = p_1 A_1 - R_z - W - p_2 A_2 \tag{4}$$

Solving Eq. 4 for the pressure drop from section (1) to section (2), $p_1 - p_2$, we obtain

$$p_1 - p_2 = \frac{\rho w_1^2}{3} + \frac{R_z}{A_1} + \frac{W}{A_1} \tag{Ans}$$

We see that the drop in pressure from section (1) to section (2) occurs because of the following:

1. The change in momentum flow between the two sections associated with going from a uniform velocity profile to a parabolic velocity profile.

2. Pipe wall friction.

3. The weight of the water column; that is hydrostatic pressure effect.

If the velocity profiles had been identically parabolic at sections (1) and (2), the momentum flowrate at each section would have been identical, a condition we call "fully developed" flow. Then, the pressure drop, $p_1 - p_2$, would be due only to pipe wall friction and the weight of the water column. If in addition to being fully developed, the flow involved negligible weight effects, for example, horizontal flow of liquids or the flow of gases in any direction, the drop in pressure between any two sections, $p_1 - p_2$, would be a result of pipe wall friction only.

EXAMPLE 5.14

A static thrust stand as sketched in Fig. E5.14 is to be designed for testing a jet engine. The following conditions are known for a typical test:

$$\text{Intake air velocity} = 200 \text{ m/s}$$
$$\text{Exhaust gas velocity} = 500 \text{ m/s}$$
$$\text{Intake cross section area} = 1 \text{ m}^2$$
$$\text{Intake static pressure} = -22.5 \text{ kPa} = 78.5 \text{ kPa (abs)}$$
$$\text{Intake static temperature} = 268 \text{ K}$$
$$\text{Exhaust static pressure} = 0 \text{ kPa} = 101 \text{ kPa (abs)}$$

Estimate the nominal thrust to design for.

FIGURE E5.14

SOLUTION

The cylindrical control volume outlined with a dashed line in Fig. E5.14 is selected. The external forces acting in the axial direction are also shown. Application of the momentum equation (Eq. 5.22) to the contents of this control volume yields

$$\int_{cs} u\rho \mathbf{V} \cdot \hat{\mathbf{n}} \, dA = p_1 A_1 + F_{th} - p_2 A_1 \tag{1}$$

or, for one-dimensional flow, Eq. 1 becomes

$$(+u_1)\rho_1(-u_1 A_1) + (+u_2)\rho_2(+u_2 A_2) = A_1(p_1 - p_2) + F_{th} \tag{2}$$

The positive direction is to the right. The conservation of mass equation (Eq. 5.12) leads to

$$\dot{m} = \rho_1 A_1 u_1 = \rho_2 A_2 u_2 \tag{3}$$

Combining Eqs. 2 and 3, we obtain

$$\dot{m}(u_2 - u_1) = A_1(p_1 - p_2) + F_{th} \tag{4}$$

Solving Eq. 4 for the thrust force, F_{th}, we obtain

$$F_{th} = -A_1(p_1 - p_2) + \dot{m}(u_2 - u_1) \tag{5}$$

We need to determine the mass flowrate, $\dot{m}$, to calculate F_{th}, and to calculate $\dot{m} = \rho_1 A_1 u_1$, we need ρ_1. From the ideal gas equation of state

$$\rho_1 = \frac{p_1}{RT_1} = \frac{(78.5 \text{ kPa})(1000 \text{ Pa/kPa})[1(\text{N/m}^2)/\text{Pa}]}{(286.9 \text{ J/kg·K})(268 \text{ K})(1 \text{ N·m/J})} = 1.02 \text{ kg/m}^3$$

Thus,

$$\dot{m} = \rho_1 A_1 u_1 = (1.02 \text{ kg/m}^3)(1 \text{ m}^2)(200 \text{ m/s}) = 204 \text{ kg/s} \tag{6}$$

Finally, combining Eqs. 5 and 6 and substituting given data, we obtain

$$F_{th} = -(1 \text{ m}^2)(-22.5 \text{ kPa} - 0 \text{ kPa})(1000 \text{ Pa/kPa})[1(\text{N/m}^2)/\text{Pa}]$$
$$+ (204 \text{ kg/s})(500 \text{ m/s} - 200 \text{ m/s})[1 \text{ N/(kg·m/s}^2)]$$

and

$$F_{th} = 22{,}500 \text{ N} + 61{,}200 \text{ N} = 83{,}700 \text{ N} \tag{Ans}$$

EXAMPLE 5.15

A sluice gate across a channel of unit width is shown in the closed and open positions in Figs. E5.15a and E5.15b. Is the anchoring force required to hold the gate in place larger when the gate is closed or when it is open?

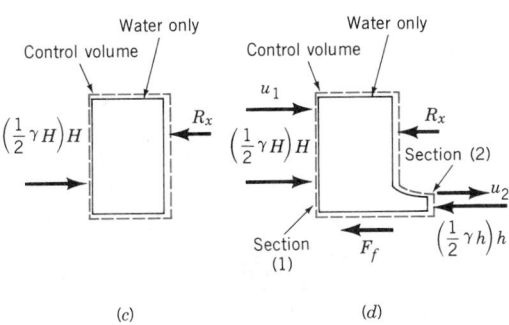

FIGURE E5.15

SOLUTION

We will answer this question by comparing expressions for the horizontal reaction force per unit width, R_x, between the gate and the water when the gate is closed and when the gate is open. The control volume used in each case is indicated with dashed lines in Figs. E5.14a and E5.14b.

When the gate is closed, the horizontal forces acting on the contents of the control volume are identified in Fig. E5.15c. Application of Eq. 5.22 to the contents of this control volume yields

$$\int_{cs} u\rho \overset{\text{0 (no flow)}}{\mathbf{V}} \cdot \hat{\mathbf{n}}\, dA = \frac{1}{2}\gamma H^2 - R_x \tag{1}$$

Note that the hydrostatic pressure force, $\gamma H^2/2$, is used. From Eq. 1, the force exerted on the water by the gate (which is equal to the force necessary to hold the gate stationary) is

$$R_x = \tfrac{1}{2}\gamma H^2 \tag{2}$$

which is equal in magnitude to the hydrostatic force exerted on the gate by the water.

When the gate is open, the horizontal forces acting on the contents of the control volume are shown in Fig. E5.15d. Application of Eq. 5.22 to the contents of this control volume leads to

$$\int_{cs} u\rho \mathbf{V} \cdot \hat{\mathbf{n}}\, dA = \frac{1}{2}\gamma H^2 - R_x - \frac{1}{2}\gamma h^2 - F_f \tag{3}$$

Note that we have assumed that the pressure distribution is hydrostatic in the water at sections (1) and (2). Also, the frictional force between the channel bottom and the water is specified as F_f. The surface integral in Eq. 3 is nonzero only where there is flow across the control surface. With the assumption of uniform velocity distributions

$$\int_{cs} u\rho \mathbf{V} \cdot \hat{\mathbf{n}}\, dA = (u_1)\rho(-u_1 H) + (+u_2)\rho(+u_2 h) \tag{4}$$

With $H \gg h$, the upstream velocity u_1, is much less than u_2 so that the contribution of the incoming momentum flow to the control surface integral can be neglected. Thus, Eqs. 3 and 4 combine to form

$$\rho u_2^2 h = \tfrac{1}{2}\gamma H^2 - R_x - \tfrac{1}{2}\gamma h^2 - F_f \tag{5}$$

Solving Eq. 5 for the reaction force, R_x, we obtain

$$R_x = \tfrac{1}{2}\gamma H^2 - \tfrac{1}{2}\gamma h^2 - F_f - \rho u_2^2 h \tag{6}$$

Comparing the expressions for R_x (Eqs. 2 and 6) we conclude that the reaction force between the gate and the water (and therefore the anchoring force required to hold the gate in place) is smaller when the gate is open than when it is closed. (Ans)

All of the linear momentum examples considered thus far have involved stationary and nondeforming control volumes which are thus inertial because there is no acceleration. A nondeforming control volume translating in a straight line at constant speed is also inertial because there is no acceleration. For a system and an inertial, moving, nondeforming control volume that are both coincident at an instant of time, the Reynolds transport theorem (Eq. 4.23) leads to

$$\frac{D}{Dt}\int_{sys} \mathbf{V}\rho\, d\Psi = \frac{\partial}{\partial t}\int_{cv} \mathbf{V}\rho\, d\Psi + \int_{cs} \mathbf{V}\rho\mathbf{V}_r \cdot \hat{\mathbf{n}}\, dA \tag{5.23}$$

When we combine Eq. 5.23 with Eqs. 5.19 and 5.20, we get

$$\frac{\partial}{\partial t} \int_{cv} \mathbf{V}\rho \, d\mathcal{V} + \int_{cs} \mathbf{V}\rho \mathbf{V}_r \cdot \hat{\mathbf{n}} \, dA = \sum \mathbf{F}_{\substack{\text{contents of the} \\ \text{control volume}}} \tag{5.24}$$

When the equation relating absolute, relative, and control volume velocities (Eq. 5.14) is used with Eq. 5.24, the result is

$$\frac{\partial}{\partial t} \int_{cv} (\mathbf{V}_r + \mathbf{V}_{cv})\rho \, d\mathcal{V} + \int_{cs} (\mathbf{V}_r + \mathbf{V}_{cv})\rho \mathbf{V}_r \cdot \hat{\mathbf{n}} \, dA = \sum \mathbf{F}_{\substack{\text{contents of the} \\ \text{control volume}}} \tag{5.25}$$

For a constant control volume velocity, $\mathbf{V}_{cv}$, and steady flow in the control volume reference frame,

$$\frac{\partial}{\partial t} \int_{cv} (\mathbf{V}_r + \mathbf{V}_{cv})\rho \, d\mathcal{V} = 0 \tag{5.26}$$

Also, for this inertial, nondeforming control volume

$$\int_{cs} (\mathbf{V}_r + \mathbf{V}_{cv})\rho \mathbf{V}_r \cdot \hat{\mathbf{n}} \, dA = \int_{cs} \mathbf{V}_r \rho \mathbf{V}_r \cdot \hat{\mathbf{n}} \, dA + \mathbf{V}_{cv} \int_{cs} \rho \mathbf{V}_r \cdot \hat{\mathbf{n}} \, dA \tag{5.27}$$

For steady flow (on an instantaneous or time-average basis) Eq. 5.15 gives

$$\int_{cs} \rho \mathbf{V}_r \cdot \hat{\mathbf{n}} \, dA = 0 \tag{5.28}$$

Combining Eqs. 5.25, 5.26, 5.27, and 5.28, we conclude that the linear momentum equation for an inertial, moving, nondeforming control volume that involves steady (instantaneously or time-average) flow is

$$\int_{cs} \mathbf{V}_r \rho \mathbf{V}_r \cdot \hat{\mathbf{n}} \, dA = \sum \mathbf{F}_{\substack{\text{contents of the} \\ \text{control volume}}} \tag{5.29}$$

Example 5.16 illustrates the use of Eq. 5.29.

EXAMPLE 5.16

An airplane moves forward on a horizontal path with a steady air speed of 500 mph. The wings of this plane deflect the air flowing near the wing surfaces by an angle of 10° as shown in Fig. E5.16. Estimate the mass flowrate of air required over the wings to produce a momentum flow change that is in itself equivalent to a vertical reaction force between air and wing surfaces equal to 5000 lb. If the effective span of the wing surfaces is 30 ft and the air density is 10^{-3} slugs/ft^3, determine the incoming height, h, of the airstream required to produce this effect.

SOLUTION

We need to relate the reaction force on the wings due to momentum flow change to the mass flowrate of air sought. To do this we select a control volume that moves with the plane as indicated in Fig. E5.16a. The control volume includes in it, at any instant, the air turned 10° by the wing surfaces. We assume that this control volume is inertial. Along the portions of the control surface not in contact with the wing, pressure forces exist. However, since we are only concerned with the reaction force between air and wing surfaces due to momentum flow change, we disregard these pressure forces. Thus, as is indicated in Fig. E5.16b, we consider the reaction force, F_L, as the only external vertical force acting on the air in the control volume. This reaction force is equal to 5000 lb.

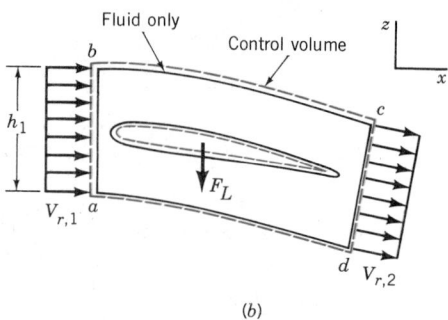

FIGURE E5.16

Applying the vertical or z component of the linear momentum equation (Eq. 5.29) to the contents of this control volume we get

$$\int_{cs} w_r \rho \mathbf{V}_r \cdot \hat{\mathbf{n}} \, dA = -F_L \tag{1}$$

where w_r is the component of the relative velocity, $\mathbf{V}_r$, in the z direction. All values of w_r and $\mathbf{V}_r$ are as measured relative to the moving control volume coordinate system. We assume that the air enters the control volume in the x direction at surface ab at a speed of 500 mph. Furthermore, we simplify the analysis by assuming that the speed of the flow leaving the control volume at surface cd is also 500 mph and that the flow is turned from its incoming direction by 10°. Thus, with $w_{r,1} = 0$ and $\mathbf{V} \cdot \hat{\mathbf{n}} = 0$ on sides bc and da of the control volume (stream surfaces), Eq. 1 becomes

$$-w_{r,2}\dot{m}_{r,2} = -F_L \tag{2}$$

where $\dot{m}_{r,2}$ is the mass flowrate relative to the control volume at section (2). Conservation of mass (Eq. 5.9) applied to this control volume leads to

$$\dot{m}_{r,2} = \dot{m}_{r,1} = \dot{m}_r \tag{3}$$

Combining Eqs. 2 and 3 we get

$$w_{r,2}\dot{m}_r = F_L$$

or

$$\dot{m}_r = \frac{F_L}{w_{r,2}} = \frac{F_L}{u_{r,1} \sin 10°}$$

and

$$\dot{m}_r = \frac{(5000 \text{ lb})(3600 \text{ s/hr})}{(500 \text{ mi/hr})(5280 \text{ ft/mi})} \frac{[1 \text{ (slug·ft)/(lb·s}^2)]}{(0.1736)} = 39.3 \text{ slugs/s} \tag{Ans}$$

To determine the entrance height of the airstream required we use

$$\dot{m}_r = \rho h_1 \ell u_{r,1} \tag{4}$$

where h_1 is the entrance height of airstream, ℓ is the effective wing span, and $u_{r,1}$ is the velocity of the flow entering the control volume. Solving Eq. 4 for entrance height, h_1, we obtain

$$h_1 = \frac{\dot{m}_r}{\rho \ell u_{r,1}}$$

or

$$h_1 = \frac{(39.3 \text{ slugs/s})(3600 \text{ s/hr})}{(10^{-3} \text{ slug/ft}^3)(30 \text{ ft})(500 \text{ mph})(5280 \text{ ft/mi})} = 1.79 \text{ ft} \tag{Ans}$$

It should be clear from what has been considered about linear momentum that fluid flows can lead to a reaction force in the following ways:

1. Linear momentum flow variation in direction and/or magnitude.
2. Fluid pressure forces.
3. Fluid friction forces.
4. Fluid weight.

The selection of a control volume is an important matter. An appropriate control volume can make a problem solution straightforward.

5.2.3 Derivation of the Moment-of-Momentum Equation[2]

In many engineering problems, the moment of a force with respect to an axis, namely, *torque,* is important. Newton's second law of motion has already led to a useful relationship between forces and linear momentum flow. The linear momentum equation can also be used to solve problems involving torques. However, by forming the moment of the linear momentum and the resultant force associated with each particle of fluid with respect to a point in an inertial coordinate system, we will develop a *moment-of-momentum equation* that relates *torques* and *angular momentum flow* for the contents of a control volume. When torques are important, the moment-of-momentum equation is often more convenient to use than the linear momentum equation.

Application of Newton's second law of motion to a particle of fluid yields

$$\frac{D}{Dt}(\mathbf{V}\rho \, \delta\Psi) = \delta\mathbf{F}_{\text{particle}} \tag{5.30}$$

where $\mathbf{V}$ is the particle velocity measured in an inertial reference system, ρ is the particle density, $\delta\Psi$ is the infinitesimally small particle volume, and $\delta\mathbf{F}_{\text{particle}}$ is the resultant external force acting on the particle. If we form the moment of each side of Eq. 5.30 with respect to the origin of an inertial coordinate system, we obtain

$$\mathbf{r} \times \frac{D}{Dt}(\mathbf{V}\rho \, \delta\Psi) = \mathbf{r} \times \delta\mathbf{F}_{\text{particle}} \tag{5.31}$$

where $\mathbf{r}$ is the position vector from the origin of the inertial coordinate system to the

---

[2] This section may be omitted, along with Sections 5.2.4 and 5.3.5, without loss of continuity in the text material.

fluid particle (Fig. 5.3). We note that

$$\frac{D}{Dt}[(\mathbf{r} \times \mathbf{V})\rho \,\delta \Psi] = \frac{D\mathbf{r}}{Dt} \times \mathbf{V}\rho \,\delta \Psi + \mathbf{r} \times \frac{D(\mathbf{V}\rho \,\delta \Psi)}{Dt} \tag{5.32}$$

and

$$\frac{D\mathbf{r}}{Dt} = \mathbf{V} \tag{5.33}$$

Thus, since

$$\mathbf{V} \times \mathbf{V} = 0 \tag{5.34}$$

by combining Eqs. 5.31, 5.32, 5.33, and 5.34, we obtain the expression

$$\frac{D}{Dt}[(\mathbf{r} \times \mathbf{V})\rho \,\delta \Psi] = \mathbf{r} \times \delta \mathbf{F}_{\text{particle}} \tag{5.35}$$

Equation 5.35 is valid for every particle of a system. For a system (collection of fluid particles), we need to use the sum of both sides of Eq. 5.35 to obtain

$$\int_{\text{sys}} \frac{D}{Dt}[(\mathbf{r} \times \mathbf{V})\rho \,d\Psi] = \sum (\mathbf{r} \times \mathbf{F})_{\text{sys}} \tag{5.36}$$

where

$$\sum \mathbf{r} \times \delta \mathbf{F}_{\text{particle}} = \sum (\mathbf{r} \times \mathbf{F})_{\text{sys}} \tag{5.37}$$

We note that

$$\frac{D}{Dt}\int_{\text{sys}} (\mathbf{r} \times \mathbf{V})\rho \,d\Psi = \int_{\text{sys}} \frac{D}{Dt}[(\mathbf{r} \times \mathbf{V})\rho \,d\Psi] \tag{5.38}$$

since the sequential order of differentiation and integration can be reversed without consequence. Thus, from Eqs. 5.36 and 5.38 we get

$$\frac{D}{Dt}\int_{\text{sys}} (\mathbf{r} \times \mathbf{V})\rho \,d\Psi = \sum (\mathbf{r} \times \mathbf{F})_{\text{sys}} \tag{5.39}$$

or

$$\begin{array}{ll} \text{the time rate of change of the} & = \text{sum of the external torques} \\ \text{moment-of-momentum of the system} & \text{acting on the system} \end{array}$$

For a control volume that is instantaneously coincident with the system, the torques acting on the system and on the control volume contents will be identical,

$$\sum (\mathbf{r} \times \mathbf{F})_{\text{sys}} = \sum (\mathbf{r} \times \mathbf{F})_{\text{cv}} \tag{5.40}$$

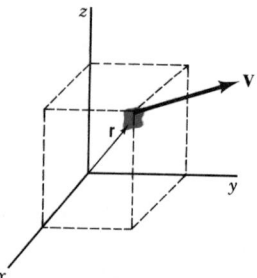

FIGURE 5.3   Inertial coordinate system.

Further, for the system and the contents of the coincident control volume that is fixed and nondeforming, the Reynolds transport theorem (Eq. 4.19) leads to

$$\frac{D}{Dt} \int_{\text{sys}} (\mathbf{r} \times \mathbf{V}) \rho \, d\mathcal{V} = \frac{\partial}{\partial t} \int_{\text{cv}} (\mathbf{r} \times \mathbf{V}) \rho \, d\mathcal{V} + \int_{\text{cs}} (\mathbf{r} \times \mathbf{V}) \rho \mathbf{V} \cdot \hat{\mathbf{n}} \, dA \qquad (5.41)$$

or

time rate of change of the moment-of-momentum of the system	=	time rate of change of the moment-of-momentum of the contents of the control volume	+	net rate of flow of moment-of-momentum through the control surface

For a control volume that is fixed (and therefore inertial) and nondeforming, we combine Eqs. 5.40 and 5.41 to obtain the moment-of-momentum equation,

$$\frac{\partial}{\partial t} \int_{\text{cv}} (\mathbf{r} \times \mathbf{V}) \rho \, d\mathcal{V} + \int_{\text{cs}} (\mathbf{r} \times \mathbf{V}) \rho \mathbf{V} \cdot \hat{\mathbf{n}} \, dA = \sum (\mathbf{r} \times \mathbf{F})_{\substack{\text{contents of the} \\ \text{control volume}}} \qquad (5.42)$$

An important category of fluid mechanical problems that are readily solved with the help of the moment-of-momentum equation (Eq. 5.42) involves machines that rotate or tend to rotate around a single axis. Examples of these machines include rotary lawn sprinklers, ceiling fans, lawn mower blades, wind turbines, turbochargers, and gas turbine engines. As a class, these devices are often called turbomachines.

5.2.4 Application of the Moment-of-Momentum Equation[3]

We simplify our use of Eq. 5.42 in several ways:

1. We assume that flows considered are one-dimensional (uniform distributions of velocity at any section).

2. We confine ourselves to steady or steady-in-the-mean cyclical flows. Thus,

$$\frac{\partial}{\partial t} \int_{\text{cv}} (\mathbf{r} \times \mathbf{V}) \rho \, d\mathcal{V} = 0$$

at any instant of time for steady flows or on a time-average basis for cyclical unsteady flows.

3. We work only with the component of Eq. 5.42 resolved along the axis of rotation.

Consider the rotating sprinkler sketched in Fig. 5.4. Because the direction of the flow through the sprinkler from the inlet [section (1)] to the outlet [section (2)] of the arm changes, the water exerts a torque on the sprinkler head causing it to tend to rotate or to actually rotate in the direction shown, much like a turbine rotor. In applying the moment-of-momentum equation (Eq. 5.42) to this flow situation, we elect to use the fixed and nondeforming control volume shown in Fig. 5.4. This disk-shaped control volume contains within its boundaries the spinning or stationary sprinkler head and the portion of the water flowing through the sprinkler contained in the control volume at an instant. The control surface cuts through the sprinkler head solid material so that the shaft torque that resists motion can be clearly identified. When the sprinkler

---

[3] This section may be omitted, along with Sections 5.2.3 and 5.3.5, without loss of continuity in the text material.

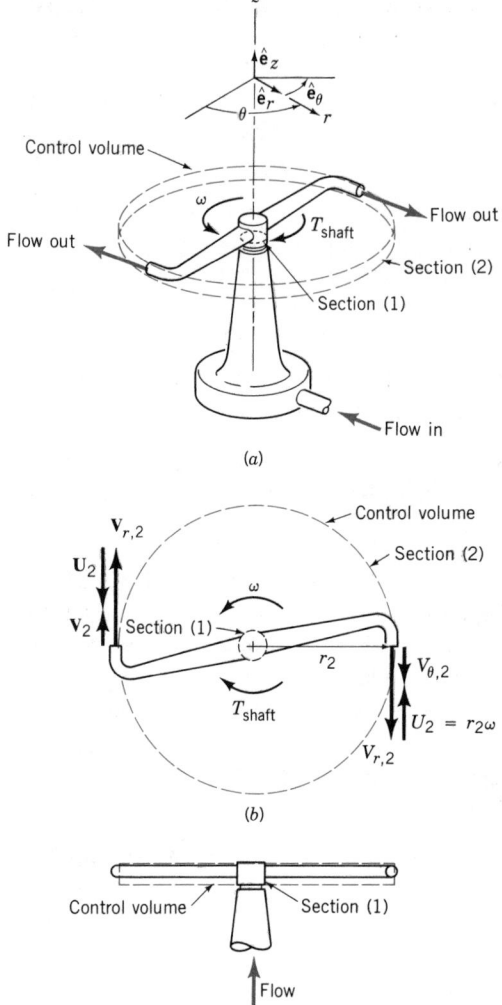

FIGURE 5.4 (*a*) Rotary water sprinkler. (*b*) Rotary water sprinkler, plan view. (*c*) Rotary water sprinkler, side view.

is rotating, the flow field in the stationary control volume is cyclical and unsteady, but steady in the mean. We proceed to use the axial component of the moment-of-momentum equation (Eq. 5.42) to analyze this flow.

The integrand of the moment-of-momentum flow term in Eq. 5.42,

$$\int_{cs} (\mathbf{r} \times \mathbf{V}) \rho \mathbf{V} \cdot \hat{\mathbf{n}} \, dA$$

can be nonzero only where fluid is crossing the control surface. Everywhere else on the control surface this term will be zero because $\mathbf{V} \cdot \hat{\mathbf{n}} = 0$. Water enters the control volume axially through the hollow stem of the sprinkler at section (1). At this portion of the control surface, the component of $\mathbf{r} \times \mathbf{V}$ resolved along the axis of rotation is zero because $\mathbf{r} \times \mathbf{V}$ and the axis of rotation are perpendicular. Thus, there is no axial moment-of-momentum flow in at section (1). Water leaves the control volume through each of the two nozzle openings at section (2). For the exiting flow, the magnitude of the axial component of $\mathbf{r} \times \mathbf{V}$ is $r_2 V_{\theta,2}$, where r_2 is the radius from the axis of rotation to the nozzle centerline and $V_{\theta,2}$ is the value of the tangential component of the velocity of the flow exiting each nozzle as observed from a frame of reference attached to the fixed and nondeforming control volume. The fluid velocity measured relative to a fixed control surface is an absolute velocity, $\mathbf{V}$. The velocity of the nozzle exit flow as viewed from the nozzle is called the relative velocity, $\mathbf{V}_r$. The absolute and relative velocities, $\mathbf{V}$ and $\mathbf{V}_r$, are related by the vector relationship

$$\mathbf{V} = \mathbf{V}_r + \mathbf{U} \tag{5.43}$$

where $\mathbf{U}$ is the velocity of the moving nozzle as measured relative to the fixed control surface. The cross product and the dot product involved in the moment-of-momentum flow term of Eq. 5.42,

$$\int_{cs} (\mathbf{r} \times \mathbf{V})\rho \mathbf{V} \cdot \hat{\mathbf{n}} \, dA$$

each result in a positive or negative value. For flow into the control volume, $\mathbf{V} \cdot \hat{\mathbf{n}}$ is negative. For flow out, $\mathbf{V} \cdot \hat{\mathbf{n}}$ is positive. The correct algebraic sign to assign the axial component of $\mathbf{r} \times \mathbf{V}$ can be ascertained by using the right-hand rule. The positive direction along the axis of rotation is the direction the thumb of the right-hand points when it is extended and the remaining fingers are curled around the rotation axis in the positive direction of rotation as illustrated in Fig. 5.5. The direction of the axial component of $\mathbf{r} \times \mathbf{V}$ is similarly ascertained by noting the direction of the cross product of the radius from the axis of rotation, $r\hat{\mathbf{e}}_r$, and the tangential component of absolute velocity, $V_\theta\hat{\mathbf{e}}_\theta$. Thus, for the sprinkler of Fig. 5.4, we can state that

$$\left[\int_{cs} (\mathbf{r} \times \mathbf{V})\rho \mathbf{V} \cdot \hat{\mathbf{n}} \, dA \right]_{\text{axial}} = (-r_2 V_{\theta,2})(+\dot{m}) \tag{5.44}$$

where, because of mass conservation, $\dot{m}$ is the total mass flowrate through both nozzles. As was demonstrated in Example 5.7, the mass flowrate is the same whether the sprinkler rotates or not.

The torque term $[\Sigma (\mathbf{r} \times \mathbf{F})_{\text{contents of the control volume}}]$ of the moment-of-momentum equation (Eq. 5.42) is analyzed next. Confining ourselves to torques acting with respect to the axis of rotation only, we conclude that the shaft torque is important. The net torque with respect to the axis of rotation associated with normal forces exerted on the contents of the control volume will be very small if not zero. The net axial torque due to fluid tangential forces is also negligibly small for the control volume of Fig. 5.4. Thus, for the sprinkler of Fig. 5.4

$$\Sigma \left[(\mathbf{r} \times \mathbf{F})_{\substack{\text{contents of the} \\ \text{control volume}}} \right]_{\text{axial}} = -T_{\text{shaft}} \tag{5.45}$$

FIGURE 5.5 Right-hand rule convention.

The shaft torque, T_{shaft}, is entered as a negative amount in Eq. 5.45 because it is shown in Fig. 5.4 as opposing the sprinkler arm motion. This is true for all turbine devices.

For the sprinkler of Fig. 5.4, the axial component of the moment-of-momentum equation (Eq. 5.42) is, from Eqs. 5.44 and 5.45

$$-r_2 V_{\theta,2} \dot{m} = -T_{shaft} \tag{5.46}$$

We could evaluate the shaft power, $\dot{W}_{shaft}$, associated with shaft torque, T_{shaft}, by forming the product of T_{shaft} and the rotational speed of the shaft, ω. Thus, from Eq. 5.46 we get

$$\dot{W}_{shaft} = T_{shaft}\omega = r_2 V_{\theta,2} \dot{m}\omega \tag{5.47}$$

Since $r_2\omega$ is the speed of each sprinkler nozzle, U, we can also state Eq. 5.47 in the form

$$\dot{W}_{shaft} = U_2 V_{\theta,2} \dot{m} \tag{5.48}$$

Shaft work per unit mass, w_{shaft}, is equal to $\dot{W}_{shaft}/\dot{m}$. Dividing Eq. 5.48 by the mass flowrate, $\dot{m}$, we obtain

$$w_{shaft} = U_2 V_{\theta,2} \tag{5.49}$$

The principles associated with this sprinkler example can be extended to handle most simplified turbomachine flows. The fundamental technique is not difficult. However, the geometry of some turbomachine flows is quite complicated.

Example 5.17 illustrates how the axial component of the moment-of-momentum equation (Eq. 5.46) can be used.

EXAMPLE 5.17

Water enters a rotating lawn sprinkler through its base at the steady rate of 1000 ml/s as sketched in Fig. E5.17. The exit area of each of the two nozzles is 30 mm^2 and the flow leaving each nozzle is in the tangential direction. The radius from the axis of rotation to the centerline of each nozzle is 200 mm.

(a) Determine the resisting torque required to hold the sprinkler head stationary.

(b) Determine the resisting torque associated with the sprinkler rotating with a constant speed of 500 rev/min.

(c) Determine the speed of the sprinkler if no resisting torque is applied.

SOLUTION

To solve parts (a), (b), and (c) of this example we can use the same fixed and nondeforming, disk shaped control volume illustrated in Fig. 5.4. As is indicated in Fig. E.5.17a the only axial torque considered is the one resisting motion, T_{shaft}.

When the sprinkler head is held stationary as specified in part (a) of this example problem, the velocities of the fluid entering and leaving the control volume are as shown in Fig. E.17b. Equation 5.46 applies to the contents of this control volume. Thus

$$T_{shaft} = r_2 V_{\theta,2} \dot{m} \tag{1}$$

Since the control volume is fixed and nondeforming, and the flow exiting from each nozzle is

(a)

(b)

(c)

FIGURE E5.17

tangential,

$$V_{\theta,2} = V_2 \tag{2}$$

Equations 1 and 2 give

$$T_{shaft} = r_2 V_2 \dot{m} \tag{3}$$

In Example 5.7, we ascertained that $V_2 = 16.7$ m/s. Thus, from Eq. 3,

$$T_{shaft} = \frac{(200 \text{ mm})(16.7 \text{ m/s})(1000 \text{ ml/s})(10^{-3} \text{ m}^3/\text{liter})(999 \text{ kg/m}^3)[1 \text{ (N/kg)}/(\text{m/s}^2)]}{(1000 \text{ mm/m})(1000 \text{ ml/liter})}$$

or

$$T_{shaft} = 3.34 \text{ N·m} \tag{Ans}$$

When the sprinkler is rotating at a constant speed of 500 rpm, the flow field in the control volume is unsteady but cyclical. Thus, the flow field is steady in the mean. The velocities of the flow entering and leaving the control volume are as indicated in Fig. E5.17c. The absolute velocity of the fluid leaving each nozzle, V_2, is, from Eq. 5.43,

$$V_2 = V_{r,2} - U_2 \tag{4}$$

where

$$V_{r,2} = 16.7 \text{ m/s}$$

as determined in Example 5.7. The speed of the nozzle, U_2, is obtained from

$$U_2 = r_2\omega \tag{5}$$

Application of the axial component of the moment-of-momentum equation (Eq. 5.46) leads again to Eq. 3. From Eqs. 4 and 5,

$$V_2 = 16.7 \text{ m/s} - r_2\omega = 16.7 \text{ m/s} - \frac{(200 \text{ mm})(500 \text{ rev/min})(2\pi \text{ rad/rev})}{(1000 \text{ mm/m})(60 \text{ s/min})}$$

or

$$V_2 = 16.7 \text{ m/s} - 10.5 \text{ m/s} = 6.2 \text{ m/s}$$

Thus, using Eq. 3, we get

$$T_{\text{shaft}} = \frac{(200 \text{ mm})(6.2 \text{ m/s})(1000 \text{ ml/s})(10^{-3} \text{ m}^3/\text{liter})(999 \text{ kg/m}^3)[1 \text{ (N/kg)/(m/s}^2)]}{(1000 \text{ mm/m})(1000 \text{ ml/liter})}$$

or

$$T_{\text{shaft}} = 1.24 \text{ N·m} \tag{Ans}$$

Note that the resisting torque associated with sprinkler head rotation is much less than the resisting torque that is required to hold the sprinkler stationary.

When no resisting torque is applied to the rotating sprinkler head, a maximum constant speed of rotation will occur as demonstrated below. Application of Eqs. 3, 4, and 5 to the contents of the control volume results in

$$T_{\text{shaft}} = r_2(V_{r,2} - r_2\omega)\dot{m} \tag{6}$$

For no resisting torque, Eq. 6 yields

$$0 = r_2(V_{r,2} - r_2\omega)\dot{m}$$

Thus,

$$\omega = \frac{V_{r,2}}{r_2} \tag{7}$$

In Example 5.7, we learned that the relative velocity of the fluid leaving each nozzle, $V_{r,2}$, is the same regardless of the speed of rotation of the sprinkler head, ω, as long as the mass flowrate of the fluid, $\dot{m}$, remains constant. Thus, by using Eq. 7 we obtain

$$\omega = \frac{V_{r,2}}{r_2} = \frac{(16.7 \text{ m/s})(1000 \text{ mm/m})}{(200 \text{ mm})} = 83.5 \text{ rad/s}$$

or

$$\omega = \frac{(83.5 \text{ rad/s})(60 \text{ s/min})}{2\pi \text{ rad/rev}} = 797 \text{ rpm} \tag{Ans}$$

For this condition ($T_{\text{shaft}} = 0$), the water both enters and leaves the control volume with zero angular momentum.

In summary, we observe that the resisting torque associated with rotation is less than the torque required to hold a rotor stationary. Even in the absence of a resisting torque, the rotor maximum speed is finite.

When the moment-of-momentum equation (Eq. 5.42) is applied to a more general, one-dimensional flow through a rotating machine, we obtain

$$T_{\text{shaft}} = (-\dot{m}_{\text{in}})(\pm r_{\text{in}} V_{\theta,\text{in}}) + \dot{m}_{\text{out}}(\pm r_{\text{out}} V_{\theta,\text{out}}) \tag{5.50}$$

by applying the same kind of analysis used with the sprinkler of Fig. 5.4. The "−" is used with mass flowrate into the control volume, $\dot{m}_{in}$, and the "+" is used with mass flowrate out of the control volume, $\dot{m}_{out}$, to account for the sign of the dot product, $\mathbf{V} \cdot \hat{\mathbf{n}}$, involved. Whether "+" or "−" is used with the rV_θ product depends on the direction of $(\mathbf{r} \times \mathbf{V})_{axial}$. A simple way to determine the sign of the rV_θ product is to compare the direction of V_θ and the blade speed, U. If V_θ and U are in the same direction, then the rV_θ product is positive. If V_θ and U are in opposite directions, the rV_θ product is negative. The sign of the shaft torque is "+" if T_{shaft} is in the same direction along the axis of rotation as ω, and "−" otherwise.

The shaft power, $\dot{W}_{shaft}$, is related to shaft torque, T_{shaft}, by

$$\dot{W}_{shaft} = T_{shaft}\omega \tag{5.51}$$

Thus, using Eq. 5.50 and 5.51 with a "+" sign for T_{shaft} in Eq. 5.50, we obtain

$$\dot{W}_{shaft} = (-\dot{m}_{in})(\pm r_{in}\omega V_{\theta,in}) + \dot{m}_{out}(\pm r_{out}\omega V_{\theta,out}) \tag{5.52}$$

or since $r\omega = U$

$$\dot{W}_{shaft} = (-\dot{m}_{in})(\pm U_{in}V_{\theta,in}) + \dot{m}_{out}(\pm U_{out}V_{\theta,out}) \tag{5.53}$$

The "+" is used for the UV_θ product when U and V_θ are in the same direction; the "−" is used when U and V_θ are in opposite directions. Also, since $+T_{shaft}$ was used to obtain Eq. 5.53, when $\dot{W}_{shaft}$ is positive, power is into the control volume (e.g., pump), and when $\dot{W}_{shaft}$ is negative, power is out of the control volume (e.g., turbine).

The shaft work per unit mass, w_{shaft}, can be obtained from the shaft power, $\dot{W}_{shaft}$, by dividing Eq. 5.53 by the mass flowrate, $\dot{m}$. By conservation of mass,

$$|\dot{m}| = |\dot{m}_{in}| = |\dot{m}_{out}|$$

From Eq. 5.53, we obtain

$$w_{shaft} = -(\pm U_{in}V_{\theta,in}) + (\pm U_{out}V_{\theta,out}) \tag{5.54}$$

The application of Eqs. 5.50, 5.53 and 5.54 is demonstrated in Examples 5.18 and 5.19.

EXAMPLE 5.18

An air fan has a bladed rotor of 12-in.-outside diameter and 10-in.-inside diameter as illustrated in Fig. E5.18a. The height of each rotor blade is constant at 1 in. from blade inlet to outlet. The flowrate is steady, on a time-average basis, at 230 ft³/min and the absolute velocity of the air at blade inlet, $\mathbf{V}_1$, is radial. The blade discharge angle is 30° from the tangential direction. If the rotor rotates at a constant speed of 1725 rpm, estimate the power required to run the fan.

SOLUTION

We select a fixed and nondeforming control volume that includes the rotating blades and the fluid within the blade row at an instant as shown with a dashed line in Fig. E5.18a. The flow within this control volume is cyclical, but steady in the mean. The only torque we consider is the driving shaft torque, T_{shaft}. This torque is provided by a motor. We assume that the entering

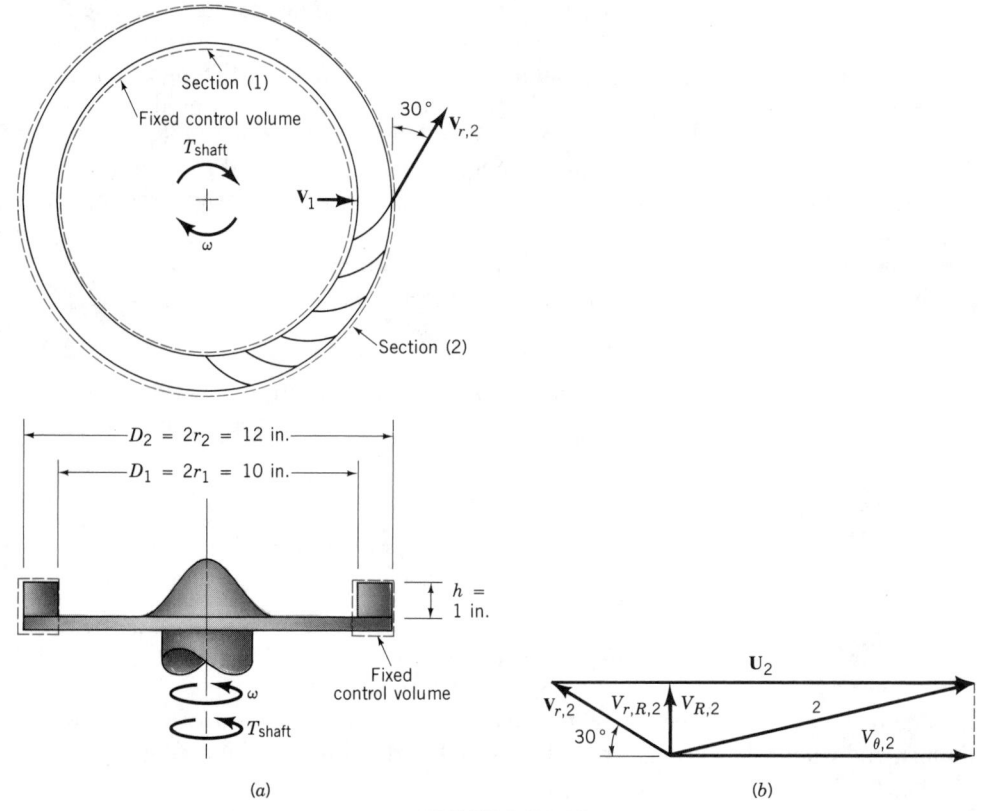

FIGURE E5.18

and leaving flows are each represented by uniformly distributed velocities and flow properties. Since shaft power is sought, Eq. 5.53 is appropriate. Application of Eq. 5.53 to the contents of the control volume in Fig. E5.18 gives

$$\dot{W}_{\text{shaft}} = (-\dot{m}_1) \; (\pm U_1 \overset{0 \;(\mathbf{V}_1 \; \text{is radial})}{\cancel{V_{\theta,1}}}) \; + \; \dot{m}_2(\pm U_2 V_{\theta,2}) \tag{1}$$

From Eq. 1 we see that to calculate fan power, we need mass flowrate, $\dot{m}$, rotor exit blade velocity, U_2, and fluid tangential velocity at blade exit, $V_{\theta,2}$. The mass flowrate, $\dot{m}$, is easily obtained from Eq. 5.6 as

$$\dot{m} = \rho Q = \frac{(2.38 \times 10^{-3} \; \text{slug/ft}^3)(230 \; \text{ft}^3/\text{min})}{(60 \; \text{s/min})} = 0.00912 \; \text{slug/s} \tag{2}$$

The rotor exit blade speed, U_2, is

$$U_2 = r_2\omega = \frac{(6 \; \text{in.})(1725 \; \text{rpm})(2\pi \; \text{rad/rev})}{(12 \; \text{in./ft})(60 \; \text{s/min})} = 90.3 \; \text{ft/s} \tag{3}$$

To determine the fluid tangential speed at the fan rotor exit, $V_{\theta,2}$, we use Eq. 5.43 to get

$$\mathbf{V}_2 = \mathbf{V}_{r,2} + \mathbf{U}_2 \tag{4}$$

The vector addition of Eq. 4 is shown in the form of a "velocity triangle" in Fig. E5.18b. From Fig. E5.18b, we can see that

$$V_{\theta,2} = U_2 - V_{r,2} \cos 30° \tag{5}$$

To solve Eq. 5 for $V_{\theta,2}$ we need a value of $V_{r,2}$, in addition to the value of U_2 already determined

(Eq. 3). To get $V_{r,2}$, we recognize that

$$V_{r,2} \sin 30° = V_{R,2} \tag{6}$$

where $V_{R,2}$ is the radial component of either $\mathbf{V}_{r,2}$ or $\mathbf{V}_2$. Also, using Eq. 5.6, we obtain

$$\dot{m} = \rho A_2 V_{R,2} \tag{7}$$

or since

$$A_2 = 2\pi r_2 h \tag{8}$$

where h is the blade height, Eqs. 7 and 8 combine to form

$$\dot{m} = \rho 2\pi r_2 h V_{R,2} \tag{9}$$

Taking Eqs. 6 and 9 together we get

$$V_{r,2} = \frac{\dot{m}}{\rho 2\pi r_2 h \sin 30°} \tag{10}$$

Substituting known values into Eq. 10, we obtain

$$V_{r,2} = \frac{(0.00912 \text{ slugs/s})(12 \text{ in./ft})(12 \text{ in./ft})}{(2.38 \times 10^{-3} \text{ slugs/ft}^3)2\pi(6 \text{ in.})(1 \text{ in.}) \sin 30°} = 29.3 \text{ ft/s}$$

By using this value of $V_{r,2}$ in Eq. 5 we get

$$V_{\theta,2} = U_2 - V_{r,2} \cos 30°$$

$$= 90.3 \text{ ft/s} - (29.3 \text{ ft/s})(0.866) = 64.9 \text{ ft/s}$$

Equation 1 can now be used to obtain

$$\dot{W}_{shaft} = \dot{m} U_2 V_{\theta,2} = \frac{(0.00912 \text{ slug/s})(90.3 \text{ ft/s})(64.9 \text{ ft/s})}{[1 \text{ (slug·ft/s}^2)/\text{lb}][550 \text{ (ft·lb)/(hp·s)}]}$$

or

$$\dot{W}_{shaft} = 0.0972 \text{ hp} \tag{Ans}$$

Note that the "+" was used with the $U_2 V_{\theta,2}$ product because U_2 and $V_{\theta,2}$ are in the same direction. This result, 0.0972 hp, is the power that needs to be delivered through the fan shaft for the given conditions. Ideally, all of this power would go into the flowing air. However, because of fluid friction, only some of this power will produce a useful effect (e.g., pressure rise) in the air. How much useful effect depends on the efficiency of the energy transfer between the fan blades and the fluid. This important aspect of energy transfer is covered in Section 5.3.5.

EXAMPLE 5.19

An air turbine is used to drive the high-speed drill used by your dentist. The turbine stage is sketched in Fig. E5.19a. Air exiting from the upstream nozzle holes forces the turbine blades to move in the direction shown. Given the following data, estimate the related shaft energy, per unit mass of air flowing through the turbine, available to drive the drill bit.

Turbine rotor outside radius, r_o	0.168 in.
Turbine rotor inside radius, r_i	0.133 in.
Turbine rotor speed	300,000 rpm
Tangential velocity of fluid out of nozzles	$2U$
Absolute tangential velocity of fluid out of rotor	0

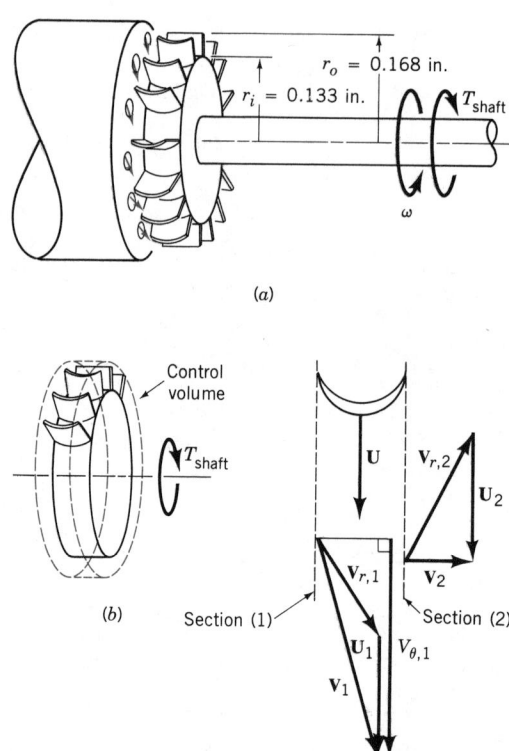

FIGURE E5.19

SOLUTION

We use a fixed and nondeforming control volume that includes the turbine rotor and the fluid in the rotor blade passages at an instant of time. The only torque we consider is the shaft torque acting on the contents of this control volume. Further, for simplicity, we analyze this problem using an arithmetic mean radius, r_m, where

$$r_m = \tfrac{1}{2}(r_o + r_i)$$

and with assumed uniform distributions of fluid velocities and other properties at rotor entrance [section (1)] and exit [section (2)]. A simplified sketch of the "velocity triangles" at rotor entrance and exit is shown in Fig. E5.19c. The application of Eq. 5.54 to the contents of this control volume gives

$$w_{\text{shaft}} = -(\pm U_{1m}V_{\theta,1}) + (\pm U_{2m}V_{\theta,2}) \tag{1}$$

or, since $V_{\theta,2} = 0$,

$$w_{\text{shaft}} = -U_{1m}V_{\theta,1} \tag{2}$$

The "+" is used with the $U_{1m}V_{\theta,1}$ product because U_{1m} and $V_{\theta,1}$ are in the same direction. Since

$$V_{\theta,1} = 2U_{1m} \tag{3}$$

combining Eqs. 2 and 3 yields

$$w_{\text{shaft}} = -2U_{1m}^2 \tag{4}$$

The arithmetic mean radius blade velocity, U_{1m}, is

$$U_{1m} = r_{1m}\,\omega = \frac{(r_o + r_i)\omega}{2}$$

or

$$U_{1m} = \frac{(0.168 \text{ in.} + 0.133 \text{ in.})(300,00 \text{ rpm})(2\pi \text{ rad/rev})}{2\,(12 \text{ in./ft})(60 \text{ s/min})} = 394 \text{ ft/s}$$

Thus, from Eq. 4,

$$w_{\text{shaft}} = -2(394 \text{ ft/s})^2[1 \text{ lb/(slug·ft/s}^2)]$$

$$= -310,000 \text{ ft·lb/slug} \qquad\qquad \text{(Ans)}$$

This amount, 310,000 ft·lb/slug, represents the amount of energy available at the shaft to drive the drill bit for every slug of air that flows through the turbine. Ideally this would also represent the amount of energy given up by each slug of air as it flows through the turbine. However, because of fluid friction, the actual amount of energy given up by each slug of air will be greater than the amount available at the shaft. How much greater depends on the efficiency of the fluid mechanical energy transfer between the fluid and the turbine blades. This important aspect of energy transfer is considered in the Section 5.3.5.

5.3   FIRST LAW OF THERMODYNAMICS—THE ENERGY EQUATION

5.3.1   Derivation of the Energy Equation

The first law of thermodynamics for a system is, in words

$$
\begin{matrix}
\text{time rate of} \\
\text{increase of the} \\
\text{total stored energy} \\
\text{of the system}
\end{matrix}
=
\begin{matrix}
\text{net time rate of} \\
\text{energy addition by} \\
\text{heat transfer into} \\
\text{the system}
\end{matrix}
+
\begin{matrix}
\text{net time rate of} \\
\text{energy addition by} \\
\text{work transfer into} \\
\text{the system}
\end{matrix}
$$

In symbolic form, this statement is

$$\frac{D}{Dt}\int_{\text{sys}} e\rho\, d\mathcal{V} = \left(\sum \dot{Q}_{\text{in}} - \sum \dot{Q}_{\text{out}}\right)_{\text{sys}} + \left(\sum \dot{W}_{\text{in}} - \sum \dot{W}_{\text{out}}\right)_{\text{sys}}$$

or

$$\frac{D}{Dt}\int_{\text{sys}} e\rho\, d\mathcal{V} = (\dot{Q}_{\underset{\text{in}}{\text{net}}} + \dot{W}_{\underset{\text{in}}{\text{net}}})_{\text{sys}} \qquad\qquad (5.55)$$

Some of these variables deserve a brief explanation before proceeding further. The total stored energy per unit mass for each particle of mass in the system, e, is related to the internal energy per unit mass, $\check{u}$, the kinetic energy per unit mass, $V^2/2$, and the potential energy per unit mass, gz, by the equation

$$e = \check{u} + \frac{V^2}{2} + gz \qquad\qquad (5.56)$$

The net rate of heat transfer into the system is denoted with "$\dot{Q}_{\text{net in}}$," and the net rate of work transfer into the system is labeled "$\dot{W}_{\text{net in}}$." Heat transfer and work transfer are considered "+" going into the system and "−" coming out.

Equation 5.55 is valid for inertial and noninertial reference systems. We proceed to develop the control volume statement of the first law of thermodynamics. For the control volume that is coincident with the system at an instant of time,

$$(\dot{Q}_{\substack{net \\ in}} + \dot{W}_{\substack{net \\ in}})_{sys} = (\dot{Q}_{\substack{net \\ in}} + \dot{W}_{\substack{net \\ in}})_{\substack{coincident \\ control \\ volume}} \tag{5.57}$$

Furthermore, for the system and the contents of the coincident control volume that is fixed and nondeforming, the Reynolds transport theorem (Eq. 4.19) allows us to conclude that

$$\frac{D}{Dt}\int_{sys} e\rho \, d\mathcal{V} = \frac{\partial}{\partial t}\int_{cv} e\rho \, d\mathcal{V} + \int_{cs} e\rho \mathbf{V} \cdot \hat{\mathbf{n}} \, dA \tag{5.58}$$

or in words,

the time rate of increase of the total stored energy of the system	=	the time rate of increase of total stored energy of the contents of the control volume	+	the net rate of flow of total stored energy out of the control volume through the control surface

Combining Eqs. 5.55, 5.57, and 5.58 we get the control volume formula for the first law of thermodynamics:

$$\frac{\partial}{\partial t}\int_{cv} e\rho \, d\mathcal{V} + \int_{cs} e\rho \mathbf{V} \cdot \hat{\mathbf{n}} \, dA = (\dot{Q}_{\substack{net \\ in}} + \dot{W}_{\substack{net \\ in}})_{cv} \tag{5.59}$$

The total stored energy per unit mass, e, in Eq. 5.59 is for particles of mass entering, leaving, and within the control volume. Further explanation of the heat transfer and work transfer involved in this equation follows.

The heat transfer rate, $\dot{Q}$, represents all of the ways in which energy is exchanged between the control volume contents and surroundings because of a temperature difference. Thus, radiation, conduction, and/or convection are possible. Heat transfer into the control volume is considered positive, heat transfer out is negative. In many engineering applications, the process is *adiabatic;* the heat transfer rate, $\dot{Q}$, is zero. The net heat transfer rate, $\dot{Q}_{net \, in}$, can also be zero when $\Sigma \dot{Q}_{in} - \Sigma \dot{Q}_{out} = 0$.

The work transfer rate, $\dot{W}$, also called *power,* is positive when work is done on the contents of the control volume by the surroundings. Otherwise, it is considered negative. Work can be transferred across the control surface in several ways. In the following paragraphs, we consider some important forms of work transfer.

In many instances, work is transferred across the control surface by a moving shaft. In rotary devices such as turbines, fans, and propellers, a rotating shaft transfers work across the control surface. Even in reciprocating machines like positive displacement internal combustion engines and compressors that utilize piston-in-cylinder arrangements, a rotating crankshaft is used. Since work is the dot product of force and related displacement, rate of work (or power) is the dot product of force and related displacement per unit time. For a rotating shaft, the power transfer, $\dot{W}_{shaft}$, is related to the shaft torque that causes the rotation, T_{shaft}, and the angular velocity of the shaft, ω, by the relationship:

$$\dot{W}_{shaft} = T_{shaft}\omega$$

When the control surface cuts through the shaft material, the shaft torque is exerted by shaft material at the control surface. To allow for consideration of problems involving more than one shaft we use the notation

$$\dot{W}_{\substack{\text{shaft} \\ \text{net in}}} = \sum \dot{W}_{\substack{\text{shaft} \\ \text{in}}} - \sum \dot{W}_{\substack{\text{shaft} \\ \text{out}}} \tag{5.60}$$

Work transfer can also occur at the control surface when a force associated with fluid normal stress acts over a distance. Consider the simple pipe flow illustrated in Fig. 5.6 and the control volume shown. For this situation, the fluid normal stress, σ, is simply equal to the negative of fluid pressure, p, in all directions; that is

$$\sigma = -p \tag{5.61}$$

This relationship can be used with varying amounts of approximation for many engineering problems (see Chapter 6). The appropriate selection of the control volume used for an analysis is important.

The power transfer associated with normal stresses acting on a single fluid particle, $\delta\dot{W}_{\text{normal stress}}$, can be evaluated as the dot product of the normal stress force, $\delta\mathbf{F}_{\text{normal stress}}$, and the fluid particle velocity, $\mathbf{V}$, as

$$\delta\dot{W}_{\text{normal stress}} = \delta\mathbf{F}_{\text{normal stress}} \cdot \mathbf{V}$$

If the normal stress force is expressed as the product of local normal stress, $\sigma = -p$, and fluid particle surface area, $\hat{\mathbf{n}}\,\delta A$, the result is

$$\delta\dot{W}_{\text{normal stress}} = \sigma\hat{\mathbf{n}}\,\delta A \cdot \mathbf{V} = -p\hat{\mathbf{n}}\,\delta A \cdot \mathbf{V} = -p\mathbf{V}\cdot\hat{\mathbf{n}}\,\delta A$$

For all fluid particles on the control surface of Fig. 5.6 at the instant considered, power transfer due to fluid normal stress, $\dot{W}_{\text{normal stress}}$, is

$$\dot{W}_{\substack{\text{normal} \\ \text{stress}}} = \int_{cs} \sigma\mathbf{V}\cdot\hat{\mathbf{n}}\,dA = \int_{cs} -p\mathbf{V}\cdot\hat{\mathbf{n}}\,dA \tag{5.62}$$

Note that the value of $\dot{W}_{\text{normal stress}}$ for particles on the wetted inside surface of the pipe is zero because $\mathbf{V}\cdot\hat{\mathbf{n}}$ is zero there. Thus, $\dot{W}_{\text{normal stress}}$ is possibly nonzero only where fluid enters and leaves the control volume. Although only a simple pipe flow was considered, Eq. 5.62 is quite general and the control volume used in this example can serve as a general model for other cases.

Work transfer can also occur at the control surface because of tangential stress forces. Rotating shaft work is transferred by tangential stresses in the shaft material. For a fluid particle, shear stress force power, $\delta\dot{W}_{\text{tangential stress}}$, can be evaluated as the dot product of tangential stress force, $\delta\mathbf{F}_{\text{tangential stress}}$, and the fluid particle velocity, $\mathbf{V}$. That is,

$$\delta\dot{W}_{\text{tangential stress}} = \delta\mathbf{F}_{\text{tangential stress}} \cdot \mathbf{V}$$

For the control volume of Fig. 5.6, the fluid particle velocity is zero everywhere on the wetted inside surface of the pipe. Thus, no tangential stress work is transferred across that portion of the control surface. Furthermore, where fluid crosses the control

$$u_1 = u_{\max}\left[1 - \left(\frac{r}{R}\right)^2\right] \qquad u_2 = u_{\max}\left[1 - \left(\frac{r}{R}\right)^2\right]$$

FIGURE 5.6 Simple fully developed pipe flow.

surface, the tangential stress force is perpendicular to the fluid particle velocity and therefore tangential stress work transfer is also zero there. In general, we select control volumes like the one of Fig. 5.6 and consider fluid tangential stress power transfer to be negligibly small.

Using the information we have developed about power, we can express the first law of thermodynamics for the contents of a control volume by combining Eqs. 5.59, 5.60, and 5.62 to obtain

$$\frac{\partial}{\partial t} \int_{cv} e\rho \, d\Psi + \int_{cs} e\rho \mathbf{V} \cdot \hat{\mathbf{n}} \, dA = \dot{Q}_{\substack{net \\ in}} + \dot{W}_{\substack{shaft \\ net \, in}} - \int_{cs} p\mathbf{V} \cdot \hat{\mathbf{n}} \, dA \qquad (5.63)$$

When the equation for total stored energy (Eq. 5.56) is considered with Eq. 5.63, we obtain the energy equation:

$$\frac{\partial}{\partial t} \int_{cv} e\rho \, d\Psi + \int_{cs} \left(\check{u} + \frac{p}{\rho} + \frac{V^2}{2} + gz \right) \rho \mathbf{V} \cdot \hat{\mathbf{n}} \, dA = \dot{Q}_{\substack{net \\ in}} + \dot{W}_{\substack{shaft \\ net \, in}} \qquad (5.64)$$

5.3.2   Application of the Energy Equation

In Eq. 5.64, the term, $\partial/\partial t \int_{cv} e\rho \, d\Psi$, represents the time rate of change of the total stored energy, e, of the contents of the control volume. This term is zero when the flow is steady. This term is also zero in the mean when the flow is steady in the mean (cyclical).

In Eq. 5.64, the integrand of

$$\int_{cs} \left(\check{u} + \frac{p}{\rho} + \frac{V^2}{2} + gz \right) \rho \mathbf{V} \cdot \hat{\mathbf{n}} \, dA$$

can be nonzero only where fluid crosses the control surface ($\mathbf{V} \cdot \hat{\mathbf{n}} \neq 0$). Otherwise, $\mathbf{V} \cdot \hat{\mathbf{n}}$ is zero and the integrand is zero for that portion of the control surface. If the properties within parentheses, $\check{u}$, p/ρ, $V^2/2$, and gz, are all assumed to be uniformly distributed over the flow cross section areas involved, the integration becomes simple and gives

$$\int_{cs} \left(\check{u} + \frac{p}{\rho} + \frac{V^2}{2} + gz \right) \rho \mathbf{V} \cdot \hat{\mathbf{n}} \, dA = \sum_{\substack{flow \\ out}} \left(\check{u} + \frac{p}{\rho} + \frac{V^2}{2} + gz \right) \dot{m}$$

$$- \sum_{\substack{flow \\ in}} \left(\check{u} + \frac{p}{\rho} + \frac{V^2}{2} + gz \right) \dot{m} \qquad (5.65)$$

Furthermore, if there is only one stream entering and leaving the control volume, then

$$\int_{cs} \left(\check{u} + \frac{p}{\rho} + \frac{V^2}{2} + gz \right) \rho \mathbf{V} \cdot \hat{\mathbf{n}} \, dA =$$

$$\left(\check{u} + \frac{p}{\rho} + \frac{V^2}{2} + gz \right)_{out} \dot{m}_{out} - \left(\check{u} + \frac{p}{\rho} + \frac{V^2}{2} + gz \right)_{in} \dot{m}_{in} \qquad (5.66)$$

Uniform flow as described above will occur in an infinitesimally small diameter streamtube as illustrated in Fig. 5.7. This kind of streamtube flow is representative of the steady flow of a particle of fluid along a pathline. We can also idealize actual conditions by disregarding nonuniformities in a finite cross section of flow. We call this one-dimensional flow and although such uniform flow rarely occurs in reality, the simplicity

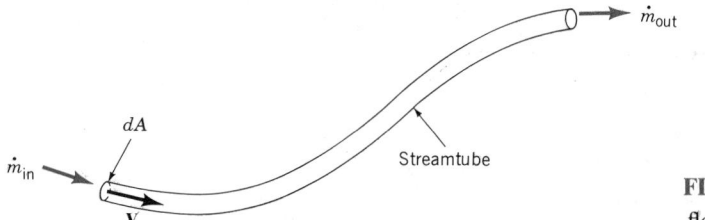

FIGURE 5.7   Streamtube flow.

achieved with the one-dimensional approximation often justifies its use. More details about the effects of nonuniform distributions of velocities and other fluid flow variables are considered in Section 5.3.4 and in Chapters 7, 8, 9, and 10.

If shaft work is involved, the flow must be unsteady, at least locally (see Refs. 1 and 2). The flow in any fluid machine that involves shaft work is unsteady within that machine. Upstream and downstream of the machine, the flow may be steady. Most often shaft work is associated with flow that is unsteady in a recurring or cyclical way. On a time-average basis for flow that is one-dimensional, cyclical, and involves only one stream of fluid entering and leaving the control volume, Eq. 5.64 can be simplified with the help of Eqs. 5.9 and 5.66 to form

$$\dot{m}\left[\check{u}_{out} - \check{u}_{in} + \left(\frac{p}{\rho}\right)_{out} - \left(\frac{p}{\rho}\right)_{in} + \frac{V^2_{out} - V^2_{in}}{2} + g(z_{out} - z_{in})\right] = \dot{Q}_{\substack{net \\ in}} + \dot{W}_{\substack{shaft \\ net\ in}} \quad (5.67)$$

We call Eq. 5.67 the *one-dimensional energy equation for steady-in-the-mean flow*. Note that Eq. 5.67 is valid for incompressible and compressible flows. Often, the fluid property called *enthalpy*, $\check{h}$, where

$$\check{h} = \check{u} + \frac{p}{\rho} \quad (5.68)$$

is used in Eq. 5.67. With enthalpy, the one-dimensional energy equation for steady-in-the-mean flow (Eq. 5.67) is

$$\dot{m}\left[\check{h}_{out} - \check{h}_{in} + \frac{V^2_{out} - V^2_{in}}{2} + g(z_{out} - z_{in})\right] = \dot{Q}_{\substack{net \\ in}} + \dot{W}_{\substack{shaft \\ net\ in}} \quad (5.69)$$

Equation 5.69 is often used for solving compressible flow problems. Examples 5.20 and 5.21 illustrate how Eqs. 5.67 and 5.69 can be used.

EXAMPLE 5.20

A pump delivers water at a steady rate of 300 gal/min as shown in Fig. E5.20. Just upstream of the pump [section (1)] where the pipe diameter is 3.5 in., the pressure is 18 psi. Just downstream of the pump [section (2)] where the pipe diameter is 1 in., the pressure is 60 psi. The change in water elevation across the pump is zero. The rise in internal energy of water, $\check{u}$, across the pump is 3000 ft·lb/slug. If the pumping process is considered to be adiabatic, determine the power (hp) required by the pump.

SOLUTION

We include in our control volume the water contained in the pump between its entrance and exit sections. Application of Eq. 5.67 to the contents of this control volume on a time-average

$$u_2 - u_1 = 3000 \text{ ft·lb/slug}$$

FIGURE E5.20

basis yields

$$\dot{m}\left[\breve{u}_2 - \breve{u}_1 + \left(\frac{p}{\rho}\right)_2 - \left(\frac{p}{\rho}\right)_1 + \frac{V_2^2 - V_1^2}{2} + \overset{\text{0 (no elevation change)}}{g(z_2 - z_1)}\right]$$

$$= \overset{\text{0 (adiabatic flow)}}{\dot{Q}_{\substack{\text{net} \\ \text{in}}}} + \dot{W}_{\substack{\text{shaft} \\ \text{net in}}} \tag{1}$$

We can solve directly for the power required by the pump, $\dot{W}_{\text{shaft net in}}$, from Eq. 1, after we first determine the value of mass flowrate, $\dot{m}$, the speed of flow into the pump, V_1, and the speed of the flow out of the pump, V_2. All other quantities in Eq. 1 are given in the problem statement. From Eq. 5.6, we get

$$\dot{m} = \rho Q = \frac{(1.94 \text{ slugs/ft}^3)(300 \text{ gal/min})}{(7.48 \text{ gal/ft}^3)(60 \text{ s/min})} = 1.30 \text{ slugs/s} \tag{2}$$

Also from Eq. 5.6,

$$V = \frac{Q}{A} = \frac{Q}{\pi D^2/4}$$

so

$$V_1 = \frac{Q}{A_1} = \frac{(300 \text{ gal/min})4 (12 \text{ in./ft})^2}{(7.48 \text{ gal/ft}^3)(60 \text{ s/min})\pi(3.5 \text{ in.})^2} = 10.0 \text{ ft/s} \tag{3}$$

and

$$V_2 = \frac{Q}{A_2} = \frac{(300 \text{ gal/min})4 (12 \text{ in./ft})^2}{(7.48 \text{ gal/ft}^3)(60 \text{ s/min})\pi(1 \text{ in.})^2} = 123 \text{ ft/s} \tag{4}$$

Substituting the values of Eqs. 2, 3, and 4 and values from the problem statement into Eq. 1 we obtain

$$\dot{W}_{\substack{\text{shaft} \\ \text{net}}} = (1.30 \text{ slugs/s}) \left[(3000 \text{ ft·lb/slug}) + \frac{(60 \text{ psi})(144 \text{ in.}^2/\text{ft}^2)}{(1.94 \text{ slugs/ft}^3)} \right.$$

$$\left. - \frac{(18 \text{ psi})(144 \text{ in.}^2/\text{ft}^2)}{(1.94 \text{ slugs/ft}^3)} + \frac{(123 \text{ ft/s})^2 - (10.0 \text{ ft/s})^2}{2[1 \text{ (slug·ft)/(lb·s}^2)]} \right]$$

$$\times \frac{1}{[550(\text{ft·lb})/(\text{s·hp})]} = 32.2 \text{ hp} \tag{Ans}$$

EXAMPLE 5.21

Steam enters a turbine with a velocity of 30 m/s and enthalpy, $\check{h}_1$, of 3348 kJ/kg (see Fig. E5.21). The steam leaves the turbine as a mixture of vapor and liquid having a velocity of 60 m/s and an enthalpy of 2550 kJ/kg. If the flow through the turbine is adiabatic and changes in elevation are negligible, determine the work output involved per unit mass of steam through-flow.

Control volume

Steam turbine

Section (1)
$V_1 = 30$ m/s
$h_1 = 3348$ kJ/kg

$w_{shaft} = ?$ Section (2)
$V_2 = 60$ m/s
$h_2 = 2550$ kJ/kg

FIGURE E5.21

SOLUTION

We use a control volume that includes the steam in the turbine from the entrance to the exit as shown in Fig. E5.21. Applying Eq. 5.69 to the steam in this control volume we get

0 (elevation change is negligible)

0 (adiabatic flow)

$$\dot{m}\left[\check{h}_2 - \check{h}_1 + \frac{V_2^2 - V_1^2}{2} + g(z_2 - z_1)\right] = \dot{Q}_{\substack{net \\ in}} + \dot{W}_{\substack{shaft \\ net \ in}} \tag{1}$$

Dividing through Eq. 1 by the mass flow rate, $\dot{m}$, we obtain

$$w_{\substack{shaft \\ net \ in}} = \frac{W_{\substack{shaft \\ net \ in}}}{\dot{m}} = \check{h}_2 - \check{h}_1 + \frac{V_2^2 - V_1^2}{2} \tag{2}$$

Since $w_{shaft \ net \ out} = -w_{shaft \ net \ in}$, we obtain

$$w_{\substack{shaft \\ net \ out}} = \check{h}_1 - \check{h}_2 + \frac{V_1^2 - V_2^2}{2}$$

or

$$w_{\substack{shaft \\ net \ out}} = 3348 \text{ kJ/kg} - 2550 \text{ kJ/kg}$$

$$+ \frac{[(30 \text{ m/s})^2 - (60 \text{ m/s})^2][1 \text{ J/(N·m)}]}{2[1 \text{ (kg·m)/(N·s}^2)](1000 \text{ J/kJ})}$$

Thus

$$w_{\substack{shaft \\ net \ out}} = 3348 \text{ kJ/kg} - 2550 \text{ kJ/kg} - 1.35 \text{ kJ/kg} = 797 \text{ kJ/kg} \tag{Ans}$$

Note that in this particular example, the change in kinetic energy is small in comparison to the difference in enthalpy involved. This is often true in applications involving steam turbines.

involved, then the shaft work is zero and the energy equation is

$$\dot{m}\left[\breve{u}_{out} - \breve{u}_{in} + \left(\frac{p}{\rho}\right)_{out} - \left(\frac{p}{\rho}\right)_{in} + \frac{V_{out}^2 - V_{in}^2}{2} + g(z_{out} - z_{in})\right] = \dot{Q}_{net\,in} \qquad (5.70)$$

We call Eq. 5.70 the *one-dimensional, steady-flow energy equation*. This equation is valid for incompressible and compressible flows. For compressible flows, enthalpy is most often used in the one-dimensional, steady-flow energy equation and, thus, we have

$$\dot{m}\left[\breve{h}_{out} - \breve{h}_{in} + \frac{V_{out}^2 - V_{in}^2}{2} + g(z_{out} - z_{in})\right] = \dot{Q}_{net\,in} \qquad (5.71)$$

An example of the application of Eq. 5.70 follows.

EXAMPLE 5.22

A 500-ft waterfall involves steady flow from one large body of water to another. Determine the temperature change associated with this flow.

SOLUTION

To solve this problem we consider a control volume consisting of a small cross section streamtube from the nearly motionless surface of the upper body of water to the nearly motionless surface of the lower body of water as is sketched in Fig. E5.22. We need to determine $T_2 - T_1$. This temperature change is related to the change of internal energy of the water, $\breve{u}_2 - \breve{u}_1$, by the relationship

$$T_2 - T_1 = \frac{\breve{u}_2 - \breve{u}_1}{\breve{c}} \qquad (1)$$

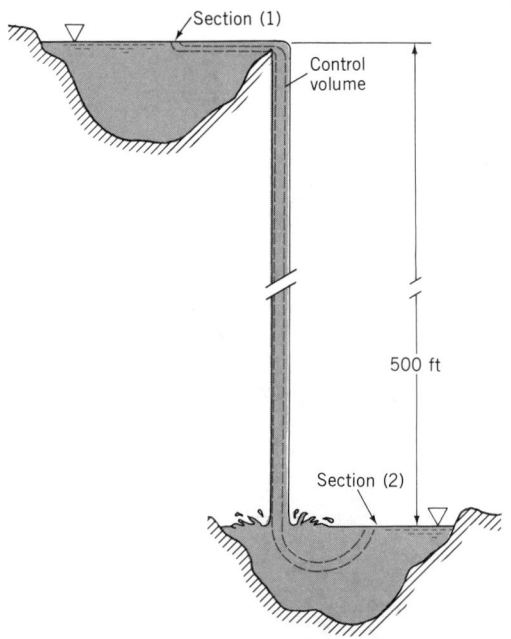

Section (1)

Control volume

500 ft

Section (2)

FIGURE E5.22

where $\check{c} = 1$ Btu/(lbm·°R) is the specific heat of water. The application of Eq. 5.70 to the contents of this control volume leads to

$$\dot{m}\left[\check{u}_2 - \check{u}_1 + \left(\frac{p}{\rho}\right)_2 - \left(\frac{p}{\rho}\right)_1 + \frac{V_2^2 - V_1^2}{2} + g(z_2 - z_1)\right] = \dot{Q}_{\substack{net \\ in}} \tag{2}$$

We assume that the flow is adiabatic. Thus $\dot{Q}_{net\ in} = 0$. Also,

$$\left(\frac{p}{\rho}\right)_1 = \left(\frac{p}{\rho}\right)_2 \tag{3}$$

because the flow is incompressible and atmospheric pressure prevails at sections (1) and (2). Furthermore,

$$V_1 = V_2 = 0 \tag{4}$$

because the surface of each large body of water is considered motionless. Thus, Eqs. 1 through 4 combine to yield

$$T_2 - T_1 = \frac{g(z_1 - z_2)}{\check{c}}$$

or

$$T_2 - T_1 = \frac{(32.2\ \text{ft/s}^2)(500\ \text{ft})}{[1\ \text{Btu/(lbm·°R)}][32.2\ \text{(lbm·ft)/(lb·s}^2)](778\ \text{ft·lb/Btu})} = 0.643\ °R \quad \text{(Ans)}$$

Note that it takes a considerable change of potential energy to produce even a small increase in temperature.

A form of the energy equation that is most often used to solve incompressible flow problems is developed in the next section.

5.3.3   Comparison of the Energy Equation with the Bernoulli Equation

When the one-dimensional energy equation for steady-in-the-mean flow, Eq. 5.67, is applied to a flow that is steady, Eq. 5.67 becomes the one-dimensional, steady-flow energy equation, Eq. 5.70. The only difference between Eq. 5.67 and Eq. 5.70 is that shaft power, $\dot{W}_{shaft\ net\ in}$, is zero if the flow is steady throughout the control volume (fluid machines involve locally unsteady flow). If in addition to being steady, the flow is incompressible, we get from Eq. 5.70,

$$\dot{m}\left[\check{u}_{out} - \check{u}_{in} + \frac{p_{out}}{\rho} - \frac{p_{in}}{\rho} + \frac{V_{out}^2 - V_{in}^2}{2} + g(z_{out} - z_{in})\right] = \dot{Q}_{\substack{net \\ in}} \tag{5.72}$$

Dividing Eq. 5.72 by the mass flowrate, $\dot{m}$, and rearranging terms we obtain

$$\frac{p_{out}}{\rho} + \frac{V_{out}^2}{2} + gz_{out} = \frac{p_{in}}{\rho} + \frac{V_{in}^2}{2} + gz_{in} - (\check{u}_{out} - \check{u}_{in} - q_{\substack{net \\ in}}) \tag{5.73}$$

where

$$q_{\substack{net \\ in}} = \frac{\dot{Q}_{net\ in}}{\dot{m}}$$

Note that Eq. 5.73 involves energy per unit mass and is applicable to one-dimensional flow of a single stream of fluid between two sections or flow along a streamline between two sections.

If the steady, incompressible flow we are considering also involves negligible viscous effects (frictionless flow), then the Bernoulli equation, Eq. 3.7, can be used to describe what happens between two sections in the flow as

$$p_{out} + \frac{\rho V_{out}^2}{2} + \gamma z_{out} = p_{in} + \frac{\rho V_{in}^2}{2} + \gamma z_{in} \qquad (5.74)$$

where $\gamma = \rho g$ is the specific weight of the fluid. To get Eq. 5.74 in terms of energy per unit mass, so that it can be compared directly with Eq. 5.73, we divide Eq. 5.75 by density, ρ, and obtain

$$\frac{p_{out}}{\rho} + \frac{V_{out}^2}{2} + g z_{out} = \frac{p_{in}}{\rho} + \frac{V_{in}^2}{2} + g z_{in} \qquad (5.75)$$

A comparison of Eqs. 5.73 and 5.75 prompts us to conclude that

$$\check{u}_{out} - \check{u}_{in} - q_{\substack{net \\ in}} = 0 \qquad (5.76)$$

when the steady, incompressible flow is frictionless. For steady, incompressible flow with friction, we learn from experience that

$$\check{u}_{out} - \check{u}_{in} - q_{\substack{net \\ in}} > 0 \qquad (5.77)$$

In Eqs. 5.73 and 5.75, we can consider the combination of variables,

$$\frac{p}{\rho} + \frac{V^2}{2} + gz$$

as equal to *useful* or *available energy*. Thus, from inspection of Eqs. 5.73 and 5.75, we can conclude that $\check{u}_{out} - \check{u}_{in} - q_{net\ in}$ represents the *loss* of useful or available energy that occurs in an incompressible fluid flow because of friction. In equation form we have

$$\check{u}_{out} - \check{u}_{in} - q_{\substack{net \\ in}} = loss \qquad (5.78)$$

For a frictionless flow, Eqs. 5.73 and 5.75 tell us that loss equals zero.

It is often convenient to express Eq. 5.73 in terms of loss as

$$\frac{p_{out}}{\rho} + \frac{V_{out}^2}{2} + g z_{out} = \frac{p_{in}}{\rho} + \frac{V_{in}^2}{2} + g z_{in} - loss \qquad (5.79)$$

An example of the application of Eq. 5.79 follows.

EXAMPLE 5.23

Compare the volume flowrates associated with two different vent configurations, a cylindrical hole in the wall having a diameter of 120 mm and the same diameter cylindrical hole in the wall but with a well-rounded entrance (see Fig. E5.23). The room pressure is held constant at 1.0 kPa above atmospheric pressure. Both vents exhaust into the atmosphere. As discussed in Section 8.4.2, the loss in available energy associated with flow through the cylindrical vent from the room to the vent exit is $0.5 V_2^2/2$ where V_2 is the uniformly distributed exit velocity of air. The loss in available energy associated with flow through the rounded entrance vent from the room to the vent exit is $0.05 V_2^2/2$ where V_2 is the uniformly distributed exit velocity of air.

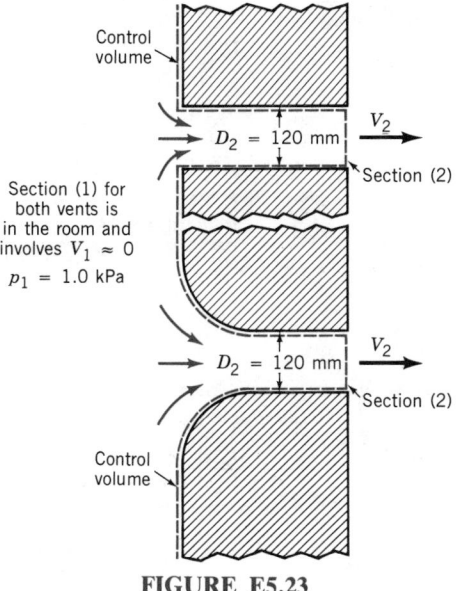

FIGURE E5.23

SOLUTION

We use the control volume for each vent sketched in Fig. E5.23. What is sought is the flowrate, $Q = A_2 V_2$, where A_2 is the vent exit cross section area and V_2 is the uniformly distributed exit velocity. For both vents, application of Eq. 5.79 leads to

$$\frac{p_2}{\rho} + \frac{V_2^2}{2} + \overbrace{gz_2}^{\text{0 (no elevation change)}} = \frac{p_1}{\rho} + \underbrace{\frac{V_1^2}{2}}_{\text{0 }(V_1 \approx 0)} + gz_1 - {_1}\text{loss}_2 \tag{1}$$

where ${_1}\text{loss}_2$ is the loss between sections (1) and (2). Solving Eq. 1 for V_2 we get

$$V_2 = \sqrt{2\left[\left(\frac{p_1 - p_2}{\rho}\right) - {_1}\text{loss}_2\right]} \tag{2}$$

Since

$${_1}\text{loss}_2 = K_L \frac{V_2^2}{2} \tag{3}$$

where K_L is the loss coefficient ($K_L = 0.5$ and 0.05 for the two vent configurations involved), we can combine Eqs. 2 and 3 to get

$$V_2 = \sqrt{2\left[\left(\frac{p_1 - p_2}{\rho}\right) - K_L \frac{V_2^2}{2}\right]} \tag{4}$$

Solving Eq. 4 for V_2 we obtain

$$V_2 = \sqrt{\frac{p_1 - p_2}{\rho\left[(1 + K_L)/2\right]}} \tag{5}$$

Therefore, for flowrate, Q, we obtain

$$Q = A_2 V_2 = \frac{\pi D_2^2}{4} \sqrt{\frac{p_1 - p_2}{\rho \, [(1 + K_L)/2]}} \tag{6}$$

For the rounded entrance cylindrical vent, Eq. 6 gives

$$Q = \frac{\pi (120 \text{ mm})^2}{4(1000 \text{ mm/m})^2} \sqrt{\frac{(1.0 \text{ kPa})(1000 \text{ Pa/kPa})[1(\text{N/m}^2)/(\text{Pa})]}{(1.23 \text{ kg/m}^3)[(1 + 0.05)/2][1(\text{N·s}^2)/(\text{kg·m})]}}$$

or

$$Q = 0.445 \text{ m}^3/\text{s} \tag{Ans}$$

For the cylindrical vent, Eq. 6 gives us

$$Q = \frac{\pi (120 \text{ mm})^2}{4(1000 \text{ mm/m})^2} \sqrt{\frac{(1.0 \text{ kPa})(1000 \text{ Pa/kPa})[1(\text{N/m}^2)/(\text{Pa})]}{(1.23 \text{ kg/m}^3)[(1 + 0.5)/2][1(\text{N.s}^2)/(\text{kg·m})]}}$$

or

$$Q = 0.372 \text{ m}^3/\text{s} \tag{Ans}$$

Note that the rounded entrance vent passes more air than the cylindrical vent because the loss associated with the rounded entrance vent is less than with the cylindrical vent.

An important group of fluid mechanics problems involves one-dimensional, incompressible, steady-in-the-mean flow with friction and shaft work. Included in this category are constant density flows through pumps, blowers, fans, and turbines. For this kind of flow, Eq. 5.67 becomes

$$\dot{m} \left[\breve{u}_{\text{out}} - \breve{u}_{\text{in}} + \frac{p_{\text{out}}}{\rho} - \frac{p_{\text{in}}}{\rho} + \frac{V_{\text{out}}^2 - V_{\text{in}}^2}{2} + g(z_{\text{out}} - z_{\text{in}}) \right] = \dot{Q}_{\substack{\text{net} \\ \text{in}}} + \dot{W}_{\substack{\text{shaft} \\ \text{net in}}} \tag{5.80}$$

Dividing Eq. 5.80 by mass flowrate using $w_{\substack{\text{shaft} \\ \text{net in}}} = \dot{W}_{\substack{\text{shaft} \\ \text{net in}}} / \dot{m}$ we obtain

$$\frac{p_{\text{out}}}{\rho} + \frac{V_{\text{out}}^2}{2} + g z_{\text{out}} = \frac{p_{\text{in}}}{\rho} + \frac{V_{\text{in}}^2}{2} + g z_{\text{in}} + w_{\substack{\text{shaft} \\ \text{net in}}} - (\breve{u}_{\text{out}} - \breve{u}_{\text{in}} - q_{\substack{\text{net} \\ \text{in}}}) \tag{5.81}$$

If the flow is steady throughout, Eq. 5.81 becomes identical to Eq. 5.73, and the previous observation that $\breve{u}_{\text{out}} - \breve{u}_{\text{in}} - q_{\text{net in}}$ equals the loss of available energy is valid. Thus, we conclude that Eq. 5.81 can be expressed as

$$\frac{p_{\text{out}}}{\rho} + \frac{V_{\text{out}}^2}{2} + g z_{\text{out}} = \frac{p_{\text{in}}}{\rho} + \frac{V_{\text{in}}^2}{2} + g z_{\text{in}} + w_{\substack{\text{shaft} \\ \text{net in}}} - \text{loss} \tag{5.82}$$

This is a form of the energy equation for steady-in-the-mean flow that is often used for incompressible flow problems. Note that Eq. 5.82 involves energy per unit mass.

According to Eq. 5.82, when the shaft work is into the control volume, as for example with a pump, a larger amount of loss will result in more shaft work being required for the same rise in available energy. Similarly, when the shaft work is out of the control volume (e.g., a turbine), a larger loss will result in less shaft work out for the same drop in available energy. Designers spend a great deal of effort on minimizing losses in fluid flow components. The following examples demonstrate why losses should be kept as small as possible in fluid systems.

EXAMPLE 5.24

Two large lakes exist as are shown in Fig. E5.24. The water level of one lake is appreciably higher than the other. During the day, water flows from the higher lake to the lower one through appropriate piping and a turbine to provide supplementary power to a nearby city. At night, water is pumped back from the lower lake to the higher one to "store" energy. Show that in each situation, minimizing losses is beneficial.

FIGURE E5.24

SOLUTION

For turbine operation, application of Eq. 5.82 to the contents of the control volume shown in Fig. E5.24 leads to

0 (motionless water surface)   0 (motionless water surface)

$$\frac{p_B}{\rho} + \frac{V_B^2}{2} + gz_B = \frac{p_A}{\rho} + \frac{V_A^2}{2} + gz_A + w_{\text{shaft}\atop\text{net in}} - {}_A\text{loss}_B \tag{1}$$

(atmospheric pressures cancel)

or

$$-w_{\text{shaft}\atop\text{net in}} = g(z_A - z_B) - {}_A\text{loss}_B = w_{\text{shaft}\atop\text{net out}} \tag{2}$$

We see clearly from Eq. 2 that the amount of work and, therefore, power obtained from the turbine is the maximum amount available diminished by the amount of loss due to frictional effects involved.

For pump operation (i.e., flow from B to A), application of Eq. 5.82 to the contents of the control volume shown in Fig. E5.24 leads to

$$gz_A = gz_B + w_{\text{shaft}\atop\text{net in}} - {}_B\text{loss}_A$$

$$w_{\text{shaft}\atop\text{net in}} = g(z_A - z_B) + {}_B\text{loss}_A$$

We note that the amount of work and, therefore, power required to move water from the lower lake to the higher one is increased above the minimum amount by the loss due to frictional effects involved.

EXAMPLE 5.25

An axial flow ventilating fan driven by a motor that delivers 0.4 kW of power to the fan blades produces a 0.6-m-diameter axial stream of air having a speed of 12 m/s. The flow upstream of the fan involves negligible speed. Determine how much of the work to the air actually produces a useful effect, that is, a rise in available energy, and estimate the fluid mechanical efficiency of this fan.

SOLUTION

We select a fixed and nondeforming control volume as is illustrated in Fig. E5.25. The application

FIGURE E5.25

of Eq. 5.82 to the contents of this control volume leads to

$$w_{\substack{shaft \\ net\ in}} - loss = \underbrace{\left(\frac{p_2}{\rho} + \frac{V_2^2}{2} + gz_2\right)}_{} - \left(\cancel{\frac{p_1}{\rho}} + \overbrace{\frac{V_1^2}{2}}^{0\ (V_1 \approx 0)} + \cancel{gz_1}\right) \tag{1}$$

$$0\ (\text{atmospheric pressures cancel}) \qquad 0\ (\text{no elevation change})$$

where $w_{\substack{shaft \\ net\ in}} - loss$ is the amount of work added to the air that produces a useful effect. Equation 1 leads to

$$w_{\substack{shaft \\ net\ in}} - loss = \frac{V_2^2}{2} = \frac{(12\ \text{m/s})^2}{2[1(\text{kg·m})/(\text{N·s}^2)]} = 72.0\ \text{N·m/kg} \tag{2} \quad \text{(Ans)}$$

A reasonable estimate of efficiency, η, would be the ratio of amount of work that produces a useful effect, Eq. 2, to the amount of work delivered to the fan blades. That is

$$\eta = \frac{w_{\substack{shaft \\ net\ in}} - loss}{w_{\substack{shaft \\ net\ in}}} \tag{3}$$

To calculate the efficiency we need a value of $w_{shaft\ net\ in}$ which is related to the power delivered to the blades, $\dot{W}_{shaft\ net\ in}$. We note that

$$w_{\substack{shaft \\ net\ in}} = \frac{\dot{W}_{\substack{shaft \\ net\ in}}}{\dot{m}} \tag{4}$$

where the mass flowrate, $\dot{m}$, is (from Eq. 5.6)

$$\dot{m} = \rho AV = \rho \frac{\pi D_2^2}{4} V_2 \qquad (5)$$

For fluid density, ρ, we use 1.23 kg/m³ (standard air) and thus from Eqs. 4 and 5 we obtain

$$w_{\substack{\text{shaft} \\ \text{net in}}} = \frac{\dot{W}_{\substack{\text{shaft} \\ \text{net in}}}}{(\rho \pi D_2^2/4)V_2} = \frac{(0.4 \text{ kW})[1000 \text{ (N·m)}/(\text{s·kW})]}{(1.23 \text{ kg/m}^3)[(\pi)(0.6 \text{ m})^2/4](12 \text{ m/s})}$$

or

$$w_{\substack{\text{shaft} \\ \text{net in}}} = 95.8 \text{ N·m/kg} \qquad (6)$$

From Eqs. 2, 3 and 6 we obtain

$$\eta = \frac{72.0 \text{ N·m/kg}}{95.8 \text{ N·m/kg}} = 0.752 \qquad \text{(Ans)}$$

Note that only 75% of the power that was delivered to the air resulted in a useful effect, and thus 25% of the shaft power is lost to air friction.

If Eq. 5.82, which involves energy per unit mass, is multiplied by fluid density, ρ, we obtain

$$p_{\text{out}} + \frac{\rho V_{\text{out}}^2}{2} + \gamma z_{\text{out}} = p_{\text{in}} + \frac{\rho V_{\text{in}}^2}{2} + \gamma z_{\text{in}} + \rho w_{\substack{\text{shaft} \\ \text{net in}}} - \rho(\text{loss}) \qquad (5.83)$$

where $\gamma = \rho g$ is the specific weight of the fluid. Equation 5.83 involves *energy per unit volume* and the units involved are identical with those used for pressure.

If Eq. 5.82 is divided by the acceleration of gravity, g, we get

$$\frac{p_{\text{out}}}{\gamma} + \frac{V_{\text{out}}^2}{2g} + z_{\text{out}} = \frac{p_{\text{in}}}{\gamma} + \frac{V_{\text{in}}^2}{2g} + z_{\text{in}} + h_s - h_L \qquad (5.84)$$

where $h_s = w_{\text{shaft net in}}/g$ and $h_L = \text{loss}/g$. Equation 5.84 involves *energy per unit weight*. In Section 3.7, we introduced the notion of "head," which is energy per unit weight. Units of length (e.g., ft, m) are used to quantify the amount of head involved. If a turbine is in the control volume, $h_s = -h_T$ is sometimes used, particularly in the field of hydraulics. For a pump in the control volume, $h_s = h_P$. The quantity h_T is termed the *turbine head* and h_P is the *pump head*. The loss term, h_L is often referred to as *head loss*.

A comparison of the energy equation and the Bernoulli equation has led to the concept of loss of available energy in incompressible fluid flows with friction. In Chapter 8, we discuss in detail some methods for estimating loss in incompressible flows with friction. In Section 5.4 and Chapter 11, we demonstrate that loss of available energy is also an important factor to consider in compressible flows with friction.

5.3.4 Application of the Energy Equation to Nonuniform Flows

The forms of the energy equation discussed in Sections 5.3.2 and 5.3.3 are applicable to one-dimensional flows, flows that are approximated with uniform velocity distributions where fluid crosses the control surface.

If the velocity profile at every section where flow crosses the control surface is not uniform, inspection of the energy equation for a control volume, Eq. 5.64, suggests that the integral

$$\int_{cs} \frac{V^2}{2} \rho \mathbf{V} \cdot \hat{\mathbf{n}} \, dA$$

will require special attention. The other terms of Eq. 5.64 can be accounted for as already discussed in Sections 5.3.2 and 5.3.3.

For one stream of fluid entering and leaving the control volume, we can define the relationship,

$$\int_{cs} \frac{V^2}{2} \rho \mathbf{V} \cdot \hat{\mathbf{n}} \, dA = \dot{m} \left(\frac{\alpha_{out} \overline{V}_{out}^2}{2} - \frac{\alpha_{in} \overline{V}_{in}^2}{2} \right) \tag{5.85}$$

where α is the *kinetic energy coefficient* and $\overline{V}$ is the average velocity defined earlier in Eq. 5.7. From Eq. 5.85, we can conclude that

$$\frac{\dot{m}\alpha \overline{V}^2}{2} = \int_A \frac{V^2}{2} \rho \mathbf{V} \cdot \hat{\mathbf{n}} \, dA$$

for flow through surface area A of the control surface. Thus,

$$\alpha = \frac{\int_A (V^2/2)\rho \mathbf{V} \cdot \hat{\mathbf{n}} \, dA}{\dot{m}\overline{V}^2/2} \tag{5.86}$$

For nonuniform velocity profiles, the energy equation on an energy per unit mass basis for the incompressible flow of one stream of fluid through a control volume that is steady in the mean is

$$\frac{p_{out}}{\rho} + \frac{\alpha_{out} \overline{V}_{out}^2}{2} + g z_{out} = \frac{p_{in}}{\rho} + \frac{\alpha_{in} \overline{V}_{in}^2}{2} + g z_{in} + w_{\substack{shaft \\ net\,in}} - loss \tag{5.87}$$

On an energy per unit volume basis we have

$$p_{out} + \frac{\rho \alpha_{out} \overline{V}_{out}^2}{2} + \gamma z_{out} = p_{in} + \frac{\rho \alpha_{in} \overline{V}_{in}^2}{2} + \gamma z_{in} + \rho w_{\substack{shaft \\ net\,in}} - \rho(loss) \tag{5.88}$$

and on an energy per unit weight or head basis we have

$$\frac{p_{out}}{\gamma} + \frac{\alpha_{out} \overline{V}_{out}^2}{2g} + z_{out} = \frac{p_{in}}{\gamma} + \frac{\alpha_{in} \overline{V}_{in}^2}{2g} + z_{in} + \frac{w_{\substack{shaft \\ net\,in}}}{g} - h_L \tag{5.89}$$

The following examples illustrate the use of the kinetic energy coefficient.

EXAMPLE 5.26

The small fan shown in Fig. E5.26 moves air at a mass flowrate of 0.1 kg/min. Upstream of the fan, the pipe diameter is 60 mm, the flow is laminar, the velocity distribution is parabolic, and the kinetic energy coefficient, α_1, is equal to 2.0. Downstream of the fan, the pipe diameter is 30 mm, the flow is turbulent, the velocity profile is quite uniform, and the kinetic energy coefficient, α_2, is equal to 1.08. If the rise in static pressure across the fan is 0.1 kPa and the fan motor draws 0.14 W, compare the value of loss calculated: (a) assuming uniform velocity distributions, (b) considering actual velocity distributions.

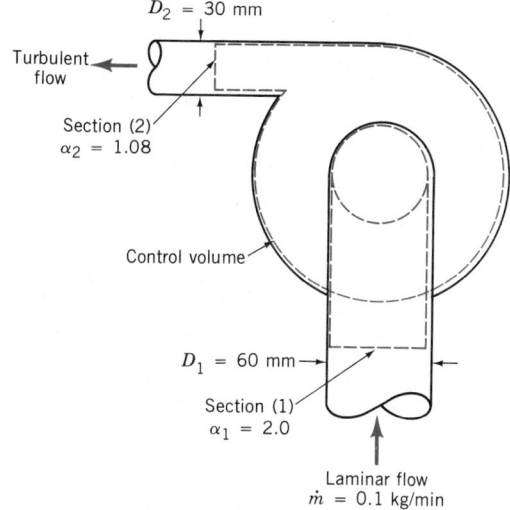

FIGURE E5.26

SOLUTION

Application of Eq. 5.87 to the contents of the control volume shown in Fig. E5.26 leads to

$$\frac{p_2}{\rho} + \frac{\alpha_2 \overline{V}_2^{\,2}}{2} + g\cancel{z_2} \overset{\text{0 (change in } gz \text{ is negligible)}}{=} \frac{p_1}{\rho} + \frac{\alpha_1 \overline{V}_1^{\,2}}{2} + g\cancel{z_1} - \text{loss} + w_{\substack{\text{shaft} \\ \text{net in}}} \tag{1}$$

or solving Eq. 1 for loss, we get

$$\text{loss} = w_{\substack{\text{shaft} \\ \text{net in}}} - \left(\frac{p_2 - p_1}{\rho}\right) + \frac{\alpha_1 \overline{V}_1^{\,2}}{2} - \frac{\alpha_2 \overline{V}_2^{\,2}}{2} \tag{2}$$

To proceed further, we need values of $w_{\text{shaft net in}}$, $\overline{V}_1$, and $\overline{V}_2$. These quantities can be obtained as follows. For shaft work

$$w_{\substack{\text{shaft} \\ \text{net in}}} = \frac{\text{power to fan motor}}{\dot{m}}$$

or

$$w_{\substack{\text{shaft} \\ \text{net in}}} = \frac{(0.14 \text{ W})([1 \text{ (N·m/s)/W}]}{0.1 \text{ kg/min}} (60 \text{ s/min})[1 \text{ (kg·m)/(N·s}^2)]$$

so that

$$w_{\substack{\text{shaft} \\ \text{net in}}} = 84.0 \text{ N·m/kg} \tag{3}$$

For the average velocity at section (1), $\overline{V}_1$, from Eq. 5.11 we obtain

$$\overline{V}_1 = \frac{\dot{m}}{\rho A_1}$$

or

$$\overline{V}_1 = \frac{\dot{m}}{\rho(\pi D_1^2/4)} = \frac{(0.1 \text{ kg/min})}{(1.23 \text{ kg/m}^3)[\pi(60 \text{ mm})^2/4][(60 \text{ s/min})/(1000 \text{ mm/m})^2]} = 0.479 \text{ m/s} \tag{4}$$

For the average velocity at section (2), $\overline{V}_2$,

$$\overline{V}_2 = \frac{(0.1 \text{ kg/min})}{(1.23 \text{ kg/m}^3)[\pi(30 \text{ mm})^2/4][(60 \text{ s/min})/(1000 \text{ mm/m})^2]} = 1.92 \text{ m/s} \qquad (5)$$

(a) For the assumed uniform velocity profiles ($\alpha_1 = \alpha_2 = 1.0$), Eq. 2 yields

$$\text{loss} = w_{\substack{\text{shaft} \\ \text{net in}}} - \left(\frac{p_2 - p_1}{\rho}\right) + \frac{\overline{V}_1^2}{2} - \frac{\overline{V}_2^2}{2} \qquad (6)$$

Using Eqs. 3, 4, and 5 and the pressure rise given in the problem statement, Eq. 6 gives

$$\text{loss} = 84.0 \frac{\text{N·m}}{\text{kg}} - \frac{(0.1 \text{ kPa})(1000 \text{ Pa/kPa})(1 \text{ N/m}^2/\text{Pa})}{1.23 \text{ kg/m}^3}$$
$$+ \frac{(0.479 \text{ m/s})^2}{2[1 \text{ (kg·m)}/(\text{N·s}^2)]}$$
$$- \frac{(1.92 \text{ m/s})^2}{2[1 \text{ (kg·m)}/(\text{N·s}^2)]}$$

or

$$\text{loss} = 84.0 \text{ N·m/kg} - 81.3 \text{ N·m/kg} + 0.115 \text{ N·m/kg} - 1.84 \text{ N·m/kg}$$
$$= 0.975 \text{ N·m/kg} \qquad \text{(Ans)}$$

(b) For the actual velocity profiles ($\alpha_1 = 2$, $\alpha_2 = 1.08$), Eq. 1 gives

$$\text{loss} = w_{\substack{\text{shaft} \\ \text{net in}}} - \left(\frac{p_2 - p_1}{\rho}\right) + \alpha_1 \frac{\overline{V}_1^2}{2} - \alpha_2 \frac{\overline{V}_2^2}{2} \qquad (7)$$

If we use Eqs. 3, 4, and 5 and the given pressure rise, Eq. 7 yields

$$\text{loss} = 84 \text{ N·m/kg} - \frac{(0.1 \text{ kPa})(1000 \text{ Pa/kPa})(1 \text{ N/m}^2/\text{Pa})}{1.23 \text{ kg/m}^3}$$
$$+ \frac{2(0.479 \text{ m/s})^2}{2[1 \text{ (kg·m)}/(\text{N·s}^2)]}$$
$$- \frac{1.08(1.92 \text{ m/s})^2}{2[1 \text{ (kg·m)}/(\text{N·s}^2)]}$$

or

$$\text{loss} = 84.0 \text{ N·m/kg} - 81.3 \text{ N·m/kg} + 0.230 \text{ N·m/kg} - 1.99 \text{ N·m/kg}$$
$$= 0.940 \text{ N·m/kg} \qquad \text{(Ans)}$$

The difference in loss calculated assuming uniform velocity profiles and actual velocity profiles is not large compared to $w_{\text{shaft net in}}$ for this fluid flow situation.

EXAMPLE 5.27

Apply Eq. 5.87 to the flow situation of Example 5.13 and develop an expression for the fluid pressure drop that occurs between sections (1) and (2). By comparing the equation for pressure drop obtained presently with the result of Example 5.13, obtain an expression for loss between sections (1) and (2).

SOLUTION

Applicaton of Eq. 5.87 to the flow of Example 5.13 (see Fig. E5.13) leads to

$$\frac{p_2}{\rho} + \frac{\alpha_2 \overline{w}_2^2}{2} + g z_2 = \frac{p_1}{\rho} + \frac{\alpha_1 \overline{w}_1^2}{2} + g z_1 - \text{loss} + \overset{\displaystyle 0 \text{ (no shaft work)}}{\cancel{w_{\substack{\text{shaft} \\ \text{net in}}}}} \tag{1}$$

Solving Eq. 1 for the pressure drop, $p_1 - p_2$, we obtain

$$p_1 - p_2 = \rho \left[\frac{\alpha_2 \overline{w}_2^2}{2} - \frac{\alpha_1 \overline{w}_1^2}{2} + g(z_2 - z_1) + \text{loss} \right] \tag{2}$$

Since the fluid velocity at section (1), w_1, is uniformly distributed over cross section area A_1, the corresponding kinetic energy coefficient, α_1, is equal to 1.0. The kinetic energy coefficient at section (2), α_2, needs to be determined from the velocity profile distribution given in Example 5.4. Using Eq. 5.86 we get

$$\alpha_2 = \frac{\displaystyle\int_{A_2} \rho w_2^3 \, dA_2}{\dot{m} \overline{w}_2^2} \tag{3}$$

Substituting the parabolic velocity profile equation into Eq. 3 we obtain

$$\alpha_2 = \frac{\rho \displaystyle\int_0^R (2w_1)^3 [1 - (r/R)^2]^3 2\pi r \, dr}{(\rho A_2 \overline{w}_2) \overline{w}_2^2}$$

From conservation of mass, since $A_1 = A_2$

$$w_1 = \overline{w}_2 \tag{4}$$

Then, substituting Eq. 4 into Eq. 3, we obtain

$$\alpha_2 = \frac{\rho 8 \overline{w}_2^3 2\pi \displaystyle\int_0^R [1 - (r/R)^2]^3 r \, dr}{\rho \pi R^2 \overline{w}_2^3}$$

or

$$\alpha_2 = \frac{16}{R^2} \int_0^R [1 - 3(r/R)^2 + 3(r/R)^4 - (r/R)^6] r \, dr = 2 \tag{5}$$

Now we combine Eqs. 2 and 5 to get

$$p_1 - p_2 = \rho \left[\frac{2.0 \overline{w}_2^2}{2} - \frac{1.0 \overline{w}_1^2}{2} + g(z_2 - z_1) + \text{loss} \right] \tag{6}$$

However, from conservation of mass $\overline{w}_2 = \overline{w}_1 = \overline{w}$ so that Eq. 6 becomes

$$p_1 - p_2 = \frac{\rho \overline{w}^2}{2} + \rho g(z_2 - z_1) + \rho(\text{loss}) \tag{7}$$

The term associated with change in elevation, $\rho g(z_2 - z_1)$, is equal to the weight per unit cross section area, W/A, of the water contained between sections (1) and (2) at any instant,

$$\rho g(z_2 - z_1) = \frac{W}{A} \tag{8}$$

Thus combining Eqs. 7 and 8 we get

$$p_1 - p_2 = \frac{\rho \overline{w}^2}{2} + \frac{W}{A} + \rho(\text{loss}) \tag{9}$$

The pressure drop between sections (1) and (2) is due to:

1. The change in kinetic energy between sections (1) and (2) associated with going from a uniform velocity profile to a parabolic velocity profile.
2. The weight of the water column; that is, hydrostatic pressure effect.
3. Viscous loss.

Comparing Eq. 9 for pressure drop with the one obtained in Example 5.13 (i.e., Eq. 7 of Example 5.13) we obtain

$$\frac{\rho \overline{w}^2}{2} + \frac{W}{A} + \rho(\text{loss}) = \frac{\rho \overline{w}^2}{3} + \frac{R_z}{A} + \frac{W}{A} \tag{10}$$

or

$$\text{loss} = \frac{R_z}{\rho A} - \frac{\overline{w}^2}{6} \tag{Ans}$$

We conclude that while some of the pipe wall friction force, R_z, resulted in loss of available energy, a portion of this friction, $\rho A \overline{w}^2/6$, led to the velocity profile change.

5.3.5 Combination of the Energy Equation and the Moment-of-Momentum Equation[3]

If Eq. 5.82 is used for one-dimensional flow through a turbomachine, we can use Eq. 5.54, developed in Section 5.2.4 from the moment-of-momentum equation (Eq. 5.42), to evaluate shaft work. This application of both Eqs. 5.54 and 5.82 allows us to ascertain the amount of loss that occurs in turbomachine flows as is demonstrated in Example 5.28.

EXAMPLE 5.28

For the fan of Example 5.18, show that only some of the shaft power into the air is converted into a useful effect. Develop a meaningful efficiency equation and a practical means for estimating lost shaft energy.

SOLUTION

We use the same control volume used in Example 5.18. Application of Eq. 5.82 to the contents of this control volume yields

$$\frac{p_2}{\rho} + \frac{V_2^2}{2} + gz_2 = \frac{p_1}{\rho} + \frac{V_1^2}{2} + gz_1 + w_{\substack{\text{shaft} \\ \text{net in}}} - \text{loss} \tag{1}$$

As in Example 5.25, we can see with Eq. 1 that a "useful effect" in this fan can be defined as

$$\text{useful effect} = w_{\substack{\text{shaft} \\ \text{net in}}} - \text{loss} = \left(\frac{p_2}{\rho} + \frac{V_2^2}{2} + gz_2 \right) - \left(\frac{p_1}{\rho} + \frac{V_1^2}{2} + gz_1 \right) \tag{2} \quad \text{(Ans)}$$

---

[3] This section may be omitted without loss of continuity in the text material. This section should not be considered without prior study of Sections 5.2.3 and 5.2.4.

In other words, only a portion of the shaft work delivered to the air by the fan blades is used to increase the available energy of the air; the rest is lost because of fluid friction.

A meaningful efficiency equation would involve the ratio of shaft work converted into a useful effect (Eq. 2) to shaft work into the air, $w_{\text{shaft net in}}$. Thus, we can express efficiency, η, as

$$\eta = \frac{w_{\substack{\text{shaft} \\ \text{net in}}} - \text{loss}}{w_{\substack{\text{shaft} \\ \text{net in}}}} \tag{3}$$

However, when Eq. 5.54, which was developed from the moment-of-momentum equation (Eq. 5.42), is applied to the contents of the control volume of Fig. E5.18, we obtain

$$w_{\substack{\text{shaft} \\ \text{net in}}} = +U_2 V_{\theta,2} \tag{4}$$

Combining Eqs. 2, 3, and 4, we obtain

$$\eta = \frac{[(p_2/\rho) + (V_2^2/2) + gz_2] - [(p_1/\rho) + (V_1^2/2) + gz_1]}{U_2 V_{\theta,2}} \tag{5} \quad \text{(Ans)}$$

Equation 5 provides us with a practical means to evaluate the efficiency of the fan of Example 5.18.

Combining Eqs. 2 and 4, we obtain

$$\text{loss} = U_2 V_{\theta,2} - \left[\left(\frac{p_2}{\rho} + \frac{V_2^2}{2} + gz_2 \right) - \left(\frac{p_1}{\rho} + \frac{V_1^2}{2} + gz_1 \right) \right] \tag{6} \quad \text{(Ans)}$$

Equation 6 provides us with a useful method of evaluating the loss due to fluid friction in the fan of Example 5.18 in terms of fluid mechanical variables that can be measured.

5.4 SECOND LAW OF THERMODYNAMICS—IRREVERSIBLE FLOW[4]

The second law of thermodynamics affords us with a means to formalize the inequality

$$\breve{u}_2 - \breve{u}_1 - q_{\substack{\text{net} \\ \text{in}}} \geq 0 \tag{5.90}$$

for steady, incompressible, one-dimensional flow with friction (see Eq. 5.73). In this section we continue to develop the notion of loss of useful or available energy for flow with friction. Minimization of loss of available energy in any flow situation is of obvious engineering importance.

5.4.1 Semi-infinitesimal Control Volume Statement of the Energy Equation

If we apply the one-dimensional, steady flow energy equation, Eq. 5.70, to the contents of a control volume that is infinitesimally thin as illustrated in Fig. 5.8, the result is

$$\dot{m} \left[d\breve{u} + d \left(\frac{p}{\rho} \right) + d \left(\frac{V^2}{2} \right) + g\,(dz) \right] = \delta \dot{Q}_{\substack{\text{net} \\ \text{in}}} \tag{5.91}$$

For all pure substances including common engineering working fluids, such as air, water, oil, and gasoline, the following relationship is valid (see, for example, Ref. 3).

$$T\,ds = d\breve{u} + pd \left(\frac{1}{\rho} \right) \tag{5.92}$$

---

[4] This entire section may be omitted without loss of continuity in the text material.

FIGURE 5.8   Semi-infinitesimal control volume.

where

$$T = \text{absolute temperature}$$

and

$$s = \textit{entropy} \text{ per unit mass}$$

Combining Eqs. 5.91 and 5.92 we get

$$\dot{m}\left(T\,ds - pd\left(\frac{1}{\rho}\right) + d\left(\frac{p}{\rho}\right) + d\left(\frac{V^2}{2}\right) + g\,dz \right) = \delta\dot{Q}_{\substack{net\\in}}$$

or, dividing through by $\dot{m}$ and letting $\delta q_{\substack{net\\in}} = \delta\dot{Q}_{\substack{net\\in}}/\dot{m}$, we obtain

$$\frac{dp}{\rho} + d\left(\frac{V^2}{2}\right) + g\,dz = -(T\,ds - \delta q_{\substack{net\\in}}) \tag{5.93}$$

5.4.2   Semi-infinitesimal Control Volume Statement of the Second Law of Thermodynamics

A general statement of the second law of thermodynamics is

$$\frac{D}{Dt}\int_{sys} s\rho\,d\Psi \geq \sum\left(\frac{\delta\dot{Q}_{\substack{net\\in}}}{T}\right)_{sys} \tag{5.94}$$

or, in words

the time rate of increase of the entropy of a system	$\geq$	sum of the ratio of net heat transfer rate into system to absolute temperature for each particle of mass in the system receiving heat from surroundings

The right-hand side of Eq. 5.94 is identical for the system and control volume at the instant when system and control volume are coincident; thus,

$$\sum\left(\frac{\delta\dot{Q}_{\substack{net\\in}}}{T}\right)_{sys} = \sum\left(\frac{\delta\dot{Q}_{\substack{net\\in}}}{T}\right)_{cv} \tag{5.95}$$

The system time derivative can be expressed for the contents of the coincident control volume that is fixed and nondeforming with the help of the Reynolds transport theorem

(Eq. 4.19). Using Eq. 4.19, we obtain

$$\frac{D}{Dt} \int_{\text{sys}} s\rho \, d\Psi = \frac{\partial}{\partial t} \int_{\text{cv}} s\rho \, d\Psi + \int_{\text{cs}} s\rho \mathbf{V} \cdot \hat{\mathbf{n}} \, dA \tag{5.96}$$

For a fixed, nondeforming control volume, Eqs. 5.94, 5.95, and 5.96 combine to give

$$\frac{\partial}{\partial t} \int_{\text{cv}} s\rho \, d\Psi + \int_{\text{cs}} s\rho \mathbf{V} \cdot \hat{\mathbf{n}} \, dA \geq \sum \left(\frac{\delta \dot{Q}_{\text{net}}}{T} \right)_{\text{cv}} \tag{5.97}$$

At any instant for steady flow

$$\frac{\partial}{\partial t} \int_{\text{cv}} s\rho \, d\Psi = 0 \tag{5.98}$$

If the flow consists of only one stream through the control volume and if the properties are uniformly distributed (one-dimensional flow), Eqs. 5.97 and 5.98 lead to

$$\dot{m}(s_{\text{out}} - s_{\text{in}}) \geq \sum \frac{\delta \dot{Q}_{\text{net}}}{T} \tag{5.99}$$

For the infinitesimally thin control volume of Fig. 5.8, Eq. 5.99 yields

$$\dot{m} \, ds \geq \sum \frac{\delta \dot{Q}_{\text{net}}}{T} \tag{5.100}$$

If all of the fluid in the infinitesimally thin control volume is considered as being at a uniform temperature, T, then from Eq. 5.100 we get

$$T \, ds \geq \delta q_{\text{net in}}$$

or

$$T \, ds - \delta q_{\text{net in}} \geq 0 \tag{5.101}$$

The equality is for any reversible (frictionless) process; the inequality is for all irreversible (friction) processes.

5.4.3 Combination of the Equations of the First and Second Laws of Thermodynamics

Combining Eqs. 5.93 and 5.101, we conclude that

$$-\left[\frac{dp}{\rho} + d\left(\frac{V^2}{2} \right) + g \, dz \right] \geq 0 \tag{5.102}$$

The equality is for any steady, reversible (frictionless) flow, an important example being flow for which the Bernoulli equation (Eq. 3.7) is applicable. The inequality is for all steady, irreversible (friction) flows. The actual amount of the inequality has physical significance. It represents the extent of loss of useful or available energy which occurs because of irreversible flow phenomena including viscous effects. Thus, Eq. 5.102 can be expressed as

$$-\left[\frac{dp}{\rho} + d\left(\frac{V^2}{2} \right) + g \, dz \right] = \delta(\text{loss}) = (T \, ds - \delta q_{\text{net in}}) \tag{5.103}$$

The irreversible flow loss is zero for a frictionless flow and greater than zero for a flow with frictional effects. Note that when the flow is frictionless, Eq. 5.103 multiplied by density, ρ, is identical to Eq. 3.5. Thus, for steady frictionless flow, Newton's second law of motion (see Section 3.1) and the first and second laws of thermodynamics lead to the same differential equation,

$$\frac{dp}{\rho} + d\left(\frac{V^2}{2}\right) + g\, dz = 0 \tag{5.104}$$

If some shaft work is involved, then the flow must be at least locally unsteady in a cyclical way and the appropriate form of the energy equation for the contents of an infinitesimally thin control volume can be developed starting with Eq. 5.67. The resulting equation is

$$-\left[\frac{dp}{\rho} + d\left(\frac{V^2}{2}\right) + g\, dz\right] = \delta(\text{loss}) - \delta w_{\substack{\text{shaft} \\ \text{net in}}} \tag{5.105}$$

Equations 5.103 and 5.105 are valid for incompressible and compressible flows. If we combine Eqs. 5.92 and 5.103, we obtain

$$d\breve{u} + pd\left(\frac{1}{\rho}\right) - \delta q_{\substack{\text{net} \\ \text{in}}} = \delta(\text{loss}) \tag{5.106}$$

For incompressible flow, $d(1/\rho) = 0$ and, thus, from Eq. 5.106,

$$d\breve{u} - \delta q_{\substack{\text{net} \\ \text{in}}} = \delta(\text{loss}) \tag{5.107}$$

Applying Eq. 5.107 to a finite control volume, we obtain

$$\breve{u}_{\text{out}} - \breve{u}_{\text{in}} - q_{\substack{\text{net} \\ \text{in}}} = \text{loss}$$

which is the same conclusion we reached earlier (see Eq. 5.78) for incompressible flows.

For compressible flow, $d(1/\rho) \neq 0$, and thus when we apply Eq. 5.106 to a finite control volume we obtain

$$\breve{u}_{\text{out}} - \breve{u}_{\text{in}} + \int_{\text{in}}^{\text{out}} pd\left(\frac{1}{\rho}\right) - q_{\substack{\text{net} \\ \text{in}}} = \text{loss} \tag{5.108}$$

indicating that $u_{\text{out}} - u_{\text{in}} - q_{\text{net in}}$ is not equal to loss.

5.4.4 Application of the Loss Form of the Energy Equation

Steady flow along a pathline in an incompressible and frictionless flow field provides a simple application of the loss form of the energy equation (Eq. 5.105). We start with Eq. 5.105 and integrate it term by term from one location on the pathline, section (1), to another one downstream, section (2). Note that because the flow is frictionless, loss = 0. Also, because the flow is steady throughout, $w_{\text{shaft net in}} = 0$. Since the flow is incompressible, the density is constant. The control volume in this case is an infinitesimally small diameter streamtube (Fig. 5.7). The resultant equation is

$$\frac{p_2}{\rho} + \frac{V_2^2}{2} + gz_2 = \frac{p_1}{\rho} + \frac{V_1^2}{2} + gz_1 \tag{5.109}$$

which is identical to the Bernoulli equation (Eq. 3.7) already discussed in Chapter 3.

If the frictionless and steady pathline flow of the fluid particle considered above were compressible, application of Eq. 5.105 would yield

$$\int_1^2 \frac{dp}{\rho} + \frac{V_2^2}{2} + gz_2 = \frac{V_1^2}{2} + gz_1 \tag{5.110}$$

To carry out the integration required, $\int_1^2 (dp/\rho)$, a relationship between fluid density, ρ, and pressure, p, must be known. If the frictionless compressible flow we are considering is adiabatic and involves the flow of an ideal gas, it is shown in Section 11.1 that

$$\frac{p}{\rho^k} = \text{constant} \tag{5.111}$$

where $k = c_p/c_v$ is the ratio of gas specific heats, c_p and c_v, which are properties of the fluid. Using Eq. 5.111 we get

$$\int_1^2 \frac{dp}{\rho} = \frac{k}{k-1}\left(\frac{p_2}{\rho_2} - \frac{p_1}{\rho_1}\right) \tag{5.112}$$

Thus, Eqs. 5.110 and 5.112 lead to

$$\frac{k}{k-1}\frac{p_2}{\rho_2} + \frac{V_2^2}{2} + gz_2 = \frac{k}{k-1}\frac{p_1}{\rho_1} + \frac{V_1^2}{2} + gz_1 \tag{5.113}$$

Note that this equation is identical to Eq. 3.24. An example application of Eqs. 5.109 and 5.113 follows.

EXAMPLE 5.29

Air steadily expands adiabatically and without friction from stagnation conditions of 100 psia and 520 °R to 14.7 psia. Determine the velocity of the expanded air assuming (a) incompressible flow; (b) compressible flow.

SOLUTION

(a) If the flow is considered incompressible, the Bernoulli equation, Eq. 5.109, can be applied to flow through an infinitesimal cross section streamtube, like the one in Fig. 5.7, from the stagnation state (1) to the expanded state (2). We get from Eq. 109,

$$\frac{p_2}{\rho} + \frac{V_2^2}{2} + g\cancel{z_2} = \frac{p_1}{\rho} + \cancel{\frac{V_1^2}{2}} + g\cancel{z_1} \tag{1}$$

0 (1 is the stagnation state)

0 (changes in gz are negligible for air flow)

or

$$V_2 = \sqrt{2\left(\frac{p_1 - p_2}{\rho}\right)}$$

We can calculate the density at state (1) by assuming that air behaves like an ideal gas,

$$\rho = \frac{p_1}{RT_1} = \frac{(100 \text{ psia})(144 \text{ in.}^2/\text{ft}^2)}{(1716 \text{ ft·lb/slug·°R})(520 \text{ °R})} = 0.0161 \text{ slug/ft}^3 \tag{2}$$

Thus,

$$V_2 = \sqrt{\frac{2(100 \text{ psia} - 14.7 \text{ psia})(144 \text{ in.}^2/\text{ft}^2)}{(0.016 \text{ slug/ft}^3)[1 \text{ (lb·s}^2)/(\text{slug·ft})]}} = 1240 \text{ ft/s} \qquad \text{(Ans)}$$

(b) If the flow is considered compressible, Eq. 5.113 can be applied to the flow through an infinitesimal cross section control volume, like the one in Fig. 5.7, from the stagnation state (1) to the expanded state (2). We obtain

$$\frac{k}{k-1}\frac{p_2}{\rho_2} + \frac{V_2^2}{2} + g z_2 = \frac{k}{k-1}\frac{p_1}{\rho_1} + \frac{V_1^2}{2} + g z_1 \qquad (3)$$

0 (1 is stagnation state)

0 (changes in gz are negligible for air flow)

or

$$V_2 = \sqrt{\frac{2k}{k-1}\left(\frac{p_1}{\rho_1} - \frac{p_2}{\rho_2}\right)} \qquad (4)$$

Given in the problem statement are values of p_1 and p_2. A value of ρ_1 was calculated earlier (Eq. 2). To determine ρ_2 we need to make use of a property relationship for reversible (frictionless) and adiabatic flow of an ideal gas that is derived in Chapter 11; namely,

$$\frac{p}{\rho^k} = \text{constant}, \qquad \text{where } k = 1.4 \text{ for air} \qquad (5)$$

Solving Eq. 5 for ρ_2 we get

$$\rho_2 = \rho_1 \left(\frac{p_2}{p_1}\right)^{1/k}$$

or

$$\rho_2 = (0.0161 \text{ slug/ft}^3)\left[\frac{14.7 \text{ psia}}{100 \text{ psia}}\right]^{1/1.4} = 0.00409 \text{ slug/ft}^3$$

Then, from Eq. 4,

$$V_2 = \sqrt{\frac{(2)(1.4)}{1.4-1}\left(\frac{100 \text{ psia}}{0.0161 \text{ slug/ft}^3} - \frac{14.7 \text{ psia}}{0.00409 \text{ slug/ft}^3}\right)\frac{(144 \text{ in.}^2/\text{ft}^2)}{1 \text{ (slug·ft)}/(\text{lb·s}^2)}}$$

or

$$V_2 = 1620 \text{ ft/s} \qquad \text{(Ans)}$$

A considerable difference exists between the air velocities calculated assuming incompressible and compressible flow.

REFERENCES

1. Eck, B., *Technische Stromungslehre*, Julius Springer, Verlag, Berlin, Germany, 1957.

2. Dean, R. C., "On the Necessity of Unsteady Flow in Fluid Machines," *ASME Journal of Basic Engineering* 81D; 24–28, March 1959.

3. Moran, M. J., and Shapiro, H. N., *Fundamentals of Engineering Thermodynamics*, Wiley, New York, 1988.

$$\boxed{2} \quad 2 \text{ ports, steady, publy}$$
$$Q + W = \dot{m} \left[\frac{(P_2 - P_1)}{g} + \frac{1}{2}(v_2{}^2 - v_1{}^2) + g(z_2 - z_1) + \text{losses} \right] \overset{h_\ell}{}$$

PROBLEMS

Note: Unless otherwise indicated, use the values of fluid properties found in the tables on the inside of the front cover. Problems designated with an (*) are intended to be solved with the aid of a programmable calculator or a computer.

5.1   A water flow situation is described by the velocity field equation

$$\mathbf{V} = (3x + 2)\hat{\mathbf{i}} + (2y - 4)\hat{\mathbf{j}} - 5z\hat{\mathbf{k}} \text{ ft/s}$$

where x, y, and z are in feet. **(a)** Determine the mass flow rate through the rectangular area in the plane corresponding to $z = 2$ feet having corners at $(x, y, z) = (0, 0, 2)$, $(5, 0, 2)$, $(5, 5, 2)$, and $(0, 5, 2)$ as shown in Fig. P5.1a. **(b)** Show that mass is conserved in the control volume having corners at $(x, y, z) = (0, 0, 2)$, $(5, 0, 2)$, $(5, 5, 2)$, $(0, 5, 2)$, $(0, 0, 0)$, $(5, 0, 0)$, $(5, 5, 0)$, and $(0, 5, 0)$ as shown in Fig. P5.1b.

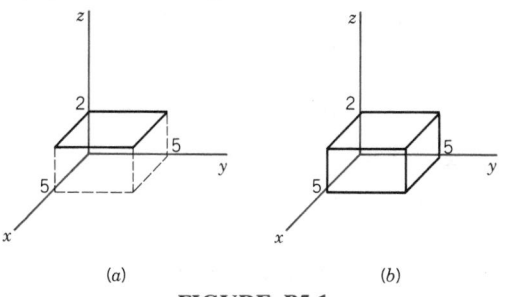

(a)   (b)

FIGURE P5.1

5.2   An incompressible flow velocity field (water) is given as

$$\mathbf{V} = -\frac{1}{r}\hat{\mathbf{e}}_r + \frac{1}{r}\hat{\mathbf{e}}_\theta \text{ m/s}$$

where r is in meters. **(a)** Calculate the mass flow-rate through the cylindrical surface at $r = 1$ m from $z = 0$ to $z = 1$ m as shown in Fig. P5.2a. **(b)** Show that mass is conserved in the annular control volume from $r = 1$ m to $r = 2$ m and $z = 0$ to $z = 1$ m as shown in Fig. P5.2b.

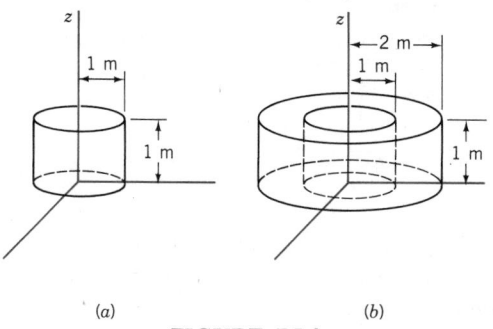

(a)   (b)

FIGURE P5.2

5.3   Seawater flows steadily through a simple conical shaped nozzle as indicated in Fig. P5.3. If the nozzle exit velocity must be at least 100 ft/s, determine the minimum pumping capacity required in gal/min.

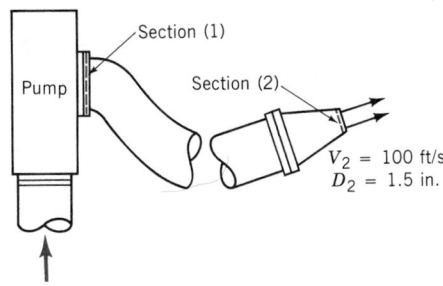

FIGURE P5.3

5.4   Air flows steadily between two cross sections in a long, straight section of 12-in.-inside diameter pipe. The static temperature and pressure at each section are indicated in Fig. P5.4. If the average air velocity at section (2) is 320 m/s, determine the average air velocity at section (1).

Section (1)	Section (2)
p_1 = 690 kPa (abs)	p_2 = 127 kPa (abs)
T_1 = 300 K	T_2 = 252 K
	V_2 = 320 m/s

FIGURE P5.4

5.5   Water enters a conical diffusing passage (see Fig. P5.5) with an average velocity of 10 ft/s. If the entrance cross section area is 1 ft², how large should the diffuser exit area be to reduce the average velocity level to 1 ft/s?

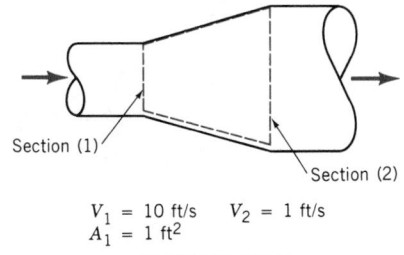

$$V_1 = 10 \text{ ft/s} \qquad V_2 = 1 \text{ ft/s}$$
$$A_1 = 1 \text{ ft}^2$$

FIGURE P5.5

5.6   A hydroelectric turbine passes 8 million gal/min through its blades. If the average velocity of the flow in the circular cross section conduit leading to the turbine is not to exceed 30 ft/s, determine the minimum allowable diameter of the conduit.

5.7 Water flows steadily through a 2-in.-inside diameter pipe at the rate of 200 gal/min. The 2-in. pipe branches into two 1-in.-inside diameter pipes. If the average velocity in one of the 1-in. pipes is 30 ft/s, what is the average velocity in the other 1-in. pipe?

5.8 Water flows along the centerline of a 50-mm-diameter pipe with an average velocity of 10 m/s and out radially between two large circular disks as shown in Fig. P5.8. The disks are parallel and spaced 10 mm apart. Determine the average velocity of the water at a radius of 300 mm in the space between the disks.

$$Q_1 = Q_2$$
$$A_1 V_1 = A_2 V_2$$
$$V_2 = \frac{A_1}{A_2} V_1 = \frac{\pi r_1^2 \, \bar{V_1}}{(4) 2\pi r_2 h}$$

FIGURE P5.8

5.9 A hydraulic jump is in place downstream from a spillway as indicated in Fig. P5.9. Upstream of the jump, the depth of the stream is 1.5 m and the average stream velocity is 20 m/s. Just downstream of the jump, the average stream velocity is 2.9 m/s. Calculate the depth of the stream, h, just downstream of the jump.

FIGURE P5.9

5.10 A water jet pump (see Fig. P5.10) involves a jet cross section area of 0.01 m², and a jet velocity of 30 m/s. The jet is surrounded by entrained water. The total cross section area associated with the jet and entrained streams is 0.075 m². These two fluid streams leave the pump thoroughly mixed with an average velocity of 6 m/s through a cross section area of 0.075 m².

Determine the pumping rate (i.e., the entrained fluid flowrate) involved in liters/s.

$$Q_1 + Q_2 = Q_3$$
$$V_1 A_1 + Q_2 = V_3 A_3 \quad \therefore Q_2 = V_3 A_3 - V_1 A_1 \Rightarrow Q_2 =$$

FIGURE P5.10

5.11 Water enters a cylindrical tank through two pipes at rates of 250 and 100 gal/min (see Fig. P5.11). If the level of the water in the tank remains constant, calculate the average velocity of the flow leaving the tank through an 8-in.-inside diameter pipe.

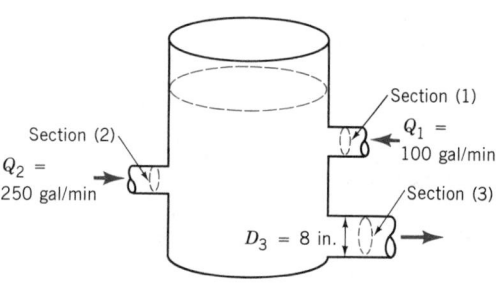

FIGURE P5.11

5.12 At cruise conditions, air flows into a jet engine at a steady rate of 65 lbm/s. Fuel enters the engine at a steady rate of 0.75 lbm/s. The average velocity of the exhaust gases is 1500 ft/s relative to the engine. If the engine exhaust effective cross section area is 3.5 ft², estimate the density of the exhaust gases in lbm/ft³.

5.13 Air at standard atmospheric conditions is drawn into a compressor at the steady rate of 30 m³/min. The compressor pressure ratio, p_{exit}/p_{inlet}, is 10 to 1. Through the compressor p/ρ^n remains constant with $n = 1.4$. If the average velocity in the compressor discharge pipe is not to exceed 30 m/s, calculate the minimum discharge pipe diameter required.

5.14 An evaporative cooling tower (see Fig. P5.14) is used to cool water from 110 to 80 °F. Water enters the tower at a rate of 250,000 lbm/hr. Dry air (no water vapor) flows into the tower at a rate of 151,000 lbm/hr. If the rate of wet air flow out of the tower is 156,900 lbm/hr, determine the rate of water evaporation in lbm/hr and the rate of cooled water flow in lbm/hr.

Wet air
$\dot{m}$ = 156,900 lbm/hr

Warm water
$\dot{m}$ = 250,000 lbm/hr

Dry air
$\dot{m}$ = 151,000 lbm/hr

Cooled
water

FIGURE P5.14

5.15 Oil having a specific gravity of 0.9 is pumped as illustrated in Fig. P5.15 with a water jet pump. The water volume flowrate is 1 m³/s. The water and oil mixture has an average specific gravity of 0.95. Calculate the rate, in m³/s, at which the pump moves oil.

Section (1)

Section (3)

Water
Q_1 =
1 m³/s

Water
and
oil
mix
(SG = 0.95)

Section (2)

Oil (SG = 0.9)
FIGURE P5.15

5.16 Water at 0.1 m³/s and alcohol (SG = 0.8) at 0.3 m³/s are mixed in a y-duct as shown in Fig. P5.16. What is the average density of the mixture of alcohol and water?

Water and
alcohol mix

Water
Q = 0.1 m³/s

Alcohol (SG = 0.8)
Q = 0.3 m³/s
FIGURE P5.16

5.17 An appropriate turbulent pipe flow velocity profile is

$$ \mathbf{V} = u_c \left(\frac{R - r}{R} \right)^{1/n} \hat{\mathbf{i}} $$

where u_c = centerline velocity, r = local radius,

R = pipe radius, and $\hat{\mathbf{i}}$ = unit vector along pipe centerline. Determine the ratio of average velocity, $\bar{u}$, to centerline velocity u_c for **(a)** n = 5; **(b)** n = 6; **(c)** n = 7; **(d)** n = 8; **(e)** n = 9; **(f)** n = 10.

5.18 The flow in an open channel has a velocity distribution

$$ \mathbf{V} = U(y/h)^{1/7} \hat{\mathbf{i}} \text{ ft/s} $$

where U = free-surface velocity, y = perpendicular distance from the channel bottom in feet, and h = depth of the channel in feet. Determine the average velocity of the channel stream as a fraction of U.

5.19 The velocity and temperature profiles for one circular cross section in laminar pipe flow of air with heat transfer are

$$ \mathbf{V} = u_c \left[1 - \left(\frac{r}{R} \right)^2 \right] \hat{\mathbf{i}} $$

where the unit vector $\hat{\mathbf{i}}$ is along the pipe axis, and

$$ T = T_c \left[1 + \frac{1}{2} \left(\frac{r}{R} \right)^2 - \frac{1}{4} \left(\frac{r}{R} \right)^4 \right] $$

The subscript c refers to centerline value, r = local radius, R = pipe radius and T = local temperature. Show how you would evaluate the mass flowrate through this cross section area.

5.20* To measure the mass flowrate of air through a 6-in.-inside diameter pipe, local velocity data are collected at different radii from the pipe axis (see Table). Determine the mass flowrate corresponding to the data listed below.

r (in.)	Axial Velocity (ft/s)
0	30
0.2	29.71
0.4	29.39
0.6	29.06
0.8	28.70
1.0	28.31
1.2	27.89
1.4	27.42
1.6	26.90
1.8	26.32
2.0	25.64
2.2	24.84
2.4	23.84
2.6	22.50
2.8	20.38
2.9	18.45
2.95	16.71
2.98	14.66
3.00	0

5.21*   Air flows steadily in the axial direction in an annulus as indicated in Fig. P5.21. The inner and outer radii of the annulus are 17.5 and 50.8 mm. Axial velocities measured at various radii are listed below. Determine the mass flowrate involved.

r (mm)	Axial Velocity (m/s)
17.5	0
18.2	32.0
18.7	33.9
19.3	35.6
20.0	36.8
20.6	37.9
21.9	39.3
23.2	40.4
25.7	41.8
28.2	42.5
30.8	42.8
33.3	42.6
35.8	42.1
38.4	41.4
40.9	40.1
43.5	38.5
46.0	36.0
47.3	34.3
48.6	31.9
49.8	27.4
50.8	0

FIGURE P5.21

5.22   Flow of a viscous fluid over a flat plate surface results in the development of a region of reduced velocity adjacent to the wetted surface as depicted in Fig. P5.22. This region of reduced flow is called a boundary layer. At the leading edge of the plate, the velocity profile may be considered uniformly distributed with a value U. All along the outer edge of the boundary layer, the fluid velocity component parallel to the plate surface is also U. If the x direction velocity profile at section (2) is

$$\frac{u}{U} = \left(\frac{y}{\delta}\right)^{1/7}$$

develop an expression for the volume flowrate through the edge of the boundary layer from the leading edge to a location downstream at x where the boundary layer thickness is δ.

FIGURE P5.22

5.23   Estimate the time required to fill with water a cone shaped container (see Fig. P5.23) 5 ft high and 5 ft across at the base if the filling rate is 20 gal/min.

FIGURE P5.23

5.24   How long would it take to fill a cylindrical shaped swimming pool having a diameter of 5 m to a depth of 1.5 m with water from a garden hose if the flowrate is 1.0 liter/s?

5.25   To determine the mass flowrate of cooling water through a test engine, the amount of water collected in a large container over a time interval is measured. Using an appropriate control volume, comment on the validity of this procedure.

5.26   Storm sewer backup causes your basement to flood at the steady rate of 1 in. of depth per hour. The basement floor area is 1000 ft². What capacity (gal/min) pump would you rent to **(a)** keep the water accumulated in your basement at a constant level until the storm sewer is blocked off; **(b)** reduce the water accumulation in your basement at a rate of 3 in./hr even while the backup problem exists?

5.27   A hypodermic syringe (see Fig. P5.27) is used to apply a vaccine. If the plunger is moved forward at the steady rate of 20 mm/s and if vaccine leaks pass the plunger at 0.1 of the volume flowrate out the needle opening, calculate the average velocity of the needle exit flow. The in-

FIGURE P5.27

side diameters of the syringe and the needle are 20 mm and 0.7 mm.

5.28 Freshwater flows steadily into an open 55-gal drum initially filled with seawater. The freshwater mixes thoroughly with the seawater and the mixture overflows out of the drum. If the freshwater flowrate is 10 gal/min, estimate the time in seconds required to decrease the difference between the density of the mixture and the density of fresh water by 50%.

5.29 Water enters a rigid, sealed, cylindrical tank at a steady rate of 100 liters/hr and forces gasoline ($SG = 0.68$) out as is indicated in Fig. P5.29. What is the time rate of change of mass of gasoline contained in the tank?

Gasoline ($SG = 0.68$)

Water

FIGURE P5.29

5.30 A gas flows steadily through a duct of varying cross section area. If the gas density is assumed to be uniformly distributed at any cross section, show that the conservation of mass principle leads to

$$\frac{d\rho}{\rho} + \frac{d\overline{V}}{V} + \frac{dA}{A} = 0$$

where ρ = gas density, $\overline{V}$ = average speed of gas, and A = cross section area.

5.31 Determine the anchoring force required to hold in place the conical nozzle attached to the end of the laboratory sink faucet shown in Fig. P5.31 when the water flowrate is 10 gal/min. The

$D_1 = 0.6$ in.

Section (1)

1.2 in.

Section (2)

$D_2 = 0.2$ in.

$Q = 10$ gal/min

FIGURE P5.31

nozzle weight is 0.2 lb. The nozzle inlet and exit inside diameters are 0.6 and 0.2 in., respectively. The nozzle axis is vertical and the axial distance between sections (1) and (2) is 1.2 in. The pressure at section (1) is 68 psi.

5.32 Water enters the horizontal, circular cross section sudden contraction nozzle sketched in Fig. P5.32 at section (1) with a uniformly distributed velocity of 25 ft/s and a pressure of 75 psi. The water exits from the nozzle into the atmosphere at section (2) where the uniformly distributed velocity is 100 ft/s. **(a)** Determine the axial component of the reaction force exerted by the contraction on the flowing water. **(b)** Determine the axial component of the anchoring force required to hold the contraction in place.

$-V_1\rho V_1 A_1 + V_2 \rho V_2 A_2 = P_1 A_1 - R_x - P_2 A_2 ; \dot{m} = \rho V_1 A_1 = \rho V_2 A_2 ;$

$\dot{m}(V_2 - V_1) = P_1 A_1 - R_x - P_2 A_2$

$R_x = P_1 A_1 - P_2 A_2 + \dot{m}(V_2 - V_1)$

$= P_1 \frac{\pi D_1^2}{4} - P_2 \frac{\pi D_2^2}{4} - \rho V_1 \pi \frac{D_1^2}{4}(V_2 - V_1)$

See P. 246

$D_1 = 3$ in.

$p_1 = 75$ psi

$V_1 = 25$ ft/s

Section (1)

Section (2)

$p_2 = 0$ psi

$V_2 = 100$ ft/s

FIGURE P5.32

5.33 Water flows through a horizontal, 180° pipe bend as is illustrated in Fig. P5.33. The flow cross section area is constant at a value of 9000 mm². The flow velocity everywhere in the bend is 15 m/s. The pressures at the entrance and exit of the bend are 210 and 165 kPa, respectively. Calculate the horizontal (x and y) components of the force exerted by the bend on the flowing water.

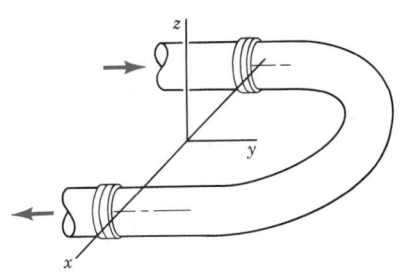

FIGURE P5.33

5.34 Determine the magnitude and direction of the x and y components of the anchoring force required to hold in place the horizontal 180° elbow and nozzle combination shown in Fig. P5.34. Also determine the magnitude and direction of the x and y components of the reaction force exerted by the 180° elbow and nozzle on the flowing water.

$$p_1 = 15 \text{ psi}$$
$$V_1 = 5 \text{ ft/s}$$

FIGURE P5.34

5.35 Determine the magnitude and direction of the x and y components of reaction force exerted on the flowing water by the horizontal elbow and nozzle combination shown in Fig. P5.35. Atmospheric pressure is 100 kPa. The gage pressure at section (1) is 100 kPa. At section (2), the water exists to the atmosphere.

$$p_1 = 100 \text{ kPa}$$
$$V_1 = 2 \text{ m/s}$$

FIGURE P5.35

5.36 Water flows through a right angle valve at the rate of 1000 lbm/s as is shown in Fig. P5.36. The pressure just upstream of the valve is 90 psi and the pressure drop across the valve is 50 psi. The inside diameters of the valve inlet and exit pipes are 12 and 24 in. If the flow through the valve occurs in a horizontal plane, determine the x and y components of the force exerted by the valve on the water.

FIGURE P5.36

5.37 A converging elbow (see Fig. P5.37) turns water through an angle of 135° in a vertical plane. The flow cross section diameter is 400 mm at the elbow inlet, section (1), and 200 mm at the elbow outlet, section (2). The elbow flow passage volume is 0.2 m³ between sections (1) and (2). The water volume flowrate is 0.4 m³/s and the elbow inlet and outlet pressures are 150 kPa and 90 kPa. The elbow mass is 12 kg. Calculate the horizontal (x direction) and vertical (z direction) anchoring forces required to hold the elbow in place.

FIGURE P5.37

5.38 For the conditions of Problem 5.4, determine the frictional force exerted by the pipe wall on the air flowing between sections (1) and (2). Assume uniform velocity distributions at each section.

5.39 Water is sprayed radially outward over 180° as indicated in Fig. P5.39. The jet sheet is in the horizontal plane. If the jet velocity at the nozzle exit is 50 ft/s, determine the direction and magnitude of the resultant horizontal anchoring force required to hold the nozzle in place.

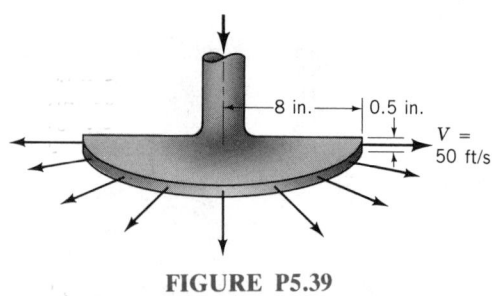

FIGURE P5.39

5.40 A circular plate having a diameter of 300 mm is held perpendicular to an axisymmetric horizontal jet of air having a velocity of 40 m/s and a diameter of 80 mm as shown in Fig. P5.40. A hole at the center of the plate results in a discharge jet of air having a velocity of 40 m/s and a diameter of 20 mm. Determine the horizontal component of force required to hold the plate stationary. $F_{A/x} = v_1^2 \rho \frac{\pi D_1^2}{4} - v_2^2 \rho \frac{\pi D_2^2}{4} = v_1^2 \rho \frac{\pi}{4}(D_1^2 - D_2^2)$

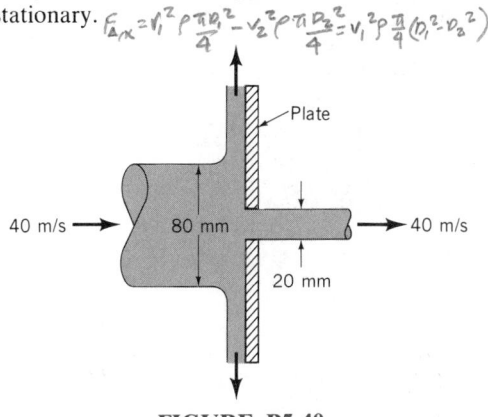

FIGURE P5.40

5.41   A nozzle is attached to an 80-mm-inside diameter flexible hose. The nozzle area is 500 mm². If the delivery pressure of water at the nozzle inlet is 700 kPa, could you hold the hose and nozzle stationary? Explain.

5.42   The results of a wind tunnel test to determine the drag on a body (see Fig. P5.42) are summarized below. The upstream [section (1)] velocity is uniform at 100 ft/s. The static pressures are given by $p_1 = p_2 = 14.7$ psia. The downstream velocity distribution which is symmetrical about the centerline is given by

$$u = 100 - 30 \left(1 - \frac{|y|}{3}\right) \qquad |y| \le 3 \text{ ft}$$

$$u = 100 \qquad |y| > 3 \text{ ft}$$

where u is the velocity in ft/s and y is the distance on either side of the centerline in feet (see Fig. P5.42). Assume that the body shape does not change in the direction normal to the paper. Calculate the drag force (reaction force in x direction) exerted on the air by the body per unit length normal to the plane of the sketch.

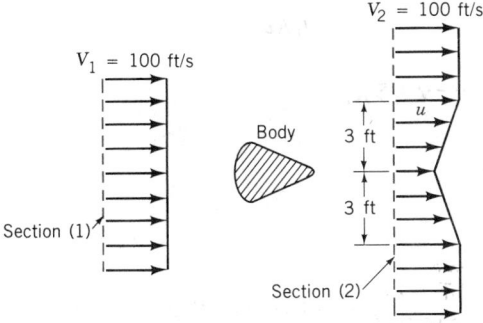

FIGURE P5.42

5.43   A variable mesh screen produces a linear and axisymmetric velocity profile as indicated in Fig. P5.43 in the air flow through a 2-ft-diameter circular cross section duct. The static pressures upstream and downstream of the screen are 0.2 and 0.15 psi and are uniformly distributed over the flow cross section area. Neglecting the force exerted by the duct wall on the flowing air, calculate the screen drag force.

FIGURE P5.43

5.44   The rocket shown in Fig. P5.44 is held stationary by the horizontal force, F_x, and the vertical force, F_z. The velocity and pressure of the exhaust gas are 5000 ft/s and 20 psia at the nozzle exit, which has a cross section area of 60 in.². The exhaust mass flowrate is constant at 21 lbm/s. Determine the value of the restraining force F_x. Assume the exhaust flow is essentially horizontal.

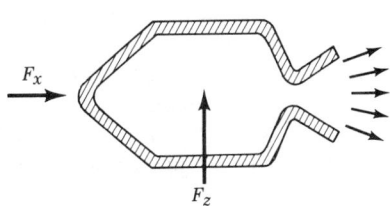

FIGURE P5.44

5.45   Water flows vertically upward in a circular cross section pipe as shown in Fig. P5.45. At section (1), the velocity profile over the cross section area is uniform. At section (2), the velocity profile is

$$\mathbf{V} = w_c \left(\frac{R - r}{R}\right)^{1/7} \hat{\mathbf{k}}$$

where $\mathbf{V}$ = local velocity vector, w_c = centerline velocity in the axial direction, R = pipe radius, and r = radius from pipe axis. Develop an expression for the fluid pressure drop that occurs between sections (1) and (2).

FIGURE P5.45

5.46 In a laminar pipe flow that is fully developed, the axial velocity profile is parabolic, that is,

$$u = u_c \left[1 - \left(\frac{r}{R} \right)^2 \right]$$

as is illustrated in Fig. P5.46. Compare the axial direction momentum flowrate calculated with the average velocity, $\bar{u}$, with the axial direction momentum flowrate calculated with the nonuniform velocity distribution taken into account.

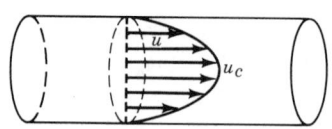

FIGURE P5.46

5.47* For the pipe (6-in.-inside diameter) air flow data of Problem 5.20, calculate the rate of flow of axial direction momentum. How large would the error be if the average axial velocity were used to calculate axial direction momentum flow?

5.48* For the air flow data of Problem 5.21, determine the rate of flow of axial direction momentum.

5.49 A static thrust stand is to be designed for testing a specific jet engine. Knowing the following conditions for a typical test,

$$
\begin{aligned}
\text{intake air velocity} &= 700 \text{ ft/s} \\
\text{exhaust gas velocity} &= 1640 \text{ ft/s} \\
\text{intake cross section area} &= 10 \text{ ft}^2 \\
\text{intake static pressure} &= 11.4 \text{ psia} \\
\text{intake static temperature} &= 480 \text{ °R} \\
\text{exhaust gas pressure} &= 0 \text{ psi}
\end{aligned}
$$

estimate a nominal thrust to design for.

5.50 Determine the magnitude of the horizontal component of the anchoring force per unit width required to hold in place the sluice gate shown in Fig. P5.50. Compare this result with the size of the horizontal component of the anchoring force required to hold in place the sluice gate when it is closed and the depth of water upstream is 10 ft.

FIGURE P5.50

5.51 A horizontal circular jet of air strikes a stationary flat plate as indicated in Fig. P5.51. The jet velocity is 40 m/s and the jet diameter is 30 mm. If the air velocity magnitude remains constant as the air flows over the plate surface in the directions shown, determine: **(a)** the magnitude of F_A, the anchoring force required to hold the plate stationary; **(b)** the fraction of mass flow along the plate surface in each of the two directions shown; **(c)** the magnitude of F_A, the anchoring force required to allow the plate to move to the right at a constant speed of 10 m/s.

$D_j = 30$ mm

$V_j = 40$ m/s

$30°$

$90°$

V_3

F_A

FIGURE P5.51

5.52 A horizontal circular cross section jet of air having a diameter of 6 in. strikes a conical deflector as shown in Fig. P5.52. A horizontal anchoring force of 5 lb is required to hold the

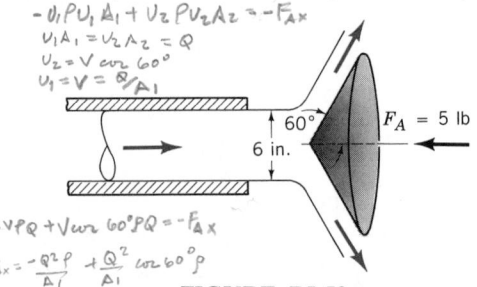

$60°$

$F_A = 5$ lb

6 in.

FIGURE P5.52

cone in place. Estimate the nozzle flowrate in ft³/s. The magnitude of the velocity of the air remains constant.

5.53 A vertical, circular cross section jet of air strikes a conical deflector as indicated in Fig. P5.53. A vertical anchoring force of 0.1 N is required to hold the deflector in the place. Determine the mass (kg) of the deflector. The magnitude of velocity of the air remains constant.

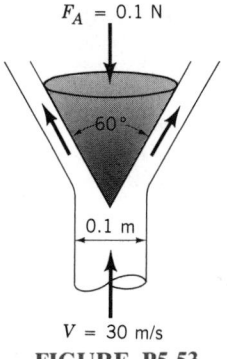

FIGURE P5.53

5.54 A horizontal air jet having a velocity of 50 m/s and a diameter of 20 mm strikes the inside surface of a hollow hemisphere as indicated in Fig. P5.54. How large is the horizontal anchoring force needed to hold the hemisphere in place? The magnitude of velocity of the air remains constant.

FIGURE P5.54

5.55 An airplane moves forward with a steady air speed of 250 km/hr. The wings of this plane deflect the air flow around each wing downward by an angle of 10°. Estimate the mass flowrate of air required over the wing surfaces if the mass of the plane and its contents is 3000 kg. Assume that air leaves the wing at the same speed it entered with and consider momentum change effects only.

5.56 A 3-in.-diameter horizontal jet of water strikes a flat plate as indicated in Fig. P5.56. Determine the jet velocity if a 10-lb horizontal force is required to: **(a)** hold the plate stationary; **(b)** allow the plate to move at a constant speed of 10 ft/s to the right.

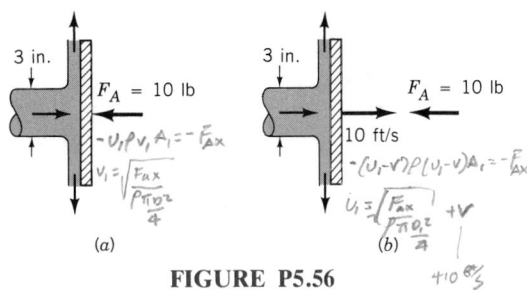

FIGURE P5.56

5.57 A vane directs a horizontal, circular cross section jet of water symmetrically as indicated in Fig. P5.57. The jet leaves the nozzle with a velocity of 100 ft/s. Determine the x direction component of anchoring force required to: **(a)** hold the vane stationary; **(b)** confine the speed of the vane to a value of 10 ft/s to the right. The fluid velocity magnitude remains constant along the vane surface.

FIGURE P5.57

5.58 A Pelton wheel turbine is illustrated in Fig. P5.58. The radius to the line of action of the tangential reaction force on each vane is 1 ft. Each

FIGURE P5.58

vane deflects fluid by an angle of 135° as indicated. Assume all of the flow occurs in a horizontal plane. Each of the four jets shown strikes a vane with a velocity of 100 ft/s and a stream diameter of 1 in. The magnitude of velocity of the jet remains constant along the vane surface. **(a)** How much torque is required to hold the wheel stationary? **(b)** How fast will the wheel rotate if friction is negligible and how much power is delivered at the shaft?

5.59   Water enters a rotating lawn sprinkler through its base at the steady rate of 16 gal/min as shown in Fig. P5.59. The exit cross section area of each of the two nozzles is 0.04 in.² and the flow leaving each nozzle is tangential. The radius from the axis of rotation to the centerline of each nozzle is 8 in. **(a)** Determine the resisting torque required to hold the sprinkler head stationary. **(b)** Determine the resisting torque associated with the sprinkler rotating with a constant speed of 500 rev/min. **(c)** Determine the angular velocity of the sprinkler if no resisting torque is applied.

sprinkler stationary. Compute the angle, θ, which the exiting water stream makes with the tangential direction. The flow leaves the nozzles in the horizontal plane.

5.61   Five liters/s of water enters the rotor shown in Fig. P5.61 along the axis of rotation. The cross section area of each of the three nozzle exits normal to the relative velocity is 18 mm². How large is the resisting torque required to hold the rotor stationary? How fast will the rotor spin steadily if the resisting torque is reduced to zero and: **(a)** $\theta = 0°$; **(b)** $\theta = 30°$; **(c)** $\theta = 60°$?

FIGURE P5.61

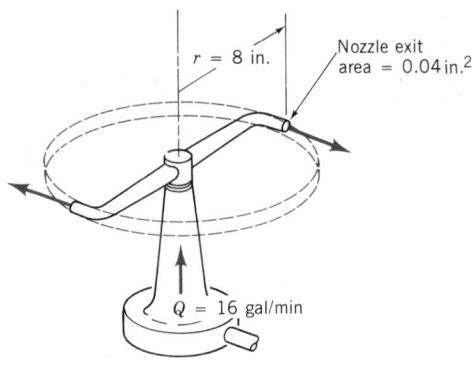

FIGURE P5.59

5.60   A lawn sprinkler is constructed from pipe with $\frac{1}{4}$-in.-inside diameter as indicated in Fig. P5.60. Each arm is 6 in. in length. Water flows through the sprinkler at the rate of 1.5 lbm/s. A force of 3 lb positioned halfway along one arm holds the

5.62   A small water turbine is designed as shown in Fig. P5.62. If the flowrate through the turbine is 0.1 lbm/s, and the rotor speed is 300 rpm, estimate the shaft torque and shaft power involved. Each nozzle exit cross section area is 3.5×10^{-5} ft².

FIGURE P5.60

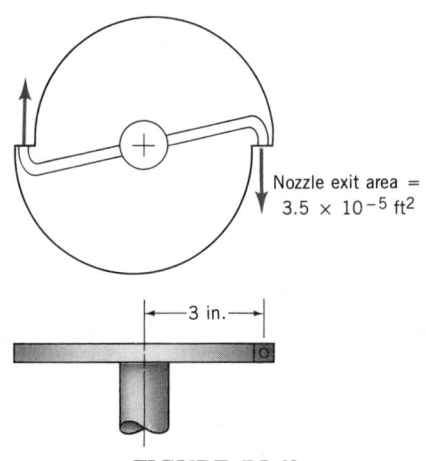

FIGURE P5.62

5.63   An inward flow radial turbine (see Fig. P5.63) involves a nozzle angle, α_1, of 60° and an inlet rotor tip speed, U_1, of 3 m/s. The ratio of rotor inlet to outlet diameters is 2.0. The absolute velocity leaving the rotor at section (2) is radial with a magnitude of 6 m/s. Determine the energy transfer per unit mass of fluid flowing through this turbine if the fluid is: **(a)** air; **(b)** water.

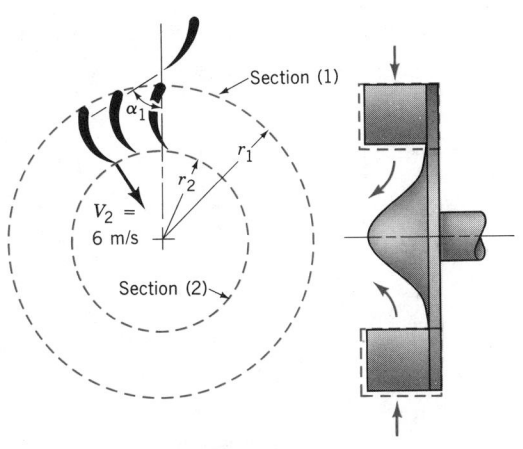

FIGURE P5.63

5.64   A simplified sketch of a hydraulic turbine runner is shown in Fig. P5.64. Relative to the rotating runner, water enters at section (1) (cylindrical cross section area A_1 at $r_1 = 1.5$ m) at an angle of 100° from the tangential direction and leaves at section (2) (cylindrical cross section area A_2 at $r_2 = 0.85$ m) at an angle of 50° from the tangential direction. The blade height at sections (1) and (2) is 0.45 m and the volume flowrate through the turbine is 30 m³/s. The runner speed is 130 rpm in the direction shown. Determine the shaft power developed.

5.65   A water turbine wheel rotates at the rate of 100 rpm in the direction shown in Fig. P5.65. The inner radius, r_2, of the blade row is 1 ft, and the outer radius, r_1, is 2 ft. The absolute velocity vector at the turbine rotor entrance makes an angle of 20° with the tangential direction. The inlet blade angle is 60° relative to the tangential direction. The blade outlet angle is 120°. The flowrate is 10 ft³/s. For the flow tangent to the rotor blade surface at inlet and outlet determine an appropriate constant blade height, b, and the corresponding power available at the rotor shaft.

FIGURE P5.65

5.66   A water turbine with radial flow has the dimensions shown in Fig. P5.66. The absolute entering velocity is 50 ft/s, and it makes an angle of 30° with the tangent to the rotor. The absolute exit velocity is directed radially inward. The angular speed of the rotor is 120 rpm. Find the power delivered to the shaft of the turbine.

FIGURE P5.64

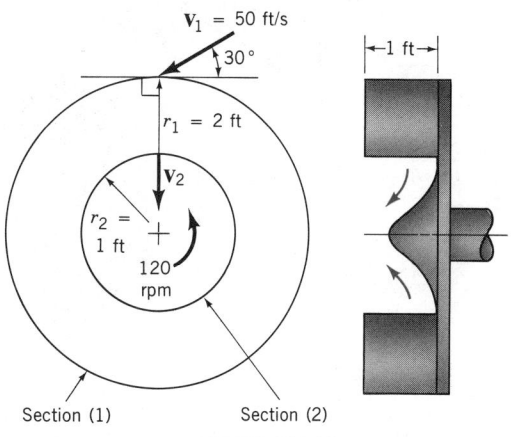

FIGURE P5.66

5.67 An incompressible fluid flows outward through a blower as indicated in Fig. P5.67. The shaft torque involved, T_{shaft}, is estimated with the following relationship:

$$T_{shaft} = \dot{m} r_2 V_{\theta,2}$$

where $\dot{m}$ = mass flowrate through the blower, r_2 = outer radius of blower, and $V_{\theta,2}$ = tangential component of absolute fluid velocity leaving the blower. State the flow conditions that make this formula valid.

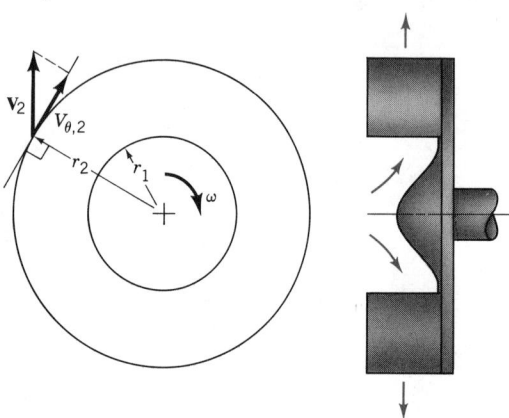

FIGURE P5.67

5.68 A centrifugal pump impeller is rotating at 1200 rpm in the direction shown in Fig. P5.68. The flow enters parallel to the axis of rotation and leaves at an angle of 30° to the radial direction. The absolute exit velocity, V_2, is 90 ft/s. **(a)** Draw the velocity triangle for the impeller exit flow. **(b)** Estimate the torque necessary to turn the impeller if the fluid density is 2.0 slugs/ft³.

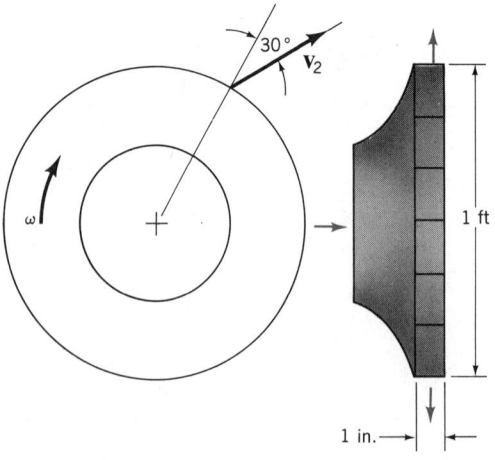

FIGURE P5.68

5.69 Shown in Fig. P5.69 are front and side views of a centrifugal pump rotor or impeller. If the pump delivers 200 liters/s of water and the blade exit angle is 35° from the tangential direction, determine the power requirement associated with flow leaving at the blade angle. The flow entering the rotor blade row is essentially radial as viewed from a stationary frame.

FIGURE P5.69

5.70 The radial component of velocity of water leaving the centrifugal pump sketched in Fig. P5.70 is 45 ft/s. The magnitude of the absolute velocity at the pump exit is 90 ft/s. The fluid enters the pump rotor radially. Calculate the shaft work required per unit mass flowing through the pump.

FIGURE P5.70

5.71 The velocity triangles for water flow through a radial pump rotor are as indicated in Fig. P5.71. **(a)** Determine the energy added to each unit mass (kg) of water as it flows through the rotor. **(b)** Sketch an appropriate blade section.

FIGURE P5.71

5.72   A fan (see Fig. P5.72) has a bladed rotor of 12-in.-outside diameter and 5-in.-inside diameter and runs at 1725 rpm. The width of each rotor blade is 1 in. from blade inlet to outlet. The volume flowrate is steady at 230 ft³/min and the absolute velocity of the air at blade inlet, V_1, is purely radial. The blade discharge angle is 30° measured with respect to the tangential direction at the outside diameter of the rotor. **(a)** What would be a reasonable blade inlet angle (measured with respect to the tangential direction at the inside diameter of the rotor)? **(b)** Find the power required to run the fan.

FIGURE P5.72

5.73   A centrifugal radial water pump has the dimensions shown in Fig. P5.73. The volume rate of flow is 0.25 ft³/s, and the absolute inlet velocity is directed radially outward. The angular velocity of the impeller is 960 rpm. The exit velocity as seen from a coordinate system attached to the

impeller can be assumed to be tangent to the vane at its trailing edge. Calculate the power required to drive the pump.

FIGURE P5.73

5.74   An axial flow gasoline pump (see Fig. P5.74) consists of a rotating row of blades (rotor) followed downstream by a stationary row of blades (stator). The gasoline enters the rotor axially (without any angular momentum) with an absolute velocity of 3 m/s. The rotor blade inlet and exit angles are 60° and 45° from the axial direction. The pump annulus passage cross section area is constant. Consider the flow as being tangent to the blades involved. Sketch velocity triangles for flow just upstream and downstream

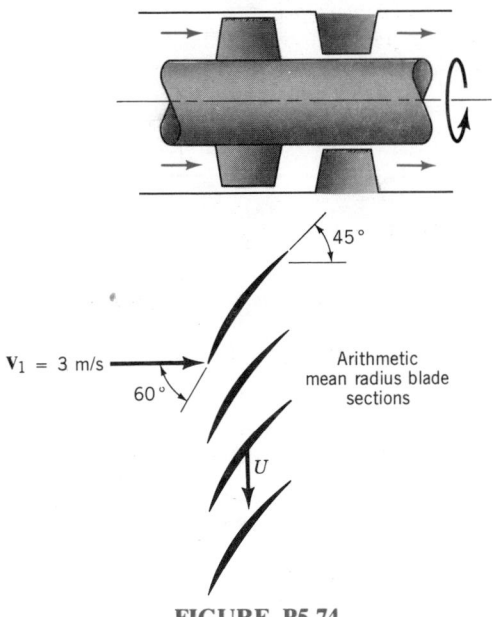

FIGURE P5.74

of the rotor and just downstream of the stator where the flow is axial. How much energy is added to each kilogram of gasoline?

5.75   The single stage, axial flow turbomachine shown in Fig. P5.75 involves water flow at a volumetric flowrate of 9 m³/s. The rotor revolves at 600 rpm. The inner and outer radii of the annular flow path through the stage are 0.46 and 0.61 m, and $\beta_2 = 60°$. The flow entering the rotor row and leaving the stator row is axial viewed from the stationary casing. Is this device a turbine or a pump? Estimate the amount of power transferred to or from the fluid.

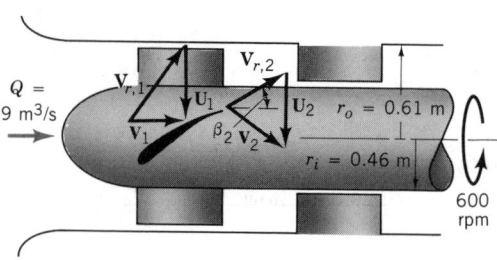

FIGURE P5.75

5.76   A sketch of the arithmetic mean radius blade sections of an axial flow water turbine stage is shown in Fig. P5.76. The rotor speed is 1500 rpm. **(a)** Sketch and label velocity triangles for the flow entering and leaving the rotor row. Use **V** for absolute velocity, V_r for relative velocity, and **U** for blade velocity. Assume flow enters and leaves each blade row at the blade angles shown. **(b)** Calculate the work per unit mass delivered at the shaft.

FIGURE P5.76

5.77   Sketched in Fig. P5.77 are the upstream [section (1)] and downstream [section (2)] velocity triangles at the arithmetic mean radius for flow through an axial flow turbomachine rotor. The axial component of velocity is 100 ft/s at sections (1) and (2). **(a)** Label each velocity vector appropriately. Use **V** for absolute velocity, V_r for relative velocity and **U** for blade velocity. **(b)** Are you dealing with a turbine or a fan? **(c)** Calculate the work per unit mass involved. **(d)** Sketch a reasonable blade section.

FIGURE P5.77

5.78   An axial flow turbomachine rotor involves the upstream (1) and downstream (2) velocity triangles shown in Fig. P5.78. Is this turbomachine a turbine or a fan? Sketch an appropriate blade section and determine energy transferred per unit mass of fluid.

FIGURE P5.78

5.79   By using velocity triangles for flow upstream (1) and downstream (2) of a turbomachine rotor, prove that the shaft work in per unit mass flowing through the rotor is

$$w_{\substack{shaft \\ net\ in}} = \frac{V_2^2 - V_1^2 + U_2^2 - U_1^2 + V_{r,1}^2 - V_{r,2}^2}{2}$$

where V = absolute flow velocity magnitude, V_r = relative flow velocity magnitude, and U = blade speed.

5.80*   Summarized below are air flow data for flow across a low-speed axial flow fan. Calculate the change in rate of flow of axial direction angular momentum across this rotor and evaluate the shaft power input involved. The inner and outer radii of the fan annulus are 142 and 203 mm. The rotor speed is 2400 rpm.

	Upstream of Rotor		Downstream of Rotor	
Radius (mm)	Axial Velocity (m/s)	Absolute Tangential Velocity (m/s)	Axial Velocity (m/s)	Absolute Tangential Velocity (m/s)
142	0	0	0	0
148	32.03	0	32.28	12.64
169	32.03	0	32.37	12.24
173	32.04	0	31.78	11.91
185	32.03	0	31.50	11.35
197	31.09	0	29.64	11.66
203	0	0	0	0

5.81 Air enters a radial blower with zero angular momentum. It leaves with an absolute tangential velocity, V_θ, of 100 ft/s. The rotor blade speed at rotor exit is 70 ft/s. If the stagnation pressure rise across the rotor is 0.1 psi, calculate the loss of available energy across the rotor.

5.82 Water enters a pump impeller radially. It leaves the impeller with a tangential component of absolute velocity of 10 m/s. The impeller exit diameter is 60 mm and the impeller speed is 1800 rpm. If the stagnation pressure rise across the impeller is 45 kPa, determine the loss of available energy across the impeller and the hydraulic efficiency of the pump.

5.83 Water enters an axial flow turbine rotor with an absolute velocity tangential component, V_θ, of 30 ft/s. The corresponding blade velocity, U, is 100 ft/s. The water leaves the rotor blade row with no angular momentum. If the stagnation pressure drop across the turbine is 45 psi, determine the hydraulic efficiency of the turbine.

5.84 An inward flow radial turbine (see Fig. P5.84) involves a nozzle angle, α_1, of 60° and an inlet rotor tip speed, U_1, of 9 m/s. The ratio of rotor inlet to outlet diameters is 2.0. The radial component of velocity remains constant at 6 m/s through the rotor and the flow leaving the rotor at section (2) is without angular momentum. If the flowing fluid is water and the stagnation pressure drop across the rotor is 110 kPa, determine the loss of available energy across the rotor and the hydraulic efficiency involved.

5.85 An inward flow radial turbine (see Fig. P5.84) involves a nozzle angle, α_1, of 60° and an inlet rotor tip speed of 9 m/s. The ratio of rotor inlet to outlet diameters is 2.0. The radial component of velocity remains constant at 6 m/s through the rotor and the flow leaving the rotor at section (2) is without angular momentum. If the flowing fluid is air and the static pressure drop across the rotor is 0.07 kPa, determine the loss of available energy across the rotor and the rotor aerodynamic efficiency.

5.86 For adiabatic flow through a turbomachine rotor, prove that the quantity

$$\breve{h} + \frac{V_r^2}{2} - \frac{U^2}{2}$$

remains constant, where

$\breve{h}$ = enthalpy
V_r = relative flow velocity
U = blade speed

5.87 A 200-m-high waterfall involves steady flow from one large body to another. Determine the temperature rise associated with this flow.

5.88 Is a necessary condition associated with application of the Bernoulli equation zero heat transfer? Explain.

5.89 Air steadily expands adiabatically and without friction from stagnation conditions of 690 kPa (abs) and 290 K to a static pressure of 101 kPa (abs). Determine the velocity of the expanded air assuming: **(a)** incompressible flow; **(b)** compressible flow.

5.90 A horizontal Venturi flow meter consists of a converging–diverging conduit as indicated in Fig. P5.90. The diameters of cross sections (1)

FIGURE P5.84

FIGURE P5.90

and (2) are 6 and 4 in. The velocity and static pressure are uniformly distributed at cross sections (1) and (2). Determine the volume flowrate (ft³/s) through the meter if $p_1 - p_2 = 3$ psi, the flowing fluid is oil ($\rho = 56$ lbm/ft³), and the loss per unit mass from (1) to (2) is negligibly small.

5.91 Oil ($SG = 0.9$) flows downward through a vertical pipe contraction as shown in Fig. P5.91. If the mercury manometer reading, h, is 120 mm, determine the volume flowrate for frictionless flow. Is the actual flowrate more or less than the frictionless value? Explain.

FIGURE P5.91

5.92 A circular disk can be lifted up by blowing on it with the device shown in Fig. P5.92. Explain why this happens.

FIGURE P5.92

5.93 A siphon is used to draw water at 20 °C from a large container as indicated in Fig. P5.93. The inside diameter of the siphon line is 1 in. and the pipe centerline rises 3 ft above the essentially constant water level in the tank. Show that by varying the length of the siphon below the water level, h, the rate of flow through the siphon can be changed. Assuming frictionless flow determine the maximum flowrate possible through the siphon. The limiting condition is the occurrence of cavitation in the siphon. Will the actual maximum

flow be more or less than the frictionless value? Explain.

FIGURE P5.93

5.94 A siphon is used to draw water at 20 °C from a large container as indicated in Fig. P5.94. Does changing the elevation, h, of the siphon centerline above the water level in the tank vary the flowrate through the siphon? Explain. What is the maximum allowable value of h?

FIGURE P5.94

5.95 A water siphon having a constant inside diameter of 3 in. is arranged as shown in Fig. P5.95. If the friction loss between A and B is

FIGURE P5.95

$0.6V^2/2$, where V is the velocity of flow in the siphon, determine the flowrate involved.

5.96   Water flows through a valve (see Fig. P5.96) at the rate of 1000 lbm/s. The pressure just upstream of the valve is 90 psi and the pressure drop across the valve is 50 psi. The inside diameters of the valve inlet and exit pipes are 12 and 24 in. If the flow through the valve occurs in a horizontal plane determine the loss in available energy across the valve.

FIGURE P5.96

5.97   Water flows steadily from one location to another in the inclined pipe shown in Fig. P5.97. At one section, the static pressure is 8 psi. At the other section, the static pressure is 5 psi. Which way is the water flowing? Explain.

FIGURE P5.97

5.98   Water flows through a vertical pipe as is indicated in Fig. P5.98. Is the flow up or down in the pipe? Explain.

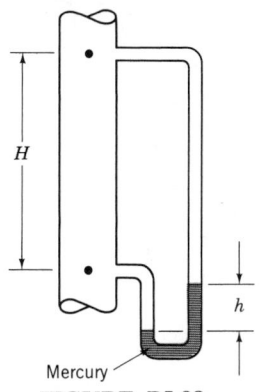

FIGURE P5.98

5.99   A fire hose nozzle is designed to deliver water that will rise 30 m vertically. Calculate the stagnation pressure required at the nozzle inlet if: **(a)** no loss is assumed; **(b)** a loss of 30 N·m/kg is assumed.

5.100   For the 180° elbow and nozzle flow shown in Fig. P5.100, determine the loss in available energy from section (1) to section (2). How much additional available energy is lost from section (2) to where the water comes to rest?

FIGURE P5.100

5.101   Compare the volume flowrates associated with two different vent configurations, a cylindrical hole in the wall having a diameter of 4 in. and the same diameter cylindrical hole in the wall but with a well-rounded entrance (see Fig. P5.101). The room is held at a constant pressure of 1.5 psi above atmospheric. Both vents exhaust into the atmosphere. The loss in available energy associated with flow through the cylindrical vent from the room to the vent exit is $0.5V_2^2/2$, where V_2 is the uniformly distributed exit velocity of air. The loss in available energy associated with flow through the rounded entrance vent from the room to the vent exit is $0.05V_2^2/2$, where V_2 is the uniformly distributed exit velocity of air.

FIGURE P5.101

5.102   An automobile engine will work best when the back pressure at the exhaust manifold, engine block interface is minimized. Show how reduction of losses in the exhaust manifold, piping, and muffler will also reduce the back pressure. How could losses in the exhaust system be reduced? What primarily limits the minimization of exhaust system losses?

5.103   Water flows vertically upward in a circular cross section pipe. At section (1), the velocity profile over the cross section area is uniform. At section (2), the velocity profile is

$$\mathbf{V} = w_c \left(\frac{R - r}{R}\right)^{1/7} \hat{\mathbf{k}}$$

where $\mathbf{V}$ = local velocity vector, w_c = centerline velocity in the axial direction, R = pipe inside radius, and, r = radius from pipe axis. Develop an expression for the loss in available energy between sections (1) and (2).

5.104   Water flowing in a pipe with a velocity of 25 m/s and a static pressure of 940 kPa is split into two branches as indicated in Fig. P5.104. If the measured static pressure at sections (2) and (3) and the velocity measured at section (3) are as shown in Fig. P5.104, determine the amount of available power lost in the horizontal y connection.

$A_1 = 0.4\ \text{m}^3$
$p_1 = 940\ \text{kPa}$
$V_1 = 25\ \text{m/s}$

$A_2 = 0.28\ \text{m}^2$
$p_2 = 1140\ \text{kPa}$

$A_3 = 0.18\ \text{m}^2$
$p_3 = 570\ \text{kPa}$
$V_3 = 30\ \text{m/s}$

FIGURE P5.104

5.105*   Air flows past an airfoil section. Upstream of the airfoil the fluid velocity and total pressure are uniformly distributed. Downstream of the airfoil, the fluid axial velocity and total pressure vary considerably. From the data provided, determine the area-averaged and mass-average total pressure loss associated with flow over the airfoil, where

$$\text{area averaged} = \frac{\int_0^Y (P_{o,\text{upstream}} - P_{o,\text{downstream}})\, dy}{Y}$$

$$\text{mass averaged} = \frac{\int_0^Y V_{\text{downstream}} (P_{o,\text{upstream}} - P_{o,\text{downstream}})\, dy}{\overline{V}_{\text{downstream}} Y}$$

where P_o = total pressure
y = tangential coordinate

$Y = y$ extent of data
V = local velocity in x (axial) coordinate direction
$\overline{V}_{\text{downstream}}$ = average velocity in x direction

$$= \frac{\int_0^Y V_{\text{downstream}}\, dy}{Y}$$

Which version of total pressure loss is correct? Why?

	Upstream of Airfoil	
y (m)	Total Pressure (Pa)	Axial Velocity (m/s)
0	572	31.8
0.00431	572	31.8
0.00844	572	31.8
0.0126	572	31.8
0.0168	572	31.8
0.0210	572	31.8
0.0252	572	31.8
0.0294	572	31.8
0.0336	572	31.8
0.0377	572	31.8

	Downstream of Airfoil	
y (m)	Total Pressure (Pa)	Axial Velocity (m/s)
0	555	32.4
0.00268	554	32.4
0.00528	559	32.5
0.00796	558	32.5
0.0106	559	32.5
0.0115	562	32.6
0.0121	559	32.5
0.0129	554	32.4
0.0136	550	32.3
0.0144	531	31.8
0.0151	463	30.0
0.0159	392	28.1
0.0167	293	25.0
0.0174	240	23.2
0.0182	321	25.9
0.0190	478	30.5
0.0197	536	32.0
0.0204	551	32.3
0.0212	552	32.4
0.0219	549	32.3
0.0229	551	32.3
0.0236	552	32.4
0.0242	551	32.4
0.0277	551	32.4
0.0310	554	32.4
0.0343	554	32.4
0.0377	551	32.4

5.106 A gas expands through a nozzle from a pressure of 300 psia to a pressure of 5 psia. The enthalpy change involved, $\check{h}_1 - \check{h}_2$, is 150 Btu/lbm. If the expansion is adiabatic but with frictional effects and the inlet gas speed is negligibly small, determine the exit gas velocity.

5.107 What is the maximum possible power output of the hydroelectric turbine shown in Fig. P5.107?

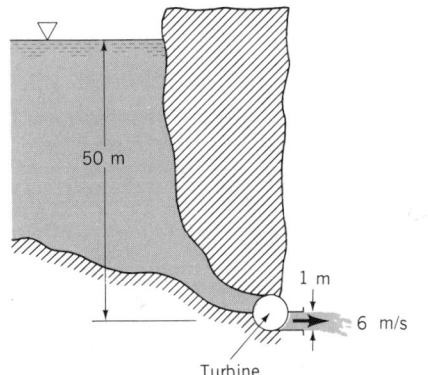

FIGURE **P5.107**

5.108 A hydroelectric turbine passes 8 million gal/min across a head of 600 ft of water. What is the maximum amount of power output possible? Why will the actual amount be less?

5.109 A hydroelectric power plant operates under the conditions illustrated in Fig. P5.109. The head loss associated with flow from the water level upstream of the dam, section (1), to the turbine discharge at atmospheric pressure, section (2), is 20 m. How much power is transferred from the water to the turbine blades?

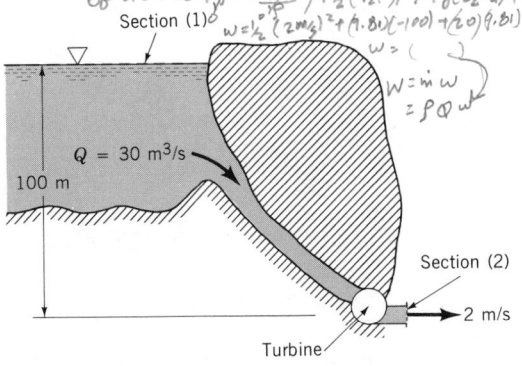

FIGURE **P5.109**

5.110 A hydraulic turbine is provided with 4.25 m³/s of water at 415 kPa. A vacuum gage in the turbine discharge 3 m below the turbine inlet centerline reads 250 mm Hg vacuum. If the turbine shaft output power is 1100 kW, calculate the power loss through the turbine. The supply and dis-

charge pipe inside diameters are identically 80 mm.

5.111 Water is supplied at 150 ft³/s and 60 psi to a hydraulic turbine through a 3-ft-inside diameter inlet pipe as indicated in Fig. P5.111. The turbine discharge pipe has a 4-ft-inside diameter. The static pressure at section (2), 10 ft below the turbine inlet, is 10 in. Hg vacuum. If the turbine develops 2500 hp, determine the rate of loss of available energy between sections (1) and (2).

FIGURE **P5.111**

5.112 A steam turbine receives steam having a static pressure, p_1, of 400 psia, an enthalpy, $\check{h}_1$, of 1407 Btu/lbm, and a velocity, V_1, of 100 ft/s. The steam leaves the turbine as a mixture of vapor and liquid having an enthalpy, $\check{h}_2$, of 1098 Btu/lbm, a pressure, p_2, of 2 psia, and a velocity, V_2, of 200 ft/s. If the flow through the turbine is essentially adiabatic and the change in elevation of the steam is negligible, calculate: **(a)** the actual work output per unit mass of steam; **(b)** the efficiency of the turbine if the ideal work output is 467 Btu/lbm.

5.113 A centrifugal air compressor stage operates between an inlet stagnation pressure of 14.7 psia and an exit stagnation pressure of 60 psia. The inlet stagnation temperature is 80 °F. If the loss of total pressure through the compressor stage associated with irreversible flow phenomena is 10 psi, calculate the actual and ideal stagnation temperature rise through the compressor. Calculate the ratio of ideal to actual temperature rise to obtain efficiency.

5.114 Explain how, in terms of the loss of available energy involved, a home water faucet valve works to vary the flow through a lawn hose from the shut-off condition to maximum flow.

5.115 When a fan or pump is tested at the factory, head curves (head across the fan or pump versus volume flowrate) are often produced. A generic fan or pump head curve is shown in Fig. P5.115a. For any piping system, the drop in pressure or head involved because of loss can be estimated as a function of volume flowrate. A generic piping system loss curve is also shown in Fig. P5.115b. When the pump or fan and piping system associated with the two curves of Fig. P5.115 are combined, what will the flowrate be? Why? How can the flowrate through this combined system be varied?

(a)   (b)

FIGURE P5.115

5.116* Total head-rise values measured for air flowing across a fan are listed below as a function of volume flowrate.

Q (m³/s)	Total Head Rise (mm H₂O)
0	79
0.14	79
0.28	76
0.42	67
0.57	65
0.71	70
0.85	76
0.99	79
1.13	75
1.27	64

Determine the flowrate that will result when this fan is connected to a piping system whose loss in total head is described by loss = $\kappa_L Q^2$ when: **(a)** κ_L = 49 mm H₂O/(m³/s)²; **(b)** κ_L = 91 mm H₂O/(m³/s)²; **(c)** κ_L = 140 mm H₂O/(m³/s)².

5.117 A pump transfers water from one large reservoir to another as shown in Fig. P5.117a. The difference in elevation between the two reservoirs is 100 ft. The friction head loss in the piping is given by $K_L \overline{V}^2/2g$, where $\overline{V}$ is the average fluid velocity in the pipe and K_L is the loss coefficient, which is considered constant. The relation between the total head rise H across the pump and the flowrate Q through the pump is given in Fig. 5.117b. If K_L = 20, and the pipe

diameter is 4 in., what is the flowrate through the pump?

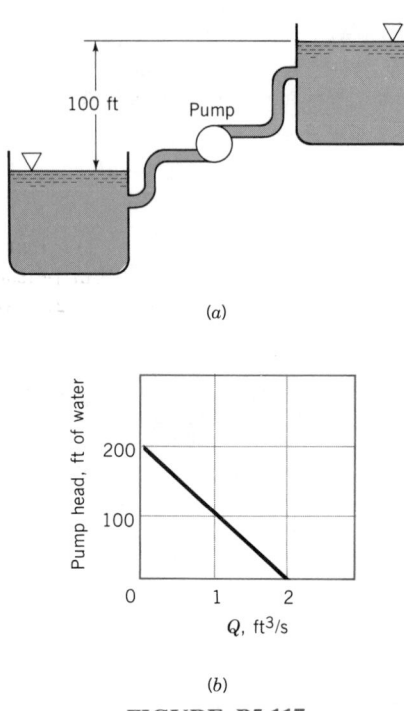

(a)

(b)

FIGURE P5.117

5.118 Water flows by gravity from one lake to another as sketched in Fig. P5.118 at the steady rate of 0.007 m³/s. What is the loss in available energy associated with this flow? If this same amount of loss is associated with pumping the fluid from the lower lake to the higher one at the same flowrate, estimate the amount of pumping power required.

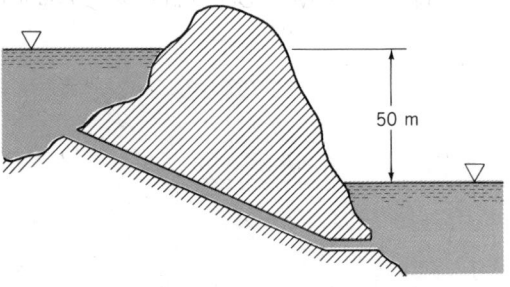

FIGURE P5.118

5.119 A ¾-hp motor is required by an air ventilating fan to produce a 24-in.-diameter stream of air having a uniform speed of 40 ft/s. Determine the aerodynamic efficiency of the fan.

5.120 A pump moves water horizontally at a rate of 0.02 m³/s. Upstream of the pump where the pipe diameter is 90 mm, the pressure is 120 kPa. Downstream of the pump where the pipe diameter is 30 mm, the pressure is 400 kPa. If the loss in energy across the pump due to fluid friction effects is 170 N·m/kg, determine the hydraulic efficiency of the pump.

5.121 Gasoline ($SG = 0.68$) flows through a pump at 0.12 m³/s as indicated in Fig. P5.121. The loss between sections (1) and (2) is equal to $0.3 V_1^2/2$. What will the difference in pressures between sections (1) and (2) be if 20 kW is delivered by the pump to the fluid?

FIGURE P5.121

5.122 Water is to be moved from one large reservoir to another at a higher elevation as indicated in Fig. P5.122. The loss in available energy associated with 2.5 ft³/s being pumped from sections (1) to (2) is $61 \overline{V}^2/2$ where $\overline{V}$ is the average velocity of water in the 8-in.-inside diameter piping involved. Determine the amount of shaft power required.

FIGURE P5.122

5.123 A liquid enters a fluid machine at sections (1) and (2) and leaves at section (3) as shown in Fig. P5.123. The density of the fluid is constant at 2 slugs/ft³. All of the flow occurs in a horizontal plane and is frictionless and adiabatic. For the above-mentioned and additional conditions indicated in Fig. 5.123, determine the amount of shaft power involved.

FIGURE P5.123

5.124 The distribution of axial direction velocity, u, in a pipe flow is linear from zero at the wall to maximum of u_c at the centerline. Determine the average velocity, $\overline{u}$, and the kinetic energy coefficient, α.

5.125 The velocity profile in a turbulent pipe flow may be approximated with the expression

$$\frac{u}{u_c} = \left(\frac{R - r}{R} \right)^{1/n}$$

where u = local velocity in the axial direction, u_c = centerline velocity in the axial direction, R = pipe inner radius from pipe axis, r = local radius from pipe axis, and n = constant. Determine the kinetic energy coefficient, α, for: **(a)** $n = 5$; **(b)** $n = 6$; **(c)** $n = 7$; **(d)** $n = 8$; **(e)** $n = 9$; **(f)** $n = 10$.

5.126 A small fan moves air at a mass flowrate of 0.004 lbm/s. Upstream of the fan, the pipe diameter is 2.5 in., the flow is laminar, the velocity distribution is parabolic, and the kinetic energy coefficient, α_1, is equal to 2.0. Downstream of the fan, the pipe diameter is 1 in., the flow is turbulent, the velocity profile is quite flat, and the kinetic energy coefficient, α_2, is equal to 1.08. If the rise in static pressure across the fan is 0.015 psi and the fan shaft draws 0.00024 hp, compare the value of loss calculated: **(a)** assuming uniform velocity distributions; **(b)** considering actual velocity distributions.

5.127 Demonstrate with the use of enthalpy versus entropy coordinates that an actual, adiabatic compressor process will always require more work than a frictionless, adiabatic process if both processes are between the same stagnation pressures. In a similar way, demonstrate that an actual, adiabatic expansion through a turbine will always produce less work than a frictionless, adiabatic process between the same stagnation pressures.

5.128 Show with the use of enthalpy versus entropy coordinates that the same amount of work input will result in less stagnation pressure rise for an actual, adiabatic compressor than a frictionless, adiabatic compressor. Also show that for the same amount of work output, an actual, adiabatic turbine requires a larger drop in stagnation pressure than a frictionless, adiabatic turbine does.

5.129 A 90° reducing bend occurs in the horizontal plane in a pipe line as shown in Fig. P5.129. The pipe diameter changes from 100 to 50 mm. The pressure upstream of the bend is 600 kPa and the flowrate is 60 liters/s. Determine the x and y components of reaction force exerted by the water on the bend. Assume frictionless flow.

$D_1 = 100$ mm
$p_1 = 600$ kPa
Section (2)
$D_2 = 50$ mm
$Q = 60$ liters/s

FIGURE P5.129

5.130 Assuming frictionless, incompressible, one-dimensional flow of water through the horizontal tee connection sketched in Fig. P5.130, estimate values of the x and y direction components of the force exerted by the water on the tee. Each pipe has an inside diameter of 2 ft.

$V_1 = 20$ ft/s
$p_1 = 35$ psia
Section (1)
Section (3)
$Q_3 = 100$ ft³/s
Section (2)

FIGURE P5.130

5.131 Assuming frictionless, incompressible, one-dimensional flow of water through the horizontal tee connection sketched in Fig. P5.131, estimate values of the x and y components of the force exerted by the tee on the water. Each pipe has an inside diameter of 1 m.

Section (3)
Section (2)
$Q_3 = 10$ m³/s
Section (1)
$V_1 = 6$ m/s
$p_1 = 200$ kPa

FIGURE P5.131

5.132 Water flows steadily down the inclined pipe as indicated in Fig. P5.132. Determine the

Flow
Section (1)
5 ft
6 in.
30°
Section (2)
6 in.
Mercury

FIGURE P5.132

following: **(a)** The difference in pressure $p_1 - p_2$. **(b)** The loss per unit mass between sections (1) and (2). **(c)** The net axial force exerted by the pipe wall on the flowing water between sections (1) and (2).

5.133 Water flows through a 2-ft-diameter pipe arranged horizontally in a circular arc as shown in Fig. P5.133. If the pipe discharges to the atmosphere ($p = 14.7$ psia) determine the x and y components of the resultant force exerted by the water on the piping between sections (1) and (2). The steady flowrate is 3000 ft^3/min. The loss in pressure due to fluid friction between sections (1) and (2) is 60 psi.

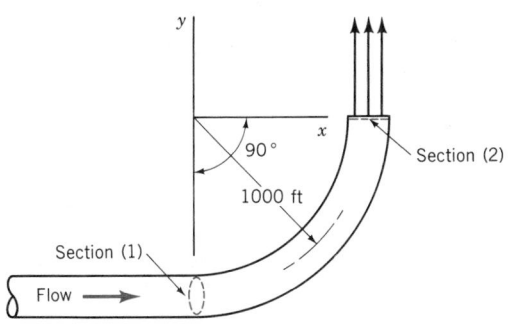

FIGURE P5.133

5.134 Eleven equally spaced turning vanes are used in the horizontal plane 90° bend as indicated in Fig. P5.134. The depth of the rectangular cross section bend remains constant at 3 in. The velocity distributions upstream and downstream of the vanes may be considered uniform. The loss in available energy across the vanes is $0.2V_1^2/2$. The required velocity and pressure downstream of the vanes, section (2), are 180 ft/s and 15 psia.

What is the average magnitude of the force exerted by the air flow on each vane? Assume the force of the air on the duct walls is equivalent to the force of the air on one vane.

5.135 When fluid flows through an abrupt expansion as indicated in Fig. P5.135, the loss in available energy across the expansion, loss$_{ex}$, is often expressed as

$$\text{loss}_{ex} = \left(1 - \frac{A_1}{A_2}\right)^2 \frac{V_1^2}{2}$$

where A_1 = cross section area upstream of expansion, A_2 = cross section area downstream of expansion, and V_1 = velocity of flow upstream of expansion. Derive this relationship.

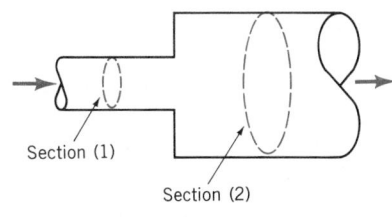

FIGURE P5.135

5.136 Near the downstream end of a river spillway, a hydraulic jump often forms, as illustrated in Fig. P5.136. The velocity of the channel flow is reduced abruptly across the jump. Using the conservation of mass and linear momentum principles, derive the expression for h_2,

$$h_2 = -\frac{h_1}{2} + \sqrt{\left(\frac{h_1}{2}\right)^2 + \frac{2V_1^2 h_1}{g}}$$

The loss of available energy across the jump can also be determined if energy conservation is considered. Derive the loss expression

$$\text{jump loss} = \frac{g(h_2 - h_1)^3}{4h_1 h_2}$$

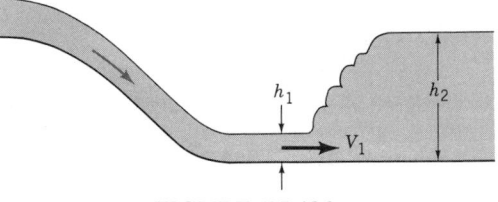

FIGURE P5.136

5.137 A vertical jet of water leaves a nozzle at a speed of 10 m/s and a diameter of 20 mm. It suspends a plate having a mass of 1.5 kg as indicated in Fig. P5.137. What is the vertical distance h?

FIGURE P5.137

5.138 A vertical jet of water having a nozzle exit velocity of 20 ft/s with a diameter of 1 in. suspends a hollow hemisphere as indicated in Fig. P5.138. If the hemisphere is stationary at an elevation of 12 in., determine its weight.

12 in.

FIGURE P5.138

Flow past an inclined plate: The streamlines of a viscous fluid flowing slowly past a two-dimensional object placed between two closely spaced plates (a Hele-Shaw cell) approximate inviscid, irrotational (potential) flow. (Dye in water between glass plates spaced 1 mm apart) (Photograph courtesy of D. H. Peregrine).

SIX

DIFFERENTIAL ANALYSIS
OF FLUID FLOW

In the previous chapter attention is focused on the use of finite control volumes for the solution of a variety of fluid mechanics problems. This approach is very practical and useful, since it does not generally require a detailed knowledge of the pressure and velocity variations within the control volume. Typically, we found that only conditions on the surface of the control volume entered the problem, and thus problems could be solved without a detailed knowledge of the flow field. Unfortunately, there are many situations that arise in which the details of the flow are important and the finite control volume approach will not yield the desired information. For example, we may need to know how the velocity varies over the cross section of a pipe, or how the pressure and shear stress vary along the surface of an airplane wing. In these circumstances we need to develop relationships that apply at a point, or at least in a very small region (infinitesimal volume) within a given flow field. This approach, which involves an *infinitesimal control volume,* as distinguished from a finite control volume, is commonly referred to as *differential analysis,* since (as we will soon discover) the governing equations are differential equations.

In this chapter we will provide an introduction to the differential equations that describe (in detail) the motion of fluids. Unfortunately, we will also find that these equations are rather complicated, partial differential equations that cannot be solved exactly except in a few cases, at least without making some simplifying assumptions. Thus, although differential analysis has the potential for supplying very detailed information about flow fields, this information is not easily extracted. Nevertheless, this approach provides a fundamental basis for the study of fluid mechanics. We do not want to be too discouraging at this point, since there are some exact solutions for laminar flow that can be obtained, and these have proved to be very useful. A few of these are included in this chapter. In addition, by making some simplifying assumptions many other analytical solutions can be obtained. For example, in some circumstances it may be reasonable to assume that the effect of viscosity is small and can be neglected. This rather drastic assumption greatly simplifies the analysis and provides the opportunity to obtain detailed solutions to a variety of complex flow problems. Some examples of these so-called *inviscid flow* solutions are also described in this chapter.

It is known that for certain types of flows the flow field can be conceptually divided into two regions—a very thin region near the boundaries of the system in which viscous effects are important, and a region away from the boundaries in which the flow is essentially inviscid. By making certain assumptions about the behavior of the fluid in the thin layer near the boundaries, and using the assumption of inviscid flow outside this layer, a large class of problems can be solved using differential analysis. These boundary layer problems are discussed in Chapter 9. Finally, it is to be noted that with the availability of powerful digital computers it is feasible to attempt to solve the differential equations using the techniques of numerical analysis. Although it is beyond the scope of this book to delve into this approach, which is generally referred to as *computational fluid dynamics* (CFD), the reader should be aware of this approach to complex flow problems. A few additional comments about CFD and other aspects of differential analysis are given in the last section of this chapter.

We begin our introduction to differential analysis by reviewing and extending some of the ideas associated with fluid kinematics that were introduced in Chapter 4. With this background the remainder of the chapter will be devoted to the derivation of the basic differential equations (which will be based on the principle of conservation of mass and Newton's second law of motion) and to some applications.

6.1 FLUID ELEMENT KINEMATICS

In this section we will be concerned with the mathematical description of the motion of fluid elements moving in a flow field. A small fluid element in the shape of a cube which is initially in one position will move to another position during a short time interval δt as illustrated in Fig. 6.1. Because of the generally complex velocity variation within the field, we expect the element to not only translate from one position but to also have its volume changed (linear deformation), to rotate, and to undergo a change in shape (angular deformation). Although these movements and deformations occur simultaneously, we can consider each one separately as illustrated in Fig. 6.1. Since element motion and deformation is intimately related to the velocity and variation of velocity throughout the flow field, we will briefly review the manner in which velocity and acceleration fields can be described.

FIGURE 6.1 Types of motion and deformation for a fluid element.

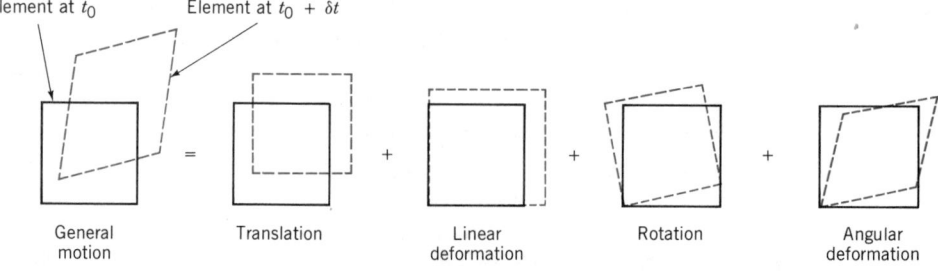

| General motion | Translation | Linear deformation | Rotation | Angular deformation |

6.1.1   Velocity and Acceleration Fields Revisited

As discussed in detail in Section 4.1 the velocity field can be described by specifying the velocity **V** at all points, and at all times, within the flow field of interest. Thus, in terms of rectangular coordinates, the notation $\mathbf{V}(x, y, z, t)$ means that the velocity of a fluid particle depends on where it is located within the flow field (as determined by its coordinates, x, y, and z) and when it occupies the particular point (as determined by the time, t). As is pointed out in Section 4.1.1, this method of describing the fluid motion is called the Eulerian method. It is also convenient to express the velocity in terms of three rectangular components so that

$$\mathbf{V} = u\hat{\mathbf{i}} + v\hat{\mathbf{j}} + w\hat{\mathbf{k}} \tag{6.1}$$

where u, v, and w are the velocity components in the x, y, and z directions, respectively, and $\hat{\mathbf{i}}, \hat{\mathbf{j}}$, and $\hat{\mathbf{k}}$ are the corresponding unit vectors. Of course, each of these components will, in general, be a function of x, y, z, and t. One of the goals of differential analysis is to determine how these velocity components specifically depend on x, y, z, and t for a particular problem.

With this description of the velocity field it was also shown in Section 4.2.1 that the acceleration of a fluid particle can be expressed as

$$\mathbf{a} = \frac{\partial \mathbf{V}}{\partial t} + u\frac{\partial \mathbf{V}}{\partial x} + v\frac{\partial \mathbf{V}}{\partial y} + w\frac{\partial \mathbf{V}}{\partial z} \tag{6.2}$$

and in component form:

$$a_x = \frac{\partial u}{\partial t} + u\frac{\partial u}{\partial x} + v\frac{\partial u}{\partial y} + w\frac{\partial u}{\partial z} \tag{6.3a}$$

$$a_y = \frac{\partial v}{\partial t} + u\frac{\partial v}{\partial x} + v\frac{\partial v}{\partial y} + w\frac{\partial v}{\partial z} \tag{6.3b}$$

$$a_z = \frac{\partial w}{\partial t} + u\frac{\partial w}{\partial x} + v\frac{\partial w}{\partial y} + w\frac{\partial w}{\partial z} \tag{6.3c}$$

The acceleration is also concisely expressed as

$$\mathbf{a} = \frac{D\mathbf{V}}{Dt} \tag{6.4}$$

where the operator

$$\frac{D(\ )}{Dt} = \frac{\partial(\ )}{\partial t} + u\frac{\partial(\ )}{\partial x} + v\frac{\partial(\ )}{\partial y} + w\frac{\partial(\ )}{\partial z} \tag{6.5}$$

is termed the material derivative, or substantial derivative. In vector notation

$$\frac{D(\ )}{Dt} = \frac{\partial(\ )}{\partial t} + (\mathbf{V} \cdot \nabla)(\ ) \tag{6.6}$$

where the gradient operator, $\nabla(\ )$, is

$$\nabla(\ ) = \frac{\partial(\ )}{\partial x}\hat{\mathbf{i}} + \frac{\partial(\ )}{\partial y}\hat{\mathbf{j}} + \frac{\partial(\ )}{\partial z}\hat{\mathbf{k}} \tag{6.7}$$

which was introduced in Chapter 2. As we will see in the following sections, the motion and deformation of a fluid element depends on the velocity field. The relationship between the motion and the forces causing the motion depends on the acceleration field.

6.1.2 Linear Motion and Deformation

The simplest type of motion that a fluid element can undergo is translation as illustrated in Fig. 6.2. In a small time interval δt a particle located at point O will move to point O' as is illustrated in the figure. If all points in the element have the same velocity (which is only true if there are no velocity gradients), then the element will simply translate from one position to another. However, because of the presence of velocity gradients, the element will generally be deformed and rotated as it moves. For example, consider the effect of a single velocity gradient, $\partial u/\partial x$, on a small cube having sides δx, δy, and δz. As is shown in Fig. 6.3a, if the x component of velocity of O and B is u, then at nearby points A and C the x component of the velocity can be expressed as $u + (\partial u/\partial x)\,\delta x$. This difference in velocity causes a "stretching" of the volume element by an amount $(\partial u/\partial x)(\delta x)(\delta t)$ during the short time interval δt in which line OA stretches to OA' and BC to BC' (Fig. 6.3b). The corresponding change in the original volume, $\delta V = \delta x\,\delta y\,\delta z$, would be

$$\text{Change in } \delta V = \left(\frac{\partial u}{\partial x}\delta x\right)(\delta y\,\delta z)(\delta t)$$

and the *rate* at which the volume δV is changing *per unit volume* due to the gradient $\partial u/\partial x$ is

$$\frac{1}{\delta V}\frac{d(\delta V)}{dt} = \lim_{\delta t \to 0}\left[\frac{(\partial u/\partial x)\,\delta t}{\delta t}\right] = \frac{\partial u}{\partial x} \tag{6.8}$$

If velocity gradients $\partial v/\partial y$ and $\partial w/\partial z$ are also present, then using a similar analysis it follows that, in the general case,

$$\frac{1}{\delta V}\frac{d V}{dt} = \frac{\partial u}{\partial x} + \frac{\partial v}{\partial y} + \frac{\partial w}{\partial z} = \nabla \cdot \mathbf{V} \tag{6.9}$$

This rate of change of the volume per unit volume is called the *volumetric dilatation rate*. Thus, we see that the volume of a fluid may change as the element moves from one location to another in the flow field. However, for an *incompressible fluid* the volumetric dilatation rate is zero, since the element volume cannot change without a change in fluid density (the element mass must be conserved). Variations in the velocity in the direction of the velocity, as represented by the derivatives $\partial u/\partial x$, $\partial v/\partial y$, and $\partial w/\partial z$, simply cause a *linear deformation* of the element in the sense that the shape of the element does not change. Cross derivatives, such as $\partial u/\partial y$, and $\partial v/\partial x$, will cause the element to rotate and generally to undergo an *angular deformation*, which changes the shape of the element.

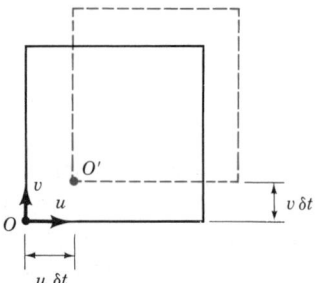

FIGURE 6.2 Translation of a fluid element.

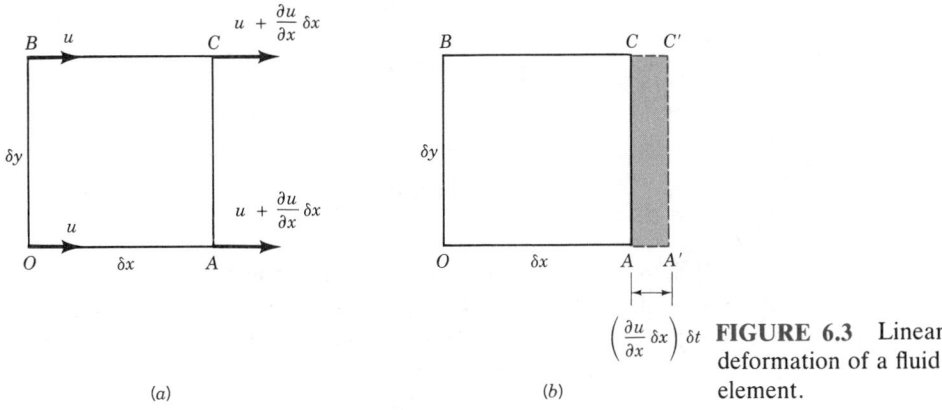

$$\left(\frac{\partial u}{\partial x}\,\delta x\right)\delta t$$

FIGURE 6.3 Linear deformation of a fluid element.

6.1.3 Angular Motion and Deformation

For simplicity we will consider motion in the x–y plane, but the results can be readily extended to the more general case. The velocity variation that causes rotation and angular deformation is illustrated in Fig. 6.4a. In a short time interval δt the line segments OA and OB will rotate through the angles $\delta\alpha$ and $\delta\beta$ to the new positions OA' and OB' as is shown in Fig. 6.4b. The angular velocity of line OA, ω_{OA}, is

$$\omega_{OA} = \lim_{\delta t \to 0} \frac{\delta\alpha}{\delta t}$$

For small angles

$$\tan\delta\alpha \approx \delta\alpha = \frac{(\partial v/\partial x)\,\delta x\,\delta t}{\delta x} = \frac{\partial v}{\partial x}\,\delta t \tag{6.10}$$

so that

$$\omega_{OA} = \lim_{\delta t \to 0}\left[\frac{(\partial v/\partial x)\,\delta t}{\delta t}\right] = \frac{\partial v}{\partial x}$$

Note that if $\partial v/\partial x$ is positive ω_{OA} will be counterclockwise. Similarly, the angular velocity of the line OB is

$$\omega_{OB} = \lim_{\delta t \to 0} \frac{\delta\beta}{\delta t}$$

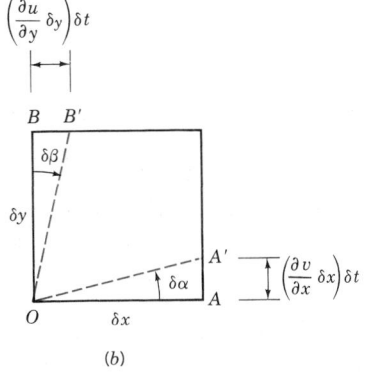

FIGURE 6.4 Angular motion and deformation of a fluid element.

and

$$\tan \delta\beta \approx \delta\beta = \frac{(\partial u/\partial y)\,\delta y\,\delta t}{\delta y} = \frac{\partial u}{\partial y}\,\delta t \tag{6.11}$$

so that

$$\omega_{OB} = \lim_{\delta t \to 0} \left[\frac{(\partial u/\partial y)\,\delta t}{\delta t} \right] = \frac{\partial u}{\partial y}$$

In this instance if $\partial u/\partial y$ is positive ω_{OB} will be clockwise. The *rotation*, ω_z, of the element about the z axis is defined as the average of the angular velocities ω_{OA} and ω_{OB} of the two mutually perpendicular lines OA and OB.[1] Thus, if counterclockwise rotation is considered to be positive, it follows that

$$\omega_z = \frac{1}{2}\left(\frac{\partial v}{\partial x} - \frac{\partial u}{\partial y}\right) \tag{6.12}$$

Rotation of the fluid element about the other two coordinate axes can be obtained in a similar manner with the result that for rotation about the x axis

$$\omega_x = \frac{1}{2}\left(\frac{\partial w}{\partial y} - \frac{\partial v}{\partial z}\right) \tag{6.13}$$

and for rotation about the y axis

$$\omega_y = \frac{1}{2}\left(\frac{\partial u}{\partial z} - \frac{\partial w}{\partial x}\right) \tag{6.14}$$

The three components ω_x, ω_y, and ω_z, can be combined to give the rotation vector, $\boldsymbol{\omega}$, in the form

$$\boldsymbol{\omega} = \omega_x \hat{\mathbf{i}} + \omega_y \hat{\mathbf{j}} + \omega_z \hat{\mathbf{k}} \tag{6.15}$$

An examination of this result reveals that $\boldsymbol{\omega}$ is equal to one-half the curl of the velocity vector. That is,

$$\boldsymbol{\omega} = \tfrac{1}{2}\,\text{curl}\,\mathbf{V} = \tfrac{1}{2}\,\nabla \times \mathbf{V} \tag{6.16}$$

since

$$\frac{1}{2}\nabla \times \mathbf{V} = \frac{1}{2}\begin{vmatrix} \hat{\mathbf{i}} & \hat{\mathbf{j}} & \hat{\mathbf{k}} \\ \dfrac{\partial}{\partial x} & \dfrac{\partial}{\partial y} & \dfrac{\partial}{\partial z} \\ u & v & w \end{vmatrix}$$

$$= \frac{1}{2}\left(\frac{\partial w}{\partial y} - \frac{\partial v}{\partial z}\right)\hat{\mathbf{i}} + \frac{1}{2}\left(\frac{\partial u}{\partial z} - \frac{\partial w}{\partial x}\right)\hat{\mathbf{j}} + \frac{1}{2}\left(\frac{\partial v}{\partial x} - \frac{\partial u}{\partial y}\right)\hat{\mathbf{k}}$$

The *vorticity*, $\boldsymbol{\zeta}$, is defined as a vector that is twice the rotation vector; that is

$$\boldsymbol{\zeta} = 2\,\boldsymbol{\omega} = \nabla \times \mathbf{V} \tag{6.17}$$

The use of the vorticity to describe the rotational characteristics of the fluid simply eliminates the $(\tfrac{1}{2})$ factor associated with the rotation vector.

---

[1] With this definition ω_z can also be interpreted to be the angular velocity of the bisector of the angle between the lines OA and OB.

We observe from Eq. 6.12 that the fluid element will rotate about the z axis as an *undeformed* block only when $\partial u/\partial y = -\partial v/\partial x$. Otherwise the rotation will be associated with an angular deformation. We also note from Eq. 6.12 that when $\partial u/\partial y = \partial v/\partial x$ the rotation around the z axis is zero. More generally if $\nabla \times \mathbf{V} = 0$ then the rotation (and the vorticity) are zero, and flow fields for which this condition applies are termed *irrotational*. We will find in Section 6.4 that the condition of irrotationality often greatly simplifies the analysis of complex flow fields. However, it is probably not immediately obvious why some flow fields would be irrotational, and we will need to examine this concept more fully in Section 6.4.

EXAMPLE 6.1

For a certain two-dimensional flow field the velocity is given by the equation

$$\mathbf{V} = 4xy\hat{\mathbf{i}} + 2(x^2 - y^2)\hat{\mathbf{j}}$$

Is this flow irrotational?

SOLUTION

For an irrotational flow the rotation vector, $\boldsymbol{\omega}$, having the components given by Eqs. 6.12, 6.13, and 6.14 must be zero. For the prescribed velocity field

$$u = 4xy \qquad v = 2(x^2 - y^2) \qquad w = 0$$

and therefore

$$\omega_x = \frac{1}{2}\left(\frac{\partial w}{\partial y} - \frac{\partial v}{\partial z}\right) = 0$$

$$\omega_y = \frac{1}{2}\left(\frac{\partial u}{\partial z} - \frac{\partial w}{\partial x}\right) = 0$$

$$\omega_z = \frac{1}{2}\left(\frac{\partial v}{\partial x} - \frac{\partial u}{\partial y}\right) = \frac{1}{2}(4x - 4x) = 0$$

Thus, the flow is irrotational.   (Ans)

It is to be noted that for a two-dimensional flow field (where the flow is in the x–y plane) ω_x and ω_y will always be zero, since by definition of two-dimensional flow u and v are not functions of z, and w is zero. In this instance the condition for irrotationality simply becomes $\omega_z = 0$ or $\partial v/\partial x = \partial u/\partial y$. (Lines OA and OB of Fig. 6.4 rotate with the same speed but in opposite directions so that there is no rotation of the fluid element.)

In addition to the rotation associated with the derivatives $\partial u/\partial y$ and $\partial v/\partial x$, it is observed from Fig. 6.4b that these derivatives can cause the fluid element to undergo an *angular deformation,* which results in a change in shape of the element. The change in the original right angle formed by the lines OA and OB is termed the *shearing strain,* $\delta\gamma$, and from Fig. 6.4b

$$\delta\gamma = \delta\alpha + \delta\beta$$

where $\delta\gamma$ is considered to be positive if the original right angle is decreasing. The rate of change of $\delta\gamma$ is called the *rate of shearing strain* or the *rate of angular deformation* and is commonly denoted with the symbol, $\dot{\gamma}$. The angles $\delta\alpha$ and $\delta\beta$ are related to the velocity gradients through Eqs. 6.10 and 6.11 so that

$$\dot{\gamma} = \lim_{\delta t \to 0} \frac{\delta \gamma}{\delta t} = \lim_{\delta t \to 0} \left[\frac{(\partial v / \partial x) \, \delta t + (\partial u / \partial y) \, \delta t}{\delta t} \right]$$

and, therefore,

$$\dot{\gamma} = \frac{\partial v}{\partial x} + \frac{\partial u}{\partial y} \tag{6.18}$$

As we will learn in Section 6.7, the rate of angular deformation is related to a corresponding shearing stress which causes the fluid element to change in shape. From Eq. 6.18 we note that if $\partial u / \partial y = -\partial v / \partial x$, the rate of angular deformation is zero, and this condition corresponds to the case in which the element is simply rotating as an undeformed block (Eq. 6.12). In the remainder of this chapter we will see how the various kinematical relationships developed in this section play an important role in the development and subsequent analysis of the differential equations that govern fluid motion.

6.2   CONSERVATION OF MASS

As is discussed in Section 5.2 conservation of mass requires that the mass, M, of a system remain constant as the system moves through the flow field. In equation form this principle is expressed as

$$\frac{DM_{\text{sys}}}{Dt} = 0$$

We found it convenient to use the control volume approach for fluid flow problems, with the control volume representation of the conservation of mass written as

$$\frac{\partial}{\partial t} \int_{\text{cv}} \rho \, d\Psi + \int_{\text{cs}} \rho \, \mathbf{V} \cdot \hat{\mathbf{n}} \, dA = 0 \tag{6.19}$$

where the equation (commonly called the continuity equation) can be applied to a finite control volume (cv), which is bounded by a control surface (cs). The first integral on the left side of Eq. 6.19 represents the rate at which the mass within the control volume is increasing, and the second integral represents the net rate at which mass is flowing out through the control surface (rate of mass outflow − rate of mass inflow). To obtain the differential form of the continuity equation, Eq. 6.19 is applied to an infinitesimal control volume.

6.2.1   Differential Form of Continuity Equation

We will take as our control volume the small cubical element shown in Fig. 6.5a. At the center of the element the fluid density is ρ and the velocity has components $u, v,$ and w. Since the element is small the volume integral in Eq. 6.19 can be expressed as

$$\frac{\partial}{\partial t} \int_{\text{cv}} \rho \, d\Psi \approx \frac{\partial \rho}{\partial t} \, \delta x \, \delta y \, \delta z \tag{6.20}$$

The rate of mass flow through the surfaces of the element can be obtained by considering the flow in each of the coordinate directions separately. For example, in Fig. 6.5b flow in the x direction is depicted. If we let ρu represent the x component of

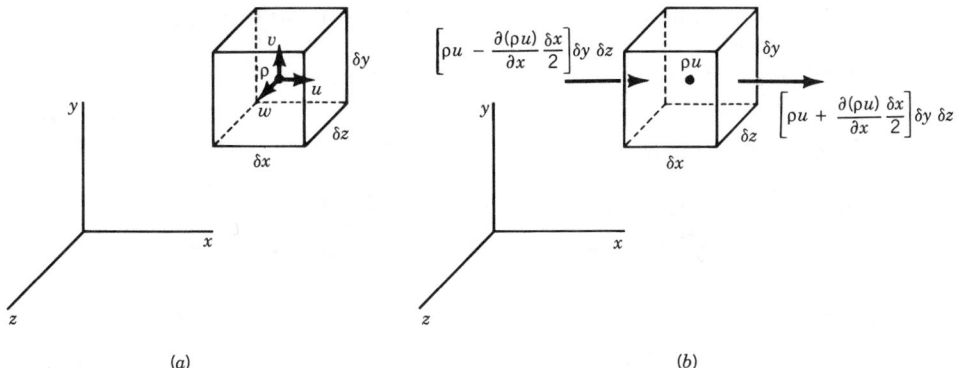

FIGURE 6.5 A differential element for the development of conservation of mass equation.

the mass rate of flow per unit area at the center of the element, then on the right face

$$\rho u \mid_{x+(\delta x/2)} = \rho u + \frac{\partial(\rho u)}{\partial x} \frac{\delta x}{2} \tag{6.21}$$

and on the left face

$$\rho u \mid_{x-(\delta x/2)} = \rho u - \frac{\partial(\rho u)}{\partial x} \frac{\delta x}{2} \tag{6.22}$$

Note that we are really using a Taylor series expansion of ρu and neglecting higher order terms such as $(\delta x)^2$, $(\delta x)^3$, and so on. When the right-hand sides of Eqs. 6.21 and 6.22 are multiplied by the area $\delta y \, \delta z$, the rate at which mass is crossing the right and left sides of the element are obtained as is illustrated in Fig. 6.5b. When these two expressions are combined, the net rate of mass flowing from the element through the two surfaces can be expressed as

$$\begin{array}{l} \text{Net rate of mass} \\ \text{outflow in } x \text{ direction} \end{array} = \left[\rho u + \frac{\partial(\rho u)}{\partial x} \frac{\delta x}{2} \right] \delta y \, \delta z$$

$$- \left[\rho u - \frac{\partial(\rho u)}{\partial x} \frac{\delta x}{2} \right] \delta y \, \delta z = \frac{\partial(\rho u)}{\partial x} \delta x \, \delta y \, \delta z \tag{6.23}$$

For simplicity, only flow in the x direction has been considered in Fig. 6.5b, but, in general, there will also be flow in the y and z directions. An analysis similar to the one used for flow in the x direction shows that

$$\begin{array}{l} \text{Net rate of mass} \\ \text{outflow in } y \text{ direction} \end{array} = \frac{\partial(\rho v)}{\partial y} \delta x \, \delta y \, \delta z \tag{6.24}$$

and

$$\begin{array}{l} \text{Net rate of mass} \\ \text{outflow in } z \text{ direction} \end{array} = \frac{\partial(\rho w)}{\partial z} \delta x \, \delta y \, \delta z \tag{6.25}$$

Thus,

$$\begin{array}{l} \text{Net rate of} \\ \text{mass outflow} \end{array} = \left[\frac{\partial(\rho u)}{\partial x} + \frac{\partial(\rho v)}{\partial y} + \frac{\partial(\rho w)}{\partial z} \right] \delta x \, \delta y \, \delta z \tag{6.26}$$

From Eqs. 6.19, 6.20, and 6.26 it now follows that the differential equation for conservation of mass is

$$\frac{\partial \rho}{\partial t} + \frac{\partial(\rho u)}{\partial x} + \frac{\partial(\rho v)}{\partial y} + \frac{\partial(\rho w)}{\partial z} = 0 \tag{6.27}$$

As previously mentioned, this equation is also commonly referred to as the continuity equation.

The continuity equation is one of the fundamental equations of fluid mechanics, and as expressed in Eq. 6.27, is valid for steady or unsteady flow, and compressible or incompressible fluids. In vector notation Eq. 6.27 can be written as

$$\frac{\partial \rho}{\partial t} + \nabla \cdot \rho \mathbf{V} = 0 \tag{6.28}$$

Two special cases are of particular interest. For *steady* flow of *compressible* fluids

$$\nabla \cdot \rho \mathbf{V} = 0$$

or

$$\frac{\partial(\rho u)}{\partial x} + \frac{\partial(\rho v)}{\partial y} + \frac{\partial(\rho w)}{\partial z} = 0 \tag{6.29}$$

This follows since by definition ρ is not a function of time for steady flow, but could be a function of position. For *incompressible* fluids the fluid density, ρ, is a constant throughout the flow field so that Eq. 6.28 becomes

$$\nabla \cdot \mathbf{V} = 0 \tag{6.30}$$

or

$$\frac{\partial u}{\partial x} + \frac{\partial v}{\partial y} + \frac{\partial w}{\partial z} = 0 \tag{6.31}$$

Equation 6.31 applies to both steady and unsteady flow of incompressible fluids. Note that Eq. 6.31 is the same as that obtained by setting the volumetric dilatation rate (Eq. 6.9) equal to zero. This result should not be surprising since both relationships are based on conservation of mass for incompressible fluids. However, the expression for the volumetric dilation rate was developed from a system approach, whereas Eq. 6.31 was developed from a control volume approach. In the former case the deformation of a particular differential mass of fluid was studied, and in the latter case mass flow through a fixed differential volume was studied.

EXAMPLE 6.2

The velocity components for a certain incompressible, steady flow field are

$$u = x^2 + y^2 + z^2$$
$$v = xy + yz + z$$
$$w = ?$$

Determine the form of the z component, w, required to satisfy the continuity equation.

SOLUTION

Any physically possible velocity distribution must for an incompressible fluid satisfy conservation of mass as expressed by the continuity equation

$$\frac{\partial u}{\partial x} + \frac{\partial v}{\partial y} + \frac{\partial w}{\partial z} = 0$$

For the given velocity distribution

$$\frac{\partial u}{\partial x} = 2x$$

and

$$\frac{\partial v}{\partial y} = x + z$$

so that the required expression for $\partial w/\partial z$ is

$$\frac{\partial w}{\partial z} = -2x - (x + z) = -3x - z$$

Integration with respect to z yields

$$w = -3xz - \frac{z^2}{2} + f(x, y) \qquad \text{(Ans)}$$

The third velocity component cannot be explicitly determined since the function $f(x, y)$ can have any form and conservation of mass will still be satisfied. The specific form of this function will be governed by the flow field described by these velocity components—that is, some additional information is needed to completely determine w.

6.2.2 Cylindrical Polar Coordinates

For some problems it is more convenient to express the various differential relationships in cylindrical polar coordinates rather than Cartesian coordinates. As is shown in Fig. 6.6, with cylindrical coordinates a point is located by specifying the coordinates r, θ, and z. The coordinate r is the radial distance from the z axis, θ is the angle measured from a line parallel to the x axis (with counterclockwise taken as positive), and z is the coordinate along the z axis. The velocity components, as are

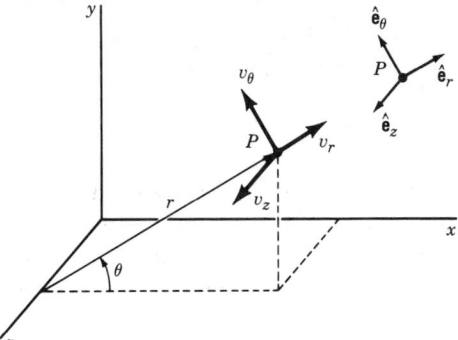

FIGURE 6.6 The representation of velocity components in cylindrical polar coordinates.

sketched in Fig. 6.6, are the radial velocity, v_r, the tangential velocity, v_θ, and the axial velocity, v_z. Thus, the velocity at some arbitrary point P can be expressed as

$$\mathbf{V} = v_r \hat{\mathbf{e}}_r + v_\theta \hat{\mathbf{e}}_\theta + v_z \hat{\mathbf{e}}_z \tag{6.32}$$

where $\hat{\mathbf{e}}_r$, $\hat{\mathbf{e}}_\theta$, and $\hat{\mathbf{e}}_z$ are the unit vectors in the r, θ, and z directions, respectively, as are illustrated in Fig. 6.6. The use of cylindrical coordinates is particularly convenient when the boundaries of the flow system are cylindrical. Several examples illustrating the use of cylindrical coordinates will be given in succeeding sections in this chapter.

The differential form of the continuity equation in cylindrical coordinates is

$$\frac{\partial \rho}{\partial t} + \frac{1}{r}\frac{\partial(r\rho v_r)}{\partial r} + \frac{1}{r}\frac{\partial(\rho v_\theta)}{\partial \theta} + \frac{\partial(\rho v_z)}{\partial z} = 0 \tag{6.33}$$

This equation can be derived by following the same procedure as used in the preceding section (see Problem 6.18). For steady, compressible flow

$$\frac{1}{r}\frac{\partial(r\rho v_r)}{\partial r} + \frac{1}{r}\frac{\partial(\rho v_\theta)}{\partial \theta} + \frac{\partial(\rho v_z)}{\partial z} = 0 \tag{6.34}$$

For incompressible fluids (for steady or unsteady flow)

$$\frac{1}{r}\frac{\partial(r v_r)}{\partial r} + \frac{1}{r}\frac{\partial v_\theta}{\partial \theta} + \frac{\partial v_z}{\partial z} = 0 \tag{6.35}$$

6.2.3 The Stream Function

Steady, incompressible, plane, two-dimensional flow represents one of the simplest types of flow of practical importance. By plane, two-dimensional flow we mean that there are only two velocity components, such as u and v, when the flow is considered to be in the x–y plane. For this flow the continuity equation, Eq. 6.31, reduces to

$$\frac{\partial u}{\partial x} + \frac{\partial v}{\partial y} = 0 \tag{6.36}$$

We still have two variables, u and v, to deal with, but they must be related in a special way as indicated by Eq. 6.36. This equation suggests that if we define a function $\psi(x, y)$, called the *stream function*, which relates the velocities as

$$u = \frac{\partial \psi}{\partial y} \qquad v = -\frac{\partial \psi}{\partial x} \tag{6.37}$$

then the continuity equation is identically satisfied. This conclusion can be verified by simply substituting the expressions for u and v into Eq. 6.36 so that

$$\frac{\partial}{\partial x}\left(\frac{\partial \psi}{\partial y}\right) + \frac{\partial}{\partial y}\left(-\frac{\partial \psi}{\partial x}\right) = \frac{\partial^2 \psi}{\partial x\, \partial y} - \frac{\partial^2 \psi}{\partial y\, \partial x} = 0$$

Thus, whenever the velocity components are defined in terms of the stream function we know that conservation of mass will be satisfied. Of course, we still do not know what $\psi(x, y)$ is for a particular problem, but at least we have simplified the analysis by having to determine only one unknown function, $\psi(x, y)$, rather than the two functions, $u(x, y)$ and $v(x, y)$.

Another particular advantage of using the stream function is related to the fact that *lines along which ψ is constant are streamlines*. Recall from Section 4.1.4 that

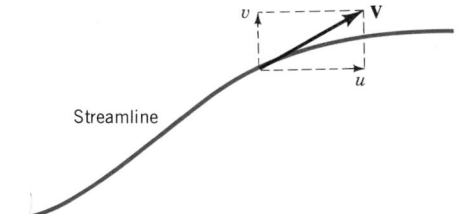

FIGURE 6.7 Velocity and velocity components along a streamline.

streamlines are lines in the flow field that are everywhere tangent to the velocities, as is illustrated in Fig. 6.7. It follows from the definition of the streamline that the slope at any point along a streamline is given by

$$\frac{dy}{dx} = \frac{v}{u}$$

The change in the value of ψ as we move from one point (x, y) to a nearby point $(x + dx, y + dy)$ is given by the relationship:

$$d\psi = \frac{\partial \psi}{\partial x} dx + \frac{\partial \psi}{\partial y} dy = -v \, dx + u \, dy$$

Along a line of constant ψ we have $d\psi = 0$ so that

$$-v \, dx + u \, dy = 0$$

and, therefore, along a line of constant ψ

$$\frac{dy}{dx} = \frac{v}{u}$$

which is the defining equation for a streamline. Thus, if we know the function $\psi(x, y)$ we can plot lines of constant ψ to provide the family of streamlines that are helpful in visualizing the pattern of flow. There are an infinite number of streamlines that can be drawn for a particular flow field, since for each constant value assigned to ψ a streamline can be drawn.

The actual numerical value associated with a particular streamline is not of particular significance, but the change in the value of ψ is related to the volume rate of flow. Consider two closely spaced streamlines, as are shown in Fig. 6.8a. The lower streamline is designated ψ and the upper one $\psi + d\psi$. Let dq represent the volume rate of flow (per unit width perpendicular to the x–y plane) passing between the two streamlines. Note that flow never crosses streamlines, since by definition the velocity

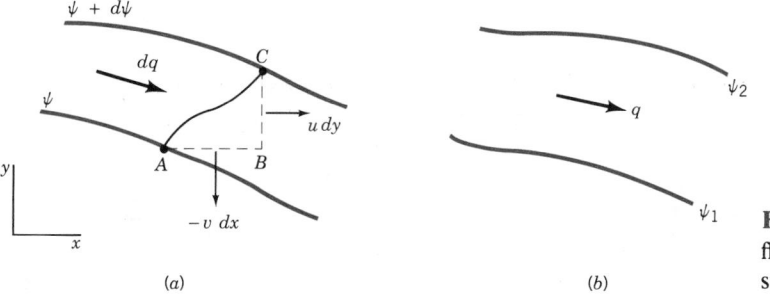

FIGURE 6.8 The flow between two streamlines.

is tangent to the streamline. From conservation of mass we know that the inflow, dq, crossing the arbitrary surface AC of Fig. 6.8a must equal the net outflow through surfaces AB and BC. Thus,

$$dq = u\,dy - v\,dx$$

or in terms of the stream function

$$dq = \frac{\partial \psi}{\partial y}\,dy + \frac{\partial \psi}{\partial x}\,dx \tag{6.38}$$

The right-hand side of Eq. 6.38 is equal to $d\psi$ so that

$$dq = d\psi \tag{6.39}$$

Thus, the volume rate of flow, q, between two streamlines such as ψ_1 and ψ_2 of Fig. 6.8b can be determined by integrating Eq. 6.39 to yield

$$q = \int_{\psi_1}^{\psi_2} d\psi = \psi_2 - \psi_1 \tag{6.40}$$

If the upper streamline, ψ_2, has a value greater than the lower streamline, ψ_1, q is positive, which indicates that the flow is from left to right. For $\psi_1 > \psi_2$ the flow is from right to left.

In cylindrical coordinates the continuity equation (Eq. 6.35) for incompressible, plane, two-dimensional flow reduces to

$$\frac{1}{r}\frac{\partial(rv_r)}{\partial r} + \frac{1}{r}\frac{\partial v_\theta}{\partial \theta} = 0 \tag{6.41}$$

and the velocity components, v_r and v_θ, can be related to the stream function, $\psi(r, \theta)$, through the equations

$$v_r = \frac{1}{r}\frac{\partial \psi}{\partial \theta} \qquad v_\theta = -\frac{\partial \psi}{\partial r} \tag{6.42}$$

Substitution of these expressions for the velocity components into Eq. 6.41 shows that the continuity equation is identically satisfied. The stream function concept can be extended to axisymmetric flows, such as flow in pipes or flow around bodies of revolution, and to two-dimensional compressible flows. However, the concept is not applicable to general three-dimensional flows.

EXAMPLE 6.3

The velocity components in a steady, two-dimensional incompressible flow field are

$$u = 2y$$

$$v = 4x$$

Determine the corresponding stream function and show on a sketch several streamlines. Indicate the direction of flow along the streamlines.

SOLUTION

From the definition of the stream function (Eqs. 6.37)

$$u = \frac{\partial \psi}{\partial y} = 2y$$

and

$$v = -\frac{\partial \psi}{\partial x} = 4x$$

The first of these equations can be integrated to give

$$\psi = y^2 + f_1(x)$$

where $f_1(x)$ is an arbitrary function of x. Similarly from the second equation

$$\psi = -2x^2 + f_2(y)$$

where $f_2(y)$ is an arbitrary function of y. It now follows that in order to satisfy both expressions for the stream function

$$\psi = -2x^2 + y^2 + C \qquad \text{(Ans)}$$

where C is an arbitrary constant.

Since the velocities are related to the derivatives of the stream function, an arbitrary constant can always be added to the function, and the value of the constant is actually of no consequence. Usually, for simplicity, we set $C = 0$ so that for this particular example the simplest form for the stream function is

$$\psi = -2x^2 + y^2 \qquad (1)(\text{Ans})$$

Either answer indicated would be acceptable.

Streamlines can now be determined by setting $\psi = $ constant and plotting the resulting curve. With the above expression for ψ (with $C = 0$) the value of ψ at the origin is zero so that the equation of the streamline passing through the origin (the $\psi = 0$ streamline) is

$$0 = -2x^2 + y^2$$

or

$$y = \pm\sqrt{2}x$$

Other streamlines can be obtained by setting ψ equal to various constants. It follows from Eq. 1 that the equations of these streamlines (for $\psi \neq 0$) can be expressed in the form

$$\frac{y^2}{\psi} - \frac{x^2}{\psi/2} = 1$$

which we recognize as the equation of a hyperbola. Thus, the streamlines are a family of hyperbolas with the $\psi = 0$ streamlines as asymptotes. Several of the streamlines are plotted in Fig. E6.3. Since the velocities can be calculated at any point, the direction of flow along a

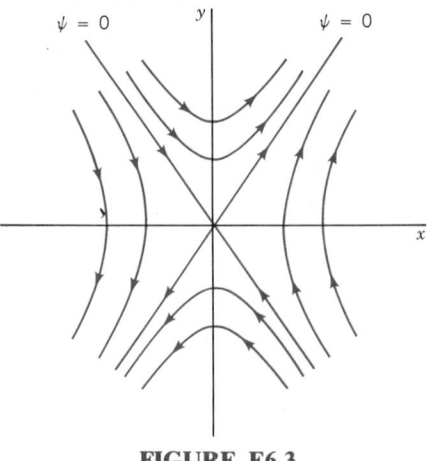

$\psi = 0$ y $\psi = 0$

x

FIGURE E6.3

given streamline can be easily deduced. For example $v = -\partial\psi/\partial x = 4x$ so that $v > 0$ if $x > 0$ and $v < 0$ if $x < 0$. The direction of flow is indicated on the figure.

6.3 CONSERVATION OF LINEAR MOMENTUM

To develop the differential, linear momentum equations we can start with the linear momentum equation

$$\mathbf{F} = \left.\frac{D\mathbf{P}}{Dt}\right|_{\text{sys}} \tag{6.43}$$

where $\mathbf{F}$ is the resultant force acting on a fluid mass, $\mathbf{P}$ is the linear momentum defined as

$$\mathbf{P} = \int_{\text{sys}} \mathbf{V}\,dm$$

and the operator $D(\ )/Dt$ is the material derivative. In the last chapter it was demonstrated how Eq. 6.43 in the form

$$\sum \mathbf{F}_{\substack{\text{contents of the} \\ \text{control volume}}} = \frac{\partial}{\partial t}\int_{\text{cv}} \mathbf{V}\rho\,d\mathcal{V} + \int_{\text{cs}} \mathbf{V}\rho\mathbf{V}\cdot\hat{\mathbf{n}}\,dA \tag{6.44}$$

could be applied to a finite control volume to solve a variety of flow problems. To obtain the differential form of the linear momentum equation, we can either apply Eq. 6.43 to a differential system, consisting of a mass, δm, or apply Eq. 6.44 to an infinitesimal control volume, $\delta\mathcal{V}$, which initially bounds the mass δm. It is probably simpler to use the system approach since application of Eq. 6.43 to the differential mass, δm, yields

$$\delta\mathbf{F} = \frac{D(\mathbf{V}\,\delta m)}{Dt}$$

where $\delta\mathbf{F}$ is the resultant force acting on δm. Using this system approach δm can be treated as a constant so that

$$\delta\mathbf{F} = \delta m\,\frac{D\mathbf{V}}{Dt}$$

But $D\mathbf{V}/Dt$ is the acceleration, $\mathbf{a}$, of the element. Thus,

$$\delta\mathbf{F} = \delta m\, \mathbf{a} \tag{6.45}$$

which is simply Newton's second law applied to the mass, δm. This is the same result that would be obtained by applying Eq. 6.44 to an infinitesimal control volume (see Ref. 1). Before we can proceed, it is necessary to examine how the force, $\delta\mathbf{F}$, can be most conveniently expressed.

6.3.1 Description of Forces Acting on Differential Element

In general, two types of forces need to be considered: *surface forces,* which act on the surface of the differential element, and *body forces,* which are distributed throughout the element. For our purpose, the only body force, $\delta\mathbf{F}_b$, of interest is the weight of the element, which can be expressed as

$$\delta\mathbf{F}_b = \delta m\, \mathbf{g} \tag{6.46}$$

where $\mathbf{g}$ is the vector representation of the acceleration of gravity. In component form

$$\delta F_{bx} = \delta m\, g_x \tag{6.47a}$$

$$\delta F_{by} = \delta m\, g_y \tag{6.47b}$$

$$\delta F_{bz} = \delta m\, g_z \tag{6.47c}$$

where g_x, g_y, and g_z are the components of the acceleration of gravity vector in the x, y, and z directions, respectively.

Surface forces act on the element as a result of its interaction with its surroundings. At any arbitrary location within a fluid mass, the force acting on a small area, δA, which lies in an arbitrary surface can be represented by $\delta\mathbf{F}_s$, as is shown in Fig. 6.9. In general, $\delta\mathbf{F}_s$ will be inclined with respect to the surface. The force $\delta\mathbf{F}_s$ can be resolved into three components, δF_n, δF_1, and δF_2, where δF_n is normal to the area, δA, and δF_1 and δF_2 are parallel to the area and orthogonal to each other. The *normal stress,* σ_n is defined as

$$\sigma_n = \lim_{\delta A \to 0} \frac{\delta F_n}{\delta A}$$

and the *shearing stresses* are defined as

$$\tau_1 = \lim_{\delta A \to 0} \frac{\delta F_1}{\delta A}$$

and

$$\tau_2 = \lim_{\delta A \to 0} \frac{\delta F_2}{\delta A}$$

We will use σ for normal stresses and τ for shearing stresses. The intensity of the

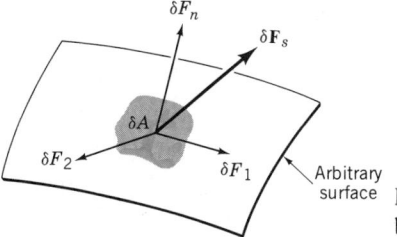

FIGURE 6.9 Components of force acting on an arbitrary differential area.

force per unit area at a point in a body can thus be characterized by a normal stress and two shearing stresses, if the orientation of the area is specified. For purposes of analysis it is usually convenient to reference the area to the coordinate system. For example, for the rectangular coordinate system shown in Fig. 6.10 we choose to consider the stresses acting on planes parallel to the coordinate planes. On the plane $ABCD$ of Fig. 6.10a, which is parallel to the y–z plane, the normal stress is denoted σ_{xx} and the shearing stresses are denoted as τ_{xy} and τ_{xz}. To easily identify the particular stress component we use a double subscript notation. The first subscript indicates the direction of the *normal* to the plane on which the stress acts, and the second subscript indicates the direction of the stress. Thus, normal stresses have repeated subscripts, whereas the subscripts for the shearing stresses are always different.

It is also necessary to establish a sign convention for the stresses. We define the positive direction for the stress as the positive coordinate direction on the surfaces for which the outward normal is in the positive coordinate direction. This is the case illustrated in Fig. 6.10a where the outward normal to the area $ABCD$ is in the positive x direction. The positive directions for σ_{xx}, τ_{xy}, and τ_{xz} are as shown in Fig. 6.10a. If the outward normal points in the negative coordinate direction, as in Fig. 6.10b for the area $A'B'C'D'$, then the stresses are considered positive if directed in the negative coordinate directions. Thus, the stresses shown in Fig. 6.10b are considered to be positive when directed as shown. Note that positive normal stresses are tensile stresses; that is, they tend to "stretch" the material.

It should be emphasized that the state of stress at a point in a material is not completely defined by simply three components of a "stress vector." This follows, since any particular stress vector depends on the orientation of the plane passing through the point. However, it can be shown that the normal and shearing stresses acting on *any* plane passing through a point can be expressed in terms of the stresses acting on three orthogonal planes passing through the point (Ref. 2).

We now can express the surface forces acting on a small cubical element of fluid in terms of the stresses acting on the faces of the element as shown in Fig. 6.11. It is expected that in general the stresses will vary from point to point within the flow field. Thus, we will express the stresses on the various faces in terms of the corresponding stresses at the center of the element of Fig. 6.11 and their gradients in the coordinate directions. For simplicity only the forces in the x direction are shown. Note that the stresses must be multiplied by the area on which they act to obtain the force. Summing all these forces in the x direction yields

$$\delta F_{sx} = \left(\frac{\partial \sigma_{xx}}{\partial x} + \frac{\partial \tau_{yx}}{\partial y} + \frac{\partial \tau_{zx}}{\partial z} \right) \delta x\, \delta y\, \delta z \qquad (6.48a)$$

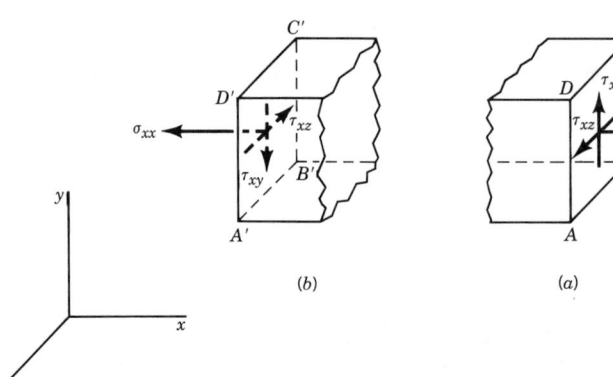

(b)   (a)

FIGURE 6.10
Double subscript notation for stresses.

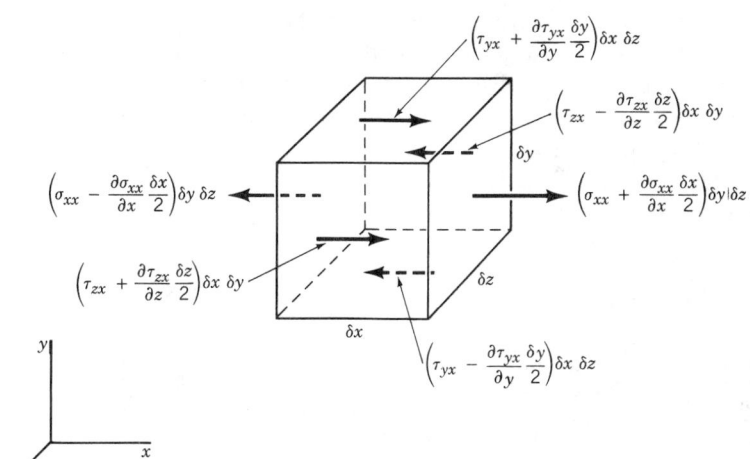

FIGURE 6.11 Surface forces in the x direction acting on a fluid element.

for the resultant surface force in the x direction. In a similar manner the resultant surface forces in the y and z direction can be obtained and expressed as

$$\delta F_{sy} = \left(\frac{\partial \tau_{xy}}{\partial x} + \frac{\partial \sigma_{yy}}{\partial y} + \frac{\partial \tau_{zy}}{\partial z} \right) \delta x \, \delta y \, \delta z \tag{6.48b}$$

and

$$\delta F_{sz} = \left(\frac{\partial \tau_{xz}}{\partial x} + \frac{\partial \tau_{yz}}{\partial y} + \frac{\partial \sigma_{zz}}{\partial z} \right) \delta x \, \delta y \, \delta z \tag{6.48c}$$

The resultant surface force can now be expressed as

$$\delta \mathbf{F}_s = \delta F_{sx} \hat{\mathbf{i}} + \delta F_{sy} \hat{\mathbf{j}} + \delta F_{sz} \hat{\mathbf{k}} \tag{6.49}$$

and this force combined with the body force, $\delta \mathbf{F}_b$, yields the resultant force, $\delta \mathbf{F}$, acting on the differential mass, δm. That is, $\delta \mathbf{F} = \delta \mathbf{F}_s + \delta \mathbf{F}_b$.

6.3.2 Equations of Motion

The expressions for the body and surface forces can now be used in conjunction with Eq. 6.45 to develop the equations of motion. In component form Eq. 6.45 can be written as

$$\delta F_x = \delta m \, a_x$$

$$\delta F_y = \delta m \, a_y$$

$$\delta F_z = \delta m \, a_z$$

where $\delta m = \rho \, \delta x \, \delta y \, \delta z$, and the acceleration components are given by Eq. 6.3. It now follows (using Eqs. 6.47 and 6.48 for the forces on the element) that

$$\rho g_x + \frac{\partial \sigma_{xx}}{\partial x} + \frac{\partial \tau_{yx}}{\partial y} + \frac{\partial \tau_{zx}}{\partial z} = \rho \left(\frac{\partial u}{\partial t} + u \frac{\partial u}{\partial x} + v \frac{\partial u}{\partial y} + w \frac{\partial u}{\partial z} \right) \tag{6.50a}$$

$$\rho g_y + \frac{\partial \tau_{xy}}{\partial x} + \frac{\partial \sigma_{yy}}{\partial y} + \frac{\partial \tau_{zy}}{\partial z} = \rho \left(\frac{\partial v}{\partial t} + u \frac{\partial v}{\partial x} + v \frac{\partial v}{\partial y} + w \frac{\partial v}{\partial z} \right) \tag{6.50b}$$

$$\rho g_z + \frac{\partial \tau_{xz}}{\partial x} + \frac{\partial \tau_{yz}}{\partial y} + \frac{\partial \sigma_{zz}}{\partial z} = \rho \left(\frac{\partial w}{\partial t} + u \frac{\partial w}{\partial x} + v \frac{\partial w}{\partial y} + w \frac{\partial w}{\partial z} \right) \tag{6.50c}$$

where the element volume $\delta x \, \delta y \, \delta z$ cancels out.

Equations 6.50 are the general differential equations of motion for a fluid. In fact, they are applicable to any continuum (solid or fluid) in motion or at rest. However, before we can use the equations to solve specific problems, some additional information about the stresses must be obtained. Otherwise, we will have more unknowns (all of the stresses and velocities and the density) than equations. It should not be too surprising that the differential analysis of fluid motion is complicated. We are attempting to describe, in detail, complex fluid motion.

6.4   INVISCID FLOW

As is discussed in Section 1.6, shearing stresses develop in a moving fluid because of the viscosity of the fluid. We know that for some common fluids, such as air and water, the viscosity is small, and therefore it seems reasonable to assume that under some circumstances we may be able to simply neglect the effect of viscosity (and thus shearing stresses). Flow fields in which the shearing stresses are assumed to be negligible are said to be *inviscid, nonviscous,* or *frictionless*. These terms are used interchangeably. As is discussed in Section 2.1, for fluids in which there are no shearing stresses the normal stress at a point is independent of direction—that is, $\sigma_{xx} = \sigma_{yy} = \sigma_{zz}$. In this instance we define the pressure, p, as the negative of the normal stress so that

$$-p = \sigma_{xx} = \sigma_{yy} = \sigma_{zz}$$

The negative sign is used so that a *compressive* normal stress (which is what we expect in a fluid) will give a *positive* value for p.

In Chapter 3 the inviscid flow concept was used in the development of the Bernoulli equation, and numerous applications of this important equation were considered. In this section we will again consider the Bernoulli equation and will show how it can be derived from the general equations of motion for inviscid flow.

6.4.1   Euler's Equations of Motion

For an inviscid flow in which all the shearing stresses are zero, and the normal stresses are replaced by $-p$, the general equations of motion (Eqs. 6.50) reduce to

$$\rho g_x - \frac{\partial p}{\partial x} = \rho \left(\frac{\partial u}{\partial t} + u \frac{\partial u}{\partial x} + v \frac{\partial u}{\partial y} + w \frac{\partial u}{\partial z} \right) \tag{6.51a}$$

$$\rho g_y - \frac{\partial p}{\partial y} = \rho \left(\frac{\partial v}{\partial t} + u \frac{\partial v}{\partial x} + v \frac{\partial v}{\partial y} + w \frac{\partial v}{\partial z} \right) \tag{6.51b}$$

$$\rho g_z - \frac{\partial p}{\partial z} = \rho \left(\frac{\partial w}{\partial t} + u \frac{\partial w}{\partial x} + v \frac{\partial w}{\partial y} + w \frac{\partial w}{\partial z} \right) \tag{6.51c}$$

These equations are commonly referred to as *Euler's equations of motion,* named in honor of Leonhard Euler (1707–1783), a famous Swiss mathematician who pioneered work on the relationship between pressure and flow. In vector notation Euler's equations can be expressed as

$$\rho \mathbf{g} - \nabla p = \rho \left[\frac{\partial \mathbf{V}}{\partial t} + (\mathbf{V} \cdot \nabla)\mathbf{V} \right] \tag{6.52}$$

Although Eqs. 6.51 are considerably simpler than the general equations of motion, they are still not amenable to a general analytical solution that would allow us to determine the pressure and velocity at all points within an inviscid flow field. The main difficulty arises from the nonlinear velocity terms (such as $u\,\partial u/\partial x$, $v\,\partial u/\partial y$, etc.), which appear in the convective acceleration. Because of these terms Euler's equations are nonlinear partial differential equations for which we do not have a general method for solving. However, under some circumstances we can use them to obtain useful information about inviscid flow fields. For example, as shown in the following section we can integrate Eq. 6.52 to obtain a relationship (the Bernoulli equation) between elevation, pressure, and velocity along a streamline.

6.4.2 The Bernoulli Equation

In Section 3.2 the Bernoulli equation was derived by a direct application of Newton's second law to a fluid particle moving along a streamline. In this section we will again derive this important equation, starting from Euler's equations. Of course, we should obtain the same result since Euler's equations simply represent a statement of Newton's second law expressed in a general form that is useful for flow problems. We will restrict our attention to steady flow so Euler's equation in vector form becomes

$$\rho\mathbf{g} - \nabla p = \rho(\mathbf{V}\cdot\nabla)\mathbf{V} \tag{6.53}$$

We wish to integrate this differential equation along some arbitrary streamline (Fig. 6.12) and select the coordinate system with the z axis vertical (with "up" being positive) so that the acceleration of gravity vector can be expressed as

$$\mathbf{g} = -g\,\nabla z$$

where g is the magnitude of the acceleration of gravity vector. Also, it will be convenient to use the vector identity

$$(\mathbf{V}\cdot\nabla)\mathbf{V} = \tfrac{1}{2}\nabla(\mathbf{V}\cdot\mathbf{V}) - \mathbf{V}\times(\nabla\times\mathbf{V})$$

Equation 6.53 can now be written in the form

$$-\rho g\,\nabla z - \nabla p = \frac{\rho}{2}\nabla(\mathbf{V}\cdot\mathbf{V}) - \rho(\mathbf{V}\times\nabla\times\mathbf{V})$$

and this equation can be rearranged to yield

$$\frac{\nabla p}{\rho} + \frac{1}{2}\nabla(V^2) + g\nabla z = \mathbf{V}\times(\nabla\times\mathbf{V})$$

We next take the dot product of each term with a differential length $d\mathbf{s}$ along a streamline (Fig. 6.12). Thus,

$$\frac{\nabla p}{\rho}\cdot d\mathbf{s} + \frac{1}{2}\nabla(V^2)\cdot d\mathbf{s} + g\,\nabla z\cdot d\mathbf{s} = [\mathbf{V}\times(\nabla\times\mathbf{V})]\cdot d\mathbf{s} \tag{6.54}$$

Since $d\mathbf{s}$ has a direction along the streamline, the vectors $d\mathbf{s}$ and $\mathbf{V}$ are parallel.

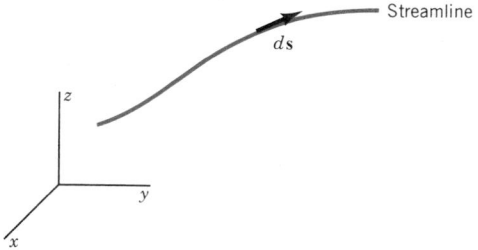

FIGURE 6.12 The notation for differential length along a streamline.

However, the vector $\mathbf{V} \times (\nabla \times \mathbf{V})$ is perpendicular to $\mathbf{V}$ (why?), so it follows that

$$[\mathbf{V} \times (\nabla \times \mathbf{V})] \cdot d\mathbf{s} = 0$$

Recall also that the dot product of the gradient of a scalar and a differential length gives the differential change in the scalar in the direction of the differential length. That is, with $d\mathbf{s} = dx\,\hat{\mathbf{i}} + dy\,\hat{\mathbf{j}} + dz\,\hat{\mathbf{k}}$ we can write $\nabla p \cdot d\mathbf{s} = (\partial p/\partial x)\,dx + (\partial p/\partial y)\,dy + (\partial p/\partial z)\,dz = dp$. Thus, Eq. 6.54 becomes

$$\frac{dp}{\rho} + \frac{1}{2}\,d(V^2) + g\,dz = 0 \tag{6.55}$$

where the change in p, V, and z is along the streamline. Equation 6.55 can now be integrated to give

$$\int \frac{dp}{\rho} + \frac{V^2}{2} + gz = \text{constant} \tag{6.56}$$

which indicates that the sum of the three terms on the left side of the equation must remain a constant along a given streamline. Equation 6.56 is valid for both compressible and incompressible inviscid flows, but for compressible fluids the variation in ρ with p must be specified before the first term in Eq. 6.56 can be evaluated.

For inviscid, incompressible fluids (commonly called *ideal fluids*) Eq. 6.56 can be written as

$$\frac{p}{\rho} + \frac{V^2}{2} + gz = \text{constant} \tag{6.57}$$

and this equation is the Bernoulli equation used extensively in Chapter 3. It is often convenient to write Eq. 6.57 between two points (1) and (2) along a streamline and to express the equation in the "head" form by dividing each term by g so that

$$\frac{p_1}{\gamma} + \frac{V_1^2}{2g} + z_1 = \frac{p_2}{\gamma} + \frac{V_2^2}{2g} + z_2 \tag{6.58}$$

It should be again emphasized that the Bernoulli equation, as expressed by Eqs. 6.57 and 6.58, is restricted to the following:

- inviscid flow
- steady flow
- incompressible flow
- flow along a streamline

You may want to go back and review some of the examples in Chapter 3 that illustrate the use of the Bernoulli equation.

6.4.3   Irrotational Flow

If we make one additional assumption—that the flow is *irrotational*—the analysis of inviscid flow problems is further simplified. Recall from Section 6.1.3 that the rotation of a fluid element is equal to $\frac{1}{2}(\nabla \times \mathbf{V})$, and an irrotational flow field is one for which $\nabla \times \mathbf{V} = 0$. Since the vorticity, ζ, is defined as $\nabla \times \mathbf{V}$, it also follows that in an irrotational flow field the vorticity is zero. The concept of irrotationality may seem to be a rather strange condition for a flow field. Why would a flow field be irrotational? To answer this question we note that if $\frac{1}{2}(\nabla \times \mathbf{V}) = 0$ then each of the components of this vector, as are give by Eqs. 6.12, 6.13, and 6.14, must be equal to zero. Since these components include the various velocity gradients in the flow field, the condition of irrotationality imposes specific relationships among these velocity gradients. For example, for rotation about the z axis to be zero, it follows from Eq. 6.12 that

$$\omega_z = \frac{1}{2}\left(\frac{\partial v}{\partial x} - \frac{\partial u}{\partial y}\right) = 0$$

and, therefore,

$$\frac{\partial v}{\partial x} = \frac{\partial u}{\partial y} \tag{6.59}$$

Similarly from Eqs. 6.13 and 6.14

$$\frac{\partial w}{\partial y} = \frac{\partial v}{\partial z} \tag{6.60}$$

and

$$\frac{\partial u}{\partial z} = \frac{\partial w}{\partial x} \tag{6.61}$$

A general flow field would not satisfy these three equations. However, a uniform flow as is illustrated in Fig. 6.13 does. Since $u = U$ (a constant), $v = 0$, and $w = 0$, it follows that Eqs. 6.59, 6.60, and 6.61 are all satisfied. Therefore, a uniform flow field (in which there are no velocity gradients) is certainly an example of an irrotational flow.

Uniform flows by themselves are not very interesting. However, many interesting and important flow problems include uniform flow in some part of the flow field. Two examples are shown in Fig. 6.14. In Fig. 6.14a a solid body is placed in a uniform stream of fluid. Far away from the body the flow remains uniform, and in this far region the flow is irrotational. In Fig. 6.14b, flow from a large reservoir enters a pipe through a streamlined entrance where the velocity distribution is essentially uniform. Thus, at the entrance the flow is irrotational.

For an inviscid fluid there are no shearing stresses—the only forces acting on a fluid element are its weight and pressure forces. Since the weight acts through the element center of gravity, and the pressure acts in a direction normal to the element surface, neither of these forces can cause the element to rotate. Therefore, for an inviscid fluid, if some part of the flow field is irrotational, the fluid elements emanating from this region will not take on any rotation as they progress through the flow field. This phenomenon is illustrated in Fig. 6.14a in which fluid elements flowing far away from the body have irrotational motion, and as they flow around the body the motion

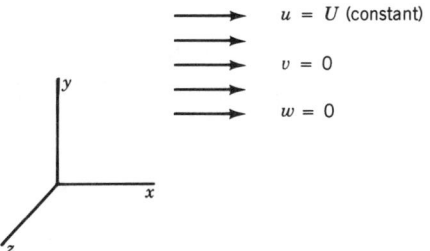

FIGURE 6.13   Uniform flow in the x direction.

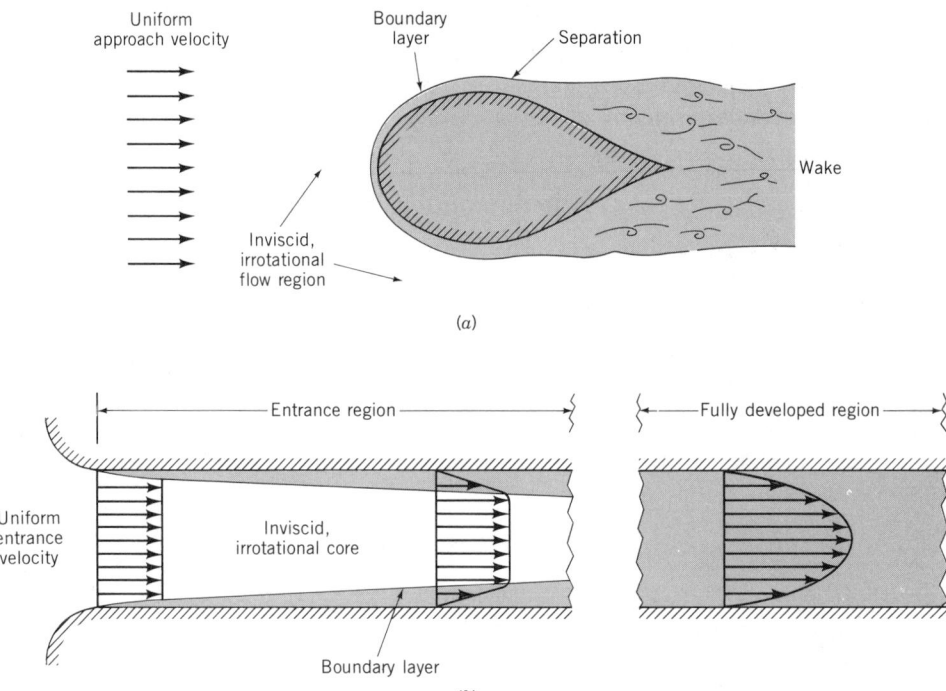

FIGURE 6.14 Various regions of flow: (*a*) around bodies; (*b*) through channels.

remains irrotational except very near the boundary. Near the boundary the velocity changes rapidly from zero at the boundary (no-slip condition) to some relatively large value in a short distance from the boundary. This rapid change in velocity gives rise to a large velocity gradient normal to the boundary and produces significant shearing stresses, even though the viscosity is small. Of course if we had a truly inviscid fluid, the fluid would simply "slide" past the boundary and the flow would be irrotational everywhere. But this is not the case for real fluids so we will typically have a layer (usually very thin) near any fixed surface in a moving stream in which shearing stresses are not negligible. This layer is called the *boundary layer*. Outside the boundary layer the flow can be treated as an irrotational flow. Another possible consequence of the boundary layer is that the main stream may "separate" from the surface and form a *wake* downstream from the body. The wake would include a region of slowly, randomly moving fluid. To completely analyze this type of problem it is necessary to consider both the inviscid, irrotational flow outside the boundary layer, and the viscous, rotational flow within the boundary layer and to somehow "match" these two regions. This type of analysis is considered in Chapter 9.

As is illustrated in Fig. 6.14*b*, the flow in the entrance to a pipe may be uniform (if the entrance is streamlined), and thus will be irrotational. In the central core of the pipe the flow remains irrotational for some distance. However, a boundary layer will develop along the wall and grow in thickness until it fills the pipe. Thus, for this type of internal flow there will be an *entrance region* in which there is a central irrotational core, followed by a so-called *fully developed region* in which viscous forces are dominant. The concept of irrotationality is completely invalid in the fully developed region. This type of internal flow problem is considered in detail in Chapter 8.

The two preceding examples are intended to illustrate the possible applicability of irrotational flow to some "real fluid" flow problems, and to indicate some limitations of the irrotationality concept. We proceed to develop some useful equations based on the assumptions of inviscid, incompressible, irrotational flow, with the admonition to use caution when applying the equations.

6.4.4   The Bernoulli Equation for Irrotational Flow

In the development of the Bernoulli equation in Section 6.4.2, Eq. 6.54 was integrated along a streamline. This restriction was imposed so the right side of the equation could be set equal to zero; that is

$$(\mathbf{V} \times \mathbf{\nabla} \times \mathbf{V}) \cdot d\mathbf{s} = 0$$

(since $d\mathbf{s}$ is parallel to $\mathbf{V}$). However, for irrotational flow, $\mathbf{\nabla} \times \mathbf{V} = 0$, so the right side of Eq. 6.54 is zero regardless of the direction of $d\mathbf{s}$. We can now follow the same procedure used to obtain Eq. 6.55, where the differential changes dp, $d(V^2)$, and dz can be taken in any direction. Integration of Eq. 6.55 again yields

$$\int \frac{dp}{\rho} + \frac{V^2}{2} + gz = \text{constant} \tag{6.62}$$

but the constant is the same throughout the flow field. Thus, for incompressible, irrotational flow the Bernoulli equation can be written as

$$\frac{p_1}{\gamma} + \frac{V_1^2}{2g} + z_1 = \frac{p_2}{\gamma} + \frac{V_2^2}{2g} + z_2 \tag{6.63}$$

between *any two points in the flow field*. Equation 6.63 is exactly the same form as Eq. 6.58 but is not limited to application along a streamline. However, Eq. 6.63 is restricted to:

- inviscid flow
- steady flow
- incompressible flow
- irrotational flow

It may be worthwhile to review the use and misuse of the Bernoulli equation for rotational flow as is illustrated in Example 3.19.

6.4.5   The Velocity Potential

For an irrotational flow the velocity gradients are related through Eqs. 6.59, 6.60, and 6.61. It follows that in this case the velocity components can be expressed in terms of a scalar function $\phi(x, y, z, t)$ as

$$u = \frac{\partial \phi}{\partial x} \qquad v = \frac{\partial \phi}{\partial y} \qquad w = \frac{\partial \phi}{\partial z} \tag{6.64}$$

where ϕ is called the *velocity potential*. Direct substitution of these expressions for the velocity components into Eqs. 6.59, 6.60, and 6.61 will verify that a velocity field defined by Eqs. 6.64 is indeed irrotational. In vector form Eqs. 6.64 can be written as

$$\mathbf{V} = \mathbf{\nabla}\phi \tag{6.65}$$

so that for an irrotational flow the velocity is expressible as the gradient of a scalar function ϕ.

The velocity potential is a consequence of the irrotationality of the flow field, whereas the stream function is a consequence of conservation of mass. It is to be noted, however, that the velocity potential can be defined for a general three-dimensional flow, whereas the stream function is restricted to two-dimensional flows.

For an incompressible fluid we know from conservation of mass that

$$\nabla \cdot \mathbf{V} = 0$$

and therefore for incompressible, irrotational flow (with $\mathbf{V} = \nabla \phi$) it follows that

$$\nabla^2 \phi = 0 \tag{6.66}$$

In Cartesian coordinates

$$\frac{\partial^2 \phi}{\partial x^2} + \frac{\partial^2 \phi}{\partial y^2} + \frac{\partial^2 \phi}{\partial z^2} = 0$$

This differential equation arises in many different areas of engineering and physics and is called *Laplace's equation*. Thus, inviscid, incompressible, irrotational flow fields are governed by Laplace's equation. This type of flow is commonly called a *potential flow*. To complete the mathematical formulation of a given problem, boundary conditions have to be specified. These are usually velocities specified on the boundaries of the flow field of interest. It follows that if the potential function can be determined, then the velocity at all points in the flow field can be determined from Eq. 6.64, and the pressure at all points can be determined from the Bernoulli equation (Eq. 6.63). Although the concept of the velocity potential is applicable to both steady and unsteady flow, we will confine our attention to steady flow.

For some problems it will be convenient to use cylindrical coordinates, r, θ, and z. In this coordinate system the gradient operator is

$$\nabla(\ ) = \frac{\partial(\ )}{\partial r} \hat{\mathbf{e}}_r + \frac{1}{r} \frac{\partial(\ )}{\partial \theta} \hat{\mathbf{e}}_\theta + \frac{\partial(\ )}{\partial z} \hat{\mathbf{e}}_z \tag{6.67}$$

so that

$$\nabla \phi = \frac{\partial \phi}{\partial r} \hat{\mathbf{e}}_r + \frac{1}{r} \frac{\partial \phi}{\partial \theta} \hat{\mathbf{e}}_\theta + \frac{\partial \phi}{\partial z} \hat{\mathbf{e}}_z \tag{6.68}$$

where $\phi = \phi(r, \theta, z)$. Since

$$\mathbf{V} = v_r \hat{\mathbf{e}}_r + v_\theta \hat{\mathbf{e}}_\theta + v_z \hat{\mathbf{e}}_z \tag{6.69}$$

it follows for an irrotational flow (with $\mathbf{V} = \nabla \phi$)

$$v_r = \frac{\partial \phi}{\partial r} \qquad v_\theta = \frac{1}{r} \frac{\partial \phi}{\partial \theta} \qquad v_z = \frac{\partial \phi}{\partial z} \tag{6.70}$$

Also, Laplace's equation in cylindrical coordinates is

$$\frac{1}{r} \frac{\partial}{\partial r} \left(r \frac{\partial \phi}{\partial r} \right) + \frac{1}{r^2} \frac{\partial^2 \phi}{\partial \theta^2} + \frac{\partial^2 \phi}{\partial z^2} = 0 \tag{6.71}$$

EXAMPLE 6.4

The two-dimensional flow of a nonviscous, incompressible fluid in the vicinity of the 90° corner of Fig. E6.4a is described by the stream function

$$\psi = 2r^2 \sin 2\theta$$

(a)

(b)

(c)

FIGURE E6.4

where ψ has units of m²/s when r is in meters. (a) Determine, if possible, the corresponding velocity potential. (b) If the pressure at point (1) on the wall is 30 kPa, what is the pressure at point (2)? Assume the fluid density is 10^3 kg/m³ and the x–y plane is horizontal—that is, there is no difference in elevation between points (1) and (2).

SOLUTION

(a) The radial and tangential velocity components can be obtained from the stream function as

$$v_r = \frac{1}{r}\frac{\partial \psi}{\partial \theta} = 4r \cos 2\theta$$

and

$$v_\theta = -\frac{\partial \psi}{\partial r} = -4r \sin 2\theta$$

Since

$$v_r = \frac{\partial \phi}{\partial r}$$

it follows that

$$\frac{\partial \phi}{\partial r} = 4r \cos 2\theta$$

and therefore by integration

$$\phi = 2r^2 \cos 2\theta + f_1(\theta) \tag{1}$$

where $f_1(\theta)$ is an arbitrary function of θ. Similarly

$$v_\theta = \frac{1}{r}\frac{\partial \phi}{\partial \theta} = -4r \sin 2\theta$$

and integration yields

$$\phi = 2r^2 \cos 2\theta + f_2(r) \tag{2}$$

where $f_2(r)$ is an arbitrary function of r. To satisfy both Eqs. 1 and 2 the velocity potential must have the form

$$\phi = 2r^2 \cos 2\theta + C \qquad \text{(Ans)}$$

where C is an arbitrary constant. As is the case for stream functions, the specific value of C is not important, and it is customary to let $C = 0$ so that the velocity potential for this corner flow is

$$\phi = 2r^2 \cos 2\theta \qquad \text{(Ans)}$$

In the statement of this problem it was implied by the wording "if possible" that we might not be able to find a corresponding velocity potential. The reason for this concern is that we can always define a stream function for two-dimensional flow, but the flow must be *irrotational* if there is a corresponding velocity potential. Thus, the fact that we were able to determine a velocity potential means that the flow is irrotational. Several streamlines and lines of constant ϕ are plotted in Fig. E6.4b. These two sets of lines are *orthogonal*. The reason why streamlines and lines of constant ϕ are always orthogonal is explained in Section 6.5.

(b)   Since we have an irrotational flow of a nonviscous, incompressible fluid, the Bernoulli equation can be applied between any two points. Thus, between points (1) and (2) with no elevation change

$$\frac{p_1}{\gamma} + \frac{V_1^2}{2g} = \frac{p_2}{\gamma} + \frac{V_2^2}{2g}$$

or

$$p_2 = p_1 + \frac{\rho}{2}(V_1^2 - V_2^2) \tag{3}$$

Since

$$V^2 = v_r^2 + v_\theta^2$$

it follows that for any point within the flow field

$$V^2 = (4r \cos 2\theta)^2 + (-4r \sin 2\theta)^2$$
$$= 16r^2(\cos^2 2\theta + \sin^2 2\theta)$$
$$= 16r^2$$

This result indicates that the square of the velocity at any point depends only on the radial distance, r, to the point. Note that the constant, 16, has units of s^{-2}. Thus,

$$V_1^2 = (16\ s^{-2})(1\ m)^2 = 16\ m^2/s^2$$

and

$$V_2^2 = (16\ s^{-2})(0.5\ m)^2 = 4\ m^2/s^2$$

Substitution of these velocities into Eq. 3 gives

$$p_2 = 30 \times 10^3\ N/m^2 + \frac{10^3\ kg/m^3}{2}\ (16\ m^2/s^2 - 4\ m^2/s^2) = 36\ kPa \qquad (Ans)$$

The stream function used in this example could also be expressed in Cartesian coordinates as

$$\psi = 2r^2 \sin 2\theta = 4r^2 \sin \theta \cos \theta$$

or

$$\psi = 4xy$$

since $x = r \cos \theta$ and $y = r \sin \theta$. However, in the cylindrical polar form the results can be generalized to describe flow in the vicinity of a corner of angle α (see Fig. E6.4c) with the equations

$$\psi = Ar^{\pi/\alpha} \sin \frac{\pi\theta}{\alpha}$$

and

$$\phi = Ar^{\pi/\alpha} \cos \frac{\pi\theta}{\alpha}$$

where A is a constant.

6.5 SOME BASIC, PLANE POTENTIAL FLOWS

A major advantage of Laplace's equation is that it is a linear partial differential equation. Since it is linear, various solutions can be added to obtain other solutions— that is, if $\phi_1(x, y, z)$ and $\phi_2(x, y, z)$ are two solutions to Laplace's equation, then $\phi_3 = \phi_1 + \phi_2$ is also a solution. The practical implication of this result is that if we have certain basic solutions we can combine them to obtain more complicated and interesting solutions. In this section several basic velocity potentials, which describe some relatively simple flows, will be determined. In the next section these basic potentials will be combined to represent more complicated flows.

For simplicity, only plane (two-dimensional) flows will be considered. In this case, by using Cartesian coordinates

$$u = \frac{\partial \phi}{\partial x} \qquad v = \frac{\partial \phi}{\partial y} \tag{6.72}$$

or by using cylindrical coordinates

$$v_r = \frac{\partial \phi}{\partial r} \qquad v_\theta = \frac{1}{r} \frac{\partial \phi}{\partial \theta} \tag{6.73}$$

Since we can define a stream function for plane flow, we can also let

$$u = \frac{\partial \psi}{\partial y} \qquad v = -\frac{\partial \psi}{\partial x} \tag{6.74}$$

or

$$v_r = \frac{1}{r} \frac{\partial \psi}{\partial \theta} \qquad v_\theta = -\frac{\partial \psi}{\partial r} \tag{6.75}$$

where the stream function was previously defined in Eqs. 6.37 and 6.42. We know that by defining the velocities in terms of the stream function, conservation of mass is identically satisfied. If we now impose the condition of irrotationality it follows from Eq. 6.59 that

$$\frac{\partial u}{\partial y} = \frac{\partial v}{\partial x}$$

and in terms of the stream function

$$\frac{\partial}{\partial y} \left(\frac{\partial \psi}{\partial y} \right) = \frac{\partial}{\partial x} \left(-\frac{\partial \psi}{\partial x} \right)$$

or

$$\frac{\partial^2 \psi}{\partial x^2} + \frac{\partial^2 \psi}{\partial y^2} = 0$$

Thus, for a plane, irrotational flow we can use either the velocity potential or the stream function—both must satisfy Laplace's equation in two dimensions. It is apparent from these results that the velocity potential and the stream function are somehow related. We have previously shown that lines of constant ψ are streamlines; that is,

$$\left. \frac{dy}{dx} \right|_{\text{along } \psi = \text{constant}} = \frac{v}{u} \tag{6.76}$$

The change in ϕ as we move from one point (x, y) to a nearby point $(x + dx, y + dy)$ is given by the relationship:

$$d\phi = \frac{\partial \phi}{\partial x} dx + \frac{\partial \phi}{\partial y} dy = u \, dx + v \, dy$$

Along a line of constant ϕ we have $d\phi = 0$ so that

$$\left. \frac{dy}{dx} \right|_{\text{along } \phi = \text{constant}} = -\frac{u}{v} \tag{6.77}$$

A comparison of Eqs. 6.76 and 6.77 shows that lines of constant ϕ (called *equipotential lines*) are orthogonal to lines of constant ψ (streamlines) at all points where they intersect. (Recall that two lines are orthogonal if the product of their slopes is minus one.) For any potential flow field a *"flow net"* can be drawn that consists of a family of streamlines and equipotential lines. The flow net is useful in visualizing flow patterns

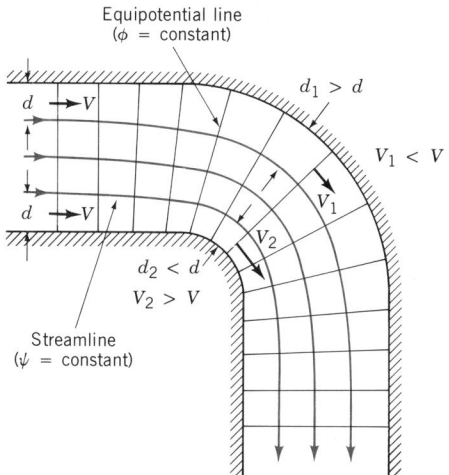

FIGURE 6.15 Flow net for a 90° bend. (From Ref. 3, used by permission.)

and can be used to obtain graphical solutions by sketching in streamlines and equipotential lines and adjusting the lines until the lines are approximately orthogonal at all points where they intersect. An example of a flow net is shown in Fig. 6.15. Velocities can be estimated from the flow net, since the velocity is inversely proportional to the streamline spacing. Thus, for example, from Fig. 6.15 we can see that the velocity near the inside corner will be higher than the velocity along the outer part of the bend.

6.5.1 Uniform Flow

The simplest plane flow is one for which the streamlines are all straight and parallel, and the magnitude of the velocity is constant. This type of flow is called a *uniform flow*. For example, consider a uniform flow in the positive x direction as is illustrated in Fig. 6.16a. In this instance, $u = U$ and $v = 0$, and in terms of the velocity potential

$$\frac{\partial \phi}{\partial x} = U \qquad \frac{\partial \phi}{\partial y} = 0$$

These two equations can be integrated to yield

$$\phi = Ux + C$$

where C is an arbitrary constant, which can be set equal to zero. Thus, for a uniform flow in the positive x direction

$$\phi = Ux \tag{6.78}$$

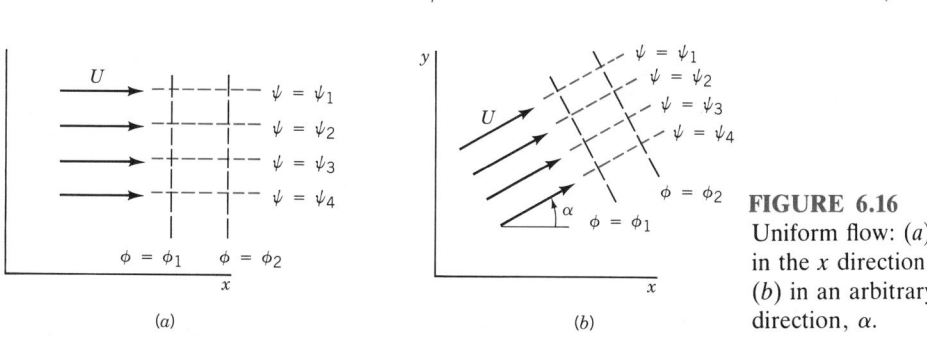

FIGURE 6.16 Uniform flow: (a) in the x direction; (b) in an arbitrary direction, α.

The corresponding stream function can be obtained in a similar manner, since

$$\frac{\partial \psi}{\partial y} = U \qquad \frac{\partial \psi}{\partial x} = 0$$

and, therefore,

$$\psi = Uy \tag{6.79}$$

These results can be generalized to provide the velocity potential and stream function for a uniform flow at an angle α with the x axis, as in Fig. 6.16b. For this case

$$\phi = U(x \cos \alpha + y \sin \alpha) \tag{6.80}$$

and

$$\psi = U(y \cos \alpha - x \sin \alpha) \tag{6.81}$$

6.5.2   Source and Sink

Consider a fluid flowing radially outward from a line through the origin perpendicular to the x–y plane as is shown in Fig. 6.17. Let m be the volume rate of flow emanating from the line (per unit length), and therefore to satisfy conservation of mass

$$(2\pi r)v_r = m$$

or

$$v_r = \frac{m}{2\pi r}$$

Also, since the flow is a purely radial flow, $v_\theta = 0$, the corresponding velocity potential can be obtained by integrating the equations:

$$\frac{\partial \phi}{\partial r} = \frac{m}{2\pi r} \qquad \frac{1}{r}\frac{\partial \phi}{\partial \theta} = 0$$

It follows that

$$\phi = \frac{m}{2\pi} \ln r \tag{6.82}$$

If m is positive, the flow is radially outward, and the flow is considered to be a *source* flow. If m is negative, the flow is toward the origin, and the flow is considered to be a *sink* flow. The flowrate, m, is the *strength* of the source or sink.

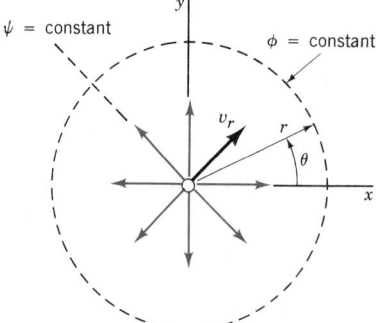

FIGURE 6.17   The streamline pattern for a source.

We note that at the origin where $r = 0$ the velocity becomes infinite, which is of course physically impossible. Thus, sources and sinks do not really exist in real flow fields, and the line representing the source or sink is a mathematical *singularity* in the flow field. However, some real flows can be approximated at points away from the origin by using sources or sinks. Also, the velocity potential representing this hypothetical flow can be combined with other basic velocity potentials to describe approximately some real flow fields. This idea is further discussed in Section 6.6

The stream function for the source can be obtained by integrating the relationships

$$\frac{1}{r}\frac{\partial \psi}{\partial \theta} = \frac{m}{2\pi r} \qquad \frac{\partial \psi}{\partial r} = 0$$

to yield

$$\psi = \frac{m}{2\pi}\theta \tag{6.83}$$

It is apparent from Eq. 6.83 that the streamlines (lines of ψ = constant) are radial lines, and from Eq. 6.82 the equipotential lines (lines of ϕ = constant) are concentric circles centered at the origin.

EXAMPLE 6.5

A nonviscous, incompressible fluid flows between wedge shaped walls into a small opening as shown in Fig. E6.5. The velocity potential (in ft²/s), which approximately describes this flow is

$$\phi = -2 \ln r$$

Determine the volume rate of flow (per unit length) into the opening.

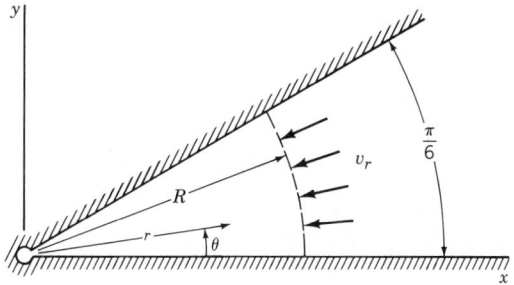

FIGURE E6.5

SOLUTION

The components of velocity are

$$v_r = \frac{\partial \phi}{\partial r} = -\frac{2}{r} \qquad v_\theta = \frac{1}{r}\frac{\partial \phi}{\partial \theta} = 0$$

which indicates we have a purely radial flow. The flowrate per unit width, q, crossing the arc of length $R\pi/6$ can thus be obtained by integrating the expression:

$$q = \int_0^{\pi/6} v_r R \, d\theta = -\int_0^{\pi/6} \left(\frac{2}{R}\right) R \, d\theta = -\frac{\pi}{3} = -1.05 \text{ ft}^2/\text{s} \tag{Ans}$$

Note that the radius R is arbitrary since the flowrate crossing any curve between the two walls must be the same. The negative sign indicates that the flow is toward the opening, that is, in the negative radial direction.

6.5.3 Vortex

We next consider a flow field in which the streamlines are concentric circles—that is, we interchange the velocity potential and stream function for the source. Thus, let

$$\phi = K\theta \tag{6.84}$$

and

$$\psi = -K \ln r \tag{6.85}$$

where K is a constant. In this case the streamlines are concentric circles as are illustrated in Fig. 6.18, with $v_r = 0$ and

$$v_\theta = \frac{1}{r}\frac{\partial \phi}{\partial \theta} = -\frac{\partial \psi}{\partial r} = \frac{K}{r} \tag{6.86}$$

This result indicates that the tangential velocity varies inversely with the distance from the origin, with a singularity occurring at $r = 0$ (where the velocity becomes infinite).

It may seem strange that this *vortex* motion is irrotational (and it is since the flow field is described by a velocity potential). However, it must be recalled that rotation refers to the orientation of a fluid element and not the path followed by the element. Thus, for an irrotational vortex, if a pair of small sticks were placed in the flow field at location A, as indicated in Fig. 6.19a, the sticks would rotate as they move to location B. One of the sticks, the one that is aligned along the streamline, would follow a circular path and rotate in a counterclockwise direction. The other stick would rotate in a clockwise direction due to the nature of the flow field—that is, the part of the stick nearest the origin moves faster than the opposite end. Although both sticks are rotating, the average angular velocity of the two sticks is zero since the flow is irrotational.

If the fluid were rotating as a rigid body, such that $v_\theta = K_1 r$ where K_1 is a constant, then sticks similarly placed in the flow field would rotate as is illustrated in Fig. 6.19b. This type of vortex motion is *rotational* and cannot be described with a velocity potential. The rotational vortex is commonly called a *forced vortex*, whereas the irrotational vortex is usually called a *free vortex*. The swirling motion of the water as

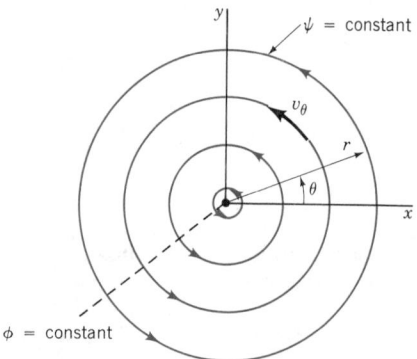

FIGURE 6.18 The streamline pattern for a vortex.

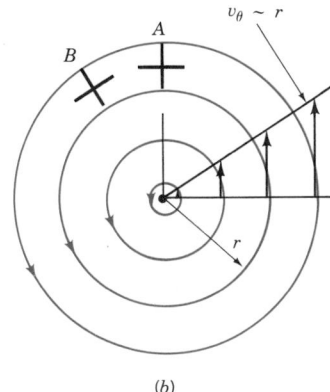

FIGURE 6.19 Motion of fluid element from A to B: (a) for irrotational (free) vortex; (b) for rotational (forced) vortex.

it drains from a bathtub is similar to that of a free vortex, whereas the motion of a liquid contained in a tank that is rotated about its axis with angular velocity ω corresponds to a forced vortex.

A *combined vortex* is one with a forced vortex as a central core and a velocity distribution corresponding to that of a free vortex outside the core. Thus, for a combined vortex

$$v_\theta = \omega r \qquad r \leq r_0 \tag{6.87}$$

and

$$v_\theta = \frac{K}{r} \qquad r > r_0 \tag{6.88}$$

where K and ω are constants and r_0 corresponds to the radius of the central core. The pressure distribution in both the free and forced vortex was previously considered in Example 3.3.

A mathematical concept commonly associated with vortex motion is that of *circulation*. The circulation, Γ, is defined as the line integral of the tangential component of the velocity taken around a closed curve in the flow field. In equation form Γ can be expressed as

$$\Gamma = \oint_C \mathbf{V} \cdot d\mathbf{s} \tag{6.89}$$

where the integral sign means that the integration is taken around a closed curve, C, in the counterclockwise direction, and $d\mathbf{s}$ is a differential length along the curve as is illustrated in Fig. 6.20. For an irrotational flow $\mathbf{V} = \nabla\phi$ so that $\mathbf{V} \cdot d\mathbf{s} = \nabla\phi \cdot d\mathbf{s} = d\phi$ and, therefore,

$$\Gamma = \oint_C d\phi = 0$$

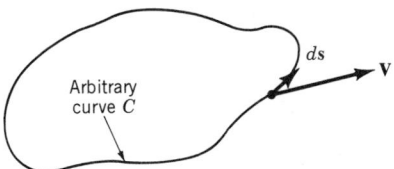

FIGURE 6.20 The notation for determining circulation around closed curve, C.

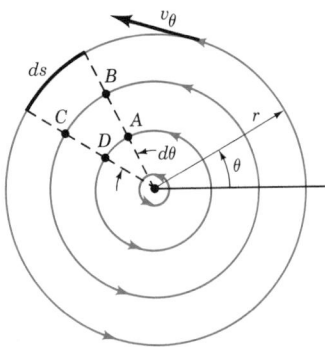

FIGURE 6.21   Circulation around various paths in a free vortex.

This result indicates that for an irrotational flow the circulation will generally be zero. However, if there are singularities enclosed within the curve the circulation may not be zero. For example, for the free vortex with $v_\theta = K/r$ the circulation around the circular path of radius r shown in Fig. 6.21 is

$$\Gamma = \int_0^{2\pi} \frac{K}{r} (r\, d\theta) = 2\pi K$$

which shows that the circulation is nonzero and the constant $K = \Gamma/2\pi$. However, the circulation around any path that does not include the singular point at the origin will be zero. This can be easily confirmed by evaluating the circulation around a closed path such as $ABCD$ of Fig. 6.21, which does not include the origin.

The velocity potential and stream function for the free vortex are commonly expressed in terms of the circulation as

$$\phi = \frac{\Gamma}{2\pi} \theta \tag{6.90}$$

and

$$\psi = -\frac{\Gamma}{2\pi} \ln r \tag{6.91}$$

The concept of circulation is often useful when evaluating the forces developed on bodies immersed in moving fluids. This application will be considered in Section 6.6.3.

EXAMPLE 6.6

A liquid drains from a large tank through a small opening as is illustrated in Fig. E6.6. A vortex forms whose velocity distribution away from the tank opening can be approximated as

FIGURE E6.6

that of a free vortex having a velocity potential

$$\phi = \frac{\Gamma}{2\pi} \theta$$

Determine an expression relating the surface shape to the strength of the vortex as specified by the circulation Γ.

SOLUTION

Since the free vortex represents an irrotational flow field the Bernoulli equation

$$\frac{p_1}{\gamma} + \frac{V_1^2}{2g} + z_1 = \frac{p_2}{\gamma} + \frac{V_2^2}{2g} + z_2$$

can be written between any two points. If the points are selected at the free surface, $p_1 = p_2 = 0$, so that

$$\frac{V_1^2}{2g} = z_s + \frac{V_2^2}{2g} \qquad (1)$$

where the free surface elevation, z_s, is measured relative to a datum passing through point (1).
The velocity is given by the equation

$$v_\theta = \frac{1}{r}\frac{\partial \phi}{\partial \theta} = \frac{\Gamma}{2\pi r}$$

We note that far from the origin at point (1) $V_1 = v_\theta \approx 0$ so that Eq. 1 becomes

$$z_s = -\frac{\Gamma^2}{8\pi^2 r^2 g} \qquad (\text{Ans})$$

which is the desired equation for the surface profile. The negative sign indicates that the surface falls as the origin is approached as shown in Fig. E6.6. This solution is not valid very near the origin since the predicted velocity becomes excessively large as the origin is approached.

6.5.4 Doublet

The final, basic potential flow to be considered is one that is formed by combining a source and sink in a special way. Consider the equal strength, source–sink pair of Fig. 6.22. The combined stream function for the pair is

$$\psi = -\frac{m}{2\pi}(\theta_1 - \theta_2)$$

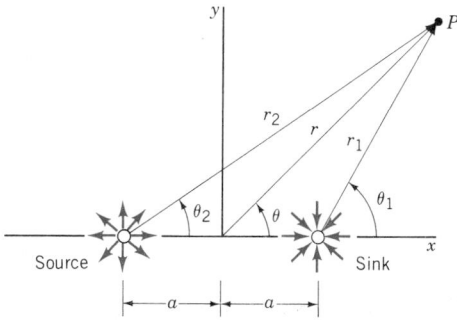

FIGURE 6.22 The combination of a source and sink of equal strength located along the x axis.

which can be rewritten as

$$\tan\left(-\frac{2\pi\psi}{m}\right) = \tan(\theta_1 - \theta_2) = \frac{\tan\theta_1 - \tan\theta_2}{1 + \tan\theta_1 \tan\theta_2} \tag{6.92}$$

From Fig. 6.22 it follows that

$$\tan\theta_1 = \frac{r\sin\theta}{r\cos\theta - a}$$

and

$$\tan\theta_2 = \frac{r\sin\theta}{r\cos\theta + a}$$

These results substituted into Eq. 6.92 give

$$\tan\left(-\frac{2\pi\psi}{m}\right) = \frac{2ar\sin\theta}{r^2 - a^2}$$

so that

$$\psi = -\frac{m}{2\pi}\tan^{-1}\left(\frac{2ar\sin\theta}{r^2 - a^2}\right) \tag{6.93}$$

For small values of a

$$\psi = -\frac{m}{2\pi}\frac{2ar\sin\theta}{r^2 - a^2} = -\frac{mar\sin\theta}{\pi(r^2 - a^2)} \tag{6.94}$$

since the tangent of an angle approaches the value of the angle for small angles.

The so-called *doublet* is formed by letting the source and sink approach one another ($a \to 0$) while increasing the strength m ($m \to \infty$) so that the product ma/π remains constant. In this case, since $r/(r^2 - a^2) \to 1/r$, Eq. 6.94 reduces to

$$\psi = -\frac{K\sin\theta}{r} \tag{6.95}$$

where K, a constant equal to ma/π, is called the *strength* of the doublet. The corresponding velocity potential for the doublet is

$$\phi = \frac{K\cos\theta}{r} \tag{6.96}$$

Plots of lines of constant ψ reveal that the streamlines for a doublet are circles through the origin tangent to the x axis as shown in Fig. 6.23. Just as sources and

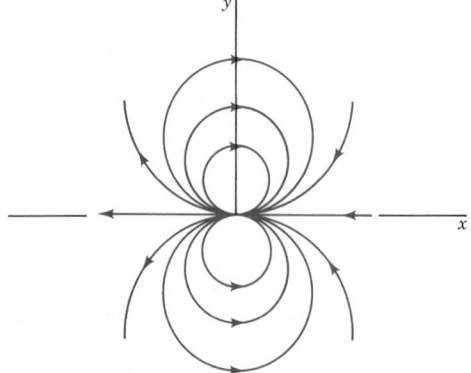

FIGURE 6.23 Streamlines for a doublet.

TABLE 6.1 Summary of Basic, Plane Potential Flows.

Description of Flow Field	Velocity Potential	Stream Function	Velocity Components[a]
Uniform flow at angle α with the x axis (see Fig. 6.16b)	$\phi = U(x \cos \alpha + y \sin \alpha)$	$\psi = U(y \cos \alpha - x \sin \alpha)$	$u = U \cos \alpha$ $v = U \sin \alpha$
Source or sink (see Fig. 6.17) $m > 0$ source $m < 0$ sink	$\phi = \dfrac{m}{2\pi} \ln r$	$\psi = \dfrac{m}{2\pi} \theta$	$v_r = \dfrac{m}{2\pi r}$ $v_\theta = 0$
Free vortex (see Fig. 6.18) $\Gamma > 0$ counterclockwise motion $\Gamma < 0$ clockwise motion	$\phi = \dfrac{\Gamma}{2\pi} \theta$	$\psi = -\dfrac{\Gamma}{2\pi} \ln r$	$v_r = 0$ $v_\theta = \dfrac{\Gamma}{2\pi r}$
Doublet	$\phi = \dfrac{K \cos \theta}{r}$	$\psi = -\dfrac{K \sin \theta}{r}$	$v_r = -\dfrac{K \cos \theta}{r^2}$ $v_\theta = -\dfrac{K \sin \theta}{r^2}$

[a] Velocity components are related to the velocity potential and stream function through the relationships:
$$u = \frac{\partial \phi}{\partial x} = \frac{\partial \psi}{\partial y} \qquad v = \frac{\partial \phi}{\partial y} = -\frac{\partial \psi}{\partial x} \qquad v_r = \frac{\partial \phi}{\partial r} = \frac{1}{r}\frac{\partial \psi}{\partial \theta} \qquad v_\theta = \frac{1}{r}\frac{\partial \phi}{\partial \theta} = -\frac{\partial \psi}{\partial r}$$

sinks are not physically realistic entities, neither are doublets. However, the doublet when combined with other basic potential flows provides a useful representation of some flow fields of practical interest. For example, we will determine in Section 6.6.3 that the combination of a uniform flow and a doublet can be used to represent the flow around a circular cylinder. Table 6.1 provides a summary of the pertinent equations for the basic, plane potential flows considered in the preceding sections.

6.6   SUPERPOSITION OF BASIC, PLANE POTENTIAL FLOWS

As was discussed in the previous section potential flows are governed by Laplace's equation, which is a linear partial differential equation. It therefore follows that the various basic velocity potentials and stream functions can be combined to form new potentials and stream functions. (Why is this true?) Whether such combinations yield useful results remains to be seen. It is to be noted that *any streamline in an inviscid flow field can be considered as a solid boundary,* since the conditions along a solid boundary and a streamline are the same—that is, there is no flow through the boundary or the streamline. Thus, if we can combine some of the basic velocity potentials or stream functions to yield a streamline that corresponds to a particular body shape of interest, that combination can be used to describe in detail the flow around that body. This method of solving some interesting flow problems, commonly called the *method of superposition,* is illustrated in the following three sections.

6.6.1   Source in a Uniform Stream—Half-Body

Consider the superposition of a source and a uniform flow as shown in Fig. 6.24a. The resulting stream function is

$$\psi = \psi_{\text{uniform flow}} + \psi_{\text{source}}$$

$$= Ur \sin \theta + \frac{m}{2\pi} \theta \qquad (6.97)$$

and the corresponding velocity potential is

$$\phi = Ur \cos \theta + \frac{m}{2\pi} \ln r \qquad (6.98)$$

It is clear that at some point along the negative x axis the velocity due to the source will just cancel that due to the uniform flow and a stagnation point will be created. For the source alone

$$v_r = \frac{m}{2\pi r}$$

so that the stagnation point will occur at $x = -b$ where

$$U = \frac{m}{2\pi b}$$

or

$$b = \frac{m}{2\pi U} \qquad (6.99)$$

The value of the stream function at the stagnation point can be obtained by evaluating ψ at $r = b$ and $\theta = \pi$, which yields from Eq. 6.97

$$\psi_{\text{stagnation}} = \frac{m}{2}$$

Since $m/2 = \pi bU$ (from Eq. 6.99) it follows that the equation of the streamline passing through the stagnation point is

$$\pi bU = Ur \sin \theta + bU\theta$$

or

$$r = \frac{b(\pi - \theta)}{\sin \theta} \qquad (6.100)$$

FIGURE 6.24   The flow around a half-body: (a) superposition of a source and a uniform flow; (b) replacement of streamline $\psi = \pi bU$ with solid boundary to form half-body.

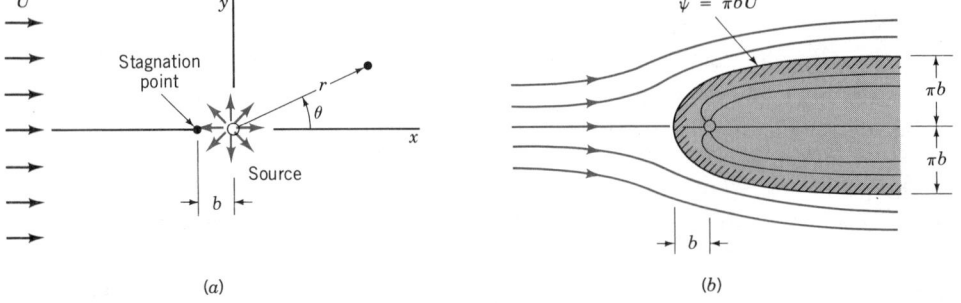

(a)   (b)

where θ can vary between 0 and 2π. A plot of this streamline is shown in Fig. 6.24b. If we replace this streamline with a solid boundary, as indicated in the figure, then it is clear that this combination of a uniform flow and a source can be used to describe the flow around a streamlined body placed in a uniform stream. The body is open at the downstream end, and thus is called a *half-body*. Other streamlines in the flow field can be obtained by setting ψ = constant in Eq. 6.97 and plotting the resulting equation. A number of these streamlines are shown in Fig. 6.24b. Although the streamlines inside the body are shown, they are actually of no interest in this case, since we are concerned with the flow field outside the body. It should be noted that the singularity in the flow field (the source) occurs inside the body, and there are no singularities in the flow field of interest (outside the body).

The width of the half-body asymptotically approaches $2\pi b$. This follows from Eq. 6.100, which can be written as

$$y = b(\pi - \theta)$$

so that as $\theta \to 0$ or $\theta \to 2\pi$ the half-width approaches $\pm b\pi$. With the stream function (or velocity potential) known, the velocity components at any point can be obtained. For the half-body, using the stream function given by Eq. 6.97,

$$v_r = \frac{1}{r}\frac{\partial \psi}{\partial \theta} = U\cos\theta + \frac{m}{2\pi r}$$

and

$$v_\theta = -\frac{\partial \psi}{\partial r} = -U\sin\theta$$

Thus, the square of the magnitude of the velocity, V, at any point is

$$V^2 = v_r^2 + v_\theta^2 = U^2 + \frac{Um\cos\theta}{\pi r} + \left(\frac{m}{2\pi r}\right)^2$$

and since $b = m/2\pi U$

$$V^2 = U^2\left(1 + 2\frac{b}{r}\cos\theta + \frac{b^2}{r^2}\right) \qquad (6.101)$$

With the velocity known, the pressure at any point can be determined from the Bernoulli equation, which can be written between any two points in the flow field since the flow is irrotational. Thus, applying the Bernoulli equation between a point far from the body, where the pressure is p_0 and the velocity is U, and some arbitrary point with pressure p and velocity V, it follows that

$$p_0 + \tfrac{1}{2}\rho U^2 = p + \tfrac{1}{2}\rho V^2 \qquad (6.102)$$

where elevation changes have been neglected. Equation 6.101 can now be substituted into Eq. 6.102 to obtain the pressure at any point in terms of the reference pressure, p_0, and velocity, U.

This relatively simple potential flow provides some useful information about the flow around the front part of a streamlined body, such as a bridge pier or strut placed in a uniform stream. An important point to be noted is that the velocity tangent to the surface of the body is not zero; that is, the fluid "slips" by the boundary. This result is a consequence of neglecting viscosity, the fluid property that causes real fluids to stick to the boundary, thus creating a "no-slip" condition. All potential flows differ from the flow of real fluids in this respect and do not accurately represent the velocity very near the boundary. However, outside this very thin boundary layer the velocity distribution will generally correspond to that predicted by potential flow theory if flow separation does not occur. Also, the pressure distribution along the surface will closely approximate that predicted from the potential flow theory, since the boundary layer is thin and there is little opportunity for the pressure to vary through the thin layer. In fact, as will be discussed in more detail in Chapter 9, the pressure distribution obtained from potential flow theory is used in conjunction with viscous flow theory to determine the nature of flow within the boundary layer.

EXAMPLE 6.7

The shape of a hill arising from a plain can be approximated with the top secton of a half-body as is illustrated in Fig. E6.7. The height of the hill approaches 200 ft as shown. (a) When a 40 mi/hr wind blows toward the hill, what is the magnitude of the air velocity at a point on the hill directly above the origin [point (2)]? (b) What is the elevation of point (2) above the plain and what is the difference in pressure between point (1) on the plain far from the hill and point (2)? Assume an air density of 0.00238 slugs/ft³.

FIGURE E6.7

SOLUTION

(a) The velocity is given by (Eq. 6.101)

$$V^2 = U^2 \left(1 + 2\frac{b}{r}\cos\theta + \frac{b^2}{r^2}\right)$$

At point (2), $\theta = \pi/2$, and since this point is on the surface (Eq. 6.100)

$$r = \frac{b(\pi - \theta)}{\sin\theta} = \frac{\pi b}{2} \tag{1}$$

Thus,

$$V_2^2 = U^2 \left[1 + \frac{b^2}{(\pi b/2)^2}\right]$$

$$= U^2 \left(1 + \frac{4}{\pi^2}\right)$$

and the magnitude of the velocity at (2) for a 40 mi/hr approaching wind is

$$V_2 = \left(1 + \frac{4}{\pi^2}\right)^{1/2} (40 \text{ mi/hr}) = 47.4 \text{ mi/hr} \tag{Ans}$$

(b) The elevation at (2) above the plain is given by Eq. 1 as

$$y_2 = \frac{\pi b}{2}$$

Since the height of the hill approaches 200 ft and this height is equal to πb it follows that

$$y_2 = \frac{200 \text{ ft}}{2} = 100 \text{ ft} \qquad \text{(Ans)}$$

From the Bernoulli equation (with the y axis the vertical axis)

$$\frac{p_1}{\gamma} + \frac{V_1^2}{2g} + y_1 = \frac{p_2}{\gamma} + \frac{V_2^2}{2g} + y_2$$

so that

$$p_1 - p_2 = \frac{\rho}{2}(V_2^2 - V_1^2) + \gamma(y_2 - y_1)$$

and with

$$V_1 = (40 \text{ mi/hr})\left(\frac{5280 \text{ ft/mi}}{3600 \text{ s/hr}}\right) = 58.7 \text{ ft/s}$$

and

$$V_2 = (47.4 \text{ mi/hr})\left(\frac{5280 \text{ ft/mi}}{3600 \text{ s/hr}}\right) = 69.5 \text{ ft/s}$$

it follows that

$$p_1 - p_2 = \frac{(0.00238 \text{ slugs/ft}^3)}{2}[(69.5 \text{ ft/s})^2 - (58.7 \text{ ft/s})^2]$$

$$+ (0.00238 \text{ slugs/ft}^3)(32.2 \text{ ft/s}^2)(100 \text{ ft} - 0 \text{ ft})$$

$$= 9.31 \text{ lb/ft}^2 = 0.0647 \text{ psi} \qquad \text{(Ans)}$$

This result indicates that the pressure on the hill at point (2) is slightly lower than the pressure on the plain at some distance from the base of the hill with a 0.0533 psi difference due to the elevation increase and a 0.0114 psi difference due the velocity increase.

The maximum velocity along the hill surface does not occur at point (2) but further up the hill at $\theta = 63°$. At this point $V_{\text{surface}} = 1.26U$ (Problem 6.60). The minimum velocity ($V = 0$) and maximum pressure occur at point (3), the stagnation point.

6.6.2   Rankine Ovals

The half-body described in the previous section is a body that is "open" at one end. To study the flow around a closed body a source and a sink of equal strength can be combined with a uniform flow as shown in Fig. 6.25a. The stream function

FIGURE 6.25   The flow around a Rankine oval: (*a*) superposition of source–sink pair and a uniform flow; (*b*) replacement of streamline $\psi = 0$ with solid boundary to form Rankine oval.

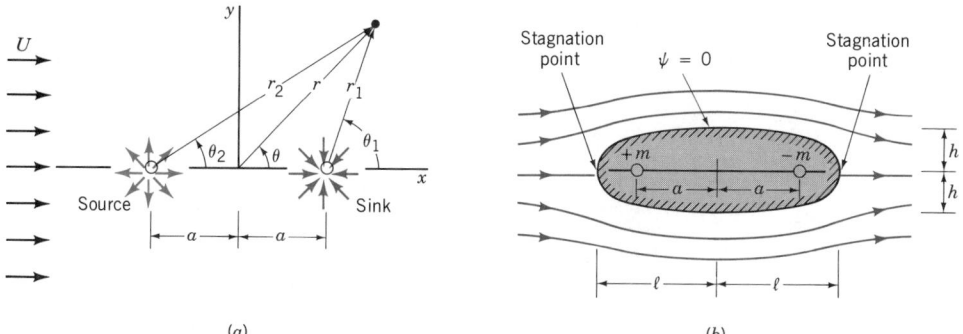

(*a*)   (*b*) for this combination is

$$\psi = Ur \sin \theta - \frac{m}{2\pi}(\theta_1 - \theta_2) \tag{6.103}$$

and the velocity potential is

$$\phi = Ur \cos \theta - \frac{m}{2\pi}(\ln r_1 - \ln r_2) \tag{6.104}$$

As is discussed in Section 6.5.4 the stream function for the source–sink pair can be expressed as in Eq. 6.93 and, therefore, Eq. 6.103 can also be written as

$$\psi = Ur \sin \theta - \frac{m}{2\pi} \tan^{-1}\left(\frac{2ar \sin \theta}{r^2 - a^2}\right)$$

or

$$\psi = Uy - \frac{m}{2\pi} \tan^{-1}\left(\frac{2ay}{x^2 + y^2 - a^2}\right) \tag{6.105}$$

The corresponding streamlines for this flow field are obtained by setting $\psi = $ constant. If several of these stream lines are plotted, it will be discovered that the streamline $\psi = 0$ will form a closed body as is illustrated in Fig. 6.25b. We can think of this streamline as forming the surface of a body of length 2ℓ and width $2h$ placed in a uniform stream. The streamlines inside the body are of no practical interest and are not shown. Note that, since the body is closed, all of the flow emanating from the source flows into the sink. These bodies have an oval shape and are termed *Rankine ovals*.

Stagnation points occur at the upstream and downstream ends of the body as are indicated in Fig. 6.25b. These points can be located by determining where along the x axis the velocity is zero. The stagnation points correspond to the points where the uniform velocity, the source velocity, and the sink velocity all combine to give a zero velocity. The location of the stagnation points depend on the value of a, m, and U. The body half-length, ℓ, can be expressed as

$$\ell = \left(\frac{ma}{\pi U} + a^2\right)^{1/2} \tag{6.106}$$

or

$$\frac{\ell}{a} = \left(\frac{m}{\pi Ua} + 1\right)^{1/2} \tag{6.107}$$

The body half-width, h, can be obtained by determining the value of y where the y axis intersects the $\psi = 0$ streamline. Thus, from Eq. 6.105 with $\psi = 0$, $x = 0$, and $y = h$, it follows that

$$h = \frac{h^2 - a^2}{2a} \tan \frac{2\pi Uh}{m} \tag{6.108}$$

or

$$\frac{h}{a} = \frac{1}{2}\left[\left(\frac{h}{a}\right)^2 - 1\right] \tan\left[2\left(\frac{\pi Ua}{m}\right)\frac{h}{a}\right] \tag{6.109}$$

Equations 6.107 and 6.109 show that both ℓ/a and h/a are functions of the dimensionless parameter, $\pi Ua/m$. Although for a given value of Ua/m the corresponding value of ℓ/a can be determined directly from Eq. 6.107, h/a must be determined by a trial and error solution of 6.109.

A large variety of body shapes with different length to width ratios can be obtained by using different values of Ua/m. As this parameter becomes large, flow around a long slender body is described, whereas for small values of the parameter, flow around a more blunt shape is obtained. Downstream from the point of maximum body width the surface pressure increases with distance along the surface. This condition (called an adverse pressure gradient) typically leads to separation of the flow from the surface, resulting in a large low pressure wake on the downstream side of the body. Separation is not predicted by potential theory (which simply indicates a symmetrical flow) and, therefore, the potential solution for the Rankine ovals will give a reasonable approximation of the velocity outside the thin, viscous boundary layer and the pressure distribution on the front part of the body only.

6.6.3 Flow Around a Circular Cylinder

As was noted in the previous section, when the distance between the source–sink pair approaches zero, the shape of the Rankine oval becomes more blunt and in fact approaches a circular shape. Since the doublet described in Section 6.5.4 was developed by letting a source–sink pair approach one another, it might be expected that a uniform flow in the positive x direction combined with a doublet could be used to represent flow around a circular cylinder. This combination gives for the stream function

$$\psi = Ur \sin \theta - \frac{K \sin \theta}{r} \tag{6.110}$$

and for the velocity potential

$$\phi = Ur \cos \theta + \frac{K \cos \theta}{r} \tag{6.111}$$

In order for the stream function to represent flow around a circular cylinder it is necessary that $\psi = $ constant for $r = a$, where a is the radius of the cylinder. Since Eq. 6.110 can be written as

$$\psi = \left(U - \frac{K}{r^2} \right) r \sin \theta$$

it follows that $\psi = 0$ for $r = a$ if

$$U - \frac{K}{a^2} = 0$$

which indicates that the doublet strength, K, must be equal to Ua^2. Thus, the stream function for flow around a circular cylinder can be expressed as

$$\psi = Ur \left(1 - \frac{a^2}{r^2} \right) \sin \theta \tag{6.112}$$

and the corresponding velocity potential is

$$\phi = Ur \left(1 + \frac{a^2}{r^2} \right) \cos \theta \tag{6.113}$$

A sketch of the streamlines for this flow field is shown in Fig. 6.26.

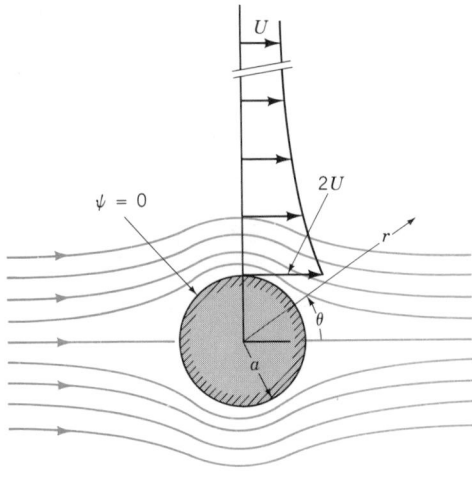

FIGURE 6.26   The flow around a circular cylinder.

The velocity components can be obtained from either Eqs. 6.112 or 6.113 as

$$v_r = \frac{\partial \phi}{\partial r} = \frac{1}{r} \frac{\partial \psi}{\partial \theta} = U \left(1 - \frac{a^2}{r^2} \right) \cos \theta \tag{6.114}$$

and

$$v_\theta = \frac{1}{r} \frac{\partial \phi}{\partial \theta} = -\frac{\partial \psi}{\partial r} = -U \left(1 + \frac{a^2}{r^2} \right) \sin \theta \tag{6.115}$$

On the surface of the cylinder ($r = a$) it follows from Eqs. 6.114 and 6.115 that $v_r = 0$ and

$$v_{\theta s} = -2U \sin \theta$$

We observe from this result that the maximum velocity occurs at the top and bottom of the cylinder ($\theta = \pm \pi/2$) and has a magnitude of twice the upstream velocity, U. As we move away from the cylinder along the ray $\theta = \pi/2$ the velocity varies as is illustrated in Fig. 6.26.

The pressure distribution on the cylinder surface is obtained from the Bernoulli equation written from a point far from the cylinder where the pressure is p_0 and the velocity is U so that

$$p_0 + \tfrac{1}{2}\rho U^2 = p_s + \tfrac{1}{2}\rho v_{\theta s}^2$$

where p_s is the surface pressure. Elevation changes are neglected. Since $v_{\theta s} = -2U \sin \theta$, the surface pressure can be expressed as

$$p_s = p_0 + \tfrac{1}{2}\rho U^2 (1 - 4 \sin^2 \theta) \tag{6.116}$$

A comparison of this theoretical, symmetrical pressure distribution expressed in dimensionless form with a typical measured distribution is shown in Fig. 6.27. This figure clearly reveals that only on the upstream part of the cylinder is there approximate agreement between the potential flow and the experimental results. Because of the viscous boundary layer that develops on the cylinder, the main flow separates from the surface of the cylinder, leading to the large difference between the theoretical, frictionless fluid solution and the experimental results on the downstream side of the cylinder (see Chapter 9).

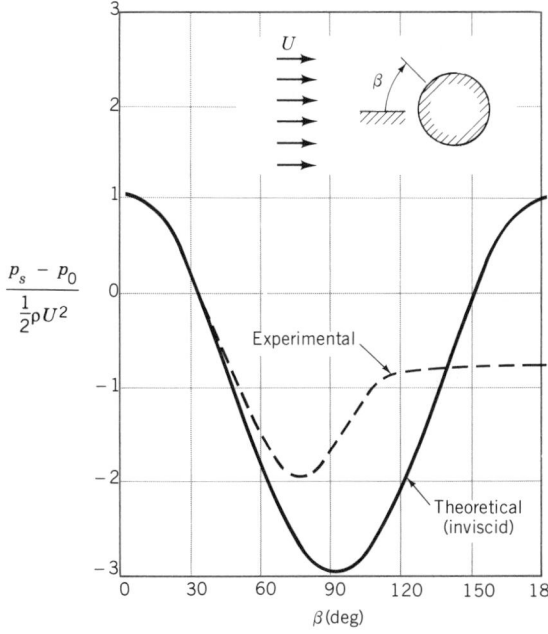

FIGURE 6.27 A comparison of theoretical (inviscid) pressure distribution on the surface of a circular cylinder with typical experimental distribution.

The resultant force (per unit length) developed on the cylinder can be determined by integrating the pressure over the surface. From Fig. 6.28 it can be seen that

$$F_x = -\int_0^{2\pi} p_s \cos\theta \, a \, d\theta \tag{6.117}$$

and

$$F_y = -\int_0^{2\pi} p_s \sin\theta \, a \, d\theta \tag{6.118}$$

where F_x is the *drag* (force parallel to direction of the uniform flow) and F_y is the *lift* (force perpendicular to the direction of the uniform flow). Substitution for p_s from Eq. 6.116 into these two equations, and subsequent integration, reveals that $F_x = 0$ and $F_y = 0$ (Problem 6.70). These results indicate that both the drag and lift as predicted by potential theory for a fixed cylinder in a uniform stream are zero. Since the pressure distribution is symmetrical around the cylinder, this is not really a surprising result. However, we know from experience that there is a significant drag developed on a cylinder when it is placed in a moving fluid. This discrepancy is known as d'Alembert's paradox. The paradox is named after Jean le Rond d'Alembert (1717–

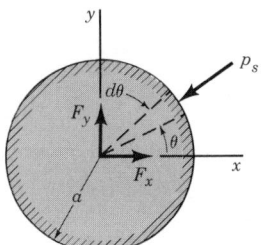

FIGURE 6.28 The notation for determining lift and drag on a circular cylinder.

1783), a French mathematician and philosopher, who first showed that the drag on bodies immersed in inviscid fluids is zero. It was not until the latter part of the nineteenth century and the early part of the twentieth century that the role viscosity plays in the steady fluid motion was understood, and d'Alembert's paradox explained (see Section 9.1).

EXAMPLE 6.8

When a circular cylinder is placed in a uniform stream, a stagnation point is created on the cylinder as is shown in Fig. E6.8a. If a small hole is located at this point, the stagnation pressure, p_{stag}, can be measured and used to determine the approach velocity, U. (a) Show how p_{stag} and U are related. (b) If the cylinder is misaligned by an angle, α (Fig. E6.8b), but the measured pressure still interpreted as the stagnation pressure, determine an expression for the ratio of the true velocity, U, to the predicted velocity, U'. Plot this ratio as a function of α for the range $-20° \leq \alpha \leq 20°$.

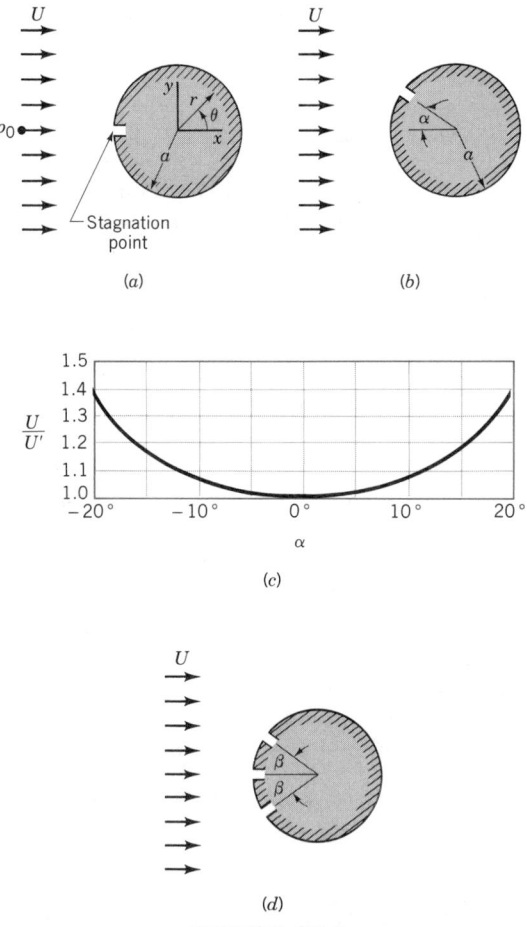

FIGURE E6.8

SOLUTION

(a)   The velocity at the stagnation point is zero so the Bernoulli equation written between a point on the stagnation streamline upstream from the cylinder and the stagnation point gives

$$\frac{p_0}{\gamma} + \frac{U^2}{2g} = \frac{p_{stag}}{\gamma}$$

Thus,

$$U = \left[\frac{2}{\rho}(p_{stag} - p_0)\right]^{1/2} \tag{Ans}$$

A measurement of the difference between the pressure at the stagnation point and the upstream pressure can be used to measure the approach velocity. This is, of course, the same result that was obtained in Section 3.5 for Pitot-static tubes.

(b)   If the direction of the fluid approaching the cylinder is not known precisely, it is possible that the cylinder is misaligned by some angle, α. In this instance the pressure actually measured, p_α, will be different from the stagnation pressure, but if the misalignment is not recognized the predicted approach velocity, U', would still be calculated as

$$U' = \left[\frac{2}{\rho}(p_\alpha - p_0)\right]^{1/2}$$

Thus,

$$\frac{U(\text{true})}{U'(\text{predicted})} = \left(\frac{p_{stag} - p_0}{p_\alpha - p_0}\right)^{1/2} \tag{1}$$

The velocity on the surface of the cylinder, v_θ, where $r = a$, is obtained from Eq. 6.115 as

$$v_\theta = -2U \sin \theta$$

If we now write the Bernoulli equation between a point upstream of the cylinder and the point on the cylinder where $r = a$, $\theta = \alpha$, it follows that

$$p_0 + \frac{1}{2}\rho U^2 = p_\alpha + \frac{1}{2}\rho(-2U \sin \alpha)^2$$

and, therefore,

$$p_\alpha - p_0 = \tfrac{1}{2}\rho U^2(1 - 4 \sin^2\alpha) \tag{2}$$

Since $p_{stag} - p_0 = \tfrac{1}{2}\rho U^2$ it follows from Eqs. 1 and 2 that

$$\frac{U(\text{true})}{U'(\text{predicted})} = (1 - 4 \sin^2\alpha)^{-1/2} \tag{Ans}$$

This velocity ratio is plotted as a function of the misalignment angle α in Fig. E6.8c.

It is clear from these results that significant errors can arise if the stagnation pressure tap is not aligned with the stagnation streamline. As is discussed in Section 3.5, if two additional, symmetrically located, holes are drilled on the cylinder, as are illustrated in Fig. E6.8d, the correct orientation of the cylinder can be determined. The cylinder is rotated until the pressure in the two symmetrically placed holes are equal, thus indicating that the center hole coincides with the stagnation streamline. For $\beta = 30°$ the pressure at the two holes theoretically corresponds to the upstream pressure, p_0. With this orientation a measurement of the difference in pressure between the center hole and the side holes can be used to determine U.

An additional, interesting potential flow can be developed by adding a free vortex to the stream function or velocity potential for the flow around a cylinder. In this case

$$\psi = Ur\left(1 - \frac{a^2}{r^2}\right)\sin\theta - \frac{\Gamma}{2\pi}\ln r \tag{6.119}$$

and

$$\phi = Ur\left(1 + \frac{a^2}{r^2}\right)\cos\theta + \frac{\Gamma}{2\pi}\theta \tag{6.120}$$

where Γ is the circulation. We note that the circle $r = a$ will still be a streamline (and thus can be replaced with a solid cylinder), since the streamlines for the added free vortex are all circular. However, the tangential velocity, v_θ, on the surface of the cylinder ($r = a$) now becomes

$$v_{\theta s} = -2U\sin\theta + \frac{\Gamma}{2\pi a} \tag{6.121}$$

This type of flow field could be approximately created by placing a rotating cylinder in a uniform stream. Because of the presence of viscosity in any real fluid, the fluid in contact with the rotating cylinder would rotate with the same velocity as the cylinder, and the resulting flow field would resemble that developed by the combination of a uniform flow past a cylinder, and a free vortex.

A variety of streamline patterns can be developed, depending on the vortex strength, Γ. For example, from Eq. 6.121 we can determine the location of stagnation points on the surface of the cylinder. These points will occur at $\theta = \theta_{\text{stag}}$ where $v_\theta = 0$ and therefore from Eq. 6.121

$$\sin\theta_{\text{stag}} = \frac{\Gamma}{4\pi Ua} \tag{6.122}$$

For $\Gamma = 0$, $\theta_{\text{stag}} = 0$, and $\pi/2$—that is, the stagnation points occur at the front and rear of the cylinder as are shown in Fig.6.29a. However, for $-1 \le \Gamma/4\pi Ua \le 1$, the

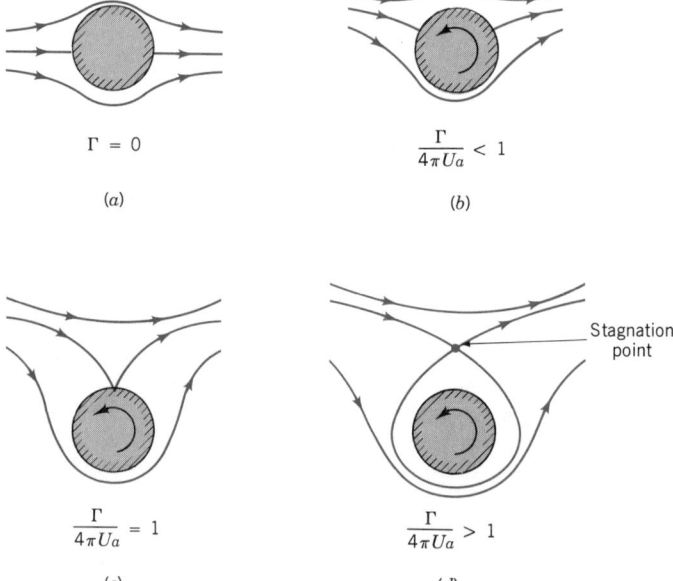

$\Gamma = 0$

(a)

$\dfrac{\Gamma}{4\pi Ua} < 1$

(b)

$\dfrac{\Gamma}{4\pi Ua} = 1$

(c)

$\dfrac{\Gamma}{4\pi Ua} > 1$

(d)

Stagnation point

FIGURE 6.29 The location of stagnation points on a circular cylinder: (a) without circulation; (b, c, d) with circulation.

stagnation points will occur at some other location on the surface as illustrated in Figs. 6.29b,c. If the absolute value of the parameter $\Gamma/4\pi Ua$ exceeds 1, Eq. 6.122 cannot be satisfied, and the stagnation point is located away from the cylinder as shown in Fig. 6.29d.

The force per unit length developed on the cylinder can again be obtained by integrating the differential pressure forces around the circumference as in Eqs. 6.117 and 6.118. For the cylinder with circulation the surface pressure, p_s, is obtained from the Bernoulli equation (with the surface velocity given by Eq. 6.121)

$$p_0 + \frac{1}{2}\rho U^2 = p_s + \frac{1}{2}\rho \left(-2U\sin\theta + \frac{\Gamma}{2\pi a}\right)^2$$

or

$$p_s = p_0 + \frac{1}{2}\rho U^2 \left(1 - 4\sin^2\theta + \frac{2\Gamma\sin\theta}{\pi aU} - \frac{\Gamma^2}{4\pi^2 a^2 U^2}\right) \tag{6.123}$$

Equation 6.123 substituted into Eq. 6.117 for the drag, and integrated, again yields (Problem 6.71)

$$F_x = 0$$

That is, even for the rotating cylinder no force in the direction of the uniform flow is developed. However, use of Eq. 6.123 with the equation for the lift, F_y (Eq. 6.118), yields (Problem 6.71)

$$F_y = -\rho U\Gamma \tag{6.124}$$

Thus, for the cylinder with circulation, lift is developed equal to the product of the fluid density, the upstream velocity, and the circulation. The negative sign means that if U is positive (in the positive x direction), and Γ is positive (a free vortex with counterclockwise rotation) the direction of the F_y is downward. Of course, if the cylinder is rotated in the clockwise direction ($\Gamma < 0$) the direction of F_y would be upward. It is this force acting in a direction perpendicular to the direction of the approach velocity that causes baseballs and golf balls to curve when they spin as they are propelled through the air. The development of this lift on rotating bodies is called the *Magnus effect*. (See Section 9.3 for further comments.) Although Eq. 6.124 was developed for a cylinder with circulation, it gives the lift per unit length for a right cylinder of any cross-sectional shape placed in a uniform, inviscid stream. The circulation is determined around any closed curve containing the body. The generalized equation relating lift to the fluid density, velocity, and circulation is called the *Kutta–Joukowski law,* and is commonly used to determine the lift on airfoils (Refs. 2–6).

6.7   OTHER ASPECTS OF POTENTIAL FLOW ANALYSIS

In the preceding section the method of superposition of basic potentials has been used to obtain detailed descriptions of irrotational flow around certain body shapes immersed in a uniform stream. For the cases considered two or more of the basic potentials were combined and the question is asked: What kind of flow does this combination represent? This approach is relatively simple and does not require the use of advanced mathematical techniques. It is, however, restrictive in its general applicability. It does not allow us to specify a priori the body shape and then determine the velocity potential or stream function that describes the flow around the particular body. Determining the velocity potential or stream function for a given body shape is a much more complicated problem.

It is possible to extend the idea of superposition by considering a *distribution* of sources and sinks, or doublets, which when combined with a uniform flow can describe the flow around bodies of arbitrary shape. Techniques are available to determine the required distribution to give a prescribed body shape. Also, for plane, potential flow problems it can be shown that complex variable theory (the use of real and imaginary numbers) can be effectively used to obtain solutions to a great variety of important flow problems. There are, of course, numerical techniques that can be used to solve not only plane two-dimensional problems, but the more general three-dimensional problems. Since potential flow is governed by Laplace's equation, any procedure that is available for solving this equation can be applied to the analysis of irrotational flow of frictionless fluids. Potential flow theory is an old and well-established discipline within the general field of fluid mechanics. The interested reader can find many detailed references on this subject, including Refs. 2, 3, 4, 5, and 6 given at the end of this chapter.

An important point to remember is that regardless of the particular technique used to obtain a solution to a potential flow problem, the solution remains approximate because of the fundamental assumption of a frictionless fluid. Thus, "exact" solutions based on potential flow theory represent, at best, only approximate solutions to real fluid problems. The applicability of potential flow theory to real fluid problems has been alluded to in a number of examples considered in the previous section. As a rule of thumb, potential flow theory will usually provide a reasonable approximation in those circumstances when we are dealing with a low viscosity fluid moving at a relatively high velocity, in regions of the flow field in which the flow is accelerating. Under these circumstances we generally find that the effect of viscosity is confined to the thin boundary layer that develops at a solid boundary. Outside the boundary layer the velocity distribution and the pressure distribution are closely approximated by the potential flow solution. However, in those regions of the flow field in which the flow is decelerating (for example, in the rearward portion of a bluff body or in the expanding region of a conduit) the pressure near a solid boundary will increase in the direction of flow. This so-called adverse pressure gradient can lead to flow separation, a phenomenon that causes dramatic changes in the flow field which are generally not accounted for by potential theory. However, as discussed in Chapter 9, in which boundary layer theory is developed, it is found that potential flow theory is used to obtain the appropriate pressure distribution that can then be combined with the viscous flow equations to obtain solutions near the boundary (and also to predict separation). The general differential equations that describe viscous fluid behavior and some simple solutions to these equations are considered in the remaining sections of this chapter.

6.8   VISCOUS FLOW

To incorporate viscous effects into the differential analysis of fluid motion we must return to the previously derived general equations of motion, Eqs. 6.50. Since these equations include both stresses and velocities, there are more unknowns than equations, and therefore before proceeding it is necessary to establish a relationship between the stresses and velocities.

6.8.1 Stress–Deformation Relationships

For incompressible, Newtonian fluids it is known that the stresses are linearly related to the rates of deformation and can be expressed in Cartesian coordinates as (for normal stresses)

$$\sigma_{xx} = -p + 2\mu \frac{\partial u}{\partial x} \tag{6.125a}$$

$$\sigma_{yy} = -p + 2\mu \frac{\partial v}{\partial y} \tag{6.125b}$$

$$\sigma_{zz} = -p + 2\mu \frac{\partial w}{\partial z} \tag{6.125c}$$

(for shearing stresses)

$$\tau_{xy} = \tau_{yx} = \mu \left(\frac{\partial u}{\partial y} + \frac{\partial v}{\partial x} \right) \tag{6.125d}$$

$$\tau_{yz} = \tau_{zy} = \mu \left(\frac{\partial v}{\partial z} + \frac{\partial w}{\partial y} \right) \tag{6.125e}$$

$$\tau_{zx} = \tau_{xz} = \mu \left(\frac{\partial w}{\partial x} + \frac{\partial u}{\partial z} \right) \tag{6.125f}$$

where p is the pressure, the negative of the average of the three normal stresses, that is $-p = (\frac{1}{3})(\sigma_{xx} + \sigma_{yy} + \sigma_{zz})$. For viscous fluids in motion the normal stresses are not necessarily the same in different directions, thus, the need to define the pressure as the average of the three normal stresses. For fluids at rest, or frictionless fluids, the normal stresses are equal in all directions. (We have made use of this fact in the chapter on fluid statics and in developing the equations for inviscid flow.) Detailed discussions of the development of these stress–velocity gradient relationships can be found in Refs. 3, 7, and 8. An important point to note is that whereas for elastic solids the stresses are linearly related to the deformation (or strain), for Newtonian fluids the stresses are linearly related to the rate of deformation (or rate of strain).

In cylindrical polar coordinates the stresses for Newtonian, incompressible fluids are expressed as (for normal stresses)

$$\sigma_{rr} = -p + 2\mu \frac{\partial v_r}{\partial r} \tag{6.126a}$$

$$\sigma_{\theta\theta} = -p + 2\mu \left(\frac{1}{r} \frac{\partial v_\theta}{\partial \theta} + \frac{v_r}{r} \right) \tag{6.126b}$$

$$\sigma_{zz} = -p + 2\mu \frac{\partial v_z}{\partial z} \tag{6.126c}$$

(for shearing stresses)

$$\tau_{r\theta} = \tau_{\theta r} = \mu \left[r \frac{\partial}{\partial r} \left(\frac{v_\theta}{r} \right) + \frac{1}{r} \frac{\partial v_r}{\partial \theta} \right] \tag{6.126d}$$

$$\tau_{\theta z} = \tau_{z\theta} = \mu \left(\frac{\partial v_\theta}{\partial z} + \frac{1}{r} \frac{\partial v_z}{\partial \theta} \right) \tag{6.126e}$$

$$\tau_{zr} = \tau_{rz} = \mu \left(\frac{\partial v_r}{\partial z} + \frac{\partial v_z}{\partial r} \right) \tag{6.126f}$$

The double subscript has a meaning similar to that of stresses expressed in Cartesian coordinates—that is, the first subscript indicates the plane on which the stress acts, and the second subscript the direction. Thus, for example, σ_{rr} refers to a stress acting on a plane perpendicular to the radial direction and in the radial direction (thus a normal stress). Similarly, $\tau_{r\theta}$ refers to a stress acting on a plane perpendicular to the radial direction but in the tangential (θ direction) and is therefore a shearing stress.

6.8.2   The Navier–Stokes Equations

The stresses as defined in the preceding section can be substituted into the differential equations of motion (Eqs. 6.50) and simplified by using the continuity equation (Eq. 6.31) to obtain (x direction)

$$\rho\left(\frac{\partial u}{\partial t} + u\frac{\partial u}{\partial x} + v\frac{\partial u}{\partial y} + w\frac{\partial u}{\partial z}\right) = -\frac{\partial p}{\partial x} + \rho g_x + \mu\left(\frac{\partial^2 u}{\partial x^2} + \frac{\partial^2 u}{\partial y^2} + \frac{\partial^2 u}{\partial z^2}\right) \quad (6.127a)$$

(y direction)

$$\rho\left(\frac{\partial v}{\partial t} + u\frac{\partial v}{\partial x} + v\frac{\partial v}{\partial y} + w\frac{\partial v}{\partial z}\right) = -\frac{\partial p}{\partial y} + \rho g_y + \mu\left(\frac{\partial^2 v}{\partial x^2} + \frac{\partial^2 v}{\partial y^2} + \frac{\partial^2 v}{\partial z^2}\right) \quad (6.127b)$$

(z direction)

$$\rho\left(\frac{\partial w}{\partial t} + u\frac{\partial w}{\partial x} + v\frac{\partial w}{\partial y} + w\frac{\partial w}{\partial z}\right) = -\frac{\partial p}{\partial z} + \rho g_z + \mu\left(\frac{\partial^2 w}{\partial x^2} + \frac{\partial^2 w}{\partial y^2} + \frac{\partial^2 w}{\partial z^2}\right) \quad (6.127c)$$

where we have rearranged the equations so the acceleration terms are on the left side and the force terms are on the right. These equations are commonly called the *Navier–Stokes* equations, named in honor of the French mathematician L. M. H. Navier (1758–1836) and the English mechanician Sir G. G. Stokes (1819–1903), who were responsible for their formulation. These three equations of motion, when combined with the conservation of mass equation (Eq. 6.31), provide a complete mathematical description of the flow of incompressible, Newtonian fluids. We have four equations and four unknowns (u, v, w, and p), and therefore the problem is "well-posed" in mathematical terms. Unfortunately, because of the general complexity of the Navier–Stokes equations (they are nonlinear, second order, partial differential equations), they are not amenable to exact mathematical solutions except in a few instances. However, in those few instances in which solutions have been obtained and compared with experimental results, the results have been in close agreement. Thus, the Navier–Stokes equations are considered to be the governing differential equations of motion for incompressible, Newtonian fluids.

In terms of cylindrical polar coordinates the Navier–Stokes equation can be written as (r direction)

$$\rho\left(\frac{\partial v_r}{\partial t} + v_r\frac{\partial v_r}{\partial r} + \frac{v_\theta}{r}\frac{\partial v_r}{\partial \theta} - \frac{v_\theta^2}{r} + v_z\frac{\partial v_r}{\partial z}\right)$$

$$= -\frac{\partial p}{\partial r} + \rho g_r + \mu\left[\frac{1}{r}\frac{\partial}{\partial r}\left(r\frac{\partial v_r}{\partial r}\right) - \frac{v_r}{r^2} + \frac{1}{r^2}\frac{\partial^2 v_r}{\partial \theta^2} - \frac{2}{r^2}\frac{\partial v_\theta}{\partial \theta} + \frac{\partial^2 v_r}{\partial z^2}\right] \quad (6.128a)$$

(θ direction)

$$\rho\left(\frac{\partial v_\theta}{\partial t} + v_r\frac{\partial v_\theta}{\partial r} + \frac{v_\theta}{r}\frac{\partial v_\theta}{\partial \theta} + \frac{v_r v_\theta}{r} + v_z\frac{\partial v_\theta}{\partial z}\right)$$

$$= -\frac{1}{r}\frac{\partial p}{\partial \theta} + \rho g_\theta + \mu\left[\frac{1}{r}\frac{\partial}{\partial r}\left(r\frac{\partial v_\theta}{\partial r}\right) - \frac{v_\theta}{r^2} + \frac{1}{r^2}\frac{\partial^2 v_\theta}{\partial \theta^2} + \frac{2}{r^2}\frac{\partial v_r}{\partial \theta} + \frac{\partial^2 v_\theta}{\partial z^2}\right] \quad (6.128b)$$

(z direction)

$$\rho\left(\frac{\partial v_z}{\partial t} + v_r\frac{\partial v_z}{\partial r} + \frac{v_\theta}{r}\frac{\partial v_z}{\partial \theta} + v_z\frac{\partial v_z}{\partial z}\right)$$

$$= -\frac{\partial p}{\partial z} + \rho g_z + \mu\left[\frac{1}{r}\frac{\partial}{\partial r}\left(r\frac{\partial v_z}{\partial r}\right) + \frac{1}{r^2}\frac{\partial^2 v_z}{\partial \theta^2} + \frac{\partial^2 v_z}{\partial z^2}\right] \quad (6.128c)$$

To provide a brief introduction to the use of the Navier–Stokes equations, a few of the simplest exact solutions are developed in the next section. Although these solutions will prove to be relatively simple, this is not the case in general. In fact, only a few other exact solutions have been obtained.

6.9 SOME SIMPLE SOLUTIONS FOR VISCOUS, INCOMPRESSIBLE FLUIDS

A principal difficulty in solving the Navier–Stokes equations is because of their non-linearity arising from the convective acceleration terms (i.e., $u\,\partial u/\partial x$, $w\,\partial v/\partial z$, etc.). There are no general analytical schemes for solving nonlinear partial differential equations (e.g., superposition of solutions cannot be used), and each problem must be considered individually. For most practical flow problems fluid particles do have accelerated motion as they move from one location to another in the flow field. Thus, the convective acceleration terms are usually important. However, there are a few special cases for which the convective acceleration vanishes because of the nature of the geometry of the flow system. In these cases exact solutions are usually possible. The Navier–Stokes equations apply to both laminar and turbulent flow, but for turbulent flow each velocity component fluctuates randomly with respect to time and this added complication makes an analytical solution intractable. Thus, the exact solutions referred to are for laminar flows in which the velocity is either independent of time (steady flow) or dependent on time in a well-defined manner (unsteady flow).

6.9.1 Steady, Laminar Flow Between Fixed Parallel Plates

We first consider flow between the two horizontal, infinite parallel plates of Fig. 6.30a. For this geometry the fluid particles move in the x direction parallel to the plates, and there is no velocity in the y or z direction—that is, $v = 0$ and $w = 0$. In this case it follows from the continuity equation (Eq. 6.31) that $\partial u/\partial x = 0$. Furthermore, there would be no variation of u in the z direction for infinite plates, and for steady flow $\partial u/\partial t = 0$ so that $u = u(y)$. If these conditions are used in the Navier–

FIGURE 6.30   The viscous flow between parallel plates: (a) coordinate system and notation used in analysis; (b) parabolic velocity distribution for flow between parallel fixed plates.

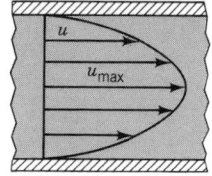

(a)   (b)

Stokes equations (Eqs. 6.127), they reduce to

$$0 = -\frac{\partial p}{\partial x} + \mu \left(\frac{\partial^2 u}{\partial y^2}\right) \tag{6.129}$$

$$0 = -\frac{\partial p}{\partial y} - \rho g \tag{6.130}$$

$$0 = -\frac{\partial p}{\partial z} \tag{6.131}$$

where we have set $g_x = 0$, $g_y = -g$, and $g_z = 0$. That is, the y axis points up. We see that for this particular problem the Navier–Stokes equations reduce to some rather simple equations.

Equations 6.130 and 6.131 can be integrated to yield

$$p = -\rho g y + f_1(x) \tag{6.132}$$

which shows that the pressure varies hydrostatically in the y direction. Equation 6.129, rewritten as

$$\frac{d^2 u}{dy^2} = \frac{1}{\mu}\frac{\partial p}{\partial x}$$

can be integrated to give

$$\frac{du}{dy} = \frac{1}{\mu}\left(\frac{\partial p}{\partial x}\right) y + c_1$$

and integrated again to yield

$$u = \frac{1}{2\mu}\left(\frac{\partial p}{\partial x}\right) y^2 + c_1 y + c_2 \tag{6.133}$$

Note that for this simple flow the pressure gradient, $\partial p/\partial x$, is treated as constant as far as the integration is concerned, since (as shown in Eq. 6.132) it is not a function of y. The two constants c_1 and c_2 must be determined from the boundary conditions. For example, if the two plates are fixed, then $u = 0$ for $y = \pm h$ (because of the no-slip condition for viscous fluids). To satisfy this condition $c_1 = 0$ and

$$c_2 = -\frac{1}{2\mu}\left(\frac{\partial p}{\partial x}\right) h^2$$

Thus, the velocity distribution becomes

$$u = \frac{1}{2\mu}\left(\frac{\partial p}{\partial x}\right)(y^2 - h^2) \tag{6.134}$$

Equation 6.134 shows that the velocity profile between the two fixed plates is parabolic as illustrated in Fig. 6.30b.

The volume rate of flow, q, passing between the plates (for a unit width in the z direction) is obtained from the relationship

$$q = \int_{-h}^{h} u\, dy = \int_{-h}^{h} \frac{1}{2\mu}\left(\frac{\partial p}{\partial x}\right)(y^2 - h^2)\, dy$$

or

$$q = -\frac{2h^3}{3\mu}\left(\frac{\partial p}{\partial x}\right) \tag{6.135}$$

The pressure gradient $\partial p/\partial x$ is negative, since the pressure decreases in the direction of flow. If we let Δp represent the pressure *drop* between two points a distance ℓ apart then

$$\frac{\Delta p}{\ell} = -\frac{\partial p}{\partial x}$$

and Eq. 6.135 can be expressed as

$$q = \frac{2h^3\,\Delta p}{3\mu\ell} \tag{6.136}$$

The flow is proportional to the pressure gradient, inversely proportional to the viscosity, and strongly dependent ($\sim h^3$) on the gap width. In terms of the mean velocity, V, where $V = q/2h$, Eq. 6.136 becomes

$$V = \frac{h^2\,\Delta p}{3\mu\ell} \tag{6.137}$$

Equations 6.136 and 6.137 provide convenient relationships for relating the pressure drop along a parallel–plate channel and the rate of flow or mean velocity. The maximum velocity, u_{max}, occurs midway ($y = 0$) between the two plates so that from Eq. 6.134

$$u_{max} = -\frac{h^2}{2\mu}\left(\frac{\partial p}{\partial x}\right)$$

or

$$u_{max} = \tfrac{3}{2}V \tag{6.138}$$

The details of the steady, laminar flow between infinite parallel plates are completely predicted by this solution to the Navier–Stokes equations. For example, if the pressure gradient, viscosity, and plate spacing are specified, then from Eq. 6.134 the velocity profile can be determined, and from Eqs. 6.136 and 6.137 the corresponding flowrate and mean velocity determined. In addition, from Eq. 6.132 it follows that

$$f_1(x) = \left(\frac{\partial p}{\partial x}\right)x + p_0$$

where p_0 is a reference pressure at $x = y = 0$, and the pressure variation throughout the fluid can be obtained from

$$p = -\rho gy + \left(\frac{\partial p}{\partial x}\right)x + p_0 \tag{6.139}$$

For a given fluid and reference pressure, p_0, the pressure at any point can be predicted. This relatively simple example of an exact solution illustrates the detailed information about the flow field which can be obtained. The flow will be laminar if the Reynolds number, $\mathrm{Re} = \rho V(2h)/\mu$, remains below about 1400. For flow with larger Reynolds numbers the flow becomes turbulent and the preceding analysis is not valid since the flow field is complex, three dimensional, and unsteady.

6.9.2 Couette Flow

Another simple parallel-plate flow can be developed by fixing one plate and letting the other plate move with a constant velocity, U, as is illustrated in Fig. 6.31a. The Navier–Stokes equations reduce to the same form as those in the preceding section

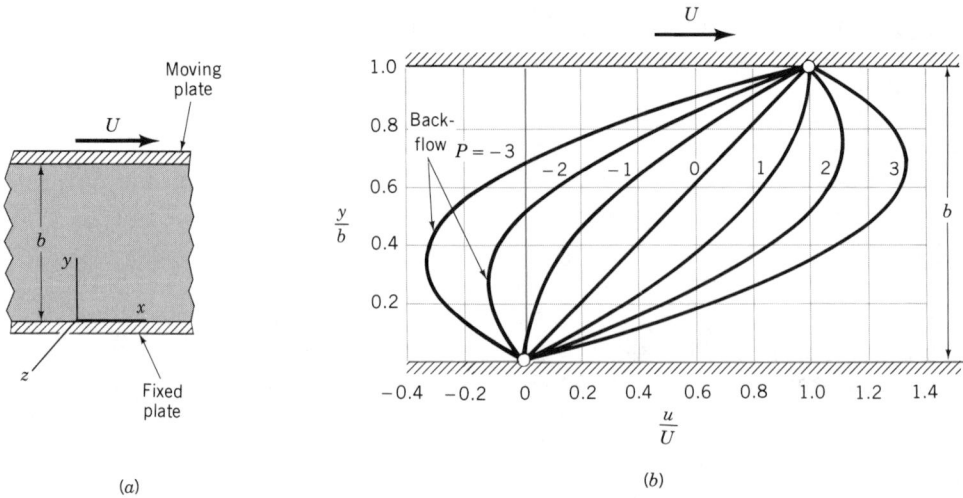

FIGURE 6.31   The viscous flow between parallel plates with bottom plate fixed and upper plate moving (Couette flow): (a) coordinate system and notation used in analysis; (b) velocity distribution as a function of parameter, P, where $P = -(b^2/2\mu U)\partial p/\partial x$. (From Ref. 8, used by permission.)

and the solution for the pressure and velocity distribution are still given by Eqs. 6.132 and 6.133, respectively. However, for the moving plate problem the boundary conditions for the velocity are different. For this case we locate the origin of the coordinate system at the bottom plate and designate the distance between the two plates as b (see Fig. 6.31a). The two constants c_1 and c_2 in Eq. 6.133 can be determined from the boundary conditions, $u = 0$ at $y = 0$ and $u = U$ at $y = 0$. It follows that

$$u = U\frac{y}{b} + \frac{1}{2\mu}\left(\frac{\partial p}{\partial x}\right)(y^2 - by) \tag{6.140}$$

or, in dimensionless form

$$\frac{u}{U} = \frac{y}{b} - \frac{b^2}{2\mu U}\left(\frac{\partial p}{\partial x}\right)\left(\frac{y}{b}\right)\left(1 - \frac{y}{b}\right) \tag{6.141}$$

The actual velocity profile will depend on the dimensionless parameter

$$P = -\frac{b^2}{2\mu U}\left(\frac{\partial p}{\partial x}\right)$$

Several profiles are shown in Fig. 6.31b. This type of flow is called *Couette flow*.

The simplest type of Couette flow is one for which the pressure gradient is zero; that is, the fluid motion is caused by the fluid being dragged along by the moving boundary. In this case, with $\partial p/\partial x = 0$, Eq. 6.140 simply reduces to

$$u = U\frac{y}{b} \tag{6.142}$$

which indicates that the velocity varies linearly between the two plates as shown in Fig. 6.31b for $P = 0$. This situation would be approximated by the flow between closely spaced concentric cylinders in which one cylinder is fixed and the other cylinder rotates with a constant angular velocity, ω. As illustrated in Fig. 6.32 the flow in an

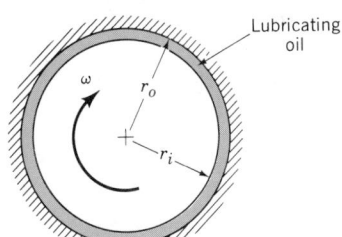

FIGURE 6.32   Flow in the narrow gap of a journal bearing.

unloaded journal bearing might be approximated by this simple Couette flow if the gap width is very small (i.e., $r_o - r_i \ll r_i$). In this case $U = r_i \omega$, $b = r_o - r_i$, and the shearing stress resisting the rotation of the shaft can be simply calculated as $\tau = \mu r_i \omega / (r_o - r_i)$. When the bearing is loaded it will no longer remain concentric with the housing and the flow cannot be treated as flow between parallel boundaries. Such problems are dealt with in lubrication theory (see, for example, Ref. 9).

EXAMPLE 6.9

A wide moving belt passes through a container of a viscous liquid. The belt moves vertically upward with a constant velocity, V_0, as illustrated in Fig. E6.9. Because of viscous forces the belt picks up a film of fluid of thickness h. Gravity tends to make the fluid drain down the belt. Use the Navier–Stokes equations to determine an expression for the average velocity of the fluid film as it is dragged up the belt. Assume that the flow is laminar, steady, and uniform.

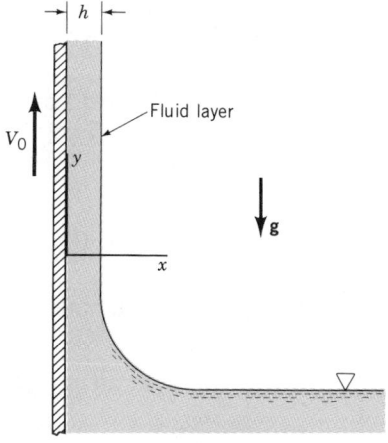

FIGURE E6.9

SOLUTION

Since the flow is assumed to be uniform, the only velocity component is in the y direction (the v component) so that $u = w = 0$. It follows from the continuity equation that $\partial v / \partial y = 0$, and for steady flow $\partial v / \partial t = 0$, so that $v = v(x)$. Under these conditions the Navier–Stokes equations for the x direction (Eq. 6.127a) and the z direction (perpendicular to the paper) (Eq. 6.127c) simply reduce to

$$\frac{\partial p}{\partial x} = 0 \qquad \frac{\partial p}{\partial z} = 0$$

This result indicates that the pressure does not vary over a horizontal plane, and since the pressure on the surface of the film is atmospheric, the pressure throughout the film must be atmospheric (or zero gage pressure). The equation of motion in the y direction (Eq. 6.127b) thus reduces to

$$0 = -\rho g + \mu \frac{d^2 v}{dx^2}$$

or

$$\frac{d^2 v}{dx^2} = \frac{\gamma}{\mu} \tag{1}$$

Integration of Eq. 1 yields

$$\frac{dv}{dx} = \frac{\gamma}{\mu} x + c_1 \tag{2}$$

On the film surface ($x = h$) we assume the shearing stress is zero—that is, the drag of the air on the film is negligible. The shearing stress at the free surface (or any interior parallel surface) is designated as τ_{xy} where from Eq. 6.125d

$$\tau_{xy} = \mu \left(\frac{dv}{dx} \right)$$

Thus, if $\tau_{xy} = 0$ at $x = h$ it follows from Eq. 2 that

$$c_1 = -\frac{\gamma h}{\mu}$$

A second integration of Eq. 2 gives the velocity distribution in the film as

$$v = \frac{\gamma}{2\mu} x^2 - \frac{\gamma h}{\mu} x + c_2$$

At the belt ($x = 0$) the fluid velocity must match the belt velocity, V_0, so that

$$c_2 = V_0$$

and the velocity distribution is therefore

$$v = \frac{\gamma}{2\mu} x^2 - \frac{\gamma h}{\mu} x + V_0$$

With the velocity distribution known we can determine the flowrate per unit width, q, from the relationship

$$q = \int_0^h v \, dx = \int_0^h \left(\frac{\gamma}{2\mu} x^2 - \frac{\gamma h}{\mu} x + V_0 \right) dx$$

and thus

$$q = V_0 h - \frac{\gamma h^3}{3\mu}$$

The average film velocity, V (where $q = Vh$), is therefore

$$V = V_0 - \frac{\gamma h^2}{3\mu} \tag{Ans}$$

It is interesting to note from this result that there will be a net upward flow of liquid (positive V) only if $V_0 > \gamma h^2 / 3\mu$.

6.9.3 Steady, Laminar Flow in Circular Tubes

Probably the best known exact solution to the Navier–Stokes equations is for steady, laminar flow through a straight circular tube of constant cross section. This type of flow is commonly called *Hagen–Poiseuille flow,* or simply *Poiseuille flow.* It is named in honor of J. L. Poiseuille (1799–1869), a French physician, and G. H. L. Hagen (1797–1884), a German hydraulic engineer. Poiseuille was interested in blood flow through capillaries and deduced experimentally the resistance laws for laminar flow through circular tubes. Hagen's investigation of flow in tubes was also experimental. It was actually after the work of Hagen and Poiseuille that the theoretical results presented in this section were determined, but their names are commonly associated with the solution of this problem.

Consider the flow through a horizontal circular tube of radius R as is shown in Fig. 6.33a. Because of the cylindrical geometry it is convenient to use cylindrical coordinates. We assume that the flow is parallel to the walls so that $v_r = 0$ and $v_\theta = 0$, and from the continuity equation $\partial v_z / \partial z = 0$. Also, for steady, axisymmetric flow, v_z is not a function of t or θ so the velocity, v_z, is only a function of the radial position within the tube—that is, $v_z = v_z(r)$. Under these conditions the Navier–Stokes equations (Eqs. 6.128) reduce to

$$0 = -\rho g \sin \theta - \frac{\partial p}{\partial r} \tag{6.143}$$

$$0 = -\rho g \cos \theta - \frac{1}{r} \frac{\partial p}{\partial \theta} \tag{6.144}$$

$$0 = -\frac{\partial p}{\partial z} + \mu \left[\frac{1}{r} \frac{\partial}{\partial r} \left(r \frac{\partial v_z}{\partial r} \right) \right] \tag{6.145}$$

where we have used the relationships $g_r = -g \sin \theta$ and $g_\theta = -g \cos \theta$ (with θ measured from the horizontal plane).

Equations 6.143 and 6.144 can be integrated to give

$$p = -\rho g (r \sin \theta) + f_1(z)$$

or

$$p = -\rho g y + f_1(z) \tag{6.146}$$

Equation 6.146 indicates that the pressure is hydrostatically distributed at any particular cross section, and the z component of the pressure gradient, $\partial p / \partial z$, is not a function of r or θ.

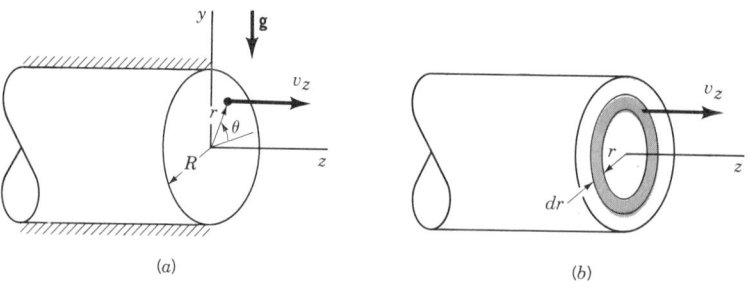

(a)

(b)

FIGURE 6.33
The viscous flow in a horizontal, circular tube: (a) coordinate system and notation used in analysis; (b) flow through differential annular ring.

The equation of motion in the z direction (Eq. 6.145) can be written in the form

$$\frac{1}{r}\frac{\partial}{\partial r}\left(r\frac{\partial v_z}{\partial r}\right) = \frac{1}{\mu}\frac{\partial p}{\partial z}$$

and integrated (using the fact that $\partial p/\partial z = $ constant) to give

$$r\frac{\partial v_z}{\partial r} = \frac{1}{2\mu}\left(\frac{\partial p}{\partial z}\right)r^2 + c_1$$

Integrating again we obtain

$$v_z = \frac{1}{4\mu}\left(\frac{\partial p}{\partial z}\right)r^2 + c_1 \ln r + c_2 \tag{6.147}$$

Since we wish v_z to be finite at the center of the tube ($r = 0$) it follows that $c_1 = 0$. At the wall ($r = R$) the velocity must be zero so that

$$c_2 = -\frac{1}{4\mu}\left(\frac{\partial p}{\partial z}\right)R^2$$

and the velocity distribution becomes

$$v_z = \frac{1}{4\mu}\left(\frac{\partial p}{\partial z}\right)(r^2 - R^2) \tag{6.148}$$

Thus, at any cross section the velocity distribution is parabolic.

To obtain a relationship between the volume rate of flow, Q, passing through the tube and the pressure gradient, we consider the flow through the differential, washer shaped ring of Fig. 6.33b. Since v_z is constant on this ring the volume rate of flow through the differential area is

$$dQ = v_z(2\pi r)\,dr$$

and therefore

$$Q = 2\pi \int_0^R v_z r\,dr \tag{6.149}$$

Equation 6.148 for v_z can be substituted into Eq. 6.149, and the resulting equation integrated to yield

$$Q = -\frac{\pi R^4}{8\mu}\left(\frac{\partial p}{\partial z}\right) \tag{6.150}$$

This relationship can be expressed in terms of the pressure *drop*, Δp, which occurs over a length, ℓ, along the tube, since

$$\frac{\Delta p}{\ell} = -\frac{\partial p}{\partial z}$$

and therefore

$$Q = \frac{\pi R^4 \Delta p}{8\mu \ell} \tag{6.151}$$

For a given pressure drop per unit length, the volume rate of flow is inversely proportional to the viscosity and proportional to the tube radius to the fourth power. A doubling of the tube radius produces a sixteenfold increase in flow! Equation 6.151 is commonly called *Poiseuille's law*.

In terms of the mean velocity, V, where $V = Q/\pi R^2$, Eq. 6.151 becomes

$$V = \frac{R^2 \, \Delta p}{8\mu \ell} \tag{6.152}$$

The maximum velocity v_{max} occurs at the center of the tube, where from Eq. 6.148

$$v_{max} = -\frac{R^2}{4\mu} \left(\frac{\partial p}{\partial z}\right) \tag{6.153}$$

so that

$$v_{max} = 2V$$

The velocity distribution can be written in terms of v_{max} as

$$\frac{v_z}{v_{max}} = 1 - \left(\frac{r}{R}\right)^2 \tag{6.154}$$

As was true for the similar case of flow between parallel plates (sometimes referred to as *plane Poiseuille flow*) a very detailed description of the pressure and velocity distribution in tube flow results from this solution to the Navier–Stokes equations. Numerous experiments performed to substantiate the theoretical results show that the theory and experiment are in agreement for the laminar flow of Newtonian fluids in circular tubes or pipes. The flow remains laminar for Reynolds numbers, $\text{Re} = \rho V(2R)/\mu$, below 2100. Turbulent flow in tubes is considered in Chapter 8.

6.9.4 Steady, Axial, Laminar Flow in an Annulus

The differential equations (Eqs. 6.143, 6.144, 6.145) used in the preceding section for flow in a tube also apply to the axial flow in the annular space between two fixed concentric cylinders (Fig. 6.34). Equation 6.147 for the velocity distribution still applies, but for the stationary annulus the boundary conditions become $v_z = 0$ at $r = r_o$ and $v_z = 0$ for $r = r_i$. With these two conditions the constants c_1 and c_2 in Eq. 6.147 can be determined and the velocity distribution becomes

$$v_z = \frac{1}{4\mu} \left(\frac{\partial p}{\partial z}\right) \left[r^2 - r_o^2 + \frac{r_i^2 - r_o^2}{\ln(r_o/r_i)} \ln \frac{r}{r_o} \right] \tag{6.155}$$

The corresponding volume rate of flow is

$$Q = \int_{r_i}^{r_o} v_z (2\pi r) \, dr = -\frac{\pi}{8\mu} \left(\frac{\partial p}{\partial z}\right) \left[r_o^4 - r_i^4 - \frac{(r_o^2 - r_i^2)^2}{\ln(r_o/r_i)} \right]$$

or in terms of the pressure drop, Δp, in length ℓ of the annulus

$$Q = \frac{\pi \, \Delta p}{8\mu \ell} \left[r_o^4 - r_i^4 - \frac{(r_o^2 - r_i^2)^2}{\ln(r_o/r_i)} \right] \tag{6.156}$$

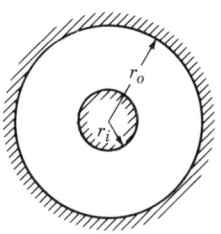

FIGURE 6.34
The viscous flow through an annulus.

The velocity at any radial location within the annular space can be obtained from Eq. 6.155. The maximum velocity occurs at the radius $r = r_m$ where $\partial v_z / \partial r = 0$. Thus,

$$r_m = \left[\frac{r_o^2 - r_i^2}{2 \ln(r_o/r_i)} \right]^{1/2} \tag{6.157}$$

An inspection of this result shows that the maximum velocity does not occur at the midpoint of the annular space, but rather it occurs nearer the inner cylinder. The specific location depends on r_o and r_i.

These results for flow through an annulus are only valid if the flow is laminar. A criterion based on the conventional Reynolds number (which is defined in terms of the tube diameter) cannot be directly applied to the annulus, since there are really "two" diameters involved. For tube cross sections other than simple circular tubes it is common practice to use an "effective" diameter, termed the *hydraulic diameter*, D_h, which is defined as

$$D_h = \frac{4 \times \text{cross-sectional area}}{\text{wetted perimeter}}$$

The wetted perimeter is the perimeter in contact with the fluid. For an annulus

$$D_h = \frac{4\pi(r_o^2 - r_i^2)}{2\pi(r_o + r_i)} = 2(r_o - r_i)$$

In terms of the hydraulic diameter the Reynolds number is $\text{Re} = \rho D_h V / \mu$ (where $V = Q/\text{cross-sectional area}$), and it is commonly assumed that if this Reynolds number remains below 2100 the flow will be laminar. A further discussion of the concept of the hydraulic diameter as it applies to other noncircular cross sections is given in Section 8.4.3.

EXAMPLE 6.10

A viscous liquid ($\rho = 1.18 \times 10^3$ kg/m³; $\mu = 0.0045$ N·s/m²) flows at a rate of 12 ml/s through a horizontal, 4-mm-diameter tube. (a) Determine the pressure drop along a 1-m length of the tube which is far from the tube entrance so that the only component of velocity is parallel to the tube axis. (b) If a 2-mm-diameter rod is placed in the 4-mm-diameter tube to form a symmetric annulus, what is the pressure drop along a 1-m length if the flowrate remains the same as in part (a)?

SOLUTION

(a)   We first calculate the Reynolds number, Re, to determine whether or not the flow is laminar. The mean velocity is

$$V = \frac{Q}{(\pi/4)D^2} = \frac{(12 \text{ ml/s})(10^{-6} \text{ m}^3/\text{ml})}{(\pi/4)(4 \text{ mm} \times 10^{-3} \text{ m/mm})^2}$$

$$= 0.955 \text{ m/s}$$

and, therefore,

$$\text{Re} = \frac{\rho V D}{\mu} = \frac{(1.18 \times 10^3 \text{ kg/m}^3)(0.955 \text{ m/s})(4 \text{ mm} \times 10^{-3} \text{ m/mm})}{0.0045 \text{ N·s/m}^2}$$

$$= 1000$$

Since the Reynolds number is well below the critical value of 2100 we can safely assume that the flow is laminar. Thus, we can apply Eq. 6.151 which gives for the pressure drop

$$\Delta p = \frac{8\mu\ell Q}{\pi R^4}$$

$$= \frac{8(0.0045 \text{ N·s/m}^2)(1 \text{ m})(12 \times 10^{-6} \text{ m}^3/\text{s})}{\pi(2 \text{ mm} \times 10^{-3} \text{ m/mm})^4}$$

$$= 8.59 \text{ kPa}$$

(Ans)

(b) For flow in the annulus the mean velocity is

$$V = \frac{Q}{\pi(r_o^2 - r_i^2)} = \frac{12 \times 10^{-6} \text{ m}^3/\text{s}}{(\pi)[(2 \text{ mm} \times 10^{-3} \text{ m/mm})^2 - (1 \text{ mm} \times 10^{-3} \text{ m/mm})^2]}$$

$$= 1.27 \text{ m/s}$$

and the Reynolds number (based on the hydraulic diameter)

$$\text{Re} = \frac{\rho 2(r_o - r_i)V}{\mu}$$

$$= \frac{(1.18 \times 10^3 \text{ kg/m}^3)(2)(2 \text{ mm} - 1 \text{ mm})(10^{-3} \text{ m/mm})(1.27 \text{ m/s})}{0.0045 \text{ N·s/m}^2}$$

$$= 666$$

This value is also well below 2100 so the flow in the annulus should also be laminar. From Eq. 6.156

$$\Delta p = \frac{8\mu\ell Q}{\pi}\left[r_o^4 - r_i^4 - \frac{(r_o^2 - r_i^2)^2}{\ln(r_o/r_i)}\right]^{-1}$$

so that

$$\Delta p = \frac{8(0.0045 \text{ N·s/m}^2)(1 \text{ m})(12 \times 10^{-6} \text{ m}^3/\text{s})}{\pi} \times \left\{(2 \times 10^{-3} \text{ m})^4 \right.$$

$$\left. - (1 \times 10^{-3} \text{ m})^4 - \frac{[(2 \times 10^{-3} \text{ m})^2 - (1 \times 10^{-3} \text{ m})^2]^2}{\ln(2 \text{ mm}/1 \text{ mm})}\right\}^{-1}$$

$$= 68.2 \text{ kPa}$$

(Ans)

The pressure drop in the annulus is much larger than that of the tube. This is not a surprising result, since to maintain the same flow in the annulus as that in the open tube the average velocity must be larger and the pressure difference along the annulus must overcome the shearing stresses that develop along both an inner and an outer wall. Even an annulus with a very small inner diameter will have a pressure drop significantly higher than that of an open tube. For example, if the inner diameter is only 1/100 of the outer diameter, Δp (annulus)/Δp (tube) = 1.28.

6.10 OTHER ASPECTS OF DIFFERENTIAL ANALYSIS

In this chapter the basic differential equations that govern the flow of fluids have been developed. The Navier–Stokes equations, which can be compactly expressed in vector notation as

$$\rho\left(\frac{\partial \mathbf{V}}{\partial t} + \mathbf{V} \cdot \nabla \mathbf{V}\right) = -\nabla p + \rho \mathbf{g} + \mu \nabla^2 \mathbf{V}$$

(6.158)

FIGURE 6.35 Flow of water from a broken dam colliding with a downstream obstacle. The dam has been removed at $t = 0$ and the fluid falls under the influence of gravity. The method used for calculating this flow is based on a finite-difference approximation of the full Navier-Stokes equations. (Courtesy of Los Alamos National Laboratory, Fluid Dynamics Group T-3.)

are the general equations of motion for incompressible, Newtonian fluids. Although we have restricted our attention to incompressible fluids these equations can be readily extended to include compressible fluids. It is well beyond the scope of this introductory text to consider in depth the variety of analytical and numerical techniques that can be used to obtain both exact and approximate solutions to the Navier–Stokes equations. Students, however, should be aware of the existence of these very general equations, which are frequently used as the basis for many advanced analyses of fluid motion. A few relatively simple solutions have been obtained and discussed in this chapter to indicate the type of detailed flow information that can be obtained by using differential analysis. However, it is hoped that the relative ease with which these solutions were obtained does not give the false impression that solutions to the Navier–Stokes equations are readily available. This is certainly not true, and as previously mentioned there are actually very few practical fluid flow problems that can be solved by using an exact analytical approach. In fact, there are no known analytical solutions to Eq. 6.158 for flow past any object such as a sphere, cube, or airplane.

Because of the difficulty in solving the Navier–Stokes equations much attention has been given to various types of approximate solutions. For example, if the viscosity is set equal to zero, the Navier–Stokes equations reduce to Euler's equations. Thus, the frictionless fluid solutions discussed previously are actually approximate solutions to the Navier–Stokes equations. At the other extreme, for problems involving slowly moving fluids, viscous effects may be dominant and the nonlinear (convective) acceleration terms can be neglected. This assumption greatly simplifies the analysis, since the equations now become linear. There are numerous analytical solutions to

these so-called *"slow flow"* or *"creeping flow"* problems. Another broad class of approximate solutions is concerned with flow in the very thin boundary layer. L. Prandtl showed in 1904 how the Navier–Stokes equations could be simplified to study flow in boundary layers. Such "boundary layer solutions" play a very important role in the study of fluid mechanics. A further discussion of boundary layers is given in Chapter 9.

Numerical techniques using digital computers are, of course, commonly used to solve a wide variety of flow problems. Numerous laminar flow solutions based on the full Navier–Stokes equations have been obtained by using both finite element and finite difference methods. As is illustrated in Fig. 6.35 the flow characteristics for highly nonlinear problems can be obtained from numerical solutions. However, these solutions do typically require the use of powerful digital computers, and the nonlinearities in the equations represent a complication that challenges the ingenuity of the numerical analyst. Practical turbulent flow problems are not amenable to a completely numerical solution because of the extreme complexity of the motion. The solution of real turbulent flows usually involves the use of some type of empirical turbulence model.

The general field of computational fluid dynamics (CFD), in which computers and numerical analysis are combined to solve fluid flow problems, represents an extremely important subject area in advanced fluid mechanics. There is a vast, continually growing literature on CFD, and a few representative references are given at the end of this chapter (Refs. 10, 11, 12, 13).

REFERENCES

1. White, F. M., *Fluid Mechanics,* Second Edition, McGraw-Hill, New York, 1986.
2. Streeter, V. L., *Fluid Dynamics,* McGraw-Hill, New York, 1948.
3. Rouse, H., *Advanced Mechanics of Fluids,* Wiley, New York, 1959.
4. Milne-Thomson, L. M., *Theoretical Hydrodynamics,* Fourth Edition, Macmillan, New York, 1960.
5. Robertson, J. M., *Hydrodynamics in Theory and Application,* Prentice-Hall, Englewood Cliffs, N.J., 1965.
6. Panton, R. L., *Incompressible Flow,* Wiley, New York, 1984.
7. Li, W. H., and Lam, S. H., *Principles of Fluid Mechanics,* Addison-Wesley, Reading, Mass., 1964.
8. Schlichting, H., *Boundary-Layer Theory,* Seventh Edition, McGraw-Hill, New York, 1979.
9. Fuller, D. D., *Theory and Practice of Lubrication for Engineers,* Wiley, New York, 1984.
10. Baker, A. J., *Finite Element Computational Fluid Mechanics,* McGraw-Hill, New York, 1983.
11. Peyret, R., and Taylor, T. D., *Computational Methods for Fluid Flow,* Springer-Verlag, New York, 1983.
12. Anderson, D. A., Tannehill, J. C., and Pletcher, R. H., *Computational Fluid Mechanics and Heat Transfer,* McGraw-Hill, New York, 1984.
13. Carey, G. F., and Oden, J. T., *Finite Elements: Fluid Mechanics,* Prentice-Hall, Englewood Cliffs, N.J., 1986.

PROBLEMS

Note: Unless otherwise indicated use the values of fluid properties found in the tables on the inside of the front cover. Problems designated with an (*) are intended to be solved with the aid of a programmable calculator or a computer.

6.1 The velocity in a certain two-dimensional flow field is given by the equation

$$V = 2xt\hat{i} - 2yt\hat{j}$$

where the velocity is in ft/s when x, y, and t are in feet and seconds, respectively. Determine expressions for the local and convective components of acceleration in the x and y directions. What is the magnitude and direction of the velocity and the acceleration at the point $x = y = 1$ ft at the time $t = 0$?

6.2 Repeat Problem 6.1 if the flow field is described by the equation

$$V = 3(x^2 - y^2)\hat{i} - 6xy\hat{j}$$

where the velocity is in ft/s when x and y are in feet.

6.3 The velocity in a certain flow field is given by the equation

$$V = 3yz^2\hat{i} + xz\hat{j} + y\hat{k}$$

Determine the expressions for the three rectangular components of acceleration.

6.4 The three components of velocity in a flow field are given by

$$u = x^2 + y^2 + z^2$$
$$v = xy + yz + z^2$$
$$w = -3xz - z^2/2 + 4$$

(a) Determine the volumetric dilatation rate, and interpret the results. **(b)** Determine an expression for the rotation vector. Is this an irrotational flow field?

6.5 Determine an expresson for the vorticity of the flow field described by

$$V = x^3y\hat{i} - xy^2\hat{j}$$

Is the flow irrotational?

6.6 A one-dimensional flow is described by the velocity field

$$u = ay + by^2$$
$$v = w = 0$$

where a and b are constants. Is the flow irrotational? For what combination of constants (if any) will the rate of angular deformation as given by Eq. 6.18 be zero?

6.7 For incompressible fluids the volumetric dilatation rate must be zero; that is, $\nabla \cdot V = 0$. For what combination of constants a, b, c, and e can the velocity components

$$u = ax + by$$
$$v = cx + ey$$
$$w = 0$$

be used to describe an incompressible flow field?

6.8 An incompressible viscous fluid is placed between two large parallel plates as shown in Fig. P6.8. The bottom plate is fixed and the upper plate moves with a constant velocity, U. For these conditions the velocity distribution between the plates is linear, and can be expressed as

$$u = U\frac{y}{b}$$

Determine: **(a)** the volumetric dilatation rate, **(b)** the rotation vector, **(c)** the vorticity, and **(d)** the rate of angular deformation.

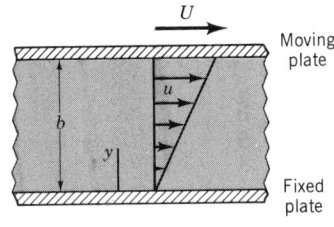

FIGURE P6.8

6.9 A two-dimensional flow field described by

$$V = (2x^2y + x)\hat{i} + (2xy^2 + y + 1)\hat{j}$$

where the velocity is in m/s when x and y are in meters. Determine the angular rotation of a fluid element located at $x = 0.5$ m, $y = 1.0$ m.

6.10 For a certain incompressible flow field it is suggested that the velocity components are given by the equations

$$u = 2xy \qquad v = -x^2y \qquad w = 0$$

Is this a physically possible flow field? Explain.

6.11 The velocity components in an incompressible, two-dimensional velocity field are given by the equations

$$u = 4xy$$
$$v = 2(x^2 - y^2)$$

Show that the flow is irrotational and satisfies conservation of mass.

6.12 For a certain incompressible, two-dimensional flow field the velocity component in the y direction is given by the equation

$$v = x^2 + 2xy$$

Determine the velocity component in the x direction so that the continuity equation is satisfied.

6.13 The velocity components in an incompressible, two-dimensional flow field are given by the equations

$$u = x^2$$
$$v = -2xy + x$$

Determine, if possible, the corresponding stream function.

6.14 The stream function for a certain incompressible flow field is given by the equation

$$\psi = 2x^2y - \tfrac{2}{3}y^3$$

Show that the velocity field represented by this stream function satisfies the continuity equation.

6.15 The stream function for an incompressible, two-dimensional flow field is

$$\psi = ay - by^3$$

where a and b are constants. Is this an irrotational flow? Explain.

6.16 The velocity components for an incompressible, plane flow are

$$v_r = Ar^{-1} + Br^{-2} \cos \theta$$
$$v_\theta = Br^{-2} \sin \theta$$

where A and B are constants. Determine the corresponding stream function.

6.17 For a certain two-dimensional flow field

$$u = 0$$
$$v = V$$

(a) What are the corresponding radial and tangential velocity components? **(b)** Determine the corresponding stream function expressed in Cartesian coordinates and in cylindrical polar coordinates.

6.18 Make use of the control volume shown in Fig. P6.18 to derive the continuity equation in cylindrical coordinates (Eq. 6.33 in text).

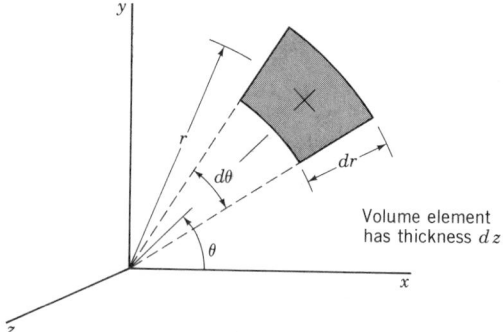

Volume element
has thickness dz

FIGURE P6.18

6.19 It is proposed that a two-dimensional, incompressible flow field be described by the velocity components

$$u = Ay$$
$$v = Bx$$

where A and B are both positive constants. **(a)** Will the continuity equation be satisfied? **(b)** Is the flow irrotational? **(c)** Determine the equation for the streamlines and show a sketch of the streamline that passes through the origin. Indicate the direction of flow along this streamline.

6.20 In a certain steady, two-dimensional flow field the fluid density varies linearly with respect to the coordinate x; that is, $\rho = Ax$ where A is a constant. If the x component of velocity u is given by the equation $u = y$, determine an expression for v.

6.21 In a two-dimensional, incompressible flow field, the x component of velocity is given by the equation $u = 2x$. **(a)** Determine the corresponding equation for the y component of velocity if $v = 0$ along the x axis. **(b)** For this flow field, what is the magnitude of the average velocity of the fluid crossing the surface OA of Fig. P6.21? Assume that the velocities are in ft/s when x and y are in feet.

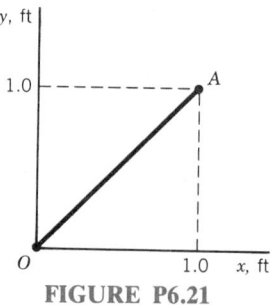

FIGURE P6.21

6.22 The stream function for a two-dimensional, incompressible flow field is given by the equation

$$\psi = 2x - 2y$$

where the stream function has the units of ft²/s with x and y in feet. **(a)** Sketch the streamlines for this flow field. Indicate the direction of flow along the streamlines. **(b)** Is this an irrotational flow field? **(c)** Determine the acceleration of a fluid particle at the point $x = 1$ ft, $y = 2$ ft.

6.23 The radial velocity component in an incompressible, two-dimensional flow field ($v_z = 0$) is

$$v_r = 2r + 3r^2 \cos \theta$$

Determine the corresponding tangential velocity component, v_θ, required to satisfy conservation of mass.

6.24 The stream function for an incompressible flow field is given by the equation

$$\psi = 3x^2y - y^3$$

where the stream function has the units of m²/s with x and y in meters. **(a)** Sketch the streamline(s) passing through the origin. **(b)** Determine the rate of flow across the straight path AB shown in Fig. P6.24.

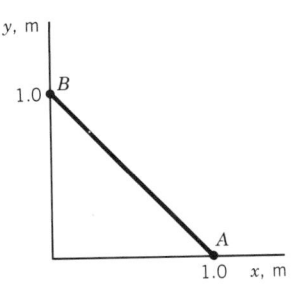

FIGURE P6.24

6.25 The streamlines in a certain incompressible, two-dimensional flow field are all concentric circles so that $v_r = 0$. Determine the stream function for **(a)** $v_\theta = Ar$ and for **(b)** $v_\theta = Ar^{-1}$, where A is a constant.

6.26* The stream function for an incompressible, two-dimensional flow field is

$$\psi = 3x^2y + y$$

For this flow field plot several streamlines.

6.27* The stream function for an incompressible, two-dimensional flow field is

$$\psi = 2r^3 \sin 3\theta$$

For this flow field plot several streamlines for $0 \le \theta \le \pi/3$.

6.28 A two-dimensional flow field for a non-viscous, incompressible fluid is described by the velocity components

$$u = U_0 + 2y$$

$$v = 0$$

where U_0 is a constant. If the pressure at the origin (Fig. P6.28) is p_0, determine an expression for the pressure at **(a)** point A, and **(b)** point B. Explain clearly how you obtained your answer. Assume the units are consistent and body forces may be neglected.

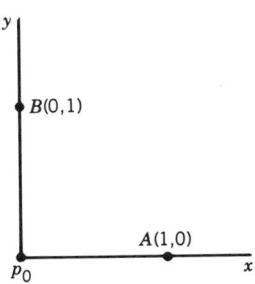

FIGURE P6.28

6.29 The velocity potential for a certain flow field is

$$\phi = 4xy$$

Determine the corresponding stream function.

6.30 The velocity potential for a given two-dimensional flow field is

$$\phi = (\tfrac{5}{3})x^3 - 5xy^2$$

Show that the continuity equation is satisfied and determine the corresponding stream function.

6.31 Determine the stream function corresponding to the velocity potential

$$\phi = x^3 - 3xy^2$$

Sketch the streamline $\psi = 0$, which passes through the origin.

6.32 A certain flow field is described by the velocity potential

$$\phi = A \ln r + Br \cos \theta$$

where A and B are positive constants. Determine the corresponding stream function and locate any stagnation points in this flow field.

6.33 It is known that the velocity distribution for two-dimensional flow of a viscous fluid between wide parallel plates (Fig. P6.33) is parabolic; that is

$$u = U_c \left[1 - \left(\frac{y}{h}\right)^2 \right]$$

with $v = 0$. Determine, if possible, the corresponding stream function and velocity potential.

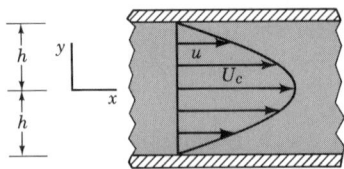

FIGURE P6.33

6.34 In a certain steady, incompressible, inviscid, two-dimensional flow field ($w = 0$, and all variables independent of z) the x component of velocity is given by the equation:

$$u = x^2 - y$$

Will the corresponding pressure gradient in the horizontal x direction be a function only of x, only of y, or of both x and y? Justify your answer.

6.35 The velocity potential for a certain inviscid flow field is

$$\phi = -(3x^2y - y^3)$$

where ϕ has the units of ft^2/s when x and y are in feet. Determine the pressure difference (in psi) between the points (1, 1) and (3, 3), where the coordinates are in feet, if the fluid is water and elevation changes are negligible.

6.36 Consider the incompressible, two-dimensional flow of a nonviscous fluid between the boundaries shown in Fig. P6.36. The velocity potential for this flow field is

$$\phi = x^2 - y^2$$

(a) Determine the corresponding stream function. **(b)** What is the relationship between the discharge, q, (per unit width normal to plane of paper) passing between the walls and the coordinates x_i, y_i of any point on the curved wall? Neglect body forces.

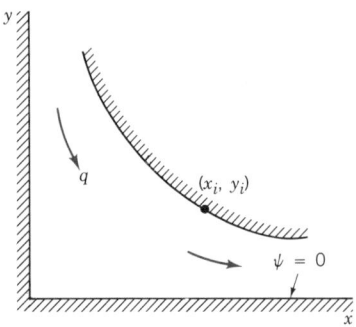

FIGURE P6.36

6.37 The stream function for a two-dimensional, nonviscous, incompressible flow field is

given by the expression

$$\psi = -x + y$$

where the stream function has the units of ft^2/s with x and y in feet. **(a)** Is the continuity equation satisfied? **(b)** Is the flow field irrotational? If so, determine the corresponding velocity potential. **(c)** Determine the pressure gradient in the horizontal x directon at the point $x = 2$ ft, $y = 2$ ft.

6.38 In a certain steady, two-dimensional flow field the fluid may be assumed to be ideal and the weight of the fluid (specific weight $= 50$ lb/ft^3) is the only body force. The x component of velocity is known to be $u = 6x$ which gives the velocity in ft/s when x is measured in feet, and the y component of velocity is known to be a function of only y. The y axis is vertical, and at the origin the velocity is zero. **(a)** Determine the y component of velocity so that the continuity equation is satisfied. **(b)** Can the difference in pressures between the points $x = 1$ ft, $y = 1$ ft and $x = 1$ ft, $y = 4$ ft be determined from the Bernoulli equation? If so, determine the value in lb/ft^2. If not, explain why not.

6.39 The velocity potential for a certain inviscid, incompressible flow field is given by the equation

$$\phi = 2x^2y - \left(\tfrac{2}{3}\right)y^3$$

where ϕ has the units of m^2/s when x and y are in meters. Determine the pressure at the point $x = 2$ m, $y = 2$ m if the pressure at $x = 1$ m, $y = 1$ m is 150 kPa. Elevation changes can be neglected and the fluid is water.

6.40 The velocity components in an ideal, two-dimensional velocity field are given by the equations

$$u = 3(x^2 - y^2)$$
$$v = -6xy$$

All body forces are negligible. **(a)** Does this velocity field satisfy the continuity equation? **(b)** Determine the equation for the pressure gradient in the x direction at any point in the field.

6.41 The streamlines for an incompressible, inviscid, two-dimensional flow field are all concentric circles and the velocity varies directly with the distance from the common center of the streamlines; that is

$$v_\theta = Kr$$

where K is a constant. **(a)** For this *rotational* flow determine, if possible, the stream function. **(b)** Can the pressure difference between the origin and any other point be determined from the Bernoulli equation? Explain.

6.42 The velocity potential

$$\phi = -k(x^2 - y^2) \quad (k = \text{constant})$$

may be used to represent the flow against an infinite plane boundary as illustrated in Fig. P6.42. For flow in the vicinity of a stagnation point it is frequently assumed that the pressure gradient along the surface is of the form

$$\frac{\partial p}{\partial x} = Ax$$

where A is a constant. Use the given velocity potential to show that this is true.

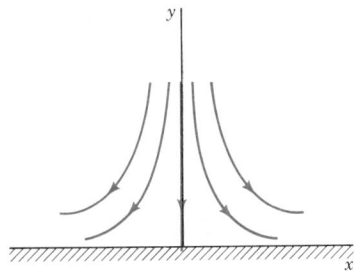

FIGURE P6.42

6.43 The stream function for the flow of a nonviscous, incompressible fluid in the vicinity of a corner (Fig. P6.43) is

$$\psi = 2r^{4/3} \sin \tfrac{4}{3}\theta$$

Determine an expression for the pressure *gradient* along the boundary, $\theta = 0$.

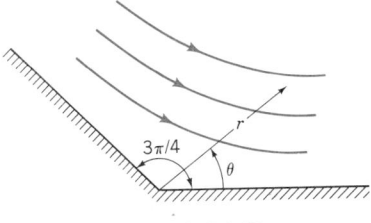

FIGURE P6.43

6.44 Water is flowing between wedge shaped walls into a small opening as shown in Fig. P6.44. The velocity potential with units m²/s for this flow is $\phi = -2 \ln r$ with r in meters. Determine the pressure differential between points A and B.

FIGURE P6.44

6.45 An ideal fluid flows between the inclined walls of a two-dimensional channel into a sink located at the origin (Fig. P6.45). The velocity potential for this flow field is

$$\phi = \frac{m}{2\pi} \ln r$$

where m is a constant. **(a)** Determine the corresponding stream function. Note that the value of the stream function along the wall OA is zero. **(b)** Determine the equation of the streamline passing through the point B, located at $x = 1$, $y = 4$.

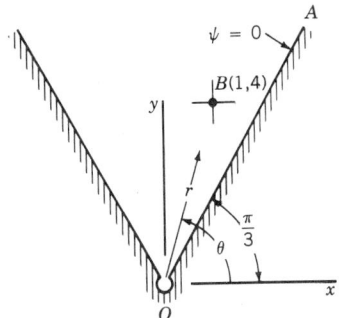

FIGURE P6.45

6.46 It is suggested that the velocity potential for the flow of an incompressible, nonviscous, two-dimensional flow along the wall shown in Fig. P6.43 is

$$\phi = r^{4/3} \cos \tfrac{4}{3}\theta$$

Is this a suitable velocity potential for flow along the wall? Explain.

6.47 As illustrated in Fig. P6.47 a tornado can be approximated by a free vortex of strength Γ for $r > R_c$, where R_c is the radius of the core. Velocity measurements at points A and B indicate that $V_A = 100$ ft/s and $V_B = 75$ ft/s. Determine the distance from point A to the center of the tornado. Why can the free vortex model not be used to approximate the tornado throughout the flow field $(r \geq 0)$?

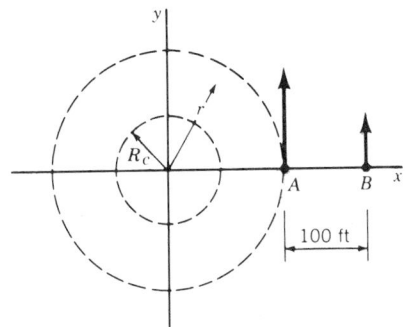

FIGURE P6.47

6.48   The velocity distribution in a horizontal, two-dimensional bend through which an ideal fluid flows can be approximated with a free vortex as shown in Fig. P6.48. Show how the discharge (per unit width normal to plane of paper) through the channel can be expressed as

$$q = C \sqrt{\frac{\Delta p}{\rho}}$$

where $\Delta p = p_B - p_A$. Determine the value of the constant C for the bend dimensions given.

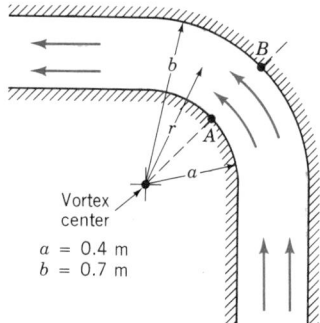

Vortex center
$a = 0.4$ m
$b = 0.7$ m

FIGURE P6.48

6.49   Water discharges from the large tank of Fig. P6.49 through an opening at A and a vortex forms above A. The velocity distribution around this vortex is approximately the same as that for a free vortex. At the same time that the fluid is being discharged from A it is necessary to discharge a small quantity of water through the pipe B. As the discharge through A is increased, the strength of the vortex, as indicated by its circulation, is increased. Determine the maximum strength that the vortex can have in order that no air is sucked in at B. Express your answer in terms of the circulation and assume that the fluid level in the tank at a large distance from the opening at A remains constant, and viscous effects are negligible.

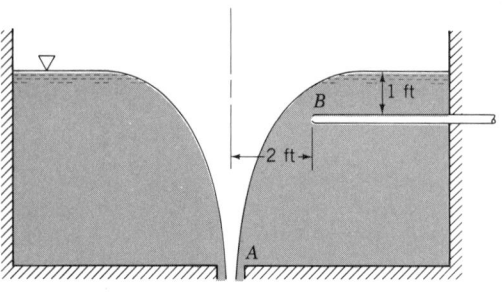

FIGURE P6.49

6.50   The streamlines in a particular two-dimensional flow field are all concentric circles, as shown in Fig. P6.50. The velocity is given by the equation $v_\theta = \omega r$ where ω is the angular velocity of the rotating mass of fluid. Determine the circulation around the path $ABCD$.

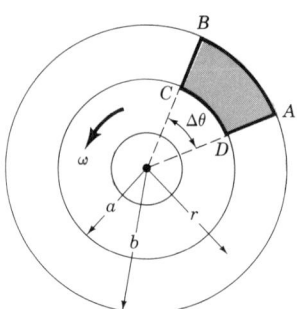

FIGURE P6.50

6.51   Water flows over a flat surface at 5 ft/s as shown in Fig. P6.51. A pump draws off water through a narrow slit at a volume rate of 0.1 ft³/s per foot length of the slit. Assume that the fluid is incompressible and inviscid and can be represented by the combination of a uniform flow and a sink. Locate the stagnation point on the wall (point A), and determine the equation for the stagnation streamline. How far above the surface, H, must the fluid be so that it does not get sucked into the slit?

5 ft/s

H

0.1 ft³/s
(per foot of length of slit)

FIGURE P6.51

6.52   A two-dimensional flow field is formed by adding a source at the origin of the coordinate system to the velocity potential

$$\phi = r^2 \cos 2\theta$$

Locate any stagnation points in the upper half of the coordinate plane ($0 \leq 0 \leq \pi$).

6.53   The velocity potential for a spiral vortex flow is given by $\phi = (\Gamma/2\pi) \theta - (m/2\pi) \ln r$, where Γ and m are constants. Show that the angle, α, between the velocity vector and the radial direction is constant throughout the flow field (see Fig. P6.53).

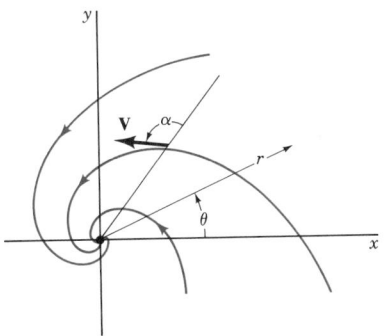

FIGURE P6.53

6.54 Consider a uniform flow in the positive x direction combined with a free vortex located at the origin of the coordinate system. The streamline $\psi = 0$ passes through the point $x = 2$, $y = 0$. Determine the equation of this streamline.

6.55 Potential flow against a flat plate (Fig. P6.55a) can be described with the stream function

$$\psi = Axy$$

where A is a constant. This type of flow is commonly called a "stagnation point" flow since it can be used to describe the flow in the vicinity of the stagnation point at O. By adding a source of strength, m, at O, stagnation point flow against a flat plate with a "bump" is obtained as illustrated in Fig. P6.55b. Determine the relationship between the bump height, h, the constant, A; and the source strength, m.

6.56* Show on a plot several "bumps" that can be generated by combining stagnation point flow against a flat plate and a source as described in Problem 6.55.

6.57 A certain body has the shape of a half-body with a thickness of 0.5 m. If this body is to be placed in an airstream moving at 10 m/s what source strength is required to simulate flow around the body?

6.58 A body having the general shape of a half-body is placed in a stream of fluid. At a great distance upstream the velocity is U as shown in Fig. P6.58. Show how a measurement of the differential pressure between the stagnation point and point A can be used to predict the free-stream velocity, U. Express the pressure differential in terms of U and fluid density. Neglect body forces and assume that the fluid is nonviscous and incompressible.

FIGURE P6.58

6.59 One end of a pond has a shoreline that resembles a half-body as shown in Fig. P6.59. A vertical porous pipe is located near the end of the pond so that water can be pumped out. When water is pumped at the rate of 0.04 m^3/s through a 3-m long pipe, what will be the velocity at point A? *Hint:* Consider the flow *inside* a half-body.

FIGURE P6.59

(a)

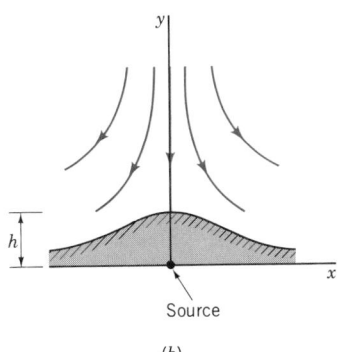

Source

(b)

FIGURE P6.55

6.60* For the half-body described in Section 6.6.1 show on a plot how the magnitude of the velocity on the surface, V_s, varies as a function

of the distance, s (measured along the surface), from the stagnation point. Use the dimensionless variables V_s/U and s/b where U and b are defined in Fig. 6.24.

6.61* Plot the shapes of half-bodies for several values of the parameter, $m/2\pi U$.

6.62 A source and a sink are located along the x axis with the source at $x = -1$ ft and the sink at $x = 1$ ft. Both the source and the sink have a strength of $10\ ft^2/s$. Determine the location of the stagnation points along the x axis when this source–sink pair is combined with a uniform velocity of 15 ft/s in the positive x direction.

6.63 A Rankine oval is formed by combining a source–sink pair, each having a strength of 36 ft^2/s, and separated by a distance of 12 ft along the x axis, with a uniform velocity of 10 ft/s (in the positive x direction). Determine the length and thickness of the oval.

6.64* Make use of Eqs. 6.107 and 6.109 to construct a table showing how ℓ/a, h/a, and ℓ/h for Rankine ovals depend on the parameter $\pi Ua/m$. Plot ℓ/h versus $\pi Ua/m$ and describe how this plot could be used to obtain the required values of m and a for a Rankine oval having a specific value of ℓ and h when placed in a uniform fluid stream of velocity, U.

6.65 Assume that the flow around the long circular cylinder of Fig. P6.65 is nonviscous and incompressible. Two pressures, p_1 and p_2, are measured on the surface of the cylinder, as illustrated. It is proposed that the free-stream velocity, U, can be related to the pressure difference $\Delta p = p_1 - p_2$ by the equation

$$U = C\sqrt{\frac{\Delta p}{\rho}}$$

where ρ is the fluid density. Determine the value of the constant C. Neglect body forces.

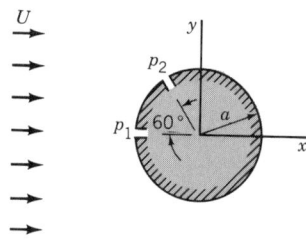

FIGURE P6.65

6.66 An ideal fluid flows past an infinitely long semicircular "hump" located along a plane boundary as shown in Fig. P6.66. Far from the hump the velocity field is uniform, and the pressure is p_0. **(a)** Determine expressions for the max-

imum and minimum values of the pressure along the hump, and indicate where these points are located. Express your answer in terms of ρ, U, and p_0. **(b)** If the solid surface is the $\psi = 0$ streamline, determine the equation of the streamline passing through the point $\theta = \pi/2$, $r = 2a$.

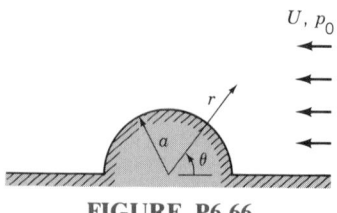

FIGURE P6.66

6.67 An ideal fluid flows around a fixed cylinder as shown in Fig. P6.67. Note that the uniform velocity is in the negative x direction. Show that the pressure gradient, $\partial p/\partial s$, is proportional to s near the stagnation point. The coordinate s is measured along the cylinder surface as shown.

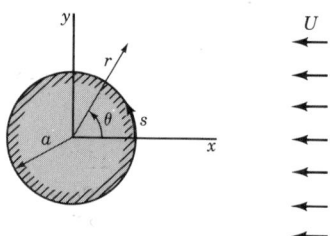

FIGURE P6.67

6.68* Consider the steady potential flow around the circular cylinder shown in Fig. 6.26. Show on a plot the variation of the magnitude of the dimensionless fluid velocity, V/U, along the positive y axis. At what distance, y/a (along the y axis), is the velocity within 1% of the free-stream velocity?

6.69 The velocity potential for a cylinder (Fig. P6.69) rotating in a uniform stream of fluid is

$$\phi = Ur\left(1 + \frac{a^2}{r^2}\right)\cos\theta + \frac{\Gamma}{2\pi}\theta$$

where Γ is the circulation. For what value of the circulation will the stagnation point be located at: **(a)** point A, **(b)** point B?

FIGURE P6.69

6.70   A fixed circular cylinder of infinite length is placed in a steady, uniform stream of an incompressible, nonviscous fluid. Assume that the flow is irrotational. Prove that the drag on the cylinder is zero. Neglect body forces.

6.71   Repeat Problem 6.70 for a rotating cylinder for which the stream function and velocity potential are given by Eqs. 6.119 and 6.120, respectively. Verify that the lift is not zero and can be expressed by Eq. 6.124.

6.72   A source of strength m is located a distance ℓ from a vertical solid wall as shown in Fig. P6.72. The velocity potential for this incompressible, irrotational flow is given by

$$\phi = \frac{m}{4\pi} \{\ln[(x - \ell)^2 + y^2]$$
$$+ \ln[(x + \ell)^2 + y^2]\}$$

(a) Show that there is no flow through the wall. **(b)** Determine the velocity distribution along the wall. **(c)** Determine the pressure distribution along the wall, assuming $p = p_0$ far from the source. Neglect the effect of the fluid weight on the pressure.

FIGURE P6.72

6.73   A long porous pipe runs parallel to a horizontal plane surface as shown in Fig. P6.73. The longitudinal axis of the pipe is perpendicular to the plane of the paper. Water flows radially from the pipe at a rate of $0.5\,\pi$ ft³/s per foot of pipe. Determine the difference in pressure (in lb/ft²) between point B and point A. The flow from the pipe may be approximated by a two-dimensional source. *Hint:* To develop the stream function or velocity potential for this type of flow, place (symmetrically) another equal source on the other side of the wall. With this combination there is no flow across the x-axis, and this axis can be replaced with a solid boundary. This technique is called the *method of images*.

FIGURE P6.73

6.74   A plane flow field is developed by the addition of a free vortex and a uniform stream in the positive x direction. If the vortex is located at the origin determine the pressure variation in this flow field. Neglect the effect of the fluid weight. Express your answer in terms of the uniform velocity, U, the strength of the vortex, Γ, and the pressure, p_0, far from the origin.

6.75   The two-dimensional velocity field for an incompressible, Newtonian fluid is described by the relationship

$$\mathbf{V} = (12xy^2 - 6x^3)\hat{\mathbf{i}} + (18x^2y - 4y^3)\hat{\mathbf{j}}$$

where the velocity has units of m/s when x and y are in meters. Determine the stresses σ_{xx}, σ_{yy}, and τ_{xy} at the point $x = 0.5$ m, $y = 1.0$ m if pressure at this point is 6 kPa and the fluid is glycerin at 20 °C. Show these stresses on a sketch.

6.76   The stream function for a certain incompressible, two-dimensional flow field is

$$\psi = 3r^3 \sin 2\theta + 2\theta$$

where ψ is in ft²/s when r is in feet and θ in radians. Determine the shearing stress, $\tau_{r\theta}$, at the point $r = 2$ ft, $\theta = \pi/3$ radians if the fluid is water.

6.77   For a two-dimensional incompressible flow in the x–y plane show that the z component of the vorticity, ζ_z, varies in accordance with the equation

$$\frac{D\zeta_z}{Dt} = \nu \nabla^2 \zeta_z$$

What is the physical interpretation of this equation for a nonviscous fluid? *Hint:* This *vorticity transport equation* can be derived from the Navier–Stokes equations by differentiating and eliminating the pressure between Eqs. 6.127a and 6.127b.

6.78   The velocity of a fluid particle moving along a horizontal streamline that coincides with the x axis in a plane, two-dimensional incompressible flow field was experimentally found to be described by the equation $u = x^2$. Along this streamline determine an expression for: **(a)** the rate of change of the v-component of velocity with

ⓐ eq 6.136 $q = \frac{2h^3 \Delta P}{3 \mu \ell}$ $\frac{\Delta P}{\ell} = 40 \frac{kP}{m}$ ⓑ $\tau_{yx} = \mu \left(\frac{\partial v}{\partial y} + \frac{\partial u}{\partial x} \right)$ $u = \frac{1}{2\mu} \frac{dP}{dx} (y^2 - h^2) \therefore \frac{\partial u}{\partial y} = \frac{1}{2\mu} \frac{dP}{dx} (2y)$

6.125d   6.136   6.136

$\tau_{yx} = \frac{dP}{dx} (h)$ ⓔ $V_{max} = \frac{3}{2} V = \frac{3}{2} \cdot \frac{5.61 \times 10^{-4}}{2(2 \times 10^{-3} m)} = .21 \frac{m}{s}$

6.138

respect to y; **(b)** the acceleration of the particle; and **(c)** the pressure gradient in the x direction. The fluid is Newtonian.

6.79   In a certain viscous, incompressible flow field with zero body forces the velocity components are

$$u = ay - b(cy - y^2)$$
$$v = w = 0$$

where a, b, and c are constants. **(a)** Use the Navier–Stokes equations to determine an expression for the pressure gradient in the x direction. **(b)** For what combination of the constants a, b, and c (if any) will the shearing stress, τ_{yx}, be zero at $y = 0$ where the velocity is zero?

6.80   Oil (SAE 30) at 15.6 °C flows steadily between fixed, horizontal, parallel plates. The pressure drop per unit length along the channel is 40 kPa/m, and the distance between the plates is 4 mm. The flow is laminar. Determine: **(a)** the volume rate of flow (per meter of width), **(b)** the magnitude and direction of the shearing stress acting on the bottom plate, and **(c)** the velocity along the centerline of the channel.

6.81   Two fixed, horizontal, parallel plates are spaced 0.2 in. apart. A viscous liquid ($\mu = 8 \times 10^{-3}$ lb·s/ft^2, $SG = 0.9$) flows between the plates with a mean velocity of 1.2 ft/s. Determine the pressure drop per unit length in the direction of flow. What is the maximum velocity in the channel?

6.82   A layer of viscous liquid of constant thickness (no velocity perpendicular to plate) flows steadily down an infinite, inclined plane. Determine, by means of the Navier–Stokes equations, the relationship between the thickness of the layer and the discharge per unit width. The flow is laminar, and assume air resistance is negligible so that the shearing stress at the free surface is zero.

6.83   A viscous, incompressible fluid flows be-tween the two infinite, vertical, parallel plates of Fig. P6.83. Determine, by use of the Navier–Stokes equations, an expression for the pressure gradient in the direction of flow. Express your answer in terms of the mean velocity. Assume that the flow is laminar, steady, and uniform.

6.84   A fluid of density ρ flows steadily *downward* between the two vertical infinite, parallel plates shown in the figure for Problem 6.83. The flow is fully developed and laminar. Make use of the Navier–Stokes equation to determine the relationship between the discharge and the other parameters involved, for the case in which the change in pressure along the channel is zero.

6.85   A viscous fluid is contained between two infinite horizontal parallel plates that are spaced 0.5 in. apart. The bottom plate is fixed and the upper plate moves with a constant velocity, U. The fluid motion is caused by the movement of the upper plate and there is no pressure gradient in the direction of flow. The flow is laminar. If the velocity of the upper plate is 2 ft/s and the fluid has a viscosity of 0.03 lb·s/ft^2 and a specific weight of 70 lb/ft^3, what is the required horizontal force per square foot on the upper plate to maintain the 2 ft/s velocity? What is the pressure differential in the fluid between the top and bottom plate?

6.86   An incompressible, viscous fluid is placed between horizontal, infinite, parallel plates as is shown in Fig. P6.86. The two plates move in opposite directions with constant velocities, U_1 and U_2, as shown. The pressure gradient in the x direction is zero and the only body force is due to the fluid weight. Use the Navier–Stokes equations to derive an expression for the velocity distribution between the plates. Assume laminar flow.

Direction of flow

FIGURE P6.83

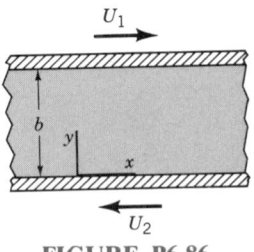

FIGURE P6.86

6.87   Two immiscible, incompressible, viscous fluids having the same densities but different viscosities are contained between two infinite, horizontal, parallel plates (Fig. P6.87). The bottom plate is fixed and the upper plate moves with a constant velocity U. Determine the velocity at the

interface. Express your answer in terms of U, μ_1, and μ_2. The motion of the fluid is caused entirely by the movement of the upper plate; that is, there is no pressure gradient in the x direction. The fluid velocity and shearing stress is continuous across the interface between the two fluids. Assume laminar flow.

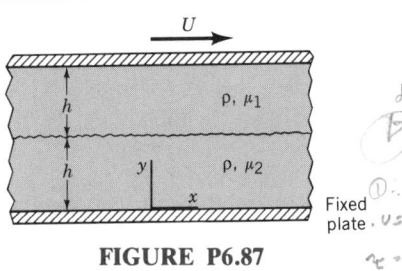

FIGURE P6.87

6.88   The viscous, incompressible flow between the parallel plates shown in Fig. P6.88 is caused by both the motion of the bottom plate and a pressure gradient, $\partial p/\partial x$. Determine the relationship between U and $\partial p/\partial x$ so that the shearing stress acting on the fixed plate is zero.

FIGURE P6.88

6.89   A viscous fluid (specific weight = 76 lb/ft³; viscosity = 0.02 lb·s/ft²) is contained between two infinite horizontal parallel plates as shown in Fig. P6.89. The fluid moves between the plates under the action of a pressure gradient, and the upper plate moves with a velocity U while the bottom plate is fixed. A U-tube manometer connected between two points along the bottom indicates a differential reading of 0.1 in. If the upper plate moves with a velocity of 0.01 ft/s, at what distance from the bottom plate does the maximum velocity in the gap between the two plates occur? Assume laminar flow.

FIGURE P6.89

6.90   A vertical shaft passes through a bearing and is lubricated with an oil having a viscosity of 0.2 N·s/m² as shown in Fig. P6.90. Assume that the flow characteristics in the gap between the shaft and bearing are the same as those for laminar flow between infinite parallel plates with zero pressure gradient in the direction of flow. Estimate the torque required to overcome viscous resistance when the shaft is turning at 100 rev/min.

FIGURE P6.90

6.91   A viscous fluid is contained between two long concentric cylinders. The geometry of the system is such that the flow between the cylinders is approximately the same as the laminar flow between two infinite parallel plates. Determine an expression for the torque required to rotate the outer cylinder with an angular velocity, ω. The inner cylinder is fixed. Express your answer in terms of the geometry of the system, the viscosity of the fluid, and the angular velocity.

6.92*   Oil (SAE 30) flows between parallel plates spaced 5 mm apart. The bottom plate is fixed but the upper plate moves with a velocity of 0.2 m/s in the positive x direction. The pressure gradient is 60 kPa/m, and is negative. Compute the velocity at various points across the channel and show the results on a plot. Assume laminar flow.

6.93   Consider the steady, laminar flow of an incompressible fluid through the horizontal rectangular channel of Fig. P6.93. Assume that the velocity components in the x and y directions are zero and the only body force is the weight. Start with the Navier–Stokes equations. **(a)** Determine the appropriate set of differential equations and boundary conditions for this problem. You need

FIGURE P6.93

not solve the equations. **(b)** Show that the pressure distribution is hydrostatic at any particular cross section.

6.94 Consider a steady, laminar flow through a straight horizontal tube having the constant elliptical cross section given by the equation:

$$\frac{x^2}{a^2} + \frac{y^2}{b^2} = 1$$

The streamlines are all straight and parallel. Investigate the possibility of using an equation for the z component of velocity of the form

$$w = A\left(1 - \frac{x^2}{a^2} - \frac{y^2}{b^2}\right)$$

as an exact solution to this problem. With this velocity distribution what is the relationship between the pressure gradient along the tube and the volume flowrate through the tube?

6.95 A fluid is initially at rest between two horizontal, infinite, parallel plates. A constant pressure gradient in a direction parallel to the plates is suddenly applied and the fluid starts to move. Determine the appropriate differential equation(s), initial condition, and boundary conditions that govern this type of flow. You need not solve the equation(s).

6.96 Ethyl alcohol flows through a horizontal tube having a diameter of 10 mm. If the mean velocity is 0.18 m/s what is the pressure drop per unit length along the tube? What is the velocity at a distance of 3 mm from the tube axis?

6.97 A viscous liquid, having a viscosity of 10^{-4} lb·s/ft² and a specific weight of 50 lb/ft³, flows steadily through the 2-in.-diameter, horizontal, smooth pipe shown in Fig. P6.97. The mean velocity in the pipe is 0.5 ft/s. Determine the differential reading, Δh, on the inclined-tube manometer.

FIGURE P6.97

6.98 A simple flow system to be used for steady-flow tests consists of a constant head tank connected to a length of 4-mm-diameter tubing as shown in Fig. P6.98. The liquid has a viscosity

of 0.015 N·s/m² and a density of 1200 kg/m³, and discharges into the atmosphere with a mean velocity of 1 m/s. **(a)** Verify that the flow will be laminar. **(b)** The flow is fully developed in the last 3 m of the tube. What is the pressure at the pressure gage? **(c)** What is the magnitude of the wall shearing stress, τ_{rz}, in the fully developed region?

FIGURE P6.98

6.99 A liquid (viscosity = 0.002 N·s/m²; density = 1000 kg/m³) is forced through the circular tube shown in Fig. P6.99. A differential manometer is connected to the tube as shown to measure the pressure drop along the tube. When the differential reading, Δh, is 6 mm, what is the mean velocity in the tube?

FIGURE P6.99

6.100 **(a)** Show that for Poiseuille flow in a tube of radius R the magnitude of the wall shearing stress, τ_{rz}, can be obtained from the relationship

$$|(\tau_{rz})_{\text{wall}}| = \frac{4\mu Q}{\pi R^3}$$

for a Newtonian fluid of viscosity μ. The volume rate of flow is Q. **(b)** Determine the magnitude of the wall shearing stress for a fluid having a viscosity of 0.003 N·s/m² flowing with an average velocity of 100 mm/s in a 2-mm-diameter tube.

6.101 An incompressible, Newtonian fluid flows steadily between two infinitely long, concentric cylinders as shown in Fig. P6.101. The

outer cylinder is fixed, but the inner cylinder moves with a longitudinal velocity V_0 as shown. For what value of V_0 will the drag on the inner cylinder be zero? Assume that the flow is laminar, axisymmetric, and fully developed.

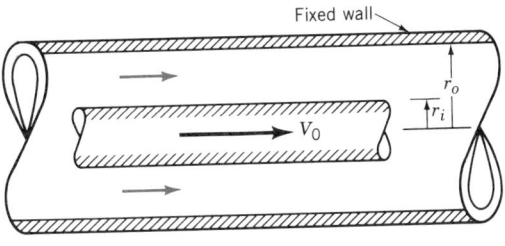

FIGURE P6.101

6.102   An infinitely long, solid, vertical cylinder of radius R is located in an infinite mass of an incompressible fluid. Start with the Navier–Stokes equation in the θ direction and derive an expression for the velocity distribution for the steady flow case in which the cylinder is rotating about a fixed axis with a constant angular velocity ω. You need not consider body forces. Assume that the flow is axisymmetric and the fluid is at rest at infinity.

6.103   A viscous fluid is contained between two infinitely long vertical concentric cylinders. The outer cylinder has a radius r_o and rotates with an angular velocity ω. The inner cylinder is fixed and has a radius r_i. Make use of the Navier–Stokes equations to obtain an exact solution for the velocity distribution in the gap. Assume that the flow in the gap is axisymmetric (neither velocity nor pressure are functions of angular position θ within gap) and that there are no velocity components other than the tangential component. The only body force is the weight.

6.104   A viscous liquid ($\mu = 0.016$ lb·s/ft², $\rho = 1.79$ slugs/ft³) flows through the annular space between two, horizontal, fixed concentric cylinders. If the radius of the inner cylinder is 1.5 in. and the radius of the outer cylinder is 2.5 in., what is the pressure drop along the axis of the annulus per foot when the volume flowrate is 0.09 ft³/s?

6.105*   Plot the velocity profile for the fluid flowing in the annular space described in Problem P6.104. Determine from the plot the radius at which the maximum velocity occurs and compare with the value predicted from Eq. 6.157.

6.106*   As is shown by Eq. 6.150 the pressure gradient for laminar flow through a tube of constant radius is given by the expression:

$$\frac{\partial p}{\partial z} = -\frac{8\mu Q}{\pi R^4}$$

For a tube whose radius is changing very gradually, such as the one illustrated in Fig. P6.106, it is expected that this equation can be used to approximate the pressure change along the tube if the actual radius, $R(z)$, is used at each cross section. The following measurements were obtained along a particular tube.

z/ℓ	0	0.1	0.2	0.3	0.4	0.5	0.6	0.7	0.8	0.9	1.0
$R(z)/R_o$	1.00	0.73	0.67	0.65	0.67	0.80	0.80	0.71	0.73	0.77	1.00

Compare the pressure drop over the length ℓ for this nonuniform tube with one having the constant radius R_o. *Hint:* To solve this problem you will need to numerically integrate the equation for the pressure gradient given above.

FIGURE P6.106

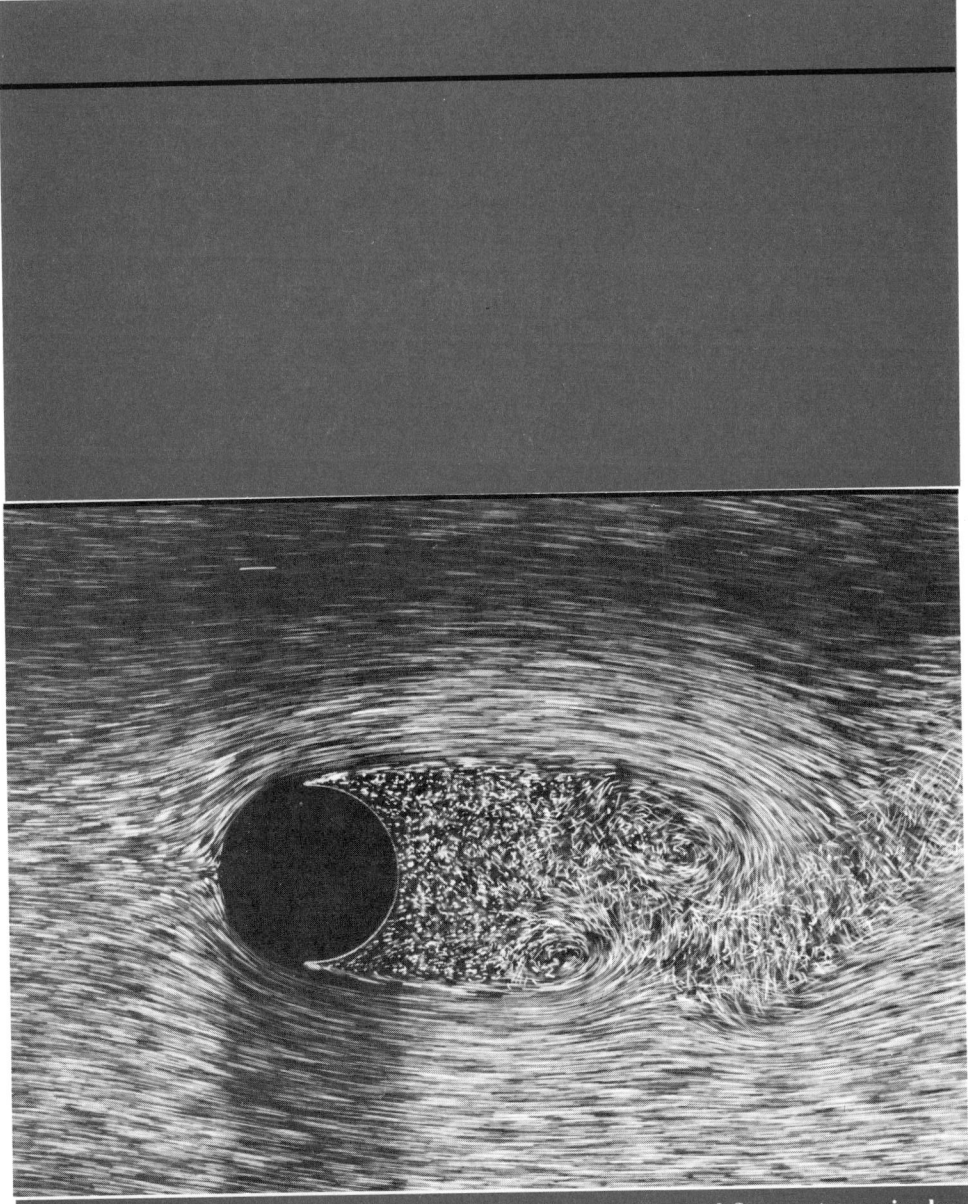

Flow past a circular cylinder with Re = 2000: The streaklines of flow past any circular cylinder (regardless of size, velocity, or fluid) are as shown provided that the dimensionless parameter called the Reynolds number, Re, is equal to 2000. For other values of Re the flow pattern will be different. (air bubbles in water) (Photograph courtesy of ONERA, France.)

___ SEVEN ___

SIMILITUDE, DIMENSIONAL ANALYSIS, AND MODELING

Although many practical engineering problems involving fluid mechanics can be solved by using the equations and analytical procedures described in the preceding chapters, there remain a large number of problems that rely on experimentally obtained data for their solution. In fact, it is probably fair to say that very few problems involving real fluids can be solved by analysis alone. The solution to many problems is achieved through the use of a combination of analysis and experimental data. Thus, engineers working on fluid mechanics problems should be familiar with the experimental approach to these problems so that they can interpret and make use of data obtained by others, such as might appear in handbooks, or be able to plan and execute the necessary experiments in their own laboratories. In this chapter we consider some techniques and ideas that are important in the planning and execution of experiments, as well as in understanding and correlating data that may have been obtained by other experimenters.

An obvious goal of any experiment is to make the results as widely applicable as possible. To achieve this end, the concept of *similitude* is often used so that measurements made on one system (for example, in the laboratory) can be used to describe the behavior of other similar systems (outside the laboratory). The laboratory systems are usually thought of as *models* and are used to study the phenomenon of interest under carefully controlled conditions. From these model studies empirical formulations can be developed, or specific predictions of one or more characteristics of some other similar system can be made. To do this, it is necessary to establish the relationship between the laboratory model and the "other" system. In the following sections, we find out how this can be accomplished in a systematic manner.

7.1   DIMENSIONAL ANALYSIS

To illustrate a typical fluid mechanics problem in which experimentation is required, consider the steady flow of an incompressible, Newtonian fluid through a long, smooth walled, horizontal, circular pipe. An important characteristic of this system, which would be of interest to an engineer designing a pipeline, is the pressure drop per unit length that develops along the pipe as a result of friction. Although this would appear to be a relatively simple flow problem, it cannot generally be solved analytically (even with the aid of large computers) without the use of experimental data.

The first step in the planning of an experiment to study this problem would be to decide on the factors, or variables, that will have an effect on the pressure drop per unit length, Δp_ℓ. We expect the list to include the pipe diameter, D, the fluid density, ρ, and fluid viscosity, μ, and the mean velocity, V, at which the fluid is flowing through the pipe. Thus, we can express this relationship as

$$\Delta p_\ell = f(D, \rho, \mu, V) \tag{7.1}$$

which simply indicates mathematically that we expect the pressure drop per unit length to be some function of the factors contained within the parentheses. At this point the nature of the function is unknown and the objective of the experiments to be performed is to determine the nature of this function.

To perform the experiments in a meaningful and systematic manner, it would be necessary to change one of the variables, such as the velocity, while holding all others constant, and measure the corresponding pressure drop. This series of tests would yield data that could be represented graphically as is illustrated in Fig. 7.1a. It is to be noted that this plot would only be valid for the specific pipe and for the specific fluid used in the tests, and certainly does not give us the general formulation we are looking for. We could repeat the process by varying each of the other variables in turn, as is illustrated in Figs. 7.1b, 7.1c, and 7.1d. This approach to determining the

(a)

(b)

(c)

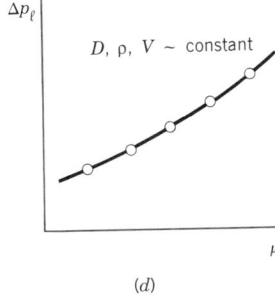

(d)

FIGURE 7.1 Illustrative plots showing how the pressure drop in a pipe may be affected by several different factors.

functional relationship between the pressure drop and the various factors that influence it, although logical in concept, is fraught with difficulties. Some of the experiments would be hard to carry out—for example, to obtain the data illustrated in Fig. 7.1c it would be necessary to vary fluid density while holding viscosity constant. How would you do this? Finally, once we obtained the various curves shown in Figs. 7.1a, 7.1b, 7.1c, and 7.1d how could we combine these data to obtain the desired general functional relationship between Δp_ℓ, D, ρ, μ, and V which would be valid for any similar pipe system?

Fortunately, there is a much simpler approach to this problem that will eliminate the difficulties described above. In the following sections we will show that rather than working with the original list of variables, as described in Eq. 7.1, we can collect these into two nondimensional combinations of variables (called *dimensionless products* or *dimensionless groups*) so that

$$\frac{D \, \Delta p_\ell}{\rho V^2} = \phi \left(\frac{\rho V D}{\mu}\right) \tag{7.2}$$

Thus, instead of having to work with five variables, we now have only two. The necessary experiment would simply consist of varying the dimensionless product $\rho V D / \mu$ and determining the corresponding value of $D \, \Delta p_\ell / \rho V^2$. The results of the experiment could then be represented by a single, universal curve as is illustrated in Fig. 7.2. This curve would be valid for any combination of smooth walled pipe and incompressible, Newtonian fluid. To obtain this curve we could choose a pipe of convenient size and a fluid that is easy to work with. Note that we wouldn't have to use different pipe sizes or even different fluids! It is clear that the experiment would be much simpler, easier to do, and less expensive (which would certainly make an impression on your boss).

The basis for this simplification lies in a consideration of the dimensions of the variable involved. As was discussed in Chapter 1, a qualitative description of physical quantities can be given in terms of basic dimensions such as mass, M, length, L, and time, T.[1] Alternatively we could use force, F, L, and T as basic dimensions, since from Newton's second law

$$F \doteq MLT^{-2}$$

(Recall from Chapter 1 that the notation $\doteq$ is used to indicate dimensional equality.) The dimensions of the variables in the pipe flow example are $\Delta p_\ell \doteq FL^{-3}$, $D \doteq L$, $\rho \doteq FL^{-4}T^2$, $\mu \doteq FL^{-2}T$, and $V \doteq LT^{-1}$. A quick check of the dimensions of the

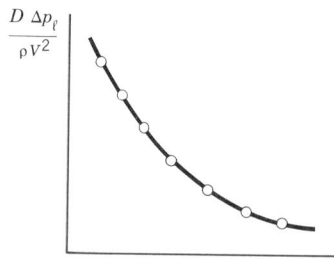

FIGURE 7.2 An illustrative plot of pressure drop data using dimensionless parameters.

---

[1] As noted in Chapter 1, we will use T to represent the basic dimension of time, although T is also used for temperature in thermodynamic relationships (such as the ideal gas law).

two groups that appear in Eq. 7.2 shows that they are in fact *dimensionless* products; that is,

$$\frac{D \, \Delta p_\ell}{\rho V^2} \doteq \frac{L(F/L^3)}{(FL^{-4}T^2)(LT^{-1})^2} \doteq F^0 L^0 T^0$$

and

$$\frac{\rho V D}{\mu} \doteq \frac{(FL^{-4}T^2)(LT^{-1})(L)}{(FL^{-2}T)} \doteq F^0 L^0 T^0$$

Not only have we reduced the numbers of variables from five to two, but the new groups are dimensionless combinations of variables, which means that the results presented in the form of Fig. 7.2 will be independent of the system of units we choose to use. This type of analysis is called *dimensional analysis,* and the basis for its application to a wide variety of problems is found in the *Buckingham pi theorem* described in the following section.

7.2   BUCKINGHAM PI THEOREM

A fundamental question we must answer is how many dimensionless products are required to replace the original list of variables? The answer to this question is supplied by the basic theorem of dimensional analysis that states the following:

If an equation involving k variables is dimensionally homogeneous, it can be reduced to a relationship among $k - r$ independent dimensionless products, where r is the minimum number of reference dimensions required to describe the variables.

The dimensionless products are frequently referred to as "pi terms," and the theorem is called the Buckingham pi theorem.[2] Buckingham used the symbol Π to represent a dimensionless product, and this notation is commonly used. Although the pi theorem is a simple one, its proof is not so simple and we will not include it here. Many entire books have been devoted to the subject of similitude and dimensional analysis, and a number of these are listed at the end of this chapter (Refs. 1–15). Students interested in pursuing the subject in more depth (including the proof of the pi theorem) can refer to one of these books.

The pi theorem is based on the idea of dimensional homogeneity which was introduced in Chapter 1. Essentially we assume that for any physically meaningful equation involving k variables, such as

$$u_1 = f(u_2, u_3, \dots, u_k)$$

the dimensions of the variable on the left side of the equal sign must be equal to the dimensions of any term that stands by itself on the right side of the equal sign. It then follows that we can rearrange the equation into a set of dimensionless products

---

[2] Although several early investigators, including Lord Rayleigh (1842–1919) in the nineteenth century, contributed to the development of dimensional analysis, Edgar Buckingham's (1867–1940) name is usually associated with the basic theorem. He stimulated interest in the subject in the United States through his publications during the early part of the twentieth century. See, for example, E. Buckingham, On Physically Similar Systems: Illustrations of the Use of Dimensional Equations, *Phys. Rev.*, 4 (1914), 345–376.

(pi terms) so that

$$\Pi_1 = \phi(\Pi_2, \Pi_3, \ldots, \Pi_{k-r})$$

The required number of pi terms is fewer than the number of original variables by r, where r is determined by the minimum number of reference dimensions required to describe the original list of variables. Usually the reference dimensions required to describe the variables will be the basic dimensions M, L, and T or F, L, and T. However, in some instances perhaps only two dimensions, such as L and T, are required, or maybe just one, such as L. Also, in a few rare cases the variables may be described by some combination of basic dimensions, such as M/T^2, and L, and in this case r would be equal to two rather than three. Although the use of the pi theorem may appear to be a little mysterious and complicated, we can actually develop a simple, systematic procedure for developing the pi terms for a given problem.

7.3 DETERMINATION OF PI TERMS

Several methods can be used to form the dimensionless products, or pi terms, that arise in a dimensional analysis. Essentially we are looking for a method that will allow us to systematically form the pi terms so that we are sure that they are dimensionless and independent, and that we have the right number. The method we will describe in detail in this section is called the *method of repeating variables*. A comparable, alternate method is described in Appendix D.

It will be helpful to break the repeating variable method down into a series of distinct steps that can be followed for any given problem. With a little practice you will be able to readily complete a dimensional analysis for your problem.

Step 1. *List all the variables that are involved in the problem.* This step is the most difficult one and it is, of course, vitally important that all pertinent variables be included. Otherwise the dimensional analysis will not be correct! We are using the term "variable" to include any quantity, including dimensional and nondimensional constants, which play a role in the phenomenon under investigation. All such quantities should be included in the list of "variables" to be considered for the dimensional analysis. The determination of the variables must be accomplished by the experimenter's knowledge of the problem and the physical laws that govern the phenomenon. Typically the variables will include those that are necessary to describe the *geometry* of the system (such as a pipe diameter), to define any *fluid properties* (such as a fluid viscosity), and to indicate *external effects* that influence the system (such as a driving pressure). These general classes of variables are intended as broad categories that should be helpful in identifying variables. It is likely, however, that there will be variables that do not fit easily into one of these categories, and each problem needs to be carefully analyzed.

Since we wish to keep the number of variables to a minimum, so that we can minimize the amount of laboratory work, it is important that all variables be independent. For example, if in a certain problem the cross-sectional area of a pipe is an important variable, either the area or the pipe diameter could be used, but not both, since they are obviously not independent. Similarly, if both fluid density, ρ, and specific weight, γ, are important variables, we could list ρ and γ, or ρ and g (acceleration of gravity), or γ

and g. However, it would be incorrect to use all three since $\gamma = \rho g$; that is, ρ, γ, and g are not independent. Note that although g would normally be constant in a given experiment, that fact is irrelevant as far as a dimensional analysis is concerned.

Step 2. ***Express each of the variables in terms of basic dimensions.*** For the typical fluid mechanics problem the basic dimensions will be either M, L, and T or F, L, and T. Dimensionally these two sets are related through Newton's second law ($\mathbf{F} = m\mathbf{a}$) so that $F \doteq MLT^{-2}$. Thus, either set can be used. The basic dimensions for typical variables found in fluid mechanics problems are listed in Table 1.1 in Chapter 1.

Step 3. ***Determine the required number of pi terms.*** This can be accomplished by means of the Buckingham pi theorem, which indicates that the number of pi terms is equal to $k - r$, where k is the number of variables in the problem (which is determined from Step 1) and r is the number of reference dimensions required to describe these variables (which is determined from Step 2). The reference dimensions usually correspond to the basic dimensions, and can be determined by an inspection of the dimensions of the variables obtained in Step 2. As previously noted, there may be occasions (usually rare) in which the basic dimensions appear in combinations so that the number of reference dimensions is less than the number of basic dimensions. This possibility is illustrated in Example 7.2.

Step 4. ***Select a number of repeating variables, where the number required is equal to the number of reference dimensions.*** Essentially what we are doing here is selecting from the original list of variables several of which can be combined with each of the remaining variables to form a pi term. All of the required reference dimensions must be included within the group of repeating variables, and each repeating variable must be dimensionally independent of the others (i.e., the dimensions of one repeating variable cannot be reproduced by some combination of products of powers of the remaining repeating variables). This means that the repeating variables cannot themselves be combined to form a dimensionless product.

For any given problem we usually are interested in determining how one particular variable is influenced by the other variables. We would consider this variable to be the dependent variable, and we would want this to appear in only one pi term. Thus, do *not* choose the dependent variable as one of the repeating variables, since the repeating variables will generally appear in more than one pi term.

Step 5. ***Form a pi term by multiplying one of the nonrepeating variables by the product of the repeating variables, each raised to an exponent that will make the combination dimensionless.*** Essentially each pi term will be of the form $u_i u_1^{a_i} u_2^{b_i} u_3^{c_i}$ where u_i is one of the nonrepeating variables, u_1, u_2, and u_3 are the repeating variables, and the exponents a_i, b_i, and c_i are determined so that the combination is dimensionless.

Step 6. ***Repeat Step 5 for each of the remaining nonrepeating variables.*** The resulting set of pi terms will correspond to the required number obtained from Step 3. If not, check your work—you have made a mistake!

Step 7. ***Check all the resulting pi terms to make sure they are dimensionless.*** It is easy to make a mistake in forming the pi terms. However, this can be checked by simply substituting the dimensions of the variables into the pi

terms to confirm that they are all dimensionless. One good way to do this is to express the variables in terms of M, L, and T if the basic dimensions F, L, and T were used initially, or vice versa, and then check to make sure the pi terms are dimensionless.

Step 8. *Express the final form as a relationship among the pi terms, and think about what it means.* Typically the final form can be written as

$$\Pi_1 = \phi(\Pi_2, \Pi_3, \ldots, \Pi_{k-r})$$

where Π_1 would contain the dependent variable in the numerator. It should be emphasized that if you started out with the correct list of variables (and the other steps were completed correctly), then the relationship in terms of the pi terms can be used to describe the problem. You need only work with the pi terms—not with the individual variables. However, it should be clearly noted that this is as far as we can go with the dimensional analysis; that is, the actual functional relationship among the pi terms must be determined by experiment.

To illustrate these various steps we will again consider the problem discussed earlier in this chapter which was concerned with the steady flow of an incompressible, Newtonian fluid through a long, smooth walled horizontal circular pipe. We are interested in the pressure drop per unit length, Δp_ℓ, along the pipe. According to Step 1 we must list all of the pertinent variables that are specified based on the experimenter's knowledge of the problem. In this problem we assume that

$$\Delta p_\ell = f(D, \rho, \mu, V)$$

where D is the pipe diameter, ρ and μ are the fluid density and viscosity, respectively, and V is the mean velocity.

Next (Step 2) we express all the variables in terms of basic dimensions. Using F, L, and T as basic dimensions it follows that

$$\Delta p_\ell \doteq FL^{-3}$$

$$D \doteq L$$

$$\rho \doteq FL^{-4}T^2$$

$$\mu \doteq FL^{-2}T$$

$$V \doteq LT^{-1}$$

We could also use M, L, and T as basic dimensions if desired—the final result will be the same! Note that for density, which is a mass per unit volume (ML^{-3}), we have used the relationship $F \doteq MLT^{-2}$ to express the density in terms of F, L, and T. Do not mix the basic dimensions; that is, use either F, L, and T or M, L, and T.

We can now apply the pi theorem to determine the required number of pi terms (Step 3). An inspection of the dimensions of the variables from Step 2 reveals that all three basic dimensions are required to describe the variables. Since there are five ($k = 5$) variables (do not forget to count the dependent variable, Δp_ℓ) and three required reference dimensions ($r = 3$), then according to the pi theorem there will be ($5 - 3$), or two pi terms required.

The repeating variables to be used to form the pi terms (Step 4) need to be selected from the list D, ρ, μ, and V. Remember, we do not want to use the dependent variable

as one of the repeating variables. Since three reference dimensions are required we will need to select three repeating variables. Generally, we would try to select for repeating variables those that are the simplest, dimensionally. For example, if one of the variables has the dimension of a length, choose it as one of the repeating variables. In this example we will use D, V, and ρ as repeating variables. Note that these are dimensionally independent, since D is a length, V involves both length and time, and ρ involves force, length, and time. This means that we cannot form a dimensionless product from this set.

We are now ready to form the two pi terms (Step 5). Typically, we would start with the dependent variable and combine it with the repeating variables to form the first pi term; that is,

$$\Pi_1 = \Delta p_\ell\, D^a V^b \rho^c$$

Since this combination is to be dimensionless it follows that

$$(FL^{-3})(L)^a(LT^{-1})^b(FL^{-4}T^2)^c \doteq F^0 L^0 T^0$$

The exponents a, b, and c must be determined such that the resulting exponent for each of the basic dimensions—F, L, and T—must be zero (so that the resulting combination is dimensionless). Thus, we can write

$$1 + c = 0 \quad \text{(for } F\text{)}$$

$$-3 + a + b - 4c = 0 \quad \text{(for } L\text{)}$$

$$-b + 2c = 0 \quad \text{(for } T\text{)}$$

The solution of this system of algebraic equations gives the desired values for a, b, and c. It follows that $a = 1$, $b = -2$, $c = -1$ and, therefore,

$$\Pi_1 = \frac{\Delta p_\ell D}{\rho V^2}$$

The process is now repeated for the remaining nonrepeating variables (Step 6). In this example there is only one additional variable (μ) so that

$$\Pi_2 = \mu D^a V^b \rho^c$$

or

$$(FL^{-2}T)(L)^a(LT^{-1})^b(FL^{-4}T^2)^c \doteq F^0 L^0 T^0$$

and, therefore,

$$1 + c = 0 \quad \text{(for } F\text{)}$$

$$-2 + a + b - 4c = 0 \quad \text{(for } L\text{)}$$

$$1 - b + 2c = 0 \quad \text{(for } T\text{)}$$

Solving these equations simultaneously it follows that $a = -1$, $b = -1$, $c = -1$ so that

$$\Pi_2 = \frac{\mu}{DV\rho}$$

Note that we end up with the correct number of pi terms as determined from Step 3.

At this point stop and check to make sure the pi terms are actually dimensionless

(Step 7). We will check using both *FLT* and *MLT* dimensions. Thus,

$$\Pi_1 = \frac{\Delta p_\ell D}{\rho V^2} \doteq \frac{(FL^{-3})(L)}{(FL^{-4}T^2)(LT^{-1})^2} \doteq F^0 L^0 T^0$$

$$\Pi_2 = \frac{\mu}{DV\rho} \doteq \frac{(FL^{-2}T)}{(L)(LT^{-1})(FL^{-4}T^2)} \doteq F^0 L^0 T^0$$

or alternatively,

$$\Pi_1 = \frac{\Delta p_\ell D}{\rho V^2} \doteq \frac{(ML^{-2}T^{-2})(L)}{(ML^{-3})(LT^{-1})^2} \doteq M^0 L^0 T^0$$

$$\Pi_2 = \frac{\mu}{DV\rho} \doteq \frac{(ML^{-1}T^{-1})}{(L)(LT^{-1})(ML^{-3})} \doteq M^0 L^0 T^0$$

Finally (Step 8), we can express the result of the dimensional analysis as

$$\frac{\Delta p_\ell D}{\rho V^2} = \tilde{\phi}\left(\frac{\mu}{DV\rho}\right)$$

This result indicates that this problem can be studied in terms of these two pi terms, rather than the original five variables we started with. However, dimensional analysis will *not* provide the form of the function $\tilde{\phi}$. This can only be obtained from a suitable set of experiments. If desired, the pi terms can be rearranged; that is, the reciprocal of $\mu/DV\rho$ could be used, and of course the order in which we write the variables can be changed. Thus, for example, Π_2 could be expressed as

$$\Pi_2 = \frac{\rho VD}{\mu}$$

and the relationship between Π_1 and Π_2 as

$$\frac{D\,\Delta p_\ell}{\rho V^2} = \phi\left(\frac{\rho VD}{\mu}\right)$$

This is the form we previously used in our initial discussion of this problem (Eq. 7.2). The dimensionless product, $\rho VD/\mu$, is a very famous one in fluid mechanics—the Reynolds number. This number has been briefly alluded to in Chapters 1 and 6 and will be further discussed in Section 7.6.

To summarize, the steps to be followed in performing a dimensional analysis using the method of repeating variables are as follows:

Step 1. List all the variables that are involved in the problem.

Step 2. Express each of the variables in terms of basic dimensions.

Step 3. Determine the required number of pi terms.

Step 4. Select a number of repeating variables, where the number required is equal to the number of reference dimensions (usually the same as the number of basic dimensions).

Step 5. Form a pi term by multiplying one of the nonrepeating variables by the product of repeating variables each raised to an exponent that will make the combination dimensionless.

Step 6. Repeat Step 5 for each of the remaining repeating variables.

Step 7. Check all the resulting pi terms to make sure they are dimensionless.

Step 8.   Express the final form as a relationship among the pi terms, and think about what it means.

EXAMPLE 7.1

A thin rectangular plate having a width w and a height h is located so that it is normal to a moving stream of fluid. Assume the drag, $\mathcal{D}$, that the fluid exerts on the plate is a function of w and h, the fluid viscosity and density, μ and ρ, respectively, and the velocity V of the fluid approaching the plate. Determine a suitable set of pi terms to study this problem experimentally.

SOLUTION

From the statement of the problem we can write

$$\mathcal{D} = f(w, h, \mu, \rho, V)$$

where this equation expresses the general functional relationship between the drag and the several variables that will affect it. The dimensions of the variables (using the MLT system) are

$$\mathcal{D} \doteq MLT^{-2}$$
$$w \doteq L$$
$$h \doteq L$$
$$\mu \doteq ML^{-1}T^{-1}$$
$$\rho \doteq ML^{-3}$$
$$V \doteq LT^{-1}$$

We see that all three basic dimensions are required to define the six variables so that the Buckingham pi theorem tells us that three pi terms will be needed (six variables minus three reference dimensions, $k - r = 6 - 3$).

We will next select three repeating variables such as w, V, ρ. A quick inspection of these three reveals that they are dimensionally independent, since each one contains a basic dimension not included in the others. Note that it would be incorrect to use both w and h as repeating variables since they have the same dimensions.

Starting with the dependent variable, $\mathcal{D}$, the first pi term can be formed by combining $\mathcal{D}$ with the repeating variables such that

$$\Pi_1 = \mathcal{D}w^a V^b \rho^c$$

and in terms of dimensions

$$(MLT^{-2})(L)^a(LT^{-1})^b(ML^{-3})^c \doteq M^0 L^0 T^0$$

Thus, for Π_1 to be dimensionless it follows that

$$1 + c = 0 \quad \text{(for } M)$$
$$1 + a + b - 3c = 0 \quad \text{(for } L)$$
$$-2 - b = 0 \quad \text{(for } T)$$

and, therefore, $a = -2$, $b = -2$, and $c = -1$. The pi term then becomes

$$\Pi_1 = \frac{\mathcal{D}}{w^2 V^2 \rho}$$

Next the procedure is repeated with the second nonrepeating variable, h, so that

$$\Pi_2 = hw^a V^b \rho^c$$

It follows that

$$(L)(L)^a(LT^{-1})^b(ML^{-3})^c \doteq M^0L^0T^0$$

and

$$c = 0 \quad \text{(for } M)$$
$$1 + a + b - 3c = 0 \quad \text{(for } L)$$
$$b = 0 \quad \text{(for } T)$$

so that $a = -1$, $b = 0$, $c = 0$, and therefore

$$\Pi_2 = \frac{h}{w}$$

The remaining nonrepeating variable is μ so that

$$\Pi_3 = \mu w^a V^b \rho^c$$

with

$$(ML^{-1}T^{-1})(L)^a(LT^{-1})^b(ML^{-3})^c \doteq M^0L^0T^0$$

and, therefore,

$$1 + c = 0 \quad \text{(for } M)$$
$$-1 + a + b - 3c = 0 \quad \text{(for } L)$$
$$-1 - b = 0 \quad \text{(for } T)$$

Solving for the exponents we obtain $a = -1$, $b = -1$, $c = -1$ so that

$$\Pi_3 = \frac{\mu}{wV\rho}$$

Now that we have the three required pi terms we should check to make sure they are dimensionless. To make this check we use F, L, and T, which will also verify the correctness of the original dimensions used for the variables. Thus,

$$\Pi_1 = \frac{D}{w^2V^2\rho} \doteq \frac{(F)}{(L)^2(LT^{-1})^2(FL^{-4}T^2)} \doteq F^0L^0T^0$$

$$\Pi_2 = \frac{h}{w} \doteq \frac{(L)}{(L)} \doteq F^0L^0T^0$$

$$\Pi_3 = \frac{\mu}{wV\rho} \doteq \frac{(FL^{-2}T)}{(L)(LT^{-1})(FL^{-4}T^2)} \doteq F^0L^0T^0$$

If these do not check, go back to the original list of variables and make sure you have the correct dimensions for each of the variables, and then check the algebra you used to obtain the exponents a, b, and c.

Finally, we can express the results of the dimensional analysis in the form

$$\frac{\mathcal{D}}{w^2V^2\rho} = \tilde{\phi}\left(\frac{h}{w}, \frac{\mu}{wV\rho}\right) \qquad \text{(Ans)}$$

Since at this stage in the analysis the nature of the function $\tilde{\phi}$ is unknown, we could rearrange the pi terms if we so desire. For example, we could express the final result in the form

$$\frac{\mathcal{D}}{w^2\rho V^2} = \phi\left(\frac{w}{h}, \frac{\rho Vw}{\mu}\right) \qquad \text{(Ans)}$$

which would be more conventional, since the ratio of the plate width to height, w/h, is called the *aspect ratio*, and $\rho Vw/\mu$ is the Reynolds number. To proceed, it would be necessary to perform a set of experiments to determine the nature of the function ϕ, as discussed in Section 7.7.

7.4 SOME ADDITIONAL COMMENTS ABOUT DIMENSIONAL ANALYSIS

The preceding section provides a systematic procedure for performing a dimensional analysis. Other methods could be used, although we think the method of repeating variables is the easiest for the beginning student to use. An alternate procedure is described in Appendix D. Pi terms can also be formed by inspection, as is discussed in Section 7.5. Regardless of the specific method used for the dimensional analysis, there are certain aspects of this important engineering tool that must seem a little baffling and mysterious to the student (and sometimes to the experienced investigator as well). In this section we will attempt to elaborate on some of the more subtle points that, based on our experience, can prove to be puzzling to students.

7.4.1 Selection of Variables

One of the most important, and difficult, steps in applying dimensional analysis to any given problem is the selection of the variables that are involved. As noted previously, for convenience we will use the term variable to indicate any quantity involved, including dimensional and nondimensional constants. There is no simple procedure whereby the variables can be easily identified. Generally, one must rely on a good understanding of the phenomenon involved and the governing physical laws. If extraneous variables are included, then too many pi terms appear in the final solution, and it may be difficult, time consuming, and expensive to eliminate these experimentally. If important variables are omitted, then an incorrect result will be obtained; and again, this may prove to be costly and difficult to ascertain. It is, therefore, imperative that sufficient time and attention be given to this first step in which the variables are determined.

Most engineering problems involve certain simplifying assumptions that have an influence on the variables to be considered. Usually we wish to keep the problem as simple as possible, perhaps even if some accuracy is sacrificed. A suitable balance between simplicity and accuracy is a desirable goal. How "accurate" the solution must be depends on the objective of the study; that is, we may be only concerned with general trends and, therefore, some variables that are thought to have only a minor influence in the problem may be neglected for simplicity.

For most engineering problems (including areas outside of fluid mechanics) pertinent variables can be classified into three general groups—geometry, material properties, and external effects.

Geometry. The geometric characteristics can usually be prescribed by a series of lengths and angles. In most problems the geometry of the system plays an important role and a sufficient number of geometric variables must be included to describe the system. These variables can usually be readily identified.

Material Properties. Since the response of a system to applied external effects such as forces, pressures, and changes in temperature is dependent on the nature of the materials involved in the system, the material properties that relate the external effects and the responses must be included as variables. For example, for Newtonian fluids the viscosity of the fluid is the property that relates the applied forces to the rates of deformation of the fluid. As the material behavior becomes more complex, such as would be true for non-Newtonian fluids, the determination of material properties becomes difficult, and this class of variables can be troublesome to identify.

External Effects.   This terminology is used to denote any variable that produces, or tends to produce, a change in the system. For example, in structural mechanics, forces (either concentrated or distributed) applied to a system tend to change its geometry, and such forces would need to be considered as pertinent variables. For fluid mechanics, variables in this class would be related to pressures, velocities, or gravity.

The above general classes of variables are intended as broad categories that should be helpful in identifying variables. It is likely, however, that there will be important variables that do not fit easily into one of the above categories and each problem needs to be carefully analyzed.

Since we wish to keep the number of variables to a minimum, it is important that all variables are independent. For example, if in a given problem we know that the moment of inertia of the area of a circular plate is an important variable, we could list either the moment of inertia or the plate diameter as the pertinent variable. However, it would be unnecessary to include both inertia and diameter, assuming that the diameter enters the problem only through the moment of inertia. In more general terms, if we have a problem in which the variables are

$$f(p, q, r, \ldots, u, v, w, \ldots) = 0 \tag{7.3}$$

and it is known that there is an additional relationship among some of the variables, for example,

$$q = f_1(u, v, w, \ldots) \tag{7.4}$$

then q is not required and can be omitted. Conversely, if it is known that the only way the variables $u, v, w, \ldots$ enter the problem is through the relationship expressed by Eq. 7.4, then the variables $u, v, w, \ldots$ can be replaced by the single variable q, therefore, reducing the number of variables.

In summary, the following points should be considered in the selection of variables:

1.   Clearly define the problem. What is the main variable of interest (the dependent variable)?

2.   Consider the basic laws that govern the phenomenon. Even a crude theory that describes the essential aspects of the system may be helpful.

3.   Start the variable selection process by grouping the variables into three broad classes: geometry, material properties, and external effects.

4.   Consider other variables that may not fall into one of the above categories. For example, time will be an important variable if any of the variables are time dependent.

5.   Be sure to include all quantities that enter the problem even though some of them may be held constant (e.g., the acceleration of gravity, g). For a dimensional analysis it is the dimensions of the quantities that are important—not specific values!

6.   Make sure that all variables are independent. Look for relationships among subsets of the variables.

7.4.2   Determination of Reference Dimensions

For any given problem it is obviously desirable to reduce the number of pi terms to a minimum and, therefore, we wish to reduce the number of variables to a minimum; that is, we certainly do not want to include extraneous variables. It is also important

to know how many reference dimensions are required to describe the variables. As we have seen in the preceding examples, F, L, and T appear to be a convenient set of basic dimensions for characterizing fluid-mechanical quantities. There is, however, really nothing "fundamental" about this set, and as previously noted M, L, and T would also be suitable. Actually any set of measurable quantities could be used as basic dimensions provided that the selected combination can be used to describe all secondary quantities. However, the use of FLT or MLT as basic dimensions is the simplest, and these dimensions can be used to describe fluid-mechanical phenomena. Of course, in some problems only one or two of these are required. In addition, we occasionally find that the number of reference dimensions needed to describe all variables is smaller than the number of basic dimensions. This point is illustrated in Example 7.2. Interesting discussions, both practical and philosophical, relative to the concept of basic dimensions can be found in the books by Huntley (Ref. 4) and by Issacson and Issacson (Ref. 12).

EXAMPLE 7.2

An open, cylindrical tank having a diameter D is supported around its bottom circumference and is filled to a depth h with a liquid having a specific weight γ. The vertical deflection, δ, of the center of the bottom is a function of D, h, d, γ, and E, where d is the thickness of the bottom and E is the modulus of elasticity of the bottom material. Perform a dimensional analysis of this problem.

SOLUTION

From the statement of the problem

$$\delta = f(D, h, d, \gamma, E)$$

and the dimensions of the variables are

$$\delta \doteq L$$
$$D \doteq L$$
$$h \doteq L$$
$$d \doteq L$$
$$\gamma \doteq FL^{-3} \doteq ML^{-2}T^{-2}$$
$$E \doteq FL^{-2} \doteq ML^{-1}T^{-2}$$

where the dimensions have been expressed in terms of both the FLT and MLT systems.

We now apply the pi theorem to determine the required number of pi terms. First, let us use F, L, and T as our system of basic dimensions. There are six variables and two reference dimensions (F and L) required so that four pi terms are needed. For repeating variables we can select D and γ so that

$$\Pi_1 = \delta D^a \gamma^b$$

$$(L)(L)^a(FL^{-3})^b \doteq F^0 L^0$$

and

$$1 + a - 3b = 0 \quad \text{(for } L)$$
$$b = 0 \quad \text{(for } F)$$

Therefore, $a = -1$, $b = 0$, and

$$\Pi_1 = \frac{\delta}{D}$$

Similarly,

$$\Pi_2 = hD^a\gamma^b$$

and following the same procedure as above $a = -1$, $b = 0$ so that

$$\Pi_2 = \frac{h}{D}$$

The remaining two pi terms can be found using the same procedure, with the result

$$\Pi_3 = \frac{d}{D} \qquad \Pi_4 = \frac{E}{D\gamma}$$

Thus, this problem can be studied by using the relationship:

$$\frac{\delta}{D} = \phi\left(\frac{h}{D}, \frac{d}{D}, \frac{E}{D\gamma}\right) \qquad\qquad \text{(Ans)}$$

Let us now solve the same problem using the MLT system. Although the number of variables is obviously the same, it would seem that there are three reference dimensions required, rather than two. If this were indeed true it would certainly be fortuitous, since we would reduce the number of required pi terms from four to three. Does this seem right? How can we reduce the number of required pi terms by simply using the MLT system of basic dimensions? The answer is that we cannot, and a closer look at the dimensions of the variables listed above reveals that actually only two reference dimensions, MT^{-2} and L, are required.

This is an example of the situation in which the number of reference dimensions differs from the number of basic dimensions. It does not happen very often and can be detected by looking at the dimensions of the variables (regardless of the systems used), and making sure how many reference dimensions are actually required to describe the variables. Once the number of reference dimensions has been determined, we can proceed as before. Since the number of repeating variables must equal the number of reference dimensions, it follows that two reference dimensions are still required and we could again use D and γ as repeating variables. The pi terms would be determined in the same manner. For example, the pi term containing E would be developed as

$$\Pi_4 = ED^a\gamma^b$$

$$(ML^{-1}T^{-2})(L)^a(ML^{-2}T^{-2})^b \doteq (MT^{-2})^0 L^0$$

$$1 + b = 0 \quad (\text{for } MT^{-2})$$

$$-1 + a - 2b = 0 \quad (\text{for } L)$$

and, therefore, $a = -1$, $b = -1$ so that

$$\Pi_4 = \frac{E}{D\gamma}$$

which is the same as Π_4 obtained using the FLT system. The other pi terms would be the same, and the final result is the same; that is,

$$\frac{\delta}{D} = \phi\left(\frac{h}{D}, \frac{d}{D}, \frac{E}{D\gamma}\right) \qquad\qquad \text{(Ans)}$$

This will always be true—you cannot affect the required number of pi terms by using M, L, and T instead of F, L, and T, or vice versa.

7.4.3   Uniqueness of Pi Terms

A little reflection on the process used to determine pi terms by the method of repeating variables reveals that the specific pi terms obtained depend on the somewhat arbitrary selection repeating variables. For example, in the problem of studying the pressure drop in a pipe, we selected D, V, and ρ as repeating variables. This led to the formulation of the problem in terms of pi terms as

$$\frac{\Delta p_\ell D}{\rho V^2} = \phi \left(\frac{\rho V D}{\mu} \right) \tag{7.5}$$

What if we had selected D, V, and μ as repeating variables? A quick check will reveal that the pi term involving Δp_ℓ is

$$\frac{\Delta p_\ell D^2}{V \mu}$$

and the second pi term remains the same. Thus, we can express the final result as

$$\frac{\Delta p_\ell D^2}{V \mu} = \phi_1 \left(\frac{\rho V D}{\mu} \right) \tag{7.6}$$

Both results are correct, and both would lead to the same final equation for Δp_ℓ. Note, however, that the functions ϕ and ϕ_1 in Eqs. 7.5 and 7.6 will be different because the dependent pi terms are different for the two relationships.

We can conclude from this illustration that there is *not* a unique set of pi terms which arise from a dimensional analysis. However, the required *number* of pi terms is fixed, and once a correct set is determined all other possible sets can be developed from this set by combinations of products of powers of the original set. Thus, if we have a problem involving, say, three pi terms,

$$\Pi_1 = \phi(\Pi_2, \Pi_3)$$

we could always form a new set from this one by combining the pi terms. For example, we could form a new pi term, Π_2', by letting

$$\Pi_2' = \Pi_2^a \, \Pi_3^b$$

where a and b are arbitrary exponents. Then the relationship could be expressed as

$$\Pi_1 = \phi_1(\Pi_2', \Pi_3)$$

or

$$\Pi_1 = \phi_2(\Pi_2, \Pi_2')$$

All of these would be correct. It should be emphasized, however, that the required number of pi terms cannot be reduced by this manipulation; only the form of the pi terms is altered. By using this technique we see that the pi terms in Eq. 7.6 could be obtained from those in Eq. 7.5; that is, we multiply Π_1 in Eq. 7.5 by Π_2 so that

$$\left(\frac{\Delta p_\ell D}{\rho V^2} \right) \left(\frac{\rho V D}{\mu} \right) = \frac{\Delta p_\ell D^2}{V \mu}$$

which is the Π_1 of Eq. 7.6.

There is no simple answer to the question: Which form for the pi terms is best? Usually our only guideline is to keep the pi terms as simple as possible. Also, it may be that certain pi terms will be easier to work with in actually performing experiments. The final choice remains an arbitrary one and generally will depend on the background

and experience of the investigator. It should again be emphasized, however, that although there is no unique set of pi terms for a given problem, the *number* required is fixed in accordance with the pi theorem. A somewhat more rigorous explanation of the nonuniqueness of pi terms is given in Appendix D.

7.5 DETERMINATION OF PI TERMS BY INSPECTION

One method for forming pi terms has been presented in Section 7.3 and an alternate method has been presented in Appendix D. These methods provide a step-by-step procedure that if executed properly will provide a correct and complete set of pi terms. Although these methods are simple and straightforward, they are rather tedious, particularly for problems in which large numbers of variables are involved. Since the only restrictions placed on the pi terms are that they be (1) correct in number, (2) dimensionless, and (3) independent, it is possible to simply form the pi terms by inspection, without resorting to one of the more formal procedures.

To illustrate this approach, we again consider the pressure drop per unit length along a smooth pipe. Regardless of the technique to be used, the starting point remains the same—determine the variables, which in the case are

$$\Delta p_\ell = f(D, \rho, \mu, V)$$

Next, the dimensions of the variables are listed:

$$\Delta p_\ell \doteq FL^{-3}$$

$$D \doteq L$$

$$\rho \doteq FL^{-4}T^2$$

$$\mu \doteq FL^{-2}T$$

$$V \doteq LT^{-1}$$

and subsequently the number of reference dimensions determined. The application of the pi theorem then tells us how many pi terms are required. In this problem, since there are five variables and three reference dimensions, two pi terms are needed. Thus, the required number of pi terms can be easily determined, and the determination of this number should always be done at the beginning of the analysis.

Once the number of pi terms is known, we can form each pi term by inspection, simply making use of the fact that each pi term must be dimensionless. We will always let Π_1 contain the dependent variable, which in this example is Δp_ℓ. Since this variable has the dimensions, FL^{-3}, we need to combine it with other variables so that a nondimensional product will result. One possibility is to first divide Δp_ℓ by ρ so that

$$\frac{\Delta p_\ell}{\rho} \doteq \frac{(FL^{-3})}{(FL^{-4}T^2)} \doteq \frac{L}{T^2}$$

The dependence on F has been eliminated but $\Delta p_\ell/\rho$ is obviously not dimensionless. To eliminate the dependence on T, we can divide by V^2 so that

$$\left(\frac{\Delta p_\ell}{\rho}\right)\frac{1}{V^2} \doteq \left(\frac{L}{T^2}\right)\frac{1}{(LT^{-1})^2} \doteq \frac{1}{L}$$

Finally, to make the combination dimensionless we multiply by D so that

$$\left(\frac{\Delta p_\ell}{\rho V^2}\right) D \doteq \left(\frac{1}{L}\right)(L) \doteq L^0$$

Thus

$$\Pi_1 = \frac{\Delta p_\ell D}{\rho V^2}$$

Next, we will form the second pi term by selecting the variable that was not used in Π_1, which in this case is μ. We simply combine μ with the other variables to make the combination dimensionless (but do not use Δp_ℓ in Π_2, since we want the dependent variable to appear only in Π_1). For example, divide μ by ρ (to eliminate F) then by V (to eliminate T), and finally by D (to eliminate L). Thus,

$$\Pi_2 = \frac{\mu}{\rho V D} \doteq \frac{(FL^{-2}T)}{(FL^{-4}T^2)(LT^{-1})(L)} \doteq F^0 L^0 T^0$$

and, therefore,

$$\frac{\Delta p_\ell D}{\rho V^2} = \phi\left(\frac{\mu}{\rho V D}\right)$$

which is, of course, the same result we obtained by using the method of repeating variables.

An additional concern, when one is forming pi terms by inspection, is to make certain that they are all independent. In the pipe flow example, Π_2 contains μ, which does not appear in Π_1, and therefore these two pi terms are obviously independent. In a more general case a pi term would not be independent of the others in a given problem if it can be formed by some combination of the others. For example, if Π_2 can be formed by a combination of say Π_3, Π_4, and Π_5 such as

$$\Pi_2 = \frac{\Pi_3^2 \Pi_4}{\Pi_5}$$

then Π_2 is not an independent pi term. We can ensure that each pi term is independent of those preceding it by incorporating a new variable in each pi term.

Although forming pi terms by inspection is essentially equivalent to the repeating variable method, it is less structured. With a little practice the pi terms can be readily formed by inspection, and this method offers an alternative to the more formal procedures.

7.6   COMMON DIMENSIONLESS GROUPS IN FLUID MECHANICS

Table 7.1 contains a list of variables that commonly arise in fluid mechanics problems. The list is obviously not exhaustive, but does indicate a broad range of variables likely to be found in a typical problem. Fortunately, not all of these variables would be encountered in all problems. However, when combinations of these variables are present, it is standard practice to combine them into some of the common dimensionless groups (pi terms) given in Table 7.1. These combinations appear so frequently that special names are associated with them as indicated in the table.

It is also often possible to provide a physical interpretation to the dimensionless groups which can be helpful in assessing their influence in a particular application.

TABLE 7.1 Some Common Variables and Dimensionless Groups in Fluid Mechanics

Variables: Acceleration of gravity, g; Bulk modulus, E_v; Characteristic length, ℓ; Density, ρ; Frequency of oscillating flow, ω; Pressure, p (or Δp); Speed of sound, c; Surface tension, σ; Velocity, V; Viscosity, μ

Dimensionless Groups	Name	Interpretation (Index of Force Ratio Indicated)	Types of Applications
$\dfrac{\rho V \ell}{\mu}$	Reynolds number, Re	$\dfrac{\text{inertia force}}{\text{viscous force}}$	Generally of importance in all types of fluid dynamics problems
$\dfrac{V}{\sqrt{g\ell}}$	Froude number, Fr	$\dfrac{\text{inertia force}}{\text{gravitational force}}$	Flow with a free surface
$\dfrac{p}{\rho V^2}$	Euler number, Eu	$\dfrac{\text{pressure force}}{\text{inertia force}}$	Problems in which pressure, or pressure differences, are of interest
$\dfrac{\rho V^2}{E_v}$	Cauchy number,[a] Ca	$\dfrac{\text{inertia force}}{\text{compressibility force}}$	Flows in which the compressibility of the fluid is important
$\dfrac{V}{c}$	Mach number,[a] Ma	$\dfrac{\text{inertia force}}{\text{compressibility force}}$	Flows in which the compressibility of the fluid is important
$\dfrac{\omega \ell}{V}$	Strouhal number, St	$\dfrac{\text{inertia (local) force}}{\text{inertia (convective) force}}$	Unsteady flow with a characteristic frequency of oscillation
$\dfrac{\rho V^2 \ell}{\sigma}$	Weber number, We	$\dfrac{\text{inertia force}}{\text{surface tension force}}$	Problems in which surface tension is important

[a] The Cauchy number and the Mach number are related and either can be used as an index of the relative effects of inertia and compressibility. See accompanying discussion.

For example, the Froude number is an index of the ratio of the force due to the acceleration of a fluid particle to the force due to gravity (weight). This can be demonstrated by considering a fluid particle moving along a streamline (Fig. 7.3). The magnitude of the component of inertia force F_I along the streamline can be expressed as $F_I = a_s m$, where a_s is the magnitude of the acceleration along the streamline for a particle having a mass, m. From our study of particle motion along

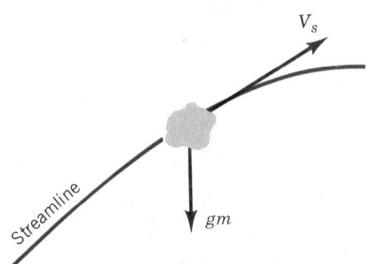

FIGURE 7.3 The force of gravity acting on a fluid particle moving along a streamline.

a curved path (see Section 3.1) we know that

$$a_s = \frac{dV_s}{dt} = V_s \frac{dV_s}{ds}$$

where s is measured along the streamline. If we write the velocity , V_s, and length, s, in dimensionless form; that is,

$$V_s^* = \frac{V_s}{V} \qquad s^* = \frac{s}{\ell}$$

where V and ℓ represent some characteristic velocity and length, respectively, then

$$a_s = \frac{V^2}{\ell} V_s^* \frac{dV_s^*}{ds^*}$$

and

$$F_I = \frac{V^2}{\ell} V_s^* \frac{dV_s^*}{ds^*} m$$

The magnitude of the weight of the particle, F_G, is $F_G = gm$, so the ratio of the inertia to the gravitational force is

$$\frac{F_I}{F_G} = \frac{V^2}{g\ell} V_s^* \frac{dV_s^*}{ds^*}$$

Thus, the force ratio F_I/F_G is proportional to $V^2/g\ell$, and the square root of this ratio, $V/\sqrt{g\ell}$, is called the *Froude number*. We see that a physical interpretation of the Froude number is that it is a measure of, or an index of, the relative importance of inertial forces acting on fluid particles to the weight of the particle. Note that the Froude number is not really *equal* to this force ratio, but simply some type of average measure of the influence of these two forces. In a problem in which gravity (or weight) is not important, the Froude number would not appear as an important pi term. A similar interpretation in terms of indices of force ratios can be given to the other dimensionless groups, as indicated in Table 7.1, and a further discussion of the basis for this type of interpretation is given in the last section in this chapter. Some additional details about these important dimensionless groups are given below, and the types of application or problem in which they arise are briefly noted in the last column of Table 7.1.

Reynolds Number   The Reynolds number is undoubtedly the most famous dimensionless parameter in fluid mechanics. It is named in honor of Osborne Reynolds (1842–1912), a British engineer who first demonstrated that this combination of variables could be used as a criterion to distinguish between laminar and turbulent flow. In most fluid flow problems there will be a characteristic length, ℓ, and a velocity, V, as well as the fluid properties of density, ρ, and viscosity, μ, which are relevant variables in the problem. Thus, with these variables the Reynolds number

$$\mathrm{Re} = \frac{\rho V \ell}{\mu}$$

arises naturally from the dimensional analysis. The Reynolds number is a measure of the ratio of the inertia force on an element of fluid to the viscous force on an element. When these two types of forces are important in a given problem, the Reynolds number will play an important role. However, if the Reynolds number is very small

(Re ≪ 1) this is an indication that the viscous forces are dominant in the problem, and it may be possible to neglect the inertial effects; that is, the density of the fluid will not be an important variable. Flows at very small Reynolds numbers are commonly referred to as "creeping flows" as discussed in Section 6.10. Conversely, for large Reynolds number flows viscous effects are small relative to inertial effects and for these cases it may be possible to neglect the effect of viscosity and consider the problem as one involving a "nonviscous" fluid. This type of problem is considered in detail in Sections 6.4 through 6.7.

Froude Number   The Froude number

$$\mathrm{Fr} = \frac{V}{\sqrt{g\ell}}$$

is distinguished from the other dimensionless groups in Table 7.1 in that it contains the acceleration of gravity, g. The acceleration of gravity becomes an important variable in a fluid dynamics problem in which the fluid weight is an important force. As discussed, the Froude number is a measure of the ratio of the inertia force on an element of fluid to the weight of the element. It will generally be important in problems involving flows with free surfaces since gravity principally affects this type of flow. Typical problems would include the study of the flow of water around ships (with the resulting wave action), or flow through rivers or open conduits. The Froude number is named in honor of William Froude (1810–1879), a British civil engineer, mathematician, and naval architect who pioneered the use of towing tanks for the study of ship design. It is to be noted that the Froude number is also commonly defined as the square of the Froude number listed in Table 7.1.

Euler Number   The Euler number

$$\mathrm{Eu} = \frac{p}{\rho V^2}$$

can be interpreted as a measure of the ratio of pressure forces to inertial forces, where p is some characteristic pressure in the flow field. Very often the Euler number is written in terms of a pressure difference, Δp, so that $\mathrm{Eu} = \Delta p/\rho V^2$. Also, this combination expressed as $\Delta p/\frac{1}{2}\rho V^2$ is called the *pressure coefficient*. Some form of the Euler number would normally be used in problems in which pressure or the pressure difference between two points is an important variable. The Euler number is named in honor of Leonhard Euler (1707–1783), a famous Swiss mathematician who pioneered work on the relationship between pressure and flow. For problems in which cavitation is of concern, the dimensionless group $(p_r - p_v)/\frac{1}{2}\rho V^2$ is commonly used, where p_v is the vapor pressure and p_r is some reference pressure. Although this dimensionless group has the same form as the Euler number, it is generally referred to as the *cavitation number*.

Cauchy Number and Mach Number   The Cauchy number

$$\mathrm{Ca} = \frac{\rho V^2}{E_v}$$

and the Mach number

$$\mathrm{Ma} = \frac{V}{c}$$

are important dimensionless groups in problems in which fluid compressibility is a significant factor. Since the speed of sound, c, in a fluid is equal to $c = \sqrt{E_v/\rho}$ (see Section 1.7.3) it follows that

$$\text{Ma} = V\sqrt{\frac{\rho}{E_v}}$$

and the square of the Mach number

$$\text{Ma}^2 = \frac{\rho V^2}{E_v} = \text{Ca}$$

is equal to the Cauchy number. Thus, either number (but not both) may be used in problems in which fluid compressibility is important. Both numbers can be interpreted as representing an index of the ratio of inertial forces to compressibility forces. When the Mach number is relatively small (say, less than 0.3), the inertial forces induced by the fluid motion are not sufficiently large to cause a significant change in the fluid density, and in this case the compressibility of the fluid can be neglected. The Mach number is the more commonly used parameter in compressible flow problems, particularly in the fields of gas dynamics and aerodynamics. The Cauchy number is named in honor of Augustin Louis de Cauchy (1789–1857), a French engineer, mathematician, and hydrodynamicist. The Mach number is named in honor of Ernst Mach (1838–1916), an Austrian physicist and philosopher.

Strouhal Number   The Strouhal number

$$\text{St} = \frac{\omega\ell}{V}$$

is a dimensionless parameter that is likely to be important in unsteady, oscillating flow problems in which the frequency of the oscillation is ω. It represents a measure of the ratio of inertial forces due to the unsteadiness of the flow (local acceleration) to the inertial forces due to changes in velocity from point to point in the flow field (convective acceleration). This type of unsteady flow may develop when a fluid flows past a solid body (such as a wire or cable) placed in the moving stream. For example, in a certain Reynolds number range, a periodic flow will develop downstream from a cylinder placed in a moving fluid due to a regular pattern of vortices that are shed from the body. This system of vortices, called a *Kármán vortex trail* [named after Theodor von Kármán (1881–1963), a famous fluid mechanician] creates an oscillating flow at a discrete frequency, ω, such that the Strouhal number can be closely correlated with the Reynolds number. When the frequency is in the audible range, a sound can be heard and the bodies appear to "sing." In fact the Strouhal number is named in honor of Vincenz Strouhal (1850–1922), who used this parameter in his study of "singing wires." The most dramatic evidence of this phenomenon occurred in 1940 with the collapse of the Tacoma Narrows bridge. The shedding frequency of the vortices coincided with the natural frequency of the bridge, thereby setting up a resonant condition that eventually led to the collapse of the bridge.

There are, of course, other types of oscillating flows. For example, blood flow in arteries is periodic and can be analyzed by breaking up the periodic motion into a series of harmonic components (Fourier series analysis), with each component having a frequency that is a multiple of the fundamental frequency, ω (the pulse rate). Rather than use the Strouhal number in this type of problem, a dimensionless group formed

by the product of St and Re is used; that is

$$St \times Re = \frac{\rho \omega \ell^2}{\mu}$$

The square root of this dimensionless group is often referred to as the *frequency parameter*.

Weber Number   The Weber number

$$We = \frac{\rho V^2 \ell}{\sigma}$$

may be important in problems in which there is an interface between two fluids. In this situation the surface tension may play an important role in the phenomenon of interest. The Weber number can be thought of as an index of the inertial force to the surface tension force acting on a fluid element. Common examples of problems in which this parameter may be important include the flow of thin films of liquid, or in the formation of droplets or bubbles. Clearly, not all problems involving flows with an interface will require the inclusion of surface tension. The flow of water in a river is not affected by surface tension, since inertial and gravitational effects are dominant (We $\gg$ 1). However, as discussed in a later section, for river models (which may have small depths) caution is required so that surface tension does not become important in the model, whereas it is not important in the actual river. The Weber number is named after Moritz Weber (1871–1951), a German professor of naval mechanics who was instrumental in formalizing the general use of common dimensionless groups as a basis for similitude studies.

7.7   CORRELATION OF EXPERIMENTAL DATA

One of the most important uses of dimensional analysis is as an aid in the efficient handling, interpretation, and correlation of experimental data. Since the field of fluid mechanics relies heavily on empirical data, it is not surprising that dimensional analysis is such an important tool in this field. As noted previously, a dimensional analysis cannot provide a complete answer to any given problem, since the analysis only provides the dimensionless groups describing the phenomenon, and not the specific relationship among the groups. To determine this relationship, suitable experimental data must be obtained. The degree of difficulty involved in this process depends on the number of pi terms, and the nature of the experiments (How hard is it to obtain the measurements?). The simplest problems are obviously those involving the fewest pi terms, and the following sections indicate how the complexity of the analysis increases with the increasing number of pi terms.

7.7.1   Problems with One Pi Term

Application of the pi theorem indicates that if the number of variables minus the number of reference dimensions is equal to unity, then only *one* pi term is required to describe the phenomenon. The functional relationship that must exist for one pi term is

$$\Pi_1 = C$$

where C is a constant. This is one situation in which a dimensional analysis reveals

the specific form of the relationship, and as is illustrated by the following example, shows how the individual variables are related. The value of the constant, however, must still be determined by experiment.

EXAMPLE 7.3

Assume that the drag, $\mathcal{D}$, acting on a spherical particle that falls very slowly through a viscous fluid is a function of the particle diameter, d, the particle velocity, V, and the fluid viscosity, μ. Determine, with the aid of dimensional analysis, how the drag depends on the particle velocity.

SOLUTION

From the information given, it follows that

$$\mathcal{D} = f(d, V, \mu)$$

and the dimensions of the variables are

$$\mathcal{D} \doteq F$$

$$d \doteq L$$

$$V \doteq LT^{-1}$$

$$\mu \doteq FL^{-2}T$$

We see that there are four variables and three reference dimensions (F, L, and T) required to describe the variables. Thus, according to the pi theorem, one pi term is required. This pi term can be easily formed by inspection and can be expressed as

$$\Pi_1 = \frac{\mathcal{D}}{\mu V d}$$

Since there is only one pi term, it follows that

$$\frac{\mathcal{D}}{\mu V d} = C$$

or

$$\mathcal{D} = C\mu V d$$

Thus, for a given particle and fluid, the drag varies directly with the velocity so that

$$\mathcal{D} \propto V \qquad \text{(Ans)}$$

Actually, the dimensional analysis reveals that the drag not only varies directly with the velocity, but it also varies directly with the particle diameter and the fluid viscosity. We could not, however, predict the value of the drag, since the constant, C, is unknown. An experiment would have to be performed in which the drag and the corresponding velocity are measured for a given particle and fluid. Although in principle we would only have to run a single test, we would certainly want to repeat it several times to obtain a reliable value for C. It should be emphasized that once the value of C is determined it is not necessary to run similar tests by using different particles and fluids; that is, C is a universal constant so long as the drag is a function only of particle diameter, velocity, and fluid viscosity.

An approximate solution to this problem can also be obtained theoretically, from which it is found that $C = 3\pi$ so that

$$\mathcal{D} = 3\pi\mu V d$$

This equation is commonly called *Stokes law* and is used in the study of the settling of particles. Our experiments would reveal that this result is only valid for small Reynolds numbers ($\rho V d / \mu \ll 1$). This follows, since in the original list of variables, we have neglected inertial effects (fluid density is not included as a variable). The inclusion of an additional variable would lead to another pi term so that there would be two pi terms rather than one.

7.7.2   Problems with Two or More Pi Terms

If a given phenomenon can be described with two pi terms such that

$$\Pi_1 = \phi(\Pi_2)$$

the functional relationship among the variables can then be determined by varying Π_2 and measuring the corresponding values of Π_1. For this case the results can be conveniently presented in graphical form by plotting Π_1 versus Π_2 as is illustrated in Fig. 7.4. It should be emphasized that the curve shown in Fig. 7.4 would be a "universal" one for the particular phenomenon studied. This means that if the variables and the resulting dimensional analysis are correct, then there is only a single relationship between Π_1 and Π_2 as illustrated in Fig. 7.4. However, since this is an empirical relationship, we can only say that it is valid over the range of Π_2 covered by the experiments. It would be unwise to extrapolate beyond this range, since as illustrated with the dashed lines in the figure, the nature of the phenomenon could dramatically change as the range of Π_2 is extended. In addition to presenting the data graphically, it may be possible (and desirable) to obtain an empirical equation relating Π_1 and Π_2 by using a standard curve fitting technique.

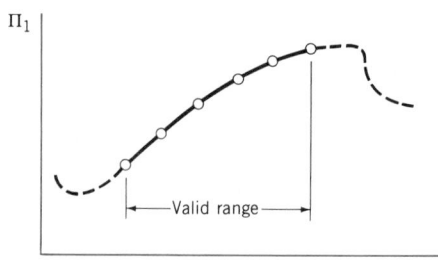

FIGURE 7.4 The graphical presentation of data for problems involving two pi terms, with an illustration of the potential danger of extrapolation of data.

EXAMPLE 7.4

The relationship between the pressure drop per unit length along a smooth walled, horizontal pipe and the variables that affect the pressure drop is to be determined experimentally. In the laboratory the pressure drop was measured over a 5-ft length of smooth walled pipe having an inside diameter of 0.496 in. The fluid used was water at 60 °F ($\mu = 2.34 \times 10^{-5}$ lb·s/ft², $\rho = 1.94$ slugs/ft³). Tests were run in which the velocity was varied and the corresponding pressure drop measured. The results of these tests are shown below:

Velocity (ft/s)	1.17	1.95	2.91	5.84	11.13	16.92	23.34	28.73
Pressure drop (lb/ft²) (for 5-ft length)	6.26	15.6	30.9	106	329	681	1200	1730

Make use of these data to obtain a general relationship between the pressure drop per unit length and the other variables.

SOLUTION

The first step is to perform a dimensional analysis during the planning stage *before* the experiments are actually run. As was discussed in Section 7.3, we will assume that the pressure drop per unit length, Δp_ℓ, is a function of the pipe diameter, D, fluid density, ρ, fluid viscosity, μ, and the velocity, V. Thus,

$$\Delta p_\ell = f(D, \rho, \mu, V)$$

and application of the pi theorem yields

$$\frac{D \, \Delta p_\ell}{\rho V^2} = \phi \left(\frac{\rho V D}{\mu}\right)$$

To determine the form of the relationship we need to vary the Reynolds number, $\rho V D / \mu$, and to measure the corresponding values of $D \, \Delta p_\ell / \rho V^2$. The Reynolds number could be varied by changing any one of the variables, ρ, V, D, or μ, or any combination of them. However, the simplest way to do this is to vary the velocity, since this will allow us to use the same fluid and pipe. Based on the data given, values for the two pi terms can be computed with the result:

$D \, \Delta p_\ell / \rho V^2$	0.0195	0.0175	0.0155	0.0132
$\rho V D / \mu$	4.01×10^3	6.68×10^3	9.97×10^3	2.00×10^4

	0.0113	0.0101	0.00939	0.00893
	3.81×10^4	5.80×10^4	8.00×10^4	9.85×10^4

These are dimensionless groups so that their values are independent of the system of units used so long as a consistent system is used. For example, if the velocity is in ft/s, then the diameter should be in feet, not inches or meters. Note that since the Reynolds numbers are all greater than 2100, the flow in the pipe is turbulent.

A plot of these two pi terms can now be made with the results shown in Fig. E7.4a. The correlation appears to be quite good, and if it was not, this would suggest that either we had large experimental measurement errors, or that we had perhaps omitted an important variable. The curve shown in Fig. E7.4a represents the general relationship between the pressure drop and the other factors in the range of Reynolds numbers between 4.01×10^3 and 9.85×10^4. Thus, for this range of Reynolds numbers it is *not* necessary to repeat the tests for other pipe sizes or other fluids provided the assumed independent variables (D, ρ, μ, V) are the only important ones.

Since the relationship between Π_1 and Π_2 is nonlinear, it is not immediately obvious what form of empirical equation might be used to describe the relationship. If, however, the same data are plotted on logarithmic graph paper, as are shown in Fig. E7.4b, the data form a straight line, suggesting that a suitable equation is of the form $\Pi_1 = A \Pi_2^n$ where A and n are empirical constants to be determined from the data by using a suitable curve fitting technique, such as a nonlinear regression program. For the data given in this example, a good fit of the data is obtained with the equation:

$$\Pi_1 = 0.150 \, \Pi_2^{-0.25} \qquad \text{(Ans)}$$

In 1911, H. Blasius (1883–1970), a German fluid mechanician, established a similar empirical equation that is widely used for predicting the pressure drop in smooth pipes in the range $4 \times 10^3 < \text{Re} < 10^5$ (Ref. 16). This equation can be expressed in the form:

$$\frac{D \, \Delta p_\ell}{\rho V^2} = 0.1582 \left(\frac{\rho V D}{\mu}\right)^{-1/4}$$

(a)

(b)

FIGURE E7.4

The so-called Blasius formula is based on numerous experimental results of the type used in this example. Flow in pipes is discussed in more detail in the next chapter, where it is shown how pipe roughness (which introduces another variable) may affect the results given in this example (which is for smooth walled pipes).

As the number of required pi terms increases, it becomes more difficult to display the results in a convenient graphical form and to determine a specific empirical equation that describes the phenomenon. For problems involving three pi terms

$$\Pi_1 = \phi(\Pi_2, \Pi_3)$$

it is still possible to show data correlations on simple graphs by plotting families of curves as are illustrated in Fig. 7.5. This is an informative and useful way of representing the data in a general way. It may also be possible to determine a suitable

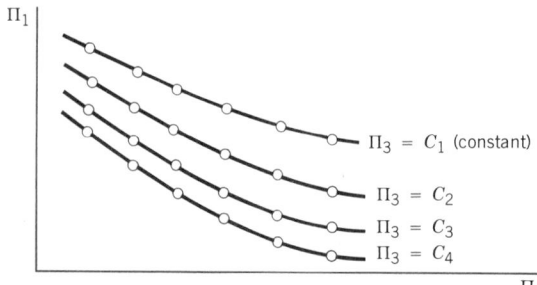

FIGURE 7.5 The graphical presentation of data for problems involving three pi terms.

empirical equation relating the three pi terms. However, as the number of pi terms continues to increase, corresponding to an increase in the general complexity of the problem of interest, both the graphical presentation and the determination of a suitable empirical equation become intractable. For these more complicated problems, it is often more feasible to use models to predict specific characteristics of the system rather than to try to develop general correlations.

7.8 MODELING AND SIMILITUDE

Models are widely used in fluid mechanics. Major engineering projects involving structures, aircraft, ships, rivers, harbors, dams, air and water pollution, and so on, frequently involve the use of models. Although the term "model" is used in many different contexts, the "engineering model" generally conforms to the following definition. *A model is a representation of a physical system that may be used to predict the behavior of the system in some desired respect.* The physical system for which the predictions are to be made is called the *prototype*. Although *mathematical* or *computer* models may also conform to this definition, our interest will be in physical models; that is, models that resemble the prototype but are generally of a different size, may involve different fluids, and often operate under different conditions (pressures, velocities, etc.). Usually a model is smaller than the prototype. Therefore, it is more easily handled in the laboratory and less expensive to construct and operate than a large prototype. Occasionally, if the prototype is very small, it may be advantageous to have a model that is larger than the prototype so that it can be more easily studied. For example, large models have been used to study the motion of red blood cells, which are approximately 8 μm in diameter. With the successful development of a valid model, it is possible to predict the behavior of the prototype under a certain set of conditions. We may also wish to examine a priori the effect of possible design changes that are proposed for a hydraulic structure or fluid-flow system. There is, of course, an inherent danger in the use of models in that predictions can be made that are in error and the error not detected until the prototype is found not to perform as predicted. It is, therefore, imperative that the model be properly designed and tested and that the results be interpreted correctly. In the following sections we will develop the procedures for designing models so that the model and prototype will behave in a similar fashion.

7.8.1 Theory of Models

The theory of models can be readily developed by using the principles of dimensional analysis. It has been shown that any given problem can be described in terms of a set of pi terms as

$$\Pi_1 = \phi(\Pi_2, \Pi_3, \ldots, \Pi_n) \tag{7.7}$$

In formulating this relationship, only a knowledge of the general nature of the physical phenomenon, and the variables involved, is required. Specific values for variables (size of components, fluid properties, and so on) are not needed to perform the dimensional analysis. Thus, Eq. 7.7 applies to any system that is governed by the same variables. If Eq. 7.7 describes the behavior of a particular prototype, a similar relationship can be written for a model of this prototype; that is,

$$\Pi_{1m} = \phi(\Pi_{2m}, \Pi_{3m}, \ldots, \Pi_{nm}) \tag{7.8}$$

where the form of the function will be the the same as long as the same phenomenon is involved in both the prototype and the model. Variables, or pi terms, without a subscript will refer to the prototype, whereas the subscript m will be used to designate the model variables or pi terms.

The pi terms can be developed so that Π_1 contains the variable that is to be predicted from observations made on the model. Therefore, if the model is designed and operated under the following conditions

$$\Pi_{2m} = \Pi_2$$

$$\Pi_{3m} = \Pi_3$$

$$\vdots \tag{7.9}$$

$$\Pi_{nm} = \Pi_n$$

then with the presumption that the form of ϕ is the same for model and prototype, it follows that

$$\Pi_1 = \Pi_{1m} \tag{7.10}$$

Equation 7.10 is the desired *prediction equation* and indicates that the measured value of Π_{1m} obtained with the model will be equal to the corresponding Π_1 for the prototype as long as the other pi terms are equal. The conditions specified by Eqs. 7.9 provide the *model design conditions,* also called *similarity requirements* or *modeling laws.*

As an example of the procedure, consider the problem of determining the drag, $\mathcal{D}$, on a thin rectangular plate ($w \times h$ in size) placed normal to a fluid with velocity, V. The dimensional analysis of this problem was performed in Example 7.1, where it was assumed that

$$\mathcal{D} = f(w, h, \mu, \rho, V)$$

Application of the pi theorem yielded

$$\frac{\mathcal{D}}{w^2 \rho V^2} = \phi\left(\frac{w}{h}, \frac{\rho V w}{\mu}\right) \tag{7.11}$$

We are now concerned with designing a model that could be used to predict the drag on a certain prototype (which presumably has a different size than the model). Since the relationship expressed by Eq. 7.11 applies to both prototype and model, Eq. 7.11 is assumed to govern the prototype, with a similar relationship

$$\frac{\mathcal{D}_m}{w_m^2 \rho_m V_m^2} = \phi\left(\frac{w_m}{h_m}, \frac{\rho_m V_m w_m}{\mu_m}\right) \tag{7.12}$$

for the model. The model design conditions, or similarity requirements, are therefore

$$\frac{w_m}{h_m} = \frac{w}{h} \qquad \frac{\rho_m V_m w_m}{\mu_m} = \frac{\rho V w}{\mu}$$

The size of the model is obtained from the first requirement which indicates that

$$w_m = \frac{h_m}{h} w \tag{7.13}$$

We are free to establish the height ratio h_m/h, but then the model plate width, w_m, is fixed in accordance with Eq. 7.13.

The second similarity requirement indicates that the model and prototype must be operated at the same Reynolds number. Thus, the required velocity for the model is obtained from the relationship:

$$V_m = \frac{\mu_m}{\mu} \frac{\rho}{\rho_m} \frac{w}{w_m} V \qquad (7.14)$$

Note that this model design requires not only geometric scaling, as specified by Eq. 7.13, but also the correct scaling of the velocity in accordance with Eq. 7.14. This result is typical of most model designs—there is more to the design than simply scaling the geometry!

With the foregoing similarity requirements satisfied, the prediction equation for the drag is

$$\frac{\mathscr{D}}{w^2 \rho V^2} = \frac{\mathscr{D}_m}{w_m^2 \rho_m V_m^2}$$

or

$$\mathscr{D} = \left(\frac{w}{w_m}\right)^2 \left(\frac{\rho}{\rho_m}\right) \left(\frac{V}{V_m}\right)^2 \mathscr{D}_m$$

Thus, a measured drag on the model, $\mathscr{D}_m$, must be multiplied by the ratio of the square of the plate widths, the ratio of the fluid densities, and the square of the velocities to obtain the predicted value of the prototype drag, $\mathscr{D}$.

Generally, as is illustrated in this example, to achieve similarity between model and prototype behavior, *all the corresponding pi terms must be equated between model and prototype*. Usually, one or more of these pi terms will involve ratios of important lengths (such as w/h in the foregoing example); that is, they are purely geometrical. Thus, when we equate the pi terms involving length ratios we are requiring that there be complete *geometric similarity* between the model and prototype. This means that the model must be a scaled version of the prototype. Geometric scaling may extend to the finest features of the system, such as surface roughness, or small protuberances on a structure, since these kinds of geometric features may significantly influence the flow. Any deviation from complete geometric similarity for a model must be carefully considered. Sometimes complete geometric scaling may be difficult to achieve, particularly when dealing with surface roughness, since roughness is difficult to characterize and control.

Another group of typical pi terms (such as the Reynolds number in the foregoing example) involves force ratios. The equality of these pi terms requires the ratio of like forces in model and prototype to be the same. Thus, for flows in which the Reynolds numbers are equal, the ratio of viscous forces in model and prototype is equal to the ratio of inertia forces. If other pi terms are involved, such as the Froude number or Weber number, a similar conclusion can be drawn; that is, the equality of these pi terms requires the ratio of like forces in model and prototype to be the same. Thus, when these types of pi terms are equal in model and prototype, we have *dynamic similarity* between model and prototype. It follows that with both geometric and dynamic similarity the streamline patterns will be the same and corresponding velocity ratios (V_m/V) and acceleration ratios (a_m/a) are the same throughout the flow field. Thus, *kinematic similarity* exists between model and prototype. To have complete similarity between model and prototype, we must maintain geometric, kinematic, and dynamic similarity between the two systems. This will automatically follow if all the important variables are included in the dimensional analysis, and if all the similarity requirements based on the resulting pi terms are satisfied.

EXAMPLE 7.5

A long structural component of a bridge has the cross section shown in Fig. E7.5. It is known that when a steady wind blows past this type of bluff body, vortices may develop on the downwind side that are shed in a regular fashion at some definite frequency. Since these vortices can create harmful periodic forces acting on the structure, it is important to determine the shedding frequency. For the specific structure of interest, $D = 0.1$ m, $H = 0.3$ m, and a representative wind velocity is 50 km/hr. Standard air can be assumed. The shedding frequency is to be determined through the use of a small-scale model that is to be tested in a water tunnel. For the model $D_m = 20$ mm and the water temperature is 20 °C. Determine the model dimension, H_m, and the velocity at which the test should be performed. If the shedding frequency for the model is found to be 49.9 Hz, what is the corresponding frequency for the prototype?

FIGURE E7.5

SOLUTION

We expect the shedding frequency, ω, to depend on the lengths, D and H, the approach velocity, V, and the fluid density, ρ, and viscosity, μ. Thus,

$$\omega = f(D, H, V, \rho, \mu)$$

where

$$\omega \doteq T^{-1}$$
$$D \doteq L$$
$$H \doteq L$$
$$V \doteq LT^{-1}$$
$$\rho \doteq ML^{-3}$$
$$\mu \doteq ML^{-1}T^{-1}$$

Since there are six variables and three reference dimensions (MLT), three pi terms are required. Application of the pi theorem yields

$$\frac{\omega D}{V} = \phi\left(\frac{D}{H}, \frac{\rho V D}{\mu}\right)$$

We recognize the pi term on the left as the Strouhal number, and the dimensional analysis indicates that the Strouhal number is a function of the geometric parameter, D/H, and the Reynolds number. Thus, to maintain similarity between model and prototype

$$\frac{D_m}{H_m} = \frac{D}{H}$$

and

$$\frac{\rho_m V_m D_m}{\mu_m} = \frac{\rho V D}{\mu}$$

From the first similarity requirement

$$H_m = \frac{D_m}{D} H$$

$$= \frac{(20 \times 10^{-3} \text{ m})}{(0.1 \text{ m})} (0.3 \text{ m})$$

$$H_m = 60 \times 10^{-3} \text{ m} = 60 \text{ mm} \qquad \text{(Ans)}$$

The second similarity requirement indicates that the Reynolds number must be the same for model and prototype so that the model velocity must satisfy the condition

$$V_m = \frac{\mu_m}{\mu} \frac{\rho}{\rho_m} \frac{D}{D_m} V \qquad (1)$$

For air at standard conditions $\mu = 1.79 \times 10^{-5}$ kg/m·s, $\rho = 1.23$ kg/m³, and for water at 20 °C, $\mu = 1.00 \times 10^{-3}$ kg/m·s, $\rho = 998$ kg/m³. The fluid velocity for the prototype is

$$V = \frac{(50 \times 10^3 \text{ m/hr})}{(3600 \text{ s/hr})} = 13.9 \text{ m/s}$$

The required velocity can now be calculated from Eq. 1 as

$$V_m = \frac{[1.00 \times 10^{-3} \text{ kg/(m·s)}](1.23 \text{ kg/m}^3)(0.1 \text{ m})}{[1.79 \times 10^{-5} \text{ kg/(m·s)}](998 \text{ kg/m}^3)(20 \times 10^{-3} \text{ m})} (13.9 \text{ m/s})$$

$$V_m = 4.79 \text{ m/s} \qquad \text{(Ans)}$$

This is a reasonable velocity that could be readily achieved in a water tunnel.

With the two similarity requirements satisfied, it follows that the Strouhal numbers for prototype and model will be the same so that

$$\frac{\omega D}{V} = \frac{\omega_m D_m}{V_m}$$

and the predicted prototype vortex shedding frequency is

$$\omega = \frac{V}{V_m} \frac{D_m}{D} \omega_m$$

$$= \frac{(13.9 \text{ m/s})}{(4.79 \text{ m/s})} \frac{(20 \times 10^{-3} \text{ m})}{(0.1 \text{ m})} (49.9 \text{ Hz})$$

$$\omega = 29.0 \text{ Hz} \qquad \text{(Ans)}$$

This same model could also be used to predict the drag per unit length, $\mathcal{D}_\ell$, on the prototype, since the drag would depend on the same variables as those used for the frequency. Thus, the similarity requirements would be the same and with these requirements satisfied it follows that the drag per unit length expressed in dimensionless form, such as $\mathcal{D}_\ell/D\rho V^2$, would be equal in model and prototype. The measured drag on the model could then be related to the corresponding drag on the prototype through the relationship

$$\mathcal{D}_\ell = \left(\frac{D}{D_m}\right) \left(\frac{\rho}{\rho_m}\right) \left(\frac{V}{V_m}\right)^2 \mathcal{D}_{\ell m}$$

7.8.2 Model Scales

It is clear from the preceding section that the ratio of like quantities for the model and prototype naturally arise from the similarity requirements. For example, if in a given problem there are two length variables ℓ_1 and ℓ_2, the resulting similarity requirement based on a pi term obtained from these two variables is

$$\frac{\ell_1}{\ell_2} = \frac{\ell_{1m}}{\ell_{2m}}$$

so that

$$\frac{\ell_{1m}}{\ell_1} = \frac{\ell_{2m}}{\ell_2}$$

We define the ratio ℓ_{1m}/ℓ_1 or ℓ_{2m}/ℓ_2 as the *length scale*. For true models there will be only one length scale, and all lengths are fixed in accordance with this scale. There are, however, other scales such as the velocity scale, V_m/V, density scale, ρ_m/ρ, viscosity scale, μ_m/μ, and so on. In fact, we can define a scale for each of the variables in the problem. Thus, it is actually meaningless to talk about a "scale" of a model without specifying which scale.

We will designate the length scale as λ_ℓ, and other scales as λ_V, λ_ρ, λ_μ, and so on, where the subscript indicates the particular scale. Also, we will take the ratio of the model value to the prototype value as the scale (rather than the inverse). Length scales are often specified, for example, as $1:10$ or as a $\frac{1}{10}$-scale model. The meaning of this specification is that the model is one-tenth the size of the prototype, and the tacit assumption is that all relevant lengths are scaled accordingly so the model is geometrically similar to the prototype.

7.8.3   Practical Aspects of Using Models

Validation of Model Design   Most model studies involve simplifying assumptions with regard to the variables to be considered. Although the number of assumptions are frequently less stringent than those required for mathematical models, they nevertheless introduce some uncertainty in the model design. It is, therefore, desirable to check experimentally the design whenever possible. In some situations the purpose of the model is to predict the effects of certain proposed changes in a given prototype, and in this instance some actual prototype data may be available. The model can be designed, constructed, and tested, and the model prediction can be compared with these data. If the agreement is satisfactory, then the model can be changed in the desired manner, and the corresponding effect on the prototype can be predicted with increased confidence.

Another useful and informative procedure is to run tests with a series of models of different sizes, where one of the models can be thought of as the prototype and the others as "models" of this prototype. With the models designed and operated on the basis of the proposed design, a necessary condition for the validity of the model design is that an accurate prediction be made between any pair of models, since one can always be considered as a model of the other. Although suitable agreement in validation tests of this type does not unequivocally indicate a correct model design (e.g., the length scales between laboratory models may be significantly different than required for actual prototype prediction), it is certainly true that if agreement between models cannot be achieved in these tests, there is no reason to expect that the same model design can be used to correctly predict prototype behavior.

Distorted Models   Although the general idea behind establishing similarity requirements for models is straightforward (we simply equate pi terms), it is not always possible to satisfy all the known requirements. If one or more of the similarity requirements are not met, for example, if $\Pi_{2m} \neq \Pi_2$, then it follows that the prediction equation $\Pi_1 = \Pi_{1m}$ is not true; that is, $\Pi_1 \neq \Pi_{1m}$. Models for which one or more of the similarity requirements are not satisfied are called *distorted models*.

Distorted models are rather commonplace, and they can arise for a variety of reasons. For example, perhaps a suitable fluid cannot be found for the model. The classic example of a distorted model occurs in the study of open channel or free-

surface flows. Typically in these problems both the Reynolds number, $\rho V \ell / \mu$, and the Froude number, $V / \sqrt{g\ell}$, are involved.

Froude number similarity requires

$$\frac{V_m}{\sqrt{g_m \ell_m}} = \frac{V}{\sqrt{g\ell}}$$

If the model and prototype are operated in the same gravitational field, then the required velocity scale is

$$\frac{V_m}{V} = \sqrt{\frac{\ell_m}{\ell}} = \sqrt{\lambda_\ell}$$

Reynolds number similarity requires

$$\frac{\rho_m V_m \ell_m}{\mu_m} = \frac{\rho V \ell}{\mu}$$

and the velocity scale is

$$\frac{V_m}{V} = \frac{\mu_m}{\mu} \frac{\rho}{\rho_m} \frac{\ell}{\ell_m}$$

Since the velocity scale must be equal to the square root of the length scale, it follows that

$$\frac{\mu_m / \rho_m}{\mu / \rho} = \frac{\nu_m}{\nu} = (\lambda_\ell)^{3/2} \tag{7.15}$$

where the ratio μ / ρ is the kinematic viscosity, ν. Although, in principle it may be possible to satisfy this design condition, it may be quite difficult, if not impossible, to find a suitable model fluid, particularly for small length scales. For problems involving rivers, spillways, and harbors, for which the prototype fluid is water, the models are also relatively large so that the only practical model fluid is water. However, in this case (with the kinematic viscosity scale equal to unity) Eq. 7.15 will not be satisfied, and a distorted model will result. Generally, hydraulic models of this type are distorted and are designed on the basis of the Froude number, with the Reynolds number different in model and prototype.

Distorted models can be successfully used, but the interpretation of results obtained with this type of model is obviously more difficult than the interpretation of results obtained with *true models* for which all similarity requirements are met. There are no general rules for handling distorted models, and essentially each problem must be considered on its own merits. The success of using distorted models depends to a large extent on the skill and experience of the investigator responsible for the design of the model, and in the interpretation of experimental data obtained from the model. Distorted models are widely used, and additional information can be found in the references at the end of the chapter. References 14 and 15 contain detailed discussions of several practical examples of distorted fluid flow and hydraulic models.

7.9 SOME TYPICAL MODEL STUDIES

Models are used to investigate many different types of fluid mechanics problems, and it is difficult to characterize in a general way all necessary similarity requirements,

since each problem is unique. We can, however, broadly classify many of the problems on the basis of the general nature of the flow, and subsequently develop some general characteristics of model designs in each of these classifications. In the following sections we will consider models for the study of: (1) flow through closed conduits, (2) flow around immersed bodies, (3) flow with a free surface, and (4) flow through turbomachinery.

7.9.1 Flow Through Closed Conduits

Common examples of this type of flow include pipe flow and flow through valves, fittings, and metering devices. Although the conduits are often circular, they could have other shapes as well and may contain expansions or contractions. Since there are no fluid interfaces or free surfaces, the dominant forces are inertial and viscous so that the Reynolds number is an important similarity parameter. For low Mach numbers ($Ma < 0.3$), compressibility effects are negligible for both the flow of liquids or gases. For this class of problems, geometric similarity between model and prototype must be maintained. Generally the geometric characteristics can be described by a series of length terms, $\ell_1, \ell_2, \ell_3, \ldots, \ell_i$, and ℓ where ℓ is some particular length dimension for the system. Such a series of length terms leads to a set of pi terms of the form

$$\Pi_i = \frac{\ell_i}{\ell}$$

where $i = 1, 2, \ldots$, and so on. In addition to the basic geometry of the system, the roughness of the internal surfaces in contact with the fluid may be important. If the average height of surface roughness elements is defined as ε, then the pi term representing roughness would be ε/ℓ. This parameter indicates that for complete geometric similarity, surface roughness would also have to be scaled. Note that this implies that for length scales less than 1, the model surfaces should be smoother than those in the prototype since $\varepsilon_m = \lambda_\ell \varepsilon$. To further complicate matters, the pattern of roughness elements in model and prototype would have to be similar. These are conditions that are virtually impossible to satisfy exactly. Fortunately, in some problems the surface roughness plays a minor role and can be neglected. However, in other problems (such as turbulent flow through pipes) roughness can be very important.

It follows from this discussion that for flow in closed conducts at low Mach numbers, any dependent pi term (the one that contains the particular variable of interest, such as pressure drop) can be expressed as

$$\text{Dependent pi term} = \phi\left(\frac{\ell_i}{\ell}, \frac{\varepsilon}{\ell}, \frac{\rho V \ell}{\mu}\right) \tag{7.16}$$

This is a general formulation for this type of problem. The first two pi terms in the right side of Eq. 7.16 lead to the requirement of geometric similarity so that

$$\frac{\ell_{im}}{\ell_m} = \frac{\ell_i}{\ell} \qquad \frac{\varepsilon_m}{\ell_m} = \frac{\varepsilon}{\ell}$$

or

$$\frac{\ell_{im}}{\ell_i} = \frac{\varepsilon_m}{\varepsilon} = \frac{\ell_m}{\ell} = \lambda_\ell$$

This result indicates that the investigator is free to choose a length scale, λ_ℓ, but once this scale is selected, all other pertinent lengths must be scaled in the same ratio.

The additional similarity requirement arises from the equality of Reynolds numbers

$$\frac{\rho_m V_m \ell_m}{\mu_m} = \frac{\rho V \ell}{\mu}$$

From this condition the velocity scale is established so that

$$\frac{V_m}{V} = \frac{\mu_m}{\mu} \frac{\rho}{\rho_m} \frac{\ell}{\ell_m} \tag{7.17}$$

and the actual value of the velocity scale depends on the viscosity and density scales, as well as the length scale. Different fluids can be used in model and prototype. However, if the same fluid is used (with $\mu_m = \mu$ and $\rho_m = \rho$), then

$$\frac{V_m}{V} = \frac{\ell}{\ell_m}$$

Thus, $V_m = V/\lambda_\ell$, which indicates that the fluid velocity in the model will be larger than that in the prototype for any length scale less than 1. Since length scales are typically much less than unity, Reynolds number similarity may be difficult to achieve because of the large model velocities required.

With these similarity requirements satisfied, it follows that the dependent pi term will be equal in model and prototype. For example, if the dependent variable of interest is the pressure differential,[3] Δp, between two points along a closed conduit, then the dependent pi term could be expressed as

$$\Pi_1 = \frac{\Delta p}{\rho V^2}$$

The prototype pressure drop would then be obtained from the relationship

$$\Delta p = \frac{\rho}{\rho_m} \left(\frac{V}{V_m}\right)^2 \Delta p_m$$

so that from a measured pressure differential in the model, Δp_m, the corresponding pressure differential for the prototype could be predicted. Note that in general $\Delta p \neq \Delta p_m$.

---

[3] In some previous examples the pressure differential *per unit length*, Δp_ℓ, was used. This is appropriate for flow in long pipes or conduits in which the pressure would vary linearly with distance. However, in the more general situation the pressure may not vary linearly with position so that it is necessary to consider the pressure differential, Δp, as the dependent variable. In this case the distance between pressure taps is an additional variable (as well as the distance of one of the taps measured from some reference point within the flow system).

EXAMPLE 7.6

Model tests are to be performed to study the flow through a large valve having a 2-ft-diameter inlet and carrying water at a flowrate of 30 cfs. The working fluid in the model is water at the same temperature as that in the prototype. Complete geometric similarity exists between model and prototype, and the model inlet diameter is 3 in. Determine the required flowrate in the model.

SOLUTION

To ensure dynamic similarity, the model tests should be run so that

$$\text{Re}_m = \text{Re}$$

or

$$\frac{V_m D_m}{v_m} = \frac{VD}{v}$$

where V and D correspond to the inlet velocity and diameter, respectively. Since the same fluid is to be used in model and prototype, $v = v_m$, and therefore

$$\frac{V_m}{V} = \frac{D}{D_m}$$

The discharge, Q, is equal to VA where A is the inlet area, so

$$\frac{Q_m}{Q} = \frac{V_m A_m}{VA} = \left(\frac{D}{D_m}\right) \frac{[(\pi/4)D_m^2]}{[(\pi/4)D^2]}$$

$$= \frac{D_m}{D}$$

and for the data given

$$Q_m = \frac{(3/12 \text{ ft})}{(2 \text{ ft})} (30 \text{ ft}^3/\text{s})$$

$$Q_m = 3.75 \text{ cfs} \tag{Ans}$$

Although this is a large flowrate to be carried through a 3-in.-diameter pipe (the corresponding velocity is 76.4 ft/s), it could be attained in a laboratory facility. However, it is to be noted that if we tried to use a smaller model, say one with $D = 1$ in., the required model velocity is 229 ft/s, a very high velocity that would be difficult to achieve. These results are indicative of one of the difficulties encountered in maintaining Reynolds number similarity—the required model velocities may be impractical to obtain.

Two additional points should be made with regard to modeling flows in closed conduits. First, for large Reynolds numbers, inertial forces are much larger than viscous forces, and in this case it may be possible to neglect viscous effects. The important practical consequence of this is that it would not be necessary to maintain Reynolds number similarity between model and prototype. However, *both* model and prototype would have to operate at large Reynolds numbers. Since we do not know, a priori, what is a "large Reynolds number," the effect of Reynolds numbers would

have to be determined from the model. This could be accomplished by varying the model Reynolds number to determine the range (if any) over which the dependent pi term ceases to be affected by changes in Reynolds number.

The second point relates to the possibility of cavitation in flow through closed conduits. For example, flow through the complex passages that may exist in valves may lead to local regions of high velocity (and thus low pressure), which can cause the fluid to cavitate. If the model is to be used to study cavitation phenomena, then the vapor pressure, p_v, becomes an important variable and an additional similarity requirement such as equality of the cavitation number $(p_r - p_v)/\frac{1}{2}\rho V^2$ is required, where p_r is some reference pressure. The use of models to study cavitation is complicated, since it is not fully understood how vapor bubbles form and grow. The initiation of bubbles seems to be influenced by the microscopic particles that exist in most liquids, and how this aspect of the problem influences model studies is not clear. Additional details can be found in Ref. 17.

7.9.2 Flow Around Immersed Bodies

Models have been widely used to study the flow characteristics associated with bodies that are completely immersed in a moving fluid. Examples include flow around aircraft, automobiles, golf balls, and buildings. (These types of models are usually tested in wind tunnels as is illustrated in Fig. 7.6.) Modeling laws for these problems are similar to those described in the preceding section; that is, geometric and Reynolds number similarity is required. Since there are no fluid interfaces, surface tension (and therefore the Weber number) is not important. Also, gravity will not affect the flow patterns so the Froude number need not be considered. The Mach number will be important for high-speed flows in which compressibility becomes an important factor, but for incompressible fluids (such as liquids or for gases at relatively low speeds) the Mach number can be omitted as a similarity requirement. In this case, a general formulation for these problems is

$$\text{Dependent pi term} = \phi \left(\frac{\ell_i}{\ell}, \frac{\varepsilon}{\ell}, \frac{\rho V \ell}{\mu} \right) \tag{7.18}$$

where ℓ is some characteristic length of the system and ℓ_i represents other pertinent lengths, ε/ℓ is the relative roughness of the surface (or surfaces), and $\rho V \ell/\mu$ is the Reynolds number.

Frequently, the dependent variable of interest for this type of problem is the drag, $\mathcal{D}$, developed on the body, and in this situation the dependent pi term would usually be expressed in the form of a *drag coefficient*, C_D, where

$$C_D = \frac{\mathcal{D}}{\frac{1}{2}\rho V^2 \ell^2}$$

FIGURE 7.6 Model of the National Bank of Commerce, San Antonio, Texas, for measurement of peak, rms, and mean pressure distributions. The model is located in a long-test-section, meteorological wind tunnel. (Photograph courtesy of Cermak Peterka Petersen, Inc.)

The numerical factor, $\frac{1}{2}$, is arbitrary but commonly included, and ℓ^2 is usually taken as some representative area of the object. Thus, drag studies can be undertaken with the formulation:

$$\frac{\mathcal{D}}{\frac{1}{2}\rho V^2 \ell^2} = C_D = \phi \left(\frac{\ell_i}{\ell}, \frac{\varepsilon}{\ell}, \frac{\rho V \ell}{\mu} \right) \tag{7.19}$$

It is clear from Eq. 7.19 that geometric similarity

$$\frac{\ell_{im}}{\ell_m} = \frac{\ell_i}{\ell} \qquad \frac{\varepsilon_m}{\ell_m} = \frac{\varepsilon}{\ell}$$

as well as Reynolds number similarity

$$\frac{\rho_m V_m \ell_m}{\mu_m} = \frac{\rho V \ell}{\mu}$$

must be maintained. If these conditions are met, then

$$\frac{\mathcal{D}}{\frac{1}{2}\rho V^2 \ell^2} = \frac{\mathcal{D}_m}{\frac{1}{2}\rho_m V_m^2 \ell_m^2}$$

or

$$\mathcal{D} = \frac{\rho}{\rho_m} \left(\frac{V}{V_m} \right)^2 \left(\frac{\ell}{\ell_m} \right)^2 \mathcal{D}_m$$

Measurements of model drag, $\mathcal{D}_m$, can then be used to predict the corresponding drag, $\mathcal{D}$, on the prototype from this relationship.

As was discussed in the previous section, one of the common difficulties with models is related to the Reynolds number similarity requirement which establishes the model velocity as

$$V_m = \frac{\mu_m}{\mu} \frac{\rho}{\rho_m} \frac{\ell}{\ell_m} V \tag{7.20}$$

or

$$V_m = \frac{\nu_m}{\nu} \frac{\ell}{\ell_m} V \tag{7.21}$$

where ν_m/ν is the ratio of kinematic viscosities. If the same fluid is used for model and prototype so that $\nu_m = \nu$, then

$$V_m = \frac{\ell}{\ell_m} V$$

and, therefore, the required model velocity will be higher than the prototype velocity for ℓ/ℓ_m greater than 1. Since this ratio is often relatively large, the required value of V_m may be large. For example, for a $\frac{1}{10}$-length scale, and a prototype velocity of 50 mph, the required model velocity is 500 mph. This is a value that is unreasonably high to achieve with liquids, and for gas flows this would be in the range where compressibility would be important in the model (but not in the prototype).

As an alternative, we see from Eq. 7.19 that V_m could be reduced by using a different fluid in the model such that $\nu_m/\nu < 1$. For example, the ratio of the kinematic viscosity of water to that of air is approximately $\frac{1}{10}$, so that if the prototype fluid were air, tests might be run on the model using water. This would reduce the required model velocity, but it still may be difficult to achieve the necessary velocity in a suitable test facility, such as a water tunnel.

Another possibility for wind tunnel tests would be to increase the air pressure in the tunnel so that $\rho_m > \rho$, thus reducing the required model velocity as specified by Eq. 7.20. Fluid viscosity is not strongly influenced by pressure. Although pressurized tunnels have been used, they are obviously more complicated and expensive.

The required model velocity can also be reduced if the length scale is modest; that is, the model is relatively large. For wind tunnel testing, this requires a large test section which greatly increases the cost of the facility. However, large wind tunnels suitable for testing very large models (or prototypes) are in use. One such tunnel, located at the NASA Ames Research Center, Moffett Field, California, has a test section that is 40 ft by 80 ft and can accommodate test speeds to 345 mph. Such a large and expensive test facility is obviously not feasible for university or industrial laboratories, so most model testing has to be accomplished with relatively small models.

EXAMPLE 7.7

The drag on an airplane cruising at 240 mph in standard air is to be determined from tests on a 1:10 scale model placed in a pressurized wind tunnel. To minimize compressibility effects, the air speed in the wind tunnel is to also be 240 mph. Determine the required air pressure in the tunnel (assuming the same air temperature for model and prototype), and the drag on the prototype corresponding to a measured force of 1 lb on the model.

SOLUTION

From Eq. 7.19 it follows that drag can be predicted from a geometrically similar model if the Reynolds numbers in model and prototype are the same. Thus,

$$\frac{\rho_m V_m \ell_m}{\mu_m} = \frac{\rho V \ell}{\mu}$$

For this example, $V_m = V$ and $\ell_m / \ell = \frac{1}{10}$ so that

$$\frac{\rho_m}{\rho} = \frac{\mu_m}{\mu} \frac{V}{V_m} \frac{\ell}{\ell_m}$$

$$= \frac{\mu_m}{\mu} (1)(10)$$

and therefore

$$\frac{\rho_m}{\rho} = 10 \frac{\mu_m}{\mu}$$

This result shows that the same fluid with $\rho_m = \rho$ and $\mu_m = \mu$ cannot be used if Reynolds number similarity is to be maintained. One possibility is to pressurize the wind tunnel to increase the density of the air. We assume that an increase in pressure does not significantly change the viscosity so that the required increase in density is given by the relationship:

$$\frac{\rho_m}{\rho} = 10$$

For an ideal gas, $p = \rho RT$ so that

$$\frac{p_m}{p} = \frac{\rho_m}{\rho}$$

for constant temperature ($T = T_m$). Therefore, the wind tunnel would need to be pressurized so that

$$\frac{p_m}{p} = 10$$

Since the prototype operates at standard atmospheric pressure, the required pressure in the wind tunnel is 10 atmospheres or

$$p_m = 10 \ (14.7 \ \text{psia})$$
$$= 147 \ \text{psia}$$
(Ans)

Thus, we see that a high pressure would be required and this could not be easily or inexpensively achieved. However, under these conditions Reynolds similarity would be attained and the drag could be obtained from Eq. 7.19 so that

$$\frac{\mathcal{D}}{\frac{1}{2}\rho V^2 \ell^2} = \frac{\mathcal{D}_m}{\frac{1}{2}\rho_m V_m^2 \ell_m^2}$$

and

$$\mathcal{D} = \frac{\rho}{\rho_m} \left(\frac{V}{V_m}\right)^2 \left(\frac{\ell}{\ell_m}\right)^2 \mathcal{D}_m$$

$$= \left(\frac{1}{10}\right) (1)^2 (10)^2 \mathcal{D}_m$$

$$= 10 \mathcal{D}_m$$

Thus, for a drag of 1 lb on the model the corresponding drag on the prototype is

$$\mathcal{D} = 10 \ \text{lb}$$
(Ans)

Fortunately, in many situations the flow characteristics are not strongly influenced by the Reynolds number over the operating range of interest. In this case we can avoid the rather stringent similarity requirement of matching Reynolds numbers. To illustrate this point, consider the variation in the drag coefficient with the Reynolds number for a smooth sphere of diameter, d, placed in a uniform stream with approach velocity, V. Some typical data are shown in Fig. 7.7. We observe that for Reynolds numbers between approximately 10^3 and 2×10^5 the drag coefficient is relatively constant and does not depend on the specific value of the Reynolds number. Thus, exact Reynolds number similarity is not required in this range. For other geometric shapes we would typically find that for high Reynolds numbers, inertial forces are dominant (rather than viscous forces), and the drag is essentially independent of the Reynolds number.

Another interesting point to note from Fig. 7.7 is the rather abrupt drop in the drag coefficient near a Reynolds number of 3×10^5. As is discussed in Section 9.3.3, this is due to a change in the flow conditions near the surface of the sphere. These changes are influenced by the surface roughness and, in fact, the drag coefficient for a sphere with a "rougher" surface will generally be less than that of the smooth sphere for high Reynolds number. For example, the dimples on a golf ball are used to reduce the drag over that which would occur for a smooth golf ball. Although this is undoubtedly of great interest to the avid golfer, it is also important to engineers responsible for fluid-flow models, since it does emphasize the potential importance of the surface roughness. However, for bodies that are sufficiently angular with sharp

FIGURE 7.7 The effect of Reynolds number on the drag coefficient, C_D, for a sphere with $C_D = \mathcal{D}/\frac{1}{2}A\rho V^2$, where A is the projected area of sphere, $\pi d^2/4$. (Data from Ref. 16, used by permission.)

corners, the actual surface roughness is likely to play a secondary role compared with the main geometric features of the body.

For problems involving high velocities in which the Mach number is greater than about 0.3, the influence of compressibility, and therefore the Mach number (or Cauchy number), becomes significant. In this case complete similarity requires not only geometric and Reynolds number similarity but also Mach number similarity so that

$$\frac{V_m}{c_m} = \frac{V}{c} \tag{7.22}$$

This similarity requirement, when combined with that for Reynolds number similarity (Eq. 7.21), yields

$$\frac{c}{c_m} = \frac{v}{v_m}\frac{\ell_m}{\ell} \tag{7.23}$$

Clearly the same fluid with $c = c_m$ and $v = v_m$ cannot be used in model and prototype unless the length scale is unity (which means that we are running tests on the prototype). In high-speed aerodynamics the prototype fluid is usually air, and it is difficult to satisfy Eq. 7.23 for reasonable length scales. Thus, models involving high-speed flows are often distorted with respect to Reynolds number similarity, but Mach number similarity is maintained.

7.9.3 Flow with a Free Surface

Flow in canals, rivers, spillways, and stilling basins, as well as flow around ships, are all examples of flow phenomena involving a free surface. For this class of problems, both gravitational and inertial forces are important and, therefore, the Froude number becomes an important similarity parameter. Also, since there is a free surface with a

liquid–air interface, forces due to surface tension may be significant, and the Weber number becomes another similarity parameter that needs to be considered along with the Reynolds number. Geometric variables will obviously still be important. Thus, a general formulation for problems involving flow with a free surface can be expressed as

$$\text{Dependent pi term} = \phi \left(\frac{\ell_i}{\ell}, \frac{\varepsilon}{\ell}, \frac{\rho V \ell}{\mu}, \frac{V}{\sqrt{g\ell}}, \frac{\rho V^2 \ell}{\sigma} \right) \tag{7.24}$$

As discussed previously, ℓ is some characteristic length of the system, ℓ_i represents other pertinent lengths, and ε/ℓ is the relative roughness of the various surfaces. Since gravity is the driving force in these problems, Froude number similarity is definitely required so that

$$\frac{V_m}{\sqrt{g_m \ell_m}} = \frac{V}{\sqrt{g\ell}}$$

The model and prototype are expected to operate in the same gravitational field ($g_m = g$), and therefore it follows that

$$\frac{V_m}{V} = \sqrt{\frac{\ell_m}{\ell}} = \sqrt{\lambda_\ell} \tag{7.25}$$

Thus, when models are designed on the basis of Froude number similarity, the velocity scale is determined by the square root of the length scale. As is discussed in Section 7.8.3, to simultaneously have Reynolds and Froude number similarity it is necessary that the kinematic viscosity scale be related to the length scale as

$$\frac{\nu_m}{\nu} = (\lambda_\ell)^{3/2} \tag{7.26}$$

The working fluid for the prototype is normally either freshwater or seawater and the length scale is small. Under these circumstances it is virtually impossible to satisfy Eq. 7.26, so models involving free-surface flows are usually distorted. The problem is further complicated if an attempt is made to model surface tension effects, since this requires equality of Weber numbers, which leads to the condition

$$\frac{\sigma_m/\rho_m}{\sigma/\rho} = (\lambda_\ell)^2 \tag{7.27}$$

for the kinematic surface tension (σ/ρ). It is again evident that the same fluid cannot be used in model and prototype if we are to have similitude with respect to surface tension effects for $\lambda_\ell \neq 1$.

Fortunately, in many problems involving free-surface flows both surface tension and viscous effects are small and consequently strict adherence to Weber and Reynolds number similarity is not required. Certainly, surface tension is not important in large hydraulic structures and rivers. Our only concern would be if in a model the depths were reduced to the point where surface tension becomes an important factor, whereas it is not in the prototype. This is of particular importance in the design of river models, since the length scales are typically small (so that the width of the model is reasonable), but with a small length scale the required model depth may be very small. To overcome this problem, different horizontal and vertical length scales are often used for river models. Although this approach eliminates surface tension effects in the model, it introduces geometric distortion that must be accounted for empirically, usually by

FIGURE 7.8 A scale hydraulic model (1:197) of the Guri Dam in Venezuela which is used to simulate the characteristics of the flow over and below the spillway and the erosion below the spillway. (Photograph courtesy of St. Anthony Falls Hydraulic Laboratory.)

increasing the model surface roughness. It is important in these circumstances that verification tests with the model be performed (if possible) in which model data are compared with available prototype river flow data. Model roughness can be adjusted to give satisfactory agreement between model and prototype, and then the model subsequently used to predict the effect of proposed changes on river characteristics (such as velocity patterns or surface elevations).

For large hydraulic structures, such as dam spillways, the Reynolds numbers are large so that viscous forces are small in comparison to the forces due to gravity and inertia. In this case Reynolds number similarity is not maintained and models are designed on the basis of Froude number similarity. Care must be taken to ensure that the model Reynolds numbers are also large, but they are not required to be equal to those of the prototype. This type of hydraulic model is usually made as large as possible so that the Reynolds number will be large. A spillway model is shown in Fig. 7.8. Also, for relatively large models the geometric features of the prototype can be accurately scaled, as well as surface roughness. Note that $\varepsilon_m = \lambda_\ell \varepsilon$, which indicates that the model surfaces must be smoother than the corresponding prototype surfaces for $\lambda_\ell < 1$.

EXAMPLE 7.8

A certain spillway for a dam is 20 m wide and is designed to carry 125 m³/s at flood stage. A 1:15 model is constructed to study the flow characteristics through the spillway. Determine the required model width and flowrate. What operating time for the model corresponds to a 24-h period in the prototype? The effects of surface tension and viscosity are to be neglected.

SOLUTION

The width, w_m, of the model spillway is obtained from the length scale, λ_ℓ, so that

$$\frac{w_m}{w} = \lambda_\ell$$

$$= \frac{1}{15}$$

and

$$w_m = \frac{20 \text{ m}}{15} = 1.33 \text{ m} \qquad \text{(Ans)}$$

Of course, all other geometric features (including surface roughness) of the spillway must be scaled in accordance with the same length scale.

With the neglect of surface tension and viscosity, Eq. 7.24 indicates that dynamic similarity will be achieved if the Froude numbers are equal between model and prototype. Thus,

$$\frac{V_m}{\sqrt{g_m \ell_m}} = \frac{V}{\sqrt{g \ell}}$$

and for $g_m = g$

$$\frac{V_m}{V} = \sqrt{\frac{\ell_m}{\ell}}$$

Since the flowrate is given by $Q = VA$, where A is an appropriate cross-sectional area, it follows that

$$\frac{Q_m}{Q} = \frac{V_m A_m}{VA} = \sqrt{\frac{\ell_m}{\ell}} \left(\frac{\ell_m}{\ell}\right)^2$$

$$= (\lambda_\ell)^{5/2}$$

where we have made use of the relationship $A_m/A = (\ell_m/\ell)^2$. For $\lambda_\ell = \frac{1}{15}$ and $Q = 125$ m³/s

$$Q_m = (\tfrac{1}{15})^{5/2}\ (125 \text{ m}^3/\text{s}) = 0.143 \text{ m}^3/\text{s} \tag{Ans}$$

The time scale can be obtained from the velocity scale, since the velocity is distance divided by time ($V = \ell/t$), and therefore

$$\frac{V}{V_m} = \frac{\ell}{t}\frac{t_m}{\ell_m}$$

or

$$\frac{t_m}{t} = \frac{V}{V_m}\frac{\ell_m}{\ell} = \sqrt{\frac{\ell_m}{\ell}} = \sqrt{\lambda_\ell}$$

This result indicates that time intervals in the model will be smaller than the corresponding intervals in the prototype if $\lambda_\ell < 1$. For $\lambda_\ell = \frac{1}{15}$ and a prototype time interval of 24 h

$$t_m = \sqrt{\tfrac{1}{15}}\ (24 \text{ hr}) = 6.20 \text{ hr} \tag{Ans}$$

The ability to scale times may be very useful, since it is possible to "speed up" events in the model which may occur over a relatively long time in the prototype.

There are, unfortunately, problems involving flow with a free surface in which viscous, inertial, and gravitational forces are all important. The drag on a ship as it moves through water is due to the viscous shearing stresses that develop along its hull, as well as a pressure-induced component of drag caused by both the shape of the hull and wave action. The shear drag is a function of the Reynolds number, whereas the pressure drag is a function of the Froude number. Since both Reynolds number and Froude number similarity cannot be simultaneously achieved by using water as the model fluid (which is the only practical fluid for ship models), some technique other than a straightforward model test must be employed. One common approach is to measure the total drag on a small, geometrically similar model as it is towed through a model basin at Froude numbers matching those of the prototype. The shear drag on the model is calculated using analytical techniques of the type described in Chapter 9. This calculated value is then subtracted from the total drag to obtain pressure drag, and using Froude number scaling the pressure drag on the prototype can then be predicted. The experimentally determined value can then be

combined with a calculated value of the shear drag (again using analytical techniques) to provide the desired total drag on the ship. Ship models are widely used to study new designs, but the tests require extensive facilities (see Fig. 7.9).

It is clear from this brief discussion of various types of models involving free-surface flows that the design and use of such models requires considerable ingenuity, as well as a good understanding of the physical phenomenon involved. This is generally true for most model studies. Modeling is both an art and a science. Motion picture producers make extensive use of model ships, fires, explosions, and the like. It is interesting to attempt to observe the flow differences between these distorted model flows and the real thing.

7.9.4   Turbomachinery

Mechanical devices are commonly used in fluid-flow systems to add or extract energy from the system. Pumps add energy, whereas turbines extract energy from the fluid. Frequently, these devices involve a rotor with blades attached and are referred to as *turbomachines*. The analytical prediction of the performance characteristics of turbomachines is complex, and often the characteristics are determined experimentally with the aid of models. For the purpose of this brief introduction to similitude considerations related to turbomachinery, we restrict the discussion to pumps. However, a similar approach can be made for other types of turbomachines.

For a pump the principal dependent variables of interest are the actual head rise, h_a, the shaft power, $\mathscr{P}$, and the efficiency, η. Note that $\mathscr{P} = \dot{W}_{\text{shaft net in}}$ of Chapter 5. The power actually delivered to the fluid is equal to the product of the specific weight of the fluid, the flowrate, and the head, so that

$$\mathscr{P}_a = \gamma Q h_a \tag{7.28}$$

The efficiency of the pump is equal to the ratio of the power actually delivered to the fluid, as given by Eq. 7.28, to the shaft power involved (sometimes called brake horsepower). We expect that these variables will depend on the geometrical configuration, which can be represented by some characteristic diameter, D, other pertinent lengths, ℓ_i, and surface roughness, ε. In addition, the other important variables are flowrate, Q, the rotational speed, ω, of the pump shaft, fluid viscosity, μ, and fluid density, ρ. We will only consider incompressible fluids so compressibility effects need not concern us. Thus, any one of the dependent variables can be expressed as

$$\text{Dependent variable} = f(D, \ell_i, \varepsilon, Q, \omega, \mu, \rho)$$

FIGURE 7.9 Instrumented, small-waterplane-area, twin hull (SWATH) model suspended from a towing carriage. (Photograph courtesy of the U.S. Navy's David W. Taylor Research Center.)

and a straightforward application of dimensional analysis leads to

$$\text{Dependent pi term} = \phi\left(\frac{\ell_i}{D}, \frac{\varepsilon}{D}, \frac{Q}{\omega D^3}, \frac{\rho \omega D^2}{\mu}\right) \qquad (7.29)$$

The dependent pi term involving the head is usually expressed as $gh_a/\omega^2 D^2$ where gh_a is the head rise in terms of energy per unit mass, rather than simply h_a, which is energy per unit weight. This dimensionless parameter is called the *head coefficient*, C_H. The dependent pi term involving the shaft power[4] is expressed as $\mathcal{P}/\rho\omega^3 D^5$, and this standard dimensionless parameter is termed the *power coefficient, $C_\mathcal{P}$*. The rotational speed, ω, which appears in these dimensionless groups, would normally be expressed in rev/s. The final dependent pi term is the efficiency, η, which is, of course, dimensionless. Thus, in terms of dimensionless parameters the performance characteristics are expressed as

$$C_H = \frac{gh_a}{\omega^2 D^2} = \phi_1\left(\frac{\ell_i}{D}, \frac{\varepsilon}{D}, \frac{Q}{\omega D^3}, \frac{\rho \omega D^2}{\mu}\right)$$

$$C_\mathcal{P} = \frac{\mathcal{P}}{\rho\omega^3 D^5} = \phi_2\left(\frac{\ell_i}{D}, \frac{\varepsilon}{D}, \frac{Q}{\omega D^3}, \frac{\rho \omega D^2}{\mu}\right)$$

$$\eta = \phi_3\left(\frac{\ell_i}{D}, \frac{\varepsilon}{D}, \frac{Q}{\omega D^3}, \frac{\rho \omega D^2}{\mu}\right)$$

The last pi term in each of the above equations is a form of Reynolds number that represents the relative influence of viscous effects. Usually, flow in turbomachines is at high Reynolds numbers, and experience has shown that the effect of Reynolds number can be neglected. The relative roughness, ε/D, is also neglected, since the highly irregular shape of the pump chamber is usually the dominant geometric factor. Thus, with these simplifications and for *geometrically similar* pumps (all pertinent dimensions, ℓ_i, scaled by a common length scale) the dependent pi terms are functions of only $Q/\omega D^3$, so that

$$\frac{gh_a}{\omega^2 D^2} = \phi_1\left(\frac{Q}{\omega D^3}\right) \qquad (7.30)$$

$$\frac{\mathcal{P}}{\rho\omega^3 D^5} = \phi_2\left(\frac{Q}{\omega D^3}\right) \qquad (7.31)$$

$$\eta = \phi_3\left(\frac{Q}{\omega D^3}\right) \qquad (7.32)$$

The dimensionless parameter $C_Q = Q/\omega D^3$ is called the *flow coefficient*. These three equations provide the desired similarity relationships among a family of geometrically similar pumps. If two pumps from the family are operated at the same value of flow coefficient

$$\left(\frac{Q}{\omega D^3}\right)_1 = \left(\frac{Q}{\omega D^3}\right)_2$$

---

[4] The power appearing in this dimensionless parameter is commonly based on the shaft (brake) horsepower, bhp, so that in BG units, $\mathcal{P} = 550 \times (\text{bhp})$.

it then follows that

$$\left(\frac{gh_a}{\omega^2 D^2}\right)_1 = \left(\frac{gh_a}{\omega^2 D^2}\right)_2$$

$$\left(\frac{\mathcal{P}}{\rho\omega^3 D^5}\right)_1 = \left(\frac{\mathcal{P}}{\rho\omega^3 D^5}\right)_2$$

$$\eta_1 = \eta_2$$

where the subscripts 1 and 2 refer to any two pumps from the family of geometrically similar pumps.

With these so-called "pump scaling laws" it is possible to experimentally determine the performance characteristics of one pump in the laboratory, and then use these data to predict the corresponding characteristics for other pumps within the family under different operating conditions. Figure 7.10a shows some typical curves obtained for a centrifugal pump. Figure 7.10b shows the results plotted in terms of the dimensionless coefficients, C_Q, C_H, $C_{\mathcal{P}}$, and η. From these curves the performance of different sized, geometrically similar pumps can be predicted, as well as the effect of changing speeds on the performance of the pump from which the curves were obtained.

From Eq. 7.32 it is noted that when a pump operates at a constant efficiency, the value of the flow coefficient is also a constant. Since the head coefficient is also a function of the flow coefficient, it follows that for constant efficiency

$$\frac{Q}{\omega D^3} = C_1$$

and

$$\frac{gh_a}{\omega^2 D^2} = C_2$$

where C_1 and C_2 are constants. Based on these two relationships, an additional useful

FIGURE 7.10 Typical performance data for a centrifugal pump. (a) Characteristic curves for a 12-in. centrifugal pump operating at 1000 rpm. (b) Dimensionless characteristic curves. (Data from Ref. 18, used by permission.)

relationship can be obtained by eliminating D between the two. This can be accomplished by raising the flow coefficient to an appropriate exponent, m, and dividing by the head coefficient raised to an appropriate exponent so that D will be eliminated. Thus,

$$\frac{(Q/\omega D^3)^m}{(gh_a/\omega^2 D^2)^{3m/2}} = \frac{C_1^m}{C_2^{3m/2}}$$

or

$$\left[\frac{Q\omega^2}{(gh_a)^{3/2}}\right]^m = \left(\frac{C_1}{C_2^{3/2}}\right)^m$$

It is conventional to let $m = \frac{1}{2}$ so that

$$\frac{\omega\sqrt{Q}}{(gh_a)^{3/4}} = \sqrt{\frac{C_1}{C_2^{3/2}}} = N_s \tag{7.33}$$

where this dimensionless parameter is called the *specific speed*, N_s. The value of the specific speed actually depends on the efficiency (since C_1 and C_2 depend on efficiency), but it is customary to specify values of specific speed at peak efficiency. For pumps with low Q and high h_a the specific speed is low compared with a pump with high Q and low h_a. Centrifugal pumps typically are low-capacity, high-head pumps, and therefore have low specific speeds. Conversely, axial flow pumps are high-capacity, low-head pumps, and have large values of specific speed. The specific speed is particularly useful to the engineer or designer, since if the required head, flowrate, and speed are specified, it is possible to select an appropriate type of pump based on the value of the specific speed.

Specific speed as defined by Eq. 7.33 is dimensionless, and therefore independent of the system of units used in its evaluation so long as a consistent system is used. However, in the United States a modified, dimensional form of specific speed, N_{sd}, is commonly used, where

$$N_{sd} = \frac{\omega \text{ (rpm)}\sqrt{Q \text{ (gpm)}}}{[h_a(\text{ft})]^{3/4}} \tag{7.34}$$

Typical values of N_{sd} are in the range $500 < N_{sd} < 4000$ for centrifugal pumps and $N_{sd} > 10,000$ for axial flow pumps. Both N_s and N_{sd} have the same physical meaning, but their magnitudes will differ by a constant conversion factor ($N_{sd} = 17,200\, N_s$).

EXAMPLE 7.9

An 8-in.-diameter centrifugal pump operating at 1200 rpm is geometrically similar to the 12-in.-diameter pump having the performance characteristics of Figs. 7.10a and 7.10b while operating at 1000 rpm. (a) For peak efficiency, predict the discharge, actual head rise, and shaft horsepower for this smaller pump. The working fluid is water at 60 °F. (b) What is the value of the specific speed, N_{sd}, for both the 12-in. pump of Fig. 7.10 and the 8-in. pump?

SOLUTION

(a)   As is indicated by Eq. 7.32, for a given efficiency the flow coefficient has the same value for a given family of pumps. From Fig. 7.10b we see that at peak efficiency $C_Q = 0.40$.

Thus, for the 8-in. pump

$$Q = C_Q \omega D^3 = (0.40) \left(\frac{1200}{60} \text{ rev/s} \right) \left(\frac{8}{12} \text{ ft} \right)^3$$

$$Q = 2.37 \text{ ft}^3/\text{s} \qquad \text{(Ans)}$$

or in terms of gpm

$$Q = (2.37 \text{ ft}^3/\text{s})(7.48 \text{ gal/ft}^3)(60 \text{ s/min}) = 1060 \text{ gpm} \qquad \text{(Ans)}$$

The actual head rise and the shaft horsepower can be determined in a similar manner since at peak efficiency $C_H = 7.7$ and $C_\wp = 3.6$, so that

$$h_a = C_H \omega^2 D^2 / g = \frac{(7.7)(1200/60 \text{ rev/s})^2 (8/12 \text{ ft})^2}{32.2 \text{ ft/s}^2}$$

$$h_a = 42.5 \text{ ft} \qquad \text{(Ans)}$$

and

$$\wp = C_\wp \rho \omega^3 D^5 = (3.6) \, (1.94 \text{ slugs/ft}^3) \left(\frac{1200}{60} \text{ rev/s} \right)^3 \left(\frac{8}{12} \text{ ft} \right)^5$$

$$= 7360 \text{ ft·lb/s}$$

$$\wp = \frac{7360 \text{ ft·lb/s}}{550 \text{ ft·lb/s/hp}} = 13.4 \text{ hp} \qquad \text{(Ans)}$$

This last result gives the shaft horsepower which is the power supplied to the pump shaft. The power actually added to the fluid is equal to $\gamma Q h_a$, which in this example is

$$\gamma Q h_a = (62.4 \text{ lb/ft}^3)(2.37 \text{ ft}^3/\text{s})(42.5 \text{ ft})$$

$$= 6290 \text{ ft·lb/s}$$

Thus, the efficiency, η, is

$$\eta = \frac{6290}{7360} = 85\%$$

which checks with the efficiency curve of Fig. 7.10b.

(b)   From Eq. 7.34

$$N_{sd} = \frac{\omega \sqrt{Q}}{(h_a)^{3/4}}$$

where ω is in rpm, Q in gpm, and h_a in feet. For the 12-in. pump, $\omega = 1000$ rpm, and from Fig. 7.10a, $Q = 3000$ gpm and $h_a = 66.5$ ft at maximum efficiency. Thus,

$$\text{(for 12-in. pump)} \qquad N_{sd} = \frac{(1000 \text{ rpm})\sqrt{3000 \text{ gpm}}}{(66.5 \text{ ft})^{3/4}} = 2350 \qquad \text{(Ans)}$$

For the 8-in. pump, $\omega = 1200$ rpm, and from part (a), $Q = 1060$ gpm and $h_a = 42.5$ ft so that

$$\text{(for 8-in. pump)} \qquad N_{sd} = \frac{(1200 \text{ rpm})\sqrt{1060 \text{ gpm}}}{(42.5 \text{ ft})^{3/4}} = 2350 \qquad \text{(Ans)}$$

The two specific speeds are the same, as expected, since the specific speed has a fixed value (at a fixed efficiency) for a given family of pumps.

7.10   SIMILITUDE BASED ON GOVERNING DIFFERENTIAL EQUATIONS

In the preceding sections of this chapter, dimensional analysis has been used to obtain similarity laws. This is a simple, straightforward approach to modeling, which is widely used. The use of dimensional analysis requires only a knowledge of the variables that influence the phenomenon of interest. Although the simplicity of this approach is attractive, it must be recognized that omission of one or more important variables may lead to serious errors in the model design. An alternative approach is available if the equations (usually differential equations) governing the phenomenon are known. In this situation similarity laws can be developed from the governing equations, even though it may not be possible to obtain analytic solutions to the equations.

To illustrate the procedure, consider the flow of an incompressible, Newtonian fluid. For simplicity we will restrict our attention to two-dimensional flow, although the results are applicable to the general three-dimensional case. From Chapter 6 we know that the governing equations are the continuity equation

$$\frac{\partial u}{\partial x} + \frac{\partial v}{\partial y} = 0 \tag{7.35}$$

and the Navier–Stokes equations

$$\rho \left(\frac{\partial u}{\partial t} + u \frac{\partial u}{\partial x} + v \frac{\partial u}{\partial y} \right) = -\frac{\partial p}{\partial x} + \mu \left(\frac{\partial^2 u}{\partial x^2} + \frac{\partial^2 u}{\partial y} \right) \tag{7.36}$$

$$\rho \left(\frac{\partial v}{\partial t} + u \frac{\partial v}{\partial x} + v \frac{\partial v}{\partial y} \right) = -\frac{\partial p}{\partial y} - \rho g + \mu \left(\frac{\partial^2 v}{\partial x^2} + \frac{\partial^2 v}{\partial y^2} \right) \tag{7.37}$$

where the y axis is vertical, so that the gravitational body force, ρg, only appears in the "y equation." To continue the mathematical description of the problem, boundary conditions are required. For example, velocities on all boundaries may be specified; that is, $u = u_B$ and $v = v_B$ at all boundary points $x = x_B$ and $y = y_B$. In some types of problems it may be necessary to specify the pressure over some part of the boundary. For time-dependent problems, initial conditions would also have to be provided, which means that the values of all dependent variables would be given at some time (usually taken at $t = 0$).

Once the governing equations, including boundary and initial conditions, are known we are ready to proceed to develop similarity requirements. The next step is to define a new set of variables that are dimensionless. To do this we select a reference quantity for each type of variable. In this problem the variables are u, v, p, x, y, and t so we will need a reference velocity, V, a reference pressure, p_0, a reference length, ℓ, and a reference time, τ. These reference quantities should be parameters that appear in the problem. For example, ℓ may be a characteristic length of a body immersed in a fluid or the width of a channel through which a fluid is flowing. The velocity, V, may be the free-stream velocity or the inlet velocity. The new dimensionless (starred) variables can be expressed as

$$u^* = \frac{u}{V} \qquad v^* = \frac{v}{V} \qquad p^* = \frac{p}{p_0}$$

$$x^* = \frac{x}{\ell} \qquad y^* = \frac{y}{\ell} \qquad t^* = \frac{t}{\tau}$$

The governing equations can now be rewritten in terms of these new variables. For example,

$$\frac{\partial u}{\partial x} = \frac{\partial V u^*}{\partial x^*}\frac{\partial x^*}{\partial x} = \frac{V}{\ell}\frac{\partial u^*}{\partial x^*}$$

and

$$\frac{\partial^2 u}{\partial x^2} = \frac{V}{\ell}\frac{\partial}{\partial x^*}\left(\frac{\partial u^*}{\partial x^*}\right)\frac{\partial x^*}{\partial x} = \frac{V}{\ell^2}\frac{\partial^2 u^*}{\partial x^{*2}}$$

The other terms that appear in the equations can be expressed in a similar fashion. Thus, in terms of the new variables the governing equations become

$$\frac{\partial u^*}{\partial x^*} + \frac{\partial v^*}{\partial y^*} = 0 \tag{7.38}$$

and

$$\left[\frac{\rho V}{\tau}\right]\frac{\partial u^*}{\partial t^*} + \left[\frac{\rho V^2}{\ell}\right]\left(u^*\frac{\partial u^*}{\partial x^*} + v^*\frac{\partial u^*}{\partial y^*}\right) = -\left[\frac{p_0}{\ell}\right]\frac{\partial p^*}{\partial x^*} + \left[\frac{\mu V}{\ell^2}\right]\left(\frac{\partial^2 u^*}{\partial x^{*2}} + \frac{\partial^2 u^*}{\partial y^{*2}}\right) \tag{7.39}$$

$$\underbrace{\left[\frac{\rho V}{\tau}\right]}_{F_{I\ell}}\frac{\partial v^*}{\partial t^*} + \underbrace{\left[\frac{\rho V^2}{\ell}\right]}_{F_{Ic}}\left(u^*\frac{\partial v^*}{\partial x^*} + v^*\frac{\partial v^*}{\partial y^*}\right)$$

$$= -\underbrace{\left[\frac{p_0}{\ell}\right]}_{F_P}\frac{\partial p^*}{\partial y^*} - \underbrace{[\rho g]}_{F_G} + \underbrace{\left[\frac{\mu V}{\ell^2}\right]}_{F_V}\left(\frac{\partial^2 v^*}{\partial x^{*2}} + \frac{\partial^2 v^*}{\partial y^{*2}}\right) \tag{7.40}$$

The terms appearing in brackets contain the reference quantities and can be interpreted as indices of the various forces (per unit volume) that are involved. Thus, as is indicated in Eq. 7.40, $F_{I\ell}$ = inertia (local) force, F_{Ic} = inertia (convective) force, F_P = pressure force, F_G = gravitational force, and F_V = viscous force. As the final step in the nondimensionalization process, we will divide each term in Eqs. 7.39 and 7.40 by one of the bracketed quantities. Although any one of these quantities could be used, it is conventional to divide by the bracketed quantity $\rho V^2/\ell$ which is the index of the convective inertia force. The final nondimensional form then becomes

$$\left[\frac{\ell}{\tau V}\right]\frac{\partial u^*}{\partial t^*} + u^*\frac{\partial u^*}{\partial x^*} + v^*\frac{\partial u^*}{\partial y^*} = -\left[\frac{p_0}{\rho V^2}\right]\frac{\partial p^*}{\partial x^*} + \left[\frac{\mu}{\rho V\ell}\right]\left(\frac{\partial^2 u^*}{\partial x^{*2}} + \frac{\partial^2 u^*}{\partial y^{*2}}\right) \tag{7.41}$$

$$\left[\frac{\ell}{\tau V}\right]\frac{\partial v^*}{\partial t^*} + u^*\frac{\partial v^*}{\partial x^*} + v^*\frac{\partial v^*}{\partial y^*} = -\left[\frac{p_0}{\rho V^2}\right]\frac{\partial p^*}{\partial y^*} - \left[\frac{g\ell}{V^2}\right] + \left[\frac{\mu}{\rho V\ell}\right]\left(\frac{\partial^2 v^*}{\partial x^{*2}} + \frac{\partial^2 v^*}{\partial y^{*2}}\right) \tag{7.42}$$

We see that bracketed terms are the standard dimensionless groups (or their reciprocals) which were developed from dimensional analysis; that is, $\ell/\tau V$ is a form of Strouhal number, $p_0/\rho V^2$ the Euler number, $g\ell/V^2$ the reciprocal of the square of the Froude number, and $\mu/\rho V\ell$ the reciprocal of the Reynolds number. From this analysis it is now clear how each of the dimensionless groups can be interpreted as the ratio of two forces, and how these groups arise naturally from the governing equations.

Although we really have not helped ourselves with regard to obtaining an analytical

solution to these equations (they are still complicated and not amenable to an analytical solution), the dimensionless forms of the equations, Eqs. 7.38, 7.41, and 7.42, can be used to establish similarity requirements. From these equations it follows that if two systems are governed by these equations, then the solutions (in terms of u^*, v^*, p^*, x^* y^*, and t^*) will be the same if the four parameters $\ell/\tau V$, $p_0/\rho V^2$, $V^2/g\ell$, and $\rho V\ell/\mu$ are equal for the two systems. The two systems will be dynamically similar. Of course, boundary and initial conditions expressed in dimensionless form must also be equal for the two systems, and this will require complete geometric similarity. These are the same similarity requirements that would be determined by a dimensional analysis if the same variables were considered. However, the advantage of working with the governing equations is that the variables appear naturally in the equations, and we do not have to worry about omitting an important one, provided the governing equations are correctly specified. We can thus use this method to deduce the conditions under which two solutions will be similar even though one of the solutions will most likely be obtained experimentally.

In the foregoing analysis we have considered a general case in which the flow may be unsteady, and both the actual pressure level, p_0, and the effect of gravity are important. A reduction in the number of similarity requirements can be achieved if one or more of these conditions is removed. For example, if the flow is steady the dimensionless group, $\ell/\tau V$, can be eliminated.

The actual pressure level will only be of importance if we are concerned with cavitation. If not, the flow patterns and the pressure differences will not depend on the pressure level. In this case, p_0 can be taken as ρV^2 (or $\frac{1}{2}\rho V^2$), and the Euler number can be eliminated as a similarity requirement. However, if we are concerned about cavitation (which will occur in the flow field if the pressure at certain points reaches the vapor pressure, p_v), then the actual pressure level is important. Usually, in this case, the characteristic pressure, p_0, is defined relative to the vapor pressure such that $p_0 = p_r - p_v$ where p_r is some reference pressure within the flow field. With p_0 defined in this manner, the similarity parameter $p_0/\rho V^2$ becomes $(p_r - p_v)/\rho V^2$. This parameter is frequently written as $(p_r - p_v)/\frac{1}{2}\rho V^2$, and in this form, as was noted previously in Section 7.6, is called the cavitation number. Thus we can conclude that if cavitation is not of concern we do not need a similarity parameter involving p_0, but if cavitation is to be modeled, then the cavitation number becomes an important similarity parameter.

The Froude number, which arises because of the inclusion of gravity, is important for problems in which there is a free surface. Examples of these types of problems include the study of rivers, flow through hydraulic structures such as spillways, and the drag on ships. In these situations the shape of the free surface is influenced by gravity, and therefore the Froude number becomes an important similarity parameter. However, if there are no free surfaces, the only effect of gravity is to superimpose a hydrostatic pressure distribution on the pressure distribution created by the fluid motion. The hydrostatic distribution can be eliminated from the governing equation (Eq. 7.37) by defining a new pressure $p' = p - \rho g y$, and with this change the Froude number does not appear in the nondimensional governing equations.

We conclude from this discussion that for the steady flow of an incompressible fluid without free surfaces, dynamic and kinematic similarity will be achieved if (for geometrically similar systems) Reynolds number similarity exists. If free surfaces are involved Froude number similarity must also be maintained. For free-surface flows we have tacitly assumed that surface tension is not important. We would find, however,

that if surface tension is included, its effect would appear in the free-surface boundary condition, and the Weber number, $\rho V^2 \ell / \sigma$, would become an additional similarity parameter. In addition, if the governing equations for compressible fluids are considered, the Mach number, V/c, would appear as an additional similarity parameter.

It is clear that all the common dimensionless groups that we previously developed by using dimensional analysis appear in the governing equation that describe fluid motion when these equations are expressed in terms of dimensionless variables. Thus, the use of the governing equations to obtain similarity laws provides an alternative to dimensional analysis. This approach has the advantage that the variables are known and the assumptions involved are clearly identified. In addition, a physical interpretation of the various dimensionless groups can often be obtained.

REFERENCES

1. Bridgman, P. W., *Dimensional Analysis,* Yale University Press, New Haven, Conn., 1922.

2. Murphy, G., *Similitude in Engineering,* Ronald Press, New York, 1950.

3. Langhaar, H. L., *Dimensional Analysis and Theory of Models,* Wiley, New York, 1951.

4. Huntley, H. E., *Dimensional Analysis,* Macdonald, London, 1952.

5. Duncan, W. J., *Physical Similarity and Dimensional Analysis: An Elementary Treatise,* Edward Arnold, London, 1953.

6. Sedov, L. I., *Similarity and Dimensional Methods in Mechanics,* Academic Press, New York, 1959.

7. Ipsen, D. C., *Units, Dimensions, and Dimensionless Numbers,* McGraw-Hill, New York, 1960.

8. Kline, S. J., *Similitude and Approximation Theory,* McGraw-Hill, New York, 1965.

9. Skoglund, V. J., *Similitude—Theory and Applications,* International Textbook, Scranton, Pa., 1967.

10. Baker, W. E., Westline, P. S., and Dodge, F. T., *Similarity Methods in Engineering Dynamics—Theory and Practice of Scale Modeling,* Hayden (Spartan Books), Rochelle Park, N.J., 1973.

11. Taylor, E. S., *Dimensional Analysis for Engineers,* Clarendon Press, Oxford, 1974.

12. Issacson, E. de St. Q, and Issacson, M. de St. Q., *Dimensional Methods in Engineering and Physics,* Wiley, New York, 1975.

13. Schuring, D. J., *Scale Models in Engineering,* Pergamon Press, New York, 1977.

14. Yalin, M. S., *Theory of Hydraulic Models,* Macmillan, London, 1971.

15. Sharp, J. J., *Hydraulic Modeling,* Butterworth, London, 1981.

16. Schlichting, H., *Boundary-Layer Theory,* 7th Ed., McGraw-Hill, New York, 1979.

17. Knapp, R. T., Daily, J. W., and Hammitt, F. G., *Cavitation,* McGraw-Hill, New York, 1970.

18. Rouse, H., *Elementary Mechanics of Fluids,* Wiley, New York, 1946.

PROBLEMS

Note: Unless otherwise indicated use the values of fluid properties found in the tables on the inside of the front cover. Problems designated with an (*) are intended to be solved with the aid of a programmable calculator or a computer.

$Re = \dfrac{\rho v \ell}{\mu} = \dfrac{(FL^{-4}T^2)(LT^{-1})(L)}{FL^{-2}T} = F^0 L^0 T^0$

7.1   The Reynolds number, $\rho VD/\mu$, is a very important parameter in fluid mechanics. Verify that the Reynolds number is dimensionless, using both the *FLT* system and the *MLT* system for basic dimensions, and determine its value for water (at 70 °C) flowing at a velocity of 3 m/s through a 1-in.-diameter pipe. *Table B-2 workup values plug in*

7.2   What are the dimensions of density, pressure, specific weight, surface tension, and dynamic viscosity in **(a)** the *FLT* system, and **(b)** the *MLT* system? Compare your results with those given in Table 1.1 in Chapter 1.

7.3   *see Table 1.1* Standard air with velocity V flows past an airfoil having a chord length, b, of 6 ft. **(a)** Determine the Reynolds number, $\rho Vb/\mu$, for $V = 200$ mph. **(b)** If this airfoil were attached to an airplane flying at the same speed in a standard atmosphere at an altitude of 10,000 ft, what would be the value of the Reynolds number?

7.4   Some common variables in fluid mechanics include: volume flowrate, Q, acceleration of gravity, g, viscosity, μ, density, ρ, and a length, ℓ. Which of the following combinations of these variables are dimensionless? **(a)** $Q^2/g\ell^2$. **(b)** $\rho Q/\mu\ell$. **(c)** $g\ell^2/Q$. *NOT* **(d)** $\rho Q\ell/\mu$. *NOT 15*

7.5   *NOT* For the flow of a thin film of a liquid with a depth h and a free surface, two important dimensionless parameters are the Froude number, $V/\sqrt{gh}$, and the Weber number, $\rho V^2 h/\sigma$. Determine the value of these two parameters for glycerin (at 20 °C) flowing with a velocity of 0.5 m/s at a depth of 3 mm.

7.6   The Mach number for a body moving through a fluid with velocity V is defined as V/c, where c is the speed of sound in the fluid. This dimensionless parameter is usually considered to be important in fluid dynamics problems when its value exceeds 0.3. What would be the velocity of a body at a Mach number of 0.3 if the fluid is: **(a)** air at standard atmospheric pressure and 20 °C, and **(b)** water at the same temperature and pressure?

7.7   At a sudden contraction in a pipe the diameter changes from D_1 to D_2. The pressure drop, Δp, which develops across the contraction

is a function of D_1 and D_2, as well as the velocity, V, in the larger pipe, and the fluid density, ρ, and viscosity, μ. Use D_1, V, and μ as repeating variables to determine a suitable set of dimensionless parameters. Why would it be incorrect to include the velocity in the smaller pipe as an additional variable?

7.8   Assume that the power, $\mathcal{P}$, required to drive a fan is a function of the fan diameter, D, the fluid density, ρ, the rotational speed, ω, and the flowrate, Q. Use D, ω, and ρ as repeating variables to determine a suitable set of pi terms.

7.9   Assume that the flowrate, Q, of a gas from a smokestack is a function of the density of the ambient air, ρ_a, the density of the gas, ρ_g, within the stack, the acceleration of gravity, g, and the height and diameter of the stack, h and d, respectively. Use ρ_a, d, and g as repeating variables to develop a set of pi terms that could be used to describe this problem.

7.10   In a fuel injection system small droplets are formed due to the breakup of the liquid jet. Assume the droplet diameter, d, is a function of the liquid density, ρ, viscosity, μ, and surface tension, σ, and the jet velocity, V, and diameter, D. Form an appropriate set of dimensionless parameters using μ, V, and D as repeating variables.

7.11   The pressure rise, Δp, across a pump can be expressed as

$$\Delta p = f(D, \rho, \omega, Q)$$

where D is the impeller diameter, ρ the fluid density, ω the rotational speed, and Q the flowrate. Determine a suitable set of dimensionless parameters.

7.12   The drag, $\mathcal{D}$, on a washer shaped plate placed normal to a stream of fluid can be expressed as

$$\mathcal{D} = f(\underset{1}{d_1}, \underset{2}{d_2}, \underset{3}{V}, \underset{4}{\mu}, \underset{5 \ 6}{\rho}) \quad \underset{6-3=3}{\text{use } d \ V \ \rho \ \text{as eq.}}$$

where d_1 is the outer diameter, d_2 the inner diameter, V the fluid velocity, μ the fluid viscosity, and ρ the fluid density. Some experiments are to be performed in a wind tunnel to determine the drag. What dimensionless parameters would you use to organize these data?

7.13   A thin elastic wire is placed between rigid supports. A fluid flows past the wire, and it is desired to study the static deflection, δ, at the center of the wire due to the fluid drag. Assume that

$$\delta = f(\ell, d, \rho, \mu, V, E)$$

where ℓ is the wire length, d the wire diameter, ρ the fluid density, μ the fluid viscosity, V the fluid velocity, and E the modulus of elasticity of the wire material. Develop a suitable set of pi terms for this problem.

7.14   A fluid flows at a velocity V through a horizontal pipe of diameter D. An orifice plate containing a hole of diameter d is placed in the pipe. It is desired to investigate the pressure drop, Δp, across the plate. Assume that

$$\Delta p = f(D, d, \rho, V)$$

where ρ is the fluid density. Determine a suitable set of pi terms.

7.15   The velocity, V, of a spherical particle falling slowly in a viscous liquid can be expressed as

$$V = f(d, \mu, \gamma, \gamma_s)$$

where d is the particle diameter, μ the liquid viscosity, and γ and γ_s the specific weight of the liquid and particle, respectively. Develop a set of dimensionless parameters that can be used to investigate this problem.

7.16   The velocity, c, at which pressure pulses travel through arteries (pulse-wave velocity) is a function of the artery diameter, D, and wall thickness, h, the density of blood, ρ, and the modulus of elasticity, E, of the arterial wall. Determine a set of nondimensional parameters that can be used to study experimentally the relationship between the pulse-wave velocity and the variables listed. Form the nondimensional parameters by inspection.

7.17   The thrust, $\mathcal{T}$, developed by a propeller of a given shape depends on its diameter, D, the fluid density, ρ, and viscosity, μ, the angular speed of rotation, ω, and the advance velocity, V. Develop a suitable set of pi terms, one of which should be $\rho D^2 \omega / \mu$. Form the pi terms by inspection.

7.18   Assume that the drag, $\mathcal{D}$, on an aircraft flying at supersonic speeds is a function of its velocity, V, fluid density, ρ, speed of sound, c, and a series of lengths, $\ell_1, \ldots, \ell_i$, which describe the geometry of the aircraft. Develop a set of pi terms that could be used to investigate experimentally how the drag is affected by the various factors listed. Form the pi terms by inspection.

7.19   A cone and plate viscometer consists of a cone with a very small angle α which rotates above a flat surface as shown in Fig. P7.19. The torque, $\mathcal{T}$, required to rotate the cone at an angular velocity ω, is a function of the radius, R,

the cone angle, α, and the fluid viscosity, μ, in addition to ω. With the aid of dimensional analysis, determine how the torque will change if both the viscosity and angular velocity are doubled.

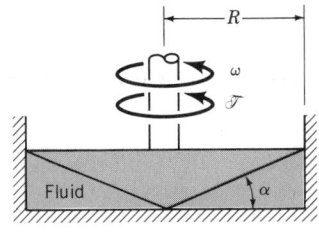

FIGURE P7.19

7.20   The pressure drop, Δp, along a straight pipe of diameter D has been experimentally studied, and it is observed that for laminar flow of a given fluid and pipe, the pressure drop varies directly with the distance, ℓ, between pressure taps. Assume that Δp is a function of D and ℓ, the velocity, V, and the fluid viscosity, μ. Use dimensional analysis to deduce how the pressure drop varies with pipe diameter.

7.21   One type of viscometer consists of an open reservoir with a small diameter tube at the bottom as illustrated in Fig. P7.21. To measure viscosity the system is filled with the liquid of interest and the time required for the liquid level to fall from level H_i to H_f is determined. Use dimensional analysis to obtain a relationship between the viscosity, μ, and the draining time, τ. Assume that the other variables involved are the initial head, H_i, the final head, H_f, the tube diameter, D, and the specific weight of the liquid, γ.

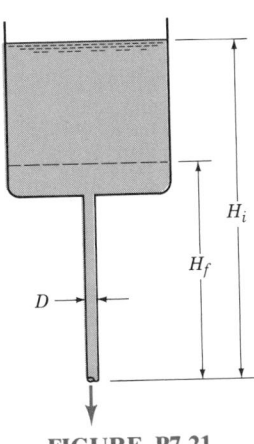

FIGURE P7.21

7.22   The buoyant force, F_B, acting on a body submerged in a fluid is a function of the specific weight, γ, of the fluid and the volume, $\mathcal{V}$, of the

body. Show, by dimensional analysis, that the buoyant force must be directly proportional to the specific weight.

7.23   Assume that the drag on a small sphere placed in a rapidly moving stream of fluid depends on the fluid density but not the fluid viscosity. Use dimensional analysis to determine how the drag is affected if the velocity of the fluid is doubled.

7.24   The height, h, that a liquid will rise in a capillary tube is a function of the tube diameter, D, the specific weight of the liquid, γ, and the surface tension, σ. Perform a dimensional analysis using both the FLT and MLT systems for basic dimensions. Note: The results should obviously be the same regardless of the system of dimensions used. If your analysis indicates otherwise, go back and check your work giving particular attention to the required number of reference dimensions.

7.25   The speed of sound in a gas, c, is a function of the gas pressure, p, and density, ρ. Determine, with the aid of dimensional analysis, how the velocity is related to the pressure and density. Be careful when you decide on how many reference dimensions are required.

7.26* The pressure drop across a short hollowed plug placed in a circular tube through which a liquid is flowing (see Fig. P7.26) can be expressed as

$$\Delta p = f(\rho, V, D, d)$$

where ρ is the fluid density, and V is the mean velocity in the tube. Some experimental data obtained with $D = 0.2$ ft, $\rho = 2.0$ slugs/ft^3, and $V = 2$ ft/s are given in the following table:

d (ft)	0.06	0.08	0.10	0.15
Δp (lb/ft^2)	493.8	156.2	64.0	12.6

Plot the results of these tests, using suitable dimensionless parameters, on log–log graph paper. Use a standard curve-fitting technique to determine a general equation for Δp. What are the limits of applicability of the equation?

7.27   A liquid flows with a velocity V through a hole in the side of a large tank. Assume that

$$V = f(h, g, \rho, \sigma)$$

where h is the depth of fluid above the hole, g is the acceleration of gravity, ρ the fluid density, and σ the surface tension. The following data were obtained by changing h and measuring V, with a fluid having a density $= 10^3$ kg/m^3 and surface tension $= 0.074$ N/m.

V (m/s)	3.13	4.43	5.42	6.25	7.00
h (m)	0.50	1.00	1.50	2.00	2.50

Plot these data by using appropriate dimensionless variables. Could any of the original variables have been omitted?

7.28* The concentric cylinder device of the type shown in Fig. P7.28 is commonly used to measure the viscosity, μ, of liquids by relating the angle of twist, θ, of the inner cylinder to the angular velocity, ω, of the outer cylinder. Assume that

$$\theta = f(\omega, \mu, K, D_1, D_2, \ell)$$

where K depends on the suspending wire properties and has the dimensions FL. The following data were obtained in a series of tests for which $\mu = 0.01$ lb·s/ft^2, $K = 10$ lb·ft, $\ell = 1$ ft, and D_1 and D_2 were constant.

θ (rad)	ω (rad/s)
0.89	0.30
1.50	0.50
2.51	0.82
3.05	1.05
4.28	1.43
5.52	1.86
6.40	2.14

FIGURE P7.26

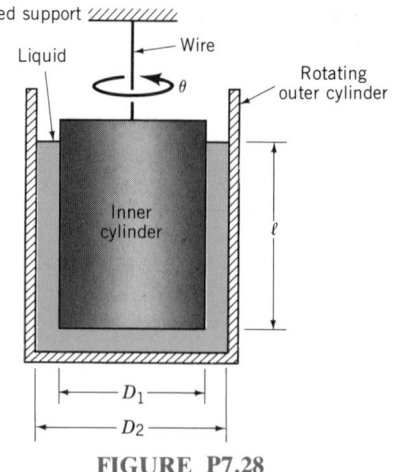

FIGURE P7.28

Determine from these data, with the aid of dimensional analysis, the relationship between θ, ω, and μ for this particular apparatus. *Hint:* Plot the data using appropriate dimensionless parameters, and determine the equation of the resulting curve using a standard curve-fitting technique. The equation should satisfy the condition that $\theta = 0$ for $\omega = 0$.

7.29   The pressure drop per unit length, Δp_ℓ, for the flow of blood through a horizontal small diameter tube is a function of the volume rate of flow, Q, the diameter, D, and the blood viscosity, μ. For a series of tests in which $d = 2$ mm, and $\mu = 0.004$ N·s/m², the following data were obtained, where the Δp listed was measured over the length, $\ell = 300$ mm.

Q (m³/s)	Δp (N/m²)
3.6×10^{-6}	1.1×10^4
4.9×10^{-6}	1.5×10^4
6.3×10^{-6}	1.9×10^4
7.9×10^{-6}	2.4×10^4
9.8×10^{-6}	3.0×10^4

Perform a dimensional analysis for this problem, and make use of the data given to determine a general relationship between Δp_ℓ and Q (one that is valid for other values of D, ℓ, and μ).

7.30*   A rectangular steel bar is rigidly attached to the floor of a channel through which a fluid is flowing with velocity V as shown in Fig. P7.30. Because of the force of the moving fluid the tip of the bar deflects a distance, δ. Assume that δ is a function of the height, h, thickness, d, modulus of elasticity, E, and kinetic energy, ρV^2, of the moving fluid. (Note the kinetic energy can be treated as a single variable.) The plate width is not an important variable. It is known from theoretical considerations that the deflection varies inversely with modulus of elasticity. In a laboratory experiment the following data were obtained when the height, h, was varied while holding the other variables constant at the values $d = 0.01$ m, $E = 2.1 \times 10^9$ N/m², $\rho = 10^3$ kg/m³, and $V = 2.0$ m/s.

δ (m)	h (m)
0.19×10^{-3}	0.10
1.52×10^{-3}	0.20
5.13×10^{-3}	0.30
12.16×10^{-3}	0.40
23.75×10^{-3}	0.50

Based on these data, predict the deflection of a similar steel bar under the following conditions: $d = 0.008$ m, $E = 2.1 \times 10^9$ N/m², $\rho = 10^3$ kg/m³, $V = 1.0$ m/s, and $h = 0.25$ m.

FIGURE P7.30

7.31   A liquid is contained in a U-tube as is shown in Fig. P7.31. When the liquid is displaced from its equilibrium position and released, it oscillates with a period, τ. Assume that τ is a function of the acceleration of gravity, g, and the column length, ℓ. Some laboratory measurements made by varying ℓ and measuring τ, with $g = 32.2$ ft/s², are given in the following table.

τ (s)	0.548	0.783	0.939	1.174
ℓ (ft)	0.49	1.00	1.44	2.25

Based on these data determine a general equation for the period.

FIGURE P7.31

7.32   The flowrate, Q, in an open canal or channel can be measured by placing a plate with a V-notch across the channel as illustrated in Fig. P7.32. This type of device is called a V-notch *weir*. The height, H, of the liquid above the crest can be used to determine Q. Assume that

$$Q = f(H, g, \theta)$$

where g is the acceleration of gravity. What are the significant dimensionless parameters for this problem?

FIGURE P7.32

7.33   From theoretical considerations it is known that for the V-notch weir described in Problem 7.32 the flowrate, Q, is directly proportional to the tan $\theta/2$. In the laboratory it was determined that for $\theta = 90°$ and $H = 0.3$ m, $Q = 0.068$ m^3/s. Based on these limited data, determine a general equation for the flowrate through a V-notch weir.

7.34   SAE 30 oil at 60 °F is pumped through a 3-ft-diameter pipeline at a rate of 6300 gal/min. A model of this pipeline is to be designed using a 2-in.-diameter pipe and water at 60 °F as the working fluid. To maintain Reynolds number similarity between these two systems what fluid velocity will be required in the model?

7.35   Glycerin at 20 °C flows with a velocity of 4 m/s through a 40-mm-diameter tube. A model of this system is to be developed using standard air as the model fluid. The air velocity is to be 2 m/s. What tube diameter is required for the model if dynamic similarity is to be maintained between model and prototype?

7.36   The drag characteristics of a torpedo are to be studied in a water tunnel using a 1:4 scale model. The tunnel operates with freshwater at 20 °C, whereas the prototype torpedo is to be used in seawater at 15.6 °C. To correctly simulate the behavior of the prototype moving with a velocity of 30 m/s, what velocity is required in the water tunnel?

7.37   The design of a river model is to be based on Froude number similarity, and a river depth of 3 m is to correspond to a model depth of 100 mm. Under these conditions what is the prototype velocity corresponding to a model velocity of 1 m/s?

7.38   For a certain fluid flow problem it is known that both the Froude number and the Weber number are important dimensionless parameters. If the problem is to be studied by using a 1:10 scale model, determine the required surface tension scale if the density scale is equal to 1. The model and prototype operate in the same gravitational field.

7.39   The fluid dynamic characteristics of an airplane flying at 280 mph at 10,000 ft are to be investigated with the aid of a 1:20 scale model. If the model tests are to be performed in a wind tunnel using standard air, what is the required air velocity in the wind tunnel? Is this a realistic velocity?

7.40   If an airplane travels at a speed of 1120 km/hr at an altitude of 15 km, what is the required speed at an altitude of 7 km to satisfy Mach number similarity? Assume the air properties correspond to those for the U.S. standard atmosphere.

7.41   The water velocity at a certain point along a 1:10 scale model of a dam spillway is 3 m/s. What is the corresponding prototype velocity if the model and prototype operate in accordance with Froude number similarity?

7.42   The flowrate over the spillway of a dam is 27,000 ft^3/min. Determine the required flowrate for a 1:25 scale model that is operated in accordance with Froude number similarity.

7.43   The lift and drag developed on a hydrofoil are to be determined through wind tunnel tests using standard air. If full scale tests are to be run, what is the required wind tunnel velocity corresponding to a hydrofoil velocity in seawater of 20 mph? Assume Reynolds number similarity is required.

7.44   Water flows at a rate of 40 m^3/s through the spillway of a dam which is 65 m wide. A model spillway, having a width of 1.5 m, is to be constructed and tested in the laboratory. What is the required flowrate in the model?

7.45   A 1:40 scale model of a ship is to be tested in a towing tank. Determine the required kinematic viscosity of the model fluid so that both the Reynolds number and the Froude number are the same for model and prototype. Assume the prototype fluid to be seawater at 60 °F. Could any of the liquids with viscosities given in Fig. B.2 in Appendix B be used as the model fluid?

7.46   The pressure drop per unit length in a 0.25-in.-diameter gasoline fuel line is to be determined from a laboratory test using the same tubing but with water as the fluid. The pressure drop at a gasoline velocity of 1.0 ft/s is of interest. **(a)** What water velocity is required? **(b)** At the properly scaled velocity from part (a) the pressure drop per unit length (using water) was found to be 0.45 psf/ft. What is the predicted pressure drop per unit length for the gasoline line?

7.47   The drag on a 2-m-diameter satellite dish due to an 80 km/hr wind is to be determined through a wind tunnel test using a geometrically similar 0.4-m-diameter model dish. Assume standard air for both model and prototype. **(a)** At what air speed should the model test be run? **(b)** With all similarity conditions satisfied, the measured drag on the model was determined to be 170 N. What is the predicted drag on the prototype dish?

7.48   The pressure drop between the entrance and exit of a 150-mm-diameter 90° elbow, through which ethyl alcohol at 20 °C is flowing, is to be determined with a geometrically similar model. The velocity of the alcohol is 5 m/s. The model fluid is to be water at 20 °C, and the model velocity is limited to 10 m/s. **(a)** What is the required diameter of the model elbow to maintain dynamic similarity? **(b)** A measured pressure drop of 15 kPa in the model will correspond to what prototype value?

7.49   For a certain model study involving a 1:5 scale model it is known that Froude number similarity must be maintained. The possibility of cavitation is also to be investigated, and it is assumed that the cavitation number must be the same for model and prototype. The prototype fluid is water at 30 °C, and the model fluid is water at 70 °C. If the prototype operates at an ambient pressure of 101 kPa (abs), what is the required ambient pressure for the model system?

7.50   A thin layer of spherical particles rests on the bottom of a horizontal tube as shown in Fig. P7.50. When an incompressible fluid flows through the tube, it is observed that at some critical velocity the particles will rise and be transported along the tube. A model is to be used to determine this critical velocity. Assume the critical velocity, V_c to be a function of the pipe diameter, D, particle diameter, d, the fluid density, ρ, and viscosity, μ, the density of the particles, ρ_p, and the acceleration of gravity, g. **(a)** Determine the similarity requirements for the model, and the relationship between the critical velocity for model and prototype (the prediction equation). **(b)** For a length scale of $\frac{1}{2}$ and a fluid density scale of 1.0, what will be the critical velocity scale (assuming all similarity requirements are satisfied)?

FIGURE P7.50

7.51   A large open tank contains an incompressible fluid and is to be drained through a long pipe connected to the bottom of the tank as shown in Fig. P7.51. It is desired to design a model of this system to predict the time required to drain the tank after the valve is opened. Surface tension effects and pipe roughness may be neglected, but inertial, gravitational, and viscous effect are all

important. **(a)** Establish the similarity requirements for the model (the model design conditions), and the relationship between the drain time for the model and that for the prototype (the prediction equation). **(b)** Can the same fluid be used for the model and prototype? Explain. **(c)** For a length scale of 1:50, what is the time scale if all similarity requirements are satisfied.

FIGURE P7.51

7.52   A model is to be used to determine the velocity, V, of liquid flow through a small-diameter passage in a wall separating two pressurized tanks as shown in Fig. P7.52. Prototype characteristics are indicated on the figure. The model is to have a length scale of $\frac{1}{4}$, a viscosity scale of 2.0, and if possible, the pressures are to be the same for model and prototype. Assume that V is a function of $p_1 - p_2$, d, ℓ, and the fluid viscosity, μ. Determine the required dimensions for the model, and the velocity scale.

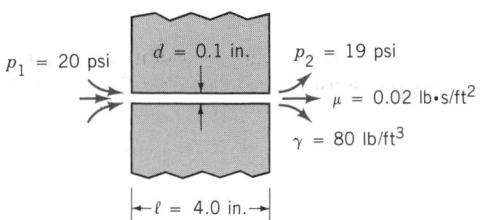

FIGURE P7.52

7.53   A thin layer of an incompressible fluid flows steadily over a horizontal smooth plate as shown in Fig. P7.53. The fluid surface is open to the atmosphere, and an obstruction having a square cross section is placed on the plate as shown. A model with a length scale of $\frac{1}{4}$ and a fluid density scale of 1.0 is to be designed to predict the depth of fluid, y, along the plate. Assume that inertial, gravitational, surface tension, and viscous effects are all important. What are the required viscosity and surface tension scales?

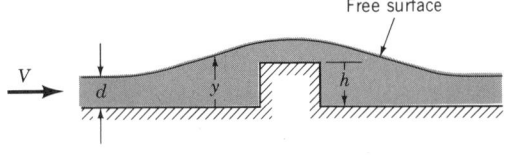

FIGURE P7.53

7.54   A thin flat plate having a diameter of 0.3 ft is towed through a tank of oil ($\gamma = 53$ lb/ft³) at a velocity of 5 ft/s. The plane of the plate is perpendicular to the direction of motion, and the plate is submerged so that wave action is negligible. Under these conditions the drag on the plate is 1.4 lb. If viscous effects are neglected, predict the drag on a geometrically similar, 2-ft-diameter plate that is towed with a velocity of 3 ft/s through water at 60 °F under conditions similar to those for the smaller plate.

7.55   A thin rectangular plate is towed through seawater at an average velocity of 5 mph. The plate is held in a vertical position and projects above the undisturbed level of the water to a height z. A 1:4 scale model is to be used to predict the drag on the plate, and the model fluid is also seawater. **(a)** Assuming that Froude number similarity must be maintained, determine the required model velocity. **(b)** What is the required value of z_m/z? **(c)** A measured drag of 1 lb on the model will correspond to what drag on the prototype?

7.56   A solid sphere having a diameter d and specific weight γ_s is immersed in a liquid having a specific weight γ_f ($\gamma_f > \gamma_s$) and then released. It is desired to use a model system to determine the maximum height, h, above the liquid surface that the sphere will rise upon release from a depth H. It can be assumed that the important liquid properties are the density, γ_f/g, specific weight, γ_f, and viscosity, μ_f. Establish the model design conditions and the prediction equation, and determine whether the same liquid can be used in both the model and prototype systems.

7.57   An incompressible fluid oscillates harmonically ($V = V_0 \sin \omega t$, where V is the velocity) with a frequency of 10 rad/s in a 4-in.-diameter pipe. A $\frac{1}{4}$ scale model is to be used to determine the pressure difference per unit length, Δp_ℓ (at any instant) along the pipe. Assume that

$$\Delta p_\ell = f(D, V_0, \omega, t, \mu, \rho)$$

where D is the pipe diameter, ω the frequency, t the time, μ the fluid viscosity, and ρ the fluid density. **(a)** Determine the similarity requirements for the model and the prediction equation for Δp_ℓ. **(b)** If the same fluid is used in the model and the prototype, at what frequency should the model operate?

7.58   The drag on a sphere moving in a fluid is known to be a function of the sphere diameter, the velocity, and the fluid viscosity and density. Laboratory tests on a 4-in.-diameter sphere were performed in a water tunnel and some model data

are plotted in Fig. P7.58. For these tests the viscosity of the water was 2.3×10^{-5} lb·s/ft² and the water density was 1.94 slugs/ft³. Estimate the drag on an 8-ft diameter balloon moving in air at a velocity of 3 ft/s. Assume the air to have a viscosity of 3.7×10^{-7} lb·s/ft² and a density of 2.38×10^{-3} slugs/ft³.

FIGURE P7.58

7.59   The drag on a small, completely submerged solid body having a characteristic length of 2.5 mm and moving with a velocity of 15 m/s through water is to be determined with the aid of a model. The length scale is to be 50, which indicates that the model is to be *larger* than the prototype. Investigate the possibility of using either an unpressurized wind tunnel or a water tunnel for this study. Determine the required velocity in both the wind and water tunnels, and the relationship between the model drag and the prototype drag for both systems. Would either type of test facility be suitable for this study?

7.60   The drag characteristics for a newly designed automobile having a maximum characteristic length of 20 ft are to be determined through a model study. The characteristics at both low speed (approximately 20 mph) and high speed (90 mph) are of interest. For a series of projected model tests an unpressurized wind tunnel that will accommodate a model with a maximum characteristic length of 4 ft is to be used. Determine the range of air velocities that would be required for the wind tunnel if Reynolds number similarity is desired. Are the velocities suitable? Explain.

7.61   If the unpressurized wind tunnel of Problem 7.60 were replaced with a tunnel in which the air can be pressurized isothermally to 8 atm (abs), what range of air velocities would be required to maintain Reynolds number similarity for the same prototype velocities given in Problem 7.60? For the pressurized tunnel the maximum characteristic model length that can be accommodated is 2 ft, whereas the maximum characteristic prototype length remains at 20 ft.

Handwritten notes at top of page:

7.68 See of - 7.32 $\dfrac{Q_m}{W_m D_m^3} = \dfrac{Q}{WD^2}$ w/ $W = W_m$ ∴ $Q_m = \dfrac{D_m^3}{D^3}$ $Q = \left(\dfrac{1}{5}\right)^3 (4.1) = 0.0328 \, m^3/s$

Similarity $\dfrac{g_m h_{am}}{W_m^2 D_m^2} = \dfrac{g h_a}{W^2 D^2}$ w/ $g = g_m$ & $W = W_m$ $h_{am} = \dfrac{D_m^2}{D^2} h_a = \left(\dfrac{1}{5}\right)^2 (200 m) = 8.0 m$

7.62   The drag characteristics of an airplane are to be determined by model tests in a wind tunnel operated at an absolute pressure of 1300 kPa. If the prototype is to cruise in standard air at 385 km/hr, and the corresponding speed of the model is not to differ by more than 20% from this (so that compressibility effects may be ignored), what range of length scales may be used if Reynolds number similarity is to be maintained? Assume the viscosity of air is unaffected by pressure, and the temperature of the air in the tunnel is equal to the temperature of the air in which the airplane will fly.

7.63   A large, rigid, rectangular billboard is supported by an elastic column as shown in Fig. P7.63. There is concern about the deflection, δ, of the top of the structure during a high wind of velocity V. A wind tunnel test is to be conducted with a 1:15 scale model. Assume the pertinent column variables are its length and cross-sectional dimensions, and the modulus of elasticity of the material used for the column. The only important "wind" variables are the air density and velocity. **(a)** Determine the model design conditions and the prediction equation for the deflection. **(b)** If the same structural materials are used for the model and prototype, and the wind tunnel operates under standard atmospheric conditions, what is the required wind tunnel velocity to match an 80 km/hr wind?

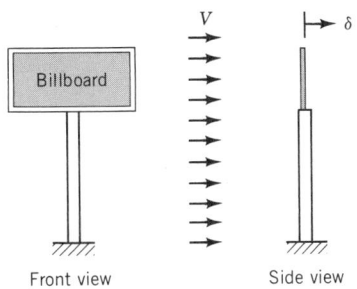

Front view   Side view

FIGURE P7.63

7.64   An open channel with a rectangular cross section has a width of 20 ft and carries water at a depth of 3 ft at a flowrate of 60 ft³/s. A model is to be designed, based on Froude number similarity, so that the discharge scale is 1/1024. At what depth and flowrate would the model operate?

7.65   An open channel, 6 ft wide, carries a highly viscous fluid ($\mu = 10^{-3}$ lb·s/ft²; $\gamma = 75$ lb/ft³) at a depth of 2 ft. A model is to be designed having a length scale of $\frac{1}{10}$. Determine the required value of the kinematic viscosity for the model fluid.

7.66   A $\frac{1}{50}$ scale model is to be used in a towing tank to determine the drag on the hull of a ship. The model is operated in accordance with the Froude number criteria for dynamic similitude. The prototype ship is designed to cruise at 18 knots. At what velocity (in m/s) should the model be towed? Under these conditions what will be the ratio of the prototype drag to the model drag? Assume the water in the towing tank to have the same properties as those for the prototype and that shear drag is negligible.

7.67   When the shaft horsepower supplied to a certain centrifugal pump is 25 hp the pump discharges 700 gpm of water while operating at 1800 rpm with a head rise of 90 ft. **(a)** If the pump speed is reduced to 1200 rpm determine the new head rise and shaft horsepower. Assume the efficiency remains the same. **(b)** What is the specific speed, N_{sd}, for this pump?

7.68   A centrifugal pump having an impeller diameter of 1 m is to be constructed so that it will supply a head rise of 200 m at a flowrate of 4.1 m³/s of water when operating at a speed of 1200 rpm. To study the characteristics of this pump, a $\frac{1}{5}$ scale, geometrically similar model operated at the same speed is to be tested in the laboratory. Determine the required model discharge and head rise. Assume both model and prototype operate with the same efficiency.

7.69   A small model of a pump is tested in the laboratory and found to have a specific speed N_{sd} equal to 1000 when operating at peak efficiency. Predict the discharge of a larger, geometrically similar pump operating at peak efficiency at a speed of 1800 rpm across a head rise of 200 ft.

7.70   Start with the two-dimensional continuity equation and the Navier–Stokes equations (Eqs. 7.35, 7.36, and 7.37) and verify the nondimensional forms of these equations (Eqs. 7.38, 7.41, and 7.42).

7.71   The flow between two concentric cylinders (see Fig. P7.71) is governed by the differential equation

$$\frac{d^2 v_\theta}{dr^2} + \frac{d}{dr}\left(\frac{v_\theta}{r}\right) = 0$$

FIGURE P7.71

where v_θ is the tangential velocity at any radial location, r. The inner cylinder is fixed and the outer cylinder rotates with an angular velocity ω. Express the equation in dimensionless form using R_o and ω as reference parameters.

7.72 An incompressible fluid is contained between two large parallel plates as shown in Fig. P7.72. The upper plate is fixed. If the fluid is initially at rest and the bottom plate suddenly starts to move with a constant velocity, U, the governing differential equation describing the fluid motion is

$$\rho \frac{\partial u}{\partial t} = \mu \frac{\partial^2 u}{\partial y^2}$$

where u is the velocity in the x direction, and ρ and μ are the fluid density and viscosity, respectively. Rewrite the equation and the inital and boundary conditions in dimensionless form using h and U as reference parameters for length and velocity, and $h^2\rho/\mu$ as a reference parameter for time.

FIGURE P7.72

7.73 The deflection of the cantilever beam of Fig. P7.73 is governed by the differential equation

$$EI \frac{d^2y}{dx^2} = P(x - \ell)$$

where E is the modulus of elasticity and I is the moment of inertia of the beam cross section. The boundary conditions are $y = 0$ at $x = 0$ and $dy/dx = 0$ at $x = 0$. **(a)** Rewrite the equation and boundary conditions in dimensionless form using the beam length, ℓ, as the reference length.

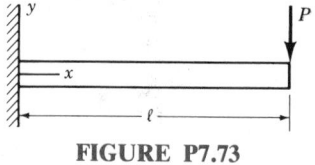

FIGURE P7.73

(b) Based on the results of part (a) what are the similarity requirements and the prediction equation for a model to predict deflections?

7.74 A liquid is contained in a pipe that is closed at one end as shown in Fig. P7.74. Initially the liquid is at rest, but if the end is suddenly opened the liquid starts to move. Assume the pressure p_1 remains constant. The differential equation that describes the resulting motion of the liquid is

$$\rho \frac{\partial v_z}{\partial t} = \frac{p_1}{\ell} + \mu \left(\frac{\partial^2 v_z}{\partial r^2} + \frac{1}{r} \frac{\partial v_z}{\partial r} \right)$$

where v_z is the velocity at any radial location, r, and t is time. Rewrite this equation in dimensionless form using the liquid density, ρ, the viscosity, μ, and the pipe radius, R, as reference parameters.

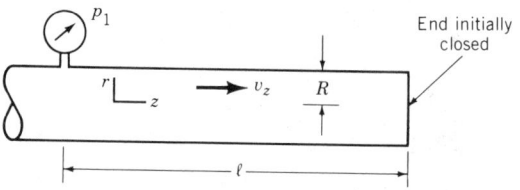

FIGURE P7.74

7.75 An incompressible fluid is contained between two infinite parallel plates as illustrated in Fig. P7.75. Under the influence of a harmonically varying pressure gradient in the x direction, the fluid oscillates harmonically with a frequency ω. The differential equation describing the fluid motion is

$$\rho \frac{\partial u}{\partial t} = X \cos \omega t + \mu \frac{\partial^2 u}{\partial y^2}$$

where X is the amplitude of the pressure gradient. Express this equation in nondimensional form using h and ω as reference parameters.

FIGURE P7.75

Turbulent jet: The jet of water from the pipe is turbulent. The complex, irregular, unsteady structure typical of turbulent flows is apparent. (Laser-induced fluorescence of dye in water.) (Photograph by P. E. Dimotakis, R. C. Lye, and D. Z. Papantoniou.)

EIGHT

VISCOUS FLOW IN PIPES

In the previous chapters we have considered a variety of topics concerning the motion of fluids. The basic governing principles concerning mass, momentum, and energy were developed and applied, in conjunction with rather severe assumptions, to numerous flow situations. In this chapter we will apply the basic principles to a specific, important topic—the flow of viscous incompressible fluids in pipes and ducts.

The transport of a fluid (liquid or gas) in a closed conduit (commonly called a *pipe* if it is of round cross section or a *duct* if it is not round) is extremely important in our daily operations. A brief consideration of the world around us will indicate that there is wide variety of applications of pipe flow. Such applications range from the large, man-made Alaskan pipeline that carries crude oil almost 800 miles across Alaska, to the more mundane (but certainly not less useful) natural systems of "pipes" that carry blood throughout our body and air into and out of our lungs. Other examples include the water pipes in our homes and the distribution system that delivers the water from the city well to the house. Numerous hoses and pipes carry hydraulic fluid or other fluids to various components of vehicles and machines. The air quality within our buildings is maintained at comfortable levels by the distribution of conditioned (heated, cooled, humidified/dehumidified) air through a maze of pipes and ducts. Although all of these systems are different, the fluid mechanics principles governing the fluid motions are common. The purpose of this chapter is to understand the basic processes involved in such flows.

Some of the basic components of a typical *pipe system* are shown in Fig. 8.1. They include the pipes themselves (perhaps of more than one diameter), the various fittings used to connect the individual pipes to form the desired system, the flowrate control devices (valves), and the pumps or turbines that add energy to or remove energy from the fluid. Even the most simple pipe systems are actually quite complex when

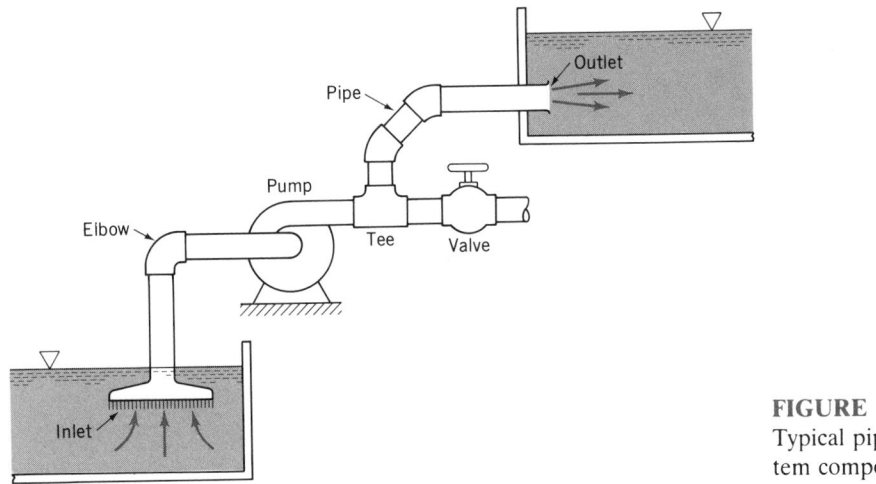

FIGURE 8.1
Typical pipe system components.

they are viewed in terms of rigorous analytical considerations. We will use an "exact" analysis of the simplest pipe flow topics (such as laminar flow in long, straight, constant diameter pipes), and dimensional analysis considerations combined with experimental results for the other pipe flow topics. Such an approach is not unusual in fluid mechanics investigations. When "real world" effects are important (such as viscous effects in pipe flows), it is often difficult or "impossible" to use only theoretical methods to obtain the desired results. A judicious combination of experimental data with theoretical considerations and dimensional analysis often provides the desired results. The flow in pipes discussed in this chapter is an example of such an analysis.

8.1   GENERAL CHARACTERISTICS OF PIPE FLOW

Before we apply the various governing equations to pipe flow examples, we will discuss some of the basic concepts of pipe flow. With these ground rules established we can then proceed to formulate and solve various important flow problems.

Most of the situations discussed in this chapter deal with fluid flowing in a round pipe of diameter D. Although not all conduits used to transport fluid from one location to another are round in cross section, most of the common ones are. These include typical water pipes, hydraulic hoses, and other conduits that are designed to withstand a considerable pressure difference across their walls without undue distortion of their shape. Typical conduits of noncircular cross section include heating and air conditioning ducts that are often of rectangular cross section. Normally the pressure difference between the inside and outside of these ducts is relatively small. Most of the basic principles involved are independent of the cross-sectional shape, although the details of the flow may be dependent on it. Unless otherwise specified, we will assume that the conduit is round, although we will show how to account for other shapes.

For all flows involved in this chapter, we assume that the pipe is completely filled with the fluid being transported as is shown in Fig. 8.2a. Thus, we will not consider a concrete pipe through which rain water flows without completely filling the pipe, as is shown in Fig. 8.2b. Such flows, called open-channel flow, are treated in Chapter 10. The difference between open-channel flow and the pipe flow of this chapter is in the fundamental mechanism that drives the flow. For open-channel flow, gravity alone

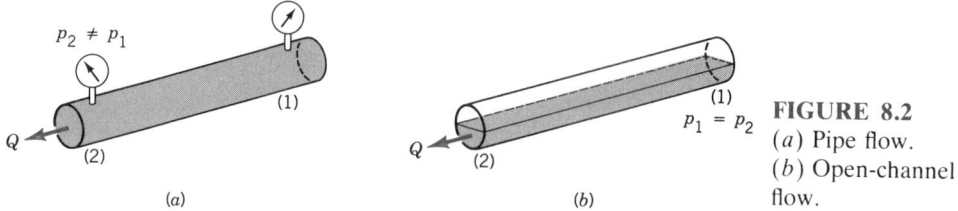

FIGURE 8.2
(*a*) Pipe flow.
(*b*) Open-channel
flow.

is the driving force—the water flows down a hill. For pipe flow, gravity may be important (the pipe need not be horizontal), but the main driving force is likely to be a pressure gradient along the pipe. If the pipe is not full, it is not possible to maintain this pressure difference, $p_1 - p_2$.

8.1.1 Laminar or Turbulent Flow

The flow of a fluid in a pipe may be laminar flow or it may be turbulent flow. Osborne Reynolds (1842–1912), a British scientist and mathematician, was the first to distinguish the difference between these two classifications of flow by using a simple apparatus as shown in Fig. 8.3a. If water runs through a pipe of diameter D with an average velocity V, the following characteristics are observed by injecting neutrally buoyant dye as shown. For "small enough flowrates" the dye streak (a streakline) will remain as a well-defined line as it flows along, with only slight blurring due to molecular diffusion of the dye into the surrounding water. For a somewhat larger "intermediate flowrate" the dye streak fluctuates in time and space, and intermittent bursts of irregular behavior appear along the streak. On the other hand, for "large enough flowrates" the dye streak almost immediately becomes blurred and spreads across the entire pipe in a random fashion. These three characteristics, denoted as *laminar, transitional,* and *turbulent* flow, respectively, are illustrated in Fig. 8.3b.

The curves shown in Fig. 8.4 represent the x component of the velocity as a function of time at a point A in the flow. The random fluctuations of the turbulent flow (with the associated particle mixing) are what disperse the dye throughout the pipe and cause the blurred appearance illustrated in Fig. 8.3b. For laminar flow in a pipe there is only one component of velocity—$\mathbf{V} = u\hat{\mathbf{i}}$. For turbulent flow the predominant component of velocity is also along the pipe, but it is unsteady (random) and accompan-

FIGURE 8.3 (*a*) Experiment to illustrate type of pipe flow. (*b*) Typical dye streaks.

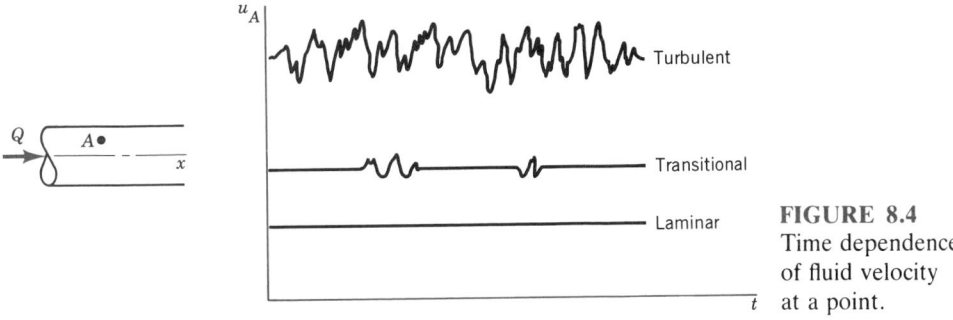

FIGURE 8.4
Time dependence of fluid velocity at a point.

ied by random components normal to the pipe axis—$\mathbf{V} = u\hat{\mathbf{i}} + v\hat{\mathbf{j}} + w\hat{\mathbf{k}}$. Such motion in a typical flow occurs too fast for our eyes to follow. Slow motion pictures of the flow can more clearly reveal the irregular, random, turbulent nature of the flow.

As was discussed in Chapter 7, we should not label dimensional quantities as being "large" or "small," such as "small enough flowrates" in the preceding paragraphs. Rather, the appropriate dimensionless quantity should be identified and the "small" or "large" character attached to it. A quantity is "large" or "small" only relative to a reference quantity. The ratio of those two quantities results in a dimensionless quantity. For pipe flow the most important dimensionless parameter is the Reynolds number, Re—the ratio of the inertia to viscous effects in the flow. Hence, in the previous paragraph the term flowrate should be replaced by Reynolds number, $\text{Re} = \rho V D/\mu$, where V is the average velocity in the pipe. That is, the flow in a pipe is laminar, transitional, or turbulent provided the Reynolds number is "small enough," "intermediate," or "large enough." It is not only the fluid velocity that determines the character of the flow—its density, viscosity, and the pipe size are of equal importance. These parameters combine to produce the Reynolds number. The distinction between laminar and turbulent pipe flow and its dependence on an appropriate dimensionless quantity was first pointed out by Osborne Reynolds in 1883.

The Reynolds number ranges for which laminar, transitional, or turbulent pipe flows are obtained cannot be precisely given. The actual transition from laminar to turbulent flow may take place at various Reynolds numbers, depending on how much the flow is "disturbed" by vibrations of the pipe, smoothness of the entrance region, and the like. For general engineering purposes (i.e., without undue precautions to eliminate such disturbances), the following values are appropriate: The flow in a round pipe is laminar if the Reynolds number is less than approximately 2100. The flow in a round pipe is turbulent if the Reynolds number is greater than approximately 4000. For Reynolds numbers between these two limits, the flow may switch between laminar and turbulent conditions in an apparently random fashion (transitional flow).

EXAMPLE 8.1

Water at a temperature of 50 °F flows through a pipe of diameter $D = 0.73$ in. (a) Determine the minimum time taken to fill a 12-oz glass (volume $= 0.0125$ ft^3) with water if the flow in the pipe is to be laminar. (b) Determine the maximum time taken to fill the glass if the flow is to be turbulent. Repeat the calculations if the water temperature is 140 °F.

SOLUTION

(a) If the flow in the pipe is to remain laminar, the minimum time to fill the glass will occur if the Reynolds number is the maximum allowed for laminar flow, typically $\text{Re} = \rho V D/\mu = 2100$. Thus, $V = 2100\,\mu/\rho D$, where from Table B.1, $\rho = 1.94$ slugs/ft³ and $\mu = 2.73 \times 10^{-5}$ lb·s/ft² at 50 °F, while $\rho = 1.91$ slugs/ft³ and $\mu = 0.974 \times 10^{-5}$ lb·s/ft² at 140 °F. Thus, the maximum average velocity for laminar flow in the pipe is

$$V = \frac{2100\mu}{\rho D} = \frac{2100(2.73 \times 10^{-5}\ \text{lb·s/ft}^2)}{(1.94\ \text{slugs/ft}^3)(0.73/12\ \text{ft})} = 0.486\ \text{lb·s/slug}$$

$$= 0.486\ \text{ft/s}$$

Similarly, $V = 0.176$ ft/s at 140 °F. With $\forall$ = volume of glass and $\forall = Qt$ we obtain

$$t = \frac{\forall}{Q} = \frac{\forall}{(\pi/4)D^2 V} = \frac{4(0.0125\ \text{ft}^3)}{(\pi[0.73/12]^2\text{ft}^2)(0.486\ \text{ft/s})}$$

$$= 8.85\ \text{s at}\ T = 50\ °\text{F}$$ (Ans)

Similarly, $t = 24.4$ s at 140 °F. To maintain laminar flow, the less viscous hot water requires a lower flowrate than the cold water.

(b) If the flow in the pipe is to be turbulent, the maximum time to fill the glass will occur if the Reynolds number is the minimum allowed for turbulent flow, $\text{Re} = 4000$. Thus, $V = 4000\mu/\rho D = 0.925$ ft/s and $t = 4.65$ s at 50 °F, while $V = 0.335$ ft/s and $t = 12.8$ s at 140 °F.

(Ans)

Note that because water is "not very viscous," the velocity must be "fairly small" to maintain laminar flow. In general, turbulent flows are encountered more often than laminar flows because of the relatively small viscosity of most common fluids (water, gasoline, air). If the flowing fluid had been honey with a kinematic viscosity ($\nu = \mu/\rho$) 3000 times greater than that of water, the above velocities would be increased by a factor of 3000 and the times reduced by the same factor. As we will see in the following sections, the pressure needed to force a very viscous fluid through a pipe at such a high velocity may be unreasonably large.

8.1.2 Entrance Region and Fully Developed Flow

Any fluid flowing in a pipe had to enter the pipe at some location. The region of flow near where the fluid enters the pipe is termed the *entrance region* and is illustrated in Fig. 8.5. It may be the first few feet of a pipe connected to a tank or the initial portion of a long run of a hot air duct coming from a furnace. In any case, the entrance region is that portion of the pipe in which the fluid makes its transition from not being influenced by the pipe to having completely lost its dependency on anything but the pipe.

As is shown in Fig. 8.5, the fluid typically enters the pipe with a nearly uniform velocity profile at section (1). As the fluid moves through the pipe, viscous effects cause it to stick to the pipe wall (the no-slip boundary condition). This is true whether the fluid is relatively inviscid air or a very viscous oil. Thus, a *boundary layer* in which viscous effects are important is produced along the pipe wall such that the initial velocity profile changes with distance along the pipe, x, until the fluid reaches the end of the entrance length, section (2), beyond which the velocity profile does not vary with x. The boundary layer has grown in thickness to completely fill the pipe.

FIGURE 8.5   Entrance region, developing flow, and fully developed flow in a pipe system.

Viscous effects are of considerable importance within the boundary layer. For fluid outside the boundary layer [within the *inviscid core* surrounding the centerline from (1) to (2)], viscous effects are negligible.

The shape of the velocity profile in the pipe depends on whether the flow is laminar or turbulent, as does the length of the entrance region, ℓ_e. As with many other properties of pipe flow, the dimensionless *entrance length*, ℓ_e/D, correlates quite well with the Reynolds number. Typical entrance lengths are given by

$$\frac{\ell_e}{D} = 0.06 \text{ Re for laminar flow} \tag{8.1}$$

and

$$\frac{\ell_e}{D} = 4.4 \, (\text{Re})^{1/6} \text{ for turbulent flow} \tag{8.2}$$

For very low Reynolds number flows the entrance length can be quite short ($\ell_e = 0.6D$ if Re = 10), whereas for large Reynolds number flows it may take a length equal to many pipe diameters before the end of the entrance region is reached ($\ell_e = 120D$ for Re = 2000). For many practical engineering problems $10^4 < \text{Re} < 10^5$ so that $20D < \ell_e < 30D$.

Calculation of the velocity profile and pressure distribution within the entrance region is quite complex. However, once the fluid reaches the end of the entrance region, section (2) of Fig. 8.5, the flow is simpler to describe because the velocity is a function of only the distance from the pipe centerline, r, and independent of x. This is true until the character of the pipe changes in some way such as a change in diameter, or the fluid flows through a bend, valve, or some other component at section (3). The flow between (2) and (3) is termed *fully developed*. Beyond the interruption of the fully developed flow [at section (4)], the flow gradually begins its return to its fully developed character [section (5)] and continues with this profile until the next pipe system component is reached [section (6)]. In many cases the pipe is long enough so that there is a considerable length of fully developed flow compared with the developing flow length [$(x_3 - x_2) \gg \ell_e$ and $(x_6 - x_5) \gg (x_5 - x_4)$]. In other cases the distances between one component (bend, tee, valve, etc.) of the pipe system and the next component is so short that fully developed flow is never achieved.

8.1.3   Pressure and Shear Stress

Fully developed steady flow in a constant diameter pipe may be driven by gravity and/or pressure forces. For horizontal pipe flow, gravity has no effect except for a hydrostatic pressure variation across the pipe, γD, that is usually negligible. It is the pressure difference, $\Delta p = p_1 - p_2$, between one section of the horizontal pipe and another which forces the fluid through the pipe. Viscous effects provide the restraining force that exactly balances the pressure force, thereby allowing the fluid to flow through the pipe with no acceleration (no convective or local acceleration). If viscous effects were absent in such flows, the pressure would be constant throughout the pipe, except for the hydrostatic variation.

In nonfully developed flow regions, such as the entrance region of a pipe, the fluid accelerates or decelerates as it flows (the velocity profile changes from a uniform profile at the entrance of the pipe to its fully developed profile at the end of the entrance region). Thus, in the entrance region there is a balance between pressure, viscous, and inertia (acceleration) forces. The result is a pressure distribution along the horizontal pipe as shown in Fig. 8.6. The magnitude of the pressure gradient, $\partial p / \partial x$, is larger in the entrance region than in the fully developed region, where it is a constant, $\partial p / \partial x = -\Delta p / \ell < 0$.

The fact that there is a nonzero pressure gradient along the horizontal pipe is a result of viscous effects. As is discussed in Chapter 3, if the viscosity were zero, the pressure would not vary with x. The need for the pressure drop can be viewed from two different standpoints. In terms of a force balance, the pressure force is needed to overcome the viscous forces generated. In terms of an energy balance, the work done by the pressure force is needed to overcome the viscous dissipation of energy throughout the fluid. If the pipe is not horizontal, the pressure gradient along it is due in part to the component of weight in that direction. As is discussed in Section 8.2.1, this contribution due to the weight either enhances or retards the flow, depending on whether the flow is downhill or uphill.

The nature of the pipe flow is strongly dependent on whether the flow is laminar or turbulent. This is a direct consequence of the differences in the nature of the shear stress in laminar and turbulent flows. As is discussed in some detail in Section 8.3.3, the shear stress in laminar flow is a direct result of momentum transfer among the randomly moving molecules (a microscopic phenomenon). The shear stress in turbulent flow is largely a result of momentum transfer among the randomly moving, finite sized bundles of fluid particles (a macroscopic phenomenon). The net result is that the physical properties of the shear stress are quite different for laminar flow

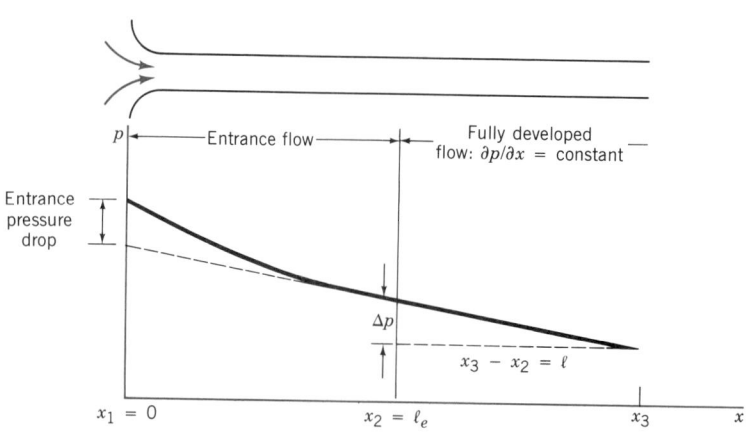

FIGURE 8.6
Pressure distribution along a horizontal pipe.

FIGURE 8.7 Flowrate as a function of pressure drop in a given horizontal pipe.

than for turbulent flow. This shows up in the plot of flowrate, Q, as a function of pressure drop, Δp, for a given length of pipe as shown in Fig. 8.7. For "small flowrates" (laminar flow) the flowrate is directly proportional to the pressure drop. In such cases, to double the flowrate the pressure drop must also be doubled. For "large flowrates" (turbulent flow) the flowrate is less sensitive to the pressure drop. For very large Reynolds number flows it is found that the flowrate is proportional to the square root of the pressure drop, $Q \sim \Delta p^{1/2}$. In such cases, to double the flowrate the pressure must be increased by a factor of 4.

Experiments show that the functional dependence among the various parameters is usually different for laminar pipe flow than it is for turbulent pipe flow. Some of these differences are indicated in Table 8.1. In some cases, the laminar and turbulent flows have the same parametric dependencies. For example, in either case the pressure drop, Δp, is proportional to the length of the pipe, ℓ, although the constant of proportionality is different for laminar flow than it is for turbulent flow. For laminar flow the pressure drop is independent of the density of the fluid, whereas for very large Reynolds number turbulent flow it is proportional to the density. For laminar flow the pressure drop is directly proportional to the viscosity, whereas for very large Reynolds number turbulent flow it is independent of the viscosity. For laminar flow the pressure drop is independent of the roughness of the pipe surface, ε, while for turbulent flow it is a function of the roughness. The reasons for such behavior are discussed in the remainder of the chapter.

TABLE 8.1 Dependence Among Various Parameters for Horizontal Pipe Flow (Other Parameters Fixed)

Parameter	Laminar flow	Very Large Re Turbulent Flow
Average velocity, V	$Q \sim V$	$Q \sim V$
Pressure drop, Δp	$Q \sim \Delta p$	$Q \sim \Delta p^{1/2}$
Density, ρ	$Q \sim \rho^0$ (independent)	$Q \sim \rho^{-1/2}$
Viscosity, μ	$Q \sim 1/\mu$	$Q \sim \mu^0$ (independent)
Pipe diameter, D	$Q \sim D^4$	$Q \sim D^{5/2}$
Pipe length, ℓ	$\Delta p \sim \ell$	$\Delta p \sim \ell$
Pipe roughness, ε	$\Delta p \sim \varepsilon^0$ (independent)	$\Delta p =$ function of ε

8.2   FULLY DEVELOPED LAMINAR FLOW

As is indicated in the previous section, the flow in long, straight, constant diameter sections of a pipe becomes fully developed. That is, the velocity profile is the same at any cross section of the pipe. Although this is true whether the flow is laminar or turbulent, the details of the velocity profile (and other flow properties) are quite different for these two types of flow. As will be seen in the remainder of this chapter, knowledge of the velocity profile can lead directly to other useful information such as pressure drop, head loss, flowrate, and the like. Thus, we begin by developing the equation for the velocity profile in fully developed laminar flow. If the flow is not fully developed, an approximate theoretical analysis becomes much more complex and is outside the scope of this text. If the flow is turbulent, a rigorous theoretical analysis is as yet not possible.

Although most flows are turbulent rather than laminar, and many pipes are not long enough to allow the attainment of fully developed flow, a theoretical treatment and full understanding of fully developed laminar flow is of considerable importance. First, it represents one of the few theoretical viscous flow analyses that can be carried out "exactly" (within the framework, of quite general assumptions) without using other "ad hoc" assumptions or approximations. An understanding of the method of analysis and the results obtained provides a foundation from which to carry out more complicated analyses. Second, there are many practical situations involving the use of fully developed laminar pipe flow.

There are numerous ways to derive important results pertaining to fully developed laminar flow. Three alternatives include: (1) from $\mathbf{F} = m\mathbf{a}$ applied directly to a fluid element, (2) from the Navier–Stokes equations of motion, and (3) from dimensional analysis methods.

8.2.1   From $\mathbf{F} = m\mathbf{a}$ Applied Directly to a Fluid Element

We consider the fluid element at time t as is shown in Fig. 8.8. It is a circular cylinder of fluid of length ℓ and radius r centered on the axis of a horizontal pipe of diameter D. Because the velocity is not uniform across the pipe, the initially flat ends of the cylinder of fluid at time t become distorted at time $t + \delta t$ when the fluid element has moved to its new location along the pipe as shown in the figure. If the flow is fully developed and steady, the distortion on each end of the fluid element is the same and no part of the fluid experiences any acceleration as it flows. The local acceleration is zero ($\partial \mathbf{V}/\partial t = 0$) because the flow is steady, and the convective acceleration is zero ($\mathbf{V} \cdot \nabla \mathbf{V} = u\, \partial u/\partial x \hat{\mathbf{i}} = 0$) because the flow is fully developed. Thus, every part of the fluid merely flows along its pathline parallel to the pipe walls with constant velocity, although neighboring particles have slightly different velocities. The velocity varies from one pathline to the next. This velocity variation, combined with the fluid viscosity, produces the shear stress.

FIGURE 8.8 Motion of a cylindrical fluid element within a pipe.

FIGURE 8.9 Free-body diagram of a cylinder of fluid.

If gravitational effects are neglected, the pressure is constant across any vertical cross section of the pipe, although it varies along the pipe from one section to the next. Thus, if the pressure is $p = p_1$ at section (1), it is $p_2 = p_1 - \Delta p$ at section (2). We have anticipated the fact that the pressure decreases in the direction of flow so that $\Delta p > 0$. A shear stress, τ, acts on the surface of the cylinder of fluid. This viscous stress is a function of the radius of the cylinder, $\tau = \tau(r)$.

As was done in fluid statics analysis (Chapter 2), we isolate the cylinder of fluid as is shown in Fig. 8.9 and apply Newton's second law, $F_x = ma_x$. In this case even though the fluid is moving, it is not accelerating, so that $a_x = 0$. Thus, fully developed horizontal pipe flow is merely a balance between pressure and viscous forces—the pressure difference acting on the end of the cylinder of area πr^2, and the shear stress acting on the lateral surface of the cylinder of area $2\pi r \ell$. This force balance can be written as

$$(p_1)\pi r^2 - (p_1 - \Delta p)\pi r^2 - (\tau)2\pi r \ell = 0$$

which can be simplified to give

$$\frac{\Delta p}{\ell} = \frac{2\tau}{r} \tag{8.3}$$

Equation 8.3 represents the basic balance in forces needed to drive each fluid particle along the pipe with constant velocity. Since neither Δp nor ℓ are functions of the radial coordinate, r, it follows that $2\tau/r$ must also be independent of r. That is, $\tau = Cr$, where C is a constant. At $r = 0$ (the centerline of the pipe) there is no shear stress ($\tau = 0$). At $r = D/2$ (the pipe wall) the shear stress is a maximum, denoted τ_w, the *wall shear stress*. Hence, $C = 2\tau_w/D$ and the shear stress distribution throughout the pipe is a linear function of the radial coordinate

$$\tau = \frac{2\tau_w r}{D} \tag{8.4}$$

as is indicated in Fig. 8.10. The linear dependence of τ on r is a result of the pressure force being proportional to r^2 (the pressure acts on the end of the fluid cylinder; area $= \pi r^2$) and the shear force being proportional to r (the shear stress acts on the

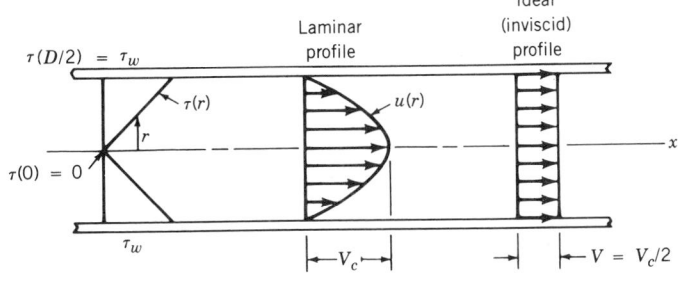

FIGURE 8.10 Shear stress distribution within the fluid in a pipe (laminar or turbulent flow) and typical velocity profiles.

lateral sides of the cylinder; area $= 2\pi r \ell$). If the viscosity were zero there would be no shear stress, and the pressure would be constant throughout the horizontal pipe ($\Delta p = 0$). As is seen from Eqs. 8.3 and 8.4, the pressure drop and wall shear stress are related by

$$\Delta p = \frac{4\ell\tau_w}{D} \tag{8.5}$$

A small shear stress can produce a large pressure difference if the pipe is relatively long ($\ell/D \gg 1$).

Although we are discussing laminar flow, a closer consideration of the assumptions involved in the derivation of Eqs. 8.3, 8.4, and 8.5 reveals that these equations are valid for both laminar and turbulent flow. To carry the analysis further we must prescribe how the shear stress is related to the velocity. This is the critical step that separates the analysis of laminar from that of turbulent flow—from being able to solve for the laminar flow properties and not being able to solve for the turbulent flow properties without additional "ad hoc" assumptions. As is discussed in Section 8.3, the shear stress dependence for turbulent flow is very complex. However, for laminar flow of a Newtonian fluid, the shear stress is simply proportional to the velocity gradient, "$\tau = \mu \, du/dy$" (see Section 1.6). In the notation associated with our pipe flow this becomes

$$\tau = -\mu\frac{du}{dr} \tag{8.6}$$

The negative sign is included to give $\tau > 0$ with $du/dr < 0$ (the velocity decreases from the pipe centerline to the pipe wall).

Equations 8.3 and 8.6 represent the two governing laws for fully developed laminar flow of a Newtonian fluid within a horizontal pipe. The one is Newton's second law of motion and the other is the definition of a Newtonian fluid. By combining these two equations we obtain

$$\frac{du}{dr} = -\left(\frac{\Delta p}{2\mu\ell}\right)r$$

which can be integrated to give the velocity profile as follows:

$$\int du = -\frac{\Delta p}{2\mu\ell}\int r \, dr$$

or

$$u = -\left(\frac{\Delta p}{4\mu\ell}\right)r^2 + C_1$$

where C_1 is a constant. Because the fluid is viscous it sticks to the pipe wall ($u = 0$ at $r = D/2$) so that $C_1 = (\Delta p/16\mu\ell)D^2$. Hence, the velocity profile can be written as

$$u(r) = \left(\frac{\Delta p D^2}{16\mu\ell}\right)\left[1 - \left(\frac{2r}{D}\right)^2\right] = V_c\left[1 - \left(\frac{2r}{D}\right)^2\right] \tag{8.7}$$

where $V_c = \Delta p D^2/(16\mu\ell)$ is the centerline velocity. An alternative expression can be written by using the relationship between the wall shear stress and the pressure

gradient (Eqs. 8.5 and 8.7) to give

$$u(r) = \frac{\tau_w D}{4\mu} \left[1 - \left(\frac{r}{R}\right)^2 \right]$$

where $R = D/2$ is the pipe radius.

This velocity profile, plotted in Fig. 8.10, is parabolic in the radial coordinate, r, has a maximum velocity, V_c, at the pipe centerline, and a minimum velocity (zero) at the pipe wall. The volume flowrate through the pipe can be obtained by integrating the velocity profile across the pipe. Since the flow is axisymmetric about the centerline, the velocity is constant on small area elements consisting of rings of radius r and thickness dr. Thus,

$$Q = \int u \, dA = \int_{r=0}^{r=R} u(r)2\pi r \, dr = 2\pi V_c \int_0^R \left[1 - \left(\frac{r}{R}\right)^2 \right] r \, dr$$

or

$$Q = \frac{\pi R^2 V_c}{2}$$

By definition, the average velocity is the flowrate divided by the cross-sectional area, $V = Q/A = Q/\pi R^2$, so that for this flow

$$V = \frac{\pi R^2 V_c}{2\pi R^2} = \frac{V_c}{2} = \frac{\Delta p D^2}{32\mu\ell} \tag{8.8}$$

and

$$Q = \frac{\pi D^4 \Delta p}{128\mu\ell} \tag{8.9}$$

As is indicated in Eq. 8.8, the average velocity is one-half of the maximum velocity. In general, for velocity profiles of other shapes (such as for turbulent pipe flow), the average velocity is not merely the average of the maximum (V_c) and minimum (0) velocities as it is for the laminar parabolic profile. The two velocity profiles indicated in Fig. 8.10 provide the same flowrate—one is the fictitious ideal ($\mu = 0$) profile; the other is the actual laminar flow profile.

The above results confirm the general properties of laminar flow indicated in Table 8.1. For a horizontal pipe the flowrate is (a) directly proportional to the pressure drop, (b) inversely proportional to the viscosity, (c) inversely proportional to the pipe length, and (d) proportional to the pipe diameter to the fourth power. With all other parameters fixed, an increase in diameter by a factor of 2 will increase the flowrate by a factor of 16—the flowrate is very strongly dependent on pipe size. A 2% error in diameter gives an 8% error in flowrate ($Q \sim D^4$ or $\delta Q \sim 4D^3 \, \delta D$, so that $\delta Q/Q = 4 \, \delta D/D$). This flow, the properties of which were first established experimentally by two independent workers, G. Hagen (1797–1884) in 1839 and J. Poiseuille (1799–1869) in 1840, is termed *Hagen–Poiseuille flow*. Equation 8.9 is commonly referred to as Poiseuille's law. Recall that all of these results are restricted to laminar flow (those with Reynolds numbers less than approximately 2100) in a horizontal pipe.

The adjustment necessary to account for nonhorizontal pipes, as shown in Fig. 8.11, can be easily included by replacing the pressure drop, Δp, by the combined effect of pressure and gravity, $\Delta p - \gamma\ell \sin\theta$, where θ is the angle between the pipe and the horizontal. (Note that $\theta > 0$ if the flow is uphill, while $\theta < 0$ if the flow is downhill.) This can be seen from the force balance in the x direction (along the pipe axis) on the cylinder of fluid shown in Fig. 8.11b. The method is exactly analogous

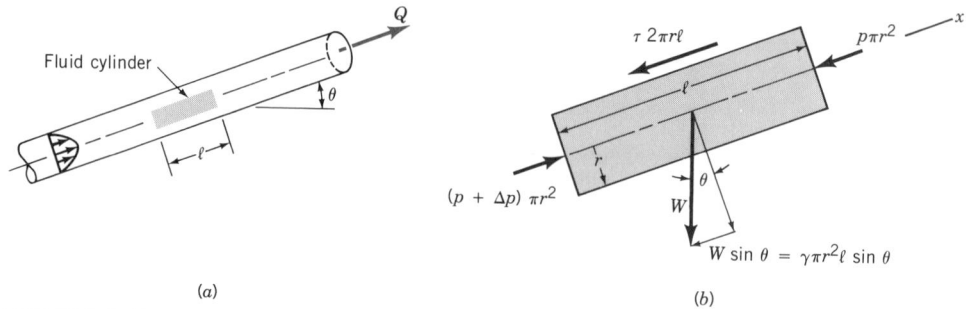

(a)   (b)

FIGURE 8.11   Free-body diagram of a fluid cylinder for flow in a nonhorizontal pipe.

to that used to obtain the Bernoulli equation (Eq. 3.6) when the streamline is not horizontal. The net force in the x direction is a combination of the pressure force in that direction, $\Delta p\pi r^2$, and the component of weight in that direction, $-\gamma\pi r^2\ell \sin\theta$. The result is a slightly modified form of Eq. 8.3 given by

$$\frac{\Delta p - \gamma\ell\sin\theta}{\ell} = \frac{2\tau}{r} \tag{8.10}$$

Thus, all of the results for the horizontal pipe are valid provided the pressure gradient is adjusted for the elevation term, that is, Δp is replaced by $\Delta p - \gamma\ell \sin\theta$ so that

$$V = \frac{(\Delta p - \gamma\ell\sin\theta)D^2}{32\mu\ell} \tag{8.11}$$

and

$$Q = \frac{\pi(\Delta p - \gamma\ell\sin\theta)D^4}{128\mu\ell} \tag{8.12}$$

It is seen that the driving force for pipe flow can be either a pressure drop in the flow direction, Δp, or the component of weight in the flow direction, $-\gamma\ell \sin\theta$. If the flow is downhill, gravity helps the flow (a smaller presure drop is required; $\sin\theta < 0$). If the flow is uphill, gravity works against the flow (a larger pressure drop is required; $\sin\theta > 0$). Note that $\gamma\ell \sin\theta = \gamma\,\Delta z$ (where Δz is the change in elevation) is a hydrostatic type pressure term. If there is no flow, $V = 0$ and $\Delta p = \gamma\ell \sin\theta = \gamma\,\Delta z$.

EXAMPLE 8.2

An oil with a viscosity of $\mu = 0.40$ N·s/m^2 and density $\rho = 900$ kg/m^3 flows in a pipe of diameter $D = 0.020$ m. (a) What pressure drop, $p_1 - p_2$, is needed to produce a flowrate of $Q = 2.0 \times 10^{-5}$ m^3/s if the pipe is horizontal with $x_1 = 0$ and $x_2 = 10$ m? (b) How steep of a hill, θ, must the pipe be on if the oil is to flow through the pipe at the same rate as in part (a), but with $p_1 = p_2$? (c) For the conditions of part (b), if $p_1 = 200$ kPa, what is the pressure at section $x_3 = 5$ m, where x is measured along the pipe?

SOLUTION

(a)   If the Reynolds number is less than 2100 the flow is laminar and the equations derived in this section are valid. Since the average velocity is $V = Q/A = 2.0 \times 10^{-5}$ m^3/s/ $[\pi(0.020)^2\text{m}^2/4] = 0.0637$ m/s, the Reynolds number is Re $= \rho VD/\mu = 2.87 < 2100$.

Hence, the flow is laminar and from Eq. 8.9 with $\ell = x_2 - x_1 = 10$ m, the pressure drop is

$$\Delta p = p_1 - p_2 = \frac{128\mu\ell Q}{\pi D^4}$$

$$= \frac{128(0.40 \text{ N·s/m}^2)(10.0 \text{ m})(2.0 \times 10^{-5} \text{ m}^3/\text{s})}{\pi(0.020 \text{ m})^4}$$

or

$$\Delta p = 20{,}400 \text{ N/m}^2 = 20.4 \text{ kPa} \qquad \text{(Ans)}$$

(b) If the pipe is on a hill of angle θ such that $\Delta p = p_1 - p_2 = 0$, Eq. 8.12 gives

$$\sin\theta = -\frac{128\mu Q}{\pi\rho g D^4} \qquad (1)$$

or

$$\sin\theta = \frac{-128(0.40 \text{ N·s/m}^2)(2.0 \times 10^{-5} \text{ m}^3/\text{s})}{\pi(900 \text{ kg/m}^3)(9.81 \text{ m/s}^2)(0.020 \text{ m})^4}$$

Thus, $\theta = -13.34°$. $\qquad$ (Ans)

This checks with the previous horizontal result as is seen from the fact that a change in elevation of $\Delta z = \ell \sin\theta = (10 \text{ m}) \sin(-13.34°) = -2.31$ m is equivalent to a pressure change of $\Delta p = \rho g\, \Delta z = (900 \text{ kg/m}^3)(9.81 \text{ m/s}^2)(2.31 \text{ m}) = 20{,}400 \text{ N/m}^2$, which is equivalent to that needed for the horizontal pipe. For the horizontal pipe it is the work done by the pressure forces that overcomes the viscous dissipation. For the zero pressure drop pipe on the hill, it is the change in potential energy of the fluid "falling" down the hill that is converted to the energy lost by viscous dissipation. Note that if it is desired to increase the flowrate to $Q = 1.0 \times 10^{-4}$ m³/s with $p_1 = p_2$, the value of θ given by Eq. 1 is $\sin\theta = -1.15$. Since the sine of an angle cannot be greater than 1, this flow would not be possible. The weight of the fluid would not be large enough to offset the viscous force generated for the flowrate desired. A larger diameter pipe would be needed.

(c) With $p_1 = p_2$ the length of the pipe, ℓ, does not appear in the flowrate equation (Eq. 1). This is a statement of the fact that for such cases the pressure is constant all along the pipe (provided the pipe lies on a hill of constant slope). This can be seen by substituting the values of Q and θ from case (b) into Eq. 8.12 and noting that $\Delta p = 0$ for any ℓ. For example, $\Delta p = p_1 - p_3 = 0$ if $\ell = x_3 - x_1 = 5$ m. Thus, $p_1 = p_2 = p_3$ so that

$$p_3 = 200 \text{ kPa} \qquad \text{(Ans)}$$

Note that if the fluid were gasoline ($\mu = 3.1 \times 10^{-4}$ N·s/m² and $\rho = 680$ kg/m³), the Reynolds number would be Re = 2790, the flow would probably not be laminar, and a use of Eqs. 8.9 and 8.12 would give incorrect results. Also note from Eq. 1 that the kinematic visosity, $v = \mu/\rho$, is the important viscous parameter. This is a statement of the fact that with constant pressure along the pipe, it is the ratio of the viscous force ($\sim\mu$) to the weight force ($\sim\gamma = \rho g$) that determines the value of θ.

8.2.2 From the Navier–Stokes Equations

In the previous section we obtained results for fully developed laminar pipe flow by applying Newton's second law and the assumption of a Newtonian fluid to a specific portion of the fluid—a cylinder of fluid centered on the axis of a long round pipe. When this governing law and assumptions are applied to a general fluid flow (not restricted to pipe flow), the result is the Navier–Stokes equations as discussed in

Chapter 6. In Section 6.9.3 these equations were solved for the specific geometry of fully developed laminar flow in a round pipe. The results are the same as those given in Eq. 8.7.

We will not repeat the detailed steps used to obtain the laminar pipe flow from the Navier–Stokes equations (see Section 6.9.3), but will indicate how the various assumptions used and steps applied in the derivation correlate with the analysis used in the previous section.

General motion of an incompressible Newtonian fluid is governed by the continuity equation (conservation of mass, Eq. 6.31) and the momentum equation (Eq. 6.127), which are rewritten here for convenience:

$$\nabla \cdot \mathbf{V} = 0 \tag{8.13}$$

$$\frac{\partial \mathbf{V}}{\partial t} + \mathbf{V} \cdot \nabla \mathbf{V} = -\frac{\nabla p}{\rho} - \mathbf{g} + \nu \nabla^2 \mathbf{V} \tag{8.14}$$

For steady, fully developed flow in a pipe, the velocity contains only an axial component, which is a function of only the radial coordinate $[\mathbf{V} = u(r)\hat{\mathbf{i}}]$. For such conditions, the left-hand side of the Eq. 8.14 is zero. This is equivalent to saying that the fluid experiences no acceleration as it flows along. The same constraint was used in the previous section when considering $\mathbf{F} = m\mathbf{a}$ for the fluid cylinder. Thus, the Navier–Stokes equations become

$$\nabla \cdot \mathbf{V} = 0$$

$$\nabla p + \rho g \hat{\mathbf{k}} = \mu \nabla^2 \mathbf{V} \tag{8.15}$$

The flow is governed by a balance of pressure, weight, and viscous forces in the flow direction, similar to that shown in Fig. 8.11 and Eq. 8.10. If the flow were not fully developed (as in an entrance region, for example), it would not be possible to simplify the Navier–Stokes equations to that form given in Eq. 8.15 (the nonlinear term $\mathbf{V} \cdot \nabla \mathbf{V}$ would not be zero), and the solution would be very difficult to obtain.

Because of the assumption that $\mathbf{V} = u(r)\hat{\mathbf{i}}$, the continuity equation, Eq. 8.13, is automatically satisfied. This conservation of mass condition was also automatically satisfied by the incompressible flow assumption in the derivation in the previous section. The fluid flows across one section of the pipe at the same rate that it flows across any other section (see Fig. 8.8).

When it is written in terms of polar coordinates (as was done in Section 6.9.3), the component of Eq. 8.15 along the pipe becomes

$$\frac{\partial p}{\partial x} + \rho g \sin \theta = \mu \frac{1}{r} \frac{\partial}{\partial r} \left(r \frac{\partial u}{\partial r} \right) \tag{8.16}$$

Since the flow is fully developed, $u = u(r)$ and the right-hand side is a function of, at most, only r. The left-hand side is a function of, at most, only x. It was shown that this leads to the condition that the pressure gradient in the x direction is a constant—$\partial p/\partial x = -\Delta p/\ell$. The same condition was used in the derivation of the previous section (Eq. 8.3).

It is seen from Eq. 8.16 that the effect of a nonhorizontal pipe enters into the Navier–Stokes equation in the same manner as was discussed in the previous section. The pressure gradient in the flow direction is coupled with the effect of the weight in that direction to produce an effective pressure gradient of $-\Delta p/\ell + \rho g \sin \theta$.

The velocity profile is obtained by integration of Eq. 8.16. Since it is a second order equation, two boundary conditions are needed—(1) the fluid sticks to the pipe

wall (as was also done in Eq. 8.7) and (2) either of the equivalent forms that the velocity remains finite throughout the flow (in particular $u < \infty$ at $r = 0$), or because of symmetry, $\partial u / \partial r = 0$ at $r = 0$. In the derivation of the previous section, only one boundary condition (the no-slip condition at the wall) was needed because the equation integrated was a first order equation. The other condition ($\partial u / \partial r = 0$ at $r = 0$) was automatically built into the analysis because of the fact that $\tau = -\mu \, du/dr$ and $\tau = 2\tau_w r / D = 0$ at $r = 0$.

The results obtained by either applying $\mathbf{F} = m\mathbf{a}$ to a fluid cylinder (Section 8.2.1) or solving the Navier–Stokes equations (Section 6.9.3) are exactly the same. Similarly, the basic assumptions regarding the flow structure are the same. This should not be surprising because the two methods are based on the same principle—Newton's second law. One is restricted to fully developed laminar pipe flow from the beginning (the drawing of the free body diagram) and the other starts with the general governing equations (the Navier–Stokes equations) with the appropriate restrictions concerning fully developed laminar flow applied as the solution process progresses.

8.2.3   From Dimensional Analysis

Although fully developed laminar pipe flow is simple enough to allow the rather straightforward solutions discussed in the previous two sections, it may be worthwhile to consider this flow from a dimensional analysis standpoint. Thus, we assume that the pressure drop in the horizontal pipe, Δp, is a function of the average velocity of the fluid in the pipe, V, the length of the pipe, ℓ, the pipe diameter, D, and the viscosity of the fluid, μ. We have not included the density or the specific weight of the fluid as parameters because for such flows they are not important parameters. There is neither mass (density) times acceleration nor a component of weight (specific weight times volume) in the flow direction involved. Thus,

$$\Delta p = F(V, \ell, D, \mu)$$

There are five variables that can be described in terms of three reference dimensions (M, L, T). According to the results of dimensional analysis (Chapter 7), this flow can be described in terms of $k - r = 5 - 3 = 2$ dimensionless groups. One such representation is

$$\frac{D \, \Delta p}{\mu V} = \phi \left(\frac{\ell}{D} \right) \tag{8.17}$$

where $\phi(\ell/D)$ is an unknown function of the length to diameter ratio of the pipe.

Although this is as far as dimensional analysis can take us, it seems reasonable to impose a further assumption that the pressure drop is directly proportional to the pipe length. That is, it takes twice the pressure drop to force fluid through a pipe if its length is doubled. The only way that this can be true is if $\phi(\ell/D) = C\ell/D$, where C is a constant. Thus, Eq. 8.17 becomes

$$\frac{D \, \Delta p}{\mu V} = \frac{C\ell}{D}$$

which can be rewritten as

$$\frac{\Delta p}{\ell} = \frac{C\mu \, V}{D^2}$$

or

$$Q = AV = \frac{(\pi/4C) \, \Delta p \, D^4}{\mu \ell} \tag{8.18}$$

The basic functional dependence for laminar pipe flow given by Eq. 8.18 is the same as that obtained by the analysis of the two previous sections. The value of C must be determined by theory (as done in the previous two sections) or experiment. For a round pipe, $C = 32$. For ducts of other cross-sectional shapes, the value of C is different (see Section 8.4.3).

It is usually advantageous to describe a process in terms of dimensionless quantities. To this end we rewrite the pressure drop equation for laminar horizontal pipe flow, Eq. 8.8, as $\Delta p = 32\mu\ell V/D^2$ and divide both sides by the dynamic pressure, $\rho V^2/2$, to obtain the dimensionless form as

$$\frac{\Delta p}{\frac{1}{2}\rho V^2} = \frac{(32\mu\ell V/D^2)}{\frac{1}{2}\rho V^2} = 64\left(\frac{\mu}{\rho VD}\right)\left(\frac{\ell}{D}\right) = \frac{64}{\mathrm{Re}}\left(\frac{\ell}{D}\right)$$

This is often written as

$$\Delta p = f\frac{\ell}{D}\frac{\rho V^2}{2}$$

where the dimensionless quantity

$$f = \Delta p(D/\ell)/(\rho V^2/2)$$

is termed the *friction factor*, or sometimes the *Darcy friction factor*. Thus, the friction factor for laminar fully developed pipe flow is simply

$$f = \frac{64}{\mathrm{Re}} \tag{8.19}$$

By substituting the pressure drop in terms of the wall shear stress (Eq. 8.5), we obtain an alternate expression for the friction factor as a dimensionless wall shear stress:

$$f = \frac{8\tau_w}{\rho V^2} \tag{8.20}$$

Knowledge of the friction factor will allow us to obtain a variety of information regarding pipe flow. For turbulent flow the dependence of the friction factor on the Reynolds number is much more complex than that given by Eq. 8.19 for laminar flow. This is discussed in detail in Section 8.4.

8.2.4   Energy Considerations

In the previous three sections we derived the basic laminar flow results from application of $\mathbf{F} = m\mathbf{a}$ or dimensional analysis considerations. It is equally important to understand the implications of energy considerations of such flows. To this end we consider the energy equation for incompressible, steady flow between two locations as is given in Eq. 5.89:

$$\frac{p_1}{\gamma} + \alpha_1\frac{V_1^2}{2g} + z_1 = \frac{p_2}{\gamma} + \alpha_2\frac{V_2^2}{2g} + z_2 + h_L \tag{8.21}$$

Recall that the kinetic energy coefficients, α_1 and α_2, compensate for the fact that the velocity profile across the pipe is not uniform. For uniform velocity profiles $\alpha = 1$, whereas for any nonuniform profile, $\alpha > 1$. The head loss term, h_L, accounts for any energy loss associated with the flow. This loss is a direct consequence of the viscous dissipation that occurs throughout the fluid in the pipe. For the ideal (inviscid) cases discussed in previous chapters, $\alpha_1 = \alpha_2 = 1$, $h_L = 0$, and the energy equation reduces to the familiar Bernoulli equation discussed in Chapter 3 (Eq. 3.7).

Even though the velocity profile in viscous pipe flow is not uniform, for fully developed flow it does not change from section (1) to section (2) so that $\alpha_1 = \alpha_2$. Thus, the kinetic energy is the same at any section ($\alpha_1 V_1^2/2 = \alpha_2 V_2^2/2$) and the energy equation becomes

$$\left(\frac{p_1}{\gamma} + z_1\right) - \left(\frac{p_2}{\gamma} + z_2\right) = h_L \tag{8.22}$$

The energy dissipated by the viscous forces within the fluid is supplied by the excess work done by the pressure and gravity forces.

A comparison of Eqs. 8.22 and 8.10 shows that the head loss is given by

$$h_L = \frac{2\tau\ell}{\gamma r}$$

(recall $p_1 = p_2 + \Delta p$ and $z_2 - z_1 = \ell \sin\theta$), which by use of Eq. 8.4 can be rewritten in the form:

$$h_L = \frac{4\ell\,\tau_w}{\gamma D} \tag{8.23}$$

It is the shear stress at the wall (which is directly related to the viscosity and the shear stress throughout the fluid) that is responsible for the head loss. A closer consideration of the assumptions involved in the derivation of Eq. 8.23 will show that it is valid for both laminar and turbulent flow.

EXAMPLE 8.3

The flowrate, Q, of corn syrup through the horizontal pipe shown in Fig. E8.3 is to be monitored by measuring the pressure difference between sections (1) and (2). It is proposed that $Q = K\,\Delta p$, where the calibration constant, K, is a function of temperature, T, because of the

(a)

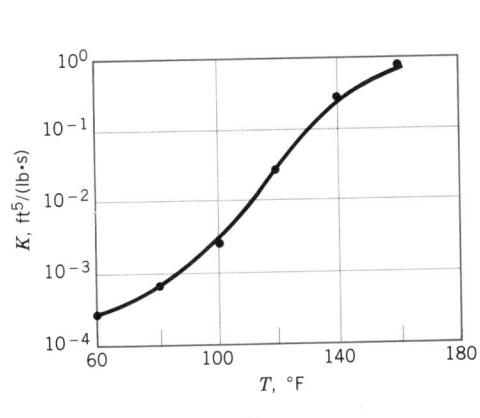

(b)

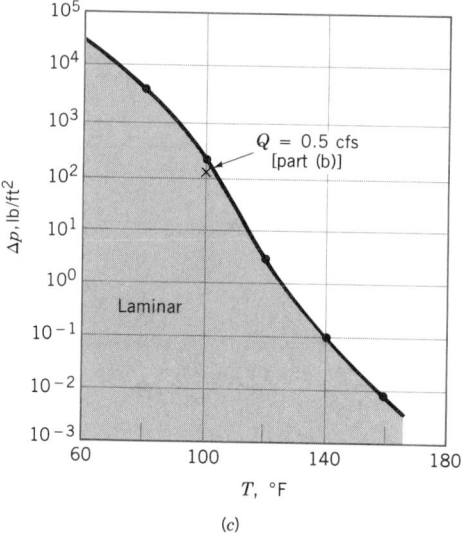

FIGURE E8.3

temperature variation of the syrup's viscosity and density. These variations are given in Table E8.3. (a) Plot $K(T)$ versus T for 60 °F $\leq T \leq$ 160 °F. (b) Determine the wall shear stress and the pressure drop, $\Delta p = p_1 - p_2$, for Q = 0.5 ft³/s and T = 100 °F. (c) For the conditions of part (b), determine the net pressure force, $(\pi D^2 / 4) \Delta p$, and the net shear force, $\pi D \ell \tau_w$, on the fluid within the pipe between the sections (1) and (2).

TABLE E8.3

T (°F)	ρ (slugs/ft³)	μ (lb·s/ft²)
60	2.07	4.0×10^{-2}
80	2.06	1.9×10^{-2}
100	2.05	3.8×10^{-3}
120	2.04	4.4×10^{-4}
140	2.03	9.2×10^{-5}
160	2.02	2.3×10^{-5}

SOLUTION

(a) If the flow is laminar it follows from Eq. 8.9 that

$$Q = \frac{\pi D^4 \, \Delta p}{128 \mu \ell} = \frac{\pi \left(\frac{3}{12} \text{ ft}\right)^4 \Delta p}{128 \mu (6 \text{ ft})}$$

or

$$Q = K \, \Delta p = \frac{1.60 \times 10^{-5}}{\mu} \Delta p \qquad (1)$$

where the units on Q, Δp, and μ are ft³/s, lb/ft², and lb·s/ft², respectively. Thus

$$K = \frac{1.60 \times 10^{-5}}{\mu} \qquad \text{(Ans)}$$

where the units of K are ft⁵/lb·s. By using values of the viscosity from Table E8.3, the calibration curve shown in Fig. E8.3b is obtained. If the upper limit on laminar flow is

Re $= \rho VD/\mu = 2100$ and the flowrate is $Q = V\pi D^2/4$, it follows that the maximum flowrate for which the above calibration is valid is

$$Q = \frac{\pi}{4}\left(\frac{2100\mu}{\rho D}\right)D^2 = \frac{\pi(2100)(\frac{3}{12}\text{ ft})\mu}{4\rho} = 412\,\frac{\mu}{\rho} \qquad (2)$$

By combining Eqs. 1 and 2 we obtain the upper limit on the value of Δp for which the calibration curve is valid as

$$\Delta p = \frac{412\mu^2}{1.6\times 10^{-5}\rho} = 2.58\times 10^7\,\frac{\mu^2}{\rho}$$

where Δp is in lb/ft^2. This limiting value of Δp that divides laminar and transitional or turbulent flow is a function of temperature because ρ and μ are functions of temperature (Table E8.3). This result is shown in Fig. E8.3c. Use of the calibration equation $Q = K\,\Delta p$ is not appropriate if the flow is turbulent. Because of the wide range in the values of K and Δp (due to the wide range in the viscosity), it would be very difficult to obtain accurate results with this device over the temperature range desired.

Note that the value of the syrup density was not needed to determine the value of K. The flow involved is governed by pressure and viscous forces. However, the density was needed to determine the upper limit for the value of Δp because the transition from laminar to turbulent flow is a process governed in part by the fluid inertia and in part by the viscous effects. That is, the Reynolds number is the governing parameter.

(b)   For $T = 100\,°F$, the viscosity is $\mu = 3.8\times 10^{-3}\,\text{lb·s/ft}^2$ so that with a flowrate of $Q = 0.5\,\text{ft}^3/\text{s}$ the pressure drop (according to Eq. 8.9) is

$$\Delta p = \frac{128\mu\ell Q}{\pi D^4} = \frac{128(3.8\times 10^{-3}\,\text{lb·s/ft}^2)(6\,\text{ft})(0.5\,\text{ft}^3/\text{s})}{\pi(\frac{3}{12}\text{ ft})^4}$$

$$= 119\,\text{lb/ft}^2 \qquad \text{(Ans)}$$

As is indicated in Fig. E8.3c, this $Q = 0.5$ cfs flowrate at $T = 100\,°F$ is near the upper limit for laminar flow conditions. From Eq. 8.5 the wall shear stress is

$$\tau_w = \frac{\Delta p D}{4\ell} = \frac{(119\,\text{lb/ft}^2)(\frac{3}{12}\text{ ft})}{4(6\,\text{ft})} = 1.24\,\text{lb/ft}^2 \qquad \text{(Ans)}$$

(c)   For the conditions of part (b), the net pressure force, F_p, on the fluid within the pipe between sections (1) and (2) is

$$F_p = \frac{\pi}{4}D^2\,\Delta p = \frac{\pi}{4}\left(\frac{3}{12}\text{ ft}\right)^2(119\,\text{lb/ft}^2) = 5.84\,\text{lb} \qquad \text{(Ans)}$$

Similarly, the net viscous force, F_v, on that portion of the fluid is

$$F_v = 2\pi\left(\frac{D}{2}\right)\ell\,\tau_w$$

$$= 2\pi\left(\frac{3}{2(12)}\text{ ft}\right)(6\,\text{ft})(1.24\,\text{lb/ft}^2) = 5.84\,\text{lb} \qquad \text{(Ans)}$$

Note that the values of these two forces are the same. The net force is zero; there is no acceleration.

8.3   FULLY DEVELOPED TURBULENT FLOW

In the previous section various properties of fully developed laminar pipe flow were discussed. Since turbulent pipe flow is actually more likely to occur than laminar flow

in practical situations, it is necessary to obtain similar information for turbulent pipe flow. However, turbulent flow is a very complex process. Numerous persons have devoted considerable effort in attempting to understand the variety of baffling aspects of turbulence. Although a considerable amount of knowledge about the topic has been developed, the field of turbulent flow still remains the least understood area of fluid mechanics. In this book we can provide only some of the very basic ideas concerning turbulence. The interested reader should consult some of the many books available for further reading (Refs. 1, 2, and 3).

8.3.1 Transition from Laminar to Turbulent Flow

Flows are classified as laminar or turbulent. For any flow situation, whether it is flow in a pipe, flow past your automobile, or flow in the kettle of soup cooking on the stove, the flow is laminar "if it is not forced too hard" and turbulent "if it is forced hard enough." The meaning of "forced hard" depends on flow situation involved. For pipe flow, "forced hard" means a large enough value of the Reynolds number, Re. If $Re = \rho VD/\mu < 2100$ the flow is laminar ("not forced too hard"), whereas if $Re > 4000$ the flow is turbulent ("forced hard enough").

For any flow geometry, there is one (or more) dimensionless parameter such that with this parameter value below a particular value the flow is laminar, whereas with the parameter value larger than a certain value the flow is turbulent. The important parameters involved (i.e., Reynolds number, Mach number, etc.) and their critical values depend on the specific flow situation involved. For example, flow in a pipe and flow along a flat plate (boundary layer flow as discussed in Section 9.2.4) can be laminar or turbulent, depending on the value of the Reynolds number involved. For pipe flow the value of the Reynolds number must be less than approximately 2100 for laminar flow and greater than approximately 4000 for turbulent flow. For flow along a flat plate the transition between laminar and turbulent flow occurs at a Reynolds number of approximately 500,000 (see Section 9.2.4), where the length term in the Reynolds number is the distance measured from the leading edge of the plate.

Consider a long section of pipe that is initially filled with a fluid at rest. As the valve is opened to start the flow, the flow velocity and, hence, the Reynolds number increases from zero (no flow) to its maximum steady state flow value, as is shown in Fig. 8.12. Assume this transient process is slow enough so that unsteady effects are negligible (quasisteady flow). For an initial time period the Reynolds number is small

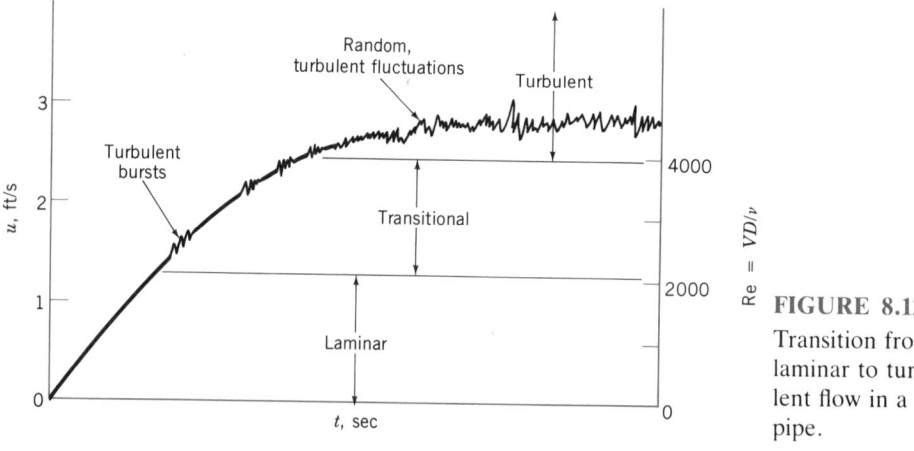

FIGURE 8.12
Transition from laminar to turbulent flow in a pipe.

enough for laminar flow to occur. At some time the Reynolds number reaches 2100, and the flow begins its transition to turbulent conditions. Intermittent spots or bursts of turbulence appear. As the Reynolds number is increased the entire flow field becomes turbulent. The flow remains turbulent as long as the Reynolds number exceeds approximately 4000.

A typical velocity trace measured at a given location in the flow, $u = u(t)$, is shown in Fig. 8.13. Its irregular, random nature is the distinguishing feature of turbulent flow. The character of many of the important properties of the flow (pressure drop, the heat transfer, etc.) depends strongly on the existence and nature of the turbulent fluctuations or randomness indicated. In previous considerations involving inviscid flow, the Reynolds number is (strictly speaking) infinite (because the viscosity is zero), and the flow most surely would be turbulent. However, reasonable results were obtained by using the inviscid Bernoulli equation as the governing equation. The reason that such simplified inviscid analyses gave reasonable results is that viscous effects were not very important and the velocity used in the calculations was actually the time-averaged velocity indicated in Fig. 8.13. Calculation of the heat transfer, pressure drop, and many other parameters would not be possible without inclusion of the seemingly "small," but very important, effects associated with the randomness of the flow.

Consider flow in a pan of water placed on a stove. With the stove turned off the fluid is stationary. The initial "sloshing" has died out because of viscous dissipation within the water. With the stove turned on, a temperature gradient in the vertical direction, $\partial T / \partial z$, is produced. The water temperature is greatest near the pan bottom and decreases toward the top of the fluid layer. If the temperature difference is very small, the water will remain stationary, even though the water density is smallest near the bottom of the pan because of the decrease in density with an increase in temperature. A further increase in the temperature gradient will cause a buoyancy driven instability that results in fluid motion—the light warm water rises to the top and the heavy cold water sinks to the bottom. This slow, regular "turning over" increases the heat transfer from the pan to the water and promotes mixing within the pan. As the temperature gradient increases still further, the fluid motion becomes more vigorous and eventually turns into a chaotic, random, turbulent flow with considerable mixing and greatly increased heat transfer rate. The flow has progressed from a stationary fluid, to laminar flow, and finally to turbulent flow.

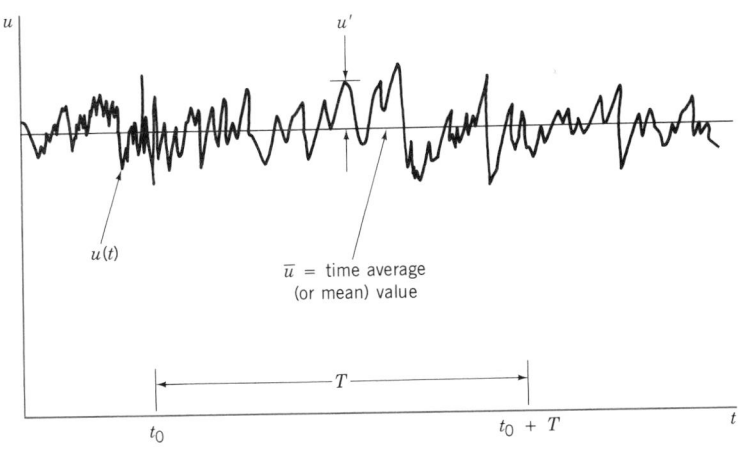

FIGURE 8.13
The time averaged, $\bar{u}$, and fluctuating, u', description of a parameter for turbulent flow.

Mixing processes and heat and mass transfer processes are considerably enhanced in turbulent flow compared to laminar flow. This is due to the macroscopic scale of the randomness in turbulent flow. We are all familiar with the "rolling," vigorous eddy type motion of the water in a pan being heated on the stove (even if it is not heated to boiling). Such finite sized random mixing is very effective in transporting energy and mass throughout the flow field, thereby increasing the various rate processes involved. Laminar flow, on the other hand, can be thought of as very small but finite sized fluid particles flowing smoothly in layers, one over another. The only randomness and mixing takes place on the molecular scale and results in relatively small heat, mass, and momentum transfer rates.

Without turbulence it would be virtually impossible to carry out life as we now know it. In some situations, turbulent flow is desirable. To transfer the required heat between a solid and an adjacent fluid (such as in the cooling coils of an air conditioner or a boiler of a power plant) would require an enormously large heat exchanger if the flow were laminar. Similarly, the required mass transfer of a liquid state to a vapor state (such as is needed in the evaporated cooling system associated with sweating) would require very large surfaces if the fluid flowing past the surface were laminar rather than turbulent.

Turbulence is also of importance in the mixing of fluids. Smoke from a stack would continue for miles as a ribbon of pollutant without rapid dispersion within the surrounding air if the flow were laminar rather than turbulent. Under certain atmospheric conditions this is observed to occur. Although there is mixing on a molecular scale (laminar flow), it is several orders of magnitude slower and less effective than the mixing on a macroscopic scale (turbulent flow). It is considerably easier to mix cream into a cup of coffee (turbulent flow) than to thoroughly mix two colors of a viscous paint (laminar flow).

In other situations laminar (rather than turbulent) flow is desirable. The pressure drop in pipes (hence, the power requirements for pumping) can be considerably lower if the flow is laminar rather than turbulent. Fortunately, the blood flow through a person's arteries is normally laminar, except in the largest arteries with high blood flowrates. The aerodynamic drag on an airplane wing can be considerably smaller with laminar flow past it than with turbulent flow.

8.3.2   Turbulent Shear Stress

The fundamental difference between laminar and turbulent flow lies in the chaotic, random behavior of the various fluid parameters. Such variations occur in the three components of velocity, the pressure, the shear stress, the temperature, and any other variable that has a field description. As is indicated in Fig. 8.13, such flows can be described in terms of their mean values (denoted with an overbar) on which are superimposed the fluctuations (denoted with a prime). Thus, if $u = u(x, y, z, t)$ is the x component of velocity, then its time mean (or *time average*) value, $\bar{u}$, is

$$\bar{u} = \frac{1}{T} \int_{t_0}^{t_0 + T} u(x, y, z, t)\, dt \tag{8.24}$$

where the time interval, T, is considerably longer than the period of the longest fluctuations, but considerably shorter than any unsteadiness of the average velocity. This is illustrated in Fig. 8.13.

The *fluctuating part* of the velocity, u', is that time-varying portion that differs from the average value:

$$u = \bar{u} + u' \quad \text{or} \quad u' = u - \bar{u} \tag{8.25}$$

Clearly, the time average of the fluctuations is zero, since

$$\overline{u'} = \frac{1}{T} \int_{t_0}^{t_0+T} (u - \overline{u}) \, dt = \frac{1}{T} \left(\int_{t_0}^{t_0+T} u \, dt - \overline{u} \int_{t_0}^{t_0+T} dt \right)$$

$$= \frac{1}{T} (T\overline{u} - T\overline{u}) = 0$$

The fluctuations "are equally distributed on either side of the average." It is also clear, as is indicated in Fig. 8.14, that since the square of a fluctuation quantity cannot be negative $[(u')^2 \geq 0]$, its average value is positive. Thus,

$$\overline{(u')^2} = \frac{1}{T} \int_{t_0}^{t_0+T} (u')^2 \, dt > 0$$

Similarly, it may be that the average of products of the fluctuations, such as $\overline{u'v'}$, are zero or nonzero (either positive or negative).

The structure and characteristics of turbulence may vary from one flow situation to another. For example, the *turbulence intensity* (or the level of the turbulence) may be larger in a very gusty wind than it is in a relatively steady (although turbulent) wind. The turbulence intensity, $\mathscr{I}$, is often defined as the square root of the mean square of the fluctuating velocity divided by the time average velocity, or

$$\mathscr{I} = \frac{\sqrt{\overline{(u')^2}}}{\overline{u}} = \frac{\left[\dfrac{1}{T} \displaystyle\int_{t_0}^{t_0+T} (u')^2 \, dt \right]^{1/2}}{\overline{u}}$$

The larger the turbulence intensity, the larger the fluctuations of the velocity (and other flow parameters). Well-designed wind tunnels have typical values of $\mathscr{I} \approx 0.01$, although with extreme care values as low as $\mathscr{I} \approx 0.0002$ have been obtained. On the other hand, values of $\mathscr{I} \gtrsim 0.1$ are found for the flow in the atmosphere and rivers.

Another turbulence parameter that is different from one flow situation to another is the period of the fluctuations—the *time scale* of the fluctuations shown in Fig. 8.13. In many flows, such as the flow of water from a faucet, typical frequencies are on the order of 10, 100, or 1000 cycles per second (cps). For other flows, such as the Gulf Stream current in the Atlantic Ocean or flow of the atmosphere of Jupiter, characteristic random oscillations may have a period on the order of hours, days, or more.

It is tempting to extend the concept of viscous shear stress for laminar flow ($\tau =$

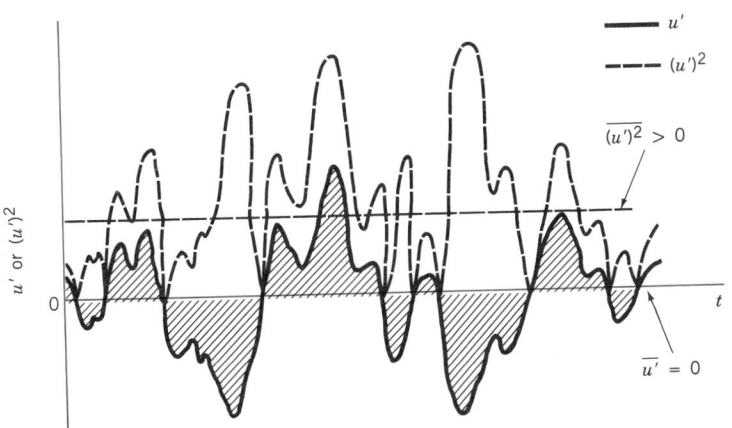

FIGURE 8.14 Average of the fluctuations and average of the square of the fluctuations.

$\mu\ du/dy$) to that of turbulent flow by replacing u, the instantaneous velocity, by $\bar{u}$, the time-average velocity. However, numerous experimental and theoretical studies have shown that such an approach will lead to completely incorrect results. That is, $\tau \neq \mu\ d\bar{u}/dy$. A physical explanation for this behavior can be found in the concept of what produces a shear stress.

Laminar flow is modeled as a set of fluid particles that flow smoothly along in layers, gliding past the slightly slower or faster ones on either side. As is discussed in Chapter 1, the fluid actually consists of numerous molecules darting about in an almost random fashion as is indicated in Fig. 8.15. The motion is not entirely random— a slight bias in one direction produces the flowrate we associate with the motion of fluid particles, $\bar{u}$. As the molecules dart across a given plane (plane A–A, for example), the ones moving upward have come from an area of smaller average x component of velocity than the ones moving downward, which have come from an area of larger velocity.

The momentum flux in the x direction across plane A–A gives rise to a drag (to the left) of the lower fluid on the upper fluid and an equal but opposite effect of the upper fluid on the lower fluid. The sluggish molecules moving upward across plane A–A must be accelerated by the fluid above this plane. The rate of change of momentum in this process produces (on the macroscopic scale) a shear force. Similarly, the more energetic molecules moving down across plane A–A must be slowed down by the fluid below that plane. This shear force is present only if there is a gradient in $u = u(y)$, otherwise the average x component of velocity (and momentum) of the upward and downward molecules is exactly the same. In addition, there are attractive forces between molecules. By combining these effects we obtain the well-known Newton viscosity law: $\tau = \mu\ du/dy$, where on a molecular basis μ is related to the mass and speed (temperature) of the random motion of the molecules.

Although the above random motion of the molecules is also present in turbulent flow, there is another factor that is generally more important. A simplistic way of thinking about turbulent flow is to consider it as consisting of a series of random, three-dimensional eddy type motions as is depicted (in one dimension only) in Fig. 8.16. These eddies range in size from very small diameter (on the order of the size of a fluid particle) to fairly large diameter (on the order of the size of the object or flow geometry considered). They move about randomly, conveying mass with an average velocity $\bar{u} = \bar{u}(y)$. This eddy structure greatly promotes mixing within the

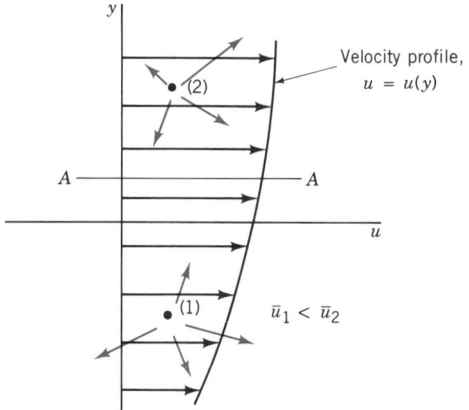

FIGURE 8.15   Laminar flow shear stress caused by random motion of molecules.

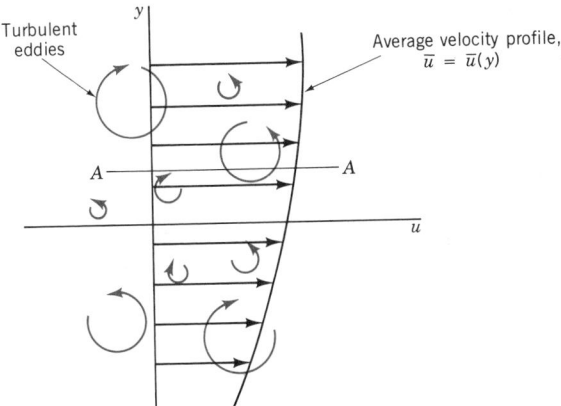

FIGURE 8.16 Turbulent flow as a series of random, three-dimensional eddies.

fluid. It also greatly increases the transport of x momentum across plane A–A. That is, finite parcels of fluid (not merely individual molecules as in laminar flow) are randomly transported across this plane, resulting in a relatively large (when compared with laminar flow) shear force.

The random velocity components that account for this momentum transfer (hence, the shear force) are u' (for the x component of velocity) and v' (for the rate of mass transfer crossing the plane). A more detailed consideration of the processes involved will show that the apparent shear stress on plane A–A is given by (Ref. 2)

$$\tau = \mu \frac{d\bar{u}}{dy} - \rho \overline{u'v'} = \tau_{\text{lam}} + \tau_{\text{turb}} \qquad (8.26)$$

Note that if the flow is laminar, $u' = v' = 0$, so that $\overline{u'v'} = 0$ and Eq. 8.26 reduces to the customary random molecule motion induced *laminar shear stress*, $\tau_{\text{lam}} = \mu \, d\bar{u}/dy$. For turbulent flow it is found that the *turbulent shear stress*, $\tau_{\text{turb}} = -\rho\overline{u'v'}$, is positive. Hence, the shear stress is greater in turbulent flow than in laminar flow. Note the units on τ_{turb} are (density)(velocity)^2 = (slugs/ft^3)(ft/s)^2 = (slugs· ft/s^2)/ft^2 = lb/ft^2, or N/m^2, as expected. Terms of the form $-\rho\overline{u'v'}$ (or $-\rho\overline{v'w'}$, etc.) are called *Reynolds stresses* in honor of Osborne Reynolds who first discussed them in 1895.

It is seen from Eq. 8.26 that the shear stress in turbulent flow is not merely proportional to the gradient of the time-average velocity, $\bar{u}(y)$. It also contains a contribution due to the random fluctuations of the x and y components of velocity. The density is involved because of the momentum transfer of the fluid within the random eddies. Although the relative magnitude of τ_{lam} compared to τ_{turb} is a complex function dependent on the specific flow involved, typical measurements indicate the structure shown in Fig. 8.17a. (Recall from Eq. 8.4 that the shear stress is proportional to the distance from the centerline of the pipe). In a very narrow region near the wall (the *viscous sublayer*), the laminar shear stress is dominant. Away from the wall (in the *outer layer*) the turbulent portion of the shear stress is dominant. The transition between these two regions occurs in the *overlap layer*. The corresponding typical velocity profile is shown in Fig. 8.17b.

The scale of the sketches shown in Fig. 8.17 is not necessarily correct. Typically the value of τ_{turb} is 100 to 1000 times greater than τ_{lam} in the outer region, while the converse is true in the viscous sublayer. A correct modeling of turbulent flow is strongly

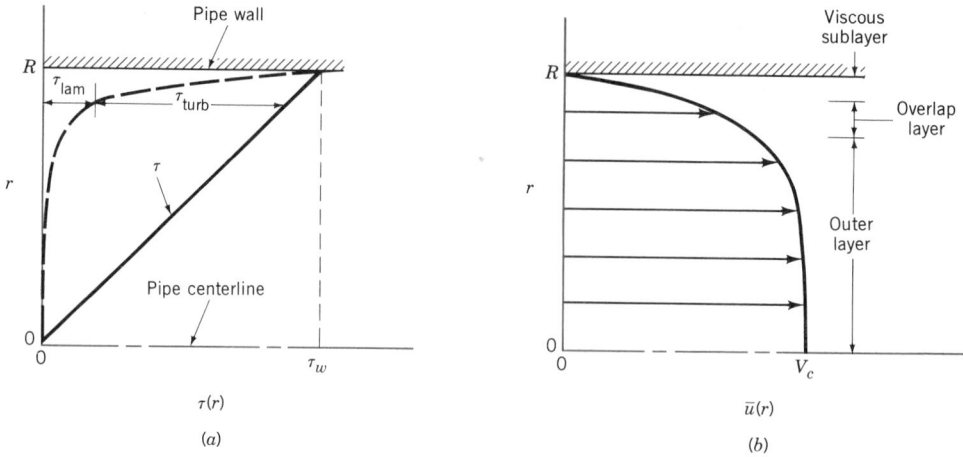

FIGURE 8.17 Structure of turbulent flow in a pipe. (*a*) Shear stress. (*b*) Average velocity.

dependent on an accurate knowledge of τ_{turb}. This, in turn, requires an accurate knowledge of the fluctuations u' and v', or $\overline{\rho u'v'}$. As yet it is not possible to solve the governing equations (the Navier–Stokes equations) for these details of the flow, although numerical techniques using the largest and fastest computers available have produced important information about some of the characteristics of turbulence. Considerable effort has gone into the study of turbulence. Much remains to be learned. Perhaps studies in the new areas of chaos and fractal geometry (Ref. 4) will provide the tools for a better understanding of turbulence.

The vertical scale of Fig. 8.17 is also distorted. The viscous sublayer is usually a very thin layer adjacent to the wall. For example, for water flow in a 3-in.-diameter pipe with an average velocity of 10 ft/s, the viscous sublayer is approximately 0.002 in. thick. Since the fluid motion within this thin layer is critical in terms of the overall flow (the no-slip condition and the wall shear stress occur in this layer), it is not surprising to find that turbulent pipe flow properties can be quite dependent on the roughness of the pipe wall, unlike laminar pipe flow which is independent of roughness. Small roughness elements (scratches, rust, sand or dirt particles, etc.) can easily disturb this viscous sublayer (see Section 8.4), thereby affecting the entire flow.

An alternate form for the shear stress for turbulent flow is given in terms of the *eddy viscosity, η,* where

$$\tau = \eta \frac{d\bar{u}}{dy} \tag{8.27}$$

This extension of laminar flow terminology was introduced by J. Boussinesq, a French scientist, in 1877. Although the concept of an eddy viscosity is intriguing, in practice it is not an easy parameter to use. Unlike the absolute viscosity, μ, which is a known value for a given fluid, the eddy viscosity is a function of both the fluid and the flow conditions. That is, the eddy viscosity of water cannot be looked up in handbooks— its value changes from one turbulent flow condition to another and from one point in a turbulent flow to another.

The inability to accurately determine the Reynolds stress, $\overline{\rho u'v'}$, is equivalent to not knowing the eddy viscosity. Several semiempirical theories have been proposed to determine approximate values of η (Ref. 3). L. Prandlt (1875–1953), a German

physicist and aerodynamicist, proposed that the turbulent process could be viewed as the random transport of bundles of fluid particles over a certain distance, ℓ_m, the *mixing length*, from a region of one velocity to another region of a different velocity. By the use of some ad hoc assumptions and physical reasoning, it was concluded that the eddy viscosity was given by

$$\eta = \rho \ell_m^2 \left| \frac{d\bar{u}}{dy} \right|$$

Thus, the turbulent shear stress is

$$\tau_{\text{turb}} = \rho \ell_m^2 \left(\frac{d\bar{u}}{dy} \right)^2 \tag{8.28}$$

The problem is thus shifted to that of determining the mixing length, ℓ_m. Further considerations indicate that ℓ_m is not a constant throughout the flow field. Near a solid surface the turbulence is dependent on the distance from the surface. Thus, additional assumptions are made regarding how the mixing length varies throughout the flow.

The net result is that as yet there is no general, all encompassing, useful model that can accurately predict the shear stress throughout a general incompressible, viscous turbulent flow. Without such information it is impossible to integrate the force balance equation to obtain the turbulent velocity profile and other useful information, as was done for laminar flow.

8.3.3 Turbulent Velocity Profile
Considerable information concerning turbulent velocity profiles has been obtained through the use of dimensional analysis, experimentation, and semiempirical theoretical efforts. As is indicated in Fig. 8.17, fully developed turbulent flow in a pipe can be broken into three regions which are characterized by their distances from the wall: the viscous sublayer very near the pipe wall, the overlap region, and the outer turbulent layer throughout the center portion of the flow. Within the viscous sublayer the viscous shear stress is dominant compared with the turbulent (or Reynolds) stress, and the random, eddying nature of the flow is essentially absent. In the outer turbulent layer the Reynolds stress is dominant, and there is considerable mixing and randomness to the flow.

The character of the flow within these two regions is entirely different. For example, within the viscous sublayer the fluid viscosity is an important parameter; the density is unimportant. In the outer layer the opposite is true. By a careful use of dimensional analysis arguments for the flow in each layer and by a matching of the results in the common overlap layer, it has been possible to obtain the following conclusions about the turbulent velocity profile in a smooth pipe (Ref. 5).

In the viscous sublayer the velocity profile can be written in dimensionless form as

$$\frac{\bar{u}}{u^*} = \frac{yu^*}{\nu} \tag{8.29}$$

where $y = R - r$ is the distance measured from the wall, $\bar{u}$ is the time average x component of velocity, and $u^* = (\tau_w/\rho)^{1/2}$ is termed the *friction velocity*. Note that u^* is not an actual velocity of the fluid—it is merely a quantity that has dimensions of velocity. As is indicated in Fig. 8.18, Eq. 8.29 (commonly called the *law of the wall*) is valid very near the smooth wall, for $0 \le yu^*/\nu \lesssim 5$.

Dimensional analysis arguments indicate that in the overlap region the velocity

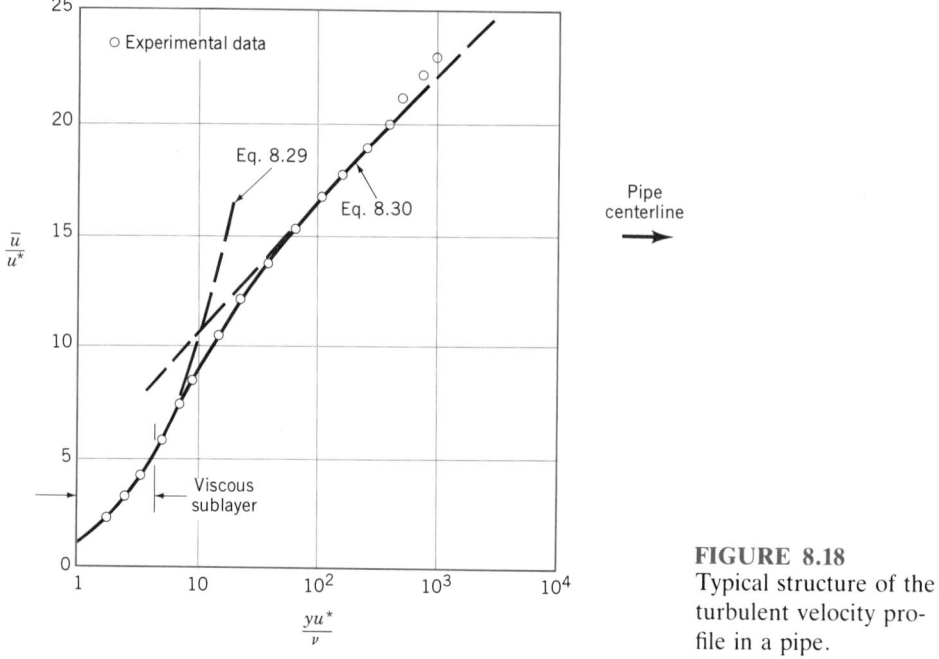

FIGURE 8.18
Typical structure of the turbulent velocity profile in a pipe.

should vary as the logarithm of y. Thus, the following expression has been proposed,

$$\frac{\bar{u}}{u^*} = 2.5 \ln \left(\frac{yu^*}{\nu}\right) + 5.0 \tag{8.30}$$

where the constants 2.5 and 5.0 have been determined experimentally. As is indicated in Fig. 8.18, for regions not too close to the smooth wall, but not all the way out to the pipe center, Eq. 8.30 gives a reasonable correlation with the experimental data. Note that the horizontal scale is a logarithmic scale. This tends to exaggerate the size of the viscous sublayer relative to the remainder of the flow. As is shown in Example 8.4, the viscous sublayer is usually quite thin. Similar results can be obtained for turbulent flow past rough walls (Ref. 17).

A number of other correlations exist for the velocity profile in turbulent pipe flow. In the central region (the outer turbulent layer) the expression $(V_c - \bar{u})/u^* = 2.5 \ln(R/y)$, where V_c is the centerline velocity, is often suggested as a good correlation with experimental data. Another often used (and relatively easy to use) correlation is the empirical *power-law velocity profile*

$$\frac{\bar{u}}{V_c} = \left(1 - \frac{r}{R}\right)^{1/n} \tag{8.31}$$

In this representation, the value of n is a function of the Reynolds number, as is indicated in Fig. 8.19. The one-seventh power law velocity profile ($n = 7$) is often used as a reasonable approximation for many practical flows. Typical turbulent velocity profiles based on this power-law representation are shown in Fig. 8.20.

A closer examination of Eq. 8.31 shows that the power-law profile cannot be valid near the wall, since the velocity gradient is infinite there. In addition, Eq. 8.31 cannot be precisely valid near the centerline because it does not give $d\bar{u}/dr = 0$ at $r = R$. However, it does provide a reasonable approximation to the measured velocity profiles across most of the pipe.

FIGURE 8.19 Exponent, n, for power-law velocity profiles (adapted from Ref. 1).

$$Re = \frac{\rho VD}{\mu}$$

Note from Fig. 8.20 that the turbulent profiles are much "flatter" than the laminar profile and that this flatness increases with Reynolds number (i.e., with n). Recall from Chapter 3 that reasonable approximate results are often obtained by using the inviscid Bernoulli equation and assuming a fictitious uniform velocity profile. Since most flows are turbulent and turbulent flows tend to have nearly uniform velocity profiles, the usefulness of the Bernoulli equation and the uniform profile assumption is not unexpected. Of course, many properties of the flow cannot be accounted for without including viscous effects.

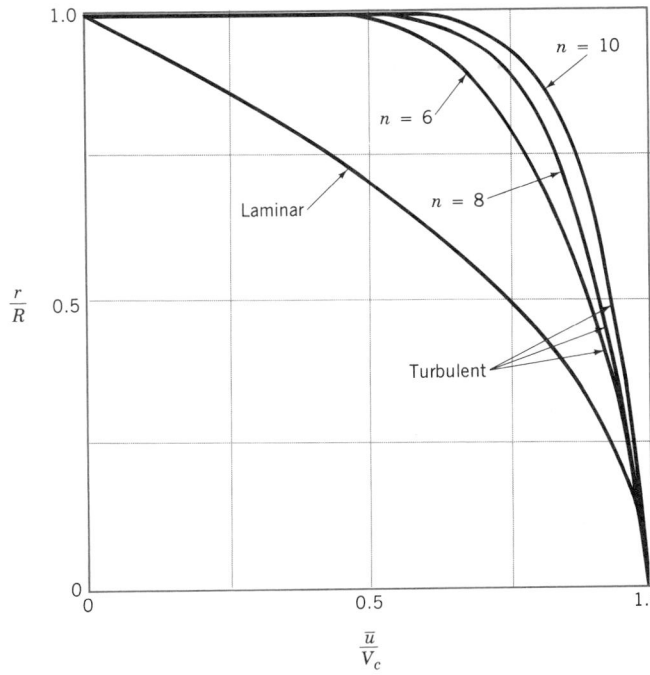

FIGURE 8.20 Typical laminar flow and turbulent flow velocity profiles.

EXAMPLE 8.4

Water at 20 °C ($\rho = 998$ kg/m^3 and $\nu = 1.004 \times 10^{-6}$ m^2/s) flows through a horizontal pipe of 0.1-m diameter with a flowrate of $Q = 4 \times 10^{-2}$ m^3/s and a pressure gradient of 2.59

kPa/m. (a) Determine the approximate thickness of the viscous sublayer. (b) Determine the approximate centerline velocity, V_c. (c) Determine the ratio of the turbulent to laminar shear stress, τ_{turb}/τ_{lam}, at a point midway between the centerline and the pipe wall (i.e., at $r = 0.025$ m).

SOLUTION

(a) According to Fig. 8.18, the thickness of the viscous sublayer, δ_s, is approximately

$$\frac{\delta_s u^*}{\nu} = 5$$

or

$$\delta_s = 5 \frac{\nu}{u^*}$$

where

$$u^* = \left(\frac{\tau_w}{\rho}\right)^{1/2} \tag{1}$$

The wall shear stress can be obtained from the pressure drop data and Eq. 8.5, which is valid for either laminar or turbulent flow. Thus,

$$\tau_w = \frac{D\,\Delta p}{4\ell} = \frac{(0.1 \text{ m})(2.59 \times 10^3 \text{ N/m}^2)}{4(1 \text{ m})} = 64.8 \text{ N/m}^2$$

Hence, from Eq. 1 we obtain

$$u^* = \left(\frac{64.8 \text{ N/m}^2}{998 \text{ kg/m}^3}\right)^{1/2} = 0.255 \text{ m/s}$$

so that

$$\delta_s = \frac{5(1.004 \times 10^{-6} \text{ m}^2/\text{s})}{0.255 \text{ m/s}} = 1.97 \times 10^{-5} \text{ m} \approx 0.02 \text{ mm} \tag{Ans}$$

As stated previously, the viscous sublayer is very thin. Minute imperfections on the pipe wall will protrude into this sublayer and likely alter some of the characteristics of the flow (i.e., wall shear stress and pressure drop).

(b) The centerline velocity can be obtained from the average velocity and the assumption of a power-law velocity profile as follows. For this flow with

$$V = \frac{Q}{A} = \frac{0.04 \text{ m}^3/\text{s}}{\pi(0.1 \text{ m})^2/4} = 5.09 \text{ m/s}$$

the Reynolds number is

$$\text{Re} = \frac{VD}{\nu} = \frac{(5.09 \text{ m/s})(0.1 \text{ m})}{(1.004 \times 10^{-6} \text{ m}^2/\text{s})} = 5.07 \times 10^5$$

Thus, from Fig. 8.19, $n = 8.4$ so that

$$\frac{\bar{u}}{V_c} \approx \left(1 - \frac{r}{R}\right)^{1/8.4}$$

   To determine the centerline velocity, V_c, we must know the relationship between V (the average velocity) and V_c. This can be obtained by integration of the power-law velocity profile as follows. Since the flow is axisymmetric,

$$Q = AV = \int \bar{u}\, dA = V_c \int_{r=0}^{r=R} \left(1 - \frac{r}{R}\right)^{1/n} (2\pi r)\, dr$$

which can be integrated to give

$$Q = 2\pi R^2 V_c \frac{n^2}{(n + 1)(2n + 1)}$$

Thus, since $Q = \pi R^2 V$, we obtain

$$\frac{V}{V_c} = \frac{2n^2}{(n + 1)(2n + 1)}$$

With $n = 8.4$ in the present case, this gives

$$V_c = \frac{(n + 1)(2n + 1)}{2n^2} V = 1.186V = 1.186 \ (5.09 \ \text{m/s})$$

$$= 6.04 \ \text{m/s} \qquad \text{(Ans)}$$

Recall that $V_c = 2V$ for laminar pipe flow.

(c) From Eq. 8.4, which is valid for laminar or turbulent flow, the shear stress at $r = 0.025$ m is

$$\tau = \frac{2\tau_w r}{D} = \frac{2(64.8 \ \text{N/m}^2)(0.025 \ \text{m})}{(0.1 \ \text{m})}$$

or

$$\tau = \tau_{\text{lam}} + \tau_{\text{turb}} = 32.4 \ \text{N/m}^2$$

where $\tau_{\text{lam}} = -\mu \ d\bar{u}/dr$. From the power-law velocity profile (Eq. 8.31) we obtain the gradient of the average velocity as

$$\frac{d\bar{u}}{dr} = -\frac{V_c}{nR}\left(1 - \frac{r}{R}\right)^{(1-n)/n}$$

which gives

$$\frac{d\bar{u}}{dr} = -\frac{(6.04 \ \text{m/s})}{8.4(0.05 \ \text{m})}\left(1 - \frac{0.025 \ \text{m}}{0.05 \ \text{m}}\right)^{(1-8.4)/8.4} = -26.5/s$$

Thus,

$$\tau_{\text{lam}} = -\mu \frac{d\bar{u}}{dr} = -(\nu\rho)\frac{d\bar{u}}{dr}$$

$$= -(1.004 \times 10^{-6} \ \text{m}^2/\text{s})(998 \ \text{kg/m}^3)(-26.5/\text{s})$$

$$= 0.0266 \ \text{N/m}^2$$

Thus, the ratio of turbulent to laminar shear stress is given by

$$\frac{\tau_{\text{turb}}}{\tau_{\text{lam}}} = \frac{\tau - \tau_{\text{lam}}}{\tau_{\text{lam}}} = \frac{32.4 - 0.0266}{0.0266} = 1220 \qquad \text{(Ans)}$$

As expected, most of the shear stress at this location in the turbulent flow is due to the turbulent shear stress.

The turbulent flow characteristics discussed in this section are not unique to turbulent flow in round pipes. Many of the characteristics introduced (i.e., the Reynolds stress, the viscous sublayer, the overlap layer, the outer layer, the general characteristics of the velocity profile, etc.) are found in other turbulent flows. In particular, turbulent pipe flow and turbulent flow past a solid wall (boundary layer flow) share many of these common traits. Such ideas are discussed more fully in Chapter 9.

8.4 DIMENSIONAL ANALYSIS OF PIPE FLOW

As was discussed in previous sections, turbulent flow can be a very complex, difficult topic—one that as yet has defied a rigorous theoretical treatment. Thus, most turbulent pipe flow analyses are based on experimental data and semiempirical formulas, even if the flow is fully developed. These results are given in dimensionless form and cover a very wide range of flow parameters, including arbitrary fluids, pipes, and flowrates. In addition to these fully-developed flow considerations, a variety of useful data is available regarding flow through pipe fittings, such as elbows, tees, valves, and the like. These data are conveniently expressed in dimensionless form.

8.4.1 The Moody Chart

A dimensional analysis treatment of pipe flow provides the most convenient base from which to consider turbulent, fully developed pipe flow. An introduction to this topic was given in Section 8.3. As is discussed in Sections 8.2.1 and 8.2.4, the pressure drop and head loss in a pipe are dependent on the wall shear stress, τ_w, between the fluid and pipe surface. A fundamental difference between laminar and turbulent flow is that the shear stress for turbulent flow is a function of the density of the fluid, ρ. For laminar flow, the shear stress is independent of the density, leaving the viscosity, μ, as the only important fluid property. This is shown by the fact that with $\tau = \tau_{lam} + \tau_{turb}$, the laminar portion is proportional to μ, and the turbulent portion, the Reynolds stress, is a function of ρ (see Eq. 8.26).

Thus, the pressure drop, Δp, for steady, incompressible turbulent flow in a horizontal round pipe of diameter D can be written in functional form as

$$\Delta p = F(V, D, \ell, \varepsilon, \mu, \rho) \tag{8.32}$$

where V is the average velocity, ℓ is the pipe length, and ε is a measure of the roughness of the pipe wall. It is clear that Δp should be a function of V, D, and ℓ. The dependence of Δp on the fluid properties μ and ρ is expected because of the dependence of τ on these parameters.

Although the pressure drop for laminar pipe flow is found to be independent of the roughness of the pipe, it is necessary to include this parameter when considering turbulent flow. As is discussed in Section 8.3.3 and illustrated in Fig. 8.21, for turbulent flow there is a relatively thin viscous sublayer formed in the fluid near the pipe wall. In many instances this layer is very thin; $\delta_s/D \ll 1$, where δ_s is the sublayer thickness. If a typical wall roughness element protrudes sufficiently far into (or even through) this layer, the structure and properties of the viscous sublayer (along with Δp and τ_w) will be different than if the wall were smooth. Thus, for turbulent flow the pressure drop is expected to be a function of the wall roughness. For laminar flow there is no thin viscous layer—viscous effects are important across the entire pipe. Thus, relatively small roughness elements have completely negligible effects on laminar pipe flow. Of course, for pipes with very large wall "roughness," ($\varepsilon/D \gtrsim 0.1$), such as that in corrugated pipes, the flowrate may be a function of the "roughness." We will consider only typical constant diameter pipes with relative roughnesses in the range $0 \leq \varepsilon/D \lesssim 0.05$. Analysis of flow in corrugated pipes does not fit into the standard constant diameter pipe category.

The list of parameters given in Eq. 8.32 is apparently a complete one. That is, experiments have shown that other parameters (such as surface tension, vapor pressure, etc.) do not affect the pressure drop for the conditions stated (steady, incom-

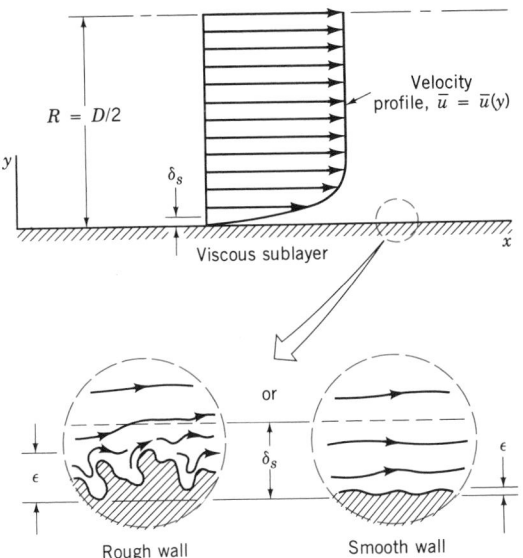

FIGURE 8.21   Flow in the viscous sublayer near rough and smooth walls.

pressible flow; round, horizontal pipe). Since there are seven variables ($k = 7$), which can be written in terms of the three reference dimensions MLT ($r = 3$), Eq. 8.32 can be written in dimensionless form in terms of $k - r = 4$ dimensionless groups. As was discussed in Section 7.9.1, one such representation is

$$\frac{\Delta p}{\frac{1}{2}\rho V^2} = \bar{\phi}\left(\frac{\rho VD}{\mu}, \frac{\ell}{D}, \frac{\varepsilon}{D}\right)$$

This result differs from that used for laminar flow (see Eq. 8.17) in two ways. First, we have chosen to make the pressure dimensionless by dividing by the dynamic pressure, $\rho V^2/2$, rather than a characteristic viscous shear stress, $\mu V/D$. This convention was chosen in recognition of the fact that the shear stress for turbulent flow is normally dominated by τ_{turb}, which is a stronger function of the density than it is of viscosity. Second, we have introduced two additional dimensionless parameters, the Reynolds number, Re $= \rho VD/\mu$, and the *relative roughness*, ε/D, which are not present in the laminar formulation because the two parameters ρ and ε are not important in fully developed laminar pipe flow.

As was done for laminar flow, the functional representation can be simplified by imposing the reasonable assumption that the pressure drop should be proportional to the pipe length. (Such a step is not within the realm of dimensional analysis. It is merely a logical assumption supported by experiments.) The only way that this can be true is if the ℓ/D dependence is factored out as

$$\frac{\Delta p}{\frac{1}{2}\rho V^2} = \frac{\ell}{D}\phi\left(\text{Re}, \frac{\varepsilon}{D}\right)$$

As was discussed in Section 8.2.3, the quantity $\Delta p D/(\ell \rho V^2/2)$ is termed the friction factor, f. Thus, for a horizontal pipe

$$\Delta p = f\frac{\ell}{D}\frac{\rho V^2}{2} \tag{8.33}$$

where

$$f = \phi\left(\text{Re}, \frac{\varepsilon}{D}\right)$$

For laminar fully developed flow the value of f is simply $f = 64/\text{Re}$, independent of ε/D. For turbulent flow, the functional dependence of the friction factor on the Reynolds number and the relative roughness, $f = \phi(\text{Re}, \varepsilon/D)$, is a rather complex one that cannot, as yet, be obtained from a theoretical analysis. The results are obtained from an exhaustive set of experiments and usually presented in terms of a curve-fitting formula or the equivalent graphical form.

From Eq. 5.89 the energy equation for steady incompressible flow is

$$\frac{p_1}{\gamma} + \alpha_1 \frac{V_1^2}{2g} + z_1 = \frac{p_2}{\gamma} + \alpha_2 \frac{V_2^2}{2g} + z_2 + h_L$$

where h_L is the head loss between sections (1) and (2). With the assumption of a constant diameter ($D_1 = D_2$ so that $V_1 = V_2$), horizontal ($z_1 = z_2$) pipe with fully developed flow ($\alpha_1 = \alpha_2$) this becomes $\Delta p = p_1 - p_2 = \gamma h_L$, which can be combined with Eq. 8.33 to give

$$h_L = f \frac{\ell}{D} \frac{V^2}{2g} \tag{8.34}$$

Equation 8.34 is valid for any fully developed, steady, incompressible pipe flow—whether the pipe is horizontal or on a hill. On the other hand, Eq. 8.33 is valid only for horizontal pipes. In general, with $V_1 = V_2$ the energy equation gives

$$p_1 - p_2 = \gamma(z_2 - z_1) + \gamma h_L = \gamma(z_2 - z_1) + f \frac{\ell}{D} \frac{\rho V^2}{2}$$

Part of the pressure change is due to the elevation change and part is due to the head loss associated with frictional effects, which are given in terms of the friction factor, f.

It is not easy to determine the functional dependence of the friction factor on the Reynolds number and relative roughness. Much of this information is a result of experiments conducted by J. Nikuradse in 1933 (Ref. 6) and amplified by many others since then. One difficulty lies in the determination of the roughness of the pipe. Nikuradse used artificially roughened pipes produced by glueing sand grains of known size onto pipe walls to produce pipes with sandpaper type surfaces. The pressure drop needed to produce a desired flowrate was measured and the data were converted into the friction factor for the corresponding Reynolds number and relative roughness. The tests were repeated numerous times for a wide range of Re and ε/D to determine the $f = \phi(\text{Re}, \varepsilon/D)$ dependence.

In commercially available pipes the roughness is not as uniform and well defined as in the artificially roughened pipes used by Nikuradse. However, it is possible to obtain a measure of the effective relative roughness of typical pipes and thus to obtain the friction factor. Such results are given in graphical form in Figs. 8.22 and 8.23. Typical roughness values for various pipe surfaces (corresponding to Fig. 8.22) are given in Table 8.2. Figure 8.23 shows the functional dependence of f on Re and ε/D and is called the *Moody chart* in honor of L. F. Moody who, along with C. F. Colebrook, correlated the original data of Nikuradse in terms of the relative roughness of commercially available pipe materials. It should be noted that the values of ε/D

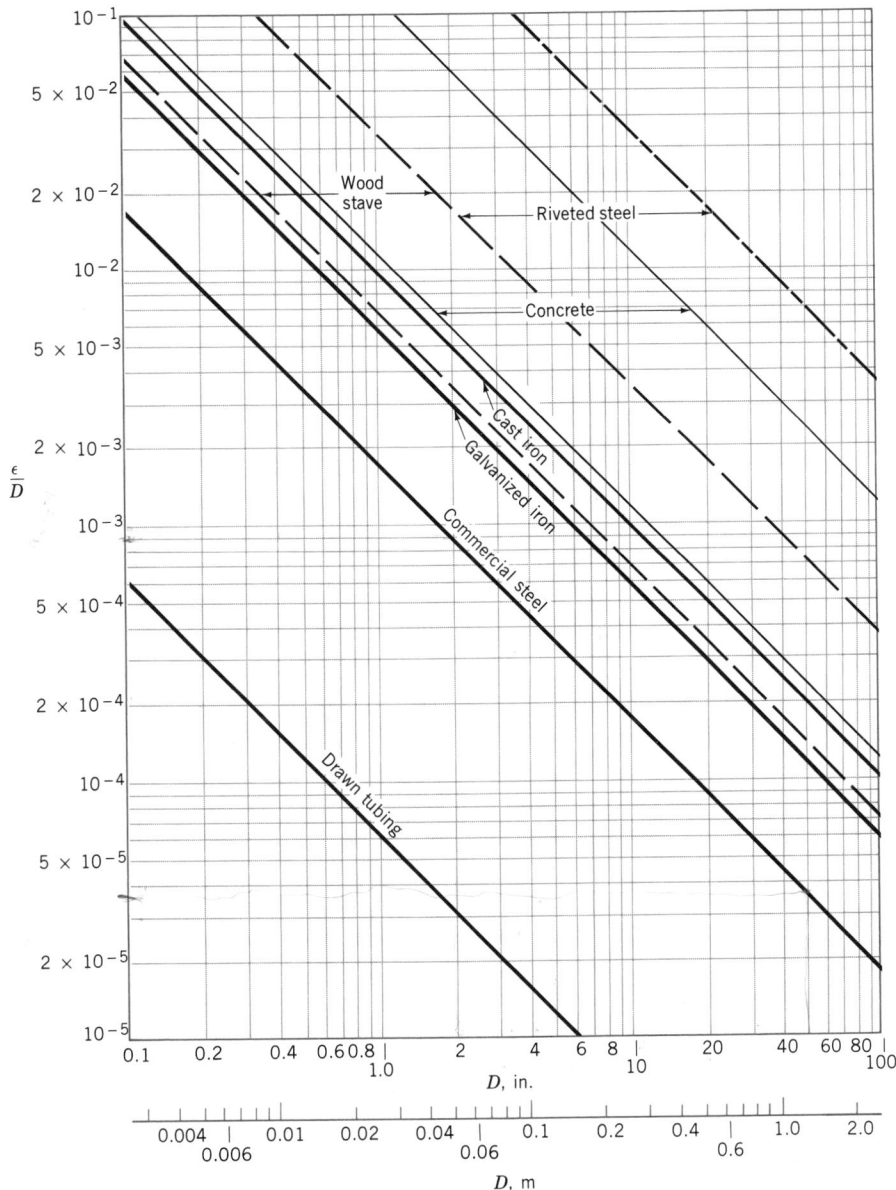

FIGURE 8.22 Relative roughness of new pipes (adapted from Ref. 7).

do not necessarily correspond to the actual values obtained by a microscopic deter-mination of the average height of the roughness of the surface. They do, however, provide the correct correlation for $f = \phi(\text{Re}, \varepsilon/D)$.

It is important to observe that the values of relative roughness given pertain to new, clean pipes. After considerable use, most pipes (because of a buildup of corrosion or scale) may have a relative roughness that is considerably larger (perhaps by an order of magnitude) than that given. Very old pipes may have enough scale buildup to not only alter the value of ε but also to change their effective diameter by a considerable amount.

FIGURE 8.23 Friction factor as a function of Reynolds number and relative roughness for round pipes—the Moody chart (Data from Ref. 7 with permission).

TABLE 8.2 Equivalent Roughness for New Pipes [from Moody (Ref. 7) and Colebrook (Ref. 8)].

Pipe	Equivalent Roughness, ε	
	Feet	Millimeters
Riveted steel	0.003–0.03	0.9–9.0
Concrete	0.001–0.01	0.3–3.0
Wood stave	0.0006–0.003	0.18–0.9
Cast iron	0.00085	0.26
Galvanized iron	0.0005	0.15
Commercial steel or wrought iron	0.00015	0.045
Drawn tubing	0.000005	0.0015
Plastic, glass	0.0 (smooth)	0.0 (smooth)

The following characteristics are observed from the data of Fig. 8.23. For laminar flow, $f = 64/\text{Re}$, independent of relative roughness. For very large Reynolds numbers, $f = \phi(\varepsilon/D)$, independent of the Reynolds number. For such flows, commonly termed *completely turbulent flow* (or *wholly turbulent flow*), the laminar sublayer is so thin (its thickness decreases with increasing Re) that the surface roughness completely dominates the character of the flow near the wall. Hence, the pressure drop required is a result of an inertia dominated turbulent shear stress rather than the viscosity dominated laminar shear stress normally found in the viscous sublayer. The characteristics indicated in Fig. 8.23 for laminar and wholly turbulent flows are consistent with the properties listed in Table 8.1 (Section 8.1.3). For flows with moderate values of Re, the friction factor is indeed dependent on both the Reynolds number and relative roughness—$f = \phi(\text{Re}, \varepsilon/D)$. The gap in the figure for which no values of f are given (the $2100 < \text{Re} < 4000$ range) is a result of the fact that the flow in this transition range may be laminar or turbulent (or an unsteady mix of both) depending on the specific circumstances involved.

Note that even for smooth pipes ($\varepsilon = 0$) the friction factor is not zero. That is, there is a head loss in any pipe, no matter how smooth the surface is made. This is a result of the no-slip boundary condition that requires any fluid to stick to any solid surface it flows over. There is always some microscopic surface roughness that produces the no-slip behavior (and thus $f \neq 0$) on the molecular level, even when the roughness is considerably less than the viscous sublayer thickness. Such pipes are called *hydraulically smooth.*

Various investigators have attempted to obtain an analytical expression for $f = \phi(\text{Re}, \varepsilon/D)$. Note that the Moody chart covers an extremely wide range in flow parameters. The nonlaminar region covers more than four orders of magnitude in Reynolds number—from $\text{Re} = 4 \times 10^3$ to $\text{Re} = 10^8$. Obviously, for a given pipe and fluid, typical values of the average velocity do not cover this range. However, because of the large variety in pipes (D), fluids (ρ and μ), and velocities (V), such a wide range in Re is needed to accommodate nearly all applications of pipe flow. In many cases the particular pipe flow of interest is confined to a relatively small region of the Moody chart and simple semiempirical expressions can be developed for those conditions. For example, a company that manufactures cast-iron water pipes with diameters between 2 and 12 in. may use a simple equation valid for their conditions only. The Moody chart, on the other hand, is universally valid for all steady, fully developed, incompressible pipe flows.

The following equation due to Colebrook is valid for the entire nonlaminar range of the Moody chart

$$\frac{1}{\sqrt{f}} = -2.0 \log \left(\frac{\varepsilon/D}{3.7} + \frac{2.51}{\text{Re}\sqrt{f}} \right) \tag{8.35}$$

In fact, the Moody chart is a graphical representation of this equation, which is an empirical fit of the pipe flow pressure drop data. Equation 8.35 is called the *Colebrook formula*. A difficulty with its use is that it is implicit in the dependence of f. That is, for given conditions (Re and ε/D), it is not possible to solve for f without some sort of iterative scheme. With the use of modern computers and calculators, such calculations are not difficult. A word of caution is in order concerning the use of the Moody chart or the equivalent Colebrook formula. Because of various inherent inaccuracies involved (uncertainty in the relative roughness, uncertainty in the experimental data used to produce the Moody chart, etc.), the use of several place accuracy in pipe flow problems is usually not justified. As a rule of thumb, a 10% accuracy is the best expected.

EXAMPLE 8.5

Air under standard conditions flows through a 4.0-mm-diameter drawn tubing with an average velocity of $V = 50$ m/s. For such conditions the flow would normally be turbulent. However, if precautions are taken to eliminate disturbances to the flow (the entrance to the tube is very smooth, the air is dust free, the tube does not vibrate, etc.), it may be possible to maintain laminar flow. (a) Determine the pressure drop in a 0.1-m section of the tube if the flow is laminar. (b) Repeat the calculations if the flow is turbulent.

SOLUTION

Under standard temperature and pressure conditions the density and viscosity are $\rho = 1.23$ kg/m³ and $\mu = 1.79 \times 10^{-5}$ N·s/m². Thus, the Reynolds number is

$$\text{Re} = \frac{\rho V D}{\mu} = \frac{(1.23 \text{ kg/m}^3)(50 \text{ m/s})(0.004 \text{ m})}{1.79 \times 10^{-5} \text{ N·s/m}^2} = 13{,}700$$

which would normally indicate turbulent flow.

(a)   If the flow were laminar, then $f = 64/\text{Re} = 64/13{,}700 = 0.00467$ and the pressure drop in a 0.1-m-long horizontal section of the pipe would be

$$\Delta p = f \frac{\ell}{D} \frac{1}{2} \rho V^2 = (0.00467) \frac{(0.1 \text{ m})}{(0.004 \text{ m})} \frac{1}{2} (1.23 \text{ kg/m}^3)(50 \text{ m/s})^2$$

or

$$\Delta p = 0.179 \text{ kPa} \tag{Ans}$$

Note that the same result is obtained from Eq. 8.8.

$$\Delta p = \frac{32\mu\ell}{D^2} V = \frac{32(1.79 \times 10^{-5} \text{ N·s/m}^2)(0.1 \text{ m})(50 \text{ m/s})}{(0.004 \text{ m})^2} = 179 \text{ N/m}^2$$

(b)   If the flow were turbulent, then $f = \phi(\text{Re}, \varepsilon/D)$, where from Table 8.2, $\varepsilon = 0.0015$ mm so that $\varepsilon/D = 0.0015$ mm/4.0 mm $= 0.000375$. (The value of ε/D can also be obtained directly from Fig. 8.22.) From the Moody chart with Re $= 1.37 \times 10^4$ and $\varepsilon/D = 0.000375$ we obtain $f = 0.028$. Thus, the pressure drop in this case would be approximately

$$\Delta p = f \frac{\ell}{D} \frac{1}{2} \rho V^2 = (0.028) \frac{(0.1 \text{ m})}{(0.004 \text{ m})} \frac{1}{2} (1.23 \text{ kg/m}^3)(50 \text{ m/s})^2$$

or

$$\Delta p = 1.076 \text{ kPa} \hspace{3cm} \text{(Ans)}$$

A considerable savings in effort to force the fluid through the pipe could be realized (0.179 kPa rather than 1.076 kPa) if the flow could be maintained as laminar flow at this Reynolds number. In general this is very difficult to do, although laminar flow in pipes has been maintained up to Re $\approx$ 100,000 in rare instances.

An alternate method to determine the friction factor for the turbulent flow would be to use the Colebrook formula, Eq. 8.35. Thus,

$$\frac{1}{\sqrt{f}} = -2.0 \log\left(\frac{\varepsilon/D}{3.7} + \frac{2.51}{\text{Re}\sqrt{f}}\right) = -2.0 \log\left(\frac{0.000375}{3.7} + \frac{2.51}{1.37 \times 10^4 \sqrt{f}}\right)$$

or

$$\frac{1}{\sqrt{f}} = -2.0 \log\left(1.01 \times 10^{-4} + \frac{1.83 \times 10^{-4}}{\sqrt{f}}\right) \hspace{2cm} (1)$$

An iterative procedure gives the solution $f = 0.0291$, in agreement (within the accuracy of reading the graph) with the Moody chart method of $f = 0.028$.

Numerous other empirical formulas can be found in the literature (Ref. 5) for portions of the Moody chart. For example, an often used equation, commonly referred to as the Blasius formula, for turbulent flow in smooth pipes ($\varepsilon/D = 0$) with Re $< 10^5$ is

$$f = \frac{0.316}{\text{Re}^{1/4}}$$

For our case this gives

$$f = 0.316(13,700)^{-0.25} = 0.0292$$

which is in agreement with the previous results. Note that the value of f is relatively insensitive to ε/D for this particular situation. Whether the tube were smooth glass ($\varepsilon/D = 0$) or the drawn tubing ($\varepsilon/D = 0.000375$) would not make much difference in the pressure drop. For this flow, an increase in relative roughness by a factor of 30 to $\varepsilon/D = 0.0113$ (equivalent to a commercial steel surface; see Table 8.2) would give $f = 0.043$. This would represent an increase in pressure drop and head loss by a factor of $0.043/0.0291 = 1.48$ compared with that for the original drawn tubing.

The pressure drop of 1.076 kPa in a length of 0.1 m of pipe corresponds to a change in absolute pressure [assuming $p = 101$ kPa (abs) at $x = 0$] of approximately $1.076/101 = 0.0107$, or about 1%. Thus, the incompressible flow assumption on which the above calculations (and all of the formulas in this chapter) are based is reasonable. However, if the pipe were 2-m long the pressure drop would be 21.5 kPa, approximately 20% of the original pressure. In this case the density would not be approximately constant along the pipe, and a compressible flow analysis would be needed. Such considerations are discussed in Chapter 11.

8.4.2   Minor Losses

As discussed in the previous section, the head loss in long, straight sections of pipe can be calculated by use of the friction factor obtained from either the Moody chart or the Colebrook equation. Most pipe systems, however, contain considerably more than straight pipes. These additional components (valves, bends, tees, and the like) add to the overall head loss of the system. Such losses are generally termed *minor losses*, with the apparent implication being that the majority of the system loss is associated with the friction in the straight portions of the pipes, the *major losses*. In many cases this is true. In other cases the minor losses are greater than the major

losses. In this section we indicate how to determine the various minor losses that commonly occur in pipe systems.

The head loss associated with flow through a valve is a common minor loss. The purpose of a valve is to provide a means to regulate the flowrate. This is accomplished by changing the geometry of the system (i.e., closing or opening the valve alters the flow pattern through the valve), which in turn alters the losses associated with the flow through the valve. The flow resistance or head loss through the valve may be a significant portion of the resistance in the system. In fact, with the valve closed, the resistance to the flow is infinite—the fluid cannot flow. Such "minor losses" may be very important indeed. With the valve wide open the extra resistance due to the presence of the valve may or may not be negligible.

The flow pattern through a typical component such as a valve is shown in Fig. 8.24. It is not difficult to realize that a theoretical analysis to predict the details of such flows to obtain the head loss for these components is not, as yet, possible. Thus, the head loss information for essentially all components is given in dimensionless form and based on experimental data. The most common method used to determine these head losses or pressure drops is to specify the *loss coefficient*, K_L, which is defined as

$$K_L = \frac{h_L}{(V^2/2g)} = \frac{\Delta p}{\frac{1}{2}\rho V^2}$$

so that

$$\Delta p = K_L \tfrac{1}{2}\rho V^2$$

or

$$h_L = K_L \frac{V^2}{2g} \tag{8.36}$$

The pressure drop across a component that has a loss coefficient of $K_L = 1$ is equal to the dynamic pressure, $\rho V^2/2$.

The actual value of K_L is strongly dependent on the geometry of the component

FIGURE 8.24   Flow through a valve.

considered. It may also be dependent on the fluid properties. That is,

$$K_L = \phi(\text{geometry, Re})$$

where Re $= \rho VD/\mu$ is the pipe Reynolds number. For many practical applications the Reynolds number is large enough so that the flow through the component is dominated by inertia effects, with viscous effects being of secondary importance. This is true because of the relatively large accelerations and decelerations experienced by the fluid as it flows along a rather curved, variable area (perhaps even torturous) path through the component (see Fig. 8.24). In a flow that is dominated by inertia effects rather than viscous effects, it is usually found that pressure drops and head losses correlate directly with the dynamic pressure. This is the reason why the friction factor for very large Reynolds number, fully developed pipe flow is independent of the Reynolds number. The same condition is found to be true for flow through pipe components. Thus, in most cases of practical interest the loss coefficients for components is a function of geometry only, $K_L = \phi(\text{geometry})$.

Minor losses are sometimes given in terms of an *equivalent length,* ℓ_{eq}. In this terminology, the head loss through a component is given in terms of the equivalent length of pipe that would produce the same head loss as the component. That is,

$$h_L = K_L \frac{V^2}{2g} = f \frac{\ell_{eq}}{D} \frac{V^2}{2g}$$

or

$$\ell_{eq} = \frac{K_L D}{f}$$

where D and f are based on the pipe containing the component. The head loss of the pipe system is the same as that produced in a straight pipe whose length is equal to the pipes of the original system plus the sum of the additional equivalent lengths of all of the components of the system. Most pipe flow analyses, including those in this book, use the loss coefficient method rather than the equivalent length method to determine the minor losses.

Many pipe systems contain various transition sections in which the pipe diameter changes from one size to another. Such changes may occur abruptly or rather smoothly through some type of gradual area change section. Any change in flow area contributes losses that are not accounted for in the fully developed head loss calculation (the friction factor). The extreme cases involve flow into a pipe from a reservoir (an entrance) or out of a pipe into a reservoir (an exit).

A fluid may flow from a reservoir into a pipe through any number of different shaped entrance regions as are sketched in Fig. 8.25. Each geometry has an associated loss coefficient. A typical flow pattern for flow entering a pipe through a square edged entrance is sketched in Fig. 8.26. As was discussed in Chapter 3, a vena contracta region may result because the fluid cannot turn a sharp right-angle corner. The flow is said to separate from the sharp corner. The maximum velocity at section (2) is greater than that in the pipe at section (3), and the pressure there is lower. If this high-speed fluid could slow down efficiently, the kinetic energy could be converted into pressure (the Bernoulli effect), and the ideal pressure distribution indicated in Fig. 8.26 would result. The head loss for the entrance would be essentially zero.

Such is not the case. Although a fluid may be accelerated very efficiently, it is very difficult to slow down (decelerate) a fluid efficiently. Thus, the extra kinetic energy of the fluid at section (2) is partially lost because of viscous dissipation, so that the

FIGURE 8.25
Entrance flow conditions and loss coefficient (Refs. 28, 29). (a) Reentrant, $K_L = 0.8$; (b) Sharp edged, $K_L = 0.5$; (c) Slightly rounded, $K_L = 0.2$ (see Fig. 8.27). (d) Well rounded, $K_L = 0.04$ (see Fig. 8.27).

pressure does not return to the ideal value. An entrance head loss (pressure drop) is produced as is indicated in Fig. 8.26. The majority of this loss is due to inertia effects that are eventually dissipated by the shear stresses within the fluid. Only a small portion of the loss is due to the wall shear stress within the entrance region. The net effect is that the loss coefficient for a square edged entrance is approximately $K_L = 0.50$. One-half of a velocity head is lost as the fluid enters the pipe. If the pipe protrudes

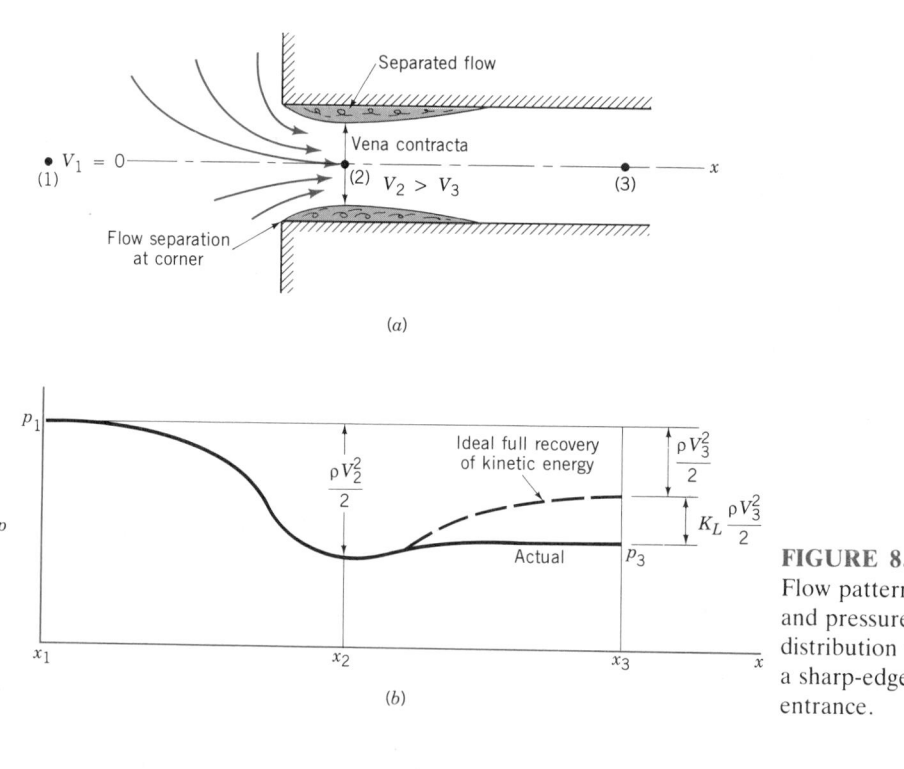

FIGURE 8.26
Flow pattern and pressure distribution for a sharp-edged entrance.

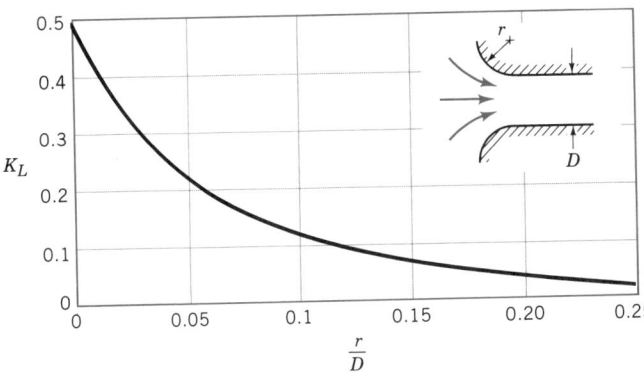

FIGURE 8.27 Entrance loss coefficient as a function of rounding of the inlet edge (Ref. 9).

into the tank (a reentrant entrance) as is shown in Fig. 8.25a, the losses are even greater.

An obvious way to reduce the entrance loss is to round the entrance region as is shown in Fig. 8.25c, thereby reducing or eliminating the vena contracta effect. Typical values for the loss coefficient for entrances with various amounts of rounding of the lip are shown in Fig. 8.27. A significant reduction in K_L can be obtained with only slight rounding.

A head loss (the exit loss) is also produced when a fluid flows from a pipe into a tank as is shown in Fig. 8.28. In these cases the entire kinetic energy of the exiting fluid (velocity V_1) is dissipated through viscous effects as the stream of fluid mixes with the fluid in the tank and eventually comes to rest ($V_2 = 0$). The exit loss from points (1) and (2) is therefore equivalent to one velocity head, or $K_L = 1$.

Losses also occur because of a change in pipe diameter as is shown in Fig. 8.29. Flows through sudden contractions or expansions produce flow patterns as shown.

FIGURE 8.28 Exit flow conditions and loss coefficient; (a) Reentrant, $K_L = 1.0$; (b) Sharp edged, $K_L = 1.0$; (c) Slightly rounded, $K_L = 1.0$; (d) Well rounded, $K_L = 1.0$.

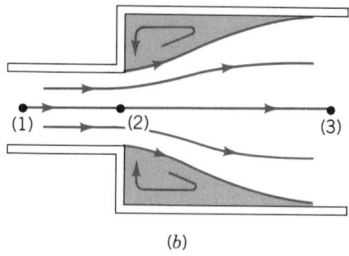

FIGURE 8.29 Flow conditions near a sudden change in diameter. (*a*) Contraction. (*b*) Expansion.

The sharp edged entrance and exit flows discussed in the previous paragraphs are limiting cases of this type of flow with either $A_1/A_2 = \infty$, or $A_1/A_2 = 0$, respectively. The loss coefficient for a sudden contraction, $K_L = h_L/(V_2^2/2g)$, is a function of the area ratio, A_2/A_1, as is shown in Fig. 8.30. The value of K_L changes gradually from one extreme of a sharp edged entrance ($A_2/A_1 = 0$ with $K_L = 0.50$) to the other extreme of no area change ($A_2/A_1 = 1$ with $K_L = 0$).

In many ways, the flow in a sudden expansion is similar to exit flow. As is indicated in Fig. 8.31, the fluid leaves the smaller pipe and initially forms a jet-type structure as it enters the larger pipe. Within a few diameters downstream of the expansion, the jet becomes dispersed across the pipe, and fully developed flow becomes established again. In this process [between sections (2) and (3)] a portion of the kinetic energy of the fluid is dissipated as a result of viscous effects. A square edged exit is the limiting case with $A_1/A_2 = 0$.

A sudden expansion is one of the few components (perhaps the only one) for which the loss coefficient can be obtained by means of a simple analysis. To do this we consider the continuity and momentum equations for the control volume shown in Fig. 8.31 and the energy equation applied between (2) and (3). We assume that the flow is uniform at sections (1), (2), and (3) and the pressure is constant across the left-hand side of the control volume ($p_a = p_b = p_c = p_1$). The resulting three governing equations (mass, momentum, and energy) are

$$A_1V_1 = A_3V_3$$

$$p_1A_3 - p_3A_3 = \rho A_3V_3(V_3 - V_1)$$

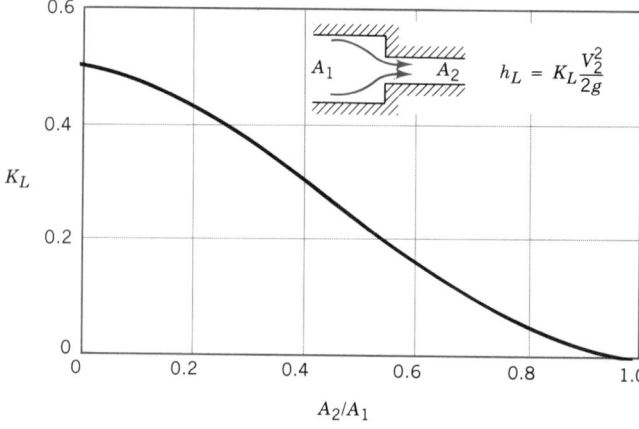

FIGURE 8.30 Loss coefficient for a sudden contraction (Ref. 10).

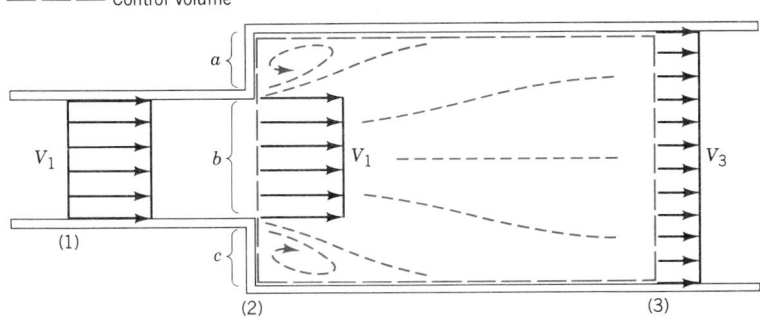

FIGURE 8.31 Control volume used to calculate the loss coefficient for a sudden expansion.

and

$$\frac{p_1}{\gamma} + \frac{V_1^2}{2g} = \frac{p_3}{\gamma} + \frac{V_3^2}{2g} + h_L$$

These can be rearranged to give the loss coefficient, $K_L = h_L/(V_1^2/2g)$, as

$$K_L = \left(1 - \frac{A_1}{A_2}\right)^2$$

where we have used the fact that $A_2 = A_3$. This result, plotted in Fig. 8.32, is in good agreement with experimental data. As with so many minor loss situations, it is not the viscous effects directly (i.e., the wall shear stress) that cause the loss. Rather, it is the dissipation of kinetic energy (another type of viscous effect) as the fluid decelerates inefficiently.

The losses may be quite different if the contraction or expansion is gradual. Typical results for a conical *diffuser* with a given area ratio, A_2/A_1, are shown in Fig. 8.33. Clearly the included angle of the diffuser, θ, is a very important parameter. For very small angles, the diffuser is excessively long and most of the head loss is due to the wall shear stress as in fully developed flow. For moderate or large angles, the flow separates from the walls and the losses are due mainly to a dissipation of the kinetic energy of the jet leaving the smaller diameter pipe. In fact, for moderate or large values of θ (i.e., $\theta > 35°$ for the case shown in Fig. 8.33), the conical diffuser is, perhaps unexpectedly, less efficient than a sharp edged expansion which has $K_L = 1$.

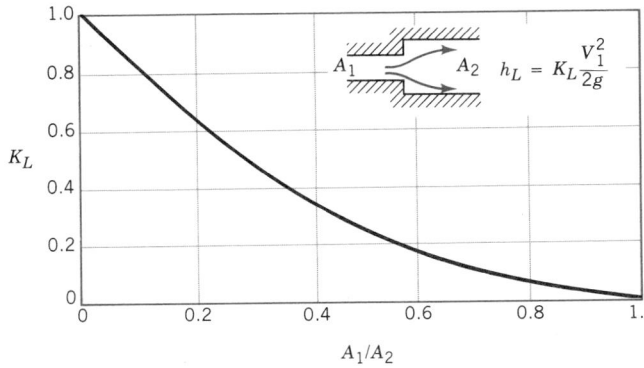

FIGURE 8.32 Loss coefficient for a sudden expansion (Ref. 10).

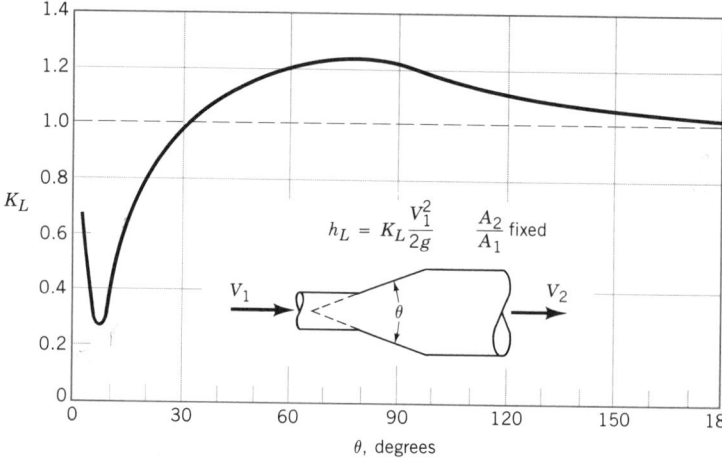

FIGURE 8.33
Loss coefficient for a typical conical diffuser (Ref. 5).

There is an optimum angle($\theta \approx 8°$ for the case illustrated) for which the loss coefficient is a minimum. The relatively small value of θ for the minimum K_L results in a long diffuser and is an indication of the fact that it is difficult to efficiently decelerate a fluid.

It must be noted that the conditions indicated in Fig. 8.33 represent typical results only. Flow through a diffuser is very complicated and may be strongly dependent on the area ratio A_2/A_1, specific details of the geometry, and the Reynolds number. The data are often presented in terms of a *pressure recovery coefficient*, $C_p = (p_2 - p_1)/(\rho V_1^2/2)$, which is the ratio of the static pressure rise across the diffuser to the inlet dynamic pressure. Considerable effort has gone into understanding this important topic (Refs. 11, 12).

Flow in a conical contraction (a nozzle; reverse the flow direction shown in Fig. 8.33) is less complex than that in a conical expansion. Typical loss coefficients based on the downstream (high-speed) velocity can be quite small, ranging from $K_L = 0.02$ for $\theta = 30°$, to $K_L = 0.07$ for $\theta = 60°$ for example. It is relatively easy to accelerate a fluid efficiently.

Bends in pipes produce a greater head loss than if the pipe were straight. The losses are due to the separated region of flow near the inside of the bend (especially if the bend is sharp) and the swirling secondary flow that occurs because of the imbalance of centripetal forces as a result of the curvature of the pipe centerline. These effects and the associated values of K_L for large Reynolds number flows through a 90° bend are shown in Fig. 8.34. The friction loss due to the axial length of the pipe bend must be calculated and added to that given by the loss coefficient of Fig. 8.34.

For situations in which space is limited, a flow direction change is often accomplished by use of miter bends, as is shown in Fig. 8.35, rather than smooth bends. The considerable losses in such bends can be reduced by the use of carefully designed guide vanes that help direct the flow with less unwanted swirl and disturbances.

Another important category of pipe system components is that of commercially available pipe fittings such as elbows, tees, reducers, valves, and filters. The values of K_L for such components depend strongly on the shape of the component and only very weakly on the Reynolds number for typical large Re flows. Thus, the loss coefficient for a 90° elbow depends on whether the pipe joints are threaded or flanged, but is, within the accuracy of the data, fairly independent of the pipe diameter (the

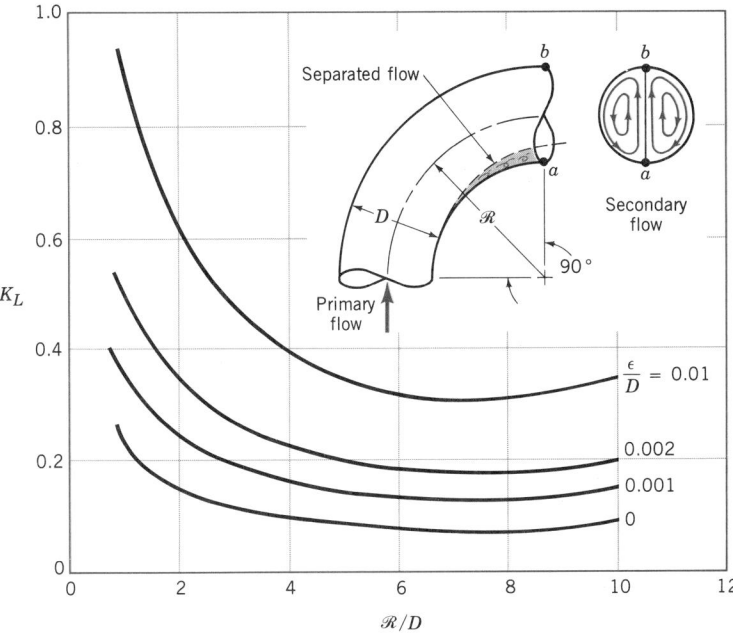

FIGURE 8.34 Character of the flow in a 90° bend and the associated loss coefficient (Ref. 5).

Reynolds number effect). Typical values of K_L for such components are given in Table 8.3. These typical components are designed more for ease of manufacturing and costs than for reduction of the head losses that they produce. The flowrate from a faucet in a typical house is sufficient whether the value of K_L for an elbow is the typical $K_L = 1.5$, or it is reduced to $K_L = 0.2$ by use of a more expensive long-radius, gradual bend (Fig. 8.34).

Valves control the flowrate by providing a means to adjust the overall system loss coefficient to the desired value. When the valve is closed, the value of K_L is infinity and no fluid flows. Opening of the valve reduces K_L, producing the desired flowrate. Typical cross sections of various types of valves are shown in Fig. 8.36. Some valves (such as the conventional globe valve) are designed for general use, providing convenient control between the extremes of fully closed and fully open. Others (such as

$K_L \approx 1.1$

(a)

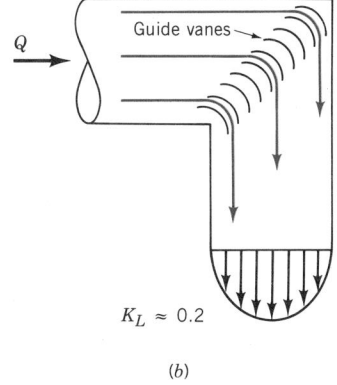

$K_L \approx 0.2$

(b)

FIGURE 8.35 Character of the flow in a 90° miter bend and the associated loss coefficient. (a) Without guide vanes. (b) With guide vanes.

TABLE 8.3 Loss Coefficients for Pipe Components $\left(h_L = K_L \dfrac{V^2}{2g}\right)$ (Data from Refs. 5, 10, 27)

Component	K_L	
a. Elbows		
Regular 90°, flanged	0.3	
Regular 90°, threaded	1.5	
Long radius 90°, flanged	0.2	
Long radius 90°, threaded	0.7	
Long radius 45°, flanged	0.2	
Regular 45°, threaded	0.4	
b. 180° return bends		
180° return bend, flanged	0.2	
180° return bend, threaded	1.5	
c. Tees		
Line flow, flanged	0.2	
Line flow, threaded	0.9	
Branch flow, flanged	1.0	
Branch flow, threaded	2.0	
d. Union, threaded	0.08	
e. Valves		
Globe, fully open	10	
Angle, fully open	2	
Gate, fully open	0.15	
Gate, $\frac{1}{4}$ closed	0.26	
Gate, $\frac{1}{2}$ closed	2.1	
Gate, $\frac{3}{4}$ closed	17	
Swing check, forward flow	2	
Swing check, backward flow	∞	
Ball valve, fully open	0.05	
Ball valve, $\frac{1}{3}$ closed	5.5	
Ball valve, $\frac{2}{3}$ closed	210	

a needle valve) are designed to provide very fine control of the flowrate. The check valve provides a diode type operation that allows fluid to flow one direction only.

Loss coefficients for typical valves are given in Table 8.3. As with many system components, the head loss in valves is mainly a result of the dissipation of kinetic energy of a high-speed portion of the flow. This is illustrated in Fig. 8.37.

FIGURE 8.36   Internal structure of various valves: (a) globe valve, (b) gate valve, (c) swing check valve, (d) stop check valve (courtesy of Crane Co., Valve Division).

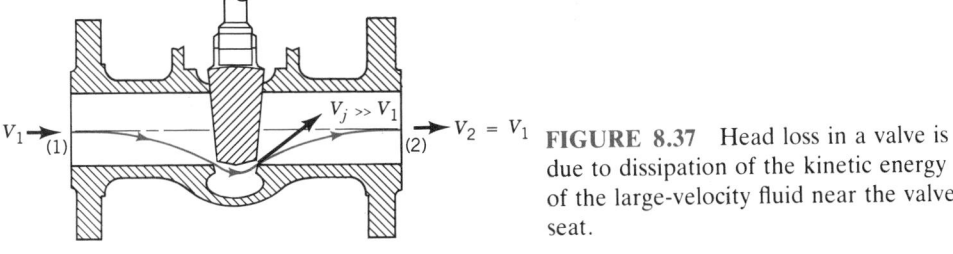

$V_1 \rightarrow$ (1) $V_j \gg V_1$ (2) $\rightarrow V_2 = V_1$

FIGURE 8.37   Head loss in a valve is due to dissipation of the kinetic energy of the large-velocity fluid near the valve seat.

EXAMPLE 8.6

Air at standard conditions is to flow through the test section [between sections (5) and (6)] of the closed circuit wind tunnel shown in Fig. E8.6 with a velocity of 200 ft/s. The flow is driven

by a fan that essentially increases the static pressure by the amount $p_1 - p_9$ that is needed to overcome the head losses experienced by the fluid as it flows around the circuit. Estimate the value of $p_1 - p_9$ and the horsepower supplied to the fluid by the fan.

FIGURE E8.6

Location	Area (ft²)	Velocity (ft/s)
1	22.0	36.4
2	28.0	28.6
3	35.0	22.9
4	35.0	22.9
5	4.0	200.0
6	4.0	200.0
7	10.0	80.0
8	18.0	44.4
9	22.0	36.4

SOLUTION

The maximum velocity within the wind tunnel occurs in the test section (smallest area). Thus, the maximum Mach number of the flow is $Ma_5 = V_5/c_5$, where $V_5 = 200$ ft/s and from Eq. 1.20 the speed of sound is $c_5 = (kRT_5)^{1/2} = \{1.4(1716 \text{ ft·lb/slug·°R})[(460 + 59)°R]\}^{1/2} = 1117$ ft/s. Thus, $Ma_5 = 200/1117 = 0.179$. As was indicated in Chapter 3 and discussed fully in Chapter 11, most flows can be considered as incompressible if the Mach number is less than about 0.3. Hence, we can use the incompressible formulas for this problem.

The purpose of the fan in the wind tunnel is to provide the necessary energy to overcome the net head loss experienced by the air as it flows around the circuit. This can be found from the energy equation between points (1) and (9) as

$$\frac{p_1}{\gamma} + \frac{V_1^2}{2g} + z_1 = \frac{p_9}{\gamma} + \frac{V_9^2}{2g} + z_9 + h_{L_{1-9}}$$

where $h_{L_{1-9}}$ is the total head loss from (1) to (9). With $z_1 = z_9$ and $V_1 = V_9$ this gives

$$\frac{p_1}{\gamma} - \frac{p_9}{\gamma} = h_{L_{1-9}} \tag{1}$$

Similarly, by writing the energy equation (Eq. 5.84) across the fan, from (9) to (1), we obtain

$$\frac{p_9}{\gamma} + \frac{V_9^2}{2g} + z_9 + h_p = \frac{p_1}{\gamma} + \frac{V_1^2}{2g} + z_9$$

where h_p is the actual head rise supplied by the pump (fan) to the air. Again since $z_9 = z_1$ and $V_9 = V_1$ this, when combined with Eq. 1, becomes

$$h_p = \frac{(p_1 - p_9)}{\gamma} = h_{L_{1-9}}$$

The actual power supplied to the air (horsepower, $\mathcal{P}_a$) is obtained from the fan head by

$$\mathcal{P}_a = \gamma Q h_p = \gamma A_5 V_5 h_p = \gamma A_5 V_5 h_{L_{1-9}} \tag{2}$$

Thus, the power that the fan must supply to the air depends on the head loss associated with the flow through the wind tunnel. To obtain a reasonable, approximate answer we make the following assumptions. We treat each of the four turning corners as a mitered bend with guide vanes so that $K_{L_{corner}} = 0.2$. Thus, for each corner

$$h_{L_{corner}} = K_L \frac{V^2}{2g} = 0.2 \frac{V^2}{2g}$$

where, because the flow is assumed incompressible, $V = V_5 A_5 / A$. The values of A and the corresponding velocities throughout the tunnel are given in Table E8.6.

We also treat the enlarging sections from the end of the test section (6) to the beginning of the nozzle (4) as a conical diffuser with a loss coefficient of $K_{L_{dif}} = 0.6$. This value is larger than that of a well-designed diffuser (see Fig. 8.33, for example). Since the wind tunnel diffuser is interrupted by the four turning corners and the fan, it may not be possible to obtain a smaller value of $K_{L_{dif}}$ for this situation. Thus,

$$h_{L_{dif}} = K_{L_{dif}} \frac{V_6^2}{2g} = 0.6 \frac{V_6^2}{2g}$$

The loss coefficients for the conical nozzle between section (4) and (5) and the flow-straightening screens are assumed to be $K_{L_{noz}} = 0.2$ and $K_{L_{scr}} = 4.0$ (Ref. 13), respectively. We neglect the head loss in the relatively short test section.

Thus, the total head loss is

$$h_{L_{1-9}} = h_{L_{corner7}} + h_{L_{corner8}} + h_{L_{corner2}} + h_{L_{corner3}} + h_{L_{dif}} + h_{L_{noz}} + h_{L_{scr}}$$

or

$$h_{L_{1-9}} = [0.2(V_7^2 + V_8^2 + V_2^2 + V_3^2) + 0.6V_6^2 + 0.2V_5^2 + 4.0V_4^2]/2g$$

$$= [0.2(80.0^2 + 44.4^2 + 28.6^2 + 22.9^2) + 0.6(200)^2$$

$$+ 0.2(200)^2 + 4.0(22.9)^2] \text{ ft}^2/\text{s}^2/[2(32.2 \text{ ft}/\text{s}^2)]$$

or

$$h_{L_{1-9}} = 560 \text{ ft}$$

Hence, from Eq. 1 we obtain the pressure rise across the fan as

$$p_1 - p_9 = \gamma h_{L_{1-9}} = (0.0765 \text{ lb}/\text{ft}^3)(560 \text{ ft})$$

$$= 42.8 \text{ lb}/\text{ft}^2 = 0.298 \text{ psi}$$

From Eq. 2 we obtain the power added to the fluid as

$$\mathcal{P}_a = (0.0765 \text{ lb}/\text{ft}^3)(4.0 \text{ ft}^2)(200 \text{ ft}/\text{s})(560 \text{ ft}) = 34,300 \text{ ft·lb}/\text{s}$$

or

$$\mathcal{P}_a = \frac{34,300 \text{ ft·lb}/\text{s}}{550 \text{ (ft·lb}/\text{s)}/\text{hp}} = 62.3 \text{ hp} \tag{Ans}$$

With a closed-return wind tunnel of this type, all of the power required to maintain the flow is dissipated through viscous effects, with the energy remaining within the closed tunnel. If heat transfer across the tunnel walls is negligible, the air temperature within the tunnel will increase in time. For steady state operations of such tunnels, it is often necessary to provide some means of cooling to maintain the temperature at acceptable levels.

It should be noted that the actual size of the motor that powers the fan must be greater than the calculated 62.3 hp because the fan is not 100% efficient. The power calculated above is that needed by the fluid to overcome losses in the tunnel, excluding those in the fan. If the fan were 60% efficient, it would require a shaft power of $\mathscr{P} = 62.3$ hp$/(0.60) = 104$ hp to run the fan. Determination of fan (or pump) efficiencies can be a complex problem which depends on the specific geometry of the fan. Considerable information concerning the design of fans and pumps is available (Refs. 14, 15, 16, for example).

It should also be noted that the above results are only approximate. Clever, careful design of the various components (corners, diffuser, etc.) may lead to improved (i.e., lower) values of the various loss coefficients, and hence lower power requirements. Since h_L is proportional to V^2, the components with the larger V tend to have the larger head loss. Thus, even though $K_L = 0.2$ for each of the four corners, the head loss for corner (7) is $(V_7/V_3)^2 = (80/22.9)^2 = 12.2$ times greater than it is for corner (3).

8.4.3 Noncircular Conduits

Many of the conduits that are used for conveying fluids are not circular in cross section. Although the details of the flows in such conduits depend on the exact cross-sectional shape, many round pipe results can be carried over, with slight modification, to flow in conduits of other shapes.

Theoretical results can be obtained for fully developed laminar flow in noncircular ducts, although the detailed mathematics often becomes rather cumbersome. For an arbitrary cross section, as is shown in Fig. 8.38, the velocity profile is a function of both y and z [$\mathbf{V} = u(y, z)\hat{\mathbf{i}}$]. This means that the governing equation from which the velocity profile is obtained (either the Navier–Stokes equations of motion or a force balance equation similar to that used for circular pipes, Eq. 8.6) is a partial differential equation rather than an ordinary differential equation. Although the equation is linear (for fully developed flow the convective acceleration is zero), its solution is not as straightforward as for round pipes. Typically the velocity profile is given in terms of an infinite series representation (Ref. 17).

Practical, easy-to-use results can be obtained as follows. Regardless of the cross-sectional shape, there are no inertia effects in fully developed laminar pipe flow. Thus, the friction factor can be written as $f = C/\text{Re}_h$, where the constant C depends on the particular shape of the duct and Re_h is the Reynolds number, $\text{Re}_h = \rho V D_h/\mu$, based on the hydraulic diameter. The *hydraulic diameter* defined as $D_h = 4A/P$ is

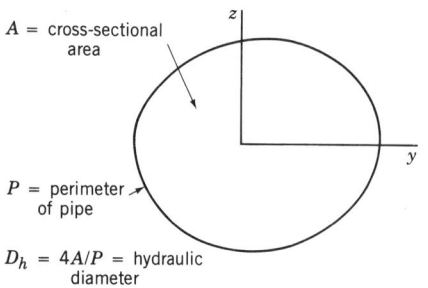

A = cross-sectional area

P = perimeter of pipe

$D_h = 4A/P$ = hydraulic diameter

(a)

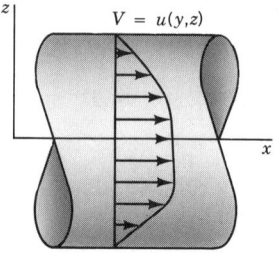

$V = u(y,z)$

(b)

FIGURE 8.38

Noncircular duct.

TABLE 8.4 Friction Factors for Laminar Flow in Noncircular Ducts (Data from Ref. 18)

Shape	Parameter	$C = f \text{Re}_h$
I. Concentric Annulus	D_1/D_2	
$D_h = D_2 - D_1$	0.0001	71.8
	0.01	80.1
	0.1	89.4
	0.6	95.6
	1.00	96.0
II. Circular segment	α (degrees)	
$D_h = D \left[1 + \dfrac{\sin(2\alpha)}{2(\pi - \alpha)} \right]$	0	64.0
	60	63.3
	90	63.1
	120	62.8
	180	62.2
III. Rectangle	a/b	
$D_h = \dfrac{2ab}{a + b}$	0	96.0
	0.05	89.9
	0.01	84.7
	0.25	72.9
	0.50	62.2
	0.75	57.9
	1.00	56.9
IV. Circular sector	α (degrees)	
$D_h = \dfrac{\alpha}{1 + \alpha} D$	0	48.0
	30	56.7
	60	60.8
	90	63.1
V. Right triangle	α (degrees)	
$D_h = \dfrac{2 b \sin \alpha}{(1 + \sin \alpha + \cos \alpha)}$	0	48.0
	10	49.9
	20	51.2
	30	52.0
	40	52.4
	45	52.5

four times the ratio of the cross-sectional flow area divided by the wetted perimeter, P, as is illustrated in Fig. 8.38. It represents a characteristic length that defines the size of a cross section of a specified shape. The factor of 4 is included in the definition of D_h so that for round pipes the diameter and hydraulic diameter are equal [$D_h = 4A/P = 4(\pi D^2/4)/(\pi D) = D$]. The hydraulic diameter is also used in the definition of the friction factor, $h_L = f(\ell/D_h)V^2/2g$, and the relative roughness, ε/D_h.

The values of $C = f\,\mathrm{Re}_h$ have been obtained from theory and/or experiment for various shapes. Typical values are given in Table 8.4 along with the hydraulic diameter. Note that the value of C is relatively insensitive to the shape of the conduit. Unless the cross section is very "thin" in some sense, the value of C is not too different from its circular pipe value, $C = 64$. Once the friction factor is obtained, the calculations for noncircular conduits are identical to those for round pipes.

Calculations for fully developed turbulent flow in ducts of noncircular cross section are usually carried out by using the Moody chart data for round pipes with the diameter replaced by the hydraulic diameter and the Reynolds number based on the hydraulic diameter. Such calculations are usually accurate to within about 15%. If greater accuracy is needed, a more detailed analysis based on the specific geometry of interest is needed.

EXAMPLE 8.7

A round pipe of diameter D is to be replaced by a square duct of width w such that for a given flowrate the head loss per foot of pipe will remain the same. Determine w in terms of D if (a) the flow is laminar, or (b) the flow is completely turbulent (i.e., large Reynolds number flow corresponding to the flat portion of the Moody chart).

SOLUTION

For either laminar or turbulent flow, $h_L = f(\ell/D_h)(V^2/2g)$. Thus, with subscripts r and s denoting the round and square pipes, we must have

$$h_L = f_r \frac{\ell_r}{D_{h_r}} \frac{V_r^2}{2g} = f_s \frac{\ell_s}{D_{h_s}} \frac{V_s^2}{2g}$$

or since $\ell_r = \ell_s$, $D_{h_r} = D$, and $D_{h_s} = 4A_s/P_s = 4w^2/(4w) = w$ this becomes

$$f_r \frac{V_r^2}{D_{h_r}} = f_s \frac{V_s^2}{D_{h_s}} \quad \text{or} \quad f_r \frac{V_r^2}{D} = f_s \frac{V_s^2}{w} \tag{1}$$

Also, $Q_r = Q_s$ or $(\pi/4)D^2 V_r = w^2 V_s$ so that

$$V_r = \frac{4w^2}{\pi D^2} V_s \tag{2}$$

Thus, Eqs. 1 and 2 combine to give

$$f_s = \frac{16}{\pi^2} \left(\frac{w}{D}\right)^5 f_r \tag{3}$$

(a) For laminar flow, $f_r = 64/\mathrm{Re}_{h_r} = 64\mu/\rho V_r D$ and $f_s = C/\mathrm{Re}_{h_s} = C\mu/\rho V_s w$, which can be combined with Eq. (2) to give

$$\frac{f_r}{f_s} = \frac{16\pi}{C} \frac{D}{w} \tag{4}$$

From Eqs. 3 and 4 we obtain (by using $C = 56.9$ from Table 8.4)

$$\left(\frac{w}{D}\right)^4 = \frac{\pi C}{256} = \frac{\pi(56.9)}{256} = 0.698$$

Thus, for the same flowrate and head loss with laminar flow the width of the square duct is

$$w = (0.698)^{1/4}D = 0.914D \qquad \text{(Ans)}$$

(b)   For completely turbulent flow the friction factor is independent of the Reynolds number. Thus, if we assume that $\varepsilon/D \approx \varepsilon/D_{hs}$, it follows that $f_r = f_s$, so that from Eq. 3

$$\left(\frac{w}{D}\right)^5 = \frac{\pi^2}{16} \quad \text{or} \quad w = 0.908D \qquad \text{(Ans)}$$

Note that although ε/D does not equal ε/D_{hs}, their values do not differ by much and the use of the same line of the Moody chart for both the round pipe and the square duct (i.e., assuming $f_r = f_s$) does not introduce significant error.

For both the laminar and turbulent cases the length of the side of the equivalent square duct is approximately 91% of the diameter of the round pipe. However, the cross-sectional area of the square duct ($A_s = w^2 \approx 0.83D^2$) is greater than that of the round pipe ($A_r = \pi D^2/4 = 0.79D^2$). It takes less material to form the round pipe (lateral area $= \pi D\ell = 3.14D\ell$) than the equivalent square duct (lateral area $= 4w\ell \approx 3.6D\ell$). Circles are very efficient shapes. They can also withstand a pressure difference without unwanted distortion of the cross-sectional shape.

8.5   PIPE FLOW EXAMPLES

In the previous sections of this chapter we discussed concepts concerning flow in pipes and ducts. The purpose of this section is to apply these ideas to the solutions of various practical problems. The application of the pertinent equations is straightforward, with rather simple calculations that give answers to problems of engineering importance. The main idea involved is to apply the energy equation between appropriate locations within the flow system, with the head loss written in terms of the friction factor and the minor loss coefficients. We will consider two classes of pipe systems: those containing a single pipe (whose length may be interrupted by various components), and those containing multiple pipes in parallel, series, or network configurations.

8.5.1   Single Pipes

The nature of the solution process for pipe flow problems can depend strongly on which of the various parameters are independent parameters (the "given") and which is the dependent parameter (the "determine"). The three most common types of problems are shown in Table 8.5 in terms of the parameters involved. We assume the pipe system is defined in terms of the length of pipe sections used and the number of elbows, bends, and valves needed to convey the fluid between the desired locations. In all instances we assume the fluid properties are given.

In a Type I problem we specify the desired flowrate or average velocity and determine the necessary pressure difference or head loss. For example, if a flowrate of 2.0 gal/min is required for a dishwasher that is connected to the water heater by a given pipe system, what pressure is needed in the water heater?

In a Type II problem we specify the applied driving pressure (or, alternatively, the

TABLE 8.5 Pipe Flow Types

Variable	Type I	Type II	Type III
a. Fluid			
Density	Given	Given	Given
Viscosity	Given	Given	Given
b. Pipe			
Diameter	Given	Given	Determine
Length	Given	Given	Given
Roughness	Given	Given	Given
c. Flow			
Flowrate or	Given	Determine	Given
Average Velocity			
d. Pressure			
Pressure Drop or	Determine	Given	Given
Head Loss			

head loss) and determine the flowrate. For example, how many gal/min of hot water are supplied to the dishwasher if the pressure within the water heater is 60 psi and the pipe system details (length, diameter, roughness of the pipe; number of elbows; etc.) are specified?

In a Type III problem we specify the pressure drop and the flowrate and determine the diameter of the pipe needed. For example, what diameter of pipe is needed between the water heater and dishwasher if the pressure in the water heater is 60 psi (determined by the city water system) and the flowrate is to be not less than 2.0 gal/min (determined by the manufacturer)?

Several examples of these types of problems follow.

EXAMPLE 8.8 (TYPE I, DETERMINE PRESSURE DROP)

Water at 60 °F flows from the basement to the second floor through the 0.75-in.- (0.0625-ft-) diameter copper pipe (a drawn tubing) at a rate of $Q = 12.0$ gal/min $= 0.0267$ ft^3/s and exits through a faucet of diameter 0.50 in. as shown in Fig. E8.8a. Determine the pressure at point (1) if: (a) viscous effects are neglected, (b) the only losses included are major losses, or (c) all losses are included.

FIGURE E8.8a

SOLUTION

Since the fluid velocity in the pipe is given by $V_1 = Q/A_1 = Q/(\pi D^2/4) = (0.0267$ ft³/s)/$[\pi(0.0625 \text{ ft})^2/4] = 8.70$ ft/s, and the fluid properties are $\rho = 1.94$ slugs/ft³, and $\mu = 2.34 \times 10^{-5}$ lb·s/ft² (see Table B.1), it follows that Re $= \rho VD/\mu = (1.94 \text{ slugs/ft}^3)(8.70$ ft/s) (0.0625 ft)/(2.34 × 10⁻⁵ lb·s/ft²) = 45,000. Thus, the flow is turbulent. The governing equation for either case (a), (b), or (c) is Eq. 8.21,

$$\frac{p_1}{\gamma} + \alpha_1 \frac{V_1^2}{2g} + z_1 = \frac{p_2}{\gamma} + \alpha_2 \frac{V_2^2}{2g} + z_2 + h_L$$

where $z_1 = 0$, $z_2 = 20$ ft, $p_2 = 0$ (free jet), $\gamma = \rho g = 62.4$ lb/ft³, and the outlet velocity is $V_2 = Q/A_2 = (0.0267 \text{ ft}^3/\text{s})/[\pi(0.50/12)^2 \text{ ft}^2/4] = 19.6$ ft/s. We have assumed that the energy correction factors α_1 and α_2 are unity. This is reasonable because turbulent velocity profiles are nearly uniform across the pipe. Thus,

$$p_1 = \gamma z_2 + \tfrac{1}{2}\rho(V_2^2 - V_1^2) + \gamma h_L \tag{1}$$

where the head loss is different for each of the three cases.

(a) If all losses are neglected ($h_L = 0$), Eq. 1 gives

$$p_1 = (62.4 \text{ lb/ft}^3)(20 \text{ ft})$$
$$+ \frac{1.94 \text{ slugs/ft}^3}{2}\left[\left(19.6^2 \frac{\text{ft}}{\text{s}}\right)^2 - \left(8.70 \frac{\text{ft}}{\text{s}}\right)^2\right]$$
$$= (1248 + 299) \text{ lb/ft}^2 = 1547 \text{ lb/ft}^2$$

or

$$p_1 = 10.7 \text{ psi} \tag{Ans}$$

Note that for this pressure drop, the amount due to elevation change (the hydrostatic effect) is $\gamma(z_2 - z_1) = 8.67$ psi and the amount due to the increase in kinetic energy is $\rho(V_2^2 - V_1^2)/2 = 2.07$ psi.

(b) If the only losses included are the major losses, the head loss is

$$h_L = f\frac{\ell}{D}\frac{V_1^2}{2g}$$

From Fig. 8.22 the relative roughness for a 0.75-in.-diameter copper pipe (drawn tubing) is $\varepsilon/D = 8 \times 10^{-5}$. With this ε/D and the calculated Reynolds number (Re = 45,000), the value of f is obtained from the Moody chart as $f = 0.0215$. Note that the Colebrook equation (Eq. 8.35) would give the same value of f. Hence, with the total length of the pipe as $\ell = (15 + 5 + 10 + 10 + 20)$ ft = 60 ft and the elevation and kinetic energy portions the same as for part (a), Eq. 1 gives

$$p_1 = \gamma z_2 + \frac{1}{2}\rho(V_2^2 - V_1^2) + \rho f\frac{\ell}{D}\frac{V_1^2}{2}$$
$$= (1248 + 299) \text{ lb/ft}^2$$
$$+ (1.94 \text{ slugs/ft}^3)(0.0215)\left(\frac{60 \text{ ft}}{0.0625 \text{ ft}}\right)\frac{(8.70 \text{ ft/s})^2}{2}$$
$$= (1248 + 299 + 1515) \text{ lb/ft}^2 = 3062 \text{ lb/ft}^2$$

or

$$p_1 = 21.3 \text{ psi} \tag{Ans}$$

Of this pressure drop, the amount due to pipe friction is 10.5 psi.

(c) If major and minor losses are included, Eq. 1 becomes

$$p_1 = \gamma z_2 + \frac{1}{2}\rho(V_2^2 - V_1^2) + f\frac{\ell}{D}\frac{V_1^2}{2g} + \sum \rho K_L \frac{V^2}{2}$$

or

$$p_1 = 21.3 \text{ psi} + \sum \rho K_L \frac{V^2}{2} \tag{2}$$

where the 21.3-psi contribution is due to elevation change, kinetic energy change, and major losses [part (b)], and the last term represents the sum of all of the minor losses. The loss coefficients of the components ($K_L = 1.5$ for each elbow and $K_L = 10$ for the wide-open globe valve) are given in Table 8.3 (except for the loss coefficient of the faucet, which is given in Fig. E8.8a as $K_L = 2$). Thus,

$$\sum \rho K_L \frac{V^2}{2} = (1.94 \text{ slugs/ft}^3)\frac{(8.70 \text{ ft})^2}{2}[10 + 4(1.5) + 2]$$

$$= 1321 \text{ lb/ft}^2$$

or

$$\sum \rho K_L \frac{V^2}{2} = 9.17 \text{ psi} \tag{3}$$

Note that we did not include an entrance or exit loss because points (1) and (2) are located within the fluid streams, not within an attaching reservoir where the kinetic energy is zero. Thus, by combining Eqs. 2 and 3 we obtain the entire pressure drop as

$$p_1 = (21.3 + 9.17) \text{ psi} = 30.5 \text{ psi} \tag{Ans}$$

This pressure drop calculated by including all losses should be the most realistic answer of the three cases considered.

More detailed calculations will show that the pressure distribution along the pipe is as illustrated in Fig. E8.8b for cases (a) and (c)—neglecting all losses or including all

FIGURE E8.8b

losses. Note that not all of the pressure drop, $p_1 - p_2$, is a "pressure loss." The pressure change due to the elevation and velocity changes are completely reversible. The portion due to the major and minor losses are irreversible.

This flow can be illustrated in terms of the energy line and hydraulic grade line concepts introduced in Section 3.7. As is shown in Fig.8.8c, for case (a) there are no losses and

FIGURE E8.8c

the energy line (EL) is horizontal, one velocity head ($V^2/2g$) above the hydraulic grade line (HGL), which is one pressure head (γz) above the pipe itself. For case (c) the energy line is not horizontal. Each bit of friction in the pipe or loss in a component reduces the available energy, thereby lowering the energy line. Thus, for case (a) the total head remains constant throughout the flow with a value of

$$H = \frac{p_1}{\gamma} + \frac{V_1^2}{2g} + z_1 = \frac{(1547 \text{ lb/ft}^2)}{(62.4 \text{ lb/ft}^3)} + \frac{(8.70 \text{ ft/s})^2}{2(32.2 \text{ ft/s}^2)} + 0$$

$$= 26.0 \text{ ft}$$

$$= \frac{p_2}{\gamma} + \frac{V_2^2}{2g} + z_2 = \frac{p_3}{\gamma} + \frac{V_3^2}{2g} + z_3 = \cdots$$

For case (c) the energy line starts at

$$H_1 = \frac{p_1}{\gamma} + \frac{V_1^2}{2g} + z_1 = \frac{(30.5 \times 144) \text{lb/ft}^2}{(62.4 \text{ lb/ft}^3)} + \frac{(8.70 \text{ ft/s})^2}{2(32.2 \text{ ft/s}^2)} + 0 = 71.6 \text{ ft}$$

and falls to a final value of

$$H_2 = \frac{p_2}{\gamma} + \frac{V_2^2}{2g} + z_2 = 0 + \frac{(19.6 \text{ ft/s})^2}{2(32.2 \text{ ft/s}^2)} + 20 \text{ ft} = 26.0 \text{ ft}$$

The elevation of the energy line can be calculated at any point along the pipe. For example, at point (7), 50 ft from point (1),

$$H_7 = \frac{p_7}{\gamma} + \frac{V_7^2}{2g} + z_7 = \frac{(9.93 \times 144) \text{ lb/ft}^2}{(62.4 \text{ lb/ft}^3)} + \frac{(8.70 \text{ ft/s})^2}{2(32.2 \text{ ft/s}^2)} + 20 \text{ ft} = 44.1 \text{ ft}$$

The head loss per foot of pipe is the same all along the pipe. That is,

$$\frac{h_L}{\ell} = f \frac{V^2}{2gD} = \frac{0.0215(8.70 \text{ ft/s})^2}{2(32.2 \text{ ft/s}^2)(0.0625 \text{ ft})} = 0.404 \text{ ft/ft}$$

Thus, the energy line is a set of straight line segments of the same slope separated by steps whose height equals the head loss of the minor component at that location. As is seen from Fig. E8.8c, the globe value produces the largest of all the minor losses.

Although the governing pipe flow equations are quite simple, they can provide very reasonable results for a variety of applications as shown in the next example.

EXAMPLE 8.9 (TYPE I, DETERMINE HEAD LOSS)

Crude oil at 140 °F with $\gamma = 53.7 \text{ lb/ft}^3$ and $\mu = 8 \times 10^{-5} \text{ lb·s/ft}^2$ (about four times the viscosity of water) is pumped across Alaska through the Alaskan pipeline, a 799-mile-long, 4-ft-diameter steel pipe, at a maximum rate of $Q = 2.4$ million barrels/day $= 117 \text{ ft}^3/\text{s}$, or $V = Q/A = 9.31 \text{ ft/s}$. Determine the horsepower needed for the pumps that drive this large system.

SOLUTION

From the energy equation (Eq. 8.21) we obtain

$$\frac{p_1}{\gamma} + \frac{V_1^2}{2g} + z_1 + h_p = \frac{p_2}{\gamma} + \frac{V_2^2}{2g} + z_2 + h_L$$

where points (1) and (2) represent locations within the large holding tanks at either end of the line and h_p is the head provided to the oil by the pumps. We assume that $z_1 = z_2$ (pumped from sea level to sea level), $p_1 = p_2 = V_1 = V_2 = 0$ (large, open tanks) and $h_L = (f\ell/D)(V^2/2g)$. Minor losses are negligible because of the large length-to-diameter ratio of the relatively straight, uninterrupted pipe; $\ell/D = (799 \text{ mi})(5280 \text{ ft/mi})/(4 \text{ ft}) = 1.05 \times 10^6$. Thus,

$$h_p = h_L = f\frac{\ell}{D}\frac{V^2}{2g}$$

where from Fig. 8.23, $f = 0.0125$ since $\varepsilon/D = 0.000036$ (see Fig. 8.22) and $\text{Re} = \rho VD/\mu = [(53.7/32.2) \text{ slugs/ft}^3](9.31 \text{ ft/s})(4.0 \text{ ft})/(8 \times 10^{-5} \text{ lb·s/ft}^2) = 7.76 \times 10^5$. Thus,

$$h_p = 0.0125(1.05 \times 10^6)\frac{(9.31 \text{ ft/s})^2}{2(32.2 \text{ ft/s}^2)} = 17{,}700 \text{ ft}$$

and the actual power supplied to the fluid, $\mathscr{P}_a$, is

$$\mathscr{P}_a = \gamma Q h_p = (53.7 \text{ lb/ft}^3)(117 \text{ ft}^3/\text{s})(17{,}700 \text{ ft})$$

$$= 1.11 \times 10^8 \text{ ft·lb/s} \left(\frac{1 \text{ hp}}{550 \text{ ft·lb/s}}\right)$$

$$= 202{,}000 \text{ hp} \qquad\qquad \text{(Ans)}$$

There are many reasons why it is not practical to drive this flow with a single pump of this size. First, there are no pumps this large! Second, the pressure at the pump outlet would need to be $p = \gamma h_L = (53.7 \text{ lb/ft}^3)(17{,}700 \text{ ft})(1 \text{ ft}^2/144 \text{ in.}^2) = 6600 \text{ psi}$. No practical 4-ft-diameter pipe would withstand this pressure. An equally unfeasible alternative would be to place the holding tank at the beginning of the pipe on top of a hill of height $h_L = 17{,}700 \text{ ft}$ and let gravity force the oil through the 799-mi pipe!

To produce the desired flow, the actual system contains 12 pumping stations positioned at strategic locations along the pipeline. Each station contains four pumps, three of which operate at any one time (the fourth is in reserve in case of emergency). Each pump is driven by a 13,500-hp motor, thereby producing a total horsepower of $\mathscr{P} = 12$ stations (3 pumps/station)(13,500 hp/pump) $= 486{,}000 \text{ hp}$. If we assume that the pump/motor combination is approximately 60% efficient, there is a total of 0.60 (486,000) hp $= 292{,}000 \text{ hp}$ available to drive the fluid. This number compares favorably with the 202,000-hp answer calculated above.

The assumption of a 140 °F oil temperature may not seem reasonable for flow across Alaska. Note, however, that the oil is warm when it is pumped from the ground and that the 202,000 hp needed to pump the oil is dissipated as a head loss (and therefore a temperature rise) along the pipe. However, if the oil temperature were 70 °F rather than 140 °F, the viscosity would

be approximately 16×10^{-5} lb·s/ft² (twice as large), but the friction factor would only increase from $f = 0.0125$ at 140 °F (Re $= 7.76 \times 10^5$) to $f = 0.0140$ at 70 °F (Re $= 3.88 \times 10^5$). This doubling of viscosity would result in only an 11% increase in power (from 202,000 to 226,000 hp). Because of the large Reynolds numbers involved, the shear stress is due mostly to the turbulent nature of the flow. That is, the value of Re for this flow is large enough (on the relatively flat part of the Moody chart) so that f is nearly independent of Re (or viscosity).

Pipe flow problems in which it is desired to determine the flowrate for a given set of conditions (Type II problems) often require trial-and-error solution techniques. This is because it is necessary to know the value of the friction factor to carry out the calculations, but the friction factor is a function of the unknown velocity (flowrate) in terms of the Reynolds number. The solution procedure is indicated in Example 8.10.

EXAMPLE 8.10 (TYPE II, DETERMINE FLOWRATE)

Water at 20 °C ($\rho = 998$ kg/m³ and $\mu = 1.00 \times 10^{-3}$ N·s/m², see Table B.2) is siphoned from one irrigation ditch to another through a 50-mm-diameter, 1.8-m-long hose ($\varepsilon = 0.20$ mm), as is shown in Fig. E8.10. The loss coefficient for each bend is $K_L = 0.4$. Determine the flowrate.

FIGURE E8.10

SOLUTION

Application of the energy equation (Eq. 8.21) between the two free surfaces, points (1) and (2) as indicated, gives

$$\frac{p_1}{\gamma} + \frac{V_1^2}{2g} + z_1 = \frac{p_2}{\gamma} + \frac{V_2^2}{2g} + z_2 + f\frac{\ell}{D}\frac{V^2}{2g} + \sum K_L \frac{V^2}{2g} \tag{1}$$

where K_L for the entrance (assumed to act as a reentrance entrance) is $K_{L_{\text{ent}}} = 0.8$, while that for each bend is given as $K_{L_{\text{bend}}} = 0.4$, and for the exit $K_{L_{\text{exit}}} = 1.0$. Note that the velocity on which the head loss is based, V, is the velocity within the hose. Thus, with $z_1 - z_2 = 0.13$ m, $\ell = 1.8$ m, $V_1 = V_2 = 0$, and $p_1 = p_2 = 0$, Eq. 1 becomes

$$0.13 \text{ m} = \left\{ f\frac{(1.8 \text{ m})}{(0.05 \text{ m})} + [0.8 + 2(0.4) + 1] \right\} \frac{V^2}{2(9.81 \text{ m/s}^2)}$$

or

$$2.55 = (36f + 2.60)V^2 \tag{2}$$

where V is in m/s.

As an alternative, we could write the energy equation between point (1) and point (3) that lies in the exit plane of the hose. In this case, $z_1 - z_3 = (0.13 + 0.07)$ m, $p_3 = 0.07\gamma$, $V_3 = V$, and $K_{L_{\text{exit}}} = 0$ (the exit kinetic energy is lost after the exit plane). The result would again be Eq. 2. Similarly, the same equation would result if points (1) and (4) were used, where

point (4) is located on the same level as the exit but far away from it. For this case, $p_4 = 0.07\gamma$, $V_4 = 0$, and $K_{L_{\text{exit}}} = 1$.

The value of f is dependent on Re, which is dependent on V, an unknown. However,

$$\text{Re} = \frac{\rho V D}{\mu} = \frac{(998 \text{ kg/m}^3)V(0.05 \text{ m})}{1.00 \times 10^{-3} \text{ N·s/m}^2}$$

or

$$\text{Re} = 49,800V \tag{3}$$

where V is in m/s.

Also, since $\varepsilon/D = (0.200 \text{ mm})/(50 \text{ mm}) = 0.004$, we know which particular curve of the Moody chart is pertinent to this flow. Thus, we have three relationships (Eqs. 2, 3, and the $\varepsilon/D = 0.04$ curve of Fig. 8.23) from which we can solve for the three unknowns f, Re, and V. This is easily done by an iterative scheme as follows.

It is usually simplest to assume a value of f, calculate V from Eq. 2, calculate Re from Eq. 3, and look up the appropriate value of f in the Moody chart for this value of Re. If the assumed f and the new f do not agree, the assumed answer is not correct—we do not have the solution to the three equations. Although values of either f, V, or Re could be assumed as starting values, it is usually simplest to assume a value of f because the correct value often lies on the relatively flat portion of the Moody chart for which f is quite insensitive to Re.

Thus, we assume $f = 0.03$, approximately the large Re limit for the given relative roughness. From Eq. 2 we obtain

$$V = \left[\frac{2.55}{36(0.03) + 2.6} \right]^{1/2} = 0.832 \text{ m/s}$$

and from Eq. 3

$$\text{Re} = 49,800(0.832) = 41,400$$

With this Re and ε/D, Fig. 8.23 gives $f = 0.031$, which is not equal to the assumed solution $f = 0.030$ (although it is close!). We try again, this time with the newly obtained value of $f = 0.031$, which gives $V = 0.828$ m/s and Re $= 41,300$. With these values, Fig. 8.23 gives $f = 0.031$, which agrees with the assumed value. Thus, the solution is $V = 0.828$ m/s, or

$$Q = AV = \frac{\pi}{4}(0.05 \text{ m})^2(0.828 \text{ m/s}) = 1.63 \times 10^{-3} \text{ m}^3/\text{s} \qquad \text{(Ans)}$$

Note that the need for the iteration scheme is because one of the equations, $f = \phi(\text{Re}, \varepsilon/D)$, is in graphical form (the Moody chart). If the dependence of f on Re and ε/D is known in equation form, this graphical dependency is eliminated, and the solution technique may be easier. Such is the case if the flow is laminar so that the friction factor is simply $f = 64/\text{Re}$. For turbulent flow, we can use the Colebrook equation rather than the Moody chart, although this would normally require an iterative scheme also because of the complexity of the equation. As is shown below, such a formulation is ideally suited for an iterative computer solution.

We keep Eqs. 2 and 3 and use the Colebrook equation (Eq. 8.35) (rather than the Moody chart) with $\varepsilon/D = 0.004$ to give

$$\frac{1}{\sqrt{f}} = -2.0 \log\left(\frac{\varepsilon/D}{3.7} + \frac{2.51}{\text{Re}\sqrt{f}} \right) = -2.0 \log\left(0.00108 + \frac{2.51}{\text{Re}\sqrt{f}} \right) \tag{4}$$

From Eq. 2 we have $V = [2.55/(36f + 2.6)]^{1/2}$, which can be combined with Eq. 3 to give

$$\text{Re} = \frac{79,500}{\sqrt{36f + 2.6}} \tag{5}$$

The combination of Eqs. 4 and 5 provides a single equation for the determination of f

$$\frac{1}{\sqrt{f}} = -2.0 \log \left(1.08 \times 10^{-3} + 3.16 \times 10^{-5} \sqrt{36 + \frac{2.6}{f}} \right) \tag{6}$$

A simple iterative solution of this equation gives $f = 0.031$, in agreement with the above solution which used the Moody chart.

Note that unlike the Alaskan pipeline example (Example 8.9) in which we assumed minor losses are negligible, minor losses are of importance in this example because of the relatively small length-to-diameter ratio; $\ell/D = 1.8/0.05 = 36$. The ratio of minor to major losses in this case is $\Sigma K_L/(f\ell/D) = 2.6/[0.031(36)] = 2.33$. A slightly smoother bend in the hose would probably increase the flowrate more than would a smoother hose.

EXAMPLE 8.11 (TYPE II, DETERMINE FLOWRATE)

The turbine shown in Fig. E8.11 extracts 50 hp from the water flowing through it. The 1-ft-diameter, 300-ft-long pipe is assumed to have a friction factor of 0.02. Minor losses are negligible. Determine the flowrate through the pipe and turbine.

FIGURE E8.11

SOLUTION

The energy equation (Eq. 8.21) can be applied between the surface of the lake and the outlet of the pipe as

$$\frac{p_1}{\gamma} + \frac{V_1^2}{2g} + z_1 = \frac{p_2}{\gamma} + \frac{V_2^2}{2g} + z_2 + h_L + h_T \tag{1}$$

where $p_1 = V_1 = p_2 = z_2 = 0$, $z_1 = 90$ ft, and $V_2 = V$, the fluid velocity in the pipe. The head loss is given by

$$h_L = f\frac{\ell}{D}\frac{V^2}{2g} = 0.02\frac{(300 \text{ ft})}{(1 \text{ ft})}\frac{V^2}{2(32.2 \text{ ft/s}^2)} = 0.0932V^2 \text{ ft}$$

where V is in ft/s. Also, the turbine head is

$$h_T = \frac{\mathscr{P}_a}{\gamma Q} = \frac{\mathscr{P}_a}{\gamma(\pi/4)D^2V} = \frac{(50 \text{ hp})[(550 \text{ ft·lb/s})/\text{hp}]}{(62.4 \text{ lb/ft}^3)[(\pi/4)(1 \text{ ft})^2V]} = \frac{561}{V} \text{ ft}$$

Thus, Eq. 1 can be written as

$$90 = \frac{V^2}{2(32.2)} + 0.0932V^2 + \frac{561}{V}$$

or

$$0.109V^3 - 90V + 561 = 0 \tag{2}$$

where V is in ft/s. The velocity of the water in the pipe is found as the solution of Eq. 2. Surprisingly, there are two real, positive roots: $V = 6.58$ ft/s or $V = 24.9$ ft/s. The third root is negative ($V = -31.4$ ft/s) and has no physical meaning for this flow. Thus, the two acceptable flowrates are

$$Q = \frac{\pi}{4} D^2 V = \frac{\pi}{4} (1 \text{ ft})^2(6.58 \text{ ft/s}) = 5.17 \text{ ft}^3/\text{s} \qquad \text{(Ans)}$$

or

$$Q = \frac{\pi}{4} (1 \text{ ft})^2(24.9 \text{ ft/s}) = 19.6 \text{ ft}^3/\text{s} \qquad \text{(Ans)}$$

Either of these two flowrates gives the same power, $\mathscr{P}_a = \gamma Q h_T$. The reason for two possible solutions can be seen from the following. With the low flowrate ($Q = 5.17$ ft^3/s), we obtain the head loss and turbine head as $h_L = 4.04$ ft and $h_T = 85.3$ ft. Because of the relatively low velocity there is a relatively small head loss and, therefore, a large head available for the turbine. With the large flowrate ($Q = 19.6$ ft^3/s), we find $h_L = 57.8$ ft and $h_T = 22.5$ ft. The high-speed flow in the pipe produces a relatively large loss due to friction, leaving a relatively small head for the turbine. However, in either case the product of the turbine head times the flowrate is the same. That is, the power extracted ($\mathscr{P}_a = \gamma Q h_T$) is identical for each case. Although either flowrate will allow the extraction of 50 hp from the water, the details of the design of the turbine itself will depend strongly on which flowrate is to be used. Such details are beyond the scope of this book. They can be found in various references about turbomachines (Refs. 14, 19, 20).

If the friction factor were not given, the solution to the problem would be much more lengthy. A trial-and-error solution similar to that in Example 8.10 would be required along with the solution of a cubic equation.

In pipe flow problems for which the diameter is the unknown (Type III), an iterative technique is required. This is, again, because the friction factor is a function of the diameter—through both the Reynolds number and the relative roughness. Thus, neither Re $= \rho V D/\mu = 4\rho Q/\pi \mu D$ nor ε/D are known unless D is known. Example 8.12 illustrates this.

EXAMPLE 8.12 (TYPE III, DETERMINE DIAMETER)

Air at standard temperature and pressure flows through a horizontal, galvanized iron pipe ($\varepsilon = 0.0005$ ft) at a rate of 2.0 ft^3/s. Determine the minimum pipe diameter if the pressure drop is to be no more than 0.50 psi per 100 ft of pipe.

SOLUTION

We assume the flow to be incompressible with $\rho = 0.00238$ slug/ft^3 and $\mu = 3.74 \times 10^{-7}$ lb·s/ft^2. Note that if the pipe were too long, the pressure drop from one end to the other, $p_1 - p_2$, would not be small relative to the pressure at the beginning, and compressible flow considerations would be required. In this case a pipe length of 200 ft gives $(p_1 - p_2)/p_1 = [(0.50 \text{ psi})/(100 \text{ ft})](200 \text{ ft})/14.7 \text{ psi} = 0.068 = 6.8\%$, which is probably small enough to justify the incompressible assumption.

With $z_1 = z_2$ and $V_1 = V_2$ the energy equation (Eq. 8.21) becomes

$$p_1 = p_2 + f \frac{\ell}{D} \frac{\rho V^2}{2} \tag{1}$$

where $V = Q/A = 4Q/(\pi D^2) = 4(2.0 \text{ ft}^3/\text{s})/\pi D^2$, or

$$V = \frac{2.55}{D^2}$$

where D is in feet. Thus, with $p_1 - p_2 = (0.5 \text{ lb/in.}^2)(144 \text{ in.}^2/\text{ft}^2)$ and $\ell = 100$ ft, Eq. 1 becomes

$$p_1 - p_2 = (0.5)(144) \text{ lb/ft}^2$$

$$= f \frac{(100 \text{ ft})}{D} (0.00238 \text{ slug/ft}^3) \frac{1}{2} \left(\frac{2.55}{D^2} \frac{\text{ft}}{\text{s}} \right)^2$$

or

$$D = 0.404 f^{1/5} \tag{2}$$

where D is in feet. Also $\text{Re} = \rho V D/\mu = (0.00238 \text{ slug/ft}^3) [(2.55/D^2) \text{ ft/s}]D/(3.74 \times 10^{-7} \text{ lb·s/ft}^2)$, or

$$\text{Re} = \frac{1.62 \times 10^4}{D} \tag{3}$$

and

$$\frac{\varepsilon}{D} = \frac{0.0005}{D} \tag{4}$$

Thus, we have four equations (Eqs. 2, 3, 4, and either the Moody chart or the Colebrook equation) and four unknowns (f, D, ε/D, and Re) from which the solution can be obtained by trial-and-error methods.

If we use the Moody chart, it is probably easiest to assume a value of f, use Eqs. 2, 3, and 4 to calculate D, Re, and ε/D, and then compare the assumed f with that from the Moody chart. If they do not agree, try again. Thus, we assume $f = 0.02$, a typical value, and obtain $D = 0.404(0.02)^{1/5} = 0.185$ ft, which gives $\varepsilon/D = 0.0005/0.185 = 0.0027$ and $\text{Re} = 1.62 \times 10^4/0.185 = 8.76 \times 10^4$. From the Moody chart we obtain $f = 0.027$ for these values of ε/D and Re. Since this is not the same as our assumed value of f, we try again. With $f = 0.027$, we obtain $D = 0.196$ ft, $\varepsilon/D = 0.0026$, and $\text{Re} = 8.27 \times 10^4$, which in turn give $f = 0.027$, in agreement with the assumed value. Thus, the diameter of the pipe should be

$$D = 0.196 \text{ ft} \qquad \text{(Ans)}$$

If we use the Colebrook equation (Eq. 8.35) with $\varepsilon/D = 0.0005/0.404 f^{1/5} = 0.00124/f^{1/5}$ and $\text{Re} = 1.62 \times 10^4/0.404 f^{1/5} = 4.01 \times 10^4/f^{1/5}$, we obtain

$$\frac{1}{\sqrt{f}} = -2.0 \log \left(\frac{\varepsilon/D}{3.7} + \frac{2.51}{\text{Re}\sqrt{f}} \right)$$

or

$$\frac{1}{\sqrt{f}} = -2.0 \log \left(\frac{3.35 \times 10^{-4}}{f^{1/5}} + \frac{6.26 \times 10^{-5}}{f^{3/10}} \right)$$

An iterative scheme to solve this equation for f gives $f = 0.027$, and hence $D = 0.196$ ft, in agreement with the Moody chart method.

In the previous example we only had to consider major losses. In some instances the inclusion of major and minor losses can cause a slightly more lengthy solution

procedure, even though the governing equations are essentially the same. This is illustrated in Example 8.13.

EXAMPLE 8.13 (TYPE III, DETERMINE DIAMETER)

Water at 10 °C ($v = 1.307 \times 10^{-6}$ m²/s, see Table B.2) is to flow from reservoir A to reservoir B through a cast-iron pipe ($\varepsilon = 0.26$ mm) of length $= 20$ m at a rate of $Q = 0.0020$ m³/s as shown in Fig. E8.13. The system contains a sharp edged entrance and six regular threaded 90° elbows. Determine the pipe diameter needed.

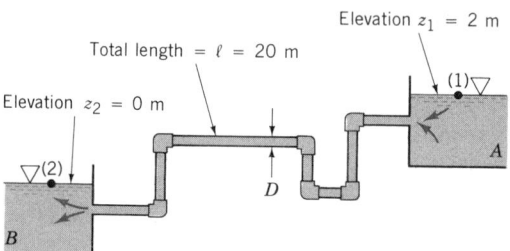

FIGURE E8.13

SOLUTION

The energy equation (Eq. 8.21) can be applied between two points on the surfaces of the reservoirs ($p_1 = p_2 = V_1 = V_2 = z_2 = 0$) as follows:

$$\frac{p_1}{\gamma} + \frac{V_1^2}{2g} + z_1 = \frac{p_2}{\gamma} + \frac{V_2^2}{2g} + z_2 + h_L$$

or

$$z_1 = \frac{V^2}{2g}\left(f\frac{\ell}{D} + \sum K_L\right) \tag{1}$$

where $V = Q/A = 4Q/\pi D^2 = 4(2 \times 10^{-3}$ m³/s$)/\pi D^2$, or

$$V = \frac{2.55 \times 10^{-3}}{D^2} \tag{2}$$

is the velocity within the pipe. (Note that the units on V and D are m/s and m, respectively.) The loss coefficients are obtained from Table 8.3 and Figs. 8.25 and 8.28 as $K_{L_{\text{ent}}} = 0.5$, $K_{L_{\text{elbow}}} = 1.5$, and $K_{L_{\text{exit}}} = 1$. Thus, Eq. 1 can be written as

$$2 \text{ m} = \frac{V^2}{2(9.81 \text{ m/s}^2)}\left\{\frac{20}{D}f + [6(1.5) + 0.5 + 1]\right\}$$

or, when combined with Eq. 2 to eliminate V,

$$6.03 \times 10^6 D^5 - 10.5D - 20f = 0 \tag{3}$$

To determine D we must know f, which is a function of Re and ε/D, where

$$\text{Re} = \frac{VD}{v} = \frac{[(2.55 \times 10^{-3})/D^2]D}{1.307 \times 10^{-6}} = \frac{1.95 \times 10^3}{D} \tag{4}$$

and

$$\frac{\varepsilon}{D} = \frac{2.6 \times 10^{-4}}{D} \tag{5}$$

where D is in meters. Again, we have four equations (Eqs. 3, 4, 5, and the Moody chart or the Colebrook equation) for the four unknowns D, f, Re, and ε/D.

Consider the solution by using the Moody chart. Although it is often easiest to assume a value of f and make calculations to determine if the assumed value is the correct one, with the inclusion of minor losses this may not be the simplest method. For example, if we assume $f = 0.02$ and calculate D from Eq. 3, we would have to solve a fifth-order equation. With only major losses (see Example 8.12), the term proportional to D in Eq. 3 is absent, and it is easy to solve for D if f is given. With both major and minor losses included (represented by the second and third terms in Eq. 3), this solution for D (given f) would require a trial-and-error or iterative technique.

Thus, for this type of problem it is perhaps easier to assume a value of D, calculate the corresponding f from Eq. 3, and with the values of Re and ε/D determined from Eqs. 4 and 5, look up the value of f in the Moody chart (or the Colebrook equation). The solution is obtained when the two values of f are in agreement. For example, assume $D = 0.05$ m, so that Eq. 3 gives $f = 0.0680$ and Eqs. 4 and 5 give Re $= 3.90 \times 10^4$ and $\varepsilon/D = 5.2 \times 10^{-3}$. With these values of Reynolds number and relative roughness, the Moody chart gives $f = 0.033$, which does not coincide with that obtained from Eq. 3 ($f = 0.0680$). Thus, $D \neq 0.05$ m.

A few more rounds of calculation will reveal that the solution is given by $D \approx 0.045$ m with $f = 0.032$.

Thus,

$$D \approx 45 \text{ mm} \qquad \text{(Ans)}$$

It is interesting to attempt to solve this example if all losses are neglected so that Eq. 1 becomes $z_1 = 0$. Clearly from Fig. E8.13 $z_1 = 2$ m. Obviously something is wrong. A fluid cannot flow from one elevation, beginning with zero pressure and velocity, and end up at a lower elevation with zero pressure and velocity unless energy is removed (i.e., a head loss or a turbine) somewhere between the two locations. If the pipe is short (negligible friction) and the minor losses are negligible, there is still the kinetic energy of the fluid as it leaves the pipe and enters the reservoir. After the fluid meanders around in the reservoir for some time, this kinetic energy is lost and the fluid is stationary. No matter how small the viscosity is, the exit loss cannot be neglected. The same result can be seen if the energy equation is written from the free surface of the upstream tank to the exit plane of the pipe, at which point the kinetic energy is still available to the fluid. In either case the energy equation becomes $z_1 = V^2/2g$ in agreement with the inviscid results of Chapter 3 (the Bernoulli equation).

8.5.2 Multiple Pipe Systems

In many pipe systems there is more than one pipe involved. The complex system of tubes in our lungs (beginning with the relatively large-diameter trachea and ending in minute bronchi after numerous branchings) and the maze of pipes in a city's water distribution system are typical of such systems. The governing mechanisms for the flow in multiple pipe systems are the same as for the single pipe systems discussed in this chapter. However, because of the numerous unknowns involved, additional complexities may arise in solving for the flow in multiple pipe systems. Some of these complexities are discussed in this section.

The simplest multiple pipe systems can be classified into series or parallel flows, as are shown in Fig. 8.39. The nomenclature is similar to that used in electrical circuits. Indeed, an analogy between fluid and electrical circuits is often made as follows. In a simple electrical circuit, there is a balance between the voltage (e), current (i), and resistance (R) as given by Ohm's law: $e = iR$. In a fluid circuit there is a balance between the pressure drop (Δp), the flowrate or velocity (Q or V), and the flow resistance as given in terms of the friction factor and minor loss coefficients (f and

(a)

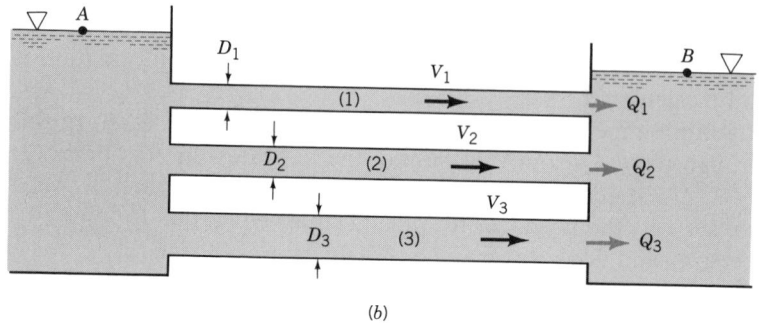

(b)

FIGURE 8.39
Series (a) and parallel (b) pipe systems.

K_L). For a simple flow [$\Delta p = f(\ell/D)(\rho V^2/2)$], it follows that $\Delta p = Q^2 \bar{R}$, where $\bar{R}$, a measure of the resistance to the flow, is proportional to f.

The main differences between the solution methods used to solve electrical circuit problems and those for fluid circuit problems lie in the fact that Ohm's law is a linear equation (doubling the voltage doubles the current), while the fluid equations are generally nonlinear (doubling the pressure drop does not double the flowrate unless the flow is laminar). Thus, although some of the standard electrical engineering methods can be carried over to help solve fluid mechanics problems, others cannot.

One of the simplest multiple pipe systems is that containing pipes in *series,* as is shown in Fig. 8.39a. Every fluid particle that passes through the system passes through each of the pipes. Thus, the flowrate (but not the velocity) is the same in each pipe, and the head loss from point A to point B is the sum of the head losses in each of the pipes. The governing equations can be written as follows

$$Q_1 = Q_2 = Q_3$$

and

$$h_{L_{A-B}} = h_{L_1} + h_{L_2} + h_{L_3}$$

where the subscripts refer to each of the pipes. In general, the friction factors will be different for each pipe because the Reynolds numbers ($\text{Re}_i = \rho V_i D_i/\mu$) and the relative roughnesses (ε_i/D_i) will be different. If the flowrate is given, it is a straightforward calculation to determine the head loss or pressure drop (Type I problem). If the pressure drop is given and the flowrate is to be calculated (Type II problem), an iteration scheme is needed. In this situation none of the friction factors, f_i, are known, so the calculations may involve more trial-and-error attempts than for corresponding single pipe systems. The same is true for problems in which the pipe diameter (or diameters) is to be determined (Type III problems).

Another common multiple pipe system contains pipes in *parallel,* as is shown in Fig. 8.39b. In this system a fluid particle traveling from A to B may take any of the paths available, with the total flowrate equal to the sum of the flowrates in each pipe. However, by writing the energy equation between points A and B it is found that the

head loss experienced by any fluid particle traveling between these locations is the same, independent of the path taken. Thus, the governing equations for parallel pipes are

$$Q = Q_1 + Q_2 + Q_3$$

and

$$h_{L_1} = h_{L_2} = h_{L_3}$$

Again, the method of solution of these equations depends on what information is given and what is to be calculated.

Another type of multiple pipe system called a *loop* is shown in Fig. 8.40. In this case the flowrate through pipe (1) equals the sum of the flowrates through pipes (2) and (3), or $Q_1 = Q_2 + Q_3$. As can be seen by writing the energy equation between the surfaces of each reservoir, the head loss for pipe (2) must equal that for pipe (3), even though the pipe sizes and flowrates may be different for each. That is,

$$\frac{p_A}{\gamma} + \frac{V_A^2}{2g} + z_A = \frac{p_B}{\gamma} + \frac{V_B^2}{2g} + z_B + h_{L_1} + h_{L_2}$$

for a fluid particle traveling through pipes (1) and (2), while

$$\frac{p_A}{\gamma} + \frac{V_A^2}{2g} + z_A = \frac{p_B}{\gamma} + \frac{V_B^2}{2g} + z_B + h_{L_1} + h_{L_2}$$

for fluid that travels through pipes (1) and (3). These can be combined to give $h_{L_2} = h_{L_3}$. This is a statement of the fact that fluid particles that travel through pipe (2) and particles that travel through pipe (3) all originate from common conditions at the junction (or node, N) of the pipes and all end up at the same final conditions.

The flow in a relatively simple looking multiple pipe system may be more complex than it appears initially. The branching system termed the *three reservoir problem* shown in Fig. 8.41 is such a system. Three reservoirs at known elevations are connected together with three pipes of known properties (lengths, diameters, and roughnesses). The problem is to determine the flowrates into or out of the reservoirs. If valve (1) were closed, the fluid would flow from reservoir B to C, and the flowrate could be easily calculated. Similar calculations could be carried out if valves (2) or (3) were closed with the others open.

With all valves open, however, it is not necessarily obvious which direction the fluid flows. For the conditions indicated in Fig. 8.41, it is clear that fluid flows from reservoir A because the other two reservoir levels are lower. Whether the fluid flows into or out of reservoir B depends on the elevation of reservoirs B and C and the

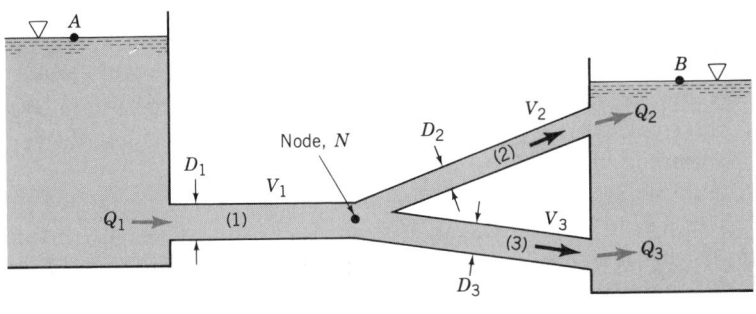

FIGURE 8.40
Multiple pipe loop system.

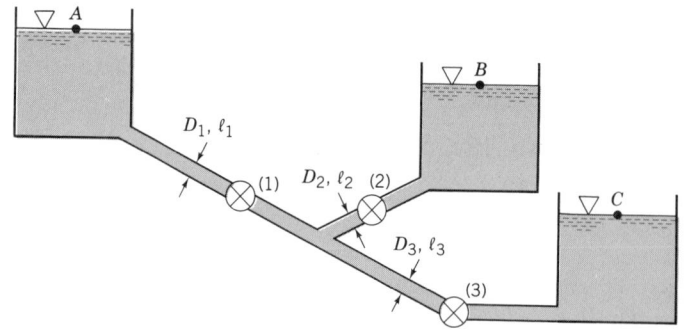

FIGURE 8.41 A three-reservoir system.

properties (length, diameter, roughness) of the three pipes. In general, the flow direction is not obvious, and the solution process must include the determination of this direction. This is illustrated in Example 8.14.

EXAMPLE 8.14

Three reservoirs are connected by three pipes as are shown in Fig. E8.14. For simplicity we assume that the diameter of each pipe is 1 ft, the friction factor for each is 0.02, and because of the large length-to-diameter ratio, minor losses are neglible. Determine the flowrate into or out of each reservoir.

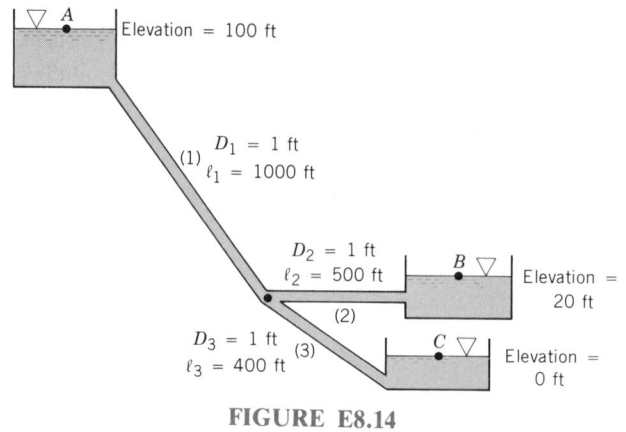

FIGURE E8.14

SOLUTION

It is not obvious which direction the fluid flows in pipe (2). However, we assume that it flows out of reservoir B, write the governing equations for this case, and check our assumption. The continuity equation requires that $Q_1 + Q_2 = Q_3$, which, since the diameters are the same for each pipe, becomes simply

$$V_1 + V_2 = V_3 \tag{1}$$

The energy equation for the fluid that flows from A to C in pipes (1) and (3) can be written as

$$\frac{p_A}{\gamma} + \frac{V_A^2}{2g} + z_A = \frac{p_C}{\gamma} + \frac{V_C^2}{2g} + z_C + f_1 \frac{\ell_1}{D_1} \frac{V_1^2}{2g} + f_3 \frac{\ell_3}{D_3} \frac{V_3^2}{2g}$$

By using the fact that $p_A = p_C = V_A = V_C = z_C = 0$, this becomes

$$z_A = f_1 \frac{\ell_1}{D_1} \frac{V_1^2}{2g} + f_3 \frac{\ell_3}{D_3} \frac{V_3^2}{2g}$$

For the given conditions of this problem we obtain

$$100 \text{ ft} = \frac{0.02}{2(32.2 \text{ ft/s}^2)} \frac{1}{(1 \text{ ft})} [(1000 \text{ ft})V_1^2 + (400 \text{ ft})V_3^2]$$

or

$$322 = V_1^2 + 0.4V_3^2 \tag{2}$$

where V_1 and V_3 are in ft/s. Similarly the energy equation for fluid flowing from B to C is

$$\frac{p_B}{\gamma} + \frac{V_B^2}{2g} + z_B = \frac{p_C}{\gamma} + \frac{V_C^2}{2g} + z_C + f_2 \frac{\ell_2}{D_2} \frac{V_2^2}{2g} + f_3 \frac{\ell_3}{D_3} \frac{V_3^2}{2g}$$

or

$$z_B = f_2 \frac{\ell_2}{D_2} \frac{V_2^2}{2g} + f_3 \frac{\ell_3}{D_3} \frac{V_3^2}{2g}$$

For the given conditions this can be written as

$$64.4 = 0.5V_2^2 + 0.4V_3^2 \tag{3}$$

Equations 1, 2, and 3 (in terms of the three unknowns V_1, V_2, and V_3) are the governing equations for this flow, provided the fluid flows from reservoir B. It turns out, however, that there is no solution for these equations with positive, real values of the velocities. Although these equations do not appear to be complicated, there is no simple way to solve them directly. Thus, a trial-and-error solution is suggested. This can be accomplished as follows. Assume a value of $V_1 > 0$, calculate V_3 from Eq. 2, and then V_2 from Eq. 3. It is found that the resulting V_1, V_2, V_2 trio does not satisfy Eq. 1 for any value of V_1 assumed. There is no solution to Eqs. 1, 2, and 3 with real, positive values of V_1, V_2, and V_3. Thus, our original assumption of flow out of reservoir B must be incorrect.

To obtain the solution, assume the fluid flows into reservoirs B and C and out of A. For this case the continuity equation becomes

$$Q_1 = Q_2 + Q_3$$

or

$$V_1 = V_2 + V_3 \tag{4}$$

Application of the energy equation between points A and B and A and C gives

$$z_A = z_B + f_1 \frac{\ell_1}{D_1} \frac{V_1^2}{2g} + f_2 \frac{\ell_2}{D_2} \frac{V_2^2}{2g}$$

and

$$z_A = z_C + f_1 \frac{\ell_1}{D_1} \frac{V_1^2}{2g} + f_3 \frac{\ell_3}{D_3} \frac{V_3^2}{2g}$$

which, with the given data, become

$$258 = V_1^2 + 0.5 \, V_2^2 \tag{5}$$

and

$$322 = V_1^2 + 0.4 \, V_3^2 \tag{6}$$

Equations (4), (5), and (6) can be solved as follows. By subtracting Eq. (5) from (6) we

obtain

$$V_3 = \sqrt{160 + 1.25V_2^2}$$

Thus, Eq. (5) can be written as

$$258 = (V_2 + V_3)^2 + 0.5V_2^2 = (V_2 + \sqrt{160 + 1.25V_2^2})^2 + 0.5V_2^2$$

or

$$2V_2\sqrt{160 + 1.25V_2^2} = 98 - 2.75V_2^2 \tag{7}$$

which, upon squaring both sides. can be written as

$$V_2^4 - 460\,V_2^2 + 3748 = 0$$

By using the quadratic formula we can solve for V_2^2 to obtain either $V_2^2 = 452$ or $V_2^2 = 8.30$. Thus, either $V_2 = 21.3$ ft/s or $V_2 = 2.88$ ft/s. The value $V_2 = 21.3$ ft/s is not a root of the original equations. It is an extra root introduced by squaring Eq. (7), which with $V_2 = 21.3$ becomes "1140 $= -1140$." Thus, $V_2 = 2.88$ ft/s and from Eq. (5), $V_1 = 15.9$ ft/s. The corresponding flowrates are

$$Q_1 = A_1V_1 = \frac{\pi}{4}\,D_1^2V_1 = \frac{\pi}{4}\,(1\text{ ft})^2\,(15.9\text{ ft/s})$$

$$= 12.5\text{ ft}^3/\text{s from }A \tag{Ans}$$

$$Q_2 = A_2V_2 = \frac{\pi}{4}\,D_2^2V_2 = \frac{\pi}{4}\,(1\text{ ft})^2\,(2.88\text{ ft/s})$$

$$= 2.26\text{ ft}^3/\text{s into }B \tag{Ans}$$

and

$$Q_3 = Q_1 - Q_2 = (12.5 - 2.26)\text{ ft}^3/\text{s} = 10.2\text{ ft}^3/\text{s into }C \tag{Ans}$$

Note the slight differences in the governing equations depending on the direction of the flow in pipe (2)—compare Eqs. (1), (2), and (3) with Eqs. (4), (5), and (6).

If the friction factors were not given, a trial-and-error procedure similar to that needed for Type II problems (see Section 8.5.1) would be required.

The ultimate in multiple pipe systems is a *network* of pipes such as that shown in Fig. 8.42. Networks like these often occur in city water distribution systems and other systems that may have multiple "inlets" and "outlets." The direction of flow in the various pipes is by no means obvious—in fact, it may vary in time, depending on how the system is used from time to time.

FIGURE 8.42 A general pipe network.

The solution for pipe network problems is often carried out by use of node and loop equations similar in many ways to that done in electrical circuits. For example, the continuity equation requires that for each *node* (the junction of two or more pipes) the net flowrate is zero. What flows into a node must flow out at the same rate. In addition, the net pressure difference completely around a *loop* (starting at one location in a pipe and returning to that location) must be zero. By combining these ideas with the usual head loss and pipe flow equations, the flow throughout the entire network can be obtained. Of course, trial-and-error solutions are usually required because the direction of flow, and the friction factors may not be known. Such a solution procedure using matrix techniques is ideally suited for computer use (Refs. 21, 22).

8.6 PIPE FLOWRATE MEASUREMENT

It is often necessary to determine experimentally the flowrate in a pipe. In Chapter 3 we introduced various types of flow-measuring devices (Venturi meter, nozzle meter, orifice meter, etc.) and discussed their operation under the assumption that viscous effects were not important. In this section we will indicate how to account for the ever-present viscous effects in these flowmeters. We will also indicate other types of commonly used flowmeters.

8.6.1 Pipe Flowrate Meters

Three of the most common devices used to measure the instantaneous flowrate in pipes are the orifice meter, the nozzle meter, and the Venturi meter. As was discussed in Section 3.6.3, each of these meters operates on the principle that a decrease in flow area in a pipe causes an increase in velocity that is accompanied by a decrease in pressure. Correlation of the pressure difference with the velocity provides a means of measuring the flowrate. In the absence of viscous effects and under the assumption of a horizontal pipe, application of the Bernoulli equation (Eq. 3.7) between points (1) and (2) shown in Fig. 8.43 gave

$$Q_{ideal} = A_2 V_2 = A_2 \sqrt{\frac{2(p_1 - p_2)}{\rho(1 - \beta^4)}} \tag{8.37}$$

where $\beta = D_2/D_1$. Based on the results of the previous sections of this chapter, we anticipate that there is a head loss between (1) and (2) so that the governing equations become

$$Q = A_1 V_1 = A_2 V_2$$

and

$$\frac{p_1}{\gamma} + \frac{V_1^2}{2g} = \frac{p_2}{\gamma} + \frac{V_2^2}{2g} + h_L$$

The ideal situation has $h_L = 0$ and results in Eq. 8.37. The difficulty in including the

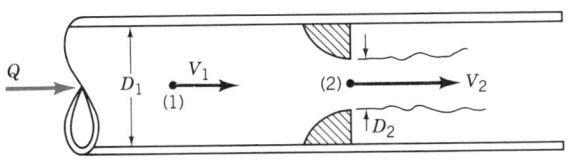

FIGURE 8.43 Typical pipe flow-meter geometry.

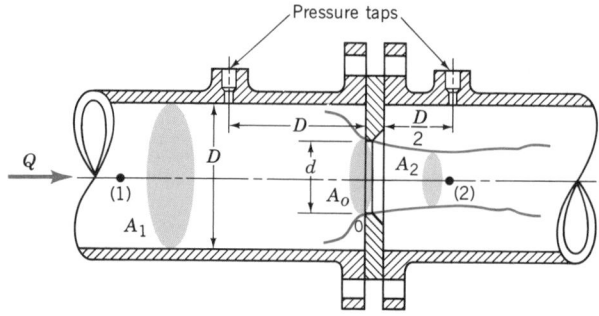

FIGURE 8.44 Typical orifice meter construction.

head loss is that there is no accurate expression for it. The net result is that empirical coefficients are used in the flowrate equations to account for the complex "real world" effects brought on by the nonzero viscosity. The coefficients are discussed below.

A typical *orifice meter* is constructed by inserting between two flanges of a pipe a flat plate with a hole, as shown in Fig. 8.44. The pressure at point (2) within the vena contracta is less than that at point (1). Nonideal effects occur for two reasons. First, the vena contracta area, A_2, is less than the area of the hole, A_o, by an unknown amount. Thus, $A_2 = C_c A_o$, where C_c is the contraction coefficient ($C_c < 1$). Second, the swirling flow and turbulent motion near the orifice plate introduce a head loss that cannot be calculated theoretically. Thus, an *orifice discharge coefficient*, C_o, is used to take these effects into account. That is,

$$Q = C_o Q_{ideal} = C_o A_o \sqrt{\frac{2(p_1 - p_2)}{\rho(1 - \beta^4)}} \tag{8.38}$$

where $A_o = \pi d^2/4$ is the area of the hole in the orifice plate. The value of C_o is a function of $\beta = d/D$ and the Reynolds number $\mathrm{Re} = \rho V D/\mu$, where $V = Q/A_1$. Typical values of C_o are given in Fig. 8.45. Note that the value of C_o depends on the specific construction of the orifice meter (i.e., the placement of the pressure taps, whether the orifice plate edge is square or beveled, etc.). Very precise conditions

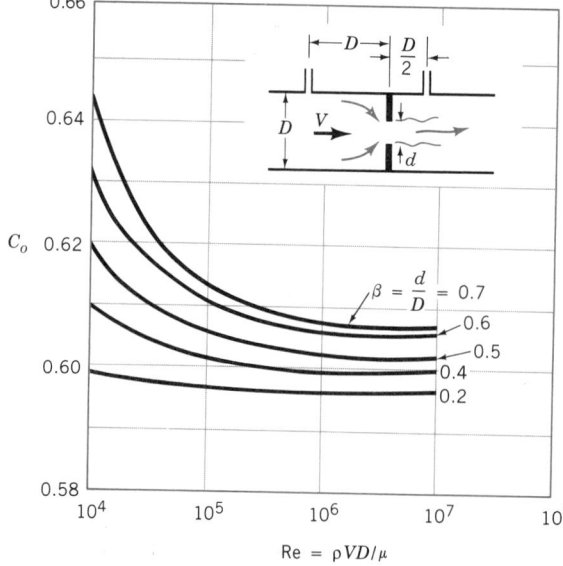

FIGURE 8.45 Orifice meter discharge coefficient (Ref. 24).

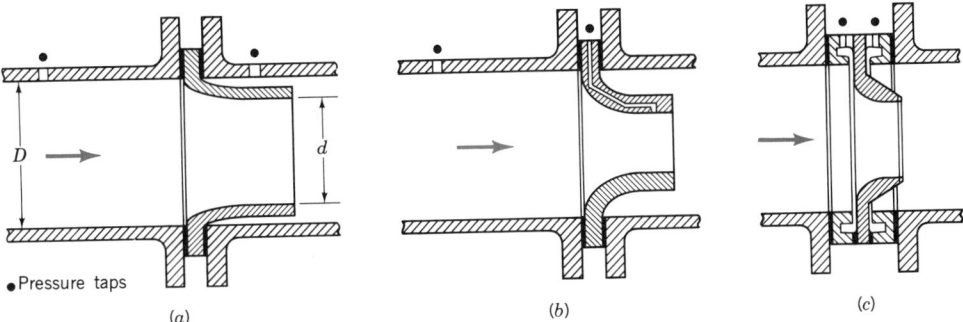

FIGURE 8.46 Typical nozzle meter construction.

governing the construction of standard orifice meters have been established to provide the greatest accuracy possible (Refs. 23, 24).

Another type of pipe flowmeter that is based on the same principles used in the orifice meter is the *nozzle meter,* three variations of which are shown in Fig. 8.46. This device uses a contoured nozzle (typically placed between flanges of pipe sections) rather than a simple (and less expensive) plate with a hole as in an orifice meter. The resulting flow pattern for the nozzle meter is closer to ideal than the orifice meter flow. There is only a slight vena contracta and the secondary flow separation is less severe, but there still are viscous effects. These are accounted for by use of the *nozzle discharge coefficient, C_n,* where

$$Q = C_n Q_{ideal} = C_n A_n \sqrt{\frac{2(p_1 - p_2)}{\rho(1 - \beta^4)}} \tag{8.39}$$

with $A_n = \pi d^2/4$. As with the orifice meter, the value of C_n is a function of the diameter ratio $\beta = d/D$ and the Reynolds number, Re $= \rho VD/\mu$. Typical values obtained from experiments are shown in Fig. 8.47. Again, precise values of C_n depend on the specific details of the nozzle design. Accepted standards have been adopted (Ref. 24). Note that $C_n > C_O$; the nozzle meter is more efficient (less energy dissipated) than the orifice meter.

The most precise and most expensive of the three obstruction type flowmeters is the *Venturi meter* shown in Fig. 8.48. Although the operating principle for this device

FIGURE 8.47 Nozzle meter discharge coefficient (Ref. 24).

FIGURE 8.48
Typical Venturi meter construction.

Re $= \rho VD/\mu$

FIGURE 8.49 Venturi meter discharge coefficient (Ref. 23).

is the same as for the orifice or nozzle meters, the geometry of the Venturi meter is designed to reduce head losses to a minimum. This is accomplished by providing a relatively streamlined contraction (which eliminates separation ahead of the throat) and a very gradual expansion downstream of the throat (which eliminates separation in this decelerating portion of the device). Most of the head loss that occurs in a well-designed Venturi meter is due to friction losses along the walls rather than losses associated with separated flows and the inefficient mixing motion that accompanies such flow.

Thus, the flowrate through a Venturi meter is given by

$$ Q = C_v Q_{ideal} = C_v A_T \sqrt{\frac{2(p_1 - p_2)}{\rho(1 - \beta^4)}} $$

where $A_T = \pi d^2/4$ is the throat area. The range of values of C_v, the *Venturi discharge coefficient,* is given in Fig. 8.49. The throat-to-pipe diameter ratio ($\beta = d/D$), the Reynolds number, and the shape of the converging and diverging sections of the meter are among the parameters that affect the value of C_v.

Again, the precise values of C_n, C_o, and C_v depend on the specific geometry of the devices used. Considerable information concerning the design and installation of standard flowmeters can be found in various books (Refs. 23, 24, 25, 26).

EXAMPLE 8.15

Ethyl alcohol flows through a pipe of diameter $D = 60$ mm in a refinery. The pressure drop across the nozzle meter used to measure the flowrate is to be $\Delta p = 4.0$ kPa when the flowrate is $Q = 0.003$ m³/s. Determine the diameter, d, of the nozzle.

SOLUTION

From Table 1.6 the properties of ethyl alcohol are $\rho = 789$ kg/m^3 and $\mu = 1.19 \times 10^{-3}$ N·s/m^2. Thus,

$$\text{Re} = \frac{\rho VD}{\mu} = \frac{4\rho Q}{\pi D \mu} = \frac{4(789 \text{ kg/m}^3)(0.003 \text{ m}^3/\text{s})}{\pi(0.06 \text{ m})(1.19 \times 10^{-3} \text{ N·s/m}^2)} = 42{,}200$$

From Eq. 8.39 the flowrate through the nozzle is

$$Q = 0.003 \text{ m}^3/\text{s} = C_n \frac{\pi}{4} d^2 \sqrt{\frac{2(4 \times 10^3 \text{ N/m}^2)}{789 \text{ kg/m}^3(1 - \beta^4)}}$$

or

$$1.20 \times 10^{-3} = \frac{C_n d^2}{\sqrt{1 - \beta^4}} \tag{1}$$

where d is in meters. Note that $\beta = d/D = d/0.06$. Equation 1 and Fig. 8.47 represent two equations for the two unknowns d and C_n that must be solved by trial and error.

As a first approximation we assume that the flow is ideal, or $C_n = 1.0$, so that Eq. 1 becomes

$$d = (1.20 \times 10^{-3} \sqrt{1 - \beta^4})^{1/2} \tag{2}$$

In addition, for many cases $1 - \beta^4 \approx 1$, so that an approximate value of d can be obtained from Eq. 2 as

$$d = (1.20 \times 10^{-3})^{1/2} = 0.0346 \text{ m}$$

Hence, with an initial guess of $d = 0.0346$ m or $\beta = d/D = 0.0346/0.06 = 0.577$, we obtain from Fig. 8.47 (using Re = 42,200) a value of $C_n = 0.972$. Clearly this does not agree with our initial assumption of $C_n = 1.0$. Thus, we do not have the solution to Eq. 1 and Fig. 8.47. Next we assume $\beta = 0.577$ and $C_n = 0.972$ and solve for d from Eq. 1 to obtain

$$d = \left(\frac{1.20 \times 10^{-3}}{0.972} \sqrt{1 - 0.577^4}\right)^{1/2}$$

or $d = 0.0341$ m. With the new value of $\beta = 0.0341/0.060 = 0.568$ and Re = 42,200, we obtain (from Fig. 8.47) $C_n \approx 0.972$ in agreement with the assumed value. Thus,

$$d = 34.1 \text{ mm} \tag{Ans}$$

If numerous cases are to be investigated, it may be much easier to replace the discharge coefficient data of Fig. 8.47 by the equivalent equation, $C_n = \phi(\beta, \text{Re})$, and use a computer to iterate for the answer. Such equations are available in the literature (Ref. 24). This would be similar to using the Colebrook equation rather than the Moody chart for pipe friction problems.

Numerous other devices are used to measure the flowrate in pipes. Many of these devices use principles other than the high-speed/low-pressure concept of the orifice, nozzle, and Venturi meters.

A quite common, accurate, and relatively inexpensive flow meter is the *rotameter*, or variable area meter as is shown in Fig. 8.50. In this device a float is contained within a tapered transparent metering tube that is attached vertically to the pipe line. As fluid flows through the meter (entering at the bottom), the float will rise within the tapered tube and reach an equilibrium height that is a function of the flowrate. This height corresponds to an equilibrium condition for which the net force on the

float (buoyancy, float weight, fluid drag) is zero. A calibration scale in the tube provides the relationship between the float position and the flowrate.

Another useful pipe flowrate meter is a *turbine meter* as is shown in Fig. 8.51. A small, freely rotating propeller or turbine within the turbine meter rotates with an angular velocity that is a function (nearly proportional to) the average fluid velocity in the pipe. This angular velocity is picked up magnetically and calibrated to provide a very accurate measure of the flowrate through the meter.

8.6.2 Volume Flowmeters

In many instances it is necessary to know the amount (volume or mass) of fluid that has passed through a pipe during a given time period, rather than the instantaneous flowrate. For example, we are interested in how many gallons of gasoline are pumped into the tank in our car rather than the rate at which it flows into the tank. There are numerous quantity-measuring devices that provide such information.

The *nutating disk meter* shown in Fig. 8.52 is widely used to measure the net amount of water used in domestic and commercial water systems as well as the amount of gasoline delivered to your gas tank. This meter contains only one essential moving part and is relatively inexpensive and accurate. Its operating principle is very simple, but it may be difficult to understand its operation without actually inspecting the device first hand. The device consists of a metering chamber with spherical sides and conical top and bottom. A disk passes through a central sphere and divides the chamber into two portions. The disk is constrained to be at an angle not normal to the axis of symmetry of the chamber. A radial plate (diaphragm) divides the chamber

FIGURE 8.50 Rotameter type flowmeter. (Courtesy of Fischer & Porter Co.).

FIGURE 8.51
Turbine type flowmeter. (Courtesy of E G & G Flow Technology, Inc.).

FIGURE 8.52 Nutating disk flowmeter. (Courtesy of Badger Meter, Inc.).

FIGURE 8.53 Bellows type flowmeter. (Courtesy of BTR—Rockwell Gas Products).
(*a*) Back case emptying, back diaphragm filling. (*b*) Front diaphragm filling, front case
emptying. (*c*) Back case filling, back diaphragm emptying. (*d*) Front diaphragm emptying, front case filling.

so that the entering fluid causes the disk to wobble (nutate), with fluid flowing alternately above or below the disk. The fluid exits the chamber after the disk has completed one wobble, which corresponds to a specific volume of fluid passing through the chamber. During each wobble of the disk, the pin attached to the tip of the center sphere, normal to the disk, completes one circle. The volume of fluid that has passed through the meter can be obtained by counting the number of revolutions completed.

Another quantity-measuring device that is used for gas flow measurements is the *bellows meter* as shown in Fig. 8.53. It contains a set of bellows that alternately fill and empty as a result of the pressure of the gas and the motion of a set of inlet and outlet valves. The common household natural gas meter is of this type. For each cycle [(a) through (d)] a known volume of gas passes through the meter.

The nutating disk meter (water meter) is an example of extreme simplicity—one cleverly designed moving part. The bellows meter (gas meter), on the other hand, is relatively complex—it contains many moving, interconnected parts. This difference is dictated by the application involved. One measures a common, safe-to-handle, relatively high-pressure liquid, whereas the other measures a relatively dangerous, low-pressure gas. Each device does its intended job very well.

There are numerous devices used to measure fluid flow, only a few of which have been discussed here. The reader is encouraged to review the literature to gain familiarity with other useful, clever devices (Refs. 25, 26).

REFERENCES

1. Hinze, J. O., *Turbulence,* 2nd Ed., McGraw-Hill, New York, 1975.

2. Panton, R. L., *Incompressible Flow,* Wiley, New York, 1984.

3. Schlichting, H., *Boundary Layer Theory,* Seventh Edition, McGraw-Hill, New York, 1979.

4. Gleick, J., *Chaos: Making a New Science,* Viking Penguin, New York, 1987.

5. White, F. M., *Fluid Mechanics,* McGraw-Hill, New York, 1979.

6. Nikurasdse, J., "Stomungsgesetz in Rauhen Rohren," *VDI-Forschungsch,* No. 361, 1933; or see NACA Tech Memo 1922.

7. Moody, L. F., "Friction Factors for Pipe Flow," *Transactions of the ASME,* Vol. 66, 1944.

8. Colebrook, C. F., "Turbulent Flow in Pipes with Particular Reference to the Transition Between the Smooth and Rough Pipe Laws," *Journal of the Institute of Civil Engineers London,* Vol. 11, 1939.

9. *ASHRAE Handbook of Fundamentals,* ASHRAE, Atlanta, 1981.

10. Streeter, V. L., ed., *Handbook of Fluid Dynamics,* McGraw-Hill, New York, 1961.

11. Sovran, G., and Klomp, E. D., "Experimentally Determined Optimum Geometries for Rectilinear Diffusers with Rectangular, Conical, or Annular Cross Sections," in *Fluid Mechanics of Internal Flow,* Sovran, G., ed., Elsevier, Amsterdam, 1967.

12. Runstadler, P. W., "Diffuser Data Book," Technical Note 186, Creare, Inc., Hanover, NH, 1975.

13. Laws, E. M., and Livesey, J. L., "Flow Through Screens," *Annual Review of Fluid Mechanics,* Vol. 10, Annual Reviews, Inc., Palo Alto, CA, 1978.

14. Balje, O. E., *Turbomachines: A Guide to Design, Selection and Theory*, Wiley, New York, 1981.

15. Wallis, R. A., *Axial Flow Fans and Ducts*, Wiley, New York, 1983.

16. Karassick, I. J. et al., *Pump Handbook*, Second Edition, McGraw-Hill, New York, 1985.

17. White, F. M., *Viscous Fluid Flow*, McGraw-Hill, New York, 1974.

18. Olson, R. M., *Essentials of Engineering Fluid Mechanics*, Fourth Edition, Harper & Row, New York, 1980.

19. Dixon, S. L., *Fluid Mechanics of Turbomachinery*, Third Edition, Pergamon, Oxford, 1978.

20. Daugherty, R. L., and Franzini, J. R., *Fluid Mechanics*, Seventh Edition, McGraw-Hill, New York, 1977.

21. Streeter, V. L., and Wylie, E. B., *Fluid Mechanics*, Eighth Edition, McGraw-Hill, New York, 1985.

22. Jepson, R. W., *Analysis of Flow in Pipe Networks*, Ann Arbor Science Publishers, Ann Arbor, Mich., 1976.

23. Bean, H. S., ed., *Fluid Meters: Their Theory and Application*, Sixth Edition, American Society of Mechanical Engineers, New York, 1971.

24. "Measurement of Fluid Flow by Means of Orifice Plates, Nozzles, and Venturi Tubes Inserted in Circular Cross Section Conduits Running Full," Int. Organ. Stand. Rep. DIS-5167, Geneva, 1976.

25. Goldstein, R. J., ed., *Fluid Mechanics Measurements*, Hemisphere Publishing, New York, 1983.

26. Benedict, R. P., *Measurement of Temperature, Pressure, and Flow*, 2nd Ed., Wiley, New York, 1977.

27. Hydraulic Institute, *Engineering Data Book*, First Edition, Cleveland Hydraulic Institute, 1979.

28. Harris, C. W., *University of Washington Engineering Experimental Station Bulletin*, 48, 1928.

29. Hamilton, J. B., *University of Washington Engineering Experimental Station Bulletin*, 51, 1929.

PROBLEMS $R_e = \frac{VD}{\nu}$; $V = \frac{Q}{A} = \frac{4Q}{\pi D^2} = \frac{4(10^{-9} \frac{m^3}{s})}{\pi (.08 \times 10^{-3} m)}$ =

Note: Unless otherwise indicated use the values of fluid properties found in the tables on the inside of the front cover. Problems designated with an (*) are intended to be solved with the aid of a programmable calculator or a computer.

8.1 Rainwater runoff from a parking lot flows through a 2 ft-diameter pipe, completely filling it. Would you expect the flow to be laminar or turbulent? Support your answer with appropriate calculations.

8.2 Air flows through a 0.08-mm-diameter passageway in a person's lungs at a rate of 10^{-9}

m³/s. Is the flow laminar or turbulent? Support your answer with appropriate calculations.

8.3 The flow of water in a 30-mm-diameter pipe is to remain laminar. Plot a graph of the maximum flowrate allowed as a function of temperature for $0 < T < 100\ °C$.

8.4 Air at 200 °F flows at standard atmospheric pressure in a pipe at a rate of 0.08 lb/s. Determine the minimum diameter allowed if the flow is to be laminar.

8.5 Carbon dioxide at 20 °C and a pressure of 550 kPa (abs) flows in a pipe at a rate of 0.04

At the top margin, handwritten:

$$\frac{p_a}{\gamma} + \frac{V_a^2}{2g} + z_a = \frac{p_b}{\gamma} + \frac{V_b^2}{2g} + z_b + h_L \quad \therefore V_a = V_b$$

$$h_L = \frac{p_a}{\gamma} + z_a - \frac{p_b}{\gamma} - z_b =$$

N/s. Determine the maximum diameter allowed if the flow is to be turbulent.

8.6 Blue and yellow streams of paint at 60 °F (each with a density of 1.6 slugs/ft³ and a viscosity 1000 times greater than water) enter a pipe with an average velocity of 4 ft/s as shown in Fig. P8.6. Would you expect the paint to exit the pipe as green paint or separate streams of blue and yellow paint? Explain. Repeat the problem if the paint were "thinned" so that it is only 10 times more viscous than water. Assume the density remains the same.

FIGURE P8.6

8.7 A long small-diameter tube is to be used as a viscometer by measuring the flowrate through the tube as a function of the pressure drop along the tube. The calibration constant, $K = Q/\Delta p$, is calculated by assuming the flow is laminar. For tubes of diameter 0.5, 1.0, and 2.0 mm, determine the maximum flowrate allowed (in cm³/s) if the fluid is **(a)** 20 °C water, or **(b)** standard air.

8.8 A soft drink with the properties of 10 °C water is sucked through a 4-mm-diameter, 0.25-m-long straw at a rate of 4 cm³/s. Is the flow at the outlet of the straw laminar? Is it fully developed? Explain. $V = \frac{Q}{A} = \frac{(4)10^{-2}}{\pi/4(.02)^2} = 0.318$.

8.9 To cool a given room it is necessary to supply 5 ft³/s of air through an 8-in.-diameter pipe. Approximately how long is the entrance length in this pipe?

8.10 The wall shear stress in a fully developed flow portion of a 12-in.-diameter pipe carrying water is 1.85 lb/ft². Determine the pressure gradient, $\partial p/\partial x$, where x is in the flow direction, if the pipe is **(a)** horizontal, **(b)** vertical with flow up, or **(c)** vertical with flow down.

8.11 The pressure drop needed to force water through a horizontal 1-in.-diameter pipe is 0.60 psi for every 12-ft length of pipe. Determine the shear stress on the pipe wall. Determine the shear stress at distances 0.3 and 0.5 in. away from the pipe wall.

8.12 Repeat Problem 8.11 if the pipe is on a 20° hill. Is the flow up or down the hill? Explain.

8.13 The pressure distribution measured along a straight, horizontal portion of a 50-mm-diam-

eter pipe attached to a tank is shown in the table below. Approximately how long is the entrance length? In the fully developed portion of the flow, what is the value of the wall shear stress?

x (m) (±0.01 m)	p (mm H₂O) (±5 mm)
0 (tank exit)	520
0.5	427
1.0	351
1.5	288
2.0	236
2.5	188
3.0	145
3.5	109
4.0	73
4.5	36
5.0 (pipe exit)	0

8.14 Water flows in a constant diameter pipe with the following conditions measured: At section (a) $p_a = 32.4$ psi and $z_a = 56.8$ ft; at section (b) $p_b = 29.7$ psi and $z_b = 58.2$ ft. Is the flow from (a) to (b) or from (b) to (a)? Explain.

8.15 Repeat Problem 8.14 if the specific gravity of the fluid is 13.6.

8.16 Some fluids behave as a non-Newtonian power-law fluid characterized by $\tau = -C(du/dr)^n$, where $n = 1, 3, 5,$ and so on, and C is a constant. (If $n = 1$, the fluid is the customary Newtonian fluid.) For flow in a round pipe of a diameter D, integrate the force balance equation (Eq. 8.3) to obtain the velocity profile

$$u(r) = \frac{-n}{(n+1)} \left(\frac{\Delta p}{2\ell C}\right)^{1/n} \left[r^{(n+1)/n} - \left(\frac{D}{2}\right)^{(n+1)/n} \right]$$

8.17* For the flow discussed in Problem 8.16, plot the dimensionless velocity profile u/V_c, where V_c is the centerline velocity (at $r = 0$), as a function of the dimensionless radial coordinate $r/(D/2)$, where D is the pipe diameter. Consider values of $n = 1, 3, 5,$ and 7.

8.18 Water at 20 °C flows through a horizontal 1-mm-diameter tube to which are attached two pressure taps a distance 1 m apart. **(a)** What is the maximum pressure drop allowed if the flow is to be laminar? **(b)** Assume the manufacturing tolerance on the tube diameter is $D = 1.0 \pm 0.1$ mm. Given this uncertainty in the tube diameter, what is the maximum pressure drop allowed if it must be assured that the flow is laminar?

8.19 A fluid of specific gravity 0.85 flows steadily in a long, vertical 1-in.-diameter pipe with an average velocity of 0.50 ft/s. If the pressure is constant throughout the fluid, what is the vis-

cosity of the fluid? Determine the shear stress on the pipe wall.

8.20 Water is pumped between two tanks as shown in Fig. P8.20. The energy line is as indicated. Is the fluid being pumped from A to B or B to A? Explain.

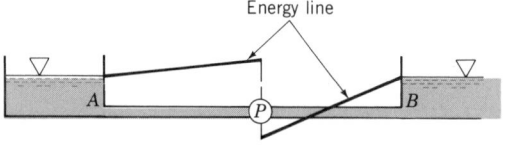

FIGURE P8.20

8.21 Oil flows through the horizontal pipe shown in Fig. P8.21 under laminar conditions. All sections are the same diameter except one. Which section of the pipe (A, B, C, D, or E) is slightly smaller diameter than the others? Explain.

FIGURE P8.21

8.22 A fluid flows through two horizontal pipes of equal length which are connected together to form a pipe of length 2ℓ. The flow is laminar and fully developed. The pressure drop for the first pipe is 1.24 times greater than it is for the second pipe. If the diameter of the first pipe is D, determine the diameter of the second pipe.

8.23 Glycerin at 20 °C flows upward in a vertical 75-mm-diameter pipe with a centerline velocity of 1.0 m/s. Determine the head loss and pressure drop in a 10-m length of the pipe.

8.24 Determine the magnitude of the velocity gradient at points 10, 20, and 30 mm from the pipe wall for the flow in Problem 8.23.

8.25 A viscous fluid flows in a 0.10-m-diameter pipe such that its velocity measured 0.012 m away from the pipe wall is 0.9 m/s. If the flow is laminar, determine the centerline velocity and the flowrate.

8.26 Laminar flow in a horizontal pipe of diameter D gives a flowrate of Q if the pressure gradient is $\partial p/\partial x = -K$. The fluid is cooled so that the density increases by a factor of 1.04 and

the viscosity increases by a factor of 3.8. Determine the new pressure gradient required (in terms of K) if the flowrate remains the same.

8.27 In Problem 8.26 if the flowrate and pressure gradient remain the same, determine the new diameter required.

8.28 Oil of $SG = 0.87$ and a kinematic viscosity $v = 1.2 \times 10^{-4}$ m²/s flows through the vertical pipe shown in Fig. P8.28 at a rate of 4×10^{-4} m³/s. Determine the manometer reading, h.

FIGURE P8.28

8.29 Determine the manometer reading, h, for Problem 8.28 if the flow is up rather than down the pipe.

8.30 For Problem 8.28, what flowrate (magnitude and direction) will cause $h = 0$?

8.31 For laminar flow in a round pipe of diameter D, at what distance from the centerline is the actual velocity equal to the average velocity?

8.32 Show that the power law approximation for the velocity profile in turbulent pipe flow (Eq. 8.31) cannot be accurate at the centerline or at the pipe wall because the velocity gradients at these locations are not correct. Explain.

8.33 The kinetic energy coefficient, α, is defined in Eq. 5.86. Show that its value for a power-law turbulent velocity profile (Eq. 8.31) is given by $\alpha = (n + 1)^3(2n + 1)^3/[4n^4(n + 3)(2n + 3)]$.

8.34 If the velocity profile for turbulent flow in a pipe is approximated by the power-law profile (Eq. 8.31), at what radial location should a Pitot tube be placed if it is to measure the average velocity in the pipe? Assume $n = 7$, 8, or 9.

8.35   Water at 80 °F flows in a 6-in.-diameter pipe with a flowrate of 2.0 cfs. What is the approximate velocity at a distance 1.5 in. away from the wall? Determine the centerline velocity.

8.36   Carbon dioxide at a temperature of 0 °C and a pressure of 600 kPa (abs) flows through a horizontal 40-mm-diameter pipe with an average velocity of 2 m/s. Determine the friction factor if the pressure drop is 235 N/m² per 10-m length of pipe.

8.37   Water flows in a cast-iron pipe of 200-mm diameter at a rate of 0.10 m³/s. Determine the friction factor for this flow.

8.38   Water flows in a galvanized iron pipe of 4-in. diameter with an average velocity of 20 ft/s. Determine the friction factor for this flow.

8.39   Water flows through a 6-in.-diameter horizontal pipe at a rate of 2.0 cfs and a pressure drop of 3.1 psi per 100 ft of pipe. Determine the friction factor.

8.40   Asphalt at 120 °F, considered to be a Newtonian fluid with a viscosity 80,000 times that of water and a specific gravity of 1.09, flows through a pipe of diameter 2.0 in. If the pressure gradient is 1.6 psi/ft determine the flowrate assuming the pipe is **(a)** horizontal; **(b)** vertical with flow up.

8.41   A fluid flows through a pipe of radius R with a Reynolds number of 100,000. At what location, r/R, does the fluid velocity equal the average velocity? Repeat if the Reynolds number is 1000. $\delta_s = \frac{5\nu}{U^*}, U^* = \left(\frac{\tau_w}{\rho}\right)^{1/2}, \tau_w = \frac{\Delta P}{4\ell}, \frac{\Delta P}{\ell} = f\frac{1}{D}\frac{v^2}{2g}\cdot\frac{\gamma}{b}=\frac{\rho v^2}{2g}f\frac{v^2}{\nu^2\rho}$

8.42   Water at 10 °C flows through a smooth 60-mm-diameter pipe with an average velocity of 8 m/s. Would a scratch of height 0.005 mm on the pipe wall protrude through the viscous sublayer? Explain. $Re=\frac{V\rho}{\nu}, f=.0138$ $ree F'9\,8.23$ $\rho lug\,in$

8.43   Water at 80 °C flows through a 120-mm-diameter pipe with an average velocity of 2 m/s. If the pipe wall roughness is small enough so that it does not protrude through the laminar sublayer, the pipe can be considered as smooth. Approximately what is the largest roughness allowed to classify this pipe as smooth? $\delta_s = \frac{5\nu}{U^*}; \tau_w=\frac{f\gamma^2\rho}{8}$

8.44   Determine the thickness of the viscous sublayer in a smooth 6-in.-diameter pipe if the Reynolds number is 25,000. $U^* = \sqrt{\frac{\tau}{\rho}}V$ $\delta_s=\frac{5\nu}{U^*}\sqrt{\frac{\rho}{8}}V$

smooth pipe Re 2×10⁶ f=0.004

8.45   Water at 60 °F flows through a 6-in.-diameter pipe with an average velocity of 15 ft/s. Approximately what is the height of the largest roughness element allowed if this pipe is to be classified as smooth?

8.46   A 50-ft-long, 0.5-in.-diameter hose with a roughness of $\varepsilon = 0.0009$ ft is fastened to a water faucet where the pressure is p_1. Determine p_1 if there is no nozzle attached and the average velocity in the hose is 6 ft/s. Neglect minor losses and elevation changes.

8.47   Repeat Problem 8.46 if there is a nozzle of diameter 0.25 in. attached to the end of the hose.

8.48*   The following equation is sometimes used in place of the Colebrook equation (Eq. 8.35):

$$f = \frac{1.325}{[\ln[(\varepsilon/3.7D) + (5.74/Re^{0.9})]]^2}$$

for $10^{-6} < \varepsilon/D < 10^{-2}$ and $5000 < Re < 10^{+8}$ (Ref. 22, pg. 220). An advantage of this equation is that given Re and ε/D, it does not require an iteration procedure to obtain f. Plot a graph of the percent difference in f as given by this equation and the original Colebrook equation for Reynolds numbers in the range of validity of the above equation, with $\varepsilon/D = 10^{-4}$.

8.49   Water flows through a horizontal 60-mm-diameter galvanized iron pipe at a rate of 0.02 m³/s. If the pressure drop is 135 kPa per 10 m of pipe, do you think this pipe is **(a)** a new pipe, **(b)** an old pipe with a somewhat increased roughness due to aging, or **(c)** a very old pipe that is partially clogged by deposits? Justify your answer.

8.50   For a given head loss per unit length, what effect on the flowrate does doubling the pipe diameter have if the flow is **(a)** laminar, or **(b)** completely turbulent?

8.51   A garden hose is attached to a faucet that is fully opened. Without a nozzle on the end of the hose, the water does not shoot very far. However, if you place your thumb over a portion of the end of the hose, it is possible to shoot the water a considerable distance. Explain this phenomena. (*Note*: The flowrate decreases as the area covered by your thumb increases.)

8.52   Air at standard temperature and pressure flows through a 1-in.-diameter galvanized iron pipe with an average velocity of 10 ft/s. What length of pipe produces a head loss equivalent to **(a)** a threaded 90° elbow, **(b)** a wide open globe valve, or **(c)** a sharp edged entrance?

8.53*   Water at 40 °C flows through drawn tubings with diameters of 0.025, 0.050, or 0.075 m. Plot the head loss in each meter length of pipe for flowrates between 5×10^{-4} m³/s and $50 \times$

$\ell_g = \frac{K_L D}{f}, Re=\frac{VD}{\nu}, \frac{\varepsilon}{D}=.0061$ $F'9\,8.22$

$\ell_g = K_L\left(\frac{1}{f}N\right)$ $\frac{}{0.043}=1.9381$ $f=.043$

$K_L=Look\,up$

[handwritten at top of page:]

8.60 ① $v_1 = \frac{Q}{A_1}$, $v_2 = \frac{Q}{A_2}$, $w/\frac{A_2}{A_1} = \left(\frac{D_2}{D_1}\right)^2 = 0.36$ Fig. 8.30 → $K_L = 0.34$

① $P_1 - P_2 = \frac{1}{2}\rho\{K_L v_2^2 + v_2^2 - v_1^2\} = 33.5 \times 10^3 N/m^2 + 84.4 \times 10^3 N/m^2 = 120 kP$

Losses · Kinetic

10^{-4} m³/s. In your solution obtain the friction factor from the Colebrook formula.

8.54 Air at standard temperature and pressure flows at a rate of 7.0 cfs through a horizontal galvanized iron duct that has a rectangular cross-sectional shape of 12 in. by 6 in. Estimate the pressure drop per 100 ft of duct.

8.55 A viscous oil with a specific gravity $SG = 0.85$ and a viscosity of 0.10 Pa·s flows from tank A to tank B through the six rectangular slots indicated in Fig. P8.55. If the total flowrate is 30 mm³/s and minor losses are negligible, determine the pressure in tank A.

FIGURE P8.55

8.56 The six rectangular channels used in Fig. P8.55 are to be replaced by a single circular tube so that the flowrate–pressure drop characteristics are the same as in Problem 8.55. Determine the diameter of this tube.

8.57 A manufacture makes two types of drinking straws: one with a square cross-sectional shape, and the other type the typical round shape. The amount of material in each straw is to be the same. That is, the length of the perimeter of the cross section of each shape is the same. For a given pressure drop, what is the ratio of the flowrates through the straws? Assume the drink is viscous enough to ensure laminar flow and neglect gravity.

8.58 To conserve water and energy, a "flow reducer" is installed in the shower head as shown in Fig. P8.58. If the pressure at point (1) remains constant and all losses except for that in the "flow

[handwritten:] W/out reducer $\frac{P_1}{\rho} + \frac{v_1^2}{2g} + z_1 = \frac{P_2}{\gamma} + \frac{v_2^2}{2g} + z_2$, $P_2 = 0$

$v_1 = \frac{Q}{A_1} = \frac{4Q}{\pi D_1^2}$

$v_2 = \frac{Q}{A_2}$

FIGURE P8.58

[handwritten:] $\therefore P_1 = \frac{1}{2}\rho(v_2^2 - v_1^2) = \frac{1}{2}\rho(147^2 Q^2 +$ ①

w/flow reducer $v_1 = \frac{1}{2}(\ )$, $v_2 = $ some ②

$\frac{P_1}{\rho} + \frac{v_1^2}{2g} = \frac{v_2^2}{2g} + K_L \frac{v_2^2}{2g}$ or $P_1 = \frac{1}{2}\rho(v_2^2 - (K_L-1)v_1^2)$

reducer" are neglected, determine the value of the loss coefficient (based on the velocity in the pipe) of the "flow reducer" if its presence is to reduce the flowrate by a factor of 2. Neglect gravity.

8.59 Air flows in a 0.50-m-diameter pipe at a rate of 10 m³/s. The pipe diameter changes to 0.75 m through a sudden expansion. Determine the pressure rise across this expansion. Explain how there can be a pressure rise across the expansion when there is an energy loss ($K_L \neq 0$).

8.60 Water flows at a rate of 0.040 m³/s in a 0.10-m-diameter pipe that contains a sudden contraction to a 0.06-m-diameter pipe. Determine the pressure drop across the contraction section. How much of this pressure difference is due to losses and how much is due to kinetic energy changes? *[handwritten:]* ① $\frac{P_1}{\gamma} + \frac{v_1^2}{2g} + z_1 = \frac{P_2}{\gamma} + \frac{v_2^2}{2g} + z_2 + K_L\frac{v_2^2}{2g}$; $z_1 = z_2$

8.61 A large artery in a person's body can be approximated by a tube of diameter 9 mm and length 0.35 m. Also assume that blood has a viscosity of approximately 4×10^{-3} N·s/m², a specific gravity of 1.0, and that the pressure at the beginning of the artery is equivalent to 120 mm Hg. If the flow were steady (it is not) with $V = 0.2$ m/s, determine the pressure at the end of the artery if it is oriented **(a)** vertically up (flow up) or **(b)** horizontal.

8.62 Blood (assume $\mu = 4.5 \times 10^{-5}$ lb·s/ft², $SG = 1.0$) flows through an artery in the neck of a giraffe from its heart to its head at a rate of 2.5×10^{-4} ft³/s. Assume the length is 10 ft and the diameter is 0.20 in. If the pressure at the beginning of the artery (outlet of the heart) is equivalent to 0.70 ft Hg, determine the pressure at the end of the artery when the head is **(a)** 8 ft above the heart, or **(b)** 6 ft below the heart. Assume steady flow. How much of this pressure difference is due to elevation effects, and how much is due to frictional effects?

8.63 At time $t = 0$ the level of water in tank A shown in Fig. P8.63 is 2 ft above that in tank

FIGURE P8.63

[handwritten at bottom right:] ② $8.07 \times 10^5 \rho Q^2 = \frac{1}{2}\rho[(\frac{14\cdot1}{2})Q^2 + (K_L-1)(\frac{73}{2}\cdot)\rho)^2]$; $K_L =$

B. Plot the elevation of the water in tank *A* as a function of time until the free surfaces in both tanks are at the same elevation. Assume quasisteady conditions—that is, the steady pipe flow equations are assumed valid at any time, even though the flowrate does change (slowly) in time. Neglect minor losses. *Note:* Verify and use the fact that the flow is laminar.

8.64* Repeat Problem 8.63 if the pipe diameter is changed to 0.1 ft rather than 0.1 in. *Note:* The flow may not be laminar for this case.

8.65 A fire protection association code requires a minimum pressure of 65 psi at the outlet end of a 200-ft-long, 4-in.- diameter hose when the flowrate is 500 gal/min. What is the minimum pressure allowed at the pumper truck that supplies water to the hose? Assume a roughness of $\varepsilon = 0.03$ in.

8.66 The relative roughness of a pipe often increases with time because of deposits of various materials. Assume the commercial steel pipe is new at $t = 0$ and that the roughness increases by a factor of 2 every 2 years. That is, $\varepsilon = \varepsilon_o[2^{t/2}]$, where ε_o is the new pipe roughness and t is in years. Plot the pressure drop needed to pump 0.20 cfs of water through a 2-in.-diameter pipe of length 200 ft for $t = 0$ to $t = 10$ years.

8.67* In some locations with very "hard" water, a scale can build up on the walls of pipes to such an extent that not only does the roughness increases with time, but the diameter significantly decreases with time. Consider a case for which the roughness and diameter vary as $\varepsilon = 0.02 + 0.01t$ mm, $D = 50 (1 - 0.02t)$ mm, where t is in years. Plot the flowrate as a function of time for $t = 0$ to $t = 10$ years if the pressure drop per 12 m of horizontal pipe remains constant at $\Delta p = 1.3$ kPa.

8.68 After a number of years of use, it is noted that to obtain a given flowrate, the head loss is increased to 1.31 times its value for the originally smooth pipe. If the Reynolds number is 10^6, determine the relative roughness of the old pipe.

8.69 Gasoline flows in a smooth pipe of 40-mm diameter at a rate of 0.001 m³/s. If it were possible to prevent turbulence from occurring, what would be the ratio of the head loss for the actual turbulent flow compared to that if it were laminar flow?

8.70 A 3-ft-diameter duct is used to carry ventilating air into a vehicular tunnel at a rate of 9000 ft³/min. Tests show that the pressure drop is 1.5 in. of water per 1500 ft of duct. What is the approximate size of the equivalent roughness of the surface of the duct?

8.71 Natural gas ($\rho = 0.0044$ slugs/ft³ and $\nu = 5.2 \times 10^{-5}$ ft²/s) is pumped through a horizontal 6-in.-diameter cast-iron pipe at a rate of 800 lb/hr. If the pressure at section (1) is 50 psi (abs), determine the pressure at section (2) 8 mi downstream if the flow is assumed incompressible. Is the incompressible assumption reasonable? Explain.

8.72* Water flows in a 20-mm-diameter galvanized iron pipe with average velocities between 0.01 and 10.0 m/s. Plot the head loss per meter of pipe length over this velocity range. Compare your plot with that shown in Fig. 8.7. Discuss.

8.73 A fluid flows through a smooth horizontal 2-m-long tube of diameter 2 mm with an average velocity of 2.1 m/s. Determine the head loss and the pressure drop if the fluid is **(a)** air, **(b)** water, or **(c)** mercury.

8.74 Air at standard temperature and pressure flows through a horizontal 2 ft by 1.3 ft rectangular galvanized iron duct with a flowrate of 8.2 cfs. Determine the pressure drop in inches of water per 100-ft length of duct.

8.75 Air flows through a rectangular galvanized iron duct of size 0.30 m by 0.15 m at a rate of 0.068 m³/s. Determine the head loss in 8 m of this duct.

8.76 Air at standard conditions flows through a horizontal 1 ft by 1.5 ft rectangular wooden duct at a rate of 5000 ft³/min. Determine the head loss, pressure drop, and power supplied by the fan to overcome the flow resistance in 500 ft of the duct.

8.77 When the valve is closed the pressure throughout the horizontal pipe shown in Fig. P8.77 is 400 kPa, and the water level in the closed surge chamber is $h = 0.4$ m. If the valve is fully opened and the pressure at point (1) remains 400 kPa, determine the new level of the water in the surge chamber. Assume the friction factor is $f = 0.02$ and the fittings are threaded fittings.

FIGURE P8.77

8.78 What horsepower is added to water to pump it vertically through a 100-ft-long, 1.0-in.-

diameter drawn tubing at a rate of 0.060 ft³/s if the pressures at the inlet and outlet are the same?

8.79   Water flows from a lake as is shown in Fig. P8.79 at a rate of 4.0 cfs. Is the device inside the building a pump or a turbine? Explain and determine the horsepower of the device. Neglect all minor losses and assume the friction factor is 0.025.

FIGURE P8.79

8.80   Repeat Problem 8.79 if the flowrate is 1.0 cfs.

8.81   At a ski resort water at 40 °F is pumped through a 3-in.-diameter, 2000-ft-long steel pipe from a pond at an elevation of 4286 ft to a snow-making machine at an elevation of 4623 ft at a rate of 0.26 ft³/s. If it is necessary to maintain a pressure of 230 psi at the snow-making machine, determine the horsepower added to the water by the pump. Neglect minor losses.

8.82*   Water flows from a large tank, through a well-rounded entrance, and into a horizontal smooth pipe of length 10 m and diameter 50 mm. It exits the pipe as a free jet. The free surface of the water in the tank remains 4.0 m above the pipe. Plot a graph of the entrance length, ℓ_e, as a function of the water temperature, T, for $0 < T < 100$ °C.

8.83   Water flows through the screen in the pipe shown in Fig. P8.83 as indicated. Determine the loss coefficient for the screen.

FIGURE P8.83

8.84   Air flows through the fine mesh gauze shown in Fig. P8.84 with an average velocity of 1.50 m/s in the pipe. Determine the loss coefficient for the gauze.

FIGURE P8.84

8.85   Ethyl alcohol flows through a filter at a rate of 2×10^{-6} m³/s as is shown in Fig. P8.85. Determine the loss coefficient (based on the fluid velocity in the 0.03-m-diameter tube) for the filter.

FIGURE P8.85

8.86   Air at 80 °F and standard atmospheric pressure flows through a furnace filter with an average velocity of 2.4 ft/s. If the pressure drop across the filter is 0.21 in. of water, what is the loss coefficient for the filter?

8.87   The exhaust from a car's engine is to be vented to the outside of a garage through a 3-in.-diameter, 40-ft-long very rough hose ($f = 0.08$) attached to the end of the exhaust pipe. If the flowrate is 0.20 cfs and the temperature of the exhaust is 150 °F, determine the backpressure imposed. That is, determine the pressure at the end of the car's exhaust pipe.

8.88   Assume a car's exhaust system can be approximated as 14 ft of 0.125-ft-diameter cast-iron pipe with the equivalent of four 90° flanged elbows and a muffler. The muffler acts as a resistor with a loss coefficient of $K_L = 8.5$. Determine the pressure at the beginning of the exhaust system if the flowrate is 0.10 cfs and the temperature is 250 °F. $z_1 = z_2$ $P_2 = 0$

$$\frac{P_1}{\delta} + \frac{V_1^2}{2\gamma} + z_1 = \frac{P_2}{\delta} + \frac{V_2^2}{2g} + z_2 + \left(f\frac{\ell}{D} + \Sigma K_L\right)\frac{V^2}{2g}$$

8.89   Air is to flow through a smooth horizontal rectangular duct at a rate of 100 m³/s with a pressure drop of not more than 40 mm of water per 50 m of duct. If the aspect ratio (width to height) is 3 to 1, determine the size of the duct. $\rho \frac{\ell}{RT}$

$V = V_1 = V_2 = \frac{Q}{A} = \frac{.1 \, ft^3/s}{\pi/4 \, (.129 ft)^2} = 8.15 \, ft/s$; $P_1 = \left(f\frac{\ell}{D} + \Sigma K_L\right)\frac{1}{2}\rho V^2$

$FIG \cdot 8.22 \quad \frac{\varepsilon}{D} = 6.5 \times 10^{-3}$; $Re = \frac{\rho V D}{N} \quad \cdot Fig.8.2 \, f = 0.041$

$\therefore P_1 =$

8.90 Air flows through the mitered bend shown in Fig. P8.90 at a rate of 5.0 cfs. To help straighten the flow after the bend, a set of 0.25-in.-diameter drinking straws is placed in the pipe as shown. Estimate the extra pressure drop between points (1) and (2) caused by these straws.

Tightly packed 0.25-in.-diameter, 12-in. long straws

FIGURE P8.90

8.91 Water at 40 °F flows through the coils of the heat exchanger as shown in Fig. P8.91 at a rate of 0.9 gal/min. Determine the pressure drop between the inlet and outlet of the horizontal device.

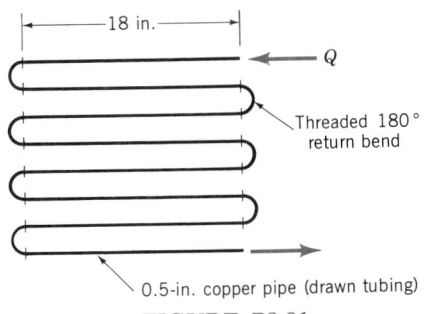

├─────18 in.─────┤

Threaded 180°
return bend

0.5-in. copper pipe (drawn tubing)

FIGURE P8.91

8.92 Water at 50 °F is pumped from a lake as shown in Fig. P8.92. What is the maximum flowrate possible without cavitation occurring?

$\frac{P_1}{\gamma} + \frac{V_1^2}{2g} + z_1 = \frac{P_2}{\gamma} + \frac{V_2^2}{2g} + z_2 + \left(\beta\frac{\ell}{D} + \Sigma K_L\right)\frac{V^2}{2g}$ $z=0, z_2=10 ft$

$P_1 = 14.7 (abs)$
$V_1 = 0$
$V_2 = V$
Table 6.1
$P_2 = 25.64 abs$
$f = .02 given$
Fig. 8.25
& Table 8.3
$k_2 = .8$
$k_{L2} = 1.5$ for $90°$ el.n.w.
$V = \frac{Q}{A} \frac{in}{fsol}$
flow

$Q = AV$
$= \frac{\pi}{4}(\frac{3}{12}ft)^2(14.4\frac{ft}{s}) = .707 ft^3/s$

├─────30 ft─────┤

10 ft

3-in.-diameter,
f = 0.02   Pump   Q

20 ft

FIGURE P8.92

8.93 Water at 10 °C is pumped from a lake as shown in Fig. P8.93. If the flowrate is 0.011 m³/s, what is the maximum length inlet pipe, ℓ, that can be used without cavitation occurring?

Length ℓ
D = 0.07 m
ε = 0.08 mm

Elevation
650 m
▽

Elevation
653 m

Q = 0.011 m³/s

FIGURE P8.93

8.94 The ½-in.-diameter hose shown in Fig. P8.94 can withstand a maximum pressure of 200 psi without rupturing. Determine the maximum length, ℓ, allowed if the friction factor is 0.015 and the flowrate is 0.010 cfs. Neglect minor losses.

$\frac{P_1}{\gamma} + \frac{V_1^2}{2g} + z_1 = \frac{P_2}{\gamma} + \frac{V_2^2}{2g} + z_2 + f\frac{\ell}{D}\frac{V^2}{2g}$ $z=0, z_2=10 ft$ $P_1=200 psi$

$P_2 = 0, V_1 = \frac{Q}{A_1} = \frac{.01}{\frac{\pi}{4}(\frac{.5}{12})} = 7.33,$

$V_2 = \frac{Q}{A_2} = \frac{.01}{\frac{\pi}{4}(\frac{.3}{12})} = 20.4$

Nozzle tip diameter = 0.30 in.   Q = 0.010 cfs

D = 0.50 in.

10 ft

├─L─┤

3 ft

plug in & solve
for ℓ

Pump

FIGURE P8.94

8.95 The hose shown in Fig. P8.94 will collapse if the pressure within it is lower than 11 psi below atmospheric pressure. Determine the maximum length, L, allowed if the friction factor is 0.015 and the flowrate is 0.010 cfs. Neglect minor losses.

8.96 A 40-m-long, 12-mm-diameter pipe with a friction factor of 0.020 is used to siphon 30 °C water from a tank as shown in Fig. P8.96. De-

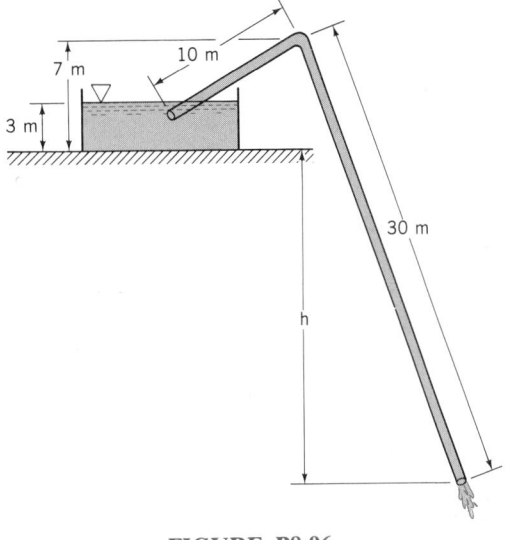

7 m

10 m

3 m

30 m

h

FIGURE P8.96

termine the maximum value of *h* allowed if there is to be no cavitation within the hose. Neglect minor losses.

8.97 The contents of the fire extinguisher shown in Fig. P8.97 are pressurized to 60 psi. Determine the flowrate if the friction factor of the tube is 0.018 and the loss coefficients for the entrance and valve/nozzle assembly (based on the velocity in the tube) are 1.0 and 7.8, respectively. Repeat the problem if pipe friction were neglected but minor losses included. Repeat the problem if minor losses were neglected but pipe friction included. Comment on the importance of the various losses for this flow.

Nozzle diameter = 0.10 in.

0.25-in.-diameter tube

p = 60 psi

1.5 ft

1.1 ft

FIGURE P8.97

8.98 Water flows from the container shown in Fig. P8.98. Determine the loss coefficient needed in the valve if the water is to "bubble up" 3 in. above the outlet pipe.

Vent

27 in.

18 in.

$\frac{1}{2}$-in.-diameter galvanized iron pipe with threaded fittings

3 in.

2 in.

32 in.

FIGURE P8.98

8.99 The pressure at section (2) shown in Fig. P8.99 is not to fall below 50 psi when the flowrate from the tank varies from 0 to 1.0 cfs and the branch line is shut off. Determine the minimum

height, *h*, of the water tank under the assumption that **(a)** minor losses are negligible; **(b)** minor losses are not negligible.

10 ft

h

All pipe is 6-in.-diameter plastic ($\epsilon/D \approx 0$), flanged fittings

6 ft

Branch line

(2) Main line

600 ft with 15 90° elbows

900 ft

FIGURE P8.99

8.100 Repeat Problem 8.99 with the assumption that the branchline is open so that half of the flow from the tank goes into the branch, and half continues in the main line.

8.101 Water flows from the nozzle attached to the spray tank shown in Fig. P8.101. Determine the flowrate if the loss coefficient for the nozzle (based on upstream conditions) is 0.75 and the friction factor for the rough hose is 0.11.

Nozzle diameter = 7.5 mm

p = 150 kPa

0.80 m

D = 15 mm
ℓ = 1.9 m

40°

FIGURE P8.101

8.102* Water flows from a large open tank through a sharp edged entrance and into a horizontal 0.5-in.-diameter copper pipe (drawn tubing) of length ℓ. The water exits the pipe as a free jet a distance of 4.0 ft below the free surface of the tank. Plot a log–log graph of the flowrate through the pipe, Q, as a function of ℓ for $10 \le \ell \le 10,000$ ft. Comment on your results.

8.103* Water flows from a large open tank through a sharp-edged entrance and into a galvanized iron pipe of length 100 m and diameter 10 mm. The water exits the pipe as a free jet at a distance *h* below the free surface of the tank.

Plot a log–log graph of the flowrate, Q, as a function of h for $0.1 \le h \le 10$ m.

8.104 Water flows at a rate of 0.50 cfs from tank A to tank B through a horizontal 3-in.-diameter cast-iron pipe of length 50 ft. If minor losses are neglected, determine the difference in elevation of the free surfaces of the tanks.

8.105 Water flows through a smooth 50-mm-diameter pipe at a rate of 0.010 m³/s and a head loss of 42 m per 50 m of horizontal pipe. An engineer claims that there is an obstruction in the pipe. Do you agree or disagree? Explain.

8.106 According to fire regulations in a town, the pressure drop in a commercial steel horizontal pipe must not exceed 1.0 psi per 150 ft of pipe for flowrates up to 500 gal/min. If the water temperature is never below 50 °F, what diameter pipe is needed?

8.107 An above ground swimming pool of 30 ft diameter and 5 ft depth is to be filled from a garden hose (smooth interior) of length 100 ft and diameter 5/8 in. If the pressure at the faucet to which the hose is attached remains at 55 psi, how long will it take to fill the pool? The water exits the hose as a free jet 6 ft above the faucet.

8.108 Water flows downward through a vertical 10-mm-diameter galvanized iron pipe with an average velocity of 5.0 m/s and exits as a free jet. There is a small hole in the pipe 4 m above the outlet. Will water leak out of the pipe through this hole, or will air enter into the pipe through the hole? Repeat the problem if the average velocity is 0.5 m/s.

8.109* Water drains from the bathtub shown in Fig. P8.109 through the drain indicated. Determine the length of time taken for the tub to drain if the water is initially 8 in. deep and the friction factor is $f = 0.02$. Assume quasisteady flow conditions and that the drain pipe remains full. See table below.

A (ft²)	h (in.)
0	0
4.1	1
6.7	2
8.5	3
9.1	4
9.4	5
9.6	6
9.8	7
10.0	8

8.110* Repeat Problem 8.109 if the "strainer" at the tub outlet is partially plugged and has a loss coefficient of 20.

8.111 Water flows through the pipe shown in Fig. P8.111. Determine the net tension in the bolts if minor losses are neglected and the wheels on which the pipe rests are frictionless.

FIGURE P8.111

8.112 Water flows through two sections of the vertical pipe shown in Fig. P8.112. The bellows connection cannot support any force in the vertical direction. The 0.4-ft-diameter pipe weighs 0.2 lb/ft and the friction factor is assumed to be 0.02. At what velocity will the force, F, required to hold the pipe be zero?

FIGURE P8.112

FIGURE P8.109

8.113 Without the pump shown in Fig. P8.113 it is determined that the flowrate is too small.

Determine the horsepower added to the fluid if the pump causes the flowrate to be doubled. Assume the friction factor remains at 0.020 in either case.

FIGURE P8.113

8.114 The pump shown in Fig. P8.114 adds 45 kW to the water and causes a flowrate of 0.04 m^3/s. Determine the flowrate expected if the pump is removed from the system. Assume $f = 0.016$ for either case and neglect minor losses.

FIGURE P8.114

8.115 A certain process requires 2.3 cfs of water to be delivered at a pressure of 30 psi. This water comes from a large diameter supply main in which the pressure remains at 60 psi. If the galvanized iron pipe connecting the two locations is 200 ft long and contains six threaded 90° elbows, determine the pipe diameter. Elevation differences are negligible.

8.116 The turbine shown in Fig. P8.116 develops 400 kW. Determine the flowrate if **(a)** head losses are negligible or **(b)** head loss due to friction in the pipe is considered. Assume $f = 0.02$. *Note*: There may be more than one solution or there may be no solution to this problem.

FIGURE P8.116

8.117 Draw the energy line and hydraulic grade line for Problem 8.83. Assume the pressure upstream of the screen is 3 psi.

8.118 Draw the energy line and hydraulic grade line for Problem 8.113.

8.119 A liquid with $SG = 0.96$, $\mu = 9.2 \times 10^{-4}$ N·s/m², and vapor pressure $p_v = 1.2 \times 10^4$ N/m²(abs) is drawn into the syringe as is indicated in Fig. P8.119. What is the maximum flowrate if cavitation is not to occur in the syringe?

FIGURE P8.119

8.120 Determine the diameter of a steel pipe that is to carry 2,000 gal/min of gasoline with a pressure drop of 5 psi per 100 ft of horizontal pipe.

8.121 Rainwater flows through the galvanized iron downspout shown in Fig. P8.121 at a rate of 0.006 m^3/s. Determine the size of the downspout cross section if it is a rectangle with an aspect ratio of 1.7 to 1 and it is completely filled with water. Neglect the velocity of the water in the gutter at the free surface and the head loss associated with the elbow.

FIGURE P8.121

8.122* Repeat Problem 8.121 if the downspout is circular.

8.123 Air, assumed incompressible, flows through the two pipes shown in Fig. P8.123. Determine the flowrate if minor losses are neglected and the friction factor in each pipe is 0.025. Determine the flowrate if the 0.5 -in.-diameter pipe were replaced by a 1-in.-diameter pipe. Comment on the assumption of incompressibility.

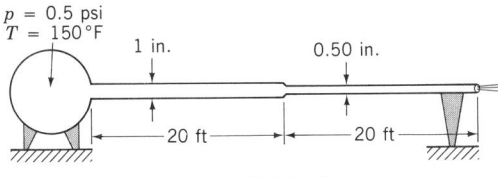

FIGURE P8.123

8.124* Repeat Problem 8.123 if the pipes are galvanized iron and the friction factors are not known a priori.

8.125 The flowrate between tank A and tank B shown in Fig. P8.125 is to be increased by 30% (i.e., from Q to $1.30Q$) by the addition of a second pipe (indicated by the dotted lines) running from node C to tank B. If the elevation of the free surface in tank A is 25 ft above that in tank B, determine the diameter, D, of this new pipe. Neglect minor losses and assume that the friction factor for each pipe is 0.02.

FIGURE P8.125

8.126 With the valve closed, water flows from tank A to tank B as shown in Fig. P8.126. What

FIGURE P8.126

is the flowrate into tank B when the valve is opened to allow water to flow into tank C also? Neglect all minor losses and assume that the friction factor is 0.02 for all pipes.

8.127* Repeat Problem 8.126 if the friction factors are not known, but the pipes are steel pipes.

8.128 The three water-filled tanks shown in Fig. P8.128 are connected by pipes as indicated. If minor losses are neglected, determine the flowrate in each pipe.

FIGURE P8.128

8.129* Repeat Problem 8.128 if the friction factors are not known, but the pipes are steel pipes.

8.130 The three tanks shown in Fig. P8.130 are connected by pipes with friction factors of 0.03 for each pipe. Determine the water velocity in each pipe. Neglect minor losses.

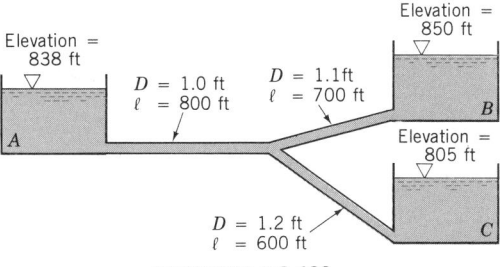

FIGURE P8.130

8.131 The surface elevations of reservoirs A, B, and C are at 2100, 1800 and 1900 ft, respectively. An 18-in.-diameter pipe 2 mi long runs from reservoir A to a node at an elevation of 2000 ft where two 12-in.-diameter pipes each of 1-mi length branch from the original pipe and connect reservoirs B and C. If the friction factor for each pipe is 0.020, determine the flowrate in each pipe.

8.132 A 2-in.-diameter orifice plate is inserted in a 3-in.-diameter pipe. If the water flowrate

through the pipe is 0.50 cfs, determine the pressure difference indicated by a manometer attached to the flow meter.

8.133 Water flows in a 0.10-m-diameter pipe at a rate of 0.02 m³/s. If the pressure difference across the orifice meter in the pipe is to be 28 kPa, what diameter orifice is needed?

8.134 Gasoline flows through a 35-mm-diameter pipe at a rate of 0.0032 m³/s. Determine the pressure drop across a flow nozzle placed in the line if the nozzle diameter is 24 mm.

8.135 Air at 200 °F and 60 psia flows in a 4-in.-diameter pipe at a rate of 0.52 lb/s. Determine the pressure at the 2-in-diameter throat of a Venturi meter placed in the pipe.

8.136 A 50-mm-diameter nozzle is installed at the end of a 80-mm-diameter pipe through which air flows. A manometer attached to the static pressure tap just upstream from the nozzle indicates a pressure of 7.3 mm of water. Determine the flowrate.

8.137 A 2.5-in.-diameter flow nozzle is installed in a 3.8-in.-diameter pipe that carries water at 160 °F. If the flowrate is 0.78 cfs, determine the reading on the air–water manometer used to measure the pressure difference across the meter.

8.138 A 2.5-in.-diameter flow nozzle is installed in a 3.8-in.-diameter pipe that carries water at 160 °F. If the air–water manometer used to measure the pressure difference across the meter indicates a reading of 3.1 ft, determine the flowrate.

8.139 Water flows through the Venturi meter shown in Fig. P8.139. The specific gravity of the manometer fluid is 1.52. Determine the flowrate.

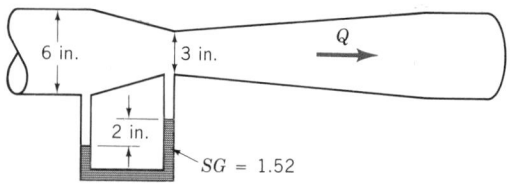

FIGURE P8.139

8.140 If the fluid flowing in Problem 8.139 were air, what would the flowrate be? Would compressibility effects be important? Explain.

8.141 Water flows through a 40-mm-diameter nozzle meter in a 75-mm-diameter pipe at a rate of 0.015 m³/s. Determine the pressure difference across the nozzle if the temperature is **(a)** 10 °C, or **(b)** 80 °C.

8.142 Water flows through the orifice meter shown in Fig. P8.142 at a rate of 0.10 cfs. If $d = 0.1$ ft, determine the value of h.

FIGURE P8.142

8.143 Water flows through the orifice meter shown in Fig. P8.142 at a rate of 0.10 cfs. If $h = 3.8$ ft, determine the value of d.

8.144 Water flows through the orifice meter shown in Fig. P8.142 such that $h = 1.6$ ft with $d = 1.5$ in. Determine the flowrate.

8.145 Air flows through a 1-in.-diameter galvanized iron pipe that contains a 0.5-in. flow nozzle as shown in Fig. P8.145. For some unexplained reason, the upstream pressure tap is located 12 ft upstream of the nozzle. Determine the manometer reading, h, if the flowrate is $Q = 0.10$ cfs.

FIGURE P8.145

8.146 Air flows through a 1-in.-diameter plastic pipe shown in Fig. P8.145. The upstream pressure tap for the nozzle flow meter is located 12 ft upstream of the nozzle and the friction factor for the pipe is $f = 0.02$. Determine the flowrate, Q, if the manometer reading is $h = 0.76$ ft. Explain why this configuration with the upstream pressure tap far upstream of the nozzle is not recommended.

Impulsive start of flow past an array of cylinders: The complex structure of laminar flow past a relatively simple geometric structure illustrates why it is often difficult to obtain exact analytical results for external flows. (Dye in water.) (Photograph courtesy of ONERA, France.)

NINE

FLOW OVER IMMERSED BODIES

In this chapter we consider various aspects of the flow over bodies that are immersed in a fluid. Examples include the flow of air around airplanes, automobiles, and falling snow flakes, or the flow of water around submarines and fish. In these situations the object is completely surrounded by the fluid and the flows are termed *external flows*.

External flows involving air are often termed aerodynamics in response to the important external flows produced when an object such as an airplane flies through the atmosphere. Although this field of external flows is extremely important, there are many other examples that are of equal importance. The fluid forces (lift and drag) on surface vehicles (cars, trucks, bicycles) have become a very important topic. By correct design of cars and trucks, it has become possible to greatly decrease the fuel consumption and improve the handling characteristics of the vehicle. Similar efforts have resulted in improved ships, whether they are surface vessels (surrounded by two fluids, air and water) or submersible vessels (surrounded completely by water).

Other applications of external flows involve objects that are not completely surrounded by fluid, although they are placed in some external type flow. For example, the proper design of a building (whether it is your house or a tall skyscraper) must include consideration of the various wind effects involved.

As with other areas of fluid mechanics, two approaches (theoretical and experimental) are used to obtain information on the fluid forces developed by external flows. Theoretical (i.e., analytical and numerical) techniques can provide much of the needed information about such flows. However, because of the complexities of the governing equations and the complexities of the geometry of the objects involved, the amount of information obtained from purely theoretical methods is limited. With current and anticipated advancements in the area of computational fluid mechanics, it is likely

that computer prediction of forces and complicated flow patterns will become more readily available.

Much of the information about external flows comes from experiments carried out, for the most part, on scale models of the actual objects. Such testing includes the obvious wind tunnel testing of model airplanes, buildings, and even entire cities. In some instances the actual device, not a model, is tested in wind tunnels. Figure 9.1 shows tests of vehicles in wind tunnels. Better performance of cars, bikes, skiers, and numerous other objects has resulted from testing in wind tunnels. The use of water tunnels and towing tanks also provides useful information about the flow around ships and other objects.

In this chapter we consider characteristics of external flows past a variety of objects. We investigate the qualitative aspects of such flows and learn how to determine the various forces on objects surrounded by a moving liquid.

9.1   GENERAL EXTERNAL FLOW CHARACTERISTICS

A body immersed in a moving fluid experiences a resultant force due to the interaction between the body and the fluid surrounding it. In some instances (such as an airplane

(a)

(b)

FIGURE 9.1
(a) Flow past a full-sized stream-lined vehicle in the GM aerodynamics laboratory wind tunnel, an 18-ft by 34-ft test section facility driven by a 4000-hp, 43-ft diameter fan. (Photograph courtesy of General Motors Corporation.) (b) Surface flow on a model vehicle as indicated by tufts attached to the surface. (Reprinted with permission from Society of Automotive Engineers, Ref. 28.)

flying through still air) the fluid far from the body is stationary and the body moves through the fluid with velocity U. In other instances (such as the wind blowing past a building) the body is stationary and the fluid flows past the body with velocity U. In some cases (such as an airplane flying into a cross wind) it is a combination of these situations. In any case, we can fix the coordinate system in the body and treat the situation as fluid flowing past a stationary body with velocity U, the *upstream velocity*. For the purposes of this book, we will assume that the upstream velocity is constant in both time and location. That is, there is a uniform, constant velocity fluid flowing past the object. In actual situations this is often not true. For example, the wind blowing past a smoke stack is nearly always turbulent and gusty (unsteady) and probably not of uniform velocity from the top to the bottom of the stack. Usually the unsteadiness and nonuniformity are of minor importance.

Even with a steady, uniform upstream flow, the flow in the vicinity of an object may be unsteady. Examples of this type of behavior include the flutter that is sometimes found in the flow past airfoils (wings), the regular oscillation of telephone wires that "sing" in a wind, and the irregular turbulent fluctuations in the wake regions behind bodies.

The structure of an external flow and the ease with which the flow can be described and analyzed often depend on the nature of the body in the flow. Three general categories of bodies are shown in Fig. 9.2. They include (a) two-dimensional objects (infinitely long and of constant cross-sectional size and shape), (b) axisymmetric bodies (formed by rotating their cross-sectional shape about the axis of symmetry), and (c) three-dimensional bodies that may or may not possess a line or plane of symmetry. In practice there can be no truly two-dimensional bodies—nothing extends to infinity. However, many objects are sufficiently long so that the end effects are negligibly small.

Another classification of body shape can be made depending on whether the body

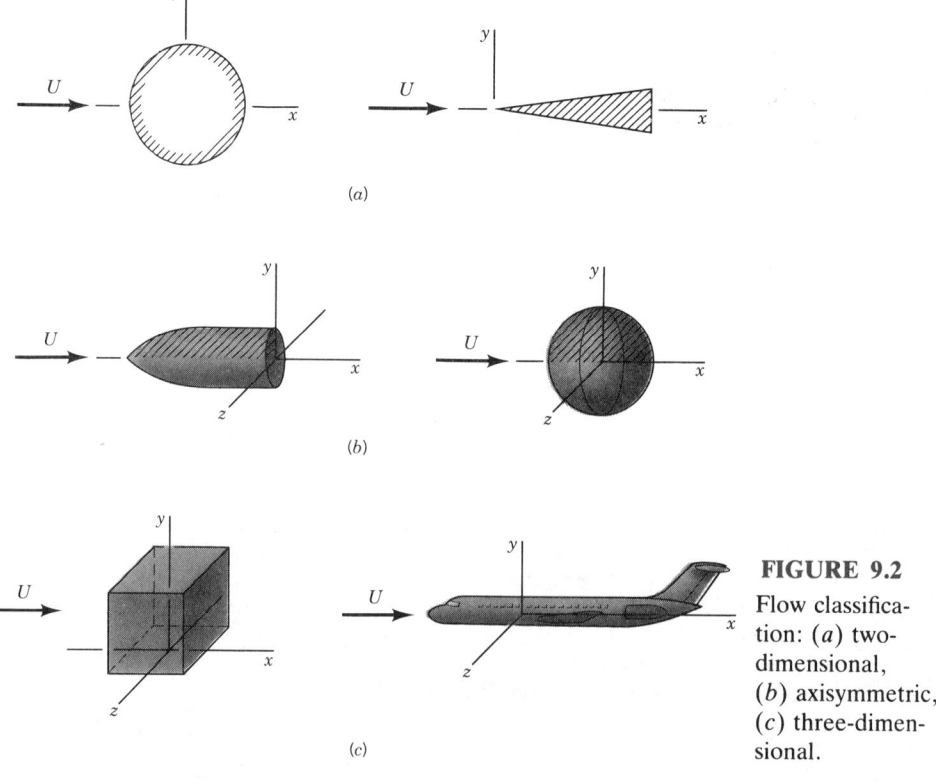

FIGURE 9.2

Flow classification: (*a*) two-dimensional, (*b*) axisymmetric, (*c*) three-dimensional.

is streamlined or blunt. The flow characteristics depend strongly on the amount of streamlining present. In general, *streamlined bodies* (i.e., airfoils, racing cars, etc.) have little effect on the surrounding fluid, compared with the effect that *blunt bodies* (i.e., parachutes, buildings, etc.) have on the fluid. Usually, but not always, it is easier to force a streamlined body through a fluid than it is to force a similar sized blunt body at the same velocity. There are important exceptions to this basic rule.

9.1.1   Lift and Drag Concepts

When any body moves through a fluid an interaction between the body and the fluid occurs; this effect can be described in terms of the forces at the fluid–body interface. This can be described in terms of the stresses—wall shear stresses, τ_w, due to viscous effects and normal stresses due to the pressure, p. Typical shear stress and pressure distributions are shown in Figs. 9.3a and 9.3b. Both τ_w and p vary in magnitude and direction along the surface.

It is often useful to know the detailed distribution of shear stress and pressure over the surface of the body, although such information is difficult to obtain. Many times, however, only the integrated or resultant effects of these distributions are needed. The resultant force in the direction of the upstream velocity is termed the *drag*, $\mathscr{D}$, and the resultant force normal to the upstream velocity is termed the *lift*, $\mathscr{L}$, as is indicated in Fig. 9.3c. For some three-dimensional bodies there may also be a side force that is perpendicular to the plane containing $\mathscr{D}$ and $\mathscr{L}$.

The resultant of the shear stress and pressure distributions can be obtained by integrating the effect of these two quantities on the body surface as is indicated in Fig. 9.4. The x and y components of the fluid force on the small area element dA are

$$dF_x = (p \, dA) \cos \theta + (\tau_w \, dA) \sin \theta$$

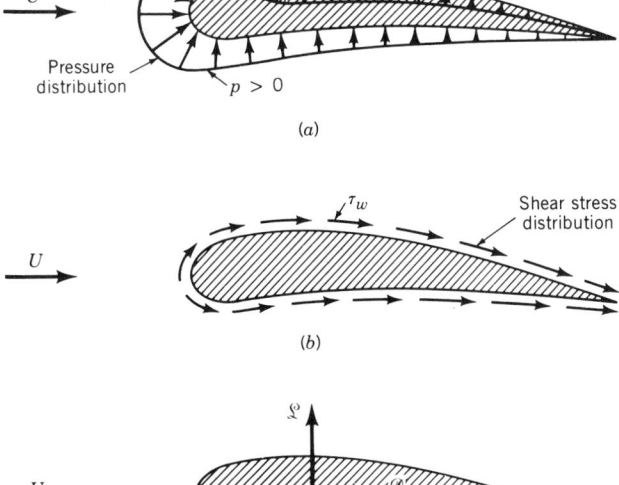

(a)

(b)

(c)

FIGURE 9.3 Forces from the surrounding fluid on a two-dimensional object: (*a*) pressure force, (*b*) viscous force, (*c*) resultant force (lift and drag).

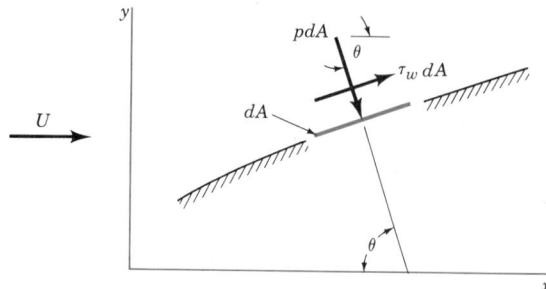

FIGURE 9.4 Pressure and shear forces on a small element of the surface of a body.

and

$$dF_y = -(p\, dA) \sin \theta + (\tau_w\, dA) \cos \theta$$

Thus, the net x and y components of the force on the object are

$$\mathcal{D} = \int dF_x = \int p \cos \theta\, dA + \int \tau_w \sin \theta\, dA \qquad (9.1)$$

and

$$\mathcal{L} = \int dF_y = -\int p \sin \theta\, dA + \int \tau_w \cos \theta\, dA \qquad (9.2)$$

Of course, to carry out the integrations and determine the lift and drag, we must know the body shape (i.e., θ as a function of location along the body) and the distribution of τ_w and p along the surface. These distributions are often extremely difficult to obtain, either experimentally or theoretically. The pressure distribution can be obtained experimentally without too much difficulty by use of a series of static pressure taps along the body surface. On the other hand, it is usually quite difficult to measure the wall shear stress distribution.

It is seen that both the shear stress and pressure force contribute to the lift and drag, since for an arbitrary body θ is neither zero nor 90° along the entire body. The exception is a flat plate aligned either parallel to the upstream flow ($\theta = 90°$) or normal to the upstream flow ($\theta = 0$) as is discussed in Example 9.1.

EXAMPLE 9.1

Air at standard conditions flows past a flat plate as is indicated in Fig. E9.1. In case (a) the plate is parallel to the upstream flow, and in case (b) it is perpendicular to the upstream flow. If the pressure and shear stress distributions on the surface are as indicated (obtained either by experiment or theory), determine the lift and drag on the plate.

$$p = p(x) = 0 \qquad b = \text{width} = 10 \text{ ft}$$

$$U = 25 \text{ ft/s}$$
$$p = 0 \text{ (gage)}$$

4 ft

$$\tau_w = \tau_w(x) = (1.24 \times 10^{-3})/\sqrt{x} \text{ lb/ft}^2$$
where x is in feet

(a)

FIGURE E9.1

(b)

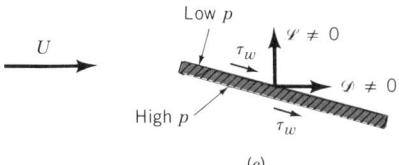

(c)

FIGURE E9.1 (*Continued*)

SOLUTION

For either orientation of the plate, the lift and drag are obtained from Eqs. 9.1 and 9.2. With the plate parallel to the upstream flow we have $\theta = 90°$ on the top surface and $\theta = 270°$ on the bottom surface so that the lift and drag are given by

$$\mathcal{L} = -\int_{\text{top}} p \, dA + \int_{\text{bottom}} p \, dA = 0$$

and

$$\mathcal{D} = \int_{\text{top}} \tau_w \, dA + \int_{\text{bottom}} \tau_w \, dA = 2\int_{\text{top}} \tau_w \, dA \tag{1}$$

where we have used the fact that because of symmetry the shear stress distribution is the same on the top and the bottom surfaces, as is the pressure also [whether we use gage ($p = 0$) or absolute ($p = p_{\text{atm}}$) pressure]. There is no lift generated—the plate does not know up from down. With the given shear stress distribution, Eq. 1 gives

$$\mathcal{D} = 2 \int_{x=0}^{4 \text{ ft}} \left(\frac{1.24 \times 10^{-3}}{x^{1/2}} \text{ lb/ft}^2 \right) (10 \text{ ft}) \, dx$$

or

$$\mathcal{D} = 0.0992 \text{ lb} \tag{Ans}$$

With the plate perpendicular to the upstream flow, we have $\theta = 0°$ on the front and $\theta = 180°$ on the back. Thus, from Eqs. 9.1 and 9.2

$$\mathcal{L} = \int_{\text{front}} \tau_w \, dA - \int_{\text{back}} \tau_w \, dA = 0$$

and

$$\mathcal{D} = \int_{\text{front}} p \, dA - \int_{\text{back}} p \, dA$$

Again there is no lift because the pressure forces act parallel to the upstream flow (in the

direction of $\mathcal{D}$ not $\mathcal{L}$) and the shear stress is symmetrical about the center of the plate. With the given relatively large pressure on the front of the plate (the center of the plate is a stagnation point) and the negative pressure (less than the upstream pressure) on the back of the plate, we obtain the following drag

$$\mathcal{D} = \int_{y=-2}^{2 \text{ ft}} \left[0.744 \left(1 - \frac{y^2}{4} \right) \text{ lb/ft}^2 - (-0.893) \text{ lb/ft}^2 \right] (10 \text{ ft}) \, dx$$

or

$$\mathcal{D} = 55.6 \text{ lb} \tag{Ans}$$

Clearly there are two mechanisms responsible for the drag. On the ultimately streamlined body (a zero thickness flat plate parallel to the flow) the drag is entirely due to the shear stress at the surface and, in this example, is relatively small. For the ultimately blunted body (a flat plate normal to the upstream flow) the drag is entirely due to the pressure difference between the front and back portions of the object and, in this example, is relatively large.

If the flat plate were oriented at an arbitrary angle relative to the upstream flow as indicated in Fig. E9.1c, there would be both a lift and a drag, each of which would be dependent on both the shear stress and the pressure. Both the pressure and shear stress distributions would be different for the top and bottom surfaces.

Although Eqs. 9.1 and 9.2 are valid for any body, the difficulty in their use lies in obtaining the appropriate shear stress and pressure distributions on the body surface. Considerable effort has gone into determining these quantities, but because of the various complexities involved, such information is available only for certain simple situations.

Without detailed information concerning the shear stress and pressure distributions on a body, Eqs. 9.1 and 9.2 cannot be used. The widely used alternative is to define dimensionless lift and drag coefficients and determine their values by means of either a simplified analysis, some numerical technique, or an appropriate experiment. The *lift coefficient, C_L,* and *drag coefficient, C_D,* are defined as

$$C_L = \frac{\mathcal{L}}{\frac{1}{2}\rho U^2 A}$$

and

$$C_D = \frac{\mathcal{D}}{\frac{1}{2}\rho U^2 A}$$

where A is a characteristic area of the object (see Chapter 7). Typically, A is taken to be *frontal area*—the projected area seen by a person looking toward the object from a direction parallel to the upstream velocity, U. It would be the area of the shadow on a screen normal to the upstream velocity as formed by a light shining along the upstream flow. In other situations A is taken to be the *planform area*—the projected area seen by an observer looking toward the object from a direction normal to the upstream velocity (i.e., from "above" it). Obviously, which characteristic area is used in the definition of the lift and drag coefficients must be clearly stated.

9.1.2 Characteristics of Flow Past an Object

External flows past objects encompass an extremely wide variety of fluid mechanics phenomena. Clearly the character of the flow field is a function of the shape of the body. Flows past relatively simple geometric shapes (i.e., a sphere or circular cylinder)

are expected to have less complex flow fields than flows past a complex shape such as an airplane or a tree. However, even the simplest shaped objects produce rather complex flows.

For a given shaped object, the characteristics of the flow depend very strongly on various parameters such as size, orientation, speed, and fluid properties. As is discussed in Chapter 7, according to dimensional analysis arguments, the character of the flow should depend on the various dimensionless parameters involved. For typical external flows the most important of these parameters are the Reynolds number, $Re = \rho U \ell / \mu = U \ell / \nu$, the Mach number, $Ma = U/c$, and for flows with a free surface (i.e., flows with an interface between two fluids, such as the flow past a surface ship), the Froude number, $Fr = U/\sqrt{g\ell}$. (Recall that ℓ is some characteristic length of the object.)

For the present, we consider how the external flow and its associated lift and drag vary as a function of Reynolds number. Recall that the Reynolds number represents the ratio of inertial effects to viscous effects. In the absence of all viscous effects ($\mu = 0$), the Reynolds number is infinite. On the other hand, in the absence of all inertial effects (negligible mass or $\rho = 0$), the Reynolds number is zero. Clearly, any actual flow will have a Reynolds number between (but not including) these two extremes. The nature of the flow past a body depends strongly on whether $Re \gg 1$ or $Re \ll 1$.

Most external flows with which we are familiar are associated with moderately sized objects with a characteristic length on the order of $0.01 \text{ m} < \ell < 10 \text{ m}$. In addition typical upstream velocities are on the order of $0.01 \text{ m/s} < U < 100 \text{ m/s}$ and the fluids involved are typically water or air. The resulting Reynolds number range for such flows is approximately $10 < Re < 10^9$. As a rule of thumb, flows with $Re > 100$ are dominated by inertial effects, whereas flows with $Re < 1$ are dominated by viscous effects. Hence, most familiar external flows are dominated by inertia.

On the other hand, there are many external flows in which the Reynolds number is considerably less than 1, indicating in some sense that viscous forces are more important than inertial forces. The gradual settling of small particles of dirt in a lake or stream is governed by low Reynolds number flow principles because of the small diameter of the particles and their small settling speed. Similarly, the Reynolds number for objects moving through large viscosity oils is small because μ is large. The general differences between small and large Reynolds number flow past streamlined and blunt objects can be illustrated by considering flows past two objects—one a flat plate parallel to the upstream velocity and the other a circular cylinder.

Flows past three flat plates of length ℓ with $Re = \rho U \ell / \mu = 0.1, 10,$ and 10^7, are shown in Fig. 9.5. If the Reynolds number is small, the viscous effects are relatively strong and the plate affects the uniform upstream flow far ahead, above, below, and behind the plate. To reach that portion of the flow field where the velocity has been altered by less than 1% of its undisturbed value (i.e., $U - u < 0.01U$) we must travel relatively far from the plate. In low Reynolds number flows the viscous effects are felt far from the object in all directions.

As the Reynolds number is increased (by increasing U, for example), the region in which viscous effects are important becomes smaller in all directions except downstream, as is shown in Fig. 9.5b. One does not need to travel very far ahead, above, or below the plate to reach areas in which the viscous effects of the plate are not felt. The streamlines are displaced from their original uniform upstream conditions, but the displacement is not as great as for the $Re = 0.1$ situation shown in Fig. 9.5a.

If the Reynolds number is large (but not infinite), the flow is dominated by inertial

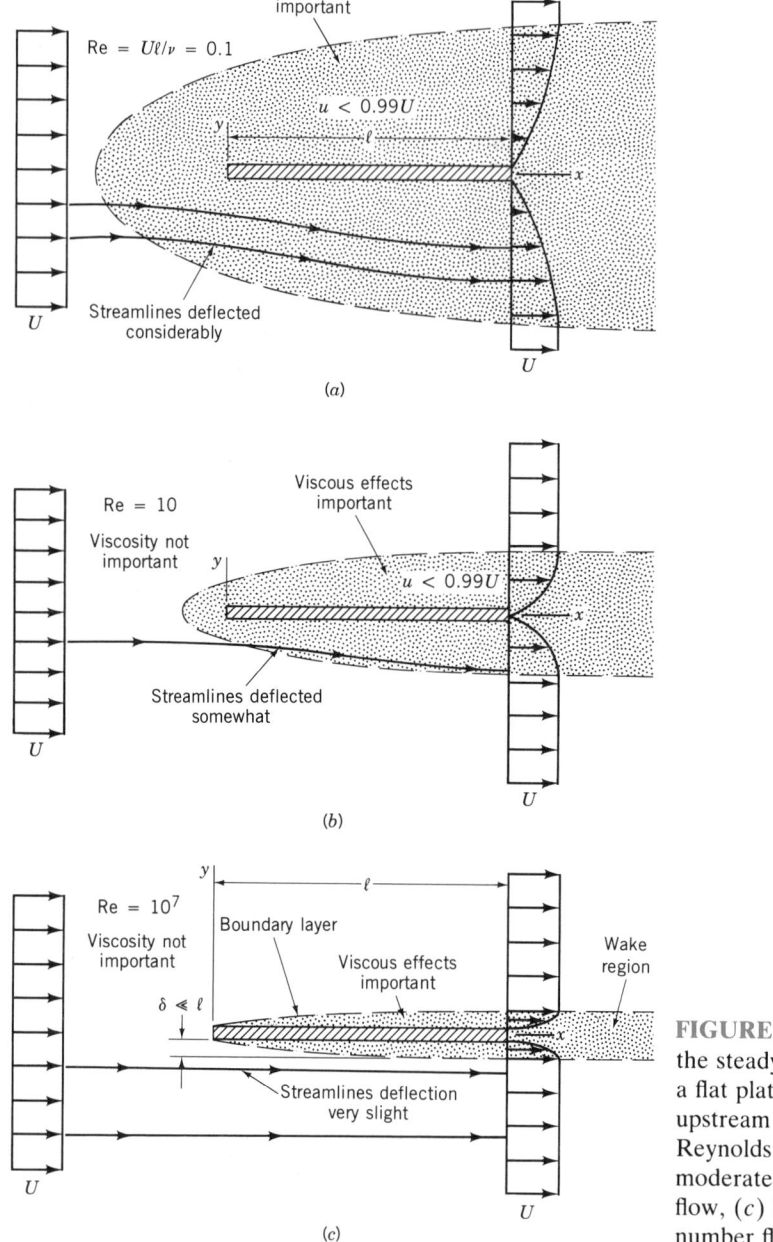

FIGURE 9.5 Character of the steady, viscous flow past a flat plate parallel to the upstream velocity: (*a*) low Reynolds number flow, (*b*) moderate Reynolds number flow, (*c*) large Reynolds number flow.

effects and the viscous effects are negligible everywhere except in a region very close to the plate and in the relatively thin *wake region* behind the plate, as shown in Fig. 9.5*c*. Since the fluid viscosity is not zero (Re $< \infty$), it follows that the fluid must stick to the solid surface (the no-slip boundary condition). There is a thin *boundary layer* region of thickness $\delta = \delta(x) \ll \ell$ (i.e., thin relative to the length of the plate) next to the plate in which the fluid velocity changes from the upstream value of $u = U$ to zero velocity on the plate. The thickness of this layer increases in the direction of flow, starting from zero at the forward or leading edge of the plate. The flow within

the boundary layer may be laminar or turbulent, depending on various parameters involved.

The streamlines of the flow outside of the boundary layer are nearly parallel to the plate. As we will see in the next section, the slight displacement of the external streamlines that are outside of the boundary layer is due to the thickening of the boundary layer in the direction of flow. The existence of the plate has very little effect on the streamlines outside of the boundary layer—either ahead, above, or below the plate. On the other hand, the wake region is due entirely to the viscous interaction between the fluid and the plate.

One of the great advancements in fluid mechanics occurred in 1904 as a result of the insight of Ludwig Prandtl (1875–1953), a German physicist and aerodynamicist. He conceived of the idea of the boundary layer—a thin region on the surface of a body in which viscous effects are very important and outside of which the fluid behaves essentially as if it were inviscid. Clearly the actual fluid viscosity is the same throughout, only the relative importance of the viscous effects is different within or outside of the boundary layer. As is discussed in the next section, by using such a hypothesis it is possible to simplify the analysis of large Reynolds number flows, thereby allowing solution to external flow problems that are otherwise still unsolvable.

As with the flow past the flat plate described above, the flow past a blunt object (such as a circular cylinder) also varies with Reynolds number. In general, the larger the Reynolds number, the smaller the region of the flow field in which viscous effects are important. For objects that are not sufficiently streamlined, however, an additional characteristic of the flow is observed. This is termed *flow separation* and is illustrated in Fig. 9.6.

Low Reynolds number flow (Re $= UD/v < 1$) past a circular cylinder is characterized by the fact that the presence of the cylinder and the accompanying viscous effects are felt throughout a relatively large portion of the flow field. As is indicated in Fig. 9.6a, for Re $= UD/v = 0.1$ the viscous effects are important several diameters in any direction from the cylinder. A somewhat surprising characteristic of this flow is that the streamlines are essentially symmetric about the center of the cylinder—the streamline pattern is the same in front of the cylinder as it is behind the cylinder.

As the Reynolds number is increased, the region ahead of the cylinder in which viscous effects are important becomes smaller, with the viscous region extending only a short distance ahead of the cylinder. The viscous effects are convected downstream and the flow loses its symmetry. Another characteristic of external flows becomes important—the flow separates from the body at the *separation location* as indicated in Fig. 9.6b. With the increase in Reynolds number, the fluid inertia becomes more important and at some location on the body, denoted the separation location, the fluid's inertia is such that it cannot follow the curved path around to the rear of the body. The result is a separation bubble behind the cylinder in which some of the fluid is actually flowing upstream, against the direction of the upstream flow.

At still larger Reynolds numbers, the area affected by the viscous forces is forced farther downstream until it involves only a thin ($\delta \ll D$) boundary layer on the front portion of the cylinder and an irregular, unsteady (perhaps turbulent) wake region that extends far downstream of the cylinder. The fluid in the region outside of the boundary layer and wake region flows as if it were inviscid. Of course, the fluid viscosity is the same throughout the entire flow field. Whether viscous effects are important or not depends on which region of the flow field we consider. The velocity gradients within the boundary layer and wake regions are much larger than those in

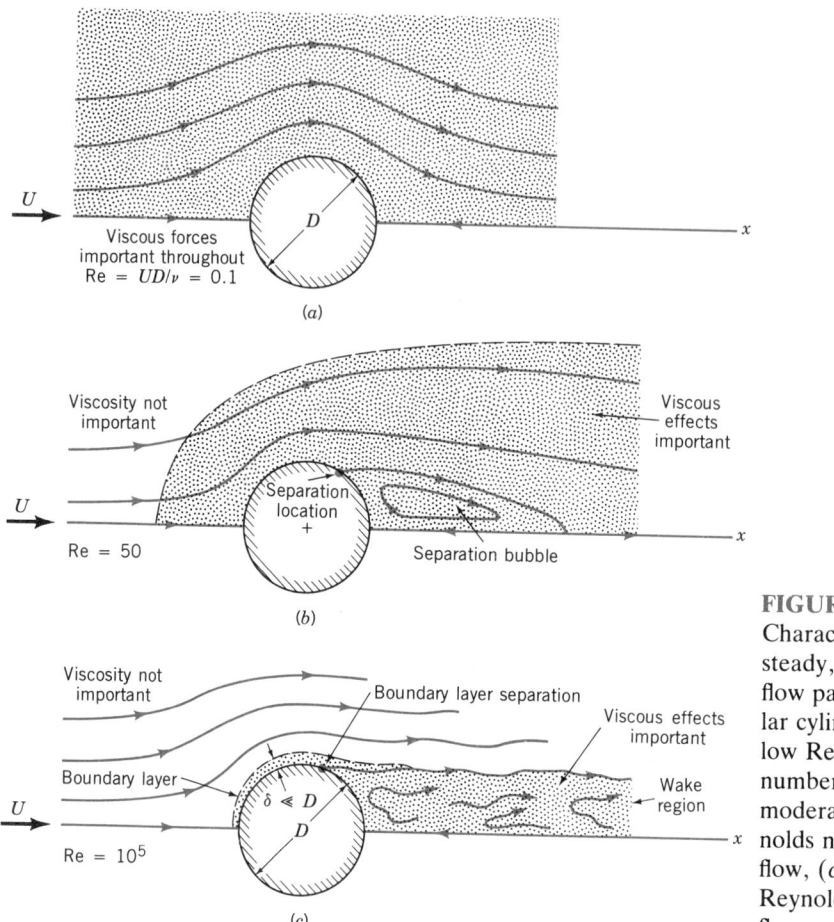

FIGURE 9.6
Character of the steady, viscous flow past a circular cylinder: (*a*) low Reynolds number flow, (*b*) moderate Reynolds number flow, (*c*) large Reynolds number flow.

the remainder of the flow field. Since the shear stress (i.e., viscous effect) is the product of the fluid viscosity and the velocity gradient, it follows that viscous effects are confined to the boundary layer and wake regions.

The characteristics described in Figs. 9.5 and 9.6 for flow past a flat plate and a circular cylinder are typical of flows past streamlined and blunt bodies, respectively. The nature of the flow depends strongly on the Reynolds number. Most familiar flows are similar to the large Reynolds number flows depicted in Figs. 9.5*c* and 9.6*c*, rather than the low Reynolds number flow situations. In the remainder of this chapter we will investigate more thoroughly these ideas and determine how to calculate the forces on immersed bodies.

EXAMPLE 9.2

It is desired to determine the various characteristics of flow past a car. The following tests could be carried out: (a) $U = 20$ mm/s flow of glycerin past a scale model that is 34 mm tall, 100 mm long, and 40 mm wide; (b) $U = 20$ mm/s air flow past the scale model; or (c) $U = 25$ m/s air flow past the actual car, which is 1.7 m tall, 5 m long, and 2 m wide. Would the flow characteristics for these three situations be similar? Explain.

SOLUTION

The characteristics of flow past an object depend on the Reynolds number. For this instance we could pick the characteristic length to be the height, h, width, b, or length, ℓ, of the car to obtain three possible Reynolds numbers, $\mathrm{Re}_h = Uh/\nu$, $\mathrm{Re}_b = Ub/\nu$, and $\mathrm{Re}_\ell = U\ell/\nu$. These numbers will be different because of the different values of h, b, and ℓ. Once we arbitrarily decide on the length we wish to use as the characteristic length, we must stick with it for all calculations when using comparisons between model and prototype.

With the values of kinematic viscosity for air and glycerin obtained from Tables 1.8 and 1.6 as $\nu_{\mathrm{air}} = 1.46 \times 10^{-5}$ m^2/s and $\nu_{\mathrm{glycerin}} = 1.19 \times 10^{-3}$ m^2/s, we obtain the following Reynolds numbers for the flows described.

Reynolds Number	(a) Model in Glycerin	(b) Model in Air	(c) Car in Air
Re_h	0.571	46.6	2.91×10^6
Re_b	0.672	54.8	3.42×10^6
Re_ℓ	1.68	137.0	8.56×10^6

Clearly, the Reynolds numbers for the three flows are quite different (regardless of which characteristic length we choose). Based on the previous discussion concerning flow past a flat plate or flow past a circular cylinder, we would expect that the flow past the actual car would behave in some way similar to the flows shown in Figs. 9.5c or 9.6c. That is, we would expect some type of boundary layer characteristic in which viscous effects would be confined to relatively thin layers near the surface of the car and the wake region behind it. Whether the car would act more like a flat plate or a cylinder would depend on the amount of streamlining incorporated into the car's design.

Because of the small Reynolds number involved, the flow past the model car in glycerin would be dominated by viscous effects, in some way reminiscent of the flows depicted in Figs. 9.5a or 9.6a. Similarly, with the moderate Reynolds number involved for the air flow past the model, a flow with characteristics similar to those indicated in Figs. 9.5b and 9.6b would be expected. Viscous effects would be important—not as important as with the glycerin flow, but more important than with the full sized car.

It would not be a wise decision to expect the flow past the full sized car to be similar to the flow past either of the models. The same conclusions result regardless of whether we use Re_h, Re_b, or Re_ℓ. As is indicated in Chapter 7, the flows past the model car and the full-sized prototype will not be similar unless the Reynolds numbers for the model and prototype are the same. It is not always an easy task to ensure this condition. One (expensive) solution is to test full-sized prototypes in very large wind tunnels (see Fig. 9.1).

9.2   BOUNDARY LAYER CHARACTERISTICS

As was discussed in the previous section, it is often possible to treat flow past an object as a combination of viscous flow in the boundary layer and inviscid flow elsewhere. If the Reynolds number is large enough, viscous effects are important only in the boundary layer regions near the object (and in the wake region behind the object). The boundary layer is needed to allow for the no-slip boundary condition that requires the fluid to cling to any solid surface that it flows past. Outside of the boundary layer the velocity gradients normal to the flow are relatively small, and the fluid acts as if

it were inviscid, even though the viscosity is not zero. A necessary condition for this structure of the flow is that the Reynolds number be large.

9.2.1   Flat Plate Boundary Layer Structure and Thickness

There can be a wide variety in the size of a boundary layer and the structure of the flow within it. Part of this variation is due to the shape of the object on which the boundary layer forms. In this section we consider the simplest situation, one in which the boundary layer is formed on an infinitely long flat plate along which flows a viscous, incompressible fluid as is shown in Fig. 9.7. If the surface were curved (i.e., a circular cylinder or an airfoil), the boundary layer structure would be more complex. Such flows are discussed in Section 9.2.6.

If the Reynolds number is sufficiently large, only the fluid in a relatively thin boundary layer on the plate will feel the effect of the plate. That is, except in the region next to the plate the flow velocity will be essentially $\mathbf{V} = u\hat{\mathbf{i}} + v\hat{\mathbf{j}} = U\hat{\mathbf{i}}$, the upstream velocity. For the infinitely long flat plate extending from $x = 0$ to $x = \infty$, it is not obvious how to define the Reynolds number because there is no characteristic length. The plate has no thickness and is not of finite length!

For a finite length plate, it is clear that the plate length, ℓ, can be used as the characteristic length. For an infinitely long plate we use x, the coordinate distance along the plate from the leading edge, as the characteristic length and define the Reynolds number as $\text{Re}_x = Ux/\nu$. Thus, for any fluid or upstream velocity the Reynolds number will be sufficiently large for boundary layer type flow (i.e., Fig. 9.5c) if the plate is long enough. Physically, this means that the flow situations illustrated in Fig. 9.5 could be thought of as occurring on the same plate, but should be viewed by looking at longer portions of the plate as we step away from the plate to see the flows in Fig. 9.5a, 9.5b, and 9.5c, respectively.

If the plate is sufficiently long, the Reynolds number $\text{Re} = U\ell/\nu$ is sufficiently large so that the flow takes on its boundary layer character (except very near the leading edge). The details of the flow field near the leading edge are lost to our eyes because we are standing so far from the plate that we cannot make out these details. On this scale (Fig. 9.5c) the plate has negligible effect on the fluid ahead of the plate. The presence of the plate is felt only in the relatively thin boundary layer and wake regions. As previously noted, Prandtl in 1904 was the first to hypothesize such a concept. It has become one of the major turning points in fluid mechanics analysis.

A better appreciation of the structure of the boundary layer flow can be obtained by considering what happens to a fluid particle that flows into the boundary layer.

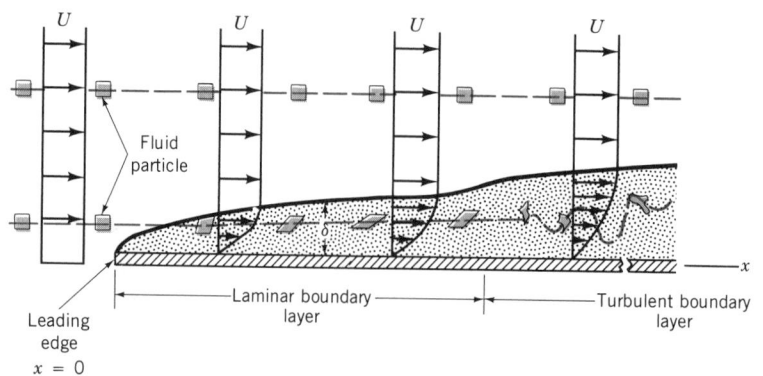

FIGURE 9.7 Distortion of a fluid particle as it flows within the boundary layer.

As is indicated in Fig. 9.7, a small rectangular particle retains its original shape as it flows in the uniform flow outside of the boundary layer. Once it enters the boundary layer, the particle begins to distort because of the velocity gradient within the boundary layer—the top of the particle has a larger speed than its bottom. The fluid particles do not rotate as they flow along outside the boundary layer, but they begin to rotate once they pass through the fictitious boundary layer surface and enter the world of viscous flow. The flow is said to be irrotational outside the boundary layer and rotational within the boundary layer. (In terms of the kinematics of fluid particles as is discussed in Section 6.1, the flow outside the boundary layer has zero vorticity, and the flow within the boundary layer has nonzero vorticity.)

At some distance downstream from the leading edge, the boundary layer flow becomes turbulent and the fluid particles become greatly distorted because of the random, irregular nature of the turbulence. One of the distinguishing features of turbulent flow is the occurrence of irregular mixing of fluid parcels that range in size from the smallest fluid particles up to those comparable in size with the object of interest. For laminar flow, mixing occurs only on the molecular scale. This molecular scale is orders of magnitude smaller in size than typical size scales for turbulent flow mixing. The transition from laminar to turbulent flow occurs at a critical value of the Reynolds number, Re_{xcr}, on the order of 2×10^5 to 3×10^6, depending on the roughness of the surface and the amount of turbulence in the upstream flow, as is discussed in Section 9.2.4.

The purpose of the boundary layer on the plate is to allow the fluid to change its velocity from the upstream value of U to zero on the plate. Thus, $\mathbf{V} = 0$ at $y = 0$ and $\mathbf{V} \approx U\hat{\mathbf{i}}$ at $y = \delta$, with the velocity profile, $u = u(x, y)$ bridging the boundary layer thickness. In actuality (both mathematically and physically), there is no sharp "edge" to the boundary layer. That is, $u \to U$ as we get farther from the plate; it is not precisely $u = U$ at $y = \delta$. We define the *boundary layer thickness*, δ, as that distance from the plate at which the fluid velocity is within some arbitrary value of the upstream velocity. Typically, as indicated in Fig. 9.8a,

$$\delta = y \quad \text{where} \quad u = 0.99U$$

To remove this arbitrariness (i.e., what is so special about 99%; why not 98%?), the following definitions are introduced. Shown in Fig. 9.8b are two velocity profiles for flow past a flat plate—one if there were no viscosity (a uniform profile) and the other if there is viscosity and zero slip at the wall (the boundary layer profile). Because of the velocity deficit, $U - u$, within the boundary layer, the flowrate across section *b–b* is less than that across section *a–a*. However, if we displace the plate at section *a–a* by an appropriate amount δ^*, the *boundary layer displacement thickness*, the

(a)

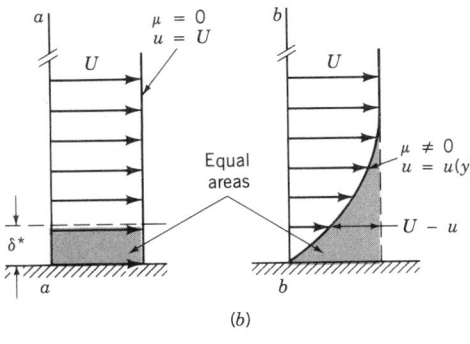

(b)

FIGURE 9.8 Boundary layer thickness: (a) standard boundary layer thickness, (b) boundary layer displacement thickness.

flowrates across each section will be identical. This is true if

$$\delta^* b U = \int_0^\infty (U - u)b \, dy$$

where b is the plate width. Thus,

$$\delta^* = \int_0^\infty \left(1 - \frac{u}{U}\right) dy \tag{9.3}$$

The displacement thickness represents the amount that the thickness of the body must be increased so that the ficticious uniform inviscid flow has the same mass flowrate properties as the actual viscous flow. It represents the outward displacement of the streamlines caused by the viscous effects on the plate. This idea allows us to simulate the presence that the boundary layer has on the flow outside of the boundary layer by adding the displacement thickness to the actual wall and treating the flow over the thickened body as an inviscid flow. The displacement thickness concept is illustrated in Example 9.3.

EXAMPLE 9.3

Air flowing into a 2-ft-square duct with a uniform velocity of 10 ft/s forms a boundary layer on the walls as shown in Fig. E9.3. The fluid within the core region (outside the boundary layers) flows as if it were inviscid. From advanced calculations it is determined that the boundary layer displacement thickness is given by

$$\delta^* = 0.0070(x)^{1/2} \tag{1}$$

where δ^* and x are in feet. Determine the velocity $U = U(x)$ of the air within the duct but outside of the boundary layer.

FIGURE E9.3

SOLUTION

If we assume incompressible flow (a reasonable assumption because of the low velocities involved), it follows that the volume flowrate across any section of the duct is equal to that at the entrance. That is,

$$Q_1 = Q_2$$

or

$$U_1 A_1 = 10 \text{ ft/s } (2 \text{ ft})^2 = 40 \text{ ft}^3/\text{s} = \int_{(2)} u \, dA$$

According to the definition of the displacement thickness, δ^*, the flowrate across section (2)

is the same as that for a uniform flow with velocity U through a duct whose walls have been moved inward by δ^*. That is,

$$40 \text{ ft}^3/\text{s} = \int_{(2)} u \, dA = U(2 \text{ ft} - 2\delta^*)^2 \tag{2}$$

By combining Eqs. 1 and 2 we obtain

$$40 \text{ ft}^3/\text{s} = 4U(1 - 0.0070x^{1/2})^2$$

or

$$U = \frac{10}{(1 - 0.0070x^{1/2})^2} \text{ ft/s} \tag{Ans}$$

Note that U increases in the downstream direction. For example, $U = 11.6$ ft/s at $x = 100$ ft. The viscous effects that cause the fluid to stick to the walls of the duct reduce the effective size of the duct, thereby (from conservation of mass principles) causing the fluid to accelerate. The pressure drop necessary to do this can be obtained by using the Bernoulli equation (Eq. 3.7) along the inviscid streamlines from section (1) to (2). (Recall that this equation is not valid for viscous flows within the boundary layer. It is, however, valid for the inviscid flow outside the boundary layer.) Thus,

$$p_1 + \tfrac{1}{2}\rho U_1^2 = p + \tfrac{1}{2}\rho U^2$$

Hence, with $\rho = 2.38 \times 10^{-3}$ slugs/ft^3 and $p_1 = 0$ we obtain

$$p = \frac{1}{2}\rho(U_1^2 - U^2)$$

$$= \frac{1}{2}(2.38 \times 10^{-3} \text{ slugs ft}^3)\left[(10 \text{ ft/s})^2 - \frac{10^2}{(1 - 0.0070x^{1/2})^4} \text{ ft}^2/\text{s}^2\right]$$

or

$$p = 0.119\left[1 - \frac{1}{(1 - 0.0070x^{1/2})^4}\right] \text{ lb/ft}^2$$

For example, $p = -0.0401$ lb/ft^2 at $x = 100$ ft.

If it were desired to maintain a constant velocity along the centerline of this entrance region of the duct, the walls could be displaced outward by an amount equal to the boundary layer displacement thickness, δ^*.

Another boundary layer thickness definition, the *boundary layer momentum thickness,* Θ, is often used when determining the drag on an object. Again because of the velocity deficit, $U - u$, in the boundary layer, the momentum flux across section b–b in Fig. 9.8 is less than that across section a–a. This deficit in momentum flux for the actual boundary layer flow is given by

$$\int \rho u(U - u) \, dA = \rho b \int_0^\infty u(U - u) \, dy$$

which by definition is the momentum flux in a layer of uniform speed U and thickness Θ. That is,

$$\rho b U^2 \Theta = \rho b \int_0^\infty u(U - u) \, dy$$

or

$$\Theta = \int_0^\infty \frac{u}{U}\left(1 - \frac{u}{U}\right)dy \tag{9.4}$$

All three boundary layer thickness definitions, δ, δ^*, and Θ, are of use in boundary layer analyses.

The boundary layer concept is based on the fact that the boundary layer is thin. For the flat plate flow this means that at any location x along the plate, $\delta \ll x$. Similarly, $\delta^* \ll x$ and $\Theta \ll x$. Again, this is true if we do not get too close to the leading edge of the plate (i.e., not closer than $Re_x = Ux/v = 1000$ or so).

The structure and properties of the boundary layer flow depend on whether the flow is laminar or turbulent. As is illustrated in Fig. 9.9 and discussed in Sections 9.2.2 through 9.2.5, both the boundary layer thickness and the wall shear stress are different in these two regimes.

9.2.2   Prandtl/Blasius Boundary Layer Solution

In theory, the details of viscous incompressible flow past any object can be obtained by solving the governing Navier–Stokes equations discussed in Section 6.8.2. For steady, two-dimensional laminar flows with negligible gravitational effects, these equations (Eqs. 6.127a, b, and c) reduce to the following

$$u\frac{\partial u}{\partial x} + v\frac{\partial u}{\partial y} = -\frac{1}{\rho}\frac{\partial p}{\partial x} + v\left(\frac{\partial^2 u}{\partial x^2} + \frac{\partial^2 u}{\partial y^2}\right) \tag{9.5}$$

$$u\frac{\partial v}{\partial x} + v\frac{\partial v}{\partial y} = -\frac{1}{\rho}\frac{\partial p}{\partial y} + v\left(\frac{\partial^2 v}{\partial x^2} + \frac{\partial^2 v}{\partial y^2}\right) \tag{9.6}$$

which express Newton's second law. In addition, the conservation of mass equation, Eq. 6.31, for incompressible flow is

$$\frac{\partial u}{\partial x} + \frac{\partial v}{\partial y} = 0 \tag{9.7}$$

The appropriate boundary conditions are that the fluid velocity is the upstream velocity

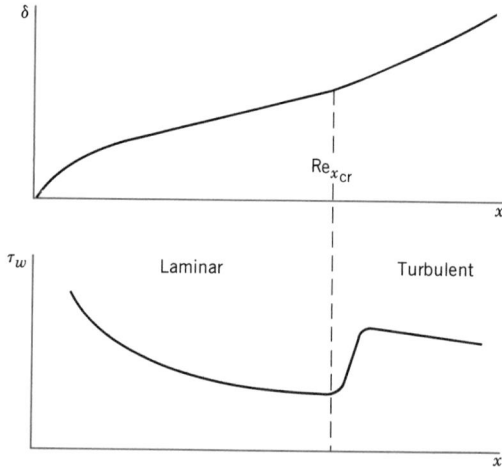

FIGURE 9.9   Typical characteristics of boundary layer thickness and wall shear stress for laminar and turbulent boundary layers.

far from the body and that the fluid sticks to the solid body surfaces. Although the mathematical problem is well posed, no one has obtained an analytical solution to these equations for flow past any shaped body! Currently much work is being done to obtain numerical solutions to these governing equations for many flow geometries.

By using boundary layer concepts introduced in the previous sections, Prandtl was able to impose certain approximations (valid for large Reynolds number flows), and thereby to simplify the governing equations. In 1908, H. Blasius (1883–1970), one of Prandtl's students, was able to solve these simplified equations for the boundary layer flow past a flat plate parallel to the flow. A brief outline of this technique and the results are presented below. Additional details may be found in the literature (Refs. 1, 2, 3).

Since the boundary layer is thin, it is expected that the component of velocity normal to the plate is much smaller than that parallel to the plate and that the rate of change of any parameter across the boundary layer should be much greater than that along the flow direction. That is,

$$v \ll u$$

and

$$\frac{\partial}{\partial x} \ll \frac{\partial}{\partial y}$$

Physically, the flow is primarily parallel to the plate and any fluid property is convected downstream much more quickly than it is diffused across the streamlines.

With these assumptions it can be shown that the governing equations, Eqs. 9.5, 9.6, and 9.7, reduce to the following boundary layer equations

$$\frac{\partial u}{\partial x} + \frac{\partial v}{\partial y} \tag{9.8}$$

$$u \frac{\partial u}{\partial x} + v \frac{\partial u}{\partial y} = v \frac{\partial^2 u}{\partial y^2} \tag{9.9}$$

Although both these boundary layer equations and the original Navier–Stokes equations are nonlinear partial differential equations, there are considerable differences between them. For one, the y momentum equation has been eliminated, leaving only the original, unaltered continuity equation and a modified x momentum equation. One of the variables, the pressure, has been eliminated, leaving only the x and y components of velocity as unknowns. For boundary layer flow over a flat plate the pressure is constant throughout the fluid. The flow represents a balance between viscous and inertial effects, with pressure playing no role.

The boundary conditions for the governing boundary layer equations are that the fluid sticks to the plate

$$u = v = 0 \quad \text{on} \quad y = 0 \tag{9.10}$$

and that outside of the boundary layer the flow is the uniform upstream flow $u = U$. That is,

$$u \to U \quad \text{as} \quad y \to \infty \tag{9.11}$$

Mathematically, the upstream velocity is approached asymptotically as one moves away from the plate. Physically, the flow velocity is within 1% of the upstream velocity at a distance of δ from the plate.

In mathematical terms, the Navier–Stokes equations (Eqs. 9.5, 9.6) and the continuity equation (Eq. 9.7) are elliptic equations, whereas the equations for boundary layer flow (Eqs. 9.8 and 9.9) are parabolic equations. The nature of the solutions to these two sets of equations, therefore, is different. Physically, this fact translates to the idea that what happens downstream of a given location in a boundary layer cannot affect what happens upstream of that point. That is, whether the plate shown in Fig. 9.5c ends with length ℓ or is extended to length 2ℓ, the flow within the first segment of length ℓ will be the same. In addition, the presence of the plate has no effect on the flow ahead of the plate.

In general, the solutions of nonlinear partial differential equations (such as the boundary layer equations, Eqs. 9.8 and 9.9) are extremely difficult to obtain. However, by applying a clever coordinate transformation and change of variables, Blasius reduced the partial differential equations to an ordinary differential equation that he was able to solve. A brief description of this process is given below. Additional details can be found in standard books dealing with boundary layer flow (Refs. 1, 2).

It can be argued that in dimensionless form the boundary layer velocity profiles on a flat plate should be similar regardless of the location along the plate. That is,

$$\frac{u}{U} = g\left(\frac{y}{\delta}\right)$$

where $g(y/\delta)$ is an unknown function to be determined. In addition, by applying an order of magnitude analysis of the forces acting on fluid within the boundary layer, it can be shown that the boundary layer thickness grows as the square root of x and inversely proportional to the square root of U. That is,

$$\delta \sim \left(\frac{\nu x}{U}\right)^{1/2}$$

Such a conclusion results from a balance between viscous and inertial forces within the boundary layer and from the fact that the velocity varies much more rapidly in the direction across the boundary layer than along it.

Thus, we introduce the dimensionless *similarity variable* $\eta = (U/\nu x)^{1/2}y$ and the stream function $\psi = (\nu x U)^{1/2} f(\eta)$, where $f = f(\eta)$ is an unknown function. Recall from Section 6.2.3 that the velocity components for two-dimensional flow are given in terms of the stream function as $u = \partial\psi/\partial y$ and $v = -\partial\psi/\partial x$, which for this flow become

$$u = Uf'(\eta) \tag{9.12}$$

and

$$v = \left(\frac{\nu U}{4x}\right)^{1/2} (\eta f' - f) \tag{9.13}$$

with the notation $(\ )' = d/d\eta$. We substitute Eqs. 9.12 and 9.13 into the governing equations, Eqs. 9.8 and 9.9, to obtain (after considerable manipulation) the following nonlinear, third order ordinary differential equation:

$$2f''' + ff'' = 0 \tag{9.14a}$$

The boundary conditions given in Eqs. 9.10 and 9.11 can be written as

$$f = f' = 0 \text{ at } \eta = 0 \quad \text{and} \quad f' \to 1 \text{ as } \eta \to \infty \tag{9.14b}$$

(a)

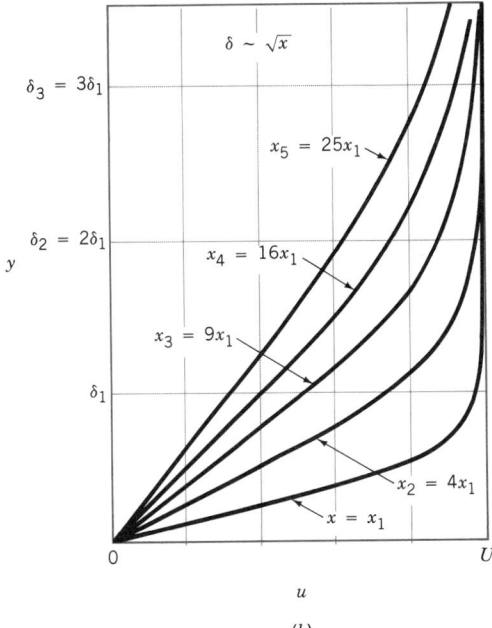

(b)

FIGURE 9.10 Blasius boundary layer profile: (a) boundary layer profile in dimensionless form using the similarity variable η, (b) similar boundary layer profiles at different locations along the flat plate.

The original partial differential equation and boundary conditions have been reduced to an ordinary differential equation by use of the similarity variable η. The two independent variables, x and y, were combined into the similarity variable in a fashion that reduced the partial differential equation (and boundary conditions) to an ordinary differential equation. This type of reduction is not generally possible. For example, this method does not work on the full Navier–Stokes equations, although it does on the boundary layer equations (Eqs. 9.8 and 9.9).

The dimensionless boundary layer profile, $u/U = f'(\eta)$, obtained by numerical solution of Eq. 9.14 (termed the Blasius solution) is sketched in Fig. 9.10a and is tabulated in Table 9.1. The velocity profiles at different x locations are similar in that

TABLE 9.1   Laminar Flow Along a Flat
Plate (the Blasius Solution)

$\eta = y(U/vx)^{1/2}$	$f'(\eta) = u/U$
0	0
0.4	0.1328
0.8	0.2647
1.2	0.3938
1.6	0.5168
2.0	0.6298
2.4	0.7290
2.8	0.8115
3.2	0.8761
3.6	0.9233
4.0	0.9555
4.4	0.9759
4.8	0.9878
5.0	0.9916
5.2	0.9943
5.6	0.9975
6.0	0.9990
∞	1.0000

there is only one curve necessary to describe the velocity at any point in the boundary layer. Because the similarity variable η contains both x and y, it is seen from Fig. 9.10b that the actual velocity profiles are a function of both x and y. The profile at location x_1 is the same as that at x_2 except that the y coordinate is stretched by a factor of $(x_2/x_1)^{1/2}$.

From the solution it is found that $u/U \approx 0.99$ when $\eta = 5.0$. Thus,

$$\delta = 5 \sqrt{\frac{vx}{U}} \tag{9.15}$$

or

$$\frac{\delta}{x} = \frac{5}{\sqrt{\mathrm{Re}_x}}$$

where $\mathrm{Re}_x = Ux/v$. It can also be shown that the displacement and momentum thicknesses are given by

$$\frac{\delta^*}{x} = \frac{1.721}{\sqrt{\mathrm{Re}_x}} \tag{9.16}$$

and

$$\frac{\Theta}{x} = \frac{0.664}{\sqrt{\mathrm{Re}_x}} \tag{9.17}$$

As postulated, the boundary layer is thin provided that Re_x is large (i.e., $\delta/x \rightarrow 0$ as $\mathrm{Re}_x \rightarrow \infty$).

With the velocity profile known it is an easy matter to determine the wall shear stress, $\tau_w = \mu(\partial u/\partial y)_{y=0}$, where the velocity gradient is evaluated at the plate. The

value of $\partial u / \partial y$ at $y = 0$ can be obtained from the Blasius solution to give

$$\tau_w = 0.332 U^{3/2} \sqrt{\frac{\rho \mu}{x}} \tag{9.18}$$

Note that the shear stress decreases with increasing x because of the increasing thickness of the boundary layer—the velocity gradient at the wall decreases with increasing x. Also, τ_w varies as $U^{3/2}$, not as U as it does for fully developed laminar pipe flow. These variations are discussed in Section 9.2.3.

9.2.3 Flat Plate Momentum Integral Boundary Layer Equation

One of the important aspects of boundary layer theory is the determination of the drag caused by shear forces on a body. As was discussed in the previous section, such results can be obtained from the governing differential equations for laminar boundary layer flow. Since these solutions are extremely difficult to obtain, it is of interest to have an alternative approximate method. The momentum integral method described in this section provides such an alternative.

We consider the uniform flow past a flat plate and the fixed control volume as shown in Fig. 9.11. In agreement with advanced theory and experiment, we assume that the pressure is constant throughout the flow field. The flow entering the control volume at the leading edge of the plate [section (1)] is uniform, while the velocity of the flow exiting the control volume [section (2)] varies from the upstream velocity at the edge of the boundary layer to zero velocity on the plate.

The fluid adjacent to the plate makes up the lower portion of the control surface. The upper surface coincides with the streamline just outside the edge of the boundary layer at section (2). It need not (in fact, does not) coincide with the edge of the boundary layer except at section (2). If we apply the x component of the momentum equation (Eq. 5.22) to the steady flow of fluid within this control volume we obtain

$$\sum F_x = \rho \int_{(1)} u \mathbf{V} \cdot \hat{\mathbf{n}} \, dA + \rho \int_{(2)} u \mathbf{V} \cdot \hat{\mathbf{n}} \, dA$$

where for a plate of width b

$$\sum F_x = -\mathcal{D} = -\int_{\text{plate}} \tau_w \, dA = -b \int_{\text{plate}} \tau_w \, dx \tag{9.19}$$

and $\mathcal{D}$ is the drag that the plate exerts on the fluid. Note that the net force caused by the uniform pressure distribution does not contribute to this flow. Since the plate is solid and the upper section of the control volume is a streamline, there is no flow

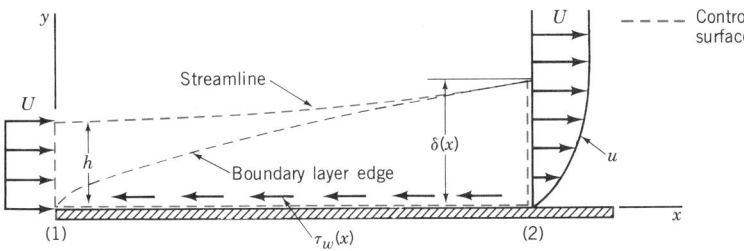

FIGURE 9.11 Control volume used in the derivation of the momentum integral equation for boundary layer flow.

through these areas. Thus,

$$- \mathcal{D} = \rho \int_{(1)} U(-U) \, dA + \rho \int_{(2)} u^2 \, dA$$

or

$$\mathcal{D} = \rho U^2 bh - \rho b \int_0^\delta u^2 \, dy \tag{9.20}$$

Although the height h is not known, it is known that for conservation of mass the flowrate through section (1) must equal that through section (2), or

$$Uh = \int_0^\delta u \, dy$$

which can be written as

$$\rho U^2 bh = \rho b \int_0^\delta Uu \, dy \tag{9.21}$$

Thus, by combining Eqs. 9.20 and 9.21 we obtain the drag in terms of the deficit of momentum flux across the outlet of the control volume as

$$\mathcal{D} = \rho b \int_0^\delta u(U - u) \, dy \tag{9.22}$$

If the flow were inviscid, the drag would be zero, since we would have $u \equiv U$ and the right-hand side of Eq. 9.22 would be zero. (This is consistent with the fact that $\tau_w = 0$ if $\mu = 0$.) Equation 9.22 points out the important fact that boundary layer flow on a flat plate is governed by a balance between shear drag (the left-hand side of Eq. 9.22) and a decrease in the momentum of the fluid (the right-hand side of Eq. 9.22). As x increases, δ increases and the drag increases. The thickening of the boundary layer is necessary to overcome the drag of the viscous shear stress on the plate. This is contrary to horizontal fully developed pipe flow in which the momentum of the fluid remains constant and the shear force is overcome by the pressure gradient along the pipe.

The development of Eq. 9.22 and its use was first put forth in 1921 by T. von Karman (1881–1963), a Hungarian/German aerodynamicist. By comparing Eqs. 9.22 and 9.4 we see that the drag can be written in terms of the momentum thickness, Θ, as

$$\mathcal{D} = \rho b U^2 \, \Theta \tag{9.23}$$

Note that this equation is valid for laminar or turbulent flows.

The shear stress distribution can be obtained from Eq. 9.23 by differentiating both sides with respect to x to obtain

$$\frac{d\mathcal{D}}{dx} = \rho b U^2 \frac{d\Theta}{dx} \tag{9.24}$$

The increase in drag per length of the plate, $d\mathcal{D}/dx$, occurs at the expense of an increase of the momentum boundary layer thickness, which represents a decrease in the momentum of the fluid.

Since $d\mathcal{D} = \tau_w\, b\, dx$ (see Eq. 9.19) it follows that

$$\frac{d\mathcal{D}}{dx} = b\tau_w \tag{9.25}$$

Hence, by combining Eqs. 9.24 and 9.25 we obtain the *momentum integral equation* for the boundary layer flow on a flat plate

$$\tau_w = \rho U^2 \frac{d\Theta}{dx} \tag{9.26}$$

The usefulness of this relationship lies in the ability to obtain approximate boundary layer results easily by using rather crude assumptions. For example, if we knew the detailed velocity profile in the boundary layer (i.e., the Blasius solution discussed in the previous section), we could evaluate either the right-hand side of Eq. 9.23 to obtain the drag, or the right-hand side of Eq. 9.26 to obtain the shear stress. Fortunately, even a rather crude guess at the velocity profile will allow us to obtain reasonable drag and shear stress results from Eq. 9.26. This method is illustrated in Example 9.4.

EXAMPLE 9.4

Consider the laminar flow of an incompressible fluid past a flat plate. The boundary layer velocity profile is approximated as $u = Uy/\delta$ for $0 \le y \le \delta$ and $u = U$ for $y > \delta$, as is shown in Fig. E9.4. Determine the shear stress by using the momentum integral equation. Compare these results with the Blasius results given by Eq. 9.18.

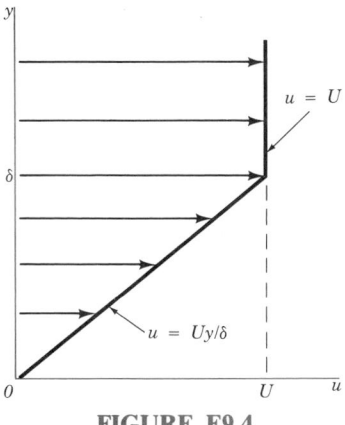

FIGURE E9.4

SOLUTION

From Eq. 9.26 the shear stress is given by

$$\tau_w = \rho U^2 \frac{d\Theta}{dx} \tag{1}$$

while for laminar flow we know that $\tau_w = \mu(\partial u/\partial y)_{y=0}$. For the assumed profile we have

$$\tau_w = \mu \frac{U}{\delta} \tag{2}$$

and from Eq. 9.4

$$\Theta = \int_0^\infty \frac{u}{U}\left(1 - \frac{u}{U}\right) dy = \int_0^\delta \frac{u}{U}\left(1 - \frac{u}{U}\right) dy = \int_0^\delta \left(\frac{y}{\delta}\right)\left(1 - \frac{y}{\delta}\right) dy$$

or

$$\Theta = \frac{\delta}{6} \tag{3}$$

Note that as yet we do not know the value of δ (we suspect that it should be a function of x). By combining Eqs. 1, 2, and 3 we obtain the following differential equation for δ.

$$\frac{\mu U}{\delta} = \frac{\rho U^2}{6}\frac{d\delta}{dx}$$

or

$$\delta\, d\delta = \frac{6\mu}{\rho U}\, dx$$

This can be integrated from the leading edge of the plate, $x = 0$, where $\delta = 0$ to an arbitrary location x where the boundary layer thickness is δ. The result is

$$\frac{\delta^2}{2} = \frac{6\mu}{\rho U}\, x$$

or

$$\delta = 3.46\sqrt{\frac{\nu x}{U}} \tag{4}$$

Note that this approximate result (i.e., the velocity profile is not actually the simple straight line we assumed) compares favorably with the (much more laborious to obtain) Blasius result given by Eq. 9.15.

The wall shear stress can also be obtained by combining Eqs. 1, 3, and 4 to give

$$\tau_w = 0.289\, U^{3/2}\sqrt{\frac{\rho\mu}{x}} \tag{Ans}$$

Again this approximate result is close (within 13%) to the Blasius value of τ_w given by Eq. 9.18.

As is illustrated in Example 9.4, the momentum integral equation, Eq. 9.26, can be used along with an assumed velocity profile to obtain reasonable approximate boundary layer results. The accuracy of these results depends on how closely the shape of the assumed velocity profile approximates the actual profile.

Thus, we consider a general velocity profile

$$\frac{u}{U} = g(Y) \quad \text{for} \quad 0 \le Y \le 1$$

and

$$\frac{u}{U} = 1 \quad \text{for} \quad Y > 1$$

where the dimensionless coordinate $Y = y/\delta$ varies from 0 to 1 across the boundary layer. The dimensionless function $g(Y)$ can be any shape we chose, although it should be a reasonable approximation to the boundary layer profile. In particular, it should

certainly satisfy the boundary conditions $u = 0$ at $y = 0$ and $u = U$ at $y = \delta$. That is,

$$g(0) = 0 \quad \text{and} \quad g(1) = 1$$

The linear function $g(Y) = Y$ used in Example 9.4 is one such possible profile. Other conditions, such as $dg/dY = 0$ at $Y = 1$ (i.e., $\partial u/\partial y = 0$ at $y = \delta$), could also be incorporated into the function $g(Y)$ to more closely approximate the actual profile.

For a given $g(Y)$, the drag can be determined from Eq. 9.22 as

$$\mathcal{D} = \rho b \int_0^\delta u(U - u)\, dy = \rho b U^2 \delta \int_0^1 g(Y)[1 - g(Y)]\, dY$$

or

$$\mathcal{D} = \rho b U^2 \delta C_1 \tag{9.27}$$

where the dimensionless constant C_1 has the value

$$C_1 = \int_0^1 g(Y)[1 - g(Y)]\, dY$$

Also, the wall shear stress can be written as

$$\tau_w = \mu \left.\frac{\partial u}{\partial y}\right|_{y=0} = \frac{\mu U}{\delta} \left.\frac{dg}{dY}\right|_{Y=0}$$

or

$$\tau_w = \frac{\mu U}{\delta} C_2 \tag{9.28}$$

where the dimensionless constant C_2 has the value

$$C_2 = \left.\frac{dg}{dY}\right|_{Y=0}$$

By combining Eqs. 9.25, 9.27 and 9.28 we obtain

$$\delta\, d\delta = \frac{\mu C_2}{\rho U C_1}\, dx$$

which can be integrated from $\delta = 0$ at $x = 0$ to give

$$\delta = \sqrt{\frac{2\nu C_2 x}{U C_1}}$$

or

$$\frac{\delta}{x} = \frac{\sqrt{2C_2/C_1}}{\sqrt{\text{Re}_x}} \tag{9.29}$$

By substituting this expression back into Eqs. 9.28 we obtain

$$\tau_w = \sqrt{\frac{C_1 C_2}{2}}\, U^{3/2} \sqrt{\frac{\rho \mu}{x}} \tag{9.30}$$

To use Eqs. 9.29 and 9.30 we must determine the values of C_1 and C_2. Several assumed velocity profiles and the resulting values of δ are given in Fig. 9.12 and Table

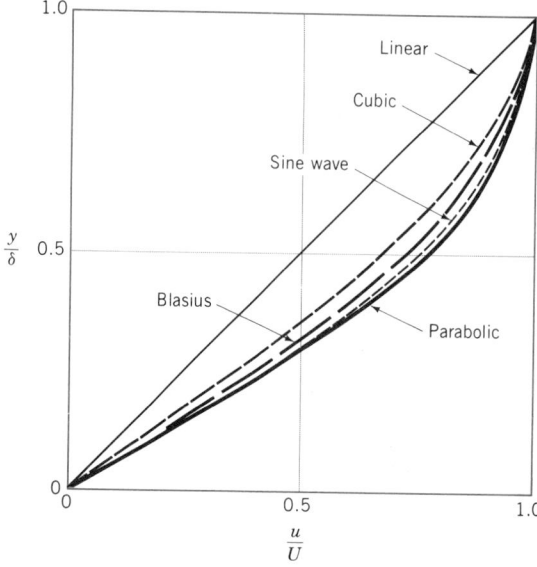

FIGURE 9.12 Typical approximate boundary layer profiles used in the momentum integral equation.

9.2. The more closely the assumed shape approximates the actual (i.e., Blasius) profile, the more accurate the final results. For any assumed profile shape, the functional dependence of δ and τ_w on the physical parameters ρ, μ, U, and x are the same. Only the constants are different. That is, $\delta \sim (\mu x/\rho U)^{1/2}$ or $\delta \mathrm{Re}_x^{1/2}/x = $ constant, and $\tau_w \sim (\rho \mu U^3/x)^{1/2}$, where $\mathrm{Re}_x = \rho U x/\mu$.

It is often convenient to use the dimensionless *local friction coefficient, c_f*, defined as

$$c_f = \frac{\tau_w}{\frac{1}{2}\rho U^2} \tag{9.31}$$

to express the wall shear stress. From Eq. 9.30 we obtain the approximate value

$$c_f = \sqrt{2C_1 C_2}\,\sqrt{\frac{\mu}{\rho U x}}$$

TABLE 9.2 Flat Plate Momentum Integral Results for Various Assumed Laminar Flow Velocity Profiles

Profile Character	$\delta\,\mathrm{Re}_x^{1/2}/x$	$c_f\mathrm{Re}_x^{1/2}$	$C_{Df}\mathrm{Re}_\ell^{1/2}$
a. Blasius solution	5.00	0.664	1.328
b. Linear $u/U = y/\delta$	3.46	0.578	1.156
c. Parabolic $u/U = 2y/\delta - (y/\delta)^2$	5.48	0.730	1.460
d. Cubic $u/U = 3(y/\delta)/2 - (y/\delta)^3/2$	4.64	0.646	1.292
e. Sine wave $u/U = \sin[\pi(y/\delta)/2]$	4.79	0.655	1.310

or

$$c_f = \frac{\sqrt{2C_1C_2}}{\sqrt{Re_x}}$$

while Blasius solution result is given by

$$c_f = \frac{0.644}{\sqrt{Re_x}} \qquad (9.32)$$

These results are also indicated in Table 9.2.

For a flat plate of length ℓ and width b, the net friction drag, $\mathcal{D}_f$, can be expressed in terms of the *friction drag coefficient, C_{Df},* as

$$C_{Df} = \frac{\mathcal{D}_f}{\frac{1}{2}\rho U^2 b\ell} = \frac{b\int_0^\ell \tau_w \, dx}{\frac{1}{2}\rho U^2 b\ell}$$

or

$$C_{Df} = \frac{1}{\ell}\int_0^\ell c_f \, dx \qquad (9.33)$$

We use the above approximate value of $c_f = (2C_1C_2\mu/\rho Ux)^{1/2}$ to obtain

$$C_{Df} = \frac{\sqrt{8C_1C_2}}{\sqrt{Re_\ell}}$$

where $Re_\ell = U\ell/\nu$ is the Reynolds number based on the plate length. The corresponding value obtained from the Blasius solution (Eq. 9.32) gives

$$C_{Df} = \frac{1.328}{\sqrt{Re_\ell}}$$

These results are also indicated in Table 9.2.

The momentum integral boundary layer method provides a relatively simple technique to obtain useful boundary layer results. As is discussed in Sections 9.2.5 and 9.2.6, this technique can be extended to boundary layer flows on curved surfaces (where the pressure and fluid velocity at the edge of the boundary layer are not constant) and to turbulent flows.

9.2.4 Transition from Laminar to Turbulent Flow

The analytical results given in Table 9.2 are restricted to laminar boundary layer flows along a flat plate with zero pressure gradient. They agree quite well with experimental results up to the point where the boundary layer flow becomes turbulent, which will occur for any free-stream velocity and any fluid provided the plate is long enough. This is true because the parameter that governs the transition to turbulent flow is the Reynolds number—in this case the Reynolds number based on the distance from the leading edge of the plate $Re_x = Ux/\nu$.

The value of the Reynolds number at the transition location is a rather complex function of various parameters involved, including the roughness of the surface, the curvature of the surface (e.g., a flat plate or a sphere), and some measure of the disturbances in the flow outside the boundary layer. On a flat plate with a sharp leading edge in a typical air stream, the transition takes place at a distance x from the lead-

FIGURE 9.13 Turbulent spots and the transition from laminar to turbulent boundary layer flow on a flat plate. (Photograph courtesy of B. Cantwell, Stanford University.)

ing edge given by $Re_{xcr} = 2 \times 10^5$ to 3×10^6. Unless otherwise stated, we will use $Re_{xcr} = 5 \times 10^5$ in our calculations.

The actual transition from laminar to turbulent boundary layer flow may occur over a region of the plate, not at a specific single location. This occurs, in part, because of the spottiness of the transition. Typically, the transition begins at random locations on the plate in the vicinity of $Re_x = Re_{xcr}$. These spots grow rapidly as they are convected downstream until the entire width of the plate is covered with turbulent flow. The photo shown in Fig. 9.13 illustrates this transition process.

FIGURE 9.14 Typical boundary layer profiles on a flat plate for laminar, transitional, and turbulent flow (Ref. 1).

The complex process of transition from laminar to turbulent flow involves the instability of the flow field. Small disturbances imposed on the boundary layer flow (i.e., from a vibration of the plate, a roughness of the surface, or a "wiggle" in the flow past the plate) will either grow (instability) or decay (stability), depending on where the disturbance is introduced into the flow. If these disturbances occur at a location with $Re_x < Re_{xcr}$ they will die out, and the boundary layer will return to laminar flow at that location. Disturbances imposed at a location with $Re_x > Re_{xcr}$ will grow and transform the boundary layer flow downstream of this location into turbulence. The study of the initiation, growth, and structure of these turbulent bursts or spots is an active area of fluid mechanics research.

Transition from laminar to turbulent flow also involves a noticeable change in the shape of the boundary layer velocity profile. Typical profiles obtained in the neighborhood of the transition location are indicated in Fig. 9.14. The turbulent profiles are flatter, have a larger velocity gradient at the wall, and produce a larger boundary layer thickness than do the laminar profiles.

EXAMPLE 9.5

A fluid flows steadily past a flat plate with a velocity of $U = 10$ ft/s. At approximately what location will the boundary layer become turbulent and how thick is the boundary layer at that point if the fluid is (a) water at 60 °F, (b) standard air, or (c) glycerin at 68 °F? Assume $Re_{xcr} = 5 \times 10^5$.

SOLUTION

For any fluid, the laminar boundary layer thickness is found from Eq. 9.15 as

$$\delta = 5\sqrt{\frac{\nu x}{U}}$$

The boundary layer remains laminar up to

$$x_{cr} = \frac{\nu Re_{xcr}}{U}$$

Thus, with $Re_{xcr} = 5 \times 10^5$ we obtain

$$x_{cr} = \frac{5 \times 10^5}{10 \text{ ft/s}} \nu = 5 \times 10^4 \nu$$

and

$$\delta_{cr} \equiv \delta \,|_{x=x_{cr}} = 5\left[\frac{\nu}{10}\,(5 \times 10^4 \,\nu)\right]^{1/2} = 354\,\nu$$

where ν is in ft²/s and x_{cr} and δ_{cr} are in feet. The values of the kinematic viscosity obtained from Tables 1.5 and 1.7 are listed in Table E9.5 along with the corresponding x_{cr} and δ_{cr}.

TABLE E9.5

Fluid	ν (ft²/s)	x_{cr} (ft)	δ_{cr} (ft)
a. Water	1.21×10^{-5}	0.605	0.00428
b. Air	1.57×10^{-4}	7.85	0.0556
c. Glycerin	1.28×10^{-2}	640.0	4.53

(Ans)

Laminar flow can be maintained on a longer portion of the plate if the viscosity is increased. However, the boundary layer flow eventually becomes turbulent, provided the plate is long enough. Similarly, the boundary layer thickness is greater if the viscosity is increased.

9.2.5 Turbulent Boundary Layer Flow

The structure of turbulent boundary layer flow is very complex, random, and irregular. It shares many of the characteristics described for turbulent pipe flow in Section 8.3. In particular, the velocity at any given location in the flow is unsteady in a random fashion. The flow can be thought of as a jumbled mix of intertwined eddies (or swirls) of different sizes (diameters and angular velocities). The various fluid quantities involved (i.e., mass, momentum, energy) are convected downstream in the free-stream direction as in a laminar boundary layer. For turbulent flow they are also convected across the boundary layer (in the direction perpendicular to the plate) by the random transport of finite-sized fluid particles associated with the turbulent eddies. There is considerable mixing involved with these finite sized eddies—considerably more than is associated with the mixing found in laminar flow where it is confined to the molecular scale. Although there is considerable random motion of fluid particles perpendicular to the plate, there is very little net transfer of mass across the boundary layer—the largest flowrate by far is parallel to the plate.

There is, however, a considerable net transfer of x component of momentum perpendicular to the plate because of the random motion of the particles. Fluid particles moving toward the plate (in the negative y direction) have some of their excess momentum (they come from areas of higher velocity) removed by the plate. Conversely, particles moving away from the plate (in the positive y direction) gain momentum from the fluid (they come from areas of lower velocity). The net result is that the plate acts as a momentum sink, continually extracting momentum from the fluid. For laminar flows, such cross-stream transfer of these properties takes place solely on the molecular scale. For turbulent flow the randomness is associated with fluid particle mixing. Consequently, the shear force for turbulent boundary layer flow is considerably greater than it is for laminar boundary layer flow.

There are no "exact" solutions for turbulent boundary layer flow. As is discussed in Section 9.2.2, it is possible to solve the Prandtl boundary layer equations for laminar flow past a flat plate to obtain the Blasius solution (which is "exact" within the framework of the assumptions involved in the boundary layer equations). Since there is no precise expression for the shear stress in turbulent flow (see Section 8.3), solutions are not available for turbulent flow. However, considerable headway has been made in obtaining numerical (computer) solutions for turbulent flow by using approximate shear stress relationships.

Approximate turbulent boundary layer results can also be obtained by use of the momentum integral equation, Eq. 9.26, which is valid for either laminar or turbulent flow. What is needed for the use of this equation are reasonable approximations to the velocity profile $u = U g(Y)$, where $Y = y/\delta$ and u is the time averaged velocity (the overbar notation, $\bar{u}$, of Section 8.3.2 has been dropped for convenience) and a functional relationship describing the wall shear stress. For laminar flow the wall shear stress was used as $\tau_w = \mu(\partial u/\partial y)_{y=0}$. In theory, such a technique should work for turbulent boundary layers also. However, as is discussed in Section 8.3, the details of the velocity gradient at the wall are not well understood for turbulent flow. Thus, it is necessary to use some empirical relationship for the wall shear stress. This is illustrated in Example 9.6.

EXAMPLE 9.6

Consider turbulent flow of an incompressible fluid past a flat plate. The boundary layer velocity profile is assumed to be $u/U = (y/\delta)^{1/7} = Y^{1/7}$ for $Y = y/\delta \le 1$ and $u = U$ for $Y > 1$ as shown in Fig. E9.6. This is a reasonable approximation of experimentally observed profiles, except very near the plate where this formula gives $\partial u/\partial y = \infty$ at $y = 0$. Note the differences between the assumed turbulent profile and the laminar profile. Also assume that the shear stress agrees with the experimentally determined formula:

$$\tau_w = 0.0225\rho U^2 \left(\frac{\nu}{U\delta}\right)^{1/4} \tag{1}$$

Determine the boundary layer thicknesses δ, δ^*, and Θ and the wall shear stress, τ_w, as a function of x. Determine the friction drag coefficient, C_{Df}.

FIGURE E9.6

SOLUTION

Whether the flow is laminar or turbulent, it is true that the drag force is accounted for by a reduction in the momentum of the fluid flowing past the plate. The shear stress is obtained from Eq. 9.26 in terms of the rate at which the momentum boundary layer thickness, Θ, increases with distance along the plate as

$$\tau_w = \rho U^2 \frac{d\Theta}{dx}$$

For the assumed velocity profile, the boundary layer momentum thickness is obtained from Eq. 9.4 as

$$\Theta = \int_0^\infty \frac{u}{U}\left(1 - \frac{u}{U}\right) dy = \delta \int_0^1 \frac{u}{U}\left(1 - \frac{u}{U}\right) dY$$

or

$$\Theta = \delta \int_0^1 Y^{1/7}(1 - Y^{1/7}) \, dY$$

where δ is an unknown function of x. This can be integrated to give

$$\Theta = \frac{7}{72}\delta \tag{2}$$

By combining the assumed shear force dependence, Eq. 1, with Eq. 2, we obtain the following differential equation for δ:

$$0.0225\rho U^2 \left(\frac{\nu}{U\delta}\right)^{1/4} = \frac{7}{72}\rho U^2 \frac{d\delta}{dx}$$

or

$$\delta^{1/4}\,d\delta = 0.231 \left(\frac{\nu}{U}\right)^{1/4} dx$$

This can be integrated from $\delta = 0$ at $x = 0$ to obtain

$$\delta = 0.370 \left(\frac{\nu}{U}\right)^{1/5} x^{4/5} \qquad\qquad (3) \quad \text{(Ans)}$$

or in dimensionless form

$$\frac{\delta}{x} = \frac{0.370}{\text{Re}_x^{1/5}}$$

Strictly speaking, the boundary layer near the leading edge of the plate is laminar, not turbulent, and the precise boundary condition should be the matching of the initial turbulent boundary layer thickness (at the transition location) with the thickness of the laminar boundary layer at that point. In practice, however, the laminar boundary layer often exists over a relatively short portion of the plate, and the error associated with starting the turbulent boundary layer with $\delta = 0$ at $x = 0$ can be negligible.

The displacement thickness, δ^*, and the momentum thickness, Θ, can be obtained from Eqs. 9.3 and 9.4 by integrating as follows.

$$\delta^* = \int_0^\infty \left(1 - \frac{u}{U}\right)dy = \delta \int_0^1 \left(1 - \frac{u}{U}\right)dY$$
$$= \delta \int_0^1 (1 - Y^{1/7})\,dY$$

or

$$\delta^* = \frac{\delta}{8}$$

Thus, by combining this with Eq. 3 we obtain

$$\delta^* = 0.0463 \left(\frac{\nu}{U}\right)^{1/5} x^{4/5} \qquad\qquad \text{(Ans)}$$

Similarly, from Eq. 2,

$$\Theta = \tfrac{7}{72}\delta$$

or

$$\Theta = 0.0360 \left(\frac{\nu}{U}\right)^{1/5} x^{4/5} \qquad\qquad (4) \quad \text{(Ans)}$$

The functional dependence for δ, δ^*, and Θ are the same, only the constants of proportionality are different. Typically, $\Theta < \delta^* < \delta$.

By combining Eqs. 1 and 3, we obtain the following result for the wall shear stress

$$\tau_w = 0.0225\rho U^2 \left[\frac{\nu}{U(0.370)(\nu/U)^{1/5}x^{4/5}}\right]^{1/4}$$

or

$$\tau_w = \frac{0.0288\rho U^2}{\text{Re}_x^{1/5}} \qquad\qquad \text{(Ans)}$$

This can be integrated over the length of the plate to obtain the friction drag on one side of the plate, $\mathcal{D}_f$, as

$$\mathcal{D}_f = \int_0^\ell b\tau_w \, dx = b(0.0288\rho U^2) \int_0^\ell \left(\frac{\nu}{Ux}\right)^{1/5} dx$$

or

$$\mathcal{D}_f = 0.0360\rho U^2 \frac{A}{\mathrm{Re}_\ell^{1/5}}$$

where $A = b\ell$ is the area of the plate. (This result can also be obtained by combining Eq. 9.23 and the expression for the momentum thickness given in Eq. 4.) The corresponding friction drag coefficient, C_{Df}, is

$$C_{Df} = \frac{\mathcal{D}_f}{\frac{1}{2}\rho U^2 A} = \frac{0.0720}{\mathrm{Re}_\ell^{1/5}} \qquad\qquad \text{(Ans)}$$

Note that for the turbulent boundary layer flow the boundary layer thickness increases as $\delta \sim x^{4/5}$ and the shear stress as $\tau_w \sim x^{-1/5}$. For laminar flow these dependencies are $x^{1/2}$ and $x^{-1/2}$, respectively. The random character of the turbulent flow causes a different structure of the flow.

Obviously the results presented in this example are valid only in the range of validity of the original data—the assumed velocity profile and shear stress. This range covers smooth flat plates with $5 \times 10^5 < \mathrm{Re}_\ell < 10^7$.

As is indicated in Example 9.6, turbulent boundary layer flow results can be obtained by using the momentum integral equation in conjunction with reasonable approximations of the velocity profile and wall shear stress. The power-law velocity profile used, $u/U = (y/\delta)^{1/n}$, is a reasonable one for such cases. As is discussed in Section 8.3 for pipe flow, it is found that the value of n is a slight function of Reynolds number. The corresponding wall shear stress relationship is of the form $\tau_w = a_1\rho U^2(\nu/U\delta)^{1/m}$, where a_1 is a constant and m is an integer dependent on the value of n. The results shown in Table 9.3 are obtained by using these relationships with the momentum integral equation as done in Example 9.6. The Reynolds number range for each situation is indicated.

TABLE 9.3 Momentum Integral Results for Various Assumed Velocity Profiles and Wall Shear Stresses (Smooth Plate, Turbulent Flow)

Profile	Wall Shear Stress	Boundary Layer Thickness	Friction Drag Coefficient	Re_x Range
$\dfrac{u}{U} = \left(\dfrac{y}{\delta}\right)^{1/n}$	$\tau_w = a_1\rho U^2\left(\dfrac{\nu}{U\delta}\right)^{1/m}$	$\delta = \dfrac{a_2 x}{\mathrm{Re}_x^{1/(m+1)}}$	$C_{Df} = \dfrac{a_3}{\mathrm{Re}_\ell^{1/(m+1)}}$	
$n = 7$	$a_1 = 0.0225$ $m = 4$	$a_2 = 0.370$	$a_3 = 0.0720$	5×10^5 to 1×10^7
$n = 8$	$a_1 = 0.0142$ $m = 5$	$a_2 = 0.252$	$a_3 = 0.0450$	1.8×10^5 to 4.5×10^7
$n = 10$	$a_1 = 0.0100$ $m = 6$	$a_2 = 0.201$	$a_3 = 0.0305$	2.9×10^7 to 5×10^8

As is indicated by the above considerations, it can become quite confusing to determine which formula to use for the drag coefficient for a flat plate. This is especially true when the effect of the plate roughness is taken into account. In general, it is easier to determine C_{Df} from a graph that covers a wide range of the parameters of interest, as is shown in Fig. 9.15. For laminar boundary layer flow the drag coefficient is a function of only the Reynolds number—surface roughness is not important. This is similar to laminar flow in a pipe. However, for turbulent flow, the surface roughness does affect the shear stress and, hence, the drag coefficient. This is similar to turbulent pipe flow in which the surface roughness may protrude into or through the viscous sublayer next to the wall and alter the flow in this thin, but very important, layer. Values of the roughness, ε, for different materials can be obtained from Table 8.2. The value of the drag coefficient in the transition region can be variable, depending on the particular situation involved.

The drag coefficient diagram of Fig. 9.15 (boundary layer flow) shares many characteristics in common with the familiar Moody diagram (pipe flow) of Fig. 8.23, even though the mechanisms governing the flow are quite different. Fully developed horizontal pipe flow is governed by a balance between pressure forces and viscous forces. The fluid inertia remains constant throughout the flow. Boundary layer flow on a horizontal flat plate is governed by a balance between inertia effects and viscous forces. The pressure remains constant throughout the flow. (As is discussed in Section 9.2.6, for boundary layer flow on curved surfaces, the pressure is not constant.)

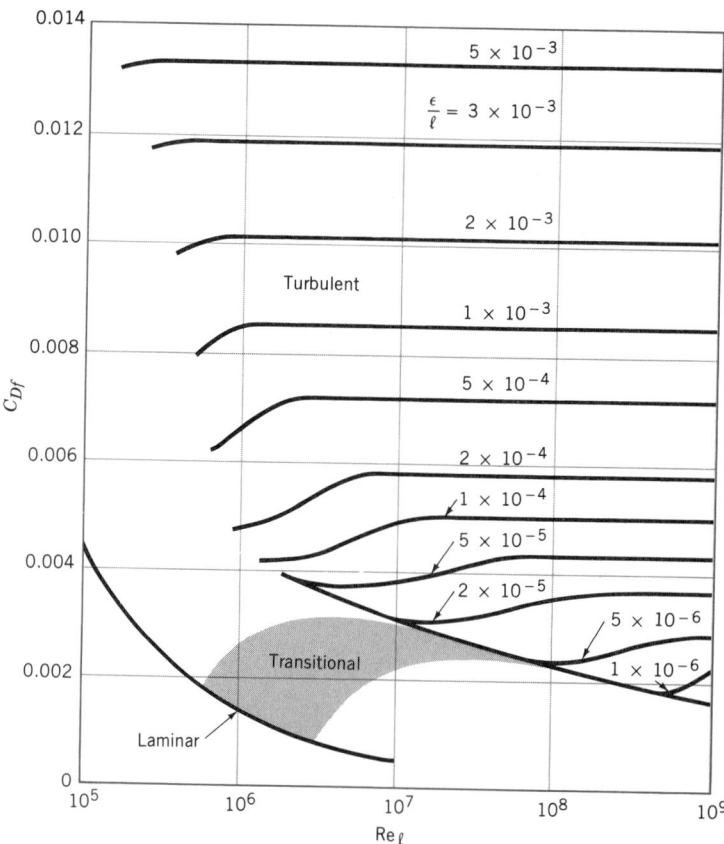

FIGURE 9.15 Friction drag coefficient for a flat plate parallel to the upstream flow (Ref. 18, with permission).

EXAMPLE 9.7

The water ski shown in Fig. E9.7a moves through 70 °F water with a velocity U. Estimate the drag caused by the shear stress on the bottom of the ski for $0 < U < 30$ ft/s.

(a)

(b)

FIGURE E9.7

SOLUTION

Clearly the ski is not a flat plate, and it is not aligned exactly parallel to the upstream flow. However, we can obtain a reasonable approximation to the shear force by using the flat plate results. That is, the friction drag, $\mathcal{D}_f$, caused by the shear stress on the bottom of the ski (the wall shear stress) can be determined as

$$\mathcal{D}_f = \tfrac{1}{2}\rho U^2 \ell b \, C_{Df}$$

With $A = \ell b = 4$ ft $\times$ 0.5 ft $= 2$ ft², $\rho = 1.94$ slugs/ft³, and $\mu = 2.04 \times 10^{-5}$ lb·s/ft² (see Table B.1) we obtain

$$\mathcal{D}_f = \tfrac{1}{2}(1.94 \text{ slugs/ft}^3)(2.0 \text{ ft}^2)U^2 C_{Df}$$

$$= 1.94 \, U^2 C_{Df} \tag{1}$$

where $\mathcal{D}_f$ and U are in pounds and ft/s, respectively.

The friction coefficient, C_{Df}, can be obtained from Fig. 9.15 or, if the ski surface is smooth and the flow entirely laminar or predominantly turbulent, from the appropriate equations given in Tables 9.2 and 9.3. However, as we will see, for this problem, much of the flow lies within the transition regime where both the laminar and turbulent portions of the boundary layer flow occupy comparable lengths of the plate. We choose to use the values of C_{Df} from the figure, since the equations mentioned are for purely laminar or purely turbulent flows.

For the given conditions we obtain

$$\mathrm{Re}_\ell = \frac{\rho U \ell}{\mu} = \frac{(1.94 \text{ slugs/ft}^3)(4 \text{ ft})U}{2.04 \times 10^{-5} \text{ lb·s/ft}^2} = 3.80 \times 10^5 \, U$$

where U is in ft/s. With $U = 10$ ft/s, or Re $= 3.80 \times 10^6$, we obtain from Fig. 9.15 $C_{Df} = 0.0024$. (We will use the middle of the shaded transition region values.) From Eq. 1 the corresponding drag is

$$\mathcal{D}_f = 1.94(10)^2(0.0024) = 0.466 \text{ lb}$$

By covering the range of upstream velocities of interest we obtain the results shown in Fig. E9.7b.

If Re < 1000, the results of boundary layer theory are not valid—inertia effects are not dominant enough and the boundary layer is not thin compared with the length of the plate. For our problem this corresponds to $U = 2.63 \times 10^{-3}$ ft/s. For all practical purposes U is greater than this value, and the flow past the ski is of the boundary layer type.

The approximate location of the transition from laminar to turbulent boundary layer flow as defined by $\mathrm{Re}_{cr} = \rho U x_{cr}/\mu = 5 \times 10^5$ is indicated in Fig. E9.7b. Up to $U = 1.31$ ft/s the entire boundary layer is laminar. The fraction of the boundary layer that is laminar decreases as U increases until only the front 0.18 ft is laminar when $U = 30$ ft/s.

For anyone who has water skied, it is clear that it can require considerably more force to be pulled along at 30 ft/s than the 2×4.53 lb $= 9.06$ lb (two skis) indicated in Fig. E9.7b. As is discussed in Section 9.3, the total drag on an object such as a water ski consists of more than just the friction drag. Other components, including pressure drag and wave-making drag, add considerably to the total resistance.

9.2.6 Effects of Pressure Gradient

The boundary layer discussions in the previous parts of Section 9.2 have dealt with flow along a flat plate in which the pressure is constant throughout the fluid. In general, when a fluid flows past an object other than a flat plate, the pressure field is not uniform. If the Reynolds number is large, relatively thin boundary layers will develop along the surfaces. Within these layers the component of the pressure gradient in the streamwise direction (i.e., along the body surface) is not zero, although the pressure gradient normal to the surface is negligibly small. That is, if we were to measure the pressure while moving across the boundary layer from the body to the boundary layer edge, we would find that the pressure is essentially constant. However, the pressure does vary in the direction along the body surface if the body is curved. The variation in the *free-stream velocity*, U_{fs}, that fluid velocity at the edge of the boundary layer, is the cause of the pressure gradient in the boundary layer. The characteristics of the entire flow (both within and outside of the boundary layer) are often highly dependent on the pressure gradient effects on the fluid within the boundary layer.

For a flat plate parallel to the upstream flow, the upstream velocity (that far ahead of the plate) and the free-stream velocity (that at the edge of the boundary layer) are equal—$U = U_{fs}$. This is a consequence of the negligible thickness of the plate. For bodies of nonzero thickness these two velocities are different. This can be seen in the flow past a circular cylinder of diameter D. The upstream velocity and pressure are U and p_0, respectively. If the fluid were completely inviscid ($\mu = 0$), the Reynolds number would be infinite (Re $= \rho U D/\mu = \infty$) and the streamlines would be symmetrical, as are shown in Fig. 9.16a. The fluid velocity along the surface would vary from $U_{fs} = 0$ at the very front and rear of the cylinder (points A and F are stagnation points) to a maximum of $U_{fs} = 2U$ at the top and bottom of the cylinder (point C). The pressure on the surface of the cylinder would be symmetrical about the verti-

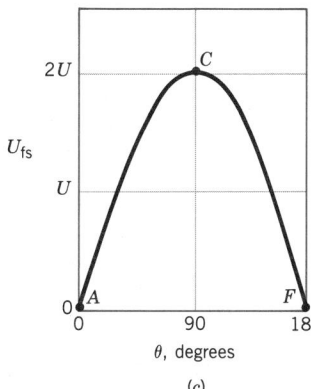

FIGURE 9.16
Inviscid flow past a circular cylinder: (*a*) streamlines for the flow if there were no viscous effects, (*b*) pressure distribution on the cylinder's surface, (*c*) free-stream velocity on the cylinder's surface.

cal midplane of the cylinder, reaching a maximum value of $p_0 + \rho U^2/2$ (the stagnation pressure) at both the front and back of the cylinder, and a minimum of $p_0 - 3\rho U^2/2$ at the top and bottom of the cylinder. The pressure and free-stream velocity distributions are shown in Figs. 9.16*b* and 9.16*c*. These characteristics can be obtained from potential flow analysis of Section 6.6.3.

Because of the absence of viscosity (therefore, $\tau_w = 0$) and the symmetry of the pressure distribution for inviscid flow past a circular cylinder, it is clear that the drag on the cylinder is zero. Although it is not obvious, it can be shown that the drag is zero for any object that does not produce a lift (symmetrical or not) in an inviscid fluid (Ref. 4). Based on experimental evidence, however, we know that there must be a net drag. Clearly, since there is no purely inviscid fluid, the reason for the observed drag must lie on the shoulders of the viscous effects.

To test this hypothesis, we could conduct an experiment by measuring the drag on an object (such as a circular cylinder) in a series of fluids with decreasing values of viscosity. To our initial surprise we would find that no matter how small we make the viscosity (provided it is not precisely zero) we would measure a finite drag, essentially independent of the value of μ. As was noted in Section 6.6.3, this leads to what has been termed *d'Alembert's paradox*—the drag on an object in an inviscid fluid is zero, but the drag on an object in a fluid with vanishingly small (but nonzero) viscosity is not zero.

The reason for the above paradox can be described in terms of the effect of the pressure gradient on boundary layer flow. Consider large Reynolds number flow past

a circular cylinder. As was discussed in Section 9.1.2, we expect the viscous effects to be confined to thin boundary layers near the surface. This allows the fluid to stick ($\mathbf{V} = 0$) to the surface—a necessary condition for any fluid, provided $\mu \neq 0$. The basic idea of boundary layer theory is that the boundary layer is thin enough so that it does not greatly disturb the flow outside the boundary layer. Based on this reasoning, for large Reynolds numbers, the flow throughout most of the flow field would be expected to be as is indicated in Fig. 9.16a, the inviscid flow field.

The pressure distribution indicated in Fig. 9.16b is imposed on the boundary layer flow along the surface of the cylinder. In fact, there is negligible pressure variation across the thin boundary layer so that the pressure within the boundary layer is that given by the inviscid flow field. This pressure distribution along the cylinder is such that the stationary fluid at the nose of the cylinder ($U_{fs} = 0$ at $\theta = 0$) is accelerated to its maximum velocity ($U_{fs} = 2U$ at $\theta = 90°$) and then is decelerated back to zero velocity at the rear of the cylinder ($U_{fs} = 0$ at $\theta = 180°$). This is accomplished by a balance between pressure and inertia effects; viscous effects are absent for the inviscid flow outside the boundary layer.

Physically, in the absence of viscous effects a fluid particle traveling from the front to the back of the cylinder coasts down the "pressure hill" from $\theta = 0$ to $\theta = 90°$ (from point A to C in Fig. 9.16b) and then back up the hill to $\theta = 180°$ (from point C to F) without any loss of energy. There is an exchange between kinetic and pressure energy, but there are no energy losses. The same pressure distribution is imposed on the viscous fluid within the boundary layer. The decrease in pressure in the direction of flow along the front half of the cylinder is termed a *favorable pressure gradient*. The increase in pressure in the direction of flow along the rear half of the cylinder is termed an *adverse pressure gradient*.

Consider a fluid particle within the boundary layer indicated in Fig. 9.17. In its attempt to flow from A to F it experiences the same pressure distribution as the particles in the free stream immediately outside the boundary layer—the inviscid flow field pressure. However, because of the viscous effects involved, the particle in the boundary layer experiences a loss of energy as it flows along. This loss means that the particle does not have enough energy to coast all of the way up the pressure hill (from C to F) and to reach point F at the rear of the cylinder. This kinetic energy deficit is seen in the velocity profile detail at point C, shown in Fig. 9.17a. Because of friction, the boundary layer fluid cannot travel from the front to the rear of the cylinder. (This conclusion can also be obtained from the concept that due to viscous effects the particle at C does not have enough momentum to allow it to coast up the pressure hill to F.)

The situation is similar to a bicyclist coasting down a hill and up the other side of the valley. If there were no friction the rider starting with zero speed could reach the same height from which he started. Clearly friction (rolling resistance, aerodynamic drag, etc.) causes a loss of energy (and momentum), making it impossible for the rider to reach the height from which he started without supplying additional energy (i.e., peddling). The fluid within the boundary layer does not have such an energy supply. Thus, the fluid flows against the increasing pressure as far as it can, at which point the boundary layer separates from (lifts off) the surface. This *boundary layer separation* is indicated in Fig. 9.17a. Typical velocity profiles at representative locations along the surface are shown in Fig. 9.17b. At the separation location (profile D), the velocity gradient at the wall and the wall shear stress are zero. Beyond that location (from D to E) there is reverse flow in the boundary layer.

As is indicated in Fig. 9.17c, because of the boundary layer separation, the average

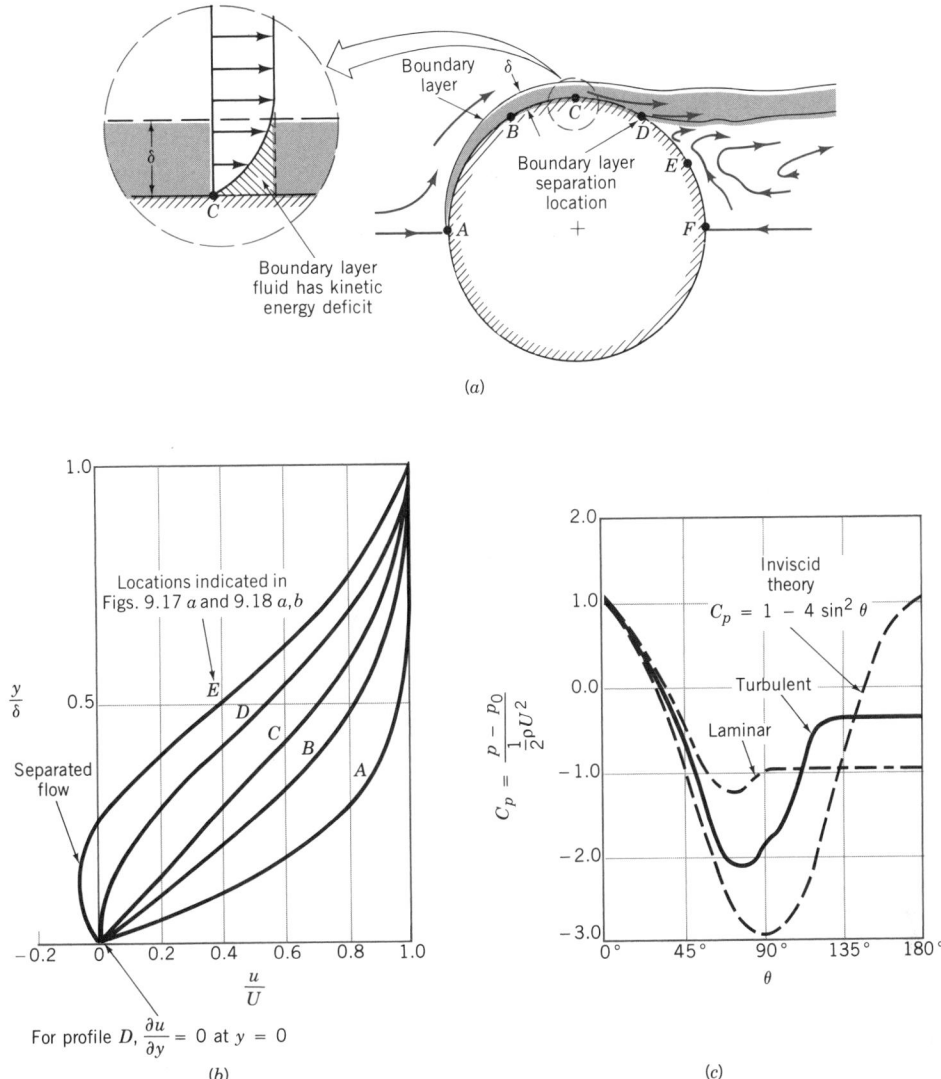

FIGURE 9.17 Boundary layer characteristics on a circular cylinder: (*a*) boundary layer separation location, (*b*) typical boundary layer velocity profiles at various locations on the cylinder, (*c*) surface pressure distributions for inviscid flow and boundary layer flow.

pressure on the rear half of the cylinder is considerably less than that on the front half. Thus, a large pressure drag is developed, even though (because of small viscosity) the viscous shear drag may be quite small. D'Alembert's paradox is explained. No matter how small the viscosity, provided it is not zero, there will be a boundary layer that separates from the surface, giving a drag that is, for the most part, independent of the value of μ.

The location of separation, the width of the wake region behind the object, and the pressure distribution on the surface depend on the nature of the boundary layer flow. Compared with a laminar boundary layer, a turbulent boundary layer flow has more kinetic energy and momentum associated with it because: (1) as is indicated in Fig. E9.6, the velocity profile is fuller, more nearly like the ideal uniform profile, and (2) there can be considerable energy associated with the swirling, random components

of the velocity that do not appear in the time averaged x component of velocity. Thus, as is indicated in Fig. 9.17c, the turbulent boundary layer can flow farther around the cylinder (farther up the pressure hill) before it separates than can the laminar boundary layer. This idea is discussed in detail in Section 9.3.2.

The structure of the flow field past a circular cylinder is completely different for a zero viscosity fluid than it is for a viscous fluid, no matter how small the viscosity is, provided it is not zero. This is due to boundary layer separation. Similar concepts hold for other shaped bodies as well. The flow past an airfoil at zero *angle of attack* (the angle between the upstream flow and the axis of the object) is shown in Fig. 9.18a; flow past the same airfoil at a 5° angle of attack is shown in Fig. 9.18b. Over the front portion of the airfoil the pressure decreases in the direction of flow—a favorable pressure gradient. Over the rear portion the pressure increases in the direction of flow—an adverse pressure gradient. The boundary layer velocity profiles at representative locations are similar to those indicated in Fig. 9.17b for flow past a circular cylinder. If the adverse pressure gradient is not too great (because the body is not too "thick" in some sense), the boundary layer fluid can flow into the slightly increasing pressure region (i.e., from C to the trailing edge in Fig. 9.18a) without separating from the surface. However, if the pressure gradient is too adverse (because the angle of attack is too large), the boundary layer will separate from the surface as indicated is in Fig. 9.18b. Such situations can lead to the catastrophic loss of lift called *stall*, which is discussed in Section 9.4.

Streamlined bodies are generally those designed to eliminate (or at least to reduce) the effects of separation, whereas nonstreamlined bodies generally have relatively large drag due to the low pressure in the separated regions (the wake). Although the boundary layer may be quite thin, it can appreciably alter the entire flow field because of boundary layer separation. These ideas are discussed in Section 9.3.

(a)

(b)

FIGURE 9.18
Flow visualization photographs of flow past an airfoil (The boundary layer velocity profiles for the points indicated are similar to those indicated in Fig. 9.17b): (a) zero angle of attack, no separation, (b) 5° angle of attack, flow separation. Dye in water. (Photograph courtesy of ONERA, France.)

9.2.7 Momentum Integral Boundary Layer Equation with Nonzero Pressure Gradient

The boundary layer results discussed in Sections 9.2.2 and 9.2.3 are valid only for boundary layers with zero pressure gradients. They correspond to the velocity profile labeled C in Fig. 9.17b. Boundary layer characteristics for flows with nonzero pressure gradients can be obtained from nonlinear partial differential boundary layer equations similar to Eqs. 9.8 and 9.9, provided the pressure gradient is appropriately accounted for. Such an approach is beyond the scope of this book (Refs. 1, 2).

An alternate approach is to extend the momentum integral boundary layer equation technique (Section 9.2.3) so that it is applicable for flows with nonzero pressure gradients. The momentum integral equation for boundary layer flows with zero pressure gradient, Eq. 9.26, is a statement of the balance between the shear force on the plate (represented by τ_w) and rate of change of momentum of the fluid within the boundary layer [represented by $\rho U^2 (d\Theta/dx)$]. For such flows the free-stream velocity is constant ($U_{fs} = U$). If the free-stream velocity is not constant [$U_{fs} = U_{fs}(x)$, where x is the distance measured along the curved body], the pressure will not be constant. This follows from the Bernoulli equation with negligible gravitational effects, since $p + \rho U_{fs}^2/2$ is constant along the streamlines outside the boundary layer. Thus,

$$\frac{dp}{dx} = -\rho U_{fs} \frac{dU_{fs}}{dx} \tag{9.34}$$

For a given body the free-stream velocity and the corresponding pressure gradient on the surface can be obtained from inviscid flow techniques (potential flow) discussed in Section 6.7. (This is how the circular cylinder results of Fig. 9.16 were obtained.)

Flow in a boundary layer with nonzero pressure gradient is very similar to that shown in Fig. 9.11, except that the upstream velocity, U, is replaced by the free-stream velocity, $U_{fs}(x)$, and the pressures at sections (1) and (2) are not necessarily equal. By using the x component of the momentum equation (Eq. 5.22) with the appropriate shear forces and pressure forces acting on the control surface indicated in Fig. 9.11, the following integral momentum equation for boundary layer flows is obtained

$$\tau_w = \rho \frac{d}{dx} (U_{fs}^2 \Theta) + \rho \delta^* U_{fs} \frac{dU_{fs}}{dx} \tag{9.35}$$

The derivation of this equation is similar to that of the corresponding equation for constant pressure boundary layer flow, Eq. 9.26, although the inclusion of the pressure gradient effect brings in additional terms (Refs. 1, 2, 3). For example, both the boundary layer momentum thickness, Θ, and the displacement thickness, δ^*, are involved.

Equation 9.35, the general momentum integral equation for two-dimensional boundary layer flow, represents a balance between viscous forces (represented by τ_w), pressure forces (represented by $\rho U_{sf} \, dU_{fs}/dx = -dp/dx$), and the fluid momentum (represented by Θ, the boundary layer momentum thickness). In the special case of a flat plate, $U_{fs} = U = $ constant, and Eq. 9.35 reduces to Eq. 9.26.

Equation 9.35 can be used to obtain boundary layer information in a manner similar to that done for the flat plate boundary layer (Section 9.2.3). That is, for a given body shape the free-stream velocity, U_{fs}, is determined, and a family of approximate boundary layer profiles is assumed. Equation 9.35 is then used to provide information about the boundary layer thickness, wall shear stress, and other properties of interest. The details of this technique are not within the scope of this book (Refs. 1, 3).

9.3   DRAG

As was discussed in Section 9.1, any object moving through a fluid will experience a drag—a net force in the direction of flow due to the pressure and shear forces on the surface of the object. This net force, a combination of flow direction components of the normal and tangential forces on the body, can be determined by uses of Eqs. 9.1 and 9.2, provided the distributions of pressure, p, and wall shear stress, τ_w, are known. Only in very rare instances can these distributions be determined analytically. The boundary layer flow past a flat plate parallel to the upstream flow as is discussed in Section 9.2 is one such case. Current advances in computational fluid dynamics (i.e., the use of computers to solve the governing equations of the flow field) have provided encouraging results for more complex shapes. However, much work in this area remains.

Most of the information pertaining to drag on objects is a result of numerous experiments with wind tunnels, water tunnels, towing tanks, and other ingenious devices that are used to measure the drag on scale models. As was discussed in Chapter 7, these data can be put into dimensionless form and the results can be appropriately ratioed for prototype calculations. Typically, the result for a given shaped object is a drag coefficient, C_D, where

$$C_D = \frac{\mathcal{D}}{\frac{1}{2}\rho U^2 A} \tag{9.36}$$

and C_D is a function of other dimensionless parameters such as Reynolds number, Re, Mach number, Ma, Froude number, Fr, and relative roughness of the surface, ε/ℓ. That is,

$$C_D = \phi(\text{shape, Re, Ma, Fr, } \varepsilon/\ell)$$

The character of C_D as a function of these parameters is discussed in this section.

9.3.1   Friction Drag

Friction drag, $\mathcal{D}_f$, is that part of the drag that is due directly to the shear stress, τ_w, on the object. It is a function of not only the magnitude of the wall shear stress, but also of the orientation of the surface on which it acts. This is indicated by the factor $\tau_w \sin \theta$ in Eq. 9.1. If the surface is parallel to the upstream velocity the entire shear force contributes directly to the drag. This is true for the flat plate parallel to the flow as was discussed in Section 9.2. If the surface is perpendicular to the upstream velocity, the shear stress contributes nothing to the drag. Such is the case for a flat plate normal to the upstream velocity as was discussed in Section 9.1.

In general, the surface of a body will contain portions parallel to and normal to the upstream flow, as well as any direction in between. A circular cylinder is such a body. Because the viscosity of most common fluids is small, the contribution of the shear force to the overall drag on a body is often quite small. Such a statement should be worded in dimensionless terms. That is, because the Reynolds number of most familiar flows is quite large, the percent of the drag caused directly by the shear stress is often quite small. For highly streamlined bodies or for low Reynolds number flow, however, most of the drag may be due to friction drag.

The friction drag on a flat plate of width b and length ℓ oriented parallel to the upstream flow can be calculated from

$$\mathcal{D}_f = \tfrac{1}{2}\rho U^2 b \ell C_{Df}$$

where C_{Df} is the friction drag coefficient. The value of C_{Df}, given as a function of Reynolds number, Re $= \rho U\ell/\mu$, and relative surface roughness, ε/ℓ, in Fig. 9.15, is a result of boundary layer analysis and experiments (see Section 9.2). Typical values of roughness, ε, for various surfaces are given in Table 8.2. As with the pipe flow discussed in Chapter 8, the flow is divided into two distinct categories—laminar or turbulent, with a transitional regime connecting them. The drag coefficient (and, hence, the drag) is not a function of the plate roughness if the flow is laminar. However, for turbulent flow the roughness does considerably affect the value of C_{Df}. As with pipe flow, this dependence is a result of the surface roughness elements protruding into or through the laminar sublayer (see Section 8.3).

Most objects are not flat plates parallel to the flow; instead, they are curved surfaces along which the pressure varies. As was discussed in Section 9.2.6, this means that the boundary layer character, including the velocity gradient at the wall, is different for most objects than that for a flat plate. This can be seen in the change of shape of the boundary layer profile along the cylinder in Fig. 9.17b.

The precise determination of the shear stress along the surface of a curved body is quite difficult to obtain. Although approximate results can be obtained by a variety of techniques (Refs. 1, 2), these are outside the scope of this text. As is shown by the following example, if the shear stress is known, its contribution to the drag can be determined.

EXAMPLE 9.8

A viscous, incompressible fluid flows past the circular cylinder shown in Fig. E9.8a. According to a more advanced theory of boundary layer flow, the boundary layer remains attached to the cylinder up to the separation location at $\theta \approx 108.8°$, with the dimensionless wall shear stress as is indicated in Fig. E9.8b (Ref. 1). The shear stress on the cylinder in the wake region, $108.8 < \theta < 180°$, is negligible. Determine C_{Df}, the drag coefficient for the cylinder based on the friction drag only.

(a)

(b)

FIGURE E9.8

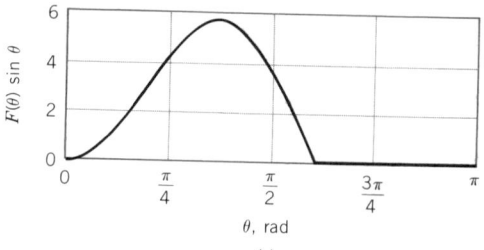

FIGURE E9.8 (*Continued*)

SOLUTION

The friction drag, $\mathcal{D}_f$, can be determined from Eq. 9.1 as

$$\mathcal{D}_f = \int \tau_w \sin \theta \, dA = 2 \left(\frac{D}{2}\right) b \int_0^\pi \tau_w \sin \theta \, d\theta$$

where b is the length of the cylinder. Thus,

$$C_{Df} = \frac{\mathcal{D}_f}{\frac{1}{2}\rho U^2 bD} = \frac{2}{\rho U^2} \int_0^\pi \tau_w \sin \theta \, d\theta$$

This can be put into dimensionless form by using the dimensionless shear stress parameter, $F(\theta) = \tau_w \sqrt{\text{Re}}/(\rho U^2/2)$, given in Fig. E9.8$b$ as follows

$$C_{Df} = \int_0^\pi \frac{\tau_w}{\frac{1}{2}\rho U^2} \sin \theta \, d\theta = \frac{1}{\sqrt{\text{Re}}} \int_0^\pi \frac{\tau_w \sqrt{\text{Re}}}{\frac{1}{2}\rho U^2} \sin \theta \, d\theta$$

where $\text{Re} = \rho U D/\mu$. Thus,

$$C_{Df} = \frac{1}{\sqrt{\text{Re}}} \int_0^\pi F(\theta) \sin \theta \, d\theta \qquad (1)$$

The function $F(\theta) \sin \theta$, obtained from Fig. E9.8b, is plotted in Fig. E9.8c. The necessary integration to obtain C_{Df} from Eq. 1 can be done by an appropriate numerical technique or by an approximate graphical method to determine the area under the given curve.

The result is $\int_0^\pi F(\theta) \sin \theta \, d\theta = 5.93$, or

$$C_{Df} = \frac{5.93}{\sqrt{\text{Re}}} \qquad \text{(Ans)}$$

Note that the total drag must include both the shear stress (friction) drag and the pressure drag. As we will see in Example 9.9, for the circular cylinder most of the drag is due to the pressure force.

The above friction drag result is valid only if the boundary layer flow on the cylinder is laminar. As is discussed in Section 9.3.3, for a smooth cylinder this means that $\text{Re} = \rho U D/\mu < 3 \times 10^5$. It is also valid only for flows that have a Reynolds number sufficiently large to ensure the boundary layer type structure to the flow. For the cylinder this means $\text{Re} > 100$.

9.3.2 Pressure Drag

Pressure drag, $\mathcal{D}_p$, is that part of the drag that is due directly to the pressure, p, on an object. It is often referred to as *form drag* because of its strong dependency on the shape or form of the object. Pressure drag is a function of the magnitude of

the pressure and the orientation of the surface element on which the pressure force acts. For example, the pressure force on either side of a flat plate parallel to the flow may be very large, but it does not contribute to the drag because it acts in the direction normal to the upstream velocity. On the other hand, the pressure force on a flat plate normal to the flow provides the entire drag.

As previously noted, for most bodies, there are portions of the surface that are parallel to the upstream velocity, others normal to the upstream velocity, and the majority of which are at some angle in between. The pressure drag can be obtained from Eq. 9.1 provided a detailed description of the pressure distribution and the body shape is given. That is,

$$\mathcal{D}_p = \int p \cos \theta \, dA$$

which can be rewritten in terms of the *pressure drag coefficient, C_{Dp}*, as

$$C_{Dp} = \frac{\mathcal{D}_p}{\frac{1}{2}\rho U^2 A} = \frac{\int p \cos \theta \, dA}{\frac{1}{2}\rho U^2 A} = \frac{\int C_p \cos \theta \, dA}{A} \tag{9.37}$$

Here $C_p = (p - p_0)/(\rho U^2/2)$ is the *pressure coefficient*, where p_0 is a reference pressure. The level of the reference pressure does not influence the drag directly because the net pressure force on a body is zero if the pressure is constant (i.e., p_0) on the entire surface.

For flows in which inertial effects are large relative to viscous effects (i.e., large Reynolds number flows), the pressure difference, $p - p_0$, scales directly with the dynamic pressure, $\rho U^2/2$, and the pressure coefficient is independent of Reynolds number. In such situations we expect the drag coefficient to be relatively independent of Reynolds number.

For flows in which viscous effects are large relative to inertial effects (i.e., very small Reynolds number flows), it is found that both the pressure difference and wall shear stress scale with the characteristic viscous stress, $\mu U/\ell$, where ℓ is a characteristic length. In such situations we expect the drag coefficient to be proportional to $1/\text{Re}$. That is, $C_D \sim \mathcal{D}/(\rho U^2/2) \sim (\mu U/\ell)/(\rho U^2/2) \sim \mu/\rho U\ell = 1/\text{Re}$. These characteristics are similar to the friction factor dependence of $f \sim 1/\text{Re}$ for laminar pipe flow and $f \sim$ constant for large Reynolds number flow (see Section 8.4).

If the viscosity were zero, the pressure drag on any shaped object (symmetrical or not) in a steady flow would be zero. There perhaps would be large pressure forces on the front portion of the object, but there would be equally large (and oppositely directed) pressure forces on the rear portion. If the viscosity is not zero, the net pressure drag may be nonzero because of boundary layer separation as is discussed in Section 9.2.6. Example 9.9 illustrates this.

EXAMPLE 9.9

A viscous incompressible fluid flows past the circular cylinder shown in Fig. E9.8a. The pressure coefficient on the surface of the cylinder (as determined from experimental measurements) is as indicated in Fig. E9.9a. Determine the pressure drag coefficient for this flow. Combine the results of Examples 9.8 and 9.9 to determine the drag coefficient for a circular cylinder. Compare your results with those given in Fig. 9.23.

(a)

(b)

(c)

FIGURE E9.9

SOLUTION

The pressure (form) drag coefficient, C_{Dp}, can be determined from Eq. 9.37 as

$$C_{Dp} = \frac{1}{A} \int C_p \cos\theta \, dA = \frac{1}{bD} \int_0^{2\pi} C_p \cos\theta \, b \left(\frac{D}{2}\right) d\theta$$

or because of symmetry

$$C_{Dp} = \int_0^\pi C_p \cos\theta \, d\theta$$

where b and D are the length and diameter of the cylinder. To obtain C_{Dp}, we must integrate the $C_{Dp} \cos \theta$ function from $\theta = 0$ to $\theta = \pi$ radians. Again, this can be done by some numerical integration scheme or by determining the area under the curve shown in Fig. E9.9b. The result is

$$C_{Dp} = 1.17 \qquad (1) \quad \text{(Ans)}$$

Note that the positive pressure on the front portion of the cylinder ($0 \leq \theta \leq 34°$) and the negative pressure (less than the upstream value) on the rear portion ($90 \leq \theta \leq 180°$) produce positive contributions to the drag. The negative pressure on the front portion of the cylinder ($34 < \theta < 90°$) reduces the drag by pulling on the cylinder in the upstream direction. The positive area under the $C_p \cos \theta$ curve is greater than the negative area—there is a net pressure drag. In the absence of viscosity, these two contributions would be equal—there would be no pressure (or friction) drag.

The net drag on the cylinder is the sum of friction and pressure drag. Thus, from Eq. 1 of Example 9.8 and Eq. 1 of this example, we obtain the drag coefficient

$$C_D = C_{Df} + C_{Dp} = \frac{5.93}{\sqrt{Re}} + 1.17 \qquad (2) \quad \text{(Ans)}$$

This result is compared with the standard experimental value (obtained from Fig. 9.23) in Fig. E9.9c. The agreement is very good over a wide range of Reynolds numbers. For $Re < 10$ the curves diverge because the flow is not a boundary layer type flow—the shear stress and pressure distributions used to obtain Eq. 2 are not valid in this range. The drastic divergence in the curves for $Re > 3 \times 10^5$ is due to the change from a laminar to turbulent boundary layer, with the corresponding change in the pressure distribution. This is discussed in Section 9.3.3.

It is of interest to compare the friction drag to the total drag on the cylinder. That is,

$$\frac{\mathcal{D}_f}{\mathcal{D}} = \frac{C_{Df}}{C_D} = \frac{5.93/\sqrt{Re}}{(5.93/\sqrt{Re}) + 1.17} = \frac{1}{1 + 0.197\sqrt{Re}}$$

For $Re = 10^3$, 10^4, and 10^5 this ratio is 0.138, 0.0483, and 0.0158, respectively. Most of the drag on the blunt cylinder is pressure drag—a result of the boundary layer separation.

9.3.3 Drag Coefficient Data and Examples

As was discussed in previous sections, the net drag is produced by both pressure and shear stress effects. In most instances these two effects are considered together and an overall drag coefficient, C_D, is used. There is an abundance of such drag coefficient data available in the literature. This information covers incompressible and compressible viscous flows past objects of almost any shape of interest—both man-made and natural objects. In this section we consider a small portion of this information for representative situations. Additional data can be obtained from various sources (Refs. 5, 6).

Shape Dependence Clearly the drag coefficient for an object depends on the shape of the object, with shapes ranging from those that are streamlined to those that are blunt. The drag on an ellipse with aspect ratio ℓ/D, where D and ℓ are the thickness and length parallel to the flow, illustrates this dependence. The drag coefficient $C_D = \mathcal{D}/(\rho U^2 bD/2)$, based on the frontal area, $A = bD$, where b is the length normal to the flow, is as is shown in Fig. 9.19. The more blunt the body, the larger the drag coefficient. With $\ell/D = 0$ (i.e., a flat plate normal to the flow) we obtain the flat plate value of $C_D = 2$. With $\ell/D = 1$ the corresponding value for a circular cylinder is obtained. As ℓ/D becomes larger the value of C_D decreases.

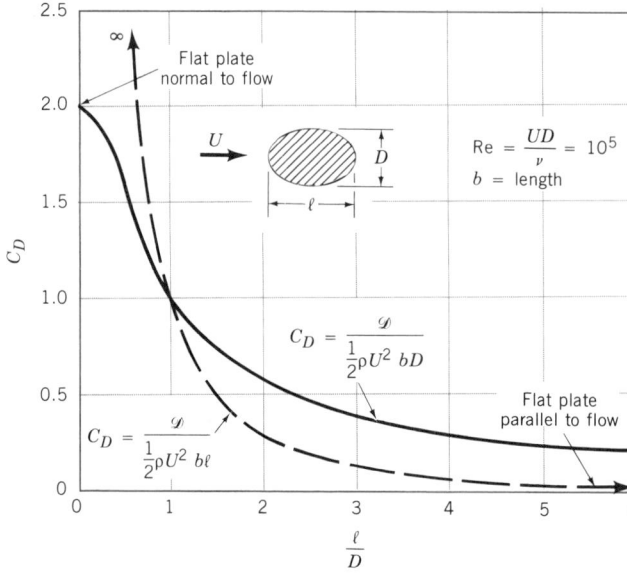

FIGURE 9.19 Drag coefficient for an ellipse with the characteristic area either the frontal area, $A = bD$, or the planform area, $A = b\ell$ (Ref. 5).

For very large aspect ratios ($\ell/D \to \infty$) the ellipse behaves as a flat plate parallel to the flow. For such cases, the friction drag is greater than the pressure drag, and the value of C_D based on the frontal area, $A = bD$, would increase with increasing ℓ/D. (This occurs for larger ℓ/D values than those shown in the figure.) For such extremely thin bodies (i.e., an ellipse with $\ell/D \to \infty$, a flat plate, or very thin airfoils) it is customary to use the planform area, $A = b\ell$, in defining the drag coefficient. After all, it is the planform area on which the shear stress acts, rather than the much smaller (for thin bodies) frontal area. The ellipse drag coefficient based on the planform area, $C_D = \mathcal{D}/(\rho U^2 b\ell/2)$, is also shown in Fig. 9.19. Clearly the drag obtained by using either of these drag coefficients would be the same. They merely represent two different ways to package the same information.

A given body can behave as a thin (streamlined) or thick (blunt) body depending on its orientation. A practical example is illustrated by the drag on an airfoil as is indicated in Fig. 9.20. If the angle of attack is small, the boundary layer remains attached to the airfoil, and the drag is relatively small (mostly friction drag). At a critical angle of attack (about 5° or 6° for this airfoil), boundary layer separation occurs on the upper surface (the surface with a large adverse pressure gradient). For angles larger than this the body appear to the flow as if it were a blunt body, and the drag increases greatly (mostly pressure drag). These characteristics are shown in the photographs of Fig. 9.18.

The amount of streamlining can have a considerable effect on the drag. Incredibly, the drag on the two two-dimensional objects drawn to scale in Fig. 9.21 is the same. The width of the wake for the streamlined strut is very thin, on the order of that for the much smaller diameter circular cylinder.

The drag on an object may depend on whether the object is two-dimensional (i.e., infinitely long normal to the flow) or three-dimensional (i.e., of finite length). The data shown in Fig. 9.22 indicate this trend for a flat plate and a circular cylinder. Apparently the finite length of the object allows the fluid to more easily flow around the object and thereby to reduce the effect of the wake, resulting in a smaller drag coefficient with $b/D < \infty$ than when $b/D = \infty$.

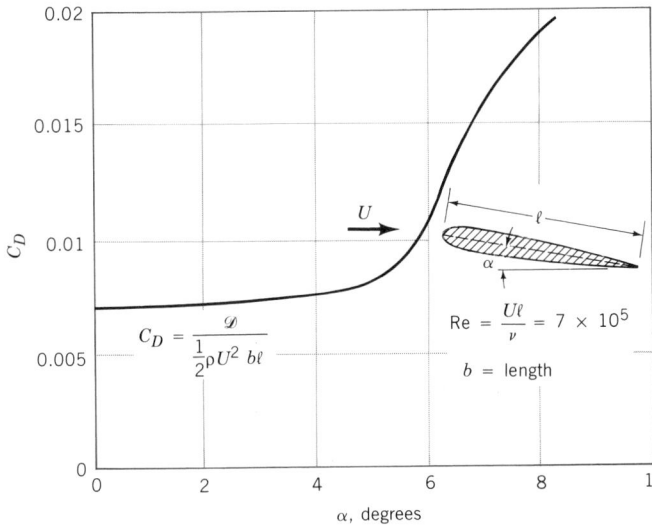

FIGURE 9.20   A typical variation of drag coefficient as a function of angle of attack for an airfoil.

Reynolds Number Dependence   Another parameter on which the drag coefficient can be very dependent is the Reynolds number. The main categories of Reynolds number dependence are: (1) very low Reynolds number flow; (2) moderate Reynolds number flow (laminar boundary layer); and (3) very large Reynolds number flow (turbulent boundary layer). Examples of these three situations are discussed below.

Low Reynolds number flows (Re < 1) are governed by a balance between viscous and pressure forces. Inertia effects are negligibly small. In such instances the drag is expected to be a function of the upstream velocity, U, the body size, ℓ, and the viscosity, μ. That is,

$$\mathcal{D} = \phi(U, \ell, \mu)$$

From dimensional considerations (see Section 7.7.1)

$$\mathcal{D} = C\mu\ell U \tag{9.38}$$

where the value of the constant C depends on the shape of the body. If we put Eq. 9.38 into dimensionless form using the standard definition of the drag coefficient, we obtain

$$C_D = \frac{\mathcal{D}}{\frac{1}{2}\rho U^2 \ell^2} = \frac{2C\mu\ell U}{\rho U^2 \ell^2} = \frac{2C}{\text{Re}}$$

FIGURE 9.21   Two objects of considerably different size that have the same drag force: (*a*) circular cylinder $C_D = 1.2$, (*b*) streamlined strut $C_D = 0.12$.

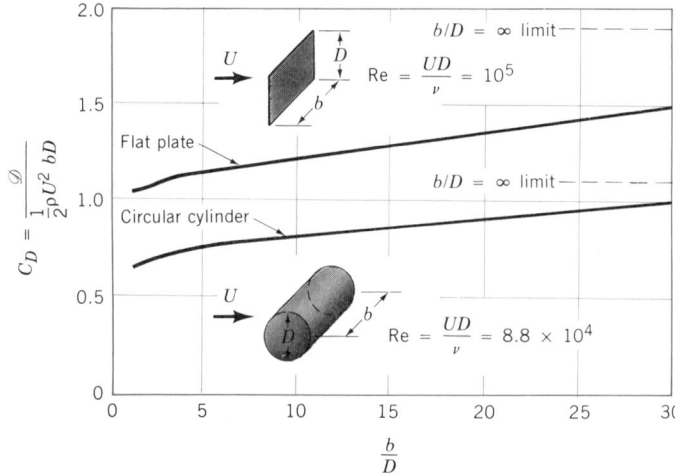

FIGURE 9.22 Drag coefficient as a function of aspect ratio for a flat plate normal to the up-stream flow and a circular cylinder (Ref. 5).

where Re $= \rho U \ell / \mu$. The use of the dynamic pressure, $\rho U^2 / 2$, in the definition of the drag coefficient is somewhat misleading in the case of creeping flows (Re < 1) because it introduces the fluid density, which is not an important parameter for such flows (inertia is not important). Use of this standard drag coefficient definition gives the 1/Re dependence for small Re drag coefficients.

Typical values of C_D for low Reynolds number flows past a variety of objects are given in Table 9.4. It is of interest that the drag on a disk normal to the flow is only

TABLE 9.4 Low Reynolds Number Drag Coefficients (Ref. 7) (Re $= \rho UD / \mu$, $A = \pi D^2 / 4$)

Object	$C_D = \mathcal{D}/(\rho U^2 A / 2)$ (for Re ≤ 1)
a. Circular disk normal to flow	20.4/Re
b. Circular disk parallel to flow	13.6/Re
c. Sphere	24.0/Re
d. Hemisphere	22.2/Re

1.5 times greater than that on a disk parallel to the flow. For large Reynolds number flows this ratio is considerably larger (see Example 9.1). Streamlining (i.e., making the body slender) can produce a considerable drag reduction for large Reynolds number flows; for very small Reynolds number flows it can actually increase the drag because of an increase in the area on which shear forces act. For most objects, the low Reynolds number flow results are valid up to a Reynolds number of about 1.

EXAMPLE 9.10

Small spherical pollen grains of diameter $D = 0.04$ mm and specific gravity $SG = 0.80$ drift down from the top of a 20-m tall pine tree. If there is a constant, uniform breeze of $U_{\text{wind}} = 1$ m/s, determine the distance from the tree that the pollen will travel before it reaches the ground. Assume the horizontal velocity of the pollen is equal to U_{wind}.

SOLUTION

In a reference frame moving horizontally with a velocity of 1 m/s, the pollen moves downward with a velocity U that is governed by a balance between the weight of the pollen, W, the

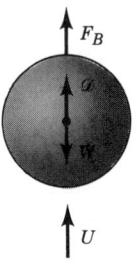

FIGURE E9.10

buoyancy force of the surrounding air, F_B, and the aerodynamic drag on the pollen, $\mathscr{D}$. From the free-body diagram of the falling pollen grain in Fig. E9.10, we obtain

$$W = \mathscr{D} + F_B$$

where

$$W = \gamma_{\text{pollen}} \mathscr{V} = SG \, \gamma_{\text{H}_2\text{O}} \, \frac{\pi}{6} \, D^3 \tag{1}$$

and

$$F_B = \gamma_{\text{air}} \mathscr{V} = \gamma_{\text{air}} \, \frac{\pi}{6} \, D^3 \tag{2}$$

Note that in the reference frame moving with the pollen grain the air appears to be moving vertically with velocity U as indicated. If we know U, we can determine the time, t, it takes to fall $h = 20$ m as

$$t = \frac{h}{U}$$

and the horizontal distance traveled, d, as

$$d = t U_{\text{wind}} = \frac{h U_{\text{wind}}}{U}$$

We assume (because of the smallness of the object) that the flow will be creeping flow (Re < 1) with $C_D = 24/\text{Re}$ (see Table 9.4) so that

$$\mathcal{D} = \frac{1}{2}\rho_{\text{air}}U^2\frac{\pi}{4}D^2C_D = \frac{1}{2}\rho_{\text{air}}U^2\frac{\pi}{4}D^2\left(\frac{24}{\rho_{\text{air}}UD/\mu}\right)$$

or

$$\mathcal{D} = 3\pi\mu UD \tag{3}$$

We must eventually check to determine if this assumption is valid or not. Equation 3 is called Stokes law in honor of G. G. Stokes (1819–1903), a British mathematician and physicist. By combining Eqs. 1, 2, and 3, we obtain

$$SG\,\gamma_{\text{H}_2\text{O}}\frac{\pi}{6}D^3 = 3\pi\mu UD + \gamma_{\text{air}}\frac{\pi}{6}D^3$$

or, since $\gamma = \rho g$,

$$U = \frac{(SG\rho_{\text{H}_2\text{O}} - \rho_{\text{air}})gD^2}{18\mu} \tag{4}$$

For standard atmospheric conditions $\rho_{\text{air}} = 1.23$ kg/m³ and $\mu = 1.79 \times 10^{-5}$ N·s/m². Thus, from Eq. 4 we obtain

$$U = \frac{(0.80 \times 1000 \text{ kg/m}^3 - 1.23 \text{ kg/m}^3)(9.81 \text{ m/s}^2)(0.04 \times 10^{-3} \text{ m})^2}{18(1.79 \times 10^{-5} \text{ N·s/m}^2)}$$

or

$$U = 3.89 \times 10^{-2} \text{ m/s}$$

Since

$$\text{Re} = \frac{\rho_{\text{air}}UD}{\mu} = \frac{(1.23 \text{ kg/m}^3)(3.89 \times 10^{-2} \text{ m/s})(0.04 \times 10^{-3} \text{ m})}{1.79 \times 10^{-5} \text{ N·s/m}^2} = 0.107$$

we see that Re < 1, and the form of the drag coefficient used is valid. Also note that since $\rho_{\text{air}} \ll \rho_{\text{pollen}}$, the buoyancy effect is negligible compared with the weight of the object. That is, $SG\rho_{\text{H}_2\text{O}} - \rho_{\text{air}} \approx SG\rho_{\text{H}_2\text{O}}$.

If there were no dispersion of the pollen because of turbulence in the prevailing breeze, the pollen would settle to the ground at a location

$$d = \frac{hU_{\text{wind}}}{U} = \frac{20 \text{ m (1 m/s)}}{3.89 \times 10^{-2} \text{ m/s}} = 514 \text{ m} \tag{Ans}$$

from the tree. In actuality, this distance can be considerably greater (perhaps many kilometers) because of the turbulent mixing of the wind. By the same argument, the distance could be considerably shorter than the calculated 514 m. Such is the random nature of turbulence.

Moderate Reynolds number flows tend to take on a boundary layer flow structure. For such flows past streamlined bodies, the drag coefficient tends to decrease slightly with Reynolds number. The $C_D \sim \text{Re}^{-1/2}$ dependence for a laminar boundary layer on a flat plate (see Table 9.1) is such an example. Moderate Reynolds number flows past blunt bodies generally produce drag coefficients that are relatively constant. The C_D values for the spheres and circular cylinders shown in Fig. 9.23a indicate this character in the range $10^3 < \text{Re} < 10^5$.

The structure of the flow field at selected Reynolds numbers indicated in Fig. 9.23a is shown in Fig. 9.23b. For a given object there is a wide variety of flow situations, depending on the Reynolds number involved. The curious reader is strongly encour-

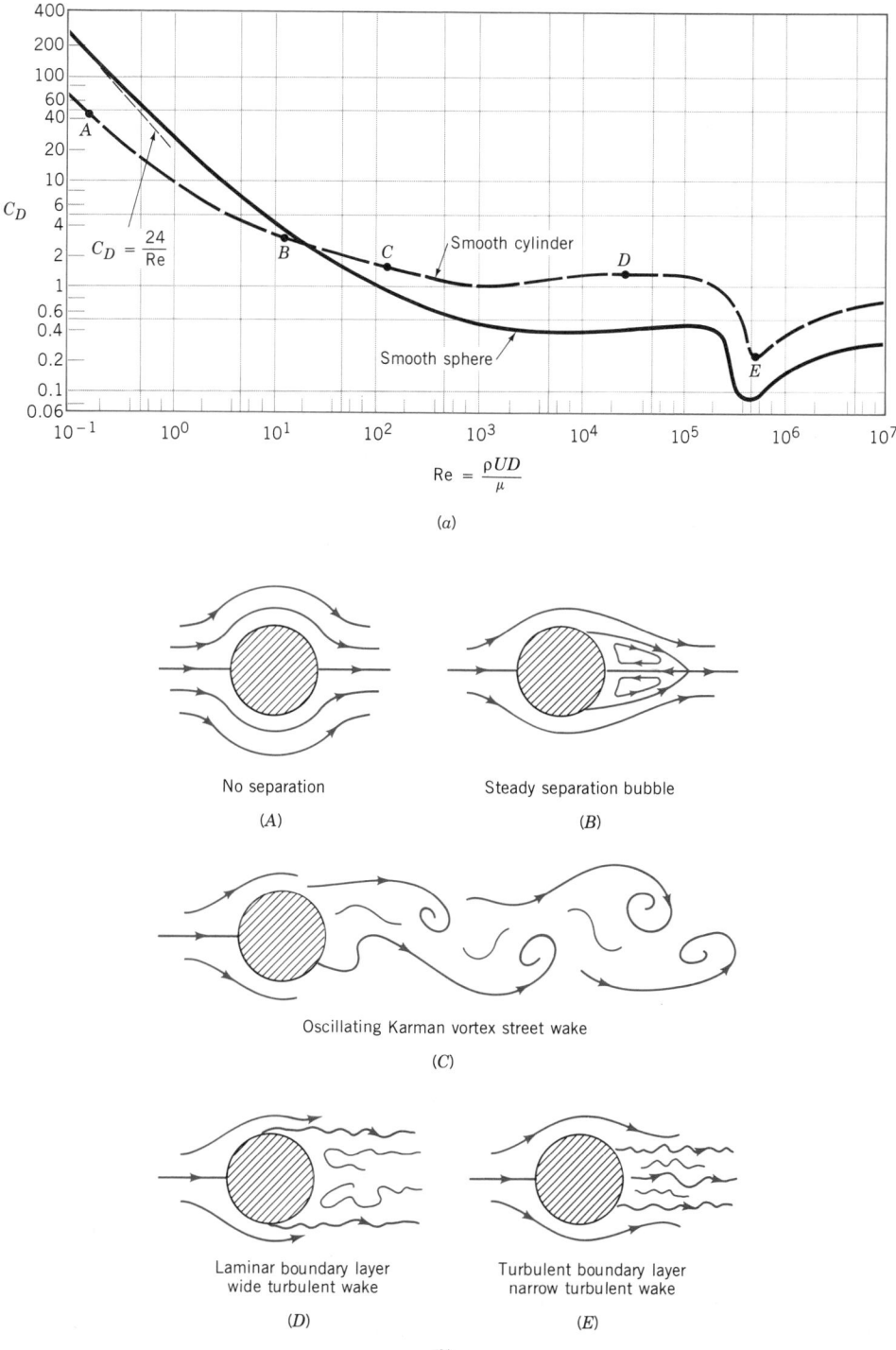

FIGURE 9.23 (a) Drag coefficient as a function of Reynolds number for a smooth circular cylinder and a smooth sphere. (b) Typical flow patterns for flow past a circular cylinder at various Reynolds numbers as indicated in (a).

aged to study the many beautiful photographs of these (and other) flow situations found in Ref. 8.

For many shapes there is a sudden change in the character of the drag coefficient when the boundary layer becomes turbulent. This is illustrated in Fig. 9.15 for the flat plate and in Fig. 9.23 for the sphere and the circular cylinder. The Reynolds number at which this transition takes place is a function of the shape of the body.

For streamlined bodies, the drag coefficient increases when the boundary layer becomes turbulent because most of the drag is due to the shear force, which is greater for turbulent flow than for laminar flow. On the other hand, the drag coefficient for a relatively blunt object, such as a cylinder or sphere, actually decreases when the boundary layer becomes turbulent. As is discussed in Section 9.2.6, a turbulent boundary layer can travel further along the surface into the adverse pressure gradient on the rear portion of the cylinder before separation occurs. The result is a thinner wake and smaller pressure drag for turbulent boundary layer flow. This is indicated in Fig. 9.23 by the sudden decrease in C_D for $10^5 < \text{Re} < 10^6$. In a portion of this range the actual drag (not just the drag coefficient) decreases with increasing speed. It would be very difficult to control the steady flight of such an object in this range—an increase in velocity requires a decrease in thrust (drag). In all other Reynolds number ranges the drag increases with an increase in the upstream velocity (even though C_D may decrease with Re).

For extremely blunt bodies, like a flat plate perpendicular to the flow, the flow separates at the edge of the plate regardless of the nature of the boundary layer flow. Thus, the drag coefficient shows very little dependence on the Reynolds number.

The drag coefficients for a series of two-dimensional bodies of varying buntness are given as a function of Reynolds number in Fig. 9.24. The characteristics described above are evident.

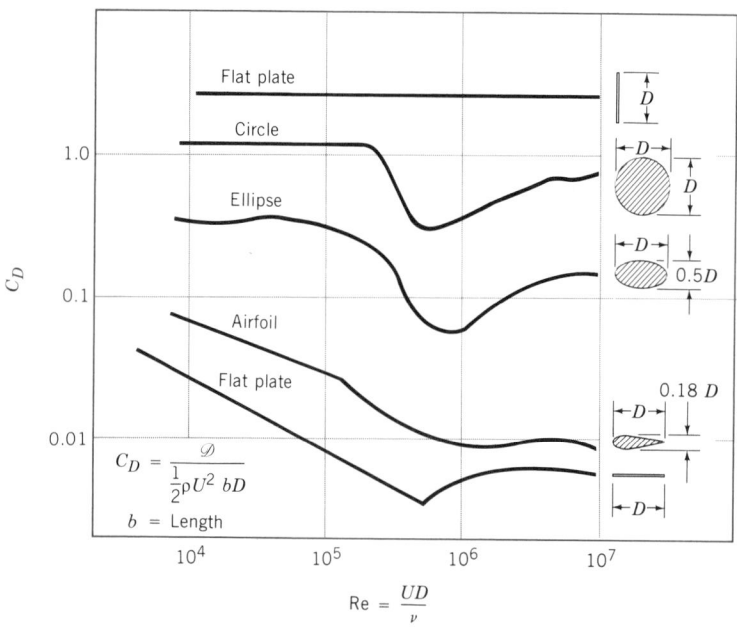

FIGURE 9.24 Character of the drag coefficient as a function of Reynolds number for objects with various degrees of streamlining, from a flat plate normal to the upstream flow to a flat plate parallel to the flow (Ref. 5).

EXAMPLE 9.11

Hail is produced by the repeated rising and falling of ice particles in the updraft of a thunderstorm, as is indicated in Fig. E9.11. When the hail becomes large enough, the aerodynamic drag from the updraft can no longer support the weight of the hail, and it falls from the storm cloud. Estimate the velocity, U, of the updraft needed to make $D = 1.5$-in. diameter (i.e., "golf ball sized") hail.

FIGURE E9.11

SOLUTION

As is discussed in Example 9.10, for steady state conditions a force balance on an object falling through a fluid gives

$$W = \mathcal{D} + F_B$$

where $F_B = \gamma_{air}\mathcal{V}$ is the buoyant force of the air on the particle, $W = \gamma_{ice}\mathcal{V}$ is the particle weight, and $\mathcal{D}$ is the aerodynamic drag. This equation can be rewritten as

$$\tfrac{1}{2}\rho_{air}U^2 \frac{\pi}{4} D^2 C_D = W - F_B \tag{1}$$

With $\mathcal{V} = \pi D^3/6$ and since $\gamma_{ice} \gg \gamma_{air}$ (i.e., $W \gg F_B$) Eq. 1 can be simplified to

$$U = \left(\frac{4}{3}\frac{\rho_{ice}}{\rho_{air}}\frac{gD}{C_D}\right)^{1/2} \tag{2}$$

By using $\rho_{ice} = 1.84$ slugs/ft³, $\rho_{air} = 2.37 \times 10^{-3}$ slug/ft³, and $D = 1.5$ in. $= 0.125$ ft, Eq. 2 becomes

$$U = \left[\frac{4(1.84 \text{ slugs/ft}^3)(32.2 \text{ ft/s}^2)(0.125 \text{ ft})}{3(2.37 \times 10^{-3} \text{ slugs/ft}^3)C_D}\right]^{1/2}$$

or

$$U = \frac{64.5}{\sqrt{C_D}} \tag{3}$$

where U is in ft/s. To determine U, we must know C_D. Unfortunately, C_D is a function of the Reynolds number (see Fig. 9.23), which is not known unless U is known. Thus, we must use an iterative technique similar to that done with the Moody chart for certain types of pipe flow problems (see Section 8.5).

From Fig. 9.23 we expect that C_D is on the order of 0.5. Thus, we assume $C_D = 0.5$ and from Eq. 3 obtain

$$U = \frac{64.5}{\sqrt{0.5}} = 91.2 \text{ ft/s}$$

The corresponding Reynolds number (assuming $v = 1.57 \times 10^{-4}$ ft²/s) is

$$\text{Re} = \frac{UD}{v} = \frac{91.2 \text{ ft/s } (0.125 \text{ ft})}{1.57 \times 10^{-4} \text{ ft}^2/\text{s}} = 7.26 \times 10^4$$

For this value of Re we obtain from Fig. 9.23, $C_D = 0.5$. Thus, our assumed value of $C_D = 0.5$ was correct. The corresponding value of U is

$$U = 91.2 \text{ ft/s} = 62.2 \text{ mph} \hspace{2cm} \text{(Ans)}$$

This result was obtained by using standard sea-level properties for the air. If conditions at 20,000 ft altitude are used (i.e., from Table C.1 $\rho_{air} = 1.267 \times 10^{-3}$ slugs/ft³ and $\mu = 3.324 \times 10^{-7}$ lb·s/ft²), the corresponding result is $U = 125$ ft/s $= 85.2$ mph.

Clearly, an airplane flying through such an updraft would feel its effects (even if it were able to dodge the hail). As seen from Eq. 2, the larger the hail, the stronger the necessary updraft. Hailstones greater than 6 in. in diameter have been reported. In reality, a hailstone is seldom spherical and often not smooth. However, the calculated updraft velocities are in agreement with measured values.

Compressibility Effects The above discussion is restricted to incompressible flows. If the velocity of the object is sufficiently large, compressibility effects become important and the drag coefficient becomes a function of the Mach number, Ma = U/c, where c is the speed of sound in the fluid. The introduction of Mach number effects complicates matters because the drag coefficient for a given object is then a function of both Reynolds number and Mach number—$C_D = \phi(\text{Re}, \text{Ma})$. The Mach number and Reynolds number effects are often closely connected because both are directly proportional to the upstream velocity. For example, both Re and Ma increase with increasing flight speed of an airplane. The changes in C_D due to a change in U are due to changes in both Re and Ma.

The precise dependence of the drag coefficient on Re and Ma is generally quite complex (Ref. 13). However, the following simplifications are often justified. For low Mach numbers, the drag coefficient is essentially independent of Ma as is indicated in Fig. 9.25. For this situation, if Ma < 0.5 or so, compressibility effects are unim-

FIGURE 9.25 Drag coefficient as a function of Mach number for two-dimensional objects in subsonic flow (Ref. 5).

portant. On the other hand, for larger Mach number flows, the drag coefficient can be strongly dependent on Ma, with only secondary Reynolds number effects.

For most objects, values of C_D increase dramatically in the vicinity of Ma = 1 (i.e., sonic flow). This change in character, indicated by Fig. 9.26, is due to the existence of shock waves (extremely narrow regions in the flow field across which the flow parameters change in a discontinuous manner), which are discussed in Chapter 11. Shock waves, which cannot exist in subsonic flows, provide a mechanism for the generation of drag that is not present in the relatively low-speed subsonic flows.

The character of the drag coefficient as a function of Mach number is different for blunt bodies than for sharp bodies. As is shown in Fig. 9.26, sharp pointed bodies develop their maximum drag coefficient in the vicinity of Ma = 1 (sonic flow), whereas the drag coefficient for blunt bodies increases with Ma far above Ma = 1. This behavior is due to the nature of the shock wave structure and the accompanying flow separation. The leading edge of wings for subsonic aircraft are usually quite rounded and blunt, while those of supersonic aircraft tend to be quite pointed and sharp. More information on these important topics can be found in standard texts about compressible flow (Refs. 9, 10).

Surface Roughness   As is indicated in Fig. 9.15, the drag on a flat plate parallel to the flow is quite dependent on the surface roughness, provided the boundary layer flow is turbulent. In such cases the surface roughness protrudes through the laminar sublayer adjacent to the surface (see Section 8.4) and alters the wall shear stress. In addition to the increased turbulent shear stress, surface roughness can alter the Rey-

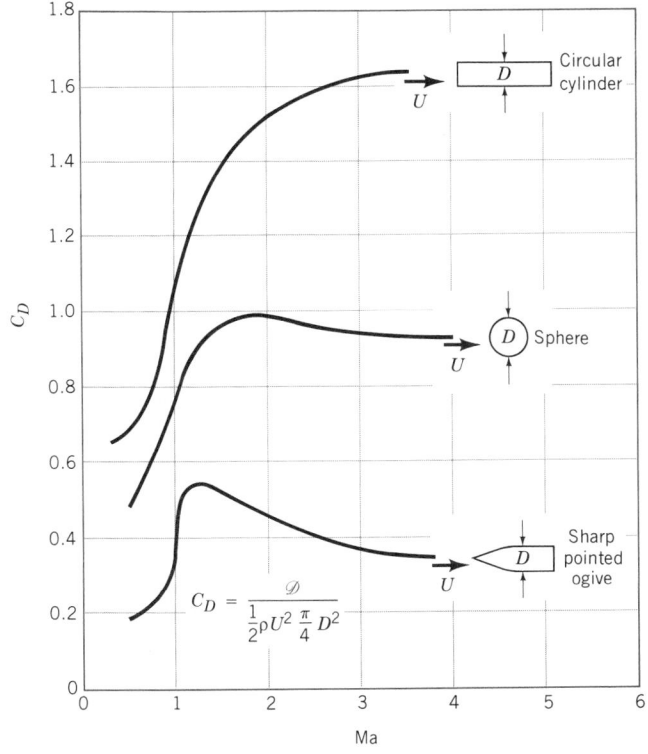

FIGURE 9.26   Drag coefficient as a function of Mach number for supersonic flow (adapted from Ref. 19).

nolds number at which the boundary layer flow becomes turbulent. Thus, a rough flat plate may have a larger portion of its length covered by a turbulent boundary layer than the corresponding smooth plate. This also acts to increase the net drag on the plate.

In general, for streamlined bodies, the drag increases with increasing surface roughness. Great care is taken to design the surfaces of airplane wings to be as smooth as possible, since protruding rivets or screw heads can cause a considerable increase in the drag. On the other hand, for an extremely blunt body, such as a flat plate normal to the flow, the drag is independent of the surface roughness, since the shear stress is not in the upstream flow direction and contributes nothing to the drag.

For blunt bodies like a circular cylinder or sphere, an increase in surface roughness can actually cause a decrease in the drag. This is illustrated for a sphere in Fig. 9.27. As is discussed in Section 9.2.6, when the Reynolds number reaches the critical value ($Re = 3 \times 10^5$ for a smooth sphere), the boundary layer becomes turbulent and the wake region behind the sphere becomes considerably narrower than if it were laminar (see Fig. 9.17). The result is a considerable drop in pressure drag with a slight increase in friction drag, combining to give a smaller overall drag (and C_D).

The boundary layer can be tripped into turbulence at a smaller Reynolds number by using a rough surfaced sphere. For example, the critical Reynolds number for a golf ball is approximately $Re = 4 \times 10^4$. In the range $4 \times 10^4 < Re < 4 \times 10^5$ the drag on the standard rough (i.e., dimpled) golf ball is considerably less ($C_{D\text{rough}}/C_{D\text{smooth}} \approx 0.25/0.5 = 0.5$) than for the smooth ball. As is shown in Example 9.12, this is precisely the Reynolds number range for well-hit golf balls—hence, the reason for dimples on golf balls. The Reynolds number range for well-hit table tennis balls is less than $Re = 4 \times 10^4$. Thus, table tennis balls are smooth.

FIGURE 9.27 The effect of surface roughness on the drag coefficient of a sphere in the Reynolds number range for which the laminar boundary layer becomes turbulent (Ref. 5).

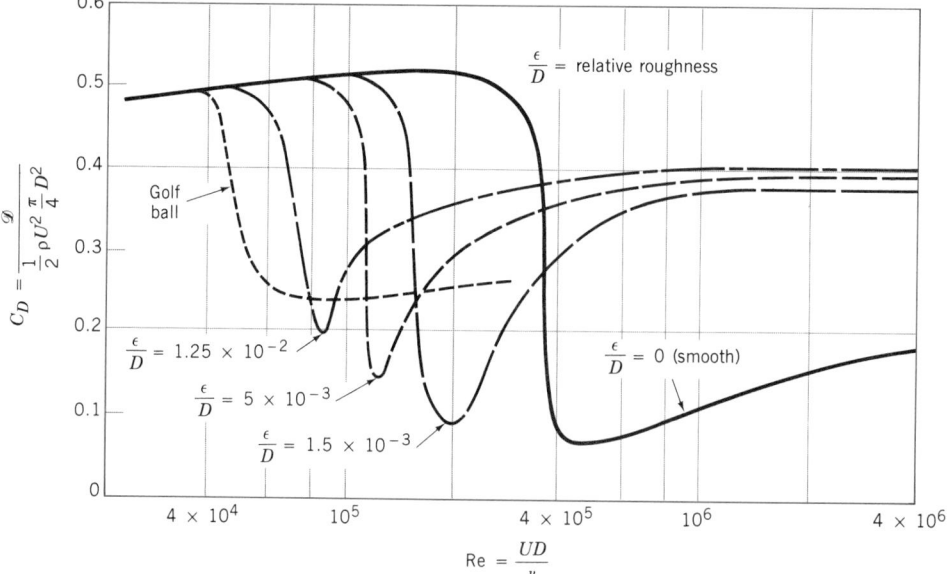

EXAMPLE 9.12

A well-hit golf ball (diameter $D = 1.69$ in., weight $W = 0.0992$ lb) can travel at $U = 200$ ft/s as it leaves the tee. A well-hit table tennis ball (diameter $D = 1.50$ in., weight $W = 0.00551$ lb) can travel at $U = 60$ ft/s as it leaves the paddle. Determine the drag on a standard golf ball, a smooth golf ball, and a table tennis ball for the conditions given. Also determine the deceleration of each ball for these conditions.

SOLUTION

For either ball, the drag can be obtained from

$$\mathcal{D} = \frac{1}{2}\rho U^2 \frac{\pi}{4}D^2 C_D \tag{1}$$

where the drag coefficient, C_D, is given in Fig. 9.27 as a function of the Reynolds number and surface roughness. For the golf ball in standard air

$$\text{Re} = \frac{UD}{\nu} = \frac{(200 \text{ ft/s})(1.69/12 \text{ ft})}{1.57 \times 10^{-4} \text{ ft}^2/\text{s}} = 1.79 \times 10^5$$

while for the table tennis ball

$$\text{Re} = \frac{UD}{\nu} = \frac{(60 \text{ ft/s})(1.50/12 \text{ ft})}{1.57 \times 10^{-4} \text{ ft}^2/\text{s}} = 4.78 \times 10^4$$

The corresponding drag coefficients are $C_D = 0.25$ for the standard golf ball, $C_D = 0.51$ for the smooth golf ball, and $C_D = 0.50$ for the table tennis ball. Hence, from Eq. 1

$$\mathcal{D} = \frac{1}{2}(2.38 \times 10^{-3} \text{ slugs/ft}^3)(200 \text{ ft/s})^2 \frac{\pi}{4}\left(\frac{1.69}{12}\text{ ft}\right)^2(0.25) = 0.185 \text{ lb} \quad \text{(Ans)}$$

for the standard golf ball

$$\mathcal{D} = \frac{1}{2}(2.38 \times 10^{-3} \text{ slugs/ft}^3)(200 \text{ ft/s})^2 \frac{\pi}{4}\left(\frac{1.69}{12}\text{ ft}\right)^2(0.51) = 0.378 \text{ lb} \quad \text{(Ans)}$$

for the smooth golf ball and

$$\mathcal{D} = \frac{1}{2}(2.38 \times 10^{-3} \text{ slugs/ft}^3)(60 \text{ ft/s})^2 \frac{\pi}{4}\left(\frac{1.50}{12}\text{ ft}\right)^2(0.50) = 0.0263 \text{ lb} \quad \text{(Ans)}$$

for the table tennis ball.

The corresponding decelerations are $a = \mathcal{D}/m = g\mathcal{D}/W$, where m is the mass of the ball. Thus, the deceleration relative to the acceleration of gravity, a/g (i.e., the number of g's deceleration) is $a/g = \mathcal{D}/W$ or

$$\frac{a}{g} = \frac{0.185 \text{ lb}}{0.0992 \text{ lb}} = 1.86 \text{ for the standard golf ball} \quad \text{(Ans)}$$

$$\frac{a}{g} = \frac{0.378 \text{ lb}}{0.0992 \text{ lb}} = 3.81 \text{ for the smooth golf ball} \quad \text{(Ans)}$$

and

$$\frac{a}{g} = \frac{0.0263 \text{ lb}}{0.00551 \text{ lb}} = 4.77 \text{ for the table tennis ball} \quad \text{(Ans)}$$

Note that there is a considerably smaller deceleration for the rough golf ball than for the smooth one. Because of its much larger drag-to-mass ratio, the table tennis ball slows down relatively

quickly and does not travel as far as the golf ball. (Note that with $U = 60$ ft/s the standard golf ball has a drag of $\mathcal{D} = 0.0200$ lb and a deceleration of $a/g = 0.202$, considerably less than the $a/g = 4.77$ of the table tennis ball. Conversely, a table tennis ball hit from a tee at 200 ft/s would decelerate at a rate of $a = 1740$ ft^3/s, or $a/g = 54.1$. It would not travel nearly as far as the golf ball.)

The Reynolds number range for which a rough golf ball has smaller drag than a smooth one (i.e., 4×10^4 to 4×10^5) corresponds to a flight velocity range of $45 < U < 450$ ft/s. This is comfortably within the range of most golfers. As is discussed in Section 9.4.4, the dimples (roughness) on a golf ball also help produce a lift (due to the spin of the ball) that allows the ball to travel farther than a smooth ball.

Froude Number Effects Another parameter on which the drag coefficient may be strongly dependent is the Froude number, Fr $= U/\sqrt{g\ell}$. As is discussed in Chapter 10, the Froude number is a ratio of the free-stream speed to a typical wave speed on the interface of two fluids, such as the surface of the ocean. An object moving on the surface, such as a ship, often produces waves that require a source of energy to generate. This energy comes from the ship and is manifest as a drag. [Recall that the rate of energy production (power) equals speed times force.] The nature of the waves produced often depends on the Froude number of the flow and the shape of the object—the waves generated by a water skier "plowing" through the water at a low speed (low Fr) are different than those generated by the skier "planing" along the surface at high speed (large Fr).

Thus, the drag coefficient for surface ships is a function of Reynolds number (viscous effects) and Froude number (wave-making effects); $C_D = \phi(\text{Re}, \text{Fr})$. As was discussed in Chapter 7, it is often quite difficult to run model tests under conditions similar to those of the prototype (i.e., same Re and Fr for surface ships). Fortunately, the viscous and wave effects can often be separated, with the total drag being the sum of the drag of these individual effects. A detailed account of this important topic can be found in standard texts (Ref. 11).

As is indicated in Fig. 9.28, the wave-making drag, $\mathcal{D}_w$, can be a complex function of the Froude number and the body shape. The rather "wiggly" dependence of wave drag coefficient, $C_{Dw} = \mathcal{D}_w/(\rho U^2 \ell^2/2)$, on the Froude number shown is typical. It results from the fact that the structure of the waves produced by the hull is a strong function of the ship speed or, in dimensionless form, the Froude number. This wave structure is also a function of the body shape. For example, the bow wave, which is often the major contributor to the wave drag, can be reduced by use of an appropriately designed bulb on the bow, as is indicated in Fig. 9.28. In this instance the streamlined body (hull without a bulb) has more drag than the less streamlined one.

Composite Body Drag Approximate drag calculations for a complex body can often be obtained by treating the body as a composite collection of its various parts. For example, the drag on an airplane can be approximated by adding the drag produced by its various components—the wings, fuselage, tail section, and so on. Considerable care must be used in such an approach because of the interactions between the various parts. The flow past the wing root (near the wing–fuselage intersection) is considerably altered by the fuselage. Hence, it may not be correct to merely add the drag of the components to obtain the drag of the entire object, although such approximations are often reasonable.

FIGURE 9.28 Typical drag coefficient data as a function of Froude number and hull characteristics for that portion of the drag due to the generation of waves (adapted from Ref. 25).

EXAMPLE 9.13

A 60-mph (i.e., 88-fps) wind blows past the water tower shown in Fig. E9.13a. Estimate the moment (torque), M, needed at the base to keep the tower from tipping over.

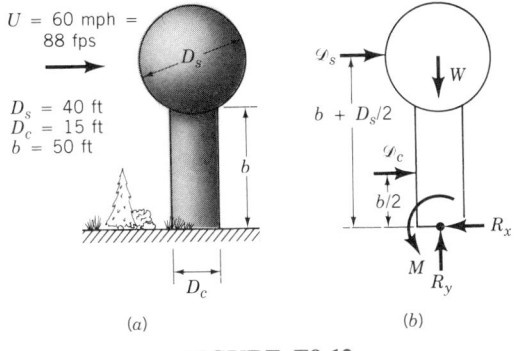

(a)   (b)

FIGURE E9.13

SOLUTION

We treat the water tower as a sphere resting on a circular cylinder and assume that the total drag is the sum of the drag from these parts. The free-body diagram of the tower is shown in Fig. E.9.13b. By summing moments about the base of the tower, we obtain

$$M = \mathcal{D}_s \left(b + \frac{D_s}{2} \right) + \mathcal{D}_c \left(\frac{b}{2} \right) \tag{1}$$

where

$$\mathcal{D}_s = \frac{1}{2}\rho U^2 \frac{\pi}{4} D_s^2 C_{Ds} \tag{2}$$

and

$$\mathcal{D}_c = \frac{1}{2}\rho U^2 b D_c C_{Dc} \tag{3}$$

are the drag on the sphere and cylinder, respectively. For standard atmospheric conditions, the Reynolds numbers are

$$\mathrm{Re}_s = \frac{U D_s}{\nu} = \frac{(88\ \mathrm{ft/s})(40\ \mathrm{ft})}{1.57 \times 10^{-4}\ \mathrm{ft^2/s}} = 2.24 \times 10^7$$

and

$$\mathrm{Re}_c = \frac{U D_c}{\nu} = \frac{(88\ \mathrm{ft/s})(15\ \mathrm{ft})}{1.57 \times 10^{-4}\ \mathrm{ft^2/s}} = 8.41 \times 10^6$$

The corresponding drag coefficients, C_{Ds} and C_{Dc}, can be approximated from Fig. 9.23 as

$$C_{Ds} \approx 0.3 \quad \text{and} \quad C_{Dc} \approx 0.7$$

Note that the value of C_{Ds} was obtained by an extrapolation of the given data to Reynolds numbers beyond those given (a potentially dangerous practice!). From Eqs. 2 and 3 we obtain

$$\mathcal{D}_s = 0.5(2.38 \times 10^{-3}\ \mathrm{slugs/ft^3})(88\ \mathrm{ft/s})^2 \frac{\pi}{4}(40\ \mathrm{ft})^2(0.3) = 3470\ \mathrm{lb}$$

and

$$\mathcal{D}_c = 0.5(2.38 \times 10^{-3}\ \mathrm{slugs/ft^3})(88\ \mathrm{ft/s})^2(50\ \mathrm{ft} \times 15\ \mathrm{ft})(0.7) = 4840\ \mathrm{lb}$$

From Eq. 1 the corresponding moment needed to prevent the tower from tipping is

$$M = 3470\ \mathrm{lb}\left(50\ \mathrm{ft} + \frac{40}{2}\ \mathrm{ft}\right) + 4840\ \mathrm{lb}\left(\frac{50}{2}\ \mathrm{ft}\right) = 3.64 \times 10^5\ \mathrm{ft \cdot lb} \qquad \text{(Ans)}$$

The above result is only an estimate because: (a) the wind is probably not uniform from the top of the tower to the ground; (b) the tower is not exactly a combination of a smooth sphere and a circular cylinder; (c) the cylinder is not of infinite length; (d) there will be some interaction between the flow past the cylinder and that past the sphere so that the net drag is not exactly the sum of the two; and (e) a drag coefficient value was obtained by extrapolation of the given data. However, such approximate results are often quite accurate.

The aerodynamic drag on automobiles provides an example of the use of composite bodies. The power required to move a car along a level street is used to overcome the rolling resistance and the aerodynamic drag. For speeds above approximately 30 mph, the aerodynamic drag becomes a significant contribution to the net propulsive force needed. The contribution of the drag due to various portions of car have been determined by numerous model and full sized tests as well as by numerical calculations. As a result it is possible to predict the aerodynamic drag on cars of a wide variety of body styles. The resulting drag coefficient is

$$C_D = 0.16 + 0.0095 \sum_{i=A}^{H} N_i$$

where the values of N_i are obtained from Table 9.5 for the given shapes of the eight

---

TABLE 9.5   Values of N_i (Ref. 12). See Fig. 9.29

A. Plan view, front end

$N_A = 1$ A-1   Approximately semicircular
2   A-2   Well-rounded outer quarters
3   A-3   Rounded corners without protuberances
4   A-4   Rounded corners with protuberances
5   A-5   Squared tapering-in corners
6   A-6   Squared constant width front

B. Plan view, windshield

$N_B = 1$ B-1   Full wraparound (approximately semicircular)
2   B-2   Wraparound ends
3   B-3   Bowed
4   B-4   Flat

C. Plan view, roof

$N_C = 1$ C-1   Well- or medium-tapered to rear
2   C-2   Tapering to front and rear or approximately constant width
3   C-3   Tapering to front (maximum width at rear)

D. Plan view, lower rear end

$N_C = 1$ D-1   Well- or medium-tapered to rear
2   D-2   Small taper to rear or constant width
3   D-3   Outward taper (or flared-out fins)

E. Side elevation, front end

$N_E = 1$ E-1   Low, rounded front, sloping up
1   E-2   High, tapered, rounded hood
2   E-3   Low, squared front, sloping up
2   E-4   High, tapered, squared hood
3   E-5   Medium-height, rounded front, sloping up
4   E-6   Medium-height, squared front, sloping up
4   E-7   High, rounded front, with horizontal hood
5   E-8   High, squared front, with horizontal hood

TABLE 9.5 **(continued)**

F. Side elevation, windshield peak

$N_F = 1$ F-1   Rounded

       2   F-2   Squared (including flanges or gutters)

       3   F-3   Forward-projecting peak

G. Side elevation, rear roof/trunk

$N_G = 1$ G-1   Fastback (roofline continuous to tail)

       2   G-2   Semifastback (with discontinuity in line to tail)

       3   G-3   Squared roof with trunk rear edge squared

       4   G-4   Rounded roof with rounded trunk

       4   G-5   Squared roof with short or no trunk

       5   G-6   Rounded roof with short or no trunk

H. Front elevation, cowl and fender cross section at windshield

$N_H = 1$ H-1   Flush hood and fenders, well-rounded body sides

       2   H-2   High cowl, low fenders

       3   H-3   Hood flush with rounded-top fenders

       3   H-4   High cowl with rounded-top fenders

       4   H-5   Hood flush with square-edged fenders

       5   H-6   Depressed hood with high squared-edged fenders

---

components identified (i.e., front end, windshield, roof, rear end, front end, windshield peak, rear roof/trunk, and cowl). The corresponding body component shapes are illustrated in Fig. 9.29.

As is indicated in Fig. 9.30, the drag coefficient for cars has decreased rather continuously over the years. This reduction is a result of careful design of the shape and the details (such as window moulding, rear view mirrors, etc.). An additional reduction in drag has been accomplished by a reduction of the projected area. The net result is a considerable increase in the gas mileage, especially at highway speeds.

The effect of several important parameters (shape, Re, Ma, Fr, and roughness) on the drag coefficient for various objects has been discussed in this section. As stated previously, drag coefficient information for a very wide range of objects is available in the literature. Some of this information is given in Figs. 9.31, 9.32, and 9.33 for a variety of two- and three-dimensional, natural and man-made objects. Recall that a drag coefficient of unity is equivalent to the drag produced by the dynamic pressure

FIGURE 9.29 Characteristics of automobile components for use with the data of Table 9.5 to determine the drag coefficient for automobiles (Ref. 12 with permission).

acting on an area of size A. That is, $\mathcal{D} = \frac{1}{2}\rho U^2 A C_D = \frac{1}{2}\rho U^2 A$ if $C_D = 1$. Typical nonstreamlined objects have drag coefficients on this order.

9.4 LIFT

As is indicated in Section 9.1, any object moving through a fluid will experience a net force of the fluid on the object. For symmetrical objects, this force will be in the

FIGURE 9.30 The historical trend of streamlining automobiles to reduce their aerodynamic drag and increase their miles per gallon (adapted from Ref. 5).

Shape	Reference area A (b = length)	Drag coefficient $C_D = \dfrac{\mathscr{D}}{\frac{1}{2}\rho U^2 A}$	Reynold number $Re = \rho UD/\nu$
Square rod with rounded corners	$A = bD$	$\begin{array}{c\|c} R/D & C_D \\ \hline 0 & 2.2 \\ 0.02 & 2.0 \\ 0.17 & 1.2 \\ 0.33 & 1.0 \end{array}$	$Re = 10^5$
Rounded equilateral triangle	$A = bD$	$\begin{array}{c\|c\|c} R/D & C_D \rightarrow & C_D \leftarrow \\ \hline 0 & 1.4 & 2.1 \\ 0.02 & 1.2 & 2.0 \\ 0.08 & 1.3 & 1.9 \\ 0.25 & 1.1 & 1.3 \end{array}$	$Re = 10^5$
Semicircular shell	$A = bD$	$\rightarrow$ 2.3 $\quad$ $\leftarrow$ 1.1	$Re = 2 \times 10^4$
Semicircular cylinder	$A = bD$	$\rightarrow$ 2.15 $\quad$ $\leftarrow$ 1.15	$Re > 10^4$
T-beam	$A = bD$	$\rightarrow$ 1.80 $\quad$ $\leftarrow$ 1.65	$Re > 10^4$
I-Beam	$A = bD$	2.05	$Re > 10^4$
Angle	$A = bD$	$\rightarrow$ 1.98 $\quad$ $\leftarrow$ 1.82	$Re > 10^4$
Hexagon	$A = bD$	1.0	$Re > 10^4$
Rectangle	$A = bD$	$\begin{array}{c\|c} \ell/D & C_D \\ \hline \leq 0.1 & 1.9 \\ 0.5 & 2.5 \\ 0.65 & 2.9 \\ 1.0 & 2.2 \\ 2.0 & 1.6 \\ 3.0 & 1.3 \end{array}$	$Re = 10^5$

FIGURE 9.31
Typical drag coefficients for regular two-dimensional objects (Refs. 5 and 6).

direction of the free stream—a drag, $\mathscr{D}$. If the object is not symmetrical (or if it does not produce a symmetrical flow field, such as the flow around a rotating sphere), there may also be a force normal to the free stream—a lift, $\mathscr{L}$. Considerable effort has been put forth to understand the various properties of the generation of lift. Some objects, such as an airfoil, are designed to generate lift. Other objects are designed to reduce the lift generated. For example, the lift on a car tends to reduce the contact force between the wheels and the ground, causing reduction in traction and cornering ability. It is desirable to reduce this lift.

9.4.1 Surface Pressure Distribution

The lift can be determined from Eq. 9.2 if the distributions of pressure and wall shear stress around the entire body are known. As is indicated in Section 9.1, such

Shape		Reference area A	Drag coefficient C_D	Reynolds number
	Solid hemisphere	$A = \frac{\pi}{4}D^2$	→ 1.17 ← 0.42	$Re > 10^4$
	Hollow hemisphere	$A = \frac{\pi}{4}D^2$	→ 1.42 ← 0.38	$Re > 10^4$
	Thin disk	$A = \frac{\pi}{4}D^2$	1.1	$Re > 10^3$
	Circular rod parallel to flow	$A = \frac{\pi}{4}D^2$	ℓ/D — C_D: 0.5 — 1.1; 1.0 — 0.93; 2.0 — 0.83; 4.0 — 0.85	$Re > 10^5$
	Cone	$A = \frac{\pi}{4}D^2$	θ, degrees — C_D: 10 — 0.30; 30 — 0.55; 60 — 0.80; 90 — 1.15	$Re > 10^4$
	Cube	$A = D^2$	1.05	$Re > 10^4$
	Cube	$A = D^2$	0.80	$Re > 10^4$
	Streamlined body	$A = \frac{\pi}{4}D^2$	0.04	$Re > 10^5$

FIGURE 9.32 Typical drag coefficients for regular three-dimensional objects (Ref. 5).

data are usually not known. Typically, the lift is given in terms of the lift coefficient

$$C_L = \frac{\mathcal{L}}{\frac{1}{2}\rho U^2 A} \tag{9.39}$$

which is obtained from experiments, advanced analysis, or numerical considerations. The lift coefficient is a function of the appropriate dimensionless parameters and, as the drag coefficient, can be written as

$$C_L = \phi(\text{shape}, Re, Ma, Fr, \varepsilon/\ell)$$

Shape	Reference area	Drag coefficient C_D
Parachute	Frontal area $A = \frac{\pi}{4}D^2$	1.4
Porous parabolic dish	Frontal area $A = \frac{\pi}{4}D^2$	Porosity: 0 | 0.2 | 0.5 → 1.42 | 1.20 | 0.82 ← 0.95 | 0.90 | 0.80 Porosity = open area/total area
Average person	Standing Sitting Crouching	$C_D A = 9$ ft^2 $C_D A = 6$ ft^2 $C_D A = 2.5$ ft^2
Fluttering flag	$A = \ell D$	ℓ/D | C_D 1 | 0.07 2 | 0.12 3 | 0.15
Empire State Building	Frontal area	1.4
Six-car passenger train	Frontal area	1.8
Bikes		
Upright commuter	$A = 5.5$ ft^2	1.1
Racing	$A = 3.9$ ft^2	0.88
Drafting	$A = 3.9$ ft^2	0.50
Streamlined	$A = 5.0$ ft^2	0.12
Tractor-trailer trucks		
Standard	Frontal area	0.96
With fairing	Frontal area	0.76
With fairing and gap seal	Frontal area	0.70
Tree $U = 10$ m/s $U = 20$ m/s $U = 30$ m/s	Frontal area	0.43 0.26 0.20
Dolphin	Wetted area	0.0036 at Re $= 6 \times 10^6$ (flat plate has $C_{D_f} = 0.0031$)
Large Birds	Frontal area	0.40

FIGURE 9.33
Typical drag coefficients for objects of interest (Refs. 5, 6, 15, and 20).

Denotes $p > p_0$

Denotes $p < p_0$

U, p_0

FIGURE 9.34
Pressure distribution on the surface of an automobile.

The Froude number, Fr, is important only if there is a free surface present, as with an underwater "wing" used to support a high-speed hydrofoil surface ship. Often the surface roughness, ε, is relatively unimportant in terms of lift—it has more of an effect on the drag. The Mach number, Ma, is of importance for relatively high-speed subsonic and supersonic flows (i.e., Ma > 0.8), and the Reynolds number effect is often not great. The most important parameter that affects the lift coefficient is the shape of the object. Considerable effort has gone into designing optimally shaped lift producing devices. We will emphasize the effect of the shape on lift—the effects of the other dimensionless parameters can be found in the literature (Refs. 13, 14).

Most common lift-generating devices (i.e., airfoils, fans, spoilers on cars, etc.) operate in the large Reynolds number range in which the flow has a boundary layer character, with viscous effects confined to the boundary layers and wake regions. For such cases the wall shear stress, τ_w, contributes little to the lift. Most of the lift comes from the surface pressure distribution. A typical pressure distribution on a moving car is shown in Fig. 9.34. The distribution, for the most part, is consistent with simple Bernoulli equation analysis. Locations with high-speed flow (i.e., over the roof and hood) have low pressure, while locations with low-speed flow (i.e., on the grill and windshield) have high pressure. It is easy to believe that the integrated effect of this pressure distribution would provide a net upward force.

For objects operating in very low Reynolds number regimes (i.e., Re < 1), viscous effects are important, and the contribution of the shear stress to the lift may be as important as that of the pressure. Such situations include the flight of minute insects and the swimming of microscopic organisms. The relative importance of τ_w and p in the generation of lift in a typical large Reynolds number flow is shown in Example 9.14.

EXAMPLE 9.14

When a uniform wind of velocity U blows past the semicircular building shown in Fig. E9.14a, the wall shear stress and pressure distributions on the outside of the building are as given previously in Figs. E9.8b and E9.9a, respectively. If the pressure in the building is atmospheric (i.e., the value, p_0, far from the building), determine the lift coefficient and the lift on the roof.

(a)

(b)

(c)

FIGURE E9.14

SOLUTION

From Eq. 9.2 we obtain the lift as

$$\mathcal{L} = -\int p \sin \theta \, dA + \int \tau_w \cos \theta \, dA \tag{1}$$

As is indicated in Fig. E9.14a, we assume that on the inside of the building the pressure is uniform $p = p_0$ and that there is no shear stress. Thus, Eq. 1 can be written as

$$\mathcal{L} = -\int_0^\pi (p - p_0) \sin \theta \, b\left(\frac{D}{2}\right) d\theta + \int_0^\pi \tau_w \cos \theta \, b\left(\frac{D}{2}\right) d\theta$$

or

$$\mathcal{L} = \frac{bD}{2} \left[-\int_0^\pi (p - p_0) \sin \theta \, d\theta + \int_0^\pi \tau_w \cos \theta \, d\theta \right] \tag{2}$$

where b and D are the length and diameter of the building, respectively, and $dA = b(D/2)d\theta$. Equation 2 can be put into dimensionless form by using the dynamic pressure, $\rho U^2/2$, planform area, $A = bD$, and dimensionless shear stress

$$F(\theta) = \tau_w (\mathrm{Re})^{1/2}/(\rho U^2/2)$$

to give

$$\mathcal{L} = \frac{1}{2} \rho U^2 A \left[-\frac{1}{2} \int_0^\pi \frac{(p - p_0)}{\frac{1}{2}\rho U^2} \sin \theta \, d\theta + \frac{1}{2\sqrt{\mathrm{Re}}} \int_0^\pi F(\theta) \cos \theta \, d\theta \right] \tag{3}$$

The values of the two integrals in Eq. 3 can be obtained by determining the area under the curves of $[(p - p_0)/(\rho U^2/2)] \sin \theta$ versus θ and $F(\theta) \cos \theta$ versus θ plotted in Figs. E9.14b and E9.14c. The results are

$$\int_0^\pi \frac{(p - p_0)}{\frac{1}{2}\rho U^2} \sin \theta \, d\theta = -1.76$$

and

$$\int_0^\pi F(\theta) \cos \theta \, d\theta = 3.92$$

Thus, the lift is

$$\mathcal{L} = \frac{1}{2} \rho U^2 A \left[\left(-\frac{1}{2} \right) (-1.76) + \frac{1}{2\sqrt{\mathrm{Re}}} (3.92) \right]$$

or

$$\mathcal{L} = \left(0.88 + \frac{1.96}{\sqrt{\mathrm{Re}}} \right) \left(\frac{1}{2} \rho U^2 A \right) \tag{Ans}$$

and

$$C_L = \frac{\mathcal{L}}{\frac{1}{2}\rho U^2 A} = 0.88 + \frac{1.96}{\sqrt{\mathrm{Re}}} \tag{4} \quad \text{(Ans)}$$

Consider a typical situation with $D = 20$ ft, $U = 30$ ft/s, $b = 50$ ft, and standard atmospheric conditions ($\rho = 2.38 \times 10^{-3}$ slugs/ft^3 and $v = 1.57 \times 10^{-4}$ ft^2/s), which gives a Reynolds number of

$$\mathrm{Re} = \frac{UD}{v} = \frac{(30 \text{ ft/s})(20 \text{ ft})}{1.57 \times 10^{-4} \text{ ft}^2/\text{s}} = 3.82 \times 10^6$$

Hence, the lift coefficient is

$$C_L = 0.88 + \frac{1.96}{(3.82 \times 10^6)^{1/2}} = 0.88 + 0.001 = 0.881$$

Note that the pressure contribution to the lift coefficient is 0.88 whereas that due to the wall shear stress is only $1.96/(\mathrm{Re}^{1/2}) = 0.001$. The Reynolds number dependency of C_L is quite minor. The lift is pressure dominated. Recall from Example 9.9 that this is also true for the drag on a similar shape.

From Eq. 4, we obtain the lift for the assumed conditions as

$$\mathcal{L} = \tfrac{1}{2}\rho U^2 A C_L = \tfrac{1}{2}(2.38 \times 10^{-3} \text{ slugs/ft}^3)(30 \text{ ft/s})^2 (20 \text{ ft} \times 50 \text{ ft})(0.881)$$

or

$$\mathcal{L} = 944 \text{ lb}$$

There is a considerable tendency for the building to lift off the ground. Clearly this is due to the object being nonsymmetrical. The lift force on a complete circular cylinder is zero, although the fluid forces do tend to pull the upper and lower halves apart.

A typical device designed to produce lift does so by generating a pressure distribution that is different on the top and bottom surfaces. For large Reynolds number flows these pressure distributions are usually directly proportional to the dynamic pressure, $\rho U^2/2$, with viscous effects being of secondary importance. Two airfoils used to produce lift are indicated in Fig. 9.35. Clearly the symmetrical one cannot produce lift unless the angle of attack, α, is nonzero. Because of the asymmetry of the non-symmetric airfoil, the pressure distributions on the upper and lower surfaces are different, and a lift is produced even with $\alpha = 0$. Of course, there will be a certain value of α (less than zero for this case) for which the lift is zero. For this situation, the pressure distributions on the upper and lower surfaces are different, but their resultant (integrated) pressure forces will be equal and opposite.

Since most airfoils are thin, it is customary to use the planform area, $A = bc$, in the definition of the lift coefficient. Here b is the length of the airfoil and c is the *chord length*—the length from the leading edge to the trailing edge as indicated in Fig. 9.35. Typical lift coefficients so defined are on the order of unity. That is, the lift force is on the order of the dynamic pressure times the planform area of the wing, $\mathcal{L} \approx (\rho U^2/2)A$. The *wing loading,* defined as the average lift per unit area of the wing, $\mathcal{L}/A$, therefore, increases with speed. For example, the wing loading of the 1903 Wright Flyer aircraft was 1.5 lb/ft^2, while for the present-day Boeing 747 aircraft it is 150 lb/ft^2. The wing loading for a bumble bee is approximately 1 lb/ft^2 (Ref. 15).

Typical lift and drag coefficient data as a function of angle of attack, α, and *aspect ratio, $\mathcal{A}$,* are indicated in Figs. 9.36a and 9.36b. The aspect ratio is defined as the ratio of the square of the wing length to the planform area, $\mathcal{A} = b^2/A$. If the chord length, c, is constant along the length of the wing this reduces to $\mathcal{A} = b/c$.

In general, the lift coefficient increases and the drag coefficient decreases with an increase in aspect ratio. Long wings are more efficient because their wing tip losses are relatively more minor than for short wings. The increase in drag due to the finite length ($\mathcal{A} < \infty$) of the wing is often termed induced drag. It is due to the interaction of the complex swirling flow structure near the wing tips (see Fig. 9.40) and the free stream (Ref. 13). High-performance soaring airplanes and highly efficient soaring

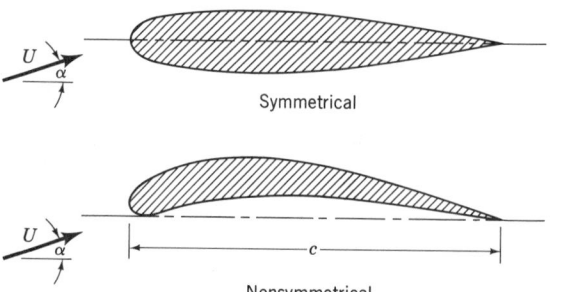

FIGURE 9.35 Symmetrical and unsymmetrical airfoils.

(a)

(b)

FIGURE 9.36 Typical lift and drag coefficient data as a function of angle of attack and the aspect ratio of the airfoil: (a) lift coefficient, (b) drag coefficient.

birds (i.e., the albatross and sea gull) have long, narrow wings. Such wings, however, have considerable inertia that inhibits rapid maneuvers. Thus, highly maneuverable fighter or acrobatic airplanes and birds (ie., the falcon) have small aspect ratio wings.

Although viscous effects and the wall shear stress contribute little to the direct generation of lift, they play an extremely important role in the design and use of lifting devices. This is because of the viscosity-induced boundary layer separation that can occur on nonstreamlined bodies, for example, airfoils, that have too large an angle of attack (see Fig. 9.18). As is indicated in Fig. 9.36, up to a certain point, the lift coefficient increases rather steadily with the angle of attack. If α is too large, the boundary layer on the upper surface separates, the flow over the wing develops a wide, turbulent wake region, the lift decreases, and the drag increases. The airfoil

stalls. Such conditions are extremely dangerous if they occur while the airplane is flying at a low altitude where there is not sufficient time and altitude to recover from the stall.

In many lift-generating devices the important quantity is the ratio of the lift to drag developed, $\mathcal{L}/\mathcal{D} = C_L/C_D$. Such information is often presented in terms of C_L/C_D versus α, as is shown in Fig. 9.37a, or in a *lift-drag polar* of C_L versus C_D with α as a parameter, as is shown in Fig. 9.37b. The most efficient angle of attack (i.e., largest C_L/C_D) can be found by drawing a line tangent to the $C_L - C_D$ curve from the origin, as is shown in Fig. 9.37b. High-performance airfoils generate lift that is perhaps 100 or more times greater than their drag. This translates into the fact that in still air they can glide a horizontal distance of 100 m for each 1 m drop in altitude.

As is indicated above, the lift and drag on an airfoil can be altered by changing the angle of attack. This actually represents a change in the shape of the object. Other shape changes can be used to alter the lift and drag when desirable. In modern airplanes it is commmon to utilize leading edge and trailing edge flaps as is shown in Fig. 9.38. To generate the necessary lift during the relatively low-speed landing and takeoff procedures, the airfoil shape is altered by extending special flaps on the front and/or rear portions of the wing. Use of the flaps considerably enhances the lift, although it is at the expense of an increase in the drag (the airfoil is in a "dirty" configuration). This increase in drag is not of much concern during landing and takeoff operations—the decrease in landing or takeoff speed is more important than is a temporary increase in drag. During normal flight with the flaps retracted (the "clean"

FIGURE 9.37 Two representations of the same lift and drag data for a typical airfoil: (a) lift-to-drag ratio as a function of angle of attack, with the onset of boundary layer separation on the upper surface indicated by the occurrence of stall, (b) the lift and drag polar diagram with the angle of attack indicated (Ref. 27).

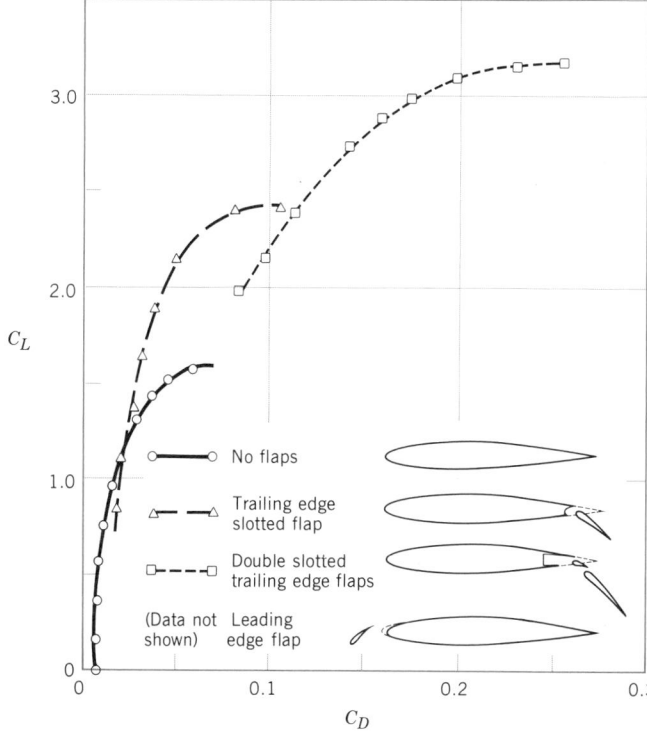

FIGURE 9.38 Typical lift and drag alterations possible with the use of various types of flap designs (Ref. 21).

configuration), the drag is relatively small, and the neeeded lift force is achieved with the smaller lift coefficient and the larger dynamic pressure (higher speed).

The use of the complex flap systems for modern aircraft has proved to be an important breakthrough of aeronautics. Actually, certain birds use the leading edge flap concept. Some species have special feathers on the leading edge of their wings that extend as a leading edge flap when low-speed flight is required (such as when their wings are fully extended during landing) (Ref. 15).

A wide variety of lift and drag information for airfoils can be found in standard aerodynamics books (Ref. 13, 14).

EXAMPLE 9.15

In 1977 the *Gossamer Condor* won the Kremer prize by being the first human-powered aircraft to complete a prescribed figure-of-eight course around two turning points 0.5 mi apart (Ref. 22). The following data pertain to this aircraft:

$$\begin{aligned}
\text{flight speed} &= U = 15 \text{ ft/s} \\
\text{wing size} &= b = 96 \text{ ft}, c = 7.5 \text{ ft (average)} \\
\text{weight (including pilot)} &= W = 210 \text{ lb} \\
\text{drag coefficient} &= C_D = 0.046 \text{ (based on planform area)} \\
\text{power train efficiency} &= \eta = \text{power to overcome drag/pilot power} = 0.8
\end{aligned}$$

Determine the lift coefficient, C_L, and the power, $\mathscr{P}$, required by the pilot.

SOLUTION

For steady flight conditions the lift must be exactly balanced by the weight, or

$$W = \mathscr{L} = \tfrac{1}{2}\rho U^2 A C_L$$

Thus,

$$C_L = \frac{2W}{\rho U^2 A}$$

where $A = bc = 96 \text{ ft} \times 7.5 \text{ ft} = 720 \text{ ft}^2$, $W = 210$ lb, and $\rho = 2.38 \times 10^{-3}$ slug/ft^3 for standard air. This gives

$$C_L = \frac{2(210 \text{ lb})}{(2.38 \times 10^{-3} \text{ slugs/ft}^3)(15 \text{ ft/s})^2(720 \text{ ft}^2)}$$

or

$$C_L = 1.09 \qquad \text{(Ans)}$$

a reasonable number. The overall lift to drag ratio for the aircraft is $C_L/C_D = 1.09/0.046 = 23.7$.

The product of the power that the pilot supplies and the power train efficiency equals the useful power needed to overcome the drag, $\mathscr{D}$. That is,

$$\eta \mathscr{P} = \mathscr{D} U$$

where

$$\mathscr{D} = \tfrac{1}{2}\rho U^2 A C_D$$

Thus,

$$\mathscr{P} = \frac{\mathscr{D} U}{\eta} = \frac{\tfrac{1}{2}\rho U^2 A C_D U}{\eta} = \frac{\rho A C_D U^3}{2\eta}$$

or

$$\mathscr{P} = \frac{(2.38 \times 10^{-3} \text{ slugs/ft}^3)(720 \text{ ft}^2)(0.046)(15 \text{ ft/s})^3}{2(0.8)}$$

$$\mathscr{P} = 166 \text{ ft·lb/s} \left(\frac{1 \text{ hp}}{550 \text{ ft·lb/s}}\right) = 0.302 \text{ hp} \qquad \text{(Ans)}$$

This power level is obtainable by a well-conditioned athlete (as is indicated by the fact that the flight was successfully completed). Note that only 80% of the pilot's power (i.e., $0.8 \times 0.302 = 0.242$ hp, which corresponds to a drag of $\mathscr{D} = 8.86$ lb) is needed to force the aircraft through the air. The other 20% is lost because of the power train inefficiency. Note also that for a constant drag coefficient, the power required increases as U^3—a doubling of the speed to 30 ft/s would require an eightfold increase in power (i.e., 2.42 hp, well beyond the range of any human).

9.4.2 Circulation

Since viscous effects are of minor importance in the generation of lift, it should be possible to calculate the lift force on an airfoil by integrating the pressure distribution obtained from the equations governing inviscid flow past the airfoil. That is, the

potential flow theory discussed in Chapter 6 should provide a method to determine the lift. Although the details are beyond the scope of this book, the following is found from such calculations (Ref. 4).

The calculation of the inviscid flow past a two-dimensional airfoil gives a flow field as indicated in Fig. 9.39. The predicted flow field past an airfoil with no lift (i.e., a symmetrical airfoil at zero angle of attack, Fig. 9.39a) appears to be quite accurate (except for the absence of thin boundary layer regions). However, as is indicated in Fig. 9.39b, the calculated flow past the same airfoil at a nonzero angle of attack (but one small enough so that boundary layer separation would not occur) is not proper near the trailing edge. In addition, the calculated lift for a nonzero angle of attack is zero—in conflict with the known fact that such airfoils produce lift.

In reality, the flow should pass smoothly over the top surface as is indicated in Fig. 9.39c, without the strange behavior indicated near the trailing edge in Fig. 9.39b. As is shown in Fig. 9.39d, the unrealistic flow situation can be corrected by adding an appropriate clockwise swirling flow around the airfoil. The results are twofold: (1) the unrealistic behavior near the trailing edge is eliminated (i.e., the flow pattern of Fig. 9.39b is changed to that of Fig. 9.39c), and (2) the average velocity on the upper surface of the airfoil is increased while that on the lower surface is decreased. From

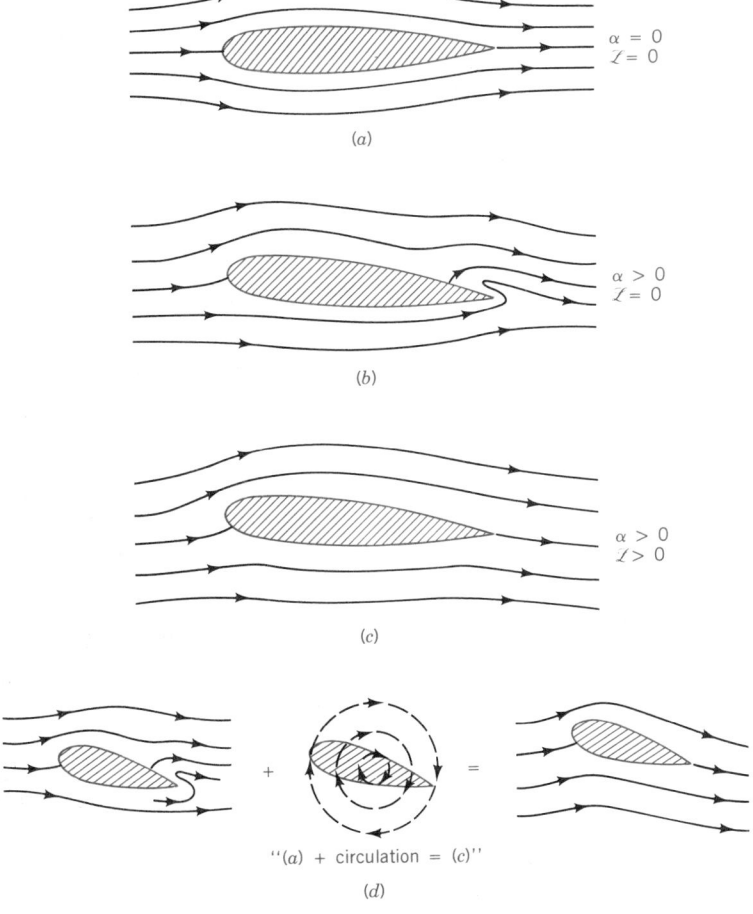

FIGURE 9.39 Inviscid flow past an airfoil: (a) symmetrical flow past the symmetrical airfoil at a zero angle of attack, (b) same airfoil at a nonzero angle of attack—no lift, flow near trailing edge not realistic, (c) same conditions as for (b) except circulation has been added to the flow—nonzero lift, realistic flow, (d) superposition of flows to produce the final flow past the airfoil.

the Bernoulli equation concepts (i.e., $p/\gamma + V^2/2g + z$ = constant), the average pressure on the upper surface is decreased and that on the lower surface is increased. The net effect is to change the original zero lift condition to that of a lift-producing airfoil.

The addition of the clockwise swirl is termed the addition of *circulation*. The amount of swirl (circulation) needed to have the flow leave the trailing edge smoothly is a function of the airfoil size and shape and can be calculated from potential flow (inviscid) theory (see Section 6.6.3). Although the addition of circulation to make the flow field physically realistic may seem artificial, it has well-founded mathematical and physical grounds. For example, consider the flow past a finite length airfoil, as is indicated in Fig. 9.40. For lift-generating conditions the average pressure on the lower surface is greater than that on the upper surface. Near the tips of the wing this pressure difference will cause some of the fluid to attempt to migrate from the lower to the upper surface, as is indicated in Fig. 9.40b. At the same time, this fluid is swept downstream, forming a *trailing vortex* (swirl) from each wing tip (see Fig. 4.3). It is speculated that the reason some birds migrate in vee-formation is to take advantage of the updraft produced by the trailing vortex of the preceding bird. [It is calculated that for a given expenditure of energy, a flock of 25 birds flying in vee-formation could travel 70% further than if each bird were to fly separately (Ref. 15.)]

The trailing vortices from the right and left wing tips are connected by the *bound vortex* along the length of the wing. It is this vortex that generates the circulation that produces the lift. The combined vortex system of the bound vortex and the trailing vortices is termed a horseshoe vortex. The strength of the trailing vortices (which is equal to the strength of the bound vortex) is proportional to the lift generated. Large

FIGURE 9.40
Flow past a finite length wing: (*a*) the horseshoe vortex system produced by the bound vortex and the trailing vorticies; (*b*) the leakage of air around the wing tips produces the trailing vorticies.

aircraft (for example, a Boeing 747) can generate very strong trailing vortices that persist for a long time before viscous effects finally cause them to die out. Such vortices are strong enough to flip smaller aircraft out of control if they follow too closely behind the large aircraft.

As is indicated above, the generation of lift is directly related to the production of a swirl or vortex flow around the object. A nonsymmetric airfoil, by design, generates its own prescribed amount of swirl and lift. A symmetric object like a circular cylinder or sphere, which normally provides no lift, can generate swirl and lift if it rotates.

As is discusssed in Section 6.6.3, the inviscid flow past a circular cylinder has the symmetrical flow pattern indicated in Fig. 9.41a. By symmetry the lift and drag are zero. However, if the cylinder is rotated about its axis in a stationary real ($\mu \neq$ 0) fluid, the rotation will drag some of the fluid around, producing circulation about the cylinder as in Fig. 9.41b. When this circulation is combined with an ideal, uniform upstream flow, the flow pattern indicated in Fig. 9.41c is obtained. The flow is no longer symmetrical about the horizontal plane through the center of the cylinder; the average pressure is greater on the lower half of the cylinder than on the upper half, and a lift is generated. This effect is called the *Magnus effect,* after Heinrich Magnus (1802–1870), a German chemist and physicist who first investigated this phenomenon. A similar lift is generated on a rotating sphere. It accounts for the various types of pitches in baseball (i.e., curve ball, floater, sinker, etc.), the ability of a soccer player to hook the ball, and the hook or slice of a golf ball.

Typical lift and drag coefficients for a smooth spinning sphere are shown in Fig. 9.42. Although the drag coefficient is fairly independent of the rate of rotation, the lift coefficient is strongly dependent on it. In addition (although not indicated in the figure), both C_L and C_D are dependent on the roughness of the surface. As was discussed in Section 9.3, in a certain Reynolds number range an increase in surface roughness actually decreases the drag coefficient. Similarly, an increase in surface roughness can increase the lift coefficient because the roughness helps drag more fluid around the sphere, increasing the circulation for a given angular velocity. Thus, a rotating rough golf ball travels farther than a smooth one because the drag is less and the lift is greater. However, do not expect a severely roughed up (cut) ball to work better—extensive testing has gone into obtaining the optimum surface roughness for golf balls.

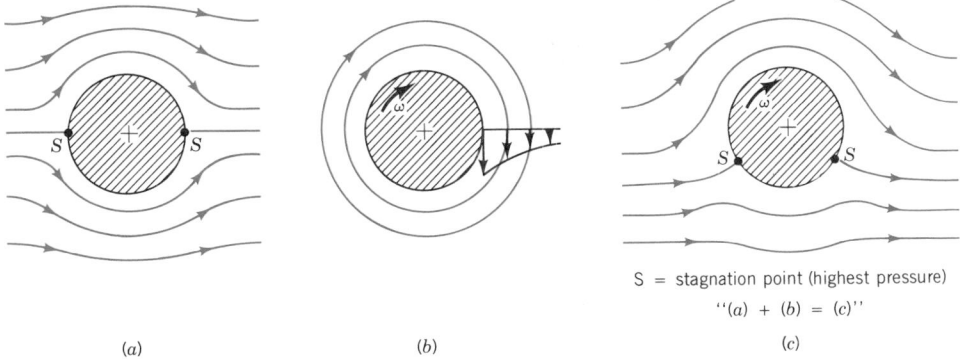

S = stagnation point (highest pressure)

"(a) + (b) = (c)"

(a)   (b)   (c)

FIGURE 9.41 Inviscid flow past a circular cylinder: (a) uniform upstream flow without circulation; (b) free vortex at the center of the cylinder, (c) combination of free vortex and uniform flow past a circular cylinder giving nonsymmetric flow and a lift.

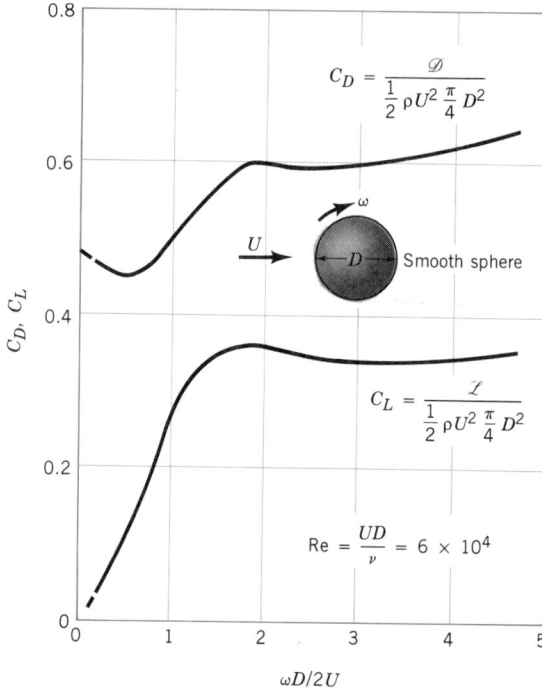

FIGURE 9.42 Lift and drag coefficients for a spinning smooth sphere (Ref. 23).

EXAMPLE 9.16

A table tennis ball weighing 2.45×10^{-2} N with diameter $D = 3.8 \times 10^{-2}$ m is hit at a velocity of $U = 12$ m/s with a back spin of angular velocity ω as is shown in Fig. E9.16. What is the value of ω if the ball is to travel on a horizontal path, not dropping due to the accleration of gravity?

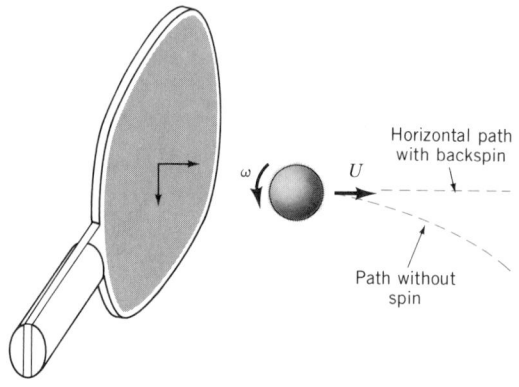

FIGURE E9.16

SOLUTION

For horizontal flight, the lift generated by the spinning of the ball must exactly balance the weight, W, of the ball so that

$$W = \mathscr{L} = \tfrac{1}{2}\rho U^2 A C_L$$

or

$$C_L = \frac{2W}{\rho U^2 (\pi/4) D^2}$$

where the lift coefficient, C_L, can be obtained from Fig. 9.42. For standard atmospheric conditions with $\rho = 1.23$ kg/m^3 we obtain

$$C_L = \frac{2(2.45 \times 10^{-2} \text{ N})}{(1.23 \text{ kg/m}^3)(12 \text{ m/s})^2(\pi/4)(3.8 \times 10^{-2} \text{ m})^2} = 0.244$$

which, according to Fig. 9.42, can be achieved if

$$\frac{\omega D}{2U} = 0.9$$

or

$$\omega = \frac{2U(0.9)}{D} = \frac{2(12 \text{ m/s})(0.9)}{3.8 \times 10^{-2} \text{ m}} = 568 \text{ rad/s}$$

Thus,

$$\omega = (568 \text{ rad/s})(60 \text{ s/min})(1 \text{ rev}/2\pi \text{ rad}) = 5420 \text{ rpm} \qquad \text{(Ans)}$$

Is it possible to impart this angular velocity to the ball? With larger angular velocities the ball will rise and follow an upward curved path. Similar trajectories can be produced by a well-hit golf ball—rather than falling like a rock, the golf ball trajectory is actually curved up and the spinning ball travels a greater distance than one without spin. However, if top spin is imparted to the ball (as in an improper tee shot) the ball will curve downward more quickly than under the action of gravity alone—the ball is "topped" and a negative lift is generated. Similarly, rotation about a vertical axis will cause the ball to hook or slice to one side or the other.

REFERENCES

1. Schlichting, H., *Boundary Layer Theory*, Seventh Edition, McGraw-Hill, New York, 1979.

2. Rosenhead, L., *Laminar Boundary Layers*, Oxford University Press, London, 1963.

3. White, F. M., *Viscous Fluid Flow*, McGraw-Hill, New York, 1974.

4. Currie, I. G., *Fundamental Mechanics of Fluids*, McGraw-Hill, New York, 1974.

5. Blevins, R. D., *Applied Fluid Dynamics Handbook*, Van Nostrand Reinhold, New York, 1984.

6. Hoerner, S. F., *Fluid-Dynamic Drag*, published by the author, Library of Congress No. 64,19666, 1965.

7. Happel, J., *Low Reynolds Number Hydrodynamics*, Prentice Hall, Englewood Cliffs, NJ, 1965.

8. Van Dyke, M., *An Album of Fluid Motion*, Parabolic Press, Stanford, Calif., 1982.

9. Thompson, P. A., *Compressible-Fluid Dynamics*, McGraw-Hill, New York, 1972.

10. Zucrow, M. J., and Hoffman, J. D., *Gas Dynamics, Vol. I*, Wiley, New York, 1976.

11. Clayton, B. R., and Bishop, R. E. D., *Mechanics of Marine Vehicles,* Gulf Publishing Co., Houston, 1982.

12. *CRC Handbook of Tables for Applied Engineering Science,* Second Edition, CRC Press, 1973.

13. Shevell, R. S., *Fundamentals of Flight,* Second Edition, Prentice-Hall, Englewood Cliffs, NJ, 1989.

14. Kuethe, A. M. and Chow, C. Y., *Foundations of Aerodynamics, Bases of Aerodynamics Design,* Fourth Edition, Wiley, 1986.

15. Vogel, J., *Life in Moving Fluids,* Willard Grant Press, Boston, 1981.

16. Kreider, J. F., *Principles of Fluid Mechanics,* Allyn and Bacon, Newton, Mass., 1985.

17. Dobrodzicki, G. A., Flow Visualization in the National Aeronautical Establishment's Water Tunnel, National Research Council of Canada, Aeronautical Report LR-557, 1972.

18. White, F. M., *Fluid Mechanics,* McGraw-Hill, New York, 1986.

19. Vennard, J. K., and Street, R. L., *Elementary Fluid Mechanics,* Sixth Edition, Wiley, New York, 1982.

20. Gross, A. C., Kyle, C. R., and Malewicki, D. J., The Aerodynamics of Human Powered Land Vehicles, *Scientific American,* Vol. 249, No. 6, 1983.

21. Abbott, I. H., and Von Doenhoff, A. E., *Theory of Wing Sections,* Dover Publications, New York, 1959.

22. MacReady, P. B., "Flight on 0.33 Horsepower: The Gossamer Condor," *Proc. AIAA 14th Annual Meeting* (Paper No. 78-308), Washington, DC, 1978.

23. Goldstein, S., *Modern Developments in Fluid Dynamics,* Oxford Press, London, 1938.

24. Achenbach, E., Distribution of Local Pressure and Skin Friction around a Circular Cylinder in Cross-Flow up to Re = 5×10^6, *Journal of Fluid Mechanics,* Vol. 34, Pt. 4, 1968.

25. Inui, T., Wave-Making Resistance of Ships, *Transactions of the Society of Naval Architects and Marine Engineers,* Vol. 70, 1962.

26. Sovran, G. et al. (ed.), *Aerodynamic Drag Mechanisms of Bluff Bodies and Road Vehicles,* Plenum Press, New York, 1978.

27. Abbott, I. H., von Doenhoff, A. E. and Stivers, L. S., Summary of Airfoil Data, NACA Report No. 824, Langley Field, Va., 1945.

28. Society of Automotive Engineers Report HSJ1566, "Aerodynamic Flow Visualization Techniques and Procedures," 1986.

PROBLEMS

Note: Unless otherwise indicated use the values of fluid properties found in the tables on the inside of the front cover. Problems designated with an (*) are intended to be solved with the aid of a programmable calculator or a computer.

9.1 Assume that water flowing past the equilateral triangular bar shown in Fig. P9.1 produces the pressure distributions indicated. Determine the lift and drag on the bar and the corresponding lift and drag coefficients (based on frontal area). Neglect shear forces.

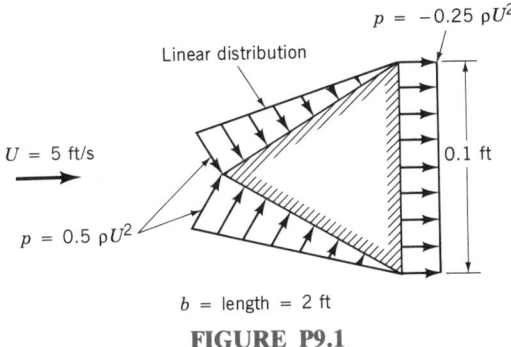

FIGURE P9.1

r (m)	p (kN/m²)
0	4.34
0.05	4.28
0.10	4.06
0.15	3.72
0.20	3.10
0.25	2.78
0.30	2.37
0.35	1.89
0.40	1.41
0.45	0.74
0.50	0.0

9.2 Determine the lift and drag coefficients (based on frontal area) for the triangular two-dimensional object shown in Fig. P9.2. Neglect shear forces.

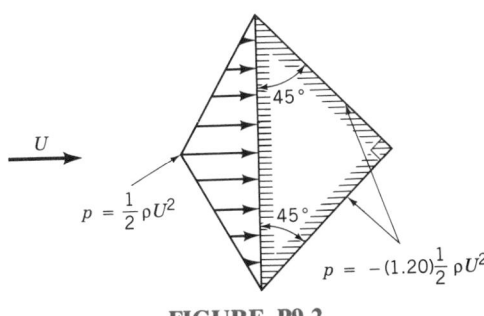

FIGURE P9.2

9.3 The average pressure and shear stress acting on the surface of the 1-m-square flat plate are as indicated in Fig. P9.3. Determine the lift and drag generated. Determine the lift and drag if the shear stress is neglected. Compare these two sets of results. $\iint p \, dA = P_{ave} A$ & $\int \tau_w \, dA = \tau_{ave} A$

$\mathcal{D} = -p_1 A_1 \sin\alpha + p_2 A_2 \sin\alpha + \tau_1 A_1 \cos\alpha + \tau_2 A_1 \cos\alpha$

$w/A_1 = A_2 \quad /\alpha = 7°$

$\mathcal{D} = A_1 \sin\alpha (p_2 - p_1) + A_1 \cos\alpha (\tau_1 + \tau_2)$

$P_{ave} = -1.2 \text{ kN/m}^2$
$\tau_{ave} = 5.8 \times 10^{-2} \text{ kN/m}^2$

if shear stress is neglected
$\mathcal{D} = .427 \quad \tau_1 = \tau_2 = 0$

$\mathcal{L} = A_1 \cos\alpha (p_1 - p_2) \quad P_{ave} = 2.3 \text{ kN/m}^2$
$\tau_{ave} = 7.6 \times 10^{-2} \text{ kN/m}^2$

$- A_1 \sin\alpha (\tau_1 - \tau_2)$

if shear stress is neglected $\tau_1 = \tau_2 = 0$
$\mathcal{L} = 0$

FIGURE P9.3

9.4* The pressure distribution on the 1-m-diameter circular disk Fig. P9.4 is given in the table below. Determine the drag on the disk.

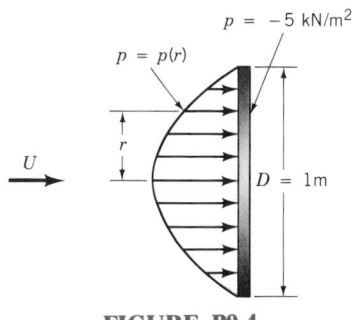

FIGURE P9.4

9.5 A small 15-mm-long fish swims with a speed of 20 mm/s. Would a boundary layer type flow be developed along the sides of the fish? Explain. $Re = \frac{U \ell}{\nu}$ or $w/\ell = 5 ft$ $U = 20 \, mm/s \, /V = 1.2 \times 10^{-5} \, ft^2/s$
$\therefore Re = 1.24 \times 10^{7}$ yes. 60°F H₂O

9.6 A 5-ft-long porpoise swims with a speed of 30 ft/s. Would a boundary layer type flow be developed along the sides of the animal? Explain.

9.7 Approximately how fast can the wind blow past a 0.25-in.-diameter twig if viscous effects are to be of importance throughout the entire flow field (i.e., Re < 1)? Explain. Repeat for a 0.004-in.-diameter hair and a 6-ft-diameter smokestack. $Re = \frac{U \ell}{\nu}$ $/V = 1.46 \times 10^{-5}$, $/V = 1 \, m/s$ assume $\ell = 1 m$
$\therefore Re = 6.85 \times 10^{9}$ all will

9.8 When you walk through still air at a rate of 1 m/s, would you expect the character of the air flow around you to be most like that depicted in Fig. 9.6a, b, or c? Explain.

9.9 A 0.23-m-diameter soccer ball moves through the air with a speed of 10 m/s. Would the flow around the ball be classified as low, moderate, or large Reynolds number flow? Explain.

9.10 The mechanism used to generate lift is different for "large" objects (i.e., airplanes, birds, butterflies, gnats, etc.) than it is for "extremely small" objects (i.e., microscopic airborne and wa-

terborne creatures). Typically large objects have wings that are planar surfaces for which viscous effects are not important, whereas microscopic objects have wings that consists of a central strut with numerous protruding hairs for which viscous effects are very important. If the condition that separates these two types of objects is whether Re > 0.1 or Re < 0.1, make an order of magnitude estimate of the maximum flight speed of the "extremely small" objects.

9.11   A viscous fluid flows past a flat plate such that the boundary layer thickness at a distance 1.3 m from the leading edge is 12 mm. Determine the boundary layer thickness at distances of 0.20, 2.0, and 20 m from the leading edge. Assume laminar flow.

9.12   If the upstream velocity of the flow in Problem 9.11 is $U = 1.5$ m/s, determine the kinematic viscosity of the fluid.

9.13   Air flows over a flat plate of length $\ell = 2$ ft such that the Reynolds number based on the plate length is Re $= 2 \times 10^5$. Plot the boundary layer thickness, δ, for $0 \le x \le \ell$.

9.14   Water flows past a flat plate with an upstream velocity of $U = 0.03$ m/s. Determine the water velocity a distance of 10 mm from the plate at distances of $x = 1.5$ m and $x = 15$ m from the leading edge.

9.15*   A Pitot tube is used to measure the total pressure within a boundary layer as indicated in Fig. P9.15. Based on the data given in the table below, determine the boundary layer thickness, δ, the displacement thickness, δ^*, and the momentum thickness, Θ.

6.4	25.6
10.7	32.5
15.0	36.9
19.3	39.4
23.6	40.5
26.8	41.0
29.3	41.0
32.7	41.0

9.16   Because of the velocity deficit, $U - u$, in the boundary layer, the streamlines for flow past a flat plate are not exactly parallel to the plate. This deviation can be determined by use of the displacement thickness, δ^*. For air blowing past the flat plate shown in Fig. P9.16, plot the streamline A–B that passes through the edge of the boundary layer ($y = \delta_B$ at $x = \ell$) at point B. That is, plot $y = y(x)$ for streamline A–B. Assume laminar boundary layer flow.

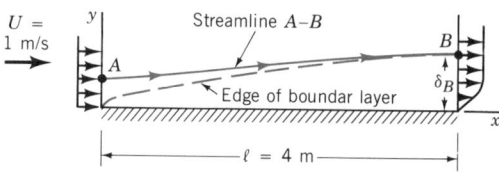

FIGURE P9.16

9.17   Air enters a square duct through a 1-ft opening as is shown in Fig. P9.17. Because the boundary layer displacement thickness increases in the direction of flow, it is necessary to increase the cross-sectional size of the duct if a constant $U = 2$ ft/s velocity is to be maintained outside the boundary layer. Plot a graph of the duct size, d, as a function of x for $0 \le x \le 10$ ft if U is to remain constant. Assume laminar flow.

FIGURE P9.17

FIGURE P9.15

y (mm)	h (mm)
0	0
2.1	10.6
4.3	21.1

9.18   A smooth flat plate of length $\ell = 4$ m and width $b = 6$ m is placed in water with an upstream velocity of $U = 0.5$ m/s. Determine the

Handwritten (top margin): $Re_{x_{cr}} = \dfrac{U x_{cr}}{v}$ whle $U = 200$ mph $(\tfrac{?}{360u})\tfrac{ft}{s} = 293\,ft/s$

, table c.1, $v = \tfrac{\mu}{\rho} = \dfrac{3.84 \times 10^{-7}}{1.756 \times 10^{-3}} = 2.01 \times 10^{-4}$

hence $w/Re_{x_{cr}} = 5 \times 10^5$ $x_{cr} = \sqrt{\dfrac{Re_{x_{cr}}}{U}} = .$

boundary layer thickness and the wall shear stress at the center and the trailing edge of the plate. Assume a laminar boundary layer.

9.19 An atmospheric boundary layer is formed when the wind blows over the earth's surface. Typically, such velocity profiles can be written as a power law: $u = ay^n$, where the constants a and n depend on the roughness of the terrain. As is indicated in Fig. P9.19, typical values are $n = 0.40$ for urban areas, $n = 0.28$ for woodland or suburban areas, and $n = 0.16$ for flat open country (Ref. 23). **(a)** If the velocity is 20 ft/s at the bottom of the sail on your boat ($y = 4$ ft), what is the velocity at the top of the mast ($y = 30$ ft)? **(b)** If the average velocity is 10 mph on the tenth floor of an urban building, what is the average velocity on the sixtieth floor?

FIGURE P9.19

9.20 A 15-story office building (each story is 12 ft tall) is built in a suburban industrial park. Plot the dynamic pressure, $\rho u^2/2$, as a function of elevation if the wind blows at hurricane strength (75 mph) at the top of the building. Use the atmospheric boundary layer information of Problem 9.19. [handwritten: Fig. 9-19 B.L. vel $u \sim y^{0.28}$ $ay^{0.28} = c + ay^{0.28}$ a constant]

9.21 The typical shape of small cumulous clouds is as indicated in Fig. P9.21. Based on boundary layer ideas, explain why it is clear that the wind is blowing from right to left as indicated.

FIGURE P9.21

9.22 Show that for any function $f = f(\eta)$ the velocity components u and v determined by Eqs. 9.12 and 9.13 satisfy the incompressible continuity equation, Eq. 9.8.

Handwritten: $\therefore \dfrac{v}{u_i} = \left(\dfrac{y}{y_i}\right)^{0.28} \Rightarrow v = 110\left(\dfrac{y}{180}\right)^{0.28}$

hen $\tfrac12 \rho v^2 = \tfrac12 (2.38 \times 10^{-3})\left[110\left(\dfrac{y}{180}\right)^{0.28}\right]^2$

$\tfrac12 \rho v^2 = 14.4 \left(\dfrac{y}{180}\right)^{0.56}$

9.23 Show that by writing the velocity in terms of the similarity variable η and the function $f(\eta)$ the momentum equation for boundary layer flow on a flat plate (Eq. 9.9) can be written as the ordinary differential equation given by Eq. 9.14.

9.24* Integrate the Blasius equation (Eq. 9.14) numerically to determine the boundary layer profile for laminar flow past a flat plate. Compare your results with those of Table 9.1.

9.25 An airplane flies at a speed of 200 mph at an altitude of 10,000 ft. If the boundary layers on the wing surfaces behave as those on a flat plate, estimate the extent of laminar boundary layer flow along the wing. Assume a transitional Reynolds number of $Re_{xcr} = 5 \times 10^5$.

9.26 If the airplane in Problem 9.25 maintains its 200-mph speed but descends to sea-level elevation, will the portion of the wing covered by a laminar boundary layer increase or decrease compared with its value at 10,000 ft? Explain.

9.27 A laminar boundary layer velocity profile is approximated by $u/U = 2(y/\delta) - 2(y/\delta)^3 + (y/\delta)^4$ for $y \le \delta$, and $u = U$ for $y > \delta$. Use the momentum integral equation to determine the boundary layer thickness, $\delta = \delta(x)$.

9.28 As is indicated in Table 9.2, the laminar boundary layer results obtained from the momentum integral equation are relatively insensitive to the shape of the assumed velocity profile. Consider the profile given by $u = U$ for $y > \delta$, and $u = U\{1 - [(y - \delta)/\delta]^2\}^{1/2}$ for $y \le \delta$ as shown in Fig. P9.28. Note that this satisfies the conditions $u = 0$ at $y = 0$ and $u = U$ at $y = \delta$. However, show that such a profile produces meaningless results when used with the momentum integral equation. Explain.

FIGURE P9.28

9.29 A laminar boundary layer velocity profile is approximated by the two straight-line segments indicated in Fig. P9.29. Use the momentum integral equation to determine the boundary layer thickness, $\delta = \delta(x)$, and wall shear stress,

Handwritten (bottom right): $v_1 = 75$ mph $\Rightarrow \tfrac{80}{60}$ $= 110$ ft/s

$y_1 = 15 \times 12$ $= 180$ ft.

$\tau_w = \tau_w(x)$. Compare these results with those in Table 9.2.

FIGURE P9.29

9.30* An assumed dimensionless laminar boundary layer profile for flow past a flat plate is given in the table below. Use the momentum integral equation to determine $\delta = \delta(x)$. Compare your result with the exact Blasius solution result (see Table 9.2).

y/δ	u/U
0	0
0.080	0.133
0.16	0.265
0.24	0.394
0.32	0.517
0.40	0.630
0.48	0.729
0.56	0.811
0.64	0.876
0.72	0.923
0.80	0.956
0.88	0.976
0.96	0.988
1.00	1.000

9.31* For a fluid of specific gravity $SG = 0.86$ flowing past a flat plate with an upstream velocity of $U = 5$ m/s, the wall shear stress on a flat plate was determined to be as indicated in the table below. Use the momentum integral equation to determine the boundary layer momentum thickness, $\Theta = \Theta(x)$. Assume $\Theta = 0$ at the leading edge, $x = 0$.

x (m)	τ_w (N/m²)
0	—
0.2	13.4
0.4	9.25
0.6	7.68
0.8	6.51

1.0	5.89
1.2	6.57
1.4	6.75
1.6	6.23
1.8	5.92
2.0	5.26

9.32 A laminar boundary layer formed on one side of a plate of length ℓ produces a drag $\mathcal{D}$. How much must the plate be shortened if the drag on the new plate is to be $\mathcal{D}/2$? Assume the upstream velocity remains the same. Explain your answer physically.

9.33 The net drag on one side of the two plates (each of size ℓ by $\ell/2$) parallel to the free stream shown in Fig. P9.33a is $\mathcal{D}$. Determine the drag (in terms of $\mathcal{D}$) on the same two plates when they are connected together as indicated in Fig. P9.33b. Assume laminar boundary flow. Explain your answer physically.

FIGURE P9.33

9.34 A plate is oriented parallel to the free stream as is indicated in Fig. 9.34. If the boundary layer flow is laminar, determine the ratio of the drag for case a to that for case b. Explain your answer physically.

FIGURE P9.34

9.35 If the drag on one side of a flat plate parallel to the upstream flow is $\mathcal{D}$ when the upstream velocity is U, what will the drag be when the upstream velocity is $2U$; or $U/2$? Assume laminar flow.

[handwritten annotations at top of page:]

9.41 $dM = $ Torque from drag, $dM = (D_{top} + D_{bot}) \cdot y = 2(\frac{1}{2}\rho U^2 C_{Df} dA) y$, $U = wy$ & toll 9.3 $C_{Df} = 0.0720 \, w / Re = \frac{Ue}{v}$

$\therefore dM = \rho U^2 \frac{.0720}{(\frac{U\ell}{v})^{1/5}} \, y \, \ell dy = .072 \rho U^{7/5} \ell^{4/5} v^{1/5} y \, dy$ w/ $U = wy$ $dM = .072 \rho w^{9/5} \ell^{4/5} v^{1/5} y^{14/5} dy$ plug in

$\frac{dm = 9.81 \times 10^{-3} y^{14/5} dy}{M = 3 \int dm = 3 \int (9.81 \times 10^{-3}) \int_0^{12} y^{14/5} dy} = 3(\ )(\frac{y^{19/5}}{5})(12)^{19/5}$

$M = 97.7 \, ft \cdot lb$

9.36 Air flows past the triangular flat plate oriented parallel to the free stream shown in Fig. P9.36. Integrate the wall shear stress over the plate to determine the friction drag on one side of the plate. Assume laminar flow.

FIGURE P9.36

U = 2 ft/s, 1.0 ft, 45°, 45°

9.37 It is relatively easy to design an efficient nozzle to accelerate a fluid. Conversely, it is very difficult to build an efficient diffuser to decelerate a fluid without boundary layer separation and its subsequent inefficient flow behavior. Use the ideas of favorable and adverse pressure gradients to explain these facts.

9.38 It is often assumed that "sharp objects can cut through the air better than blunt ones." Based on this assumption, the drag on the object shown in Fig. P9.38 should be less when the wind blows from right to left than when it blows from left to right. Experiments show that the opposite is true. Explain.

U? ... *U?*

FIGURE P9.38

9.39* The device shown in Fig. P9.39 is to be designed to measure the wall shear stress as air flows over the smooth surface with an upstream velocity U. It is proposed that τ_w can be obtained

U, ℓ, 5 mm, Square, 10 mm, (1)

FIGURE P9.39

[handwritten at bottom left:]

$D = C_{Df2} \frac{1}{2} \rho U^2 A_2 - C_{Df1} \frac{1}{2} \rho U^2 A_1$

$D = 2.75 (100 C_{Df2} - 20 C_{Df1}) \ell b$

since $Re_{f_1} = \frac{U \ell_1}{v} = \frac{81.7)(5)}{1.57 \times 10^{-4}} = 2.57 \times 10^6$

$Re_{f_2} = \frac{U \ell_2}{v} = 1.29 \times 10^7$

by measuring the bending moment, M, at the base [point (1)] of the support that holds the small surface element which is free from contact with the surrounding surface. Plot a graph of M as a function of U for $5 \le U \le 50$ m/s, with $\ell = 2$, 3, 4, and 5 m.

9.40 Repeat Problem 9.39 if the fluid is water, the upstream velocity is $U = 5$ m/s and $\ell = 2$ m.

9.41 A three-bladed helicopter blade rotates at 200 rpm. If each blade is 12 ft long and 1.5 ft wide, estimate the torque needed to overcome the friction on the blades. Assume they act as flat plates and that the drag coefficient data associated with the $\frac{1}{7}$th power velocity profile (see Table 9.3) is accurate for this problem.

9.42 A ceiling fan consists of four blades of 0.80-m length and 0.10-m width which rotate at 100 rpm. Estimate the torque needed to overcome the friction on the blades if they act as flat plates.

9.43 A thin smooth sign is attached to the side of a truck as is indicated in Fig. P9.43. Estimate the friction drag on the sign when the truck is driven at 55 mph. *U = 55 mph (7(')= 80.7 ft/s*

[handwritten:] $D = D_{f2} - D_{f1}$ where area 1 & 2 —> 5 ft, 20 ft, A_1, A_2

SIGN, 4 ft

FIGURE P9.43

9.44 A sphere of diameter D and density ρ_s falls at a steady rate through a liquid of density ρ and viscosity μ. If the Reynolds number, $Re = \rho D U/\mu$, is less than 1, show that the speed of fall is $U = gD^2(\rho_s - \rho)/18 \mu$.

9.45 Determine the drag on a small circular disk of 0.01 ft diameter moving 0.10 ft/s through oil with a specific gravity of 0.87 and a viscosity 10,000 times that of water. The disk is oriented normal to the upstream velocity. By what percent is the drag reduced if the disk is oriented parallel to the flow?

9.46 How fast do small water droplets of 0.06-μm (6 × 10^{-8} m) diameter fall through the air under standard sea-level conditions? Assume the drops do not evaporate. Repeat the problem for standard conditions at 5000-m altitude.

[handwritten at bottom right:] Fig 9.15 $C_{Df_1} \approx .002$ $C_{Df_2} \approx .0028$

$D = $ plug in $\approx 1.86 \, \ell b$

9.47 A small spherical water drop of diameter 0.002 in. exists in the atmosphere at 5000-ft altitude. Will the drop rise or fall if it is in a thermal (an upward flowing column of air) having a speed of 4 ft/s? Repeat for speeds of 1 ft/s and 0.1 ft/s?

9.48 Compare the rise velocity of an $\frac{1}{8}$-in.-diameter air bubble in water to the fall velocity of an $\frac{1}{8}$-in.-diameter water drop in air. Assume each to behave as a solid sphere.

9.49 A 1.51-in.-diameter, 0.0882 oz table tennis ball is released from the bottom of a swimming pool. With what velocity does it rise to the surface. Assume it has reached its terminal velocity.

9.50 How fast will a toy balloon filled with helium rise through still air? List all of your assumptions.

9.51 A hot air balloon roughly spherical in shape has a volume of 70,000 ft³ and a weight of 500 lb (including passengers, basket, balloon fabric, etc.). If the outside air temperature is 80 °F and the temperature within the balloon is 165 °F, estimate the rate at which it will rise under steady state conditions if the atmospheric pressure is 14.7 psi.

9.52 A 200-N rock (roughly spherical in shape) of specific gravity $SG = 1.93$ falls at a constant speed U. Determine U if the rock falls through **(a)** air; **(b)** water.

9.53* A 3-m-diameter meteorological balloon that weighs 6 N when deflated is filled with helium having $\gamma = 1.63$ N/m³. Plot a graph of the rate at which it rises as a function of altitude for altitudes from sea level to its maximum altitude. Assume standard atmospheric conditions (Table C.2) and that the diameter remains constant.

9.54* For the balloon of Problem 9.53, plot a graph of the altitude of the balloon as a function of time. Note any assumptions made.

9.55* A 50-lb box shaped as a 1-ft cube falls from the cargo hold of an airplane at an altitude of 30,000 ft. If the drag coefficient of the falling box is 1.2, determine the time it takes for the box to hit the ocean. Assume that it falls at the terminal velocity corresponding to its current altitude and use a standard atmosphere (see Table C.1).

9.56 A 50-mph wind blows against an outdoor movie screen that is 70 ft wide and 20 ft tall. Estimate the wind force on the screen.

9.57 Determine the moment needed at the base of 30-m-tall, 0.12-m-diameter flag pole to keep it in place in a 20 m/s wind.

9.58 Repeat Problem 9.57 if a 2 m by 2.5 m flag is attached to the top of the pole. See Fig. 9.33 for drag coefficient data for flags.

9.59 A rectangular car-top carrier of 1.6-ft height, 5.0-ft length (front to back), and 4.2-ft width is attached to the top of a car. Estimate the additional power required to drive the car with the carrier at 60 mph through still air compared with the power required to driving only the car at 60 mph.

9.60 If for a given vehicle it takes 30 hp to overcome aerodynamic drag while being driven at 55 mph, estimate the horsepower required at 65 mph.

9.61 How much more power is required to peddle a bicycle at 15 mph into a 20-mph headwind than at 15 mph through still air? Assume a frontal area of 3.9 ft² and a drag coefficient of $C_D = 0.88$.

9.62 Two bicycle racers ride 30 km/hr through still air. By what percentage is the power required to overcome aerodynamic drag for the second cyclist reduced if she drafts closely behind the first cyclist rather than riding alongside her? Neglect any forces other than aerodynamic drag. (See Fig. 9.33).

9.63 To reduce aerodynamic drag on a bicycle, it is proposed that the cross-sectional shape of the handlebar tubes be made "teardrop" shape rather than circular. Make a rough estimate of the reduction in aerodynamic drag for a bike with this type of handlebars compared with the standard handlebars. List all assumptions.

9.64 It is suggested that the power, $\mathcal{P}$, needed to overcome the aerodynamic drag on a vehicle traveling at a speed U varies as $\mathcal{P} \sim U^n$. What is an appropriate value for the constant n? Explain.

9.65 Using the data in Fig. 9.29, determine the drag coefficient for a car of your choice.

9.66 Estimate the wind velocity necessary to knock over a 10-lb garbage can that is 3 ft tall and 2 ft in diameter. List your assumptions.

9.67 Estimate the wind velocity necessary to blow over the 250-kN boxcar shown in Fig. P9.67.

hence $F_p = 0A(\frac{1}{2})(1.23)(30)^2\frac{1}{2}0(.012)$ the $C_D = 0.4$

$= 133 N$

$F_p =$ force on one pole $= \mathcal{D}$; $\mathcal{D} = C_D' \frac{1}{2} \rho U^2 A$

$Re = \frac{U \cdot D}{\nu} = \frac{(30)(.012)}{1.46 \times 10^{-5}} = 2.47 \times 10^4$

From Fig 9.23

FIGURE P9.67

9.68 Estimate the maximum wind velocity in which you can stand without holding on to something. List your assumptions.

9.69 Estimate the velocity with which you would contact the ground if you jumped from an airplane at an altitude of 1 km and **(a)** air resistance is negligible, **(b)** air resistance is important, but you forgot your parachute, or **(c)** you use an 8-m-diameter parachute.

9.70* The helium-filled balloon shown in Fig. P9.70 is to be used as a wind speed indicator. The specific weight of the helium is $\gamma = 0.011$ lb/ft³, the weight of the balloon material is 0.20 lb, and the weight of the anchoring cable is negligible. Plot a graph of θ as a function of U for $1 \leq U \leq 50$ mph. Would this be an effective device over the range of U indicated? Explain.

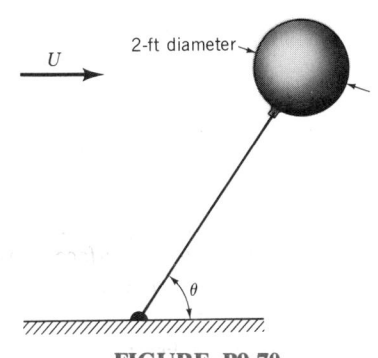

FIGURE P9.70

9.71 A 0.30-m-diameter cork ball ($SG = 0.21$) is tied to an object on the bottom of a river as is shown in Fig. P9.71. Estimate the speed of the river current.

FIGURE P9.71

9.72 A 12-mm-diameter cable is strung between a series of poles that are 50 m apart. Determine the horizontal force this cable puts on each pole if the wind velocity is 30 m/s.

9.73 Air flows past two equal sized spheres (one rough, one smooth) that are attached to the arm of a balance as is indicated in Fig. P9.73. With $U = 0$ the beam is balanced. What is the minimum air velocity for which the balance arm will rotate clockwise?

FIGURE P9.73

9.74 A radio antenna on a car consists of a circular cylinder $\frac{1}{4}$ in. in diameter and 4 ft long. Determine the bending moment at the base of the antenna if the car is driven 55 mph through still air.

9.75 Determine the power needed to overcome the aerodynamic drag of the antenna in Problem 9.74.

9.76 Estimate the wind force on your hand when you hold it out of your car window while driving 55 mph. Repeat your calculations if you were to hold your hand out of the window of an airplane flying 550 mph.

9.77* Let $\mathcal{P}_0$ be the power required to fly a particular airplane at 500 mph at sea-level conditions. Plot a graph of the ratio $\mathcal{P}/\mathcal{P}_0$, where $\mathcal{P}$ is the power required at a speed of U, for 500 mph $\leq U \leq$ 3000 mph at altitudes of sea level, 10,000 ft, 20,000 ft, and 30,000 ft. Assume that the drag coefficient for the aircraft behaves similarly to that of the sharp-pointed ogive indicated in Fig. 9.26.

9.78 A 2-mm-diameter meteor of specific gravity 2.9 has a speed of 6 km/s at an altitude of 50,000 m where the air density is 1.03×10^{-3} kg/m³. If the drag coefficient at this large Mach number condition is 1.5, determine the deceleration of the meteor.

9.79 Estimate the deceleration (in terms of the number of g's) of a 10-ft-diameter, 12,500-lb blunt reentry vehicle traveling 15,000 mph through the atmosphere at an altitude of 30 mi.

9.80 A shortwave radio antenna is constructed from circular tubing, as is illustrated in

Fig. P9.80. Estimate the wind force on the antenna in a 100 km/hr wind.

10-mm diameter
1 m long

0.6 m

0.5 m

20-mm diameter
1.5 m long

40-mm diameter
5 m long

0.25 m

FIGURE P9.80

9.81 A 50-ft-tall tower is constructed of equal 1-ft segments as is indicated in Fig. P9.81. Each of the four sides is similar. Estimate the drag on the tower when a 75-mph wind blows against it.

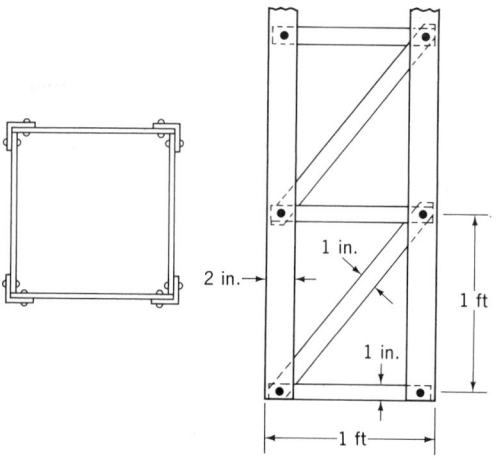

2 in.

1 in.

1 in.

1 ft

1 ft

FIGURE P9.81

9.82 The bearings on the four-cup anemometer (wind speed indicator) shown in top view in Fig. P9.82 are rusty. Determine the starting torque if it takes a $U = 3$ m/s wind to start the cups

80 mm

90 mm

U

A

O

B

Hollow hemisphere

FIGURE P9.82

rotating about their shaft, O. Neglect the effects of cups A and B.

9.83 If the anemometer of Problem 9.82 were placed in a stream of water, what upstream velocity would be needed to start the cups rotating about their shaft? Assume the starting torque is independent of the fluid in which the anemometer is immersed.

9.84 The United Nations Building in New York is a rectangle approximately 87.5 m wide and 154 m tall. Determine the drag on this building if the drag coefficient is 1.3 and the wind speed is 20 m/s.

9.85 Repeat Problem 9.84 if the velocity profile against the building is a typical profile for an urban area (see Problem 9.19) and the wind speed half way up the building is 20 m/s.

9.86 A 1.2-lb kite with an area of 6 ft² flies in a 20 ft/s wind such that the weightless string makes an angle of 55° relative to the horizontal. If the pull on the string is 1.5 lb, determine the lift and drag coefficients based on the kite area.

9.87 A regulation football is 6.78 in. in diameter and weighs 0.91 lb. If its drag coefficient is $C_D = 0.2$, determine its deceleration if it has a speed of 35 ft/s at the top of its trajectory.

9.88 Explain how the drag on a given smoke stack could be the same in a 1 m/s wind as in a 2 m/s wind. Assume the values of ρ and μ are the same for each case.

9.89 As is discussed in Section 9.3, the drag on a rough golf ball is less than that on an equal-sized smooth ball. Does it follow that a 10-m-diameter spherical water tank resting on a 20-m-tall support should have a rough surface so as to reduce the moment needed at the base of the support when a wind blows? Explain.

9.90 If golf balls are deliberately made rough so as to decrease the drag and increase the lift, why are airplane surfaces made smooth to accomplish the same thing?

9.91 The paint stirrer shown in Fig. P9.91 consists of two circular discs attached to the end

80 rpm

$\frac{7}{8}$ in.

1.5 in.

FIGURE P9.91

of a thin rod that rotates at 80 rpm. The specific gravity of the paint is $SG = 1.1$ and its viscosity is $\mu = 2 \times 10^{-2}$ lb·s/ft². Estimate the power required to drive the mixer if the induced motion of the liquid is neglected.

9.92 An airplane tows a banner that is $b = 0.8$ m tall and $\ell = 25$ m long at a speed of 150 km/hr. If the drag coefficient based on the area $b\ell$ is $C_D = 0.06$, estimate the power required to tow the banner. Compare the drag force on the banner with that on a rigid flat plate of the same size. Which has the larger drag force and why?

9.93 By appropriate streamlining the drag coefficient for an airplane is reduced by 18% while the frontal area remains the same. For the same power output, by what percentage is the flight speed increased? $Power = \mathscr{D} v \; : \; \mathscr{D} = C_D \frac{1}{2} \rho u^2 A$

                                  $v = 37.5 \, m/s$

9.94 The dirigible Akron had a length of 239 m and a maximum diameter of 40.2 m. Estimate the power required at its maximum speed of 135 km/hr if the drag coefficient based on frontal area is 0.060. $P = C_D \frac{1}{2} \rho U^3 A = 2.47 \times 10^3 \, kw \left(\frac{1.34 hp}{kw} \right)$

                      $= 3310 \, hp.$

9.95 Estimate the power needed to overcome the aerodynamic drag of a person who runs at a rate of 100 yds in 10 s in still air. Repeat the calculations if the race is run into a 10-mph headwind.

9.96 A vertical wind tunnel built with upward flow is to be used for the training of sky divers. That is, the wind velocity is to be such that the sky divers remain suspended in the air stream without use of a parachute. Estimate the wind velocity needed.

9.97 A fishnet consists of 0.10-in.-diameter strings tied into squares 4 in. per side. Estimate the force needed to tow a 15 ft by 30 ft section of this net through seawater at 5 ft/s.

9.98 An iceberg floats with approximately $\frac{1}{7}$ of its volume in the air as is shown in Fig. P9.98.

If the wind velocity is U and the water is stationary, estimate the speed at which the wind forces the iceberg through the water.

9.99 A Piper Cub airplane has a gross weight of 1750 lb, a cruising speed of 115 mph, and a wing area of 179 ft². Determine the lift coefficient of this airplane for these conditions.

9.100 The wing area of a small airplane weighing 6.22 kN is 10.2 m². **(a)** If the cruising speed of the plane is 210 km/hr, determine the lift coefficient of the wing. **(b)** If the engine delivers 150 kW at this speed, and if 60% of this power represents propeller loss and body resistance, what is the drag coefficient of the wing?

9.101 Show that for level flight the drag on a given airplane is independent of altitude if the lift and drag coefficients remain constant. Note that with C_L constant the airplane must fly faster at a higher altitude.

9.102 Show that for level flight at a given speed, the power required to overcome aerodynamic drag decreases as the altitude increases. Assume that the drag coefficient remains constant. This is one reason why airlines fly at high altitudes.

9.103 For a given airplane, compare the power to maintain level flight at 5000-ft altitude with that at 30,000 ft at the same velocity. Assume C_D remains constant.

9.104 A wing generates a lift $\mathscr{L}$ when moving through sea-level air with a velocity U. How fast must the wing move through the air at an altitude of 10,000 m with the same lift coefficient if it is to generate the same lift?

9.105* When air flows past the airfoil shown in Fig. P9.105 the velocity just outside the boundary layer, u, is as indicated. Estimate the lift coefficient for these conditions.

FIGURE P9.98

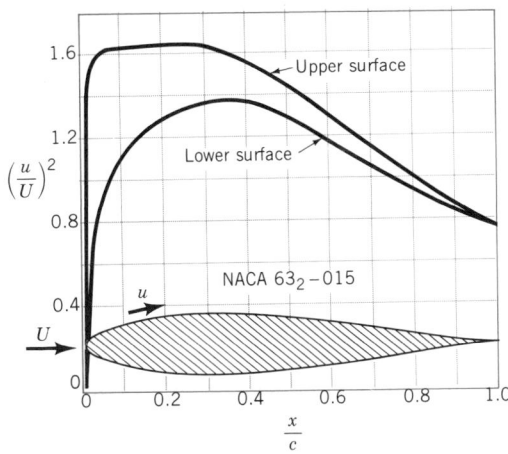

FIGURE P9.105

Handwritten top notes:
$$107 \quad \mathscr{L} = C_L \tfrac{1}{2} \rho v^2 A = W \qquad \frac{\mathscr{L}_{100}}{\mathscr{L}_{372}} = \frac{v^2_{100}}{v^2_{372}} \quad \text{so} \quad v_{372} = v_{100}\left\{ \quad \right\} = 140$$
$$\text{also} \; v_{372} = 196\,mph$$

9.106 A certain airfoil has the measured lift and drag coefficient characteristics listed in the table below.

α, deg	C_L	C_D
-3.3	-0.182	0.0098
-1.8	-0.025	0.0094
-0.5	0.135	0.0087
1.2	0.286	0.0087
2.6	0.441	0.0093
5.8	0.746	0.0104
8.6	1.05	0.0141
11.8	1.33	0.0203
13.6	1.46	0.0264
15.3	1.58	0.0361
15.9	1.60	0.0436
19.7	1.33	0.231
25.9	1.03	0.426

Plot a graph of C_L and C_D as a function of angle of attack. Plot the polar diagram for this airfoil. Determine the stall angle.

9.107 A Boeing 747 aircraft weighing 580,000 lb when loaded with fuel and 100 passengers takes off with an airspeed of 140 mph. With the same configuration (i.e., angle of attack, flap settings, etc.) what is its takeoff speed if it is loaded with 372 passengers. Assume each passenger with luggage weighs 200 lb.

9.108 Show that for unpowered flight (for which the lift, drag, and weight forces are in equilibrium) the glide slope angle, θ, is given by $\tan\theta = C_D/C_L$.

9.109 If the lift coefficient for a Boeing 727 aircraft is 16 times greater than its drag coefficient, how far horizontally can it glide from an altitude of 30,000 ft if it looses power from its engines? (See Problem 9.108.)

9.110 On its final approach to the airport an airplane flies on a flight path that is 3.0° relative to the horizontal. What lift-to-drag ratio is needed if the airplane is to land with its engines idled back to zero power? (See Problem 9.108.)

9.111 Explain why aircraft and birds take off and land into the wind?

9.112 The stall speed, U_{stall}, is the minimum landing speed for an aircraft. Show that $U_{stall} = [2W/(\rho C_{Lmax} A)]^{1/2}$, where W is the weight of the airplane. (Normally, aircraft land at a higher speed to allow a margin of safety.) By what percentage can the landing speed be reduced by using double slotted trailing edge flaps (see Fig. 9.38) rather than no flaps?

9.113 If the takeoff speed of a particular airplane is 200 km/hr at sea level, what will it be at Denver (elevation 1500 m)? Use properties of the U.S. Standard Atmosphere.

9.114* Air blows over the flat-bottomed, two-dimensional object shown in Fig. P9.114. The shape of the object, $y = y(x)$, and the fluid speed along the surface, $u = u(x)$, are given in the table below. Determine the lift coefficient for this object.

x (% c)	y (% c)	u/U
0	0	0
2.5	3.72	0.971
5.0	5.30	1.232
7.5	6.48	1.273
10	7.43	1.271
20	9.92	1.276
30	11.14	1.295
40	11.49	1.307
50	10.45	1.308
60	9.11	1.195
70	6.46	1.065
80	3.62	0.945
90	1.26	0.856
100	0	0.807

FIGURE P9.114

9.115 A 5.25-oz, 2.9-in.-diameter baseball is thrown toward home plate with an initial speed of 90 mph. The ball is released 6.2 ft above home plate with a horizontal velocity directed toward the outside edge (for a right-handed batter) of home plate. The ball spins clockwise (when viewed from above) at 2500 rpm about a vertical axis. If the ball travels 60 ft to a 17-in.-wide home plate, determine the elevation and location relative to home plate when the ball crosses the plate. Neglect aerodynamic drag and assume the angular velocity of the ball remains constant.

9.116 Repeated controversy regarding the ability of a baseball to curve appeared in the literature for years. According to a test (*Life*, July 27, 1953) a baseball (assume the diameter is 2.9 in. and weight 5.25 oz) spinning 1400 rpm while traveling 43 mph was observed to follow a path with an 800-ft horizontal radius of curvature. Based on the data of Fig. 9.42 do you agree with this test result? Explain.

Handwritten bottom notes:
$$30,000 \cdot \tan\theta = \frac{C_D}{C_L} = \frac{1}{16}$$
$$\text{here} \; \frac{3000}{16} = \frac{x}{16} \quad \text{so} \; d =$$

Motion of water induced by surface waves: As a wave passes along the surface of the water, the water particles follow elliptical paths. There is no net motion of the water, just a periodic, cyclic trajectory. (Neutrally buoyant particles in water.) (Photograph by A. Wallet and F. Ruellan, Ref. 13, courtesy of M. C. Vasseur, Sogreah.)

TEN

OPEN-CHANNEL FLOW

Open-channel flow involves the flow of a liquid in a channel or conduit that is not completely filled. There exists a free surface between the flowing fluid (usually water) and fluid above it (usually the atmosphere). The main driving force for such flows is the fluid weight—gravity forces the fluid to flow down hill. Under steady, fully-developed flow conditions the component of the weight force in the direction of flow is balanced by the equal and opposite shear force between the fluid and the channel surfaces. For unsteady or nonfully developed situations, the inertia of the flowing fluid is also important.

Such flows are different from the pipe flows discussed in Chapter 8 in that there can be no pressure force driving the fluid through the channel or conduit. Any attempt to impose a pressure gradient in the flow direction is met with failure because of the negligible inertial and viscous effects of the gas (atmosphere) above the flowing fluid. For steady, fully developed channel flow, the pressure distribution within the fluid is merely hydrostatic.

Open-channel flows are essential to the world as we know it. The natural drainage of water through the numerous creek and river systems is a complex example of open-channel flow. Although the flow geometry for these systems is extremely complex, the resulting flow properties are of considerable economic and recreational importance. Other examples of open-channel flows include the flow of rain water in the gutters of our houses; the flow in canals, drainage ditches, sewers, and gutters along roads; the flow of small rivulets and sheets of water across fields or parking lots; and the flow in the chutes of water rides in amusement parks.

Clearly the character, description, and complexity of open-channel flow geometry is quite variable. The bounding geometry for flow in a sewer pipe laid on a constant slope and running half-full is much simpler than the geometry of the Mississippi River

with its variable cross-sectional shape, bends, bottom slope variation, and character of its bounding surfaces. Because of complexities like these, most open-channel flow results are based on correlations obtained from model and full-scale experiments. Additional information can be gained from various analytical and numerical efforts.

The purpose of this chapter is to investigate the concepts of open-channel flow. Because of the amount and variety of material available, only a brief introduction to the topic can be presented. Further information can be obtained from the references indicated.

10.1 GENERAL CHARACTERISTICS OF OPEN-CHANNEL FLOW

In our study of pipe flow (Chapter 8) we found that there are many ways to classify a flow—developing, fully developed, laminar, turbulent, and so on. For open-channel flow the existence of a free surface allows additional types of flow. The extra degree of freedom because the fluid is free to select its free surface location and configuration (because it does not completely fill a pipe or conduit) allows important phenomena in open-channel flow that cannot occur in pipe flow. Some of the classifications of the flows are described below.

The manner in which the fluid depth, y, varies with time, t, and distance along the channel, x, is used to partially classify a flow. For example, the flow is *unsteady* or *steady* depending on whether the depth at a given location does or does not change with time. Some unsteady flows can be viewed as steady flows if the reference frame of the observer is changed. For example, a tidal bore moving up a river is unsteady to an observer standing on the bank, but steady to an observer moving along the bank with the speed of the wave front of the bore. Other flows are unsteady regardless of the reference frame used. The complex, time-dependent wind generated waves on a lake are in this category. Another example of this type of unsteady flow is the flood surge downstream of a dam just after the dam fails (see Fig. 6.35). In this book we will consider only steady open-channel flows.

An open-channel flow is classified as *uniform flow* (UF) if the depth of flow does not vary along the channel ($dy/dx = 0$). Conversely, it is *nonuniform flow* or *varied flow* if the depth varies with distance ($dy/dx \neq 0$). Nonuniform flows are further classified as *rapidly varying flow* (RVF) if the flow depth changes considerably over a relatively short distance; $dy/dx \sim 1$. *Gradually varying flows* (GVF) are those in which the flow depth changes slowly with distance along the channel; $dy/dx \ll 1$. Examples of these types of flow are illustrated in Fig. 10.1. The relative importance

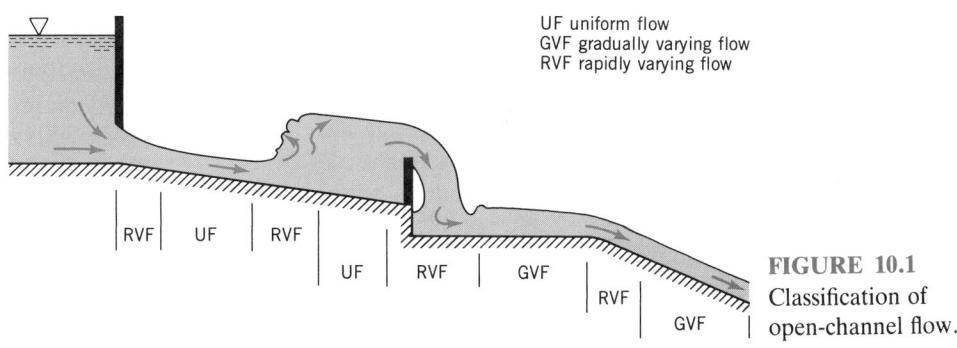

UF uniform flow
GVF gradually varying flow
RVF rapidly varying flow

FIGURE 10.1
Classification of open-channel flow.

of the various types of forces involved (pressure, weight, shear, inertia) is different for the different types of flows.

As for any flow geometry, open-channel flow may be *laminar, transitional,* or *turbulent,* depending on various conditions involved. Which type of flow occurs depends on the Reynolds number, Re $= \rho V R_h / \mu$, where V is the average velocity of the fluid and R_h is the hydraulic radius of the channel (see Section 10.4). A general rule is that open-channel flow is laminar if Re < 500, turbulent if Re $> 12{,}500$, and transitional otherwise. The values of these dividing Reynolds numbers are only approximate—a precise knowledge of the channel geometry is necessary to obtain specific values. Since many open-channel flows involve water (which has a fairly small viscosity) and have relatively large characteristic lengths, it is uncommon to have laminar open-channel flows. For example, flow of 50 °F water ($\nu = 1.41 \, 10^{-5} \, \text{ft}^2/\text{s}$) with an average velocity of $V = 1$ ft/s in a river with a hydraulic radius of $R_h = 10$ ft has Re $= V R_h / \nu = 7.1 \times 10^5$. The flow is turbulent. However, flow of a thin sheet of water down a driveway with an average velocity of $V = 0.25$ ft/s such that $R_h = 0.02$ ft (in such cases the hydraulic radius is approximately equal to the fluid depth, see Section 10.4) has Re $= 355$. The flow is laminar.

All of the open-channel flows considered in this book are *homogeneous flows.* That is, the fluid has uniform properties throughout. In some cases *stratified flows* are important. In such situations layers of two or more fluids of different densities flow in a channel. A layer of oil on water is one example of this type of flow.

Open-channel flows involve a free surface that can deform from its undisturbed relatively flat configuration to form waves. Such waves move across the surface at speeds that depend on their size (height, length) and properties of the channel (depth, fluid velocity, etc.). The character of an open-channel flow may depend strongly on how fast the fluid is flowing relative to how fast a typical wave moves relative to the fluid. The dimensionless parameter that describes this behavior is termed the Froude number, Fr $= V/(g\ell)^{1/2}$, where ℓ is an appropriate characteristic length of the flow. This dimensionless parameter was introduced in Chapter 7 and is discussed more fully in Section 10.2. The special case of a flow with a Froude number of unity, Fr $= 1$, is termed a *critical flow.* If the Froude number is less than 1, the flow is *subcritical* (or *tranquil*). A flow with the Froude number greater than 1 is termed *supercritical* (or *rapid*).

10.2 SURFACE WAVES

The distinguishing feature of flows involving a free surface (as in open-channel flows) is the opportunity for the free surface to distort into various shapes. The surface of a lake or the ocean is seldom "smooth as a mirror." It is usually distorted into ever-changing patterns associated with surface waves. Some of these waves are very high, some barely ripple the surface; some waves are very long (the distance between wave crests), some are short; some are breaking waves that form white caps, others are quite smooth. Although a general study of this wave motion is beyond the scope of this book, an understanding of certain fundamental properties of simple waves is necessary for open-channel flow considerations. The interested reader is encouraged to use some of the excellent references available for further study about wave motion (Refs. 1, 2, 3).

10.2.1 Wave Speed

Consider the situation illustrated in Fig. 10.2a in which a single elementary wave of small height, δy, is produced on the surface of a channel by suddenly moving the

FIGURE 10.2
(*a*) Production of a single elementary wave in a channel as seen by a stationary observer. (*b*) Wave as seen by an observer moving with a speed equal to the wave speed.

initially stationary end wall with speed δV. The water in the channel was stationary at the initial time, $t = 0$. A stationary observer will observe a single wave move down the channel with a *wave speed c*, with no fluid motion ahead of the wave and a fluid velocity of δV behind the wave. The motion is unsteady for such an observer. For an observer moving along the channel with speed c, the flow will appear steady as shown in Fig. 10.2*b*. To this observer, the fluid velocity will be $\mathbf{V} = -c\hat{\mathbf{i}}$ on the observer's right and $\mathbf{V} = (-c + \delta V)\hat{\mathbf{i}}$ to the left of the observer.

The relationship between the various parameters involved for this flow can be obtained by application of the continuity and momentum equations to the control volume shown in Fig. 10.2*b* as follows. With the assumption of uniform one-dimensional flow, the continuity equation (Eq. 5.12) becomes

$$-cyb = (-c + \delta V)(y + \delta y)b$$

where b is the channel width. This simplifies to

$$c = \frac{(y + \delta y)\,\delta V}{\delta y}$$

or in the limit of small amplitude waves with $\delta y \ll y$

$$c = y\,\frac{\delta V}{\delta y} \tag{10.1}$$

Similarly, the momentum equation (Eq. 5.22) is

$$\tfrac{1}{2}\gamma y^2 b - \tfrac{1}{2}\gamma(y + \delta y)^2 b = \rho bcy[(c - \delta V) - c]$$

where we have written the mass flowrate as $\dot{m} = \rho bcy$ and have assumed that the pressure variation is hydrostatic within the fluid. That is, the pressure forces on the channel cross sections (1) and (2) are $F_1 = \gamma y_{c1} A_1 = \gamma(y + \delta y)^2 b/2$ and $F_2 = \gamma y_{c2} A_2 = \gamma y^2 b/2$, respectively. If we again impose the assumption of small amplitude waves [that is, $(\delta y)^2 \ll y\,\delta y$], the momentum equation reduces to

$$\frac{\delta V}{\delta y} = \frac{g}{c} \tag{10.2}$$

Combination of Eqs. 10.1 and 10.2 gives the wave speed as

$$c = \sqrt{gy} \tag{10.3}$$

The speed of a small amplitude solitary wave as is indicated in Fig. 10.2 is proportional to the square root of the fluid depth, y, and independent of the wave amplitude, δy. The fluid density is not an important parameter, although the acceleration of gravity is. This is a result of the fact that such wave motion is a balance between inertial effects (proportional to ρ) and weight or hydrostatic pressure effects (proportional to $\gamma = \rho g$). A ratio of these forces eliminates the common factor ρ but retains g.

The wave speed can also be calculated by using the energy and continuity equations rather than the momentum and continuity equations as is done above. A simple wave on the surface is shown in Fig. 10.3. As seen by an observer moving with the wave speed, c, the flow is steady. Since the pressure is constant at any point on the free surface, the Bernoulli equation for this frictionless flow is simply

$$\frac{V^2}{2g} + y = \text{constant}$$

or by differentiating

$$\frac{V\,\delta V}{g} + \delta y = 0$$

Also, by differentiating the continuity equation, $Vy = \text{constant}$, we obtain

$$y\,\delta V + V\,\delta y = 0$$

We combine these two equations to eliminate δV and δy and use the fact that $V = c$ for this situation (the observer moves with speed c) to obtain the wave speed given by Eq. 10.3. The above results are restricted to waves of small amplitude because we have assumed one-dimensional flow. That is, $\delta y/y \ll 1$. More advanced analysis and experiments show that the wave speed for finite size solitary waves exceeds that given by Eq. 10.3. To a first approximation one obtains (Ref. 4)

$$c \approx \sqrt{gy}\left(1 + \frac{\delta y}{y}\right)^{1/2}$$

The larger the amplitude, the faster the wave travels.

A more general description of wave motion can be obtained by considering continuous (not solitary) waves of sinusoidal shape as is shown in Fig. 10.4. By combining waves of various wavelengths, λ, and amplitudes, δy, it is possible to describe very complex surface patterns found in nature, such as the wind-driven waves on a lake. Mathematically, such a process consists of using a Fourier series (each term of the series represented by a wave of different wavelength and amplitude) to represent an arbitrary function (the free-surface shape).

A more advanced analysis of such sinusoidal surface waves of small amplitude

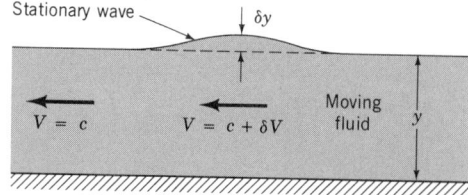

FIGURE 10.3 Stationary simple wave in a flowing fluid.

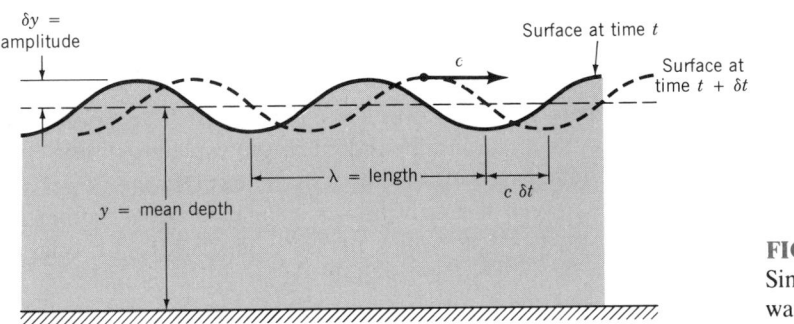

FIGURE 10.4
Sinusoidal surface wave.

shows that the wave speed varies with both the wavelength and fluid depth as (Ref. 1)

$$c = \left[\frac{g\lambda}{2\pi} \tanh\left(\frac{2\pi y}{\lambda} \right) \right]^{1/2} \tag{10.4}$$

where $\tanh(2\pi y/\lambda)$ is the hyperbolic tangent of the argument $2\pi y/\lambda$. This result is plotted in Fig. 10.5. For layers that are very deep compared with the wavelength ($y \gg \lambda$, as in the ocean), the wave speed is independent of y and given by

$$c = \sqrt{\frac{g\lambda}{2\pi}}$$

This result follows from Eq. 10.4, since $\tanh(2\pi y/\lambda) \to 1$ as $y/\lambda \to \infty$. On the other hand, if the fluid layer is shallow ($y \ll \lambda$, as often happens in open channels), the wave speed is given by $c = (gy)^{1/2}$, as derived for the solitary wave in Fig. 10.2. This result also follows from Eq. 10.4, since $\tanh(2\pi y/\lambda) \to 2\pi y/\lambda$ as $y/\lambda \to 0$. These two limiting cases are shown in Fig. 10.5. For moderate depth layers ($y \sim \lambda$), the results are given by the complete Eq. 10.4. Note that for a given fluid depth, the long wave

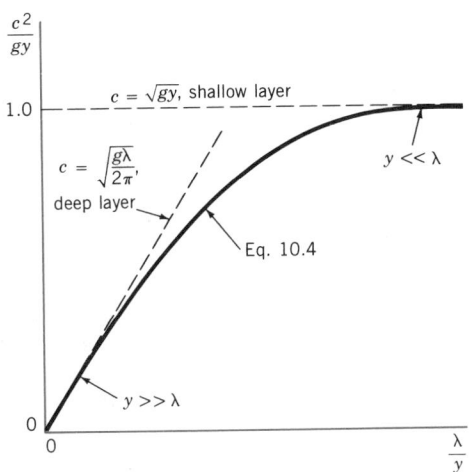

FIGURE 10.5 Wave speed as a function of wavelength.

travels the fastest. Hence, for our purposes we will consider the wave speed to be this limiting situation, $c = (gy)^{1/2}$.

10.2.2 Froude Number Effects

Consider an elementary wave traveling on the surface of a fluid, as is shown in Fig. 10.2a. If the fluid layer is stationary, the wave moves to the right with speed c relative to the fluid and the stationary observer. If the fluid is flowing to the left with velocity $V < c$, the wave (which travels with speed c relative to the fluid) will travel to the right with a speed of $c - V$ relative to a fixed observer. If the fluid flows to the left with $V = c$, the wave will remain stationary, but if $V > c$ the wave will be washed to the left with speed $V - c$.

The above ideas can be expressed in dimensionless form by use of the Froude number, Fr $= V/(gy)^{1/2}$, where we take the characteristic length to be the fluid depth, y. Thus, the Froude number, Fr $= V/(gy)^{1/2} = V/c$, is the ratio of the fluid velocity to the wave speed.

The following characteristics are observed when a wave is produced on the surface of a moving stream, as happens when a rock is thrown into a river. If the stream is not flowing, the wave spreads equally in all directions. If the stream is nearly stationary or moving in a tranquil manner (i.e., $V < c$), the wave can move upstream. Upstream locations are said to be in hydraulic commuication with the downstream locations. That is, an observer upstream of a disturbance can tell that there has been a disturbance on the surface because that disturbance can propagate upstream to the observer. Viscous effects, which have been neglected in this discussion, will eventually damp out such waves far upstream. Such flow conditions, $V < c$, or Fr < 1, are termed subcritical.

On the other hand, if the stream is moving rapidly so that the flow velocity is greater than the wave speed (i.e., $V > c$), no upstream communication with downstream locations is possible. Any disturbance on the surface downstream from the observer will be washed further downstream. Such conditions, $V > c$ or Fr > 1, are termed supercritical. For the special case of $V = c$ or Fr $= 1$, the wave remains stationary and the flow is termed critical.

The character of an open-channel flow may depend strongly on whether the flow is subcritical or supercritical. The characteristics of the flow may be completely opposite for subcritical flow than for supercritical flow. For example, as is discussed in Section 10.3, a "bump" on the bottom of a river (such as a submerged log) may cause the surface of the river to dip below the level it would have had if the log were not there, or it may cause the surface level to rise above its undisturbed level. Which situation will happen depends on the value of Fr. Similarly, for supercritical flows it is possible to produce steplike discontinuities in the fluid depth (called a hydraulic jump; see Section 10.6.1). For subcritical flows, however, changes in depth must be smooth and continuous. Certain open-channel flows, such as the broad-crested weir (Section 10.6.3), depend on the existence of critical flow conditions for their operation.

As strange as it may seem, there exist many similarities between the open-channel flow of a liquid and the compressible flow of a gas. The governing dimensionless parameter in each case is the fluid velocity, V, divided by a wave speed; the surface wave speed for open-channel flow or sound wave speed for compressible flow. Many of the differences between subcritical (Fr < 1) and supercritical (Fr > 1) open-channel flows have analogs in subsonic (Ma < 1) and supersonic (Ma > 1) compressible gas

flow, where Ma is the Mach number. Some of these similarities are discussed in this chapter and in Chapter 11.

10.3 ENERGY CONSIDERATIONS

A typical segment of an open-channel flow is shown in Fig. 10.6. The slope of the channel bottom (or *bottom slope*), $S_0 = (z_1 - z_2)/\ell$, is assumed constant over the segment shown. The fluid depths and velocities are y_1, y_2, V_1, and V_2 as indicated. Note that the fluid depth is measured in the vertical direction and the distance x is horizontal. For most open-channel flows the value is S_0 is very small (the bottom is nearly horizontal). For example, the Mississippi River drops a distance of 1470 ft in its 2350-mi length to give an average value of $S_0 = 0.000118$. In such circumstances the values of x and y are often taken as the distance along the channel bottom and the depth normal to the bottom, with negligibly small differences introduced by the two coordinate schemes.

With the assumption of a unform velocity profile across any section of the channel, the one-dimensional energy equation for this flow (Eq. 5.84) becomes

$$\frac{p_1}{\gamma} + \frac{V_1^2}{2g} + z_1 = \frac{p_2}{\gamma} + \frac{V_2^2}{2g} + z_2 + h_L \tag{10.5}$$

where h_L is the head loss due to viscous effects between sections (1) and (2) and $z_1 - z_2 = S_0\ell$. Since the pressure is essentially hydrostatic at any cross section, we find that $p_1/\gamma = y_1$ and $p_2/\gamma = y_2$ so that Eq. 10.5 becomes

$$y_1 + \frac{V_1^2}{2g} + S_0\ell = y_2 + \frac{V_2^2}{2g} + h_L \tag{10.6}$$

One of the difficulties of analyzing open-channel flow, similar to that discussed in Chapter 8 for pipe flow, is associated with the determination of the head loss in terms of other physical parameters. Without getting into such details at present, we write the head loss in terms of the slope of the energy line, $S_f = h_L/\ell$ (often termed the *friction slope*), as is indicated in Fig. 10.6. Recall from Chapter 3 that the energy line is located a distance z (the elevation from some datum to the channel bottom) plus the pressure head (p/γ) plus the velocity head ($V^2/2g$) above the datum. Therefore,

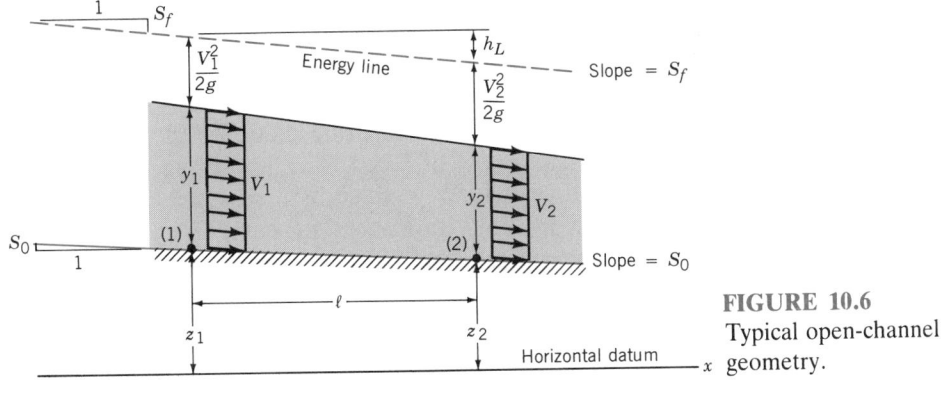

FIGURE 10.6
Typical open-channel geometry.

Eq. 10.6 can be written as

$$y_1 - y_2 = \frac{(V_2^2 - V_1^2)}{2g} + (S_f - S_0)\ell \qquad (10.7)$$

If there is no head loss, the energy line is horizontal ($S_f = 0$), and the total energy of the flow is free to shift between kinetic energy and potential energy in a conservative fashion. In the specific instance of a horizontal channel bottom ($S_0 = 0$) and negligible head loss ($S_f = 0$), Eq. 10.7 simply becomes

$$y_1 - y_2 = \frac{(V_2^2 - V_1^2)}{2g}$$

10.3.1 Specific Energy

The concept of the *specific energy*, E, defined as

$$E = y + \frac{V^2}{2g} \qquad (10.8)$$

is often useful in open-channel flow considerations. The energy equation, Eq. 10.7, can be written in terms of E as

$$E_1 = E_2 + (S_f - S_0)\ell \qquad (10.9)$$

If head losses are negligible, the sum of the specific energy and the elevation of the channel bottom remains constant (i.e., $E_1 + z_1 = E_2 + z_2$, a statement of the Bernoulli equation).

If we consider a simple channel whose cross-sectional shape is a rectangle of width b, the specific energy can be written in terms of the flowrate per unit width, $q = Q/b = Vyb/b = Vy$, as

$$E = y + \frac{q^2}{2gy^2} \qquad (10.10)$$

For a given channel of constant width, the value of q remains constant along the channel, although the depth, y, may vary. To gain insight into the flow processes involved, we consider the *specific energy diagram*, a graph of $E = E(y)$ with q fixed, as shown in Fig. 10.7.

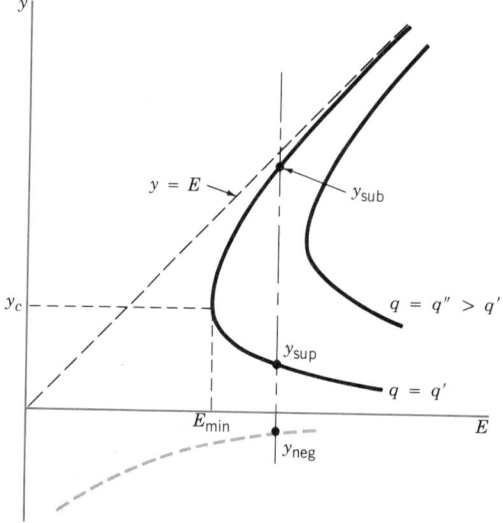

FIGURE 10.7 Specific energy diagram.

For given q and E, Eq. 10.10 is a cubic equation with three solutions, y_{sup}, y_{sub}, and y_{neg}. If the specific energy is large enough (i.e., $E > E_{min}$, where E_{min} is a function of q), two of the solutions are positive and the other, y_{neg}, is negative. The negative root, represented by the curved dashed line in Fig. 10.7, has no physical meaning and can be ignored. Thus, for a given flowrate and specific energy there are two possible depths, unless the vertical line from the E axis does not intersect the specific energy curve corresponding to the value of q given (i.e., $E < E_{min}$). These two depths are termed *alternate depths*.

For large values of E the upper and lower branches of the specific energy diagram (y_{sub} and y_{sup}) approach $y = E$ and $y = 0$, respectively. These limits correspond to a very deep channel flowing very slowly ($E = y + V^2/2g \rightarrow y$ as $y \rightarrow \infty$ with $q = Vy$ fixed), or a very high-speed flow in a shallow channel ($E = y + V^2/2g \rightarrow V^2/2g$ as $y \rightarrow 0$).

As is indicated in Fig. 10.7, $y_{sup} < y_{sub}$. Thus, since $q = Vy$ is constant along the curve, it follows that $V_{sup} > V_{sub}$, where the subscripts "sub" and "sup" on the velocities correspond to the depths so labeled. The specific energy diagram consists of two portions divided by the E_{min} "nose" of the curve. We will show that the flow conditions at this location correspond to critical conditions (Fr = 1); those on the upper portion of the curve correspond to subcritical conditions (hence, the "sub" subscript); and those on the lower portion of the curve correspond to supercritical conditions (hence, the "sup" subscript).

To determine the value of E_{min}, we use Eq. 10.10 and set $dE/dy = 0$ to obtain

$$\frac{dE}{dy} = 1 - \frac{q^2}{gy^3} = 0$$

or

$$y_c = \left(\frac{q^2}{g}\right)^{1/3} \tag{10.11}$$

where the subscript "c" denotes conditions at E_{min}. By substituting this back into Eq. 10.10 we obtain

$$E_{min} = \frac{3y_c}{2}$$

By combining Eq. 10.11 and $V_c = q/y_c$ we obtain

$$V_c = \frac{q}{y_c} = \frac{(y_c^{3/2} g^{1/2})}{y_c} = \sqrt{gy_c}$$

or $Fr_c \equiv V_c/(gy_c)^{1/2} = 1$. Thus, critical conditions (Fr = 1) occur at the location of E_{min}. Since the layer is deeper and the velocity smaller for the upper part of the specific energy diagram (compared with the conditions at E_{min}), such flows are subcritical (Fr < 1). Conversely, flows for the lower part of the diagram are supercritical. Thus, for a given flowrate, q, if $E > E_{min}$ there are two possible depths of flow, one subcritical and the other supercritical.

EXAMPLE 10.1

Water flows under the sluice gate in a constant width rectangular channel as shown in Fig. E10.1a. Describe this flow in terms of the specific energy diagram. Assume inviscid flow.

(a)   (b)

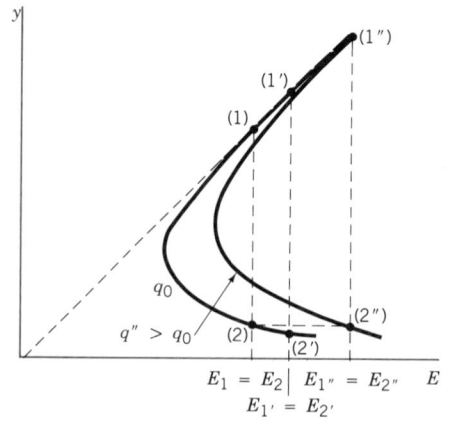

(c)

FIGURE E10.1

SOLUTION

For this inviscid ($S_f = 0$) flow the channel bottom is horizontal, $z_1 = z_2$ (or $S_0 = 0$), so that the energy equation (Eq. 10.9) reduces to $E_1 = E_2$. Although the flowrate is not given, we do know that $q_1 = q_2$ and that the specific energy diagram for this flow is as shown in Fig. E10.1b. The flow upstream of the gate is subcritical while the flow downstream is supercritical. The particular value of q obtained, $q = q_0$, and hence the specific curve of Fig. E10.1b, which is valid for this flow, is illustrated by the one that gives $E_1 = E_2$ for the given y_1 and y_2.

The flowrate can remain the same for this channel even if the upstream depth is increased. This is indicated by depths $y_{1'}$ and $y_{2'}$ in Fig. E10.1c. Of course, to do this the distance between the bottom of the gate and the channel bottom must be decreased to give a smaller flow area ($y_{2'} < y_2$), and the upstream depth must be increased to give a bigger head ($y_{1'} > y_1$). On the other hand, if the gate remains fixed so that the downstream depth remains fixed ($y_{2''} = y_2$), the flowrate will increase as the upstream depth increases to $y_{1''} > y_1$. This is indicated in Fig. E10.1c by the curve with flowrate $q'' > q_0$.

It is often possible to determine various characteristics of a flow by considering the specific energy diagram. Example 10.2 illustrates this for a situation in which the channel bottom elevation is not constant.

EXAMPLE 10.2

Water flows up a 0.5-ft-tall ramp in a constant width rectangular channel at a rate $q = 5.75$ ft²/s as is shown in Fig. E10.2a. (For now disregard the "bump" shown by the dashed line.) If the upstream depth is 2.3 ft, determine the elevation of the water surface downstream of the ramp, $y_2 + z_2$. Neglect viscous effects.

FIGURE E10.2

SOLUTION

With $S_0\ell = z_1 - z_2$ and $h_L = 0$, conservation of energy (Eq. 10.6 which, under these conditions, is actually the Bernoulli equation) requires that

$$y_1 + \frac{V_1^2}{2g} + z_1 = y_2 + \frac{V_2^2}{2g} + z_2$$

For the conditions given ($z_1 = 0$, $z_2 = 0.5$ ft, $y_1 = 2.3$ ft, and $V_1 = q/y_1 = 2.5$ ft/s), this becomes

$$1.90 = y_2 + \frac{V_2^2}{64.4} \tag{1}$$

where V_2 and y_2 are in ft/s and feet, respectively. The continuity equation provides the second equation

$$y_2V_2 = y_1V_1$$

or

$$y_2V_2 = 5.75 \text{ ft}^2/\text{s} \tag{2}$$

Equations 1 and 2 can be combined to give

$$y_2^3 - 1.90y_2^2 + 0.513 = 0$$

which has solutions

$$y_2 = 1.72 \text{ ft}, \quad y_2 = 0.638 \text{ ft}, \quad \text{or} \quad y_2 = -0.466 \text{ ft}$$

Note that two of these solutions are physically realistic, but the negative solution is meaningless. This is consistent with the previous discussions concerning the specific energy (recall the three roots indicated in Fig. 10.7). The corresponding elevations of the free surface are either

$$y_2 + z_2 = 1.72 \text{ ft} + 0.50 \text{ ft} = 2.22 \text{ ft}$$

or

$$y_2 + z_2 = 0.638 \text{ ft} + 0.50 \text{ ft} = 1.14 \text{ ft}$$

The question is which of these two flows is to be expected? This can be answered by use of the specific energy diagram obtained from Eq. 10.10, which for this problem is

$$E = y + \frac{0.513}{y^2}$$

where E and y are in feet. The diagram is shown in Fig. E10.2b. The upstream condition corresponds to subcritical flow; the downstream condition is either subcritical or supercritical, corresponding to points 2 or 2'. Note that since $E_1 = E_2 + (z_2 - z_1) = E_2 + 0.5$ ft, it follows that the downstream conditions are located 0.5 ft to the left of the upstream conditions on the diagram.

With a constant width channel, the value of q remains the same for any location along the channel. That is, all points for the flow from (1) to (2) or (2') must lie along the $q = 5.75$ ft^2/s curve shown. Any deviation from this curve would imply either a change in q or a relaxation of the one-dimensional flow assumption. To stay on the curve and go from (1) around the critical point (point c) to point (2') would require a reduction in specific energy to $E_{\min}$. As is seen from Fig. E10.2a, this would require a specified elevation (bump) in the channel bottom so that critical conditions would occur above this bump. In particular, since $E_1 = y_1 + 0.513/y_1^2 = 2.40$ ft and $E_{\min} = 3y_c/2 = 3(q^2/g)^{1/3}/2 = 1.51$ ft, the top of this bump would need to be $z_c - z_1 = E_1 - E_{\min} = 2.40$ ft $- 1.51$ ft $= 0.89$ ft above the channel bottom at section (1). The flow could then accelerate to supercritical conditions (Fr$_{2'} > 1$) as is shown by free surface represented by the dashed line in Fig. E10.2a.

Since the actual elevation change (a ramp) shown in Fig. E10.2a does not contain a bump, the downstream conditions will correspond to the subcritical flow denoted by (2), not the supercritical condition (2'). Without a bump on the channel bottom, the state (2') is inaccessible from the upstream condition state (1). Such considerations are often termed the *accessibility of flow regimes*. Thus, the surface elevation is

$$y_2 + z_2 = 2.22 \text{ ft} \qquad \text{(Ans)}$$

Note that since $y_1 + z_1 = 2.30$ ft and $y_2 + z_2 = 2.22$ ft, the elevation of the free surface decreases as it goes across the ramp.

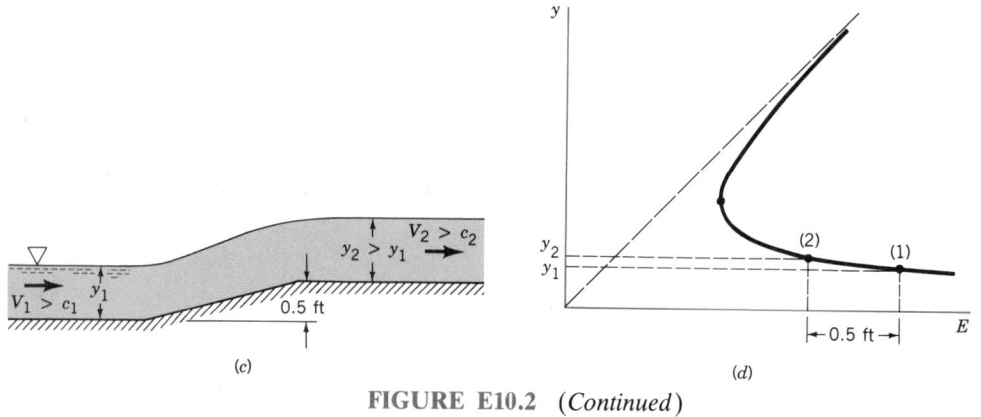

FIGURE E10.2 *(Continued)*

If the flow conditions upstream of the ramp were supercritical, the free-surface elevation and fluid depth would increase as the fluid flows up the ramp. This is indicated in Fig. E10.2c along with the corresponding specific energy diagram, as is shown in Fig. E10.2d. For this case the flow starts at (1) on the lower (supercritical) branch of the specific energy curve and ends at (2) on the same branch with $y_2 > y_1$. Since both y and z increase from (1) to (2), the surface elevation, $y + z$, also increases. Thus, flow up a ramp is different for subcritical than it is for supercritical conditions.

10.3.2   Channel Depth Variations

By using the concepts of the specific energy and critical flow conditions (Fr = 1), it is possible to determine how the depth of a flow in an open channel changes with distances along the channel. In some situations the depth change is very rapid so that the value of dy/dx is of the order one. Complex effects involving two- or three-dimensional flow phenomena are often involved in such flows.

In this section we consider only gradually varying flows. For such flows, $dy/dx \ll 1$ and it is reasonable to impose the one-dimensional velocity assumption. At any section the total head is $H = V^2/2g + y + z$ and the energy equation (Eq. 10.5) becomes

$$H_1 = H_2 + h_L$$

where h_L is the head loss between sections (1) and (2).

As is discussed in the previous section, the slope of the energy line is $dH/dx = dh_L/dx = S_f$ and the slope of the channel bottom is $dz/dx = S_0$. Thus, since

$$\frac{dH}{dx} = \frac{d}{dx}\left(\frac{V^2}{2g} + y + z\right) = \frac{V}{g}\frac{dV}{dx} + \frac{dy}{dx} + \frac{dz}{dx}$$

we obtain

$$\frac{dh_L}{dx} = \frac{V}{g}\frac{dV}{dx} + \frac{dy}{dx} + S_0$$

or

$$\frac{V}{g}\frac{dV}{dx} + \frac{dy}{dx} = S_f - S_0 \tag{10.12}$$

For a given flowrate per unit width, q, in a rectangular channel of constant width b, we have $V = q/y$ or

$$\frac{dV}{dx} = -\frac{q}{y^2}\frac{dy}{dx} = -\frac{V}{y}\frac{dy}{dx}$$

so that the kinetic energy term in Eq. 10.12 becomes

$$\frac{V}{g}\frac{dV}{dx} = -\frac{V^2}{gy}\frac{dy}{dx} = -\mathrm{Fr}^2\frac{dy}{dx} \tag{10.13}$$

where $\mathrm{Fr} = V/(gy)^{1/2}$ is the local Froude number of the flow. Substituting Eq. 10.13 into Eq. 10.12 and simplifying gives

$$\frac{dy}{dx} = \frac{(S_f - S_0)}{(1 - \mathrm{Fr}^2)} \tag{10.14}$$

It is seen that the rate of change of fluid depth, dy/dx, depends on the local slope of the channel bottom, the slope of the energy line, and the Froude number. The value of dy/dx can be either negative, zero, or positive, depending on the values of these three parameters. That is, the channel flow depth may be constant or it may increase or decrease in the flow direction, depending on the values of S_0, S_f, and Fr. The behavior of subcritical flow may be the opposite of that for supercritical flow, as seen by the denominator, $(1 - Fr^2)$, of Eq. 10.14.

Equation 10.14 is valid for channels of any constant cross-sectional shape, not just rectangular shapes, provided the Froude number is interpreted properly (Ref. 3). In this book we will consider only rectangular cross section channels when using this equation.

10.4 UNIFORM DEPTH CHANNEL FLOW

Many channels are designed to carry fluid at a uniform depth all along their length. Irrigation canals are frequently of uniform depth and cross section for considerable lengths. Natural channels such as rivers and creeks are seldom of uniform shape, although a reasonable approximation to the flowrate in such channels can often be obtained by assuming uniform flow. In this section we will discuss various aspects of such flows.

Uniform depth flow ($dy/dx = 0$) can be accomplished by adjusting the bottom slope, S_0, so that it precisely equals the slope of the energy line, S_f. That is, $S_0 = S_f$. This can be seen from Eq. 10.14. From an energy point of view, uniform depth flow is achieved by a balance between the potential energy gained by the fluid as it coasts downhill and the energy that is dissipated by viscous effects (head loss) associated with shear stresses throughout the fluid. Similar conclusions can be reached from a force balance analysis as discussed in the following section.

10.4.1 Uniform Flow Approximations

We consider fluid flowing in an open channel of constant cross-sectional size and shape such that the depth of flow remains constant as is indicated in Fig. 10.8. The area of the section is A and the *wetted perimeter* (i.e., the length of the perimeter of the cross section in contact with the fluid) is P. The interaction between the fluid and the atmosphere at the free surface is assumed negligible so that this portion of the perimeter is not included in the definition of the wetted perimeter.

Since the fluid must adhere to the solid surfaces, the actual velocity distribution in an open channel is not uniform. Some typical velocity profiles measured in channels

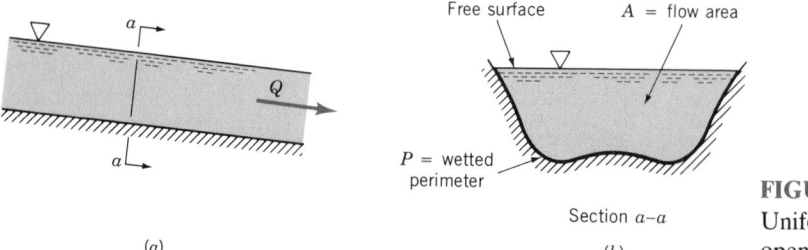

FIGURE 10.8
Uniform flow in an open channel.

of various shapes are indicated in Fig. 10.9a. The maximum velocity is often found somewhat below the free surface, and the fluid velocity is zero on the wetted perimeter, where a wall shear stress, τ_w, is developed. This shear stress is seldom uniform along the wetted perimeter, with typical variations as are indicated in Fig. 10.9b.

Fortunately, reasonable analytical results can be obtained by assuming a uniform velocity profile, V, and a constant wall shear stress, τ_w. Similar assumptions were made for pipe flow situations (Chapter 8), with the friction factor being used to obtain the head loss.

10.4.2 The Chezy and Manning Equations

The basic equations used to determine the uniform flowrate in open channels were derived many years ago. Continual refinements have taken place to obtain better values of the empirical coefficients involved. The result is a semiempirical equation that provides reasonable engineering results. A more refined analysis is perhaps not warranted because of the complexity and uncertainty of the flow geometry (i.e., channel shape and the irregular makeup of the wetted perimeter, particularly for natural channels).

Under the assumptions of steady, uniform flow, the x component of the momentum equation (Eq. 5.22) applied to the control volume indicated in Fig. 10.10 simply reduces to

$$\Sigma F_x = \rho Q(V_2 - V_1) = 0$$

since $V_1 = V_2$. There is no acceleration of the fluid, and the momentum flux across section (1) is the same as that across section (2). The flow is governed by a simple balance between the forces in the direction of the flow. Thus, $\Sigma F_x = 0$, or

$$F_1 - F_2 - \tau_w P\ell + W \sin \theta = 0 \tag{10.15}$$

Lines of constant velocity

Centerline velocity profiles

Actual
Uniform

τ_w

τ_w = shear stress distribution

(a)

(b)

FIGURE 10.9 Typical velocity and shear stress distributions in an open channel: (a) velocity distribution throughout the cross section, (b) shear stress distribution on the wetted perimeter.

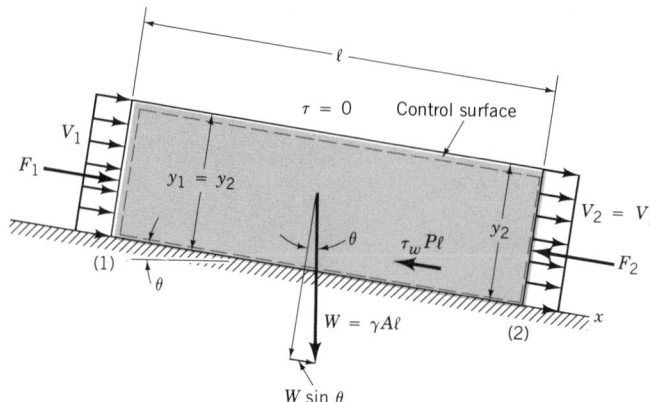

FIGURE 10.10
Control volume for uniform flow in an open channel.

where F_1 and F_2 are the hydrostatic pressure forces across either end of the control volume. Because the flow is at a uniform depth ($y_1 = y_2$), it follows that $F_1 = F_2$ so that these two forces do not contribute to the force balance. The term $W \sin \theta$ is the component of the fluid weight that acts down the slope, and $\tau_w P\ell$ is the shear force acting up the slope as a result of the interaction of the water and the channel wetted perimeter. Thus, Eq. 10.15 becomes

$$\tau_w = \frac{W \sin \theta}{P\ell} = \frac{W S_0}{P\ell}$$

where we have used the approximation that $\sin \theta \approx \tan \theta = S_0$, since the bottom slope is typically very small (i.e., $S_0 \ll 1$). Since $W = \gamma A\ell$ and the *hydraulic radius* is defined as $R_h = A/P$, the force balance equation becomes

$$\tau_w = \frac{\gamma A\ell S_0}{P\ell} = \gamma R_h S_0 \tag{10.16}$$

Most open-channel flows are turbulent rather than laminar. In fact, typical Reynolds numbers are quite large, well above the transitional value and into the wholly turbulent regime. As was discussed in Chapter 8, for very large Reynolds number pipe flows (wholly turbulent flows), the friction factor, f, is found to be independent of Reynolds number, dependent only on the relative roughness of the pipe surface. For such cases, the wall shear stress is proportional to the dynamic pressure, $\rho V^2/2$, and independent of the viscosity. That is,

$$\tau_w = K\rho \frac{V^2}{2}$$

where K is a constant dependent upon the roughness of the pipe.

It is not unreasonable that similar shear stress dependencies occur for the large Reynolds number open-channel flows. In such situations, Eq. 10.16 becomes

$$K\rho \frac{V^2}{2} = \gamma R_h S_0$$

or

$$V = C \sqrt{R_h S_0} \tag{10.17}$$

where the constant C is termed the Chezy coefficient and Eq. 10.17 is termed the *Chezy equation*. This equation, one of the oldest in the area of fluid mechanics, was developed in 1768 by A. Chezy (1718–1798), a French engineer who designed a canal for the Paris water supply. The value of the Chezy coefficient, which must be determined by experiments, is not dimensionless but has the dimensions of (length)$^{1/2}$ per time (i.e., the square root of the units of acceleration).

After a series of experiments, it was found that the slope dependence of Eq. 10.16 ($V \sim S_0^{1/2}$) is reasonable, but that the dependence on the hydraulic radius is actually not as given. In 1889, R. Manning (1816–1897), an Irish engineer, developed the following somewhat modified equation for open-channel flow to more accurately describe the R_h dependence:

$$V = \frac{R_h^{2/3} S_0^{1/2}}{n} \qquad (10.18)$$

Equation 10.18 is termed the *Manning equation,* and the parameter n is the *Manning resistance coefficient.* Its value is dependent on the surface material of the channel's wetted perimeter and is obtained from experiments. It is not dimensionless, having the units of s/m$^{1/3}$ or s/ft$^{1/3}$.

As is discussed in Chapter 7, any correlation should be expressed in dimensionless form, with the coefficients that appear being dimensionless coefficients, such as the friction factor for pipe flow or the drag coefficient for flow past objects. Thus, Eq. 10.18 should be expressed in dimensionless form. Unfortunately, the Manning equation is so widely used and has been used for so long that it will continue to be used in its dimensional form with a coefficient, n, that is not dimensionless. The values of n found in the literature (such as Table 10.1) were developed for SI units. Standard practice is to use the same value of n when using the BG system of units, and to insert a conversion factor into the equation.

Thus, uniform flow in an open channel is obtained from the Manning equation written as

$$V = \frac{\kappa}{n} R_n^{2/3} S_0^{1/2} \qquad (10.19)$$

and

$$Q = \frac{\kappa}{n} A R_h^{2/3} S_0^{1/2} \qquad (10.20)$$

where $\kappa = 1$ if SI units are used, and $\kappa = 1.49$ if BG units are used. The value 1.49 is the cube root of the number of feet per meter: $(3.281 \text{ ft/m})^{1/3} = 1.49$. Thus, by using R_h in meters, A in m^2, and $\kappa = 1$, the average velocity is m/s and the flowrate m^3/s. By using R_h in feet, A in ft^2, and $\kappa = 1.49$, the average velocity is ft/s and the flowrate ft^3/s.

Typical values of the Manning coefficient are indicated in Table 10.1. As expected, the rougher the wetted perimeter, the larger the value of n. For example, the roughness of floodplain surfaces increases from pasture to brush to tree conditions. So does the corresponding value of the Manning coefficient. Precise values of n are often difficult to obtain. Except for artificially lined channel surfaces like those found in new canals or flumes, the channel surface structure may be quite complex and variable. There are various methods used to obtain a reasonable estimate of the value of n for a given situation (Ref. 5). For the purpose of this book, the values from Table 10.1 are

TABLE 10.1 Values of the Manning Coefficient, n (Ref. 6)	
Wetted Perimeter	n
A. Natural channels	
Clean and straight	0.030
Sluggish with deep pools	0.040
Major rivers	0.035
B. Floodplains	
Pasture, farmland	0.035
Light brush	0.050
Heavy brush	0.075
Trees	0.15
C. Excavated earth channels	
Clean	0.022
Gravelly	0.025
Weedy	0.030
Stony, cobbles	0.035
D. Artificially lined channels	
Glass	0.010
Brass	0.011
Steel, smooth	0.012
Steel, painted	0.014
Steel, riveted	0.015
Cast iron	0.013
Concrete, finished	0.012
Concrete, unfinished	0.014
Planed wood	0.012
Clay tile	0.014
Brickwork	0.015
Asphalt	0.016
Corrugated metal	0.022
Rubble masonry	0.025

sufficient. Note that the error in Q is directly proportional to the error in n. A 10% error in the value of n produces a 10% error in the flowrate. Considerable effort has been put forth to obtain the best estimate of n, with extensive tables of values covering a wide variety of surfaces (Ref. 7). It should be noted that the values of n given in Table 10.1 are valid only for water as the flowing fluid.

Both the friction factor for pipe flow and the Manning coefficient for channel flow are parameters that relate the wall shear stress to the makeup of the bounding surface. Thus, various results are available that describe n in terms of the equivalent pipe friction factor, f, and the surface roughness, ε (Ref. 8). For our purposes we will use the values of n from Table 10.1.

10.4.3 Uniform Depth Examples

A variety of interesting and useful results can be obtained from the Manning equation. The following examples illustrate some of the typical considerations.

The main parameters involved in uniform depth open-channel flow are the size and shape of the channel cross section (A, R_h), the slope of the channel bottom (S_0),

the character of the material lining the channel bottom and walls (n), and the average velocity of flowrate (V or Q). Although the Manning equation is a rather simple equation, the ease of using it depends in part on which variables are given and which are to be determined.

Determination of the flowrate for a given channel with flow at a given depth (often termed the *normal flowrate* or *normal depth*, sometimes denoted y_n) is obtained from a straightforward calculation as is shown in Example 10.3.

EXAMPLE 10.3

Water flows in the canal of trapezoidal cross section shown in Fig. E10.3. The bottom drops 1.4 ft per 1000 ft of length. Determine the flowrate if the canal is lined with new smooth concrete, or if weeds cover the wetter perimeter. Determine the Froude number for each of these flows.

FIGURE E10.3

SOLUTION

From Eq. 10.20,

$$Q = \frac{1.49}{n} A R_h^{2/3} S_0^{1/2} \tag{1}$$

where we have used $\kappa = 1.49$, since the dimensions are given in BG units. For a depth of $y = 5$ ft, the flow area is

$$A = 12 \text{ ft } (5 \text{ ft}) + 5 \text{ ft} \left(\frac{5}{\tan 40°} \text{ ft} \right) = 89.8 \text{ ft}^2$$

so that with a wetted perimeter of $P = 12 \text{ ft} + 2(5/\sin 40° \text{ ft}) = 27.6$ ft, the hydraulic radius is determined to be $R_h = A/P = 3.25$ ft. Note that even though the channel is quite wide (the free-surface width is 23.9 ft), the hydraulic radius is only 3.25 ft, which is less than the depth.

Thus, with $S_0 = 1.4 \text{ ft}/1000 \text{ ft} = 0.0014$, Eq. 1 becomes

$$Q = \frac{1.49}{n} (89.8)(3.25)^{2/3}(0.0014)^{1/2} = \frac{10.98}{n}$$

where Q is in ft³/s.

From Table 10.1, the values of n are estimated to be $n = 0.012$ for the smooth concrete and $n = 0.030$ for the weedy conditions. Thus,

$$Q = \frac{10.98}{0.012} = 915 \text{ cfs} \tag{Ans}$$

for the new concrete lining and

$$Q = \frac{10.98}{0.030} = 366 \text{ cfs} \tag{Ans}$$

for the weedy lining. The corresponding average velocities, $V = Q/A$, are 10.2 ft/s and 4.08 ft/s, respectively. It does not take a very steep slope ($S_0 = 0.0014$ or $\theta = \tan^{-1}(0.0014) = 0.080°$) for this velocity.

Note that the increased roughness causes a decrease in the flowrate. This is an indication that for the turbulent flows involved, the wall shear stress increases with surface roughness. [For water at 50 °F, the Reynolds number based on the 3.25-ft hydraulic radius of the channel is Re $= R_h V/\nu = 3.25$ ft $(4.08$ ft/s$)/(1.41 \times 10^{-5}$ ft^2/s$) = 9.40 \times 10^5$, well into the turbulent regime.]

The Froude numbers based on the maximum depths for the two flows can be determined from Fr $= V/(gy)^{1/2}$. For the new concrete case,

$$Fr = \frac{10.2 \text{ ft/s}}{(32.2 \text{ ft/s}^2 \times 5 \text{ ft})^{1/2}} = 0.804 \qquad \text{(Ans)}$$

while for the weedy case

$$Fr = \frac{4.08 \text{ ft/s}}{(32.2 \text{ ft/s}^2 \times 5 \text{ ft})^{1/2}} = 0.322 \qquad \text{(Ans)}$$

In either case the flow is subcritical.

The same results would be obtained if the channel size were given in meters. We would use the same value of n but set $\kappa = 1$ for this SI units situation.

In some instances a trial-and-error or iteration method must be used to solve for the dependent variable. This is often encountered when the flowrate, channel slope, and channel material are given, and the flow depth is to be determined as illustrated in the following examples.

EXAMPLE 10.4

Water flows in the channel shown in Fig. E10.3 at a rate of $Q = 10.0$ m^3/s. If the canal lining is weedy, determine the depth of the flow.

SOLUTION

In this instance neither the flow area nor the hydraulic radius are known, although they can be written in terms of the depth, y, as

$$A = 1.19y^2 + 3.66y$$

where the bottom width is $(12 \text{ ft})(1 \text{ m}/3.281 \text{ ft}) = 3.66$ m and A and y are in square meters and meters, respectively. Also, the wetted perimeter is

$$P = 3.66 + 2\left(\frac{y}{\sin 40°}\right) = 3.11y + 3.66$$

so that

$$R_h = \frac{A}{P} = \frac{1.19y^2 + 3.66y}{3.11y + 3.66}$$

where R_h and y are in meters. Thus, with $n = 0.030$ (from Table 10.1), Eq. 10.20 can be written as

$$Q = 10 = \frac{\kappa}{n} A R_h^{2/3} S_0^{1/2}$$

$$= \frac{1.0}{0.030}(1.19y^2 + 3.66y)\left(\frac{1.19y^2 + 3.66y}{3.11y + 3.66}\right)^{2/3}(0.0014)^{1/2}$$

which can be rearranged into the form

$$(1.19y^2 + 3.66y)^5 - 515(3.11y + 3.66)^2 = 0 \tag{1}$$

where y is in meters. The solution of Eq. 1 can be easily obtained by use of a simple root-finding numerical technique or by trial-and-error methods. The only physically meaningful root of Eq. 1 (i.e., a positive, real number) gives the solution for the normal flow depth at this flowrate as

$$y = 1.50 \text{ m} \tag{Ans}$$

In Example 10.4 we found the flow depth for a given flowrate. Since the equation for this depth is a nonlinear equation, it may be that there is more than one solution to the problem. For a given channel there may be two or more depths that carry the same flowrate. Although this is not normally so, it can and does happen, as is illustrated by Example 10.5.

EXAMPLE 10.5

Water flows in a round pipe of diameter D at a depth of $0 \le y \le D$, as is shown in Fig. E10.5a. The pipe is laid on a constant slope of S_0, and the Manning coefficient is n. At what depth does the maximum flowrate occur? Show that for certain flowrates there are two depths possible with the same flowrate. Explain this behavior.

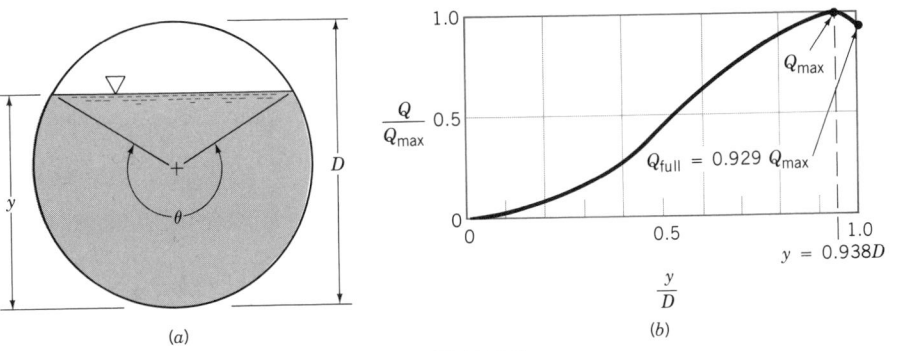

FIGURE E10.5

SOLUTION

According to the Manning equation (Eq. 10.20) the flowrate is

$$Q = \frac{\kappa}{n} A R_h^{2/3} S_0^{1/2} \tag{1}$$

where S_0, n, and κ are constants for this problem. From geometry it can be shown that

$$A = \frac{D^2}{8}(\theta - \sin\theta)$$

where θ, the angle indicated in Fig. E10.5a, is in radians. Similarly, the wetted perimeter is

$$P = \frac{D\theta}{2}$$

so that the hydraulic radius is

$$R_h = \frac{A}{P} = \frac{D(\theta - \sin \theta)}{4\theta}$$

Therefore, Eq. 1 becomes

$$Q = \frac{\kappa}{n} S_0^{1/2} \frac{D^{8/3}}{8(4)^{2/3}} \left[\frac{(\theta - \sin \theta)^{5/3}}{\theta^{2/3}} \right]$$

This can be written in terms of the flow depth by using $y = (D/2)[1 - \cos(\theta/2)]$.

A graph of flowrate versus flow depth, $Q = Q(y)$, has the characteristic indicated in Fig. E10.5b. In particular, the maximum flowrate, Q_{max}, does not occur when the pipe is full; $Q_{full} = 0.929 Q_{max}$. It occurs when $y = 0.938D$, or $\theta = 5.28$ rad $= 303°$. Thus,

$$Q = Q_{max} \text{ when } y = 0.938D \qquad \text{(Ans)}$$

For any $0.929 < Q/Q_{max} < 1$ there are two possible depths that give the same Q. The reason for this behavior can be seen by considering the gain in flow area, A, compared to the increase in wetted perimeter, P, for $y \approx D$. The flow area increase for an increase in y is very slight in this region, whereas the increase in wetted perimeter, and hence the increase in shear force holding back the fluid, is relatively large. The net result is a decrease in flowrate as the depth increases. For most practical problems, the slight difference between the maximum flowrate and full pipe flowrates is negligible, particularly in light of the usual inaccuracy of the value of n.

Sometimes it is necessary to determine the slope needed to produce a desired flowrate in a channel of specified shape, size, and material. Such a calculation is a straightforward application of Eq. 10.20. In addition, it is often of use to know whether the flow obtained will be subcritical, critical, or supercritical. Example 10.6 illustrates these ideas.

EXAMPLE 10.6

Water flows in a rectangular channel of width $b = 10$ m that has a Manning coefficient of $n = 0.025$. Plot a graph of flowrate, Q, as a function of slope, S_0, indicating lines of constant depth and lines of constant Froude number.

SOLUTION

For this channel the flow area is $A = by = 10y$, and the hydraulic radius is $R_h = A/P = by/(b + 2y)$, where y is the flow depth. Thus, the Manning equation (Eq. 10.19) becomes

$$V = \frac{\kappa}{n} R_h^{2/3} S_0^{1/2} = \frac{1.0}{0.025} \left(\frac{10y}{10 + 2y} \right)^{2/3} S_0^{1/2} \qquad (1)$$

where we have set $\kappa = 1$ because we are using SI units. Since we wish to plot Q versus S_0 at constant values of Fr, we use the fact that for a rectangular channel $\mathrm{Fr} = V/(gy)^{1/2}$ and write Eq. 1 as

$$(gy)^{1/2}\mathrm{Fr} = \frac{1}{0.025}\left(\frac{10y}{10 + 2y}\right)^{2/3} S_0^{1/2}$$

which simplifies to

$$S_0 = 0.00613(\mathrm{Fr})^2\, y\left(\frac{5 + y}{5y}\right)^{4/3} \tag{2}$$

For a given value of Fr, we pick various values of y, determine the corresponding values of S_0 from Eq. 2, and then calculate $Q = VA$, with V from either Eq. 1 or $V = (gy)^{1/2}$ Fr. The results are indicated in Fig. E10.6.

FIGURE E10.6

Note that for a given flowrate, there is a specific value of the slope that gives critical condition, $\mathrm{Fr} = 1$. This slope, denoted S_{0c}, is called the *critical slope*. The critical slope line divides the graph into two regions—one subcritical and the other supercritical. The dependence of S_{0c} on Q is rather weak over a large range of Q. If the slope is such that the flow is critical or nearly critical, it will remain so over a wide range of flowrates (and depths).

Lines of constant depth, y, are also indicated in Fig. E10.6. A figure like this allows one to easily see what effects are to be expected by varying the parameters involved.

In many man-made channels and in most natural channels, the surface roughness (and hence the Manning coefficient) varies along the wetted perimeter of the channel. A drainage ditch, for example, may have a rocky bottom surface with concrete side walls to prevent erosion. Thus, the effective n will be different for shallow depths than for deep depths of flow. Similarly, a river channel may have one value of n appropriate for its normal channel and another very different value of n during its flood stage where a portion of the flow occurs across fields or through floodplain woods. An ice-covered channel usually has a different value of n for the ice than for the remainder

of the wetted perimeter (Ref. 7). (Strictly speaking, such ice-covered channels are not "open" channels, although analysis of their flow is often based on open-channel flow equations. This is acceptable, since the ice covering is often thin enough so that it represents a fixed boundry in terms of the shear stress resistance, but it cannot support a significant pressure differential as in pipe flow situations.)

A variety of methods has been used to determine an appropriate value of the effective roughness of channels that contain subsections with different values of n. Which method gives the most accurate, easy-to-use results is not firmly established, since the results are nearly the same for each method (Ref. 5). A reasonable approximation is to divide the channel cross section into N subsections, each with its own wetted perimeter, P_i, area, A_i, and Manning coefficient, n_i. The P_i values do not include the imaginary boundaries between the different subsections. The total flowrate is assumed to be the sum of the flowrates through each section. This technique is illustrated by Example 10.7.

EXAMPLE 10.7

Water flows along the drainage canal having the properties shown in Fig. E10.7. If the bottom slope is $S_0 = 1 \text{ ft}/500 \text{ ft} = 0.002$, estimate the flowrate.

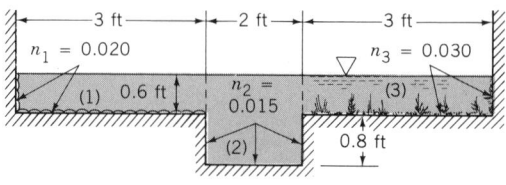

FIGURE E10.7

SOLUTION

We divide the cross section into three subsections as is indicated in Fig. E10.7 and write the flowrate as $Q = Q_1 + Q_2 + Q_3$, where for each section

$$Q_i = \frac{1.49}{n_i} A_i R_{h_i}^{2/3} S_0^{1/2}$$

The appropriate values of A_i, P_i, R_{hi}, and n_i are listed in Table E10.7. Note that the imaginary

TABLE E10-7

i	A_i (ft²)	P_i (ft)	R_{hi} (ft)	n_i
1	1.8	3.6	0.500	0.020
2	2.8	3.6	0.778	0.015
3	1.8	3.6	0.500	0.030

portions of the perimeters between sections (denoted by the dashed lines in Fig. E10.7) are not included in the P_i. That is, for section (2)

$$A_2 = 2 \text{ ft} (0.8 + 0.6) \text{ ft} = 2.8 \text{ ft}^2$$

and

$$P_2 = 2 \text{ ft} + 2(0.8 \text{ ft}) = 3.6 \text{ ft}$$

so that

$$R_{h_2} = \frac{A_2}{P_2} = \frac{2.8 \text{ ft}^2}{3.6 \text{ ft}} = 0.778 \text{ ft}$$

Thus, the total flowrate is

$$Q = Q_1 + Q_2 + Q_3 = 1.49(0.002)^{1/2}$$
$$\times \left[\frac{1.8(0.500)^{2/3}}{0.020} + \frac{2.8(0.778)^{2/3}}{0.015} + \frac{1.8(0.500)^{2/3}}{0.030} \right]$$

or

$$Q = 16.8 \text{ ft}^3/\text{s} \qquad \text{(Ans)}$$

If the entire channel cross section were considered as one flow area, then $A = A_1 + A_2 + A_3 = 6.4 \text{ ft}^2$ and $P = P_1 + P_2 + P_3 = 10.8 \text{ ft}$, or $R_h = A/P = 6.4 \text{ ft}^2/10.8 \text{ ft} = 0.593 \text{ ft}$. The flowrate is given by Eq. 10.20, which can be written as

$$Q = \frac{1.49}{n_{\text{eff}}} A R_h^{2/3} S_0^{1/2}$$

where n_{eff} is the effective value of n for this channel. With $Q = 16.8 \text{ ft}^3/\text{s}$ as determined above, the value of n_{eff} is found to be

$$n_{\text{eff}} = \frac{1.49 A R_h^{2/3} S_0^{1/2}}{Q}$$
$$= \frac{1.49(6.4)(0.593)^{2/3}(0.002)^{1/2}}{16.8} = 0.0179$$

As expected, the effective roughness (Manning n) is between the minimum ($n_2 = 0.015$) and maximum ($n_3 = 0.030$) values for the individual subsections.

One type of problem often encountered in open-channel flows is that of determining the *best hydraulic cross section* defined as the section of the minimum area for a given flowrate, Q, slope, S_0, and roughness coefficient, n. By using $R_h = A/P$ we can write Eq. 10.20 as

$$Q = \frac{\kappa}{n} A \left(\frac{A}{P} \right)^{2/3} S_0^{1/2} = \frac{\kappa}{n} \frac{A^{5/3} S_0^{1/2}}{P^{2/3}}$$

which can be rearranged as

$$A = \left(\frac{nQ}{\kappa S_0^{1/2}} \right)^{3/5} P^{2/5}$$

where the quantity in the parentheses is a constant. Thus, a channel with minimum A is one with a minimum P, so that both the amount of excavation needed and the amount of material to line the surface are minimized by the best hydraulic cross section.

The best hydraulic cross section possible is that of a semicircular channel. No other shape has as small a wetted perimeter for a given area. It is often desired to determine the best shape for a class of cross sections such as rectangles, trapezoids, or triangles. Example 10.8 illustrates the concept of the best hydraulic cross section for rectangular channels.

EXAMPLE 10.8

Water flows uniformly in a rectangular channel of width b and depth y. Determine the aspect ratio, b/y, for the best hydraulic cross section.

SOLUTION

The uniform flow is given by Eq. 10.20 as

$$Q = \frac{\kappa}{n} A R_h^{2/3} S_0^{1/2} \tag{1}$$

where $A = by$ and $P = b + 2y$, so that $R_h = A/P = by/(b + 2y)$. We rewrite the hydraulic radius in terms of A as

$$R_h = \frac{A}{(2y + b)} = \frac{A}{(2y + A/y)} = \frac{Ay}{(2y^2 + A)}$$

so that Eq. 1 becomes

$$Q = \frac{\kappa}{n} A \left(\frac{Ay}{2y^2 + A} \right)^{2/3} S_0^{1/2}$$

This can be rearranged to give

$$A^{5/2} y = K(2y^2 + A) \tag{2}$$

where $K = (nQ/\kappa S_0^{1/2})^{3/2}$ is a constant. The best hydraulic section is the one that gives the minimum A for all y. That is, $dA/dy = 0$. By differentiating Eq. 2 we obtain

$$\frac{5}{2} A^{3/2} \frac{dA}{dy} y + A^{5/2} = K \left(4y + \frac{dA}{dy} \right)$$

which, with $dA/dy = 0$, reduces to

$$A^{5/2} = 4Ky \tag{3}$$

With $K = A^{5/2} y/(2y^2 + A)$ from Eq. 2, Eq. 3 can be written in the form

$$A^{5/2} = \frac{4A^{5/2} y^2}{(2y^2 + A)}$$

which simplifies to $y = (A/2)^{1/2}$. Thus, since $A = by$, the best hydraulic cross section for a rectangular shape has a width b and a depth

$$y = \left(\frac{A}{2} \right)^{1/2} = \left(\frac{by}{2} \right)^{1/2}$$

or

$$2y^2 = by$$

That is, the rectangle with the best hydraulic cross section is twice as wide as it is deep, or

$$b/y = 2 \tag{Ans}$$

A rectangular channel with $b/y = 2$ will give the smallest area (and smallest wetted perimeter) for a given flowrate. Conversely, for a given area, the largest flowrate in a rectangular channel will occur when $b/y = 2$. For $A = by = $ constant, if $y \to 0$ then $b \to \infty$, and the flowrate is small because of the large wetted perimeter $P = b + 2y \to \infty$. Similarly, if $y \to \infty$ then $b \to 0$, and the flowrate is small because $P = b + 2y \to \infty$. The maximum Q occurs when $y = b/2$. However, as is seen in Fig. E10.8a, the maximum represented by this optimal configuration is a rather weak one. For example, for aspect ratios between 1 and 4, the flowrate is within 96% of the maximum flowrate obtained with the same area and $b/y = 2$.

An alternate but equivalent method to obtain the above answer is to use the fact that $dR_h/dy = 0$, which follows from Eq. 1 using $dQ/dy = 0$ (constant flowrate) and $dA/dy = 0$ (best hydraulic cross section has minimum area). Differentiation of $R_h = Ay/(2y^2 + A)$ with constant A gives $b/y = 2$ when $dR_h/dy = 0$.

The best hydraulic cross section can be calculated for other shapes in a similar fashion. The results (given here without proof) for circular, rectangular, trapezoidal (with 60° sides), and triangular shapes are shown in Fig. E10.8b.

FIGURE E10.8

10.5 GRADUALLY VARIED FLOW

In many situations the flow in an open channel is not of uniform depth (y = constant) along the channel. This can occur because of several reasons: the bottom slope is not constant, the cross-sectional shape and area varies in the flow direction, or there is some obstruction across a portion of the channel. Such flows are classified as gradually varying flows if $dy/dx \ll 1$ or rapidly varying flows if $dy/dx \sim 1$. In this section we consider some of the properties of gradually varied flows.

In a complex open-channel flow situation, it is common practice to divide the channel into specific segments or reaches possessing certain qualities. One such method is to use a classification scheme denoting how the flow depth varies with distance along the channel.

As discussed in the previous section, if the channel bottom slope is equal to the slope of the energy line, $S_0 = S_f$, the flow depth is constant, $dy/dx = 0$. Physically, the gain in potential energy of the fluid as it flows downhill is exactly balanced by the dissipation of energy through viscous effects.

If the bottom slope and the energy line slope are not equal, the flow depth will vary along the channel, either increasing or decreasing in the flow direction. In such cases $dy/dx \neq 0$, and the left-hand side of Eq. 10.14 is not zero. Physically, the difference between the component of weight and the shear forces in the direction of flow produces a change in the fluid momentum which requires a change in velocity and, from continuity considerations, a change in depth. Whether the depth increases or decreases depends on various parameters of the flow, with 12 types of surface profile configurations [flow depth as a function of distance, $y = y(x)$] possible.

The sign of dy/dx, that is, whether the flow depth increases or decreases with distance along the channel, depends on the sign of both the numerator and denominator of Eq. 10.14. The sign of the denominator clearly depends on whether the flow is subcritical or supercritical. For a given channel (i.e., size, shape, Manning coefficient, and flowrate) there is a *critical slope*, $S_0 = S_{0c}$, and a corresponding critical depth, $y = y_c$, such that Fr $= 1$ under conditions of uniform flow.

For uniform depth flow, it follows that $S_0 = S_f$ and the numerator of Eq. 10.14 is zero. However, if S_0 is greater than S_f, the bottom slope is steeper than is required for uniform flow, and the numerator is negative. If S_0 is less than S_f the numerator is positive. Thus, there are a number of combinations of numerator and denominator of Eq. 10.14 that give positive, zero, or negative values of dy/dx. These conditions result in the following flow configurations.

10.5.1 Classification of Surface Shapes

The character of a gradually varying flow is often classified in terms of the actual channel slope, S_0, compared with the slope required to produce uniform critical flow, S_{0c}. In addition, the character of the flow depends on whether the fluid depth is less than or greater than the uniform normal depth, y_n, that would occur in the channel with the given slope and flowrate. The 12 configurations indicated in Table 10.2 are possible. We will merely indicate the possible types of surface configurations without becoming involved in the details of how to calculate their shape. Additional discussion

TABLE 10.2 Possible Free-Surface Configurations

Slope Type	Slope Notation	Froude No.	Surface Shape Designation
$S_0 < S_{0c}$	Mild (M)	Fr < 1	M-1
		Fr < 1	M-2
		Fr > 1	M-3
$S_0 = S_{0c}$	Critical (C)	Fr < 1	C-1
		Fr > 1	C-3
$S_0 > S_{0c}$	Steep (S)	Fr < 1	S-1
		Fr > 1	S-2
		Fr > 1	S-3
$S_0 = 0$	Horizontal (H)	Fr < 1	H-2
		Fr > 1	H-3
$S_0 < 0$	Adverse (A)	Fr < 1	A-2
		Fr > 1	A-3

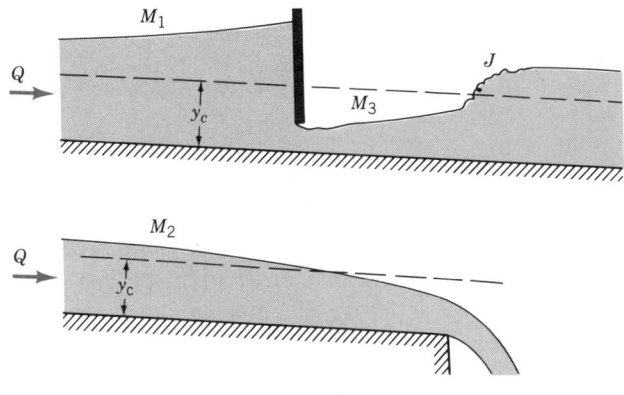

FIGURE 10.11 Typical surface configurations for nonuniform depth flow with a mild slope; $S_0 < S_{0c}$.

Mild slope

of this complex topic can be found in standard references about open-channel flows (Refs. 3, 9).

The five types of slopes indicated include: (1) mild slope with $S_0 < S_{0c}$ (the flow would be subcritical if it were uniform depth), (2) critical slope with $S_0 = S_{0c}$ (the flow would have Fr = 1 if it were uniform depth), (3) steep slope with $S_0 > S_{0c}$ (the flow would be supercritical if it were uniform depth), (4) horizontal slope with $S_0 = 0$, and (5) adverse slope with $S_0 < 0$ (flow uphill). Since we are considering nonuniform depth flows, the flows within these categories may have Fr < 1 or Fr > 1, depending on whether $y < y_c$ or $y < y_c$, respectively. Here, y_c is the critical depth, the depth of flow for the given conditions that would give critical uniform depth flow. The standard designation attached to these flows (i.e., *M*-1, *A*-2, etc.) are indicated in the table.

If the flow were uniform depth flow along the channel, the determination of whether the flow is sub- or supercritical would be dependent solely on whether $S_0 < S_{0c}$ or $S_0 > S_{0c}$, respectively. For gradually varying flow, inertia of the fluid becomes a governing mechanism (in addition to weight and shear force), and it is possible to have a variety of additional flow situations. Whether the actual flow is subcritical or supercritical depends on additional conditions, not just the value of S_0. For example, with $S_0 < S_{0c}$ (a mild slope indicating Fr < 1 if the flow were uniform depth), it is possible to have either Fr < 1 or Fr > 1 depending on the depth. That is, a balance of weight, shear, and inertia effects provides an additional freedom to the flow that is not present in uniform flow where inertial effects are absent.

10.5.2 Examples of Gradually Varied Flows

Although calculation of the actual surface shapes is beyond the scope of this book, flow situations involving the 12 types of surface profiles possible are indicated in Figs. 10.11 through 10.15. In each instance the depth variation is caused by some obstruction

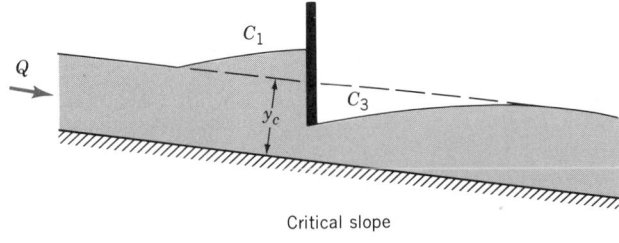

FIGURE 10.12 Typical surface configurations for nonuniform depth flow with a critical slope; $S_0 = S_{0c}$.

Critical slope

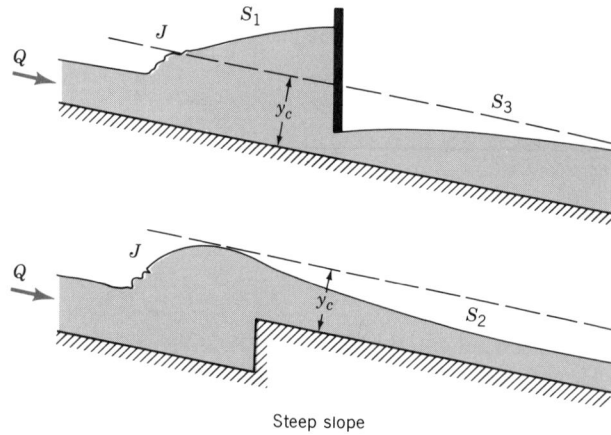

FIGURE 10.13 Typical surface configurations for nonuniform depth flow with a steep slope; $S_0 > S_{0c}$.

or change in the channel geometry, such as a dam, a sluice gate, or a sudden change in the channel elevation. Gradually varied flows can also occur in the vicinity of the location where the channel bottom changes from one slope to another slope, either steeper or shallower.

Note that the horizontal length scale of Figs. 10.11 through 10.15 has been reduced. They should have $dy/dx \ll 1$ and $S_0 \ll 1$. Generally, the bottom slope is so small that it would not be visible on a true scale drawing and the distances over which the depth changes occur are very large compared with the depth change. For rapidly varying flows, such as the hydraulic jumps indicated by the regions J in the figures and discussed in Section 10.6.1, the depth variation occurs in a relatively short distance.

The most common of the gradually varying flows is the M-1 type indicated in Fig. 10.11. Such flows occur upstream of a dam or sluice gate where the subcritical flow is slowed down by the obstruction across the channel. It is often of considerable importance to determine the extent to which the water level is raised above what it would assume if the dam were not present. The lake surface immediately upstream of a dam is horizontal—the water is stationary. However, the surface profile in the transition region between the lake and the unaffected flowing river is neither horizontal nor at an elevation it would have if the dam were not present. Such water surface profiles are called the *backwater curve*.

The M-2 type curve is obtained by the reduction of the downstream resistance in a channel such as that produced by the outflow condition shown in Fig. 10.11. The subcritical flow accelerates as it approaches the end of the channel and produces the *drop-down* profile as indicated.

The M-3 type curve indicated in Fig. 10.11 results from the fact that the mild slope is not sufficient to maintain the supercritical conditions after the sluice gate. At a certain point downstream the supercritical flow passes through a hydraulic jump, J, and changes to subcritical flow.

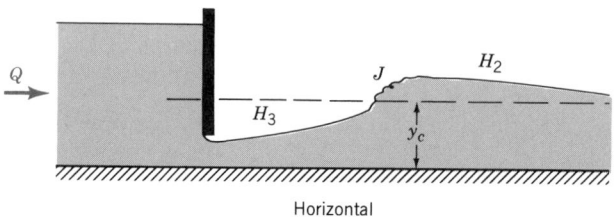

FIGURE 10.14 Typical surface configurations for nonuniform depth flow with a horizontal slope; $S_0 = 0$.

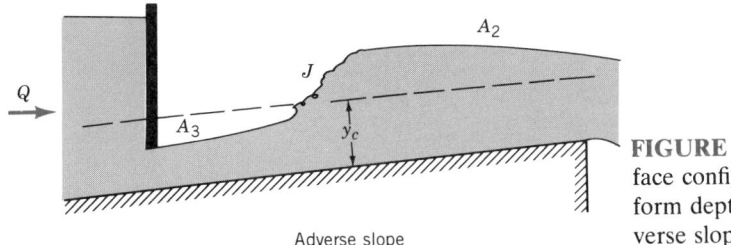

FIGURE 10.15 Typical surface configurations for nonuniform depth flow with an adverse slope; $S_0 < 0$.

The C-1 and C-3 critical slope profiles indicated in Fig. 10.12 can be produced by a sluice gate across the channel. However, they are infrequently encountered because the precise Fr = 1 or $S_0 = S_{0c}$ conditions are not often found.

The steep slope S-1 and S-3 conditions indicated in Fig. 10.13 can be produced by an obstruction such as a dam or sluice gate in a manner very similar to the M-1 and M-3 mild slope flows. Note that the sign of the free-surface curvature is opposite for the steep slope case compared with the mild slope case. The S-2 profile is obtained when a channel receives flow from a critical depth situation, as often occurs for flow from a reservoir into a channel.

The horizontal or adverse slope situations are unique in that it is impossible to maintain uniform flow under such conditions. The fluid inertia is clearly an important mechanism in these flows, since the component of weight along the channel is either zero (horizontal slope) or in the direction opposite to the flow (adverse slope). Typical flows are indicated in Figs. 10.14 and 10.15.

As is discussed above, a variety of surface profiles is possible in gradually varied open-channel flow. The free surface is relatively free to conform to the shape that satisfies the governing mass, momentum, and energy equations. The actual shape of the surface is often very important in the design of open-channel devices or in the prediction of flood levels in natural channels. The surface shape, $y = y(x)$, can be calculated by solving the governing differential equation obtained from a combination of the Manning equation (Eq. 10.20) and the energy equation (Eq. 10.14). The result is a complex nonlinear differential equation that has been the subject of many investigations over the past century. With the use of modern computers, a variety of numerical techniques has been developed and is used with considerable success in determining the various gradually varied flow surface profiles. Such results are beyond the scope of this book, but are readily available in the literature (Refs. 5, 9).

10.6   RAPIDLY VARIED FLOW

In many open channels flow depth changes occur over a relatively short distance so that $dy/dx \sim 1$. Such rapidly varied flow conditions are often quite complex and difficult to analyze in a precise fashion. Fortunately, many useful approximate results can be obtained by using a simple one-dimensional model along with appropriate experimentally determined coefficients when necessary. In this section we discuss several of these flows.

Some rapidly varied flows occur in constant area channels for reasons that are not immediately obvious. The hydraulic jump is one such case. As is indicated in Fig.

FIGURE 10.16 A hydraulic jump at the base of a model spillway. (Photograph courtesy of U. S. Army Corps of Engineers, Waterways Experiment Station.)

10.16, the flow may change from a relatively shallow, high-speed condition into a relatively deep, low-speed condition within a horizontal distance of just a few channel depths. Other rapidly varied flows may be due to a sudden change in the channel geometry such as the flow between bridge piers as is indicated in Fig. 10.17 or the flow in an expansion or contraction section of a channel as is indicated in Fig. 10.18.

In such situations the flow field is often two- or three-dimensional in character. There may be regions of flow separation, flow reversal, or unsteady oscillations of the free surface. For the purpose of some analyses, these complexities can be neglected and a simplified analysis can be undertaken. In other cases, however, it is the complex details of the flow that are the most important property of the flow; any analysis must include their effects. The scouring of a river bottom in the neighborhood of a bridge pier, as is indicated in Fig. 10.19, is such an example. A one- or two-dimensional

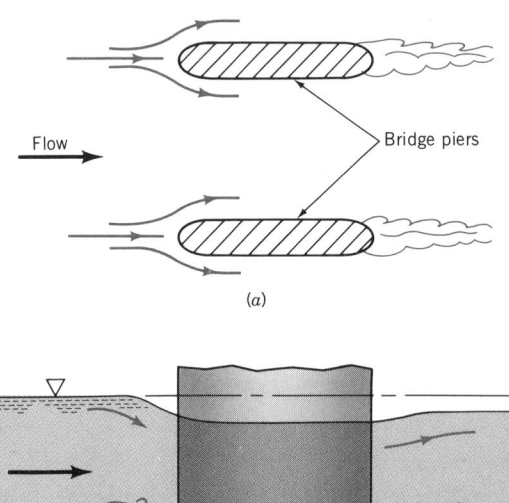

FIGURE 10.17 Rapidly varied flow in the vicinity of a bridge pier: (*a*) top view, (*b*) side view.

FIGURE 10.18 Rapidly varied flow may occur in a channel transition section.

model of this flow would not be sufficient to describe the complex structure of the flow that is responsible for the erosion near the foot of the bridge pier.

Many open-channel flow-measuring devices are based on principles associated with rapidly varied flows. Among these devices are broad crested weirs, sharp crested weirs, critical flow flumes, and sluice gates. The operation of such devices is discussed in the following sections.

10.6.1 The Hydraulic Jump

Observations of flows in open channels show that under certain conditions it is possible that the fluid depth will change very rapidly over a short length of the channel without any change in the channel configuration. Such changes in depth can be approximated as a discontinuity in the free-surface elevation ($dy/dx = \infty$). For reasons discussed below, this step change in depth is always from a shallow to a deeper depth—always a step up, never a step down.

Physically, this near discontinuity, called a *hydraulic jump,* may result when there is a conflict between the upstream and downstream influences that control a particular section (or reach) of a channel. For example, a sluice gate may require that the conditions at the upstream portion of the channel (downstream of the gate) be supercritical flow, while obstructions in the channel on the downstream end of the reach may require that the flow be subcritical. The hydraulic jump provides the mechanism (a nearly discontinuous one at that) to make the transition between the two types of flow.

The simplest type of hydraulic jump occurs in a horizontal, rectangular channel as is indicated in Fig. 10.20. Although the flow within the jump itself is extremely complex and agitated, it is reasonable to assume that the flow at sections (1) and (2) is nearly uniform and one-dimensional. In addition, we neglect any wall shear stresses, τ_w,

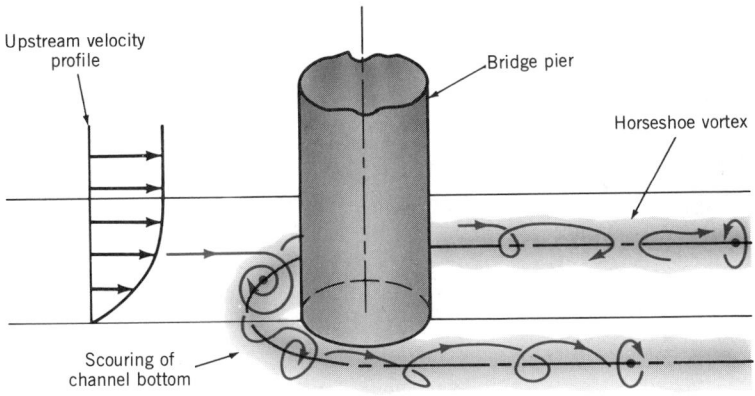

FIGURE 10.19 The complex three-dimensional flow structure around a bridge pier.

FIGURE 10.20
Hydraulic jump geometry.

within the relatively short segment between these two sections. Under these conditions the x component of the momentum equation (Eq. 5.22) for the control volume indicated can be written as

$$F_1 - F_2 = \rho Q(V_2 - V_1) = \rho V_1 y_1 b(V_2 - V_1)$$

where the pressure force at either section is hydrostatic. That is, $F_1 = p_{c_1} A_1 = \gamma y_1^2\, b/2$ and $F_2 = p_{c_2} A_2 = \gamma y_2^2 b/2$, where $p_{c_1} = \gamma y_1/2$ and $p_{c_2} = \gamma y_2/2$ are the pressures at the centroids of the channel cross sections and b is the channel width. Thus, the momentum equation becomes

$$\frac{y_1^2}{2} - \frac{y_2^2}{2} = \frac{V_1 y_1}{g}(V_2 - V_1) \tag{10.21}$$

In addition to the momentum equation, we have the conservation of mass equation (Eq. 5.12)

$$y_1 b V_1 = y_2 b V_2 = Q \tag{10.22}$$

and the energy equation (Eq. 5.84)

$$y_1 + \frac{V_1^2}{2g} = y_2 + \frac{V_2^2}{2g} + h_L \tag{10.23}$$

The head loss, h_L, in Eq. 10.23 is due to the violent turbulent mixing and dissipation that occur within the jump itself. We have neglected any head loss due to wall shear stresses.

Clearly Eqs. 10.21, 10.22, and 10.23 have a solution $y_1 = y_2$, $V_1 = V_2$, and $h_L = 0$. This represents the trivial case of no jump. Since these are nonlinear equations, it may be possible that more than one solution exists. The other solutions can be obtained as follows. By combining Eqs. 10.21 and 10.22 to eliminate V_2 we obtain

$$\frac{y_1^2}{2} - \frac{y_2^2}{2} = \frac{V_1 y_1}{g}\left(\frac{V_1 y_1}{y_2} - V_1\right) = \frac{V_1^2 y_1}{g y_2}(y_1 - y_2)$$

which can be simplified by factoring out a common nonzero factor $y_1 - y_2$ from each side to give

$$\left(\frac{y_2}{y_1}\right)^2 + \left(\frac{y_2}{y_1}\right) - 2\,\mathrm{Fr}_1^2 = 0$$

where $\mathrm{Fr}_1 = V_1/\sqrt{g y_1}$ is the upstream Froude number. By using the quadratic formula we obtain

$$\frac{y_2}{y_1} = \frac{1}{2}\left(-1 \pm \sqrt{1 + 8\mathrm{Fr}_1^2}\right)$$

Clearly the solution with the minus sign is not possible (it would give a negative y_2/y_1). Thus,

$$\frac{y_2}{y_1} = \frac{1}{2}\left(-1 + \sqrt{1 + 8\text{Fr}_1^2}\right) \qquad (10.24)$$

This depth ratio, y_2/y_1, across the hydraulic jump is shown as a function of the upstream Froude number in Fig. 10.21. The portion of the curve for $\text{Fr}_1 < 1$ is dashed in recognition of the fact that to have a hydraulic jump the flow must be supercritical. That is, the solution as given by Eq. 10.24 must be restricted to $\text{Fr}_1 \geq 1$, for which $y_2/y_1 \geq 1$. This can be shown by consideration of the energy equation, Eq. 10.23, as follows. The dimensionless head loss, h_L/y_1, can be obtained from Eq. 10.23 as

$$\frac{h_L}{y_1} = 1 - \frac{y_2}{y_1} + \frac{\text{Fr}_1^2}{2}\left[1 - \left(\frac{y_1}{y_2}\right)^2\right] \qquad (10.25)$$

where, for given values of Fr_1, the values of y_2/y_1 are obtained from Eq. 10.24. As is indicated in Fig. 10.22, the head loss is negative if $\text{Fr}_1 < 1$. Since negative head losses violate the second law of thermodynamics (viscous effects dissipate energy, they cannot create energy; see Section 5.3), it is not possible to produce a hydraulic jump with $\text{Fr}_1 < 1$. The head loss is indicated by the drop of the energy line shown in Fig. 10.20.

A flow must be supercritical (Froude number > 1) to produce the discontinuity called a hydraulic jump. This is analogous to the compressible flow ideas discussed in Chapter 11 in which it is shown that the flow of a gas must be supersonic (Mach number > 1) to produce the discontinuity called a normal shock wave. However, the fact that a flow is supercritical (or supersonic) does not guarantee the production of a hydraulic jump (or shock wave). The trivial solution $y_1 = y_2$ and $V_1 = V_2$ is also possible.

The fact that there is an energy loss across a hydraulic jump is useful in many situations. For example, the relatively large amount of energy contained in the fluid

FIGURE 10.21   Depth ratio across a hydraulic jump as a function of upstream Froude number.

FIGURE 10.22 Dimensionless head loss across a hydraulic jump as a function of upstream Froude number.

flowing down the spillway of a dam could cause damage to the channel below the dam. By placing suitable flow control objects in the channel downstream of the spillway, it is possible (if the flow is supercritical) to produce a hydraulic jump on the apron of the spillway and thereby dissipate a considerable portion of the energy of the flow. That is, the dam spillway produces supercritical flow, and the channel downstream of the dam requires subcritical flow. The resulting hydraulic jump allows the needed change in flow character.

EXAMPLE 10.9

Water on the horizontal apron of the 100-ft-wide spillway shown in Fig. E10.9a has a depth of 0.60 ft and a velocity of 18 ft/s. Determine the depth, y_2, after the jump, the Froude numbers before and after the jump, Fr_1 and Fr_2, and the power dissipated, $\mathcal{P}_d$, within the jump.

SOLUTION

FIGURE E10.9

Conditions across the jump are determined by the upstream Froude number

$$Fr_1 = \frac{V_1}{\sqrt{gy_1}} = \frac{18 \text{ ft/s}}{[(32.2 \text{ ft/s}^2)(0.60 \text{ ft})]^{1/2}} = 4.10 \qquad \text{(Ans)}$$

Thus, the upstream flow is supercritical, and it is possible to generate a hydraulic jump as sketched.

From Eq. 10.24 we obtain the depth ratio across the jump as

$$\frac{y_2}{y_1} = \frac{1}{2}(-1 + \sqrt{1 + 8\, \text{Fr}_1^2}) = \frac{1}{2}(-1 + \sqrt{1 + 8(4.10)^2}) = 5.32$$

or

$$y_2 = 5.32\,(0.60\ \text{ft}) = 3.19\ \text{ft} \qquad\qquad \text{(Ans)}$$

Since $Q_1 = Q_2$, or $V_2 = (y_1 V_1)/y_2 = 0.60\ \text{ft}\,(18\ \text{ft/s})/3.19\ \text{ft} = 3.39\ \text{ft/s}$, it follows that

$$\text{Fr}_2 = \frac{V_2}{\sqrt{gy_2}} = \frac{3.39\ \text{ft/s}}{[(32.2\ \text{ft/s}^2)(3.19\ \text{ft})]^{1/2}} = 0.334 \qquad\qquad \text{(Ans)}$$

As is true for any hydraulic jump, the flow changes from supercritical to subcritical flow across the jump.

The power (energy per unit time) dissipated, $\mathscr{P}_d$, by viscous effects within the jump can be determined from the head loss as

$$\mathscr{P}_d = \gamma Q h_L = \gamma b y_1 V_1 h_L \qquad\qquad (1)$$

where h_L is obtained from Eqs. 10.23 or 10.25 as

$$h_L = \left(y_1 + \frac{V_1^2}{2g}\right) - \left(y_2 + \frac{V_2^2}{2g}\right) = \left[0.60\ \text{ft} + \frac{(18.0\ \text{ft/s})^2}{2(32.2\ \text{ft/s}^2)}\right]$$
$$- \left[3.19\ \text{ft} + \frac{(3.39\ \text{ft/s})^2}{2(32.2\ \text{ft/s}^2)}\right]$$

or

$$h_L = 2.26\ \text{ft}$$

Thus, from Eq. 1,

$$\mathscr{P}_d = (62.4\ \text{lb/ft}^3)(100\ \text{ft})(0.60\ \text{ft})(18.0\ \text{ft/s})(2.26\ \text{ft})$$
$$= 1.52 \times 10^5\ \text{ft·lb/s}$$

or

$$\mathscr{P}_d = \frac{1.52 \times 10^5\ \text{ft·lb/s}}{550[(\text{ft·lb/s})/\text{hp}]} = 277\ \text{hp} \qquad\qquad \text{(Ans)}$$

This power, which is dissipated within the highly turbulent motion of the jump, is converted into an increase in water temperature, T. That is, $T_2 > T_1$. Although the power dissipated is considerable, the difference in temperature is not great because the flowrate is quite large.

The hydraulic jump flow process can be illustrated by use of the specific energy concept introduced in Section 10.3 as follows. Equation 10.23 can be written in terms of the specific energy, $E = y + V^2/2g$, as $E_1 = E_2 + h_L$, where $E_1 = y_1 + V_1^2/2g = 5.63$ ft and $E_2 = y_2 + V_2^2/2g = 3.37$ ft. As is discussed in Section 10.3, the specific energy diagram for this flow can be obtained by using $V = q/y$, where

$$q = q_1 = q_2 = \frac{Q}{b} = y_1 V_1 = 0.60\ \text{ft}\,(18.0\ \text{ft/s}) = 10.8\ \text{ft}^2/\text{s}$$

Thus,

$$E = y + \frac{q^2}{2gy^2} = y + \frac{(10.8\ \text{ft}^2/\text{s})^2}{2(32.2\ \text{ft/s}^2)y^2}$$

or

$$E = y + \frac{1.81}{y^2}$$

where y and E are in feet. The resulting specific energy diagram is shown in Fig. E10.9b. Because of the head loss across the jump, the upstream and downstream values of E are different. In going from state (1) to state (2) the fluid does not proceed along the specific energy

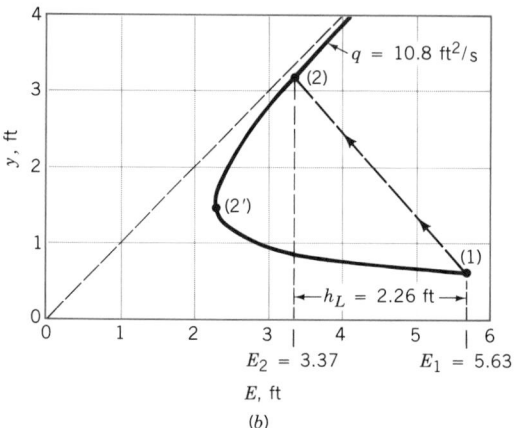

FIGURE E10.9   (*Continued*)

curve and pass through the critical condition at state 2′. Rather, it jumps from (1) to (2) as is represented by the dashed line in the figure. From a one-dimensional consideration, the jump is a discontinuity. In actuality, the jump is a complex three-dimensional flow incapable of being represented on the one-dimensional specific energy diagram.

The actual structure of a hydraulic jump is a complex function of Fr_1, even though the depth ratio and head loss are given quite accurately by a simple one-dimensional flow analysis (Eqs. 10.24 and 10.25). A detailed investigation of the flow indicates that there are essentially five types of surface and jump conditions. The classification of these jumps is indicated in Table 10.3, along with sketches of the structure of the jump. For flows that are barely supercritical the jump is more like a standing wave, without a nearly step change in depth. In some Froude number ranges the jump is unsteady, with regular periodic oscillations traveling downstream. (Recall that the wave cannot travel upstream against the supercritical flow.)

The length of a hydraulic jump (the distance between the nearly uniform upstream and downstream flows) may be of importance in the design of channels. Although its value cannot be determined theoretically, experimental results indicate that over a wide range of Froude numbers, the jump is approximately seven downstream depths long (Ref. 5).

Hydraulic jumps can occur in a variety of channel flow configurations, not just in horizontal rectangular channels as discussed above. Jumps in nonrectangular channels (i.e., circular pipes, trapezoidal canals, etc.) behave in a manner quite like those in rectangular channels, although the details of the depth ratio and head loss are somewhat different than for jumps in rectangular channels.

TABLE 10.3   Classification of Hydraulic Jumps (Ref. 12)

Fr_1	y_2/y_1	Classification	Sketch
< 1	1	Jump impossible	
1 to 1.7	1 to 2.0	Standing wave or undular jump	
1.7 to 2.5	2.0 to 3.1	Weak jump	
2.5 to 4.5	3.1 to 5.9	Oscillating jump	
4.5 to 9.0	5.9 to 12	Stable, well-balanced steady jump; insensitive to downstream conditions	
> 9.0	> 12	Rough, somewhat intermittent strong jump	

Other common types of hydraulic jumps include those that occur in sloping channels as is indicated in Fig. 10.23a and the submerged hydraulic jumps that can occur just downstream of a sluice gate as is indicated in Fig. 10.23b. Details of these and other jumps can be found in standard open-channel flow references (Refs. 3 and 5).

10.6.2   Sharp Crested Weirs

A weir is an obstruction on a channel bottom over which the fluid must flow. It provides a convenient method of determining the flowrate in an open channel in terms of a single depth measurement. A *sharp crested weir* is essentially a vertical sharp edged flat plate placed across the channel in a way such that the fluid must flow across the sharp edge and drop into the pool downstream of the weir plate, as is shown in Fig. 10.24. The specific shape of the flow area in the plane of the weir plate is used to designate the type of weir. Typical shapes include the rectangular weir, the triangular weir, and the trapezoidal weir, as indicated in Fig. 10.25.

The complex nature of the flow over a weir makes it impossible to obtain precise analytical expressions for the flow as a function of other parameters, such as the weir height, P_w, the head of the weir, H, the fluid depth upstream, and the geometry of the weir plate (angle θ for triangular weirs or aspect ratio, b/H, for rectangular weirs). The flow structure is far from one-dimensional, with a variety of interesting flow phenomena obtained. Another parameter that can affect the flowrate for the rectangular weir is the pressure in the air pocket between the nappe and the weir plate (see Fig. 10.24). For accurate flowrate measurements, it is often necessary to provide vent tubes to maintain this pressure at atmospheric pressure.

The main mechanisms governing flow over a weir are gravity and inertia. From a

(a)

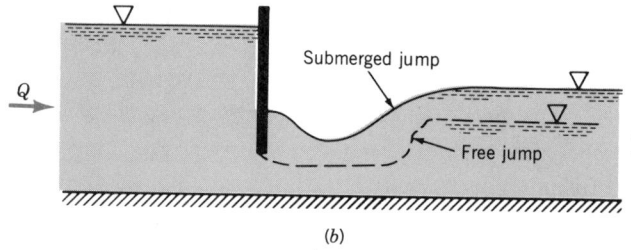

(b)

FIGURE 10.23 Hydraulic jump variations: (a) jump caused by a change in channel slope, (b) submerged jump.

highly simplified point of view, gravity accelerates the fluid from its free-surface elevation upstream of the weir to larger velocity as it flows down the hill formed by the napp. Although viscous and surface tension effects are usually of secondary importance, such effects cannot be entirely neglected. Generally, appropriate experimentally determined coefficients are used to account for these effects.

As a first approximation, we assume that the velocity profile upstream of the weir plate is uniform and that the pressure within the nappe is atmospheric. In addition, we assume that the fluid flows horizontally over the weir plate with a nonuniform velocity profile, indicated in Fig. 10.26. With $p_B = 0$ the Bernoulli equation for flow along the arbitrary streamline A–B indicated can be written as

$$\frac{p_A}{\gamma} + \frac{V_1^2}{2g} + z_A = (H + P_w - h) + \frac{u_2^2}{2g} \qquad (10.26)$$

where h is the distance that point B is below the free surface. We do not know the location of point A from which came the fluid that passes over the weir at point B. However, since the total head for any particle along the vertical section (1) is the same, $z_A + p_A/\gamma + V_1^2/2g = H + P_w + V_1^2/2g$, the specific location of A is not

FIGURE 10.24 Sharp crested weir geometry.

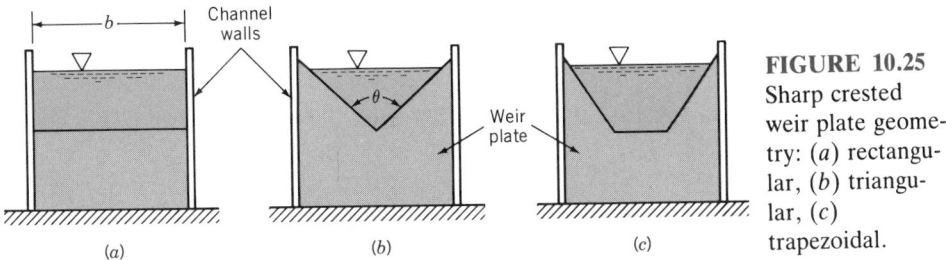

FIGURE 10.25
Sharp crested weir plate geometry: (*a*) rectangular, (*b*) triangular, (*c*) trapezoidal.

needed, and the velocity of the fluid over the weir plate is obtained from Eq. 10.26 as

$$u_2 = \sqrt{2g\left(h + \frac{V_1^2}{2g}\right)}$$

The flowrate can be calculated from

$$Q = \int_{(2)} u_2 \, dA = \int_{h=0}^{h=H} u_2 \ell \, dh \tag{10.27}$$

where $\ell = \ell(h)$ is the cross channel width of a strip of the weir area, as is indicated in Fig. 10.26*b*. For a rectangular weir ℓ is constant. For other weirs, such as triangular or circular weirs, the value of ℓ is known as a function of h.

For a rectangular weir, $\ell = b$, and the flowrate becomes

$$Q = \sqrt{2g} \, b \int_0^H \left(h + \frac{V_1^2}{2g}\right)^{1/2} dh$$

or

$$Q = \frac{2}{3}\sqrt{2g} \, b \left[\left(H + \frac{V_1^2}{2g}\right)^{3/2} - \left(\frac{V_1^2}{2g}\right)^{3/2}\right] \tag{10.28}$$

Equation 10.28 is a rather cumbersome expression that can be simplified by using the fact that with $P_w \gg H$ (as often happens in practical situations) the upstream velocity is negligibly small. That is, $V_1^2/2g \ll H$ and Eq. 10.28 simplifies to the basic rectangular

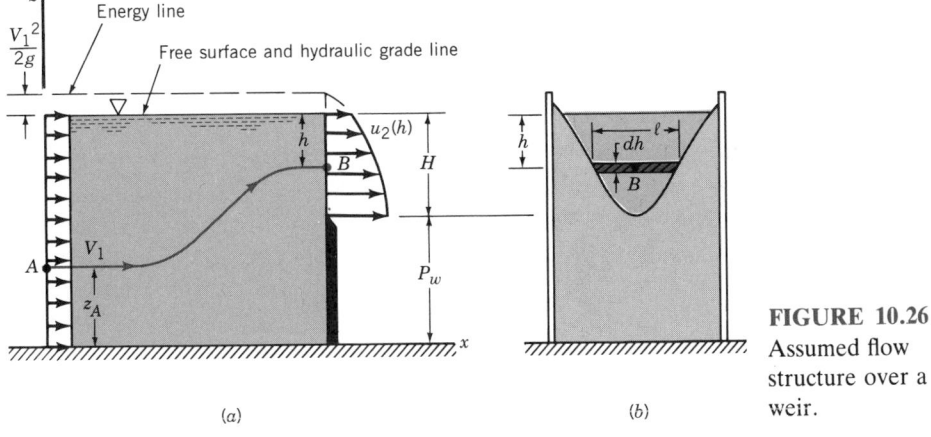

FIGURE 10.26
Assumed flow structure over a weir.

weir equation:

$$Q = \tfrac{2}{3} \sqrt{2g} \, b \, H^{3/2} \tag{10.29}$$

Note that the weir head, H, is the height of the upstream free surface above the crest of the weir. As is indicated in Fig. 10.24, because of the drawdown effect, H is not the distance of the free surface above the weir crest as is measured directly above the weir plate.

Because of the numerous approximations made to obtain Eq. 10.29, it is not unexpected that an experimentally determined correction factor must be used to obtain the actual flowrate as a function of weir head. Thus, the final form is

$$Q = C_{wr} \tfrac{2}{3} \sqrt{2g} \, b \, H^{3/2} \tag{10.30}$$

where C_{wr} is the rectangular weir coefficient. From dimensional analysis arguments, it is expected that C_{wr} is a function of Reynolds number (viscous effects), Weber number (surface tension effects), and H/P_w (geometry). In most practical situations, the Reynolds and Weber number effects are negligible, and the following correlation can be used (Refs. 4, and 7)

$$C_{wr} = 0.611 + 0.075 \left(\frac{H}{P_w} \right) \tag{10.31}$$

More precise values of C_{wr} can be found in the literature, if needed (Ref. 3).

The triangular sharp crested weir is often used for flow measurements, particularly for measuring flowrates over a wide range of values. For small flowrates, the head, H, for a rectangular weir would be very small and the flowrate could not be measured accurately. However, with the triangular weir, the flow area decreases as H decreases so that even for small flowrates, reasonable heads are developed. Accurate results can be obtained over a wide range of Q.

The triangular weir equation can be obtained from Eq. 10.27 by using

$$\ell = 2(H - h) \tan \left(\frac{\theta}{2} \right)$$

where θ is the angle of the vee-notch (see Figs. 10.25 and 10.26). After carrying out the integration and again neglecting the upstream velocity ($V_1^2/2g \ll H$), we obtain

$$Q = \frac{8}{15} \tan \left(\frac{\theta}{2} \right) \sqrt{2g} \, H^{5/2}$$

An experimentally determined triangular weir coefficient, C_{wt}, is used to account for the real-world effects neglected in the analysis to give

$$Q = C_{wt} \frac{8}{15} \tan \left(\frac{\theta}{2} \right) \sqrt{2g} \, H^{5/2} \tag{10.32}$$

Typical values of C_{wt} for triangular weirs are in the range of 0.58 to 0.62, as is shown in Fig. 10.27. Note that although C_{wt} and θ are dimensionless, the value of C_{wt} is given as a function of the weir head, H, which is a dimensional quantity. Although using dimensional parameters is not recommended (see the dimensional analysis discussion in Chapter 7), such parameters are often used for open-channel flow.

The above results for sharp crested weirs are valid provided the area under the nappe is ventilated to atmospheric pressure. Although this is not a problem for triangular weirs, for rectangular weirs it is sometimes necessary to provide ventilation tubes to ensure atmospheric pressure in this region. In addition, depending on down-

FIGURE 10.27 Weir coefficient for triangular sharp crested weirs (Ref. 10).

stream conditions it is possible to obtain submerged weir operation, as is indicated in Fig. 10.28. Clearly the flowrate will be different for these situations than that given by Eqs. 10.30 and 10.32.

10.6.3 Broad Crested Weirs

A *broad crested weir* is a structure in an open channel that has a horizontal crest above which the fluid pressure may be considered hydrostatic. A typical configuration is shown in Fig. 10.29. Generally, to ensure proper operation, these weirs are restricted to the range $0.08 < H/L_w < 0.50$. For long weir blocks (H/L_w less than 0.08) head losses across the weir cannot be neglected. On the other hand, for short weir blocks (H/L_w greater than 0.50) the streamlines of the flow over the weir block are not horizontal. Although broad crested weirs can be used in channels of any cross-sectional shape, we restrict our attention to rectangular channels.

The operation of a broad crested weir is based on the fact that nearly uniform critical flow is achieved in the short reach above the weir block. (If $H/L_w < 0.08$ viscous effects are important, and the flow is subcritical over the weir.) If the kinetic energy of the upstream flow is negligible, then $V_1^2/2g \ll y_1$ and the upstream specific energy is $E_1 = V_1^2/2g + y_1 \approx y_1$. Observations show that as the flow passes over the weir block, it accelerates and reaches critical conditions, $y_2 = y_c$ and $\mathrm{Fr}_2 = 1$ (i.e., $V_2 = c_2$), corresponding to the nose of the specific energy curve (see Fig. 10.7). The

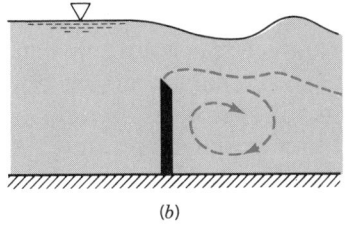

FIGURE 10.28 Flow conditions over a weir without a free nappe: (*a*) plunging nappe, (*b*) submerged nappe.

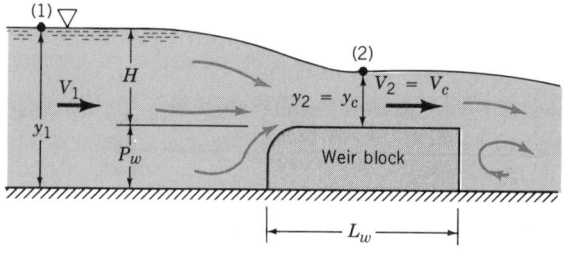

FIGURE 10.29 Broad crested weir geometry.

flow does not accelerate to supercritical conditions ($Fr_2 > 1$). To do so would require the ability of the downstream fluid to communicate with the upstream fluid to let it know that there is an end of the weir block. Since waves cannot propagate upstream against a critical flow, this information cannot be transmitted. The flow remains critical, not supercritical, across the weir block.

The Bernoulli equation can be applied between point (1) upstream of the weir and point (2) over the weir where the flow is critical to obtain

$$H + P_w + \frac{V_1^2}{2g} = y_c + P_w + \frac{V_c^2}{2g}$$

or, if the upstream velocity head is negligible

$$H - y_c = \frac{(V_c^2 - V_1^2)}{2g} \approx \frac{V_c^2}{2g}$$

However, since $V_2 = V_c = (gy_c)^{1/2}$, we find that $V_c^2 = gy_c$ so that we obtain

$$H - y_c = \frac{y_c}{2}$$

or

$$y_c = \frac{2H}{3}$$

Thus, the flowrate is

$$Q = by_2V_2 = by_cV_c = by_c(gy_c)^{1/2} = b \sqrt{g} \, y_c^{3/2}$$

or

$$Q = b \sqrt{g} \left(\frac{2}{3}\right)^{3/2} H^{3/2}$$

Again an empirical weir coefficient is used to account for the various real-world effects not included in the above simplified analysis. That is

$$Q = C_{wb} \, b \sqrt{g} \left(\frac{2}{3}\right)^{3/2} H^{3/2} \tag{10.33}$$

where typical values of C_{wb}, the broad crested weir coefficient, can be obtained from the equation (Ref. 6)

$$C_{wb} = \frac{0.65}{(1 + H/P_w)^{1/2}} \tag{10.34}$$

EXAMPLE 10.10

Water flows in a rectangular channel of width $b = 2$ m with flowrates between $Q_{min} = 0.02$ m³/s and $Q_{max} = 0.60$ m³/s. This flowrate is to be measured by using either (a) a rectangular sharp crested weir, (b) a triangular sharp crested weir with $\theta = 90°$, or (c) a broad crested weir. In all cases the bottom of the flow area over the weir is a distance $P_w = 1$ m above the channel bottom. Plot a graph of $Q = Q(H)$ for each weir and comment on which weir would be best for this application.

SOLUTION

For the rectangular weir with $P_w = 1$ m, Eqs. 10.30 and 10.31 give

$$Q = C_{wr} \frac{2}{3} \sqrt{2g}\, bH^{3/2} = \left(0.611 + 0.075 \frac{H}{P_w}\right) \frac{2}{3} \sqrt{2g}\, bH^{3/2}$$

Thus,

$$Q = (0.611 + 0.075H) \frac{2}{3} \sqrt{2\,(9.81 \text{ m/s}^2)}\,(2 \text{ m})\, H^{3/2}$$

or

$$Q = 5.91(0.611 + 0.075H)H^{3/2} \tag{1}$$

where H and Q are in meters and m³/s, respectively. The results from Eq. 1 are plotted in Fig. E10.10.

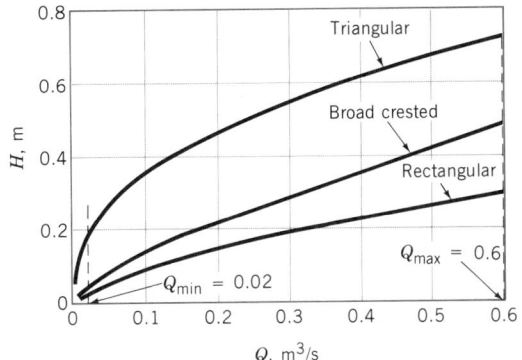

FIGURE E10.10

Similarly, for the triangular weir Eq. 10.32 gives

$$Q = C_{wt} \frac{8}{15} \tan\left(\frac{\theta}{2}\right) \sqrt{2g}\, H^{5/2}$$

$$= C_{wt} \frac{8}{15} \tan(45°) \sqrt{2(9.81 \text{ m/s}^2)}\, H^{5/2}$$

or

$$Q = 2.36 C_{wt}\, H^{5/2} \tag{2}$$

where H and Q are in meters and m³/s and C_{wt} is obtained from Fig. 10.27. For example, with $H = 0.20$ m, we find $C_{wt} = 0.60$, or $Q = 2.36\,(0.60)(0.20)^{5/2} = 0.0253$ m³/s. The triangular weir results are also plotted in Fig. E10.10.

For the broad crested weir, Eqs. 10.33 and 10.34 give

$$Q = C_{wb} \, b \sqrt{g} \left(\frac{2}{3}\right)^{3/2} H^{3/2} = \frac{0.65}{(1 + H/P_w)^{1/2}} \, b \sqrt{g} \left(\frac{2}{3}\right)^{3/2} H^{3/2}$$

Thus, with $P_w = 1$ m

$$Q = \frac{0.65}{(1 + H)^{1/2}} \, (2 \text{ m}) \sqrt{9.81 \text{ m/s}^2} \left(\frac{2}{3}\right)^{3/2} H^{3/2}$$

or

$$Q = \frac{2.22}{(1 + H)^{1/2}} H^{3/2} \tag{3}$$

where, again, H and Q are in meters and m³/s. This result is also plotted in Fig. E10.10.

Although it appears as though any of the three weirs would work well for the upper portion of the flowrate range, neither the rectangular nor the broad crested weir would be very accurate for small flowrates near $Q = Q_{min}$ because of the small head, H, at these conditions. The triangular weir, however, would allow reasonably large values of H at the lowest flowrates.

In addition, as is discussed in this section, for proper operation the broad crested weir geometry is restricted to $0.08 < H/L_w < 0.50$, where L_w is the weir block length. From Eq. 3 with $Q_{max} = 0.60$ m³/s, we obtain $H_{max} = 0.476$. Thus, we must have $L_w > H_{max}/0.5 = 0.952$ m to maintain proper critical flow conditions at the largest flowrate in the channel. On the other hand, with $Q = Q_{min} = 0.02$ m³/s, we obtain $H_{min} = 0.0439$ m. Thus, we must have $L_w < H_{min}/0.08 = 0.549$ m to ensure that frictional effects are not important. Clearly, these two constraints on the geometry of the weir block, L_w, are incompatible.

A broad crested weir will not function properly under the wide range of flowrates considered in this example. The sharp crested triangular weir would be the best of the three types considered.

10.6.4 Underflow Gates

A variety of gate type structures is available for flowrate control at the crest of an overflow spillway, or at the entrance of an irrigation canal or river from a lake. Three typical types are illustrated in Fig. 10.30. Each has certain advantages and disadvantages in terms of costs of construction, ease of use, and the like, although the basic fluid mechanics involved are the same in all instances.

The flow under a gate is said to be free outflow when the fluid issues as a jet of supercritical flow with a free surface open to the atmosphere as shown in Fig. 10.30. In such cases it is customary to write this flowrate as the product of the distance, a, between the channel bottom and the bottom of the gate times the convenient reference velocity $(2gy_1)^{1/2}$. That is

$$q = C_d a \sqrt{2gy_1} \tag{10.35}$$

where q is the flowrate per unit width. The discharge coefficient, C_d, is a function of the contraction coefficient, $C_c = y_2/a$, and the depth ratio y_1/a. Typical values of the

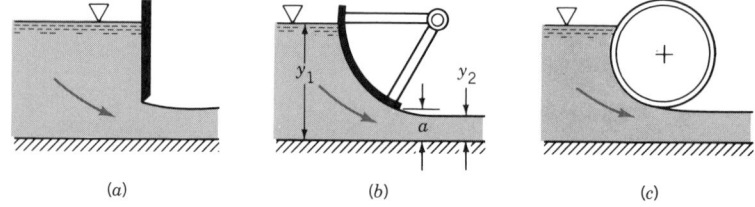

FIGURE 10.30
Three variations of underflow gates: (a) vertical gate, (b) radial gate, (c) drum gate.

(a) (b) (c)

FIGURE 10.31 Typical discharge coefficients for underflow gates (Ref. 3).

discharge coefficient for free outflow (or free discharge) from a vertical sluice gate are on the order of 0.55 to 0.60 as indicated by the top line in Fig. 10.31 (Ref. 3).

As is indicated in Fig. 10.32, in certain situations the depth downstream of the gate is controlled by some downstream obstacle and the jet of water issuing from under the gate is overlaid by a mass of water that is quite turbulent.

The flowrate for a submerged (or drowned) gate can be obtained from the same equation that is used for free outflow (Eq. 10.35), provided the discharge coefficient is modified appropriately. Typical values of C_d for drowned outflow cases are indicated as the series of lower curves in Fig. 10.31. Consider flow for a given gate and upstream conditions (i.e., given y_1/a) corresponding to a vertical line in the figure. With $y_3/a = y_1/a$ (i.e., $y_3 = y_1$) there is no head to drive the flow so that $C_d = 0$ and the fluid is stationary. For a given upstream depth (y_1/a fixed), the value of C_d increases with decreasing y_3/a until the maximum value of C_d is reached. This maximum corresponds to the free discharge conditions and is represented by the free outflow line so labeled in Fig. 10.31. For values of y_3/a that give C_d values between zero and its maximum, the jet from the gate is overlaid (drowned) by the downstream water and the flowrate is therefore reduced when compared with a free discharge situation. Similar results are obtained for the radial gate and drum gate.

FIGURE 10.32 Drowned outflow from a sluice gate.

EXAMPLE 10.11

Water flows under the sluice gate shown in Fig. E10.11. The channel width is $b = 20$ ft, the upstream depth is $y_1 = 6$ ft, and the gate is $a = 1.0$ ft off the channel bottom. Plot a graph of flowrate, Q, as a function of y_3.

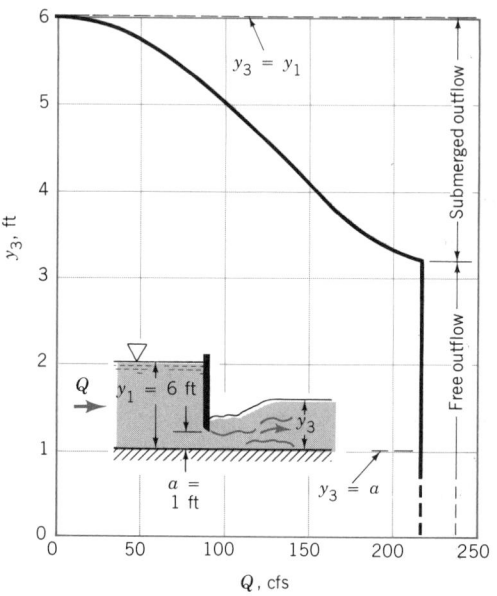

FIGURE E10.11

SOLUTION

From Eq. 10.35 we have

$$Q = bq = baC_d\sqrt{2gy_1} = 20 \text{ ft } (1.0 \text{ ft}) \, C_d\sqrt{2(32.2 \text{ ft/s}^2)(6.0 \text{ ft})}$$

or

$$Q = 393C_d \text{ cfs} \qquad (1)$$

The value of C_d is obtained from Fig. 10.31 along the vertical line $y_1/a = 6 \text{ ft}/1 \text{ ft} = 6$. For $y_3 = 6$ ft (i.e., $y_3/a = 6 = y_1/a$) we obtain $C_d = 0$ indicating that there is no flow when there is no head difference across the gate. The value of C_d increases as y_3/a decreases, reaching a maximum of $C_d = 0.56$ when $y_3/a = 3.2$. Thus, with $y_3 = 3.2a = 3.2$ ft

$$Q = 393 \, (0.56) \text{ cfs} = 220 \text{ cfs}$$

For $y_3 < 3.2$ ft the flowrate is independent of y_3, and the outflow is a free (not submerged) outflow. For such cases the inertia of the water flowing under the gate is sufficient to produce free outflow even with $y_3 > a$.

The flowrate for $3.2 \text{ ft} \leq y_3 \leq 6$ ft is obtained from Eq. 1 and the C_d values of Fig. 10.30 with the results as indicated in Fig. E10.11.

REFERENCES

1. Currie, I. G., *Fundamental Mechanics of Fluids,* McGraw-Hill, New York, 1974.

2. Stoker, J. J., *Water Waves,* Interscience, New York, 1957.

3. Henderson, F. M., *Open Channel Flow,* Macmillan, New York, 1966.

4. Rouse, H., *Elementary Fluid Mechanics,* Wiley, New York, 1946.

5. French, R. H., *Open Channel Hydraulics,* McGraw-Hill, New York, 1985.

6. Chow, V. T., *Open Channel Hydraulics,* McGraw-Hill, New York, 1959.

7. Blevins, R. D., *Applied Fluid Dynamics Handbook,* Van Nostrand Reinhold, New York, 1984.

8. Daugherty, R. L., and Franzini, J. B., *Fluid Mechanics with Engineering Applications,* McGraw-Hill, New York, 1977.

9. Vennard, J. K., and Street, R. L., *Elementary Fluid Mechanics,* Wiley, New York, 1976.

10. Lenz, A. T., "Viscosity and Surface Tension Effects on V–Notch Weir Coefficients," *Transactions of the American Society of Chemical Engineers,* Vol. 108, 759–820, 1943.

11. White, F. M., *Fluid Mechanics,* McGraw-Hill, New York, 1986.

12. U.S. Bureau of Reclamation, Research Studies on Stilling Basins, Energy Dissipators, and Associated Appurtenances, Hydraulic Lab Report Hyd.-399, June 1, 1955.

13. Wallet, A., and Ruellan, F., *Houille Blanche,* Vol. 5, 1950.

PROBLEMS

Note: Unless otherwise indicated, use the values of fluid properties found in the tables on the inside of the front cover. Problems designated with an (*) are intended to be solved with the aid of a programmable calculator or a computer.

10.1 The flowrate in a 50-ft-wide, 2-ft-deep river is $Q = 190$ cfs. Is the flow subcritical or supercritical?

10.2 The flowrate per unit width in a wide channel is $q = 2.3$ m²/s. Is the flow subcritical or supercritical if the depth is **(a)** 0.2 m, **(b)** 0.8m, or **(c)** 2.5 m?

10.3 Water flows in a canal at a depth of 2.8 ft and a velocity of 5.3 ft/s. Will waves produced by throwing a stick into the canal travel both upstream and downstream, or will they all be washed downstream? Explain.

10.4 A trout jumps, producing waves on the surface of a 1.3-m-deep mountain stream. If it is observed that the waves do not travel upstream, what is the minimum velocity of the current?

10.5 The average velocity and average depth of a river from its beginning in the mountains to its discharge into the ocean are given in the table below. Plot a graph of the Froude number as a function of distance along the river.

Distance (mi)	Average Velocity (ft/s)	Average Depth (ft)
0	13	1.5
5	10	2.0
10	9	2.3
30	5	3.7
50	4	5.4
80	4	6.0
90	3	6.2

10.6 A river barge often produces sizable bow waves because it is not well "streamlined." If the barge travels 10 mph in a 12-ft-deep river channel that has negligible current, would you expect the bow waves to be observed on the shore ahead of the boat? Explain. Repeat the problem for a speed boat towing a water skier at 20 mph.

10.7   Observations at a shallow sandy beach show that even though the waves several hundred yards out from the shore are not parallel to the beach, the waves often "break" on the beach nearly parallel to the shore as is indicated in Fig. P10.7. Explain this behavior based on the wave speed $c = (gy)^{1/2}$.

FIGURE P10.7

10.8   Waves on the surface of a tank containing water are observed to move with a velocity of 1.8 m/s. If the water is replaced by mercury, with all other conditions the same, determine the wave speed expected. Determine the wave speed if the tank were in a laboratory on the surface of the moon where the acceleration of gravity is $\frac{1}{6}$ that on earth.

10.9   Often when an earthquake shifts a segment of the ocean floor, a relatively small amplitude wave of very long wavelength is produced. Such waves go unnoticed as they move across the open ocean; only when they approach the shore do they become dangerous (a tsunami or "tidal wave"). Determine the wave speed if the wavelength, λ, is 6000 ft and the ocean depth is 15,000 ft.

10.10   If the water depth in a pond is 15 ft, determine the speed of small-amplitude, long-wavelength ($\lambda \gg y$) waves on the surface.

10.11   A small amplitude wave travels across a pond with a speed of 9.6 ft/s. Determine the water depth.

10.12   Plot the specific energy diagram for a wide channel carrying $q = 50$ ft²/s. Determine **(a)** the critical depth, **(b)** the minimum specific energy, **(c)** the alternate depth corresponding to a depth of 2.5 ft, and **(d)** the possible flow velocities if $E = 10$ ft.

10.13*   Water flows in a rectangular channel with a specific energy of $E = 5$ ft. If the flowrate per unit width is $q = 30$ ft²/s, determine the two possible flow depths and the corresponding Froude

numbers. Plot the specific energy diagram for this flow. Repeat the problem for $E = 1, 2, 3$, and 4 ft.

10.14   Water flows in a rectangular channel with a flowrate per unit width of $q = 2.5$ m²/s. Plot the specific energy diagram for this flow. Determine the two possible depths of flow if $E = 2.2$ m?

10.15   Water flows in a rectangular channel at a rate of $q = 20$ cfs/ft. When a Pitot tube is placed in the stream, water in the tube rises to a level of 4.5 ft above the channel bottom. Determine the two possible flow depths in the channel. Illustrate this flow on a specific energy diagram.

10.16   Water flows in a $b = 10$-ft-wide rectangular channel with a flowrate of $Q = 60$ ft³/s and an upstream depth of $y_1 = 2$ ft as is shown in Fig. P10.16. Determine the flow depth and the surface elevation at section (2).

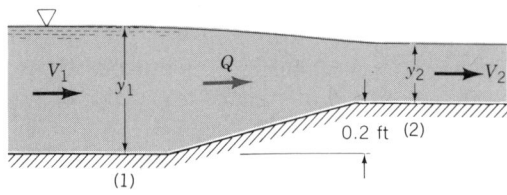

FIGURE P10.16

10.17   Repeat Problem 10.16 if the upstream depth is $y_1 = 0.5$ ft.

10.18*   Water flows over the bump in the bottom of the rectangular channel shown in Fig. P10.18 with a flowrate per unit width of $q = 4$ m²/s. The channel bottom contour is given by $z_B = 0.2e^{-x^2}$, where z_B and x are in meters. The water depth far upstream of the bump is $y_1 = 2$ m. Plot a graph of the water depth, $y = y(x)$, and the surface elevation, $z = z(x)$, for -4 m $\leq x \leq 4$ m. Assume one-dimensional flow.

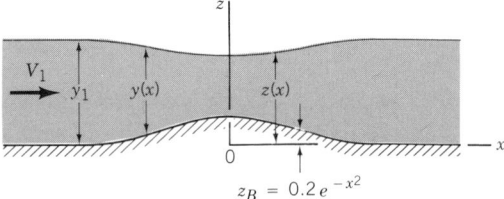

$$z_B = 0.2e^{-x^2}$$

FIGURE P10.18

10.19*   Repeat Problem 10.18 if the upstream depth is 0.4 m.

10.20   Water in a rectangular channel flows into a gradual contraction section as is indicated

in Fig. P10.20. If the flowrate is $Q = 25$ ft^3/s and the upstream depth is $y_1 = 2$ ft, determine the downstream depth, y_2.

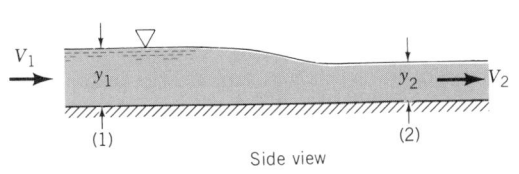

FIGURE P10.20

10.21 Sketch the specific energy diagram for the flow of Problem 10.20 and indicate its important characteristics. Note that $q_1 \neq q_2$.

10.22 Repeat Problem 10.20 if the upstream depth is $y_1 = 0.5$ ft. Assume that there are no losses between sections (1) and (2).

10.23 Water flows in a rectangular channel with a flowrate per unit width of $q = 1.5$ m^2/s and a depth of 0.5 m at section (1). The head loss between sections (1) and (2) is 0.02 m. Plot the specific energy diagram for this flow and locate states (1) and (2) on this diagram. Is it possible to have a head loss of 0.10 m? Explain.

10.24 Water flows in a horizontal rectangular channel with a flowrate per unit width of $q = 10$ ft^2/s and a depth of 1.0 ft at the downstream section (2). The head loss between section (1) upstream and section (2) is 0.2 ft. Plot the specific energy diagram for this flow and locate states (1) and (2) on this diagram.

10.25 A rectangular channel 3 m wide carries 10 m^3/s at a depth of 2 m. Is the flow subcritical or supercritical? For the same flowrate, what depth will give critical flow?

10.26 Water flows at a rate of 1000 ft^3/s in a horizontal rectangular channel 30 ft wide with a 2-ft depth. Determine the depth and the surface elevation if the channel contracts to a width of 25 ft? Explain.

10.27 A smooth transition section connects two rectangular channels as shown in Fig. P10.27. The channel width increases from 6.0 to 7.0 ft and the water surface elevation is the same in each channel. If the upstream depth of flow is 3.0

ft, determine h, the amount the channel bed needs to be raised across the transition section to maintain the same surface elevation.

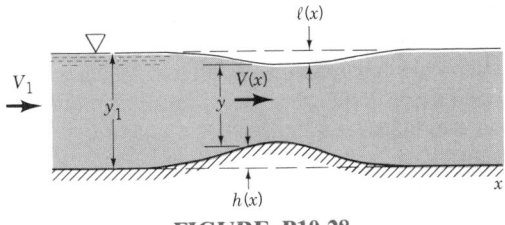

FIGURE P10.27

10.28 Water flows over a bump of height $h = h(x)$ on the bottom of a wide rectangular channel as is indicated in Fig. P10.28. If energy losses are negligible, show that the slope of the water surface is given by $dy/dx = -(dh/dx)/[1 - (V^2/gy)]$, where $V = V(x)$ and $y = y(x)$ are the local velocity and depth of flow. Comment on the sign (i.e., <0, $=0$, or >0) of dy/dx relative to the sign of dh/dx.

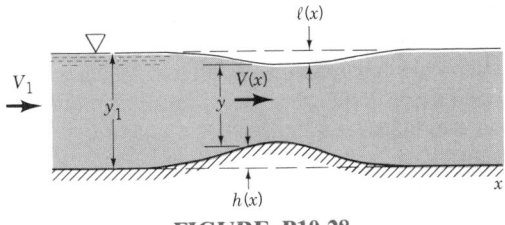

FIGURE P10.28

10.29 Integrate the differential equation obtained in Problem 10.28 to determine the "drawdown" distance, $\ell = \ell(x)$, indicated in Fig. P10.28. Comment on your results.

10.30 Determine the minimum depth in a 3-m-wide rectangular channel if the flow is to be subcritical with a flowrate of $Q = 30$ m/s.

10.31 Water flows in a rectangular channel at a depth of 4 ft and a flowrate of $Q = 150$ cfs. Determine the maximum channel width if the flow is to be supercritical.

10.32 Fluid properties such as viscosity or density do not appear in the Manning equation (Eq. 10.20). Does this mean that this equation is valid for any open-channel flow such as that in-

volving mercury, water, oil, or molasses? Explain.

10.33   The following data are taken from measurements on Indian Fork Creek: $A = 26 \text{ m}^2$, $P = 16$ m, and $S_0 = 0.02$ m/62 m. Determine the average shear stress on the wetted perimeter of this channel.

10.34   A viscous oil flows down a wide plate with a uniform depth of 8 mm and an average velocity of 50 mm/s. The plate is on a 5° hill and the specific gravity of the oil is 0.85. Determine the average shear stress between the oil and the plate.

10.35   The following data are obtained for a particular reach of the Provo River in Utah: $A = 183 \text{ ft}^2$, free-surface width = 55 ft, average depth = 3.3 ft, $R_h = 3.22$ ft, $V = 6.56$ ft/s, length of reach = 116 ft, and elevation drop of reach = 1.04 ft. Determine the **(a)** average shear stress on the wetted perimeter, **(b)** the Manning coefficient, n, and **(c)** the Froude number of the flow.

10.36   Water flows in a 10-ft-wide rectangular channel with a flowrate of 100 cfs and a depth of 3 ft. If the slope is 0.005, determine the Manning coefficient, n, and the average shear stress at the sides and bottom of the channel.

10.37   By what percent is the flowrate reduced in the rectangular channel shown in Fig. P10.37 because of the addition of the thin center board? All surfaces are of the same material.

FIGURE P10.37

10.38   Water flows in a weedy earthen channel at a rate of 30 m³/s. What flowrate can be expected if the weeds are removed and the depth remains constant?

10.39   Show that the triangular channel with the best hydraulic cross section (i.e., minimum area, A, for a given flowrate) is a right triangle as is shown in Fig. E10.8b.

10.40   Show that for a trapezoidal channel with 60° angle sides, the best hydraulic cross section (i.e., minimum area, A, for a given flowrate) is as shown in Fig. E10.8b.

10.41   At a particular location the cross section of the Columbia River is as indicated in Fig. P10.41. If on a day without wind it takes 5 min to float 0.5 mi along the river, which drops 0.46 ft in that distance, determine the value of the Manning coefficient, n.

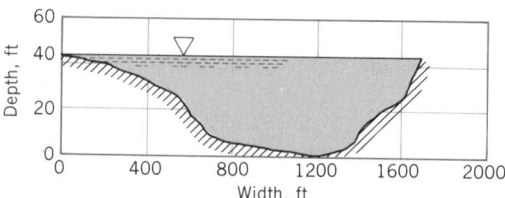

FIGURE P10.41

10.42   If the free surface of the Columbia River shown in Fig. P10.41 were 20 ft above the bottom rather than 44 ft, as is indicated in the figure, how long would it take to float the 0.5-mi stretch considered in Problem 10.41? Assume the elevation change remains 0.46 ft.

10.43   Rain water runoff from a 200 ft by 500 ft parking lot is to drain through a circular concrete pipe that is laid on a slope of 5 ft/mi. Determine the pipe diameter if it is to be half-full with a steady rainfall of 0.5 in./hr.

10.44   A weedy irrigation canal of trapezoidal cross section is to carry 20 m³/s when built on a slope of 0.56 m/km. If the sides are at a 45° angle and the bottom is 8 m wide, determine the width of the waterline at the free surface.

10.45   To prevent weeds from growing in a clean earthen lined canal it is recommended that the velocity be no less than 2.5 ft/s. For the symmetrical canal shown in Fig. P10.45, determine the minimum slope needed.

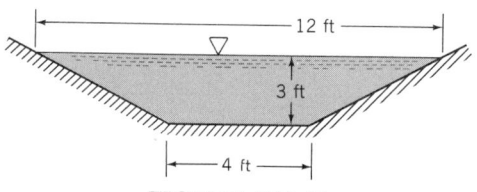

FIGURE P10.45

10.46   The smooth concrete-lined symmetrical channel shown in Fig. P10.45 carries water from the silt-laden Colorado River. If the velocity must be 5.0 ft/s to prevent the silt from settling out (and eventually clogging the channel), determine the minimum slope needed.

10.47   The symmetrical channel shown in Fig. P10.45 is dug in sandy loam soil with $n = 0.020$.

For such surface material it is recommended that to prevent scouring of the surface the average velocity be no more than 1.75 ft/s. Determine the maximum slope allowed.

10.48 The flowrate in the clay-lined channel ($n = 0.025$) shown in Fig. P10.48 is to be 300 ft³/s. To prevent erosion of the sides, the velocity must not exceed 4 ft/s. For this maximum velocity determine the width of the bottom, b, and the slope, S_0.

FIGURE P10.48

10.49 A trapezoidal channel with a bottom width of 3.0 m and sides with a slope of 2:1 (horizontal:vertical) is lined with fine gravel ($n = 0.020$) and is to carry 10 m³/s. Can this channel be built with a slope of $S_0 = 0.00010$ if it is necessary to keep the velocity below 0.75 m/s to prevent scouring of the bottom? Explain.

10.50 Water flows in a 2-m-diameter finished concrete pipe so that it is completely full and the pressure is constant all along the pipe. If the slope is $S_0 = 0.005$, determine the flowrate by using open-channel flow methods. Compare this result with that obtained by using pipe flow methods of Chapter 8.

10.51 Water flows along the curbing of a street as is shown in Fig. P10.51. If the slope of the street is $S_0 = 0.01$ would you expect the flow to be laminar or turbulent? Explain.

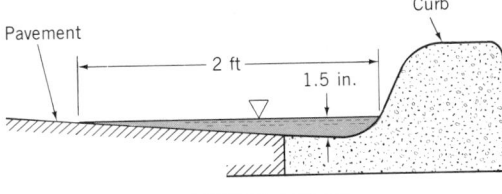

FIGURE P10.51

10.52 A natural riverbed has an approximate parabolic cross section such that the free-surface width is 200 ft when the center depth is 5 ft. Determine the flowrate if $n = 0.035$ and the slope is 2.5 ft/mi.

10.53 The triangular flume shown in Fig. P10.53 is built to carry its design flowrate, Q_0, at a depth of 0.90 m as is indicated. If the flume is to be able to carry up to twice its design flowrate, $Q = 2Q_0$, determine the freeboard, ℓ, needed.

FIGURE P10.53

10.54 A circular finished concrete culvert is to carry a discharge of 50 ft³/s on a slope of 0.0010. It is to flow not more than half-full. The culvert pipes are available from the manufacture with diameters that are multiples of 1 ft. Determine the smallest suitable culvert diameter.

10.55 A rectangular unfinished concrete channel of 28-ft-width is laid on a slope of 8 ft/mi. Determine the flow depth and Froude number of the flow if the flowrate is 400 ft³/s.

10.56 Repeat Problem 10.55 if the flowrate is 20 ft³/s.

10.57 An engineer is to design a channel lined with planed wood to carry water at a flowrate of 2 m³/s on a slope of 10 m/800 m. The channel cross section can be either a 90° triangle or a square. Which would require less wood and by what percent?

10.58 Water flows in a rectangular channel of width b at a depth of $b/3$. Determine the diameter of a circular channel (in terms of b) that carries the same flowrate when it is half-full. Both channels have the same Manning coefficient, n, and slope.

10.59 Water flows in a channel with an equilateral triangle cross section as shown in Fig. P10.59. For a given Manning coefficient, n, and channel slope, determine the depth that gives the maximum flowrate.

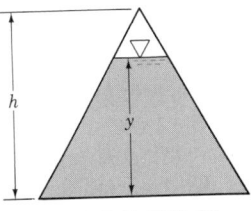

FIGURE P10.59

10.60 Water flows in a channel with an equilateral triangle cross section as is shown in Fig. P10.60. Let Q_{full} denote the flowrate when $y = h$. By what percent is Q_{full} less than Q when $y = h - \delta y$, where $\delta y \ll h$? That is, placing a lid on

this channel reduces the flowrate by what percent?

FIGURE P10.60

10.61 Water flows in the symmetrical, unfinished concrete trapezoidal channel shown in Fig. P10.61 at a rate of 120 ft³/s. The slope is 4.2 ft/2000 ft. Determine the number of cubic yards of concrete needed to line each mile of the channel.

FIGURE P10.61

10.62 Determine the critical depth for a flow of 200 m³/s through a rectangular channel of 10-m width. If the water flows 3.8 m deep, is the flow supercritical? Explain.

10.63 The depth downstream of a sluice gate in a rectangular wooden channel of width 5 m is 0.60 m. If the flowrate is 18 m³/s, determine the channel slope needed to maintain this depth. Will the depth increase or decrease in the flow direction if the slope is (a) 0.02; (b) 0.01?

10.64 A smooth steel water slide at an amusement park is of semicircular cross section with a diameter of 2.5 ft. The slide descends a vertical distance of 35 ft in its 420-ft length. If pumps supply water to the slide at a rate of 4 cfs, determine the depth of flow. Neglect the effects of the curves and bends of the slide.

10.65 Four sewer pipes of 0.5-m diameter join to form one pipe of diameter D. If the Manning coefficient, n, and the slope are the same for all of the pipes, and if each pipe flows half-full, determine D.

10.66* Water flows in the painted steel rectangular channel with rounded corners shown in Fig. P10.66. The bottom slope is 1 ft/200 ft. Plot a graph of flowrate as a function of water depth for $0 \le y \le 1$ ft with corner radii of $r = 0, 0.2, 0.4, 0.6, 0.8,$ and 1.0 ft.

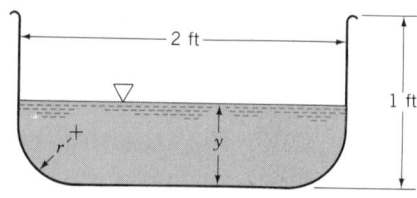

FIGURE P10.66

10.67* Water flows in the fiberglass ($n = 0.014$) triangular channel with a round bottom shown in Fig. P10.67. The channel slope is 0.1 m/90 m. Plot a graph of flowrate as a function of water depth for $0 \le y \le 0.50$ m with bottom radii of $r = 0, 0.25, 0.50, 0.75,$ and 1.0 m.

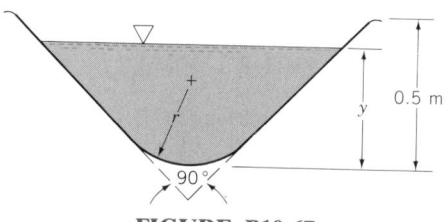

FIGURE P10.67

10.68* The cross section of a long tunnel carrying water through a mountain is as indicated in Fig. P10.68. Plot a graph of flowrate as a function of water depth, y, for $0 \le y \le 18$ ft. The slope is 2 ft/mi and the surface of the tunnel is rough rock (equivalent to rubble masonry). At what depth is the flowrate maximum? Explain.

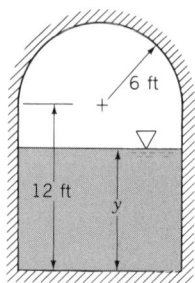

FIGURE P10.68

10.69 The smooth concrete lined channel shown in Fig. P10.69 is built on a slope of 3 m/km. Determine the flowrate if the depth is $y = 1.5$ m.

FIGURE P10.69

10.70 Determine the flow depth for the channel shown in Fig. P10.69 if the flowrate is 15 m³/s.

10.71* The cross section of a creek valley is shown in Fig. P10.71. Plot a graph of flowrate as a function of depth, y, for $0 \leq y \leq 10$ ft. The slope is 5 ft/mi.

FIGURE P10.71

10.72 Repeat Problem 10.69 if the surfaces are smooth concrete as is indicated except for the diagonal surface, which is grass with $n = 0.030$.

10.73* Water flows through the storm sewer shown in Fig. P10.73. The slope of the bottom is 2 m/400 m. Plot a graph of the flowrate as a function of depth for $0 \leq y \leq 1.7$ m. On the same graph plot the flowrate expected if the entire surface were lined with material similar to that of a clay tile.

FIGURE P10.73

10.74 Determine the maximum flowrate possible for the creek shown in Fig. P10.74 if it is not to overflow onto the flood plain. The creek bed drops an average of 3 ft/half mile of length. Determine the flowrate during a flood if the depth is 8 ft.

FIGURE P10.74

10.75 Determine the flowrate for the symmetrical channel shown in Fig. P10.45 if the bottom is smooth concrete and the sides are masonry rubble. The bottom slope is $S_0 = 0.001$.

10.76 Water in a rectangular painted steel channel of width $b = 1$ ft and depth y is to flow at critical conditions, Fr = 1. Plot a graph of the critical slope, S_{0c}, as a function of y for 0.05 ft $\leq y \leq 5$ ft. What is the maximum slope allowed if critical flow is not to occur regardless of the depth?

10.77 Water flows in a rectangular channel of width b and depth y with a Froude number of unity. The slope, S_{0c}, of the channel needed to produce this critical flow is a function of y. Show that as $y \rightarrow \infty$ the slope becomes proportional to y (i.e., $S_{0c} = C_1 y$, where C_1 is a constant) and that as $y \rightarrow 0$ the slope becomes proportional to $y^{-1/3}$ (i.e., $S_{0c} = C_2 / y^{1/3}$, where C_2 is a constant). Show that the channel with an aspect ratio of $b/y = 6$ gives the minimum value of S_{0c}.

10.78 Water flows in a rectangular channel with a bottom slope of 2.2 ft/mi and a head loss of 4.3 ft/mi. At a section where the depth is 5.8 ft and the average velocity 5.9 ft/s, does the flow depth increase or decrease in the direction of flow? Explain.

10.79 Water flows in the river shown in Fig. P10.79 with a uniform bottom slope. The total head at each section is measured by using Pitot tubes as indicated. Determine the value of dy/dx at a location where the Froude number is 0.357.

$z_1 = 620.1$ ft
$z_3 = 628.3$ ft
$x_2 - x_1 = 4100$ ft
$z_2 = 618.7$ ft
$z_4 = 625.0$ ft

FIGURE P10.79

10.80 Repeat Problem 10.79 if the Froude number is 1.75.

10.81 Assume that the conditions given in Fig. P10.79 are as indicated except that the value of z_4 is not known. Determine the value of z_4 if the flow is uniform depth.

10.82 At the bottom of a water ride in an amusement park, the water in the rectangular channel has a depth of 1.2 ft and a velocity of

15.6 ft/s. Determine the height of the "standing wave" (a hydraulic jump) that the boat passes through for its final "splash."

10.83 The water depths upstream and downstream of a hydraulic jump are 0.3 and 1.2 m, respectively. Determine the upstream velocity and the power dissipated if the channel is 50 m wide.

10.84 Show that the Froude number downstream of a hydraulic jump in a rectangular channel is $(y_1/y_2)^{3/2}$ times the Froude number upstream of the jump, where (1) and (2) denote the upstream and downstream conditions, respectively.

10.85 Show that for a hydraulic jump in a rectangular channel, the Froude number upstream, Fr_1, and the Froude number downstream, Fr_2, are related by

$$Fr_2^2 = \frac{8Fr_1^2}{[(1 + 8\ Fr_1^2)^{1/2} - 1]^3}$$

Plot Fr_2 as a function of Fr_1 and show that the flow downstream of a jump is subcritical.

10.86 Water flows in a 2-ft-wide rectangular channel at a rate of 10 ft³/s. If the water depth downstream of a hydraulic jump is 2.5 ft, determine **(a)** the water depth upstream of the jump, **(b)** the upstream and downstream Froude numbers, and **(c)** the head loss across the jump.

10.87 A hydraulic jump at the base of a spillway of a dam is such that the depths upstream and downstream of the jump are 0.70 and 3.6 m, respectively. If the spillway is 50 m wide, what is the flowrate over the spillway?

10.88 Determine the head loss and power dissipated by the hydraulic jump of Problem 10.87.

10.89 At a given location in a 12-ft-wide rectangular channel the flowrate is 900 ft³/s and the depth is 4 ft. Is this location upstream or downstream of the hydraulic jump that occurs in this channel? Explain.

10.90 Water flows in a wide finished concrete channel as is shown in Fig. P10.90 such that a hydraulic jump occurs at the transition of the change in slope of the channel bottom. If the upstream Froude number and depth are 4.0 and

0.2 ft, respectively, determine the slopes upstream, S_{01}, and downstream, S_{02}, of the jump to maintain uniform flows in those regions. The jump can be treated as a jump on a horizontal surface.

10.91* A rectangular channel of width b is to carry water at flowrates from $30 \le Q \le 600$ cfs. The water depth upstream of the hydraulic jump that occurs (if one does occur) is to remain 1.5 ft for all cases. Plot the power dissipated in the jump as a function of flowrate for channels of width $b = 10$, 20, 30, and 40 ft.

10.92 Water flows in a rectangular channel at a depth of $y = 1$ ft and a velocity of $V = 20$ ft/s. When a gate is suddenly placed across the end of the channel, a wave (a moving hydraulic jump) travels upstream with velocity V_w as is indicated in Fig. P10.92. Determine V_w. Note that this is an unsteady problem for a stationary observer. However, for an observer moving to the left with velocity V_w, the flow appears as a steady hydraulic jump.

FIGURE P10.92

10.93 Water flows in a rectangular channel with velocity $V = 6$ m/s. A gate at the end of the channel is suddenly closed so that a wave (a moving hydraulic jump) travels upstream with velocity $V_w = 2$ m/s as is indicated in Fig. P10.92. Determine the depths ahead of and behind the wave. Note that this is an unsteady problem for a stationary observer. However, for an observer moving to the left with velocity V_w, the flow appears as a steady hydraulic jump.

10.94 Water flows over a 5-ft-wide rectangular sharp crested weir that is $P_w = 4$ ft tall. If the depth upstream is 5 ft, determine the flowrate.

10.95 A rectangular sharp crested weir is used to measure the flowrate in a channel of width 10 ft. It is desired to have the channel flow depth be 6 ft when the flowrate is 50 cfs. Determine the height, P_w, of the weir plate.

10.96 Determine the head, H, required to allow a flowrate of 500 m³/hr over a sharp crested triangular weir with $\theta = 60°$.

10.97 Water flows over a sharp crested triangular weir with $\theta = 90°$. The head range covered is $0.2 \le H \le 1.0$ ft and the accuracy in the

FIGURE P10.90

measurement of the head, H, is $\delta H = \pm 0.01$ ft. Plot a graph of the percent error expected in Q as a function of Q.

10.98 Water flows over a broad crested weir that has a width of 2 m and a height of $P_w = 1.5$ m. The free-surface well upstream of the weir is at a height of 0.5 m above the surface of the weir. Determine the flowrate in the channel and the minimum depth of the water above the weir block.

10.99 The top of a broad crested weir block is at an elevation of 724.5 ft, 4 ft above the channel bottom. If the weir is 20 ft wide and the flowrate is 400 cfs, determine the elevation of the reservoir upstream of the weir.

10.100 Determine the flowrate per unit width, q, over a broad crested weir that is 1.5 m tall if the head, H, is 0.50 m.

10.101 Water flows under a sluice gate in a channel of 10-ft width. If the upstream depth remains constant at 5 ft, plot a graph of flowrate as a function of the distance between the gate and channel bottom as the gate is slowly opened. Assume free outflow.

10.102 Water flows under a sluice gate in a 60-ft-wide finished concrete channel as is shown in Fig. P10.102. Determine the flowrate. If the slope of the channel is 2.5 ft/200 ft, will the water depth increase or decrease downstream of the gate? Assume $C_c = y_2/a = 0.65$. Explain.

FIGURE P10.102

10.103 Water flows over the rectangular sharp crested weir in a wide channel as shown in Fig. P10.103. If the channel is lined with unfinished concrete with a bottom slope of 2 m/300 m, will it be possible to produce a hydraulic jump in the channel downstream of the weir? Explain.

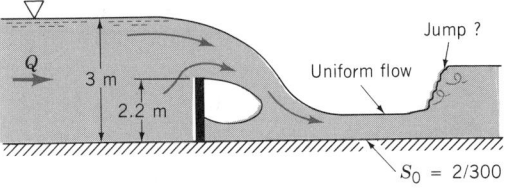

FIGURE P10.103

10.104 Water flows in a rectangular channel of width $b = 20$ ft at a rate of 100 ft³/s. The flowrate is to be measured by using either a rectangular weir of height $P_w = 4$ ft or a triangular ($\theta = 90°$) sharp crested weir. Determine the head, H, necessary. If measurement of the head is accurate to only ± 0.04 ft, determine the accuracy of the measured flowrate expected for each of the weirs. Which weir would be the most accurate? Explain.

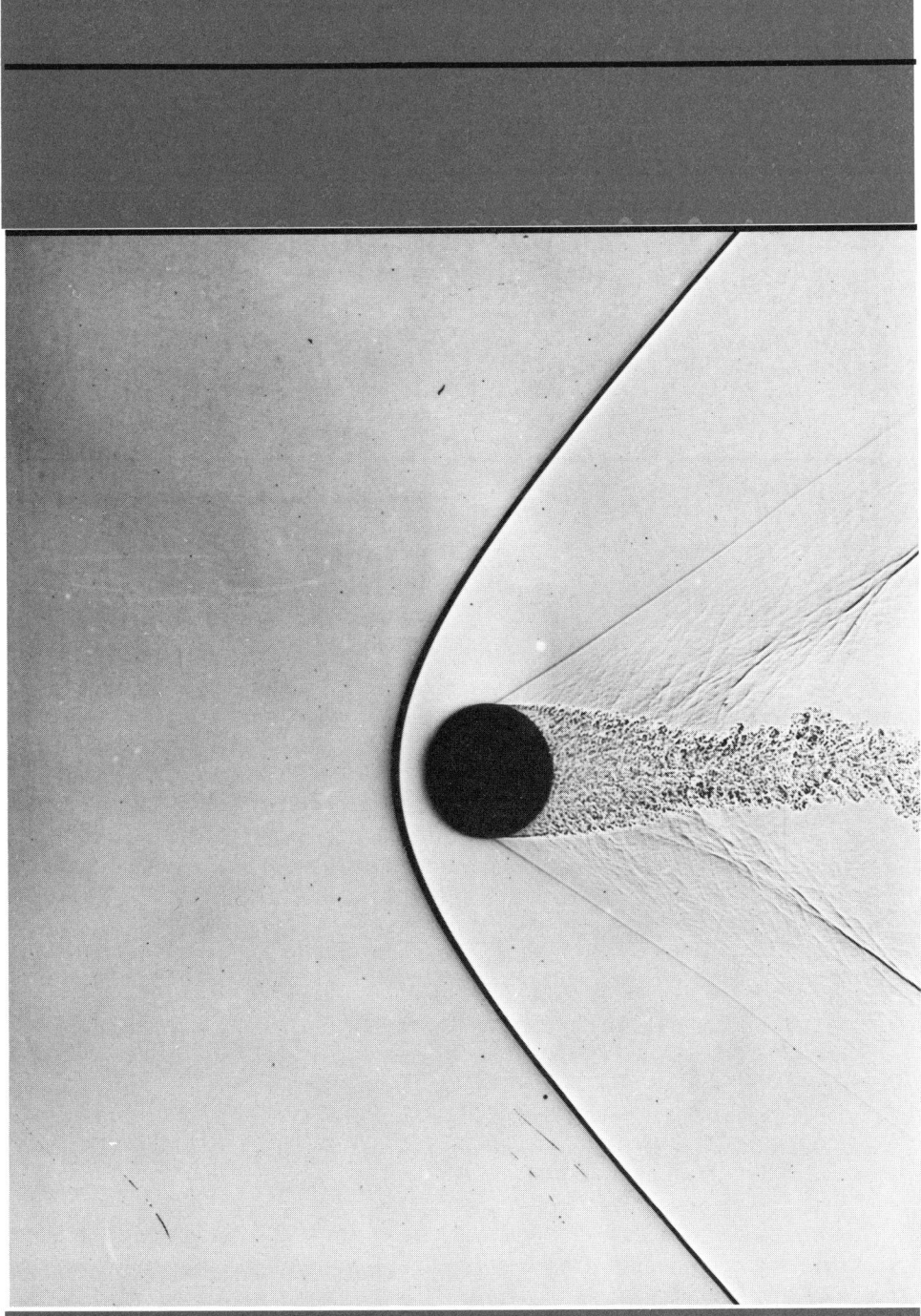

Flow past a sphere at Mach 1.53: An object moving through a fluid at supersonic speed (Mach number greater than one) creates a shock wave (a discontinuity in flow conditions shown by the dark curved line), which is heard as a sonic boom as the object passes overhead. The turbulent wake is also shown. (Shadowgraph technique used in air) (Photograph courtesy of A. C. Charters)

ELEVEN

COMPRESSIBLE FLOW

Most first courses in fluid mechanics concentrate on constant density (incompressible) flows. In earlier chapters of this book, we mainly considered incompressible flow behavior. In a few instances, variable density (compressible) flow effects were covered briefly. The notion of an incompressible fluid is convenient because when constant density and constant (including zero) viscosity are assumed, problem solutions are greatly simplified. Also, fluid incompressibility allows us to build on the Bernoulli equation as was done, for example, in Chapter 5. Preceding examples should have convinced us that nearly incompressible flows are common in everyday experiences.

Any study of fluid mechanics would, however, be incomplete without a brief introduction to compressible flow behavior. Fluid compressibility is a very important consideration in numerous engineering applications of fluid mechanics. For example, the measurement of high-speed flow velocities requires compressible flow theory. The flows in gas turbine engine components are generally compressible. Many aircraft fly fast enough to involve a compressible flow field.

The variation of fluid density for compressible flows requires attention to density and other fluid property relationships. The fluid equation of state, often unimportant for incompressible flows, is vital in the analysis of compressible flows. Also, temperature variations for compressible flows are usually significant and thus the energy equation is important. Curious phenomena can occur with compressible flows. For example, with compressible flows we can have fluid acceleration because of friction, fluid deceleration in a converging duct, fluid temperature decrease with heating, and the formation of abrupt discontinuities in flows across which fluid properties change appreciably.

For simplicity in this introductory study of compressibility effects we mainly consider the steady, one-dimensional, constant (including zero) viscosity, compressible flow of

an ideal gas. In this chapter, one-dimensional flow refers to flow involving uniform distributions of fluid properties over any flow cross section area. Both frictionless ($\mu = 0$) and frictional ($\mu \neq 0$) compressible flows are considered. If the change in volume associated with a change of pressure is considered a measure of compressibility, our experience suggests that gases and vapors are much more compressible than liquids. We focus our attention on the compressible flow of a gas because such flows occur often. We limit our discussion to ideal gases, since the equation of state for an ideal gas is uncomplicated, yet representative of actual gases at pressures and temperatures of engineering interest, and because the flow trends associated with an ideal gas are generally applicable to other compressible fluids.

An excellent film about compressible flow is available (see Ref. 1). This resource is a useful supplement to the material covered in this chapter.

11.1 IDEAL GAS RELATIONSHIPS

Before we can proceed to develop compressible flow equations, we need to become more familiar with the fluid we will work with, the ideal gas. Specifically, we must learn how to evaluate ideal gas property changes. The equation of state for an ideal gas is

$$p = \rho R T \tag{11.1}$$

We have already discussed fluid pressure, p, density, ρ, and temperature, T, in earlier chapters. The gas constant, R, represents a constant for each distinct ideal gas or mixture of ideal gases, where

$$R = \frac{\lambda}{M_{\text{gas}}} \tag{11.2}$$

With this notation, λ is the universal gas constant and M_{gas} is the molecular weight of the ideal gas or gas mixture. Listed in Tables 1.7 and 1.8 are values of the gas constants of some commonly used gases. Nonideal gas state equations are beyond the scope of this text, and those interested in this topic are directed to texts on engineering thermodynamics, for example, Refs. 2 and 3. Note that the trends of ideal gas flows are generally good indicators of what nonideal gas flow behavior is like. Nonideal gas flows are also discussed briefly in Refs. 2 and 3.

For an ideal gas, internal energy, $\breve{u}$, is considered to be a function of temperature only (Refs. 2 and 3). Thus, the ideal gas specific heat at constant volume, c_v, can be expressed as

$$c_v = \left(\frac{\partial \breve{u}}{\partial T}\right)_v = \frac{d\breve{u}}{dT} \tag{11.3}$$

where the subscript v on the partial derivative refers to differentiation at constant specific volume, $v = 1/\rho$. From Eq. 11.3 we conclude that for a particular ideal gas, c_v is a function of temperature only. Equation 11.3 can be rearranged to yield

$$d\breve{u} = c_v \, dT$$

Thus,

$$\breve{u}_2 - \breve{u}_1 = \int_{T_1}^{T_2} c_v \, dT \tag{11.4}$$

Equation 11.4 is useful because it allows us to evaluate the change in internal energy, $\breve{u}_2 - \breve{u}_1$, associated with ideal gas flow from section (1) to section (2) in a flow. For simplicity, we can assume that c_v is constant for a particular ideal gas and obtain from Eq. 11.4

$$\breve{u}_2 - \breve{u}_1 = c_v(T_2 - T_1) \qquad (11.5)$$

Actually, c_v for a particular gas varies with temperature (see Refs. 2 and 3). However, for moderate changes in temperature, the constant c_v assumption is reasonable.

The fluid property enthalpy, $\breve{h}$, is defined as

$$\breve{h} = \breve{u} + \frac{p}{\rho} \qquad (11.6)$$

For an ideal gas, we have already stated that

$$\breve{u} = \breve{u}(T)$$

From the equation of state (Eq. 11.1)

$$\frac{p}{\rho} = RT$$

Thus, it follows that

$$\breve{h} = \breve{h}(T)$$

Since for an ideal gas enthalpy is a function of temperature only, then the ideal gas specific heat at constant pressure, c_p, can be expressed as

$$c_p = \left(\frac{\partial \breve{h}}{\partial T}\right)_p = \frac{d\breve{h}}{dT} \qquad (11.7)$$

where the subscript p on the partial derivative refers to differentiation at constant pressure, and c_p is a function of temperature only. The rearrangement of Eq. 11.7 leads to

$$d\breve{h} = c_p \, dT$$

and

$$\breve{h}_2 - \breve{h}_1 = \int_{T_1}^{T_2} c_p \, dT \qquad (11.8)$$

Equation 11.8 is useful because it allows us to evaluate the change in enthalpy, $\breve{h}_2 - \breve{h}_1$, associated with ideal gas flow from section (1) to section (2) in a flow. For simplicity, we can assume that c_p is constant for a specific ideal gas and obtain from Eq. 11.8

$$\breve{h}_2 - \breve{h}_1 = c_p(T_2 - T_1) \qquad (11.9)$$

As is true for c_v, c_p for a given gas varies with temperature. Nevertheless, for moderate changes in temperature, the constant c_p assumption is reasonable.

From Eqs. 11.5 and 11.9 we see that changes in internal energy and enthalpy are related to changes in temperature by values of c_v and c_p. We turn our attention now to developing useful relationships for determining c_v and c_p. Combining Eqs. 11.6 and 11.1 we get

$$\breve{h} = \breve{u} + RT \qquad (11.10)$$

Differentiating Eq. 11.10 leads to

$$d\check{h} = d\check{u} + R\,dT$$

or

$$\frac{d\check{h}}{dT} = \frac{d\check{u}}{dT} + R \tag{11.11}$$

From Eqs. 11.3, 11.7, and 11.11 we conclude that

$$c_p - c_v = R \tag{11.12}$$

Equation 11.12 indicates that the difference between c_p and c_v is constant for each ideal gas regardless of temperature. Also $c_p > c_v$. If the specific heat ratio, k, is defined as

$$k = \frac{c_p}{c_v} \tag{11.13}$$

then combining Eqs. 11.12 and 11.13 leads to

$$c_p = \frac{Rk}{k-1} \tag{11.14}$$

and

$$c_v = \frac{R}{k-1} \tag{11.15}$$

Actually, c_p, c_v, and k are all temperature dependent for any ideal gas. We will assume constant values for these variables in this book. Values of k and R for some commonly used gases at nominal temperatures are listed in Tables 1.7 and 1.8. These tabulated values can be used with Eqs. 11.13 and 11.14 to determine the values of c_p and c_v. Example 11.1 demonstrates how internal energy and enthalpy changes can be calculated for a flowing ideal gas having constant c_p and c_v.

EXAMPLE 11.1

Air flows steadily between two sections in a long straight portion of 4-in. diameter pipe as is indicated in Fig. E11.1. The uniformly distributed temperature and pressure at each section are $T_1 = 540\ °R$, $p_1 = 100$ psia, and $T_2 = 453\ °R$, $p_2 = 18.4$ psia. Calculate the (a) change in internal energy between sections (1) and (2), (b) change in enthalpy between sections (1) and (2), and (c) change in density between sections (1) and (2).

$D_1 = D_2 = 4\ \text{in.}$

FIGURE E11.1

SOLUTION

(a) Assuming air behaves as an ideal gas, we can use Eq. 11.5 to evaluate the change in internal energy between sections (1) and (2). Thus

$$\check{u}_2 - \check{u}_1 = c_v(T_2 - T_1) \tag{1}$$

From Eq. 11.15 we have

$$c_v = \frac{R}{k - 1} \tag{2}$$

From Table 1.7

$$R = 1716 \text{ (ft·lb)/(slug· °R)} \tag{3}$$

and

$$k = 1.4 \tag{4}$$

Throughout this book, we use the nominal values of k for common gases listed in Tables 1.7 and 1.8 and consider these values as being representative. From Eqs. 2, 3, and 4 we obtain

$$c_v = \frac{1716}{(1.4 - 1)} \text{ (ft·lb)/(slug· °R)} = 4290 \text{ (ft·lb)/(slug· °R)} \tag{5}$$

Combining Eqs. 1 and 5 yields

$$\begin{aligned} \breve{u}_2 - \breve{u}_1 &= c_v(T_2 - T_1) = 4290 \text{ (ft·lb)/(slug· °R)} \\ &\times (453 \text{ °R} - 540 \text{ °R}) = -3.73 \times 10^5 \text{ ft·lb/slug} \end{aligned} \tag{Ans}$$

(b) For enthalpy change we use Eq. 11.9. Thus

$$\breve{h}_2 - \breve{h}_1 = c_p(T_2 - T_1) \tag{6}$$

We need a value of c_p for air to solve for $\breve{h}_2 - \breve{h}_1$. Since

$$k = \frac{c_p}{c_v} \tag{7}$$

substituting Eq. 5 into Eq. 7 gives

$$c_p = kc_v = (1.4)[4290 \text{ (ft·lb)/(slug· °R)}] = 6006 \text{ (ft·lb)/(slug· °R)} \tag{8}$$

From Eqs. 6 and 8 we obtain

$$\begin{aligned} \breve{h}_2 - \breve{h}_1 &= c_p(T_2 - T_1) = 6006 \text{ (ft·lb)/(slug· °R)} \\ &\times (453 \text{ °R} - 540 \text{ °R}) = -5.22 \times 10^5 \text{ ft·lb/slug} \end{aligned} \tag{Ans}$$

(c) For density change we use the ideal gas equation of state (Eq. 11.1) to get

$$\rho_2 - \rho_1 = \frac{p_2}{RT_2} - \frac{p_1}{RT_1} = \frac{1}{R}\left(\frac{p_2}{T_2} - \frac{p_1}{T_1}\right) \tag{9}$$

Using the pressures and temperatures given in the problem statement we calculate from Eq. 9

$$\begin{aligned} \rho_2 - \rho_1 &= \frac{1}{1716 \text{ (ft·lb)/(slug· °R)}} \\ &\times \left(\frac{(18.4 \text{ psia})(144 \text{ in.}^2/\text{ft}^2)}{453 \text{ °R}} - \frac{(100 \text{ psia})(144 \text{ in.}^2/\text{ft}^2)}{540 \text{ °R}}\right) \end{aligned}$$

or

$$\rho_2 - \rho_1 = -0.0121 \text{ slug/ft}^3 \tag{Ans}$$

This is a significant change in density when compared with the upstream density

$$\rho_1 = \frac{p_1}{RT_1} = \frac{(100 \text{ psia})(144 \text{ in.}^2/\text{ft}^2)}{[1716 \text{ (ft·lb)/(slug· °R)}](540 \text{ °R})} = 0.0155 \text{ slug/ft}^3$$

For compressible flows, changes in the thermodynamic property *entropy, s*, are important. For any pure substance including ideal gases, the "first *T ds* equation" is (see Ref. 2 or Ref. 3)

$$T \, ds = d\breve{u} + p \, d\left(\frac{1}{\rho}\right) \tag{11.16}$$

where T is absolute temperature, s is entropy, $\breve{u}$ is internal energy, p is absolute pressure, and ρ is density. Differentiating Eq. 11.6 leads to

$$d\breve{h} = d\breve{u} + p \, d\left(\frac{1}{\rho}\right) + \left(\frac{1}{\rho}\right) dp \tag{11.17}$$

By combining Eqs. 11.16 and 11.17, we obtain

$$T \, ds = d\breve{h} - \left(\frac{1}{\rho}\right) dp \tag{11.18}$$

Equation 11.18 is often referred to as the "second *T ds* equation." For an ideal gas, Eqs. 11.1, 11.3, and 11.16 can be combined to yield

$$ds = c_v \frac{dT}{T} + \frac{R}{1/\rho} \, d\left(\frac{1}{\rho}\right) \tag{11.19}$$

and Eqs. 11.1, 11.7, and 11.18 can be combined to yield

$$ds = c_p \frac{dT}{T} - R \frac{dp}{p} \tag{11.20}$$

If c_p and c_v are assumed to be constant for a given gas, Eqs. 11.19 and 11.20 can be integrated to get

$$s_2 - s_1 = c_v \ln \frac{T_2}{T_1} + R \ln \frac{\rho_1}{\rho_2} \tag{11.21}$$

and

$$s_2 - s_1 = c_p \ln \frac{T_2}{T_1} - R \ln \frac{p_2}{p_1} \tag{11.22}$$

Equations 11.21 and 11.22 allow us to calculate the change of entropy of an ideal gas flowing from one section to another with constant specific heat values (c_p and c_v).

EXAMPLE 11.2

For the air flow of Example 11.1, calculate the change in entropy, $s_2 - s_1$, between sections (1) and (2).

SOLUTION

Assuming that the flowing air in Fig. E11.1 behaves as an ideal gas, we can calculate the entropy change between sections by using either Eq. 11.21 or Eq. 11.22. We use both to demonstrate that the same result is obtained either way.

From Eq. 11.21,

$$s_2 - s_1 = c_v \ln \frac{T_2}{T_1} + R \ln \frac{\rho_1}{\rho_2} \tag{1}$$

To evaluate $s_2 - s_1$ from Eq. 1 we need the density ratio, ρ_1/ρ_2, which can be obtained from the ideal gas equation of state (Eq. 11.1) as

$$\frac{\rho_1}{\rho_2} = \left(\frac{p_1}{T_1}\right)\left(\frac{T_2}{p_2}\right) \tag{2}$$

and thus from Eqs. 1 and 2,

$$s_2 - s_1 = c_v \ln \frac{T_2}{T_1} + R \ln \left[\left(\frac{p_1}{T_1}\right)\left(\frac{T_2}{p_2}\right)\right] \tag{3}$$

By substituting values already identified in the Example 11.1 problem statement and solution into Eq. 3 we get

$$s_2 - s_1 = [4290 \text{ (ft·lb)/(slug· °R)}] \ln \left(\frac{453 \text{ °R}}{540 \text{ °R}}\right)$$

$$+ [1716 \text{ (ft·lb)/(slug· °R)}] \ln \left[\left(\frac{100 \text{ psia}}{540 \text{ °R}}\right)\left(\frac{453 \text{ °R}}{18.4 \text{ psia}}\right)\right]$$

or

$$s_2 - s_1 = 1850 \text{ (ft·lb)/(slug· °R)} \tag{Ans}$$

From Eq. 11.22,

$$s_2 - s_1 = c_p \ln \frac{T_2}{T_1} - R \ln \frac{p_2}{p_1} \tag{4}$$

By substituting known values into Eq. 3 we obtain

$$s_2 - s_1 = [6006 \text{ (ft·lb)/(slug· °R)}] \ln \left(\frac{453 \text{ °R}}{540 \text{ °R}}\right)$$

$$- [1716 \text{ (ft·lb)/(slug· °R)}] \ln \left(\frac{18.4 \text{ psia}}{100 \text{ psia}}\right)$$

or

$$s_2 - s_1 = 1850 \text{ (ft·lb)/(slug· °R)} \tag{Ans}$$

As anticipated both Eqs. 11.21 and 11.22 yield the same result for the entropy change $(s_2 - s_1)$.

Note that since the ideal gas equation of state was used in the derivation of the entropy difference equations, both the pressures and temperatures used must be absolute.

If internal energy, enthalpy, and entropy changes for ideal gas flow with variable specific heats are desired, Eqs. 11.4, 11.8, and 11.19 or 11.20 must be used as explained in Refs. 2 and 3. Detailed tables (see for example Ref. 4) are available for variable specific heat calculations.

The second law of thermodynamics requires that the adiabatic and frictionless flow of any fluid results in

$$ds = 0$$

or

$$s_2 - s_1 = 0$$

Constant entropy flow is called *isentropic* flow. For the isentropic flow of an ideal gas with constant c_p and c_v, we get from Eqs. 11.21 and 11.22

$$c_v \ln \frac{T_2}{T_1} + R \ln \frac{\rho_1}{\rho_2} = c_p \ln \frac{T_2}{T_1} - R \ln \frac{p_2}{p_1} = 0 \tag{11.23}$$

By combining Eq. 11.23 with Eqs. 11.14 and 11.15 we obtain

$$\left(\frac{T_2}{T_1}\right)^{k/(k-1)} = \left(\frac{p_2}{\rho_1}\right)^k = \left(\frac{p_2}{p_1}\right) \tag{11.24}$$

which is a useful relationship between temperature, density, and pressure for the isentropic flow of an ideal gas. From Eq. 11.24 we can conclude that

$$\frac{p}{\rho^k} = \text{constant} \tag{11.25}$$

for an ideal gas with constant c_p and c_v flowing isentropically, a result already used without proof earlier in Chapters 1, 3, and 5.

11.2 MACH NUMBER AND SPEED OF SOUND

The Mach number, Ma, was introduced in Chapters 1 and 7 as a dimensionless measure of compressibility in a fluid flow. In this and subsequent sections, we develop some useful relationships involving the Mach number. The Mach number is defined as the ratio of the value of the local flow velocity, V, to the local speed of sound, c. In other words,

$$\text{Ma} = \frac{V}{c}$$

What we perceive as sound generally consists of weak pressure pulses that move through air. When our ear drums respond to a succession of moving pressure pulses, we hear sounds.

To better understand the notion of speed of sound, we analyze the one-dimensional fluid mechanics of an infinitesimally thin, weak pressure pulse moving at the speed of sound through a fluid at rest (see Fig. 11.1a). Ahead of the pressure pulse, the fluid velocity is zero and the fluid pressure and density are p and ρ. Behind the pressure pulse, the fluid velocity has changed by an amount δV, and the pressure and density of the fluid have also changed by amounts δp and $\delta \rho$. We select an infinitesimally thin control volume that moves with the pressure pulse as is sketched in Fig. 11.1a. The speed of the weak pressure pulse is considered constant and in one direction only; thus, our control volume is inertial.

For an observer moving with this control volume (Fig. 11.1b), it appears as if fluid is entering the control volume through surface area A with speed c at pressure p and density ρ and leaving the control volume through surface area A with speed $c - \delta V$,

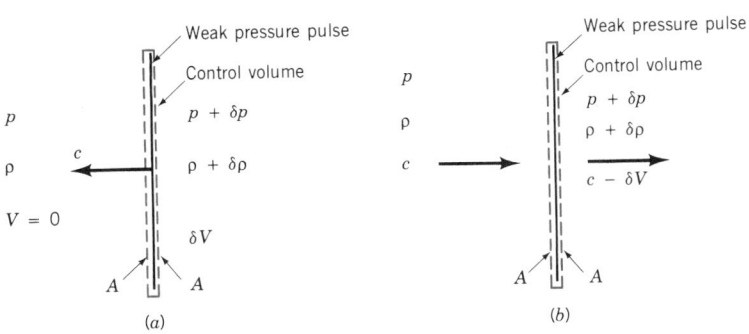

FIGURE 11.1 **(a)** Weak pressure pulse moving through a fluid at rest. **(b)** The flow relative to a control volume containing a weak pressure pulse.

pressure $p + \delta p$, and density $\rho + \delta \rho$. When the continuity equation (Eq. 5.16) is applied to the flow through this control volume, the result is

$$\rho A c = (\rho + \delta \rho) A (c - \delta V) \tag{11.26}$$

or

$$\rho c = \rho c - \rho\, \delta V + c\, \delta \rho - (\delta \rho)(\delta V) \tag{11.27}$$

Since $(\delta \rho)(\delta V)$ is much smaller than the other terms in Eq. 11.27, we drop it from further consideration and keep

$$\rho\, \delta V = c\, \delta \rho \tag{11.28}$$

The linear momentum equation (Eq. 5.29) can also be applied to the flow through the control volume of Fig. 11.1b. The result is

$$-c\rho c A + (c - \delta V)(\rho + \delta \rho)(c - \delta V)A = pA - (p + \delta p)A \tag{11.29}$$

Note that any frictional forces are considered as being negligibly small. By combining Eqs. 11.26 and 11.29, we get

$$-c\rho c A + (c - \delta V)\rho A c = -\delta p A$$

or

$$\rho \delta V = \frac{\delta p}{c} \tag{11.30}$$

From Eqs. 11.28 (continuity) and 11.30 (linear momentum) we obtain

$$c^2 = \frac{\delta p}{\delta \rho}$$

or

$$c = \sqrt{\frac{\delta p}{\delta \rho}} \tag{11.31}$$

This expression for the speed of sound results from application of the conservation of mass and conservation of linear momentum principles to the flow through the control volume of Fig. 11.1b. These principles were used in Section 10.2.1 to obtain an expression for the speed of surface waves traveling on the surface of fluid in a channel.

The conservation of energy principle can also be applied to the flow through the control volume of Fig. 11.1b. If the energy equation (Eq. 5.103) is used for the flow through this control volume, the result is

$$\frac{\delta p}{\rho} + \delta \left(\frac{V^2}{2} \right) + g\, \delta z = \delta (\text{loss}) \tag{11.32}$$

For gas flow we can consider $g\, \delta z$ as being negligibly small in comparison to the other terms in the equation. Also, if we assume that the flow is frictionless, then $\delta(\text{loss}) = 0$ and Eq. 11.32 becomes

$$\frac{\delta p}{\rho} + \frac{(c - \delta V)^2}{2} - \frac{c^2}{2} = 0$$

or, neglecting $(\delta V)^2$ compared to $c \, \delta V$, we obtain

$$\rho \, \delta V = \frac{\delta p}{c} \tag{11.33}$$

By combining Eqs. 11.28 (continuity) and 11.33 (energy) we again find that

$$c = \sqrt{\frac{\delta p}{\delta \rho}}$$

which is identical to Eq. 11.31. Thus, the conservation of linear momentum and the conservation of energy principles lead to the same result. If we further assume that the frictionless flow through the control volume of Fig. 11.1b is adiabatic (no heat transfer), then the flow is isentropic. In the limit, as δp becomes vanishingly small $(\delta p \to \partial p \to 0)$

$$c = \sqrt{\left(\frac{\partial p}{\partial \rho}\right)_s} \tag{11.34}$$

where the subscript s is used to designate that the partial differentiation occurs at constant entropy.

Equation 11.34 suggests to us that we can calculate the speed of sound by determining the partial derivative of pressure with respect to density at constant entropy. For the isentropic flow of an ideal gas (with constant c_p and c_v), we learned earlier (Eq. 11.25) that

$$p = (\text{constant})(\rho^k)$$

and thus

$$\left(\frac{\partial p}{\partial \rho}\right)_s = (\text{constant}) \, k\rho^{k-1} = \frac{p}{\rho^k} \, k\rho^{k-1} = \frac{p}{\rho} \, k = RTk \tag{11.35}$$

Thus, for an ideal gas

$$c = \sqrt{RTk} \tag{11.36}$$

More generally, the bulk modulus of elasticity, E_v, of any fluid including liquids is defined as

$$E_v = \frac{dp}{d\rho/\rho} = \rho \left(\frac{\partial p}{\partial \rho}\right)_s \tag{11.37}$$

Thus, in general, from Eqs. 11.34 and 11.37,

$$c = \sqrt{\frac{E_v}{\rho}} \tag{11.38}$$

Values of the speed of sound are tabulated in Tables B.1 and B.2 for water and in Tables B.3 and B.4 for air. From experience we know that air is more easily compressed than water. Note from the values of c in Tables B.1 through B.4 that the speed of sound in air is much less than it is in water. From Eq. 11.37, we can conclude that if a fluid is truly incompressible, its bulk modulus would be infinitely large, as would be the speed of sound in that fluid. Thus, an incompressible flow must be considered an idealized approximation of reality.

EXAMPLE 11.3

Verify the speed of sound for air at 0 °C listed in Table B.4.

SOLUTION

In Table B.4, we find the speed of sound of air at 0 °C given as 331.4 m/s. Assuming that air behaves as an ideal gas, we can calculate the speed of sound from Eq. 11.36 as

$$c = \sqrt{RTk} \tag{1}$$

The value of the gas constant is obtained from Table 1.8 as

$$R = 286.9 \ J/(kg \cdot K)$$

and the specific heat ratio is listed in Table B.4 as

$$k = 1.401$$

By substituting values of R, k, and T into Eq. 1 we obtain

$$c = \sqrt{[286.9) \ J/(kg \cdot K)](273.15K)(1.401)[1(kg \cdot m)/(N \cdot s^2)][1(N \cdot m)/J]} = 331.4 \ m/s \quad \text{(Ans)}$$

The value of the speed of sound calculated with Eq. 11.36 agrees very well with the value of c listed in Table B.4.

11.3 CATEGORIES OF COMPRESSIBLE FLOW

In Section 3.8.1, we learned that the effects of compressibility become more significant as the Mach number increases. For example, the error associated with calculating the stagnation pressure of an ideal gas with $\rho V^2/2$ increases at larger Mach numbers. From Fig. 3.23 we can conclude that incompressible flows can only occur at low Mach numbers.

Experience has also demonstrated that compressibility can have a large influence on other important flow variables. For example, in Fig. 11.2 the variation of drag coefficient with Reynolds number and Mach number is shown for air flow over a sphere. Compressibility effects can be of considerable importance.

To further illustrate some curious features of compressible flow, a simplified example is considered. Imagine the emission of weak pressure pulses from a point source. These pressure waves are spherical and expand radially outward from the point source at the speed of sound, c. If a pressure wave is emitted at different times, t_{wave}, we can determine where several waves will be at a common instant of time, t, by using the relationship

$$r = (t - t_{wave})c$$

where r is the radius of the sphere shaped wave emitted at time $= t_{wave}$. For a stationary point source, the symmetrical wave pattern shown in Fig. 11.3a is involved.

When the point source moves to the left with a constant velocity, V, the wave pattern is no longer symmetrical. In Figs. 11.3b, 11.3c, and 11.3d are illustrated the wave patterns at $t = 3$ s for different values of V. Also shown with a "$+$" are the positions of the moving point source at values of time, t, equal to 0 s, 1 s, 2 s, and

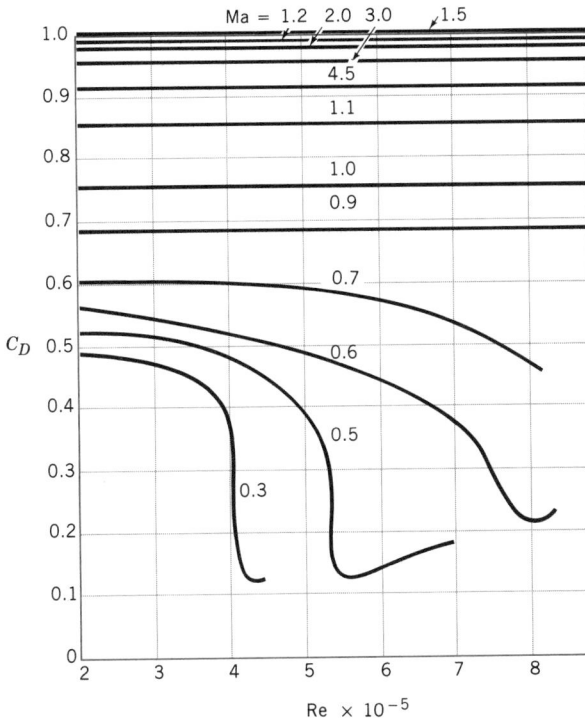

FIGURE 11.2 The variation of the drag coefficient of a sphere with Reynolds number and Mach number. (Adapted from Fig. 1.8 in Ref. 1 of Chapter 9)

3 s. Knowing where the point source has been at different instances is important because it indicates to us where the different waves originated.

From the pressure wave patterns of Fig. 11.3, we can draw some useful conclusions. Before doing this we should recognize that if instead of moving the point source to the left, we held the point source stationary and moved the fluid to the right with velocity, V, the resulting pressure wave patterns would be identical to those indicated in Fig. 11.3.

When the point source and the fluid are stationary, the pressure wave pattern is symmetrical (Fig. 11.3a) and an observer anywhere in the pressure field would hear the same sound frequency from the point source. When the velocity of the point source (or the fluid) is very small in comparison with the speed of sound, the pressure wave pattern will still be nearly symmetrical. The speed of sound in an incompressible fluid is infinitely large. Thus, the stationary point source and stationary fluid situation are representative of incompressible flows. For truly incompressible flows, the communication of pressure information throughout the flow field is unrestricted and instantaneous ($c = \infty$).

When the point source moves in fluid at rest (or when fluid moves past a stationary point source) the pressure wave patterns vary in asymmetry, with the extent of asymmetry depending on the ratio of the point source (or fluid) velocity and the speed of sound. When $V/c < 1$, the wave pattern is similar to the one shown in Fig. 11.3b. This flow is considered *subsonic* and compressible. A stationary observer will hear a different sound frequency coming from the point source depending on where the observer is relative to the source because the wave pattern is asymmetrical. We call this phenomenon the Doppler effect. Pressure information can still travel unrestricted throughout the flow field, but not symmetrically or instantaneously.

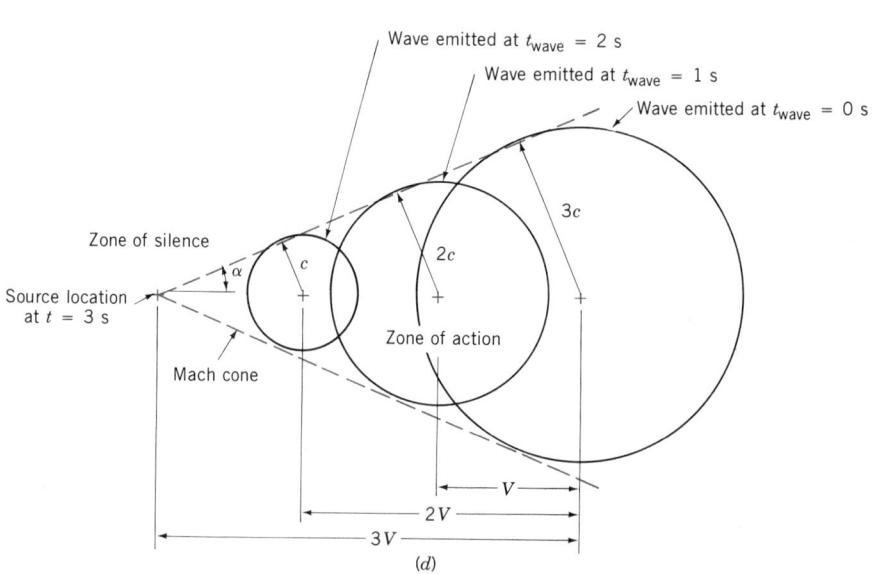

FIGURE 11.3 **(a)** Pressure waves at $t = 3$ s; $V = 0$. **(b)** Pressure waves at $t = 3$ s; $V < c$.
(c) Pressure waves at $t = 3$ s; $V = c$. **(d)** Pressure waves at $t = 3$ s; $V > c$.

When $V/c = 1$, pressure waves are not present ahead of the moving point source. The flow is *sonic*. If you were positioned to the left of the moving point source, you would not hear the point source until it was coincident with your location. For flow moving past a stationary point source at the speed of sound ($V/c = 1$), the pressure waves are all tangent to a plane that is perpendicular to the flow and that passes through the point source. The concentration of pressure waves in this tangent plane suggests the formation of a significant pressure variation across the plane. This plane is often called a Mach wave. Note that communication of pressure information is restricted to the region of flow downstream of the Mach wave. The region of flow upstream of the Mach wave is called the *zone of silence* and the region of flow downstream of the tangent plane is called the *zone of action*.

When $V > c$, the flow is *supersonic* and the pressure wave pattern resembles the one depicted in Fig. 11.3d. A cone (*Mach cone*) that is tangent to the pressure waves can be constructed to represent the Mach wave that separates the zone of silence from the zone of action in this case. The communication of pressure information is restricted to the zone of action. From the sketch of Fig. 11.3d, we can see that the angle of this cone, α, is given by

$$\sin \alpha = \frac{c}{V} = \frac{1}{\text{Ma}} \qquad (11.39)$$

Equation 11.39 is often used to relate the Mach cone angle, α, and the flow Mach number, Ma, when studying flows involving $V/c > 1$. The concentration of pressure waves at the surface of the Mach cone suggests a significant pressure, and thus density, variation across the cone surface. An abrupt density change can be visualized in a flow field by using special optics. Examples of flow visualization methods include the schlieren, shadowgraph, and interferometer techniques (see Ref. 5). A schlieren photo of a flow for which $V > c$ is shown in Fig. 11.4. The air flow through the row of compressor blade airfoils is as shown with the arrow. The flow enters supersonically ($\text{Ma}_1 = 1.14$) and leaves subsonically ($\text{Ma}_2 = 0.86$). The center two airfoils have pressure tap hoses connected to them. Regions of significant changes in fluid density appear in the supersonic portion of the flow. Also, the region of separated flow on each airfoil is visible.

This discussion about pressure wave patterns suggests the following categories of fluid flow,

1. Incompressible flow: $\text{Ma} \leq 0.3$. Unrestricted, nearly symmetrical and instantaneous pressure communication.

FIGURE 11.4 The schlieren visualization of flow (supersonic to subsonic) through a row of compressor airfoils. (Photograph provided by Dr. Hans Starken of the DLR Köln-Porz, FRG.)

2. Compressible subsonic flow: $0.3 < \mathrm{Ma} < 1.0$. Unrestricted but asymmetrical pressure communication.

3. Compressible supersonic flow: $\mathrm{Ma} \geq 1.0$. Formation of Mach wave; pressure communication restricted to zone of action.

In addition to the above-mentioned categories of flows, two other regimes are commonly referred to: namely, transonic flows ($0.9 \leq \mathrm{Ma} \leq 1.2$) and hypersonic flows ($\mathrm{Ma} > 5$). Modern aircraft are mainly powered by gas turbine engines that involve transonic flows. When a space shuttle reenters the earth's atmosphere, the flow is hypersonic. Future aircraft may be expected to operate from subsonic to hypersonic flow conditions.

EXAMPLE 11.4

An aircraft cruising at 1000-m elevation, z, above you moves past in a flyby. How many seconds after the plane passes overhead do you expect to elapse before you hear the aircraft if it is moving with a Mach number equal to 1.5 and the ambient temperature is 20 °C?

SOLUTION

Since the aircraft is moving supersonically ($\mathrm{Ma} > 1$), we can imagine a Mach cone originating from the forward tip of the craft as is illustrated in Fig. E11.4. When the surface of the cone

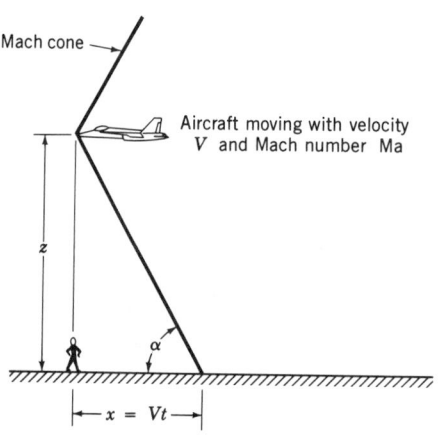

Mach cone

Aircraft moving with velocity V and Mach number Ma

z

α

$x = Vt$

FIGURE E11.4

reaches the observer, the "sound" of the aircraft is perceived. The angle α in Fig. E11.4 is related to the elevation of the plane, z, and the ground distance, x, by

$$\alpha = \tan^{-1} \frac{z}{x} = \tan^{-1} \frac{1000}{Vt} \qquad (1)$$

Also, assuming negligible change of Mach number with elevation, we can use Eq. 11.39 to relate Mach number to the angle α. Thus

$$\mathrm{Ma} = \frac{1}{\sin \alpha} \qquad (2)$$

Combining Eqs. 1 and 2 we obtain

$$\mathrm{Ma} = \frac{1}{\sin \left[\tan^{-1} (1000/Vt) \right]} \qquad (3)$$

The speed of the aircraft can be related to the Mach number with

$$V = (\text{Ma})c \qquad (4)$$

where c is the speed of sound. From Table B.4, $c = 343.3$ m/s. Using Ma $= 1.5$, we get from Eqs. 3 and 4

$$1.5 = \frac{1}{\sin\left\{\tan^{-1}\left[\dfrac{1000 \text{ m}}{(1.5)(343.3 \text{ m/s})t}\right]\right\}}$$

or

$$t = 2.17 \text{ s} \qquad \text{(Ans)}$$

11.4 ISENTROPIC FLOW OF AN IDEAL GAS

In this section, we consider in further detail the steady, one-dimensional, isentropic flow of an ideal gas with constant specific heat values (c_p and c_v). Because the flow is steady throughout, shaft work cannot be involved. Also, as explained earlier, the one-dimensionality of flows we discuss in this chapter implies velocity and fluid property changes in the streamwise direction only. We consider flows through finite control volumes with uniformly distributed velocities and fluid properties at each section of flow. Much of what we develop can also apply to the flow of a fluid particle along its pathline.

Isentropic flow involves constant entropy and was discussed earlier in Section 11.1, where we learned that adiabatic and frictionless (reversible) flow is one form of isentropic flow. Some ideal gas relationships for isentropic flows were developed in Section 11.1. An isentropic flow is not achievable with actual fluids because of friction. Nonetheless, the study of isentropic flow trends is useful because it helps us to gain an understanding of actual compressible flow phenomena.

11.4.1 Effect of Flow Cross Section Area Variations

When fluid flows steadily through a conduit that has a flow cross section area that varies with axial distance, the conservation of mass (continuity) equation

$$\dot{m} = \rho A V = \text{constant} \qquad (11.40)$$

can be used to relate the flow rates at different sections. For incompressible flow, the fluid density remains constant and the flow velocity from section to section varies inversely with cross section area. However, when the flow is compressible, density, cross section area, and flow velocity can all vary from section to section. We proceed to determine how fluid density and flow velocity change with axial location in a variable area duct when the fluid is an ideal gas and the flow through the duct is steady and isentropic.

In Chapter 3, Newton's second law was applied to the inviscid (frictionless) and steady flow of a fluid particle. For the streamwise direction, the result (Eq. 3.5) for either compressible or incompressible flows is

$$dp + \tfrac{1}{2}\rho d(V^2) + \gamma \, dz = 0 \qquad (11.41)$$

The frictionless flow from section to section through a finite control volume is also governed by Eq. 11.41 if the flow is one-dimensional because every particle of fluid

involved will have the same experience. For ideal gas flow, the potential energy difference term, $\gamma\,dz$, can be dropped because of its small size in comparison to the other terms, namely, dp and $d(V^2)$. Thus, an appropriate equation of motion in the streamwise direction for the steady, one-dimensional, and isentropic (adiabatic and frictionless) flow of an ideal gas is obtained from Eq. 11.41 as

$$\frac{dp}{\rho V^2} = -\frac{dV}{V} \tag{11.42}$$

If we form the logarithm of both sides of the continuity equation (Eq. 11.40), the result is

$$\ln\rho + \ln A + \ln V = \text{constant} \tag{11.43}$$

Differentiating Eq. 11.43 we get

$$\frac{d\rho}{\rho} + \frac{dA}{A} + \frac{dV}{V} = 0$$

or

$$-\frac{dV}{V} = \frac{d\rho}{\rho} + \frac{dA}{A} \tag{11.44}$$

Now we combine Eqs. 11.42 and 11.44 to obtain

$$\frac{dp}{\rho V^2}\left(1 - \frac{V^2}{dp/d\rho}\right) = \frac{dA}{A} \tag{11.45}$$

Since the flow being considered is isentropic, the speed of sound is related to variations of pressure with density by Eq. 11.34, repeated here for convenience as

$$c = \sqrt{\left(\frac{\partial p}{\partial \rho}\right)_s}$$

Equation 11.34, combined with the definition of Mach number

$$\text{Ma} = \frac{V}{c} \tag{11.46}$$

and Eq. 11.45 yields

$$\frac{dp}{\rho V^2}(1 - \text{Ma}^2) = \frac{dA}{A} \tag{11.47}$$

Equations 11.42 and 11.47 merge to form

$$\frac{dV}{V} = -\frac{dA}{A}\frac{1}{(1 - \text{Ma}^2)} \tag{11.48}$$

We can use Eq. 11.48 to conclude that when the flow is subsonic (Ma < 1), velocity and section area changes are in opposite directions. In other words, the area increase associated with subsonic flow through a diverging duct like the one shown in Fig. 11.5a is accompanied by a velocity decrease. Subsonic flow through a converging duct (see Fig. 11.5b) involves an increase of velocity. These trends are consistent with incompressible flow behavior, which we described earlier in this book, for instance, in Chapters 3 and 8.

Equation 11.48 also serves to show us that when the flow is supersonic (Ma > 1), velocity and area changes are in the same direction. A diverging duct (Fig. 11.5a)

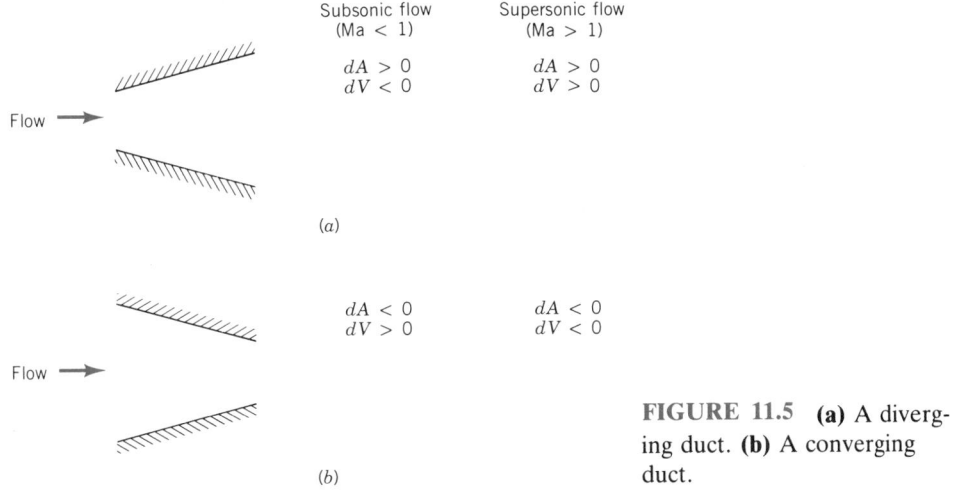

FIGURE 11.5 **(a)** A diverging duct. **(b)** A converging duct.

will accelerate a supersonic flow. A converging duct (Fig. 11.5b) will decelerate a supersonic flow. These trends are the opposite of what happens for incompressible and subsonic compressible flows.

To better understand why subsonic and supersonic duct flows are so different we combine Eqs. 11.44 and 11.48 to form

$$\frac{d\rho}{\rho} = \frac{dA}{A} \frac{\text{Ma}^2}{(1 - \text{Ma}^2)} \tag{11.49}$$

Using Eq. 11.49, we can conclude that for subsonic flows (Ma < 1), density and area changes are in the same direction, whereas for supersonic flows (Ma > 1), density and area changes are in opposite directions. Since ρAV must remain constant (Eq. 11.40), when the duct diverges and the flow is subsonic, density and area both increase and thus flow velocity must decrease. However, for supersonic flow through a diverging duct, when the area increases, the density decreases enough so that the flow velocity has to increase to keep ρAV constant.

By rearranging Eq. 11.48, we can obtain

$$\frac{dA}{dV} = -\frac{A}{V}(1 - \text{Ma}^2) \tag{11.50}$$

Equation 11.50 gives us some insight into what happens when Ma = 1. For Ma = 1, Eq. 11.50 requires that $dA/dV = 0$. This result suggests that the area associated with Ma = 1 is either a minimum or a maximum amount.

A converging–diverging duct (Fig. 11.6a) involves a minimum area. If the flow

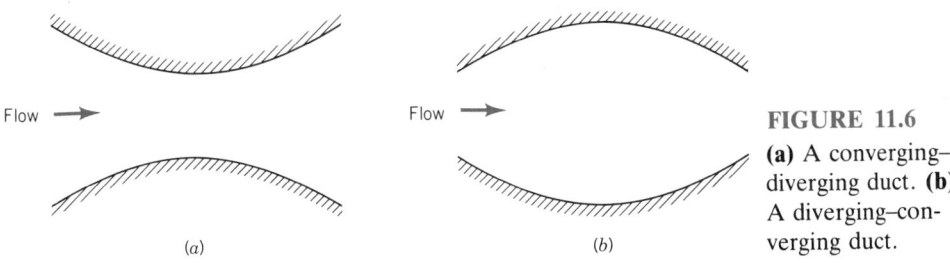

FIGURE 11.6 **(a)** A converging–diverging duct. **(b)** A diverging–converging duct.

entering such a duct were subsonic, Eq. 11.48 discloses that the fluid velocity would increase in the converging portion of the duct, and achievement of a sonic condition (Ma = 1) at the minimum area location appears possible. If the flow entering the converging–diverging duct is supersonic, Eq. 11.48 states that the fluid velocity would decrease in the converging portion of the duct and the sonic condition at the minimum area is possible.

A diverging–converging duct (Fig. 11.6b), on the other hand, would involve a maximum area. If the flow entering this duct were subsonic, the fluid velocity would decrease in the diverging portion of the duct and the sonic condition could not be attained at the maximum area location. For supersonic flow in the diverging portion of the duct, the fluid velocity would increase and thus Ma = 1 at the maximum area is again impossible.

For the steady isentropic flow of an ideal gas, we conclude that the sonic condition (Ma = 1) can be attained in a converging–diverging duct at the minimum area location. This minimum area location is often called the *throat* of the converging–diverging duct. Furthermore, to achieve supersonic flow from a subsonic state in a duct, a converging–diverging area variation is necessary. For this reason, we often refer to such a duct as a *converging–diverging nozzle*. Note that a converging–diverging duct can also decelerate a supersonic flow to subsonic conditions. Thus, a converging–diverging duct can be a nozzle or a diffuser depending on whether the flow in the converging portion of the duct is subsonic or supersonic. A supersonic wind tunnel test section is generally preceded by a converging–diverging nozzle and followed by a converging–diverging diffuser (see Ref. 1). Further details about steady, isentropic, ideal gas flow through a converging–diverging duct are discussed in the next section.

11.4.2 Converging–Diverging Duct Flow

In the preceding section, we discussed the variation of density and velocity of the steady isentropic flow of an ideal gas through a variable area duct. We proceed now to develop equations that help us determine how other important flow properties vary in these flows.

It is convenient to use the stagnation state of the fluid as a reference state for compressible flow calculations. The stagnation state is associated with zero flow velocity and an entropy value that corresponds to the entropy of the flowing fluid. The subscript 0 is used to designate the stagnation state. Thus, stagnation temperature and pressure are T_0 and p_0. For example, if the fluid flowing through the converging–diverging duct of Fig. 11.6a were drawn isentropically from the atmosphere, the atmospheric pressure and temperature would represent the stagnation state of the flowing fluid. The stagnation state can also be achieved by isentropically decelerating a flow to zero velocity. This can be accomplished with a diverging duct for subsonic flows or a converging–diverging duct for supersonic flows. Also, as discussed earlier in Chapter 3, an approximately isentropic deceleration can be accomplished with a Pitot-static tube (see Fig. 3.6). It is thus possible to measure, with only a small amount of uncertainty, values of stagnation pressure, p_0, and stagnation temperature, T_0, of a flowing fluid.

In Section 11.1, we demonstrated that for the isentropic flow of an ideal gas (see Eq. 11.25)

$$\frac{p}{\rho^k} = \text{constant} = \frac{p_0}{\rho_0^k}$$

The streamwise equation of motion for steady and frictionless flow (Eq. 11.41) can be expressed for an ideal gas as

$$\frac{dp}{\rho} + d\left(\frac{V^2}{2}\right) = 0 \tag{11.51}$$

since the potential energy term ($\gamma\, dz$) can be considered as being negligibly small in comparison with the other terms involved.

By incorporating Eq. 11.25 into Eq. 11.51 we obtain

$$\frac{p_0^{1/k}}{\rho_0} \frac{dp}{(p)^{1/k}} + d\left(\frac{V^2}{2}\right) = 0 \tag{11.52}$$

Consider the steady, one-dimensional, isentropic flow of an ideal gas with constant c_p and c_v through the converging–diverging nozzle of Fig. 11.6a. Equation 11.52 is valid for this flow and can be integrated between the common stagnation state of the flowing fluid to the state of the gas at any location in the converging–diverging duct to give

$$\frac{k}{k-1}\left(\frac{p_0}{\rho_0} - \frac{p}{\rho}\right) - \frac{V^2}{2} = 0 \tag{11.53}$$

By using the ideal gas equation of state (Eq. 11.1) with Eq. 11.53 we obtain

$$\frac{kR}{k-1}(T_0 - T) - \frac{V^2}{2} = 0 \tag{11.54}$$

It is of interest to note that combining Eqs. 11.14 and 11.54 leads to

$$c_p(T_0 - T) - \frac{V^2}{2} = 0$$

which, when merged with Eq. 11.9, results in

$$\check{h}_0 - \left(\check{h} + \frac{V^2}{2}\right) = 0 \tag{11.55}$$

where $\check{h}_0$ is the stagnation enthalpy. If the steady flow energy equation (Eq. 5.69) is applied to the flow situation we are presently considering, the resulting equation would be identical to Eq. 11.55. Further, we would conclude that the stagnation enthalpy is constant. The conservation of momentum and energy principles lead to the same equation (Eq. 11.55) for steady isentropic flows.

The definition of Mach number (Eq. 11.46) and the speed of sound relationship for ideal gases (Eq. 11.36) can be combined with Eq. 11.54 to yield

$$\frac{T}{T_0} = \frac{1}{1 + [(k-1)/2]\mathrm{Ma}^2} \tag{11.56}$$

With Eq. 11.56 we can calculate the temperature of an ideal gas anywhere in the converging–diverging duct of Fig. 11.6a if the flow is steady, one-dimensional, and isentropic, provided we know the value of the local Mach number and the stagnation temperature.

We can also develop an equation for pressure variation. Since $p/\rho = RT$, then

$$\left(\frac{p}{p_0}\right)\left(\frac{\rho_0}{\rho}\right) = \frac{T}{T_0} \tag{11.57}$$

From Eqs. 11.57 and 11.25 we obtain

$$\left(\frac{p}{p_0}\right) = \left(\frac{T}{T_0}\right)^{k/(k-1)}$$

(11.58)

Combining Eqs. 11.58 and 11.56 leads to

$$\frac{p}{p_0} = \left\{\frac{1}{1 + [(k-1)/2]\mathrm{Ma}^2}\right\}^{k/(k-1)}$$

(11.59)

For density variation we consolidate Eqs. 11.56, 11.57, and 11.59 to get

$$\frac{\rho}{\rho_0} = \left\{\frac{1}{1 + [(k-1)/2]\mathrm{Ma}^2}\right\}^{1/(k-1)}$$

(11.60)

A very useful means of keeping track of the states of an isentropic flow of an ideal gas involves a temperature–entropy $(T - s)$ diagram as is shown in Fig. 11.7. Experience has shown (see, for example, Refs. 2 and 3), that lines of constant pressure are generally as are sketched in Fig. 11.7. An isentropic flow is confined to a vertical line on a $T - s$ diagram. The vertical line in Fig. 11.7 is representative of flow between the stagnation state and any state within the converging–diverging nozzle. Equation 11.56 shows that fluid temperature decreases with an increase in Mach number. Thus, the lower temperature levels on a $T - s$ diagram correspond to higher Mach numbers. Equation 11.59 suggests that fluid pressure also decreases with an increase in Mach number. Thus, lower fluid temperatures and pressures are associated with higher Mach numbers in our isentropic converging–diverging duct example.

One way to produce flow through a converging–diverging duct like the one in Fig. 11.6a is to connect the downstream end of the duct to a vacuum pump. When the pressure at the downstream end of the duct (the back pressure) is decreased slightly, air will flow from the atmosphere through the duct and vacuum pump. Neglecting friction and heat transfer and considering the air to act as an ideal gas, Eqs. 11.56, 11.59, and 11.60 and a $T - s$ diagram can be used to describe steady flow through the converging–diverging duct.

If the pressure in the duct is only slightly less than atmospheric pressure, we predict with Eq. 11.59 that the Mach numbers levels in the duct will be low. Thus, with Eq. 11.60 we conclude that the variation of fluid density in the duct is also small. The continuity equation (Eq. 11.40) leads us to state that there is a small amount of fluid flow acceleration in the converging portion of the duct followed by flow deceleration in the diverging portion of the duct. We considered this type of flow when we discussed

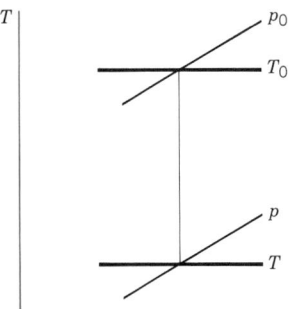

FIGURE 11.7 The T–s diagram relating stagnation and static states.

the Venturi meter in Section 3.6.3. The $T - s$ diagram for this flow is sketched in Fig. 11.8.

We next consider what happens when the back pressure is lowered further. Since the flow starts from rest upstream of the converging portion of the duct of Fig. 11.6a, Eqs. 11.48 and 11.50 reveal to us that flow up to the nozzle throat can be accelerated to a maximum allowable Mach number of 1 at the throat. Thus, when the duct back pressure is lowered sufficiently, the Mach number at the throat of the duct will be 1. Any further decrease of the back pressure will not affect the flow in the converging portion of the duct because, as is discussed in Section 11.3, information about pressure cannot move upstream when Ma = 1. When Ma = 1 at the throat of the converging–diverging duct, we have a condition called *choked flow*. Some useful equations for choked flow are developed below.

We have already used the stagnation state for which Ma = 0 as a reference condition. It will prove helpful to us to use the state associated with Ma = 1 and the same entropy level as the flowing fluid as another reference condition we shall call the *critical state*, denoted ( )*.

The ratio of pressure at the converging–diverging duct throat for choked flow, p^*, to stagnation pressure, p_0, is referred to as the *critical pressure ratio*. By substituting Ma = 1 into Eq. 11.59 we obtain

$$\frac{p^*}{p_0} = \left(\frac{2}{k + 1}\right)^{k/(k-1)} \tag{11.61}$$

For $k = 1.4$, the nominal value of k for air, Eq. 11.61 yields

$$\left(\frac{p^*}{p_0}\right)_{k=1.4} = 0.528 \tag{11.62}$$

Because the stagnation pressure for our converging–diverging duct example is the atmospheric pressure, p_{atm}, the throat pressure for choked air flow is, from Eq. 11.62

$$p^*_{k=1.4} = 0.528 p_{atm}$$

We can get a relationship for the critical temperature ratio, T^*/T_0, by substituting Ma = 1 into Eq. 11.56. Thus,

$$\frac{T^*}{T_0} = \frac{2}{k + 1} \tag{11.63}$$

or for $k = 1.4$

$$\left(\frac{T^*}{T_0}\right)_{k=1.4} = 0.833 \tag{11.64}$$

For the duct of Fig. 11.6a, Eq. 11.64 yields

$$T^*_{k=1.4} = 0.833 T_{atm}$$

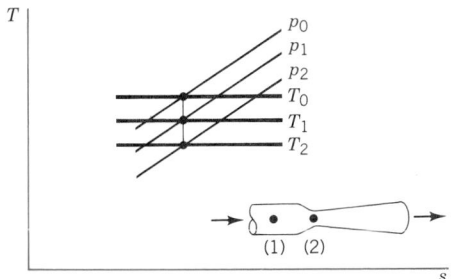

FIGURE 11.8   The T–s diagram for Venturi meter flow.

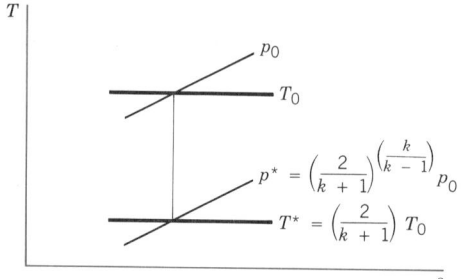

FIGURE 11.9 The relationship between the stagnation and critical states.

The stagnation and critical pressures and temperatures are shown on the $T - s$ diagram of Fig. 11.9.

When we combine the ideal gas equation of state (Eq. 11.1) with Eqs. 11.61 and 11.63, for Ma = 1 we get

$$\frac{\rho^*}{\rho_0} = \left(\frac{p^*}{T^*}\right)\left(\frac{T_0}{p_0}\right) = \left(\frac{2}{k+1}\right)^{k/(k-1)} \left(\frac{k+1}{2}\right) = \left(\frac{2}{k+1}\right)^{1/(k-1)} \qquad (11.65)$$

For air ($k = 1.4$), Eq. 11.65 leads to

$$\left(\frac{\rho^*}{\rho_0}\right)_{k=1.4} = 0.634 \qquad (11.66)$$

and we see that when the converging–diverging duct flow is choked, the density of the air at the duct throat is 63.4% of the density of atmospheric air.

EXAMPLE 11.5

A converging duct passes air steadily from standard atmospheric conditions to a receiver pipe as illustrated in Fig. E11.5a. The throat (minimum) flow cross section area of the converging

(a)

(b)

duct is 1×10^{-4} m². Determine the mass flowrate through the duct if the receiver pressure is (a) 80 kPa (abs); (b) 40 kPa (abs). Sketch temperature–entropy diagrams for situations (a) and (b).

FIGURE E11.5   (*Continued*)

SOLUTION

To determine the mass flowrate through the converging duct we use Eq. 11.40. Thus,

$$\dot{m} = \rho A V = \text{constant}$$

or in terms of the given throat area, A_{th},

$$\dot{m} = \rho_{th} A_{th} V_{th} \qquad (1)$$

We assume that the flow through the converging duct is isentropic and that the air behaves as an ideal gas with constant c_p and c_v. Then, from Eq. 11.60

$$\frac{\rho_{th}}{\rho_0} = \left\{ \frac{1}{1 + [(k-1)/2]\text{Ma}_{th}^2} \right\}^{1/(k-1)} \qquad (2)$$

The stagnation density, ρ_0, for standard atmosphere is 1.23 kg/m³ and the specific heat ratio is 1.4. To determine the throat Mach number, Ma_{th}, we can use Eq. 11.59,

$$\frac{p_{th}}{p_0} = \left\{ \frac{1}{1 + [(k-1)/2]\text{Ma}_{th}^2} \right\}^{k/(k-1)} \qquad (3)$$

The critical pressure, p^*, is obtained from Eq. 11.62 as

$$p^* = 0.528 p_0 = 0.528 p_{atm} = (0.528)[101 \text{ kPa(abs)}] = 53.3 \text{ kPa(abs)}$$

If the receiver pressure, p_{re}, is greater than or equal to p^*, then $p_{th} = p_{re}$. If $p_{re} < p^*$, then $p_{th} = p^*$ and the flow is choked. With p_{th}, p_0, and k known, Ma_{th} can be obtained from Eq. 3, and ρ_{th} can be determined from Eq. 2.

The flow velocity at the throat can be obtained from Eqs. 11.36 and 11.46 as

$$V_{th} = \text{Ma}_{th} c_{th} = \text{Ma}_{th} \sqrt{R T_{th} k} \qquad (4)$$

The value of temperature at the throat, T_{th}, can be calculated from Eq. 11.56,

$$\frac{T_{th}}{T_0} = \frac{1}{1 + [(k-1)/2]\text{Ma}_{th}^2} \qquad (5)$$

Since the flow through the converging duct is assumed to be isentropic, the stagnation

temperature is considered constant at the standard atmosphere value of $T_0 = 15$ K + 273 K = 288 K. Note that absolute pressures and temperatures are used.

(a) For $p_{re} = 80$ kPa(abs) > 53.3 kPa(abs) = p^*, we have $p_{th} = 80$ kPa(abs). Then from Eq. 3,

$$\frac{80 \text{ kPa(abs)}}{101 \text{ kPa(abs)}} = \left\{ \frac{1}{1 + [(1.4 - 1)/2]\text{Ma}_{th}^2} \right\}^{1.4/(1.4-1)}$$

or

$$\text{Ma}_{th} = 0.587$$

From Eq. 2,

$$\frac{\rho_{th}}{1.23 \text{ kg/m}^3} = \left\{ \frac{1}{1 + [(1.4 - 1)/2](0.587)^2} \right\}^{1/(1.4-1)}$$

or

$$\rho_{th} = 1.04 \text{ kg/m}^3$$

From Eq. 5,

$$\frac{T_{th}}{288 \text{ K}} = \frac{1}{1 + [(1.4 - 1)/2](0.587)^2}$$

or

$$T_{th} = 269 \text{ K}$$

Substituting $\text{Ma}_{th} = 0.587$ and $T_{th} = 269$ K into Eq. 4 we obtain

$$V_{th} = 0.587 \sqrt{[286.9 \text{ J/(kg·K)}](269 \text{ K})(1.4)[1(\text{kg·m})/(\text{N·s}^2)][1(\text{N·m})/\text{J}]}$$

or

$$V_{th} = 193 \text{ m/s}$$

Finally from Eq. 1 we have

$$\dot{m} = (1.04 \text{ kg/m}^3)(1 \times 10^{-4} \text{ m}^2)(193 \text{ m/s}) = 0.0201 \text{ kg/s} \qquad \text{(Ans)}$$

(b) For $p_{re} = 40$ kPa(abs) < 53.3 kPa(abs) = p^*, we have $p_{th} = p^* = 53.3$ kPa(abs) and $\text{Ma}_{th} = 1$. The converging duct is choked. From Eq. 2 (see also Eq. 11.66)

$$\frac{\rho_{th}}{1.23 \text{ kg/m}^3} = \left\{ \frac{1}{1 + 1.4 - 1/2(1)^2} \right\}^{1/(1.4-1)}$$

or

$$\rho_{th} = 0.780 \text{ kg/m}^3$$

From Eq. 5 (see also Eq. 11.64),

$$\frac{T_{th}}{288 \text{ K}} = \frac{1}{1 + [(1.4 - 1)/2](1)^2}$$

or

$$T_{th} = 240 \text{ K}$$

From Eq. 4,

$$V_{th} = (1) \sqrt{[286.9 \text{ J/(kg·K)}](240 \text{ K})(1.4)[1(\text{kg·m})/(\text{N·s}^2)][1(\text{N·m})/\text{J}]} = 310 \text{ m/s}$$

Finally from Eq. 1,

$$\dot{m} = (0.780 \text{ kg/m}^3)(1 \times 10^{-4} \text{ m}^2)(310 \text{ m/s}) = 0.0242 \text{ kg/s} \qquad \text{(Ans)}$$

From the values of throat temperature and throat pressure calculated above for flow situations (a) and (b) we can construct the temperature–entropy diagram shown in Fig. E11.5b.

Note that the flow from standard atmosphere to the receiver for receiver pressure (p_{re}) greater than or equal to the critical pressure (p^*) is isentropic. When the receiver pressure is less than the critical pressure as in situation (b) above, what is the flow like downstream from the exit of the converging duct? Experience suggests that this flow, when $p_{re} < p^*$, is three-dimensional and nonisentropic and involves a drop in pressure from p_{th} to p_{re}, a drop in temperature, and an increase of entropy as are indicated in Fig. E11.5c.

Isentropic flow Eqs. 11.56, 11.59, and 11.60 have been used to construct Table E.1 in Appendix E for air ($k = 1.4$). Examples 11.6 and 11.7 illustrate how these tabulated values of T/T_0, p/p_0, and ρ/ρ_0 as a function of Mach number, Ma, can be used to solve compressible flow problems.

EXAMPLE 11.6

Solve Example 11.5 using the values listed in Table E.1 of Appendix E.

SOLUTION

We still need the density and velocity of the air at the converging duct throat to solve for mass flowrate from

$$\dot{m} = \rho_{th} A_{th} V_{th} \tag{1}$$

(a) Since the receiver pressure, $p_{re} = 80$ kPa (abs), is greater than the critical pressure, $p^* = 53.3$ kPa (abs) the throat pressure, p_{th}, is equal to the receiver pressure. Thus

$$\frac{p_{th}}{p_0} = \frac{80 \text{ kPa(abs)}}{101 \text{ kPa(abs)}} = 0.792$$

From Table E.1, the closest entry for p/p_0 is 0.79012. We will use Table E.1 without interpolation to illustrate how the table can be used. For $p/p_0 = 0.790 \approx p_{th}/p_0$, we get from the table

$$\text{Ma}_{th} = 0.59$$

$$\frac{T_{th}}{T_0} = 0.935 \tag{2}$$

$$\frac{\rho_{th}}{\rho_0} = 0.8451 \tag{3}$$

Thus, from Eqs. 2 and 3,

$$T_{th} = (0.935)(288 \text{ K}) = 269 \text{ K}$$

and

$$\rho_{th} = (0.8451)(1.23 \text{ kg/m}^3) = 1.04 \text{ kg/m}^3$$

Furthermore, using Eqs. 11.36 and 11.46 we get

$$V_{th} = \text{Ma}_{th} \sqrt{R T_{th} k}$$
$$= (0.59) \sqrt{[286.9 \text{ J/(kg·K)}](269 \text{ K})(1.4)[1(\text{kg·m})/(\text{N·s}^2)][1(\text{N·m})/\text{J}]} = 194 \text{ m/s}$$

Finally, from Eq. 1

$$\dot{m} = (1.04 \text{ kg/m}^3)(1 \times 10^{-4} \text{ m}^2)(194 \text{ m/s}) = 0.0202 \text{ kg/s} \tag{Ans}$$

(b) For p_{re} = 40 kPa (abs) < 53.3 kPa (abs) = p^*, the throat pressure is equal to 53.3 kPa (abs) and the duct is choked with Ma_{th} = 1. From Table E.1, for Ma = 1, we get

$$\frac{T_{th}}{T_0} = 0.8333 \tag{4}$$

and

$$\frac{\rho_{th}}{\rho_0} = 0.6339 \tag{5}$$

From Eqs. 4 and 5 we obtain

$$T_{th} = (0.8333)(288 \text{ K}) = 240 \text{ K}$$

and

$$\rho_{th} = (0.6339)(1.23 \text{ kg/m}^3) = 0.780 \text{ kg/m}^3$$

Also, from Eqs. 11.36 and 11.46 we conclude that

$$V_{th} = Ma_{th} \sqrt{RT_{th}k}$$

$$= (1) \sqrt{[286.9 \text{ J/(kg·K)}](240 \text{ K})(1.4)[1(\text{kg·m})/(\text{N·s}^2)][1(\text{N·m})/\text{J}]} = 310 \text{ m/s}$$

Then, from Eq. 1,

$$m = (0.78 \text{ kg/m}^3)(1 \times 10^{-4} \text{ m}^2)(310 \text{ m/s}) = 0.0242 \text{ kg/s} \tag{Ans}$$

The values from Table E.1 resulted in answers for mass flowrate that are essentially the same as those using the ideal gas equations (see Example 11.5).

The temperature–entropy diagrams remain the same as those provided in the solution of Example 11.5.

EXAMPLE 11.7

The static pressure to stagnation pressure ratio at a point in a flow stream is measured with a Pitot-static tube (see Fig. 3.6) as being equal to 0.82. The stagnation temperature of the fluid is 68 °F. Determine the flow velocity if the fluid is (a) air; (b) helium.

SOLUTION

We consider both air and helium, flowing as described above, to act as ideal gases with constant specific heats. Then, we can use any of the ideal gas relationships developed in this chapter. To determine the flow velocity, we can combine Eqs. 11.36 and 11.46 to obtain

$$V = Ma \sqrt{RTk} \tag{1}$$

By knowing the value of static to stagnation pressure ratio, p/p_0, and specific heat ratio we can obtain the corresponding Mach number from Eq. 11.59, or for air, from Table E.1. Table E.1 cannot be used for helium, since k for helium is 1.66 and Table E.1 is for k = 1.4 only. With Mach number, specific heat ratio, and stagnation temperature known, the value of static temperature can be subsequently ascertained from Eq. 11.56 (or Table E.1 for air).

(a) For air, p/p_0 = 0.82; thus from Table E.1, without interpolation

$$Ma = 0.54 \tag{2}$$

and

$$\frac{T}{T_0} = 0.9449 \tag{3}$$

Then, from Eq. 3,

$$T = (0.9449)[(68 + 460) \, ^\circ R] = 499 \, ^\circ R \tag{4}$$

and using Eqs. 1, 2, and 4 we get

$$V = (0.54) \sqrt{[1.716 \times 10^3 \, (\text{ft·lb})/(\text{slug·}^\circ R)](499 \, ^\circ R)(1.4)[1 \, (\text{slug·ft})/(\text{lb·s}^2)]}$$

or

$$V = 591 \, \text{ft/s} \tag{Ans}$$

(b) For helium, $p/p_0 = 0.82$ and $k = 1.66$. By substituting these values into Eq. 11.59 we get

$$0.82 = \left\{ \frac{1}{1 + [(1.66 - 1)/2] \, \text{Ma}^2} \right\}^{1.66/(1.66-1)}$$

or

$$\text{Ma} = 0.499$$

From Eq. 11.56 we obtain

$$\frac{T}{T_0} = \frac{1}{1 + [(k - 1)/2]\text{Ma}^2}$$

Thus,

$$T = \left\{ \frac{1}{1 + [(1.66 - 1)/2](0.499)^2} \right\} [(68 + 460) \, ^\circ R] = 488 \, ^\circ R$$

From Eq. 1 we obtain

$$V = (0.499) \sqrt{[1.242 \times 10^4 (\text{ft·lb})/(\text{slug·}^\circ R)](488 ^\circ R)(1.66)[1(\text{slug·ft})/(\text{lb·s}^2)]}$$

or

$$V = 1580 \, \text{ft/s} \tag{Ans}$$

Note that the isentropic flow equations and Table E.1 for $k = 1.4$ were used presently to describe fluid particle isentropic flow along a pathline in a stagnation process. Even though these equations and table were developed for one-dimensional duct flows, they can be used for frictionless adiabatic pathline flows also.

Furthermore, while the Mach numbers calculated above are of similar size for the air and helium flows, the flow speed is much larger for helium than for air because the speed of sound in helium is much larger than it is in air.

Also included in Table E.1 are values of the ratio of local area, A, to critical area, A^*, for different values of local Mach number. The importance of this area ratio is clarified below.

For choked flow through the converging–diverging duct of Fig. 11.6a, the conservation of mass equation (Eq. 11.40) yields

$$\rho A V = \rho^* A^* V^*$$

or

$$\frac{A}{A^*} = \left(\frac{\rho^*}{\rho} \right) \left(\frac{V^*}{V} \right) \tag{11.67}$$

From Eqs. 11.36 and 11.46, we obtain

$$V^* = \sqrt{RT^*k} \tag{11.68}$$

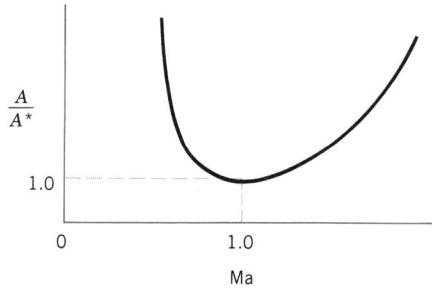

FIGURE 11.10 The variation of area ratio with Mach number for isentropic flow of an ideal gas ($k = 1.4$).

and

$$V = \text{Ma } \sqrt{RTk} \tag{11.69}$$

By combining Eqs. 11.67, 11.68, and 11.69 we get

$$\frac{A}{A^*} = \frac{1}{\text{Ma}} \left(\frac{\rho^*}{\rho_0}\right)\left(\frac{\rho_0}{\rho}\right) \sqrt{\frac{(T^*/T_0)}{(T/T_0)}} \tag{11.70}$$

The incorporation of Eqs. 11.56, 11.60, 11.63, 11.65, and 11.70 results in

$$\frac{A}{A^*} = \frac{1}{\text{Ma}} \left\{ \frac{1 + [(k-1)/2]\text{Ma}^2}{1 + [(k-1)/2]} \right\}^{(k+1)/[2(k-1)]} \tag{11.71}$$

Equation 11.71 was used to generate the values of A/A^* for air ($k = 1.4$) in Table E.1. These values of A/A^* are graphed as a function of Mach number in Fig. 11.10. As is demonstrated in the following examples, whether or not the critical area, A^*, is physically present in the flow, the area ratio, A/A^*, is still a useful concept for the isentropic flow of an ideal gas through a converging–diverging duct.

EXAMPLE 11.8

Air enters subsonically from standard atmosphere and flows isentropically through a choked converging–diverging duct having a circular cross section area, A, that varies with axial distance from the throat, x, according to the formula

$$A = 0.1 + x^2$$

where A is in square meters and x is in meters. The duct extends from $x = -0.5$ m to $x = +0.5$ m. For this flow situation, sketch the side view of the duct and graph the variation of Mach number, static temperature to stagnation temperature ratio, T/T_0, and static pressure to stagnation pressure ratio, p/p_0, through the duct from $x = -0.5$ m to $x = +0.5$ m. Also show the possible fluid states at $x = -0.5$ m, 0 m, and $+0.5$ m using temperature–entropy coordinates.

SOLUTION

The side view of the converging–diverging duct is a graph of radius from the duct axis, r, as a function of axial distance. For a circular flow cross section we have

$$A = \pi r^2 \tag{1}$$

where

$$A = 0.1 + x^2 \tag{2}$$

Thus, combining Eqs. 1 and 2, we have

$$r = \left(\frac{0.1 + x^2}{\pi}\right)^{1/2} \tag{3}$$

and a graph of radius as a function of axial distance can be easily constructed (see Fig. E11.8a).

Since the converging–diverging duct in this example is choked, the throat area is also the critical area (A^*). From Eq. 2 we see that

$$A^* = 0.1 \text{ m}^2 \tag{4}$$

For any axial location, from Eqs. 2 and 4 we get

$$\frac{A}{A^*} = \frac{0.1 + x^2}{0.1} \tag{5}$$

Values of A/A^* from Eq. 5 can be used in Eq. 11.71 to calculate corresponding values of Mach number, Ma. For air with $k = 1.4$, we could enter Table E.1 with values of A/A^* and read off values of Mach number. With values of Mach number ascertained, we could use Eqs. 11.56

(a)

(b)

(c)

FIGURE E11.8

(d)

FIGURE E11.8 (*Continued*)

and 11.59 to calculate related values of T/T_0 and p/p_0. For air with $k = 1.4$, Table E.1 could be entered with A/A^* or Ma to get values of T/T_0 and p/p_0. To solve this example, we elect to use the closest uninterpolated values from Table E.1.

The table below was constructed by using Eqs. 3 and 5 and Table E.1.

With the air entering the choked converging–diverging duct subsonically, only one isentropic solution exists for the converging portion of the duct. This solution involves an accelerating flow that becomes sonic (Ma = 1) at the throat of the passage. Two isentropic flow solutions are possible for the diverging portion of the duct, one subsonic, the other supersonic. If the pressure ratio, p/p_0, is set at 0.98003 at $x = +0.5$ m (the outlet), the subsonic flow will occur. Alternatively, if p/p_0 is set at 0.03685 at $x = +0.5$ m, the supersonic flow field will exist. These conditions are illustrated in Fig. 11.8. An unchoked subsonic flow through the converging–diverging duct of this example is discussed in Example 11.10. Choked flows involving other than the two isentropic flows in the diverging portion of the duct of this example are discussed after Example 11.10.

x (m)	From Eq. 3, r (m)	From Eq. 5, A/A^*	From Table E.1 A/A^*	Ma	T/T_0	p/p_0	State
			Subsonic solution				
-0.5	0.334	3.5	3.4635	0.17	0.99425	0.98003	a
-0.4	0.288	2.6	2.5968	0.23	0.98953	0.96383	
-0.3	0.246	1.9	1.9218	0.32	0.97993	0.93150	
-0.2	0.211	1.4	1.4018	0.47	0.95769	0.85958	
-0.1	0.187	1.1	1.1018	0.69	0.91306	0.72735	
0	0.178	1	1.0000	1.00	0.83333	0.52828	b
$+0.1$	0.187	1.1	1.1018	0.69	0.91306	0.72735	
$+0.2$	0.211	1.4	1.4018	0.47	0.95769	0.85958	
$+0.3$	0.246	1.9	1.9218	0.32	0.97993	0.93150	
$+0.4$	0.288	2.6	2.5968	0.23	0.98953	0.96883	
$+0.5$	0.344	3.5	3.4635	0.17	0.99425	0.98003	c
			Supersonic solution				
$+0.1$	0.187	1.1	1.0990	1.37	0.72707	0.32774	
$+0.2$	0.211	1.4	1.3967	1.76	0.61747	0.18499	
$+0.3$	0.246	1.9	1.9018	2.14	0.52194	0.10272	
$+0.4$	0.288	2.6	2.5880	2.48	0.44841	0.06038	
$+0.5$	0.334	3.5	3.5001	2.80	0.38941	0.03685	d

EXAMPLE 11.9

Air enters supersonically with T_0 and p_0 equal to standard atmosphere values and flows isentropically through the choked converging–diverging duct described in Example 11.8. Graph the variation of Mach number, Ma, static temperature to stagnation temperature ratio, T/T_0, and static pressure to stagnation pressure ratio, p/p_0, through the duct from $x = -0.5$ m to $x = +0.5$ m. Also show the possible fluid states at $x = -0.5$ m, 0 m, and $+0.5$ m by using temperature–entropy coordinates.

SOLUTION

With the air entering the converging–diverging duct of Example 11.8 supersonically instead of subsonically, a unique isentropic flow solution is obtained for the converging portion of the duct. Now, however, the flow decelerates to the sonic condition at the throat. The two solutions obtained previously in Example 11.8 for the diverging portion are still valid. Since the area variation in the duct is symmetrical with respect to the duct throat, we can use the supersonic flow values obtained from Example 11.8 in Table E.1 for the supersonic flow in the converging portion of the duct. The supersonic flow solution for the converging passage is summarized in the table below. The solution values for the entire duct are graphed in Fig E11.9.

| x (m) | A/A^* | From Table E.1 | | | | State |
		A/A^*	Ma	T/T_0	p/p_0	
-0.5	3.5	3.5001	2.80	0.38941	0.03685	e
-0.4	2.6	2.5880	2.48	0.44841	0.06038	
-0.3	1.9	1.9018	2.14	0.52194	0.10272	
-0.2	1.4	1.3967	1.76	0.61747	0.18499	
-0.1	1.1	1.0990	1.37	0.72707	0.32744	
0	1	1.0000	1.00	0.83333	0.52828	b

(a)

(b)

FIGURE E11.9

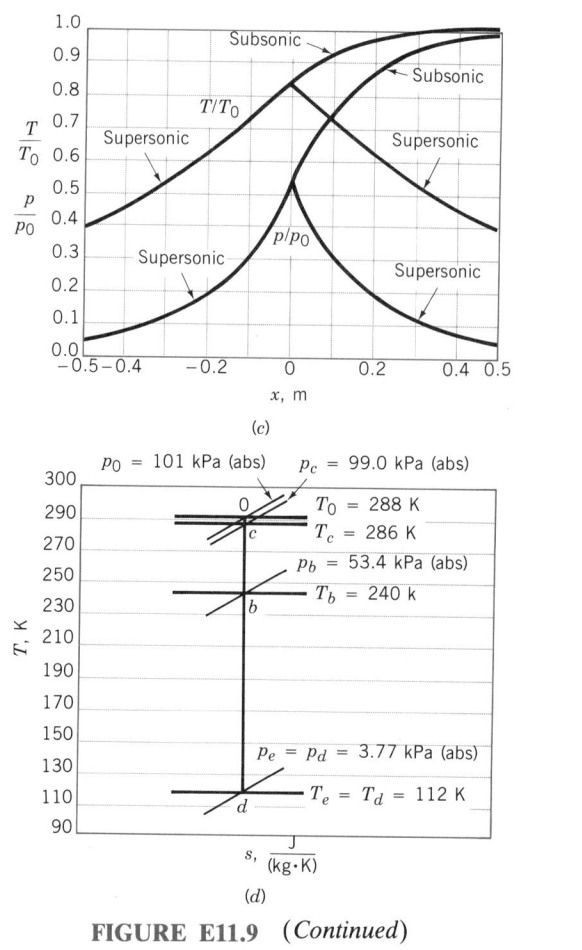

FIGURE E11.9   (*Continued*)

EXAMPLE 11.10

Air flows subsonically and isentropically through the converging–diverging duct of Example 11.8. Graph the variation of Mach number, Ma, static temperature to stagnation temperature ratio, T/T_0, and static pressure to stagnation pressure ratio, p/p_0, through the duct from $x = -0.5$ m to $x = +0.5$ m for $p/p_0 = 0.99$ at $x = +0.5$ m. Show the corresponding temperature–entropy diagram.

SOLUTION

Since for this example, $p/p_0 = 0.99$ at $x = +0.5$ m is greater than the corresponding value for choked flow ($p/p_0 = 0.98003$ at $x = +0.5$ m, see Example 11.8), the isentropic flow through the converging–diverging duct will be entirely subsonic and not choked. For air ($k = 1.4$) flowing isentropically through the duct, we can use Table E.1 for flow field quantities. Entering Table E.1 with $p/p_0 = 0.99$ we read off (without interpolation) the closest values $p/p_0 = 0.98998$, Ma $= 0.12$, $T/T_0 = 0.99714$, and $A/A^* = 4.8643$. Even though the duct flow is not choked in this example and A^* does not therefore exist physically, it still represents a valid reference. For a given isentropic flow, p_0, T_0, and A^* are constants. Since A at $x = +0.5$ m

is equal to 0.350 m² (from Eq. 2 of Example 11.8), A^* for this example is

$$A^* = \frac{A}{(A/A^*)} = \frac{0.35 \text{ m}^2}{4.8643} = 0.07195 \text{ m}^2 \tag{1}$$

With known values of duct area at different axial locations, we can calculate corresponding area ratios, A/A^*, knowing $A^* = 0.07195$ m². Having values of area ratio, we can enter Table E.1 and obtain related values of Ma, T/T_0, and p/p_0. The table below summarizes flow quantities obtained in this manner. The results are graphed in Fig. E11.10.

(a)

(b)

(c)

(d)

FIGURE E11.10

A more precise solution for the flow of this example could have been obtained with isentropic flow equations by following the steps outlined below.

1. Use Eq. 11.59 to get Ma at $x = +0.5$ m knowing k and p/p_0.
2. From Eq. 11.71, obtain value of A/A^* at $x = +0.5$ m having k and Ma.
3. Determine A^* knowing A and A/A^* at $x = +0.5$ m.
4. Determine A/A^* at different axial locations, x.
5. Use Eq. 11.71 and A/A^* from step 4 above to get values of Mach number at different axial locations.
6. Use Eqs. 11.56 and 11.59 and Ma from step 5 above to obtain T/T_0 and p/p_0 at different axial locations, x.

There are an infinite number of subsonic isentropic flow solutions for the converging–diverging duct considered in this example. For all of these flows, the pressure ratio, p/p_0, at $x = +0.5$ m must be larger than the pressure ratio there for choked flow, namely, 0.98003.

x (m)	Calculated, A/A^*	From Table E.1 A/A^*	Ma	T/T_0	p/p_0	State
−0.5	4.864	4.8643	0.12	0.99714	0.98998	a
−0.4	3.614	3.6727	0.16	0.99490	0.98228	
−0.3	2.641	2.5968	0.23	0.98953	0.96383	
−0.2	1.946	1.9218	0.32	0.97993	0.93150	
−0.1	1.529	1.5289	0.42	0.96592	0.88572	
0	1.390	1.3801	0.48	0.95595	0.85413	b
+0.1	1.529	1.5289	0.42	0.96592	0.88572	
+0.2	1.946	1.9218	0.32	0.97993	0.93150	
+0.3	2.641	2.5968	0.23	0.98953	0.96383	
+0.4	3.614	3.6727	0.16	0.99490	0.98228	
+0.5	4.864	4.8643	0.12	0.99714	0.98998	c

The isentropic flow behavior for the converging–diverging duct discussed in Examples 11.8, 11.9, and 11.10 is summarized on the area ratio–Mach number graphs sketched in Fig. 11.11. The points a, b, and c represent states at axial distance $x = -0.5$ m, 0 m, and $+0.5$ m. In Fig. 11.11a, the isentropic flow through the converging–diverging duct is subsonic without choking at the throat. This situation was discussed in Example 11.10. Figure 11.11b represents subsonic to subsonic choked flow (Example 11.8) and Fig. 11.11c is for subsonic to supersonic choked flow (also Example 11.8). The states in Fig. 11.11d are related to the supersonic to supersonic choked flow of Example 11.9; the states in Fig. 11.11e are for the supersonic to subsonic choked flow of Example 11.9. Not covered by an example but also possible are the isentropic flow states a, b, and c shown in Fig. 11.11f for supersonic to supersonic flow without choking. These six categories generally represent the possible kinds of isentropic, ideal gas flow through a converging–diverging duct.

For a given stagnation state (i.e., T_0 and p_0 fixed), ideal gas (k = constant), and converging–diverging duct geometry, an infinite number of isentropic subsonic to subsonic (not choked) and supersonic to supersonic (not choked) flow solutions exist. In contrast, the isentropic subsonic to supersonic (choked), subsonic to subsonic (choked), supersonic to subsonic (choked), and supersonic to supersonic (choked) flow solutions are each unique. The above-mentioned isentropic flow solutions are represented in Fig. 11.12. When the pressure at $x = +0.5$ (exit) is greater than or equal to p_I indicated in Fig. 11.12d, an isentropic flow is possible. When the pressure

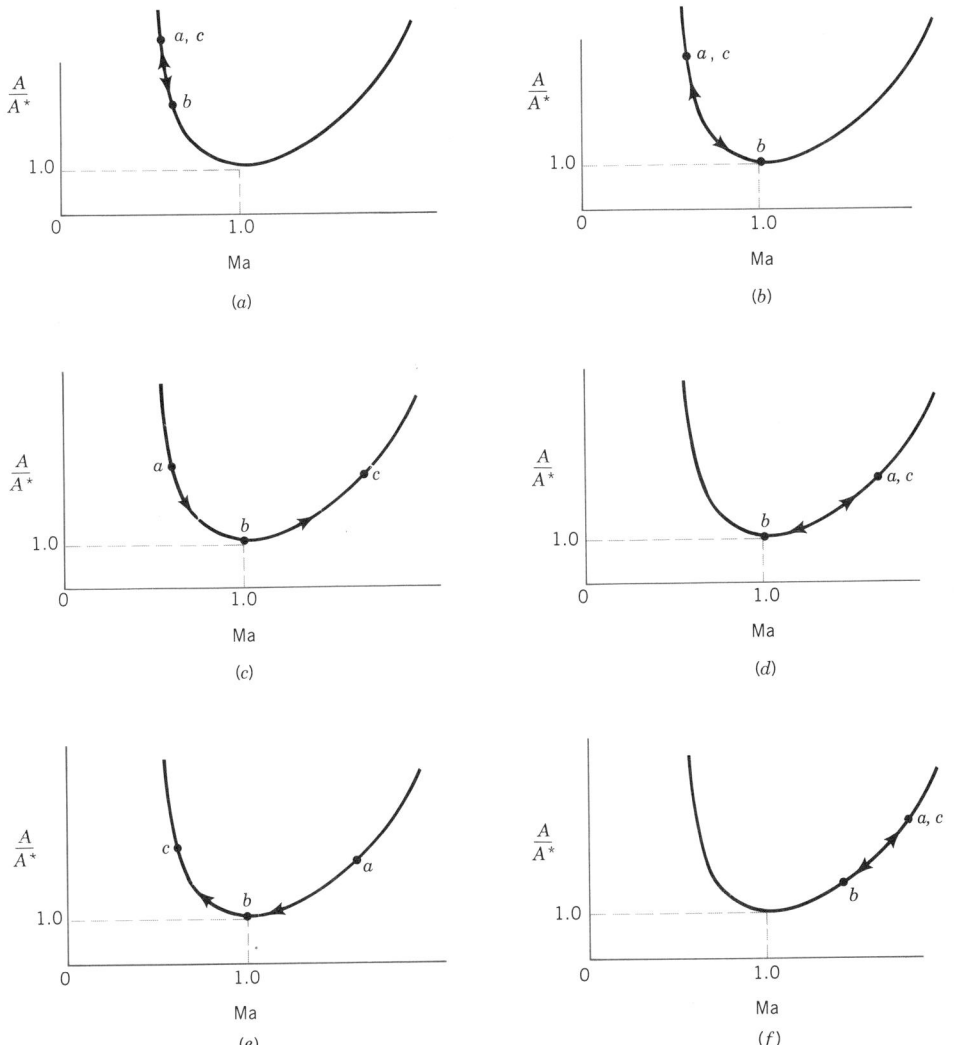

FIGURE 11.11 **(a)** Subsonic to subsonic isentropic flow (not choked). **(b)** Subsonic to subsonic isentropic flow (choked). **(c)** Subsonic to supersonic isentropic flow (choked). **(d)** Supersonic to supersonic isentropic flow (choked). **(e)** Supersonic to subsonic isentropic flow (choked). **(f)** Supersonic to supersonic isentropic flow (not choked).

at $x = +0.5$ is equal to or less than p_{II}, isentropic flows in the duct are possible. However, when the exit pressure is less than p_I and greater than p_{III} indicated in Fig. 11.13, isentropic flows are no longer possible in the duct. Determination of the value of P_{III} is discussed in Example 11.20.

Some possible nonisentropic choked flows through our converging–diverging duct are represented in Fig. 11.13. Each abrupt pressure rise shown within and at the exit of the flow passage occurs across a very thin discontinuity in the flow called a *normal shock wave*. Except for flow across the normal shock wave, the flow is isentropic. The nonisentropic flow equations that describe the changes in fluid properties that take place across a normal shock wave are developed in Section 11.5.3. The less abrupt pressure rise or drop that occurs after the flow leaves the duct is nonisentropic and attributable to three-dimensional *oblique shock waves*. If the pressure rises down-

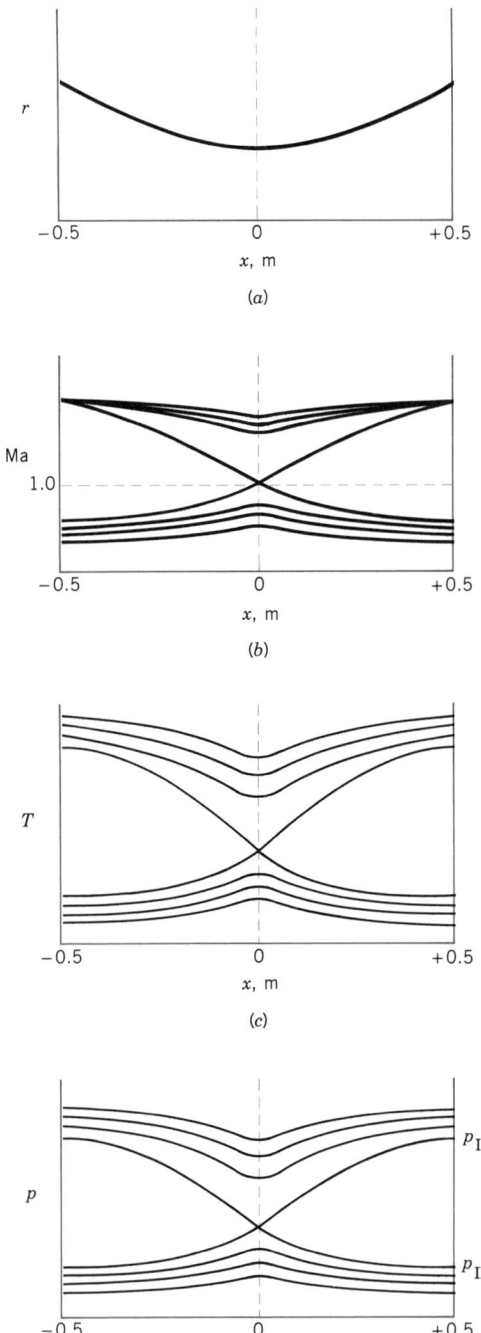

(a)

(b)

(c)

(d)

FIGURE 11.12 **(a)** The variation of duct radius with axial distance. **(b)** The variation of Mach number with axial distance. **(c)** The variation of temperature with axial distance. **(d)** The variation of pressure with axial distance.

stream of the duct exit, the flow is considered *overexpanded*. If the pressure drops downstream of the duct exit, the flow is called *underexpanded*. Further details about over- and underexpanded flows and oblique shock waves are beyond the scope of this text. Interested readers are referred to texts on compressible flows and gas dynamics, for example, Refs. 5, 6, 7, and 8, for additional material on this subject.

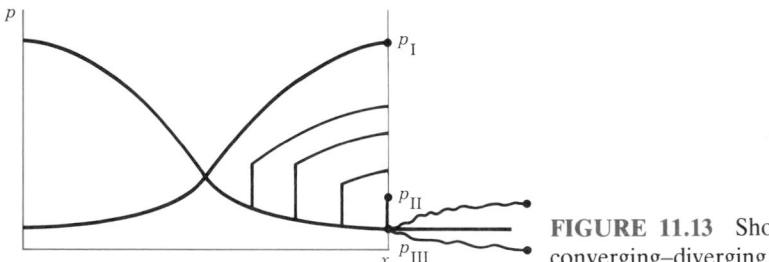

FIGURE 11.13 Shock formation in converging–diverging duct flows.

11.4.3 Constant Area Duct Flow

For steady, one-dimensional, isentropic flow of an ideal gas through a constant area duct (see Fig. 11.14), Eq. 11.50 suggests that $dV = 0$ or that flow velocity remains constant. With the energy equation (Eq. 5.69) we can conclude that since flow velocity is constant, the fluid enthalpy and thus temperature are also constant for this flow. This information and Eqs. 11.36 and 11.46 indicate that the Mach number is constant for this flow also. This being the case, Eqs. 11.59 and 11.60 tell us that fluid pressure and density also remains unchanged. Thus, we see that a steady, one-dimensional, isentropic flow of an ideal gas does not involve varying velocity or fluid properties unless the flow cross section area changes.

In Section 11.5 we discuss nonisentropic, steady, one-dimensional flows of an ideal gas through a constant area duct and also a normal shock wave. We learn that friction and/or heat transfer can also accelerate or decelerate a fluid.

11.5 NONISENTROPIC FLOW OF AN IDEAL GAS

Actual fluid flows are generally nonisentropic. An important example of nonisentropic flow involves adiabatic (no heat transfer) flow with friction. Flows with heat transfer (diabatic flows) are generally nonisentropic also. In this section we consider the adiabatic flow of an ideal gas through a constant cross section area duct with friction. This kind of flow is often referred to as *Fanno flow*. We also analyze the diabatic flow of an ideal gas through a constant cross section area duct without friction (*Rayleigh flow*). The concepts associated with Fanno and Rayleigh flows lead to further discussion of normal shock waves.

11.5.1 Adiabatic Constant Area Duct Flow with Friction (Fanno Flow)

Consider the steady, one-dimensional, and adiabatic flow of an ideal gas through the constant area duct shown in Fig. 11.15. This is Fanno flow. For the control volume indicated, the energy equation (Eq. 5.69) leads to

$$\dot{m} \left[\check{h}_2 - \check{h}_1 + \frac{V_2^2 - V_1^2}{2} + g(z_2 - z_1) \right] = \dot{Q}_{\substack{net \\ in}} + \dot{W}_{\substack{shaft \\ net \, in}}$$

where, above the terms: 0(negligibly small for gas flow) over $g(z_2 - z_1)$; 0(flow is adiabatic) over $\dot{Q}_{net \, in}$; 0(flow is steady throughout) over $\dot{W}_{shaft \, net \, in}$.

Constant area duct

Fluid flow →

FIGURE 11.14 Constant area duct flow.

FIGURE 11.15 Adiabatic constant area flow.

or

$$\check{h} + \frac{V^2}{2} = \check{h}_0 = \text{constant} \tag{11.72}$$

where h_0 is the stagnation enthalpy. For an ideal gas we gather from Eq. 11.9 that

$$\check{h} - \check{h}_0 = c_p(T - T_0) \tag{11.73}$$

so that by combining Eqs. 11.72 and 11.73 we get

$$T + \frac{V^2}{2c_p} = T_0 = \text{constant}$$

or

$$T + \frac{(\rho V)^2}{2c_p \rho^2} = T_0 = \text{constant} \tag{11.74}$$

By substituting the ideal gas equation of state (Eq. 11.1) into Eq. 11.74 we obtain

$$T + \frac{(\rho V)^2 T^2}{2c_p (p^2/R^2)} = T_0 = \text{constant} \tag{11.75}$$

From the continuity equation (Eq. 11.40) we can conclude that the density–velocity product, ρV, is constant for a given Fanno flow since the area, A, is constant. Also, for a particular Fanno flow, the stagnation temperature, T_0, is fixed. Thus, Eq. 11.75 allows us to calculate values of fluid temperature corresponding to values of fluid pressure in the Fanno flow. We postpone our discussion of how pressure is determined until later.

As with earlier discussions in this chapter, it is helpful to describe Fanno flow with a temperature–entropy diagram. From the second $T \, ds$ relationship, an expression for entropy variation was already derived (Eq. 11.22). If the temperature, T_1, pressure, p_1, and entropy, s_1, at the entrance of the Fanno flow duct are considered as reference values, then Eq. 11.22 yields

$$s - s_1 = c_p \ln \frac{T}{T_1} - R \ln \frac{p}{p_1} \tag{11.76}$$

Equations 11.75 and 11.76 taken together result in a curve with $T - s$ coordinates as is illustrated in Fig. 11.16 for a given gas (c_p and R) and fixed values of stagnation temperature, density–velocity product, and inlet temperature, pressure, and entropy. Curves like the one sketched in Fig. 11.16 are called Fanno lines.

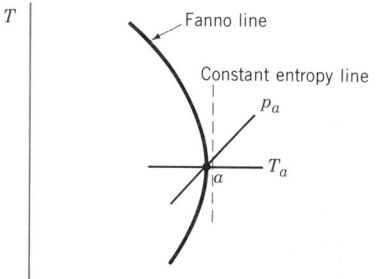

FIGURE 11.16 The T–s diagram for Fanno flow.

EXAMPLE 11.11

Air ($k = 1.4$) enters [section (1)] an insulated, constant flow cross section area duct with the following properties:

$$T_0 = 518.67 \text{ °R}$$

$$T_1 = 514.55 \text{ °R}$$

$$p_1 = 14.3 \text{ psia}$$

For Fanno flow, determine corresponding values of fluid temperature and entropy change for various values of downstream pressures and plot the related Fanno line.

SOLUTION

To plot the Fanno line we can use Eq. 11.75

$$T + \frac{(\rho V)^2 T^2}{2 c_p p^2 / R^2} = T_0 = \text{constant} \tag{1}$$

and Eq. 11.76

$$s - s_1 = c_p \ln \frac{T}{T_1} - R \ln \frac{p}{p_1} \tag{2}$$

to construct a table of values of temperature and entropy change corresponding to different levels of pressure in the Fanno flow.

We need values of the ideal gas constant and the specific heat at constant pressure to use in Eqs. 1 and 2. From Table 1.7 we read for air,

$$R = 1716 \text{ (ft·lb)/(slug·°R)}$$

From Eq. 11.14 we obtain

$$c_p = \frac{Rk}{k - 1} \tag{3}$$

or

$$c_p = \frac{[1716 \text{ (ft·lb)/(slug·°R)}](1.4)}{1.4 - 1} = 6006 \text{ (ft·lb)/(slug·°R)}$$

From Eqs. 11.1 and 11.69 we obtain

$$\rho V = \frac{p}{RT} \text{Ma} \sqrt{RTk}$$

and ρV is constant for this flow

$$\rho V = \rho_1 V_1 = \frac{p_1}{RT_1} \text{Ma}_1 \sqrt{RT_1 k} \tag{4}$$

But

$$\frac{T_1}{T_0} = \frac{514.55 \text{ °R}}{518.67 \text{ °R}} = 0.99206$$

and from Table E.1

$$\text{Ma}_1 = 0.2$$

Thus with Eq. 4

$$\rho V = \frac{(14.3 \text{ psia})(144 \text{ in.}^2/\text{ft}^2)(0.2)\sqrt{(1.4)[1716(\text{ft·lb})/(\text{slug·°R})](514.55 \text{ °R})[1(\text{slug·ft})/(\text{lb·s}^2)]}}{[1716(\text{ft·lb})/(\text{slug·°R})](514.55 \text{ °R})}$$

or

$$\rho V = 0.519 \text{ slug/(ft}^2\cdot\text{s)}$$

For $p = 7$ psia we have from Eq. 1

$$T + \frac{[0.519 \text{ slug/(ft}^2\cdot\text{s)}]^2 T^2}{2[6006 \text{ (ft}\cdot\text{lb)/(slug}\cdot\text{°R)}] \dfrac{(7 \text{ psia})^2 (144 \text{ in.}^2/\text{ft}^2)^2}{[1716 \text{ (ft}\cdot\text{lb)/(slug}\cdot\text{°R)}]^2}} = 518.67 \text{ °R}$$

or

$$6.5 \times 10^{-5} T^2 + T - 518.67 = 0$$

Thus,

$$T = 502.3 \text{ °R} \tag{Ans}$$

From Eq. 2, we obtain

$$s - s_1 = [6006 \text{ (ft}\cdot\text{lb)/(slug}\cdot\text{°R)}] \ln \left(\frac{502.3 \text{ °R}}{514.55 \text{ °R}} \right)$$

$$- [1716 \text{ (ft}\cdot\text{lb)/(slug}\cdot\text{°R)}] \ln \left(\frac{7 \text{ psia}}{14.3 \text{ psia}} \right)$$

or

$$s - s_1 = 1081 \text{ (ft}\cdot\text{lb)/(slug}\cdot\text{°R)} \tag{Ans}$$

Proceeding as outlined above, we construct the table of values shown below and graphed in Fig. E11.11. The maximum entropy difference occurs at a pressure of 2.62 psia and a temperature of 432.1 °R.

FIGURE E11.11

p (psia)	T (°R)	$s - s_1$ [(ft·lb)/(slug·°R)]
7	502.3	1081
6	496.8	1280
5	488.3	1489
4	474.0	1693
3	447.7	1844
2.62	432.1	1863
2	394.7	1783
1.8	378.1	1706
1.5	347.6	1513
1.4	335.6	1421

We can learn more about Fanno lines by further analyzing the equations that describe the physics involved. For example, the second $T\,ds$ equation (Eq. 11.18) is

$$T\,ds = d\breve{h} - \frac{dp}{\rho} \tag{11.18}$$

For an ideal gas,

$$d\breve{h} = c_p\,dT \tag{11.7}$$

and

$$p = \rho RT \tag{11.1}$$

or

$$\frac{dp}{p} = \frac{d\rho}{\rho} + \frac{dT}{T} \tag{11.77}$$

Thus, consolidating Eqs. 11.1, 11.7, 11.18, and 11.77 we obtain

$$T\,ds = c_p\,dT - RT\left(\frac{d\rho}{\rho} + \frac{dT}{T}\right) \tag{11.78}$$

Also, from the continuity equation (Eq. 11.40), we get for Fanno flow

$$\rho V = \text{constant}$$

or

$$\frac{d\rho}{\rho} = -\frac{dV}{V} \tag{11.79}$$

Substituting Eq. 11.79 into Eq. 11.78 yields

$$T\,ds = c_p\,dT - RT\left(-\frac{dV}{V} + \frac{dT}{T}\right)$$

or

$$\frac{ds}{dT} = \frac{c_p}{T} - R\left(-\frac{1}{V}\frac{dV}{dT} + \frac{1}{T}\right) \tag{11.80}$$

By differentiating the energy equation (11.74) obtained earlier, we obtain

$$\frac{dV}{dT} = -\frac{c_p}{V} \tag{11.81}$$

which, when substituted into Eq. 11.80, results in

$$\frac{ds}{dT} = \frac{c_p}{T} - R\left(\frac{c_p}{V^2} + \frac{1}{T}\right) \tag{11.82}$$

The Fanno line in Fig. 11.16 goes through a state (labeled state a) for which $ds/dT = 0$. At this state, we can conclude from Eqs. 11.14 and 11.82 that

$$V_a = \sqrt{RT_a k} \tag{11.83}$$

However, by comparing Eqs. 11.83 and 11.36 we see that the Mach number at state a is 1. Since the stagnation temperature is the same for all points on the Fanno line [see energy equation (Eq. 11.74)], the temperature at point a is the critical temperature, T^*, for the entire Fanno line. Thus, Fanno flow corresponding to the portion

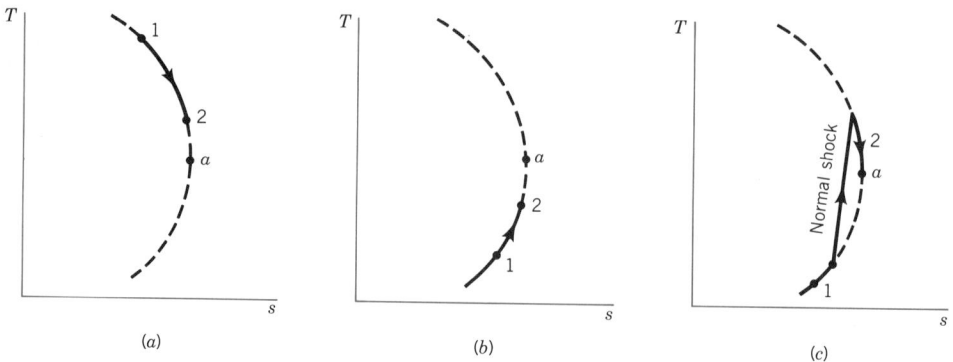

FIGURE 11.17 **(a)** Subsonic Fanno flow. **(b)** Supersonic Fanno flow. **(c)** Normal shock occurrence in Fanno flow.

of the Fanno line above the critical temperature must be subsonic, and Fanno flow on the line below T^* must be supersonic.

The second law of thermodyamics states that, based on all past experience, entropy can only remain constant or increase for adiabatic flows. For Fanno flow to be consistent with the second law of thermodynamics, flow can only proceed along the Fanno line toward state a, the critical state. The critical state may or may not be reached by the flow. If it is, the Fanno flow is *choked*. Some examples of Fanno flow behavior are summarized in Fig. 11.17. A case involving subsonic Fanno flow that is accelerated by friction to a higher Mach number without choking is illustrated in Fig. 11.17a. A supersonic flow that is decelerated by friction to a lower Mach number without choking is illustrated in Fig. 11.17b. In Fig. 11.17c, an abrupt change from supersonic to subsonic flow in the Fanno duct is represented. This sudden deceleration occurs across a standing *normal shock wave* that is described in more detail in Section 11.5.3.

A summary of the qualitative aspects of Fanno flow that we have already discussed is outlined in Table 11.1 and Fig. 11.18. To quantify Fanno flow behavior we need to combine a relationship that represents the linear momentum law with the set of equations already derived in this chapter.

If the linear momentum equation (Eq. 5.22) is applied to the Fanno flow through the control volume sketched in Fig. 11.19a, the result is

$$p_1 A_1 - p_2 A_2 - R_x = \dot{m}(V_2 - V_1)$$

where R_x is the frictional force exerted by the inner pipe wall on the fluid. Since

TABLE 11.1 Summary of Fanno Flow Behavior

Parameter	Flow	
	Subsonic Flow	Supersonic Flow
Stagnation temperature	Constant	Constant
Ma	Increases	Decreases
	(maximum is 1)	(miminum is 1)
Friction	Accelerates flow	Decelerates flow
Pressure	Decreases	Increases
Temperature	Decreases	Increases

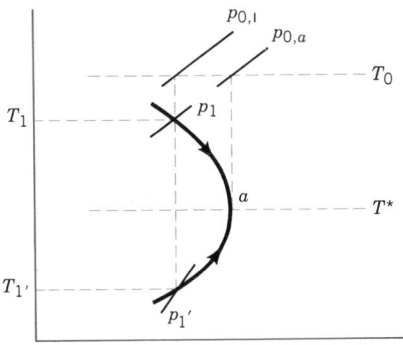

FIGURE 11.18   Fanno flow.

$A_1 = A_2 = A$ and $\dot{m} = \rho A V$ = constant

$$p_1 - p_2 - \frac{R_x}{A} = \rho V(V_2 - V_1) \tag{11.84}$$

The differential form of Eq. 11.84, which is valid for Fanno flow through the semi-infinitesimal control volume shown in Fig. 11.19b, is

$$-dp - \frac{\tau_w \pi D \, dx}{A} = \rho V \, dV \tag{11.85}$$

The wall shear stress, τ_w, is related to the wall friction factor, f, by Eq. 8.20 as

$$f = \frac{8\tau_w}{\rho V^2} \tag{11.86}$$

By substituting Eq. 11.86 and $A = \pi D^2/4$ into Eq. 11.85, we obtain

$$-dp - f\rho \frac{V^2}{2} \frac{dx}{D} = \rho V \, dV \tag{11.87}$$

or

$$\frac{dp}{p} + \frac{f}{p} \frac{\rho V^2}{2} \frac{dx}{D} + \frac{\rho}{p} \frac{d(V^2)}{2} = 0 \tag{11.88}$$

(a)

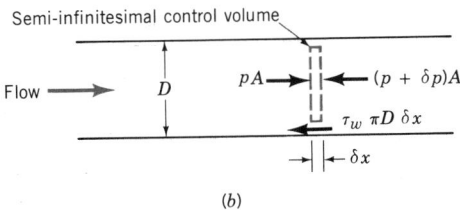

(b)

FIGURE 11.19   **(a)** Finite control volume. **(b)** Semi-infinitesimal control volume.

Combining the ideal gas equation of state (Eq. 11.1), the ideal gas speed of sound equation (Eq. 11.36), and the Mach number definition (Eq. 11.46) with Eq. 11.88 leads to

$$\frac{dp}{p} + \frac{fk}{2} \text{Ma}^2 \frac{dx}{D} + k \frac{\text{Ma}^2}{2} \frac{d(V^2)}{V^2} = 0 \tag{11.89}$$

Since $V = \text{Ma } c = \text{Ma}\sqrt{RTk}$, then

$$V^2 = \text{Ma}^2 RTk$$

or

$$\frac{d(V^2)}{V^2} = \frac{d(\text{Ma}^2)}{\text{Ma}^2} + \frac{dT}{T} \tag{11.90}$$

The application of the energy equation (Eq. 5.69) to Fanno flow gave Eq. 11.74. If Eq. 11.74 is differentiated and divided by temperature, the result is

$$\frac{dT}{T} + \frac{d(V^2)}{2c_p T} = 0 \tag{11.91}$$

Substituting Eqs. 11.14, 11.36, and 11.46 into Eq. 11.91 yields

$$\frac{dT}{T} + \frac{k-1}{2} \text{Ma}^2 \frac{d(V^2)}{V^2} = 0 \tag{11.92}$$

which can be combined with Eq. 11.90 to form

$$\frac{d(V^2)}{V^2} = \frac{d(\text{Ma}^2)/\text{Ma}^2}{1 + [(k-1)/2]\text{Ma}^2} \tag{11.93}$$

We can merge Eqs. 11.77, 11.79, and 11.90 to get

$$\frac{dp}{p} = \frac{1}{2} \frac{d(V^2)}{V^2} - \frac{d(\text{Ma}^2)}{\text{Ma}^2} \tag{11.94}$$

Consolidating Eqs. 11.94 and 11.89 leads to

$$\frac{1}{2}(1 + k\text{Ma}^2) \frac{d(V^2)}{V^2} - \frac{d(\text{Ma}^2)}{\text{Ma}^2} + \frac{fk}{2} \text{Ma}^2 \frac{dx}{D} = 0 \tag{11.95}$$

Finally, incorporating Eq. 11.93 into Eq. 11.95 yields

$$\frac{(1 - \text{Ma}^2)\, d(\text{Ma}^2)}{\{1 + [(k-1)/2]\text{Ma}^2\}k\text{Ma}^4} = f\frac{dx}{D} \tag{11.96}$$

Equation 11.96 can be integrated from one section to another in a Fanno flow duct. We elect to use the critical (*) state as a reference and to integrate Eq. 11.96 from an upstream state to the critical state. Thus

$$\int_{\text{Ma}}^{\text{Ma}^*=1} \frac{(1 - \text{Ma}^2)\, d(\text{Ma}^2)}{\{1 + [(k-1)/2]\,\text{Ma}^2\}k\text{Ma}^4} = \int_{\ell}^{\ell^*} f\frac{dx}{D} \tag{11.97}$$

where ℓ is length measured from an arbitrary but fixed upstream reference location to a section in the Fanno flow. For an approximate solution, we can assume that the friction factor is constant at an average value over the integration length, $\ell^* - \ell$. We

also consider a constant value of k. Thus, we obtain from Eq. 11.97

$$\frac{1}{k}\frac{(1 - \text{Ma}^2)}{\text{Ma}^2} + \frac{k + 1}{2k}\ln\left\{\frac{[(k + 1)/2]\text{Ma}^2}{1 + [(k - 1)/2]\text{Ma}^2}\right\} = \frac{f(\ell^* - \ell)}{D} \qquad (11.98)$$

For a given gas, values of $f(\ell^* - \ell)/D$ can be tabulated as a function of Mach number for Fanno flow. For example, values of $f(\ell^* - \ell)/D$ for air ($k = 1.4$) Fanno flow are listed as a function of Mach number in Table E.2 in Appendix E. Note that the critical state does not have to exist in the actual Fanno flow being considered, since for any two sections in a given Fanno flow

$$\frac{f(\ell^* - \ell_2)}{D} - \frac{f(\ell^* - \ell_1)}{D} = \frac{f}{D}(\ell_1 - \ell_2) \qquad (11.99)$$

The sketch in Fig. 11.20 illustrates the physical meaning of Eq. 11.99.

For a given Fanno flow (constant specific heat ratio, duct diameter, and friction factor) the length of duct required to change the Mach number from Ma_1 to Ma_2 can be determined from Eqs. 11.98 and 11.99 or a table such as Table E.2. To get the values of other fluid properties in the Fanno flow field we need to develop more equations.

By consolidating Eqs. 11.90 and 11.92 we obtain

$$\frac{dT}{T} = -\frac{(k - 1)}{2\{1 + [(k - 1)/2]\text{Ma}^2\}}d(\text{Ma}^2) \qquad (11.100)$$

Integrating Eq. 11.100 from any state upstream in a Fanno flow to the critical (*)

FIGURE 11.20 (a) Unchoked Fanno flow. **(b)** Choked Fanno flow.

(a)

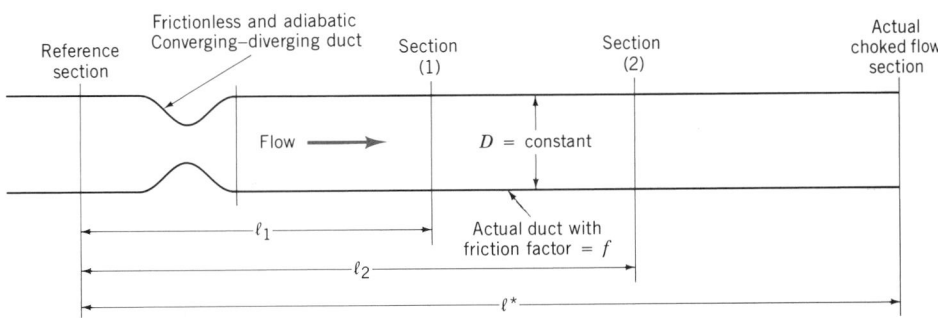

(b)

state leads to

$$\frac{T}{T^*} = \frac{(k + 1)/2}{1 + [(k - 1)/2]\text{Ma}^2} \tag{11.101}$$

Equations 11.68 and 11.69 allow us to write

$$\frac{V}{V^*} = \frac{\text{Ma}\sqrt{RTk}}{\sqrt{RT^*k}} = \text{Ma}\sqrt{\frac{T}{T^*}} \tag{11.102}$$

Substituting Eq. 11.101 into Eq. 11.102 yields

$$\frac{V}{V^*} = \left\{\frac{[(k + 1)/2]\text{Ma}^2}{1 + [(k - 1)/2]\text{Ma}^2}\right\}^{1/2} \tag{11.103}$$

From the continuity equation (Eq. 11.40) we get for Fanno flow

$$\frac{\rho}{\rho^*} = \frac{V^*}{V} \tag{11.104}$$

Combining Eqs. 11.104 and 11.103 results in

$$\frac{\rho}{\rho^*} = \left\{\frac{1 + [(k - 1)/2]\text{Ma}^2}{[(k + 1)/2]\text{Ma}^2}\right\}^{1/2} \tag{11.105}$$

The ideal gas equation of state (Eq. 11.1) leads to

$$\frac{p}{p^*} = \frac{\rho}{\rho^*}\frac{T}{T^*} \tag{11.106}$$

and merging Eqs. 11.106, 11.105, and 11.101 gives

$$\frac{p}{p^*} = \frac{1}{\text{Ma}}\left\{\frac{(k + 1)/2}{1 + [(k - 1)/2]\text{Ma}^2}\right\}^{1/2} \tag{11.107}$$

Finally, the stagnation pressure ratio can be written as

$$\frac{p_0}{p_0^*} = \left(\frac{p_0}{p}\right)\left(\frac{p}{p^*}\right)\left(\frac{p^*}{p_0^*}\right) \tag{11.108}$$

which by use of Eqs. 11.59 and 11.107 yields

$$\frac{p_0}{p_0^*} = \frac{1}{\text{Ma}}\left[\left(\frac{2}{k + 1}\right)\left(1 + \frac{k - 1}{2}\text{Ma}^2\right)\right]^{[(k+1)/2(k-1)]} \tag{11.109}$$

Values of $f(\ell^* - \ell)/D$, T/T^*, V/V^*, p/p^*, and p_0/p_0^* for Fanno flow of air ($k = 1.4$) are tabulated as a function of Mach number (using Eqs. 11.99, 11.101, 11.103, 11.107, and 11.109) in Table E.2 of Appendix E. The usefulness of Table E.2 is illustrated in Examples 11.12, 11.13, and 11.14.

EXAMPLE 11.12

Standard atmospheric air [$T_0 = 288$ K, $p_0 = 101$ kPa (abs)] is drawn steadily through a frictionless adiabatic converging nozzle into an adiabatic, constant cross section area duct as shown in Fig. E11.12a. The duct is 2 m long and has an inside diameter of 0.1 m. The average friction factor for the duct is estimated as being equal to 0.02. What is the maximum mass flowrate through the duct? For this maximum flowrate determine the values of static temper-

ature, static pressure, stagnation temperature, stagnation pressure, and velocity at the inlet [section (1)] and exit [section (2)] of the constant area duct. Sketch a temperature–entropy diagram for this flow.

(a)

(b)

FIGURE E11.12

SOLUTION

We consider the flow through the converging nozzle to be isentropic, and the flow through the constant area duct to be Fanno flow. A decrease in the pressure at the exit of the constant area duct (back pressure) causes the mass flowrate through the nozzle and the duct to increase. The flow throughout is subsonic. The maximum flowrate will occur when the back pressure is lowered to the extent that the constant area duct chokes and the Mach number at the duct exit is equal to 1. Any further decrease of back pressure will not affect the flowrate through the nozzle–duct combination.

For the maximum flowrate condition, the constant area duct must be choked, and

$$\frac{f(\ell^* - \ell_1)}{D} = \frac{f(\ell_2 - \ell_1)}{D} = \frac{(0.02)(2 \text{ m})}{(0.1 \text{ m})} = 0.4 \tag{1}$$

With $k = 1.4$ for air and the above calculated value of $f(\ell_1 - \ell^*)/D = 0.4$, we could use Eq. 11.98 to determine a value of Mach number at the entrance of the duct [section (1)]. With $k = 1.4$ and Ma_1 known, we could then rely on Eqs. 11.101, 11.103, 11.107, and 11.109 to obtain values of T_1/T^*, V_1/V^*, p_1/p^*, and $p_{0,1}/p_0^*$. Alternatively, for air ($k = 1.4$), we can enter Table E.2 with $f(\ell^* - \ell_1)/D = 0.4$ and read off values of Ma_1, T_1/T^*, V_1/V^*, p_1/p^*, and $p_{0,1}/p_0^*$. We simplify this example by using the closest values Table E.2 without interpolation.

The pipe entrance Mach number, Ma_1, also represents the Mach number at the throat (and exit) of the isentropic, converging nozzle. Thus, the isentropic flow equations of Section 11.4 or Table E.1 can be used with Ma_1. We use Table E.1 in this example.

With Ma_1 known, we can enter Table E.1 and get values of T_1/T_0, p_1/p_0, and ρ_1/ρ_0. Through the isentropic nozzle, T_0, p_0, and ρ_0 are each constant, and thus T_1, p_1, and ρ_1 can be readily obtained.

Since T_0 also remains constant through the constant area duct (see Eq. 11.75), we can use Eq. 11.63 to get T^*. Thus,

$$\frac{T^*}{T_0} = \frac{2}{k+1} = \frac{2}{1.4+1} = 0.8333 \tag{2}$$

Since $T_0 = 288$ K, we get from Eq. 2,

$$T^* = (0.8333)(288 \text{ K}) = 240 \text{ K} = T_2 \tag{3}\quad\text{(Ans)}$$

With T^* known, we can calculate V^* from Eq. 11.36 as

$$V^* = \sqrt{RT^*k}$$

$$= \sqrt{[(286.9 \text{ J/(kg·K)}](240 \text{ K})(1.4)[1(\text{kg·m})/(\text{N·s}^2)][1(\text{N·m})/\text{J}]} = 310 \text{ m/s} = V_2 \tag{4}\quad\text{(Ans)}$$

Now V_1 can be obtained from V^* and V_1/V^*. Having A_1, ρ_1, and V_1 we can get the mass flowrate from

$$\dot{m} = \rho_1 A_1 V_1 \tag{5}$$

Values of the other variables asked for can be obtained from the ratios mentioned above.
Entering Table E.2 with $f(\ell^* - \ell)/D = 0.4$ we read closest values

$$\frac{f(\ell^* - \ell_1)}{D} = 0.38411 \tag{6}$$

$$Ma_1 = 0.63 \tag{7}$$

$$\frac{T_1}{T^*} = 1.1118 \tag{8}$$

$$\frac{V_1}{V^*} = 0.66427 \tag{9}$$

$$\frac{p_1}{p^*} = 1.6737 \tag{10}$$

$$\frac{p_{0.1}}{p_0^*} = 1.1551 \tag{11}$$

Entering Table E.1 with $Ma_1 = 0.63$ we read

$$\frac{T_1}{T_0} = 0.92646 \tag{12}$$

$$\frac{p_1}{p_{0.1}} = 0.76540 \tag{13}$$

$$\frac{\rho_1}{\rho_{0.1}} = 0.82616 \tag{14}$$

Thus, from Eqs. 4 and 9 we obtain

$$V_1 = (0.66427)(310 \text{ m/s}) = 206 \text{ m/s} \quad\text{(Ans)}$$

From Eq. 14 we get

$$\rho_1 = 0.82616\rho_0 = (0.82616)(1.23 \text{ kg/m}^3) = 1.016 \text{ kg/m}^3$$

and from Eq. 5 we conclude that

$$\dot{m} = (1.016 \text{ kg/m}^3) \left[\frac{\pi (0.1 \text{ m})^2}{4} \right] (206 \text{ m/s}) = 1.64 \text{ kg/s} \qquad \text{(Ans)}$$

From Eq. 12, it follows that

$$T_1 = (0.92646)(288 \text{ K}) = 267 \text{ K} \qquad \text{(Ans)}$$

Equation 13 yields

$$p_1 = (0.76540)[101 \text{ kPa (abs)}] = 77.3 \text{ kPa (abs)} \qquad \text{(Ans)}$$

The stagnation temperature, T_0, remains constant through this adiabatic flow at a value of

$$T_{0,1} = T_{0,2} = 288 \text{ K} \qquad \text{(Ans)}$$

The stagnation pressure, p_0, at the entrance of the constant area duct is the same as the constant value of stagnation pressure through the isentropic nozzle. Thus

$$p_{0,1} = 101 \text{ kPa (abs)} \qquad \text{(Ans)}$$

To obtain the duct exit pressure ($p_2 = p^*$) we can use Eqs. 10 and 13. Thus,

$$p_2 = \left(\frac{p^*}{p_1} \right)\left(\frac{p_1}{p_{0,1}} \right) (p_{0,1}) = \left(\frac{1}{1.6737} \right) (0.76540)[101 \text{ kPa (abs)}] = 46.2 \text{ kPa (abs)} \qquad \text{(Ans)}$$

For the duct exit stagnation pressure ($p_{0,2} = p_0^*$) we can use Eq. 11 as

$$p_{0,2} = \left(\frac{p_0^*}{p_{0,1}} \right) (p_{0,1}) = \left(\frac{1}{1.1551} \right) [101 \text{ kPa (abs)}] = 87.4 \text{ kPa (abs)} \qquad \text{(Ans)}$$

The stagnation pressure, p_0, decreases in a Fanno flow because of friction.

Use of tables such as Tables E.1 and E.2 greatly simplifies the solution of a problem involving Fanno flow. The $T - s$ diagram for this flow is shown in Fig. E.11.12b, where the entropy difference, $s_2 - s_1$, is obtained from Eq. 11.22.

EXAMPLE 11.13

The duct in Example 11.12 is shortened by 50%, but the duct discharge pressure is maintained at the choked flow value for Example 11.12, namely,

$$p_d = 46.2 \text{ kPa (abs)}$$

Will shortening the duct cause the mass flowrate through the duct to increase or decrease? Assume that the average friction factor for the duct remains constant at a value of $f = 0.02$.

SOLUTION

We guess that the shortened duct will still choke and check our assumption by comparing p_d with p^*. If $p_d < p^*$, the flow is choked; if not, another assumption has to be made. For choked flow we can calculate the mass flowrate just as we did for Example 11.12. For unchoked flow, we will have to devise another strategy.

For choked flow,

$$\frac{f(\ell^* - \ell_1)}{D} = \frac{(0.02)(1 \text{ m})}{0.1 \text{ m}} = 0.2$$

and from Table E.2, we read the closest values

$$\frac{f(\ell^* - \ell_1)}{D} = 0.20814$$

$$Ma_1 = 0.70$$

$$\frac{p_1}{p^*} = 1.4934$$

With $Ma_1 = 0.70$, we enter Table E.1 and get

$$\frac{p_1}{p_0} = 0.72092$$

Now the duct exit pressure ($p_2 = p^*$) can be obtained from

$$p_2 = p^* = \left(\frac{p^*}{p_1}\right)\left(\frac{p_1}{p_{0.1}}\right)(p_{0.1})$$

$$= \left(\frac{1}{1.4934}\right)(0.72092)[101 \text{ kPa (abs)}] = 48.8 \text{ kPa (abs)}$$

and we see that $p_d < p^*$. Our assumption of choked flow is justified. The pressure at the exit plane is greater than the surrounding pressure outside the duct exit. The final drop of pressure from 48.8 kPa (abs) to 46.2 kPa (abs) involves complicated three-dimensional flow downstream of the exit.

To determine the mass flowrate we use

$$\dot{m} = \rho_1 A_1 V_1 \tag{1}$$

The density at section (1) is obtained from

$$\frac{\rho_1}{\rho_{0.1}} = 0.79158 \tag{2}$$

which is read in Table E.1 for $Ma_1 = 0.7$. Thus

$$\rho_1 = (0.79158)(1.23 \text{ kg/m}^3) = 0.974 \text{ kg/m}^3 \tag{3}$$

We get V_1 from

$$\frac{V_1}{V^*} = 0.73179 \tag{4}$$

out of Table E.2 for $Ma_1 = 0.7$. The value of V^* is the same as it was in Example 11.12, namely,

$$V^* = 310 \text{ m/s} \tag{5}$$

Thus, from Eqs. 4 and 5 we obtain

$$V_1 = (0.73179)(310) = 227 \text{ m/s} \tag{6}$$

and from Eqs. 1, 3, and 6 we get

$$m = (0.974 \text{ kg/m}^3)\left[\frac{\pi(0.1)^2}{4}\right](227 \text{ m/s}) = 1.74 \text{ kg/s} \tag{Ans}$$

The mass flowrate associated with a shortened tube is larger than the mass flowrate for the longer tube. This trend is general for subsonic Fanno flow. For the same upstream stagnation state and downstream pressure, the mass flowrate for the Fanno flow will decrease with increase in length of duct for subsonic flow. Equivalently, if the length of the duct remains the same but the wall friction is increased, the mass flowrate will decrease.

EXAMPLE 11.14

If the same flowrate obtained in Example 11.12 ($\dot{m} = 1.64$ kg/s) is desired through the shortened duct of Example 11.13 ($\ell_2 - \ell_1 = 1$ m), determine the Mach number at the exit of the duct, M_2, and the back pressure, p_2, required. Assume f remains constant at a value of 0.02.

SOLUTION

Since the mass flowrate of Example 11.12 is desired, the Mach number and other properties at the entrance of the constant area duct remain at the values determined in Example 11.12. Thus, from Example 11.12

$$\text{Ma}_1 = 0.63$$

and

$$\frac{f(\ell^* - \ell_1)}{D} = 0.4 \ (\approx 0.38411, \text{ the closest value in Table E.2})$$

For this example,

$$\frac{f(\ell_2 - \ell_1)}{D} = \frac{f(\ell^* - \ell_1)}{D} - \frac{f(\ell^* - \ell_2)}{D}$$

or

$$\frac{(0.02)(1 \text{ m})}{(0.1 \text{ m})} = 0.4 - \frac{f(\ell^* - \ell_2)}{D}$$

so that

$$\frac{f(\ell^* - \ell_2)}{D} = 0.2 \tag{1}$$

By using the value from Eq. 1 and Table E.2, we get for the closest values

$$\frac{f(\ell^* - \ell_2)}{D} = 0.20814$$

$$\text{Ma}_2 = 0.70 \tag{Ans}$$

and

$$\frac{p_2}{p^*} = 1.4934 \tag{2}$$

We obtain p_2 from

$$p_2 = \left(\frac{p_2}{p^*}\right)\left(\frac{p^*}{p_1}\right)\left(\frac{p_1}{p_{0,1}}\right)(p_{0,1})$$

where p_2/p^* is given in Eq. 2 and p^*/p_1, $p_1/p_{0,1}$, and $p_{0,1}$ are the same as they were in Example 11.12. Thus,

$$p_2 = (1.4934)\left(\frac{1}{1.6737}\right)(0.76540)[101 \text{ kPa (abs)}] = 69.0 \text{ kPa (abs)} \tag{Ans}$$

A larger back pressure [69.0 kPa (abs)] than the one associated with choked flow through a Fanno duct [46.2 kPa (abs)] will maintain the same flowrate through a shorter Fanno duct with the same friction coefficient. The flow through the shorter duct is not choked. It would not be possible to maintain the same flowrate through a Fanno duct longer than the choked one with the same friction coefficient, regardless of what back pressure is used.

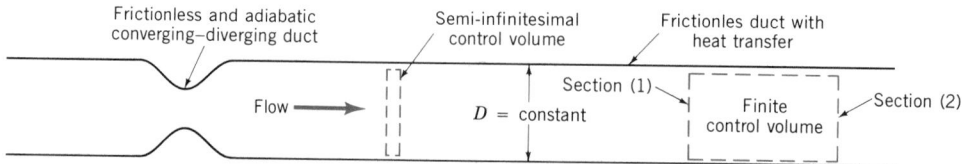

FIGURE 11.21 Rayleigh flow.

11.5.2 Frictionless Constant Area Duct Flow with Heat Transfer (Rayleigh Flow)

Consider the steady, one-dimensional, and frictionless flow of an ideal gas through the constant area duct with heat transfer illustrated in Fig. 11.21. This is *Rayleigh flow*. Application of the linear momentum equation (Eq. 5.22) to the Rayleigh flow through the finite control volume sketched in Fig. 11.21 results in

$$p_1 A_1 + \dot{m} V_1 = p_2 A_2 + \dot{m} V_2 + \overset{0(\text{frictionless flow})}{\cancel{R_x}}$$

or

$$p + \frac{(\rho V)^2}{\rho} = \text{constant} \tag{11.110}$$

Use of the ideal gas equation of state (Eq. 11.1) in Eq. 11.110 leads to

$$p + \frac{(\rho V)^2 RT}{p} = \text{constant} \tag{11.111}$$

Since the flow cross section area remains constant for Rayleigh flow, from the continuity equation (Eq. 11.40) we conclude that

$$\rho V = \text{constant}$$

For a given Rayleigh flow, the constant in Eq. 11.111, the density–velocity product, ρV, and the ideal gas constant are all fixed. Thus, Eq. 11.111 can be used to determine values of fluid temperature corresponding to the local pressure in a Rayleigh flow.

To construct a temperature–entropy diagram for a given Rayleigh flow, we can use Eq. 11.76, which was developed earlier from the second $T\, ds$ relationship. Equations 11.111 and 11.76 can be solved simultaneously to obtain the curve sketched in Fig. 11.22. Curves like the one in Fig. 11.22 are called *Rayleigh lines*.

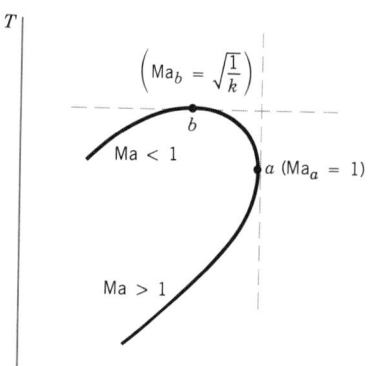

s **FIGURE 11.22** Rayleigh line.

EXAMPLE 11.15

Air ($k = 1.4$) enters [section (1)] a frictionless, constant flow cross section area duct with the following properties:

$$T_0 = 518.67 \,°R$$

$$T_1 = 514.55 \,°R$$

$$p_1 = 14.3 \text{ psia}$$

For Rayleigh flow, determine corresponding values of fluid temperature and entropy change for various levels of downstream pressure and plot the related Rayleigh line.

SOLUTION

To plot the Rayleigh line asked for we use Eq. 11.111

$$p + \frac{(\rho V)^2 \, RT}{p} = \text{constant} \tag{1}$$

and Eq. 11.76

$$s - s_1 = c_p \ln \frac{T}{T_1} - R \ln \frac{p}{p_1} \tag{2}$$

to construct a table of values of temperature and entropy change corresponding to different levels of pressure downstream in a Rayleigh flow.

We use the value of ideal gas constant for air from Table 1.7

$$R = 1716 \text{ (ft·lb)/(slug·°R)}$$

and the value of specific heat at constant pressure for air from Example 11.11, namely,

$$c_p = 6006 \text{ (ft·lb)/(slug·°R)}$$

Also, from Example 11.11, $\rho V = 0.519$ slug/(ft²·s). For the given inlet [section (1)] conditions, we get from Eq. 1

$$p + \frac{(\rho V)^2 \, RT}{p} = 14.3 \text{ psia}$$

$$+ \frac{[0.519 \text{ slug/(ft}^2\cdot\text{s)}]^2[1716 \text{ (ft·lb)/(slug·°R)}](514.55 \,°R)}{(144 \text{ in.}^2/\text{ft}^2)^2 \, 14.3 \text{ psia}} = \text{constant}$$

or

$$p + \frac{(\rho V)^2 \, RT}{p} = 15.10 \text{ psia} = \text{constant} \tag{3}$$

From Eq. 3, for downstream pressure $p = 13.5$ psia, we obtain

$$13.5 \text{ psia} + \frac{[0.519 \text{ slug/(ft}^2\cdot\text{s)}]^2[1716 \text{ (ft·lb)/(slug·°R)}]T}{(144 \text{ in.}^2/\text{ft}^2)^2 \, 13.5 \text{ psia}} = 15.10 \text{ psia}$$

or

$$T = 969 \,°R$$

Then for downstream pressure $p = 13.5$ psia and $T = 969 \,°R$, from Eq. 2 we get

$$s - s_1 = [6006 \text{ (ft·lb)/(slug·°R)}] \ln \left(\frac{969 \,°R}{514.55 \,°R}\right) - [1716 \text{ (ft·lb)/(slug·°R)}] \ln \left(\frac{13.5 \text{ psia}}{14.3 \text{ psia}}\right)$$

$$s - s_1 = 3900 \text{ (ft·lb)/(slug·°R)}$$

By proceeding as outlined above, we can construct the table of values shown below and graph the Rayleigh line of Fig. E11.15.

p (psia)	T (°R)	$s - s_1$ [(ft·lb)/(slug·°R)]
13.5	969	3,900
12.5	1459	6,490
11.5	1859	8,089
10.5	2168	9,168
9.0	2464	10,202
8.0	2549	10,607
7.6	2558	10,716
7.5	2558	10,739
7.0	2544	10,825
6.3	2488	10,872
6.0	2450	10,863
5.5	2369	10,810
5.0	2266	10,707
4.5	2140	10,544
4.0	1992	10,315
2.0	1175	8,335
1.0	633	5,809

$$s - s_1, \frac{(ft \cdot lb)}{(slug \cdot °R)}$$

FIGURE E11.15

At point a on the Rayleigh line of Fig. 11.22, $ds/dT = 0$. To determine the physical importance of point a, we analyze further some of the governing equations. By differentiating the linear momentum equation for Rayleigh flow (Eq. 11.110) we obtain

$$dp = -\rho V \, dV$$

or

$$\frac{dp}{\rho} = -V \, dV \tag{11.112}$$

Combining Eq. 11.112 with the second $T \, ds$ equation (Eq. 11.18) leads to

$$T \, ds = d\check{h} + V \, dV \tag{11.113}$$

For an ideal gas (Eq. 11.7)

$$d\check{h} = c_p \, dT$$

Thus, substituting Eq. 11.7 into Eq. 11.113 gives

$$T \, ds = c_p \, dT + V \, dV$$

or

$$\frac{ds}{dT} = \frac{c_p}{T} + \frac{V}{T}\frac{dV}{dT} \tag{11.114}$$

Consolidation of Eqs. 11.114, 11.112 (linear momentum), 11.1 and 11.77 (differentiated equation of state), and 11.79 (continuity) leads to

$$\frac{ds}{dT} = \frac{c_p}{T} + \frac{V}{T}\frac{1}{[(T/V) - (V/R)]} \tag{11.115}$$

Hence, at state a where $ds/dT = 0$, Eq. 11.115 reveals that

$$V_a = \sqrt{RT_a k} \tag{11.116}$$

Comparison of Eqs. 11.116 and 11.36 tells us that the Mach number at state a is equal to 1,

$$\text{Ma}_a = 1 \tag{11.117}$$

At point b on the Rayleigh line of Fig. 11.22, $dT/ds = 0$. From Eq. 11.115 we get

$$\frac{dT}{ds} = \frac{1}{ds/dT} = \frac{1}{(c_p/T) + (V/T)[(T/V) - (V/R)]^{-1}}$$

which for $dT/ds = 0$ (point b) gives,

$$\text{Ma}_b = \sqrt{\frac{1}{k}} \tag{11.118}$$

The flow at point b is subsonic ($\text{Ma}_b < 1.0$).

To learn more about Rayleigh flow, we need to consider the energy equation, in addition to the equations already used. Application of the energy equation (Eq. 5.69) to the Rayleigh flow through the finite control volume of Fig. 11.21 yields

$$\dot{m}\left[\check{h}_2 - \check{h}_1 + \frac{V_2^2 - V_1^2}{2} + g(z_2 \overset{\displaystyle 0\text{(negligibly small}}{\underset{\text{for gas flow)}}{\diagup} z_1)\right] = \dot{Q}_{\substack{\text{net} \\ \text{in}}} + \overset{\displaystyle 0\text{(flow is steady}}{\underset{\text{throughout)}}{\diagup} \dot{W}_{\substack{\text{shaft} \\ \text{net in}}}$$

or in differential form for Rayleigh flow through the semi-infinitesimal control volume of Fig. 11.21,

$$d\check{h} + V \, dV = \delta q \tag{11.119}$$

where δq is the heat transfer per unit mass of fluid in the semi-infinitesimal control volume.

By using $d\check{h} = c_p \, dT = Rk \, dT/(k - 1)$ in Eq. 11.119, we obtain

$$\frac{dV}{V} = \frac{\delta q}{c_p T}\left[\frac{V}{T}\frac{dT}{dV} + \frac{V^2(k - 1)}{kRT}\right]^{-1} \tag{11.120}$$

Thus, by combining Eqs. 11.36 (ideal gas speed of sound), 11.46 (Mach number), 11.1 and 11.77 (ideal gas equation of state), 11.79 (continuity) and 11.112 (linear

TABLE 11.2 Summary of Rayleigh Flow
Characteristics

	Heating	Cooling
Ma < 1	Acceleration	Deceleration
Ma > 1	Deceleration	Acceleration

momentum) with Eq. 11.120 (energy) we get

$$\frac{dV}{V} = \frac{\delta q}{c_p T} \frac{1}{(1 - Ma^2)} \tag{11.121}$$

With the help of Eq. 11.121, we see clearly that when the Rayleigh flow is subsonic (Ma < 1), fluid heating ($\delta q > 0$) increases fluid velocity while fluid cooling ($\delta q < 0$) decreases fluid velocity. When Rayleigh flow is supersonic (Ma > 1), fluid heating decreases fluid velocity and fluid cooling increases fluid velocity.

The second law of thermodynamics states that, based on experience, entropy increases with heating and decreases with cooling. With this additional insight provided by the conservation of energy principle and the second law of thermodynamics we can say more about the Rayleigh line in Fig. 11.22. A summary of the qualitative aspects of Rayleigh flow is outlined in Table 11.2 and Fig. 11.23. Along the upper portion of the line, which includes point b, the flow is subsonic. Heating the fluid results in flow acceleration to a maximum Mach number of 1 at point a. Note that between points b and a along the Rayleigh line, heating the fluid results in a temperature decrease and cooling the fluid leads to a temperature increase. This trend is not surprising if we consider the stagnation temperature and fluid velocity changes that occur between points a and b when the fluid is heated or cooled. Along the lower portion of the Rayleigh curve the flow is supersonic. Rayleigh flows may or may not be choked. The amount of heating or cooling involved determines what will happen in a specific instance. As with Fanno flows, an abrupt deceleration from supersonic flow to subsonic flow across a normal shock wave can also occur in Rayleigh flows.

To quantify Rayleigh flow behavior we need to develop appropriate forms of the governing equations. We elect to use the state of the Rayleigh flow fluid at point a of Fig. 11.22 as the reference state. As shown earlier, the Mach number at point a is 1. Even though the Rayleigh flow being considered may not choke and state a is not achieved by the flow, this reference state is useful.

FIGURE 11.23 **(a)** Subsonic Rayleigh flow. **(b)** Supersonic Rayleigh flow. **(c)** Normal shock in a Rayleigh flow.

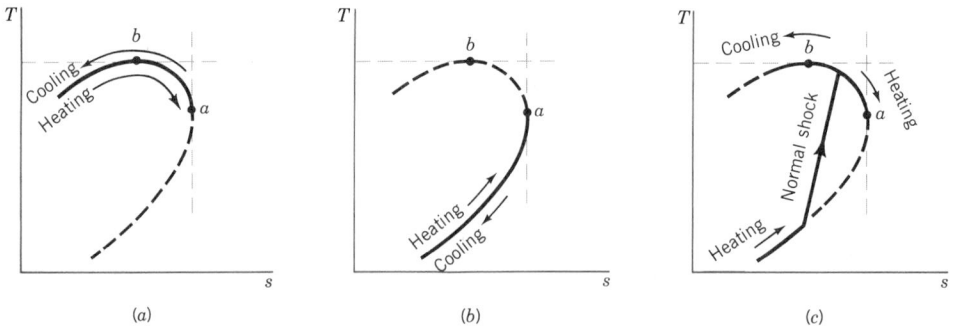

If we apply the linear momentum equation (Eq. 11.110) to Rayleigh flow between any upstream section and the section, actual or imagined, where state a is attained, we get

$$p + \rho V^2 = p_a + \rho_a V_a^2$$

or

$$\frac{p}{p_a} + \frac{\rho V^2}{p_a} = 1 + \frac{\rho_a}{p_a} V_a^2 \qquad (11.122)$$

By substituting the ideal gas equation of state (Eq. 11.1) into Eq. 11.122 and making use of the ideal gas speed of sound equation (Eq. 11.36) and the definition of Mach number (Eq. 11.46), we obtain

$$\frac{p}{p_a} = \frac{1 + k}{1 + k\,\mathrm{Ma}^2} \qquad (11.123)$$

From the ideal gas equation of state (Eq. 11.1) we conclude that

$$\frac{T}{T_a} = \frac{p}{p_a}\frac{\rho_a}{\rho} \qquad (11.124)$$

Conservation of mass (Eq. 11.40) with constant A gives

$$\frac{\rho_a}{\rho} = \frac{V}{V_a} \qquad (11.125)$$

which when combined with Eqs. 11.36 (ideal gas speed of sound) and 11.46 (Mach number definition) gives

$$\frac{\rho_a}{\rho} = \mathrm{Ma}\sqrt{\frac{T}{T_a}} \qquad (11.126)$$

Combining Eqs. 11.124 and 11.126 leads to

$$\frac{T}{T_a} = \left(\frac{p}{p_a}\,\mathrm{Ma}\right)^2 \qquad (11.127)$$

which when combined with Eq. 11.123 gives

$$\frac{T}{T_a} = \left[\frac{(1 + k)\mathrm{Ma}}{1 + k\,\mathrm{Ma}^2}\right]^2 \qquad (11.128)$$

From Eqs. 11.125, 11.126, and 11.128 we see that

$$\frac{\rho_a}{\rho} = \frac{V}{V_a} = \mathrm{Ma}\left[\frac{(1 + k)\mathrm{Ma}}{1 + k\,\mathrm{Ma}^2}\right] \qquad (11.129)$$

The energy equation (Eq. 5.69) tells us that because of the heat transfer involved in Rayleigh flows, the stagnation temperature varies. We note that

$$\frac{T_0}{T_{0,a}} = \left(\frac{T_0}{T}\right)\left(\frac{T}{T_a}\right)\left(\frac{T_a}{T_{0,a}}\right) \qquad (11.130)$$

We can use Eq. 11.56 (developed earlier for steady, isentropic, ideal gas flow) to evaluate T_0/T and $T_a/T_{0,a}$ because these two temperature ratios, by definition of the

stagnation state, involve isentropic processes. Equation 11.128 can be used for T/T_a. Thus, consolidating Eqs. 11.130, 11.56, and 11.128 we obtain

$$\frac{T_0}{T_{0,a}} = \frac{2(k + 1)\text{Ma}^2\left(1 + \dfrac{k - 1}{2}\text{Ma}^2\right)}{(1 + k\text{Ma}^2)^2} \tag{11.131}$$

Finally, we observed that

$$\frac{p_0}{p_{0,a}} = \left(\frac{p_0}{p}\right)\left(\frac{p}{p_a}\right)\left(\frac{p_a}{p_{0,a}}\right) \tag{11.132}$$

We can use Eq. 11.59 developed earlier for steady, isentropic, ideal gas flow to evaluate p_0/p and $p_a/p_{0,a}$ because these two pressure ratios, by definition, involve isentropic processes. Equation 11.123 can be used for p/p_a. Together, Eqs. 11.59, 11.123, and 11.132 give

$$\frac{p_0}{p_{0,a}} = \frac{(1 + k)}{(1 + k\text{Ma}^2)}\left[\left(\frac{2}{k + 1}\right)\left(1 + \frac{k - 1}{2}\text{Ma}^2\right)\right]^{k/(k-1)} \tag{11.133}$$

Values of p/p_a, T/T_a, ρ_a/ρ, or V/V_a, $T_0/T_{0,a}$, and $p_0/p_{0,a}$ are given in Table E.3 of Appendix E as a function of Mach number for Rayleigh flow of air ($k = 1.4$). The values in Table E.3 were calculated from Eqs. 11.123, 11.128, 11.129, 11.131, and 11.133. The usefulness of Table E.3 is illustrated in Examples 11.16 and 11.17.

EXAMPLE 11.16

The information in Table 11.2 tells us that subsonic Rayleigh flow accelerates when heated and decelerates when cooled. Supersonic Rayleigh flow behaves just opposite to subsonic Rayleigh flow; it decelerates when heated and accelerates when cooled. Using Table E.3 for air ($k = 1.4$) state whether velocity, Mach number, static temperature, stagnation temperature, static pressure, and stagnation pressure increase or decrease as subsonic and supersonic Rayleigh flow is (a) heated; (b) cooled.

SOLUTION

Acceleration occurs when V/V_a in Table E.3 increases. For deceleration, V/V_a decreases. From Table E. 3 and Table 11.2 the following chart can be constructed.

	Heating		Cooling	
	Subsonic	Supersonic	Subsonic	Supersonic
V	Increase	Decrease	Decrease	Increase
Ma	Increase	Decrease	Decrease	Increase
T	Increase for $0 \leq \text{Ma} \leq \sqrt{1/k} = 0.845$ Decrease for $\sqrt{1/k} \leq \text{Ma} \leq 1$	Increase	Decrease for $0 \leq \text{Ma} \leq \sqrt{1/k} = 0.845$ Increase for $\sqrt{1/k} \leq \text{Ma} \leq 1$	Decrease
T_0	Increase	Increase	Decrease	Decrease
p	Decrease	Increase	Increase	Decrease
p_0	Decrease	Decrease	Increase	Increase

From the Rayleigh flow trends summarized in the table above, we note that heating affects Rayleigh flows much like friction affects Fanno flows. Heating and friction both accelerate subsonic flows and decelerate supersonic flows. More importantly, both heating and friction cause the stagnation pressure to decrease. Since stagnation pressure loss is considered undesirable in terms of fluid mechanical efficiency, heating a fluid flow must be accomplished with this loss in mind.

EXAMPLE 11.17

Standard atmospheric air ($T_0 = 519$ °R, $p_0 = 14.7$ psia) is drawn steadily through a frictionless and adiabatic converging nozzle into a frictionless and diabatic ($q = 200$ Btu/lbm) constant cross section area duct as is shown in Fig. E11.17a. For maximum flowrate determine the values of static temperature, static pressure, stagnation temperature, stagnation pressure, and flow velocity at the inlet [section(1)] and exit [section (2)] of the constant area duct. Sketch a temperature–entropy diagram for this flow.

(a)

(b)

FIGURE E11.17

SOLUTION

As the back pressure, p_d, is decreased, the flowrate of air through the isentropic converging nozzle and the constant area duct will increase. When the back pressure is lowered sufficiently, the duct will choke with an exit Mach number of 1. Any further reduction in back pressure will not affect the flow. In the choked condition, the duct will pass the maximum flowrate possible.

Since this flow is subsonic and heating is involved, the flow is represented by the upper portion of the Rayleigh line (see Figs. 11.22 and 11.23a with states downstream to the right on the curve). For choked flow, the condition at the exit of the constant area duct will correspond to state *a* on the Rayleigh line (Figs. 11.22 and 11.23).

For the flow through the isentropic nozzle, the stagnation temperature remains constant at a value

$$T_{0,1} = 519 \; °R \qquad \text{(Ans)}$$

Application of the energy equation (Eq. 5.69) to the flow through the constant area duct yields

$$\check{h}_{0,2} - \check{h}_{0,1} = q \qquad (1)$$

For an ideal gas with constant c_p, Eq. 11.9 allows us to state Eq. 1 as

$$T_{0,2} - T_{0,1} = \frac{q}{c_p} \qquad (2)$$

With Eq. 2 and known values of $T_{0,1}$, q, and c_p we can calculate $T_{0,2}$, which for choked flow is $T_{0,a}$. For air ($k = 1.4$), Eq. 11.14 allows us to determine c_p as

$$c_p = \frac{Rk}{k-1} = \frac{[1716(\text{ft·lb})/(\text{slug·}°R)](1.4)}{1.4 - 1} = 6006 \; (\text{ft·lb})/(\text{slug·}°R)$$

Then, with Eq. 2 we get

$$T_{0,2} = 519 + \frac{(200 \; \text{Btu/lbm})[778 \; (\text{ft·lb})/\text{Btu}](32.2 \; \text{lbm/slug})}{6006 \; (\text{ft·lb})/(\text{slug·}°R)} = 1353 \; °R = T_{0,a} \qquad \text{(Ans)}$$

Thus

$$\frac{T_{0,1}}{T_{0,a}} = \frac{519 \; °R}{1353 \; °R} = 0.3836$$

and we enter Table E.3 for Rayleigh flow of air ($k = 1.4$) and read off the following closest values.

$$\frac{T_{0,1}}{T_{0,a}} = 0.38369$$

$$\text{Ma}_1 = 0.32$$

$$\frac{T_1}{T_a} = 0.45119 \qquad (3)$$

$$\frac{p_1}{p_a} = 2.0991 \qquad (4)$$

$$\frac{V_1}{V_a} = 0.21494 \qquad (5)$$

and

$$\frac{p_{0,1}}{p_{0,a}} = 1.1904 \qquad (6)$$

For the flow through the isentropic nozzle, the stagnation pressure remains constant. Thus,

$$p_{0,1} = 14.7 \; \text{psia} \qquad \text{(Ans)}$$

and with Eq. 6,

$$p_{0,a} = \frac{14.7 \; \text{psia}}{1.1904} = 12.35 \; \text{psia} = p_{0,2} \qquad \text{(Ans)}$$

For the isentropic nozzle flow, with $\text{Ma}_1 = 0.32$, we read in Table E.1,

$$\frac{p_1}{p_{0,1}} = 0.93150 \qquad (7)$$

and

$$\frac{T_1}{T_{0.1}} = 0.97993 \qquad\qquad (8)$$

Thus, from Eqs. 7 and 8 we obtain

$$p_1 = (0.93150)(14.7 \text{ psia}) = 13.7 \text{ psia} \qquad\qquad \text{(Ans)}$$

and

$$T_1 = (0.97993)(519 \text{ °R}) = 509 \text{ °R} \qquad\qquad \text{(Ans)}$$

With the above values of p_1 and T_1 and Eqs. 3 and 4 we get

$$T_2 = T_a = \frac{509 \text{ °R}}{0.45119} = 1128 \text{ °R} \qquad\qquad \text{(Ans)}$$

and

$$p_2 = p_a = \frac{13.7 \text{ psia}}{2.0991} = 6.53 \text{ psia} \qquad\qquad \text{(Ans)}$$

We can use 11.36 and 11.46 to determine the velocity at section (1) as,

$$V_1 = \text{Ma}_1 \sqrt{RT_1 k}$$

or

$$V_1 = 0.32 \sqrt{[1716 \text{ (ft·lb)/(slug·°R)}](509 \text{ °R})(1.4)[1(\text{slug·ft})/(\text{lb·s}^2)]} = 354 \text{ ft/s} \quad \text{(Ans)}$$

With $V_1 = 354$ ft/s and Eq. 5 we obtain

$$V_a = \frac{354 \text{ ft/s}}{0.21494} = 1650 \text{ ft/s} \qquad\qquad \text{(Ans)}$$

The temperature–entropy diagram for this flow is sketched in Fig. E11.17b.

Note that even though the nozzle and duct involved frictionless flow, the heat addition in the duct resulted in a substantial loss of stagnation pressure from $p_{0.1} = 14.7$ psia to $p_{0.2} = 12.4$ psia. The back pressure needed to be lowered to 6.53 psia to choke the flow. Any further reduction in the back pressure to values less than 6.53 psia would not affect the flow through the nozzle and duct.

11.5.3   Normal Shock Waves

As mentioned earlier, normal shock waves can occur in supersonic flows through converging–diverging and constant area ducts. Past experience suggests that normal shock waves involve deceleration from a supersonic flow to a subsonic flow, a pressure rise, and an increase of entropy. To develop the equations that verify this observed behavior of flows across a normal shock, we apply first principles to the flow through a control volume that completely surrounds a normal shock wave (see Fig. 11.24). We consider the normal shock and thus the control volume to be infinitesimally thin and stationary.

For steady flow through the control volume of Fig. 11.24, the conservation of mass principle yields

$$\rho V = \text{constant} \qquad\qquad (11.134)$$

because the flow cross section area remains essentially constant within the infinitesimal thickness of the normal shock. Note that Eq. 11.134 is identical to the continuity equation used for Fanno and Rayleigh flows considered earlier.

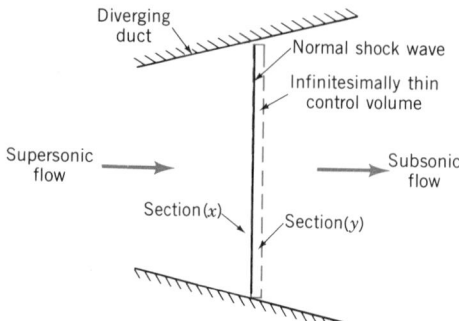

FIGURE 11.24 Normal shock control volume.

The friction force acting on the contents of the infinitesimally thin control volume surrounding the normal shock is considered to be negligibly small. Also for ideal gas flow, the effect of gravity is neglected. Thus, the linear momentum equation (Eq. 5.22) describing steady gas flow through the control volume of Fig. 11.24 is

$$p + \rho V^2 = \text{constant}$$

or for an ideal gas for which $p = \rho RT$,

$$p + \frac{(\rho V)^2 RT}{p} = \text{constant} \qquad (11.135)$$

Equation 11.135 is the same as the linear momentum equation for Rayleigh flow, which was derived earlier (Eq. 11.111).

For the control volume containing the normal shock, no shaft work is involved and the heat transfer is assumed negligible.

The energy equation (Eq. 5.69) can be applied to steady gas flow through the control volume of Fig. 11.24 to obtain

$$\check{h} + \frac{V^2}{2} = \check{h}_0 = \text{constant}$$

or, for an ideal gas, since $\check{h} - \check{h}_0 = c_p(T - T_0)$ and $p = \rho RT$,

$$T + \frac{(\rho V)^2 T^2}{2c_p(p^2/R^2)} = T_0 = \text{constant} \qquad (11.136)$$

Equation 11.136 is identical to the energy equation for Fanno flow analyzed earlier (Eq.11.75).

The T ds relationship previously used for ideal gas flow (Eq. 11.22) is valid for the flow through the normal shock (Fig. 11.24) because it (Eq. 11.22) is an ideal gas property relationship.

From the analyses of the previous paragraphs, it is apparent that the steady flow of an ideal gas across a normal shock is governed by some of the same equations used for describing Fanno and Rayleigh flows (energy equation for Fanno flows and momentum equation for Rayleigh flow). Thus, for a given density–velocity product (ρV), gas (R,k), and conditions at the inlet of the normal shock $(T_x, p_x, \text{and } s_x)$, the conditions downstream of the shock (state y) will be on both a Fanno line and a Rayleigh line that pass through the inlet state (state x), as is illustrated in Fig. 11.25. To conform with common practice we designate the states upstream and downstream of the normal shock with an x and a y instead of numerals 1 and 2. The Fanno and

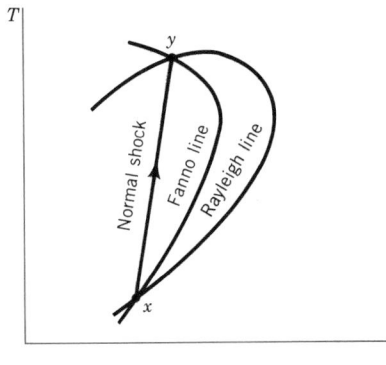

FIGURE 11.25   The relationship between a normal shock and Fanno and Rayleigh lines.

Rayleigh lines describe more of the flow field than just in the vicinity of the normal shock when Fanno and Rayleigh flows are actually involved (solid lines in Fig. 11.26*a* and 11.26*b*). Otherwise, these lines (dashed lines in Figs. 11.26*a*, 11.26*b*, and 11.26*c*) are useful mainly as a way to better visualize how the governing equations combine to yield a solution to the normal shock flow problem.

The second law of thermodynamics requires that entropy must increase across a normal shock wave. This law and sketches of the Fanno line and Rayleigh line intersections like those of Figs. 11.25 and 11.26 persuade us to conclude that flow across

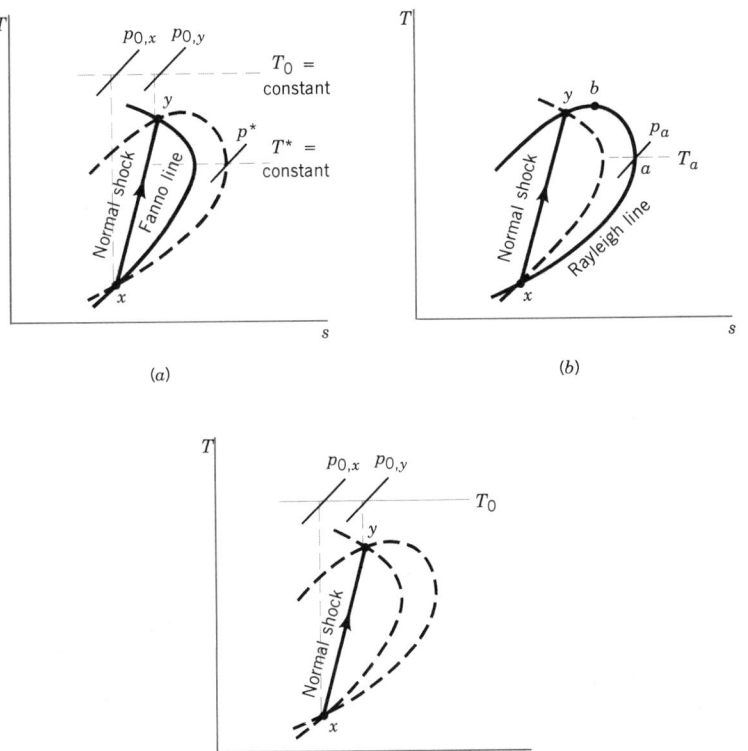

FIGURE 11.26
(a) The normal shock in a Fanno flow. **(b)** The normal shock in a Rayleigh flow. **(c)** The normal shock in a frictionless and adiabatic flow.

a normal shock can only proceed from supersonic to subsonic flow. Similarly, in open channel flows (see Chapter 10), the flow across a hydraulic jump proceeds from supercritical to subcritical conditions.

Since the states upstream and downstream of a normal shock wave are represented by the supersonic and subsonic intersections of actual and/or imagined Fanno and Rayleigh lines, we should be able to use equations developed earlier for Fanno and Rayleigh flows to quantify normal shock flow. For example, for the Rayleigh line of Fig. 11.26*b*

$$\frac{p_y}{p_x} = \left(\frac{p_y}{p_a}\right)\left(\frac{p_a}{p_x}\right) \tag{11.137}$$

But from Eq. 11.123 for Rayleigh flow we get

$$\frac{p_y}{p_a} = \frac{1 + k}{1 + k\text{Ma}_y^2} \tag{11.138}$$

and

$$\frac{p_x}{p_a} = \frac{1 + k}{1 + k\text{Ma}_x^2} \tag{11.139}$$

Thus, by combining Eqs. 11.137, 11.138, and 11.139 we get

$$\frac{p_y}{p_x} = \frac{1 + k\text{Ma}_x^2}{1 + k\text{Ma}_y^2} \tag{11.140}$$

Equation 11.140 can also be derived starting with

$$\frac{p_y}{p_x} = \left(\frac{p_y}{p^*}\right)\left(\frac{p^*}{p_x}\right)$$

and using the Fanno flow equation (Eq. 11.107)

$$\frac{p}{p^*} = \frac{1}{\text{Ma}} \left\{ \frac{(k + 1)/2}{1 + [(k - 1)/2]\text{Ma}^2} \right\}^{1/2}$$

As might be expected, Eq. 11.140 can be obtained directly from the linear momentum equation

$$p_x + \rho_x V_x^2 = p_y + \rho_y V_y^2$$

since $\rho V^2/p = V^2/RT = kV^2/RTk = k\,\text{Ma}^2$.

For the Fanno flow of Fig. 11.26*a*

$$\frac{T_y}{T_x} = \left(\frac{T_y}{T^*}\right)\left(\frac{T^*}{T_x}\right) \tag{11.141}$$

From Eq. 11.101 for Fanno flow we get

$$\frac{T_y}{T^*} = \frac{(k + 1)/2}{1 + [(k - 1)/2]\text{Ma}_y^2} \tag{11.142}$$

and

$$\frac{T_x}{T^*} = \frac{(k + 1)/2}{1 + [(k - 1)/2]\text{Ma}_x^2} \tag{11.143}$$

A consolidation of Eqs. 11.141, 11.142, and 11.143 gives

$$\frac{T_y}{T_x} = \frac{1 + [(k-1)/2]\text{Ma}_x^2}{1 + [(k-1)/2]\text{Ma}_y^2} \qquad (11.144)$$

We seek next to develop an equation that will allow us to determine the Mach number downstream of the normal shock, Ma_y, when the Mach number upstream of the normal shock, Ma_x, is known. From the ideal gas equation of state (Eq. 11.1), we can form

$$\frac{p_y}{p_x} = \left(\frac{T_y}{T_x}\right)\left(\frac{\rho_y}{\rho_x}\right) \qquad (11.145)$$

Using the continuity equation

$$\rho_x V_x = \rho_y V_y$$

with Eq. 11.145 we obtain

$$\frac{p_y}{p_x} = \left(\frac{T_y}{T_x}\right)\left(\frac{V_x}{V_y}\right) \qquad (11.146)$$

When combined with the Mach number definition (Eq. 11.46) and the ideal gas speed of sound equation (Eq. 11.36), Eq. 11.146 becomes

$$\frac{p_y}{p_x} = \left(\frac{T_y}{T_x}\right)^{1/2}\left(\frac{\text{Ma}_x}{\text{Ma}_y}\right) \qquad (11.147)$$

Thus, Eqs. 11.147 and 11.144 lead to

$$\frac{p_y}{p_x} = \left\{\frac{1 + [(k-1)/2]\text{Ma}_x^2}{1 + [(k-1)/2]\text{Ma}_y^2}\right\}^{1/2} \frac{\text{Ma}_x}{\text{Ma}_y} \qquad (11.148)$$

which can be merged with Eq. 11.140 to yield

$$\text{Ma}_y^2 = \frac{\text{Ma}_x^2 + [2/(k-1)]}{[2k/(k-1)]\text{Ma}_x^2 - 1} \qquad (11.149)$$

Thus, we can use Eq. 11.149 to calculate values of Mach number downstream of a normal shock from a known Mach number upstream of the shock.

If we combine Eqs. 11.149 and 11.140, we get

$$\frac{p_y}{p_x} = \frac{2k}{k+1}\text{Ma}_x^2 - \frac{k-1}{k+1} \qquad (11.150)$$

which allows us to calculate the pressure ratio across a normal shock from a known upstream Mach number. Similarly, taking Eqs. 11.149 and 11.144 together we obtain

$$\frac{T_y}{T_x} = \frac{\{1 + [(k-1)/2]\text{Ma}_x^2\}\{[2k/(k-1)]\text{Ma}_x^2 - 1\}}{\{(k+1)^2/[2(k-1)]\}\text{Ma}_x^2} \qquad (11.151)$$

From the continuity equation (Eq. 11.40) we have for flow across a normal shock

$$\frac{\rho_y}{\rho_x} = \frac{V_x}{V_y} \qquad (11.152)$$

TABLE 11.3 Summary of Normal Shock Wave Characteristics

Variable	Change Across Normal Shock Wave
Mach number	Decrease
Static pressure	Increase
Stagnation pressure	Decrease
Static temperature	Increase
Stagnation temperature	Constant
Density	Increase
Velocity	Decrease

and from the ideal gas equation of state (Eq. 11.1)

$$\frac{\rho_y}{\rho_x} = \left(\frac{p_y}{p_x}\right)\left(\frac{T_x}{T_y}\right) \tag{11.153}$$

Thus, by combining Eqs. 11.152, 11.153, 11.150, and 11.151, we get

$$\frac{\rho_y}{\rho_x} = \frac{V_x}{V_y} = \frac{(k + 1)\text{Ma}_x^2}{(k - 1)\text{Ma}_x^2 + 2} \tag{11.154}$$

The stagnation pressure ratio across the shock can be determined by combining

$$\frac{p_{0,y}}{p_{0,x}} = \left(\frac{p_{0,y}}{p_y}\right)\left(\frac{p_y}{p_x}\right)\left(\frac{p_x}{p_{0,x}}\right) \tag{11.155}$$

with Eqs. 11.59, 11.149, and 11.150 to get

$$\frac{p_{0,y}}{p_{0,x}} = \frac{\left(\dfrac{k + 1}{2}\text{Ma}_x^2\right)^{k/(k-1)}\left(1 + \dfrac{k-1}{2}\text{Ma}_x^2\right)^{k/(1-k)}}{\left(\dfrac{2k}{k + 1}\text{Ma}_x^2 - \dfrac{k - 1}{k + 1}\right)^{1/(k-1)}} \tag{11.156}$$

Table E.4 in Appendix E lists values of downstream Mach numbers, Ma_y, pressure ratio, p_y/p_x, temperature ratio, T_y/T_x, density ratio, ρ_y/ρ_x or velocity ratio, V_x/V_y, and stagnation pressure ratio, $p_{0,y}/p_{0,x}$, as a function of upstream Mach number, Ma_x, for the steady flow across a normal shock wave of an ideal gas having a specific heat ratio $k = 1.4$. These values were calculated from Eqs. 11.149, 11.150, 11.151, 11.154, and 11.156.

Important trends associated with the steady flow of an ideal gas across a normal shock wave can be determined by studying Table E.4. These trends are summarized in Table 11.3.

Examples 11.18, 11.19, and 11.20 illustrate how Table E.4 can be used to solve fluid flow problems involving normal shock waves.

EXAMPLE 11.18

Designers involved with fluid mechanics work hard at minimizing loss of available energy in their designs. Adiabatic frictionless flows involve no loss in available energy. Entropy remains

constant for these idealized flows. Adiabatic flows with friction involve available energy loss and entropy increase. Generally, larger entropy increases imply larger losses. For normal shocks, show that entropy rise and thus loss are larger for higher Mach numbers. Confirm your conclusion about loss by examining corresponding stagnation pressure changes across a normal shock.

SOLUTION

We assume that air ($k = 1.4$) behaves as a typical gas and use Table E.4 to respond to the above stated requirements. We use Eq. 11.76 to determine entropy rise across the normal shock as

$$s_y - s_x = c_p \ln \frac{T_y}{T_x} - R \ln \frac{p_y}{p_x} \tag{1}$$

For the gas constant we use

$$R = 286.9 \text{ J/(kg·K)}$$

from Table 1.8. For c_p we use Eq. 11.14 to get

$$c_p = \frac{Rk}{k-1} = [286.9 \text{ J/(kg·K)}] \frac{1.4}{(1.4-1)} = 1004 \text{ J/(kg·K)}$$

The values of T_x/T_y and p_x/p_y to use with Eq. 1 are obtained from Table E.4 for selected upstream Mach numbers. Also from Table E.4 we get corresponding values of stagnation pressure ratio, $p_{0,y}/p_{0,x}$.

For example, for $Ma_x = 1.1$ from Eq. 1 and Table E.4

$$s_y - s_x = [1004 \text{ J/(kg·K)}] \ln(1.06494) - [286.9 \text{ J/(kg·K)}] \ln(1.2450)$$

or

$$s_y - s_x = 0.300 \text{ J/(kg·K)}$$

Also from Table E.4

$$\frac{p_{0,y}}{p_{0,x}} = 0.99892$$

The following table was constructed using the preceding procedure.

Ma_x	$s_y - s_x$ [J/(kg·K)]	$p_{0,y}/p_{0,x}$
1.0	0	1.0
1.1	0.300	0.99892
1.2	2.07	0.99280
1.5	20.8	0.92978
2.0	93.8	0.72088
2.5	199.3	0.49902
3.0	319.4	0.32834
3.5	443.6	0.21295
4.0	566.4	0.13876
5.0	798.8	0.06172

When the Mach number of the flow entering the shock is low, say $Ma_x = 1.1$, the flow across the shock is nearly isentropic and the loss in stagnation pressure is small. However, at larger Mach numbers, the entropy change across the normal shock rises dramatically and the

stagnation pressure drop across the shock is appreciable. If a shock occurs at $Ma_x = 2.5$, only about 50% of the upstream stagnation pressure is recovered.

In devices where supersonic flows occur, for example, high-performance aircraft engine inlet ducts and high-speed wind tunnels, designers attempt to prevent shock formation, or if shocks must occur, they design the flow path so that shocks are positioned where they are weak (small Mach number).

Of interest also is the static pressure rise that occurs across a normal shock. These static pressure ratios, p_y/p_x, obtained from Table E.4 are shown in the table below for a few Mach numbers. For a developing boundary layer any pressure rise in the flow direction is considered as an adverse pressure gradient that can possibly cause flow separation (see Section 9.2.6). Thus, shock-boundary layer interactions are of great concern to designers of high-speed flow devices.

Ma_x	p_y/p_x
1.0	1.00
1.2	1.51
1.5	2.46
2.0	4.50
3.0	10.3
4.0	18.5
5.0	29.0

EXAMPLE 11.19

A total pressure probe is inserted into a supersonic air flow. A shock wave forms just upstream of the impact hole and head as illustrated in Fig. E11.19. The probe measures a total pressure of 60 psia. The stagnation temperature at the probe head is 1000 °R. The static pressure upstream of the shock is measured with a wall tap to be 12 psia. From these data determine the Mach number and velocity of the flow.

FIGURE E11.19

SOLUTION

We assume that the flow along the stagnation pathline is isentropic except across the shock. Also, the shock is treated as a normal shock. Thus, in terms of the data we have

$$\frac{p_{0,y}}{p_x} = \left(\frac{p_{0,y}}{p_{0,x}}\right)\left(\frac{p_{0,x}}{p_x}\right) \tag{1}$$

where $p_{0,y}$ is the stagnation pressure measured by the probe, and p_x is the static pressure measured by the wall tap. The stagnation pressure upstream of the shock, $p_{0,x}$, is not measured.

Combining Eqs. 1, 11.156, and 11.59 we obtain

$$\frac{p_{0,y}}{p_x} = \frac{\{[(k+1)/2]\text{Ma}_x^2\}^{k/(k-1)}}{\{[2k/(k+1)]\text{Ma}_x^2 - [(k-1)/(k+1)]\}^{1/(k-1)}} \tag{2}$$

which is called the *Rayleigh pitot-tube formula.* Values of $p_{0,y}/p_x$ from Eq. 2 are considered important enough to be included in Table E.4 for $k = 1.4$. Thus, for $k = 1.4$ and

$$\frac{p_{0,y}}{p_x} = \frac{60 \text{ psia}}{12 \text{ psia}} = 5$$

we use Table E.4 to ascertain that

$$\text{Ma}_x = 1.87 \tag{Ans}$$

To determine the flow velocity we need to know the static temperature upstream of the shock, since Eqs. 11.36 and 11.46 can be used to yield

$$V_x = \text{Ma}_x \sqrt{RT_x k} \tag{3}$$

The stagnation temperature downstream of the shock was measured and found to be

$$T_{0,y} = 1000 \text{ °R}$$

Since the stagnation temperature remains constant across a normal shock (see Eq. 11.136)

$$T_{0,x} = T_{0,y} = 1000 \text{ °R}$$

For the flow upstream of the shock, Eq. 11.56 or Table E.1 can be used. For $\text{Ma}_x = 1.87$,

$$\frac{T_x}{T_{0,x}} = 0.58845$$

or

$$T_x = (0.58845)(1000 \text{ °R}) = 588.45 \text{ °R}$$

With Eq. 3 we obtain

$$V_x = 1.87 \sqrt{[1716(\text{ft·lb})/(\text{slug·°R})](588.45 \text{ °R})(1.4)[1(\text{slug·ft})/(\text{lb·s}^2)]} = 2220 \text{ ft/s} \quad (\text{Ans})$$

EXAMPLE 11.20

Determine, for the converging–diverging duct of Example 11.8, the ratio of back presure to inlet stagnation pressure, $p_{\text{III}}/p_{0,x}$, see Fig. 11.13, that will result in a standing normal shock at the exit ($x = +0.5$ m) of the duct. What value of the ratio of back pressure to inlet stagnation pressure would be required to position the shock at $x = +0.3$ m? Show related temperature–entropy diagrams for these flows.

SOLUTION

For supersonic, isentropic flow through the nozzle to just upstream of the standing normal shock at the duct exit, we have from the table of Example 11.8 at $x = +0.5$ m

$$\text{Ma}_x = 2.8$$

and

$$\frac{p_x}{p_{0,x}} = 0.03685$$

From Table E.4 for $Ma_x = 2.8$, we obtain

$$\frac{p_y}{p_x} = 8.9800$$

Thus,

$$\frac{p_y}{p_{0,x}} = \left(\frac{p_y}{p_x}\right)\left(\frac{p_x}{p_{0,x}}\right) = (8.9800)(0.03685) = 0.331 = \frac{p_{\mathrm{III}}}{p_{0,x}} \qquad \text{(Ans)}$$

When the duct back pressure to inlet stagnation pressure ratio, $p_{\mathrm{III}}/p_{0,x}$, is set equal to 0.331, the air will accelerate through the converging–diverging duct to a Mach number of 2.8 at the duct exit. The air will subsequently decelerate to a subsonic flow across a normal shock at the duct exit. The stagnation pressure ratio across the normal shock, $p_{0,y}/p_{0,x}$, is 0.38946 (Table E.4 for $M_x = 2.8$). A considerable amount of available energy is lost across the shock.

For a normal shock at $x = +0.3$ m, we note from the table of Example 11.8 that

$$Ma_x = 2.14$$

and

$$\frac{p_x}{p_{0,x}} = 0.10272 \qquad (1)$$

From Table E.4 for $Ma_x = 2.14$, we obtain

$$\frac{p_y}{p_x} = 5.1762$$

$$Ma_y = 0.55538$$

and

$$\frac{p_{0,y}}{p_{0,x}} = 0.65567 \qquad (2)$$

From Table E.1, for $Ma_y = 0.55538 \approx 0.56$ we get

$$\frac{A_y}{A^*} = 1.2403 \qquad (3)$$

For $x = +0.3$ m, the ratio of duct exit area to local area (A_2/A_y) is, using the area equation from Example 11.8,

$$\frac{A_2}{A_y} = \frac{0.1 + (0.5)^2}{0.1 + (0.3)^2} = 1.842 \qquad (4)$$

Using Eqs. 3 and 4 we get

$$\frac{A_2}{A^*} = \left(\frac{A_y}{A^*}\right)\left(\frac{A_2}{A_y}\right) = (1.2403)(1.842) = 2.285$$

Note that for the isentropic flow upstream of the shock, $A^* = 0.10$ m² (the actual throat area), while for the isentropic flow downstream of the shock, $A^* = A_2/2.285 = 0.35$ m²/2.285 = 0.15 m². With $A_2/A^* = 2.285$ we enter Table E.1 and without interpolating get for closest values

$$Ma_2 = 0.26$$

and

$$\frac{p_2}{p_{0,y}} = 0.95408 \qquad (5)$$

Combining Eqs. 2 and 5 we obtain

$$\frac{p_2}{p_{0,x}} = \left(\frac{p_2}{p_{0,y}}\right)\left(\frac{p_{0,y}}{p_{0,x}}\right) = (0.95408)(0.65567) = 0.626 \qquad \text{(Ans)}$$

When the back pressure, p_2, is set equal to 0.626 times the inlet stagnation pressure, $p_{0,x}$, the normal shock will be positioned at $x = +0.3$ m. Note that $p_2/p_{0,x} = 0.626$ is less than the value

(a)

(b)

FIGURE E11.20

of this ratio for subsonic isentropic flow through the converging–diverging duct, $p_2/p_0 = 0.98003$ (from Example 11.8), and is larger than $p_{III}/p_{0.x} = 0.331$, for duct flow with a normal shock at the exit (see Fig. 11.13). Also the stagnation pressure ratio with the shock at $x = +0.3$ m, $p_{0.y}/p_{0.x} = 0.65567$, is much greater than the stagnation pressure ratio, 0.38946, when the shock occurs at the exit ($x = +0.5$ m) of the duct. The corresponding $T - s$ diagrams are shown in Figs. E11.20a and E11.20b.

EXAMPLE 11.21

Supersonic air flow enters an adiabatic, 25 m long constant cross section area pipe from the converging–diverging duct of Example 11.8 (see Fig. E11.21a). The flow through the converging–diverging nozzle is isentropic. The cross section area of the adiabatic pipe is identical to the discharge area of the converging–diverging nozzle. The adiabatic pipe friction factor is $f = 0.02$. What ratio of back pressure at the adiabatic pipe discharge to the pipe inlet stagnation pressure, $p_2/p_{0.1}$, results in a standing normal shock wave located one-tenth the length of the pipe downstream from the pipe inlet? Sketch the related temperature–entropy diagram.

(a)

(b)

FIGURE E11.21

SOLUTION

The isentropic supersonic flow solution of Example 11.8 is unique. Thus, from the table of Example 11.8, the Mach number at the entrance of the adiabatic pipe is,

$$Ma_1 = 2.80$$

With $Ma_1 = 2.80$ we enter Table E.2 for Fanno flow of air ($k = 1.4$) and get

$$\frac{f(\ell^* - \ell_1)}{D} = 0.48976$$

But

$$\frac{f(\ell^* - \ell_1)}{D} = \frac{f(\ell^* - \ell_x)}{D} + \frac{f(\ell_x - \ell_1)}{D} \tag{1}$$

Thus, from Eq. 1

$$\frac{f(\ell^* - \ell_x)}{D} = 0.48976 - \frac{(0.02)(2.5 \text{ m})}{(0.668 \text{ m})} = 0.4149 \tag{2}$$

With the result of Eq. 2 we use Table E.2 to determine (closest values)

$$\frac{f(\ell^* - \ell_x)}{D} = 0.41442$$

and

$$Ma_x = 2.42 \tag{3}$$

With $Ma_x = 2.42$ we go into Table E.4 for a normal shock wave and get

$$Ma_y = 0.521 \tag{4}$$

Now with $Ma_y = 0.521$, we obtain from Table E.2 for Fanno flow (closest values)

$$Ma_y = 0.52$$

and

$$\frac{f(\ell^* - \ell_y)}{D} = 0.91741$$

Since

$$\frac{f(\ell^* - \ell_2)}{D} = \frac{f(\ell^* - \ell_y)}{D} - \frac{f(\ell_2 - \ell_y)}{D}$$

we get

$$\frac{f(\ell^* - \ell_2)}{D} = 0.91741 - \frac{(0.02)(22.5 \text{ m})}{(0.668 \text{ m})} = 0.2438 \tag{5}$$

We enter Table E.2 with the result of Eq. 5 and ascertain that (closest values)

$$\frac{f(\ell^* - \ell_2)}{D} = 0.24978$$

and

$$Ma_2 = 0.68 \tag{6}$$

Now we form

$$\frac{p_2}{p_{0,1}} = \left(\frac{p_2}{p^*}\right)\left(\frac{p^*}{p_y}\right)\left(\frac{p_y}{p_x}\right)\left(\frac{p_x}{p^*}\right)\left(\frac{p^*}{p_1}\right)\left(\frac{p_1}{p_{0,1}}\right) \tag{7}$$

Knowing $Ma_2 = 0.68$, we get from Table E.2 (Fanno)

$$\frac{p_2}{p^*} = 1.5413 \tag{8}$$

With $Ma_y = 0.52$, we obtain from Table E.2 (Fanno)

$$\frac{p_y}{p^*} = 2.0519 \tag{9}$$

For $Ma_x = 2.42$, Table E.4 (shock) gives

$$\frac{p_y}{p_x} = 6.6658 \tag{10}$$

and Table E.2 (Fanno) yields

$$\frac{p_x}{p^*} = 0.30720 \tag{11}$$

For $Ma_1 = 2.8$, Table E.2 (Fanno) gives

$$\frac{p_1}{p^*} = 0.24414 \tag{12}$$

and Table E.1 (isentropic) yields

$$\frac{p_1}{p_{0,1}} = 0.03685 \tag{13}$$

Consolidating Eqs. 7 through 13 we obtain

$$\frac{p_2}{p_{0,1}} = (1.5413)\left(\frac{1}{2.0519}\right)(6.6658)$$

$$\times \ (0.30720)\left(\frac{1}{0.24414}\right)(0.03685) = 0.232 \tag{Ans}$$

When the back pressure is set to a value that is 23% of the inlet stagnation pressure, the flow through the converging–diverging nozzle and the adiabatic pipe will involve a normal shock at 2.5 m downstream from the pipe entrance. The flow in the adiabatic pipe will decelerate from $Ma_1 = 2.8$ to $Ma_x = 2.42$. Then, across the normal shock, the flow will further decelerate to $Ma_y = 0.52$. From the normal shock on downstream to the adiabatic pipe exit, the flow accelerates from $Ma_y = 0.52$ to $Ma_2 = 0.68$. The temperature–entropy diagram for this flow is sketched in Fig. E11.21b.

11.6 ANALOGY BETWEEN COMPRESSIBLE AND OPEN-CHANNEL FLOWS

During a first course in fluid mechanics, students rarely study both open-channel flows (Chapter 10) and compressible flows. This is unfortunate because these two kinds of flows are strikingly similar in several ways. Furthermore, the analogy between open-channel and compressible flows is useful because important two-dimensional compressible flow phenomena can be simply and inexpensively demonstrated with a shallow, open-channel flow field in a *ripple tank* or *water table*.

The propagation of weak pressure pulses (sound waves) in a compressible flow can be considered to be comparable to the movement of small amplitude waves on the surface of an open-channel flow. In each case, two-dimensional compressible flow and open-channel flow, the influence of flow velocity on wave pattern is similar. When

the flow velocity is less than the wave speed, wave fronts can move upstream of the wave source and the flow is subsonic (compressible flow) or subcritical (open-channel flow). When the flow velocity is equal to the wave speed, wave fronts cannot move upstream of the wave source and the flow is sonic (compressible flow) or critical (open-channel flow). When the flow velocity is greater than the wave speed, the flow is supersonic (compressible flow) or supercritical (open-channel flow). Normal shocks can occur in supersonic compressible flows. Hydraulic jumps can occur in supercritical open-channel flows. Comparison of the characteristics of normal shocks (Section 11.5.3) and hydraulic jumps (Section 10.6.1) suggest a strong resemblance and thus analogy between the two phenomena.

For compressible flows a meaningful dimensionless variable is the Mach number, where

$$\text{Ma} = \frac{V}{c} \tag{11.46}$$

In open-channel flows, an important dimensionless variable is the Froude number, where

$$\text{Fr} = \frac{V_{oc}}{\sqrt{gy}} \tag{11.157}$$

The velocity of the channel flow is V_{oc}, the acceleration of gravity is g, and the depth of the flow is y. Since the speed of a small amplitude wave on the surface of an open-channel flow, c_{oc}, is (see Section 10.2.1)

$$c_{oc} = \sqrt{gy} \tag{11.158}$$

we conclude that

$$\text{Fr} = \frac{V_{oc}}{c_{oc}} \tag{11.159}$$

From Eqs. 11.46 and 11.159 we see the similarity between Mach number (compressible flow) and Froude number (open-channel flow).

For compressible flow, the continuity equation is

$$\rho A V = \text{constant} \tag{11.160}$$

where V is the flow velocity, ρ is the fluid density, and A is the flow cross section area. For an open-channel flow, conservation of mass leads to

$$y b V_{oc} = \text{constant} \tag{11.161}$$

where V_{oc} is the flow velocity, and y and b are the depth and width of the open-channel flow. Comparing Eqs. 11.160 and 11.161 we note that if flow velocities are considered similar and flow area, A, and channel width, b, are considered similar, then compressible flow density, ρ, is analogous to open-channel flow depth, y.

It should be pointed out that the similarity between Mach number and Froude number is generally not exact. If compressible flow and open-channel flow velocities are considered to be similar, then it follows that for Mach number and Froude number similarity the wave speeds c and c_{oc} must also be similar.

From the development of the equation for the speed of sound in an ideal gas (see Eqs. 11.34 and 11.35) we have for the compressible flow

$$c = \sqrt{(\text{constant}) k \rho^{k-1}} \tag{11.162}$$

From Eqs. 11.162 and 11.158, we see that if y is to be similar to ρ as suggested by comparing Eq. 11.160 and 11.161, then k should be equal to 2. Typically $k = 1.4$ or 1.67, not 2. This limitation to exactness is, however, usually not serious enough to compromise the benefits of the analogy between compressible and open-channel flows.

REFERENCES

1. Coles, D., "Channel Flow of a Compressible Fluid," Summary description of film in *Illustrated Experiments in Fluid Mechanics, The NCFMF Book of Film Notes,* MIT Press, Cambridge, Mass., 1972.

2. Jones, J. B., and Hawkins, G. A., *Engineering Thermodynamics,* Second Edition, Wiley, New York, 1986.

3. Moran, M. J., and Shapiro, H. N., *Fundamentals of Engineering Thermodynamics,* Wiley, New York, 1988.

4. Keenan, J. H., Chao, J., and Kaye, J., *Gas Tables,* Second Edition, Wiley, New York, 1980.

5. Shapiro, A. H., *The Dynamics and Thermodynamics of Compressible Fluid Flow,* Vol. 1, Ronald Press, New York, 1953.

6. Thompson, P. A., *Compressible-Fluid Dynamics,* McGraw-Hill, New York, 1972.

7. Zuchrow, M. J., and Hoffman, J. D., *Gas Dynamics,* Vol. 1, Wiley, New York, 1976.

8. Saad, M. A., *Compressible Fluid Flow,* Prentice-Hall, Englewood Cliffs, N.J., 1985.

PROBLEMS

Note: Unless otherwise indicated, use the values of fluid properties found in the tables on the inside of the front cover. Problems designated with an (*) are intended to be solved with the aid of a programmable calculator or a computer. If $k = 1.4$ and the tables of Appendix E can be used to simplify a problem solution, use the closest tabulated value, do not interpolate, unless instructed differently.

11.1 Determine, in SI units, nominal values of c_p and c_v for: **(a)** air; **(b)** carbon dioxide; **(c)** helium; **(d)** hydrogen; **(e)** methane; **(f)** nitrogen; **(g)** oxygen. Use information provided in Table 1.8.

11.2 Determine, in BG units, nominal values of c_p and c_v for: **(a)** air; **(b)** carbon dioxide; **(c)** helium; **(d)** hydrogen; **(e)** methane; **(f)** nitrogen; **(g)** oxygen. Use information provided in Table 1.7.

11.3 Nitrogen flows steadily between two sections in a duct. At section (1), the temperature and pressure are $T_1 = 80$ °C, $p_1 = 301$ kPa (abs), and at section (2), the temperature and pressure are $T_2 = 180$ °C, $p_2 = 181$ kPa (abs). Calculate the **(a)** change in internal energy between sections (1) and (2); **(b)** change in enthalpy between sections (1) and (2); **(c)** change in density between sections (1) and (2); **(d)** change in entropy between sections (1) and (2).

11.4 Three kilograms of hydrogen contained in a nondeforming sealed vessel are cooled from 400 °C, 400 kPa (abs) until the hydrogen pressure is 100 kPa (abs). Calculate the change in internal energy, enthalpy, and entropy associated with this cooling process.

11.5 Five pounds mass of oxygen are heated in a closed, rigid container from 80 °F, 15 psia to 500 °F. Determine the final pressure of the oxygen and the entropy rise involved.

11.6 Helium is compressed isothermally from 121 kPa (abs) to 301 kPa (abs). Determine the entropy change associated with this process.

11.7   Air at 14.7 psia and 70 °F is compressed adiabatically by a centrifugal compressor to a pressure of 60 psia. What is the minimum temperature rise possible? Explain.

11.8   Methane is compressed adiabatically from 100 kPa (abs), 25 °C to 200 kPa (abs). What is the minimum compressor exit temperature possible? Explain.

11.9   Air expands adiabatically through a turbine from a pressure and temperature of 80 psia, 1600 °R to a pressure of 14.7 psia. If the actual temperature change is 80% of the ideal temperature change, determine the actual temperature of the expanded air and the actual enthalpy and entropy differences across the turbine.

11.10   An expression for the value of c_p for carbon dioxide as a function of temperature is

$$c_p = 9210 - \frac{3.71 \times 10^6}{T} + \frac{8.02 \times 10^8}{T^2}$$

where c_p is in (ft · lb)/(slug · °R) and where T is in °R. Compare the change in enthalpy of carbon dioxide using the constant value of c_p from Problem 11.2 with the change in enthalpy of carbon dioxide using the expression above, for $T_2 - T_1$ equal to **(a)** 10 °R; **(b)** 1000 °R and **(c)** 3000 °R. Set $T_1 = 540$ °R.

11.11   Confirm the speed of sound for air at 70 °F listed in Table B.3.

11.12   From Table B.1 we can conclude that the speed of sound in water at 60 °F is 4814 ft/s. Is this value of c consistent with the value of bulk modulus, E_v, listed in Table 1.5?

11.13   An airplane moves forward in air with a speed of 500 mph at an altitude of 40,000 ft. Determine the Mach number involved if the air is considered as U.S. standard atmosphere (see Table C.1).

11.14   Determine the Mach number of a car moving in standard air at a speed of **(a)** 25 mph, **(b)** 55 mph, and **(c)** 100 mph.

11.15   Using information provided in Table C.2, develop a table of speed of sound in m/s as a function of elevation for U.S. standard atmosphere.

11.16   Using information provided in Table C.1, develop a table of speed of sound in ft/s as a function of elevation for U.S. standard atmosphere.

11.17   If a high-performance aircraft is able to cruise at a Mach number of 3.0 at an altitude of 80,000 ft, how fast is this in: **(a)** mph; **(b)** ft/s; **(c)** m/s?

11.18   Compare values of the speed of sound in ft/s at 68 °F in the following gases: **(a)** air; **(b)** carbon dioxide; **(c)** helium; **(d)** hydrogen; **(e)** methane.

11.19   Compare values of the speed of sound in m/s in the following liquids at 20 °C: **(a)** ethyl alcohol; **(b)** glycerin; **(c)** mercury.

11.20   If the observed speed of sound in steel is 5300 m/s, determine the bulk modulus of elasticity of steel in N/m³. The density of steel is nominally 7790 kg/m³. How does your value of E_v for steel compare with E_v for water at 15.6 °C? Compare the speed of sound in steel and in water and comment on what you observe.

11.21   The flow of an ideal gas may be considered incompressible if the Mach number is less than 0.3. Determine the velocity level in ft/s and in m/s for Ma = 0.3 in the following gases: **(a)** standard air; **(b)** hydrogen at 68 °F.

11.22   At the seashore, you observe a high-speed aircraft moving overhead at an elevation of 10,000 ft. You hear the plane 8 s after it passes directly overhead. Using a nominal air temperature of 40 °F, estimate the Mach number and speed of the aircraft.

11.23   A schlieren photo of a bullet moving through air at 101 kPa (abs) and 15 °C indicates a Mach cone angle of 28°. How fast was the bullet moving in: **(a)** m/s; **(b)** ft/s; **(c)** mph?

11.24   At a given instant of time, two of the pressure waves, each moving at the speed of sound, emitted by a point source moving with constant

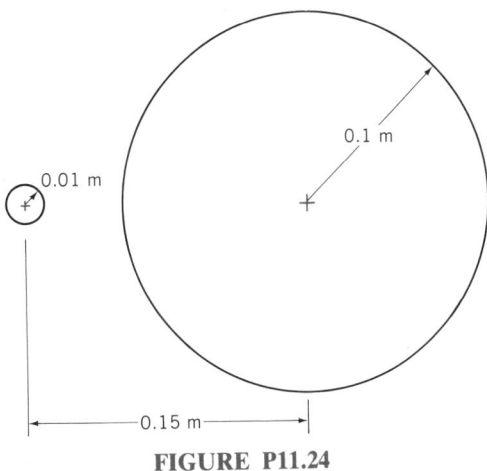

0.1 m

0.01 m

0.15 m

FIGURE P11.24

velocity in a fluid at rest are shown in Fig. P11.24. Determine the Mach number involved and indicate with a sketch the instantaneous location of the point source.

11.25   At a given instant of time, two of the pressure waves, each moving at the speed of sound, emitted by a point source moving with constant velocity in a fluid at rest are shown in Fig. P11.25. Determine the Mach number involved and indicate with a sketch the instantaneous location of the point source.

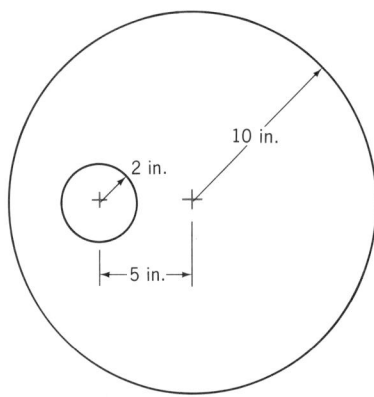

FIGURE P11.25

11.26   At a given instant of time, a pressure wave moving at the speed of sound which is emitted by a point source moving with constant velocity along the line *A–B* in a fluid at rest is shown in Fig. P11.26. If the point source Mach number is 0.8, sketch, for the given instant, the pressure wave emitted by the point source when it was at location *B*.

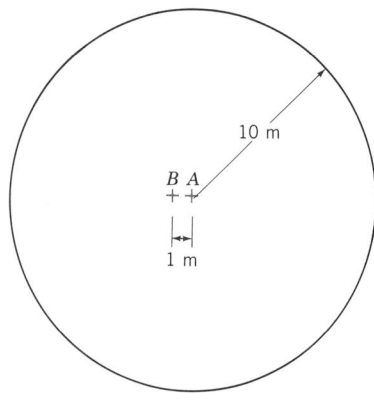

FIGURE P11.26

11.27   A stationary point source emits weak pressure pulses in a flow that moves uniformly

from left to right with a Mach number of 0.5. Sketch the instantaneous outline at time $= 10/c$ of pressure waves emitted earlier at time $= 5/c$ and time $= 8/c$. The time is in seconds and the speed of sound is denoted with *c*.

11.28   Starting with Eq. 11.52, prove that the stagnation enthalpy (and temperature) of an ideal gas remains constant during isentropic flow. Does this result fix the stagnation state for the flow? Explain. Using Eq. 5.69, comment on heat transfer and shaft work during any constant stagnation state process.

11.29   Explain how fluid pressure varies with cross section area change for the isentropic flow of an ideal gas when the flow is **(a)** subsonic; **(b)** supersonic.

11.30   For any ideal gas, prove that the slope of constant pressure lines on a temperature–entropy diagram is positive and that higher pressure lines are above lower pressure lines.

11.31   Determine the critical pressure and temperature ratios for: **(a)** air; **(b)** carbon dioxide; **(c)** helium; **(d)** hydrogen; **(e)** methane; **(f)** nitrogen; **(g)** oxygen.

11.32   Air flows steadily and isentropically from standard atmospheric conditions to a receiver pipe through a converging duct. The cross section area of the throat of the converging duct is 0.05 ft². Determine the mass flowrate through the duct if the receiver pressure is **(a)** 10 psia; **(b)** 5 psia. Sketch temperature–entropy diagrams for situations (a) and (b). Verify results obtained with values from the appropriate table in Appendix E with calculations involving ideal gas equations.

11.33   Helium at 68 °F and 14.7 psia in a large tank flows steadily and isentropically through a converging nozzle to a receiver pipe. The cross section area of the throat of the converging passage is 0.05 ft². Determine the mass flowrate through the duct if the receiver pressure is **(a)** 10 psia; **(b)** 5 psia. Sketch temperature–entropy diagrams for situations (a) and (b).

11.34   What is the static pressure to stagnation pressure ratio associated with the following motion in standard air: **(a)** a runner moving at the rate of 20 mph; **(b)** a cyclist moving at the rate of 40 mph; **(c)** a car moving at the rate of 65 mph; **(d)** an airplane moving at the rate of 500 mph.

11.35   The static pressure to stagnation pressure ratio at a point in an ideal gas flow field is

measured with a Pitot-static probe as being equal to 0.6. The stagnation temperature of the gas is 20 °C. Determine the flow speed in m/s and Mach number if the gas is **(a)** air; **(b)** carbon dioxide; **(c)** hydrogen.

11.36 The stagnation pressure and temperature of air flowing past a probe are 120 kPa (abs) and 100 °C, respectively. The air pressure is 80 kPa (abs). Determine the air speed and Mach number considering the flow to be **(a)** incompressible; **(b)** compressible.

11.37 The stagnation pressure indicated by a Pitot tube mounted on an airplane in flight is 45 kPa (abs). If the aircraft is cruising in standard atmosphere at an altitude of 10,000 m, determine the speed and Mach number involved.

11.38 Graph, for the isentropic flow of an ideal gas having a constant specific heat ratio, k, of 1.4, the variation of critical area ratio, A/A^*, as a function of Mach number for $0.1 \leq \text{Ma} \leq 3.0$.

11.39 An ideal gas enters subsonically and flows isentropically through a choked converging–diverging duct having a circular cross section area, A, that varies with axial distance from the throat, x, according to the formula

$$A = 0.1 + x^2$$

where A is in square feet and x is in feet. For this flow situation, sketch the side view of the duct and graph the variation of Mach number, static temperature to stagnation temperature ratio, T/T_0, and static pressure to stagnation pressure ratio, p/p_0, through the duct from $x = -1.0$ ft to $x = +1.0$ ft. Also show the possible fluid states at $x = -1.0$ ft, 0 ft, and $+1.0$ ft using temperature–entropy coordinates. Consider the gas as being **(a)** air; **(b*)** helium (use $0.051 \leq \text{Ma} \leq 5.193$).

11.40 An ideal gas enters supersonically and flows isentropically through the choked converging–diverging duct described in Problem 11.39. Graph the variation of Ma, T/T_0, and p/p_0 from the entrance to the exit sections of the duct for **(a)** air; **(b*)** helium (use $0.051 \leq \text{Ma} \leq 5.193$). Show the possible fluid states at $x = -1.0$ ft, 0 ft, and $+1.0$ ft using temperature–entropy coordinates.

11.41* Helium enters supersonically and flows isentropically through the choked converging–diverging duct described in Example 11.8. Compare the variation of Ma, T/T_0, and p/p_0 for helium with the variation of these parameters for air throughout the duct. Use $0.163 \leq \text{Ma} \leq 3.221$.

11.42* Helium enters subsonically and flows isentropically through the converging–diverging duct of Example 11.8. Compare the values of Ma, T/T_0, and p/p_0 for helium with those for air at several locations in the duct. Use $0.163 \leq \text{Ma} \leq 3.221$.

11.43* Helium flows subsonically and isentropically through the converging–diverging duct of Example 11.8. Graph the variation of Ma, T/T_0, and p/p_0 through the duct from $x = -0.5$ m to $x = +0.5$ m for $p/p_0 = 0.99$ at $x = -0.5$ m. Sketch the corresponding $T - s$ diagram. Use $0.110 \leq \text{Ma} \leq 0.430$.

11.44 An ideal gas flows subsonically and isentropically through the converging–diverging duct described in Problem 11.39. Graph the variation of Ma, T/T_0, and p/p_0 from the entrance to the exit sections of the duct for **(a)** air; **(b*)** helium (use $0.047 \leq \text{Ma} \leq 0.722$). The value of p/p_0 is 0.6708 at $x = 0$ ft. Sketch important states on a $T - s$ diagram.

11.45 An ideal gas contained in a large storage container at a constant temperature and pressure of 59 °F and 25 psia is to be expanded isentropically through a duct to standard atmospheric discharge conditions. Describe in general terms the kind of duct required and determine the duct exit cross section area if the discharge mass flowrate required is 1.0 lbm/s and the gas is **(a)** air; **(b)** carbon dioxide; **(c)** helium.

11.46 An ideal gas is to flow isentropically from a large tank where the air is maintained at a temperature and pressure of 59 °F and 80 psia to standard atmospheric discharge conditions. Describe in general terms the kind of duct involved and determine the duct exit Mach number and velocity in ft/s if the gas is **(a)** air; **(b)** methane; **(c)** helium.

11.47 At section (1) in the isentropic flow of carbon dioxide, $p_1 = 40$ kPa (abs), $T_1 = 60$ °C, and $V_1 = 350$ m/s. Determine the flow velocity, V_2, in m/s, at another section, section (2), where the Mach number is 2.0. Also calculate the section area ratio, A_2/A_1.

11.48 An ideal gas flows isentropically through a converging–diverging nozzle. At a section in the converging portion of the nozzle, $A_1 = 0.1$ m², $p_1 = 600$ kPa (abs), $T_1 = 20$ °C, and $\text{Ma}_1 = 0.6$. For section (2) in the diverging part of the nozzle

determine A_2, p_2, and T_2 if $Ma_2 = 3.0$ and the gas is **(a)** air; **(b)** helium.

11.49   Upstream of the throat of an isentropic converging–diverging nozzle at section (1), $V_1 = 150$ m/s, $p_1 = 100$ kPa (abs), and $T_1 = 20$ °C. If the discharge flow is supersonic and the throat area is 0.1 m², determine the mass flowrate in kg/s for the flow of **(a)** air; **(b)** methane; **(c)** helium.

11.50   An ideal gas in a large storage tank at 100 psia and 60 °F flows isentropically through a converging duct to the atmosphere. The throat area of the duct is 0.1 ft². Determine the pressure, temperature, velocity, and mass flowrate of the gas at the duct throat if the gas is **(a)** air; **(b)** carbon dioxide; **(c)** helium.

11.51   The flow blockage associated with the use of an intrusive probe can be important. Determine the percentage increase in section velocity corresponding to a 0.5% reduction in flow area due to probe blockage for air flow if the section area is 1.0 m², $T_0 = 20$ °C, and the unblocked flow Mach numbers are **(a)** Ma = 0.2; **(b)** Ma = 0.8; **(c)** Ma = 1.5; **(d)** Ma = 3.0.

11.52   An ideal gas enters [section (1)] an insulated, constant flow cross section area duct with the following properties:

$$T_0 = 293 \text{ K}$$

$$p_0 = 101 \text{ kPa (abs)}$$

$$Ma_1 = 0.2$$

For Fanno flow, determine corresponding values of fluid temperature and entropy change for various levels of pressure and plot the Fanno line if the gas is **(a)** air; **(b)** helium.

11.53   For Fanno flow, prove that

$$\frac{dV}{V} = \frac{fk(Ma^2/2)(dx/D)}{1 - Ma^2}$$

and in so doing show that when the flow is subsonic, friction accelerates the fluid, and when the flow is supersonic, friction decelerates the fluid.

11.54   Standard atmospheric air ($T_0 = 59$ °F, $p_0 = 14.7$ psia) is drawn steadily through a frictionless and adiabatic converging nozzle into an adiabatic, constant cross section area duct. The duct is 10 ft long and has an inside diameter of 0.5 ft. The average friction factor for the duct may be estimated as being equal to 0.03. What is the maximum mass flowrate in slugs/s through the duct? For this maximum flowrate determine

the values of static temperature, static pressure, stagnation temperature, stagnation pressure, and velocity at the inlet [section (1)] and exit [section (2)] of the constant area duct. Sketch a temperature–entropy diagram for this flow.

11.55   An ideal gas [$T_0 = 288$ K and $p_0 = 101$ kPa (abs)] is drawn steadily through the flow configuration of Example 11.12. Compare the maximum mass flowrate, and constant area duct inlet [section (1)] and exit [section (2)] values of static temperature, static pressure, stagnation temperature, stagnation pressure, and velocity for **(a)** helium and **(b)** carbon dioxide with the quantities for air.

11.56   The duct in Problem 11.54 is shortened by 50%. The duct discharge pressure is maintained at the choked flow value determined in Problem 11.54. Determine the change in mass flowrate through the duct associated with the 50% reduction in length. The average friction factor remains constant at a value of 0.03.

11.57   If the same mass flowrate of air obtained in Problem 11.54 is desired through the shortened duct of Problem 11.56, determine the back pressure, p_2, required. Assume f remains constant at a value of 0.03.

11.58   If the average friction factor of the duct of Example 11.12 is changed to **(a)** 0.01 or **(b)** 0.03, determine the maximum mass flowrate of air through the duct associated with each new friction factor and compare with the maximum mass flowrate value of Example 11.12.

11.59   If the length of the constant area duct of Example 11.12 is changed to **(a)** 1 m or **(b)** 3 m, and all other specifications in the problem statement remain the same, determine the maximum mass flowrate of air through the duct associated with each new length and compare with the maximum mass flowrate of Example 11.12.

11.60   The duct of Example 11.12 is lengthened by 50%. If the duct discharge pressure is set at a value of $p_d = 46.2$ kPa (abs), determine the mass flowrate of air through the lengthened duct. The average friction factor for the duct remains constant at a value of 0.02.

11.61   A long smooth wall pipe ($f = 0.01$) is to deliver 8000 ft³/min of air at 60 °F and 14.7 psia. The inside diameter of the pipe is 0.5 ft and the length of the pipe is 100 ft. Determine the static temperature and pressure required at the pipe entrance if the flow through the pipe is adiabatic.

11.62   An ideal gas flows adiabatically with friction through a long, constant cross section area pipe. At upstream section (1), $p_1 = 60$ kPa (abs), $T_1 = 60\,°C$ and $V_1 = 200$ m/s. At downstream section (2), $T_2 = 30\,°C$. Determine p_2, V_2, and the stagnation pressure ratio $p_{0,2}/p_{0,1}$ if the gas is **(a)** air; **(b)** helium.

11.63   For the air flow of Problem 11.62, determine T, p, and V for the section halfway between sections (1) and (2).

11.64   An ideal gas flows adiabatically between two sections in a constant cross section area pipe. At upstream section (1), $p_{0,1} = 100$ psia, $T_{0,1} = 600\,°R$, and $Ma_1 = 0.5$. At downstream section (2), the flow is choked. Determine the magnitude of the force per unit cross section area exerted by the inside wall of the pipe on the fluid between sections (1) and (2) if the gas is **(a)** air; **(b)** helium.

11.65   An ideal gas enters [section (1)] a frictionless, constant flow cross section area duct with the following properties:

$$T_0 = 293\ K$$

$$p_0 = 101\ \text{kPa (abs)}$$

$$Ma_1 = 0.2$$

For Rayleigh flow, determine corresponding values of fluid temperature and entropy change for various levels of pressure and plot the Rayleigh line if the gas is **(a)** air; **(b)** helium.

11.66   Standard atmospheric air [$T_0 = 288$ K, $p_0 = 101$ kPa (abs)] is drawn steadily through an isentropic converging nozzle into a frictionless and diabatic ($q = 500$ kJ/kg) constant cross section area duct. For maximum flow determine the values of static temperature, static pressure, stagnation temperature, stagnation pressure, and flow velocity at the inlet [section (1)] and exit [section (2)] of the constant area duct. Sketch a temperature–entropy diagram for this flow.

11.67   Air enters a constant cross section area duct that may be considered frictionless with $T_1 = 300$ K and $V_1 = 300$ m/s. Determine the amount of heat transfer in kJ/kg required to choke the Rayleigh flow involved.

11.68   An ideal gas enters a 0.5-ft inside diameter duct with $p_1 = 20$ psia, $T_1 = 80\,°F$, and $V_1 = 200$ ft/s. What frictionless heat addition rate in Btu/s is necessary for an exit gas temperature, $T_2 = 1500\,°F$? Determine p_2, V_2, and Ma_2 also. The gas is **(a)** air; **(b)** helium.

11.69   Air enters a length of constant cross section area pipe with $p_1 = 200$ kPa (abs), $T_1 = 500$ K, and $V_1 = 400$ m/s. If 500 kJ/kg of energy is removed from the air by frictionless heat transfer between sections (1) and (2), determine p_2, T_2, and V_2. Sketch a temperature–entropy diagram for the flow between sections (1) and (2).

11.70   Air flows through a constant cross section area pipe. At an upstream section (1), $p_1 = 15$ psia, $T_1 = 530\,°R$, and $V_1 = 200$ ft/s. Downstream at section (2), $p_2 = 10$ psia and $T_2 = 1760\,°R$. For this flow, determine the stagnation temperature and pressure ratios, $T_{0,2}/T_{0,1}$ and $p_{0,2}/p_{0,1}$, and the heat transfer per unit mass of air flowing between sections (1) and (2). Is the flow between sections (1) and (2) frictionless? Explain.

11.71   The Mach number and stagnation pressure of an ideal gas are 2.0 and 200 kPa (abs) just upstream of a normal shock. Determine the stagnation pressure loss across the shock for the following gases: **(a)** air; **(b)** helium. Comment on the effect of specific heat ratio, k, on shock loss.

11.72   The stagnation pressure ratio across a normal shock in an ideal gas flow is 0.5. Determine the Mach number of the flow entering the shock if the gas is air.

11.73   Just upstream of a normal shock in an ideal gas flow, $Ma = 3.0$, $T = 600\,°R$, and $p = 30$ psia. Determine values of Ma, T_0, T, p_0, p, and V downstream of the shock if the gas is **(a)** air; **(b)** helium.

11.74   A total pressure probe is inserted into a supersonic air flow. A shock wave forms just upstream of the impact hole. The probe measures a total pressure of 600 kPa (abs). The stagnation temperature at the probe head is 500 K. The static pressure upstream of the shock is measured with a wall tap to be 100 kPa (abs). From these data determine the Mach number and velocity of the flow.

11.75   The Pitot tube on a supersonic aircraft cruising at an altitude of 50,000 ft senses a stagnation pressure of 12 psia. If the atmosphere is considered standard, determine the air speed and Mach number of the aircraft. A shock wave is present just upstream of the probe impact hole.

11.76   An aircraft cruises at a Mach number of 2.0 at an altitude of 15 km. Inlet air is decelerated to a Mach number of 0.4 at the engine compressor inlet. A normal shock occurs in the inlet diffuser upstream of the compressor inlet at

a section where the Mach number is 1.2. For isentropic diffusion, except across the shock, and for standard atmosphere determine the stagnation temperature and pressure of the air entering the engine compressor.

11.77 Standard atmospheric air enters subsonically and accelerates isentropically to supersonic flow in a duct. If the ratio of duct exit to throat cross section areas is 3.0, determine the ratio of back pressure to inlet stagnation pressure that will result in a standing normal shock at the duct exit. Determine also the stagnation pressure loss across the normal shock in kPa.

11.78 Determine, for the air flow through the frictionless and adiabatic converging–diverging duct of Example 11.8, the ratio of duct exit pressure to duct inlet stagnation pressure that will result in a standing normal shock at: **(a)** $x = +0.1$ m; **(b)** $x = +0.2$ m; **(c)** $x = +0.4$ m. How large is the stagnation pressure loss in each case?

11.79 A normal shock is positioned in the diverging portion of a frictionless, adiabatic, converging–diverging air flow duct where the cross section area is 0.1 ft² and the local Mach number is 2.0. Upstream of the shock, $p_0 = 200$ psia and $T_0 = 1200$ °R. If the duct exit area is 0.15 ft², determine the exit area temperature and pressure and the duct mass flowrate.

11.80 Supersonic air flow enters an adiabatic, constant cross section area (inside diameter = 1 ft) pipe 30 ft long with $Ma_1 = 3.0$. The pipe friction factor is estimated to be 0.02. What ratio of pipe exit pressure to pipe inlet stagnation pressure would result in a normal shock wave standing at **(a)** $x = 5$ ft, or **(b)** $x = 10$ ft, where x is the distance downstream from the pipe entrance?

Determine also the duct exit Mach number and sketch the temperature–entropy diagram for each situation.

11.81 Supersonic ideal gas flow enters an adiabatic, constant cross section area pipe (inside diameter = 0.1 m) with $Ma_1 = 2.0$. The pipe friction factor is 0.02. If a standing normal shock is located right at the pipe exit, and the Mach number just upstream of the shock is 1.2, determine the length of the pipe if the gas is **(a)** air; **(b)** helium.

11.82 Air enters a frictionless, constant cross section area duct with $Ma_1 = 2.0$, $T_{0,1} = 59$ °F, and $p_{0,1} = 14.7$ psia. The air is decelerated by heating until a normal shock wave occurs where the local Mach number is 1.5. Downstream of the normal shock, the subsonic flow is accelerated with heating until it chokes at the duct exit. Determine the static temperature and pressure, the stagnation temperature and pressure, and the fluid velocity at the duct entrance, just upstream and downstream of the normal shock and at the duct exit. Sketch the temperature–entropy diagram for this flow.

11.83 An ideal gas enters a frictionless, constant cross section area duct with $Ma = 2.5$, $T_0 = 20$ °C, and $p_0 = 101$ kPa (abs). The gas is decelerated by heating until a normal shock occurs where the local Mach number is 1.3. Downstream of the shock, the subsonic flow is accelerated with heating until it exits with a Mach number of 0.9. Determine the static temperature and pressure, the stagnation temperature and pressure, and the fluid velocity at the duct entrance, just upstream and downstream of the normal shock and at the duct exit if the gas is **(a)** air or **(b)** helium. Sketch the temperature–entropy diagram for each flow.

APPENDICES

APPENDIX A

UNIT CONVERSION TABLES[1]

The following tables express the definitions of miscellaneous units of measure as exact numerical multiples of coherent SI units, and provide multiplying factors for converting numbers and miscellaneous units to corresponding new numbers and SI units.

Conversion factors are expressed using computer exponential notation, and an asterisk follows each number which expresses an exact definition. For example, the entry "2.54 E − 2*" expresses the fact that 1 inch = 2.54×10^{-2} meter, exactly by definition. Numbers not followed by an asterisk are only approximate representations of definitions, or are the results of physical measurements. In these tables pound-force is designated as lbf, whereas in the text pound-force is designated as lb.

TABLE A.1 Listing by Physical Quantity

To convert from	to	Multiply by
Acceleration		
foot/second^2	meter/second^2	3.048 E − 1*
free fall, standard	meter/second^2	9.806 65 E + 0*
gal (galileo)	meter/second^2	1.00 E − 2*
inch/second^2	meter/second^2	2.54 E − 2*
Area		
acre	meter^2	4.046 856 422 4 E + 3*
are	meter^2	1.00 E + 2*

---

[1] These tables abridged from Mechtly, E. A., *The International System of Units, 2nd Revision*, NASA SP-7012, 1973.

TABLE A.1 (continued)

To convert from	to	Multiply by
barn	meter^2	1.00 E − 28*
foot^2	meter^2	9.290 304 E − 2*
hectare	meter^2	1.00 E + 4*
inch^2	meter^2	6.4516 E − 4*
mile^2 (U.S. statute)	meter^2	2.589 988 110 336 E + 6*
section	meter^2	2.589 988 110 336 E + 6*
township	meter^2	9.323 957 2 E + 7
yard^2	meter^2	8.361 273 6 E − 1*

Density

gram/centimeter^3	kilogram/meter^3	1.00 E + 3*
lbm/inch^3	kilogram/meter^3	2.767 990 5 E + 4
lbm/foot^3	kilogram/meter^3	1.601 846 3 E + 1
slug/foot^3	kilogram/meter^3	5.153 79 E + 2

Energy

British thermal unit:		
(IST after 1956)	joule	1.055 056 E + 3
British thermal unit (thermochemical)	joule	1.054 350 E + 3
calorie (International Steam Table)	joule	4.1868 E + 0
calorie (thermochemical)	joule	4.184 E + 0*
calorie (kilogram, International Steam Table)	joule	4.1868 E + 3
calorie (kilogram, thermochemical)	joule	4.184 E + 3*
electron volt	joule	1.602 191 7 E − 19
erg	joule	1.00 E − 7*
foot lbf	joule	1.355 817 9 E + 0
foot poundal	joule	4.214 011 0 E − 2
joule (international of 1948)	joule	1.000 165 E + 0
kilocalorie (International Steam Table)	joule	4.1868 E + 3
kilocalorie (thermochemical)	joule	4.184 E + 3*
kilowatt hour	joule	3.60 E + 6*
watt hour	joule	3.60 E + 3*

Force

dyne	newton	1.00 E − 5*
kilogram force (kgf)	newton	9.806 65 E + 0*
kilopond force	newton	9.806 65 E + 0*
kip	newton	4.448 221 615 260 5 E + 3*
lbf (pound force, avoirdupois)	newton	4.448 221 615 260 5 E + 0*
ounce force (avoirdupois)	newton	2.780 138 5 E − 1
pound force, lbf (avoirdupois)	newton	4.448 221 615 260 5 E + 0*
poundal	newton	1.382 549 543 76 E − 1*

Length

angstrom	meter	1.00 E − 10*
astronomical unit (IAU)	meter	1.496 00 E + 11
cubit	meter	4.572 E − 1*
fathom	meter	1.8288 E + 0*
foot	meter	3.048 E − 1*
furlong	meter	2.011 68 E + 2*
hand	meter	1.016 E − 1*
inch	meter	2.54 E − 2*
league (international nautical)	meter	5.556 E + 3*
light year	meter	9.460 55 E + 15
meter	wavelengths Kr 86	1.650 763 73 E + 6*
micron	meter	1.00 E − 6*
mil	meter	2.54 E − 5*
mile (U.S. statute)	meter	1.609 344 E + 3*

TABLE A.1   (continued)

To convert from	to	Multiply by
nautical mile (U.S.)	meter	1.852 E + 3*
rod	meter	5.0292 E + 0*
yard	meter	9.144 E − 1*

Mass

carat (metric)	kilogram	2.00 E − 4*
grain	kilogram	6.479 891 E − 5*
gram	kilogram	1.00 E − 3*
ounce mass (avoirdupois)	kilogram	2.834 952 312 5 E − 2*
pound mass, lbm (avoirdupois)	kilogram	4.535 923 7 E − 1*
slug	kilogram	1.459 390 29 E + 1
ton (long)	kilogram	1.016 046 908 8 E + 3*
ton (metric)	kilogram	1.00 E + 3*
ton (short, 2000 pound)	kilogram	9.071 847 4 E + 2*
tonne	kilogram	1.00 E + 3*

Power

Btu (thermochemical)/second	watt	1.054 350 264 488 E + 3
calorie (thermochemical)/second	watt	4.184 E + 0*
foot lbf/second	watt	1.355 817 9 E + 0
horsepower (550 foot lbf/second)	watt	7.456 998 7 E + 2
kilocalorie (thermochemical)/second	watt	4.184 E + 3*
watt (international of 1948)	watt	1.000 165 E + 0

Pressure

atmosphere	newton/meter^2	1.013 25 E + 5*
bar	newton/meter^2	1.00 E + 5*
barye	newton/meter^2	1.00 E − 1*
centimeter of mercury (0° C)	newton/meter^2	1.333 22 E + 3
centimeter of water (4° C)	newton/meter^2	9.806 38 E + 1
dyne/centimeter^2	newton/meter^2	1.00 E − 1*
foot of water (39.2° F)	newton/meter^2	2.988 98 E + 3
inch of mercury (32° F)	newton/meter^2	3.386 389 E + 3
inch of mercury (60° F)	newton/meter^2	3.376 85 E + 3
inch of water (39.2° F)	newton/meter^2	2.490 82 E + 2
inch of water (60° F)	newton/meter^2	2.4884 E + 2
kgf/centimeter^2	newton/meter^2	9.806 65 E + 4*
kgf/meter^2	newton/meter^2	9.806 65 E + 0*
lbf/foot^2	newton/meter^2	4.788 025 8 E + 1
lbf/inch^2 (psi)	newton/meter^2	6.894 757 2 E + 3
millibar	newton/meter^2	1.00 E + 2*
millimeter of mercury (0° C)	newton/meter^2	1.333 224 E + 2
pascal	newton/meter^2	1.00 E + 0*
psi (lbf/inch^2)	newton/meter^2	6.894 757 2 E + 3
torr (0° C)	newton/meter^2	1.333 22 E + 2

Speed

foot/second	meter/second	3.048 E − 1*
inch/second	meter/second	2.54 E − 2*
kilometer/hour	meter/second	2.777 777 8 E − 1
knot (international)	meter/second	5.144 444 444 E − 1
mile/hour (U.S. statute)	meter/second	4.4704 E − 1*

Temperature

Celsius	kelvin	$t_K = t_C + 273.15$
Fahrenheit	kelvin	$t_K = (5/9)(t_F + 459.67)$
Fahrenheit	Celsius	$t_C = (5/9)(t_F − 32)$
Rankine	kelvin	$t_K = (5/9)t_R$

TABLE A.1 (continued)

To convert from	to	Multiply by
	Time	
day (mean solar)	second (mean solar)	8.64 E + 4*
hour (mean solar)	second (mean solar)	3.60 E + 3*
minute (mean solar)	second (mean solar)	6.00 E + 1*
year (calendar)	second (mean solar)	3.1536 E + 7*
	Viscosity	
centistoke	meter^2/second	1.00 E − 6*
stoke	meter^2/second	1.00 E − 4*
foot^2/second	meter^2/second	9.290 304 E − 2*
centipoise	newton second/meter^2	1.00 E − 3*
lbm/foot second	newton second/meter^2	1.488 163 9 E + 0
lbf second/foot^2	newton second/meter^2	4.788 025 8 E + 1
poise	newton second/meter^2	1.00 E − 1*
poundal second/foot^2	newton second/meter^2	1.488 163 9 E + 0
slug/foot second	newton second/meter^2	4.788 025 8 E + 1
rhe	meter^2/newton second	1.00 E + 1*
	Volume	
acre foot	meter^3	1.233 481 837 547 52 E + 3*
barrel (petroleum, 42 gallons)	meter^3	1.589 873 E − 1
board foot	meter^3	2.359 737 216 E − 3*
bushel (U.S.)	meter^3	3.523 907 016 688 E − 2*
cord	meter^3	3.624 556 3 E + 0
cup	meter^3	2.365 882 365 E − 4*
dram (U.S. fluid)	meter^3	3.696 691 195 312 5 E − 6*
fluid ounce (U.S.)	meter^3	2.957 352 956 25 E − 5*
foot^3	meter^3	2.831 684 659 2 E − 2*
gallon (U.K. liquid)	meter^3	4.546 087 E − 3
gallon (U.S. liquid)	meter^3	3.785 411 784 E − 3*
inch^3	meter^3	1.638 706 4 E − 5*
liter	meter^3	1.00 E − 3*
ounce (U.S. fluid)	meter^3	2.957 352 956 25 E − 5*
peck (U.S.)	meter^3	8.809 767 541 72 E − 3*
pint (U.S. liquid)	meter^3	4.731 764 73 E − 4*
quart (U.S. liquid)	meter^3	9.463 529 5 E − 4
stere	meter^3	1.00 E + 0*
tablespoon	meter^3	1.478 676 478 125 E − 5*
teaspoon	meter^3	4.928 921 593 75 E − 6*
yard^3	meter^3	7.645 548 579 84 E − 1*

APPENDIX B

PHYSICAL PROPERTIES OF FLUIDS

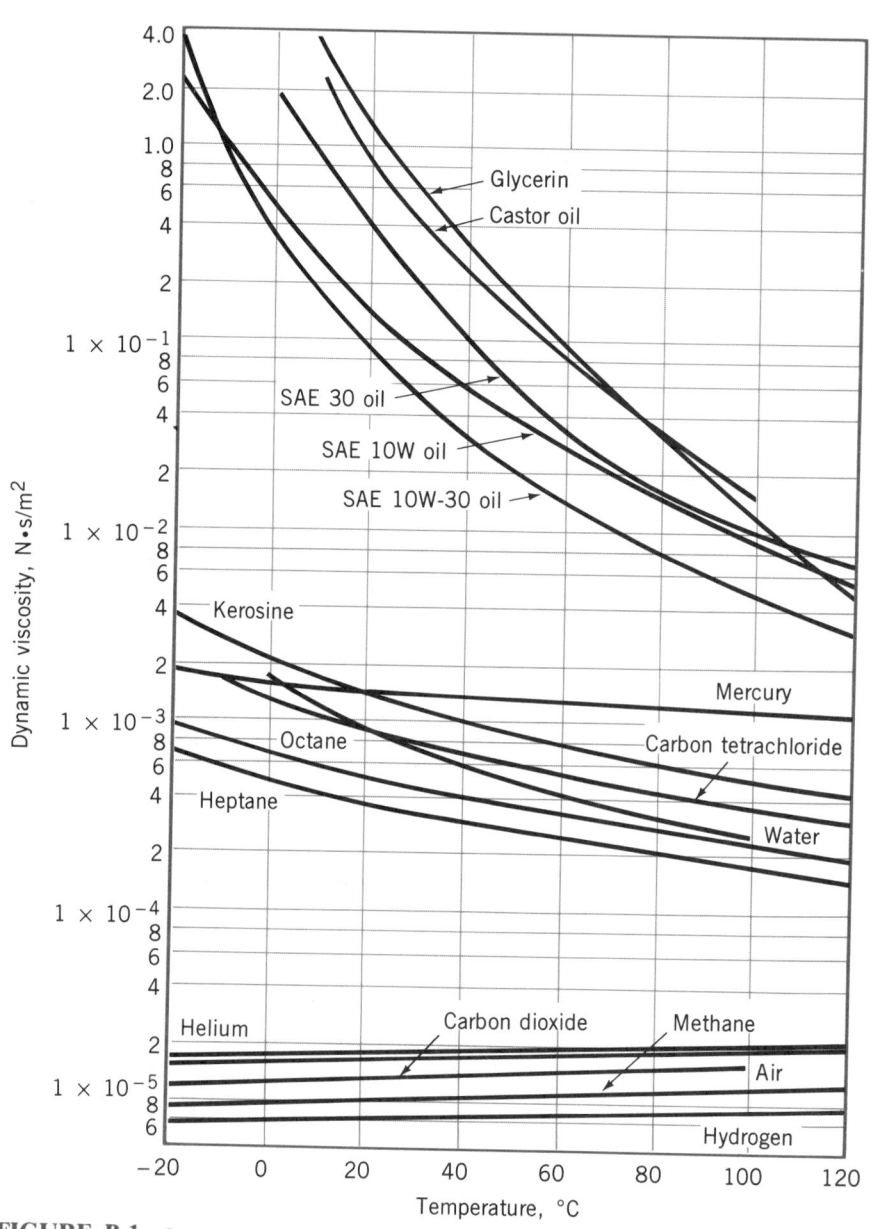

FIGURE B.1 Dynamic (absolute) viscosity of common fluids as a function of temperature. To convert to BG units of lb·s/ft² multiply N·s/m² by 2.089×10^{-2}. Curves from R. W. Fox, and A. T. McDonald, *Introduction to Fluid Mechanics,* Third Edition, Wiley, New York, 1985. Used by permission.

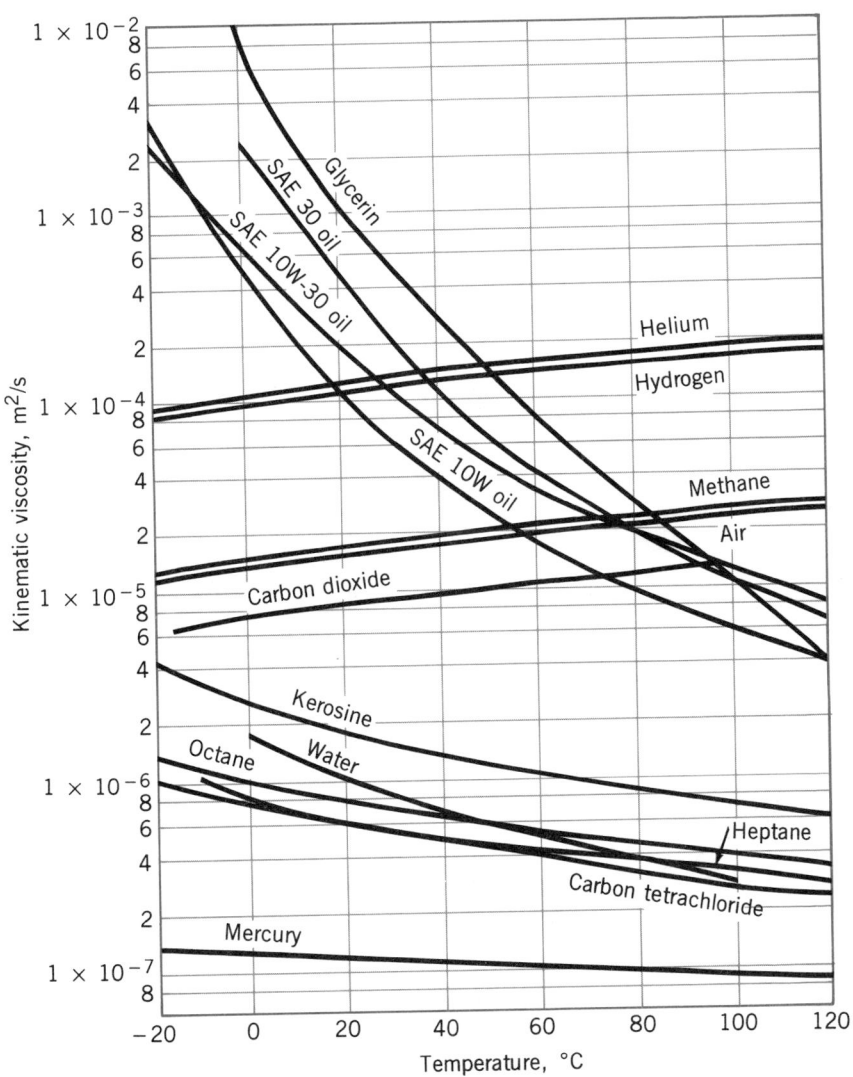

FIGURE B.2 Kinematic viscosity of common fluids (at atmospheric pressure) as a function of temperature. To convert to BG units of ft²/s multiply m²/s by 10.76. Curves from R. W. Fox, and A. T. McDonald, *Introduction to Fluid Mechanics,* Third Edition, Wiley, New York, 1985. Used by permission.

TABLE B.1 Physical Properties of Water (BG Units)[a]

Temperature (°F)	Density, ρ (slugs/ft^3)	Specific Weight[b], γ (lb/ft^3)	Dynamic Viscosity, μ (lb·s/ft^2)			Kinematic Viscosity, ν (ft^2/s)			Surface Tension[c], σ (lb/ft)			Vapor Pressure, p_v [lb/in.^{2}(abs)]			Speed of Sound[d], c (ft/s)
32	1.940	62.42	3.732	E	− 5	1.924	E	− 5	5.18	E	− 3	8.854	E	− 2	4603
40	1.940	62.43	3.228	E	− 5	1.664	E	− 5	5.13	E	− 3	1.217	E	− 1	4672
50	1.940	62.41	2.730	E	− 5	1.407	E	− 5	5.09	E	− 3	1.781	E	− 1	4748
60	1.938	62.37	2.344	E	− 5	1.210	E	− 5	5.03	E	− 3	2.563	E	− 1	4814
70	1.936	62.30	2.037	E	− 5	1.052	E	− 5	4.97	E	− 3	3.631	E	− 1	4871
80	1.934	62.22	1.791	E	− 5	9.262	E	− 6	4.91	E	− 3	5.069	E	− 1	4819
90	1.931	62.11	1.500	E	− 5	8.233	E	− 6	4.86	E	− 3	6.979	E	− 1	4960
100	1.927	62.00	1.423	E	− 5	7.383	E	− 6	4.79	E	− 3	9.493	E	− 1	4995
120	1.918	61.71	1.164	E	− 5	6.067	E	− 6	4.67	E	− 3	1.692	E	+ 0	5049
140	1.908	61.38	9.743	E	− 6	5.106	E	− 6	4.53	E	− 3	2.888	E	+ 0	5091
160	1.896	61.00	8.315	E	− 6	4.385	E	− 6	4.40	E	− 3	4.736	E	+ 0	5101
180	1.883	60.58	7.207	E	− 6	3.827	E	− 6	4.26	E	− 3	7.507	E	+ 0	5195
200	1.869	60.12	6.342	E	− 6	3.393	E	− 6	4.12	E	− 3	1.152	E	+ 1	5089
212	1.860	59.83	5.886	E	− 6	3.165	E	− 6	4.04	E	− 3	1.469	E	+ 1	5062

[a] Based on data from *Handbook of Chemistry and Physics*, 69th Ed., CRC Press, 1988. Where necessary, values obtained by interpolation.

[b] Density and specific weight are related through the equation $\gamma = \rho g$. For this table, $g = 32.174$ ft/s^2.

[c] In contact with air.

[d] From R. D. Blevins, *Applied Fluid Dynamics Handbook*, Van Nostrand Reinhold Co., Inc., New York, 1984.

TABLE B.2 Physical Properties of Water (SI Units)[a]

Temperature (°C)	Density, ρ (kg/m^3)	Specific Weight[b], γ (kN/m^3)	Dynamic Viscosity, μ (N·s/m^2)			Kinematic Viscosity, ν (m^2/s)			Surface Tension[c], σ (N/m)			Vapor Pressure, p_v [N/m^2(abs)]			Speed of Sound[d], c (m/s)
0	999.9	9.806	1.787	E	− 3	1.787	E	− 6	7.56	E	− 2	6.105	E	+ 2	1403
5	1000.0	9.807	1.519	E	− 3	1.519	E	− 6	7.49	E	− 2	8.722	E	+ 2	1427
10	999.7	9.804	1.307	E	− 3	1.307	E	− 6	7.42	E	− 2	1.228	E	+ 3	1447
20	998.2	9.789	1.002	E	− 3	1.004	E	− 6	7.28	E	− 2	2.338	E	+ 3	1481
30	995.7	9.765	7.975	E	− 4	8.009	E	− 7	7.12	E	− 2	4.243	E	+ 3	1507
40	992.2	9.731	6.529	E	− 4	6.580	E	− 7	6.96	E	− 2	7.376	E	+ 3	1526
50	988.1	9.690	5.468	E	− 4	5.534	E	− 7	6.79	E	− 2	1.233	E	+ 4	1541
60	983.2	9.642	4.665	E	− 4	4.745	E	− 7	6.62	E	− 2	1.992	E	+ 4	1552
70	977.8	9.589	4.042	E	− 4	4.134	E	− 7	6.44	E	− 2	3.116	E	+ 4	1555
80	971.8	9.530	3.547	E	− 4	3.650	E	− 7	6.26	E	− 2	4.734	E	+ 4	1555
90	965.3	9.467	3.147	E	− 4	3.260	E	− 7	6.08	E	− 2	7.010	E	+ 4	1550
100	958.4	9.399	2.818	E	− 4	2.940	E	− 7	5.89	E	− 2	1.013	E	+ 5	1543

[a] Based on data from *Handbook of Chemistry and Physics*, 69th Ed., CRC Press, 1988.

[b] Density and specific weight are related through the equation $\gamma = \rho g$. For this table, $g = 9.807$ m/s^2.

[c] In contact with air.

[d] From R. D. Blevins, *Applied Fluid Dynamics Handbook*, Van Nostrand Reinhold Co., Inc., New York, 1984.

TABLE B.3 Physical Properties of Air at Standard Atmospheric Pressure (BG Units)[a]

Temperature ($°F$)	Density, ρ (slugs/ft^3)			Specific Weight[b], γ (lb/ft^3)			Dynamic Viscosity, μ (lb·s/ft^2)			Kinematic Viscosity, ν (ft^2/s)			Specific Heat Ratio, k (—)	Speed of Sound, c (ft/s)
−40	2.939	E	− 3	9.456	E	− 2	3.29	E	− 7	1.12	E	− 4	1.401	1004
−20	2.805	E	− 3	9.026	E	− 2	3.34	E	− 7	1.19	E	− 4	1.401	1028
0	2.683	E	− 3	8.633	E	− 2	3.38	E	− 7	1.26	E	− 4	1.401	1051
10	2.626	E	− 3	8.449	E	− 2	3.44	E	− 7	1.31	E	− 4	1.401	1062
20	2.571	E	− 3	8.273	E	− 2	3.50	E	− 7	1.36	E	− 4	1.401	1074
30	2.519	E	− 3	8.104	E	− 2	3.58	E	− 7	1.42	E	− 4	1.401	1085
40	2.469	E	− 3	7.942	E	− 2	3.60	E	− 7	1.46	E	− 4	1.401	1096
50	2.420	E	− 3	7.786	E	− 2	3.68	E	− 7	1.52	E	− 4	1.401	1106
60	2.373	E	− 3	7.636	E	− 2	3.75	E	− 7	1.58	E	− 4	1.401	1117
70	2.329	E	− 3	7.492	E	− 2	3.82	E	− 7	1.64	E	− 4	1.401	1128
80	2.286	E	− 3	7.353	E	− 2	3.86	E	− 7	1.69	E	− 4	1.400	1138
90	2.244	E	− 3	7.219	E	− 2	3.90	E	− 7	1.74	E	− 4	1.400	1149
100	2.204	E	− 3	7.090	E	− 2	3.94	E	− 7	1.79	E	− 4	1.400	1159
120	2.128	E	− 3	6.846	E	− 2	4.02	E	− 7	1.89	E	− 4	1.400	1180
140	2.057	E	− 3	6.617	E	− 2	4.13	E	− 7	2.01	E	− 4	1.399	1200
160	1.990	E	− 3	6.404	E	− 2	4.22	E	− 7	2.12	E	− 4	1.399	1220
180	1.928	E	− 3	6.204	E	− 2	4.34	E	− 7	2.25	E	− 4	1.399	1239
200	1.870	E	− 3	6.016	E	− 2	4.49	E	− 7	2.40	E	− 4	1.398	1258
300	1.624	E	− 3	5.224	E	− 2	4.97	E	− 7	3.06	E	− 4	1.394	1348
400	1.435	E	− 3	4.616	E	− 2	5.24	E	− 7	3.65	E	− 4	1.389	1431
500	1.285	E	− 3	4.135	E	− 2	5.80	E	− 7	4.51	E	− 4	1.383	1509
750	1.020	E	− 3	3.280	E	− 2	6.81	E	− 7	6.68	E	− 4	1.367	1685
1000	8.445	E	− 4	2.717	E	− 2	7.85	E	− 7	9.30	E	− 4	1.351	1839
1500	6.291	E	− 4	2.024	E	− 2	9.50	E	− 7	1.51	E	− 3	1.329	2114

[a] Based on data from R. D. Blevins, *Applied Fluid Dynamics Handbook,* Van Nostrand Reinhold Co., Inc., New York, 1984.

[b] Density and specific weight are related through the equation $\gamma = \rho g$. For this table $g = 32.174$ ft/s^2.

TABLE B.4   Physical Properties of Air at Standard Atmospheric Pressure (SI Units)[a]

Temperature (°C)	Density, ρ (kg/m³)	Specific Weight[b], γ (N/m³)	Dynamic Viscosity, μ (N·s/m²)	Kinematic Viscosity, ν (m²/s)	Specific Heat Ratio, k (−)	Speed of Sound, c (m/s)
−40	1.514	14.85	1.57 E − 5	1.04 E − 5	1.401	306.2
−20	1.395	13.68	1.63 E − 5	1.17 E − 5	1.401	319.1
0	1.292	12.67	1.71 E − 5	1.32 E − 5	1.401	331.4
5	1.269	12.45	1.73 E − 5	1.36 E − 5	1.401	334.4
10	1.247	12.23	1.76 E − 5	1.41 E − 5	1.401	337.4
15	1.225	12.01	1.80 E − 5	1.47 E − 5	1.401	340.4
20	1.204	11.81	1.82 E − 5	1.51 E − 5	1.401	343.3
25	1.184	11.61	1.85 E − 5	1.56 E − 5	1.401	346.3
30	1.165	11.43	1.86 E − 5	1.60 E − 5	1.400	349.1
40	1.127	11.05	1.87 E − 5	1.66 E − 5	1.400	354.7
50	1.109	10.88	1.95 E − 5	1.76 E − 5	1.400	360.3
60	1.060	10.40	1.97 E − 5	1.86 E − 5	1.399	365.7
70	1.029	10.09	2.03 E − 5	1.97 E − 5	1.399	371.2
80	0.9996	9.803	2.07 E − 5	2.07 E − 5	1.399	376.6
90	0.9721	9.533	2.14 E − 5	2.20 E − 5	1.398	381.7
100	0.9461	9.278	2.17 E − 5	2.29 E − 5	1.397	386.9
200	0.7461	7.317	2.53 E − 5	3.39 E − 5	1.390	434.5
300	0.6159	6.040	2.98 E − 5	4.84 E − 5	1.379	476.3
400	0.5243	5.142	3.32 E − 5	6.34 E − 5	1.368	514.1
500	0.4565	4.477	3.64 E − 5	7.97 E − 5	1.357	548.8
1000	0.2772	2.719	5.04 E − 5	1.82 E − 4	1.321	694.8

[a] Based on data from R. D. Blevins, *Applied Fluid Dynamics Handbook,* Van Nostrand Reinhold Co., Inc., New York, 1984.

[b] Density and specific weight are related through the equation $\gamma = \rho g$. For this table $g = 9.807$ m/s².

APPENDIX C

PROPERTIES OF THE U.S. STANDARD ATMOSPHERE

TABLE C.1 Properties of the U.S. Standard Atmosphere (BG Units)[a]

Altitude (ft)	Temperature (°F)	Acceleration of Gravity, g (ft/s²)	Pressure, p [lb/in.²(abs)]	Density, ρ (slugs/ft³)	Dynamic Viscosity, μ (lb·s/ft²)
−5,000	76.84	32.189	17.554	2.745 E − 3	3.836 E − 7
0	59.00	32.174	14.696	2.377 E − 3	3.737 E − 7
5,000	41.17	32.159	12.228	2.048 E − 3	3.637 E − 7
10,000	23.36	32.143	10.108	1.756 E − 3	3.534 E − 7
15,000	5.55	32.128	8.297	1.496 E − 3	3.430 E − 7
20,000	− 12.26	32.112	6.759	1.267 E − 3	3.324 E − 7
25,000	− 30.05	32.097	5.461	1.066 E − 3	3.217 E − 7
30,000	− 47.83	32.082	4.373	8.907 E − 4	3.107 E − 7
35,000	− 65.61	32.066	3.468	7.382 E − 4	2.995 E − 7
40,000	− 69.70	32.051	2.730	5.873 E − 4	2.969 E − 7
45,000	− 69.70	32.036	2.149	4.623 E − 4	2.969 E − 7
50,000	− 69.70	32.020	1.692	3.639 E − 4	2.969 E − 7
60,000	− 69.70	31.990	1.049	2.256 E − 4	2.969 E − 7
70,000	− 67.42	31.959	0.651	1.392 E − 4	2.984 E − 7
80,000	− 61.98	31.929	0.406	8.571 E − 5	3.018 E − 7
90,000	− 56.54	31.897	0.255	5.610 E − 5	3.052 E − 7
100,000	− 51.10	31.868	0.162	3.318 E − 5	3.087 E − 7
150,000	19.40	31.717	0.020	3.658 E − 6	3.511 E − 7
200,000	− 19.78	31.566	0.003	5.328 E − 7	3.279 E − 7
250,000	− 88.77	31.415	0.000	6.458 E − 8	2.846 E − 7

[a] Data abridged from *U.S. Standard Atmosphere*, 1976, U.S. Government Printing Office, Washington, D.C.

TABLE C.2 Properties of the U.S. Standard Atmosphere (SI Units)[a]

Altitude (m)	Temperature (°C)	Acceleration of Gravity, g (m/s²)	Pressure, p [N/m²(abs)]		Density, ρ (kg/m³)		Dynamic Viscosity, μ (N·s/m²)	
−1,000	21.50	9.810	1.139	E + 5	1.347	E + 0	1.821	E − 5
0	15.00	9.807	1.013	E + 5	1.225	E + 0	1.789	E − 5
1,000	8.50	9.804	8.988	E + 4	1.112	E + 0	1.758	E − 5
2,000	2.00	9.801	7.950	E + 4	1.007	E + 0	1.726	E − 5
3,000	−4.49	9.797	7.012	E + 4	9.093	E − 1	1.694	E − 5
4,000	−10.98	9.794	6.166	E + 4	8.194	E − 1	1.661	E − 5
5,000	−17.47	9.791	5.405	E + 4	7.364	E − 1	1.628	E − 5
6,000	−23.96	9.788	4.722	E + 4	6.601	E − 1	1.595	E − 5
7,000	−30.45	9.785	4.111	E + 4	5.900	E − 1	1.561	E − 5
8,000	−36.94	9.782	3.565	E + 4	5.258	E − 1	1.527	E − 5
9,000	−43.42	9.779	3.080	E + 4	4.671	E − 1	1.493	E − 5
10,000	−49.90	9.776	2.650	E + 4	4.135	E − 1	1.458	E − 5
15,000	−56.50	9.761	1.211	E + 4	1.948	E − 1	1.422	E − 5
20,000	−56.50	9.745	5.529	E + 3	8.891	E − 2	1.422	E − 5
25,000	−51.60	9.730	2.549	E + 3	4.008	E − 2	1.448	E − 5
30,000	−46.64	9.715	1.197	E + 3	1.841	E − 2	1.475	E − 5
40,000	−22.80	9.684	2.871	E + 2	3.996	E − 3	1.601	E − 5
50,000	−2.50	9.654	7.978	E + 1	1.027	E − 3	1.704	E − 5
60,000	−26.13	9.624	2.196	E + 1	3.097	E − 4	1.584	E − 5
70,000	−53.57	9.594	5.221	E + 0	8.283	E − 5	1.438	E − 5
80,000	−74.51	9.564	1.052	E + 0	1.846	E − 5	1.321	E − 5

[a] Data abridged from *U.S. Standard Atmosphere,* 1976, U.S. Government Printing Office, Washington, D.C.

APPENDIX D

ALTERNATE METHOD FOR DETERMINATION OF PI TERMS

A different approach to the determination of pi terms can be developed by recognizing that the pi terms always consist of products of the variables raised to appropriate powers. If in a given problem there are k variables, $u_1, u_2, \ldots, u_k$, then any pi term is of the form $u_1^{x_1} u_2^{x_2} \cdots u_k^{x_k}$, where the x's are chosen so that the combination is dimensionless. Also, each of the variables can be expressed in terms of their basic dimensions (such as FLT) in the form

$$u^i = F^{a_i} L^{b_i} T^{c_i}$$

where a_i, b_i, and c_i are known for any variable. Thus, the dimensions for any pi term can be expressed as

$$(F^{a_1} L^{b_1} T^{c_1})^{x_1} (F^{a_2} L^{b_2} T^{c_2})^{x_2} \cdots (F^{a_k} L^{b_k} T^{c_k})^{x_k}$$

To make the product dimensionless, the x's must be selected so that the dependence of the product on each basic dimension is eliminated. For example, to eliminate F

$$a_1 x_1 + a_2 x_2 + \cdots + a_k x_k = 0 \tag{D.1}$$

Similarly, to eliminate L

$$b_1 x_1 + b_2 x_2 + \cdots + b_k x_k = 0 \tag{D.2}$$

and to eliminate T

$$c_1 x_1 + c_2 x_2 + \cdots + c_k x_k = 0 \tag{D.3}$$

Thus, to make the product dimensionless these three algebraic equations must be satisfied simultaneously.

There will be as many equations as basic dimensions, and the number of unknown

x's will equal the number of variables. Any set of x's that satisfies this system of equations will yield a dimensionless product. Since there will always be more unknowns than independent equations, a unique solution will not exist. However, from the theory of linear algebraic equations it is known that the number of linearly independent solutions is equal to the number of unknown x's minus the rank of the matrix of coefficients, where this matrix is of the form

$$\begin{bmatrix} a_1 & a_2 & \cdots & a_k \\ b_1 & b_2 & \cdots & b_k \\ c_1 & c_2 & \cdots & c_k \end{bmatrix}$$

and the *rank* is the highest order, nonzero determinant which is contained within the matrix. This matrix is called the *dimensional matrix*. The rank is usually equal to the number of basic dimensions. However, in those few instances in which the number of required dimensions (reference dimensions) is smaller than the number of basic dimensions, the rank of the dimensional matrix will be less than the number of basic dimensions. The rank will equal the number of reference dimensions required to describe the variables. Since each independent solution will give a pi term, it can be concluded that the number of independent dimensionless products obtained from this method is equal to the number of variables minus the rank of the dimensional matrix (number of reference dimensions). This result is, of course, consistent with the pi theorem.

To illustrate this procedure with a specific example, the problem of pressure drop along a tube described in Section 7.3 will again be considered. Since the variables are Δp_{ℓ}, ρ, D, V, and μ, the pi terms are of the form:

$$(\Delta p_{\ell})^{x_1}(\rho)^{x_2}(D)^{x_3}(V)^{x_4}(\mu)^{x_5}$$

Note that the order in which we list the variables is not important. In terms of basic dimensions

$$(FL^{-3})^{x_1}(FL^{-4}T^2)^{x_2}(L)^{x_3}(LT^{-1})^{x_4}(FL^{-2}T)^{x_5}$$

For this combination to be dimensionless, three auxiliary equations must be satisfied:

$$x_1 + x_2 + 0 + 0 + x_5 = 0 \qquad \text{(for } F) \tag{D.4}$$

$$-3x_1 - 4x_2 + x_3 + x_4 - 2x_5 = 0 \qquad \text{(for } L) \tag{D.5}$$

$$0 + 2x_2 + 0 - x_4 + x_5 = 0 \qquad \text{(for } T) \tag{D.6}$$

The coefficients of the dimensional matrix can, therefore, be written as:

	Δp_{ℓ}	ρ	D	V	μ
F	1	1	0	0	1
L	-3	-4	1	1	-2
T	0	2	0	-1	1

The rank of the matrix is three and there will be two linearly independent solutions for the x's.

There are three equations (Eqs. D.4, D.5, D.6) available for solving for five unknowns, and arbitrary values can be assigned to two of the unknowns. Many

combinations are possible. For example, let

$$x_1 = 1$$

$$x_5 = 0$$

Equations D.4, D.5, and D.6 then become

$$1 + x_2 = 0$$

$$-3 - 4x_2 + x_3 + x_4 = 0$$

$$2x_2 - x_4 = 0$$

and it then follows that $x_2 = -1$, $x_3 = 1$, and $x_4 = -2$. From this result

$$\Pi_1 = (\Delta p_\ell)^1 (\rho)^{-1} (D)^1 (V)^{-2} (\mu)^0$$

and, therefore,

$$\Pi_1 = \frac{D \, \Delta p_\ell}{\rho V^2}$$

For the second pi term, let $x_1 = 0$ and $x_5 = 1$ and it follows that $x_2 = -1$, $x_3 = -1$, and $x_4 = -1$ so that

$$\Pi_2 = \frac{\mu}{\rho D V}$$

This result is the same as that previously obtained in Section 7.3 using the method of repeating variables. However, the same pi terms were obtained by judicious selection of the assumed values for two of the x's. An infinite number of correct pi terms exist since there is an infinite number of combinations of x's that could be assumed. The only guide at this point is to keep the pi terms as simple as possible. This can be achieved by setting assumed values for the x's equal to unity or zero.

Although there are an infinite number of pi terms possible, once a suitable set is found, all the rest can be formed by products of this set raised to various powers. This can be demonstrated as follows. Pi terms obtained by the method described in this appendix are based on obtaining a solution to a set of equations such as Eqs. D.1, D.2, and D.3. Let one solution be designated as x_1', x_2', . . . , x_k', and another solution as x_1'', x_2'', . . . , x_k''. Each solution yields a pi term of the form $\Pi_1 = u_1^{x_1'} u_2^{x_2'} \cdots u_k^{x_k'}$ and $\Pi_2 = u_1^{x_1''} u_2^{x_2''} \cdots u_k^{x_k''}$. As noted the number of linearly independent solutions is equal to the number of unknown x's minus the rank of the dimensional matrix. Let us assume that the number of independent solutions is two, and these are indicated above. The linear independence concept means that if we find a third solution, it will be related to the two independent solutions in the form

$$x_1''' = \alpha x_1' + \beta x_1'' \tag{D.7}$$

$$x_2''' = \alpha x_2' + \beta x_2'' \tag{D.8}$$

$$\vdots$$

$$x_k''' = \alpha x_k' + \beta x_k'' \tag{D.9}$$

where α and β are constants. Thus, once we have found the required number of independent solutions, all other solutions can be obtained from a linear combination of these solutions.

The new pi term formed by this new solution will be of the form:

$$\Pi_3 = u_1^{x_1'''} u_2^{x_2'''} \cdots u_k^{x_k'''}$$

We now make use of Eqs. D.7, D.8, and D.9 and write

$$\Pi_3 = u_1^{\alpha x_1' + \beta x_1''} u_2^{\alpha x_2' + \beta x_2''} \cdots u_k^{\alpha x_k' + \beta x_k''}$$

$$= (u_1^{x_1'} u_2^{x_2'} \cdots u_k^{x_k'})^\alpha (u_1^{x_1''} u_2^{x_2''} \cdots u_k^{x_k''})^\beta$$

$$= (\Pi_1)^\alpha (\Pi_2)^\beta \tag{D.10}$$

Thus, once we have found a set of independent solutions (where each solution yields a pi term), all other solutions will give pi terms that can be formed by products of the original pi terms raised to appropriate powers as shown in Eq. D.10. It is clear that although there is not a unique set of pi terms for any given problem, once an independent set is found, any other set can be obtained from the first set. The practical significance of this result is that new pi terms can be easily formed by simply taking products of the original pi terms each raised to any desired exponent. A further discussion of this point is given in Section 7.4.3.

APPENDIX E

COMPRESSIBLE FLOW TABLES FOR AN IDEAL GAS ($k = 1.4$)
(values obtained from Ref. 4 of Chapter 11)

TABLE E.1 Isentropic Flow of an Ideal Gas ($k = 1.4$)

Ma	T/T_0	p/p_0	ρ/ρ_0	A/A^*	Ma	T/T_0	p/p_0	ρ/ρ_0	A/A^*
0	1.00000	1.00000	1.00000	∞	0.24	0.98861	0.96070	0.97177	2.4956
0.01	0.99998	0.99993	0.99995	57.874	0.25	0.98765	0.95745	0.96942	2.4027
0.02	0.99992	0.99972	0.99980	28.942	0.26	0.98666	0.95408	0.96699	2.3173
0.03	0.99982	0.99937	0.99955	19.300	0.27	0.98563	0.95060	0.96446	2.2385
0.04	0.99968	0.99888	0.99920	14.482	0.28	0.98456	0.94700	0.96185	2.1656
0.05	0.99950	0.99825	0.99875	11.592	0.29	0.98346	0.94329	0.95916	2.0979
0.06	0.99928	0.99748	0.99820	9.6659	0.30	0.98232	0.93947	0.95638	2.0351
0.07	0.99902	0.99658	0.99755	8.2915	0.31	0.98114	0.93554	0.95352	1.9765
0.08	0.99872	0.99553	0.99680	7.2616	0.32	0.97993	0.93150	0.95058	1.9218
0.09	0.99838	0.99435	0.99596	6.4613	0.33	0.97868	0.92736	0.94756	1.8707
0.10	0.99800	0.99303	0.99502	5.8218	0.34	0.97740	0.92312	0.94446	1.8229
0.11	0.99758	0.99157	0.99398	5.2992	0.35	0.97608	0.91877	0.94128	1.7780
0.12	0.99714	0.98998	0.99284	4.8643	0.36	0.97473	0.91433	0.93803	1.7358
0.13	0.99664	0.98826	0.99160	4.4968	0.37	0.97335	0.90979	0.93470	1.6961
0.14	0.99610	0.98640	0.99027	4.1824	0.38	0.97193	0.90516	0.93129	1.6587
0.15	0.99552	0.98441	0.98884	3.9103	0.39	0.97048	0.90044	0.92782	1.6234
0.16	0.99490	0.98228	0.98731	3.6727	0.40	0.96899	0.89562	0.92428	1.5901
0.17	0.99425	0.98003	0.98569	3.4635	0.41	0.96747	0.89071	0.92066	1.5587
0.18	0.99356	0.97765	0.98398	3.2779	0.42	0.96592	0.88572	0.91697	1.5289
0.19	0.99283	0.97514	0.98217	3.1122	0.43	0.96434	0.88065	0.91322	1.5007
0.20	0.99206	0.97250	0.98027	2.9635	0.44	0.96272	0.87550	0.90940	1.4740
0.21	0.99125	0.96973	0.97828	2.8293	0.45	0.96108	0.87027	0.90552	1.4487
0.22	0.99041	0.96685	0.97621	2.7076	0.46	0.95940	0.86496	0.90157	1.4246
0.23	0.98953	0.96383	0.97403	2.5968	0.47	0.95769	0.85958	0.89756	1.4018

TABLE E.1 (continued)

Ma	T/T_0	p/p_0	ρ/ρ_0	A/A^*	Ma	T/T_0	p/p_0	ρ/ρ_0	A/A^*
0.48	0.95595	0.85413	0.89349	1.3801	0.99	0.83611	0.53446	0.63923	1.00008
0.49	0.95418	0.84861	0.88936	1.3594	1.00	0.83333	0.52828	0.63394	1.00000
0.50	0.95238	0.84302	0.88517	1.3398	1.01	0.83055	0.52213	0.62866	1.00008
0.51	0.95055	0.83737	0.88092	1.3212	1.02	0.82776	0.51602	0.62339	1.00033
0.52	0.94869	0.83166	0.87662	1.3034	1.03	0.82496	0.50994	0.61813	1.00074
0.53	0.94681	0.82589	0.87227	1.2864	1.04	0.82215	0.50389	0.61288	1.00130
0.54	0.94489	0.82005	0.86788	1.2703	1.05	0.81933	0.49787	0.60765	1.00202
0.55	0.94295	0.81416	0.86342	1.2550	1.06	0.81651	0.49189	0.60243	1.00290
0.56	0.94098	0.80822	0.85892	1.2403	1.07	0.81368	0.48595	0.59722	1.00394
0.57	0.93898	0.80224	0.85437	1.2263	1.08	0.81084	0.48005	0.59203	1.00512
0.58	0.93696	0.79621	0.84977	1.2130	1.09	0.80800	0.47418	0.58685	1.00645
0.59	0.93491	0.79012	0.84513	1.2003	1.10	0.80515	0.46835	0.58169	1.00793
0.60	0.93284	0.78400	0.84045	1.1882	1.11	0.80230	0.46256	0.57655	1.00955
0.61	0.93074	0.77784	0.83573	1.1766	1.12	0.79944	0.45682	0.57143	1.01131
0.62	0.92861	0.77164	0.83096	1.1656	1.13	0.79657	0.45112	0.56632	1.01322
0.63	0.92646	0.76540	0.82616	1.1551	1.14	0.79370	0.44545	0.56123	1.01527
0.64	0.92428	0.75913	0.82132	1.1451	1.15	0.79083	0.43983	0.55616	1.01746
0.65	0.92208	0.75283	0.81644	1.1356	1.16	0.78795	0.43425	0.55112	1.01978
0.66	0.91986	0.74650	0.81153	1.1265	1.17	0.78507	0.42872	0.54609	1.02224
0.67	0.91762	0.74014	0.80659	1.1178	1.18	0.78218	0.42323	0.54108	1.02484
0.68	0.91535	0.73376	0.80162	1.1096	1.19	0.77929	0.41778	0.53610	1.02757
0.69	0.91306	0.72735	0.79662	1.1018	1.20	0.77640	0.41238	0.53114	1.03044
0.70	0.91075	0.72092	0.79158	1.09437	1.21	0.77350	0.40702	0.52620	1.03344
0.71	0.90842	0.71448	0.78652	1.08729	1.22	0.77061	0.40171	0.52129	1.03657
0.72	0.90606	0.70802	0.78143	1.08057	1.23	0.76771	0.39645	0.51640	1.03983
0.73	0.90368	0.70155	0.77632	1.07419	1.24	0.76481	0.39123	0.51154	1.04323
0.74	0.90129	0.69507	0.77119	1.06814	1.25	0.76190	0.38606	0.50670	1.04676
0.75	0.89888	0.68857	0.76603	1.06242	1.26	0.75900	0.38094	0.50189	1.05041
0.76	0.89644	0.68207	0.76086	1.05700	1.27	0.75610	0.37586	0.49710	1.05419
0.77	0.89399	0.67556	0.75567	1.05188	1.28	0.75319	0.37083	0.49234	1.05810
0.78	0.89152	0.66905	0.75046	1.04705	1.29	0.75029	0.36585	0.48761	1.06214
0.79	0.88903	0.66254	0.74524	1.04250	1.30	0.74738	0.36092	0.48291	1.06631
0.80	0.88652	0.65602	0.74000	1.03823	1.31	0.74448	0.35603	0.47823	1.07060
0.81	0.88400	0.64951	0.73474	1.03422	1.32	0.74158	0.35119	0.47358	1.07502
0.82	0.88146	0.64300	0.72947	1.03046	1.33	0.73867	0.34640	0.46895	1.07957
0.83	0.87890	0.63650	0.72419	1.02696	1.34	0.73577	0.34166	0.46436	1.08424
0.84	0.87633	0.63000	0.71890	1.02370	1.35	0.73287	0.33697	0.45980	1.08904
0.85	0.87374	0.62351	0.71361	1.02067	1.36	0.72997	0.33233	0.45527	1.09397
0.86	0.87114	0.61703	0.70831	1.01787	1.37	0.72707	0.32774	0.45076	1.09902
0.87	0.86852	0.61057	0.70300	1.01530	1.38	0.72418	0.32319	0.44628	1.10420
0.88	0.86589	0.60412	0.69769	1.01294	1.39	0.72128	0.31869	0.44183	1.10950
0.89	0.86324	0.59768	0.69237	1.01080	1.40	0.71839	0.31424	0.43742	1.1149
0.90	0.86058	0.59126	0.68704	1.00886	1.41	0.71550	0.30984	0.43304	1.1205
0.91	0.85791	0.58486	0.68171	1.00713	1.42	0.71261	0.30549	0.42869	1.1262
0.92	0.85523	0.57848	0.67639	1.00560	1.43	0.70973	0.30119	0.42436	1.1320
0.93	0.85253	0.57212	0.67107	1.00426	1.44	0.70685	0.29693	0.42007	1.1379
0.94	0.84982	0.56578	0.66575	1.00311	1.45	0.70397	0.29272	0.41581	1.1440
0.95	0.84710	0.55946	0.66044	1.00214	1.46	0.70110	0.28856	0.41158	1.1502
0.96	0.84437	0.55317	0.65513	1.00136	1.47	0.69823	0.28445	0.40738	1.1565
0.97	0.84162	0.54691	0.64982	1.00076	1.48	0.69537	0.28039	0.40322	1.1629
0.98	0.83887	0.54067	0.64452	1.00033	1.49	0.69251	0.27637	0.39909	1.1695

TABLE E.1 (continued)

Ma	T/T_0	p/p_0	ρ/ρ_0	A/A^*	Ma	T/T_0	p/p_0	ρ/ρ_0	A/A^*
1.50	0.68965	0.27240	0.39498	1.1762	2.01	0.55310	0.12583	0.22751	1.7017
1.51	0.68680	0.26848	0.39091	1.1830	2.02	0.55064	0.12389	0.22499	1.7160
1.52	0.68396	0.26461	0.38687	1.1899	2.03	0.54819	0.12198	0.22250	1.7305
1.53	0.68112	0.26078	0.38287	1.1970	2.04	0.54576	0.12009	0.22004	1.7452
1.54	0.67828	0.25700	0.37890	1.2042	2.05	0.54333	0.11823	0.21760	1.7600
1.55	0.67545	0.25326	0.37496	1.2115	2.06	0.54091	0.11640	0.21519	1.7750
1.56	0.67262	0.24957	0.37105	1.2190	2.07	0.53850	0.11460	0.21281	1.7902
1.57	0.66980	0.24593	0.36717	1.2266	2.08	0.53611	0.11282	0.21045	1.8056
1.58	0.66699	0.24233	0.36332	1.2343	2.09	0.53373	0.11107	0.20811	1.8212
1.59	0.66418	0.23878	0.35951	1.2422	2.10	0.53135	0.10935	0.20580	1.8369
1.60	0.66138	0.23527	0.35573	1.2502	2.11	0.52898	0.10766	0.20352	1.8529
1.61	0.65858	0.23181	0.35198	1.2583	2.12	0.52663	0.10599	0.20126	1.8690
1.62	0.65579	0.22839	0.34826	1.2666	2.13	0.52428	0.10434	0.19902	1.8853
1.63	0.65301	0.22501	0.34458	1.2750	2.14	0.52194	0.10272	0.19681	1.9018
1.64	0.65023	0.22168	0.34093	1.2835	2.15	0.51962	0.10113	0.19463	1.9185
1.65	0.64746	0.21839	0.33731	1.2922	2.16	0.51730	0.09956	0.19247	1.9354
1.66	0.64470	0.21515	0.33372	1.3010	2.17	0.51499	0.09802	0.19033	1.9525
1.67	0.64194	0.21195	0.33016	1.3099	2.18	0.51269	0.09650	0.18821	1.9698
1.68	0.63919	0.20879	0.32664	1.3190	2.19	0.51041	0.09500	0.18612	1.9873
1.69	0.63645	0.20567	0.32315	1.3282	2.20	0.50813	0.09352	0.18405	2.0050
1.70	0.63372	0.20259	0.31969	1.3376	2.21	0.50586	0.09207	0.18200	2.0229
1.71	0.63099	0.19955	0.31626	1.3471	2.22	0.50361	0.09064	0.17998	2.0409
1.72	0.62827	0.19656	0.31286	1.3567	2.23	0.50136	0.08923	0.17798	2.0592
1.73	0.62556	0.19361	0.30950	1.3665	2.24	0.49912	0.08784	0.17600	2.0777
1.74	0.62286	0.19070	0.30617	1.3764	2.25	0.49689	0.08648	0.17404	2.0964
1.75	0.62016	0.18782	0.30287	1.3865	2.26	0.49468	0.08514	0.17211	2.1154
1.76	0.61747	0.18499	0.29959	1.3967	2.27	0.49247	0.08382	0.17020	2.1345
1.77	0.61479	0.18220	0.29635	1.4071	2.28	0.49027	0.08252	0.16830	2.1538
1.78	0.61211	0.17944	0.29314	1.4176	2.29	0.48809	0.08123	0.16643	2.1734
1.79	0.60945	0.17672	0.28997	1.4282	2.30	0.48591	0.07997	0.16458	2.1931
1.80	0.60680	0.17404	0.28682	1.4390	2.31	0.48374	0.07873	0.16275	2.2131
1.81	0.60415	0.17140	0.28370	1.4499	2.32	0.48158	0.07751	0.16095	2.2333
1.82	0.60151	0.16879	0.28061	1.4610	2.33	0.47944	0.07631	0.15916	2.2537
1.83	0.59888	0.16622	0.27756	1.4723	2.34	0.47730	0.07513	0.15739	2.2744
1.84	0.59626	0.16369	0.27453	1.4837	2.35	0.47517	0.07396	0.15564	2.2953
1.85	0.59365	0.16120	0.27153	1.4952	2.36	0.47305	0.07281	0.15391	2.3164
1.86	0.59105	0.15874	0.26857	1.5069	2.37	0.47095	0.07168	0.15220	2.3377
1.87	0.58845	0.15631	0.26563	1.5188	2.38	0.46885	0.07057	0.15052	2.3593
1.88	0.58586	0.15392	0.26272	1.5308	2.39	0.46676	0.06948	0.14885	2.3811
1.89	0.58329	0.15156	0.25984	1.5429	2.40	0.46468	0.06840	0.14720	2.4031
1.90	0.58072	0.14924	0.25699	1.5552	2.41	0.46262	0.06734	0.14557	2.4254
1.91	0.57816	0.14695	0.25417	1.5677	2.42	0.46056	0.06630	0.14395	2.4479
1.92	0.57561	0.14469	0.25138	1.5804	2.43	0.45851	0.06527	0.14235	2.4706
1.93	0.57307	0.14247	0.24862	1.5932	2.44	0.45647	0.06426	0.14078	2.4936
1.94	0.57054	0.14028	0.24588	1.6062	2.45	0.45444	0.06327	0.13922	2.5168
1.95	0.56802	0.13813	0.24317	1.6193	2.46	0.45242	0.06229	0.13768	2.5403
1.96	0.56551	0.13600	0.24049	1.6326	2.47	0.45041	0.06133	0.13616	2.5640
1.97	0.56301	0.13390	0.23784	1.6461	2.48	0.44841	0.06038	0.13465	2.5880
1.98	0.56051	0.13184	0.23522	1.6597	2.49	0.44642	0.05945	0.13316	2.6122
1.99	0.55803	0.12981	0.23262	1.6735	2.50	0.44444	0.05853	0.13169	2.6367
2.00	0.55556	0.12780	0.23005	1.6875	2.51	0.44247	0.05763	0.13023	2.6615

TABLE E.1 (continued)

Ma	T/T_0	p/p_0	ρ/ρ_0	A/A^*	Ma	T/T_0	p/p_0	ρ/ρ_0	A/A^*
2.52	0.44051	0.05674	0.12879	2.6865	2.90	0.37286	0.03165	0.08489	3.8498
2.53	0.43856	0.05586	0.12737	2.7117	2.91	0.37125	0.03118	0.08398	3.8866
2.54	0.43662	0.05500	0.12597	2.7372	2.92	0.36965	0.03071	0.08308	3.9238
2.55	0.43469	0.05415	0.12458	2.7630	2.93	0.36806	0.03025	0.08218	3.9614
2.56	0.43277	0.05332	0.12321	2.7891	2.94	0.36648	0.02980	0.08130	3.9993
2.57	0.43085	0.05250	0.12185	2.8154	2.95	0.36490	0.02935	0.08043	4.0376
2.58	0.42894	0.05169	0.12051	2.8420	2.96	0.36333	0.02891	0.07957	4.0763
2.59	0.42705	0.05090	0.11918	2.8689	2.97	0.36177	0.02848	0.07872	4.1153
2.60	0.42517	0.05012	0.11787	2.8960	2.98	0.36022	0.02805	0.07788	4.1547
2.61	0.42330	0.04935	0.11658	2.9234	2.99	0.35868	0.02764	0.07705	4.1944
2.62	0.42143	0.04859	0.11530	2.9511	3.00	0.35714	0.02722	0.07623	4.2346
2.63	0.41957	0.04784	0.11403	2.9791	3.10	0.34223	0.02345	0.06852	4.6573
2.64	0.41772	0.04711	0.11278	3.0074	3.20	0.32808	0.02023	0.06165	5.1210
2.65	0.41589	0.04639	0.11154	3.0359	3.30	0.31466	0.01748	0.05554	5.6287
2.66	0.41406	0.04568	0.11032	3.0647	3.40	0.30193	0.01512	0.05009	6.1837
2.67	0.41224	0.04498	0.10911	3.0938	3.50	0.28986	0.01311	0.04523	6.7896
2.68	0.41043	0.04429	0.10792	3.1233	3.60	0.27840	0.01138	0.04089	7.4501
2.69	0.40863	0.04361	0.10674	3.1530	3.70	0.26752	0.00990	0.03702	8.1691
2.70	0.40684	0.04295	0.10557	3.1830	3.80	0.25720	0.00863	0.03355	8.9506
2.71	0.40505	0.04230	0.10442	3.2133	3.90	0.24740	0.00753	0.03044	9.7990
2.72	0.40327	0.04166	0.10328	3.2440	4.00	0.23810	0.00658	0.02766	10.719
2.73	0.40151	0.04102	0.10215	3.2749	4.10	0.22925	0.00577	0.02516	11.715
2.74	0.39976	0.04039	0.10104	3.3061	4.20	0.22085	0.00506	0.02292	12.792
2.75	0.39801	0.03977	0.09994	3.3376	4.30	0.21286	0.00445	0.02090	13.955
2.76	0.39627	0.03917	0.09885	3.3695	4.40	0.20525	0.00392	0.01909	15.210
2.77	0.39454	0.03858	0.09777	3.4017	4.50	0.19802	0.00346	0.01745	16.562
2.78	0.39282	0.03800	0.09671	3.4342	4.60	0.19113	0.00305	0.01597	18.018
2.79	0.39111	0.03742	0.09566	3.4670	4.70	0.18457	0.00270	0.01463	19.583
2.80	0.38941	0.03685	0.09462	3.5001	4.80	0.17832	0.00240	0.01343	21.264
2.81	0.38771	0.03629	0.09360	3.5336	4.90	0.17235	0.00213	0.01233	23.067
2.82	0.38603	0.03574	0.09259	3.5674	5.00	0.16667	$189(10)^{-5}$	0.01134	25.000
2.83	0.38435	0.03520	0.09158	3.6015	6.00	0.12195	$633(10)^{-6}$	0.00519	53.180
2.84	0.38268	0.03467	0.09059	3.6359	7.00	0.09259	$242(10)^{-6}$	0.00261	104.143
2.85	0.38102	0.03415	0.08962	3.6707	8.00	0.07246	$102(10)^{-6}$	0.00141	190.109
2.86	0.37937	0.03363	0.08865	3.7058	9.00	0.05814	$474(10)^{-7}$	0.000815	327.189
2.87	0.37773	0.03312	0.08769	3.7413	10.00	0.04762	$236(10)^{-7}$	0.000495	535.938
2.88	0.37610	0.03262	0.08674	3.7771	∞	0	0	0	∞
2.89	0.37448	0.03213	0.08581	3.8133					

TABLE E.2   Fanno Flow on an Ideal Gas ($k = 1.4$)

Ma	$\dfrac{f(l^* - l)}{D}$	T/T^*	V/V^*	p/p^*	p_0/p_0^*	Ma	$\dfrac{f(l^* - l)}{D}$	T/T^*	V/V^*	p/p^*	p_0/p_0^*
0	∞	1.2000	0	∞	∞	0.07	140.66	1.1988	0.07664	15.642	8.2915
0.01	7134.40	1.2000	0.01095	109.544	57.874	0.08	106.72	1.1985	0.08758	13.684	7.2616
0.02	1778.45	1.1999	0.02191	54.770	28.942	0.09	83.496	1.1981	0.09851	12.162	6.4614
0.03	787.08	1.1998	0.03286	36.511	19.300	0.10	66.922	1.1976	0.10943	10.9435	5.8218
0.04	440.35	1.1996	0.04381	27.382	14.482	0.11	54.688	1.1971	0.12035	9.9465	5.2992
0.05	280.02	1.1994	0.05476	21.903	11.5914	0.12	45.408	1.1966	0.13126	9.1156	4.8643
0.06	193.03	1.1991	0.06570	18.251	9.6659	0.13	38.207	1.1960	0.14216	8.4123	4.4968

TABLE E.2 (continued)

Ma	$\dfrac{f(\ell^* - \ell)}{D}$	T/T^*	V/V^*	p/p^*	p_0/p_0^*	Ma	$\dfrac{f(\ell^* - \ell)}{D}$	T/T^*	V/V^*	p/p^*	p_0/p_0^*
0.14	32.511	1.1953	0.15306	7.8093	4.1824	0.75	0.12728	1.07865	0.77893	1.3848	1.06242
0.15	27.932	1.1946	0.16395	7.2866	3.9103	0.76	0.11446	1.07573	0.78825	1.3647	1.05700
0.16	24.198	1.1939	0.17482	6.8291	3.6727	0.77	0.10262	1.07279	0.79753	1.3451	1.05188
0.17	21.115	1.1931	0.18568	6.4252	3.4635	0.78	0.09167	1.06982	0.80677	1.3260	1.04705
0.18	18.543	1.1923	0.19654	6.0662	3.2779	0.79	0.08159	1.06684	0.81598	1.3074	1.04250
0.19	16.375	1.1914	0.20739	5.7448	3.1123	0.80	0.07229	1.06383	0.82514	1.2892	1.03823
0.20	14.533	1.1905	0.21822	5.4555	2.9635	0.81	0.06375	1.06080	0.83426	1.2715	1.03422
0.21	12.956	1.1895	0.22904	5.1936	2.8293	0.82	0.05593	1.05775	0.84334	1.2542	1.03047
0.22	11.596	1.1885	0.23984	4.9554	2.7076	0.83	0.04878	1.05468	0.85239	1.2373	1.02696
0.23	10.416	1.1874	0.25063	4.7378	2.5968	0.84	0.04226	1.05160	0.86140	1.2208	1.02370
0.24	9.3865	1.1863	0.26141	4.5383	2.4956	0.85	0.03632	1.04849	0.87037	1.2047	1.02067
0.25	8.4834	1.1852	0.27217	4.3546	2.4027	0.86	0.03097	1.04537	0.87929	1.1889	1.01787
0.26	7.6876	1.1840	0.28291	4.1850	2.3173	0.87	0.02613	1.04223	0.88818	1.1735	1.01529
0.27	6.9832	1.1828	0.29364	4.0280	2.2385	0.88	0.02180	1.03907	0.89703	1.1584	1.01294
0.28	6.3572	1.1815	0.30435	3.8820	2.1656	0.89	0.01793	1.03589	0.90583	1.1436	1.01080
0.29	5.7989	1.1802	0.31504	3.7460	2.0979	0.90	0.014513	1.03270	0.91459	1.12913	1.00887
0.30	5.2992	1.1788	0.32572	3.6190	2.0351	0.91	0.011519	1.02950	0.92332	1.11500	1.00714
0.31	4.8507	1.1774	0.33637	3.5002	1.9765	0.92	0.008916	1.02627	0.93201	1.10114	1.00560
0.32	4.4468	1.1759	0.34700	3.3888	1.9219	0.93	0.006694	1.02304	0.94065	1.08758	1.00426
0.33	4.0821	1.1744	0.35762	3.2840	1.8708	0.94	0.004815	1.01978	0.94925	1.07430	1.00311
0.34	3.7520	1.1729	0.36822	3.1853	1.8229	0.95	0.003280	1.01652	0.95782	1.06129	1.00215
0.35	3.4525	1.1713	0.37880	3.0922	1.7780	0.96	0.002056	1.01324	0.96634	1.04854	1.00137
0.36	3.1801	1.1697	0.38935	3.0042	1.7358	0.97	0.001135	1.00995	0.97481	1.03605	1.00076
0.37	2.9320	1.1680	0.39988	2.9209	1.6961	0.98	0.000493	1.00664	0.98324	1.02379	1.00033
0.38	2.7055	1.1663	0.41039	2.8420	1.6587	0.99	0.000120	1.00333	0.99164	1.01178	1.00008
0.39	2.4983	1.1646	0.42087	2.7671	1.6234	1.00	0	1.00000	1.00000	1.00000	1.00000
0.40	2.3085	1.1628	0.43133	2.6958	1.5901	1.01	0.000114	0.99666	1.00831	0.98844	1.00008
0.41	2.1344	1.1610	0.44177	2.6280	1.5587	1.02	0.000458	0.99331	1.01658	0.97711	1.00033
0.42	1.9744	1.1591	0.45218	2.5634	1.5289	1.03	0.001013	0.98995	1.02481	0.96598	1.00073
0.43	1.8272	1.1572	0.46257	2.5017	1.5007	1.04	0.001771	0.98658	1.03300	0.95506	1.00130
0.44	1.6915	1.1553	0.47293	2.4428	1.4739	1.05	0.002712	0.98320	1.04115	0.94435	1.00203
0.45	1.5664	1.1533	0.48326	2.3865	1.4486	1.06	0.003837	0.97982	1.04925	0.93383	1.00291
0.46	1.4509	1.1513	0.49357	2.3326	1.4246	1.07	0.005129	0.97642	1.05731	0.92350	1.00394
0.47	1.3442	1.1492	0.50385	2.2809	1.4018	1.08	0.006582	0.97302	1.06533	0.91335	1.00512
0.48	1.2453	1.1471	0.51410	2.2314	1.3801	1.09	0.008185	0.96960	1.07331	0.90338	1.00645
0.49	1.1539	1.1450	0.52433	2.1838	1.3595	1.10	0.009933	0.96618	1.08124	0.89359	1.00793
0.50	1.06908	1.1429	0.53453	2.1381	1.3399	1.11	0.011813	0.96276	1.08913	0.88397	1.00955
0.51	0.99042	1.1407	0.54469	2.0942	1.3212	1.12	0.013824	0.95933	1.09698	0.87451	1.01131
0.52	0.91741	1.1384	0.55482	2.0519	1.3034	1.13	0.015949	0.95589	1.10479	0.86522	1.01322
0.53	0.84963	1.1362	0.56493	2.0112	1.2864	1.14	0.018187	0.95244	1.11256	0.85608	1.01527
0.54	0.78662	1.1339	0.57501	1.9719	1.2702	1.15	0.02053	0.94899	1.1203	0.84710	1.01746
0.55	0.72805	1.1315	0.58506	1.9341	1.2549	1.16	0.02298	0.94554	1.1280	0.83827	1.01978
0.56	0.67357	1.1292	0.59507	1.8976	1.2403	1.17	0.02552	0.94208	1.1356	0.82958	1.02224
0.57	0.62286	1.1268	0.60505	1.8623	1.2263	1.18	0.02814	0.93862	1.1432	0.82104	1.02484
0.58	0.57568	1.1244	0.61500	1.8282	1.2130	1.19	0.03085	0.93515	1.1508	0.81263	1.02757
0.59	0.53174	1.1219	0.62492	1.7952	1.2003	1.20	0.03364	0.93168	1.1583	0.80436	1.03044
0.60	0.49081	1.1194	0.63481	1.7634	1.1882	1.21	0.03650	0.92820	1.1658	0.79623	1.03344
0.61	0.45270	1.1169	0.64467	1.7325	1.1766	1.22	0.03942	0.92473	1.1732	0.78822	1.03657
0.62	0.41720	1.1144	0.65449	1.7026	1.1656	1.23	0.04241	0.92125	1.1806	0.78034	1.03983
0.63	0.38411	1.1118	0.66427	1.6737	1.1551	1.24	0.04547	0.91777	1.1879	0.77258	1.04323
0.64	0.35330	1.1091	0.67402	1.6456	1.1451	1.25	0.04858	0.91429	1.1952	0.76495	1.04676
0.65	0.32460	1.10650	0.68374	1.6183	1.1356	1.26	0.05174	0.91080	1.2025	0.75743	1.05041
0.66	0.29785	1.10383	0.69342	1.5919	1.1265	1.27	0.05494	0.90732	1.2097	0.75003	1.05419
0.67	0.27295	1.10114	0.70306	1.5662	1.1179	1.28	0.05820	0.90383	1.2169	0.74274	1.05809
0.68	0.24978	1.09842	0.71267	1.5413	1.1097	1.29	0.06150	0.90035	1.2240	0.73556	1.06213
0.69	0.22821	1.09567	0.72225	1.5170	1.1018	1.30	0.06483	0.89686	1.2311	0.72848	1.06630
0.70	0.20814	1.09290	0.73179	1.4934	1.09436	1.31	0.06820	0.89338	1.2382	0.72152	1.07060
0.71	0.18949	1.09010	0.74129	1.4705	1.08729	1.32	0.07161	0.88989	1.2452	0.71465	1.07502
0.72	0.17215	1.08727	0.75076	1.4482	1.08057	1.33	0.07504	0.88641	1.2522	0.70789	1.07957
0.73	0.15606	1.08442	0.76019	1.4265	1.07419	1.34	0.07850	0.88292	1.2591	0.70123	1.08424
0.74	0.14113	1.08155	0.76958	1.4054	1.06815	1.35	0.08199	0.87944	1.2660	0.69466	1.08904

TABLE E.2   (continued)

Ma	$\dfrac{f(\ell^* - \ell)}{D}$	T/T^*	V/V^*	p/p^*	p_0/p_0^*	Ma	$\dfrac{f(\ell^* - \ell)}{D}$	T/T^*	V/V^*	p/p^*	p_0/p_0^*
1.36	0.08550	0.87596	1.2729	0.68818	1.09397	1.97	0.29599	0.67561	1.6193	0.41724	1.6461
1.37	0.08904	0.87249	1.2797	0.68180	1.09902	1.98	0.29901	0.67262	1.6239	0.41421	1.6597
1.38	0.09259	0.86901	1.2864	0.67551	1.10419	1.99	0.30201	0.66964	1.6284	0.41121	1.6735
1.39	0.09616	0.86554	1.2932	0.66931	1.10948	2.00	0.30499	0.66667	1.6330	0.40825	1.6875
1.40	0.09974	0.86207	1.2999	0.66320	1.1149	2.01	0.30796	0.66371	1.6375	0.40532	1.7017
1.41	0.10333	0.85860	1.3065	0.65717	1.1205	2.02	0.31091	0.66076	1.6420	0.40241	1.7160
1.42	0.10694	0.85514	1.3131	0.65122	1.1262	2.03	0.31384	0.65783	1.6465	0.39954	1.7305
1.43	0.11056	0.85168	1.3197	0.64536	1.1320	2.04	0.31675	0.65491	1.6509	0.39670	1.7452
1.44	0.11419	0.84822	1.3262	0.63958	1.1379	2.05	0.31965	0.65200	1.6553	0.39389	1.7600
1.45	0.11782	0.84477	1.3327	0.63387	1.1440	2.06	0.32253	0.64910	1.6597	0.39110	1.7750
1.46	0.12146	0.84133	1.3392	0.62824	1.1502	2.07	0.32528	0.64621	1.6640	0.38834	1.7902
1.47	0.12510	0.83788	1.3456	0.62269	1.1565	2.08	0.32822	0.64333	1.6683	0.38562	1.8056
1.48	0.12875	0.83445	1.3520	0.61722	1.1629	2.09	0.33104	0.64047	1.6726	0.38292	1.8212
1.49	0.13240	0.83101	1.3583	0.61181	1.1695	2.10	0.33385	0.63762	1.6769	0.38024	1.8369
1.50	0.13605	0.82759	1.3646	0.60648	1.1762	2.11	0.33664	0.63478	1.6811	0.37760	1.8528
1.51	0.13970	0.82416	1.3708	0.60122	1.1830	2.12	0.33940	0.63195	1.6853	0.37498	1.8690
1.52	0.14335	0.82075	1.3770	0.59602	1.1899	2.13	0.34215	0.62914	1.6895	0.37239	1.8853
1.53	0.14699	0.81734	1.3832	0.59089	1.1970	2.14	0.34488	0.62633	1.6936	0.36982	1.9018
1.54	0.15063	0.81394	1.3894	0.58583	1.2043	2.15	0.34760	0.62354	1.6977	0.36728	1.9185
1.55	0.15427	0.81054	1.3955	0.58084	1.2116	2.16	0.35030	0.62076	1.7018	0.36476	1.9354
1.56	0.15790	0.80715	1.4015	0.57591	1.2190	2.17	0.35298	0.61799	1.7059	0.36227	1.9525
1.57	0.16152	0.80376	1.4075	0.57104	1.2266	2.18	0.35564	0.61523	1.7099	0.35980	1.9698
1.58	0.16514	0.80038	1.4135	0.56623	1.2343	2.19	0.35828	0.61249	1.7139	0.35736	1.9873
1.59	0.16876	0.79701	1.4195	0.56148	1.2422	2.20	0.36091	0.60976	1.7179	0.35494	2.0050
1.60	0.17236	0.79365	1.4254	0.55679	1.2502	2.21	0.36352	0.60704	1.7219	0.35254	2.0228
1.61	0.17595	0.79030	1.4313	0.55216	1.2583	2.22	0.36611	0.60433	1.7258	0.35017	2.0409
1.62	0.17953	0.78695	1.4371	0.54759	1.2666	2.23	0.36868	0.60163	1.7297	0.34782	2.0592
1.63	0.18311	0.78361	1.4429	0.54308	1.2750	2.24	0.37124	0.59895	1.7336	0.34550	2.0777
1.64	0.18667	0.78028	1.4487	0.53862	1.2835	2.25	0.37378	0.59627	1.7374	0.34319	2.0964
1.65	0.19022	0.77695	1.4544	0.53421	1.2922	2.26	0.37630	0.59361	1.7412	0.34091	2.1154
1.66	0.19376	0.77363	1.4601	0.52986	1.3010	2.27	0.37881	0.59096	1.7450	0.33865	2.1345
1.67	0.19729	0.77033	1.4657	0.52556	1.3099	2.28	0.38130	0.58833	1.7488	0.33641	2.1538
1.68	0.20081	0.76703	1.4713	0.52131	1.3190	2.29	0.38377	0.58570	1.7526	0.33420	2.1733
1.69	0.20431	0.76374	1.4769	0.51711	1.3282	2.30	0.38623	0.58309	1.7563	0.33200	2.1931
1.70	0.20780	0.76046	1.4825	0.51297	1.3376	2.31	0.38867	0.58049	1.7600	0.32983	2.2131
1.71	0.21128	0.75718	1.4880	0.50887	1.3471	2.32	0.39109	0.57790	1.7637	0.32767	2.2333
1.72	0.21474	0.75392	1.4935	0.50482	1.3567	2.33	0.39350	0.57532	1.7673	0.32554	2.2537
1.73	0.21819	0.75067	1.4989	0.50082	1.3665	2.34	0.39589	0.57276	1.7709	0.32342	2.2744
1.74	0.22162	0.74742	1.5043	0.49686	1.3764	2.35	0.39826	0.57021	1.7745	0.32133	2.2953
1.75	0.22504	0.74419	1.5097	0.49295	1.3865	2.36	0.40062	0.56767	1.7781	0.31925	2.3164
1.76	0.22844	0.74096	1.5150	0.48909	1.3967	2.37	0.40296	0.56514	1.7817	0.31720	2.3377
1.77	0.23183	0.73774	1.5203	0.48527	1.4070	2.38	0.40528	0.56262	1.7852	0.31516	2.3593
1.78	0.23520	0.73453	1.5256	0.48149	1.4175	2.39	0.40760	0.56011	1.7887	0.31314	2.3811
1.79	0.23855	0.73134	1.5308	0.47776	1.4282	2.40	0.40989	0.55762	1.7922	0.31114	2.4031
1.80	0.24189	0.72816	1.5360	0.47407	1.4390	2.41	0.41216	0.55514	1.7956	0.30916	2.4254
1.81	0.24521	0.72498	1.5412	0.47042	1.4499	2.42	0.41442	0.55267	1.7991	0.30720	2.4479
1.82	0.24851	0.72181	1.5463	0.46681	1.4610	2.43	0.41667	0.55021	1.8025	0.30525	2.4706
1.83	0.25180	0.71865	1.5514	0.46324	1.4723	2.44	0.41891	0.54776	1.8059	0.30332	2.4936
1.84	0.25507	0.71551	1.5564	0.45972	1.4837	2.45	0.42113	0.54533	1.8092	0.30141	2.5168
1.85	0.25832	0.71238	1.5614	0.45623	1.4952	2.46	0.42333	0.54291	1.8126	0.29952	2.5403
1.86	0.26156	0.70925	1.5664	0.45278	1.5069	2.47	0.42551	0.54050	1.8159	0.29765	2.5640
1.87	0.26478	0.70614	1.5714	0.44937	1.5188	2.48	0.42768	0.53810	1.8192	0.29579	2.5880
1.88	0.26798	0.70304	1.5763	0.44600	1.5308	2.49	0.42983	0.53571	1.8225	0.29395	2.6122
1.89	0.27116	0.69995	1.5812	0.44266	1.5429	2.50	0.43197	0.53333	1.8257	0.29212	2.6367
1.90	0.27433	0.69686	1.5861	0.43936	1.5552	2.51	0.43410	0.53097	1.8290	0.29031	2.6615
1.91	0.27748	0.69379	1.5909	0.43610	1.5677	2.52	0.43621	0.52862	1.8322	0.28852	2.6865
1.92	0.28061	0.69074	1.5957	0.43287	1.5804	2.53	0.43831	0.52627	1.8354	0.28674	2.7117
1.93	0.28372	0.68769	1.6005	0.42967	1.5932	2.54	0.44040	0.52394	1.8386	0.28498	2.7372
1.94	0.28681	0.68465	1.6052	0.42651	1.6062	2.55	0.44247	0.52163	1.8417	0.28323	2.7630
1.95	0.28989	0.68162	1.6099	0.42339	1.6193	2.56	0.44452	0.51932	1.8448	0.28150	2.7891
1.96	0.29295	0.67861	1.6146	0.42030	1.6326	2.57	0.44655	0.51702	1.8479	0.27978	2.8154

TABLE E.2 (continued)

Ma	$\frac{f(\ell^* - \ell)}{D}$	T/T^*	V/V^*	p/p^*	p_0/p_0^*	Ma	$\frac{f(\ell^* - \ell)}{D}$	T/T^*	V/V^*	p/p^*	p_0/p_0^*
2.58	0.44857	0.51474	1.8510	0.27808	2.8420	2.85	0.49828	0.45723	1.9271	0.23726	3.6707
2.59	0.45059	0.51247	1.8541	0.27640	2.8689	2.86	0.49995	0.45525	1.9297	0.23592	3.7058
2.60	0.45259	0.51020	1.8571	0.27473	2.8960	2.87	0.50161	0.45328	1.9322	0.23458	3.7413
2.61	0.45457	0.50795	1.8602	0.27307	2.9234	2.88	0.50326	0.45132	1.9348	0.23326	3.7771
2.62	0.45654	0.50571	1.8632	0.27143	2.9511	2.89	0.50489	0.44937	1.9373	0.23196	3.8133
2.63	0.45850	0.50349	1.8662	0.26980	2.9791	2.90	0.50651	0.44743	1.9398	0.23066	3.8498
2.64	0.46044	0.50127	1.8691	0.26818	3.0074	2.91	0.50812	0.44550	1.9423	0.22937	3.8866
2.65	0.46237	0.49906	1.8721	0.26658	3.0359	2.92	0.50973	0.44358	1.9448	0.22809	3.9238
2.66	0.46429	0.49687	1.8750	0.26499	3.0647	2.93	0.51133	0.44167	1.9472	0.22682	3.9614
2.67	0.46619	0.49469	1.8779	0.26342	3.0938	2.94	0.51291	0.43977	1.9497	0.22556	3.9993
2.68	0.46807	0.49251	1.8808	0.26186	3.1234	2.95	0.51447	0.43788	1.9521	0.22431	4.0376
2.69	0.46996	0.49035	1.8837	0.26032	3.1530	2.96	0.51603	0.43600	1.9545	0.22307	4.0763
2.70	0.47182	0.48820	1.8865	0.25878	3.1830	2.97	0.51758	0.43413	1.9569	0.22185	4.1153
2.71	0.47367	0.48606	1.8894	0.25726	3.2133	2.98	0.51912	0.43226	1.9592	0.22063	4.1547
2.72	0.47551	0.48393	1.8922	0.25575	3.2440	2.99	0.52064	0.43041	1.9616	0.21942	4.1944
2.73	0.47734	0.48182	1.8950	0.25426	3.2749	3.0	0.52216	0.42857	1.9640	0.21822	4.2346
2.74	0.47915	0.47971	1.8978	0.25278	3.3061	3.5	0.58643	0.34783	2.0642	0.16850	6.7896
2.75	0.48095	0.47761	1.9005	0.25131	3.3376	4.0	0.63306	0.28571	2.1381	0.13363	10.719
2.76	0.48274	0.47553	1.9032	0.24985	3.3695	4.5	0.66764	0.23762	2.1936	0.10833	16.562
2.77	0.48452	0.47346	1.9060	0.24840	3.4017	5.0	0.69380	0.20000	2.2361	0.08944	25.000
2.78	0.48628	0.47139	1.9087	0.24697	3.4342	6.0	0.72987	0.14634	2.2953	0.06376	53.180
2.79	0.48803	0.46933	1.9114	0.24555	3.4670	7.0	0.75280	0.11111	2.3333	0.04762	104.14
2.80	0.48976	0.46729	1.9140	0.24414	3.5001	8.0	0.76819	0.08696	2.3591	0.03686	190.11
2.81	0.49148	0.46526	1.9167	0.24274	3.5336	9.0	0.77898	0.06977	2.3772	0.02935	327.19
2.82	0.49321	0.46324	1.9193	0.24135	3.5674	10.00	0.78683	0.05714	2.3905	0.02390	535.94
2.83	0.49491	0.46122	1.9220	0.23997	3.6015	∞	0.82153	0	2.4495	0	∞
2.84	0.49660	0.45922	1.9246	0.23861	3.6359						

TABLE E.3 Rayleigh Flow of an Ideal Gas ($k = 1.4$)

Ma	p/p_a	T/T_a	V/V_a or ρ_a/ρ	$T_0/T_{0,a}$	$p_0/p_{0,a}$	Ma	p/p_a	T/T_a	V/V_a or ρ_a/ρ	$T_0/T_{0,a}$	$p_0/p_{0,a}$
0	2.4000	0	0	0	1.2679	0.25	2.2069	0.30440	0.13793	0.25684	1.2177
0.01	2.3997	0.000576	0.000240	0.000480	1.2678	0.26	2.1925	0.32496	0.14821	0.27446	1.2140
0.02	2.3987	0.00230	0.000959	0.00192	1.2675	0.27	2.1777	0.34573	0.15876	0.29231	1.2102
0.03	2.3970	0.00516	0.00216	0.00431	1.2671	0.28	2.1626	0.36667	0.16955	0.31035	1.2064
0.04	2.3946	0.00917	0.00383	0.00765	1.2665	0.29	2.1472	0.38773	0.18058	0.32855	1.2025
0.05	2.3916	0.01430	0.00598	0.01192	1.2657	0.30	2.1314	0.40887	0.19183	0.34686	1.1985
0.06	2.3880	0.02053	0.00860	0.01712	1.2647	0.31	2.1154	0.43004	0.20329	0.36525	1.1945
0.07	2.3837	0.02784	0.01168	0.02322	1.2636	0.32	2.0991	0.45119	0.21494	0.38369	1.1904
0.08	2.3787	0.03621	0.01522	0.03021	1.2623	0.33	2.0825	0.47228	0.22678	0.40214	1.1863
0.09	2.3731	0.04562	0.01922	0.03807	1.2608	0.34	2.0657	0.49327	0.23879	0.42057	1.1821
0.10	2.3669	0.05602	0.02367	0.04678	1.2591	0.35	2.0487	0.51413	0.25096	0.43894	1.1779
0.11	2.3600	0.06739	0.02856	0.05630	1.2573	0.36	2.0314	0.53482	0.26327	0.45723	1.1737
0.12	2.3526	0.07970	0.03388	0.06661	1.2554	0.37	2.0140	0.55530	0.27572	0.47541	1.1695
0.13	2.3445	0.09290	0.03962	0.07768	1.2533	0.38	1.9964	0.57553	0.28828	0.49346	1.1652
0.14	2.3359	0.10695	0.04578	0.08947	1.2510	0.39	1.9787	0.59549	0.30095	0.51134	1.1609
0.15	2.3267	0.12181	0.05235	0.10196	1.2486	0.40	1.9608	0.61515	0.31372	0.52903	1.1566
0.16	2.3170	0.13743	0.05931	0.11511	1.2461	0.41	1.9428	0.63448	0.32658	0.54651	1.1523
0.17	2.3067	0.15377	0.06666	0.12888	1.2434	0.42	1.9247	0.65345	0.33951	0.56376	1.1480
0.18	2.2959	0.17078	0.07438	0.14324	1.2406	0.43	1.9065	0.67205	0.35251	0.58075	1.1437
0.19	2.2845	0.18841	0.08247	0.15814	1.2377	0.44	1.8882	0.69025	0.36556	0.59748	1.1394
0.20	2.2727	0.20661	0.09091	0.17355	1.2346	0.45	1.8699	0.70803	0.37865	0.61393	1.1351
0.21	2.2604	0.22533	0.09969	0.18943	1.2314	0.46	1.8515	0.72538	0.39178	0.63007	1.1308
0.22	2.2477	0.24452	0.10879	0.20574	1.2281	0.47	1.8331	0.74228	0.40493	0.64589	1.1266
0.23	2.2345	0.26413	0.11820	0.22244	1.2248	0.48	1.8147	0.75871	0.41810	0.66139	1.1224
0.24	2.2209	0.28411	0.12792	0.23948	1.2213	0.49	1.7962	0.77466	0.43127	0.67655	1.1182

TABLE E.3 (continued)

Ma	p/p_a	T/T_a	V/V_a or ρ_a/ρ	$T_0/T_{0,a}$	$p_0/p_{0,a}$	Ma	p/p_a	T/T_a	V/V_a or ρ_a/ρ	$T_0/T_{0,a}$	$p_0/p_{0,a}$
0.50	1.7778	0.79012	0.44445	0.69136	1.1140	1.11	0.88075	0.95577	1.08518	0.99274	1.00588
0.51	1.7594	0.80509	0.45761	0.70581	1.1099	1.12	0.87078	0.95115	1.09230	0.99148	1.00699
0.52	1.7410	0.81955	0.47075	0.71990	1.1059	1.13	0.86094	0.94646	1.09933	0.99013	1.00820
0.53	1.7226	0.83351	0.48387	0.73361	1.1019	1.14	0.85123	0.94169	1.10626	0.98871	1.00951
0.54	1.7043	0.84695	0.49696	0.74695	1.0979	1.15	0.84166	0.93685	1.1131	0.98721	1.01092
0.55	1.6860	0.85987	0.51001	0.75991	1.09397	1.16	0.83222	0.93195	1.1198	0.98564	1.01243
0.56	1.6678	0.87227	0.52302	0.77248	1.09010	1.17	0.82292	0.92700	1.1264	0.98400	1.01403
0.57	1.6496	0.88415	0.53597	0.78467	1.08630	1.18	0.81374	0.92200	1.1330	0.98230	1.01572
0.58	1.6316	0.89552	0.54887	0.79647	1.08255	1.19	0.80468	0.91695	1.1395	0.98054	1.01752
0.59	1.6136	0.90637	0.56170	0.80789	1.07887	1.20	0.79576	0.91185	1.1459	0.97872	1.01941
0.60	1.5957	0.91670	0.57447	0.81892	1.07525	1.21	0.78695	0.90671	1.1522	0.97685	1.02140
0.61	1.5780	0.92653	0.58716	0.82956	1.07170	1.22	0.77827	0.90153	1.1584	0.97492	1.02348
0.62	1.5603	0.93585	0.59978	0.83982	1.06821	1.23	0.76971	0.89632	1.1645	0.97294	1.02566
0.63	1.5427	0.94466	0.61232	0.84970	1.06480	1.24	0.76127	0.89108	1.1705	0.97092	1.02794
0.64	1.5253	0.95298	0.62477	0.85920	1.06146	1.25	0.75294	0.88581	1.1764	0.96886	1.03032
0.65	1.5080	0.96081	0.63713	0.86833	1.05820	1.26	0.74473	0.88052	1.1823	0.96675	1.03280
0.66	1.4908	0.96816	0.64941	0.87709	1.05502	1.27	0.73663	0.87521	1.1881	0.96461	1.03536
0.67	1.4738	0.97503	0.66159	0.88548	1.05192	1.28	0.72865	0.86988	1.1938	0.96243	1.03803
0.68	1.4569	0.98144	0.67367	0.89350	1.04890	1.29	0.72078	0.86453	1.1994	0.96022	1.04080
0.69	1.4401	0.98739	0.68564	0.90117	1.04596	1.30	0.71301	0.85917	1.2050	0.95798	1.04365
0.70	1.4235	0.99289	0.69751	0.90850	1.04310	1.31	0.70535	0.85380	1.2105	0.95571	1.04661
0.71	1.4070	0.99796	0.70927	0.91548	1.04033	1.32	0.69780	0.84843	1.2159	0.95341	1.04967
0.72	1.3907	1.00260	0.72093	0.92212	1.03764	1.33	0.69035	0.84305	1.2212	0.95108	1.05283
0.73	1.3745	1.00682	0.73248	0.92843	1.03504	1.34	0.68301	0.83766	1.2264	0.94873	1.05608
0.74	1.3585	1.01062	0.74392	0.93442	1.03253	1.35	0.67577	0.83227	1.2316	0.94636	1.05943
0.75	1.3427	1.01403	0.75525	0.94009	1.03010	1.36	0.66863	0.82698	1.2367	0.94397	1.06288
0.76	1.3270	1.01706	0.76646	0.94546	1.02776	1.37	0.66159	0.82151	1.2417	0.94157	1.06642
0.77	1.3115	1.01971	0.77755	0.95052	1.02552	1.38	0.65464	0.81613	1.2467	0.93915	1.07006
0.78	1.2961	1.02198	0.78852	0.95528	1.02337	1.39	0.64778	0.81076	1.2516	0.93671	1.07380
0.79	1.2809	1.02390	0.79938	0.95975	1.02131	1.40	0.64102	0.80540	1.2564	0.93425	1.07765
0.80	1.2658	1.02548	0.81012	0.96394	1.01934	1.41	0.63436	0.80004	1.2612	0.93178	1.08159
0.81	1.2509	1.02672	0.82075	0.96786	1.01746	1.42	0.62779	0.79469	1.2659	0.92931	1.08563
0.82	1.2362	1.02763	0.83126	0.97152	1.01569	1.43	0.62131	0.78936	1.2705	0.92683	1.08977
0.83	1.2217	1.02823	0.84164	0.97492	1.01399	1.44	0.61491	0.78405	1.2751	0.92434	1.09400
0.84	1.2073	1.02853	0.85190	0.97807	1.01240	1.45	0.60860	0.77875	1.2796	0.92184	1.0983
0.85	1.1931	1.02854	0.86204	0.98087	1.01091	1.46	0.60237	0.77346	1.2840	0.91933	1.1028
0.86	1.1791	1.02826	0.87206	0.98363	1.00951	1.47	0.59623	0.76819	1.2884	0.91682	1.1073
0.87	1.1652	1.02771	0.88196	0.98607	1.00819	1.48	0.59018	0.76294	1.2927	0.91431	1.1120
0.88	1.1515	1.02690	0.89175	0.98828	1.00698	1.49	0.58421	0.75771	1.2970	0.91179	1.1167
0.89	1.1380	1.02583	0.90142	0.99028	1.00587	1.50	0.57831	0.75250	1.3012	0.90928	1.1215
0.90	1.1246	1.02451	0.91097	0.99207	1.00486	1.51	0.57250	0.74731	1.3054	0.90676	1.1264
0.91	1.1114	1.02297	0.92039	0.99366	1.00393	1.52	0.56677	0.74215	1.3095	0.90424	1.1315
0.92	1.09842	1.02120	0.92970	0.99506	1.00310	1.53	0.56111	0.73701	1.3135	0.90172	1.1367
0.93	1.08555	1.01921	0.93889	0.99627	1.00237	1.54	0.55553	0.73189	1.3175	0.89920	1.1420
0.94	1.07285	1.01702	0.94796	0.99729	1.00174	1.55	0.55002	0.72680	1.3214	0.89669	1.1473
0.95	1.06030	1.01463	0.95692	0.99814	1.00121	1.56	0.54458	0.72173	1.3253	0.89418	1.1527
0.96	1.04792	1.01205	0.96576	0.99883	1.00077	1.57	0.53922	0.71669	1.3291	0.89167	1.1582
0.97	1.03570	1.00929	0.97449	0.99935	1.00043	1.58	0.53393	0.71168	1.3329	0.88917	1.1639
0.98	1.02364	1.00636	0.98311	0.99972	1.00019	1.59	0.52871	0.70669	1.3366	0.88668	1.1697
0.99	1.01174	1.00326	0.99161	0.99993	1.00004	1.60	0.52356	0.70173	1.3403	0.88419	1.1756
1.00	1.00000	1.00000	1.00000	1.00000	1.00000	1.61	0.51848	0.69680	1.3439	0.88170	1.1816
1.01	0.98841	0.99659	1.00828	0.99993	1.00004	1.62	0.51346	0.69190	1.3475	0.87922	1.1877
1.02	0.97697	0.99304	1.01644	0.99973	1.00019	1.63	0.50851	0.68703	1.3511	0.87675	1.1939
1.03	0.96569	0.98936	1.02450	0.99940	1.00043	1.64	0.50363	0.68219	1.3546	0.87429	1.2002
1.04	0.95456	0.98553	1.03246	0.99895	1.00077	1.65	0.49881	0.67738	1.3580	0.87184	1.2066
1.05	0.94358	0.98161	1.04030	0.99838	1.00121	1.66	0.49405	0.67259	1.3614	0.86940	1.2131
1.06	0.93275	0.97755	1.04804	0.99769	1.00175	1.67	0.48935	0.66784	1.3648	0.86696	1.2197
1.07	0.92206	0.97339	1.05567	0.99690	1.00238	1.68	0.48471	0.66312	1.3681	0.86453	1.2264
1.08	0.91152	0.96913	1.06320	0.99600	1.00311	1.69	0.48014	0.65843	1.3713	0.86211	1.2332
1.09	0.90112	0.96477	1.07062	0.99501	1.00394	1.70	0.47563	0.65377	1.3745	0.85970	1.2402
1.10	0.89086	0.96031	1.07795	0.99392	1.00486	1.71	0.47117	0.64914	1.3777	0.85731	1.2473

TABLE E.3 **(continued)**

Ma	p/p_a	T/T_a	V/V_a or ρ_a/ρ	$T_0/T_{0,a}$	$p_0/p_{0,a}$	Ma	p/p_a	T/T_a	V/V_a or ρ_a/ρ	$T_0/T_{0,a}$	$p_0/p_{0,a}$
1.72	0.46677	0.64455	1.3809	0.85493	1.2545	2.33	0.27905	0.42276	1.5150	0.73482	1.9320
1.73	0.46242	0.63999	1.3840	0.85256	1.2618	2.34	0.27695	0.41999	1.5165	0.73327	1.9476
1.74	0.45813	0.63546	1.3871	0.85020	1.2692	2.35	0.27487	0.41724	1.5180	0.73173	1.9634
1.75	0.45390	0.63096	1.3901	0.84785	1.2767	2.36	0.27281	0.41451	1.5195	0.73020	1.9794
1.76	0.44972	0.62649	1.3931	0.84551	1.2843	2.37	0.27077	0.41181	1.5209	0.72868	1.9955
1.77	0.44559	0.62205	1.3960	0.84318	1.2920	2.38	0.26875	0.40913	1.5223	0.72718	2.0118
1.78	0.44152	0.61765	1.3989	0.84087	1.2998	2.39	0.26675	0.40647	1.5237	0.72569	2.0283
1.79	0.43750	0.61328	1.4018	0.83857	1.3078	2.40	0.26478	0.40383	1.5252	0.72421	2.0450
1.80	0.43353	0.60894	1.4046	0.83628	1.3159	2.41	0.26283	0.40122	1.5266	0.72274	2.0619
1.81	0.42960	0.60463	1.4074	0.83400	1.3241	2.42	0.26090	0.39863	1.5279	0.72129	2.0789
1.82	0.42573	0.60036	1.4102	0.83174	1.3324	2.43	0.25899	0.39606	1.5293	0.71985	2.0961
1.83	0.42191	0.59612	1.4129	0.82949	1.3408	2.44	0.25710	0.39352	1.5306	0.71842	2.1135
1.84	0.41813	0.59191	1.4156	0.82726	1.3494	2.45	0.25523	0.39100	1.5320	0.71700	2.1311
1.85	0.41440	0.58773	1.4183	0.82504	1.3581	2.46	0.25337	0.38850	1.5333	0.71559	2.1489
1.86	0.41072	0.58359	1.4209	0.82283	1.3669	2.47	0.25153	0.38602	1.5346	0.71419	2.1669
1.87	0.40708	0.57948	1.4235	0.82064	1.3758	2.48	0.24972	0.38356	1.5359	0.71280	2.1850
1.88	0.40349	0.57540	1.4261	0.81846	1.3848	2.49	0.24793	0.38112	1.5372	0.71142	2.2033
1.89	0.39994	0.57135	1.4286	0.81629	1.3940	2.50	0.24616	0.37870	1.5385	0.71005	2.2218
1.90	0.39643	0.56734	1.4311	0.81414	1.4033	2.51	0.24440	0.37630	1.5398	0.70870	2.2405
1.91	0.39297	0.56336	1.4336	0.81200	1.4127	2.52	0.24266	0.37392	1.5410	0.70736	2.2594
1.92	0.38955	0.55941	1.4360	0.80987	1.4222	2.53	0.24094	0.37157	1.5422	0.70603	2.2785
1.93	0.38617	0.55549	1.4384	0.80776	1.4319	2.54	0.23923	0.36923	1.5434	0.70471	2.2978
1.94	0.38283	0.55160	1.4408	0.80567	1.4417	2.55	0.23754	0.36691	1.5446	0.70340	2.3173
1.95	0.37954	0.54774	1.4432	0.80359	1.4516	2.56	0.23587	0.36461	1.5458	0.70210	2.3370
1.96	0.37628	0.54391	1.4455	0.80152	1.4616	2.57	0.23422	0.36233	1.5470	0.70081	2.3569
1.97	0.37306	0.54012	1.4478	0.79946	1.4718	2.58	0.23258	0.36007	1.5482	0.69953	2.3770
1.98	0.36988	0.53636	1.4501	0.79742	1.4821	2.59	0.23096	0.35783	1.5494	0.69825	2.3972
1.99	0.36674	0.53263	1.4523	0.79540	1.4925	2.60	0.22936	0.35561	1.5505	0.69699	2.4177
2.00	0.36364	0.52893	1.4545	0.79339	1.5031	2.61	0.22777	0.35341	1.5516	0.69574	2.4384
2.01	0.36057	0.52526	1.4567	0.79139	1.5138	2.62	0.22620	0.35123	1.5527	0.69450	2.4593
2.02	0.35754	0.52161	1.4589	0.78941	1.5246	2.63	0.22464	0.34906	1.5538	0.69327	2.4804
2.03	0.35454	0.51800	1.4610	0.78744	1.5356	2.64	0.22310	0.34691	1.5549	0.69205	2.5017
2.04	0.35158	0.51442	1.4631	0.78549	1.5467	2.65	0.22158	0.34478	1.5560	0.69084	2.5233
2.05	0.34866	0.51087	1.4652	0.78355	1.5579	2.66	0.22007	0.34267	1.5571	0.68964	2.5451
2.06	0.34577	0.50735	1.4673	0.78162	1.5693	2.67	0.21857	0.34057	1.5582	0.68845	2.5671
2.07	0.34291	0.50386	1.4694	0.77971	1.5808	2.68	0.21709	0.33849	1.5593	0.68727	2.5892
2.08	0.34009	0.50040	1.4714	0.77781	1.5924	2.69	0.21562	0.33643	1.5603	0.68610	2.6116
2.09	0.33730	0.49697	1.4734	0.77593	1.6042	2.70	0.21417	0.33439	1.5613	0.68494	2.6342
2.10	0.33454	0.49356	1.4753	0.77406	1.6161	2.71	0.21273	0.33236	1.5623	0.68378	2.6571
2.11	0.33181	0.49018	1.4773	0.77221	1.6282	2.72	0.21131	0.33035	1.5633	0.68263	2.6802
2.12	0.32912	0.48683	1.4792	0.77037	1.6404	2.73	0.20990	0.32836	1.5644	0.68150	2.7035
2.13	0.32646	0.48351	1.4811	0.76854	1.6528	2.74	0.20850	0.32638	1.5654	0.68038	2.7270
2.14	0.32383	0.48022	1.4830	0.76673	1.6653	2.75	0.20712	0.32442	1.5663	0.67926	2.7508
2.15	0.32122	0.47696	1.4849	0.76493	1.6780	2.76	0.20575	0.32248	1.5673	0.67815	2.7748
2.16	0.31864	0.47373	1.4867	0.76314	1.6908	2.77	0.20439	0.32055	1.5683	0.67704	2.7990
2.17	0.31610	0.47052	1.4885	0.76137	1.7037	2.78	0.20305	0.31864	1.5692	0.67595	2.8235
2.18	0.31359	0.46734	1.4903	0.75961	1.7168	2.79	0.20172	0.31674	1.5702	0.67487	2.8482
2.19	0.31110	0.46419	1.4921	0.75787	1.7300	2.80	0.20040	0.31486	1.5711	0.67380	2.8731
2.20	0.30864	0.46106	1.4939	0.75614	1.7434	2.81	0.19909	0.31299	1.5721	0.67273	2.8982
2.21	0.30621	0.45796	1.4956	0.75442	1.7570	2.82	0.19780	0.31114	1.5730	0.67167	2.9236
2.22	0.30381	0.45489	1.4973	0.75271	1.7707	2.83	0.19652	0.30931	1.5739	0.67062	2.9493
2.23	0.30143	0.45184	1.4990	0.75102	1.7846	2.84	0.19525	0.30749	1.5748	0.66958	2.9752
2.24	0.29908	0.44882	1.5007	0.74934	1.7986	2.85	0.19399	0.30568	1.5757	0.66855	3.0013
2.25	0.29675	0.44582	1.5024	0.74767	1.8128	2.86	0.19274	0.30389	1.5766	0.66752	3.0277
2.26	0.29445	0.44285	1.5040	0.74602	1.8271	2.87	0.19151	0.30211	1.5775	0.66650	3.0544
2.27	0.29218	0.43990	1.5056	0.74438	1.8416	2.88	0.19029	0.30035	1.5784	0.66549	3.0813
2.28	0.28993	0.43698	1.5072	0.74275	1.8562	2.89	0.18908	0.29860	1.5792	0.66449	3.1084
2.29	0.28771	0.43409	1.5088	0.74114	1.8710	2.90	0.18788	0.29687	1.5801	0.66350	3.1358
2.30	0.28551	0.43122	1.5104	0.73954	1.8860	2.91	0.18669	0.29515	1.5809	0.66252	3.1635
2.31	0.28333	0.42837	1.5119	0.73795	1.9012	2.92	0.18551	0.29344	1.5818	0.66154	3.1914
2.32	0.28118	0.42555	1.5134	0.73638	1.9165	2.93	0.18435	0.29175	1.5826	0.66057	3.2196

TABLE E.3 (continued)

Ma	p/p_a	T/T_a	V/V_a or ρ_a/ρ	$T_0/T_{0,a}$	$p_0/p_{0,a}$	Ma	p/p_a	T/T_a	V/V_a or ρ_a/ρ	$T_0/T_{0,a}$	$p_0/p_{0,a}$
2.94	0.18320	0.29007	1.5834	0.65961	3.2481	4.50	0.08177	0.13540	1.6559	0.56983	12.502
2.95	0.18205	0.28841	1.5843	0.65865	3.2768	5.00	0.06667	0.11111	1.6667	0.55555	18.634
2.96	0.18091	0.28676	1.5851	0.65770	3.3058	6.00	0.04669	0.07849	1.6809	0.53633	38.946
2.97	0.17978	0.28512	1.5859	0.65676	3.3351	7.00	0.03448	0.05826	1.6897	0.52437	75.414
2.98	0.17867	0.28349	1.5867	0.65583	3.3646	8.00	0.02649	0.04491	1.6954	0.51646	136.62
2.99	0.17757	0.28188	1.5875	0.65490	3.3944	9.00	0.02098	0.03565	1.6993	0.51098	233.88
3.00	0.17647	0.28028	1.5882	0.65398	3.4244	10.00	0.01702	0.02897	1.7021	0.50702	381.61
3.50	0.13223	0.21419	1.6198	0.61580	5.3280	∞	0	0	1.7143	0.48980	∞
4.00	0.10256	0.16831	1.6410	0.58909	8.2268						

TABLE E.4 Normal Shock Flow of an Ideal Gas ($k = 1.4$)

Ma_x	Ma_y	$\dfrac{p_y}{p_x}$	T_y/T_x	$\rho_y/\rho_x =$ V_x/V_y	$\dfrac{p_{0,y}}{p_{0,x}}$	$p_{0,y}/p_x$	Ma_x	Ma_y	$\dfrac{p_y}{p_x}$	T_y/T_x	$\rho_y/\rho_x =$ V_x/V_y	$\dfrac{p_{0,y}}{p_{0,x}}$	$p_{0,y}/p_x$
1.00	1.00000	1.00000	1.00000	1.00000	1.00000	1.8929	1.49	0.70466	2.4234	1.3136	1.8449	0.93292	3.3756
1.01	0.99013	1.02345	1.00665	1.01669	0.99999	1.9152	1.50	0.70109	2.4583	1.3202	1.8621	0.92978	3.4133
1.02	0.98052	1.04713	1.01325	1.03344	0.99998	1.9379	1.51	0.69758	2.4934	1.3269	1.8792	0.92658	3.4512
1.03	0.97115	1.07105	1.01981	1.05024	0.99997	1.9610	1.52	0.69413	2.5288	1.3336	1.8962	0.92331	3.4894
1.04	0.96202	1.09520	1.02634	1.06709	0.99994	1.9845	1.53	0.69073	2.5644	1.3403	1.9133	0.91999	3.5279
1.05	0.95312	1.1196	1.03284	1.08398	0.99987	2.0083	1.54	0.68739	2.6003	1.3470	1.9303	0.91662	3.5667
1.06	0.94444	1.1442	1.03931	1.10092	0.99976	2.0325	1.55	0.68410	2.6363	1.3538	1.9473	0.91319	3.6058
1.07	0.93598	1.1690	1.04575	1.11790	0.99962	2.0570	1.56	0.68086	2.6725	1.3606	1.9643	0.90970	3.6451
1.08	0.92772	1.1941	1.05217	1.13492	0.99944	2.0819	1.57	0.67768	2.7090	1.3674	1.9812	0.90615	3.6847
1.09	0.91965	1.2194	1.05856	1.15199	0.99921	2.1072	1.58	0.67455	2.7458	1.3742	1.9981	0.90255	3.7245
1.10	0.91177	1.2450	1.06494	1.1691	0.99892	2.1328	1.59	0.67147	2.7828	1.3811	2.0149	0.89889	3.7645
1.11	0.90408	1.2708	1.07130	1.1862	0.99858	2.1588	1.60	0.66844	2.8201	1.3880	2.0317	0.89520	3.8049
1.12	0.89656	1.2968	1.07764	1.2034	0.99820	2.1851	1.61	0.66545	2.8575	1.3949	2.0485	0.89144	3.8456
1.13	0.88922	1.3230	1.08396	1.2206	0.99776	2.2118	1.62	0.66251	2.8951	1.4018	2.0652	0.88764	3.8866
1.14	0.88204	1.3495	1.09027	1.2378	0.99726	2.2388	1.63	0.65962	2.9330	1.4088	2.0820	0.88380	3.9278
1.15	0.87502	1.3762	1.09657	1.2550	0.99669	2.2661	1.64	0.65677	2.9712	1.4152	2.0986	0.87992	3.9693
1.16	0.86816	1.4032	1.10287	1.2723	0.99605	2.2937	1.65	0.65396	3.0096	1.4228	2.1152	0.87598	4.0111
1.17	0.86145	1.4304	1.10916	1.2896	0.99534	2.3217	1.66	0.65119	3.0482	1.4298	2.1318	0.87201	4.0531
1.18	0.85488	1.4578	1.11544	1.3069	0.99455	2.3499	1.67	0.64847	3.0870	1.4369	2.1484	0.86800	4.0954
1.19	0.84846	1.4854	1.12172	1.3243	0.99371	2.3786	1.68	0.64579	3.1261	1.4440	2.1649	0.86396	4.1379
1.20	0.84217	1.5133	1.1280	1.3416	0.99280	2.4075	1.69	0.64315	3.1654	1.4512	2.1813	0.85987	4.1807
1.21	0.83601	1.5414	1.1343	1.3590	0.99180	2.4367	1.70	0.64055	3.2050	1.4583	2.1977	0.85573	4.2238
1.22	0.82998	1.5698	1.1405	1.3764	0.99073	2.4662	1.71	0.63798	3.2448	1.4655	2.2141	0.85155	4.2672
1.23	0.82408	1.5984	1.1468	1.3938	0.98975	2.4961	1.72	0.63545	3.2848	1.4727	2.2304	0.84735	4.3108
1.24	0.81830	1.6272	1.1531	1.4112	0.98835	2.5263	1.73	0.63296	3.3250	1.4800	2.2467	0.84312	4.3547
1.25	0.81264	1.6562	1.1594	1.4286	0.98706	2.5568	1.74	0.63051	3.3655	1.4873	2.2629	0.83886	4.3989
1.26	0.80709	1.6855	1.1657	1.4460	0.98568	2.5876	1.75	0.62809	3.4062	1.4946	2.2791	0.83456	4.4433
1.27	0.80165	1.7150	1.1720	1.4634	0.98422	2.6187	1.76	0.62570	3.4472	1.5019	2.2952	0.83024	4.4880
1.28	0.79631	1.7448	1.1782	1.4808	0.98268	2.6500	1.77	0.62335	3.4884	1.5093	2.3113	0.82589	4.5330
1.29	0.79108	1.7748	1.1846	1.4983	0.98106	2.6816	1.78	0.62104	3.5298	1.5167	2.3273	0.82152	4.5783
1.30	0.78596	1.8050	1.1909	1.5157	0.97935	2.7135	1.79	0.61875	3.5714	1.5241	2.3433	0.81711	4.6238
1.31	0.78093	1.8354	1.1972	1.5331	0.97758	2.7457	1.80	0.61650	3.6133	1.5316	2.3592	0.81268	4.6695
1.32	0.77600	1.8661	1.2035	1.5505	0.97574	2.7783	1.81	0.61428	3.6554	1.5391	2.3751	0.80823	4.7155
1.33	0.77116	1.8970	1.2099	1.5680	0.97382	2.8112	1.82	0.61209	3.6978	1.5466	2.3909	0.80376	4.7618
1.34	0.76641	1.9282	1.2162	1.5854	0.97181	2.8444	1.83	0.60993	3.7404	1.5542	2.4067	0.79926	4.8083
1.35	0.76175	1.9596	1.2226	1.6028	0.96972	2.8778	1.84	0.60780	3.7832	1.5617	2.4224	0.79474	4.8551
1.36	0.75718	1.9912	1.2290	1.6202	0.96756	2.9115	1.85	0.60570	3.8262	1.5694	2.4381	0.79021	4.9022
1.37	0.75269	2.0230	1.2354	1.6376	0.96534	2.9455	1.86	0.60363	3.8695	1.5770	2.4537	0.78567	4.9498
1.38	0.74828	2.0551	1.2418	1.6550	0.96304	2.9798	1.87	0.60159	3.9130	1.5847	2.4693	0.78112	4.9974
1.39	0.74396	2.0874	1.2482	1.6723	0.96065	3.0144	1.88	0.59957	3.9568	1.5924	2.4848	0.77656	5.0453
1.40	0.73971	2.1200	1.2547	1.6896	0.95819	3.0493	1.89	0.59758	4.0008	1.6001	2.5003	0.77197	5.0934
1.41	0.73554	2.1528	1.2612	1.7070	0.95566	3.0844	1.90	0.59562	4.0450	1.6079	2.5157	0.76735	5.1417
1.42	0.73144	2.1858	1.2676	1.7243	0.95306	3.1198	1.91	0.59368	4.0894	1.6157	2.5310	0.76273	5.1904
1.43	0.72741	2.2190	1.2742	1.7416	0.95039	3.1555	1.92	0.59177	4.1341	1.6236	2.5463	0.75812	5.2394
1.44	0.72345	2.2525	1.2807	1.7589	0.94765	3.1915	1.93	0.58988	4.1790	1.6314	2.5615	0.75347	5.2886
1.45	0.71956	2.2862	1.2872	1.7761	0.94483	3.2278	1.94	0.58802	4.2242	1.6394	2.5767	0.74883	5.3381
1.46	0.71574	2.3202	1.2938	1.7934	0.94196	3.2643	1.95	0.58618	4.2696	1.6473	2.5919	0.74418	5.3878
1.47	0.71198	2.3544	1.3004	1.8106	0.93901	3.3011	1.96	0.58437	4.3152	1.6553	2.6070	0.73954	5.4378
1.48	0.70829	2.3888	1.3070	1.8278	0.93600	3.3382	1.97	0.58258	4.3610	1.6633	2.6220	0.73487	5.4880

TABLE E.4 (continued)

Ma$_x$	Ma$_y$	$\dfrac{p_y}{p_x}$	T_y/T_x	$\rho_y/\rho_x =$ V_x/V_y	$\dfrac{p_{0,y}}{p_{0,x}}$	$p_{0,y}/p_x$	Ma$_x$	Ma$_y$	$\dfrac{p_y}{p_x}$	T_y/T_x	$\rho_y/\rho_x =$ V_x/V_y	$\dfrac{p_{0,y}}{p_{0,x}}$	$p_{0,y}/p_x$
1.98	0.58081	4.4071	1.6713	2.6369	0.73021	5.5385	2.55	0.50831	7.4196	2.1875	3.3918	0.47927	8.8505
1.99	0.57907	4.4534	1.6794	2.6518	0.72554	5.5894	2.56	0.50740	7.4792	2.1976	3.4034	0.47540	8.9162
2.00	0.57735	4.5000	1.6875	2.6666	0.72088	5.6405	2.57	0.50651	7.5391	2.2077	3.4149	0.47155	8.9821
2.01	0.57565	4.5468	1.6956	2.6814	0.71619	5.6918	2.58	0.50562	7.5992	2.2179	3.4263	0.46772	9.0482
2.02	0.57397	4.5938	1.7038	2.6962	0.71152	5.7434	2.59	0.50474	7.6595	2.2281	3.4376	0.46391	9.1146
2.03	0.57231	4.6411	1.7120	2.7109	0.70686	5.7952	2.60	0.50387	7.7200	2.2383	3.4489	0.46012	9.1813
2.04	0.57068	4.6886	1.7203	2.7255	0.70218	5.8473	2.61	0.50301	7.7808	2.2486	3.4602	0.45636	9.2481
2.05	0.56907	4.7363	1.7286	2.7400	0.69752	5.8997	2.62	0.50216	7.8418	2.2589	3.4714	0.45262	9.3154
2.06	0.56747	4.7842	1.7369	2.7545	0.69284	5.9523	2.63	0.50132	7.9030	2.2693	3.4825	0.44891	9.3829
2.07	0.56589	4.8324	1.7452	2.7690	0.68817	6.0052	2.64	0.50048	7.9645	2.2797	3.4936	0.44522	9.4507
2.08	0.56433	4.8808	1.7536	2.7834	0.68351	6.0584	2.65	0.49965	8.0262	2.2901	3.5047	0.44155	9.5187
2.09	0.56280	4.9295	1.7620	2.7977	0.67886	6.1118	2.66	0.49883	8.0882	2.3006	3.5157	0.43791	9.5869
2.10	0.56128	4.9784	1.7704	2.8119	0.67422	6.1655	2.67	0.49802	8.1504	2.3111	3.5266	0.43429	9.6553
2.11	0.55987	5.0275	1.7789	2.8216	0.66957	6.2194	2.68	0.49722	8.2128	2.3217	3.5374	0.43070	9.7241
2.12	0.55830	5.0768	1.7874	2.8402	0.66492	6.2736	2.69	0.49642	8.2754	2.3323	3.5482	0.42713	9.7932
2.13	0.55683	5.1264	1.7960	2.8543	0.66029	6.3280	2.70	0.49563	8.3383	2.3429	3.5590	0.42359	9.8625
2.14	0.55538	5.1762	1.8046	2.8683	0.65567	6.3827	2.71	0.49485	8.4014	2.3536	3.5697	0.42007	9.9320
2.15	0.55395	5.2262	1.8132	2.8823	0.65105	6.4377	2.72	0.49408	8.4648	2.3643	3.5803	0.41657	10.002
2.16	0.55254	5.2765	1.8219	2.8962	0.64644	6.4929	2.73	0.49332	8.5284	2.3750	3.5909	0.41310	10.072
2.17	0.55114	5.3270	1.8306	2.9100	0.64185	6.5484	2.74	0.49256	8.5922	2.3858	3.6014	0.40965	10.142
2.18	0.54976	5.3778	1.8393	2.9238	0.63728	6.6042	2.75	0.49181	8.6562	2.3966	3.6119	0.40622	10.212
2.19	0.54841	5.4288	1.8481	2.9376	0.63270	6.6602	2.76	0.49107	8.7205	2.4074	3.6224	0.40282	10.283
2.20	0.54706	5.4800	1.8569	2.9512	0.62812	6.7163	2.77	0.49033	8.7850	2.4183	3.6328	0.39945	10.354
2.21	0.54572	5.5314	1.8657	2.9648	0.62358	6.7730	2.78	0.48960	8.8497	2.4292	3.6431	0.39610	10.426
2.22	0.54440	5.5831	1.8746	2.9783	0.61905	6.8299	2.79	0.48888	8.9147	2.4402	3.6533	0.39276	10.498
2.23	0.54310	5.6350	1.8835	2.9918	0.61453	6.8869	2.80	0.48817	8.9800	2.4512	3.6635	0.38946	10.569
2.24	0.54182	5.6872	1.8924	3.0052	0.61002	6.9442	2.81	0.48746	9.0454	2.4622	3.6737	0.38618	10.641
2.25	0.54055	5.7396	1.9014	3.0186	0.60554	7.0018	2.82	0.48676	9.1111	2.4733	3.6838	0.38293	10.714
2.26	0.53929	5.7922	1.9104	3.0319	0.60106	7.0597	2.83	0.48607	9.1770	2.4844	3.6939	0.37970	10.787
2.27	0.53805	5.8451	1.9194	3.0452	0.59659	7.1178	2.84	0.48538	9.2432	2.4955	3.7039	0.37649	10.860
2.28	0.53683	5.8982	1.9285	3.0584	0.59214	7.1762	2.85	0.48470	9.3096	2.5067	3.7139	0.37330	10.933
2.29	0.53561	5.9515	1.9376	3.0715	0.58772	7.2348	2.86	0.48402	9.3762	2.5179	3.7238	0.37013	11.006
2.30	0.53441	6.0050	1.9468	3.0846	0.58331	7.2937	2.87	0.48334	9.4431	2.5292	3.7336	0.36700	11.080
2.31	0.53322	6.0588	1.9560	3.0976	0.57891	7.3529	2.88	0.48268	9.5102	2.5405	3.7434	0.36389	11.154
2.32	0.53205	6.1128	1.9652	3.1105	0.57452	7.4123	2.89	0.48203	9.5775	2.5518	3.7532	0.36080	11.228
2.33	0.53089	6.1670	1.9745	3.1234	0.57015	7.4720	2.90	0.48138	9.6450	2.5632	3.7629	0.35773	11.302
2.34	0.52974	6.2215	1.9838	3.1362	0.56580	7.5319	2.91	0.48074	9.7127	2.5746	3.7725	0.35469	11.377
2.35	0.52861	6.2762	1.9931	3.1490	0.56148	7.5920	2.92	0.48010	9.7808	2.5860	3.7821	0.35167	11.452
2.36	0.52749	6.3312	2.0025	3.1617	0.55717	7.6524	2.93	0.47946	9.8491	2.5975	3.7917	0.34867	11.527
2.37	0.52638	6.3864	2.0119	3.1743	0.55288	7.7131	2.94	0.47883	9.9176	2.6090	3.8012	0.34570	11.603
2.38	0.52528	6.4418	2.0213	3.1869	0.54862	7.7741	2.95	0.47821	9.986	2.6206	3.8106	0.34275	11.679
2.39	0.52419	6.4974	2.0308	3.1994	0.54438	7.8354	2.96	0.47760	10.055	2.6322	3.8200	0.33982	11.755
2.40	0.52312	6.5533	2.0403	3.2119	0.54015	7.8969	2.97	0.47699	10.124	2.6438	3.8294	0.33692	11.831
2.41	0.52206	6.6094	2.0499	3.2243	0.53594	7.9587	2.98	0.47638	10.194	2.6555	3.8387	0.33404	11.907
2.42	0.52100	6.6658	2.0595	3.2366	0.53175	8.0207	2.99	0.47578	10.263	2.6672	3.8479	0.33118	11.984
2.43	0.51996	6.7224	2.0691	3.2489	0.52758	8.0830	3.00	0.47519	10.333	2.6790	3.8571	0.32834	12.061
2.44	0.51894	6.7792	2.0788	3.2611	0.52344	8.1455	3.50	0.45115	14.125	3.3150	4.2608	0.21295	16.242
2.45	0.51792	6.8362	2.0885	3.2733	0.51932	8.2083	4.00	0.43496	18.500	4.0469	4.5714	0.13876	21.068
2.46	0.51691	6.8935	2.0982	3.2854	0.51521	8.2714	4.50	0.42355	23.458	4.8751	4.8119	0.09170	26.539
2.47	0.51592	6.9510	2.1080	3.2975	0.51112	8.3347	5.00	0.41523	29.000	5.8000	5.0000	0.06172	32.654
2.48	0.51493	7.0088	2.1178	3.3095	0.50706	8.3983	6.00	0.40416	41.833	7.941	5.2683	0.02965	46.815
2.49	0.51395	7.0668	2.1276	3.3214	0.50303	8.4622	7.00	0.39736	57.000	10.469	5.4444	0.01535	63.552
2.50	0.51299	7.1250	2.1375	3.3333	0.49902	8.5262	8.00	0.39289	74.500	13.387	5.5652	0.00849	82.865
2.51	0.51204	7.1834	2.1474	3.3451	0.49502	8.5904	9.00	0.38980	94.333	16.693	5.6512	0.00496	104.753
2.52	0.51109	7.2421	2.1574	3.3569	0.49104	8.6549	10.00	0.38757	116.50	20.388	5.7143	0.00304	129.217
2.53	0.51015	7.3010	2.1674	3.3686	0.48709	8.7198	∞	0.37796	∞	∞	6.000	0	∞
2.54	0.50923	7.3602	2.1774	3.3802	0.48317	8.7850							

ANSWERS TO SELECTED EVEN NUMBERED HOMEWORK PROBLEMS

CHAPTER 1

1.4 (a) FL^{-2}; (b) FL^{-3}; (c) FL

1.6 (b)

1.8 (a) L^4T^{-2}; (b) $F^2L^{-1}T$; (c) FL^{-2}; (d) L

1.10 $A \doteq T^{-1}; B \doteq T^{-2}$

1.12 No

1.16 (a) 4.66×10^4 ft; (b) 5.18×10^{-2} lb/ft³; (c) 3.12×10^{-3} slugs/ft³; (d) 2.36×10^{-2} ft · lb/s; (e) 5.17×10^{-6} ft/s

1.18 (a) 3.43×10^2 ft · lb/s; (b) 67.9 ft · lb; (c) 11.2 lb/ft²; (d) 3.03×10^{-6} ft³; (e) 2.46 slugs/ft²

1.22 7.57×10^{-2} m³/s; 4.54×10^3 liters/min

1.24 40.8 kg; 49.7 N

1.26 2.18 slugs/ft³; 1.12

1.28 0.998 at 22 °C; 0.966 at 89 °C; 1.002×10^{-3} m³/kg at 22 °C; 1.035×10^{-3} m³/kg at 89 °C

1.30 12.0 kN/m³; 1.22×10^3 kg/m³; 1.22

1.32* $\rho = 1001 - 0.05333\,T - 0.004095\,T^2$; 991.5 kg/m³

1.34 1.72×10^{-2} slugs/ft³; 0.554 lb/ft³

1.36 64 °C

1.38 0.314 lb/ft³

1.40 1.26 m³

1.44 0.6 N · s/m²; 1.3×10^{-2} lb · s/ft²

1.46 0.287 N · s/m²

1.48 4×10^{-5} N · s/m²; 8.36×10^{-7} lb · s/ft²

1.50 (kerosene) $\mu = 1.1 \times 10^{-3}$ N · s/m²; $\mu = 2.3 \times 10^{-5}$ lb · s/ft²; $v = 1.4 \times 10^{-6}$ m²/s; $v = 1.5 \times 10^{-5}$ ft²/s; (mercury) $\mu = 1.4 \times 10^{-3}$ N · s/m²; $\mu = 2.9 \times 10^{-5}$ lb · s/ft²; $v = 1.2 \times 10^{-7}$ m²/s; $v = 1.3 \times 10^{-6}$ ft²/s

1.52 7490 (water); 376 (air)

1.54* $C = 1.43 \times 10^{-6}$ kg/(m · s · K$^{1/2}$); $S = 107$ K

1.56* $D = 1.767 \times 10^{-6}$ N · s/m²; $B = 1.870 \times 10^3$ K; 5.76×10^{-4} N · s/m²

1.58 $\tau = 0.690\,U/\delta$ acting to right on plate (τ in N/m²)

1.60 0.300 N · s/m²

1.62 1.28×10^{-4} ft

1.64 13.3 N/m² (same value for either wall); stresses act on fixed walls in direction of moving plate

1.66 $\Im = 2\pi R_i^3 \ell \mu \omega / (R_o - R_i)$

1.68 0.589 N · m; 92.0%

1.70 286 N

1.72 4.14×10^3 psi

1.74 (a) 1.38 km/s; (b) 1.45 km/s; (c) 1.51 km/s

1.76 147 psi (gage)

1.78 3.83×10^{-3} slugs/ft³; 268 °F

1.80 5600 lb/ft²

1.82 (a) 343 m/s; (b) 1010 m/s; (c) 446 m/s

1.84 70.1 kPa (abs)

1.86 5.9 kPa (abs)

1.88 0.85 psi (abs)

1.90 538 Pa

1.92 0.186 in.; 4.65 in.

1.94 2.25 mm

1.96* $\sigma \cos \theta = 4.37 \times 10^{-3}$ lb/ft; $\theta = 29.7°$; $\sigma = 4.37 \times 10^{-3}$ lb/ft

CHAPTER 2

2.2 10.4 m

2.4 1940 lb/ft²

2.6 (a) 13.5 ft; (b) 6.73 psi; (c) 19.7 psia

2.8 (a) $p = -E_v \ln (1 - \rho_0 gh/E_v)$;
(b) 61.4 MPa; 60.6 MPa (constant density)

2.10 1890 lb

2.14 20,300 N

2.16 464 mm

2.18 63.8 psi

2.20 (a) 0.759 m; 0.759 m (without vapor pressure); (b) 10.1 m; 10.3 m (without vapor pressure); (c) 12.3 m; 13.0 m (without vapor pressure)

2.22 (a) 1240 lb/ft² (abs); (b) 1040 lb/ft² (abs); (c) 1270 lb/ft² (abs)

2.24 12.1 kPa; 0.195 kg/m³

2.26 (a) 4.00 ft; (b) 2.08 ft

2.28 17.4 psi

2.30 0.449 m

2.32 1.45 ft

2.34 0.424 psi

2.36 $h = (p_1 - p_2)/(\gamma_2 - \gamma_1)$

2.38 18.2 psi

2.40 −3.32 kPa

2.42* $\Delta h = [h_i + (p_g + p_{atm})/2\gamma_{Hg}] - \{[h_i + (p_g + p_{atm})/2\gamma_{Hg}]^2 - 2 p_g h_i/\gamma_{Hg}\}^{1/2}$

2.44 2.90 lb

2.46 11.7

2.48 0.244 m

2.52 $F_R = 1.16$ MN; $y_R = 10.6$ m

2.54 2.18 ft

2.56 7.11 m; No

2.58 $F_R = 33,900$ lb; 2.49 ft above base and 0.07 ft to right of centroid

2.60 (a) 2.11 m; (b) 941 kN

2.62 373 kN

2.64 $F_R = 78.5$ kN; 2.03 m below oil free surface

2.66 6330 lb

2.68 180 kN

2.70* 426 kN; 2.46 m below liquid free surface

2.72 0.146

2.74 $F_H = 294$ kN; $F_V = 328$ kN; Yes

2.76 20,200 lb

2.78 2880 psi

2.80 60.8 kN; 0.100 m below center of tank end wall

2.82 203 kN

2.84 2300 lb acting to right

2.86 (b) 735 lb

2.88 6.27 kN/m³; 707 N

2.90 0.0522 m³

2.92 59,200 ft³; 48.3 ft

2.94 2480 kg

2.96 (a) 18.9 kPa; (b) 0.208 m³

2.98 $dz/dy = 0.592$

2.100 4.91 m/s²

2.102 158 lb/ft²

2.104 $h = a\ell/g$

2.106 141 rad/s

2.108 6.04 rad/s

CHAPTER 3

3.2 (a) $-194 (1 + x) + 62.4$ lb/ft³; (b) 41.2 psi

3.4 −9.99 kPa/m

3.6 −0.703 psi/ft; 0.164 psi/ft

3.10 12.0 kPa; −20.1 kPa

3.12 20.1 kPa; −0.979 kPa

3.14 3.21 N/m²

3.16 Fig. (a)

3.18 11.5 m

3.20 61.9 psi

3.24 7.75 lb/ft²; 124 lb/ft²

3.26 −76.0 lb/ft²; 88.0 lb/ft²

3.28 115 ft; 0.119 ft; 135 ft

3.30 0 Pa; 0.210 m

3.32 104 ft/s; 72.8 psi; 304 ft/s; 620 psi

3.34 $0.0156 D^2/(10^{-4} - D^4)^{1/2}$ m³/s

3.36 203 ft/s

3.38 45.7 ft

3.40 2.35 m; 11.9 m

3.42 $D = D_1/[1 - (\pi^2 g D_1^4 z)/(8Q^2)]^{1/4}$

3.44 3.13 ft

3.46 0.158 ft³/s

3.48 $V_2 = \{2 RT \ln[g(\rho_m - \rho)h/p_2 + 1]/[1 - (D_2/D_1)^4]\}^{1/2}$

3.50 5.12×10^{-4} m³/s

3.52 19.56 psi

3.54 0.0252 m³/s; 8.40 m

3.56 2.94 m

3.58 −7.03 psi

3.60 29.5 Pa

3.62 0.0863 in.

3.66 13.0 psi

3.68   0.346 ft³/s; 7.91 × 10⁻⁴ slugs/s; 0.0255 lb/s

3.70   2.14 ft

3.72 $h = 0$

3.74   2.45 hr

3.76   0.303 ft³/s; 499 lb/ft²; 312 lb/ft²; − 187 lb/ft²

3.78   0.37 m

3.80   7.53 ft

3.82   9.05 m³/s

3.84   0.379 ft³/s

3.86   8.65 × 10⁻³ m³/s

3.94   404.5 kPa

3.96   9.10 × 10⁻³ m³/s; 57.9 kPa

3.98   2.00 × 10⁻⁴ m³/s; 0.129 m

3.100   − 4.04 × 10⁻⁵ psi

3.102   3120 ft³/s; 2080 ft³/s; 2200 ft³/s

3.104   9.89 ft/s

3.106   3.46 Q_o

3.108   0.630 ft; 4.48 ft

3.110   0.0434 m³/s

3.112   83.6 ft³/s

3.114   6.51 m; 25.4 m; 6.51 m; − 9.59 m

CHAPTER 4

4.2 $2\sqrt{2}\, x^2 t$ m/s; 45 deg from x axis

4.4   3.61 m/s; 3.24 m/s; 3 m/s; 0 s; ∞ s

4.6 $y = e^{(x^2/2 - x)} - 1$

4.8   Both the x and y axes; $x = y = 0$

4.16 $2\, c^2 x^3$; $2\, c^2 y^3$; $x = y = 0$

4.18   0.275 ft/s²; 0.476 ft/s²; 0.825 ft/s²

4.20   − 10 m/s²; − 2.5 m/s²; − 0.10 m/s²

4.26 $-x/t^2$; x/t^2

4.28   40 m/s²; 4.08

4.30   0; $(100\, x + 50)\hat{i}$ m/s; $50\hat{i}$ m/s; $150\hat{i}$ m/s

4.32 $(20 + 50)\hat{i}$ m/s²; $(60 + 150)\hat{i}$ m/s²

4.38   8.02 ft/s

4.42   4180 m/s²; 4990 m/s

4.44 $4V_0^2 \sin^2\theta/a$; $4V_0^2 \sin\theta \cos\theta/a$

4.46 $2\, c^2 x(x^2 + y^2)$; $2\, c^2 y\,(x^2 + y^2)$

4.48   7.20 m/s²; 3.33 m/s²

4.50   − 225 ft/s²; − 28.1 ft/s²; − 8.33 ft/s²

4.52   5.0 m³/s

4.54   132 ft³/s

4.56 $M = \dfrac{\pi}{4}\, D^2[\rho_1 \ell + \rho_2 \ell/2 + V_s t(\rho_2 - \rho_1)]$

4.58   0; 10⁴ particles/s

CHAPTER 5

5.2   (a) 6280 kg/s

5.4   70.1 m/s

5.6   27.5 ft

5.8   1.04 m/s

5.10   150 liters/s

5.12   0.0125 lbm/ft³

5.14   5900 lbm/hr (water evaporation); 244,000 lbm/hr (cooled water)

5.16   849 kg/m³

5.18   7/8

5.20   0.0114 slug/s

5.22 $(7/8)\, U\ell\delta$

5.24   491 min

5.26   (a) 10.4 gal/min;   (b) 41.6 gal/min

5.28   229 s

5.32   (a) 352 lb to left;   (b) 352 lb to left

5.34 $F_{A,x} = 1890$ lb to left; $F_{A,y} = 0$; $R_x = 1890$ lb to left; $R_y = 0$

5.36 $R_x = 18{,}200$ lb to left; $R_y = 10{,}800$ lb up

5.38   30,800 N against flow

5.40   9.27 N to left

5.42   17.1 lb/ft

5.44   3580 lb

5.46   axial-direction momentum flowrate with nonuniform velocity profile is 4/3 times the axial direction momentum flowrate with uniform velocity profile at $\bar{u}$ level

5.48   13.0 (kg · m)/s²

5.50   1290 lb (gate open); 3120 lb (gate closed)

5.52   28.7 ft³/s

5.54   1.93 N

5.56   (a) 10.2 ft/s;   (b) 20.2 ft/s

5.58   (a) 722 ft · lb;   (b) 100 rad/s; 0 hp

5.60   23.9°

5.62   0.0117 ft · lb opposing rotation; 0.368 (ft · lb)/s

5.64   12.8 MW

5.66   1200 hp

5.68   (b) 918 ft · lb in rotation direction

5.70 1.22×10^4 (ft · lb)/ slug

5.72 (a) 43°; (b) 53.4 (ft · lb)/s

5.74 11.4 (N · m)/kg

5.76 (b) 7010 (ft · lb)/slug

5.78 turbine; 36.8 (ft · lb)/slug

5.80 4.79 (kg · m²)/s²; 1.20 kW

5.82 11.6 (N · m)/kg; 79.6%

5.84 16.5 (N · m)/kg; 85.0%

5.88 No

5.90 2.17 ft³/s

5.94 $h_{max} = 9.06$ m

5.96 3910 (ft · lb)/slug

5.98 flow is up

5.100 926 (ft · lb)/slug [from (1) to (2)]; 200 (ft · lb)/slug [from (2) on]

5.104 1.16 MW

5.106 2740 ft/s

5.108 1.22×10^6 hp

5.110 930 kW

5.112 308 Btu/lbm; 66%

5.116 (a) 1.20 m³/s; (b) 0.928 m³/s; (c) 0.705 m³/s

5.118 490 (N · m)/kg; 6860 W

5.120 79.9%

5.122 28 hp

5.124 $\bar{u} = u_c/3$; $\alpha = 2.7$

5.126 (a) 3.43 (ft · lb)/lbm; (b) 3.36 (ft · lb)/lbm

5.130 $R_x = 10{,}800$ lb in flow direction; $R_y = 1870$ lb opposite to flow direction

5.132 (a) 1.65 psi; (b) 203 (ft · lb)/slug; (c) 7.72 lb opposite to flow direction

5.134 4.60 lb

5.138 8.12 lb

CHAPTER 6

6.2 a_x (local) = 0; a_x (conv) = $18(x^3 + xy^2)$; a_y (local) = 0; a_y (conv) = $18(x^2y + y^3)$; $\mathbf{V} = -6\hat{\mathbf{j}}$; $|\mathbf{V}| = 6$ ft/s; $\mathbf{a} = 36\hat{\mathbf{i}} + 36\hat{\mathbf{j}}$; $|\mathbf{a}| = 50.9$ ft/s²

6.4 (a) 0; (b) $\boldsymbol{\omega} = -(y/2 + z)\hat{\mathbf{i}} + (5z/2)\hat{\mathbf{j}} - (y/2)\hat{\mathbf{k}}$; No

6.6 No; None (except both equal to zero)

6.8 (a) 0; (b) $\boldsymbol{\omega} = -(U/2b)\hat{k}$; (c) $\zeta = -(U/b)\hat{k}$; (d) $\dot{\gamma} = U/b$

6.10 No

6.12 $u = -x^2 + f(y)$

6.16 $\psi = A\theta + Br^{-1}\sin\theta + C$

6.20 $v = -y^2/2x + f(x)$

6.22 (b) Yes; (c) 0

6.24 (b) 1 m³/s (flow right to left)

6.28 (a) $p_A = p_0$; (b) $p_B = p_0$

6.30 $\psi = 5x^2y - (5/3)y^3 + C$

6.32 $\psi = A\theta + Br\sin\theta + C$; $\theta = \pi$, $r = A/B$

6.34 Function only of x

6.36 (a) $\psi = 2xy$; (b) $q = 2x_iy_i$

6.38 (a) $v = -6y$; (b) Yes; 569 lb/ft²

6.40 (a) Yes; (b) $\partial p/\partial x = -18\rho(x^3 + xy^2)$

6.44 $p_A - p_B = -7.10$ kPa

6.46 Yes

6.48 $C = 0.386$ m

6.50 $\Gamma = \omega\Delta\theta(b^2 - a^2)$

6.52 $\theta_{stag} = \pi/2$, $r_{stag} = \sqrt{m/4\pi}$

6.54 $\sin\theta = (\Gamma/2\pi rU) \ln(r/2)$

6.58 $p_{stag} - p_A = 0.703 \rho U^2$

6.62 $x_{stag} = \pm 1.10$ ft

6.66 (a) $p_{max} = p_0 + \rho U^2/2$ (at $\theta = 0$ and π); $p_{min} = p_0 - 3\rho U^2/2$ (at $\theta = \pi/2$); (b) $(2r/3a)(1 - a^2/r^2)\sin\theta = 1$

6.68* $y/a \geqslant 10$

6.72 (b) $V_{wall} = (m/\pi)[y/(\ell^2 + y^2)]$; (c) $p_{wall} = p_0 - (\rho m^2/2\pi^2)[y/(\ell^2 + y^2)]^2$

6.74 $p = p_0 + (\rho U^2/2)[(\Gamma/\pi Ur)\sin\theta - (\Gamma/2\pi Ur)^2]$

6.76 -85.1×10^{-5} lb/ft²

6.78 (a) $\partial v/\partial y = -2x$; (b) $\mathbf{a} = 2x^3\hat{\mathbf{i}}$; (c) $\partial p/\partial x = 2\mu - 2\rho x^3$

6.80 (a) 5.61×10^{-4} m²/s; (b) 80 N/m² (acting in direction of flow); (c) 0.210 m/s

6.82 $q = (\rho g h^3 \sin\alpha)/3\mu$

6.84 $q = -2\rho g h^3/3\mu$

6.86 $u = [(U_1 + U_2)/b]y - U_2$

6.88 $U = (b^2/2\mu) \partial p/\partial x$

6.90 0.444 N · m

6.94 $\partial p/\partial z = -(4\mu Q/\pi ab)(1/a^2 + 1/b^2)$

6.96 68.5 N/m² per meter; 0.230 m/s

6.98 (a) Re = 320 < 2100; (b) 90.0 kPa; (c) 30.0 N/m²

6.100 (b) 1.20 Pa

6.102 $v_\theta = R^2\omega/r$

6.104 28.4 lb/ft² per ft

6.106* Δp (nonuniform)/Δp (constant radius) = 3.57

CHAPTER 7

7.4 (b)

7.6 (a) 103 m/s;   (b) 444 m/s

7.8 $\mathcal{P}/\rho D^5 \omega^3 = \phi(Q/D^3\omega)$

7.10 $d/D = \phi(\rho V D/\mu, \sigma/\mu V)$

7.12 $\mathcal{D}/d_i^2 V^2 \rho = \phi(d_2/d_1, \rho V d_1/\mu)$

7.14 $\Delta p/\rho V^2 = \phi(d/D)$

7.16 $c\sqrt{\rho/E} = \phi(h/D)$

7.18 $\mathcal{D}/\rho V^2 \ell_1^2 = \phi(V/c, \ell_i/\ell_1)$

7.20 $\Delta p \propto 1/D^2$ (for a given velocity)

7.24 $h/D = \phi(\sigma/\gamma D^2)$

7.26* $\Delta p/\rho V^2 = 0.505\,(D/d)^{3.99}$ for $1.33 \le D/d \le 3.33$

7.28* $\theta = 298\,\omega\mu$

7.30* 1.16 mm

7.32 $Q/\sqrt{gH^5} = \phi(\theta)$

7.34 9.63×10^{-2} ft/s

7.36 103 m/s

7.38 $\sigma_m/\sigma = 0.010$

7.40 1180 km/hr

7.42 8.64 ft³/min

7.44 3.24×10^{-3} m³/s

7.46 (a) 2.45 ft/s;   (b) 0.0510 psf/ft

7.48 (a) 49.9 mm;   (b) 2.96 kPa

7.50 (b) $V_{cm}/V_c = 0.707$

7.52 $\ell_m = 1.00$ in.; $d_m = 0.025$ in; $V_m/V = 0.125$

7.54 26.4 lb

7.56 Same liquid cannot be used if $d_m/d \ne 1$

7.58 0.274 lb

7.60 100 to 450 mph; No.

7.62 0.0647 to 0.0971

7.64 0.188 ft; 0.0586 ft³/s

7.66 1.31 m/s; 1.25×10^5

7.68 0.0328 m³/s; 8.00 m

CHAPTER 8

8.2 Laminar

8.4 3.36 ft

8.6 Blue and yellow; green

8.8 Laminar; fully developed

8.10 (a) -7.40 lb/ft³;   (b) -69.8 lb/ft³; (c) 55.0 lb/ft³

8.12 Downhill; 0.596 lb/ft²; 0.238 lb/ft²; 0

8.14 (a) to (b)

8.18 (a) 6.72×10^4 Pa;   (b) 5.05×10^4 Pa

8.20 B to A

8.22 1.055 D

8.24 -39.1 s⁻¹; -24.9 s⁻¹; -10.7 s⁻¹

8.26 -3.80 K

8.28 10.1 m

8.30 0

8.34 0.758; 0.760; 0.762

8.36 0.0404

8.38 0.0220

8.40 (a) 4.69×10^{-3} ft³/s; 3.30×10^{-3} ft³/s

8.42 No.

8.44 0.00183 ft

8.46 15.1 psi

8.50 (a) 16;   (b) 6.36

8.52 (a) 2.91 ft;   (b) 19.4 ft;   (c) 0.97 ft

8.54 0.00551 psi

8.56 2.63 mm

8.58 9.00

8.60 120 kPa; 33.8 kPa; 86.4 kPa

8.62 34.5 lb/ft²; 908 lb/ft²

8.68 0.00025

8.70 0.0132 ft

8.74 0.00360 in.

8.76 440 ft; 0.233 psi; 5.11 hp

8.78 0.973 hp

8.80 1.20 hp

8.84 56.7

8.86 166

8.88 0.865 lb/ft²

8.90 0.223 lb/ft²

8.92 0.707 ft³/s

8.94 1490 ft

8.96 19.6 m

8.98 5.68

8.100 (a) 112 ft;   (b) 113 ft

8.104 8.89 ft

8.106 0.513 ft

8.108 Water out; air in

8.110* 144 s

8.112 5.73 ft/s

8.114 0.0144 m³/s

8.116 (a) 2.08 m³/s or 14.4 m³/s;   (b) no solution

8.120	0.535 ft
8.122	0.0445 m
8.124	0.0746 cfs; 0.339 cfs
8.126	0.0180 m³/s
8.128	0.0284 m³/s; 0.0143 m³/s; 0.0141 m³/s
8.130	4.73 ft/s; 8.35 ft/s; 10.3 ft/s
8.132	7.63 psi
8.134	13.7 kPa
8.136	0.0221 m³/s
8.138	0.528 ft³/s
8.140	5.69 ft³/s
8.142	5.77 ft
8.144	0.0936 ft³/s
8.146	0.254 ft³/s

CHAPTER 9

9.2	0; 1.7
9.4*	5.44 kN
9.6	Yes.
9.8	Figure 9.6c
9.12	6.65×10^{-6} m²/s
9.14	0.0131 m/s; 0.00420 m/s
9.18	0.0106 m; 0.0877 N/m²; 0.0150 m; 0.0620 N/m²
9.26	Decrease
9.30*	$\delta/x = 5.03/\mathrm{Re}_x^{1/2}$
9.32	$\ell/4$
9.34	2
9.36	2.64×10^{-5} lb
9.40	7.35×10^{-6} N · m
9.42	0.0350 N · m
9.46	1.10×10^{-7} m/s; 1.20×10^{-7} m/s
9.48	0.946 ft/s; 30.2 ft/s
9.52	(a) 149 m/s; (b) 4.57 m/s
9.56	10,300 lb
9.58	18,000 N · m
9.60	49.5 hp
9.62	43.2%
9.64	3
9.66	30.6 ft/s
9.72	133 N
9.74	1.68 ft · lb
9.76	1.29 lb; 129 lb
9.78	7220 m/s²

9.80	180 N
9.82	3.76×10^{-3} N · m
9.84	4.31 MN
9.86	0.851; 0.301
9.92	53.5 kW; 4.46 kW
9.94	2.47×10^3 kW
9.96	83.2 mph
9.98	0.0187 U
9.100	0.292; 0.0483
9.104	1.72 U
9.110	19.1
9.112	29.3%
9.114*	0.327

CHAPTER 10

10.2	(a) supercritical; (b) supercritical; (c) subcritical
10.4	3.57 m/s
10.6	Yes; no.
10.8	0.735 m/s
10.10	22.0 ft/s
10.12	(a) 4.27 ft; (b) 6.41 ft; (c) 8.12 ft
10.14	2.12 m; 0.426 m
10.16	1.77 ft; 1.97 ft
10.20	1.83 ft
10.22	0.694 ft
10.26	2.57 ft
10.30	2.17 m
10.34	5.81 N/m²
10.36	0.0481; 0.585 lb/ft²
10.38	40.9 m³/s
10.42	7.42 min
10.44	12.0 m
10.46	0.000791
10.48	34.8 ft; 0.00206
10.50	11.7 m³/s; 12.2 m³/s
10.52	1380 ft³/s
10.54	6 ft
10.56	0.352 ft; 0.603
10.58	0.889b
10.60	23.7%
10.62	3.44 m; subcritical
10.64	0.303 ft
10.68	17.1 ft

10.70   1.22 m

10.72   20.8 m³/s

10.74   142 ft³/s; 1180 ft³/s

10.76   0.00757

10.78   Increase

10.80   − 0.000225

10.82   2.50 ft

10.86   (a) 0.228 ft;   (b) 8.10, 0.223;
(c) 5.15 ft

10.88   2.41 m; 8600 kW

10.90   0.0577; 0.000240

10.92   4.36 ft/s

10.94   16.9 ft³/s

10.96   0.500 m

10.98   0.679 m³/s; 0.333 m

10.100   0.339 m/s

10.102   1670 ft³/s; increase

10.104   1.29 ft; 4.39 ft; triangular

CHAPTER 11

11.2   (a) c_p = 6006 (ft · lb)/(slug · °R);
c_v = 4290 (ft · lb)/(slug · °R) (b) c_p = 4897
(ft · lb)/(slug · °R); c_v = 3767 (ft · lb)/(slug ·
°R) (c) c_p = 31,240 (ft · lb)/(slug · °R); c_v =
18,820 (ft · lb)/(slug · °R) (d) c_p = 84,810 (ft ·
lb)/(slug · °R); c_v = 60,150 (ft · lb)/(slug · °R)
(e) c_p = 13,100 (ft · lb)/(slug · °R); c_v = 9,997
(ft · lb)/(slug · °R) (f) c_p = 6212 (ft · lb)/
(slug · °R); c_v = 4438 (ft · lb)/(slug · °R)
(g) c_p = 5439 (ft · lb)/(slug · °R); c_v = 3885
(ft · lb)/(slug · °R)

11.4 $\breve{u}_2 - \breve{u}_1 = -5.08 \times 10^6$ (N · m)/kg;
$\breve{h}_2 - \breve{h}_1 = -7.16 \times 10^6$ (N · m)/kg; $s_2 -$
$s_1 = -1.40 \times 10^4$ (N · m)/(kg · K)

11.6   − 1890 J/(kg · K)

11.8   351 K

11.10   (a) constant c_p, $\breve{h}_2 - \breve{h}_1$ = 49,000
(ft · lb)/slug; varying c_p, $\breve{h}_2 - \breve{h}_1$ = 51,000 (ft ·
lb)/slug; (b) constant c_p, $\breve{h}_2 - \breve{h}_1$ = 4.9 × 10⁶
(ft · lb)/slug; varying c_p, $\breve{h}_2 - \breve{h}_1$ = 6.29 × 10⁶
(ft · lb)/slug; (c) constant c_p, $\breve{h}_2 - \breve{h}_1$ =
1.47 × 10⁷ (ft · lb)/slug; varying c_p, $\breve{h}_2 - \breve{h}_1$ =
2.19 × 10⁷ (ft · lb)/slug

11.14   (a) 0.0328;   (b) 0.0722;   (c) 0.131

11.18   (a) 1130 ft/s;   (b) 881 ft/s;   (c) 3300
ft/s;   (d) 4280 ft/s;   (e) 1460 ft/s

11.20   2.19 × 10¹¹ N/m²; 1470 m/s

11.22   2.08; 2280 ft/s

11.24   1.67

11.32   (a) 0.0727 slug/s;   (b) 0.0769

11.34   (a) 0.99937;   (b) 0.99825;
(c) 0.99435;   (d) 0.7465 (all using closest values
from Table E.1)

11.36   (a) 267 m/s; 0.725;   (b) 285 m/s; 0.78

11.46   converging-diverging nozzle;
(a) 1.76; 1540 ft/s;   (b) 1.78, 2120 ft/s;
(c) 1.71; 3990 ft/s

11.48   (a) 0.356 m²; 20.8 kPa (abs); 112 K;
(b) 0.257 m²; 24.8 kPa (abs); 82.6 K

11.50   (a) 52.8 psia; 433 °R; 1020 ft/s; 1.04
slugs/s;   (b) 54.6 psia; 452 °R; 815 ft/s; 1.26
slugs/s;   (c) 48.8 psia; 391 °R; 2840 ft/s; 0.411
slug/s

11.54 $\dot{m}$ = 0.246 slug/s; T_1 = 487 °R; p_1 =
11.8 psia; $T_{0,1}$ = 519 °R; $p_{0,1}$ = 14.7 psia; V_1 =
617 ft/s; T_2 = 432 °R; p_2 = 6.34 psia; $T_{0,2}$ =
519 °R, $p_{0,2}$ = 12.0 psia; V_2 = 1020 ft/s

11.56   from 0.246 slug/s to 0.268 slug/s with
reduction

11.58   (a) 1.73 kg/s;   (b) 1.54 kg/s

11.60   1.54 kg/s

11.62   (a) 34.6 kPa (abs); 318 m/s; 0.803;
(b) 18.4 kPa (abs); 593 m/s; 0.390

11.64   (a) 2730 lb/ft²; 2550 lb/ft²

11.66 T_1 = 283 K; p_1 = 94.5 kPa (abs);
$T_{0,1}$ = 288 K; $p_{0,1}$ = 101 kPa (abs); V_1 = 104
m/s; T_2 = 658 K; p_2 = 44.7 kPa; $T_{0,2}$ = 786
K; $p_{0,2}$ = 84.6 kPa (abs); V_2 = 512 m/s

11.68   (a) 1390 Btu/s; 17.1 psia; 868 ft/s;
0.400;   (b) 969 Btu/s; 19.7 psia; 737 m/s; 0.116

11.70 $(T_{0,2}/T_{0,1})$ = 3.45; $(p_{0,2}/p_{0,1})$ = 0.763;
q = 7.84 × 10⁶ (ft · lb)/slug; flow is not frictionless

11.72   2.5

11.74   2.07; 680 m/s

11.76   390 K; 94.1 psia

11.78   (a) 0.946; 3.5 kPa;   (b) 0.805; 17.2
kPa;   (c) 0.466; 49.8 kPa

11.80   (a) $(p_{exit}/p_{0,inlet})$ = 0.213; Ma_{exit} =
0.62;   (b) $(p_{exit}/p_{0,inlet})$ = 0.190; Ma_{exit} = 0.69

11.82 T_1 = 288 °R; p_1 = 1.88 psia; $T_{0,1}$ =
519 °R; $p_{0,1}$ = 14.7 psia; V_1 = 1660 ft/s; T_x =
410 °R; p_x = 3.00 psia; $T_{0,x}$ = 595 °R; $p_{0,x}$ =
11.0 psia; V_x = 1490 ft/s; T_y = 541 °R; p_y =
7.38 psia; $T_{0,y}$ = 595 °R; $p_{0,y}$ = 10.2 psia; V_y =
800 ft/s; T_2 = 544 °R; p_2 = 5.17 psia; $T_{0,2}$ =
654 °R; $p_{0,2}$ = 9.78 psia; V_2 = 1140 ft/s

— INDEX —

TABLE 1.3 Conversion Factors from BG and EE Units to SI Units[a]

	To Convert from	to	Multiply by
Acceleration	ft/s^2	m/s^2	3.048 E − 1
Area	ft^2	m^2	9.290 E − 2
Density	lbm/ft^3	kg/m^3	1.602 E + 1
	slugs/ft^3	kg/m^3	5.154 E + 2
Energy	Btu	J	1.055 E + 3
	ft·lb	J	1.356
Force	lb	N	4.448
Length	ft	m	3.048 E − 1
	in.	m	2.540 E − 2
	mile	m	1.609 E + 3
Mass	lbm	kg	4.536 E − 1
	slug	kg	1.459 E + 1
Power	ft·lb/s	W	1.356
	hp	W	7.457 E + 2
Pressure	in. Hg (60 °F)	N/m^2	3.377 E + 3
	lb/ft^2 (psf)	N/m^2	4.788 E + 1
	lb/in.^{2} (psi)	N/m^2	6.895 E + 3
Specific weight	lb/ft^3	N/m^3	1.571 E + 2
Temperature	°F	°C	$T_C = (5/9)(T_F - 32°)$
	°R	K	5.556 E − 1
Velocity	ft/s	m/s	3.048 E − 1
	mi/hr (mph)	m/s	4.470 E − 1
Viscosity (dynamic)	lb·s/ft^2	N·s/m^2	4.788 E + 1
Viscosity (kinematic)	ft^2/s	m^2/s	9.290 E − 2
Volume flowrate	ft^3/s	m^3/s	2.832 E − 2
	gal/min (gpm)	m^3/s	6.309 E − 5

[a] If more than four-place accuracy is desired, refer to Appendix A.